Mathematics with Applications

in Business and Social Sciences

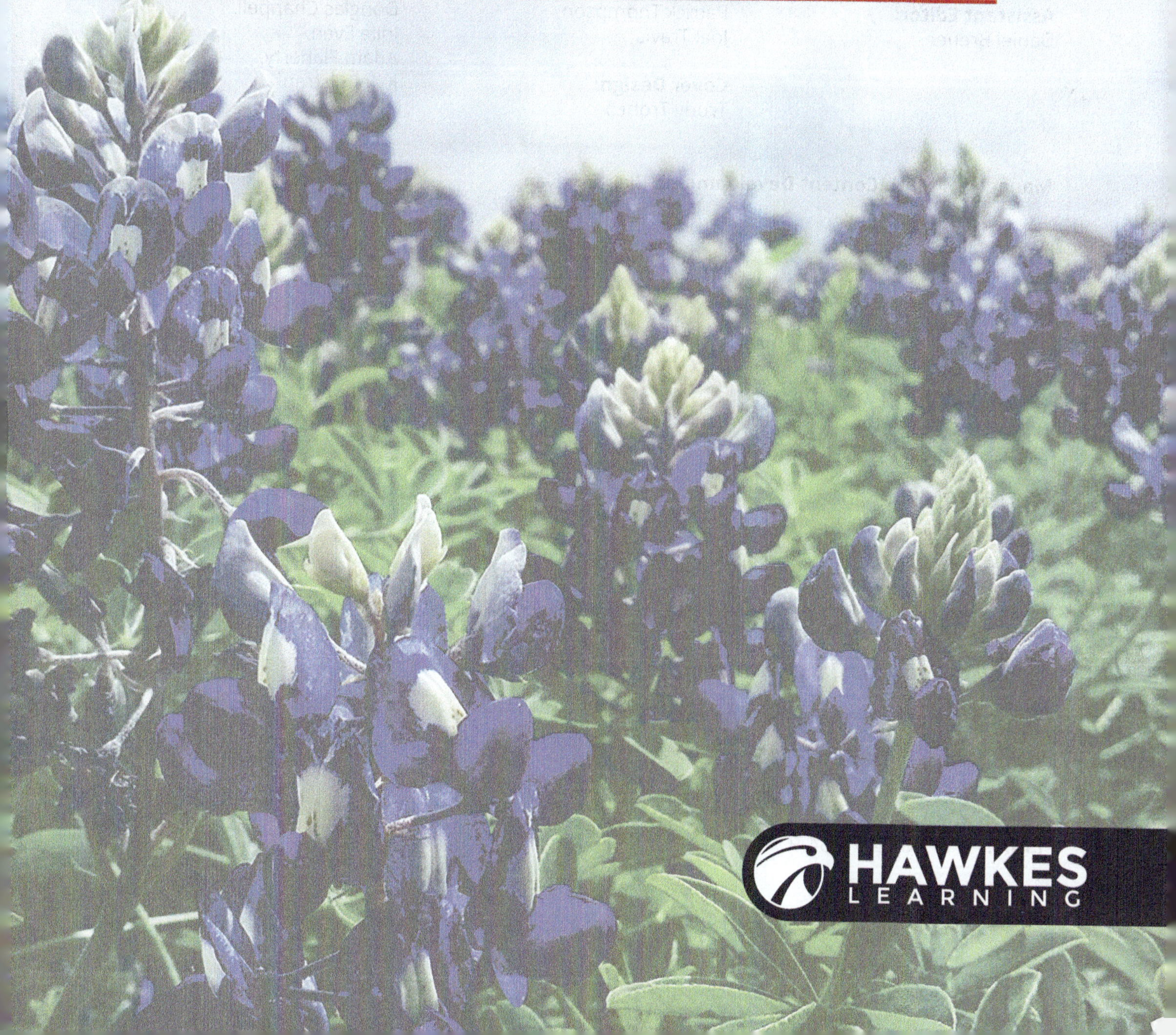

HAWKES LEARNING

Editors:
Danielle C. Bess,
Marvin Glover,
Claudia Vance

Assistant Editor:
Daniel Breuer

Creative Services Manager:
Trudy Tronco

Designers:
Lizbeth Mendoza,
Patrick Thompson,
Joel Travis

Cover Design:
Trudy Tronco

**Composition and
Answer Key Assistance:**
Quant Systems India Pvt. Ltd.

Courseware Developers:
Douglas Chappell,
Jolie Even,
Adam Flaherty,
Kyle Gilstrap

Manager of Math Content Development: Blair Dunivan

A division of Quant Systems, Inc.

546 Long Point Road
Mount Pleasant, SC 29464

Library of Congress Control Number   2021917837

Printed in the United States of America

10 9 8 7 6 5 4 3 2 1

ISBN: 978-1-64277-484-9

Table of Contents

5 Mathematics of Finance

6 Systems of Linear Equations; Matrices

7 Inequalities and Linear Programming

8 Probability

9 Statistics

15 Additional Integration Topics

16 Multivariable Calculus

PREFACE

Features

Definitions, Theorems, Formulas, Properties, and Procedures

Definitions, theorems, and formulas are clearly set apart in highly visible green boxes for easy reference, and properties and procedures are similarly set apart in distinctive blue boxes. All formally identified terms appear in bold print when first defined, and other useful terms appear in italic font.

Linear Functions

A **linear function** f in the variable x is any function that can be written in the form

Solving Elementary Exponential Equations

To solve an elementary exponential equation, complete the following steps.

Step 1: Isolate the exponential. Move the exponential containing x to one side of the equation and any constants or other variables in the expression to the other side. Simplify, if necessary.

Step 2: Find a base that can be used to rewrite both sides of the equation.

Step 3: Equate the powers, and solve the resulting equation.

Cautions

Many common errors are pointed out, along with how to correct them. These are set apart in red boxes.

Notes and Helpful Hints

These green and red boxes in the margins help clarify subtle details and provide problem-solving tips.

✎ NOTE

A function cannot represent a vertical line (since it fails the vertical line test). Vertical lines can represent the graphs of equations, but not functions.

☞ HELPFUL HINT

If calculating large formulas in pieces on a calculator, round each step to at least six decimal places to avoid rounding errors in the subsequent calculations.

⚠ CAUTION

It doesn't matter how you assign the labels (x_1, y_1) and (x_2, y_2) to the two points you are using to calculate slope, but it *is* important that you are consistent as you apply the formula. You cannot change the order in which you are subtracting as you determine the numerator and denominator in the slope formula.

Correct	Incorrect
$\dfrac{y_2 - y_1}{x_2 - x_1}$ ✓	$\dfrac{y_1 - y_2}{x_2 - x_1}$ ✗
$\dfrac{y_1 - y_2}{x_1 - x_2}$ ✓	$\dfrac{y_2 - y_1}{x_1 - x_2}$ ✗

Examples

Examples are presented in a step-by-step manner that is easy for students to follow. Each example has a title indicating the problem-solving skill being presented. Examples make use of tables, diagrams, and graphs for additional clarity where applicable.

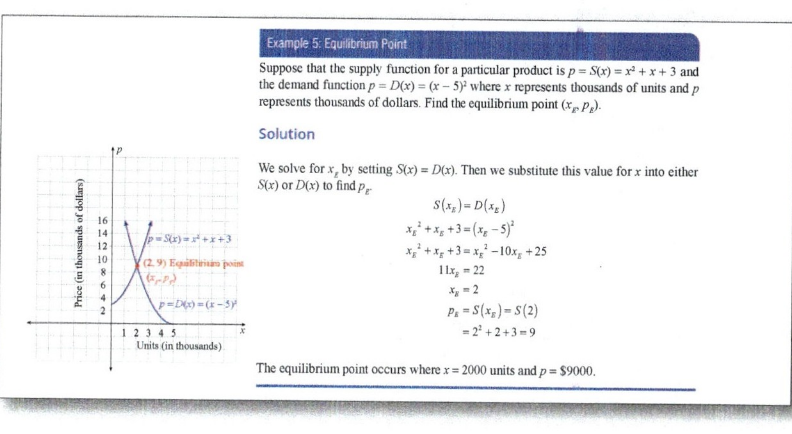

Example 5: Equilibrium Point

Suppose that the supply function for a particular product is $p = S(x) = x^2 + x + 3$ and the demand function $p = D(x) = (x - 5)^2$ where x represents thousands of units and p represents thousands of dollars. Find the equilibrium point (x_E, p_E).

Solution

We solve for x_E by setting $S(x) = D(x)$. Then we substitute this value for x into either $S(x)$ or $D(x)$ to find p_E.

$$S(x_E) = D(x_E)$$
$$x_E^2 + x_E + 3 = (x_E - 5)^2$$
$$x_E^2 + x_E + 3 = x_E^2 - 10x_E + 25$$
$$11x_E = 22$$
$$x_E = 2$$
$$p_E = S(x_E) = S(2)$$
$$= 2^2 + 2 + 3 = 9$$

The equilibrium point occurs where $x = 2000$ units and $p = \$9000$.

Technology Instructions

Technology notes and screenshots are included throughout the text to highlight ways that graphing utilities and other forms of technology can help solve problems or explain concepts. Step-by-step instructions for using a TI-84 Plus are given in many cases.

Solving Systems of Equations Using Technology

The solution to a consistent pair of linear equations is the point common to both equations. Graphically speaking, the solution is the point where the graphs of the two equations intersect. We can use a graphing calculator to find this point. Consider the following system of equations: $\begin{cases} 2x - 3y = -13 \\ x = y - 6 \end{cases}$. One way to solve this system using a calculator is to graph each equation. Remember to solve for y before entering the equation in `Y=` and selecting `graph`.

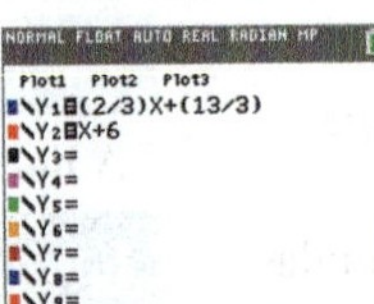

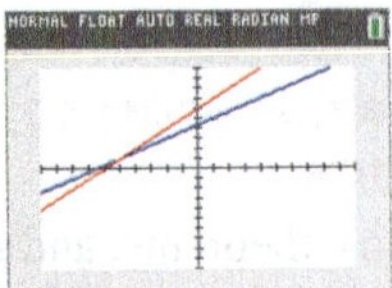

Once the graph of the two lines is displayed, press `2nd` `trace` to access the CALC menu and select `intersect`. The phrase "First curve?" should appear. Use the arrows to move the cursor along the first line to where it appears to intersect the other line and press `enter`. When the phrase "Second curve?" appears, press `enter` again (as the cursor should now be on the second line, still near the point of intersection). Now the word "Guess?" should appear. Press `enter` a final time and the x- and y-values of the point of intersection will appear at the bottom.

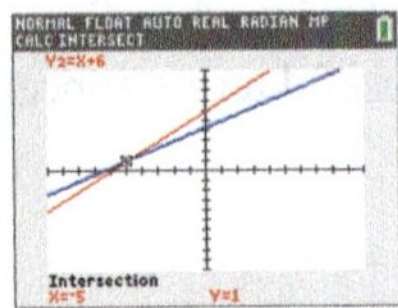

To evaluate the function $C(x)$ in Example 1 with a TI-83/84 Plus calculator, perform the following steps:

1. Enter the function $C(x)$ that was found into **Y1**.

2. Now, on the main screen, press **VARS**, scroll right to **Y-VARS**, select **1:Function**, and then **1:Y1** to display **Y1** on the main screen.

3. With your cursor after **Y1**, type an opening parenthesis and then the x-value we are evaluating, 500. Type a closing parenthesis.

4. Press `enter` to calculate the corresponding y-value.

5. This process can be repeated for the next x-value, 1.

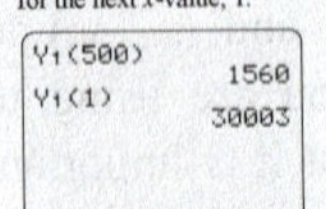

Solution

Using the given information, we can determine a function for the inventory costs, $C(x)$. Let $x =$ lot size. Then $\dfrac{500}{x}$ is the number of orders per year, and $\dfrac{x}{2}$ is the average inventory.

$$C(x) = (\text{storage cost per item}) \cdot \left(\frac{x}{2}\right) + (\text{cost per order}) \cdot \left(\frac{500}{x}\right)$$

$$C(x) = 6\left(\frac{x}{2}\right) + 60\left(\frac{500}{x}\right)$$

The number of desks ordered is between 1 and 500. At the extremes, one order for 500 desks would cost

$$C(500) = 6\left(\frac{500}{2}\right) + 60\left(\frac{500}{500}\right) = \$1560,$$

and 500 orders for one desk at a time would cost

$$C(1) = 6\left(\frac{1}{2}\right) + 60\left(\frac{500}{1}\right) = \$30,003.$$

Now we need to differentiate $C(x)$ so we can determine the local minima.

$$C(x) = 6\left(\frac{x}{2}\right) + 60\left(\frac{500}{x}\right)$$

$$= 3x + 30,000x^{-1} \qquad \text{Rewrite } C(x) \text{ using exponents.}$$

$$C'(x) = 3 - 30,000x^{-2}$$

$$= 3 - \frac{30,000}{x^2}$$

We set $C'(x) = 0$ and solve for x.

$$\frac{30,000}{\qquad}$$

Applications

Many examples and exercises illustrate practical applications, keeping students engaged.

Example 5: How Much House Can You Afford?

Suppose you have recently graduated from college and want to purchase a house. Your take-home pay is \$3220 per month and you wish to stay within the recommended guidelines for mortgage amounts by only spending $\dfrac{1}{4}$ of your take-home pay on a house payment. You have \$15,300 saved for a down payment. With your good credit and the down payment you can get an APR from your bank of 3.37% compounded monthly.

a. What is the total cost of a house you could afford with a 15-year mortgage?

b. What is the most that you could afford with a traditional 30-year mortgage instead of a 15-year?

Solution

a. The first thing to do is to calculate the size of the monthly mortgage payment you are willing to spend. Since you have \$3220 per month in take-home pay, multiply this by 25% to find your maximum monthly payment.

$$\text{advised monthly payment} = \$3220 \cdot 0.25 = \$805$$

We know that $r = 0.0337$ and that because this is a 15-year mortgage, $n = 12$ and $t = 15$. Substituting these values in the formula, we have the following.

$$\text{maximum purchase price} = PMT \cdot \dfrac{\left[1 - \left(1 + \frac{r}{n}\right)^{-nt}\right]}{\left(\frac{r}{n}\right)}$$

$$= 805 \cdot \dfrac{\left[1 - \left(1 + \frac{0.0337}{12}\right)^{-12 \cdot 15}\right]}{\left(\frac{0.0337}{12}\right)}$$

Round your answer to the nearest cent, if necessary.

7. Given the chart below, solve the following problems.

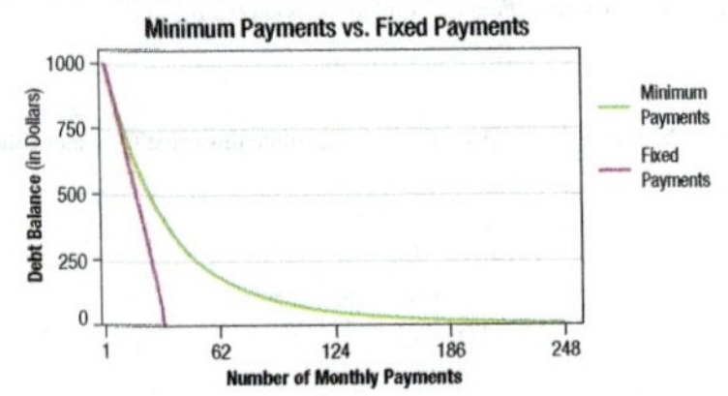

a. Estimate the total amount paid when a debt balance was paid using a fixed monthly payment of \$40.

b. Estimate the total amount paid when a debt balance was paid using the minimum monthly payment of \$18.

Categorized Exercises

Each section concludes with a selection of exercises designed to allow the student to practice skills and master concepts. Many levels of difficulty exist within each exercise set, providing instructors with flexibility in assigning exercises and allowing students to practice elementary skills or stretch themselves, as appropriate. Exercise sets are organized into categories such as Concept Check, Practice, Applications, Writing & Thinking, and Technology.

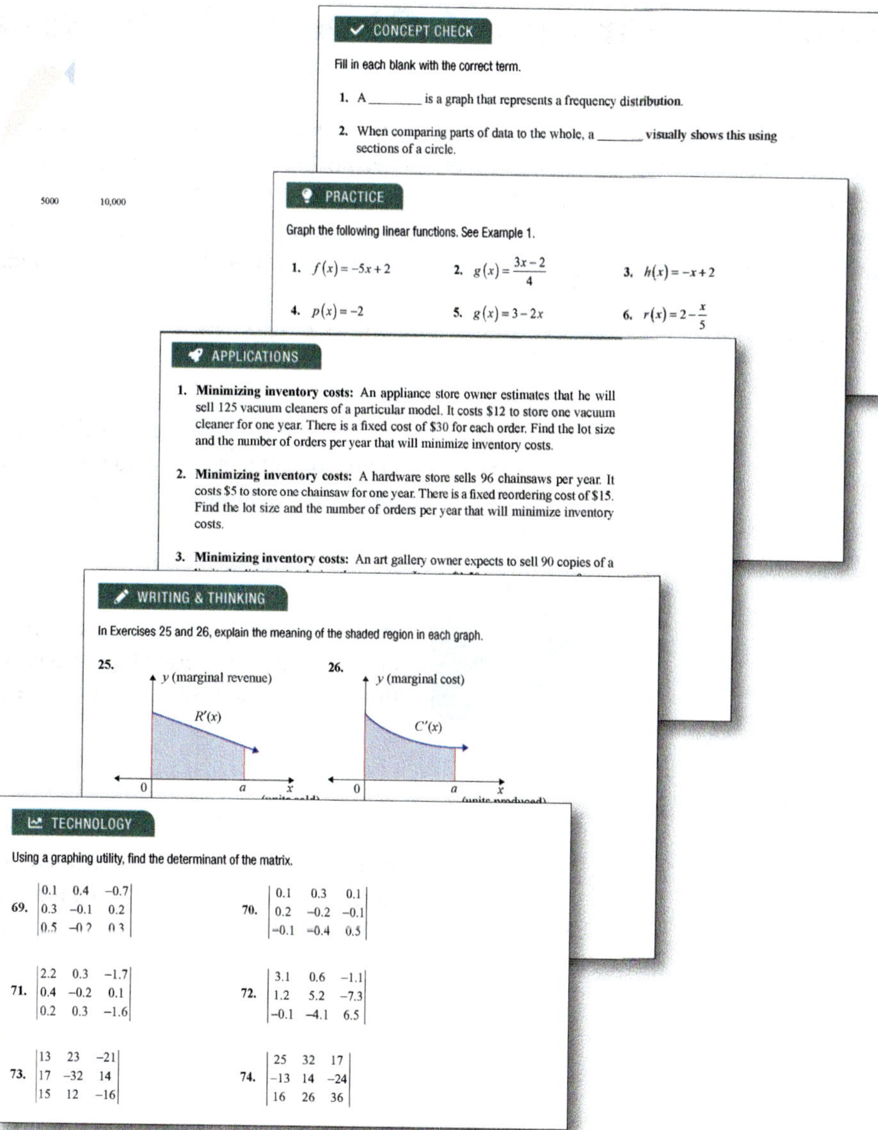

Answer Key

The Answer Key in the back of the book contains the answers for odd-numbered exercises. This allows students to check their work to ensure they are accurately applying the methods and skills that they have learned.

Chapter 3: Functions and Their Graphs

3.1 EXERCISES

1. a. 3 b. -11
   c. $2a - 5$ d. $2a - 6$
3. a. 9 b. 4
   c. a^2 d. $a^2 - 2a + 2$
5. a. 4 b. -8
   c. $a^3 + 4a^2 + 2a$
   d. $a^3 + a^2 - 3a + 2$
7. a. 35
   b. $4a^2 + 16a + 15$
   c. $4x^2 + 8xh + 4h^2 - 1$
   d. 12
9. a. 2 b. $\sqrt{a + 7}$
   c. $\sqrt{x + h + 5}$ d. $3 - \sqrt{6}$
11. a. -5 b. -2
    c. 0.25 d. 3
13. $3h$ 15. $h^2 + 2hx$
17. $2h^2 + 4hx$ 19. $h^2 + 2hx - h$
21. $3h - h^2 - 2hx$
23. $2h^2 + 4xh - 3h$

17. a
19. a. $C(x) = 135 + 0.5x$
    b. $385
21. a. $P(x) = 4.5x - 135$
    b. $2115
23. $P(x) = 100(4.5x - 135)$
25. 0.75 atm
27. 56.03 atm
29. a. $R(x) = 6.5x$
    b. $C(x) = 1.1x + 378$
    c. $P(x) = 5.4x - 378$
    d. $x = 70$ pies
31. a. $R(x) = 243x$
    b. $C(x) = 73x + 5780$
    c. $P(x) = 170x - 5780$
    d. $x = 34$ sets of clubs
33. a. $0.15/pen
    b. $C(x) = 0.15x + 260$
    c. $260
35. a. $R(x) = 31x - 0.5x^2$

3.3 EXERCISES

1.
3.
5.
7.

Hawkes Learning: A Clear Path to Mastery

Hawkes' software employs an adaptive, competency-based approach to knowledge mastery supported by a user-friendly interface. The student-centric platform promotes positive active learning by adapting to each student's needs through algorithmically generated questions based on an individual learner's pace, skill, and knowledge level. The real-time adaptive feedback addresses errors immediately, so that students learn from their mistakes when they make them. For each topic, the Hawkes Learning path to content mastery engages students through three simple modes: Learn, Practice, and Certify.

Competency-based learning made simple in three steps:

LEARN offers a multimedia-rich presentation of the lesson content. It includes instructional videos, interactive examples, and more.

PRACTICE engages students with algorithmically generated questions and intelligent tutoring in an ungraded, penalty-free environment.

CERTIFY requires students to demonstrate mastery of the material at a defined proficiency level without access to tutoring aids.

Support

If you have questions or comments, we can be contacted as follows:

24/7 Chat: chat.hawkeslearning.com

Phone: 1-800-426-9538

Email: support@hawkeslearning.com

Web: support.hawkeslearning.com

Real Numbers ($\mathbb{R}$)

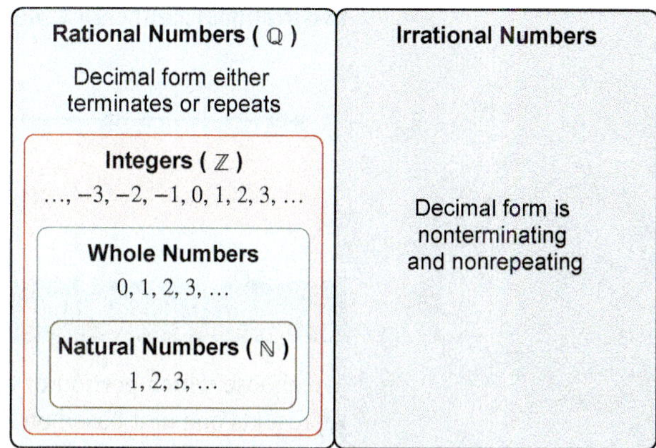

FIGURE 1: The Real Numbers

Example 1: Types of Real Numbers

Consider the set $S = \left\{ -15, -7.5, -\dfrac{7}{3}, 0, \sqrt{2}, 1.\overline{6}, \sqrt{9}, \pi, 10^{17} \right\}$.

a. The natural numbers in S are $\sqrt{9}$ and 10^{17}. Note that $\sqrt{9}$ is a natural number since $\sqrt{9} = 3$.

b. The whole numbers in S are 0, $\sqrt{9}$, and 10^{17}.

c. The integers in S are -15, 0, $\sqrt{9}$, and 10^{17}.

d. The rational numbers in S are -15, -7.5, $-\dfrac{7}{3}$, 0, $1.\overline{6}$, $\sqrt{9}$, and 10^{17}. The numbers -7.5 and $1.\overline{6}$ are both rational numbers since $-7.5 = \dfrac{-15}{2}$ and $1.\overline{6} = \dfrac{5}{3}$ (the bar over the last digit indicates that the digit repeats indefinitely). Note that any integer p is also a rational number, since it can be written as $\dfrac{p}{1}$.

e. The only irrational numbers in S are $\sqrt{2}$ and π. Although well known now, the irrationality of $\sqrt{2}$ came as a bit of a surprise to the early Greek mathematicians who discovered this fact. The irrationality of π was not proven until 1767.

The Real Number Line

Mathematicians often depict the set of real numbers as a horizontal line, with each point on the line representing a unique real number (so each real number is associated with a unique point on the line). The real number corresponding to a given point is called the **coordinate** of that point. Thus one (and only one) point on the real number line represents the number 0, and this point is called the **origin**. Points to the right of the origin represent positive real numbers, while points to the left of the origin represent negative real numbers.

Figure 2 is an illustration of the real number line with several points plotted. Note that two irrational numbers are plotted, though their locations on the line are approximations.

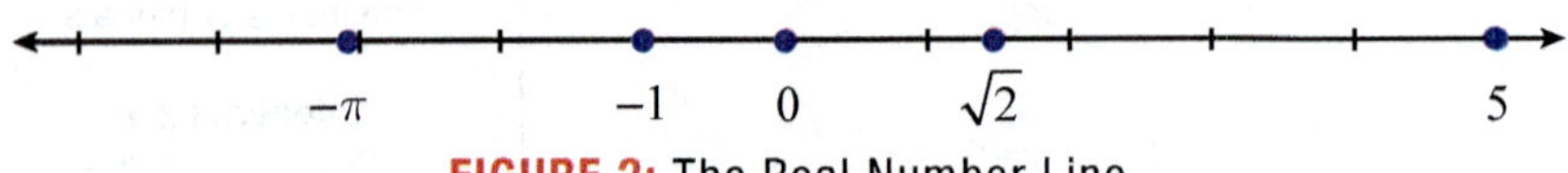

FIGURE 2: The Real Number Line

Example 2: Drawing the Real Number Line

We choose which portion of the real number line to show and the physical length that represents one unit based on the numbers that we wish to plot.

a. If we want to plot the numbers 101, 106, and 107, we might construct the graph below.

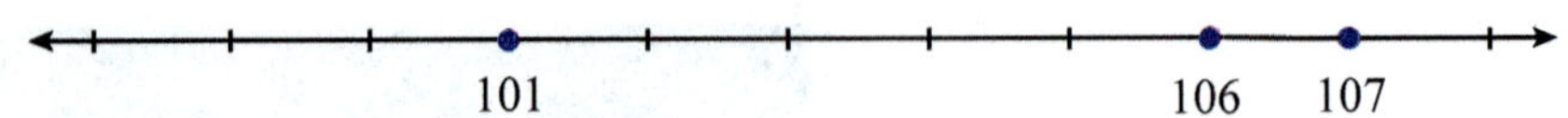

b. If we want to plot the numbers $-\dfrac{3}{4}$, $-\dfrac{1}{2}$, and $\dfrac{1}{4}$, we might make the unit interval longer.

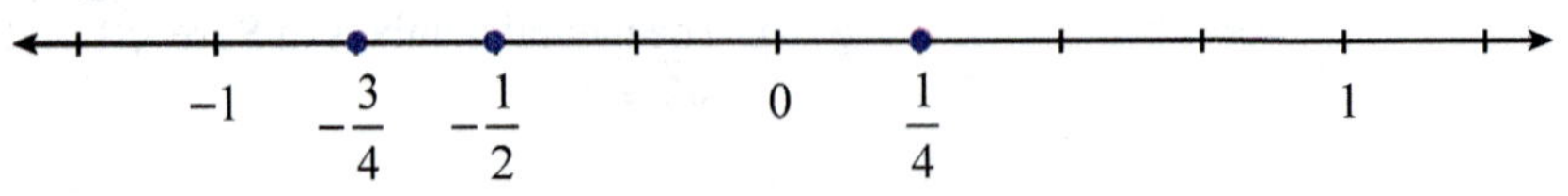

Order on the Real Number Line

Representing the real numbers as a line leads naturally to the idea of *ordering* the real numbers. We say that the real number a is **less than** the real number b (in symbols, $a < b$) if a lies to the left of b on the real number line. This is equivalent to saying that b is **greater than** a (in symbols, $b > a$). The following definition gives the meaning of these and two other symbols indicating order.

Inequality Symbols (Order)

Symbol	Reading	Meaning
$a < b$	"a is **less than** b"	a lies to the left of b on the number line
$a \le b$	"a is **less than or equal to** b"	a lies to the left of b or is equal to b
$b > a$	"b is **greater than** a"	b lies to the right of a on the number line
$b \ge a$	"b is **greater than or equal to** a"	b lies to the right of a or is equal to a

The two symbols $<$ and $>$ are called *strict* inequality signs, while the symbols $\le$ and $\ge$ are *nonstrict* inequality signs.

⚠ CAUTION

Remember that order is defined by the placement of real numbers on the number line, *not* by magnitude (its distance from zero). For instance, $-36 < 5$ because -36 lies to the left of 5 on the number line. Also, be aware that the negation of the statement $a \le b$ is the statement $a > b$. Furthermore, if $a \le b$ and $a \ge b$, then it must be the case that $a = b$.

Example 3: Working with Order

a. $5 \le 9$, since 5 lies to the left of 9.

b. $5 \le 5$, since 5 is equal to 5. Note that for every real number a, we have $a \le a$ and $a \ge a$.

c. $-7 > -163$, since -7 lies to the right of -163.

d. The statement "5 is greater than -2" can be written $5 > -2$.

e. The statement "a is less than or equal to $b + c$" can be written $a \le b + c$.

f. The statement "x is strictly less than y" can be written $x < y$.

g. The negation of the statement $a \le b$ is the statement $a > b$.

h. If $a \le b$ and $a \ge b$, then it must be the case that $a = b$.

Set-Builder Notation and Interval Notation

To describe the solutions to equations and inequalities, we need a precise, consistent way of expressing sets of real numbers. **Set-builder notation** is a general method of describing the elements that belong to a given set. **Interval notation** is a way of describing certain subsets of the real line.

Set-Builder Notation

The notation $\{x \mid x$ has property $P\}$ is used to describe a set of real numbers, all of which have the property P. This can be read "the set of all real numbers x having property P."

The symbol $\mid$ is also read as "such that," so the above notation can also be read "the set of all real numbers x, *such that* x has property P."

Example 4: Set-Builder Notation

a. $\{x \mid x$ is an even integer$\}$ is another way of describing the set $\{\dots, -4, -2, 0, 2, 4, \dots\}$. We could also describe this set as $\{2n \mid n$ is an integer$\}$, since every even integer is a multiple of 2.

b. $\{x \mid x$ is an integer such that $-3 \le x < 2\}$ describes the set $\{-3, -2, -1, 0, 1\}$.

The Empty Set

A set with no elements is called the **empty set** or the **null set**, and is denoted by the symbol $\emptyset$.

The empty set can arise from a set defined using set-builder notation; for example, the set $\{y \mid y > 1 \text{ and } y \leq -4\}$ is equivalent to the empty set, since no real number y satisfies the stated property.

Sets that consist of all real numbers bounded by two endpoints, possibly including those endpoints, are called **intervals**. Intervals can also consist of a portion of the real line extending indefinitely in either direction from just one endpoint.

We can describe such sets with set-builder notation, but intervals occur frequently enough that special notation has been devised to define them succinctly.

Interval Notation

Interval Notation	Set-Builder Notation	Meaning
(a, b)	$\{x \mid a < x < b\}$	all real numbers strictly between a and b
$[a, b]$	$\{x \mid a \leq x \leq b\}$	all real numbers between a and b, including both a and b
$(a, b]$	$\{x \mid a < x \leq b\}$	all real numbers between a and b, including b but not a
$(-\infty, b)$	$\{x \mid x < b\}$	all real numbers less than b
$[a, \infty)$	$\{x \mid x \geq a\}$	all real numbers greater than or equal to a

Intervals of the form (a, b) are called **open** intervals, while those of the form $[a, b]$ are **closed** intervals. The interval $(a, b]$ is **half-open** (or **half-closed**). Of course, a half-open interval may be open at either endpoint, as long as it is closed at the other. The symbols $-\infty$ and ∞ indicate that the interval extends indefinitely in the left and the right directions, respectively. Note that $(-\infty, b)$ excludes the endpoint b, while $[a, \infty)$ includes the endpoint a.

> **⚠ CAUTION**
>
> The symbols $-\infty$ and ∞ are just that: symbols! They are not real numbers, so they cannot be solutions to a given equation. The fact that they are symbols, and not numbers, means that they can never be included in a set of real numbers. For this reason, a parenthesis always appears next to either $-\infty$ or ∞; a bracket should never appear next to either infinity symbol.

Example 5: Intervals of Real Numbers

a. The interval $(2, 8)$ represents the set $\{x \mid 2 < x < 8\}$. This interval is open at both endpoints, so neither 2 nor 8 is included in the set.

b. The interval $[-5, -1]$ is another way to write the set $\{x \mid -5 \leq x \leq -1\}$. This interval is closed at both endpoints, so both -5 and -1 are included in the set.

c. The interval $[-3, 10)$ stands for the set $\{x \mid -3 \leq x < 10\}$. This interval is closed at the left endpoint, -3, and open at the right endpoint, 10.

d. The interval $(4, \infty)$ stands for the set $\{x \mid x > 4\}$. Since the interval is open on the left endpoint, it is the set of numbers greater than (but not equal to) 4.

e. The interval $(-\infty, \infty)$ is just another way of describing the entire set of real numbers.

Absolute Value and Distance

In addition to order, the depiction of the set of real numbers as a line leads to the notion of *distance*. Physically we understand distance as a number, indicating how close two objects are to one another. The idea of *absolute value* gives us a way to define distance in a mathematical setting.

Absolute Value

The **absolute value** of a real number a, denoted as $|a|$, is defined by:

$$|a| = \begin{cases} a & \text{if } a \geq 0 \\ -a & \text{if } a < 0 \end{cases}$$

The absolute value of a number is also referred to as its **magnitude**; it is the nonnegative number corresponding to its distance from the origin. Note that 0 is the only real number whose absolute value is 0.

Distance on the Real Number Line

Given two real numbers a and b, the **distance** between them is defined to be $|a - b|$. In particular, the distance between a and 0 is $|a - 0|$ or just $|a|$.

The distance from a to b should be the same as the distance from b to a. The definition confirms this intuition; note that it does not state whether a or b is smaller. This is because $|a - b| = |b - a|$. No matter which number we call a, the distance is the same.

Example 6: Absolute Value

a. $|17 - 3| = |3 - 17| = 14$. 17 and 3 are 14 units apart.

b. $|-\pi| = |\pi| = \pi$. Both $-\pi$ and π are π units from 0.

c. $\dfrac{|7|}{7} = \dfrac{7}{7} = 1$

d. $\dfrac{|-7|}{-7} = \dfrac{7}{-7} = -1$

e. $-|-5| = -5$. Note that the negative sign outside the absolute value symbol is not affected by the absolute value. Compare this with the fact that $-(-5) = 5$.

f. $\left|\sqrt{7} - 2\right| = \sqrt{7} - 2$. Even without a calculator, we know $\sqrt{7}$ is larger than 2 (since $2 = \sqrt{4}$), so $\sqrt{7} - 2$ is positive and hence $\left|\sqrt{7} - 2\right| = \sqrt{7} - 2$.

g. $\left|\sqrt{7} - 19\right| = 19 - \sqrt{7}$. In contrast to the last example, we know $\sqrt{7} - 19$ is negative, so its absolute value is $-\left(\sqrt{7} - 19\right) = 19 - \sqrt{7}$.

The following properties can all be derived from the definition of absolute value.

Properties of Absolute Value

In these properties, a and b represent arbitrary real numbers.

1. $|a| \geq 0$ (The absolute value of a number is never negative.)

2. $|-a| = |a|$

3. $a \leq |a|$

4. $|ab| = |a||b|$

5. $\left|\dfrac{a}{b}\right| = \dfrac{|a|}{|b|}$, $b \neq 0$

6. $|a + b| \leq |a| + |b|$ (This is called the **triangle inequality**, as it is a reflection of the fact that one side of a triangle is never longer than the sum of the other two sides.)

Example 7: Using Absolute Value Properties

a. $|(-3)(5)| = |-15| = 15 = |-3|\,|5|$

b. $1 = |-3 + 4| \leq |-3| + |4| = 7$

c. $7 = |-3 - 4| \leq |-3| + |-4| = 7$

d. $\left|\dfrac{-3}{7}\right| = \dfrac{|-3|}{|7|} = \dfrac{3}{7}$

0.1 EXERCISES

PRACTICE

Which elements of the following sets are **a.** natural numbers, **b.** whole numbers, **c.** integers, **d.** rational numbers, **e.** irrational numbers, **f.** real numbers, **g.** undefined? See Example 1.

1. $\left\{19,\ -4.3,\ -\sqrt{3},\ \dfrac{15}{0},\ \dfrac{0}{15},\ 2^5,\ -33\right\}$

2. $\left\{5\sqrt{7},\ 4\pi,\ \sqrt{16},\ 3.\bar{3},\ -1,\ \dfrac{22}{7},\ |-8|\right\}$

3. $\left\{5.41, |-16|,\ \dfrac{12}{3},\ 0,\ \sqrt{4},\ 2.\overline{145},\ \dfrac{1}{4}\right\}$

4. $\left\{2\sqrt{25},\ -4,\ 0.125,\ |32|,\ 2.1563,\ 6,\ \sqrt[3]{8}\right\}$

Plot the real numbers in the following sets on a number line. Choose the unit length appropriately for each set. See Example 2.

5. $\{-4.5, -1, 2.5\}$

6. $\{5.1, 5.2, 5.8\}$

7. $\{-24, 2, 15\}$

8. $\left\{0, \dfrac{1}{2}, \dfrac{5}{6}\right\}$

Select all of the symbols from the set $\{<, \le, >, \ge\}$ that can be placed in the blank to make each statement true. See Example 3.

9. 12 ____ 14

10. -3.4 ____ -3.5

11. -102 ____ 9

12. 3 ____ 3

13. -50 ____ -45

14. $-\dfrac{1}{4}$ ____ $-\dfrac{1}{3}$

15. 0.0087 ____ -42.9

16. $\dfrac{2}{16}$ ____ 0.125

17. -7 ____ -9

18. -8 ____ 2

Write each statement as an inequality, using the appropriate inequality symbol. See Example 3.

19. "$2a + b$ is strictly greater than c"

20. "2 is less than or equal to x"

21. "9 is greater than or equal to 7"

22. "7 is less than or equal to 9"

23. "$x + 5$ is strictly less than 3"

24. "$2c$ is no more than $3d$"

25. "9 is no less than 8"

26. "$6 + x$ is greater than or equal to $4x$"

Describe each of the following sets using set-builder notation. There may be more than one correct way to do this. See Example 4.

27. $\{-6, -3, 0, 3, 6, 9\}$

28. $\{5, 6, 7, \ldots, 105\}$

29. $\{2, 3, 5, 7, 11, 13, 17, \ldots\}$

30. $\{1, 2, 4, 8, 16, 32, \ldots\}$

31. $\left\{\ldots, \dfrac{1}{3}, \dfrac{1}{5}, \dfrac{1}{7}, \dfrac{1}{9}, \ldots\right\}$

32. $\{0, 1, 2, 3, 4, 5, \ldots\}$

Write each set as an interval using interval notation. See Example 5.

33. $x < 15$

34. $-9 \le x \le 6$

35. $2.5 < x \le 3.7$

36. $\{x \mid -3 \le x < 19\}$

37. $\{x \mid x < 4\}$

38. The positive real numbers

39. $\left\{x \mid -\dfrac{1}{2} < x < \dfrac{2}{5}\right\}$

40. $\{x \mid 1 \le x \le 2\}$

41. The nonnegative real numbers

Graph the following intervals.

42. $[5, 14)$

43. $[-9, -1]$

44. $(0, 2)$

45. $(-3, 18]$

46. $(-\infty, 7]$

47. $(25, \infty)$

Evaluate the absolute value expressions. See Examples 6 and 7.

48. $-\left|-11\right|$

49. $\left|3-7\right|$

50. $-\left|4-9\right|$

51. $\left|\sqrt{3}-\sqrt{5}\right|$

52. $\sqrt{\left|-4\right|}$

53. $-\left|-4-\left|-11\right|\right|$

54. $\left|-\sqrt{2}\right|$

55. $\dfrac{\left|-x\right|}{\left|x\right|}\ (x\neq 0)$

56. $\left|(-7)(-5)\right|$

57. $-\left|\sqrt{16}-5\right|$

58. $\left|2-\sqrt{7}\right|$

59. $-\left|-\sqrt{\left|-9\right|}-\left|-9\right|\right|$

Find the distance on the real number line between each pair of numbers given. See Example 6.

60. $a=8,\ b=3$

61. $a=6,\ b=14$

62. $a=5,\ b=5$

63. $a=4,\ b=-2$

64. $a=-7,\ b=7$

65. $a=-12,\ b=-1$

🚀 APPLICATIONS

66. Jess, Stan, Nina, and Michele are in a marathon. Twenty-five minutes after beginning, Jess has run 3.4 miles, Stan has run 4 miles, Nina has run 2.25 miles, and Michele has walked 1.6 miles. Using 0 as the beginning point, plot each competitor's location on a real number line using an appropriate interval.

67. Freddie, Sarah, Elizabeth, JR, and Aubrey are trying to line up by height for a photo shoot. JR is the tallest and Elizabeth is the shortest. Freddie is taller than Sarah, and Sarah is taller than Aubrey. Express their line-up using appropriate inequality symbols.

68. Sue boards an eastbound train in Center Station at the same time Joy boards a westbound train in Center Station. After riding the Straight Line for 20 minutes, Sue's train has traveled 13 miles east, while Joy's train (also on the Straight Line) has traveled 7 miles west. Find the distance between the two trains at this time. (Assume the Straight Line is true to its name and that the tracks lie literally along a straight line.)

69. The admission prices at the local zoo are as follows.

Admission Prices	
Children under 2	free
Children under 12	$3
Adults	$7
Seniors (65 and up)	$5

Express the age range for each of these prices in set-builder notation and interval notation.

70. A particular fudge recipe calls for at least 3 but no more than 4 cups of sugar and at least $\dfrac{1}{2}$ but no more than $\dfrac{2}{3}$ of a cup of walnuts. Express the amount of sugar and nuts needed in both set-builder and interval notation.

✏ WRITING & THINKING

71. Can a natural number be irrational? Explain.

72. Are all whole numbers also integers? Are all integers also whole numbers? Explain your answers.

73. In your own words, define absolute value.

74. Write a short paragraph explaining the similarities and differences between $>$ and $\geq$.

📈 TECHNOLOGY

Select all of the symbols from the set $\{<, \leq, >, \geq\}$ that can be placed in the blank to make each statement true. Use a graphing utility to check your answers.

75. -2.9 _____ -3.1

76. 2.1 _____ -5.5

77. 100 _____ -4

78. 0.001 _____ -99.8

79. $\dfrac{1}{3}$ _____ $\dfrac{1}{4}$

80. $-\dfrac{1}{5}$ _____ $-\dfrac{3}{4}$

0.2 THE ARITHMETIC OF ALGEBRAIC EXPRESSIONS

■ TOPICS

- Components and Terminology of Algebraic Expressions
- The Field Properties and Their Use in Algebra
- Order of Mathematical Operations
- Basic Set Operations and Venn Diagrams

Components and Terminology of Algebraic Expressions

Algebraic expressions are made up of constants and variables, combined by the operations of addition, subtraction, multiplication, division, exponentiation and the taking of roots. **Constants** like 6 and −3, are fixed numbers, while **variables** like x and y are usually letters that represent unspecified numbers. To **evaluate** a given expression means to replace the variables (if there are any) with specific numbers, perform the indicated mathematical operations and simplify the result.

The **terms** of an algebraic expression are those parts joined by addition, while the **factors** of a term are the individual parts of the term that are joined by multiplication. In this context, addition also covers subtraction (as $a - b$ can be thought of as $a + (-b)$) and multiplication covers division (as $\dfrac{a}{b}$ can be thought of as $a \cdot \dfrac{1}{b}$). The **coefficient** of a term is the constant factor of the term, while the remaining part of the term is the **variable factor**.

Example 1: Terminology of Algebraic Expressions

Consider the algebraic expression $-17x\left(x^2 + 4y\right) + 5\sqrt{x} - 13$.

a. This expression contains three terms: $-17x\,(x^2 + 4y)$, $5\sqrt{x}$, and -13. The terms are combined by addition and subtraction to form the whole expression.

b. The factors of the term $-17x(x^2 + 4y)$ are -17, x, and $(x^2 + 4y)$. The factors are combined by multiplication to form the whole term. The coefficient of $-17x(x^2 + 4y)$ is -17, and the variable part is $x(x^2 + 4y)$.

c. The factor $(x^2 + 4y)$ itself consists of the two terms x^2 and $4y$, but these two terms are not terms of the original algebraic expression $-17x\left(x^2 + 4y\right) + 5\sqrt{x} - 13$.

Example 2: Evaluating Algebraic Expressions

Evaluate the following algebraic expressions.

a. $5x^3 - 16$ for $x = 4$

b. $-3x^2 - 2(x + y)$ for $x = -2$ and $y = 3$

Solution

In both cases, we simply "plug in" the given values for each variable, then simplify.

a.
$$5(4)^3 - 16 = 5(64) - 16$$
$$= 320 - 16$$
$$= 304$$

b.
$$-3(-2)^2 - 2(-2 + 3) = -3(4) - 2(1)$$
$$= -12 - 2$$
$$= -14$$

The Field Properties and Their Use in Algebra

The following properties of addition and multiplication on the set of real numbers are probably familiar. You have likely used many of them in the past, though you may not have known the technical names of the properties. These properties and the few that follow them form the basis of the logical steps we use to solve equations and inequalities in algebra.

The set of real numbers forms what is known mathematically as a *field*, and consequently, the following properties are called *field properties*.

Field Properties

In these properties, a, b, and c represent arbitrary real numbers. The first five properties apply to addition and multiplication, while the last combines the two.

Name of Property	Additive Version	Multiplicative Version
Closure	$a + b$ is a real number	ab is a real number
Commutative	$a + b = b + a$	$ab = ba$
Associative	$a + (b + c) = (a + b) + c$	$a(bc) = (ab)c$
Identity	$a + 0 = 0 + a = a$	$a \cdot 1 = 1 \cdot a = a$
Inverse	$a + (-a) = 0$	$a \cdot \dfrac{1}{a} = 1$ (for $a \neq 0$)
Distributive		$a(b + c) = ab + ac$

Example 3: Applying the Field Properties

a. $3(2 + (-8)) = 3 \cdot (-6) = -18$ and $3 \cdot 2 + 3 \cdot (-8) = 6 - 24 = -18$. This demonstrates the distributive property: $3(2 + (-8)) = 3 \cdot 2 + 3 \cdot (-8)$.

b. $3 - 4 = 3 + (-4) = -4 + 3 = -1$. While subtraction is not commutative, we can rewrite any difference as a sum (with the sign changed on the second term) and then apply the commutative property.

c. $\dfrac{-x}{y} = (-x)\left(\dfrac{1}{y}\right) = \left(\dfrac{1}{y}\right)(-x)$. Division can be restated as multiplication by the reciprocal of the denominator, and multiplication is commutative.

d. $\left(x^2 + y\right)\left(\dfrac{1}{x^2 + y}\right) = 1$, provided that $x^2 + y \neq 0$. Any nonzero expression, when multiplied by its reciprocal, yields the multiplicative identity 1. The expression $\dfrac{1}{x^2 + y}$ is the multiplicative inverse of $x^2 + y$.

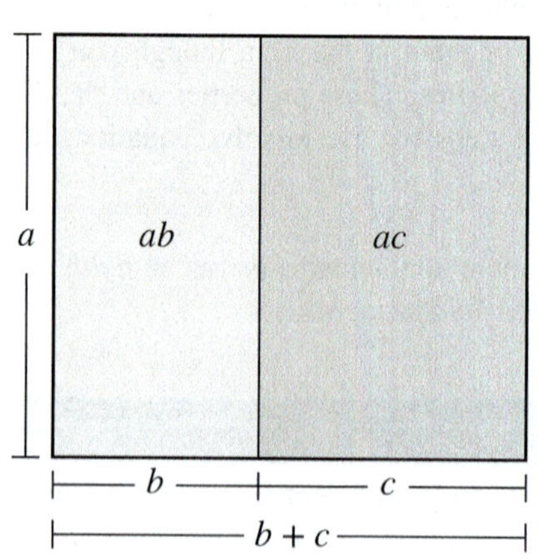

FIGURE 1

Example 4: Visualizing the Distributive Property

Consider the equation $a(b + c) = ab + ac$, which demonstrates the distributive property. We can represent the equation geometrically to understand why the distributive property works.

Figure 1 shows that the area of the shaded region represents the product $a(b + c)$, and that total area is the sum of two smaller areas that represent the products ab and ac.

While the field properties are of fundamental importance to algebra, they imply further properties that are often of more immediate use.

Cancellation Properties

Let A, B, and C be algebraic expressions.

Additive Cancellation: Adding the same quantity to both sides of an equation results in an equivalent equation.

$$\text{If } A = B, \text{ then } A + C = B + C.$$

Multiplicative Cancellation: Multiplying both sides of an equation by the same *nonzero* quantity results in an equivalent equation.

$$\text{If } A = B \text{ and } C \neq 0, \text{ then } A \cdot C = B \cdot C.$$

Zero-Factor Property

Let A and B represent algebraic expressions. If the product of A and B is 0, then at least one of A and B is itself 0.

$$AB = 0 \text{ implies that } A = 0 \text{ or } B = 0 \text{ (or both).}$$

Example 5: Properties of Real Numbers

a.

$$y + 12 = 18$$
$$y + 12 + (-12) = 18 + (-12)$$
$$y = 6$$

Using additive cancellation, we add -12 to both sides, then simplify.

This shows that the equation $y + 12 = 18$ is equivalent to the equation $y = 6$.

b.

$$-6x = 30$$
$$-6x\left(-\frac{1}{6}\right) = 30\left(-\frac{1}{6}\right)$$
$$x = -5$$

Using multiplicative cancellation, we multiply both sides by $-\frac{1}{6}$, then simplify.

Thus, the equation $-6x = 30$ is equivalent to $x = -5$. We can see how cancellation properties can help us *solve* equations for variables.

c. Multiplying both sides of the equation $x^2 - x = 2$ by 0 leads to the equation $0 = 0$, a true statement. However, these two equations are not equivalent!

While replacing x in the first equation by -1 or 2 leads to a true statement, any other value for x leads to a false statement. By contrast, the equation $0 = 0$, is true for all values of x, as there is no x in the equation to replace with a number. This example illustrates why we must multiply both sides of an equation by a nonzero quantity to apply multiplicative cancellation.

d. The equation $(x - y)(x + y) = 0$ means that either $x - y = 0$ or $x + y = 0$, by the Zero-Factor Property. Remember that the only way for a product of two (or more) factors to be 0 is for *at least* one of the factors to be 0 itself. For instance, in this example it might be that *both* $x - y = 0$ and $x + y = 0$. If $x = 0$ and $y = 0$, this is indeed the case.

Order of Mathematical Operations

Consider the two arithmetic expressions $4 - \dfrac{6}{2}$ and -3^2. Both expressions contain two operations: subtraction and division in the first one and multiplication and exponentiation in the second. Two reasonable people, lacking any indication of which operation is to be performed first, might very well proceed to simplify these expressions in two different ways and consequently arrive at different answers. It is important, therefore, that we decide the order in which the various mathematical operations are to be performed. The following list is the order which has evolved over time and which is assumed to be understood and applied in mathematics.

Order of Operations

Step 1: If the expression is a fraction, simplify the numerator and denominator individually, according to the guidelines in the following steps.

Step 2: Parentheses, braces and brackets are all used as grouping symbols. Simplify expressions within each set of grouping symbols, if any are present, working from the innermost outward.

Step 3: Simplify all powers (exponents) and roots.

Step 4: Perform all multiplications and divisions in the expression in the order they occur, working from left to right.

Step 5: Perform all additions and subtractions in the expression in the order they occur, working from left to right.

Example 6: Order of Operations

Simplify the following expressions using the correct order of operations.

a. $4 - \dfrac{6}{2}$

b. -3^2

c. $\dfrac{3 - \left(3\sqrt{4} - 2^2\right)\left(\dfrac{6}{-2}\right)}{3 + 2(-2)}$

Solution

a. $4 - \dfrac{6}{2} = 4 - 3$ 　　　　　　　　Perform the division first.

$\qquad\quad = 1$ 　　　　　　　　　　　Then subtract.

Now observe what happens if we do not follow the order of operations:

$4 - \dfrac{6}{2} = \dfrac{-2}{2}$ 　　　　　We subtract first, against order of operations. The result is now incorrect.

$\qquad\quad = -1$

b. $-3^2 = (-1)\left(3^2\right)$ 　　　We simplify the power before multiplying by -1.

$\qquad\ = (-1)(9)$

$\qquad\ = -9$

c. $\dfrac{3 - \left(3\sqrt{4} - 2^2\right)\left(\dfrac{6}{-2}\right)}{3 + 2(-2)} = \dfrac{3 - (3\cdot 2 - 4)(-3)}{3 - 4}$ 　　We simplify within each grouping symbol, following order of operations.

$\qquad\qquad\qquad\qquad\quad = \dfrac{3 - (2)(-3)}{-1}$ 　　Recall that fractions also act as a grouping symbol.

$\qquad\qquad\qquad\qquad\quad = \dfrac{3 + 6}{-1}$

$\qquad\qquad\qquad\qquad\quad = -9$

Basic Set Operations and Venn Diagrams

The sets that arise most frequently in algebra are sets of real numbers, and these sets are often the solutions of equations or inequalities. We will need to combine two or more such sets through the set operations of *union* and *intersection*. These operations are defined on sets in general, not just sets of real numbers, and can be illustrated by means of Venn diagrams.

A **Venn diagram** is a pictorial representation of a set or sets, and it indicates, through shading, the outcome of set operations such as union and intersection. In the following definitions, these two operations are first defined with set-builder notation and then demonstrated with a Venn diagram. The symbol $\in$ is read "is an element of".

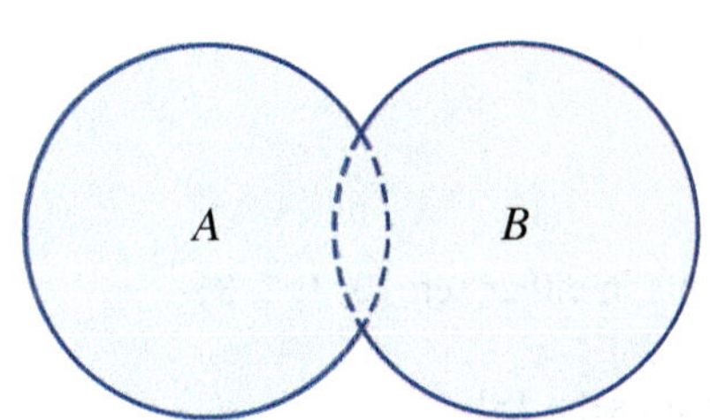

Union

In this definition, A and B denote two sets, and are represented in the Venn diagram by circles. The operation of union is depicted in the diagram by shading.

The **union** of A and B, denoted $A \cup B$, is the set $\{x \mid x \in A \text{ or } x \in B\}$. That is, an element x is in $A \cup B$ if it is in the set A, the set B, or both. Note that the union of A and B contains both individual sets.

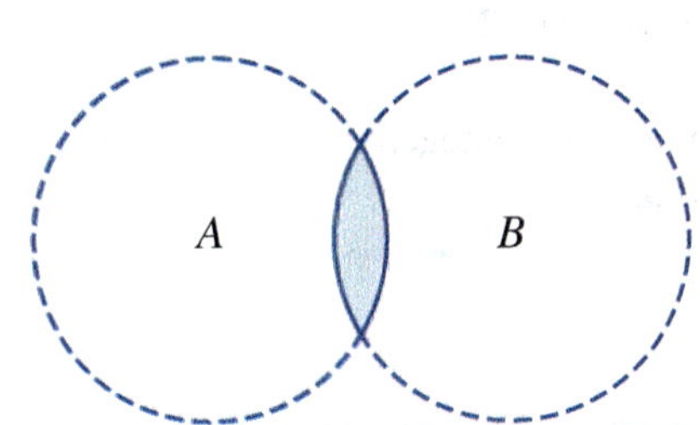

Intersection

In this definition, A and B denote two sets, and are represented in the Venn diagram by circles. The operation of intersection is depicted in the diagram by shading.

The **intersection** of A and B, denoted $A \cap B$, is the set $\{x \mid x \in A \text{ and } x \in B\}$. That is, an element x is in $A \cap B$ if it is in both A and B. Note that the intersection of A and B is contained in each individual set.

Example 7: Union and Intersection of Intervals

Simplify the following unions and intersections of intervals.

a. $(-2, 4] \cup [0, 9]$ **b.** $(-2, 4] \cap [0, 9]$

c. $[3, 4) \cap (4, 9)$ **d.** $(-\infty, 4] \cup (-1, \infty)$

Solutions

a. $(-2, 4] \cup [0, 9] = (-2, 9]$ Since these two intervals overlap, their union is described with a single interval.

b. $(-2, 4] \cap [0, 9] = [0, 4]$ This intersection of two intervals can also be described with a single interval.

    c. $[3, 4) \cap (4, 9) = \varnothing$

These two intervals have no elements in common, so their intersection is the empty set.

    d. $(-\infty, 4] \cup (-1, \infty) = (-\infty, \infty)$

The union of these two intervals is the entire set of real numbers.

Example 8: Union and Intersection

Simplify each of the following set expressions.

    a. $\{1, 2\} \cup \{0, 3\}$ **b.** $\{x, y, z\} \cap \{w, x\}$

    c. $\mathbb{Z} \cup \mathbb{R}$ **d.** $\mathbb{Z} \cap \mathbb{R}$

Solutions

a. The union of the two sets consists of all elements in either set: $\{0, 1, 2, 3\}$.

b. The intersection consists only of elements in both sets: $\{x\}$.

c. Since the integers are all also real numbers, the union of these two sets is simply the set of real numbers $\mathbb{R}$. We say that $\mathbb{Z}$ is *contained* in $\mathbb{R}$.

d. Similarly, since all integers are also real numbers, the integers are the elements contained in both sets. Thus, the intersection is $\mathbb{Z}$.

0.2 EXERCISES

PRACTICE

Identify the components of the algebraic expressions, as indicated. See Example 1.

1. Identify the terms in the expression $3x^2y^3 - 2\sqrt{x+y} + 7z$.

2. Identify the coefficients in the expression $3x^2y^3 - 2\sqrt{x+y} + 7z$.

3. Identify the factors in the term $-2\sqrt{x+y}$.

4. Identify the terms in the expression $x^2 + 8.5x - 14y^3$.

5. Identify the coefficients in the expression $x^2 + 8.5x - 14y^3$.

6. Identify the factors in the term $8.5x$.

7. Identify the terms in the expression $\dfrac{-5x}{2yz} - 8x^5y^3 + 6.9z$.

8. Identify the coefficients in the expression $\dfrac{-5x}{2yz} - 8x^5y^3 + 6.9z$.

9. Identify the factors in the term $\dfrac{-5x}{2yz}$.

Evaluate the following algebraic expressions for the given values of the variables. See Example 2.

10. $3x^3 + 5x - 2$ for $x = -3$

11. $-8(2x - y) + 4x^2$ for $x = 3$ and $y = 4$

12. $\sqrt{2x} + \dfrac{3x}{4}$ for $x = 8$

13. $3x^2 y^3 - 2\sqrt{x+y} + 7z$ for $x = -1$, $y = 2$, and $z = -2$

14. $-3\pi y + 8x + y^3$ for $x = 2$ and $y = -2$

15. $\dfrac{|x|\sqrt{2}}{x^3 y^2} - \dfrac{3y}{x}$ for $x = -3$ and $y = 2$

16. $y\sqrt{x^3 - 2} + \sqrt{x - 2y} - 3y$ for $x = 3$ and $y = -\dfrac{1}{2}$

17. $\left| -x^2 + 2xy - y^2 \right|$ for $x = -3$ and $y = -5$

18. $\dfrac{1}{32} x^2 y^3 + y\sqrt{x} - 7y$ for $x = 4$ and $y = 2$

19. $6x^2 + 3\pi y + y^2$ for $x = 3$ and $y = 2$

20. $|x - 9y| - (8z - 8)$ for $x = -3$, $y = 1$, and $z = 5$

21. $\dfrac{x^2 y^3}{8z} - \dfrac{|2xy|}{8z}$ for $x = 2$, $y = -1$, and $z = 3$

22. $5\sqrt{x+6} - 8y^2$ for $x = 10$ and $y = -2$

Identify the property that justifies each of the following statements. If one of the cancellation properties is being used to transform an equation, identify the quantity that is being added to both sides or the quantity by which both sides are being multiplied. See Examples 3 and 5.

23. $(x - y)(z^2) = (z^2)(x - y)$

24. $3 - 7 = -7 + 3$

25. $(3x + 2) + z = 3x + (2 + z)$

26. $4(y - 3) = 4y - 12$

27. $-3(4x^6 z) = (-3)(4)(x^6 z) = -12x^6 z$

28. $4 + (-3 + x) = (4 - 3) + x = 1 + x$

29. $-2(4 - x) = -8 + 2x$

30. $(x + y)\left(\dfrac{1}{x + y} \right) = 1$

31. $(-5 + 1)(7^7) = (7^7)(-5 + 1)$

32. $-5(-7x^8 y^4 z) = [(-5)(-7)](x^8 y^4 z)$

33. $25x^3 = 10y \Leftrightarrow 5x^3 = 2y$

34. $-14y = 7 \Leftrightarrow y = -\dfrac{1}{2}$

35. $14 - x = 2x \Leftrightarrow 14 = 3x$

36. $5 + 3x - y = 2x - y \Leftrightarrow 5 + x = 0$

37. $x^2 z = 0 \Rightarrow x^2 = 0$ or $z = 0$

38. $(a + b)(x) = 0 \Rightarrow a + b = 0$ or $x = 0$

39. $\dfrac{x}{6} + \dfrac{y}{3} - 2 = 0 \Leftrightarrow x + 2y - 12 = 0$

40. $(x - 3)(x + 2) = 0 \Rightarrow x - 3 = 0$ or $x + 2 = 0$

41. $21x^4 = 15y^4z \Leftrightarrow 7x^4 = 5y^4z$

42. $6x + \dfrac{25}{4}y^9 - z = \dfrac{1}{4}y^9 - z \Leftrightarrow 6x + 6y^9 = 0$

Evaluate each of the following expressions. Be sure to use the correct order of operations. See Example 6.

43. $2 + 3 - 4 \div 8 + (-1)^2$

44. $\dfrac{-2\left(13 - \sqrt{9} + 2\right)}{14 - 4 \div 2}$

45. $-3^2 - 2 \div 2$

46. $\left(-3^2 - 2\right) \div 2$

47. $\dfrac{\sqrt{\sqrt{81} + 4^2}}{10(4 - 7 \div 2)}$

48. $4\pi + 6^{\sqrt{5 - \frac{2}{2}}} - 3\pi\left[8 - 15 \div (2 + 3)\right]$

49. $4 - 10 \cdot (-1) \div 5 + (-8)^2$

50. $-3^2 + 2 \cdot \sqrt{2 + 1 \cdot 2} - 7\pi$

51. $1 \div 6 + 3^{\sqrt{2^2}} - (-4 \cdot 2)$

52. $\dfrac{8 - 9 \cdot 5 - 7}{-4\left(-9 - 5 \div (2 + 4)\right)}$

53. $-3 + 6 \cdot 1 \div 5 + (-3)^3$

54. $-5^2 + 4 \cdot \sqrt{2 + 7 \cdot 2} - 2\pi$

55. $9 \div 2 + 2^{\sqrt{2^4}} - (1 \cdot 2)$

56. $\dfrac{4 + 3 \cdot 8 - 6}{-5\left(3 - 8 \div (2 + 5)\right)}$

Use a calculator to evaluate each of the following expressions. Be sure to use the correct order of operations. Round your answers to two decimal places. See Example 6.

57. $(-3.28)^2 + 4 \cdot \sqrt{2 + 7 \cdot 3} - 2\pi$

58. $2.66 - 7 \cdot 4 \div 5 + (2 \div 3)^2$

59. $\dfrac{7.6 - 5.2 \cdot 9.8 - 8.1}{-3.22\left(11 - 6 \div (-1.45 + 6.32)\right)}$

60. $7 \div 4.6 + 2.4^{\sqrt{3}} - (1.23 \cdot 2)^4$

Translate each of the following directions into an algebraic expression.

61. Begin with 3. Add 7, and multiply the result by 3. Subtract 5. Take the square root, raise the result to the 3rd power, and then multiply by $-\dfrac{1}{5}$.

62. Begin with −6. Add 4, raise the result to the 3rd power, multiply by −2, and take the fourth root of the result.

63. Begin with x. Subtract 4, and take the third root of the result. Divide by 2, and square the result.

Simplify the following set expressions. See Examples 7 and 8.

64. $[-7, 7) \cup (2, 5)$

65. $(-5, 2] \cup (2, 4]$

66. $(-5, 2] \cap (2, 4]$

67. $[3, 5] \cap [2, 4]$

68. $(-\infty, 4] \cup (0, \infty)$

69. $(-\infty, \infty) \cap [-\pi, 21)$

70. $[2, \infty) \cap (-4, 7) \cap (-3, 2]$

71. $(3, 5] \cup [5, 9]$

72. $[-\pi, 2\pi) \cap [0, 4\pi]$

73. $\mathbb{Q} \cap \mathbb{Z}$

74. $\mathbb{N} \cup \mathbb{R}$

75. $\mathbb{N} \cup \mathbb{Z} \cap \mathbb{Q}$

76. $(-4.8, -3.5) \cap \mathbb{Z}$

77. At the beginning of the month, your checking account contains $128. For your birthday, your mother deposits $50 and your grandmother deposits $25. After you write three checks for $17, $23, and $62, you make a deposit of $41. At the end of the month, your bank removes half of the balance to put in your savings account and then charges you a $5 fee for doing so. How much do you have remaining in your checking account?

78. A particular liquid boils at 268 °F. Given the formula $C = \dfrac{5}{9}(F - 32)$ for converting temperatures from Celsius (C) to Fahrenheit (F), find the boiling point of this liquid in the Celsius scale. Round your answer to two decimal places.

79. Stephen received $75 as a gift from his aunt. With this money, he decided to start saving to buy the newest gaming console, which costs $398 after tax. After working two weeks at his part-time job, he got one check for $123 and a second check for $98. How much more does Stephen need to save to buy his gaming console?

80. Body mass index, abbreviated BMI, is one way doctors determine an adult's weight status. A BMI below 18.5 is considered underweight, the range 18.5–24.9 is normal, the range 25.0–29.9 is overweight, and a BMI above 30.0 indicates obesity. The formula used to determine BMI is $\text{BMI} = 703\left(\dfrac{\text{weight in pounds}}{\left(\text{height in inches}\right)^2}\right)$.

Derek weighs 180 lbs and is 73 inches tall. Use this formula to determine Derek's BMI and weight status. Round your answer to one decimal place.

81. The Du Bois Method provides a formula used to estimate your body's surface area in meters squared: $\text{BSA} = 0.007184 h^{0.725} w^{0.425}$, where h is height in centimeters and w is weight in kilograms. Assume Juan is 193 cm tall and weighs 88 kg. Use the Du Bois Method to estimate his body's surface area in square meters. Round your answer to two decimal places.

82. Samantha drops a tennis ball from the top of the mathematics building. If it takes the ball 3.42 seconds to hit the ground, use the formula $\text{distance} = \dfrac{1}{2}\left(\text{acceleration}\right)\left(\text{time}\right)^2$ to find the height of the building, which is equivalent to the distance the ball falls. Use the value of $32\,\text{ft/s}^2$ for the acceleration of a falling object. Round your answer to the nearest foot.

83. Choose a number. Multiply it by 3 and then add 4. Now multiply by 2 and subtract 8. Finally divide by 6. What do you notice about your final answer? Explain why you got this as a result.

84. Use your knowledge of the order of operations to check the following problem for accuracy. Explain any errors you find.

$$-8 \div 4 + 2^3 - (3 \cdot 2) = -8 \div 4 + 2^3 - (6)$$
$$= -8 \div 4 + 8 - 6$$
$$= -8 \div 4 + 2$$
$$= -8 \div 6$$
$$= \frac{-4}{3}$$

85. A mnemonic is a device used to recall particular information. For example, "*My Very Educated Mother Just Served Us Nachos*" is often used to recall the order of the planets in our solar system: *My* = Mercury, *Very* = Venus, *Educated* = Earth, and so on. Come up with your own mnemonic for remembering the order of operations.

86. After taking a poll in her town, Sally began grouping the citizens into various sets. One set contained all the citizens with brown hair and another set contained all the citizens with blue eyes. What do you know about the citizens who would be listed in the union of these two sets? What do you know about the citizens who would be listed in the intersection of these two sets?

87. In your own words, explain the difference between a union and an intersection of two sets.

Use a graphing utility to evaluate the following algebraic expressions.

88. $\sqrt{x^4 y - z} + \dfrac{x - y^3}{z^2}$ for $x = -3$, $y = 2$, and $z = -2$

89. $\dfrac{\left(x - pq^2\right)^3}{2q^3}$ for $x = -5$, $p = 2$, and $q = -3$

90. $\dfrac{\left| x^2 - y^3 \right| - 4x}{3y^5}$ for $x = 2$ and $y = 3$

91. $\sqrt{p^3 q - q^3} - \left| p + q^2 \right|$ for $p = -5$ and $q = 2$

0.3 INTEGER EXPONENTS

▌ TOPICS

- Natural Number Exponents
- Integer Exponents
- Properties of Exponents

Natural Number Exponents

As we progress, we will encounter a variety of algebraic expressions. Algebraic expressions consist of constants and variables combined by the basic operations of addition, subtraction, multiplication, and division, along with exponentiation and the taking of roots. In this section, we will explore the meaning of exponentiation and the properties of exponents.

We will begin with the most basic type of exponent: an exponent consisting of a natural number.

Natural Number Exponents

If a is any real number and if n is any natural number, then $a^n = \underbrace{a \cdot a \cdot \ldots \cdot a}_{n \text{ factors}}$.

That is, a^n is just a shorter, more precise way of denoting the product of n factors of a.

In the expression a^n, a is called the **base**, and n is the **exponent**. The process of multiplying n factors of a is called "raising a to the nth power," and the expression a^n may be referred to as "the nth power of a" or "a to the nth power." Note that a^1 is simply a.

Other phrases are also commonly used to denote the raising of something to a power, especially when the power is 2 or 3. For instance, a^2 is often referred to as "a squared" and a^3 is often referred to as "a cubed." These phrases have their basis in geometry, as the area of a square region with side length a is a^2 and the volume of a three-dimensional cube with side length a is a^3.

Example 1: Natural Number Exponents

a. $4^3 = 4 \cdot 4 \cdot 4 = 64$. Thus "four cubed is sixty-four."

b. $(-3)^2 = (-3)(-3) = 9$. Thus "negative three, squared, is nine."

c. $-3^4 = -(3 \cdot 3 \cdot 3 \cdot 3) = -81$. Note that, by order of operations, the exponent 4 applies only to the number 3. After raising 3 to the 4th power, the result is multiplied by -1.

d. $-(-2)^3 \cdot 5^2 = -((-2) \cdot (-2) \cdot (-2))(5 \cdot 5) = -(-8)(25) = 200$.

e. $x^3 \cdot x^4 = (x \cdot x \cdot x)(x \cdot x \cdot x \cdot x) = x^7$. Even though x is a variable, preventing us from writing the expression as a number, we can use the definition of natural number exponents to write the product in a simpler way.

f. $\dfrac{7^6}{7^4} = \dfrac{7 \cdot 7 \cdot \cancel{7} \cdot \cancel{7} \cdot \cancel{7} \cdot \cancel{7}}{\cancel{7} \cdot \cancel{7} \cdot \cancel{7} \cdot \cancel{7}} = 7 \cdot 7 = 49$. We can cancel four factors of 7 from the numerator and denominator of the original fraction, leaving us with $7^2 = 49$.

Examples 1e and 1f illustrate two basic properties of exponents that we will shortly state more generally. For the moment, however, we will use similar examples to guide our extension of the definition of exponents to include integer exponents.

Integer Exponents

The ultimate goal is to give meaning to the expression a^n for any real number n and to do so in such a way that the properties of exponents hold consistently. For example, analysis of Example 1e leads to the observation that if n and m are natural numbers, then

$$a^n \cdot a^m = \underbrace{a \cdot a \cdot \ldots \cdot a}_{n \text{ factors}} \cdot \underbrace{a \cdot a \cdot \ldots \cdot a}_{m \text{ factors}} = \underbrace{a \cdot \ldots \cdot a \cdot a \cdot \ldots \cdot a}_{n+m \text{ factors}} = a^{n+m}.$$

To extend the meaning of a^n to the case where $n = 0$, we might start by noting that the following statement should be true:

$$a^0 \cdot a^m = a^{0+m} = a^m$$

In order for this specific property to hold in the case when one exponent is 0, $a^0 \cdot a^m$ must be equal to a^m. This suggests the following:

0 as an Exponent

For any real number $a \neq 0$, we define $a^0 = 1$. 0^0 is **undefined**, just as division by 0 is undefined.

With this small extension of the meaning of exponents, let us continue. Consider the following powers of 3:

$$3^3 = 27$$
$$3^2 = 9$$
$$3^1 = 3$$
$$3^0 = 1$$
$$3^{-1} = ?$$
$$3^{-2} = ?$$
$$3^{-3} = ?$$

Notice the exponent is decreased by one at each step and the result is $\dfrac{1}{3}$ of the result from the previous line. In order to maintain the pattern that has begun to emerge, we are led to $3^{-1} = \dfrac{1}{3}$, $3^{-2} = \dfrac{1}{9}$, and $3^{-3} = \dfrac{1}{27}$. In general, negative integer exponents are defined as follows.

Negative Integer Exponents

For any real number $a \neq 0$ and for any natural number n, $a^{-n} = \dfrac{1}{a^n}$. (We don't allow a to be 0 simply to avoid the possibility of division by 0.) Since any negative integer is the negative of a natural number, this defines exponentiation by negative integers.

We now have a definition for a^n when n is any integer. Note that this definition is consistent; the properties of exponentiation do not depend on whether n is positive, negative, or zero. Later, we will see that this is true even when n is not an integer.

Example 2: Simplifying Exponents

a. $\dfrac{y^2}{y^7} = \dfrac{\cancel{x} \cdot \cancel{x}}{y \cdot y \cdot y \cdot y \cdot y \cdot \cancel{x} \cdot \cancel{x}} = \dfrac{1}{y \cdot y \cdot y \cdot y \cdot y} = \dfrac{1}{y^5} = y^{-5}$

b. $\dfrac{6x^2}{-3x^2} = \dfrac{6}{-3} = -2$

Note that the variable x cancels out entirely, if $x \neq 0$.

c. $5^0 \cdot 5^{-3} = 5^{0-3} = 5^{-3} = \dfrac{1}{5^3} = \dfrac{1}{125}$

Note that $5^0 = 1$, as does a^0 for any $a \neq 0$.

d. $\dfrac{1}{t^{-3}} = \dfrac{1}{\dfrac{1}{t^3}} = 1 \cdot \dfrac{t^3}{1} = t^3$

e. $(x^2 y)^3 = (x^2 y)(x^2 y)(x^2 y) = x^2 \cdot x^2 \cdot x^2 \cdot y \cdot y \cdot y = x^6 y^3$

Properties of Exponents

The following table lists the properties of exponents that are used frequently in algebra. Most of these properties have been illustrated already in Examples 1 and 2. All of them can be readily demonstrated by applying the definition of integer exponents.

Properties of Exponents

Throughout this table, a and b may be taken to represent constants, variables, or more complicated algebraic expressions. The letters n and m represent integers.

Property	Example
1. $a^n \cdot a^m = a^{n+m}$	$3^3 \cdot 3^{-1} = 3^{3+(-1)} = 3^2 = 9$
2. $\dfrac{a^n}{a^m} = a^{n-m}$	$\dfrac{7^9}{7^{10}} = 7^{9-10} = 7^{-1}$
3. $a^{-n} = \dfrac{1}{a^n}$	$5^{-2} = \dfrac{1}{5^2} = \dfrac{1}{25}$ and $x^3 = \dfrac{1}{x^{-3}}$
4. $\left(a^n\right)^m = a^{nm}$	$\left(2^3\right)^2 = 2^{3 \cdot 2} = 2^6 = 64$
5. $(ab)^n = a^n b^n$	$(7x)^3 = 7^3 x^3 = 343x^3$ and $\left(-2x^5\right)^2 = (-2)^2 \left(x^5\right)^2 = 4x^{10}$
6. $\left(\dfrac{a}{b}\right)^n = \dfrac{a^n}{b^n}$	$\left(\dfrac{3}{x}\right)^2 = \dfrac{3^2}{x^2} = \dfrac{9}{x^2}$ and $\left(\dfrac{1}{3z}\right)^2 = \dfrac{1^2}{(3z)^2} = \dfrac{1}{9z^2}$

Here we assume every expression is defined. That is, if an exponent is 0, then the base is nonzero, and if an expression appears in the denominator of a fraction, then that expression is nonzero. Remember that $a^0 = 1$ for every $a \neq 0$.

NOTE

There are often many ways to simplify an expression; the order in which you apply the properties of exponents will not change the result.

Example 3: Properties of Exponents

Simplify the following expressions by using the properties of exponents. Write the final answers with only positive exponents. (As in the table of properties, it is assumed that every expression is defined.)

a. $(17x^4 + 5x^2 + 2)^0$

b. $\dfrac{\left(x^2 y^3\right)^{-1} z^{-2}}{x^3 z^{-3}}$

c. $\dfrac{\left(-2x^3 y^{-1}\right)^{-3}}{\left(18x^{-3}\right)^0 (xy)^{-2}}$

d. $\left(7xz^{-2}\right)^2 \left(5x^2 y\right)^{-1}$

Solution

a. $(17x^4 + 5x^2 + 2)^0 = 1$

Any nonzero expression with an exponent of 0 is equal to 1.

b. $\dfrac{\left(x^2 y^3\right)^{-1} z^{-2}}{x^3 z^{-3}} = \dfrac{x^{-2} y^{-3} z^{-2}}{x^3 z^{-3}}$

Apply Property 4.

$= \dfrac{z^3}{x^3 x^2 y^3 z^2}$

Apply Property 3 several times to reach this point.

$= \dfrac{z}{x^5 y^3}$

Then apply Properties 1 and 2.

> ⚠ **CAUTION**
>
> Many errors can be made in applying the properties of exponents as a result of forgetting the exact form of the properties. The first column below contains examples of some common errors. The second column contains the corrected statements.
>
> | Incorrect | Correct |
> |---|---|
> | $x^2x^5 = x^{10}$ | $x^2x^5 = x^{2+5} = x^7$ |
> | $2^4 2^3 = 4^7$ | $2^4 2^3 = 2^{4+3} = 2^7$ |
> | $(3+4)^2 = 3^2 + 4^2$ | $(3+4)^2 = 7^2$ |
> | $(x^2 + 3y)^{-1} = \dfrac{1}{x^2} + \dfrac{1}{3y}$ | $(x^2 + 3y)^{-1} = \dfrac{1}{x^2 + 3y}$ |
> | $(3x)^2 = 3x^2$ | $(3x)^2 = 3^2 x^2 = 9x^2$ |
> | $\dfrac{x^5}{x^{-2}} = x^3$ | $\dfrac{x^5}{x^{-2}} = x^{5-(-2)} = x^7$ |

c.
$$\frac{\left(-2x^3 y^{-1}\right)^{-3}}{\left(18x^{-3}\right)^0 (xy)^{-2}} = \frac{(-2)^{-3} x^{-9} y^3}{x^{-2} y^{-2}}$$

$$= (-2)^{-3} x^{-9-(-2)} y^{3-(-2)}$$

$$= (-2)^{-3} x^{-7} y^5$$

$$= \frac{y^5}{-8x^7}$$

$$= -\frac{y^5}{8x^7}$$

Begin by applying Property 4 in the numerator, Property 5 in the denominator.

Then simplify using Property 2.

Unlike the previous example, Property 3 gets applied at the very end.

d.
$$\left(7xz^{-2}\right)^2 \left(5x^2 y\right)^{-1} = \frac{49x^2 z^{-4}}{5x^2 y}$$

$$= \frac{49}{5yz^4}$$

Note that the variable x no longer appears in the expression.

0.3 EXERCISES

💡 PRACTICE

Simplify each of the following expressions, writing your answer with only positive exponents. See Examples 1 and 2.

1. $(-2)^4$

2. -2^4

3. -3^2

4. $(-3)^2$

5. $3^2 \cdot 3^2$

6. $2^3 \cdot 3^2$

7. $4 \cdot 4^2$

8. $(-3)^3$

9. $\dfrac{8^2}{4^3}$

10. $2^2 \cdot 2^3$

11. $\dfrac{7^4}{7^5}$

12. $n^2 \cdot n^5$

13. $\dfrac{x^5}{x^2}$

14. $\dfrac{y^3 \cdot y^8}{y^2}$

15. $\dfrac{3^7}{3^4 s^{-10}}$

Use the properties of exponents to simplify each of the following expressions, writing your answer with only positive exponents. See Examples 1, 2, and 3.

16. $\dfrac{3t^{-2}}{t^3}$

17. $-2y^0$

18. $\dfrac{1}{7x^{-5}}$

19. $9^0 x^3 y^0$

20. $\dfrac{2n^3}{n^{-5}}$

21. $\dfrac{11^{21}}{11^{19} x^{-7}}$

22. $\dfrac{x^7 y^{-3} z^{12}}{x^{-1} z^9}$

23. $\dfrac{x^4 \left(-x^{-3}\right)}{-y^0}$

24. $\dfrac{s^3}{s^{-2}}$

25. $\dfrac{x^{-1}}{x}$

26. $x^{\left(y^0\right)} x^9$

27. $\dfrac{x^2 y^{-2}}{x^{-1} y^{-5}}$

28. $\dfrac{s^5 y^{-5} z^{-11}}{s^8 y^{-7}}$

29. $\dfrac{2^7 s^{-3}}{2^3}$

30. $\dfrac{3^{-5}}{\left(3^{-4} x^5 y^4\right)^2}$

31. $\dfrac{-9^0 \left(x^2 y^{-2}\right)^{-3}}{3x^{-4} y}$

32. $\left[\left(2x^{-1} z^3\right)^{-2}\right]^{-1}$

33. $\dfrac{\left(3yz^{-2}\right)^0}{3y^2 z}$

34. $\left(12a^2 - 3b^4\right)^0$

35. $\dfrac{3^{-1}}{\left(3^2 xy^2\right)^{-2}}$

36. $\left[9m^2 - \left(2n^2\right)^3\right]^{-1}$

37. $\left[\left(12x^{-6} y^4 z^3\right)^5\right]^0$

38. $\dfrac{x\left(x^{-2} y^3\right)^3}{\left(2x^4\right)^{-2} y}$

39. $\dfrac{\left(-3a\right)^{-2}\left(bc^{-2}\right)^{-3}}{a^5 c^4}$

40. $\left[\left(5m^4 n^{-2}\right)^{-1}\right]^{-2}$

41. $\left(9x^{-1} z\right)^2 \left(2xy^{-3}\right)^{-1}$

42. $\left(4^{-2} x^5 y^{-3} z^4\right)^{-2}$

43. $\left[\left(4a^2 b^{-5}\right)^{-1}\right]^{-3}$

44. $\left[\left(2^{-3} m^{-6} n^3\right)^3\right]^{-1}$

45. $\left[\left(3^{-1} x^{-1} y\right)\left(x^2 y\right)^{-1}\right]^{-3}$

46. $\left[\dfrac{100^0 \left(x^{-1} y^3\right)^{-1}}{x^2 y}\right]^{-3}$

47. $\left(5z^6 - \left(3x^3\right)^4\right)^{-1}$

48. $\left[\dfrac{y^6 \left(xy^2\right)^{-3}}{3x^{-3} z}\right]^{-2}$

✎ WRITING & THINKING

49. In your own words, explain why $a^0 = 1$.

Apply the definition of integer exponents to demonstrate the following properties.

50. $a^n \cdot a^m = a^{n+m}$

51. $\left(a^n\right)^m = a^{nm}$

52. $\left(ab\right)^n = a^n b^n$

0.4 RADICALS

■ TOPICS

- ■ Roots and Radical Notation
- ■ Simplifying Radical Expressions

Roots and Radical Notation

Taking the nth root of an expression (where n is a natural number) is the opposite operation of exponentiation by n. For example, the equation $4^3 = 64$ (which can be read as "four cubed is sixty-four") implies that the cube root of 64 is 4 (written $\sqrt[3]{64} = 4$). Similarly, the equation $2^4 = 16$ leads us to write $\sqrt[4]{16} = 2$ (read "the fourth root of sixteen is two"). At first glance, then, the definition of nth roots appears to be a simple matter, but there are some important difficulties to resolve. For instance, the statement $(-2)^4 = 16$ is also true, which might lead us to write $\sqrt[4]{16} = -2$. Can $\sqrt[4]{16}$ equal 2 and -2? Also, since any real number can be raised to any natural number power, we might be led to think that for any natural number n and any real number a, the nth root of a should be defined. What about $\sqrt[4]{-16}$? In order to evaluate $\sqrt[4]{-16}$ we seek a number whose fourth power is -16. Is there such a number? To answer the previous question, we need an understanding of complex numbers, but we can begin now by defining $\sqrt[n]{a}$ in the cases where $\sqrt[n]{a}$ is a real number.

> ### Radical Notation
>
> **Case 1: n is an even natural number.** If a is a nonnegative real number and n is an even natural number, $\sqrt[n]{a}$ is the nonnegative real number b with the property that $b^n = a$. That is, $\sqrt[n]{a} = b$ if and only if $a = b^n$. Note that $\left(\sqrt[n]{a}\right)^n = a$ and $\sqrt[n]{a^n} = a$.
>
> **Case 2: n is an odd natural number.** If a is any real number and n is an odd natural number, $\sqrt[n]{a}$ is the real number b (whose sign will be the same as the sign of a) with the property that $b^n = a$. Again, $\sqrt[n]{a} = b$ if and only if $a = b^n$, $\left(\sqrt[n]{a}\right)^n = a$, and $\sqrt[n]{a^n} = a$.

The expression $\sqrt[n]{a}$ gives the nth root of a in **radical notation**. The natural number n is called the **index**, a is the **radicand**, and $\sqrt{}$ is called a **radical sign**. By convention, $\sqrt[2]{}$ is usually simply written as $\sqrt{}$.

The important distinction between the two cases is that when n is even, $\sqrt[n]{a}$ is defined only when a is nonnegative, whereas if n is odd, $\sqrt[n]{a}$ is defined for all real numbers a. Mathematicians prevent any ambiguity in the meaning of $\sqrt[n]{a}$ when n is even and a is nonnegative by defining $\sqrt[n]{a}$ to be the *nonnegative number* whose n^{th} power is a. For example, $\sqrt{4}$ is equal to 2, not -2. Similarly, $\sqrt[4]{16}$ equals 2, not -2. Note that just as in the case of powers, alternative phrases are commonly used when the index is 2 or 3. We usually read $\sqrt{a}$ as the "square root of a" and $\sqrt[3]{a}$ as the "cube root of a."

Perfect Powers

A **perfect square** is an integer equal to the square of another integer. The square root of a perfect square is always an integer.

A **perfect cube** is an integer equal to the cube of another integer. The cube root of a perfect cube is always an integer.

Example 1: Radical Notation

a. $\sqrt[5]{-32} = -2$ because $(-2)^5 = -32$.

b. $\sqrt[4]{-16}$ is not a real number, as no real number raised to the fourth power is -16.

c. $-\sqrt[4]{16} = -2$. Note that the fourth root of 16 is a real number, which is then multiplied by -1.

d. $\sqrt{0} = 0$. In fact, $\sqrt[n]{0} = 0$ for any natural number n.

e. $\sqrt[n]{1} = 1$ for any natural number n. $\sqrt[n]{-1} = -1$ for any odd natural number n.

f. $\sqrt[3]{-\dfrac{27}{64}} = -\dfrac{3}{4}$ because $\left(-\dfrac{3}{4}\right)^3 = -\dfrac{27}{64}$.

g. $\sqrt[5]{-\pi^5} = -\pi$ because $(-\pi)^5 = (-1)^5\pi^5 = -\pi^5$.

h. $\sqrt[4]{(-3)^4} = \sqrt[4]{81} = 3$. In general, if n is an even natural number, $\sqrt[n]{a^n} = |a|$ for any real number a. Remember, though, that $\sqrt[n]{a^n} = a$ if n is an odd natural number.

Example 2: The Pythagorean Theorem

Given a right triangle with sides of length a and b, the Pythagorean theorem states that the length of the hypotenuse c is given by $c = \sqrt{a^2 + b^2}$. In this formula from the Pythagorean theorem, find the following.

a. The radicand

b. The index

c. The value of c if $a = 5$ and $b = 12$

d. The value of c if $a = 1$ and $b = 2$

Solution

a. The radicand is the quantity beneath the radical sign, $a^2 + b^2$.

b. Because no index is indicated, the index is 2.

c. $c = \sqrt{a^2 + b^2}$

$c = \sqrt{(5)^2 + (12)^2}$ Substitute.

$c = \sqrt{169}$ Since 169 is a perfect square, the solution is an integer.

$c = 13$

d. $c = \sqrt{a^2 + b^2}$

$c = \sqrt{(1)^2 + (2)^2}$ Substitute.

$c = \sqrt{5}$ Since 5 is not a perfect square, the solution is not an integer.

Simplifying Radical Expressions

When solving equations that contain radical expressions, it is often helpful to simplify the expressions first. The following definition establishes clear rules for simplifying radical expressions.

Simplified Form of Radical Expressions

A radical expression is in **simplified form** when the following conditions are met:

1. The radicand contains no factor with an exponent greater than or equal to the index of the radical.

2. The radicand contains no fractions.

3. The denominator, if there is one, contains no radical.

4. The greatest common factor of the index and any exponent occurring in the radicand is 1. That is, the index and any exponent in the radicand have no common factor other than 1.

Why do mathematicians adopt these particular conditions?

Condition 1 is a reflection of the fact that roots and powers undo one another. Specifically, $\sqrt[n]{a^n} = \begin{cases} |a| & \text{if } n \text{ is even} \\ a & \text{if } n \text{ is odd} \end{cases}$, so if the radicand contains an exponent greater than or equal to the index, a factor can be brought out from under the radical.

Conditions 2 and 3 both aim to remove radicals from the denominators of fractions, which is useful when solving equations that contain radicals. The process of removing radicals from the denominator is called **rationalizing the denominator**.

Condition 4 is again a reflection of the fact that roots and powers undo one another, as in the following example:

$$\sqrt[3]{a^6} = \sqrt[3]{a^2 a^2 a^2} = \sqrt[3]{(a^2)^3} = a^2$$

This simplification was possible because the exponent in the radicand and the index had a common factor of 3. This sort of simplification must be approached with caution, however, as we will see in a later example.

Before illustrating the process of simplifying radicals, we will review a few useful properties of radicals. These properties can be proved using nothing more than the definition of roots, but their validity will be more clear once we have discussed rational exponents later.

Properties of Radicals

Let a and b represent constants, variables, or more complicated algebraic expressions. The letters n and m represent natural numbers. Assume that all expressions are defined and are real numbers.

Property	Example
1. $\sqrt[n]{ab} = \sqrt[n]{a}\sqrt[n]{b}$	$\sqrt[3]{3x^6 y^2} = \sqrt[3]{3} \cdot \sqrt[3]{x^3} \cdot \sqrt[3]{x^3} \cdot \sqrt[3]{y^2}$
	$= \sqrt[3]{3} \cdot x \cdot x \cdot \sqrt[3]{y^2} = x^2 \sqrt[3]{3y^2}$
2. $\sqrt[n]{\dfrac{a}{b}} = \dfrac{\sqrt[n]{a}}{\sqrt[n]{b}}$	$\sqrt[4]{\dfrac{x^4}{16}} = \dfrac{\sqrt[4]{x^4}}{\sqrt[4]{16}} = \dfrac{|x|}{2}$
3. $\sqrt[m]{\sqrt[n]{a}} = \sqrt[mn]{a}$	$\sqrt[3]{\sqrt{64}} = \sqrt[3]{\sqrt[2]{64}} = \sqrt[6]{64} = 2$

Example 3: Simplifying Radical Expressions

Simplify the following radical expressions:

  a. $\sqrt[3]{-16x^8 y^4}$ **b.** $\sqrt{8z^6}$ **c.** $\sqrt[3]{\dfrac{72x^2}{y^3}}$

Begin by factoring the radicand, looking for perfect powers that match the index. This makes it easier to recognize what terms can be "pulled out" of the radical.

Solution

a. $\sqrt[3]{-16x^8 y^4} = \sqrt[3]{(-2)^3 \cdot 2 \cdot x^3 \cdot x^3 \cdot x^2 \cdot y^3 \cdot y}$

      $= -2x^2 y\sqrt[3]{2x^2 y}$

Factor out perfect cubes inside the radical.

b. $\sqrt{8z^6} = \sqrt{2^2 \cdot 2 \cdot \left(z^3\right)^2}$

      $= \left|2z^3\right|\sqrt{2}$

      $= 2\left|z^3\right|\sqrt{2}$

Factor out perfect squares.

Since the index, 2, is even, absolute value signs are needed around the factor of z^3.

c. $\sqrt[3]{\dfrac{72x^2}{y^3}} = \dfrac{\sqrt[3]{8 \cdot 9 \cdot x^2}}{\sqrt[3]{y^3}}$

      $= \dfrac{2\sqrt[3]{9x^2}}{y}$

Note that, in this case, simplifying rationalizes the denominator.

> **⚠ CAUTION**
>
> As with the properties of exponents, many mistakes arise from forgetting the properties of radicals. One common error is to rewrite $\sqrt{a+b}$ as $\sqrt{a}+\sqrt{b}$. These two expressions are not equal! To convince yourself of this, evaluate the two expressions with actual constants in place of a and b. Using $a = 9$ and $b = 16$, we see that $5 = \sqrt{9+16} \neq \sqrt{9}+\sqrt{16} = 7$.

Rationalizing denominators sometimes requires more effort than in Example 3c, while sometimes it is impossible! The following methods will, however, take care of two common cases.

Case 1: Denominator is a single term containing a root.

If the denominator is a single term containing a factor of $\sqrt[n]{a^m}$, we take advantage of the fact that $\sqrt[n]{a^m} \cdot \sqrt[n]{a^{n-m}} = \sqrt[n]{a^m \cdot a^{n-m}} = \sqrt[n]{a^n}$ and that this last expression is either a or $|a|$, depending on whether n is odd or even. Now, if we multiply the denominator by a factor of $\sqrt[n]{a^{n-m}}$ we must also multiply the numerator by the same factor. Thus in this case we multiply the fraction by $\dfrac{\sqrt[n]{a^{n-m}}}{\sqrt[n]{a^{n-m}}}$. For example, $\dfrac{1}{\sqrt{a}} = \dfrac{1}{\sqrt{a}} \cdot \dfrac{1}{1} = \dfrac{1}{\sqrt{a}} \cdot \dfrac{\sqrt{a}}{\sqrt{a}} = \dfrac{\sqrt{a}}{a}$.

Case 2: Denominator consists of two terms, one or both of which are square roots.

Let $A + B$ represent the denominator of the fraction under consideration, where at least one of A and B stands for a square root term. We will take advantage of the fact that $(A + B)(A - B) = A^2 - B^2$ and that the exponents of 2 will eliminate the square root (or roots) initially in the denominator. Just as in Case 1, we can't multiply the denominator by $A - B$ unless we multiply the numerator by this same factor. The method is thus to multiply the fraction by $\dfrac{A-B}{A-B}$. The factor $A - B$ is called the **conjugate radical expression** of $A + B$.

For example, $\dfrac{1}{\sqrt{a}+\sqrt{b}} = \dfrac{1}{\sqrt{a}+\sqrt{b}} \cdot \dfrac{1}{1} = \dfrac{1}{\sqrt{a}+\sqrt{b}} \cdot \dfrac{\sqrt{a}-\sqrt{b}}{\sqrt{a}-\sqrt{b}} = \dfrac{\sqrt{a}-\sqrt{b}}{a-b}$.

> **✎ NOTE**
>
> The first two examples follow Case 1. In general, it still helps to factor the radicands before further simplifying. The second pair of examples have two terms in the denominator, and thus follow Case 2. Begin by multiplying the numerator and denominator by the conjugate radical of the denominator.

Example 4: Rationalizing the Denominator

Simplify the following radical expressions:

a. $\dfrac{1}{\sqrt{x}}$ **b.** $\sqrt[5]{\dfrac{-4x^6}{8y^2}}$ **c.** $\dfrac{4}{\sqrt{7}+\sqrt{3}}$ **d.** $\dfrac{-\sqrt{5x}}{5-\sqrt{x}}$

Solution

a. $\dfrac{1}{\sqrt{x}} = \dfrac{1}{\sqrt{x}} \cdot \dfrac{\sqrt{x}}{\sqrt{x}}$ Multiply the numerator and denominator by $\sqrt{x}$.

$\qquad = \dfrac{\sqrt{x}}{x}$

b. $\sqrt[5]{\dfrac{-4x^6}{8y^2}} = \dfrac{\sqrt[5]{-4x \cdot x^5}}{\sqrt[5]{8y^2}}$

$= \dfrac{-x\sqrt[5]{4x}}{\sqrt[5]{2^3 y^2}} \cdot \dfrac{\sqrt[5]{2^2 y^3}}{\sqrt[5]{2^2 y^3}}$

$= \dfrac{-x\sqrt[5]{16xy^3}}{\sqrt[5]{2^5 y^5}}$

$= \dfrac{-x\sqrt[5]{16xy^3}}{2y}$

Since $2^3 y^2 \cdot 2^2 y^3 = 2^5 y^5$, a perfect fifth power, multiply the numerator and denominator by $\sqrt[5]{2^2 y^3}$.

c. $\dfrac{4}{\sqrt{7}+\sqrt{3}} = \left(\dfrac{4}{\sqrt{7}+\sqrt{3}}\right)\left(\dfrac{\sqrt{7}-\sqrt{3}}{\sqrt{7}-\sqrt{3}}\right)$

$= \dfrac{4\left(\sqrt{7}-\sqrt{3}\right)}{7-3}$

$= \dfrac{4\left(\sqrt{7}-\sqrt{3}\right)}{4}$

$= \sqrt{7}-\sqrt{3}$

Multiply the numerator and denominator by the conjugate of the denominator, then simplify.

d. $\dfrac{-\sqrt{5x}}{5-\sqrt{x}} = \left(\dfrac{-\sqrt{5x}}{5-\sqrt{x}}\right)\left(\dfrac{5+\sqrt{x}}{5+\sqrt{x}}\right)$

$= \dfrac{-5\sqrt{5x}-\sqrt{5x^2}}{25-x}$

$= \dfrac{-5\sqrt{5x}-x\sqrt{5}}{25-x}$

Multiply the numerator and denominator by the conjugate of the denominator. Simplify. Note that we can write $-x\sqrt{5}$ instead of $-|x|\sqrt{5}$ since the original expression is not real if x is negative.

There are occasions when rationalizing the numerator instead of the denominator is desirable. For instance, some problems in calculus (which the author encourages all college students to take!) are much easier to solve after rationalizing the numerator of a given fraction. This is accomplished by the same methods, as seen in the following example.

Example 5: Rationalizing the Numerator

Rationalize the numerator of the fraction $\dfrac{\sqrt{4x}-\sqrt{6y}}{2x-3y}$.

Solution

$$\frac{\sqrt{4x}-\sqrt{6y}}{2x-3y}=\left(\frac{\sqrt{4x}-\sqrt{6y}}{2x-3y}\right)\left(\frac{\sqrt{4x}+\sqrt{6y}}{\sqrt{4x}+\sqrt{6y}}\right)$$

Multiply both the numerator and denominator by the conjugate of the numerator.

$$=\frac{4x-6y}{(2x-3y)\left(\sqrt{4x}+\sqrt{6y}\right)}$$

$$=\frac{2(2x-3y)}{(2x-3y)\left(2\sqrt{x}+\sqrt{6y}\right)}$$

Note that we could have begun by simplifying the term $\sqrt{4x}$. The final answer is the same.

$$=\frac{2}{2\sqrt{x}+\sqrt{6y}}$$

0.4 EXERCISES

PRACTICE

Evaluate the following radical expressions. See Example 1.

1. $-\sqrt{9}$ **2.** $\sqrt[3]{-27}$ **3.** $\sqrt{-25}$ **4.** $\sqrt[6]{-64}$

5. $-\sqrt[6]{64}$ **6.** $-\sqrt{169}$ **7.** $\sqrt[3]{-125}$ **8.** $\sqrt{-49}$

9. $\sqrt[4]{-256}$ **10.** $-\sqrt[3]{-64}$ **11.** $\sqrt[3]{-\dfrac{27}{125}}$ **12.** $\sqrt{\dfrac{25}{121}}$

13. $\sqrt[3]{\dfrac{-8}{64}}$ **14.** $\sqrt{\dfrac{1}{4}}$ **15.** $-\sqrt[3]{-8}$

16. $\sqrt[4]{\sqrt{16}-\sqrt[3]{-27}+\sqrt{81}}$ **17.** $\sqrt{\dfrac{\sqrt[3]{-64}}{-\sqrt{144}-\sqrt{169}}}$

18. $\sqrt{\sqrt[3]{64}+\sqrt[4]{81}+\sqrt[5]{32}}$

Simplify the following radical expressions. See Example 3.

19. $\sqrt{9x^2}$ **20.** $\sqrt[3]{-8x^6y^9}$ **21.** $\sqrt[4]{\dfrac{x^8z^4}{16}}$ **22.** $\sqrt{2x^6y}$

23. $\sqrt[7]{x^{14}y^{49}z^{21}}$ **24.** $\sqrt{\dfrac{x^2}{4x^4y^6}}$ **25.** $\sqrt[3]{\dfrac{a^3b^{12}}{27c^6}}$ **26.** $\sqrt[3]{-125x^{12}y^9}$

27. $\sqrt[4]{\dfrac{x^{12}y^8}{16}}$ **28.** $\sqrt[3]{81m^4n^7}$ **29.** $\sqrt[5]{\dfrac{y^{30}z^{25}}{32x^{35}}}$ **30.** $\sqrt[5]{32x^7y^{10}}$

Simplify the following radicals by rationalizing the denominators. See Example 4.

31. $\sqrt[3]{\dfrac{4x^2}{3y^4}}$

32. $\dfrac{-\sqrt{3a^3}}{\sqrt{6a}}$

33. $\dfrac{3}{\sqrt{2}-\sqrt{5}}$

34. $\dfrac{10}{\sqrt{7}-\sqrt{2}}$

35. $\dfrac{3}{\sqrt{6}-\sqrt{3}}$

36. $\dfrac{5}{6-\sqrt{5}}$

37. $\dfrac{\sqrt{x}}{\sqrt{x}-\sqrt{2}}$

38. $\dfrac{x-y}{\sqrt{x}+\sqrt{y}}$

39. $\dfrac{\sqrt{x}+\sqrt{y}}{\sqrt{x}-\sqrt{y}}$

40. $\dfrac{1}{2-\sqrt{x}}$

41. $\dfrac{\sqrt{y}}{\sqrt{y}+2}$

42. $-\dfrac{\sqrt{6y^7}}{\sqrt{5y}}$

Rationalize the numerators of the following expressions. See Example 5.

43. $\dfrac{\sqrt{5}-3}{-4}$

44. $\dfrac{\sqrt{7}-6}{7}$

45. $\dfrac{3+\sqrt{y}}{6}$

46. $\dfrac{\sqrt{x}+\sqrt{y}}{\sqrt{x}}$

47. $\dfrac{\sqrt{13}+\sqrt{t}}{13-t}$

48. $\dfrac{2\sqrt{x}+\sqrt{y}}{\sqrt{x}-\sqrt{y}}$

49. $\dfrac{\sqrt{6}+\sqrt{y}}{\sqrt{6}-\sqrt{y}}$

50. $\dfrac{4\sqrt{xy}+y}{x-y}$

🚀 APPLICATIONS

51. The prism shown below is a right triangular cylinder, where the base is a right triangle. Find the surface area of the prism in terms of b, h, and l.

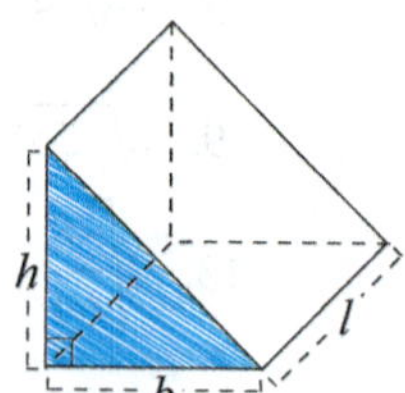

52. Terri has made a home for her pet guinea pig (Ralph) in the shape of a right triangular cylinder. Before she can put the new home in Ralph's cage, she must paint it with a nontoxic outer coat. If the front of the home has a base of 17.5 cm and a height of 15 cm and the length of the home is 25 cm, what is the surface area of Ralph's home, rounded to the nearest square centimeter? The small bottle of nontoxic coating will cover up to 1500 cm^2. Will the small bottle contain enough nontoxic coating to cover Ralph's home?

✏️ WRITING & THINKING

53. Explain, in your own words, why the square root of a negative number is not a real number.

0.5 RATIONAL EXPONENTS

■ TOPICS

- ■ Combining Radical Expressions
- ■ Rational Number Exponents

Combining Radical Expressions

Frequently, a sum of two or more radical expressions can be combined into one. This can be done if the radical expressions are **like radicals**, meaning that they have the same index and the same radicand. Often, you may have to simplify the radical expressions before determining if they are like or not.

Example 1: Combining Radical Expressions

Combine the radical expressions, if possible.

a. $-3\sqrt{8x^5} + \sqrt{18x}$ **b.** $\sqrt[3]{54x^3} + \sqrt{50x^2}$ **c.** $\sqrt{\dfrac{1}{12}} - \sqrt{\dfrac{25}{48}}$

Solution

a. $-3\sqrt{8x^5} + \sqrt{18x} = -3\sqrt{2^2 \cdot 2 \cdot x^4 \cdot x} + \sqrt{2 \cdot 3^2 \cdot x}$ Simplify each radical separately.

$\qquad\qquad\qquad\; = -6x^2\sqrt{2x} + 3\sqrt{2x}$ Now we can see that the radicals have the same index and radicand.

$\qquad\qquad\qquad\; = \left(-6x^2 + 3\right)\sqrt{2x}$

b. $\sqrt[3]{54x^3} + \sqrt{50x^2} = \sqrt[3]{2 \cdot 3^3 \cdot x^3} + \sqrt{2 \cdot 5^2 \cdot x^2}$ The radicands are the same, but the indices are not, so the terms cannot be combined.

$\qquad\qquad\qquad\; = 3x\sqrt[3]{2} + 5|x|\sqrt{2}$

c. $\sqrt{\dfrac{1}{12}} - \sqrt{\dfrac{25}{48}} = \dfrac{1}{\sqrt{2^2 \cdot 3}} - \dfrac{\sqrt{5^2}}{\sqrt{4^2 \cdot 3}}$ Simplify the radicals.

$\qquad\qquad\; = \dfrac{1}{2\sqrt{3}} \cdot \dfrac{\sqrt{3}}{\sqrt{3}} - \dfrac{5}{4\sqrt{3}} \cdot \dfrac{\sqrt{3}}{\sqrt{3}}$ Rationalize denominators.

$\qquad\qquad\; = \dfrac{2\sqrt{3}}{4 \cdot 3} - \dfrac{5 \cdot \sqrt{3}}{4 \cdot 3}$ Multiply the first term by $\dfrac{2}{2}$ to get a common denominator.

$\qquad\qquad\; = -\dfrac{3\sqrt{3}}{12} = -\dfrac{\sqrt{3}}{4}$

Rational Number Exponents

We can now return to defining exponentiation and give meaning to a^r when r is a rational number.

Rational Number Exponents

Meaning of $a^{\frac{1}{n}}$: If n is a natural number and if $\sqrt[n]{a}$ is a real number, then $a^{\frac{1}{n}} = \sqrt[n]{a}$.

Meaning of $a^{\frac{m}{n}}$: If m and n are natural numbers with $n \neq 0$, if m and n have no common factors greater than 1, and if $\sqrt[n]{a}$ is a real number, then $a^{\frac{m}{n}} = \sqrt[n]{a^m} = \left(\sqrt[n]{a}\right)^m$. Either $\sqrt[n]{a^m}$ or $\left(\sqrt[n]{a}\right)^m$ can be used to evaluate $a^{\frac{m}{n}}$, as they are equal. $a^{-\frac{m}{n}}$ is defined to be $\dfrac{1}{a^{\frac{m}{n}}}$.

In addition to giving meaning to rational exponentiation, this definition describes how to convert between radical notation and exponential notation. Often, one notation is much more convenient than the other, so converting between the two can be a crucial step in solving problems.

Although originally stated only for integer exponents, the properties of exponents listed earlier also hold for rational exponents (and for real exponents as well). Further, since we defined rational exponentiation using radical notation, we can now better understand the properties of radicals mentioned earlier. For instance,

$$\sqrt[m]{\sqrt[n]{a}} = \left(a^{\frac{1}{n}}\right)^{\frac{1}{m}} = a^{\frac{1}{n} \cdot \frac{1}{m}} = a^{\frac{1}{mn}} = \sqrt[mn]{a}.$$

The following examples illustrate radical notation, exponential notation, and the properties of each.

When simplifying, it is good practice to write the final answer in the same form as the original expression. However, it is often useful to convert between notations to make available a particular method of simplification. The first two examples exhibit this type of strategy.

Example 2: Simplifying Expressions

Simplify each of the following expressions, writing your answer using the same notation as the original expression.

a. $27^{-\frac{2}{3}}$

b. $\sqrt[9]{-8x^6}$

c. $\left(5x^2 + 3\right)^{\frac{8}{3}} \left(5x^2 + 3\right)^{-\frac{2}{3}}$

d. $\sqrt[5]{\sqrt[3]{x^2}}$

e. $\dfrac{5x - y}{\left(5x - y\right)^{\frac{-1}{3}}}$

Solution

a. $27^{-\frac{2}{3}} = \left(27^{\frac{1}{3}}\right)^{-2}$

$\qquad = 3^{-2}$

$\qquad = \dfrac{1}{3^2}$

$\qquad = \dfrac{1}{9}$

Writing $27^{-\frac{2}{3}} = \left(27^{-2}\right)^{\frac{1}{3}}$ is also a valid first step, but it leads to a messier calculation.

b. $\sqrt[9]{-8x^6} = -\sqrt[9]{2^3 x^6}$

$$= -2^{\frac{3}{9}} x^{\frac{6}{9}}$$

$$= -2^{\frac{1}{3}} x^{\frac{2}{3}}$$

$$= -\sqrt[3]{2x^2}$$

Rewrite the expression using rational exponents.

Note that the exponents in the radicand and the index now have no common factors other than 1.

c. $\left(5x^2 + 3\right)^{\frac{8}{3}} \left(5x^2 + 3\right)^{-\frac{2}{3}} = \left(5x^2 + 3\right)^{\left(\frac{8}{3}\right) + \left(-\frac{2}{3}\right)}$

$$= \left(5x^2 + 3\right)^{\frac{6}{3}}$$

$$= \left(5x^2 + 3\right)^2$$

The bases are the same, so we add the exponents.

d. $\sqrt[5]{\sqrt[3]{x^2}} = \sqrt[15]{x^2}$

Apply the property $\sqrt[m]{\sqrt[n]{a}} = \sqrt[mn]{a}$.

e. $\dfrac{5x - y}{\left(5x - y\right)^{\frac{-1}{3}}} = \left(5x - y\right)^{1 - \left(\frac{-1}{3}\right)}$

$$= \left(5x - y\right)^{\frac{4}{3}}$$

Apply the property $\dfrac{a^n}{a^m} = a^{n-m}$ to write the expression as a single term.

The following examples are a bit more complex.

Example 3: Simplifying Radical Expressions

a. Simplify the expression $\sqrt[4]{x^2}$.   **b.** Write $\sqrt[3]{2} \cdot \sqrt{3}$ as a single radical.

Solution

a. $\sqrt[4]{x^2} = \left(x^2\right)^{\frac{1}{4}}$

$$= |x|^{\frac{1}{2}}$$

$$= \sqrt{|x|}$$

Since the original expression is defined for all real numbers, but $\sqrt{x}$ is defined only for nonnegative real numbers, we need absolute value bars.

b. $\sqrt[3]{2} \cdot \sqrt{3} = 2^{\frac{1}{3}} \cdot 3^{\frac{1}{2}}$

$$= 2^{\frac{2}{6}} \cdot 3^{\frac{3}{6}}$$

$$= \left(2^2\right)^{\frac{1}{6}} \left(3^3\right)^{\frac{1}{6}}$$

$$= 4^{\frac{1}{6}} \cdot 27^{\frac{1}{6}}$$

$$= 108^{\frac{1}{6}}$$

$$= \sqrt[6]{108}$$

Rewrite the expression using rational exponents.

Write the two exponents with a common denominator.

By applying $a^{nm} = (a^n)^m$ we can give both factors the same exponent.

This allows us to combine the factors using the property $a^n b^n = (ab)^n$.

Heron's Formula

The area of a triangle with sides of length a, b, and c is

$$A = \sqrt{s(s-a)(s-b)(s-c)},$$

where $s = \dfrac{a+b+c}{2}$.

Example 4: Finding Areas with Heron's Formula

A regular hexagon is a six-sided plane figure whose sides are all the same length and whose interior angles are all the same. Use Heron's formula to derive a formula for the area of a regular hexagon with side-length d.

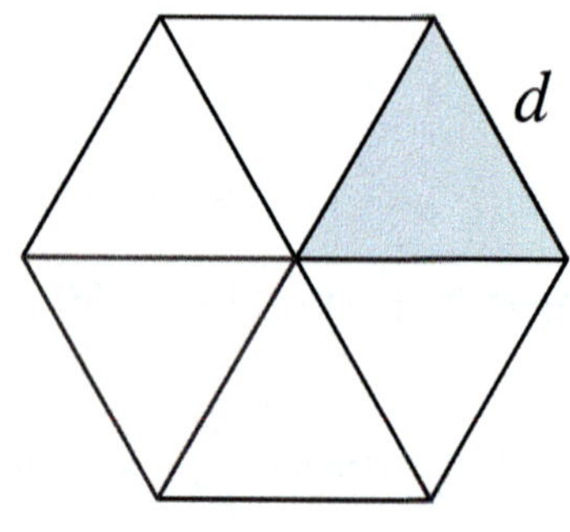

Solution

Begin by drawing three line segments joining each vertex to the vertex diagonally opposite it. The sum of the six angles meeting in the middle must be 360 degrees and all of the angles are equal, so each one must be a 60 degree angle. By symmetry, each of the six triangles making up the hexagon is isosceles, so each of the remaining two angles in any one triangle must measure 60 degrees as well. Thus each triangle is equilateral (equal sides and equal angles).

Heron's formula tells us that the area of an equilateral triangle of side-length d is $\sqrt{s(s-d)^3}$ where $s = \dfrac{3d}{2}$. Expressing this in terms of d alone, each triangle has area $\sqrt{\left(\dfrac{3d}{2}\right)\left(\dfrac{d}{2}\right)^3}$ or $\sqrt{\dfrac{3d^4}{16}}$. Simplifying this radical, we obtain $\dfrac{d^2\sqrt{3}}{4}$. Since the hexagon is made up of six of these triangles, the total area A of the hexagon is

$$A = 6\left(\frac{d^2\sqrt{3}}{4}\right) = \frac{3d^2\sqrt{3}}{2}.$$

0.5 EXERCISES

💡 PRACTICE

Combine the radical expressions, if possible. See Example 1.

1. $\sqrt[3]{-16x^4} + 5x\sqrt[3]{2x}$

2. $\sqrt{27xy^2} - 4\sqrt{3xy^2}$

3. $\sqrt{7x} - \sqrt[3]{7x}$

4. $|x|\sqrt{8xy^2z^3} - |yz|\sqrt{18x^3z}$

5. $-x^2\sqrt[3]{54x} + 3\sqrt[3]{2x^7}$

6. $\sqrt[5]{32x^{13}} + 3x\sqrt[5]{x^8}$

7. $\sqrt[3]{-16z^4} + 6z\sqrt[3]{2z}$

8. $\sqrt[3]{7y} - \sqrt[4]{7y}$

9. $-x^2\sqrt[3]{16x} + 2\sqrt[3]{2x^7}$

Simplify the following expressions, writing your answer using the same notation as the original expression. See Example 2.

10. $\sqrt[3]{\sqrt[4]{x^{36}}}$

11. $\left(3x^2 - 4\right)^{\frac{1}{3}}\left(3x^2 - 4\right)^{\frac{5}{3}}$

12. $32^{\frac{-3}{5}}$

13. $81^{\frac{3}{4}}$

14. $\dfrac{(x-z)^y}{(x-z)^4}$

15. $\sqrt[7]{n^9} \cdot \sqrt[7]{n^5}$

16. $(-8)^{\frac{2}{3}}$

17. $\dfrac{x^{\frac{1}{5}} y^{\frac{-2}{3}}}{x^{\frac{-3}{5}} y}$

18. $(1024)^{-\frac{2}{5}}$

19. $(625)^{-\frac{3}{4}}$

20. $\sqrt[8]{49a^2}$

21. $\sqrt[3]{\sqrt[5]{y^{25}}}$

22. $\dfrac{(a-b)^{-\frac{2}{3}}}{(a-b)^{-2}}$

23. $\left(ax^2 + by\right)^{\frac{3}{4}}\left(by + ax^2\right)^{-\frac{2}{3}}$

24. $\dfrac{\sqrt[3]{a^2}}{\sqrt[3]{a^5}}$

Convert the following expressions from radical notation to exponential notation, or vice versa. Simplify each expression in the process, if possible.

25. $\sqrt[4]{a^3} \cdot \sqrt[3]{a^9}$

26. $256^{-\frac{3}{4}}$

27. $\sqrt[12]{x^3}$

28. $\left(9y^2\right)^{\frac{3}{2}}\left(y^6\right)^{\frac{5}{3}}$

29. $\sqrt[6]{\dfrac{2}{72}}$

30. $\left(36n^4\right)^{\frac{5}{6}}$

Simplify the following expressions. See Example 3.

31. $\sqrt{5} \cdot \sqrt[4]{5}$

32. $\sqrt[4]{25}$

33. $\sqrt[16]{y^4}$

34. $\sqrt[4]{36}$

35. $\sqrt[3]{x^7} \cdot \sqrt[9]{x^6}$

36. $\sqrt[5]{y^{16}} \cdot \sqrt[25]{y^{20}}$

37. $\sqrt[4]{7} \cdot \sqrt[16]{7}$

38. $\sqrt{y^4} \cdot \sqrt[6]{y^3}$

39. A jeweler decides to construct a pendant for a necklace by simply attaching equilateral triangles to each edge of a regular hexagon. The edge length of one of the points of the resulting star is $d = 0.8$ cm. Find the formula for the area of the star in terms of d and then evaluate for $d = 0.8$ cm (rounding to three decimal places). Remember that the area of an equilateral triangle of side length d is $A = \dfrac{d^2 \sqrt{3}}{4}$. See Example 4.

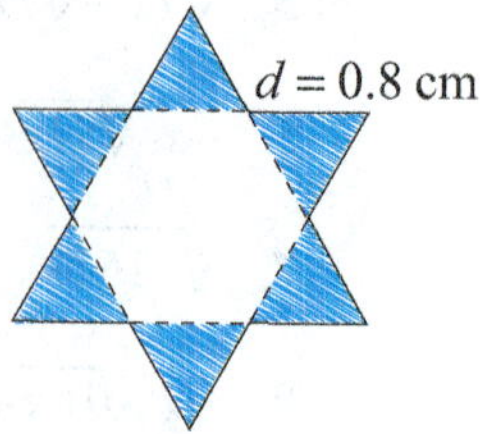

40. The pyramids in Egypt each consist of a square base and four triangular sides. For a class project, Karim constructs a model pyramid with equilateral triangles as sides. The side length is $d = 43$ cm. Find the total surface area of the pyramid (rounding to the nearest square centimeter). See Example 4.

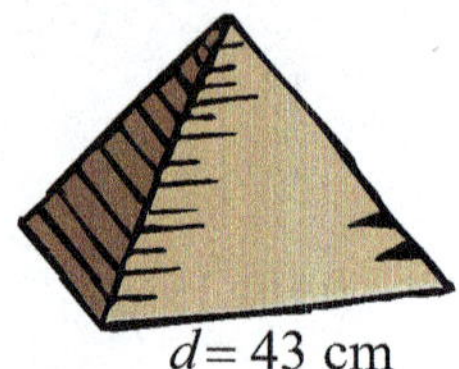

41. Explain, in your own words, why exponents and roots are evaluated at the same time in the order of operations.

Apply the definition of rational exponents to demonstrate the following properties.

42. $\sqrt[n]{ab} = \sqrt[n]{a} \cdot \sqrt[n]{b}$

43. $\sqrt[n]{\dfrac{a}{b}} = \dfrac{\sqrt[n]{a}}{\sqrt[n]{b}}$

44. $\sqrt[m]{\sqrt[n]{a}} = \sqrt[mn]{a}$

0.6 POLYNOMIALS AND FACTORING

■ TOPICS

- ■ Polynomials
- ■ Addition and Subtraction
- ■ Multiplication
- ■ Factoring

Polynomials

Constants are fixed numbers and **variables** are letters that represent unspecified numbers. A **monomial** is a single term with whole number exponents on its variables and no variable in the denominator. The general form of a **monomial in** x is kx^n, where n is a whole number and k is a constant. The constant k is called the **coefficient** and n is called the **degree**. Any monomial or algebraic sum of monomials is called a **polynomial**.

Classification of Polynomials		Example
Monomial	(polynomial with one term)	$5x$
Binomial	(polynomial with two terms)	$7x - 8$
Trinomial	(polynomial with three terms)	$x^3 + 9x - 4$
Polynomial	(any sum or difference of a set of monomials)	$7x - 8, x^3 + 9x - 4$

The **degree of a polynomial** is the largest of the degrees of all its terms. Although the emphasis in this review will be on polynomials in one variable, polynomials may contain more than one variable. The degree of a term in more than one variable is the sum of the exponents on its variables. Thus $-4x^2y$ is a third-degree term in x and y. The expression $-4x^2y + 5xy - 6y^2$ is a third-degree polynomial in x and y, and $5x^3 + 8x^2$ is a third-degree polynomial in x.

Addition and Subtraction

Like terms (or **similar terms**) are terms that contain the same variable factors with the same exponents. Constants are like terms. We combine like terms by using the distributive property and adding (or subtracting) the coefficients.

Example 1: Simplify Polynomials

Simplify the following polynomials by combining like terms.

a. $5x^2 + 7x^2$

Solution

$$5x^2 + 7x^2 = (5 + 7)x^2 = 12x^2$$

b. $3x + 4x - 13x + 1$

Solution

$$3x + 4x - 13x + 1 = (3 + 4 - 13)x + 1 = -6x + 1$$

c. $4x^3 - x^3 - 5 + 3 - 3x^3$

Solution

$$4x^3 - x^3 - 5 + 3 - 3x^3 = (4 - 1 - 3)x^3 - 5 + 3 = 0 \cdot x^3 - 2 = -2$$

The **sum** of two or more polynomials is found by combining like terms. The **difference** of two polynomials is found by adding the opposite of each term being subtracted.

Example 2: Addition with Polynomials

Find the sum $(2x^2 + 5x - 7) + (8x^2 - 6x + 1)$.

Solution

$$\begin{aligned}
(2x^2 + 5x - 7) + (8x^2 - 6x + 1) &= 2x^2 + 8x^2 + 5x - 6x - 7 + 1 \\
&= (2 + 8)x^2 + (5 - 6)x + (-7 + 1) \\
&= 10x^2 - x - 6
\end{aligned}$$

Example 3: Subtraction with Polynomials

Find the difference $(9x^3 - 4x^2 + 3x) - (8x^3 + 4x^2 - 5)$.

Solution

$$\begin{aligned}
(9x^3 - 4x^2 + 3x) - (8x^3 + 4x^2 - 5) &= 9x^3 - 4x^2 + 3x - 8x^3 - 4x^2 + 5 \\
&= 9x^3 - 8x^3 - 4x^2 - 4x^2 + 3x + 5 \\
&= x^3 - 8x^2 + 3x + 5
\end{aligned}$$

Multiplication

The product of a monomial with a polynomial of two or more terms can be found by using the distributive property and the rules of exponents.

Example 4: Multiplication with Polynomials

Use the distributive property to find the products.

a. $5x^2 (2x^3 + 3)$

Solution

$$5x^2(2x^3 + 3) = 5x^2 \cdot 2x^3 + 5x^2 \cdot 3$$
$$= 10x^5 + 15x^2$$

b. $3x (x^2 - 6x + 2)$

Solution

$$3x(x^2 - 6x + 2) = 3x \cdot x^2 + 3x \cdot (-6x) + 3x \cdot 2$$
$$= 3x^3 - 18x^2 + 6x$$

We can find the product of two binomials by using the **FOIL method**. For example,

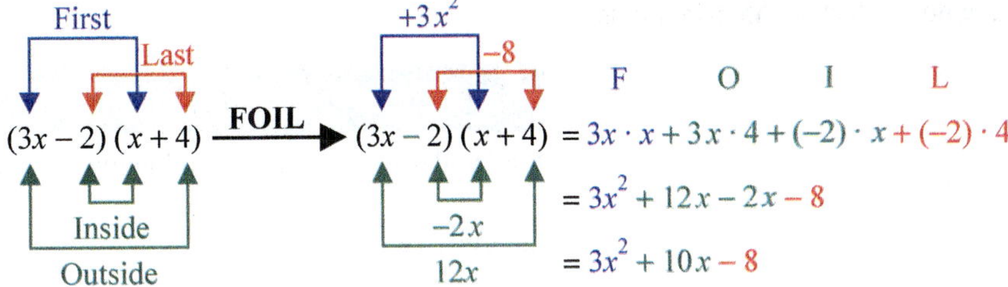

Example 5: Using the FOIL Method

Use the FOIL method to find the products.

a. $(x + 5)(x + 6)$

Solution

b. $(3x - 2)(7x + 5)$

Solution

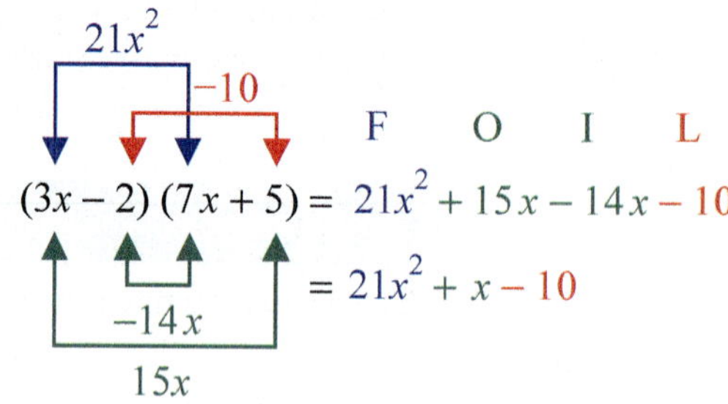

Certain products occur so frequently that they are stated as general formulas and are given names:

	Special Products	Names
I	$X^2 - A^2 = (X + A)(X - A)$	Difference of two squares
II	$(X + A)^2 = X^2 + 2AX + A^2$	Perfect square trinomial
III	$(X - A)^2 = X^2 - 2AX + A^2$	Perfect square trinomial
IV	$X^3 + A^3 = (X + A)(X^2 - AX + A^2)$	Sum of two cubes
V	$X^3 - A^3 = (X - A)(X^2 + AX + A^2)$	Difference of two cubes

⚠ CAUTION

Be careful to note that the two trinomials $X^2 - AX + A^2$ and $X^2 + AX + A^2$ in Formulas IV and V are not perfect square trinomials.

Example 6: Using Special Products

Find and name each of the following products.

a. $9x^2 - 16$

Solution

$$9x^2 - 16 = (3x)^2 - (4)^2$$
$$= (3x + 4)(3x - 4) \qquad \text{Difference of two squares}$$

b. $(x + 5)^2$

Solution

$$(x + 5)^2 = x^2 + 2 \cdot 5 \cdot x + 5^2$$
$$= x^2 + 10x + 25 \qquad \text{Perfect square trinomial}$$

c. $(x + 3)(x^2 - 3x + 9)$

Solution

$$(x+3)(x^2-3x+9)=x^3+3^3$$
$$=x^3+27 \qquad \text{Sum of two cubes}$$

This product can also be found by using the distributive property and combining like terms as follows:

$$(x+3)(x^2-3x+9)=(x+3)(x^2)+(x+3)(-3x)+(x+3)(9)$$
$$=x^3+3x^2-3x^2-9x+9x+27$$
$$=x^3+27$$

Factoring

Since factoring polynomials requires evidence of multiplication, our previous work with multiplication gives clues to factoring successfully. The following sequence of steps is recommended for factoring polynomials.

Steps for Factoring Polynomials

1. Look for any common factors (usually monomials or binomials).

2. See if the product fits any of the special forms just listed.

3. Try the reverse of the FOIL method if the product is a trinomial.

Example 7: Factoring Polynomials

Factor out any common factors in each of the following polynomials.

a. $3x^3 + 15x^2 + 21x$

Solution

$$3x^3+15x^2+21x=3x\cdot x^2+3x\cdot 5x+3x\cdot 7 \qquad \text{The largest common factor is } 3x.$$
$$=3x\left(x^2+5x+7\right)$$

b. $x^2(x^2 + 2) + 5x(x^2 + 2) + 3(x^2 + 2)$

Solution

$$x^2(x^2+2)+5x(x^2+2)+3(x^2+2)=(x^2+2)(x^2+5x+3) \qquad \text{The binomial } (x^2+2) \text{ is a common factor.}$$

Example 8: Using the FOIL Method

Use the FOIL method to factor the following polynomials.

a. $x^2 + 8x + 12$

Solution

For $x^2 + 8x + 12$, we have $F = x^2$ and $L = 12$. Thus, in the following diagram, we want to find the two missing numbers such that $L = 12$ and $O + I = 8x$.

$$x^2 + 8x + 12 = (x + \ ?)(x + \ ?)$$

Since $6 \cdot 2 = 12$ and $6 + 2 = 8$, we have $x^2 + 8x + 12 = (x + 6)(x + 2)$.

b. $2x^2 - 13x - 7$

Solution

For $2x^2 - 13x - 7$, we have $F = 2x^2$ and $L = -7$. In the following diagram we need factors of -7 such that $O + I = -13x$.

$$2x^2 - 13x - 7 = (2x + \ ?)(x + \ ?)$$

We want to fill in the question marks with $+1$ and -7 in that order. Doing so will result in $O = 2x(-7) = -14x$, $I = 1x$, and $O + I = -14x + 1x = -13x$. Thus

$$2x^2 - 13x - 7 = (2x + 1)(x - 7).$$

Example 9: Using Special Products

Factor each of the following polynomials by using the list of special products.

a. $4x^2 - 25$

Solution

$4x^2 - 25$ is the difference of two squares: $4x^2 - 25 = (2x + 5)(2x - 5)$.

b. $x^2 + 12x + 36$

Solution

$x^2 + 12x + 36$ is a perfect square trinomial with $A = \dfrac{1}{2}(12) = 6$ and $A^2 = 6^2 = 36$:

$$x^2 + 12x + 36 = (x + 6)^2.$$

c. $x^3 - 125$

Solution

$x^3 - 125$ is the difference of two cubes with $A^3 = 5^3$:

$$x^3 - 125 = (x - 5)(x^2 + 5x + 25).$$

0.6 EXERCISES

○ PRACTICE

In Exercises 1–14, perform the indicated operation and simplify the expressions.

1. $\left(x + 3x^2\right) + \left(5 - x^2\right)$ 　　　　**2.** $\left(x^2 + 2x - 4\right) + \left(x^2 - 4\right)$

3. $\left(8a^2 + 5a + 2\right) + \left(-3a^2 + 9a - 4\right)$ 　　**4.** $\left(3x^2 + 5x - 4\right) + \left(2x^2 - 2x + 4\right)$

5. $\left(2x^2 + 3x + 8\right) - \left(x^2 - 5x + 6\right)$ 　　**6.** $\left(4x^3 - 7x^2 + 3x\right) - \left(-2x^3 + 5x - 1\right)$

7. $\left(8x^2 + 9\right) - \left(4x^2 - 3x - 2\right)$ 　　**8.** $\left(y^3 + 4y^2 - 7\right) - \left(3y^3 + y^2 + 2y + 1\right)$

9. $\left(a^2 - 3ab + b^2\right) + \left(2a^2 - 5ab - b^2\right)$

10. $\left(7x^2 - 2xy + 3y^2\right) + \left(-3x^2 - 2xy + 5y^2\right)$

11. $\left(-3x^2 - 2xy + 5y^2\right) - \left(4x^2 + 3xy\right)$

12. $\left(5x^2 - 3xy + 7y^2\right) - \left(6x^2 - 9xy + 8y^2\right)$

13. $2x^2\left(3x^2 + 5x - 1\right)$ 　　　　**14.** $-4y^2\left(2y^2 + 5y - 4\right)$

Fill in the missing expressions in Exercises 15–17.

15. $\left(2x + 3y\right)^2 = 4x^2 + \underline{\qquad} + 9y^2$

16. $\left(9x - 5y\right)^3 = 729x^3 - 3\left(\underline{\qquad}\right)x^2y + 3\left(\underline{\qquad}\right)xy^2 - 125y^3$

17. $\left(3x^2 + 8y\right)^2 = 9\left(\underline{\qquad}\right) + 48\left(\underline{\qquad}\right) + 64\left(\underline{\qquad}\right)$

In Exercises 18–33, find the products.

18. $(3x-8)(x-5)$

19. $(7x+6)(2x-3)$

20. $(5x+11)(3x-4)$

21. $(3x-4)(4x-3)$

22. $(3x+1)^2$

23. $(4x-3)^2$

24. $(7x-4y)^2$

25. $(3x+2y)^2$

26. $(4x-5)(4x+5)$

27. $(6x+y)(6x-y)$

28. $3x^2(1+3x)$

29. $2x(x^2+3x-4)$

30. $(x+2)(x^2-2x+4)$

31. $(x+3)(x^2-3x+9)$

32. $(y-5)(y^2+5y+25)$

33. $(x+2y)(x^2-2xy+4y^2)$

In Exercises 34–39, simplify each expression.

34. $5a+2(a-3)-(3a+7)$

35. $11+\left[3x-2(1+5x)\right]$

36. $3y-\left[5-7(y+2)-6y\right]$

37. $10t-\left[8-5(3-2t)-7t\right]$

38. $x(x-5)+\left[6x-x(4-x)\right]$

39. $x(2x+1)-\left[5x-x(2x+3)\right]$

Fill in the missing expressions in Exercises 40 and 41.

40. $11x^2-99y^2=11\left(x-\underline{\quad}\right)\left(x+\underline{\quad}\right)$

41. $16x^3+54y^3=2\left(2x+\underline{\quad}\right)\left(\underline{\quad}x^2-\underline{\quad}xy+9y^2\right)$

In Exercises 42–60, factor each expression completely. (Each factor should have integer coefficients.)

42. $x^2+6x-27$

43. $s^2-5s-14$

44. $x^2+27x+50$

45. $x^2+11x-26$

46. $2x^2-98$

47. $4b^3-64b$

48. $9y^3-16y$

49. $27a^2-12$

50. $x^2+6xy+9y^2$

51. x^6-1

52. x^5-x^3

53. $25x^8-16$

54. $125y^6-27z^3$

55. $2t^3+16y^3$

56. $1600x^2+880xy+121y^2$

57. s^4-1

58. $3a^2+12ab+12b^2$

59. $100xy^2+200xy+100x$

60. $x^{21}-x^{19}$

61. If you were teaching Algebra I to ninth grade students, how would you explain the difference between a variable and a constant in an algebraic expression?

# Chapter 1

EQUATIONS AND INEQUALITIES IN ONE VARIABLE

1.1 LINEAR EQUATIONS IN ONE VARIABLE

■ TOPICS

- Linear Equations in One Variable: $ax + b = c$
- Conditional Equations, Identities, and Contradictions

Linear Equations in One Variable: $ax + b = c$

An **algebraic expression** is a combination of variables and numbers using any of the operations of addition, subtraction, multiplication, or division as well as exponents. An **equation** is a statement that two algebraic expressions are equal. That is, both expressions represent the same number. If an equation contains a variable, any number that gives a true statement when substituted for the variable is called a **solution** to the equation. For example, replacing x with 5 in the equation $3x + 4 = 10$ gives the false statement $3 \cdot 5 + 4 = 10$. Therefore, 5 *is not a solution* to the equation. However, replacing x with 2 gives the true statement $3 \cdot 2 + 4 = 10$. Therefore, 2 *is a solution* to the equation.

The solutions to an equation are said to form a **solution set**. The process of finding the solution set is called **solving the equation**. In this course we will study various types of equations that have more than one solution. However, *equations of the form $ax + b = c$ (linear equations in one variable) have exactly one solution.*

Linear Equations in x

If a, b, and c are constants and $a \neq 0$, then a **linear equation in x** is an equation that can be written in the form

$$ax + b = c.$$

Note: A linear equation in x is also called a **first-degree equation in x** because the variable x can be written with the exponent 1. That is, $x = x^1$.

To **solve** (or **find the solution set of**) a linear equation, we need the following two properties of equality.

Addition Property of Equality

If the same algebraic expression is added to both sides of an equation, the new equation has the same solutions as the original equation. Symbolically, if A, B, and C are algebraic expressions, then the equations

$$A = B$$
$$\text{and } A + C = B + C$$

have the same solutions. Equations with the same solutions are said to be **equivalent**.

Multiplication (or Division) Property of Equality

If both sides of an equation are multiplied by (or divided by) the same nonzero constant, the new equation has the same solutions as the original equation. Symbolically, if A and B are algebraic expressions and C is any nonzero constant, then the equations

$$A = B$$
$$\text{and } AC = BC \text{ where } C \neq 0$$
$$\text{and } \frac{A}{C} = \frac{B}{C} \text{ where } C \neq 0$$

have the same solutions and are equivalent.

The basic strategy in solving linear equations in one variable is to find equivalent equations until an equation is found with a single variable on one side and a constant on the other side. We use the Addition Property and the Multiplication Property of Equality in this process. Then a simplified equation such as $x = 5$ or $x = 7$, in which the variable has a coefficient of $+1$, gives the solution to the original equation.

Solving Linear Equations

1. Simplify each side of the equation by removing any grouping symbols and combining like terms. (In some cases, you may want to multiply both sides of the equation by a constant to clear fractional or decimal coefficients.)

2. Use the addition property of equality to add the opposites of constants or variable expressions so that variable expressions are on one side of the equation and constants on the other.

3. Use the multiplication property of equality to multiply both sides by the reciprocal of the coefficient of the variable (that is, divide both sides by the coefficient) so that the new coefficient is 1.

4. Check your answer by substituting it into the *original* equation.

Example 1: Solving a Linear Equation

Solve the linear equation $5 + 8x - 4 - 3x + 2 = -8 + 1$.

Solution

NOTE

To avoid errors and to help make your work easy to read and understand, try to align the $=$ signs in a vertical format so that each new equation is directly below the previous equation.

$5 + 8x - 4 - 3x + 2 = -8 + 1$	Write the equation.
$5x + 3 = -7$	Combine like terms.
$5x + 3 - 3 = -7 - 3$	Add -3 to both sides of the equation.
$5x = -10$	Simplify.
$\dfrac{5x}{5} = \dfrac{-10}{5}$	Divide both sides of the equation by 5.
$x = -2$	Simplify.

Check

$$5 + 8(-2) - 4 - 3(-2) + 2 \overset{?}{=} -8 + 1$$

$$5 - 16 - 4 + 6 + 2 \overset{?}{=} -7$$

$$-7 = -7$$

The solution is –2. We usually write just $x = -2$ to indicate the solution to the original equation. But, writing $\{-2\}$ as the solution set is also acceptable.

Many of the steps shown in Example 1 can be done mentally. Also, there is generally more than one correct way to proceed. In Example 1, you may choose to add +7 to both sides of the equation instead of adding –3 to both sides. In this case, the steps that follow will be different, too. However, the solution will be the same.

Example 2: Solving a Linear Equation

Solve $2(y - 7) = 4(y + 1) - 26$.

Solution

$$2(y-7) = 4(y+1) - 26$$

$2y - 14 = 4y + 4 - 26$	Use the distributive property.
$2y - 14 = 4y - 22$	Combine like terms.
$2y - 14 + 22 = 4y - 22 + 22$	Add 22 to both sides.
$2y + 8 = 4y$	Simplify.
$-2y + 2y + 8 = -2y + 4y$	Add $-2y$ to both sides. Here we put the variables on the right side to get a positive coefficient of y.
$8 = 2y$	Simplify.
$\dfrac{8}{2} = \dfrac{2y}{2}$	Divide both sides by 2.
$4 = y$	Simplify.

Check

$$2(y-7) = 4(y+1) - 26$$

$2(4-7) \overset{?}{=} 4(4+1) - 26$	Substitute $y = 4$.
$2(-3) \overset{?}{=} 4(5) - 26$	
$-6 \overset{?}{=} 20 - 26$	
$-6 = -6$	True statement.

Thus, the solution to $2(y - 7) = 4(y + 1) - 26$ is $y = 4$.

Example 3: Solving a Linear Equation

Solve $\dfrac{x-4}{10}+\dfrac{7}{2}=\dfrac{x+1}{5}$.

Solution

$$\dfrac{x-4}{10}+\dfrac{7}{2}=\dfrac{x+1}{5}$$

Write the equation.

$$10\left(\dfrac{x-4}{10}\right)+10\left(\dfrac{7}{2}\right)=10\left(\dfrac{x+1}{5}\right)$$

Multiply both sides by 10, the LCM of the denominators.

$$(x-4)+5(7)=2(x+1)$$

$$x-4+35=2x+2$$

Use the distributive property.

$$x+31=2x+2$$

Combine like terms.

$$x+31-x=2x+2-x$$

Add $-x$ to both sides.

$$31=x+2$$

Simplify.

$$31-2=x+2-2$$

Add -2 to both sides.

$$29=x$$

Simplify.

Check

$$\dfrac{29-4}{10}+\dfrac{7}{2}\overset{?}{=}\dfrac{29+1}{5}$$

$$\dfrac{25}{10}+\dfrac{35}{10}\overset{?}{=}\dfrac{30}{5}$$

$$\dfrac{60}{10}\overset{?}{=}\dfrac{30}{5}$$

$$6=6$$

Example 4: Solving a Linear Equation

Solve $16.53-18.2z=7.43$.

Solution

$$16.53-18.2z=7.43$$

$$16.53-18.2z+(-16.53)=7.43+(-16.53)$$

Add -16.53 to both sides.

$$-18.2z=-9.1$$

Simplify.

$$\dfrac{-18.2z}{-18.2}=\dfrac{-9.1}{-18.2}$$

Divide both sides by -18.2.

$$z=0.5$$

$$\text{or } z=\dfrac{1}{2}$$

NOTE

Checking can be quite time-consuming and need not be done for every problem. This is particularly important on exams. You should check only if you have time after the entire exam is completed.

Check

$$16.53 - 18.2z = 7.43$$

$$16.53 - 18.2(0.5) \overset{?}{=} 7.43 \qquad \text{Substitute } z = 0.5.$$

$$16.53 - 9.1 \overset{?}{=} 7.43 \qquad \text{Simplify.}$$

$$7.43 = 7.43 \qquad \text{True statement.}$$

Thus, the solution to $16.53 - 18.2z = 7.43$ is $z = 0.5$.

Conditional Equations, Identities, and Contradictions

Type of Equation	Number of Solutions
Conditional	Finite number of solutions
Identity	Infinite number of solutions
Contradiction	No solutions

TABLE 1

When solving equations, there are times that we are concerned with the number of solutions that an equation has. If an equation has a finite number of solutions (the number of solutions is a countable number), the equation is said to be a **conditional equation**. As stated earlier, every linear equation has exactly one solution. Thus *every linear equation is a conditional equation.* However, in some cases, simplifying an equation will lead to a statement that is always true, such as $0 = 0$. In these cases the original equation is called an **identity** and has an infinite number of solutions which can be written as all real numbers or $\mathbb{R}$. If the equation simplifies to a statement that is never true, such as $0 = 2$, then the original equation is called a **contradiction** and its solution set is the empty set, $\varnothing$. Table 1 summarizes these ideas.

Example 5: Solutions of Equations

Determine whether each of the following equations is a conditional equation, an identity, or a contradiction.

a. $0.7 + 0.9x = 16$

Solution

$$0.7 + 0.9x = 16$$

$$0.9x = 15.3 \qquad \text{Add } -0.7 \text{ to both sides.}$$

$$x = 17 \qquad \text{Solve for } x.$$

The equation has one solution and it is a conditional equation.

b. $3x + 2(x + 5) = 5(x - 1)$

Solution

$$3x + 2(x + 5) = 5(x - 1) \qquad \text{Use the distributive property.}$$

$$3x + 2x + 10 = 5x - 5 \qquad \text{Simplify.}$$

$$5x + 10 = 5x - 5 \qquad \text{Add } -5x \text{ to both sides.}$$

$$10 = -5$$

The last equation is never true. Therefore, the original equation is a contradiction and has no solution.

c. $5(x + 1) - 6x = -x + 5$

Solution

$$5(x+1) - 6x = -x + 5 \qquad \text{Use the distributive property.}$$
$$5x + 5 - 6x = -x + 5 \qquad \text{Simplify.}$$
$$-x + 5 = -x + 5 \qquad \text{Add } x \text{ to both sides.}$$
$$5 = 5$$

The last equation is always true. Therefore, the original equation is an identity and has an infinite number of solutions. Every real number is a solution.

1.1 EXERCISES

⚲ PRACTICE

Solve each equation. See Examples 1 through 4.

1. $3x + 11 = 2$

2. $3x + 10 = -5$

3. $5x - 4 = 6$

4. $4y - 8 = -12$

5. $6x + 10 = 22$

6. $3n + 7 = 19$

7. $9x - 5 = 13$

8. $2x - 4 = 12$

9. $1 - 3y = 4$

10. $5 - 2x = 9$

11. $14 + 9t = 5$

12. $5 + 2x = -7$

13. $-5x + 2.9 = 3.5$

14. $3x + 2.7 = -2.7$

15. $10 + 3x - 4 = 18$

16. $5 + 5x - 6 = 9$

17. $15 = 7x + 7 + 8$

18. $14 = 9x + 5 + 8$

19. $5y - 3y + 2 = 2$

20. $6y + 8y - 7 = -7$

21. $x - 4x + 25 = 31$

22. $3y + 9y - 13 = 11$

23. $-20 = 7y - 3y + 4$

24. $-20 = 5y + y + 16$

25. $4n - 10n + 35 = 1 - 2$

26. $-5n - 3n + 2 = 34$

27. $3n - 15 - n = 1$

28. $2n + 12 + n = 0$

29. $5.4x - 0.2x = 0$

30. $0 = 5.1x + 0.3x$

31. $\dfrac{1}{2}x + 7 = \dfrac{7}{2}$

32. $\dfrac{3}{5}x + 4 = \dfrac{9}{5}$

33. $\dfrac{1}{2} - \dfrac{8}{3}x = \dfrac{5}{6}$

34. $\dfrac{2}{5} - \dfrac{1}{2}x = \dfrac{7}{4}$

35. $\dfrac{3}{2} = \dfrac{1}{3}x + \dfrac{11}{3}$

36. $\dfrac{11}{8} = \dfrac{1}{5}x + \dfrac{4}{5}$

37. $\dfrac{7}{2} - 5 - \dfrac{5}{2}x = 9$

38. $\dfrac{8}{3} + 2 - \dfrac{7}{3}x = 6$

39. $\dfrac{5}{8}x - \dfrac{1}{4}x + \dfrac{1}{2} = \dfrac{3}{10}$

40. $\dfrac{1}{2}x + \dfrac{3}{4}x - \dfrac{5}{3} = \dfrac{5}{6}$

41. $\dfrac{y}{2} + \dfrac{1}{5} = 3$

42. $\dfrac{y}{3} - \dfrac{2}{3} = 7$

43. $\dfrac{7}{8} = \dfrac{3}{4}x - \dfrac{5}{8}$

44. $\dfrac{1}{10} = \dfrac{4}{5}x + \dfrac{3}{10}$

45. $\dfrac{y}{7} + \dfrac{y}{28} + \dfrac{1}{2} = \dfrac{3}{4}$

46. $\dfrac{5y}{6} - \dfrac{7y}{8} - \dfrac{1}{12} = \dfrac{1}{3}$

47. $x + 1.2x + 6.9 = -3.0$

48. $3x - 0.75x - 1.72 = 3.23$

49. $10 = x - 0.5x + 32$

50. $33 = y + 3 - 0.4y$

51. $2.5x + 0.5x - 3.5 = 2.5$

52. $4.7 - 0.5x - 0.3x = -0.1$

53. $6.4 + 1.2x + 0.3x = 0.4$

54. $5.2 - 1.3x - 1.5x = -0.4$

55. $-12.13 = 2.42y + 0.6y - 13.64$

56. $-7.01 = 1.75x + 3.05x - 8.45$

57. $-0.4x + x + 17.2 = 18.1$

58. $y - 0.75y + 13.76 = 14.66$

59. $0 = 17.3x - 15.02x - 0.456$

60. $0 = 20.5x - 16.35x + 0.1245$

61. $3x + 2 = x - 8$

62. $5x + 1 = 2x - 5$

63. $4n - 3 = n + 6$

64. $6y + 3 = y - 7$

65. $3y + 18 = 7y - 6$

66. $2y + 5 = 8y + 10$

67. $3x + 11 = 8x - 4$

68. $9x + 3 = 5x - 9$

69. $14n = 3n$

70. $1.6x = 0.8x$

71. $6y - 2.1 = y - 2.1$

72. $13x + 5 = 2x + 5$

73. $2(z + 1) = 3z + 3$

74. $6x - 3 = 3(x + 2)$

75. $16y + 23y - 3 = 16y - 2y + 2$

76. $5x - 2x + 4 = 3x + x - 1$

77. $0.25 + 3x + 6.5 = 0.75x$

78. $0.9y + 3 = 0.4y + 1.5$

79. $6.5 + 1.2x = 0.5 - 0.3x$

80. $x - 0.1x + 0.8 = 0.2x + 0.1$

81. $\dfrac{2}{3}x + 1 = \dfrac{1}{3}x - 6$

82. $\dfrac{4}{5}n + 2 = \dfrac{2}{5}n - 4$

83. $\dfrac{y}{5} + \dfrac{3}{4} = \dfrac{y}{2} + \dfrac{3}{4}$

84. $\dfrac{5n}{6} + \dfrac{1}{9} = \dfrac{3n}{2} + \dfrac{1}{9}$

85. $\dfrac{3}{8}\left(y - \dfrac{1}{2}\right) = \dfrac{1}{8}\left(y + \dfrac{1}{2}\right)$

86. $\dfrac{1}{2}\left(\dfrac{x}{2} + 1\right) = \dfrac{1}{3}\left(\dfrac{x}{2} - 1\right)$

87. $\dfrac{2x}{3} + \dfrac{x}{3} = -\dfrac{3}{4} + \dfrac{x}{2}$

88. $\dfrac{3}{4}x + \dfrac{1}{5}x = \dfrac{1}{2}x - \dfrac{3}{10}$

89. $x + \dfrac{2}{3}x - 2x = \dfrac{x}{6} - \dfrac{1}{8}$

90. $3x + \dfrac{1}{2}x - \dfrac{2}{5}x = \dfrac{x}{10} + \dfrac{7}{20}$

91. $3(1 + 9x) = 6(2 - 4x)$

92. $4(5 - x) = 8(3x + 10)$

93. $3(4x - 1) = 4(2x - 3) + 8$

94. $7(2x - 1) = 5(x + 6) - 13$

95. $5 - 3(2x + 1) = 4(x - 5) + 6$

96. $-2(y + 5) - 4 = 6(y - 2) + 2$

97. $8 + 4(2x - 3) = 5 - (x + 3)$

98. $8(3x + 5) - 9 = 9(x - 2) + 14$

99. $4.7 - 0.3x = 0.5x - 0.1$

100. $5.8 - 0.1x = 0.2x - 0.2$

101. $0.2(x + 3) = 0.1(x - 5)$

102. $0.4(x + 3) = 0.3(x - 6)$

103. $\dfrac{1}{2}(4 - 8x) = \dfrac{1}{3}(4x + 7) - 3$

104. $3 + \dfrac{1}{4}(x - 4) = \dfrac{2}{5}(2 + 3x)$

105. $0.6x - 22.9 = 1.5x - 18.4$

106. $0.1y + 3.8 = 5.72 - 0.3y$

107. $0.12n + 0.25n - 5.895 = 4.3n$

108. $0.15n + 32n - 21.0005 = 10.5n$

109. $0.7(x + 14.1) = 0.3(x + 32.9)$

110. $0.8(x - 6.21) = 0.2(x - 24.84)$

Determine whether each equation is a conditional equation, an identity, or a contradiction. See Example 5.

111. $2(3x - 1) + 5 = 3$

112. $-2x + 13 = -2(x - 7)$

113. $5x + 13 = -2(x - 7) + 3$

114. $3x + 9 = -3(x - 3) + 6x$

115. $7(x - 1) = -3(3 - x) + 4x$

116. $3(x - 2) + 4x = 6(x - 1) + x$

117. $5(x + 1) = 3(x + 1) + 2(x + 1)$

118. $8x - 20 + x = -3(5 - 2x) + 3(x - 4)$

119. $2x + 3x = 5.2(3 - x)$

120. $5.2x + 3.4x = 0.2(x - 0.42)$

121. Find the error(s) made in solving each equation and give the correct solution.

a.
$$\frac{1}{3}x + 4 = 9$$
$$3 \cdot \frac{1}{3}x + 4 = 3 \cdot 9$$
$$x + 4 = 27$$
$$x + 4 - 4 = 27 - 4$$
$$x = 23$$

b.
$$5x + 3 = 11$$
$$(5x - 3) + (3 - 3) = 11 - 3$$
$$2x + 0 = 8$$
$$\frac{2x}{2} = \frac{8}{2}$$
$$x = 4$$

122. Answer each question.

  a. Simplify the expression $3(x+5) + 2(x-7)$.

  b. Solve the equation $3(x+5) + 2(x-7) = 31$.

  c. How are the methods you used to answer parts **a.** and **b.** similar? How are they different?

123. Write an equation to represent each situation, using x to represent Ryan's current age. Determine whether each equation is a conditional equation, an identity, or a contradiction, and explain why that makes sense for the situation represented.

  a. In 6 years, Ryan will be 20 years old.

  b. In 6 years, Ryan will be 8 years older than he is now.

  c. In 6 years, Ryan will be 3 years older than he will be 3 years from now.

1.2 APPLICATIONS OF LINEAR EQUATIONS IN ONE VARIABLE

■ TOPICS

- ■ Solving Linear Equations for One Variable
- ■ Distance and Interest Problems

Solving Linear Equations for One Variable

One common task in applied mathematics is to solve a given equation in two or more variables for one of the variables. **Solving for a variable** means to transform the equation into an equivalent one in which the specified variable is isolated on one side of the equation. For linear equations we accomplish this by the same methods we have used previously.

Example 1: Solving Linear Equations for One Variable

Solve each of the following equations for the specified variable. All of the equations are formulas that arise in various applications, and they are linear in the specified variable.

a. $P = 2l + 2w$. Solve for w.

b. $A = P\left(1 + \dfrac{r}{m}\right)^{mt}$. Solve for P.

c. $S = 2\pi r^2 + 2\pi rh$. Solve for h.

Solution

a.
$$P = 2l + 2w$$
$$P - 2l = 2w$$
$$\frac{P - 2l}{2} = w$$
$$w = \frac{P - 2l}{2}$$

This is the formula for the perimeter P of a rectangle of length l and width w.

The last equation is equivalent to the preceding one, but it is conventional to put the specified variable on the left side of the equation.

b.
$$A = P\left(1 + \frac{r}{m}\right)^{mt}$$
$$\frac{A}{\left(1 + \dfrac{r}{m}\right)^{mt}} = P$$
$$P = A\left(1 + \frac{r}{m}\right)^{-mt}$$

This is the formula for compound interest. If principal P is invested at an annual rate r for t years, compounded m times a year, the value of the investment at time t is A. This formula is linear in the variables P and A, but not in m, t, or r.

We use the properties of exponents to find a cleaner solution.

c.
$$S = 2\pi r^2 + 2\pi rh$$
$$S - 2\pi r^2 = 2\pi rh$$
$$\frac{S - 2\pi r^2}{2\pi r} = h$$
$$h = \frac{S - 2\pi r^2}{2\pi r}$$

This is the formula for the surface area of a right circular cylinder of radius r and height h. It is linear in the variables S and h, but not in r.

Distance and Interest Problems

Many applications lead to equations more complicated than those that we have studied so far, but good examples of linear equations arise from distance and simple interest problems. This is because the basic distance and simple interest formulas are linear in all of their variables.

Distance: $d = rt$, where d is the distance traveled at rate r for time t

Simple Interest: $I = Prt$, where I is the interest earned on principal P invested at rate r for time t

Example 2: Calculating Average Speed

The distance from Shreveport, LA to Austin, TX by one route is 325 miles. If Kevin made the trip in five and a half hours, what was his average speed?

Solution

We know that $d = 325$ miles and $t = 5\frac{1}{2}$ hours. After substituting these values in the formula $d = rt$ we need to solve the linear equation $325 = \frac{11}{2}r$ for r (note that we have written five and a half as $\frac{11}{2}$). We do this by multiplying both sides by $\frac{2}{11}$.

$$\frac{2}{11}(325) = \frac{2}{11}\left(\frac{11}{2}r\right)$$
$$\frac{650}{11} = r$$
$$r \approx 59.1 \, \text{mph}$$

Alternatively, the time can be expressed in decimal form as 5.5 hours.

$$325 = 5.5r$$
$$\frac{325}{5.5} = r$$
$$r \approx 59.1 \, \text{mph}$$

NOTE

With application or mathematical modeling problems, it often helps to list the variables in the problem. As you read through the problem statement, fill in this list and determine which variable to solve for.

Example 3: Calculating Average Interest Rate

Julie invested \$1500 in a risky high-tech stock on January 1st. On July 1st, her stock is worth \$2100. She knows that her investment does not earn interest at a constant rate, but she wants to determine her average annual rate of return at this point in the year. What is the average annual rate of return she has earned so far?

Solution

The interest that Julie has earned in half a year is \$600 (or \$2100 − \$1500). Replacing P with 1500, t with $\dfrac{1}{2}$, and I with 600 in the formula $I = Prt$, we have:

$$600 = (1500)\left(\frac{1}{2}\right)r$$

$$\frac{1200}{1500} = r$$

$$r = 0.8$$

$$r = 80\% \text{ average rate of return per year}$$

1.2 EXERCISES

PRACTICE

Solve each of the following equations for the indicated variable. See Example 1.

1. Circumference of a circle: $C = 2\pi r$; solve for r

2. Ideal Gas Law: $PV = nRT$; solve for T

3. Velocity: $v^2 = v_0^2 + 2ax$; solve for a

4. Area of a trapezoid: $A = \dfrac{1}{2}h(b+c)$; solve for h

5. Temperature conversions: $C = \dfrac{5}{9}(F-32)$; solve for F

6. Volume of a right circular cone: $V = \dfrac{1}{3}\pi r^2 h$; solve for h

7. Surface area of a rectangular prism: $A = 2lw + 2wh + 2hl$; solve for h

8. Distance: $d = rt_1 + rt_2$; solve for r

9. Kinetic energy of protons: $K = \dfrac{1}{2}mv^2$; solve for m

10. Finance: $A = P(1+rt)$; solve for t

11. A riverboat leaves port and proceeds to travel downstream at an average speed of 15 miles per hour. How long will it take for the boat to arrive at the next port, 95 miles downstream?

12. Two trucks leave a warehouse at the same time. One travels due east at an average speed of 45 miles per hour, and the other travels due west at an average speed of 55 miles per hour. After how many hours will they be 450 miles apart?

13. Two cars leave a rest stop at the same time and proceed to travel down the highway in the same direction. One travels at an average rate of 62 miles per hour, and the other at an average rate of 59 miles per hour. How far apart are the two cars after four and a half hours?

14. Two trains are 630 miles apart, heading directly toward each other on parallel tracks. The first train is traveling at 95 mph, and the second train is traveling at 85 mph. How long will it be before the trains pass each other?

15. Two brothers, Rick and Tom, each inherit $10,000. Rick invests his inheritance in a savings account with an annual return of 2.25%, while Tom invests his in a CD paying 6.15% annually. How much more money does Tom have than Rick after 1 year?

16. Sarah, sister to Rick and Tom in the previous problem, also inherits $10,000, but she invests her inheritance in a global technology mutual fund. At the end of 1 year, her investment is worth $12,800. What has her effective annual rate of return been?

17. An industrial acid-etching procedure calls for 3 gallons of a 46% hydrofluoric acid solution, but the supplier currently only has 44% solution and 50% solution. How many gallons of each should be mixed for the procedure?

18. An agricultural stress test calls for soaking seeds in 8% saline solution. The scientist running the test wants to make use of 1 liter of 20% saline solution that is already made up. How much pure water should she add to the 20% solution to obtain an 8% solution?

19. A total of 39 tickets were sold for a puppet show, with child tickets selling for $7.50 and adult tickets selling for $10.00. The ticket sales raised $330.00 in all. How many child tickets and how many adult tickets were sold?

20. Joe's Java Joint wants to make a blend of two coffees that can be sold for $15 per pound. The first of the two types of coffee costs $18 per pound, while the second costs $13 per pound. How many pounds of each should be mixed to get 10 pounds of the desired blend?

21. Bob buys a large screen digital TV priced at $9500, but pays $10,212.50 with tax. What is the rate of tax where Bob lives?

22. Will and Matt are brothers. Will is 6 feet, 4 inches tall, and Matt is 6 feet, 7 inches tall. How tall is Will as a percentage of Matt's height? How tall is Matt as a percentage of Will's height?

23. A farmer wants to fence in three square garden plots situated along a road, as shown, and he decides not to install fencing along the edge of the road. If he has 182 feet of fencing material total, what dimensions should he make each square plot?

24. Find three consecutive integers whose sum is 288. (**Hint:** If n represents the smallest of the three, then $n+1$ and $n+2$ represent the other two numbers.)

25. Find three consecutive odd integers whose sum is 165. (**Hint:** If n represents the smallest of the three, then $n+2$ and $n+4$ represent the other two numbers.)

26. Kathy buys last year's best-selling novel, in hardcover, for $15.05. This is a 30% discount from the original price. What was the original price?

27. The highest point on Earth is the peak of Mount Everest. If you climbed to the top, you would be approximately 29,035 feet above sea level. Remembering that a mile is 5280 feet, what percentage of the height of the mountain would you have to climb to reach a point two miles above sea level?

☐ TECHNOLOGY

Use a graphing utility to solve the following equations. Round your answers to two decimal places if necessary.

28. $453x = 95(34x + 291)$

29. $-0.23 = 0.79x - 0.47(x + 0.98)$

30. $254 + 0.98(x - 124) = 0$

31. $323x - 1745 = 531(68x - 887)$

1.3 LINEAR INEQUALITIES IN ONE VARIABLE

■ TOPICS

- Solving Linear Inequalities
- Solving Compound Linear Inequalities
- Solving Absolute Value Inequalities
- Applications of Linear Inequalities

Solving Linear Inequalities

If the equality symbol in a linear equation is replaced with $<$, $\leq$, $>$, or $\geq$, the result is a **linear inequality**. One difference between linear equations and linear inequalities is the way in which the solutions are described. Typically, the solution of a linear inequality consists of some interval of real numbers; such solutions can be described graphically or with interval notation. The process of obtaining the solution, however, is much the same as the process for solving linear equations, with the one important difference discussed next.

When solving linear inequalities, the field properties outlined earlier all still apply, and we often use the distributive and commutative properties in order to simplify one or both sides of an inequality. The additive version of the two cancellation properties is also used in the same way as in solving equations. The one difference lies in applying the multiplicative version of cancellation. When dealing with linear *equations* we can multiply both sides by a positive or negative value and obtain an equivalent equation. When dealing with linear *inequalities* some problems arise when multiplying both sides by a negative value.

Example 1: Multiplying Inequalities by Negative Numbers

Consider the following two inequalities: $-3 < 2$ and $x < 0$. Observe what happens if we multiply both sides of each inequality by -1.

1. The statement $-3 < 2$ is clearly true, but if we multiply both sides by -1, we obtain the false statement $3 < -2$.

2. Now consider the inequality $x < 0$. If we multiply both sides by -1, we have the inequality $-x < 0$. But these two statements can't both be true!

These examples show that multiplicative cancellation must behave a bit differently for linear inequalities. Note that if we reverse the inequality sign in our results, we actually get true statements. This provides a clue to how we approach multiplicative cancellation in the case of linear inequalities.

Cancellation Properties for Inequalities

In these properties, A, B, and C represent algebraic expressions and D represents a nonzero constant. Each of the properties is stated for the inequality symbol $<$, but they are also true for the other three symbols (when substituted below).

Property	Description
If $A < B$, then $A + C < B + C$.	Adding the same quantity to both sides of an inequality results in an equivalent inequality.
If $A < B$ and $D > 0$, then $A \cdot D < B \cdot D$.	If both sides of an inequality are multiplied by a positive constant, the sense of the inequality is unchanged.
If $A < B$ and $D < 0$, then $A \cdot D > B \cdot D$.	If both sides are multiplied by a negative constant, the sense of the inequality is reversed.

Keep in mind that multiplying (or dividing) both sides of an inequality by a negative quantity requires reversing, or "flipping" the inequality symbol. We will see this several times in the examples to follow.

Example 2: Solving Linear Inequalities

Solve the following inequalities, using interval notation to describe the solution set.

a. $5 - 2(x - 3) \le -(1 - x)$

b. $\dfrac{3(a-2)}{2} < \dfrac{5a}{4}$

Solution

a. $5 - 2(x-3) \le -(1-x)$

$\qquad 5 - 2x + 6 \le -1 + x$

$\qquad -2x + 11 \le -1 + x$

Begin by using the distributive property, then combine like terms.

$\qquad\qquad -3x \le -12$

Now, all we need to do is divide by -3.

$\qquad\qquad\quad x \ge 4$

Note the reversal of the inequality symbol.

In interval notation, the solution is $[4, \infty)$.

b. $\dfrac{3(a-2)}{2} < \dfrac{5a}{4}$

Just as with equations, fractions in inequalities can be eliminated by multiplying both sides by the least common denominator.

$\quad 4\left(\dfrac{3(a-2)}{2}\right) < 4\left(\dfrac{5a}{4}\right)$

$\qquad\quad 6(a-2) < 5a$

Since we do not need to multiply or divide by a negative value, the sense of the inequality does not change.

$\qquad\quad 6a - 12 < 5a$

$\qquad\qquad\quad a < 12$

Thus, in interval notation, the solution is $(-\infty, 12)$.

The solutions in Example 2 were described using interval notation, but solutions can also be described by set-builder notation or by graphing. Graphing a solution to an inequality can lead to a better understanding of which real numbers solve the inequality.

The symbols used for graphing intervals are the same as the symbols in interval notation. Parentheses are used to indicate excluded endpoints of intervals and brackets are used when the endpoints are included in the interval. The portion of the number line that constitutes the interval is then shaded. (Other commonly used symbols in graphing are open circles for parentheses and filled-in circles for brackets.)

For example, the two solutions from Example 2, $[4, \infty)$ and $(-\infty, 12)$, are graphed as follows.

Example 3: Graphing Intervals of Real Numbers

Graph the following intervals.

a. $[-3, 6]$ **b.** $(-\infty, 5]$ **c.** $[2, 9)$

Solution

Both endpoints are included in the interval.

The left-hand side of the graph extends to negative infinity.

The left endpoint is included in the interval, while the right endpoint is excluded.

Solving Compound Linear Inequalities

A **compound inequality** is a statement containing two inequality symbols and can be interpreted as two distinct inequalities joined by the word "and" (this means both inequalities must be true). We'll explore how to solve such inequalities with an application example you are likely to encounter at some point.

Example 4: Calculating Final Grades

The final grade in a class depends on the grades of 5 exams, each worth a maximum of 100 points. Suppose Janice's scores on the first four tests are 67, 82, 73 and 85. What scores can she make on the fifth test to get a B in the class?

We follow the same process in solving compound inequalities as we do for standard ones. The only difference is that operations are applied to all three "sides" of the statements.

Solution

If the B range corresponds to an average greater than or equal to 80 and less than 90, and if x represents the fifth test score, we need to solve the following compound inequality:

$$80 \le \frac{67+82+73+85+x}{5} < 90$$

This could be solved by breaking it into the two inequalities

$$80 \le \frac{67+82+73+85+x}{5} \quad \text{and} \quad \frac{67+82+73+85+x}{5} < 90,$$

but it is more efficient to solve both at the same time as a compound inequality.

$$80 \le \frac{67+82+73+85+x}{5} < 90 \qquad \text{First, combine like terms in the numerator of the fraction.}$$

$$80 \le \frac{307+x}{5} < 90$$

$$400 \le 307+x < 450 \qquad \text{Just as if we were working with a single inequality, we begin by multiplying by 5, then subtract 307 from all three parts.}$$

$$93 \le x < 143$$

Thus, the mathematical solution to the compound inequality is [93, 143). However, this is not the solution to our application problem!

Why not? There is an additional restriction on the solution set based on the context of the problem; each exam is worth a maximum of 100 points. This means that any value in the calculated solution set greater than 100 does not apply, making the actual solution [93, 100].

Example 5: Compound Linear Inequalities

Solve the following compound inequalities.

a. $-1 < 3 - 2x \le 5$ **b.** $2(2x-1) \le 4x+2 \le 4(x+1)$

Solution

a. $-1 < 3 - 2x \le 5$ Begin by subtracting 3 from all three expressions.

$$-4 < -2x \le 2$$

Since we divide each expression by -2, we must reverse each inequality symbol.

$$2 > x \ge -1$$

The final compound inequality is identical to the one before it, but has been written so that the smaller number appears first.

$$-1 \le x < 2$$

In interval notation, the solution is [−1, 2).

b. $2(2x-1) \le 4x+2 \le 4(x+1)$ First, apply the distributive property to simplify.

$$4x-2 \le 4x+2 \le 4x+4$$

$$-2 \le 2 \le 4$$

The variable disappears from the inequality, and we are left to assess whether the statement is true.

Since we are left with a true statement, the compound inequality is true for all values of the variable x, and the solution set is $(-\infty, \infty)$.

Solving Absolute Value Inequalities

An **absolute value inequality** is an inequality in which some variable expression appears inside absolute value symbols. In the problems that we will study, the inequality would be linear if the absolute value symbols were not there.

The geometric meaning of absolute value provides the method by which absolute value inequalities are solved. Recall that $|x|$ represents the distance between x and 0 on the real number line. If a is a positive real number, the inequality $|x| < a$ means that x is less than a units from 0, and the inequality $|x| > a$ means that x is greater than a units from 0 (similar interpretations hold for the symbols $\le$ and $\ge$). This means that absolute value inequalities can be written without absolute values as follows:

$$|x| < a \quad \Leftrightarrow \quad -a < x < a$$

and

$$|x| > a \quad \Leftrightarrow \quad x < -a \text{ or } x > a$$

The two absolute value inequalities can serve as a template for rewriting more complicated inequalities without absolute values. Take note of the fact that the $<$ symbol leads to a set of two inequalities that must *both* be true, while the $>$ symbol leads to a solution in which *either* of two inequalities must hold.

Example 6: Absolute Value Inequalities

Solve the following absolute value inequalities.

a. $|4 - 2x| > 6$

b. $2|3y - 2| + 3 \le 11$

c. $|5 + 2s| \le -3$

d. $|5 + 2s| \ge -3$

Solution

a. $|4-2x| > 6$

$$4-2x < -6 \text{ or } 4-2x > 6$$

$$-2x < -10 \text{ or } -2x > 2$$

$$x > 5 \text{ or } x < -1$$

We can immediately rewrite the inequality without absolute values and begin solving the two independent inequalities.

Once again, we need to reverse the sense of the inequality after dividing by -2.

The solution is $(-\infty, -1) \cup (5, \infty)$. The graph of the solution is:

b. $2|3y-2|+3 \leq 11$

$|3y-2| \leq 4$

$-4 \leq 3y-2 \leq 4$

$-2 \leq 3y \leq 6$

$-\dfrac{2}{3} \leq y \leq 2$

Isolate the term containing absolute values by subtracting 3 from both sides, then dividing both sides by 2.

After rewriting the inequality as described earlier, we have a compound inequality to solve.

Thus, the solution is $\left[-\dfrac{2}{3}, 2\right]$. Graphically, the solution can be written as:

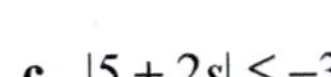

c. $|5 + 2s| \leq -3$

We conclude that the solution set is the empty set, as it is impossible for the absolute value of any expression to be negative.

The solution is $\varnothing$.

d. $|5 + 2s| \geq -3$

Since every absolute value is greater than or equal to 0, the inequality is true for all s.

The solution is $\mathbb{R}$.

Applications of Linear Inequalities

Many real-world applications leading to inequalities involve notions such as "is no greater than", "at least as large as", "does not exceed", and so on. Phrases such as these all have precise mathematical translations that use one of the four inequality symbols.

Let's look at how the phrases translate when variables are added.

"x is no greater than y"

This means that x is not greater than y, which is the same as saying x is less than or equal to y, so this translates to $x \leq y$.

"x is at least as large as y"

If x is at least as large as y, then it can either be as large as (equal to) y or larger (greater) than y, so this phrase translates to $x \geq y$.

"x does not exceed y"

Compare this to the first phrase; the words "is no" carry the same meaning as "does not", and "greater than" is a synonym for "exceed". The two phrases have the same meaning, and so "x does not exceed y" also translates to $x \leq y$.

While it is important to be able to reason out what inequality a particular phrase represents, it is also useful to have a reliable technique for doing these translations.

Given a statement like "x (phrase) y", one method is to ask whether the statement makes sense if x is less than y, if x is equal to y, and if x is greater than y. The answers to these three questions uniquely determine the appropriate inequality symbol.

Applying the process to the first two phrases from Example 7 and one new phrase, we have the following results.

Phrase	Can x be less than y?	Can x equal y?	Can x be greater than y?	Inequality
"x is no greater than y"	Yes	Yes	No	$x \leq y$
"x is at least as large as y"	No	Yes	Yes	$x \geq y$
"y exceeds x"	Yes	No	No	$x < y$

Example 8: Applications of Inequalities

Express each of the following problems as an inequality, and then solve the inequality.

a. The average daily high temperature in Santa Fe, NM over the course of three days exceeded 75. Given that the high on the first day was 72 and the high on the third day was 77, what can we say about the high temperature on the second day?

b. As a test for quality at a plant manufacturing silicon wafers for computer chips, a random sample of 10 batches of 1000 wafers each must not detect more than 5 defective wafers per batch on average. In the first 9 batches tested, the average number of defective wafers per batch is found to be 4.78 (to the nearest hundredth). What is the maximum number of defective wafers that can be found in the 10th batch for the plant to pass the quality test?

Solution

a. We'll begin building the inequality with an expression for calculating the average. Let x represent the high temperature on the second day.

$$\frac{72 + x + 77}{3}$$

What inequality symbol do we use? The problem states the average *exceeded* 75. To exceed is to be greater than, so we use the $>$ symbol.

$$\frac{72 + x + 77}{3} > 75$$

We then proceed to solve the inequality.

$$\frac{72 + x + 77}{3} > 75$$
$$149 + x > 225$$
$$x > 76$$

Thus, the high temperature on the second day exceeded 76 degrees.

b. The phrase "must not detect more than 5 defective wafers per batch on average" means the average number must be less than or equal to 5. Let x denote the maximum number of defective wafers in the last batch.

$$\frac{(9)(4.78) + x}{10} \le 5 \qquad \text{The number of defective wafers found in the first 9 batches is } (9)(4.78) = 43.02.$$
$$43.02 + x \le 50$$
$$x \le 6.98$$

Since it is not possible to have a fractional number of wafers, there must have been 43 defective wafers in the first 9 batches, so the maximum allowable number of defective wafers in the final batch is 7.

1.3 EXERCISES

💡 PRACTICE

Determine which elements of $S = \{12, -9, 3.14, -2.83, 1, 5.24, 8, -3, 4\}$ satisfy each inequality below.

1. $7y - 33.6 < -8.6 + 2y$

2. $-2.2y - 18.8 \ge 5.2(1 - y)$

3. $-40 < 4y - 8 \le 4$

4. $-4 < -2(z - 2) \le 2$

Solve the following linear inequalities. Describe the solution set using interval notation and by graphing. See Examples 2 and 3.

5. $4 + 3t \le t - 2$

6. $x - 7 \ge 5 + 3x$

7. $5y - 24 < -9.6 + 2y$

8. $-\dfrac{v + 2}{3} > \dfrac{5 - v}{2}$

9. $4.2x - 5.6 < 1.6 + x$

10. $8.5y - 3.5 \ge 2.5(3 - y)$

11. $-2(3 - x) < -2x$

12. $\dfrac{1 - x}{5} > \dfrac{-x}{10}$

13. $4w + 7 \le -7w + 4$

14. $-5(p - 3) > 19.8 - p$

15. $\dfrac{6f - 2}{5} < \dfrac{5f - 3}{4}$

16. $\dfrac{u - 6}{7} \ge \dfrac{2u - 1}{3}$

17. $0.04n + 1.7 < 0.13n - 1.45$

18. $2k + \dfrac{3}{2} < 5k - \dfrac{7}{3}$

19. $\dfrac{4x+4}{5} > \dfrac{3x+2.6}{4}$

20. $-1.4z-19.6 \geq 4.4(1-z)$

21. $6m+\dfrac{7}{4} > \dfrac{4m+5.8}{5}$

22. $-3.9n-5.4 \geq 6.2(2-3n)$

Solve the following compound inequalities. Describe the solution set using interval notation and by graphing. See Examples 4 and 5.

23. $-4 < 3x-7 \leq 8$

24. $5 \leq 2m-3 \leq 13$

25. $-36 < 3x-6 \leq 12$

26. $2 < 3(x+2) \leq 21$

27. $-8 \leq \dfrac{z}{2}-4 < -5$

28. $6(x-1) < 2(3x+5) \leq 6x+10$

29. $3 < \dfrac{w+3}{8} \leq 9$

30. $4 \leq \dfrac{p+7}{-2} < 9$

31. $\dfrac{1}{3} < \dfrac{7}{6}(l-3) < \dfrac{2}{3}$

32. $-10 < -2(4+y) \leq 9$

33. $\dfrac{1}{4} \leq \dfrac{g}{2}-3 < 5$

34. $-1.2 \leq \dfrac{x+3}{-5} \leq 0.2$

35. $0.08 < 0.03c+0.13 \leq 0.16$

Solve the following absolute value inequalities. Describe the solution set using interval notation and by graphing. See Example 6.

36. $|x-2| \geq 5$

37. $|4-2x| > 11$

38. $4+|3-2y| \leq 6$

39. $4+|3-2y| > 6$

40. $2|z+5| < 12$

41. $7-\left|\dfrac{q}{2}+3\right| \geq 12$

42. $4|z+3| \leq 28$

43. $-3|4-t| < -6$

44. $-3|4-t| > -6$

45. $3|4-t| < -6$

46. $7-|4-2y| \leq -5$

47. $11-\left|\dfrac{w}{4}+1\right| \geq 12$

48. $5.5+|x-7.2| \leq 3.5$

49. $6-5|x+2| \geq -4$

50. $|2x-1| < x+4$

51. $|3t+4| > -8$

52. $2 < |6w-2|+7$

The words "and" and "or" can appear explicitly between two inequalities, and their meaning in such cases is the same as in absolute value inequalities. If two inequalities are joined by the word "and," the solution set consists of all those real numbers that satisfy both inequalities; that is, the solution set overall is the intersection of the two individual solution sets. If the word "or" appears between two inequalities, the solution set consists of all those real numbers that satisfy at least one of the two inequalities; in other words, the solution set overall is the union of the two individual solution sets.

Guided by the above paragraph, solve the following inequality problems. Describe the solution set using interval notation and by graphing.

53. $t < 2t - 3$ and $-3(t+4) > -57$

54. $7 - \dfrac{3x}{5} < \dfrac{2}{5}$ or $2 - 3x \geq 5$

55. $-2(a-1) < 4$ and $6 + a \leq 9$

56. $-2(a-1) < 4$ and $6 - a \leq 9$

57. $-2(a-1) < 4$ or $6 + a \leq 9$

58. $\dfrac{5n+6}{3} < -10$ and $-3(n-1) < -6$

59. $\dfrac{23x-3}{-7} \leq 7$ and $-x < -(4x-9)$

60. $7 - \dfrac{x}{3} \leq 14 + \dfrac{x}{2}$ or $-3x < 15$

🚀 APPLICATIONS

61. In a class in which the final course grade depends entirely on the average of four equally weighted 100-point tests, Cindy has scored 96, 94, and 97 on the first three. The professor has announced that there will be a 15-point bonus problem on the fourth test, and anyone who finishes the semester with an average of more than 100 will receive an A+. What interval of scores on the fourth test will give Cindy an A for the semester (an average between 90 and 100, inclusive), and what interval will give Cindy an A+?

62. In a series of 30 racquetball games played to date, Larry has won 10, giving him a winning average so far of 33.3% (to the nearest tenth of a percent). If he continues to play, what interval describes the number of games he must now win in a row to have an overall winning average greater than 50%?

63. Assume that the national average SAT score for high school seniors is 1020 out of 1600. A group of seven students receive their scores in the mail, and six of them look at their scores. Two students scored 1090, one got an 1120, two others each got a 910, and the sixth student received an 880. What interval of scores can the seventh student receive to pull the group's average above the national average?

64. The central bank of a certain country tries to keep the inflation rate below 5.0% on an annual basis. Assume that inflation rates for the first three quarters of a given year are as follows: 5.2%, 4.3%, and 4.7%. What interval of inflation rates for the final quarter would satisfy the government's goal?

1.4 QUADRATIC EQUATIONS IN ONE VARIABLE

■ TOPICS

- ■ Quadratic Equations

Quadratic Equations

Previously, we studied first-degree polynomial equations in one variable. We will now expand the class of one-variable polynomial equations that we can solve to include *quadratic* or *second-degree* equations.

Recall that the method of solving linear equations was particularly straightforward, and that the method always works for *any* such equation. We will, by the end of this lesson, develop a method for solving one-variable second-degree equations that is also guaranteed to work. This is in contrast to polynomial equations in general. In fact, it can be shown that for polynomial equations of degree five and higher there is *no* method that always works.

Our development will begin with a formal definition of quadratic equations, and we will then proceed to study those quadratic equations that can be solved by factoring.

Quadratic Equations

A **quadratic equation in one variable**, say the variable x, is an equation that can be transformed into the form $ax^2 + bx + c = 0$, where a, b, and c are real numbers and $a \neq 0$. Such equations are also called **second-degree equations**, as x appears to the second power. The name *quadratic* comes from the Latin word *quadrus*, meaning square.

The key to using factoring to solve a quadratic equation, or indeed any polynomial equation, is to rewrite the equation so that 0 appears by itself on one side of the equation. This often allows us to use the Zero-Factor Property.

Zero-Factor Property

Let A and B represent algebraic expressions. If the product of A and B is 0, then at least one of A and B is itself 0. That is,

$$AB = 0 \quad \Rightarrow \quad A = 0 \text{ or } B = 0.$$

If the trinomial $ax^2 + bx + c$ can be factored, it can be written as a product of two linear factors A and B. The Zero-Factor Property then implies that the only way for $ax^2 + bx + c$ to be 0 is if one (or both) of A and B is 0. This is all we need to solve the equation.

Example 1: Solving Quadratic Equations by Factoring

Solve the following quadratic equations by factoring.

a. $x^2 + \dfrac{11x}{3} = \dfrac{4}{3}$

Solution

$$x^2 + \frac{11x}{3} = \frac{4}{3}$$

$$3x^2 + 11x = 4$$

$$3x^2 + 11x - 4 = 0$$

$$(3x - 1)(x + 4) = 0$$

$$3x - 1 = 0 \quad x + 4 = 0$$

$$x = \frac{1}{3} \qquad x = -4$$

To make the polynomial easier to factor, we multiply both sides by the LCD.

Although we could factor $3x^2 + 11x$, this would not do us any good. We must have 0 on one side in order to apply the Zero-Factor Property.

After factoring, we have two linear equations to solve.

The solution set is $\left\{ \dfrac{1}{3}, -4 \right\}$.

b. $s^2 + 25 = 10s$

Solution

$$s^2 + 25 = 10s$$

$$s^2 - 10s + 25 = 0$$

$$(s - 5)^2 = 0$$

$$s - 5 = 0 \text{ or } s - 5 = 0$$

$$s = 5$$

Again, we rewrite the equation with 0 on one side, and then factor the quadratic.

In this example two linear factors are the same. In such cases, the single solution is called a *double solution* or *double root*.

c. $4x^2 + 28x = 0$

Solution

$$4x^2 + 28x = 0$$

$$4x(x + 7) = 0$$

$$4x = 0 \quad x + 7 = 0$$

$$x = 0 \qquad x = -7$$

An alternative approach in this example would be to divide both sides by 4 at the very beginning. This would lead to the equation $x(x + 7) = 0$, which gives us the same solution set of $\{0, -7\}$.

The factoring method is fine when it works, but there are two potential problems with the method: (1) the second-degree polynomial in question might not factor over the integers, and (2) even if the polynomial does factor, the factored form may not be obvious.

In some cases where the factoring method is unsuitable, the solution can be obtained by using our knowledge of square roots. If A is an algebraic expression and if c is a constant, the equation $A^2 = c$ means $A = \sqrt{c}$ or $A = -\sqrt{c}$. We will find it convenient to summarize this as follows:

$$A^2 = c \text{ implies } A = \pm\sqrt{c}.$$

If a given quadratic equation can be written in the form $A^2 = c$, we can use the above observation to obtain two linear equations that can be easily solved.

Example 2: "Perfect Square" Quadratic Equations

Solve the following quadratic equations by taking square roots.

a. $(2x + 5)^2 = 12$

Solution

$$(2x+5)^2 = 12$$
$$2x+5 = \pm\sqrt{12}$$
$$2x+5 = \pm 2\sqrt{3}$$

We begin by taking the square root of each side, keeping in mind that there are two numbers whose square is 12.

$$2x = -5 \pm 2\sqrt{3}$$
$$x = \frac{-5 \pm 2\sqrt{3}}{2}$$

We solve the two linear equations at once by subtracting 5 from both sides and then dividing both sides by 2.

The solution set is $x = \dfrac{-5 + 2\sqrt{3}}{2}$, $x = \dfrac{-5 - 2\sqrt{3}}{2}$.

b. $(x - 6)^2 - 7 = 0$

Solution

$$(x-6)^2 - 7 = 0$$
$$(x-6)^2 = 7$$
$$x-6 = \pm\sqrt{7}$$
$$x = 6 \pm \sqrt{7}$$

Before taking square roots, we isolate the perfect square algebraic expression on one side, and put the constant on the other.

There are potential pitfalls, once again, with the method just developed. If the quadratic equation under consideration appears in the form $A^2 = c$, the method works well. But what if the equation doesn't have the form $A^2 = c$?

The method of **completing the square** allows us to write an arbitrary quadratic equation $ax^2 + bx + c = 0$ in the desired form.

Method of Completing the Square

Step 1: Write the equation $ax^2 + bx + c = 0$ in the form $ax^2 + bx = -c$.

Step 2: Divide by a, if $a \neq 1$, so that the coefficient of x^2 is 1: $x^2 + \dfrac{b}{a}x = -\dfrac{c}{a}$.

Step 3: Divide the coefficient of x by 2, square the result, and add this to both sides.

Step 4: The trinomial on the left side will now be a perfect square. That is, it can be written as the square of an algebraic expression.

At this point, the equation will have the form $A^2 = c$ and can be solved by taking the square root of both sides.

Example 3: Completing the Square

Solve the following quadratic equations by completing the square.

a. $x^2 - 6x - 2 = 0$

Solution

$$x^2 - 6x - 2 = 0$$
$$x^2 - 6x = 2$$
$$x^2 - 6x + 9 = 2 + 9$$
$$(x - 3)^2 = 11$$
$$x - 3 = \pm\sqrt{11}$$
$$x = 3 \pm \sqrt{11}$$

After moving the constant term to the right-hand side, we divide -6 (the coefficient of x) by 2 to get -3, and add $(-3)^2$ to both sides of the equation.

The trinomial on the left can now be factored.

Taking square roots leads to two easily solved linear equations.

b. $25x^2 + 5x = 6$

Solution

$$25x^2 + 5x = 6$$
$$x^2 + \frac{1}{5}x = \frac{6}{25}$$
$$x^2 + \frac{1}{5}x + \frac{1}{100} = \frac{6}{25} + \frac{1}{100}$$
$$\left(x + \frac{1}{10}\right)^2 = \frac{1}{4}$$
$$x + \frac{1}{10} = \pm\frac{1}{2}$$
$$x = \frac{-1}{10} \pm \frac{1}{2}$$
$$x = \frac{-3}{5}, \frac{2}{5}$$

The constant term is already isolated on the right-hand side, so our first step is to divide by 25 (and simplify the resulting fractions, if possible).

Half of the coefficient of x is $\frac{1}{10}$, and the square of this is $\frac{1}{100}$.

After simplifying the sum of the fractions on the right, we take the square root of each side.

Since the answer of $\frac{-1}{10} \pm \frac{1}{2}$ can be simplified, we do so to obtain the final answer.

It is time to mention an incidental benefit of the method that we have devised. The polynomial in Example 3a, $x^2 - 6x - 2$, does not factor over the integers. That is, it cannot be written as a product of two first-degree polynomials with integer coefficients. Nevertheless, it can be factored as a product of two first-degree polynomials:

$$x^2 - 6x - 2 = \left(x - 3 - \sqrt{11}\right)\left(x - 3 + \sqrt{11}\right).$$

We know this because we know that the two solutions of the equation $x^2 - 6x - 2 = 0$ are $3 - \sqrt{11}$ and $3 + \sqrt{11}$, and we know that there is a close relationship between factors of a quadratic polynomial and the solutions of the equation in which that quadratic polynomial is equal to 0. Specifically, if a quadratic polynomial can be factored as $(x - p)(x - q)$, then p and q solve the equation $(x - p)(x - q) = 0$.

The quadratic equation $25x^2 + 5x = 6$ can be rewritten as $25x^2 + 5x - 6 = 0$. What are the factors of the quadratic polynomial $25x^2 + 5x - 6$? Based on the solutions we found in Example 3b, we might guess factors of $x - \dfrac{2}{5}$ and $x + \dfrac{3}{5}$.

But,

$$\left(x - \frac{2}{5}\right)\left(x + \frac{3}{5}\right) = x^2 + \frac{1}{5}x - \frac{6}{25}.$$

It shouldn't be surprising that the product of these two factors has a leading coefficient of 1, since each of them individually has a leading coefficient of 1. To get the correct leading coefficient, we need to multiply by 25:

$$25\left(x - \frac{2}{5}\right)\left(x + \frac{3}{5}\right) = 25\left(x^2 + \frac{1}{5}x - \frac{6}{25}\right) = 25x^2 + 5x - 6.$$

And to make the factors look better, we can factor 25 into two factors of 5 and rearrange the products:

$$25\left(x - \frac{2}{5}\right)\left(x + \frac{3}{5}\right) = 5\left(x - \frac{2}{5}\right) \cdot 5\left(x + \frac{3}{5}\right) = (5x - 2)(5x + 3).$$

The method of completing the square will always serve to solve any equation of the form $ax^2 + bx + c = 0$. But this begs the question: why not just solve $ax^2 + bx + c = 0$ once and for all? Since a, b, and c represent arbitrary constants, the ideal situation would be to find a formula for the solutions of $ax^2 + bx + c = 0$ based on a, b, and c. That is exactly what the quadratic formula is: a formula that solves *any* equation of the form $ax^2 + bx + c = 0$. We will derive the formula now, using what we have learned.

$$ax^2 + bx + c = 0$$

We begin, as always in completing the square, by moving the constant to the right-hand side and dividing by a.

$$x^2 + \frac{b}{a}x = -\frac{c}{a}$$

$$x^2 + \frac{b}{a}x + \frac{b^2}{4a^2} = -\frac{c}{a} + \frac{b^2}{4a^2}$$

We next divide $\dfrac{b}{a}$ by 2 to get $\dfrac{b}{2a}$, and add $\left(\dfrac{b}{2a}\right)^2 = \dfrac{b^2}{4a^2}$ to both sides of the equation. Note that to add the fractions on the right, we need a common denominator of $4a^2$.

$$\left(x + \frac{b}{2a}\right)^2 = -\frac{4ac}{4a^2} + \frac{b^2}{4a^2}$$

$$\left(x + \frac{b}{2a}\right)^2 = \frac{b^2 - 4ac}{4a^2}$$

$$x + \frac{b}{2a} = \pm \frac{\sqrt{b^2 - 4ac}}{2a}$$

Taking square roots leads to two linear equations, which we then solve for x.

$$x = \frac{-b}{2a} \pm \frac{\sqrt{b^2 - 4ac}}{2a}$$

$$x = \frac{-b \pm \sqrt{b^2 - 4ac}}{2a}$$

Since the fractions have the same denominator, they are easily added to obtain the final formula.

The solutions of the general quadratic equation $ax^2 + bx + c = 0$, where $a \neq 0$ are

$$x = \frac{-b \pm \sqrt{b^2 - 4ac}}{2a}.$$

The expression $b^2 - 4ac$ is called the **discriminant**. The discriminant determines the number of solutions to the given quadratic equation.

- If the discriminant is *positive*, $b^2 - 4ac > 0$, there are two real solutions to the quadratic equation. This occurs because $\sqrt{b^2 - 4ac}$ and $-\sqrt{b^2 - 4ac}$ are two different numbers.

- If the discriminant is *zero*, $b^2 - 4ac = 0$, there is one real solution to the quadratic equation. This occurs because $\sqrt{0} = -\sqrt{0} = 0$, and therefore the only solution is $\dfrac{-b}{2a}$.

- If the discriminant is *negative*, $b^2 - 4ac < 0$, there are no real solutions to the quadratic equation. This occurs because the square root of a negative number is not a real number.

Example 4: The Quadratic Formula

Solve the following quadratic equations by using the quadratic formula.

a. $6y^2 - 5y = 1$

Solution

$$6y^2 - 5y = 1$$
$$6y^2 - 5y - 1 = 0$$

Before applying the quadratic formula, move all terms to one side so that a, b, and c can be identified correctly.

$$y = \frac{-(-5) \pm \sqrt{(-5)^2 - (4)(6)(-1)}}{(2)(6)}$$

Applying the quadratic formula by making the appropriate replacements for a, b, and c.

$$y = \frac{5 \pm \sqrt{25 + 24}}{12}$$

$$y = \frac{5 \pm 7}{12}$$

$$y = \frac{-1}{6}, 1$$

The two solutions are $\dfrac{-1}{6}$ and 1.

b. $t^2 + 8t - 12 = 0$

Solution

$$t^2 + 8t - 12 = 0$$

The equation is already in the proper form to apply the quadratic formula.

$$t = \frac{-8 \pm \sqrt{(8)^2 - (4)(1)(-12)}}{2(1)}$$

$$t = \frac{-8 \pm \sqrt{64 + 48}}{2}$$

$$t = \frac{-8 \pm \sqrt{112}}{2}$$

$$t = \frac{-8 \pm 4\sqrt{7}}{2}$$

$$t = -4 - 2\sqrt{7}, -4 + 2\sqrt{7}$$

1.4 EXERCISES

PRACTICE

Solve the following quadratic equations by factoring. See Example 1.

1. $2x^2 - x = 3$

2. $3x^2 - 7x = 0$

3. $x^2 - 14x + 49 = 0$

4. $9x - 5x^2 = -2$

5. $y(2y + 9) = -9$

6. $2x^2 - 3x = x^2 + 18$

7. $(3x + 2)(x - 1) = 7 - 7x$

8. $3x^2 + 33 = 2x^2 + 14x$

9. $5x^2 + 2x + 3 = 4x^2 + 6x - 1$

10. $15x^2 + x = 2$

11. $(x - 7)^2 = 16$

12. $4x^2 - 9 = 0$

Solve the following quadratic equations by taking square roots. See Example 2.

13. $(x - 3)^2 = 9$

14. $(8t - 3)^2 = 0$

15. $(2x + 1)^2 - 7 = 0$

16. $(y - 18)^2 - 1 = 0$

17. $9 = (3s + 2)^2$

18. $(2x - 1)^2 = 8$

19. $x^2 - 4x + 4 = 49$

20. $-3(n + 7)^2 = -27$

21. $(3x - 6)^2 = 4x^2$

Solve the following quadratic equations by completing the square. See Example 3.

22. $x^2 + 8x + 7 = -8$

23. $2x^2 + 6x - 10 = 10$

24. $2x^2 + 7x - 15 = 0$

25. $4x^2 - 4x - 63 = 0$

26. $u^2 + 10u + 9 = 0$

27. $4x^2 - 56x + 195 = 0$

28. $4x^2 + 32x - 260 = 0$

29. $z^2 + 26z + 2 = -23$

30. $y^2 + 22y + 96 = 0$

Solve the following quadratic equations using the quadratic formula. See Example 4.

31. $3x^2 - 4 = -x$

32. $2.1y^2 - 3.5y = 4$

33. $a(a + 2) = -1$

34. $3x^2 - 2x = 0$

35. $6x^2 + 5x - 4 = 3x - 2$

36. $7x^2 - 4x = 51$

37. $4x^2 - 14x - 27 = 3$

Solve the following quadratic equations using any appropriate method.

38. $(z - 11)^2 = 9$

39. $x^2 + 20x + 36 = -48$

40. $256t^2 - 324 = 0$

41. $(y - 8)^2 = 36$

42. $(9y - 6)^2 = 121y^2$

43. $2x^2 + 8x - 3 = 6x$

44. $x^2 - 6x = 27$

45. $3a^2 + 12a - 576 = 0$

46. $-3(b + 5)^2 = -768$

47. $y^2 + 13y + 42 = 0$

48. $3x^2 - 6x = 0$

49. $7x^2 - 42x = 0$

50. $y^2 + 24y + 23 = 0$

51. $5x^2 - 5x - 10 = 0$

52. $4w^2 + 10w + 5 = 3w^2 + 18w - 10$

53. $\left| x^2 - 3x \right| = 2$ (**Hint:** Replace $\left| x^2 - 3x \right|$ first with $x^2 - 3x$ and solve the resulting equation, then replace it with $-\left(x^2 - 3x \right)$ and solve the resulting equation.)

54. $\left| x^2 - 8 \right| = 1$

✏ WRITING & THINKING

55. Factor the quadratic $9x^2 - 6x - 4$.

56. Factor the quadratic $4x^2 + 12x + 1$.

57. Determine b and c so that the equation $x^2 + bx + c = 0$ has the solution set $\{-3, 8\}$.

📈 TECHNOLOGY

Use a graphing utility to solve the following quadratic equations.

58. $5x^2 - 3x = 17$

59. $(a+4)(4a-3) = 5$

60. $10\pi r + \pi r^2 = 107$

61. $4.8x^2 + 3.5x - 9.2 = 0$

1.5 HIGHER DEGREE POLYNOMIAL EQUATIONS

■ TOPICS

- Solving Quadratic-Like Equations
- Solving General Polynomial Equations by Factoring
- Solving Polynomial-Like Equations by Factoring

Solving Quadratic-Like Equations

A polynomial equation of degree n in one variable, say x, is an equation that can be written in the form $a_n x^n + a_{n-1} x^{n-1} + \cdots + a_1 x + a_0 = 0$, where each a_i is a constant and $a_n \neq 0$. As we have seen, such equations can always be solved if $n = 1$ or $n = 2$, but in general there is no method for solving polynomial equations that is guaranteed to find all solutions. There are formulas, called the *cubic* and *quartic* formulas, that solve third- and fourth-degree polynomial equations, but there are no formulas to solve polynomial equations of degree five (or higher)! Moreover, many nonpolynomial equations have no solution method that is guaranteed to work.

However, even though no formula exists for solving these equations, it may still be possible to solve them! In this section, we'll use factoring, the Zero-Factor Property, and our knowledge of quadratic equations to solve higher-degree polynomial equations.

The Zero-Factor Property applies whenever a product of any finite number of factors is equal to 0; if $A_1 \cdot A_2 \cdot \cdots \cdot A_n = 0$, then at least one of the A_i's must equal 0. Recall that we used the Zero-Factor Property to solve quadratic equations. This means that if we can rewrite an equation in a quadratic form, we can use the Zero-Factor Property to solve this equation as well.

Quadratic-Like Equations

An equation is **quadratic-like**, or **quadratic in form**, if it can be written in the form

$$aA^2 + bA + c = 0,$$

where a, b, and c are constants, $a \neq 0$, and A is an algebraic expression. Such equations can be solved by first solving for A and then solving for the variable in the expression A. This method of solution is called **substitution**.

Example 1: Quadratic-Like Equations

Solve the quadratic-like equations.

a. $(x^2 + 2x)^2 - 7(x^2 + 2x) - 8 = 0$

b. $y^{\frac{2}{3}} + 4y^{\frac{1}{3}} - 5 = 0$

Solution

a.
$$\left(x^2 + 2x\right)^2 - 7\left(x^2 + 2x\right) - 8 = 0$$
$$A^2 - 7A - 8 = 0$$
$$(A - 8)(A + 1) = 0$$
$$A = 8, -1$$

Making the substitution $A = x^2 + 2x$ transforms the quadratic-like equation into a quadratic equation that can be solved by factoring.

$$
\begin{array}{ll}
A = 8 \quad \text{or} & A = -1 \\
x^2 + 2x = 8 & x^2 + 2x = -1 \\
x^2 + 2x - 8 = 0 & x^2 + 2x + 1 = 0 \\
(x + 4)(x - 2) = 0 & (x + 1)^2 = 0
\end{array}
$$

Once we have solved for A, we replace A with $x^2 + 2x$ and solve for x.

$$x = -4 \text{ or } x = 2 \text{ or } x = -1$$

Note that -1 is a double root, while -4 and 2 are single roots.

While the substitution method does not necessarily introduce extraneous solutions, you should still check that each solution solves the original quadratic-like equation.

b.
$$y^{\frac{2}{3}} + 4y^{\frac{1}{3}} - 5 = 0$$
$$\left(y^{\frac{1}{3}}\right)^2 + 4\left(y^{\frac{1}{3}}\right) - 5 = 0$$
$$A^2 + 4A - 5 = 0$$
$$(A + 5)(A - 1) = 0$$
$$A = -5, 1$$

Using properties of exponents, we can see that the substitution $A = y^{\frac{1}{3}}$ will make this equation quadratic.

Now that we've solved for A, we reverse the substitution and solve for y in each case.

$$
\begin{array}{ll}
A = -5 \quad \text{or} & A = 1 \\
y^{\frac{1}{3}} = -5 & y^{\frac{1}{3}} = 1 \\
y = (-5)^3 & y = (1)^3 \\
y = -125 & y = 1
\end{array}
$$

Once again, you should confirm that both values do indeed solve the original equation $y^{\frac{2}{3}} + 4y^{\frac{1}{3}} - 5 = 0$.

Solving General Polynomial Equations by Factoring

If an equation consists of a polynomial on one side and 0 on the other, and if the polynomial can be factored completely, then the equation can be solved by using the Zero-Factor Property. If the coefficients in the polynomial are all real, the polynomial can, in principle, be factored into a product of first-degree and second-degree factors. In practice, however, this may be difficult to accomplish unless the degree of the polynomial is small or the polynomial is easily recognizable as a special product. For higher-degree polynomials, the GCF factoring method and factoring by grouping can be very effective.

Example 2: Solving Equations by Factoring

Solve the equations by factoring.

a. $x^4 = 9$

b. $y^3 + y^2 - 4y - 4 = 0$

c. $8t^3 - 27 = 0$

d. $3x^4 + 18x^3 - 21x^2 = 0$

Solution

> **NOTE**
>
> While checking solutions is always a good practice, solving by factoring does not produce any extraneous solutions.

a.
$$x^4 = 9$$
$$x^4 - 9 = 0$$
$$\left(x^2 - 3\right)\left(x^2 + 3\right) = 0$$

After isolating 0 on one side, we see the polynomial is a difference of two squares, which can always be factored.

The Zero-Factor Property gives us two equations, both of which can be solved by taking square roots.

$$x^2 = 3 \quad \text{or} \quad x^2 = -3$$
$$x = \pm\sqrt{3} \qquad x = \pm\sqrt{-3}$$
$$x = \pm\sqrt{3} \qquad x = \pm i\sqrt{3}$$

b.
$$y^3 + y^2 - 4y - 4 = 0$$
$$y^2(y+1) - 4(y+1) = 0$$
$$(y+1)(y^2 - 4) = 0$$
$$(y+1)(y-2)(y+2) = 0$$

We factor the initial equation by grouping.

We can factor $y^2 - 4$ further, since it is a difference of squares.

There are three solutions by the Zero-Factor Property.
$$y = -1 \text{ or } y = 2 \text{ or } y = -2$$

c.
$$8t^3 - 27 = 0$$
$$(2t)^3 - 3^3 = 0$$
$$(2t - 3)\left(4t^2 + 6t + 9\right) = 0$$

The polynomial in this case is a difference of two cubes, which can always be factored.

The Zero-Factor Property gives us two equations to solve. One of the equations is quadratic, and we can use the quadratic formula to solve it.

$$2t - 3 = 0 \quad \text{or} \quad 4t^2 + 6t + 9 = 0$$

$$2t = 3$$
$$t = \frac{3}{2}$$

$$t = \frac{-(6) \pm \sqrt{6^2 - 4(4)(9)}}{2(4)}$$
$$t = \frac{-6 \pm \sqrt{36 - 144}}{8}$$
$$t = \frac{-6 \pm \sqrt{-108}}{8}$$
$$t = \frac{-6 \pm 6i\sqrt{3}}{8}$$
$$t = \frac{-3 \pm 3i\sqrt{3}}{4}$$

Thus, we have three solutions to the original equation.

d. $3x^4 + 18x^3 - 21x^2 = 0$

    $3x^2\left(x^2 + 6x - 7\right) = 0$

    $3x^2\left(x - 1\right)\left(x + 7\right) = 0$

    $x^2\left(x - 1\right)\left(x + 7\right) = 0$

Factoring out the GCF of $3x^2$ yields a quadratic polynomial that we can factor.

The Zero-Factor Property yields three solutions to the original equation.

$$x = 0,\ 1,\ \text{or} -7$$

Solving Polynomial-Like Equations by Factoring

The last equations that we will consider in this section are equations that are not polynomials, but which can be solved using the methods we have developed so far. We have already seen one such equation in Example 1b: the equation $y^{\frac{2}{3}} + 4y^{\frac{1}{3}} - 5 = 0$ is quadratic-like, and can be solved using polynomial methods. Like the equation in Example 1b, some polynomial-like equations can be solved by substitution, transforming them into polynomial equations. Other equations can be solved by rewriting the equation so that 0 appears on one side, factoring the equation, and then applying the Zero-Factor Property. Often, equations involving rational exponents can be solved by factoring out a common factor, as in the following examples.

Example 3: Solving Equations by Factoring

Solve the equations by factoring.

a. $x^{\frac{7}{3}} + x^{\frac{4}{3}} - 2x^{\frac{1}{3}} = 0$

b. $(x-1)^{\frac{1}{2}} - (x-1)^{-\frac{1}{2}} = 0$

Solution

a. $x^{\frac{7}{3}} + x^{\frac{4}{3}} - 2x^{\frac{1}{3}} = 0$

    $x^{\frac{1}{3}}\left(x^2 + x - 2\right) = 0$

    $x^{\frac{1}{3}}\left(x + 2\right)\left(x - 1\right) = 0$

Recall that in cases like this, we factor out x raised to the lowest exponent. In this case, the remaining factor is a factorable trinomial.

The Zero-Factor Property leads to three simple equations.

$$x^{\frac{1}{3}} = 0 \quad \text{or} \quad x + 2 = 0 \quad \text{or} \quad x - 1 = 0$$
$$x = 0 \qquad\qquad x = -2 \qquad\qquad x = 1$$

b. $(x-1)^{\frac{1}{2}} - (x-1)^{-\frac{1}{2}} = 0$

    $(x-1)^{-\frac{1}{2}}\left((x-1) - 1\right) = 0$

    $(x-1)^{-\frac{1}{2}}\left(x - 2\right) = 0$

Again, we factor out the common algebraic expression raised to the lowest exponent.

This equation leads to two equations, only one of which has a solution. (Note that there is no value for x which would solve the first of the two equations.)

$$(x-1)^{-\frac{1}{2}} = 0 \quad \text{or} \quad x-2 = 0$$

$$\frac{1}{(x-1)^{\frac{1}{2}}} = 0 \qquad\qquad x = 2$$

The original equation has only one solution: $x = 2$

1.5 EXERCISES

PRACTICE

Solve the following quadratic-like equations. See Example 1.

1. $(x-1)^2 + (x-1) - 12 = 0$

2. $(z-8)^2 - 7(z-8) + 12 = 0$

3. $(y-5)^2 - 11(y-5) + 24 = 0$

4. $(x^2-1)^2 + (x^2-1) - 12 = 0$

5. $(x^2+1)^2 + (x^2+1) - 12 = 0$

6. $(x^2-13)^2 + (x^2-13) - 12 = 0$

7. $(x^2-2x+1)^2 + (x^2-2x+1) - 12 = 0$

8. $2y^{\frac{2}{3}} + y^{\frac{1}{3}} - 1 = 0$

9. $2x^{\frac{2}{3}} - 7x^{\frac{1}{3}} + 3 = 0$

10. $(x^2-6x)^2 + 4(x^2-6x) - 5 = 0$

11. $(y^2-5)^2 + 5(y^2-5) - 36 = 0$

12. $(x^2+7)^2 + 8(x^2+7) + 12 = 0$

13. $(t^2-t)^2 - 8(t^2-t) + 12 = 0$

14. $2x^{\frac{1}{2}} - 5x^{\frac{1}{4}} + 2 = 0$

15. $3x^{\frac{2}{3}} - x^{\frac{1}{3}} - 2 = 0$

16. $y^{\frac{1}{2}} - 5y^{\frac{1}{4}} + 6 = 0$

17. $(z^2+4z)^2 + 7(z^2+4z) + 12 = 0$

18. $5y^{\frac{2}{3}} + 33y^{\frac{1}{3}} + 18 = 0$

Solve the following polynomial equations by factoring. See Example 2.

19. $a^3 - 3a^2 = a - 3$

20. $2x^3 + x^2 + 2x + 1 = 0$

21. $2x^3 - x^2 = 15x$

22. $x^4 + 5x^2 - 36 = 0$

23. $y^4 + 21y^2 - 100 = 0$

24. $y^3 + 8 = 0$

25. $5s^3 + 6s^2 - 20s = 24$

26. $8a^3 - 27 = 0$

27. $16a^4 = 81$

28. $6x^3 + 8x^2 = 14x$

29. $14x^3 + 27x^2 - 20x = 0$

30. $5z^3 + 28z^2 = 49z$

31. $27x^3 + 64 = 0$

32. $x^3 - 4x^2 + x = 4$

33. $x^3 + 27 = 0$

Solve the following equations by factoring. See Example 3.

34. $3x^{\frac{11}{3}} + 2x^{\frac{8}{3}} - 5x^{\frac{5}{3}} = 0$

35. $(x-3)^{-\frac{1}{2}} + 2(x-3)^{\frac{1}{2}} = 0$

36. $(y-6)^{-\frac{5}{2}} + 7(y-6)^{-\frac{3}{2}} = 0$

37. $y^{-2} - 2y^{-1} + 1 = 0$

38. $2x^{\frac{13}{5}} - 5x^{\frac{8}{5}} + 2x^{\frac{3}{5}} = 0$

39. $(2x-5)^{\frac{1}{3}} - 3(2x-5)^{\frac{-2}{3}} = 0$

40. $x^{-4} - 13x^{-2} + 36 = 0$

41. $y^{\frac{7}{2}} - 5y^{\frac{5}{2}} + 6y^{\frac{3}{2}} = 0$

42. $(t+4)^{\frac{2}{3}} + 2(t+4)^{\frac{8}{3}} = 0$

43. $y^{-2} - 2y^{-1} - 35 = 0$

44. $x^{\frac{11}{2}} - 6x^{\frac{9}{2}} + 9x^{\frac{7}{2}} = 0$

45. $5y^{\frac{11}{3}} + 3y^{\frac{8}{3}} - 2y^{\frac{5}{3}} = 0$

46. $5y^{\frac{12}{5}} - 43y^{\frac{7}{5}} + 24y^{\frac{2}{5}} = 0$

47. $(3x-3)^{\frac{-1}{3}} - 5(3x-3)^{\frac{-4}{3}} = 0$

48. $x^{-2} + 8x^{-1} + 15 = 0$

49. $(y+3)^{\frac{2}{5}} + 4(y+3)^{\frac{7}{5}} = 0$

✏ WRITING & THINKING

50. Find b, c, and d so the equation $x^3 + bx^2 + cx + d = 0$ has solutions of -3, -1, and 5.

51. Find b, c, and d so the equation $x^3 + bx^2 + cx + d = 0$ has solutions of -2, 0, and 6.

52. Find b and c so the equation $x^3 + bx^2 + cx = 0$ has solutions of 0, 1, and -7.

53. Find a, c, and d so the equation $ax^3 + 4x^2 + cx + d = 0$ has solutions of -4, 6, and -6.

54. Find a, b, and d so the equation $ax^3 + bx^2 + 3x + d = 0$ has solutions of -3, $-\dfrac{1}{2}$, and 0.

55. Find a, b, and c so the equation $ax^3 + bx^2 + cx + 6 = 0$ has solutions of $-\dfrac{3}{5}$, $\dfrac{2}{3}$, and 1.

1.6 RATIONAL AND RADICAL EQUATIONS

■ TOPICS

- ■ Proportions
- ■ Solving Equations with Rational Expressions
- ■ Solving Radical Equations

Proportions

A **ratio** is a comparison of two numbers by division. Ratios are written in the form

$$a : b \quad \text{or} \quad \frac{a}{b} \quad \text{or} \quad a \text{ to } b.$$

For example, suppose the ratio of pages in the first two chapters of a text is 4 to 3. We can also write this ratio in the form $4 : 3$ or in the fraction form $\frac{4}{3}$. This ratio does not mean that there are only 4 pages in one chapter and 3 pages in the other. There are many fractions that reduce to $\frac{4}{3}$. If there are 105 pages total in both chapters, then there are 60 pages in the first chapter and 45 pages in the second because $60 + 45 = 105$ and $\frac{60}{45} = \frac{4}{3}$.

Proportion

A **proportion** is an equation stating that two ratios are equal.

Proportions may involve only numbers as in $\frac{2}{3} = \frac{10}{15}$. However, proportions can also be used to find unknown quantities in applications, and in such cases, will involve variables. One method of solving proportions with variables is to "clear" the equation of fractions by first multiplying both sides of the equation by the least common multiple (LCM) of the denominators. This method is illustrated in Example 1.

✐ NOTE

Proportions can be used to solve many everyday types of word problems. Using the correct ratios of units on both sides of the proportion is critical. One of the following conditions must be true:

1. The numerators agree in type and the denominators agree in type.

2. The numerators correspond and the denominators correspond.

Example 1: Proportions

Solve the following proportions.

a. $\dfrac{x-2}{5x} = \dfrac{6}{3x}$

Solution

$$\overset{3}{\cancel{15x}} \cdot \left(\frac{x-2}{\cancel{5x}} \right) = \overset{5}{\cancel{15x}} \cdot \left(\frac{6}{\cancel{3x}} \right) \qquad x \neq 0 \, ; \, \text{LCM} = 15x$$

$$3(x-2) = 5(6)$$

$$3x - 6 = 30$$

$$3x = 36$$

$$x = 12$$

Check

$$\frac{12-2}{5\cdot12} \overset{?}{=} \frac{6}{3\cdot12}$$

$$\frac{10}{60} \overset{?}{=} \frac{6}{36}$$

$$\frac{1}{6} = \frac{1}{6}$$

Thus, the solution is $x = 12$.

b. $\dfrac{7}{x-2} = \dfrac{5}{x}$

Solution

$$x(x-2)\cdot\frac{7}{x-2} = x(x-2)\cdot\frac{5}{x} \qquad x \neq 0, 2;$$
$$\text{LCM} = x(x-2)$$
$$7x = 5(x-2)$$
$$7x = 5x-10$$
$$2x = -10$$
$$x = -5$$

Check

$$\frac{7}{-5-2} \overset{?}{=} \frac{5}{-5}$$

$$\frac{7}{-7} \overset{?}{=} \frac{5}{-5}$$

$$-1 = -1$$

Thus, the solution is $x = -5$.

Solving Equations with Rational Expressions

An equation such as

$$\frac{3}{x}+\frac{1}{8} = \frac{13}{4x}$$

that involves the sum of rational expressions is *not a proportion*. However, the method of finding the solution to the equation is similar in that we "clear" the fractions by multiplying both sides of the equation by the LCM of the denominators. In this example the LCM of the denominators is $8x$ and we can proceed as follows.

$$\frac{3}{x} + \frac{1}{8} = \frac{13}{4x}$$

Note $x \neq 0$.

$$8x\left(\frac{3}{x} + \frac{1}{8}\right) = 8x \cdot \frac{13}{4x}$$

Multiply both sides by $8x$, the LCM of the denominators.

$$8x \cdot \frac{3}{x} + 8x \cdot \frac{1}{8} = \overset{2}{8x} \cdot \frac{13}{4x}$$

Use the distributive property.

$$24 + x = 26$$

Simplify.

$$x = 2$$

As we have seen, rational expressions may contain variables in either the numerator or denominator or both. In any case, a general approach to solving equations that contain rational expressions is as follows.

Solving Equations Containing Rational Expressions

Step 1: Find the LCM of the denominators.

Step 2: Multiply both sides of the equation by this LCM and simplify.

Step 3: Solve the resulting equation. (This equation will have only polynomials on both sides.)

Step 4: Check each solution in the **original equation**. (Remember that no denominator can be 0 and any solution that gives a 0 denominator is to be discarded.)

Checking is particularly important when equations have rational expressions. Multiplying by the LCM may introduce solutions that are not solutions to the original equation. Such solutions are called **extraneous solutions** or **extraneous roots** and occur because multiplication by a variable expression may in effect be multiplying the original equation by 0.

Example 2: Solving Equations Involving Rational Expressions

State any restrictions on the variable, and then solve the equation.

a. $\dfrac{1}{x+5} = \dfrac{2}{x^2 - x}$

Solution

The restrictions on the variable occur for values of the variable that will make the rational expressions undefined.

Setting each denominator equal to zero, we have

$$x + 5 = 0 \quad \text{and} \quad x^2 - x = 0$$
$$x = -5 \qquad \quad x(x-1) = 0$$
$$x = 0, 1 \quad \text{By the zero-factor property.}$$

Thus $x \neq -5, 0, 1$.

Next, find the LCM of the denominators and then multiply each term on both sides of the equation by the LCM.

$$\left.\begin{array}{l} x+5 \\ x^2 - x = x(x-1) \end{array}\right\} \text{LCM} = x(x-1)(x+5)$$

$$x(x-1)(x+5) \cdot \frac{1}{x+5} = x(x-1)(x+5) \cdot \frac{2}{x(x-1)} \qquad x \neq -5, 0, 1$$

$$x(x-1) = 2(x+5)$$

$$x^2 - x = 2x + 10$$

$$x^2 - 3x - 10 = 0$$

$$(x-5)(x+2) = 0$$

$$x - 5 = 0 \quad \text{or} \quad x + 2 = 0$$

$$x = 5 \qquad \qquad x = -2$$

Since 5 and −2 are not restrictions, there are two solutions, $x = 5$ and $x = -2$.

b. $\dfrac{1}{x} = \dfrac{2x}{x^2 - 16} - \dfrac{1}{x-4}$

Solution

The restrictions on the variable occur for values of the variable that will make the rational expressions undefined.

Setting each denominator equal to zero, we have

$$x = 0 \quad \text{and} \qquad x^2 - 16 = 0 \qquad \text{and} \ \ x - 4 = 0$$

$$(x+4)(x-4) = 0 \qquad \qquad x = 4$$

$$x = -4, 4$$

Thus $x \neq -4, 0, 4$.

Next, find the LCM of the denominators and then multiply each term on both sides of the equation by the LCM.

$$\left.\begin{array}{l} x \\ x^2 - 16 = (x+4)(x-4) \\ x - 4 \end{array}\right\} \text{LCM} = x(x+4)(x-4)$$

$$x(x+4)(x-4) \cdot \frac{1}{x} = x(x+4)(x-4) \cdot \frac{2x}{(x+4)(x-4)} - x(x+4)(x-4) \cdot \frac{1}{x-4}$$

$$(x+4)(x-4) = x \cdot 2x - x(x+4) \qquad\qquad x \neq -4, 0, 4$$

$$x^2 - 16 = 2x^2 - x^2 - 4x$$

$$4x - 16 = 0$$

$$4x = 16$$

$$x = 4$$

Note that 4 is one of the restrictions, so there is no solution. The solution set is the empty set, ∅. The original equation is a contradiction.

Solving Radical Equations

The last one-variable equations that we will discuss are those that contain radical expressions. A **radical equation** is an equation that has at least one radical expression containing a variable, while any non-radical expressions are polynomial terms. As with the rational equations discussed earlier, we will develop a general method of solution that converts a given radical equation into a polynomial equation. We will see, however, that just as with rational equations, we must check our potential solutions carefully to see if they actually solve the original radical equation.

Our method of solving rational equations involved multiplying both sides of the equation by an algebraic expression (the LCM of the denominators of all the rational expressions), and in some cases potential solutions had to be discarded because they led to division by 0 in one or more of the rational expressions. Something similar can happen with radical equations. Since our goal is to convert a given radical equation into a polynomial equation, one reasonable approach is to raise both sides of the equation to whatever power is necessary to "undo" the radical (or radicals). The problem is that this does *not* result in an equivalent equation; the only means we have of transforming an equation into an equivalent equation is to add the same quantity to both sides or to multiply both sides by a non-zero quantity.

We won't *lose* any solutions by raising both sides of an equation to the same power, but we may *gain* some extraneous solutions. We identify these and discard them by simply checking all of our eventual solutions in the original equation.

A simple (indeed trivial) example will make this clear. Consider the equation

$$x = -5.$$

This equation is so simple that it is its own solution. But if, solely for the purposes of demonstration, we square both sides, we obtain the equation

$$x^2 = 25.$$

This second-degree equation can be solved by factoring the polynomial $x^2 - 25$ or by taking the square root of both sides, and in either case we obtain the solution set $\{-5, 5\}$. That is, by squaring both sides of the original equation, we gained a second (and false) solution, 5.

The specifics of our method are as follows.

Solving Radical Equations

Step 1: Begin by isolating the radical expression on one side of the equation. If there is more than one radical expression, choose one to isolate on one side.

Step 2: Raise both sides of the equation by the power necessary to "undo" the isolated radical. That is, if the radical is an nth root, raise both sides to the nth power.

Step 3: If any radical expressions remain, simplify the equation if possible and then repeat steps 1 and 2 until the result is a polynomial equation. When a polynomial equation has been obtained, solve the equation using polynomial methods.

Step 4: Check your solutions in the original equation! Any extraneous solutions must be discarded.

If the equation contains many radical expressions, and especially if they are of differing indices, eliminating all the radicals may be a long process! The equations that we will solve will not require more than a few repetitions of steps 1 and 2.

Example 3: Radical Equations

Solve the following radical equations.

a. $\sqrt{4-x}-2=x$

Solution

$$\sqrt{4-x}-2=x$$
$$\sqrt{4-x}=x+2$$

There is only one radical expression, so we isolate it and proceed to square both sides.

$$\left(\sqrt{4-x}\right)^2=(x+2)^2$$
$$4-x=x^2+4x+4$$
$$0=x^2+5x$$
$$0=x(x+5)$$

The result is a second-degree polynomial equation that can be easily solved by factoring.

$$x=0,\;\cancel{-5}$$
$$x=0$$

Note, though, that $\sqrt{4-(-5)}\neq-5+2$. So -5 must be discarded. The solution is the single number 0.

b. $\sqrt[3]{x^2+5x+21}-3=0$

Solution

$$\sqrt[3]{x^2+5x+21}-3=0$$
$$\sqrt[3]{x^2+5x+21}=3$$

We first isolate the radical (a third root in this case) and then raise both sides to the third power.

$$\left(\sqrt[3]{x^2+5x+21}\right)^3=3^3$$
$$x^2+5x+21=27$$
$$x^2+5x-6=0$$
$$(x+6)(x-1)=0$$

The resulting second-degree equation can again be solved by factoring.

$$x=-6,1$$

Both roots solve the original equation.

In the problems that we will consider here, equations containing terms with positive rational exponents can be viewed as radical equations. Rewriting each term that has a positive rational exponent as a radical will allow us to use the method previously developed to solve rational equations.

Example 4: Equations with Positive Rational Exponents

Solve the following equations with rational exponents.

a. $x^{\frac{3}{4}} - 8 = 0$

Solution

$$x^{\frac{3}{4}} - 8 = 0$$

$$x^{\frac{3}{4}} = 8$$

$$\sqrt[4]{x^3} = 8$$

$$x^3 = 8^4$$

$$x = 8^{\frac{4}{3}}$$

$$x = \left(8^{\frac{1}{3}}\right)^4$$

$$x = (2)^4$$

$$x = 16$$

Since the term containing the rational exponent can be rewritten as a radical expression, we will begin by isolating that term.

Raising both sides to the fourth power eliminates the fourth root.

Raising both sides to the $\frac{1}{3}$ power solves the equation for x, but we can evaluate the expression on the right-hand side.

Verify that this number solves the original equation.

b. $\left(18x^2 - 54x - 8\right)^{\frac{1}{6}} = 2$

Solution

$$\left(18x^2 - 54x - 8\right)^{\frac{1}{6}} = 2$$

$$\sqrt[6]{18x^2 - 54x - 8} = 2$$

$$18x^2 - 54x - 8 = 2^6$$

$$18x^2 - 54x - 8 = 64$$

$$18x^2 - 54x - 72 = 0$$

$$18\left(x^2 - 3x - 4\right) = 0$$

$$x^2 - 3x - 4 = 0$$

$$(x-4)(x+1) = 0$$

$$x = 4, -1$$

The exponent of $\frac{1}{6}$ indicates we should raise both sides to the sixth power in order to eliminate the radical.

We are left with a second-degree polynomial equation that can be solved by factoring.

Note that both solutions solve the original equation.

1.6 EXERCISES

☿ PRACTICE

State any restrictions on x, then solve the proportions. See Example 1.

1. $\dfrac{4x}{7} = \dfrac{x+5}{3}$

2. $\dfrac{3x+1}{4} = \dfrac{2x+1}{3}$

3. $\dfrac{10}{x} = \dfrac{5}{x-2}$

4. $\dfrac{8}{x-3} = \dfrac{12}{2x-3}$

5. $\dfrac{4}{x-4} = \dfrac{2}{x+3}$

6. $\dfrac{3}{x+5} = \dfrac{6}{x-2}$

7. $\dfrac{x+2}{5x} = \dfrac{x-6}{3x}$

8. $\dfrac{x-4}{3x} = \dfrac{x-2}{5x}$

9. $\dfrac{5x+2}{x-6} = \dfrac{11}{4}$

10. $\dfrac{x+9}{3x+2} = \dfrac{5}{8}$

State any restrictions on *x*, and then solve the equations. See Example 2.

11. $\dfrac{5x}{4} - \dfrac{1}{2} = -\dfrac{3}{16}$

12. $\dfrac{x}{6} - \dfrac{1}{42} = \dfrac{1}{7}$

13. $\dfrac{3x-1}{6} - \dfrac{x+3}{4} = \dfrac{7}{12}$

14. $\dfrac{x-2}{3} - \dfrac{x-3}{5} = \dfrac{13}{15}$

15. $\dfrac{2+x}{4} - \dfrac{5x-2}{12} = \dfrac{8-2x}{5}$

16. $\dfrac{4x+1}{5} = \dfrac{2x+3}{2} - \dfrac{x+2}{4}$

17. $\dfrac{2}{3x} = \dfrac{1}{4} - \dfrac{1}{6x}$

18. $\dfrac{1}{x} - \dfrac{8}{21} = \dfrac{3}{7x}$

19. $\dfrac{3}{5x} - \dfrac{1}{5} = \dfrac{3}{4x}$

20. $\dfrac{3}{8x} - \dfrac{7}{10} = \dfrac{1}{5x}$

21. $\dfrac{3}{4x} - \dfrac{1}{2} = \dfrac{7}{8x} + \dfrac{1}{6}$

22. $\dfrac{5}{3x} + \dfrac{1}{2} = \dfrac{7}{9x} - \dfrac{5}{6}$

23. $\dfrac{2}{4x+1} = \dfrac{4}{x^2+9x}$

24. $\dfrac{3}{4x-1} = \dfrac{4}{x^2+x}$

25. $\dfrac{9}{x^2-6x} = \dfrac{5}{2x-3}$

26. $\dfrac{-9}{x^2+5x} = \dfrac{8}{4-9x}$

27. $\dfrac{x}{x-4} - \dfrac{4}{2x-1} = 1$

28. $\dfrac{x}{x+3} + \dfrac{1}{x+2} = 1$

29. $\dfrac{x+2}{x+1} + \dfrac{x+2}{x+4} = 2$

30. $\dfrac{3x-2}{x+4} + \dfrac{2x+5}{x-1} = 5$

31. $\dfrac{2}{4x-1} + \dfrac{1}{x+1} = \dfrac{3}{x+1}$

32. $\dfrac{x-2}{x+4} - \dfrac{3}{2x+1} = \dfrac{x-7}{x+4}$

33. $\dfrac{x-2}{x-3} + \dfrac{x-3}{x-2} = \dfrac{2x^2}{x^2-5x+6}$

34. $\dfrac{x}{x-4} - \dfrac{12x}{x^2+x-20} = \dfrac{x-1}{x+5}$

35. $\dfrac{3x+5}{3x+2} + \dfrac{8x+16}{3x^2-4x-4} = \dfrac{x+2}{x-2}$

36. $\dfrac{3x+5}{3x+2} - \dfrac{4-2x}{3x^2+8x+4} = \dfrac{x+4}{x+2}$

37. $\dfrac{3}{3x-1} + \dfrac{1}{x+1} = \dfrac{4}{2x-1}$

38. $\dfrac{2}{x+1} + \dfrac{4}{2x-3} = \dfrac{4}{x-5}$

Solve the following radical equations. See Example 3.

39. $\sqrt{4-x} - x = 2$

40. $\sqrt{x^2 - 4x + 5} - x + 2 = 0$

41. $\sqrt{x^2 - 4x + 4} + 2 = 3x$

42. $\sqrt{50 + 7s} - s = 8$

43. $\sqrt[4]{2x+3} = -1$

44. $\sqrt{11x+3} + 4x = 18$

45. $\sqrt{x+10} + 1 = x - 1$

46. $\sqrt{x+1} + 10 = x - 1$

47. $\sqrt{x^2 - 10} - 1 = x + 1$

48. $\sqrt[3]{5x^2 - 14x} = -2$

49. $\sqrt{4z+41} + 3 = z + 2$

50. $\sqrt[3]{3 - 2x} - \sqrt[3]{x+1} = 0$

51. $\sqrt[4]{x^2 - x} = \sqrt[4]{x-1}$

52. $\sqrt[5]{7t^2 + 2t} = \sqrt[5]{5t^2 + 4}$

53. $\sqrt[3]{y^3 - 7y + 2} = \sqrt[3]{2 - 3y}$

54. $\sqrt{3y+4} + \sqrt{5y+6} = 2$

55. $\sqrt{3 - 3x} - 3 = \sqrt{3x+2}$

56. $\sqrt{2b-1} + 3 = \sqrt{10b-6}$

57. $\sqrt{5x+5} = \sqrt{4x-7} + 2$

58. $\sqrt{14y^2 - 18y + 4} + 2 = 2y$

59. $\sqrt{9x+4} = \sqrt{7x+1} + 1$

Solve the following equations. See Example 4.

60. $(x+3)^{\frac{1}{4}} + 2 = 0$

61. $(2x-5)^{\frac{1}{4}} = (x-1)^{\frac{1}{4}}$

62. $(2x-1)^{\frac{2}{3}} = x^{\frac{1}{3}}$

63. $(3y^2 + 9y - 5)^{\frac{1}{2}} = y + 3$

64. $(3x-5)^{\frac{1}{5}} = (x+1)^{\frac{1}{5}}$

65. $w^{\frac{3}{5}} + 8 = 0$

66. $z^{\frac{4}{3}} - \dfrac{16}{81} = 0$

67. $x^{\frac{2}{3}} - \dfrac{25}{49} = 0$

68. $(x-2)^{\frac{2}{3}} = (14-x)^{\frac{1}{3}}$

69. $(y-2)^{\frac{2}{3}} = (13y - 66)^{\frac{1}{3}}$

70. $(x^2 + 21)^{\frac{-3}{2}} = \dfrac{1}{125}$

71. $(x^2 + 7)^{\frac{-3}{2}} = \dfrac{1}{64}$

🚀 APPLICATIONS

72. Computers: Making a statistical analysis, Ana found 3 defective computers in a sample of 20 computers. If this ratio is consistent, how many defective computers does she expect to find in a batch of 2400 computers?

73. Manufacturing: At the Bright-As-Day light bulb plant, 3 out of each 100 bulbs produced are defective. If the daily production is 4800 bulbs, how many are defective?

74. Education: The University of Arizona has a ratio of 1 professor for every 23 students. If there are 1600 faculty members at the university, how many students are enrolled there?

75. **Baseball:** New York Yankees player Didi Gregorius has a recorded batting average of 15 hits for every 50 times at bat. If he maintains this average, how many at bats will he need to achieve 111 hits? (Round to the nearest whole number.)

76. **Cartography:** On a map of Maryland, one inch represents 4 miles. If there are 8.5 inches between Baltimore, MD and Washington, DC, how far are the two cities from each other?

77. **Architecture:** A floor plan is drawn to scale in which 1 inch represents 4 feet. What size will the drawing be for a room that is 30 feet by 40 feet? (**Hint:** Set up two proportions.)

78. **Baking:** The recipe for Nestle Tollhouse Chocolate Chip Cookies calls for 2 cups of chocolate chips to make 5 dozen cookies. If you want to bake 17 dozen cookies, how many cups of chocolate chips do you need?

79. **Car maintenance:** In the instructions for Never-Ice Antifreeze it states that 4 quarts of antifreeze are needed for every 10 quarts of radiator capacity. If Sal's car has a 22-quart radiator, how many quarts of antifreeze will it need?

80. **Landscape architecture:** An architect is to draw plans for a city park. He intends to use a scale of $\frac{1}{2}$ inch to represent 25 feet. How many inches will be needed to use for the length and width of a rectangular playing field that is 50 yards by 125 yards? (1 yard = 3 feet)

81. **Testing cars:** A test driver wants to increase the speed of the car he is driving by 3 miles per hour every 2 seconds. But he can only check his speed every 5 seconds because he is busy with other items during the test drive.
    a. By how much should he increase his speed in 5 seconds?
    b. If he starts checking his speed at 40 miles per hour, how fast should he be going after 10 seconds?

82. **Decorating:** Jack and Diane are decorating a nursery room for their baby, which will be born in a few months. In one hour, Jack can get $\frac{1}{6}$ of the nursery done and Diane can get $\frac{1}{12}$ of the nursery done. If they work together, they can get $\frac{1}{x}$ of the nursery done in one hour. Determine how many hours it will take Jack and Diane to decorate the nursery if they work together by solving the equation $\frac{1}{6}+\frac{1}{12}=\frac{1}{x}$ for x.

83. **Printing:** A local print shop has a big order of pamphlets to print, so they decide to use two of their printers for the one job. The newer printer can print the pamphlets four times as fast as the older printer. That means in one hour, the newer printer can complete $\frac{1}{x}$ of the print job and the older printer can complete $\frac{1}{4x}$ of the print job. Working together, the printers can complete the job in 4 hours. Determine how many hours it would take the newer printer to print all of the pamphlets by itself by solving the equation $\frac{1}{x}+\frac{1}{4x}=\frac{1}{4}$ for x.

84. Construction: Two groups of civil engineers are surveying an area to prepare for the construction of a shopping center. The first group is full of new college graduates and it will take them four more hours than it takes the second group, which is full of seasoned professionals. The second group can complete the job in x hours. This means that in one hour, the first group can complete $\dfrac{1}{x+4}$ of the job and the second group can complete $\dfrac{1}{x}$ of the job. Working together, they can complete the surveying job in $\dfrac{15}{4}$ hours. Determine how many hours it would take each team to complete the job individually by solving the equation $\dfrac{1}{x+4}+\dfrac{1}{x}=\dfrac{4}{15}$ for x.

85. Running: Terrence and Alicia are competing in a marathon where the average running speed is x kilometers per hour. Terrence is running 2 kilometers per hour slower than the average running speed. Alicia is running 2 kilometers per hour faster than the average running speed. After a certain amount of time, Terrence ran 4 kilometers and Alicia ran 6 kilometers.

a. Determine the speed of the average runner by solving the equation $\dfrac{4}{x-2}=\dfrac{6}{x+2}$ for x.

b. What was Terrence's average running speed?

c. What was Alicia's average running speed?

d. How long did it take Terrence to run 4 kilometers and Alicia to run 6 kilometers?

✏ WRITING & THINKING

In simplifying rational expressions, the result is a rational or polynomial expression. However, in solving equations with rational expressions, the goal is to find a value (or values) for the variable that will make the equation a true statement. Many students confuse these two ideas. To avoid confusing the techniques for adding and subtracting rational expressions with the techniques for solving equations, simplify the expression in part **a.** and solve the equation in part **b.** Explain, in your own words, the differences in your procedures. Assume no denominator has a value of 0.

86. a. $\dfrac{10}{x}+\dfrac{31}{x-1}+\dfrac{4x}{x-1}$

b. $\dfrac{10}{x}+\dfrac{31}{x-1}=\dfrac{4x}{x-1}$

87. a. $\dfrac{-4}{x^2-16}+\dfrac{x}{2x+8}-\dfrac{1}{4}$

b. $\dfrac{-4}{x^2-16}+\dfrac{x}{2x+8}=\dfrac{1}{4}$

88. a. $\dfrac{3x}{x^2-4}+\dfrac{5}{x+2}+\dfrac{2}{x-2}$

b. $\dfrac{3x}{x^2-4}+\dfrac{5}{x+2}=\dfrac{2}{x-2}$

89. a. $\dfrac{7}{5x}+\dfrac{2}{x-4}-\dfrac{3}{5x}$

b. $\dfrac{7}{5x}+\dfrac{2}{x-4}=\dfrac{3}{5x}$

90. a. $\dfrac{2}{x+9}-\dfrac{2}{x-9}+\dfrac{1}{2}$

b. $\dfrac{2}{x+9}-\dfrac{2}{x-9}=\dfrac{1}{2}$

Chapter 2

LINEAR EQUATIONS IN TWO VARIABLES

2.1 THE CARTESIAN COORDINATE SYSTEM

■ TOPICS

- The Cartesian Coordinate System
- The Graph of an Equation
- The Distance and Midpoint Formulas
- Graphing an Equation Using Technology

The Cartesian Coordinate System

Previously, we studied the algebra of a single variable. We learned how to solve equations and inequalities in one variable, and how to isolate a single variable in equations with more than one variable.

While many important problems can be studied and solved using one variable, many more problems require two or more variables. In this chapter, you will learn how to write and solve equations in two variables.

The first question we need to answer is how to express a solution to an equation in two variables. Recall that a solution to an equation in one variable (for example, x) is any value of x that when substituted in the equation results in a true statement. Consider an equation in two variables x and y. A particular solution of the equation, if there is one, must consist of a value for x and a *corresponding* value for y. The solution consists of a *pair* of numbers, called an ordered pair.

Ordered Pairs

An **ordered pair** (a, b) consists of two real numbers a and b such that the order of a and b matters. That is, $(a, b) = (b, a)$ if and only if $a = b$. The number a is called the **first coordinate** and the number b is called the **second coordinate**.

We can then write a solution to an equation in two variables as an ordered pair (x, y).

The ordered pair notation is also very useful for graphing the solutions to equations in two variables. For one-variable problems, we used the real number line, a one-dimensional coordinate system, to graph solutions. Think about how difficult it would be to interpret solutions if we plotted the values of both variables on a single number line. Because the two variables are linked in each solution, a two-dimensional coordinate system is a more natural place to graph solutions of two-variable equations. The coordinate system we use is named after René Descartes (pronounced "day-cart"), the 17th century French mathematician largely responsible for its development.

The Cartesian Coordinate System

The **Cartesian coordinate system** (also called the **Cartesian plane**) consists of two perpendicular real number lines (each called an **axis**) intersecting at the 0 point of each line. The point of intersection is called the **origin** of the system, and the four quarters defined by the two lines are called the **quadrants** of the plane, numbered as indicated below in Figure 1. Because the Cartesian plane consists of two crossed real lines, it is often given the symbol $\mathbb{R} \times \mathbb{R}$, or $\mathbb{R}^2$. Each point P in the plane is identified by an ordered pair. The first coordinate indicates the horizontal displacement of the point from the origin, and the second coordinate indicates the vertical displacement. Figure 1 is an example of a Cartesian coordinate system and illustrates how several ordered pairs are **graphed**, or **plotted**.

> ### ⚠ CAUTION
>
> Unfortunately, mathematics uses parentheses to denote ordered pairs as well as open intervals, which sometimes leads to confusion. Context is the key to interpreting notation correctly. For instance, in the context of solving a one-variable inequality, the notation $(-2, 5)$ most likely refers to the open interval with endpoints at -2 and 5, while in the context of solving an equation in two variables, $(-2, 5)$ probably refers to a point in the Cartesian plane.

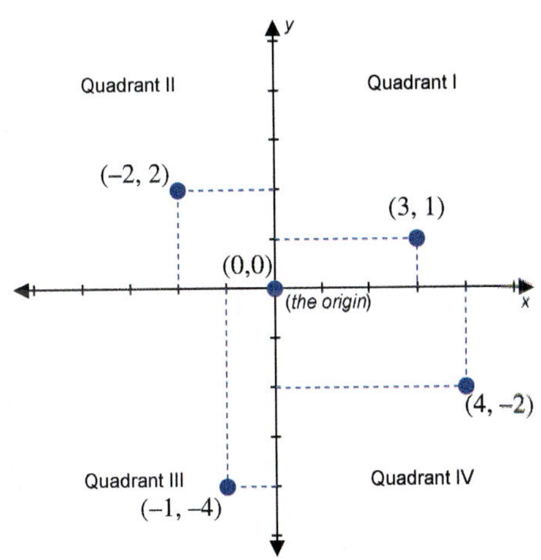

FIGURE 1: The Cartesian Plane

Example 1: Plotting Points in the Cartesian Coordinate System

Plot the following ordered pairs on the Cartesian plane, and identify which quadrant they lie in (or which axis they lie on).

a. $(2, 3)$ **b.** $(-5, 0)$ **c.** $(1, -3)$

d. $(-2, 4)$ **e.** $(-6, -6)$ **f.** $(0, 5)$

> ### 📝 NOTE
>
> While there is no required method when plotting points, it is helpful to establish a set pattern that you follow; for example, you may always count the horizontal value first, then the vertical displacement.

Solution

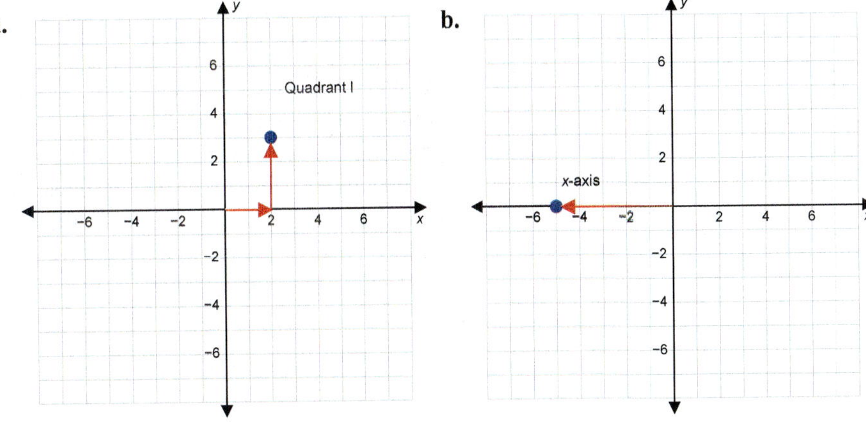

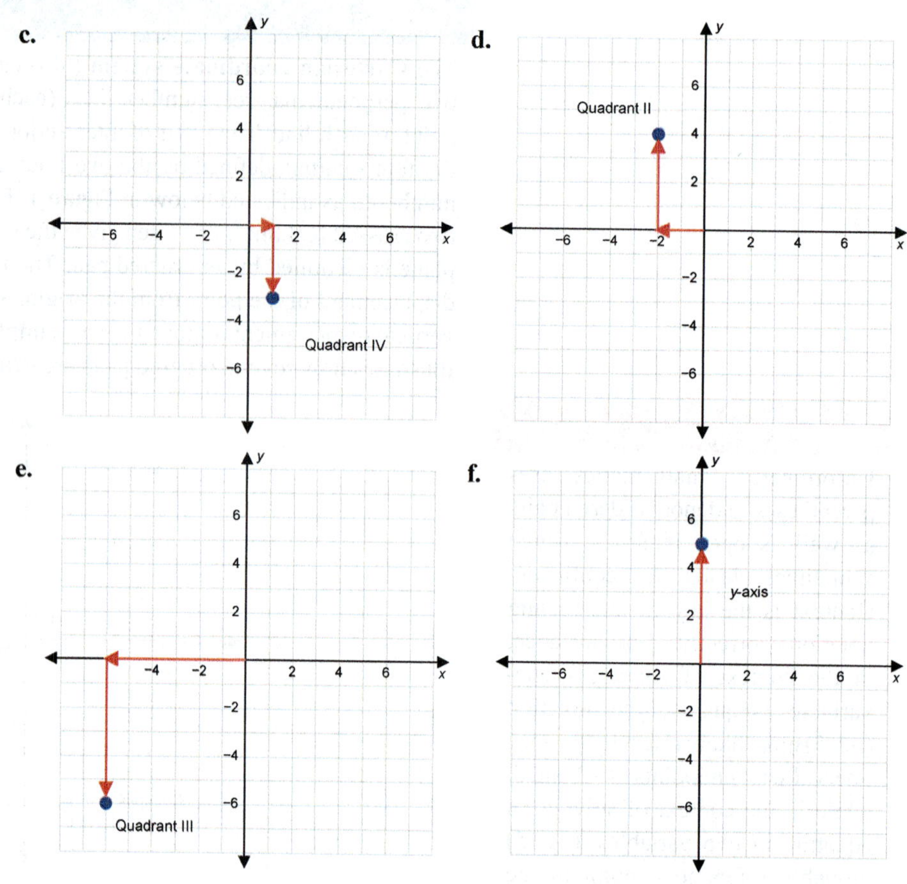

The Graph of an Equation

A solution of an equation in x and y must consist of a value a for x and a corresponding value b for y. It is natural to write such a solution as an ordered pair (a, b), and equally natural to graph it as a point in the plane whose coordinates are a and b. In this context we refer to the horizontal number line as the **x-axis**, the vertical number line as the **y-axis**, and the two coordinates of the ordered pair (a, b) as the **x-coordinate** and the **y-coordinate**.

As we will see, an equation in x and y usually consists of far more than one ordered pair (a, b). The **graph of an equation** is a plot in the Cartesian plane of *all* of the ordered pairs that make up the solution set of the equation.

We can make rough sketches of the graphs of many equations just by plotting enough solutions to give us a sense of the entire solution set. We can find individual ordered pair solutions of a given equation by selecting numbers that seem appropriate for one of the variables and then solving the equation for the other variable. This changes the task of solving a two-variable equation into that of solving a one-variable equation, and we have all the previous methods at our disposal to accomplish this. The process is illustrated in Example 2.

Example 2: Graphing Equations in Two Variables

Sketch graphs of the following equations by plotting points.

a. $2x - 5y = 10$ **b.** $x^2 + y^2 - 6x = 0$ **c.** $y = x^2 - 2x$

Solution

a. In each row in the first table, we select a value for one of the two variables.

x	y
−3	?
0	?
?	0
?	5
1	?

Once we substitute a value, the equation $2x - 5y = 10$ can be solved for the other variable. An example of this is shown below.

$$2(-3) - 5y = 10$$
$$-6 - 5y = 10$$
$$-5y = 16$$
$$y = -\frac{16}{5}$$

This gives us a list of 5 ordered pairs that can be plotted, though the ordered pair $\left(\dfrac{35}{2}, 5\right)$ is off the coordinate system we draw.

x	y
−3	$-\dfrac{16}{5}$
0	−2
5	0
$\dfrac{35}{2}$	5
1	$-\dfrac{8}{5}$

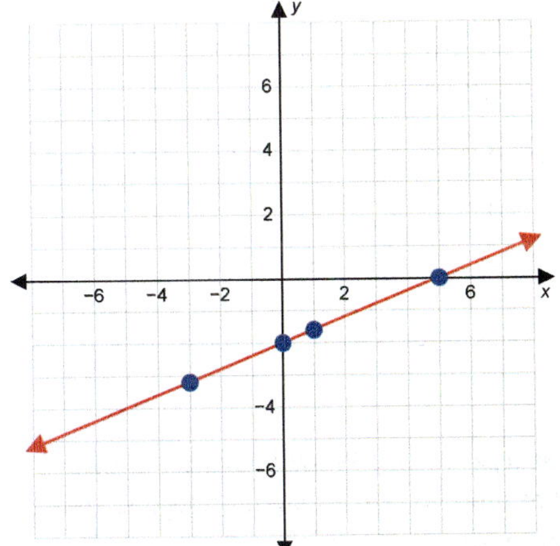

The four ordered pairs appear to lie on a straight line, and this is indeed the case. To gain more confidence in this fact, we could continue to plot more solutions of the equation, and we would find they all lie along the line that has been drawn through the four plotted ordered pairs. The infinite number of solutions of the equation are depicted by the line drawn through the plotted points.

> **NOTE**
>
> Rather than solving a new equation each time you substitute a value, it is more efficient to solve the equation for one variable before making substitutions. This method is shown in Example 2b.

b. For this example, we solve the original equation for y, then substitute several values for x to generate a series of y-values.

$$x^2 + y^2 - 6x = 0$$
$$y^2 = 6x - x^2$$
$$y = \pm\sqrt{6x - x^2}$$

Again, we plot enough solutions to feel confident in sketching the entire solution set.

x	y
0	0
1	$\pm\sqrt{5} \approx \pm2.2$
2	$\pm2\sqrt{2} \approx \pm2.8$
3	± 3
4	$\pm2\sqrt{2} \approx \pm2.8$
5	$\pm\sqrt{5} \approx \pm2.2$
6	0

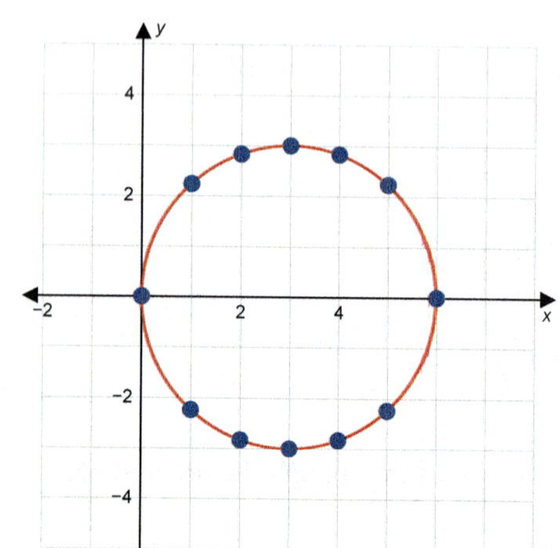

Note that for $x < 0$ and $x > 6$, the corresponding y would be imaginary, and thus irrelevant when graphing the equation. Similarly, for $y < -3$ and for any $y > 3$, the corresponding x would be a complex number (you can use the quadratic formula to verify this).

Once we plot enough solutions, the graph of the equation begins to take the shape of a circle. In fact, the graph is a circle, with center $(3, 0)$ and a radius of 3, but we will not be able to prove this claim until later.

c. Since this equation is already solved for y, we use a table of x-values and substitute them in the given equation. Again, enough points should be plotted to give some idea of the nature of the entire solution set of the equation.

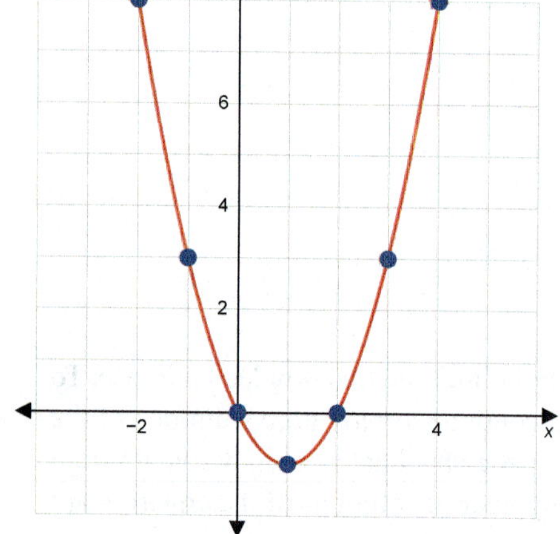

x	y
0	0
2	0
1	-1
-1	3
3	3
-2	8
4	8

The graph of $y = x^2 - 2x$ is a shape known as a *parabola*. We will encounter these shapes again later, and we will be able, at that time, to prove that our rough sketch at the left is indeed the graph of the solution set of $y = x^2 - 2x$.

Plotting points is, for the most part, easily accomplished, but it is also a rather crude and tedious method, and there are a few concerns about graphing by plotting points:

Have we really plotted enough points to accurately "fill in" the gaps and sketch the entire solution set of each equation? Is filling in the gaps justified in the first place? What proof do we have that *all* the ordered pairs along our sketches actually solve the corresponding equation? Finally, is there a faster and more sophisticated way to determine the graph of an equation?

These concerns are not trivial. Later, we will address them for linear equations. Much of the rest of this course works to answer these questions for more complicated equations and graphs. Regardless, plotting points is an important skill since it is so useful when dealing with new or unfamiliar situations.

The Distance and Midpoint Formulas

Throughout the rest of this text, we will have reasons for wanting to know, on occasion, the *distance* between two points in the Cartesian plane. We already have the tools necessary to answer this question, and we will now derive a formula that we can apply whenever necessary.

Let (x_1, y_1) and (x_2, y_2) be the coordinates of two points in the plane. By drawing the dotted lines parallel to the coordinate axes as shown in Figure 2, we can form a right triangle. Note that we are able to determine the coordinates of the vertex at the right angle from the other two vertices (x_1, y_1) and (x_2, y_2).

The lengths of the two legs of the triangle are easy to find, as they are just distances between numbers on real number lines. (The absolute value symbols are present since in general, $x_2 - x_1$ and $y_2 - y_1$ could be positive or negative.) Recall that the Pythagorean Theorem states that for a right triangle the sum of the squares of its legs is equal to the square of the hypotenuse ($a^2 + b^2 = c^2$). We can apply the Pythagorean Theorem to determine the distance labeled in Figure 2.

$$d^2 = (|x_2 - x_1|)^2 + (|y_2 - y_1|)^2, \text{ so}$$

$$d = \sqrt{(x_2 - x_1)^2 + (y_2 - y_1)^2}$$

Notice that the absolute value symbols are not necessary in the final formula, as any quantity squared is automatically nonnegative.

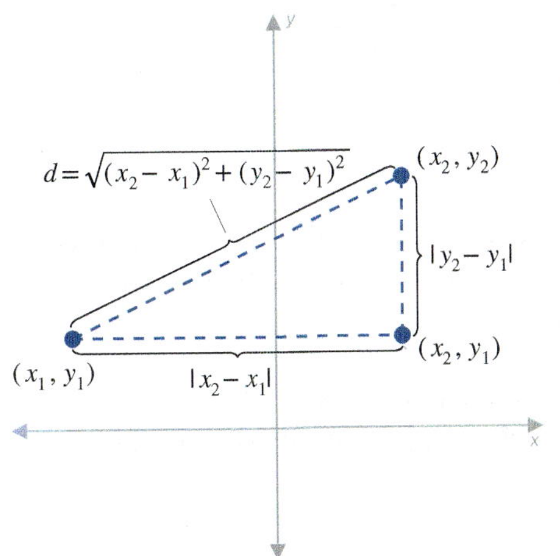

FIGURE 2: The Distance Formula

Distance Formula

The **distance** between two points (x_1, y_1) and (x_2, y_2) in the Cartesian plane is given by the following.

$$d = \sqrt{(x_2 - x_1)^2 + (y_2 - y_1)^2}$$

Example 3: Using the Distance Formula

Calculate the distance between the following pairs of points.

a. $(-4, -2)$ and $(-7, 2)$ **b.** $(5, 1)$ and $(-1, 3)$

Solutions

a. $d = \sqrt{\left((-4)-(-7)\right)^2 + \left((-2)-2\right)^2}$ Substitute the coordinates of each point into the Distance Formula.

$= \sqrt{3^2 + (-4)^2}$ Simplify.

$= \sqrt{9+16}$

$= \sqrt{25}$

$= 5$ Note that we only take the positive square root, since we are calculating a distance.

b. $d = \sqrt{\left(5-(-1)\right)^2 + \left(1-3\right)^2}$ Again, substitute the coordinates into the Distance Formula, then simplify.

$= \sqrt{36+4}$

$= \sqrt{40}$ Simplify the radical by factoring out $2^2 = 4$.

$= 2\sqrt{10}$

We will also want to be able to determine the midpoint of a line segment in the plane. That is, given two points (x_1, y_1) and (x_2, y_2), we want to know the coordinates of the point exactly halfway between the two given points.

Consider the points plotted in Figure 3. The x-coordinate of the midpoint is the average of the two x-coordinates of the given points, and the y-coordinate of the midpoint is the average of the two y-coordinates.

Since x_1 and x_2 are numbers on a real number line, and y_1 and y_2 are numbers on a (different) real number line, determining the averages of these two pairs of numbers is straightforward: the average of the x-coordinates is $\dfrac{x_1+x_2}{2}$ and the average of the y-coordinates is $\dfrac{y_1+y_2}{2}$. Putting these two coordinates together gives us the desired formula.

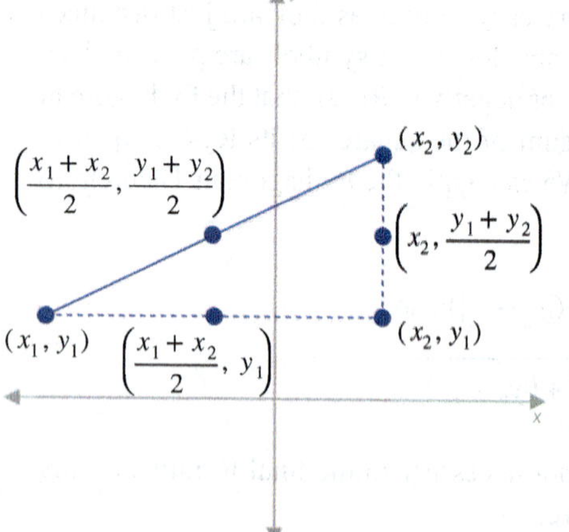

FIGURE 3: The Midpoint Formula

Midpoint Formula

The **midpoint** between two points (x_1, y_1) and (x_2, y_2) in the Cartesian plane has the following coordinates:

$$\left(\frac{x_1+x_2}{2}, \frac{y_1+y_2}{2}\right)$$

Example 4: Using the Midpoint Formula

Calculate the midpoint of the line connecting each pair of points.

a. $(5, 1)$ and $(-1, 3)$

b. $(3, 0)$ and $(-6, 11)$

Solution

a. $\left(\dfrac{5+(-1)}{2}, \dfrac{1+3}{2} \right) = (2, 2)$

In each case, we simply substitute the coordinates of each point into the midpoint formula. This has the effect of averaging both x-coordinates and both y-coordinates.

b. $\left(\dfrac{3+(-6)}{2}, \dfrac{0+11}{2} \right) = \left(-\dfrac{3}{2}, \dfrac{11}{2} \right)$

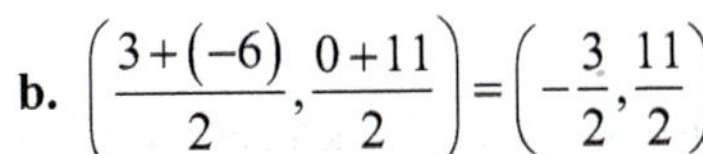

Graphing an Equation Using Technology

If we were trying to sketch the graph of the equation $y = 0.1x^4 - 2.2x^2 + 2.4x + 4.5$, the method of plotting enough ordered pairs that solve the equation would not be very efficient, or accurate. We can use a calculator to graph this equation.

Press **Y=** and type in the equation next to Y1. To type in the variable, x, press **X,T,θ,n**.

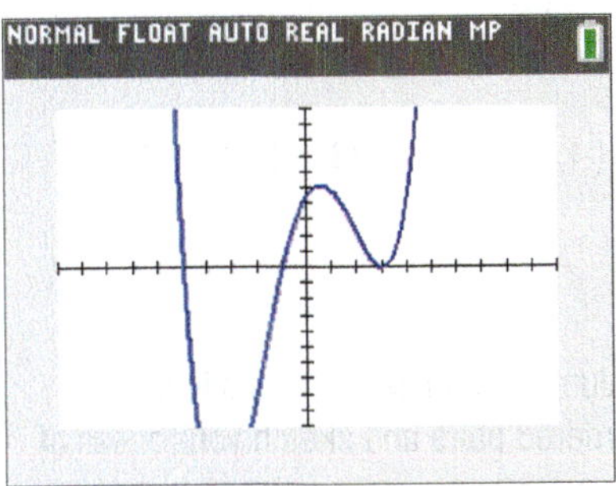

Press **graph** and the graph should appear.

Notice that you can't see the very bottom of the curve. Oftentimes when we use a calculator to graph equations, we have to adjust the viewing window to see the whole graph. To do so, press **window**. The default window displays the graph with x- and y-values ranging from -10 to 10. This window can be changed by changing the values for Xmin, Xmax, Ymin, and Ymax. Since the graph that appears descends below our viewing screen, we need to change the Ymin to something smaller, like -20.

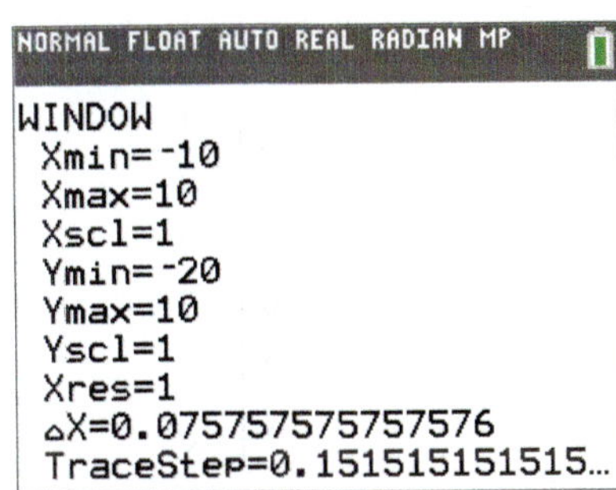

Press **graph** again and the calculator will display the graph again with the new window settings.

Keep in mind that the screen is not square: one unit on the x-axis looks longer than one unit on the y-axis, so the picture will not be an accurate representation unless the window is set to a ratio of about 3:2. One way to attain a window with this ratio is to press **zoom** and select 5:ZSquare. This will change the values in the Window screen.

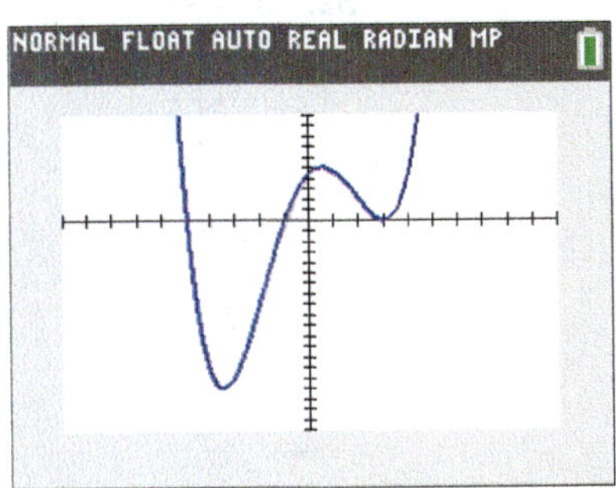

Finally, notice that the equation we graphed has two variables, specifically x and y, and is solved for y. An equation must be in this form in order to graph it on a calculator. For example, in order to graph the equation $4x + 2y = 1$, we would first have to solve the equation for y: $y = -2x + \dfrac{1}{2}$.

2.1 EXERCISES

⚙ PRACTICE

Plot the following sets of points in the Cartesian plane. See Example 1.

1. $\{(-3, 2), (5, -1), (0, -2), (3, 0)\}$ 2. $\{(-4, 0), (0, -4), (-3, -3), (3, -3)\}$

3. $\{(3, 4), (-2, -1), (-1, -3), (-3, 0)\}$ 4. $\{(2, 2), (0, 3), (4, -5), (-1, 3)\}$

5. $\{(0, 5), (-3, 2), (2, 4), (1, 1)\}$ 6. $\{(8, 3), (-3, 4), (-4, -6), (3, -4)\}$

7. $\{(-5, -4), (3, 2), (4, 5), (-2, -1), (-4, -4), (1, 1)\}$

8. $\{(-2, 5), (0, 1), (1, -1), (1, -3), (0, 0), (-1, 2), (0, -2)\}$

Identify the quadrant in which each point lies, if possible. If a point lies on an axis, specify which part (positive or negative) of which axis (x or y). See Example 1.

9. $(-2, -4)$ 10. $(0, -12)$ 11. $(4, -7)$ 12. $(-2, 0)$

13. $(9, 0)$ 14. $(3, 26)$ 15. $(-4, -7)$ 16. $(0, 1)$

17. $(17, -2)$ 18. $\left(-\sqrt{2}, 4\right)$ 19. $(-1, 1)$ 20. $(-4, 0)$

21. $(3, -9)$ 22. $(0, 0)$ 23. $(4, 3)$ 24. $(-3, -11)$

25. $(0, -97)$ 26. $\left(\dfrac{1}{3}, 0\right)$

For each of the following equations, determine the value of the missing entries in the accompanying table of ordered pairs. Then plot the ordered pairs and sketch your guess of the complete graph of the equation. See Example 2.

27. $6x - 4y = 12$

x	y
0	?
?	0
3	?
?	3

28. $y = x^2 + 2x + 1$

x	y
?	0
1	?
?	1
2	?
-3	?

29. $x = y^2$

x	y
0	?
1	?
4	?
9	?
?	$-\sqrt{2}$

30. $5x - 2 = -y$

x	y
?	0
0	?
1	?
?	7
-2	?

31. $x^2 + y^2 = 9$

x	y
0	?
?	0
-1	?
1	?
?	2

32. $y = -x^2$

x	y
0	?
-1	?
1	?
-2	?
2	?

Determine **a.** the distance between the following pairs of points, and **b.** the midpoint of the line segment joining each pair of points. See Examples 3 and 4.

33. $(-2, 3)$ and $(-5, -2)$

34. $(-1, -2)$ and $(2, 2)$

35. $(0, 7)$ and $(3, 0)$

36. $\left(-\dfrac{1}{2}, 5\right)$ and $\left(\dfrac{9}{2}, -7\right)$

37. $(-2, 0)$ and $(0, -2)$

38. $(5, 6)$ and $(-3, -2)$

39. $(13, -14)$ and $(-7, -2)$

40. $(-8, 3)$ and $(2, 11)$

41. $(-3, -3)$ and $(5, -9)$

42. $(7, -7)$ and $(-7, -6)$

43. $(5, -4)$ and $(-1, 5)$

44. $(4, 6)$ and $(2, -7)$

45. $(8, 8)$ and $(-2, -2)$

46. $\left(3, \dfrac{26}{5}\right)$ and $\left(9, -\dfrac{14}{5}\right)$

47. Given $(10, 4)$ and $(x, -2)$, find x such that the distance between these two points is 10.

48. Given $(1, y)$ and $(13, -3)$, find y such that the distance between these two points is 15.

49. Given $(x, 3)$ and $(-6, y)$, find x and y such that the midpoint between these two points is $(2, 2)$.

Find the perimeter of the triangle whose vertices are the specified points in the plane.

50. $(-2, 3), (-2, 1)$, and $(-5, -2)$

51. $(-1, -2), (2, -2)$, and $(2, 2)$

52. $(6, -1), (-6, 4)$, and $(9, 3)$

53. $(3, -4), (-7, 0)$, and $(-2, -5)$

54. $(-3, 7), (5, 1)$, and $(-3, -14)$

55. $(-12, -3), (-7, 9)$, and $(9, -3)$

🚀 **APPLICATIONS**

56. Two college friends are taking a weekend road trip. Friday they leave home and drive 87 miles north for a night of dinner and dancing in the city. The next morning they drive 116 miles east to spend a day at the beach. If they drive straight home from the beach the next day, how far do they have to travel on Sunday?

57. Your backpacker's guide contains a grid map of Paris, with each unit on the grid representing 0.25 kilometers. If the Eiffel Tower is located at $(-8, -1)$ and the Arc de Triomphe is located at $(-8, 4)$, what is the direct distance (not walking distance, which would have to account for bridges and roadways) between the two monuments in kilometers?

58. Your hotel, located at $(-1, -2)$ on the map from Exercise 57, is advertised as exactly halfway between the Eiffel Tower and Notre Dame. What are the grid coordinates of Notre Dame on your map? Find the direct distance from the Eiffel Tower to Notre Dame, rounded to the nearest hundredth of a kilometer.

59. The navigator of a submarine plots the position of the submarine and surrounding objects using a Cartesian coordinate system, where each block is one square meter.
   a. If his submarine is located at $(50, 231)$ and the mobile base to which he is heading is located at $(83, 478)$, how far is he from the mobile base?
   b. Suppose there is another submarine located halfway between the first submarine and the mobile base. What is the position of the second sub?

60. At the entrance to Paradise Island Theme Park you are given a map of the park that is in the form of a grid, with the park entrance located at $(-5, -5)$. After walking past three rides and the restrooms, you arrive at the Tsunami Water Ride, which is located at $(-3, -1)$ on the grid. If you have traveled halfway along a straight line to your favorite ride, Thundering Tower, where on the grid is your favorite ride located? How far is Thundering Tower from the park entrance on the map?

✎ WRITING & THINKING

61. Use the distance formula to prove that the triangle with vertices at the points $(1, 1)$, $(-2, -5)$, and $(3, 0)$ is a right triangle. Then determine the area of the triangle.

62. Use the distance formula to prove that the triangle with vertices at the points $(-2, 2)$, $(1, -2)$, and $(2, 5)$ is isosceles. Then determine the area of the triangle. (**Hint:** Make use of the midpoint formula.)

63. Use the distance formula to prove that the triangle with vertices at the points $(5, 1)$, $(-3, 7)$, and $(8, 5)$ is a right triangle. Then determine the area of the triangle.

64. Use the distance formula to prove that the triangle with vertices at the points $(1, 2)$, $(-2, 0)$, and $(3, 5)$ is isosceles. Then determine the area of the triangle. (**Hint:** Make use of the midpoint formula.)

65. Use the distance formula to prove that the triangle with vertices at the points $(2, 2)$, $(6, 3)$, and $(4, 11)$ is a right triangle. Then determine the area of the triangle.

66. Use the distance formula to prove that the triangle with vertices at the points $(2, -1)$, $(4, 3)$, and $(-2, -3)$ is isosceles. Then determine the area of the triangle. (**Hint:** Make use of the midpoint formula.)

67. Use the distance formula to prove that the polygon with vertices at the points $(-2, -1)$, $(6, 5)$, $(-2, 5)$, and $(6, -1)$ is a rectangle. Then determine the area of the rectangle. (**Hint:** It may help to plot the points before you begin.)

68. Plot the points $(-3, 3)$, $(-5, -2)$, $(3, -2)$, and $(1, 3)$ to demonstrate they are the vertices of a trapezoid. Then determine the area of the trapezoid.

Determine appropriate settings on a graphing utility so that each of the given points will lie within the viewing window. Answers will vary slightly.

69. $\{(-4, 1), (2, 8), (5, 7)\}$

70. $\{(12, 3), (5, -11), (-9, 6)\}$

71. $\{(3, 2), (-2, 4), (5, -3)\}$

72. $\{(30, 55), (40, 25), (-80, -10)\}$

73. $\{(3.75, -8.5), (-5.25, 6.0), (7.5, -2.25)\}$

74. $\{(63, 99), (-87, 34), (45, -22)\}$

2.2 LINEAR EQUATIONS IN TWO VARIABLES

■ TOPICS

- Recognizing Linear Equations in Two Variables
- Intercepts of the Coordinate Axes
- Horizontal and Vertical Lines
- Finding Intercepts Using Technology

Recognizing Linear Equations in Two Variables

Our first goal in this section is to recognize when an equation in two variables is linear; we want to know when the solution set of an equation is a straight line in the Cartesian plane.

Linear Equations in Two Variables

A **linear equation in two variables**, say the variables x and y, is an equation that can be written in the form $ax + by = c$, where a, b, and c are constants and a and b are not both zero. This form of such an equation is called the **standard form**.

Of course, an equation may be linear but not appear in standard form. Some algebraic manipulation is often necessary in order to determine if a given equation is linear. We will see in the next section that there are other forms of linear equations that are useful in different situations. For now, we will focus on the standard form when identifying linear equations.

Example 1: Linear Equations in Two Variables

Determine if the following equations are linear equations.

a. $3x - (2 - 4y) = x - y + 1$

b. $3x + 2(x + 7) - 2y = 5x$

c. $\dfrac{x+2}{3} - y = \dfrac{y}{5}$

d. $7x - (4x - 2) + y = y + 3(x - 1)$

e. $4x^3 - 2y = 5x$

f. $x^2 - (x - 3)^2 = 3y$

Solution

a.
$$3x - \left(2 - 4y\right) = x - y + 1$$
$$3x - 2 + 4y = x - y + 1$$
$$3x - x + 4y + y = 1 + 2$$
$$2x + 5y = 3$$

First, apply the distributive property.

Arrange the variables on one side.

Combine like terms. The equation is linear.

b. $3x + 2(x+7) - 2y = 5x$

     $3x + 2x + 14 - 2y = 5x$

       $5x - 5x - 2y = -14$

             $-2y = -14$

                $y = 7$

Begin, again, with the distributive property.

Move the variables to one side.

Combine like terms. The x variable disappears, indicating a coefficient of 0, but the coefficient on y is nonzero, so the equation is still linear.

c. $\dfrac{x+2}{3} - y = \dfrac{y}{5}$

    $\dfrac{1}{3}x + \dfrac{2}{3} - y = \dfrac{1}{5}y$

    $\dfrac{1}{3}x - y - \dfrac{1}{5}y = -\dfrac{2}{3}$

      $\dfrac{1}{3}x - \dfrac{6}{5}y = -\dfrac{2}{3}$

For this equation, we need to separate the fraction into a variable part and a constant part.

Once again, we move all the variables to one side, then combine like terms.

The equation is linear. Note that we could also have begun by clearing the fractions.

d. $7x - (4x - 2) + y = y + 3(x - 1)$

    $7x - 4x + 2 + y = y + 3x - 3$

    $3x - 3x + y - y = -3 - 2$

            $0 = -5$

After simplifying this equation, we see that the coefficients on x and y are both 0. Thus, the equation is not linear.

Further, the equation simplifies to a false statement, so it actually has no solutions!

e. $4x^3 - 2y = 5x$

The presence of the cubed term in this already simplified equation makes it not linear.

f. $x^2 - (x - 3)^2 = 3y$

   $x^2 - x^2 + 6x - 9 = 3y$

        $6x - 3y = 9$

First, expand the squared binomial term.

In contrast to the last equation, when we simplify this equation the result is linear.

Intercepts of the Coordinate Axes

Often, the goal in working with a given linear equation is to graph its solution set. Since two points determine a line, all we need to do to graph the solution set of a linear equation is to find two different solutions.

If the equation under consideration is in the two variables x and y, it is natural to call the point where the graph crosses the x-axis the **x-intercept** and the point where it crosses the y-axis the **y-intercept**. If the line does indeed cross both axes, the two intercepts are easy to find: the y-coordinate of the x-intercept is 0, and the x-coordinate of the y-intercept is 0.

The *x*- and *y*-Intercepts

Given a graph in the Cartesian plane, any point where the graph intersects the *x*-axis is called an **x-intercept** and any point where the graph intersects the *y*-axis is called a *y***-intercept**.

All *x*-intercepts are of the form $(c, 0)$ and all *y*-intercepts are of the form $(0, c)$.

Example 2: Finding Intercepts and Graphing Linear Equations

Find the *x*- and *y*-intercepts of the following equations, and sketch their graphs.

a. $3x - 4y = 12$

b. $4x - (3 - x) + 2y = 7$

Solutions

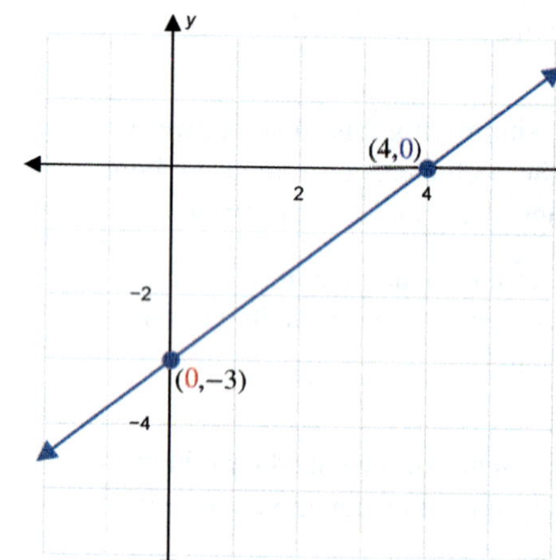

a. To find the two intercepts, first set x equal to 0 and solve for y.

$$3x - 4y = 12$$
$$3(0) - 4y = 12$$
$$y = -3$$

Then set y equal to 0 and solve for x.

$$3x - 4y = 12$$
$$3x - 4(0) = 12$$
$$x = 4$$

This gives us the coordinates of the two intercepts, *y*-intercept: $(0, -3)$ and *x*-intercept: $(4, 0)$. Once we have plotted the intercepts, drawing a straight line through them gives us the graph of the equation.

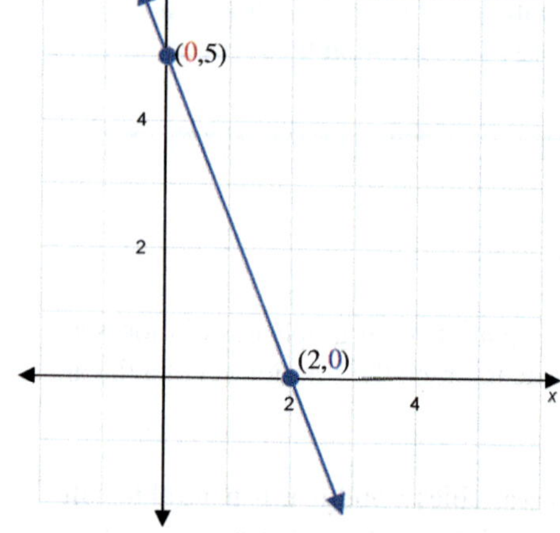

b. Again, find the two intercepts by setting the appropriate variables equal to 0.

$$4x - (3 - x) + 2y = 7 \qquad 4x - (3 - x) + 2y = 7$$
$$5x + 2y = 10 \qquad\qquad 5x + 2y = 10$$
$$5(0) + 2y = 10 \qquad\qquad 5x + 2(0) = 10$$
$$y = 5 \qquad\qquad\qquad x = 2$$

Solving the resulting equations in one variable yields the intercept solutions, *y*-intercept: $(0, 5)$ and *x*-intercept: $(2, 0)$. Plot the two intercepts, then draw the line passing through these two points. Note that in both graphs, the location of the origin has been chosen in order to conveniently plot the intercepts.

Horizontal and Vertical Lines

If a linear equation doesn't have two intercepts, the graphing process in Example 2 does not work. When might this happen? One case is when the line passes through the origin, (0, 0). Then, the x-intercept and y-intercept are the same point, instead of two distinct points. The other possibility is that the graph may be parallel to an axis (and thus not have one intercept); this happens when the equation is a horizontal or vertical line. In order to graph these equations, a second point (not an intercept) must be found in order to have two points to connect with a line.

Equations of horizontal or vertical lines are missing one of the two variables. In the absence of any other information, it is impossible to know if the solutions of equations like $x = 4$ or $y = -3$ consist of a point on the real number line or a line in the Cartesian plane. You must rely on the context of the problem to know how many variables should be considered. Throughout this chapter, all equations are assumed to be in two variables unless otherwise stated, so an equation of the form $ax = d$ (with $a \neq 0$) or $by = d$ (with $b \neq 0$) should be thought of as representing a line in the plane.

Consider an equation of the form $ax = d$. The variable y is absent, so *any* value for y will give a solution as long as we pair it with $x = \dfrac{d}{a}$. Thinking of the solution set as a set of ordered pairs, the solution consists of ordered pairs with a fixed first coordinate and arbitrary second coordinate. This describes, geometrically, a vertical line with an x-intercept of $\left(\dfrac{d}{a}, 0 \right)$. Similarly, the equation $by = d$ represents a horizontal line with y-intercept equal to $\left(0, \dfrac{d}{b} \right)$.

Example 3: Graphing Horizontal and Vertical Lines

Graph the following equations.

   a. $5x = 0$ **b.** $2x - 2 = 3$ **c.** $3x + 2(x + 7) - 2y = 5x$

Solution

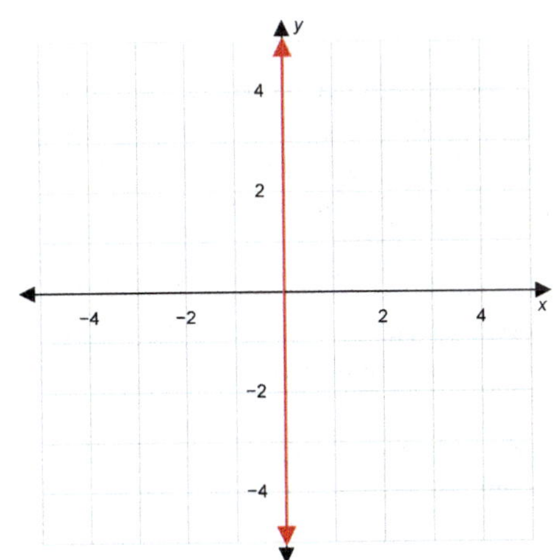

a. The first step is to divide both sides by 5, leaving the simple equation $x = 0$.

$$5x = 0$$
$$x = 0$$

The graph of this equation is the y-axis, as all ordered pairs on the y-axis have an x-coordinate of 0.

This equation is unique in that it has an infinite number of y-intercepts (since each point on the graph is on the y-axis) and one x-intercept (the origin).

Similarly, the equation $y = 0$ has an infinite number of x-intercepts and one y-intercept.

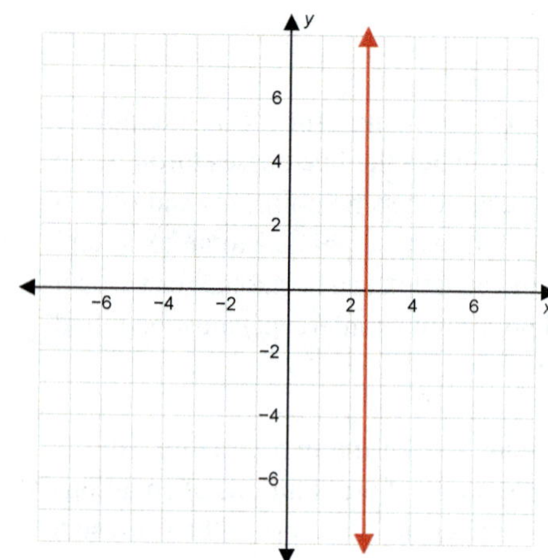

b. Upon simplifying, it is apparent that this equation also represents a vertical line, this time passing through $\dfrac{5}{2}$ on the x-axis.

$$2x - 2 = 3$$

$$x = \frac{5}{2}$$

c. We encountered this equation in Example 1b, and have already written it in standard form as shown.

$$3x + 2(x + 7) - 2y = 5x$$

$$y = 7$$

The graph of this equation is the horizontal line consisting of all those ordered pairs whose y-coordinate is 7.

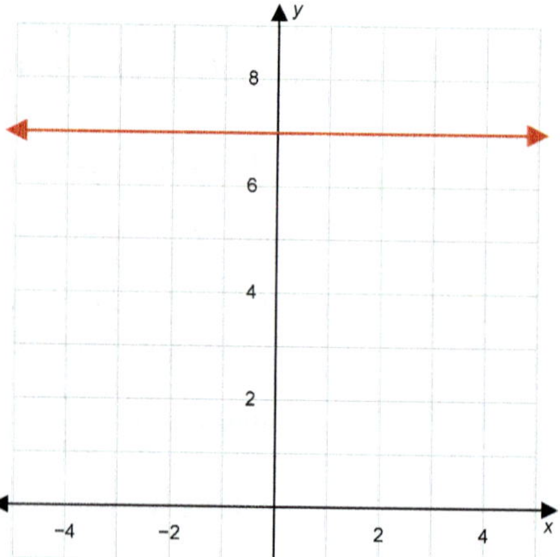

Finding Intercepts Using Technology

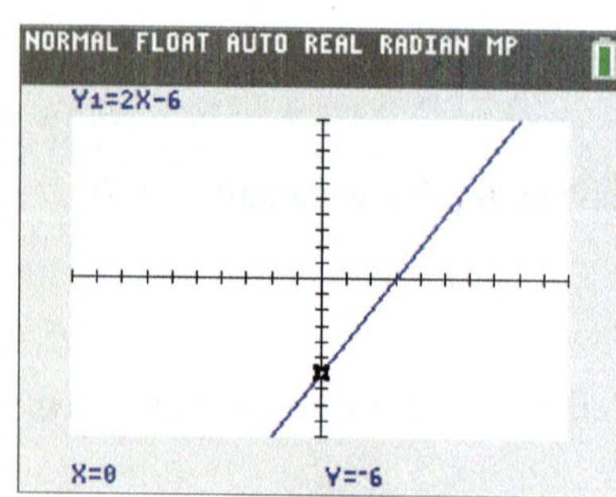

Since all linear equations can be solved for the variable y, we can use the calculator to graph them. Doing so enables us to use the calculator to find the x- and y-intercepts. Suppose we've graphed the equation $y = 2x - 6$.

To find the y-intercept, press **trace**. Since the y-intercept occurs where $x = 0$, use the arrows to move the cursor along the line until the x-value is zero. Alternatively, just press 0 and ENTER to place the cursor at that point. The corresponding y-value, -6, is shown at the bottom of the screen. So the point $(0, -6)$ is the y-intercept.

We will find the x-intercept using a different method. Press **2nd** **trace** to access the CALC menu and select 2:zero and press ENTER. The screen should now display the graph with the words "Left Bound?" shown at the bottom.

Use the arrows to move the cursor anywhere to the left of where the line crosses the x-axis and press ENTER. The screen should now say "Right Bound?".

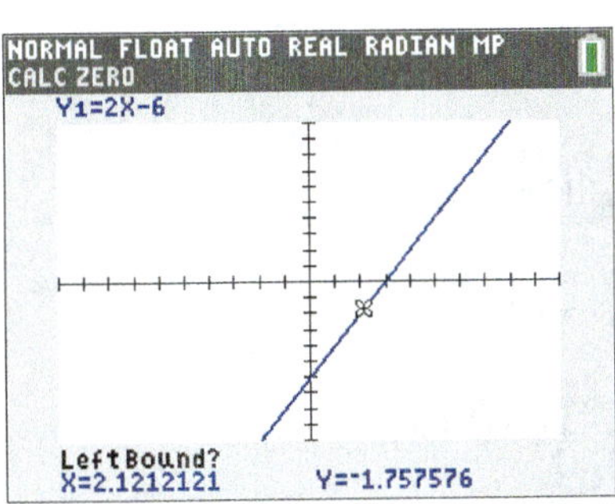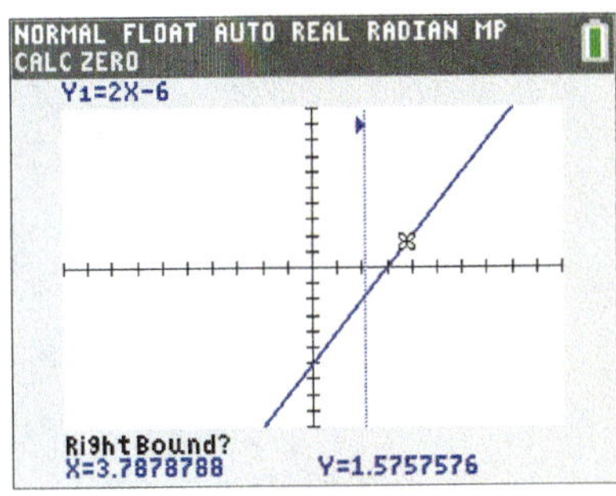

Use the right arrow to move the cursor to the right of where the line crosses the x-axis and press ENTER again. The text should now read "Guess?".

Press ENTER a third time and the x- and y-values of the x-intercept will appear at the bottom of the screen.

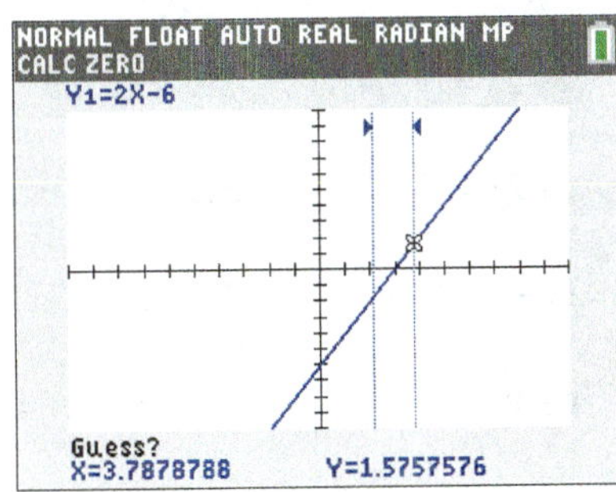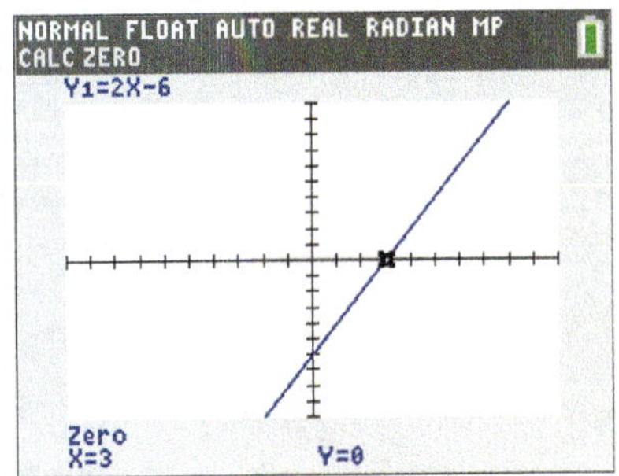

So the x-intercept is (3, 0). Both of these techniques can be used to find the x- and y-intercepts of any equation graphed with a calculator, not just linear equations.

2.2 EXERCISES

PRACTICE

Determine if the following equations are linear. See Example 1.

1. $3x + 2(x - 4y) = 2x - y$

2. $9x + 4(y - x) = 3$

3. $9x^2 - (x + 1)^2 = y - 3$

4. $3x + xy = 2y$

5. $8 - 4xy = x - 2y$

6. $\dfrac{x - y}{2} + \dfrac{7y}{3} = 5$

7. $\dfrac{6}{x} - \dfrac{5}{y} = 2$

8. $3x - 3(x - 2y) = y + 1$

9. $2y - (x + y) = y + 1$

10. $(3 - y)^2 - y^2 = x + 2$

11. $x^2 - (x - 1)^2 = y$

12. $(x + y)^2 - (x - y)^2 = 1$

13. $x(y + 1) = 16 - y(1 - x)$

14. $\dfrac{x - 3}{2} = \dfrac{4 + y}{5}$

15. $x - 2x^2 + 3 = \dfrac{x - 7}{2}$

16. $x - 3 = \dfrac{4x + 17}{5}$

17. $13x - 17y = y(7 - 2x)$

18. $y^2 - 3y = (1 + y)^2 - 2x$

19. $x - 1 = \dfrac{2y}{x} - x$

20. $3x - 4 = 89(x - y) - y$

21. $x - x(1 + x) = y - 3x$

22. $x^2 - 2x = 3 - x^2 + y$

23. $\dfrac{2y - 5}{14} = \dfrac{x - 3}{9}$

24. $16x = y(4 + (x - 3)) - xy$

Find the *x*- and *y*-intercepts of the given equations, if possible, and then sketch their graphs. See Examples 2 and 3.

25. $4x - 3y = 12$

26. $y - 3x = 9$

27. $5 - y = 10x$

28. $y - 2x = y - 4$

29. $3y = 9$

30. $2x - (x + y) = x + 1$

31. $x + 2y = 7$

32. $y - x = x - y$

33. $y = -x$

34. $2x - 3 = 1 - 4y$

35. $3y + 7x = 7(3 + x)$

36. $4 - 2y = -2 - 6x$

37. $x + y = 1 + 2y$

38. $3y + x = 2x + 3y + 4$

39. $3(x + y) + 1 = x - 5$

Match each equation to the correct graph.

40. $y = 2x + 3$

41. $2x + 3y = 4$

42. $2x - 1 = 5$

43. $y + 3 - x = 3$

44. $4y + 3 = 11$

45. $5y - x - 1 = 4y + 3x + 5$

a.

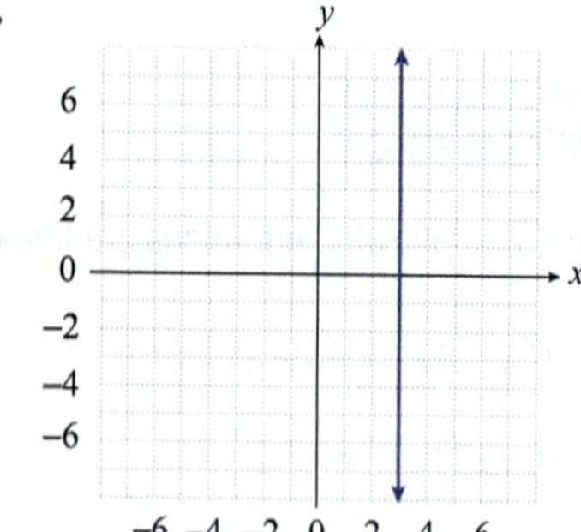

b.

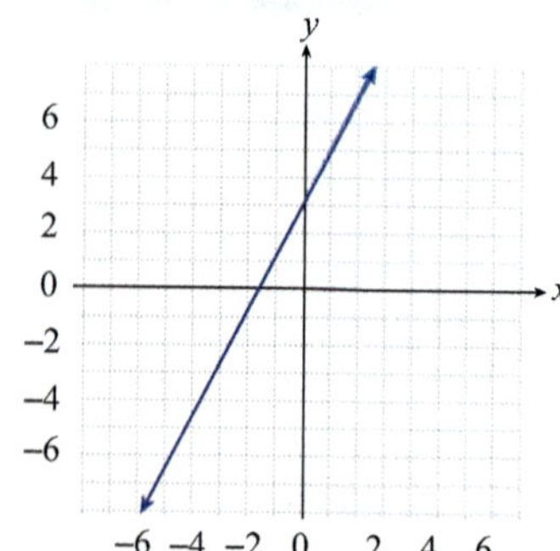

c.

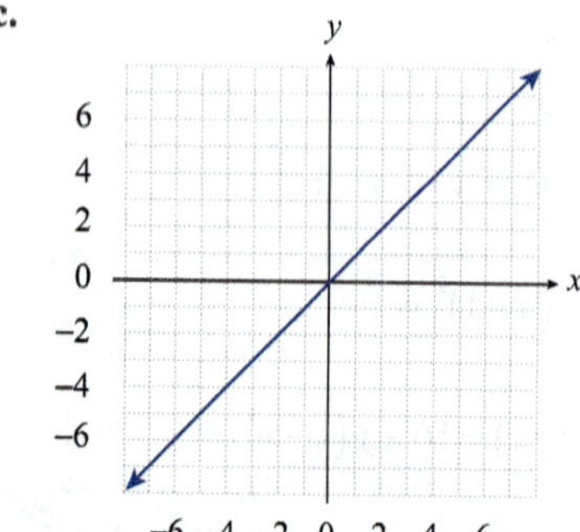

d.

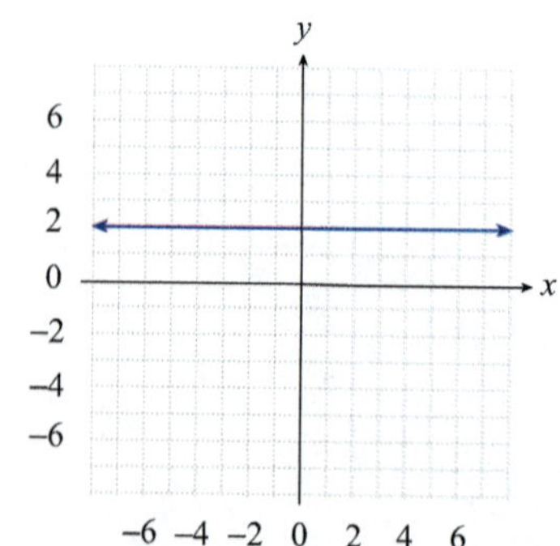

e.

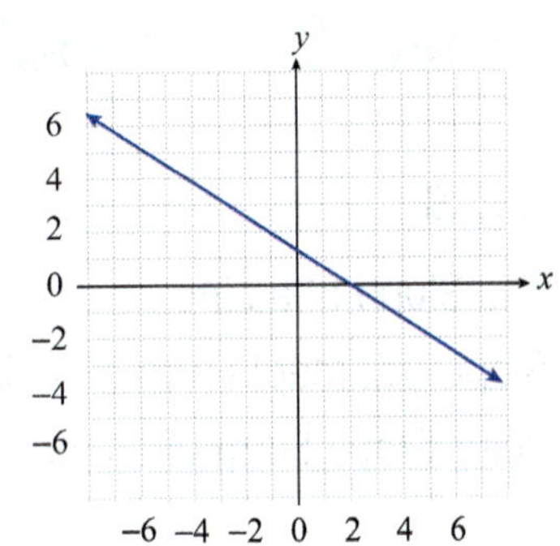

f.

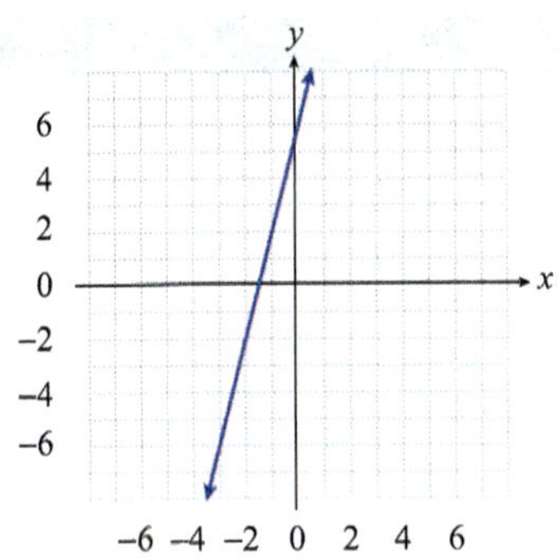

Solve each equation for the specified variable.

46. Standard form of a line: $ax + by = c$; solve for y

47. Perimeter of a triangle: $P = a + b + c$; solve for a

48. Surface area of a rectangular solid: $S = 2lw + 2wh + 2lh$; solve for w

🚀 APPLICATIONS

49. In your history class, you were told that the current population of Jamaica is approximately 24,000 more than 9 times the population of the Bahamas. Using j to represent the population of Jamaica and b to represent the population of the Bahamas, write this in the form of an equation. Then solve your equation for b to find an equation representing the population of the Bahamas. Are these equations linear?

50. The lowest point in the ocean, the bottom of the Mariana Trench, is about 1100 feet deeper than 26 times the depth of the lowest point on land, the Dead Sea. Find an equation to express the depth of the Mariana Trench, m, in terms of the depth of the Dead Sea, d. Then solve your equation for d to find the depth of the Dead Sea in terms of the depth of the Mariana Trench. Are these equations linear?

2.3 FORMS OF LINEAR EQUATIONS

■ TOPICS

- The Slope of a Line
- Slope-Intercept Form of a Line
- Point-Slope Form of a Line

The Slope of a Line

There are several ways to characterize a given line in the plane. We have already used one way repeatedly: two distinct points in the Cartesian plane determine a line. Another, often more useful, approach is to identify just one point on the line and to indicate how "steeply" the line is rising or falling as we scan the plane from left to right. It turns out that a single number is sufficient to convey this notion of "steepness."

> **The Slope of a Line**
>
> Let L stand for a given line in the Cartesian plane, and let (x_1, y_1) and (x_2, y_2) be the coordinates of any two distinct points on L. The **slope** of the line L is the ratio $\dfrac{y_2 - y_1}{x_2 - x_1}$ which can be described in words as "change in y over change in x" or "rise over run."

> **⚠ CAUTION**
>
> It doesn't matter how you assign the labels (x_1, y_1) and (x_2, y_2) to the two points you are using to calculate slope, but it *is* important that you are consistent as you apply the formula. You cannot change the order in which you are subtracting as you determine the numerator and denominator in the slope formula.
>
> | Correct | Incorrect |
> |---|---|
> | $\dfrac{y_2 - y_1}{x_2 - x_1}$ ✓ | $\dfrac{y_1 - y_2}{x_2 - x_1}$ ✗ |
> | $\dfrac{y_1 - y_2}{x_1 - x_2}$ ✓ | $\dfrac{y_2 - y_1}{x_1 - x_2}$ ✗ |

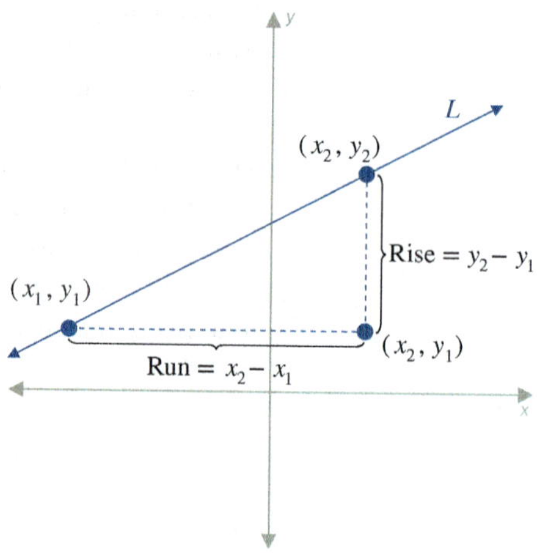

FIGURE 1: Rise and Run between Two Points

In the line drawn in Figure 1, the ratio $\dfrac{y_2 - y_1}{x_2 - x_1}$ is positive, the line rises from the lower left to the upper right, and we say that the line has a positive slope. If the rise and run have opposite signs, the slope of the line is negative and the line under consideration would fall from the upper left to the lower right.

Example 1: Calculating the Slope of a Line

Determine the slopes of the lines passing through the following pairs of points in $\mathbb{R}^2$.

a. $(-4, -3)$ and $(2, -5)$　　　**b.** $\left(\dfrac{3}{2}, 1\right)$ and $\left(1, -\dfrac{4}{3}\right)$　　　**c.** $(-2, 7)$ and $(1, 7)$

Solution

a. $\dfrac{-3-(-5)}{-4-2} = \dfrac{2}{-6} = -\dfrac{1}{3}$

We calculate the slope in two ways; first set $(x_1, y_1) = (-4, -3)$ and $(x_2, y_2) = (2, -5)$.

$\dfrac{-5-(-3)}{2-(-4)} = \dfrac{-2}{6} = -\dfrac{1}{3}$

We get the same result by setting $(x_1, y_1) = (2, -5)$ and $(x_2, y_2) = (-4, -3)$.

b. $\dfrac{1-\left(-\dfrac{4}{3}\right)}{\dfrac{3}{2}-1} = \dfrac{\dfrac{7}{3}}{\dfrac{1}{2}} = \dfrac{7}{3}\cdot\dfrac{2}{1} = \dfrac{14}{3}$

The final answer tells us that the line through the two points rises 14 units for every run of 3 units horizontally.

c. $\dfrac{7-7}{-2-1} = \dfrac{0}{-3} = 0$

These two points have the same y-coordinate and thus lie on a horizontal line.

The formula $\dfrac{y_2 - y_1}{x_2 - x_1}$ is only valid if $x_2 - x_1 \neq 0$, since division by zero is undefined. What lines have undefined slope? If $x_2 - x_1 = 0$, then the line has two points with the same x-coordinate, which defines a *vertical* line. The other extreme is that the numerator is 0 (and the denominator is nonzero), in which case the slope is 0. For what sorts of lines will this happen? If two points on a line have the same y-coordinate, that is if $y_1 = y_2$, the line must be *horizontal*.

Slopes of Horizontal and Vertical Lines

Horizontal lines, which can be written in the form $y = c$, have a **slope of 0**.

Vertical lines, which can be written in the form $x = c$, have an **undefined slope**.

Example 2: Calculating the Slope of a Line

Determine the slopes of the lines defined by the following equations.

a. $4x - 3y = 12$　　　**b.** $2x + 7y = 9$

c. $x = -\dfrac{3}{4}$　　　**d.** $y = 9$

✐ NOTE

Intercepts are often good points to use in calculating the slope, since they have at least one coordinate equal to zero.

Solution

a. First, we find two points on the line by calculating the intercepts.

$$4x - 3y = 12$$

$$
\begin{array}{ll}
4(0) - 3y = 12 & 4x - 3(0) = 12 \\
-3y = 12 & 4x = 12 \\
y = -4 & x = 3
\end{array}
$$

$$y\text{-intercept:} (0, -4) \quad x\text{-intercept:} (3, 0)$$

$$\text{slope} = \frac{-4 - 0}{0 - 3} = \frac{-4}{-3} = \frac{4}{3}$$

Recall that the x-intercept is found by setting y equal to 0 and solving for x, and vice versa for the y-intercept.

Once we have two points, we apply the slope formula.

b. x-intercept: $\left(\dfrac{9}{2}, 0\right)$

second point on the line: $(1, 1)$

$$\text{slope} = \frac{1 - 0}{1 - \dfrac{9}{2}} = \frac{1}{-\dfrac{7}{2}} = -\frac{2}{7}$$

In this example, we have found the x-intercept. We do not have to find both intercepts; the point $(1, 1)$ is clearly on the line and is simple to use in calculation.

c. The equation is of the form $x = c$, and is a vertical line. Therefore, the slope is undefined.

d. This equation is of the form $y = c$, and is a horizontal line. Therefore, it has a slope of 0.

Note that the line in Example 2a has a positive slope and the line in Example 2b has a negative slope. Without graphing these lines, we know that the first line will rise from the lower left to the upper right part of the plane, while the second line will fall from the upper left to the lower right. You should practice your graphing skills and verify that these observations are indeed correct.

Slope-Intercept Form of a Line

Example 2 illustrates the most elementary way of determining the slope of a line from an equation. With a little work, we can develop a faster method for determining not only the slope of a line, but also the y-intercept.

Consider a nonvertical line in the plane. The variable y must appear in the linear equation that describes the line (otherwise the line would be vertical), so the equation can be solved for y. The result will be an equation of the form $y = mx + b$, where m and b are constants, and it turns out that these constants provide a lot of information about the graph of the line.

Suppose that (x_1, y_1) and (x_2, y_2) are two points that lie on the line $y = mx + b$. Then, it must be the case that $y_1 = mx_1 + b$ and $y_2 = mx_2 + b$. If we use these two points to determine the slope of the line, we obtain:

$$\text{slope} = \frac{y_2 - y_1}{x_2 - x_1} = \frac{(mx_2 + b) - (mx_1 + b)}{x_2 - x_1} = \frac{m(x_2 - x_1)}{x_2 - x_1} = m$$

Now, let's calculate the y-intercept of this line. As usual, we substitute 0 for x and then solve for y.

$$y = mx + b$$
$$y = m(0) + b$$
$$y = b$$

So, the y-intercept is $(0, b)$. Thus, the two constants m and b describe the slope and y-intercept of the line. As such, we call this the *slope-intercept* form of a linear equation.

Slope-Intercept Form of a Line

If the equation of a nonvertical line in x and y is solved for y, the result is an equation in **slope-intercept form**.

$$y = mx + b$$

The constant m is the slope of the line, and the y-intercept of the line is $(0, b)$. If the variable x does not appear in the equation, the slope is 0 and the equation is simply of the form $y = b$ and is a horizontal line.

We can make use of the slope-intercept form of a line to graph the line, as illustrated in the following example.

Example 3: Slope-Intercept Form of a Line

Use the slope-intercept form of the line to graph the equation $4x - 3y = 6$.

Solution

Solving the equation for y puts it in slope-intercept form.

$$4x - 3y = 6$$
$$-3y = -4x + 6$$
$$y = \frac{4}{3}x - 2$$

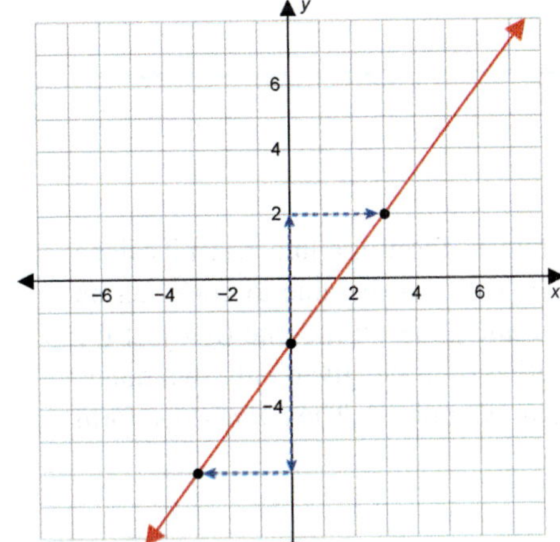

Once we have done this, we know that the line has a slope of $\frac{4}{3}$ and crosses the y-axis at -2. Immediately, we can plot the y-intercept.

A second point can now be found by using the fact that slope is "rise over run." This means a second point must lie 4 units up and 3 units to the right, i.e. at $(3, 2)$.

Alternatively, we could locate a second point by moving down 4 units and moving to the left 3 units.

In some cases, we can also make use of the slope-intercept form to find the equation of a line that has certain properties.

Find the equation of the line that passes through the point (0, 3) and has a slope of $-\dfrac{3}{5}$. Then graph the line.

Solution

NOTE

If a line is already in slope-intercept form, it's usually easier to graph the line by plotting the y-intercept and then using the slope to find a second point.

First, we write the equation of this line in slope-intercept form. We are given the y-intercept of (0, 3) and the slope of $-\dfrac{3}{5}$. We can immediately write down the equation since the only information we need is the y-intercept and the slope.

$$y = mx + b$$

$$y = -\frac{3}{5}x + 3$$

We first plot the y-intercept, then move down 3 units and to the right 5 units to find a second point.

Or, we could have plotted a second point by moving up 3 units and to the left 5 units. Both methods make use of the fact that the slope is $-\dfrac{3}{5}$.

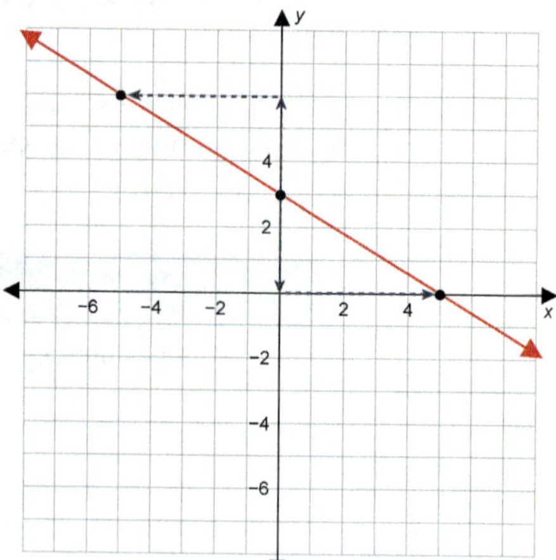

Point-Slope Form of a Line

As in Example 4, we can easily find the slope-intercept form of a line given its slope and y-intercept. In the most general case, we would like to be able to construct an equation of a line given the slope of the line and *any* point on the line (not just the y-intercept). This would allow us to find the equation of a line given only two points on that line (since we can determine the slope from two points). The *point-slope* form of a line meets these requirements.

Suppose we know that a given line has slope m and passes through the point (x_1, y_1). If we plug any two points on the line into the slope formula, we must get a result of m. Choosing (x_1, y_1) and the generic point (x, y), the definition of slope states that

$$\frac{y - y_1}{x - x_1} = m.$$

A simple rearrangement of the slope equation leads to a linear equation defined by the slope and the coordinates of a single, arbitrary point.

$$y - y_1 = m(x - x_1)$$

This is the point-slope form.

Point-Slope Form of a Line

The **point-slope form** of the equation for the line passing through the point (x_1, y_1) with slope m is

$$y - y_1 = m(x - x_1).$$

Note that $m, x_1,$ and y_1 are all constants, while x and y are variables. Note also that since the line, by definition, has slope m, vertical lines cannot be described in this form.

Example 5: Point-Slope Form of a Line

Find the equation, in slope-intercept form, of the line that passes through the point $(-2, 5)$ with slope 3.

NOTE

When asked for an equation in either standard form or slope-intercept form, it's frequently easiest to write the point-slope form and then convert the equation to the desired form.

Solution

Since we are given the slope of the line and a point on the line, we can substitute directly into the point-slope form, then solve for y to obtain the slope-intercept form.

$$y - y_1 = m\,(x - x_1) \quad \text{The point-slope form}$$
$$y - 5 = 3(x - (-2))$$
$$y - 5 = 3(x + 2)$$
$$y - 5 = 3x + 6$$
$$y = 3x + 11 \quad \text{The slope-intercept form}$$

We know that two distinct points in the plane are sufficient to determine a line. With our knowledge of the point-slope form of a line, we can now deduce the equation for the line determined by two points.

Example 6: Point-Slope Form of a Line

Find the equation, in slope-intercept form, of the line that passes through the two points $(-3, -2)$ and $(1, 6)$.

Solution

We already have a point (actually, two) on the line, but we still need the slope to use the point-slope form. We can calculate this using the two points and the slope formula.

$$m = \frac{-2-6}{-3-1} = \frac{-8}{-4} = 2$$

Now we can substitute into the point-slope form, then solve for y to obtain the desired slope-intercept equation.

$$y - y_1 = m\,(x - x_1)$$
$$y - 6 = 2(x - 1)$$
$$y - 6 = 2x - 2$$
$$y = 2x + 4$$

Note that no matter which point we substitute into the point-slope form, the resulting slope-intercept equation is the same.

We close this section with a summary of the different forms of linear equations, what information we need to write them, and what they are each most useful for.

Standard Form: $ax + by = c$

Information Required: Typically, we arrive at the standard form when given a linear equation in another form.

Potential Uses: The standard form is most useful for easily calculating the x- and y-intercepts.

Slope-Intercept Form: $y = mx + b$

Information Required: The slope m and the y-intercept $(0, b)$.

Potential Uses: The slope-intercept form makes it very easy to find the y-intercept and slope, and therefore to graph the line.

Point-Slope Form: $y - y_1 = m(x - x_1)$

Information Required: The slope m and a point on the line (x_1, y_1) or two points on the line (x_1, y_1) and (x_2, y_2).

Potential Uses: The point-slope form allows us to find the equation for a line when the y-intercept is unknown.

2.3 EXERCISES

⚡ PRACTICE

Determine the slope of the line passing through the specified points. See Example 1.

1. $(0, -3)$ and $(-2, 5)$

2. $(-3, 2)$ and $(7, -10)$

3. $(4, 5)$ and $(-1, 5)$

4. $(3, -1)$ and $(-7, -1)$

5. $(3, -5)$ and $(3, 2)$

6. $(0, 0)$ and $(-2, 5)$

7. $(-2, 1)$ and $(-5, -1)$

8. $\left(\dfrac{1}{2}, -7\right)$ and $\left(\dfrac{3}{4}, -5\right)$

9. $\left(10, \dfrac{1}{5}\right)$ and $\left(4, -\dfrac{4}{5}\right)$

10. $(-2, 4)$ and $(6, 9)$

11. $(0, -21)$ and $(-3, 0)$

12. $(-3, -5)$ and $(-2, 8)$

13. $\left(\dfrac{1}{3}, 9\right)$ and $(2, 4)$

14. $(29, -17)$ and $(31, -29)$

15. $(7, 4)$ and $(-6, 13)$

Determine the slopes of the lines defined by the following equations. See Example 2.

16. $8x - 2y = 11$

17. $2x + 8y = 11$

18. $12x - 4y = -9$

19. $4y = 13$

20. $\dfrac{x - y}{3} + 2 = 4$

21. $7x = 2$

22. $3y - 2 = \dfrac{x}{5}$

23. $3 - y = 2(5 - x)$

24. $3(2y - 1) = 5(2 - x)$

25. $\dfrac{x + 2}{3} + 2(1 - y) = -2x$

26. $2y - 7x = 4y + 5x$

27. $x - 7 = \dfrac{2y - 1}{-5}$

Use the slope-intercept form to graph the equations. See Example 3.

28. $6x - 2y = 4$

29. $3y + 2x - 9 = 0$

30. $5y - 15 = 0$

31. $x + 4y = 20$

32. $\dfrac{x - y}{2} = -1$

33. $3x + 7y = 8y - x$

34. $-4x - 4y = 8$

35. $-5x + 3y + 16 = 0$

36. $3x = 3y - 21$

Find the equation, in slope-intercept form, of the line with the given y-intercept and slope. See Example 4.

37. y-intercept $(0, -3)$; slope of $\dfrac{3}{4}$

38. y-intercept $(0, 5)$; slope of -3

39. y-intercept $(0, -7)$; slope of $-\dfrac{5}{2}$

40. y-intercept $(0, 6)$; slope of 4

41. y-intercept $(0, -9)$; slope of -5

42. y-intercept $(0, 2)$; slope of $\dfrac{1}{2}$

Find the equation, in standard form, of the line passing through the given point with the given slope.

43. point $(-1, -3)$; slope of $\dfrac{3}{2}$

44. point $(6, 0)$; slope of $\dfrac{5}{4}$

45. point $(-3, 5)$; slope of 0

46. point $(-2, -13)$; undefined slope

47. point $(3, -1)$; slope of 10

48. point $(-1, 3)$; slope of $-\dfrac{2}{7}$

49. point $(5, 11)$; slope of -3

50. point $(5, -9)$; slope of $-\dfrac{1}{2}$

Find the equation, in standard form, of the line passing through the specified points.

51. $(-1, 3)$ and $(2, -1)$

52. $(1, 3)$ and $(-2, 3)$

53. $(2, -2)$ and $(2, 17)$

54. $(-9, 2)$ and $(1, 5)$

55. $(3, -1)$ and $(8, -1)$

56. $\left(\dfrac{4}{3}, 1\right)$ and $\left(\dfrac{2}{5}, \dfrac{3}{7}\right)$

57. $(-2, 8)$ and $(5, 6)$

58. $(8, -10)$ and $(8, 0)$

59. $(7, 5)$ and $(-9, 5)$

60. $(7, 7)$ and $(9, -8)$

61. $\left(\dfrac{2}{3}, \dfrac{5}{4}\right)$ and $\left(\dfrac{3}{5}, \dfrac{9}{8}\right)$

62. $(-5, -5)$ and $(10, -11)$

Match each equation or description to the correct graph.

63. $-3x - 2y = 17$

64. $-4y + 10 = -4x$

65. $-6y + 9 = \dfrac{x}{-2}$

66. point $(-9, 7)$; slope $\dfrac{4}{3}$

67. point $(-2, 4)$; slope -2

68. point $(0, -5)$; slope -9

a.
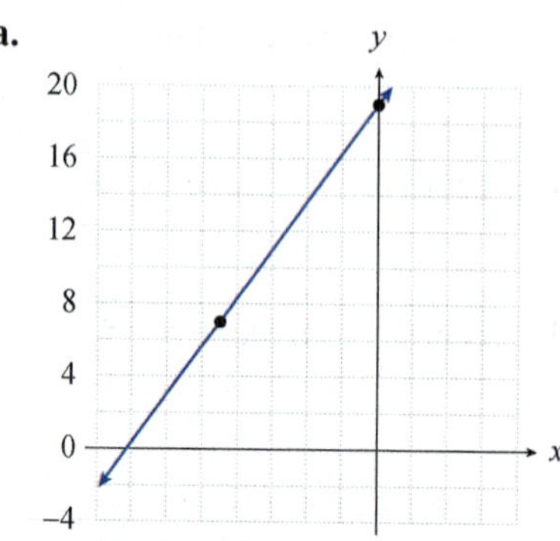

b.
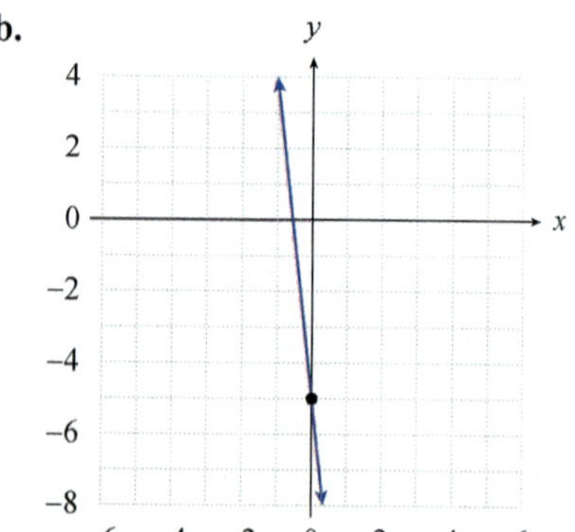

c. 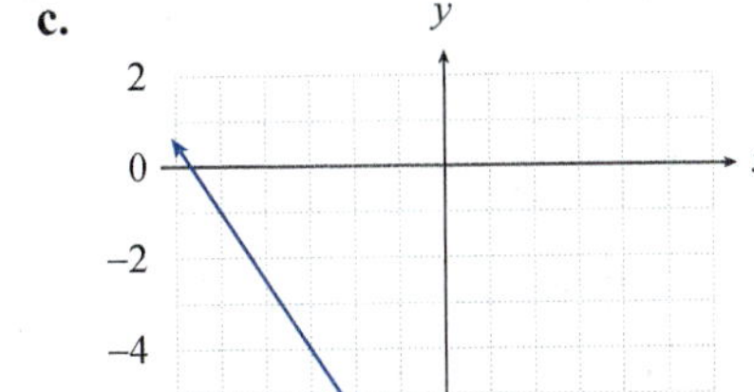

d. 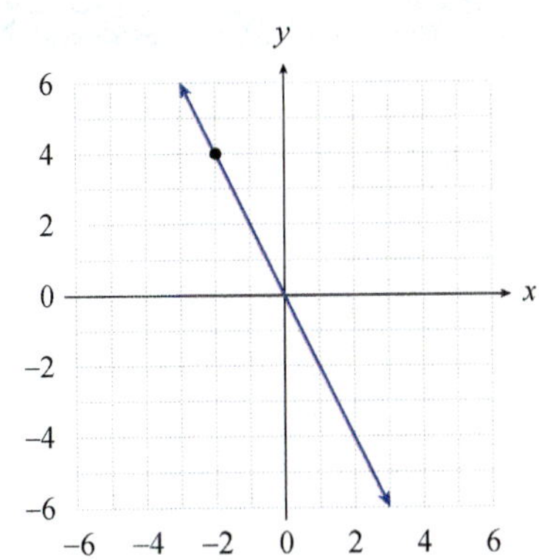

e. 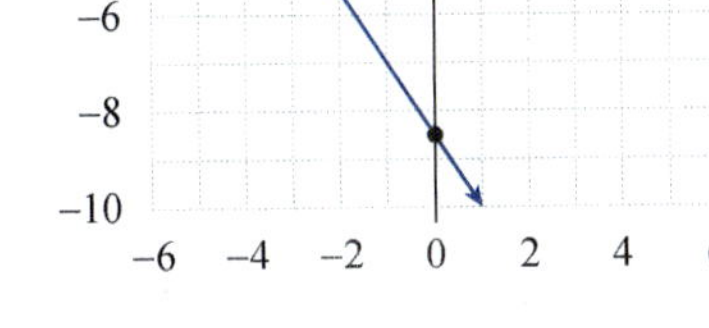

f. 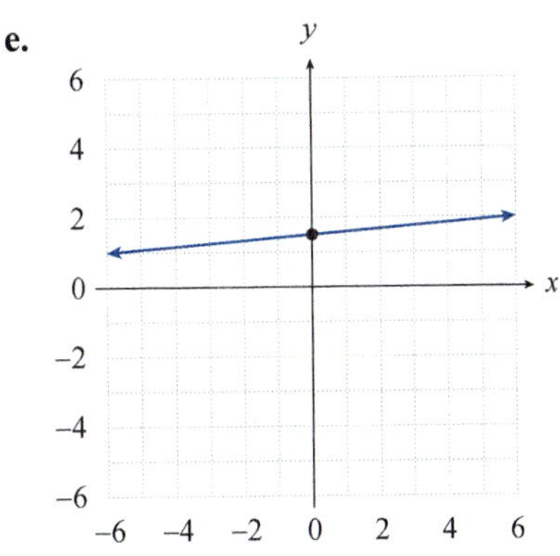

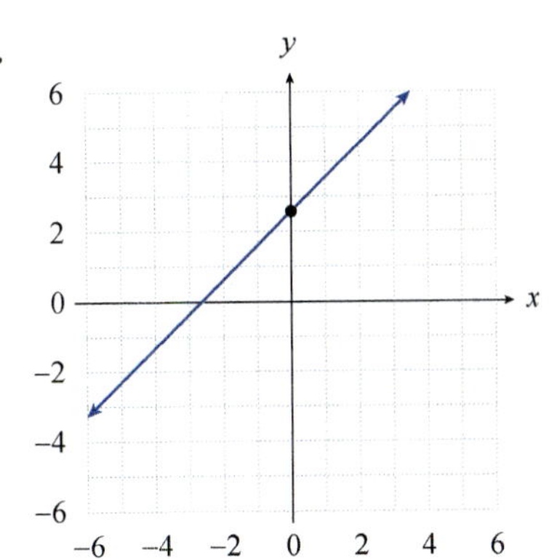

🚀 APPLICATIONS

69. A bottle manufacturer has determined that the total cost (C) in dollars of producing x bottles is $C = 0.25x + 2100$.
  a. What is the cost of producing 500 bottles?
  b. What are the fixed costs (costs incurred even when 0 bottles are produced)?
  c. What is the increase in cost for each bottle produced?

70. Sales at Glover's Golf Emporium have been increasing linearly for the past couple of years. Last year, sales were \$163,000. This year, sales were \$215,000. If sales continue to increase at this linear rate, predict the sales for next year.

71. Amy owns stock in a company. If the stock had a value of \$2500 in 2018 when she purchased it, what has been the average change in value per year if in 2020 the stock was worth \$3150?

72. For tax and accounting purposes, businesses often have to depreciate equipment values over time. One method of depreciation is the straight-line method. Three years ago Hilde Construction purchased a bulldozer for \$51,500. Using the straight-line method, the bulldozer has now depreciated to a value of \$43,200. If V equals the value at the end of year t, write a linear equation expressing the value of the bulldozer over time. How many years from the purchase date will the value equal \$0? Round your answer to two decimal places.

2.4 PARALLEL AND PERPENDICULAR LINES

■ TOPICS

- ■ Slopes of Parallel Lines
- ■ Slopes of Perpendicular Lines

Slopes of Parallel Lines

In this section, we will explore the relationship between slope and the geometric concepts of parallel and perpendicular lines. This will allow us to use algebra to construct lines parallel or perpendicular to a given line. We will begin with parallel lines; consider the following figure showing "rise" and "run" of two parallel lines:

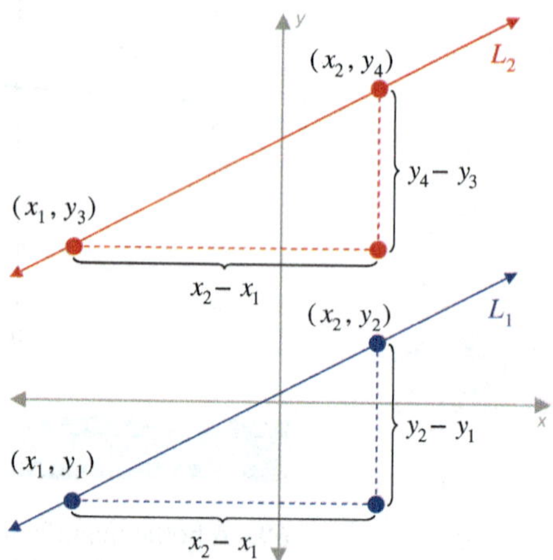

FIGURE 1: Slopes of Parallel Lines

In Figure 1, we can see that if we choose the right points on each line, the rise and run are equal. Further, no matter which points we choose, the ratio of rise to run (which we know as the slope) is equal for parallel lines. Thus, we have a very simple algebraic definition for parallel lines.

> **Slopes of Parallel Lines**
>
> Two nonvertical lines with slopes m_1 and m_2 are **parallel** if and only if $m_1 = m_2$. Also, two vertical lines (with undefined slopes) are always parallel to each other.

This fact gives us a straightforward way of finding the equations of lines parallel to a given line. First, we calculate the slope of the given line, and then construct new lines using that same slope. Usually, it will be easier to use the slope-intercept or point-slope form of a linear equation, since the slope appears directly.

Example 1: Finding Equations of Parallel Lines

Find equations for two lines parallel to each of the given lines.

a. $y = -\dfrac{2}{3}x + 4$ **b.** $10x - 2y = 14$

Solution

When given a line in standard form, it is usually easier to find its slope by rewriting it in slope-intercept form than to find two points on the line and calculate the slope directly.

a. This line is already in slope-intercept form, so we immediately know the slope of the line is $-\dfrac{2}{3}$. Any line parallel to this one must also have a slope of $-\dfrac{2}{3}$. To find two parallel lines, we can simply change the value of the y-intercept.

$$y = -\frac{2}{3}x + 1 \ \text{ and } \ y = -\frac{2}{3}x - 10$$

b. This line is in standard form. Our first step is to rewrite it in slope-intercept form.

$$10x - 2y = 14$$
$$-2y = -10x + 14 \qquad \text{Subtract } 10x \text{ from both sides.}$$
$$y = \frac{-10}{-2}x + \frac{14}{-2} \qquad \text{Divide each term by } -2.$$
$$y = 5x - 7 \qquad \text{Simplify.}$$

Again, once the line is in slope-intercept form, we can change the y-intercept to find two lines parallel to the original line.

$$y = 5x \ \text{ and } \ y = 5x + 8$$

Example 2: Finding Equations of Parallel Lines

Find the equation, in slope-intercept form, for the line which is parallel to the line $3x + 5y = 23$ and which passes through the point $(-2, 1)$.

Solution

Again, our first step is to write the initial equation in slope-intercept form.

$$3x + 5y = 23$$
$$5y = -3x + 23$$
$$y = -\frac{3}{5}x + \frac{23}{5}$$

This tells us that the slope of the line whose equation we seek is $-\dfrac{3}{5}$. We also know that the line is to pass through $(-2, 1)$, so we can use the point-slope form to obtain the desired equation.

$$y - y_1 = m(x - x_1)$$

$$y - 1 = -\frac{3}{5}\left(x - (-2)\right)$$

$$y - 1 = -\frac{3}{5}(x + 2)$$

$$y - 1 = -\frac{3}{5}x - \frac{6}{5}$$

$$y = -\frac{3}{5}x - \frac{1}{5}$$

Begin by substituting our known information into the point-slope form: $m = -\frac{3}{5}$, $(x_1, y_1) = (-2, 1)$.

The instructions asked for the equation in slope-intercept form, so we solve for y to obtain the final answer.

We can also use the knowledge that parallel lines have the same slope to answer questions that are more geometric in nature.

Example 3: Identifying a Quadrilateral

Determine if the graphed quadrilateral (four-sided figure) is a parallelogram (a quadrilateral in which both pairs of opposite sides are parallel).

Solution

The four vertices are plotted, and the sides of the quadrilateral are drawn in the given graph. The figure is a parallelogram if the left and right sides are parallel and the top and bottom sides are parallel.

The slopes of the left and right sides are, respectively,

$$\frac{4-2}{-3-(-4)} = 2 \text{ and } \frac{3-1}{2-1} = 2,$$

and the slopes of the top and bottom sides are, respectively,

$$\frac{4-3}{-3-2} = -\frac{1}{5} \text{ and } \frac{2-1}{-4-1} = -\frac{1}{5}.$$

Thus the figure is indeed a parallelogram.

Slopes of Perpendicular Lines

The relationship between the slopes of perpendicular lines is a bit less obvious. Consider a nonvertical line L_1 and two points (x_1, y_1) and (x_2, y_2) on the line, as shown in Figure 2. These two points can be used to calculate the slope m_1 of L_1 with the result that $m_1 = \frac{a}{b}$, where $a = y_2 - y_1$ and $b = x_2 - x_1$.

If we now draw a line L_2 perpendicular to L_1, we can use a and b to determine the slope m_2 of line L_2. There are an infinite number of lines that are perpendicular to L_1; one of them is drawn in Figure 3.

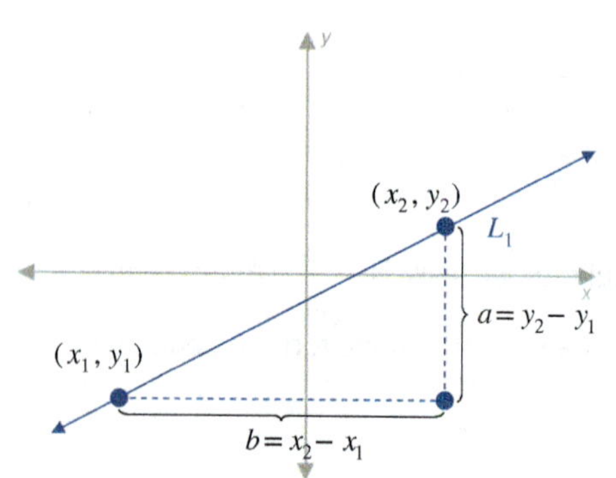

FIGURE 2: Definition of a and b

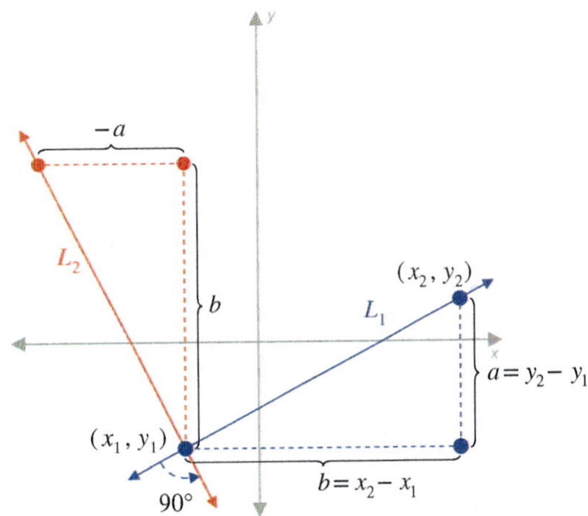

FIGURE 3: Perpendicular Lines

Note that in rotating the line L_1 by 90 degrees to obtain L_2, we have also rotated the right triangle drawn with dashed lines, so the sides of the triangle are the same length. But to travel along the line L_2 from the point (x_1, y_1) to the second point drawn requires a positive rise and a negative run, whereas the rise and run between (x_1, y_1) and (x_2, y_2) are both positive. In other words, $m_2 = -\dfrac{b}{a}$, the negative reciprocal of the slope m_1.

This relationship always exists between the slopes of two perpendicular lines, assuming neither one is vertical. Of course, if one line is vertical, any line perpendicular to it will be horizontal with a slope of zero, while if one line is horizontal, any line perpendicular to it will be vertical with undefined slope. This is summarized next.

Slopes of Perpendicular Lines

Suppose m_1 and m_2 represent the slopes of two lines, neither of which is vertical. The two lines are **perpendicular** if and only if $m_1 = -\dfrac{1}{m_2}$ (equivalently, $m_2 = -\dfrac{1}{m_1}$ and $m_1 m_2 = -1$). If one of two perpendicular lines is vertical, the other is horizontal, and the slopes are, respectively, undefined and zero.

The following examples illustrate how we can use the relationship between slopes of perpendicular lines to solve problems.

Example 4: Finding Equations of Perpendicular Lines

For each given line, find the equation of a perpendicular line.

a. $y = -\dfrac{4}{9}x + 2$

b. The line passing through the points $(-1, 3)$ and $(4, 1)$

Solution

a. This line is in slope-intercept form, so we immediately identify the slope of $-\dfrac{4}{9}$.

The slope of any perpendicular line must equal $\dfrac{9}{4}$, the negative reciprocal of the original slope. Thus, one solution is $y = \dfrac{9}{4}x$.

b. Since we only need the slope of the original line, there is no need to find its equation; we can calculate the slope directly from the given points.

$$m = \frac{1-3}{4-(-1)} = -\frac{2}{5}$$

Again, the slope of a line perpendicular to the line through the given points must have a slope equal to the negative reciprocal of $-\frac{2}{5}$, which is $\frac{5}{2}$. One perpendicular line is $y = \frac{5}{2}x + 6$.

Example 5: Finding Equations of Perpendicular Lines

Find the equation, in standard form, of the line that passes through the point $(-3, 13)$ and that is perpendicular to the line $y = -7$.

Solution

> **NOTE**
>
> Remember that if you encounter a horizontal or vertical line, you cannot use the slope formulas to find a perpendicular line.

The line $y = -7$ is a horizontal line, and hence any line perpendicular to it must be a vertical line, having the form $x = c$. Since the perpendicular line must pass through the point $(-3, 13)$, the desired solution is $x = -3$.

Given a pair of lines, we can use their slopes to determine if they are parallel, perpendicular, or neither. Note that the only information we need is the slope! The equations do not have to be written in the same form, and we do not need to know anything about their intercepts. Find the most efficient way to calculate the slope of each line to avoid any unnecessary work.

Example 6: Identifying Parallel and Perpendicular Lines

For each pair of lines, determine if the lines are parallel, perpendicular, or neither.

a. $3x - 7y = 12$ and $14x + 6y = -5$

b. $y - \dfrac{263}{4} = 9\left(x + \dfrac{77}{13}\right)$ and the line passing through the points $(0, 4)$ and $(2, 22)$

c. $y = \dfrac{3}{4}x + 1$ and $y = \dfrac{4}{3}x - 5$

Solution

> **NOTE**
>
> A pair of lines can not be *both* parallel and perpendicular.

a. Both equations are in standard form, so our first step is to rewrite them in slope-intercept form to identify the slopes.

$$
\begin{array}{ll}
3x - 7y = 12 & 14x + 6y = -5 \\
-7y = -3x + 12 & 6y = -14x - 5 \\
y = \dfrac{3}{7}x - \dfrac{12}{7} & y = -\dfrac{7}{3}x - \dfrac{5}{6}
\end{array}
$$

Are the lines parallel?

No, the slopes are not equal.

Are the lines perpendicular?

Yes, the slopes are negative reciprocals of each other.

Thus, the lines are perpendicular.

b. One line is in point-slope form, so we can see its slope is **9**. We calculate the slope of the other line using the two points given.

$$m = \frac{22-4}{2-0} = \frac{18}{2} = 9$$

Are the lines parallel?

Yes, the slopes are equal. Thus, the lines are parallel. Note that we didn't need to find the equation of the second line.

c. Both lines are in slope-intercept form, so we can read off the slopes: $\frac{3}{4}$ and $\frac{4}{3}$.

Are the lines parallel?

No, the slopes are not equal.

Are the lines perpendicular?

No, the slopes are reciprocals, not *negative* reciprocals.

Thus, the lines are neither parallel nor perpendicular.

2.4 EXERCISES

PRACTICE

Find the equation, in slope-intercept form, for the line parallel to the given line and passing through the indicated point. See Examples 1 and 2.

1. $y - 4x = 7;\quad (-1, 5)$

2. $6x + 2y = 19;\quad (-6, -13)$

3. $3x + 2y = 3y - 7;\quad (3, -2)$

4. $2 - \dfrac{y-3x}{3} = 5;\quad (0, -2)$

5. $y - 4x = 7 - 4x;\quad (23, -9)$

6. $2(y-1) + \dfrac{x+3}{5} = -7;\quad (-5, 0)$

7. $6y - 4 = -3(1 - 2x);\quad (-2, -2)$

8. $5 - \dfrac{7y+5x}{2} = 1;\quad (4, 1)$

9. $2(y-1) - \dfrac{7x+1}{3} = -3;\quad (1, 10)$

10. $8y - 6 = -3(4 - x);\quad (11, -5)$

Each set of four ordered pairs defines the vertices, in counterclockwise order, of a quadrilateral. Determine if the quadrilateral is a parallelogram. See Example 3.

11. $\{(-2,2),(-5,-2),(2,-3),(5,1)\}$ **12.** $\{(-1,6),(-4,7),(-2,3),(1,1)\}$

13. $\{(-3,3),(-2,-2),(3,-1),(2,4)\}$ **14.** $\{(-2,-3),(-3,-6),(1,-2),(2,1)\}$

15. $\{(-6,-2),(-1,0),(-3,4),(-8,2)\}$ **16.** $\{(-3,-2),(3,-3),(5,2),(-1,3)\}$

17. $\{(-1,-1),(5,1),(3,5),(-2,3)\}$ **18.** $\{(0,1),(6,0),(7,4),(1,6)\}$

Determine if the two lines are parallel. See Example 6.

19. $y = 8x + 7$ and $y = -8x + 7$

20. $x - 5y = 2$ and $5x - y = 2$

21. $2x - 3y = (x-1) - (y-x)$ and $-2y - x = 9$

22. $3 - (2y + x) = 7(x - y)$ and $\dfrac{5y+1}{4} = 3 + 2x$

23. $6 = -12(x-y) + y$ and $13y = -12x + 3$

24. $\dfrac{2x-3y}{3} = \dfrac{x-1}{6}$ and $2y - x = 3$

25. $\dfrac{x-y}{2} = \dfrac{x+y}{3}$ and $\dfrac{2x+3}{5} - 4y = 1 + 2y$

26. $5 - (4y + 3x) = 5(x - y)$ and $y + 4 = 5 + 8x$

27. $7x - 2(x+3) = 5y - x$ and $-6x = 1 - 5y$

28. $\dfrac{2y+11x}{3} = x + 1$ and $7x - 8y = 9x + 7$

29. $\dfrac{x-y}{5} = \dfrac{x+y}{3} - 1$ and $7 = -2(x-y) + 6y$

30. $2x + 5y = 14$ and the line passing through the points $(8,-5)$ and $(3,-3)$

Find the equation, in slope-intercept form, for the line perpendicular to the given line and passing through the indicated point. See Examples 4 and 5.

31. $3x + 2y = 3y - 7;\quad (3,-2)$ **32.** $6y + 2x = 1;\quad (-4,-12)$

33. $-y + 3x = 5 - y;\quad (-2,7)$ **34.** $x + y = 5;\quad$ origin

35. $x = \dfrac{1}{4}y - 3;\quad (1,-1)$ **36.** $2(y+x) - 3(x-y) = -9;\quad (2,5)$

37. $4x + 8y = 4y - 3;\quad (-2,1)$ **38.** $\dfrac{3x-y}{4} = \dfrac{4x-5}{2};\quad (8,5)$

39. $4(y+x) - 8(x-y) = -1;\quad (6,10)$ **40.** $\dfrac{3x+4}{3} - 3y = 1 - 4y;\quad (2,-8)$

Determine if the two lines are perpendicular. See Example 6.

41. $x - 5y = 2$ and $5x - y = 2$

42. $y = 5x + 4$ and $y = -\dfrac{1}{5}x - 9$

43. $3x + y = 2$ and $x + 3y = 2$

44. $\dfrac{3x - y}{3} = x + 2$ and $x = 9$

45. $5x - 6(x + 1) = 2y - x$ and $2y - (x + y) = 4y + x$

46. $-6y + 3x = 7$ and $8x - 3(x + 1) = 3y - x$

47. $-x = -\dfrac{2}{5}y + 2$ and $5y = 2x$

48. $\dfrac{7x - 5y}{4} = x + 2$ and $-3y - 3x = 2x + 4$

49. $3(4 - x) = 6y + 3$ and $-3y - 2x = 3 - 8x$

50. $\dfrac{x - 1}{2} + \dfrac{3y + 2}{3} = -9$ and $3y - 5x = x + 5$

51. $1 - \dfrac{2y - 5x}{2} = 7x + 4$ and $9x - 2y = 11$

52. $y - \dfrac{2}{3} = 4\left(x + \dfrac{7}{11}\right)$ and the line passing through the points $(-2, 4)$ and $(7, -14)$

Each set of four ordered pairs defines the vertices, in counterclockwise order, of a quadrilateral. Use the ideas in this section to determine if the quadrilateral is a rectangle.

53. $\{(-2, 2), (-5, -2), (2, -3), (5, 1)\}$ **54.** $\{(2, -1), (-2, 1), (-3, -1), (1, -3)\}$

55. $\{(1, 2), (3, -3), (9, -1), (7, 4)\}$ **56.** $\{(5, -7), (1, -13), (28, -31), (32, -25)\}$

57. $\{(-5, -1), (0, -6), (5, -1), (0, 4)\}$ **58.** $\{(-3, -3), (3, -2), (1, 2), (-5, 1)\}$

🚀 APPLICATIONS

59. A construction company is building a new suspension bridge that has support cables attached to a center tower at various heights. One cable is attached at a height of 30 feet and connects to the roadbed 50 feet from the base of the tower. If the support cables should run parallel to each other, how far from the base should the company attach a cable whose other end is connected to the tower at a height of 25 feet?

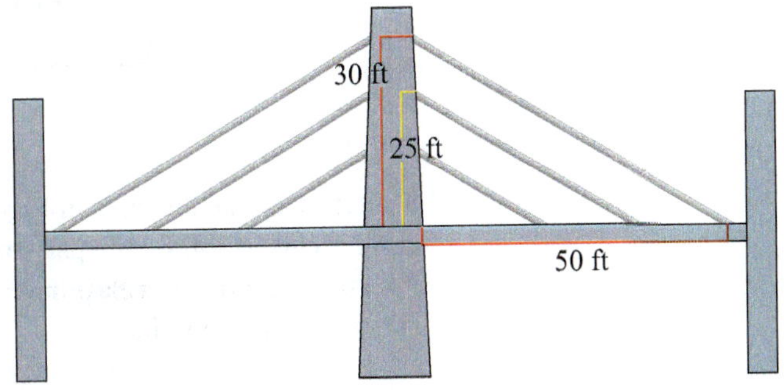

60. A light beam hits a mirror and is reflected off the mirror at a right angle. If the line formed by the original beam of light can be described by an equation of the form $y = -3.2x + b$ (for some constant b), write the form of an equation that describes the line of the reflected beam (use an arbitrary constant c in your answer).

2.5 LINEAR REGRESSION

■ TOPICS

- ■ Scatter Plots
- ■ Correlation Coefficient
- ■ Significance
- ■ Regression Lines

The data analysis we're going to look at involves evaluating the relationship between two variables. For example, we could compare the heights and weights of a sample of teenage boys, or the crime rate of small communities to the size of local police stations in those communities. Let's start by visually considering the data.

Scatter Plots

Scatter Plot

A **scatter plot** is a graphical display that is most commonly used to show two variables and how they might relate to one another. It is a graph on the coordinate plane that contains one point for each pair of data values.

Consider the scatter plots shown in Figure 1. Scatter plots **A**, **B**, and **C** have visible relationships while **D** and **E** don't display any obvious patterns.

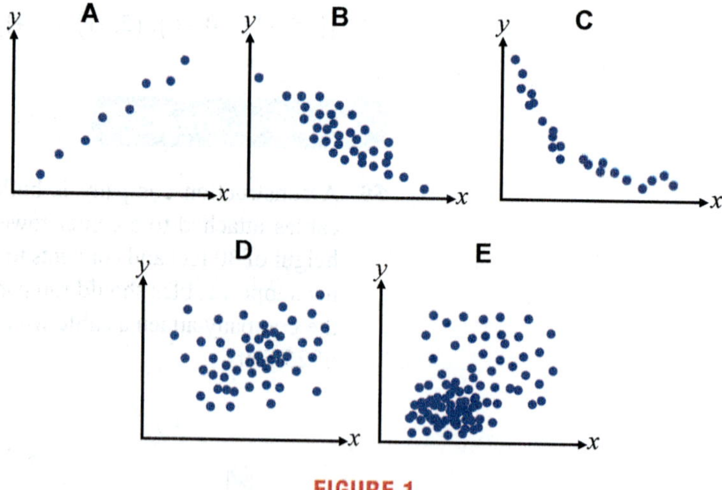

FIGURE 1

Scatter plots like the ones in Figure 1 help us to identify any trends in the data. If the trend seems to follow the pattern of a straight line, as in scatter plots **A** and **B**, there is said to be a **linear relationship** between the variables. Scatter plot **C** has a visible pattern; it's just not linear.

Consider the relationship between the number of children in a household and the number of bedrooms in the house. We can make a scatter plot that shows one point for each household, with the horizontal axis representing the number of children and the vertical axis the number of bedrooms, as shown in Figure 2.

Number of Children vs. Number of Bedrooms in Household

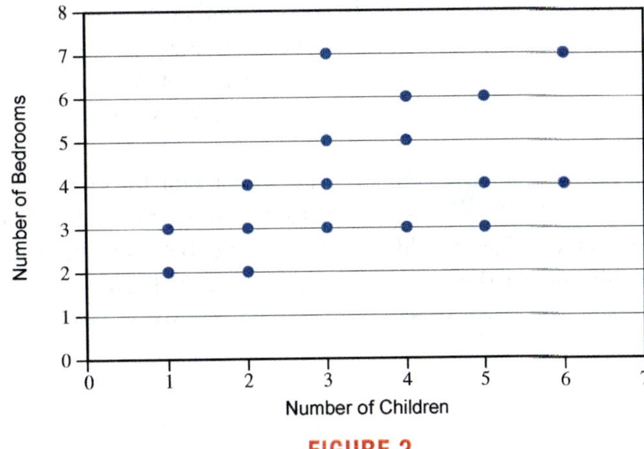

FIGURE 2

Although there seems to be a linear relationship between the number of children in a household and the number of bedrooms, it doesn't seem to be a very strong one. Both the direction and the position of the points, that is, how close the points are to lying in a straight line, tell us information about the relationship. When data are plotted in a scatter plot and the points seem to lie in a linear pattern, we say variables represented by the data are **correlated**.

Correlation

When data are plotted in a scatter plot and there appears to be an upward trend, that is, as one variable increases the other increases as well, we say there is a **positive correlation** between the variables. If the scatter plot trends downward, that is, as one variable increases the other decreases, we say there is a **negative correlation** between the variables.

Example 1: Identifying Correlations

Consider the relationship between the following variables and what kind of correlation might show up in a scatter plot of the data. Decide if the variables would likely have a positive correlation, negative correlation, or no linear correlation.

a. The number of cigarettes smoked and the probability of lung cancer

b. The number of minutes spent on social media sites by college students and their first semester grades

c. The amount of credit card debt incurred by college freshmen and their IQ score

Solution

a. As the number of cigarettes smoked increases, so does the chance of lung cancer. Thus, the scatter plot is likely to have upward-trending data points. The variables would be positively correlated.

b. As the number of minutes (or hours) spent on social media sites increases, your grades are likely to decrease. This would result in a downward-trending scatter plot and a negative correlation between the amount of time spent on social media sites and grade point average.

c. The scatter plot for these variables would likely contain a wide range of credit card debt and a wide range of IQ scores. It would be unlikely that there is a linear relationship between these two variables. Thus, they are neither positively or negatively correlated.

Identify two variables that would likely have a positive correlation.

Correlation Coefficient

As well as the direction, we can talk about the strength of a linear correlation by calculating what is called the **Pearson correlation coefficient**.

The **Pearson correlation coefficient**, rounded to the nearest thousandth, is a value between -1 and 1 that measures the strength of a linear correlation. For a sample of data, it is represented by the variable r.

The stronger the correlation, the closer the correlation coefficient is to either -1 or 1. If there is a very strong positive correlation, r will be close to 1. Conversely, if there is a strong negative correlation r will be close to -1. The closer r is to 0, the less correlation between the variables. A correlation coefficient of 0 means that there is no linear relationship between the variables at all.

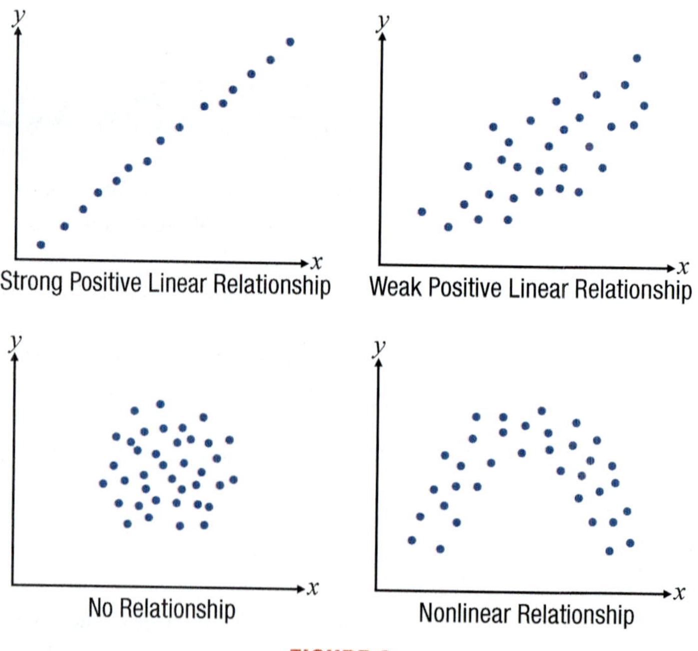

FIGURE 3

Using a TI-84 Plus calculator, we can easily calculate r, as well as some other variables we'll look at later in the section, to help us build a prediction model. Because the calculator does this simultaneously, we'll point these out as we go along. Begin by pressing `stat`; select `1:Edit`, and enter the values for one variable in L1 and the values for the other variable in L2. Then press `stat`, scroll over to CALC, and select option `4:LinReg(ax+b)`. Press `enter` twice. The output will include the correlation coefficient r.

Let's try an example with actual data.

Example 2: Using a TI-84 Calculator to Find the Pearson Correlation Coefficient

The following is a small sample of data collected from male participants in a 161 km trail ultramarathon. The survey collected the BMI (Body Mass Index) of each participant and their age. A table of the data, along with the scatter plot of the data, is given. Use your calculator to find the correlation coefficient r between the variables.

BMI	Age
23.4	53
25.7	47
23.3	28
25.9	52
24.1	60
23.0	34
22.4	39
21.6	36
23.4	52
24.8	54
24.3	53
24.7	53

TABLE 1: BMI vs. Age for Men in a 161 km Trail Ultramarathon

Solution

Begin by pressing `stat` and then select `1:Edit`. Enter the values for BMI in L1 and the values for the ages in L2. Then press `stat` and choose CALC and option `4:LinReg(ax+b)`. Press `enter` twice. The output should appear as shown in the screenshot.

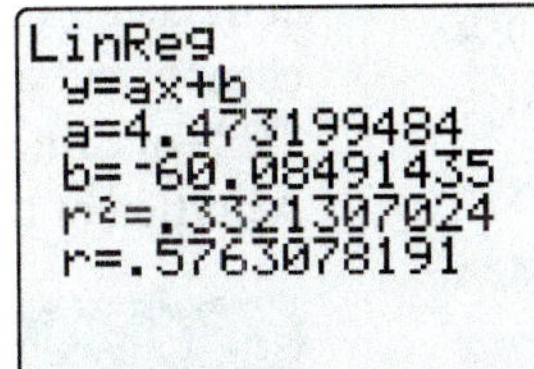

Since $r = 0.5763078191$, we can conclude that there is a positive correlation between the male participants' BMI and their age, but not a very strong one.

🖐 HELPFUL HINT

In order to make sure that the output on a TI-84 Plus will contain the correlation values, begin by pressing `2nd` and `0` which will bring up the Catalog. Scroll down to DiagnosticOn. Press `enter` twice. You need to perform this step only once.

The most recent TI-84 Plus calculators contain a new feature called STAT WIZARDS. This feature is not used in the directions in this text and is turned on by default. STAT WIZARDS can be turned off under the second page of `mode` options.

> **TECH TRAINING**
>
> A TI-30XIIS/B can be used to calculate the correlation coefficient from Example 2. Begin by pressing `2nd` `stat`, select `2-Var`, then press `enter`. Press `data`. Enter the first x-value for X_1, press the down arrow key, and then enter the first y-value for Y_1. Continue to press the down arrow key and enter the values until all of the data are entered, then press `enter`.
>
> Press `statvar` and scroll to the right until the value for r is displayed.

Significance

The correlation coefficient r can also help us determine if the linear relationship is strong enough that one variable may be used to reasonably approximate the value of the other variable. This means that the relationship is *statistically significant*. In general, we say that if the correlation coefficient for a particular set of data is less than 0.5, then there is no significant correlation between the two variables.

So how do we know? By using critical values for the Pearson correlation coefficient, we can determine how large r needs to be for the relationship to be statistically significant. All we need is the sample size and a level of confidence we want to achieve. The **level of confidence** c is the probability that the assertions made about the data are in fact correct. The **level of significance** α determines the probability that we are wrong in our assertions made about the data. (Note: Probabilities fall between 0 and 1, inclusively, and typically are converted into percentages. For example, if $c = 0.95$, we interpret that as the equivalent of 95% probability.)

Level of Confidence and Level of Significance

The **level of confidence** c is the probability that the assertions made about the data are correct.

The **level of significance** α is the probability that the assertions made about the data are incorrect.

$$c + \alpha = 1$$

In other words, if $\alpha = 0.05$, then $c = 1 - \alpha = 0.95$. This means there is a 5% probability that although we think there is a correlation between the two variables, they are actually not related at all. It also implies that there is a 95% probability that the variables are correlated. If $\alpha = 0.01$, then there is only a 1% probability that the results occurred by chance, and our level of confidence is $c = 0.99$. The Pearson Correlation Coefficient Table of Critical Values gives us the critical values for r for both $\alpha = 0.05$ and $\alpha = 0.01$. A portion of this table is shown in Table 2.

Statistically Significant

If $|r|$ is greater than the critical value listed in the table, then r is **statistically significant**, which means it is unlikely to have occurred by chance.

n	$\alpha = 0.05$	$\alpha = 0.01$
4	0.950	0.990
5	0.878	0.959
6	0.811	0.917
7	0.754	0.875
8	0.707	0.834
9	0.666	0.798
10	0.632	0.765
11	0.602	0.735
12	0.576	0.708

TABLE 2: Critical Values of the Pearson Correlation Coefficient

Use the critical values in Table 2 to determine if the correlation between BMI and age in Example 2 is statistically significant. Recall that $r = 0.5763078191$. Use a 0.05 level of significance.

Solution

There are twelve pairs of data in Example 2, so $n = 12$. By looking along the row where $n = 12$ and down the column where $\alpha = 0.05$, we see the critical value is 0.576. Comparing this critical value to the correlation coefficient we found for the data in Example 2, we have $|r| \approx 0.5763 > 0.576$. So the linear relationship between the variables is statistically significant at the 0.05 level of significance. Therefore, we have enough evidence to conclude that a linear relationship exists between BMI and age for male ultramarathoners.

When we do find data that prove to have a statistically significant r, even at the 0.01 level of significance, we need to be careful to not assign causation to the situation. When two variables are statistically correlated, we are often tempted to infer that one thing caused the other to happen. For instance, we might find that the number of watermelons consumed is positively correlated with the number of drownings that occur. This in no way means that eating a watermelon causes you to drown. Instead it might indicate that during the hotter months of the year, more people eat watermelons (which are in season) while at the same time the number of drownings increases simply because more people go swimming when it's hot outside.

Regression Lines

Once we've established that there actually *is* a statistically significant correlation between two variables, we can use the **regression line**, or **line of best fit**, to help us make predictions.

The **regression line**, also known as the **line of best fit**, is a particular line that most closely "fits" the data points on the scatter plot. The regression line can be represented by

$$\hat{y} = ax + b,$$

where a is the slope of the line and b is the y-intercept.

You have seen the equation of a line in slope-intercept form, such as $y = mx + b$, where m is the slope and b is the y-intercept. The regression line is similar in form. You can determine a regression line using a graphing calculator, such as a TI-83/84, or a statistical software package, such as that in Microsoft Excel. The letters a and b are used to represent the slope and y-intercept, respectively.

Consider the following data for newborn male babies. The table shows each baby boy's birth weight (in grams) versus his gestational age (in completed weeks).

Gestational Age (Weeks)	Birth Weight (Grams)
22	401
26	908
26	686
31	1259
31	1698
31	2209
33	2127
37	2384
37	2552
37	3080
38	3665
39	2701
39	4049
39	3465
39	2942
40	3613
41	4328
41	3179
41	3733
42	3851

TABLE 3: Birth Weight and Gestational Age for Baby Boys

```
LinReg
y=ax+b
a=187.9457364
b=-4030.573643
r²=.8796609889
r=.937902441
```

Be sure to enter the value of the gestational ages in the L1 column (inputs) and the birth weights in the L2 column (outputs) when you enter the data values in the calculator. Then, we will run the linear regression to find the line of best fit. In the calculator screenshot, we can see the values of r, a, and b that were calculated from the data.

Using the Pearson Correlation Coefficient Table of Critical Values, you can see that r is most definitely statistically significant. In fact, as long as the sample size is bigger than 5, r is significant at both the 95% and 99% confidence levels. Check for yourself that the r we calculated is larger than any critical r in the table, as long as the sample size is larger than 5.

Knowing that r is significant, we can write the regression line by simply substituting the calculated values for the slope, $a = 187.9457364$, and y-intercept, $b = -4030.573643$, to get

$$\hat{y} = 187.946x - 4030.574 .$$ Rounded to thousandths

So, given a value for x in this example, we can predict what the value for y will be by substituting the given value for the variable x into the regression line equation. Let's try it. Predict the weight of a baby boy if he is born after 40 complete weeks; that is, find the value of $\hat{y}$ when x is 40 and use it to predict y.

$$\hat{y} = 187.946x - 4030.574$$
$$= 187.946(40) - 4030.574$$
$$= 7517.84 - 4030.574$$
$$= 3487.266$$

So, when x is 40, we can predict that y will be 3487.266. In other words, we can predict that a baby boy born at 40 weeks will weigh about 3487 grams.

As with all predictions, you should be careful not to get too carried away. We've already stated that you should only use the regression line for predictions if you find r to be statistically significant. You should also make sure that you are predicting for values that are within the range and population of the original sample data. In other words, don't try to predict something that is either way out of the scope of the original data or not from the same type of population the sample was drawn from. For instance, it would not be wise to try to predict a baby's weight for a gestational age of 19 weeks. Nor would it be reasonable to apply these results to baby girls, which are a different population.

Weight	Self-Esteem
54.3	2.6
62.0	4.6
88.1	1.0
61.8	3.7
69.0	4.7
55.8	2.9
77.2	1.5
66.4	4.8
63.3	3.9
75.2	1.1
79.0	1.9
68.3	2.9

TABLE 4: Self-Esteem in Children

A recent study sought to find if any correlation existed between childhood weight and self-esteem. Children between the ages of 9 and 11 were surveyed. Weight was measured in pounds, while self-esteem was measured based on the average of the answers to 15 questions asking the participant to rate themselves on a scale of 1 to 5—where higher scores mean higher self-esteem. Table 4 contains the data for 12 children.

a. Is r statistically significant at the 0.05 level? How about the 0.01 level?

b. Write down the linear regression line in the form $\hat{y} = ax + b$.

c. If appropriate, predict the value for the level of self-esteem given that the child weighs 64.1 pounds.

d. If appropriate, predict the value for the level of self-esteem given that the child weighs 45.0 pounds.

e. If appropriate, predict the value for the level of self-esteem given that the 13-year-old weighs 88.0 pounds.

Solution

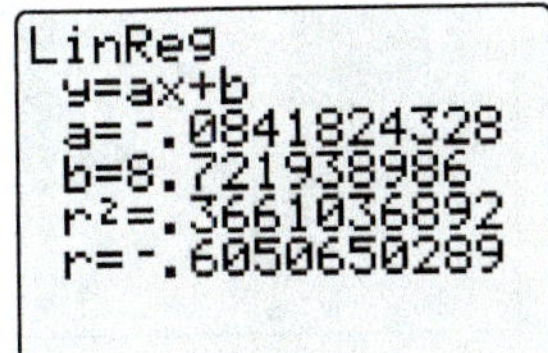

a. Begin by putting both sets of data into your calculator. Input the weights in L1 and the self-esteem score in L2. After having the calculator compute the statistics for us, we can see that $r = -0.6051$.

Looking at the Critical Values of the Pearson Correlation Coefficient table (Table 2) with $n = 12$ and $\alpha = 0.05$, we see that $|r| = 0.6051 > 0.576$, and is therefore statistically significant. However, with $\alpha = 0.01$, $|r| = 0.6051 < 0.708$, and hence is not statistically significant.

b. Using the values from the calculator, we know that the slope is $a = -0.084$ and the y-intercept is $b = 8.722$. So, the linear regression line, or line of best fit, is $\hat{y} = -0.084x + 8.722$.

c. Because the data are statistically significantly correlated at the 0.05 level, it is appropriate for us to consider the linear regression for predictions. Substituting $x = 64.1$ into the equation of the line written in the previous step, we have the following.

$$\hat{y} = -0.084x + 8.722$$
$$= -0.084(64.1) + 8.722$$
$$= -5.3844 + 8.722$$
$$= 3.3376$$

So, when a child's weight is 64.1 pounds, we can predict that his or her self-esteem score would be around 3.3.

d. The weight of 45.0 pounds is outside of the range of the original data since it is smaller than any of the other data pieces, so it is not appropriate to use the regression line for prediction.

e. Once again it is not appropriate to use the regression line for predictions in this case. The study included only children between the ages of 9 and 11. A 13-year-old is not in the same population as the study and cannot be assumed to have the same characteristics.

2.5 EXERCISES

💡 PRACTICE

In each scatter plot, determine whether there appears to be a positive linear correlation, a negative linear correlation, or no linear correlation.

1.
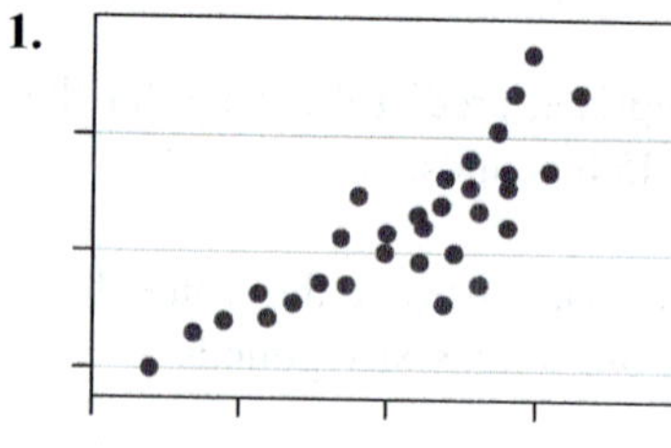

2.
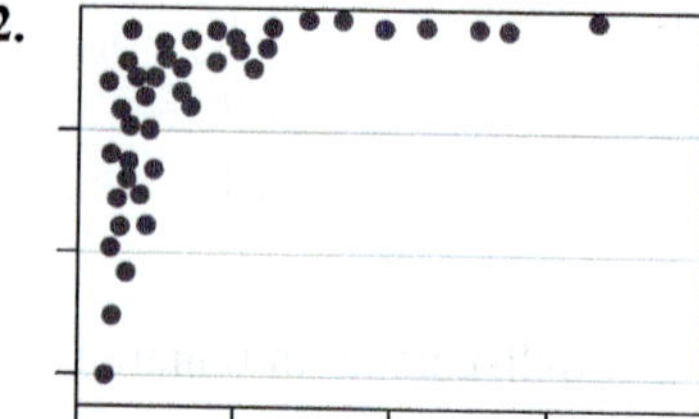

3.
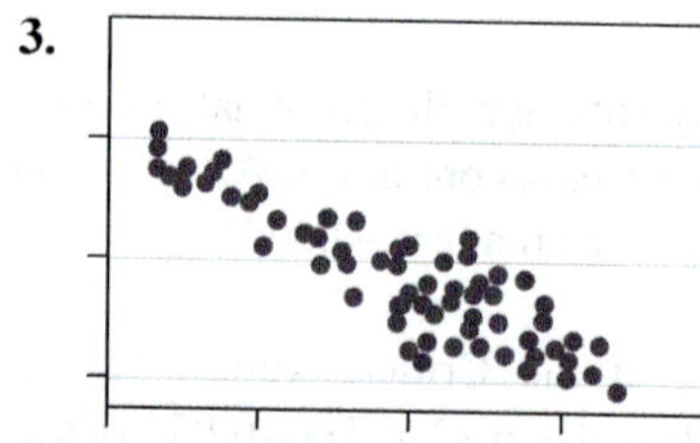

4.
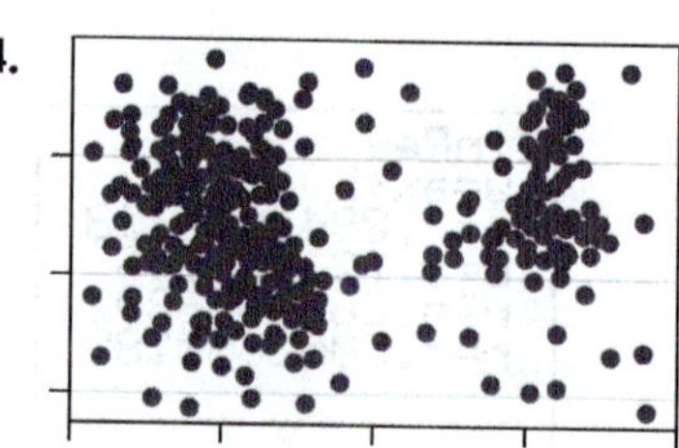

Consider each set of variables and predict whether the variables would have a weak negative relationship, a strong negative relationship, a weak positive relationship, a strong positive relationship, or no relationship at all.

5. Body weight and hours of exercise per week

6. A person's height and their self-esteem

7. Vision ability and IQ

8. Number of hours spent studying for a test and the grade on the test

Determine whether each correlation coefficient is statistically significant at the specified level of significance for the given sample size.

9. $r = 0.703,\ \alpha = 0.01,\ n = 12$ 　　　　**10.** $r = 0.403,\ \alpha = 0.05,\ n = 25$

11. $r = 0.378,\ \alpha = 0.05,\ n = 29$ 　　　　**12.** $r = 0.809,\ \alpha = 0.01,\ n = 8$

Use the linear regression model $\hat{y} = ax + b$, to predict the y-value for each value of x.

13. $\hat{y} = 28.01x + 17.83$ 　　　　　　　**14.** $\hat{y} = -16.5x + 230.55$

    a. $x = 21$ 　　　　　　　　　　　　　**a.** $x = 5$

    b. $x = 31$ 　　　　　　　　　　　　　**b.** $x = 13$

    c. $x = 40$ 　　　　　　　　　　　　　**c.** $x = 35$

🚀 APPLICATIONS

For each data set, find the following.

    a. Estimate the correlation in words as positive, negative, or no correlation.

    b. Calculate the correlation coefficient r. Round your answer to the nearest thousandth.

    c. Determine whether r is statistically significant at the 0.01 level of significance.

15. The following table gives the number of hours a student watches TV per week and his or her overall GPA.

Hours of TV per Week and Overall GPA								
TV Hours	20	10	25	15	14	13	21	9	5
GPA	2.0	2.46	2.3	2.9	3.0	3.2	3.5	3.3	3.7

16. The following table gives a sample of annual income and number of years of education.

Annual Income and Years of Education						
Annual Income	$21,000	$39,000	$40,000	$39,500	$42,000	$55,500
Years of Education	12	12	14	16	16	16
Annual Income	$61,000	$45,000	$100,000	$142,000	$240,000	$205,000
Years of Education	17	16	16	20	22	21

17. The following table shows the diastolic blood pressure reading and the stress test score for 20 adults.

Diastolic Blood Pressure Reading and Stress Test Score			
Stress Test Score	**Diastolic Blood Pressure Reading**	**Stress Test Score**	**Diastolic Blood Pressure Reading**
51	67	78	79
59	66	79	83
62	71	83	81
63	76	84	83
64	73	88	85
68	77	87	90
71	77	89	82
70	76	91	80
72	80	90	86
82	82	90	88

18. The following table shows the heights of identical twins in centimeters.

Heights of Identical Twins	
Sibling 1	**Sibling 2**
110.5	109.5
116.6	115.6
122.6	121.6
128.2	127.4
133.5	133.5
138.8	140.2
145.0	146.7
152.3	151.9
159.6	155.0
165.1	156.6
168.3	157.1
169.9	157.6
170.7	158.0

Solve each problem.

19. The following table gives the data for the number of cigarettes women smoked in their third trimester of pregnancy and the number of nonviolent crime arrests for their male babies.

Number of Cigarettes and Number of Arrests for Sons										
# of Cigarettes	0	5	3	10	22	19	30	15	8	12
# of Arrests	1	4	0	5	9	12	10	0	4	9

   a. Determine the regression line $\hat{y} = ax + b$. Round the slope and y-intercept to the nearest thousandth.

   b. Determine if the regression equation is appropriate, at the 0.05 level of significance, to use for making predictions. If so, answer part **c.**

   c. If a mother smokes eight cigarettes in her third trimester, make a prediction for the number of times her son will be arrested for a nonviolent crime, if appropriate.

20. The following table shows students' test grades on the first two tests in an introductory literature class.

Test Grades in Introductory Literature Class											
Test 1 (x)	61	45	71	81	89	55	84	91	95	59	77	88
Test 2 (y)	67	79	68	80	87	68	87	90	97	71	77	74

   a. Determine the regression line $\hat{y} = ax + b$. Round the slope and y-intercept to the nearest thousandth.

   b. Determine if the regression equation is appropriate, at the 0.05 level of significance, to use for making predictions. If so, answer part **c.**

   c. If a student scored a 70 on his first test, make a prediction for his score on the second test, if appropriate.

21. The following table shows the results on evaluations measuring self-esteem and perceived family support from 10 adolescents.

Self-Esteem and Perceived Family Support Evaluation Results										
Self-Esteem	30	31	31	28	27	26	15	32	27	33
Family Support	13	13	19	21	8	4	10	12	7	17

   a. Determine the regression line $\hat{y} = ax + b$. Round the slope and y-intercept to the nearest thousandth.

   b. Determine if the regression equation is appropriate, at the 0.05 level of significance, to use for making predictions. If so, answer part **c.**

   c. If an adolescent had a self-esteem score of 22, make a prediction for his perceived family support score, if appropriate.

22. A medical equipment company wishes to show that a new device works with the same degree of accuracy and precision as an earlier model to perform an electrocardiogram. One of the measurements tested was the change in radio electric waves during a cardiac cycle. The following results were collected from both healthy adults and those with cardiovascular problems.

Change in Radio Electric Waves during Cardiac Cycle	
# of 5 mm Squares between R Waves	
Old	**New**
2	2
3	3
4	4.5
3	3
6	6
4	4.5
3	3
5	5
3	3.5
2	2
6	6
4	4
6	6
5	5
3	3
2	2

a. Determine the regression line $\hat{y} = ax + b$. Round the slope and y-intercept to the nearest thousandth.

b. Determine if the regression equation is appropriate, at the 0.01 level of significance, to use for making predictions. If so, answer part **c.**

c. If the old machine had a reading of 5.5, make a prediction for the new machine reading, if appropriate.

Chapter 3

FUNCTIONS AND THEIR GRAPHS

3.1 INTRODUCTION TO FUNCTIONS

■ TOPICS

- Definition of a Function
- More on the Domain of a Function
- The Vertical Line Test

Definition of a Function

Suppose that the labor costs to produce a book are $5 per book and there are fixed costs of $1000 (rent, light, heat, etc.) regardless of whether or not any book is produced. Then, for x books, the costs in dollars can be expressed as

$$C = 5x + 1000.$$

We say that the cost is a **function** of the number of books produced. In function notation, we write

$$C(x) = 5x + 1000 \qquad \text{$C(x)$ is read ``C of x''}$$

and $C(2000) = 5 \cdot 2000 + 1000 = 11{,}000$ 2000 is the x-value input and 11,000 is the y-value output.

Thus "C of 2000 equals 11,000." Or the cost of producing 2000 books is $11,000.

For example, when we write that $f(4) = 2$, the point $(x, y) = (4, 2)$ is a point on the graph of $f(x)$. It may be helpful to read $f(4) = 2$ as "the value of f at $x = 4$ is 2."

The following formal definition of a **function** includes two important terms, **domain** and **range**. In this course, the domain and range will always be sets of real numbers.

> ### Function, Domain, and Range
>
> Let D and R be two sets of real numbers. A **function** f is a rule that matches each number x in D with exactly one number y (or $f(x)$) in R. D is called the **domain of** f, and R is called the **range** of f.

Note: In a function, there is only one y-value (or $f(x)$ value) for each value of x.

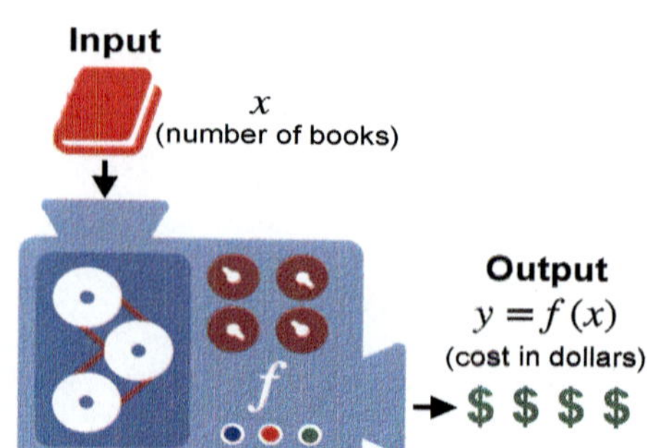

FIGURE 1: "A Function Machine"

Intuitively, a function can be thought of as an input-output machine. Values of x are input (such as numbers of books), and values of y (or $f(x)$) are output (such as dollars representing costs). For our purposes, we can think of the machine as an algebraic expression, such as $5x + 1000$. (See Figure 1.)

We say that x is the **independent variable** and that y (or $f(x)$) is the **dependent variable** because we choose the value of x but the value of y depends on our independent choice.

Example 1: Function Evaluation

For the function $f(x) = x^2$, evaluate the following.

a. $f(3)$ Read "f of 3."

Solution

$$f(3) = 3^2 = 9$$ Substitute 3 for x.

b. $f(5)$

Solution

$$f(5) = 5^2 = 25$$ Substitute 5 for x.

c. $f(-2)$

Solution

$$f(-2) = (-2)^2 = 4$$ Substitute -2 for x.

Example 2: Function Evaluation

For $f(x) = x^3 - 2x^2 + 3x + 100$, evaluate the following.

a. $f(2)$

Solution

$$f(2) = 2^3 - 2 \cdot 2^2 + 3 \cdot 2 + 100$$ Substitute 2 for x.
$$= 8 - 8 + 6 + 100 = 106$$

b. $f(a)$

Solution

$$f(a) = a^3 - 2a^2 + 3a + 100$$ Substitute a for x.

c. $f(a + 1)$

Solution

$$f(a+1) = (a+1)^3 - 2(a+1)^2 + 3(a+1) + 100$$ Substitute $a + 1$ for x.

Expanding and combining like terms, we have:

$$f(a+1) = a^3 + 3a^2 + 3a + 1 - 2(a^2 + 2a + 1) + 3a + 3 + 100$$
$$= a^3 + 3a^2 + 3a + 1 - 2a^2 - 4a - 2 + 3a + 3 + 100$$
$$= a^3 + a^2 + 2a + 102$$

⚠ CAUTION

As illustrated in Examples 2c and 2d, $f(a + 1) \neq f(a) + 1$.

d. $f(a) + 1$

Solution

$$f(a)+1 = a^3 - 2a^2 + 3a + 100 + 1 \qquad \text{Add 1 to } f(a).$$
$$= a^3 - 2a^2 + 3a + 101$$

Example 3: Function Evaluation

For $F(x) = x^2 - 8x$, evaluate the following.

a. $F(5) - F(2)$

Solution

We want the difference between the two functional values $F(5)$ and $F(2)$.

$$F(5) = 5^2 - 8 \cdot 5 = 25 - 40 = -15 \qquad \text{Substitute 5 for } x.$$
$$F(2) = 2^2 - 8 \cdot 2 = 4 - 16 = -12 \qquad \text{Substitute 2 for } x.$$
$$F(5) - F(2) = -15 - (-12) = -15 + 12 = -3 \qquad \text{Find the difference.}$$

b. $F(x + h)$

Solution

$$F(x+h) = (x+h)^2 - 8(x+h) \qquad \text{Substitute } x + h \text{ for } x.$$
$$= x^2 + 2xh + h^2 - 8x - 8h$$

c. $F(x + h) - F(x)$

Solution

$$F(x+h) - F(x) = \left[(x+h)^2 - 8(x+h) \right] - \left[x^2 - 8x \right] \quad \text{Find the difference between}$$
$$= x^2 + 2xh + h^2 - 8x - 8h - x^2 + 8x \qquad F(x+h) \text{ and } F(x).$$
$$= 2xh + h^2 - 8h$$

The following example shows how a single function can be represented by different algebraic expressions, each to be used for different parts of the domain. We say that the function is **defined in pieces** or **defined piecewise**.

Example 4: Piecewise Function

A real estate broker charges a commission of 6% on sales valued up to \$300,000. For sales valued at more than \$300,000, the commission is \$6000 plus 4% of the sale price.

a. Represent the commission earned as a function R.

Solution

$$R(x) = \begin{cases} 0.06x & \text{for } 0 \le x \le 300{,}000 \\ 0.04x + 6000 & \text{for } x > 300{,}000 \end{cases}$$

b. Find $R(200{,}000)$.

Solution

$$R(200{,}000) = 0.06(200{,}000) = \$12{,}000 \qquad \text{Use } R(x) = 0.06x \text{ since } 200{,}000 \le 300{,}000.$$

c. Find $R(500{,}000)$.

Solution

$$R(500{,}000) = 0.04(500{,}000) + 6000 \qquad \text{Use } R(x) = 0.04x + 6000 \text{ since } 500{,}000 > 300{,}000.$$
$$= 20{,}000 + 6000$$
$$= \$26{,}000$$

The letters f, g, and h (as well as F, G, and H) are commonly used to represent functions. In some cases where specific meanings are intended, we use other letters, such as C (for cost), R (for revenue), and P (for profit).

More on the Domain of a Function

The **domain** of a function f is the set of all possible values for x. The domain may be stated to be a specific set of real numbers; or, if not stated explicitly, the domain is understood to be the set of real numbers such that the corresponding functional values (or y-values) are real numbers. For polynomial functions, the domain is the set of all real numbers (the whole x-axis) unless specifically limited in a particular application.

Example 5: Domain

Find the domain of the following functions.

a. $f(x) = \dfrac{1}{x - 2}$

Solution

Since no denominator can be 0, $x - 2 \neq 0$. Thus the domain consists of all real numbers but 2. We indicate this by writing $x \neq 2$.

b. $h(x) = \sqrt{x-2}$

Solution

Since we are only interested in real numbers, $x - 2$ must be nonnegative. Thus the domain is indicated by $x - 2 \geq 0$ or $x \geq 2$.

The Vertical Line Test

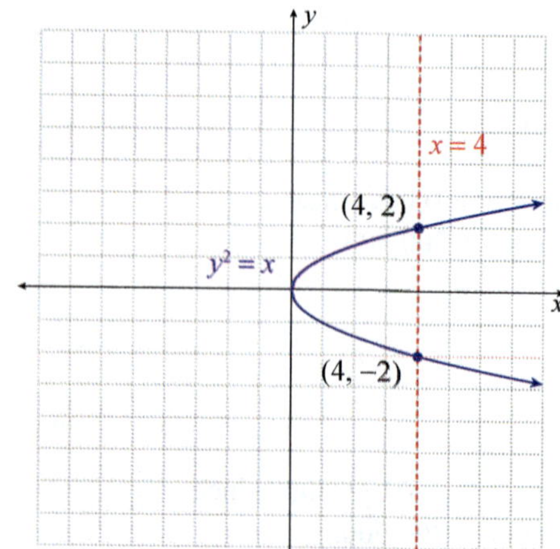

The vertical line $x = 4$ intersects the graph in two points. Therefore, the graph is **not** a function.

FIGURE 2

Many familiar formulas or equations in mathematics have useful or informative graphs. But not every such graph is the graph of a function. Remember that a function can be viewed as a rule, which, when one number is the input, then only one number is the output. If a graph shows that two or more y-values correspond to a single x input, then the graph is not that of a function.

We can test (usually visually) whether or not a graph represents a function by using the **vertical line test.**

Vertical Line Test

If any vertical line intersects a graph at more than one point, then the graph does not represent a function.

As an example, the graph of $y^2 = x$ is shown in Figure 2. Since there is a vertical line that intersects the graph in more than one point, the graph does not represent a function. For $x = 4$, there are two corresponding y-values, $y = 2$ and $y = -2$.

Example 6: Vertical Line Test

Use the vertical line test to determine whether or not each of the following graphs is a function.

a.

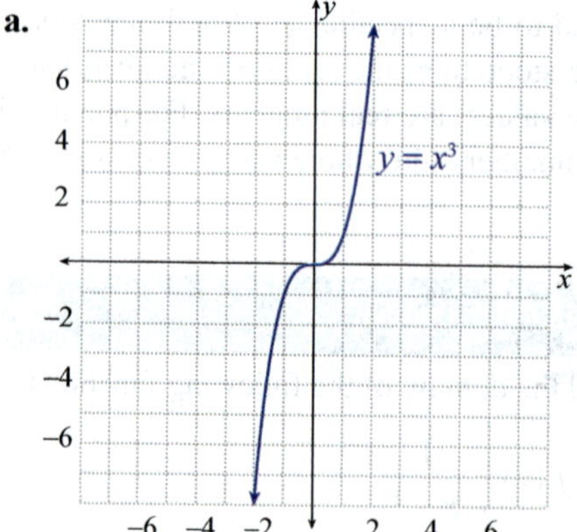

b.

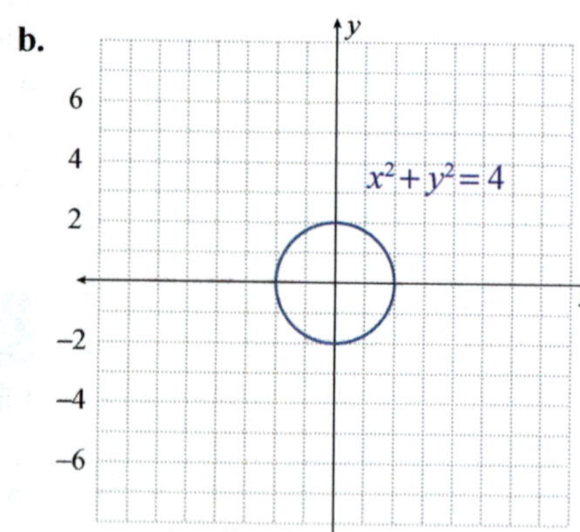

Solution

a.

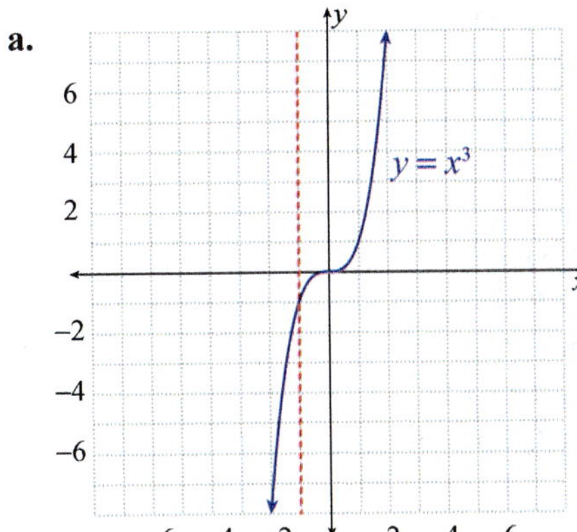

The equation $y = x^3$ represents a function (which can also be written as $f(x) = x^3$). No vertical line can intersect the graph in more than one point.

b.

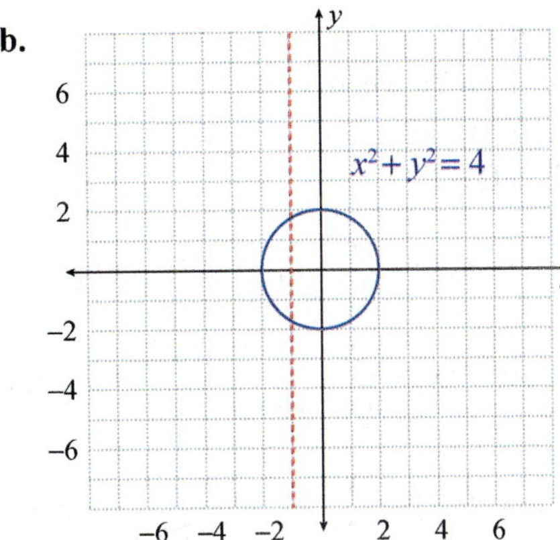

The graph of $x^2 + y^2 = 4$ is a circle. The graph shows that the equation does not represent a function. Vertical lines can be drawn that intersect the graph in more than one point.

3.1 EXERCISES

☉ PRACTICE

In Exercises 1–9, evaluate the given function for parts **a.–d.**

1. $f(x) = 2x - 7$
  a. $f(5)$
  b. $f(-2)$
  c. $f(a + 1)$
  d. $f(a) + 1$

2. $f(x) = 3x + 5$
  a. $f(2)$
  b. $f(-1)$
  c. $f(a + 1)$
  d. $f(a) + 1$

3. $f(x) = x^2 - 2x + 1$
  a. $f(-2)$
  b. $f(3)$
  c. $f(a + 1)$
  d. $f(a) + 1$

4. $f(x) = 3x^2 - x + 2$
  a. $f(-3)$
  b. $f(2)$
  c. $f(a + 1)$
  d. $f(a) + 1$

5. $f(x) = x^3 + x^2 - 3x + 1$
  a. $f(-1)$
  b. $f(-3)$
  c. $f(a + 1)$
  d. $f(a) + 1$

6. $f(x) = 2x^3 - 4x^2 + x - 6$
  a. $f(-2)$
  b. $f(4)$
  c. $f(a + 1)$
  d. $f(a) + 1$

7. $f(x) = 4x^2 - 1$
  a. $f(3)$
  b. $f(a + 2)$
  c. $f(x + h)$
  d. $f(-2) - f(-1)$

8. $f(x) = 2 - 3x^2$
  a. $f(5)$
  b. $f(a - 3)$
  c. $f(x + h)$
  d. $f(3) - f(2)$

9. $f(x) = \sqrt{x + 5}$
  a. $f(-1)$
  b. $f(a + 2)$, where $a \geq -7$
  c. $f(x + h)$
  d. $f(4) - f(1)$

10. Let $f(x) = \sqrt{x^2 + 1}$. Find **a.** $f\left(\sqrt{3}\right)$, and **b.** $f(a + 1)$.

11. Let $f(x) = \begin{cases} x - 4 & \text{if } x \leq 2 \\ x^2 - 6 & \text{if } x > 2 \end{cases}$. Find **a.** $f(-1)$, **b.** $f(2)$, **c.** $f(2.5)$, and **d.** $f(3)$.

12. Let $f(x) = \begin{cases} x^2 & \text{if } x < 0 \\ 3x - 2 & \text{if } x \geq 0 \end{cases}$. Find **a.** $f(0)$, **b.** $f(-2)$, **c.** $f(1.5)$, and **d.** $f(3)$.

In Exercises 13–27, find $f(x + h) - f(x)$.

13. $f(x) = 3x - 1$ **14.** $f(x) = 5x - 2$ **15.** $f(x) = x^2 + 4$

16. $f(x) = x^2 - 3$ **17.** $f(x) = 2x^2 + 1$ **18.** $f(x) = 5 + 3x^2$

19. $f(x) = x^2 - x$ **20.** $f(x) = x^2 + 2x$ **21.** $f(x) = 3x - x^2$

22. $f(x) = 4x^2 - x$ **23.** $f(x) = 2x^2 - 3x$ **24.** $f(x) = x^3$

25. $f(x) = x^3 - 1$ **26.** $f(x) = x^3 + 7$ **27.** $f(x) = x^3 + 5$

In Exercises 28–39, determine the domain of each function.

28. $f(x) = \dfrac{3x + 1}{(x - 5)(x - 6)}$ **29.** $f(x) = \sqrt{2x + 10}$

30. $f(x) = \sqrt{x^2 + 2}$ **31.** $f(x) = \dfrac{5}{\sqrt{x + 10}}$

32. $f(x) = \dfrac{2x}{x - 2}$ **33.** $f(x) = \dfrac{x - 3}{x + 1}$

34. $f(x) = \dfrac{4}{x^2 - x - 12}$ **35.** $f(x) = x - 3$

36. $f(x) = 4 - 3x$

37. $f(x) = \dfrac{1}{\sqrt{2x+5}}$

38. $f(x) = \begin{cases} 3x+1 & \text{if } 0 \le x < 4 \\ 5x-2 & \text{if } x \ge 4 \end{cases}$

39. $f(x) = \begin{cases} 2-x^2 & \text{if } x \le 2 \\ x-4 & \text{if } x > 2 \end{cases}$

In Exercises 40–51, use the vertical line test to determine whether or not each graph represents a function.

40.

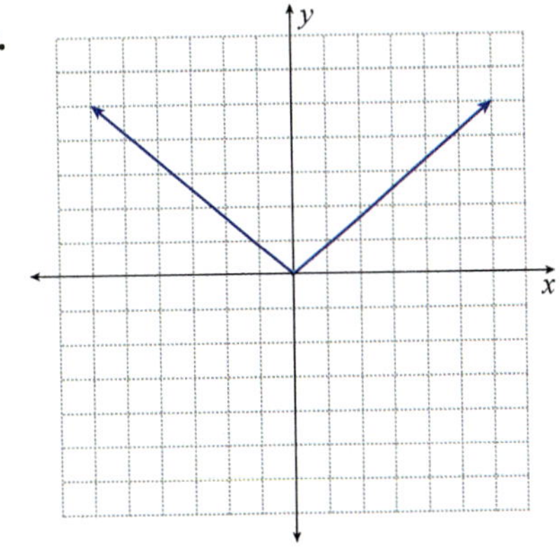

41.

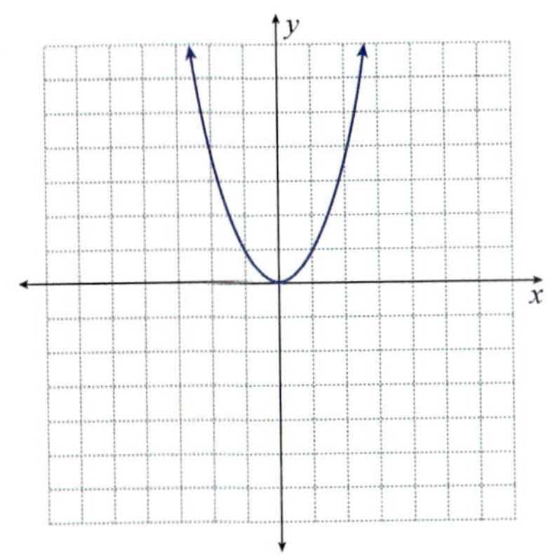

42.

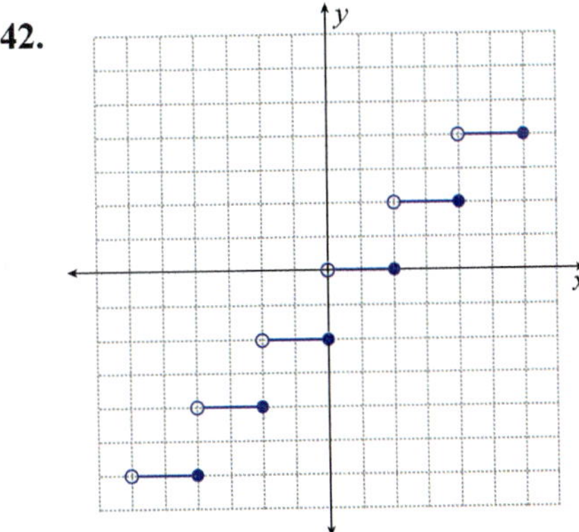

43.

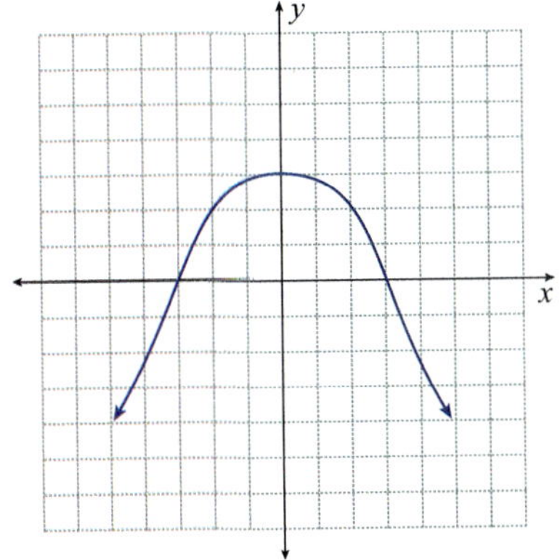

44.

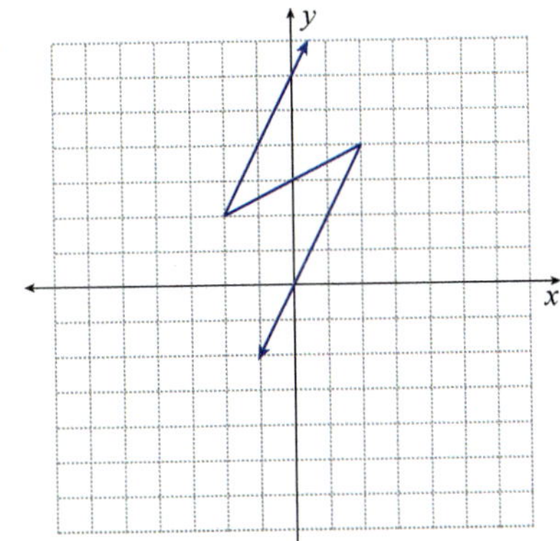

45.

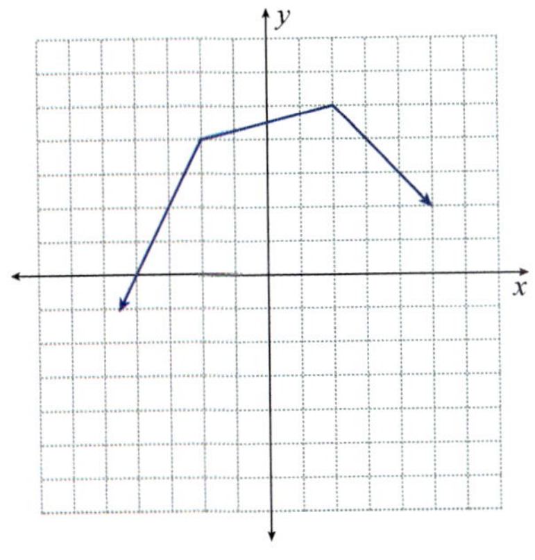

46.

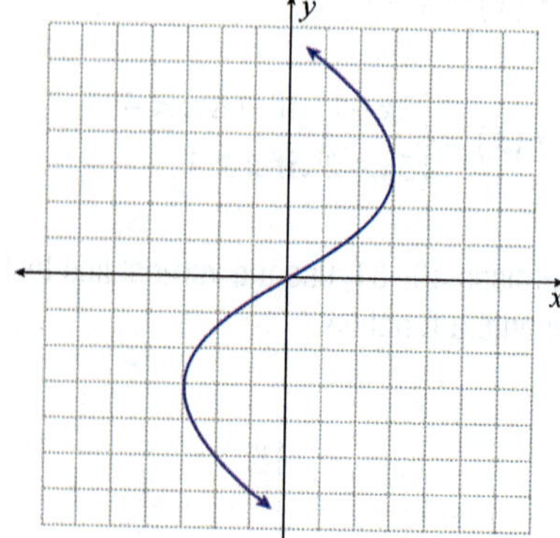

47.

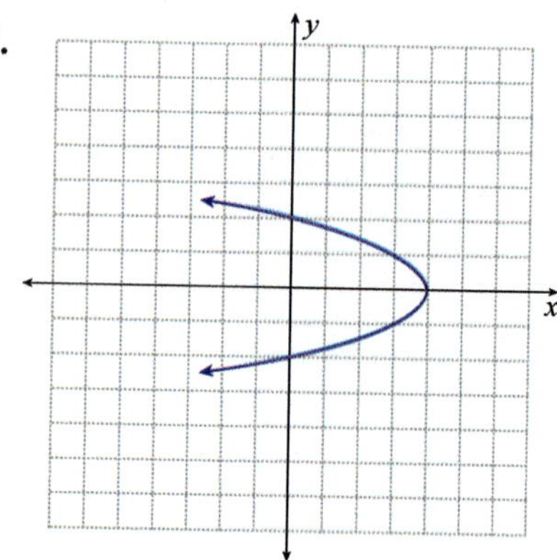

48.

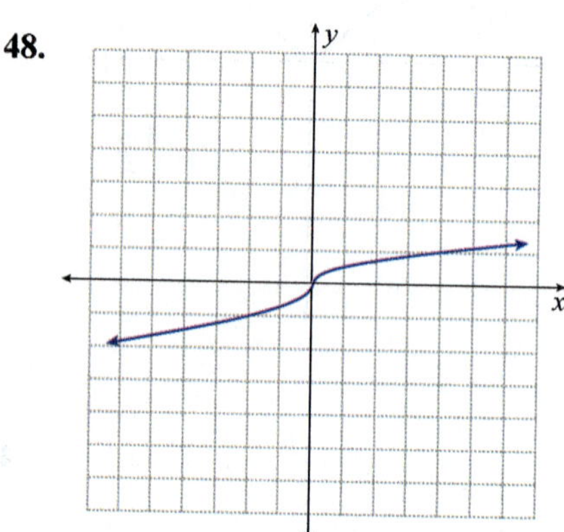

49.

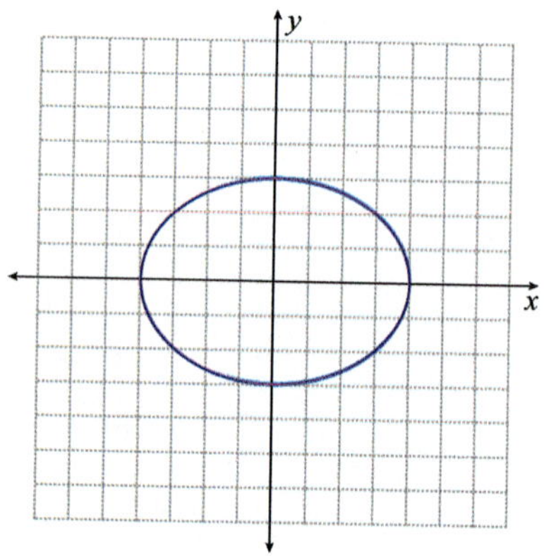

50.

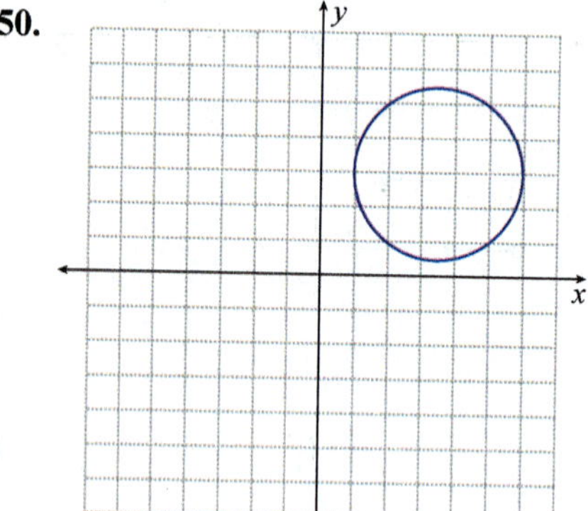

51.

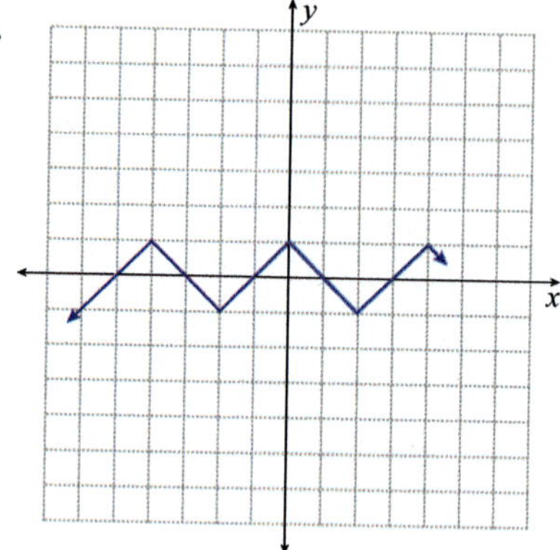

3.2 FUNCTIONS AND MODELS

■ TOPICS

- ■ Mathematical Modeling in Business and Economics
- ■ A General Comment about Price
- ■ Other Applications

Mathematical Modeling in Business and Economics

In this section we will introduce terms and ideas from business and economics that will be used throughout the remainder of the course. Most people are familiar with terms such as **profit** and **loss**, and many are familiar with the concepts of **supply** and **demand**. We want to represent these ideas and others with mathematical expressions. Creating a mathematical formula that describes a real-world problem is called **mathematical modeling**. The formula itself is called a **model**. Some models describe the problem accurately with a great deal of precision while others yield only approximations or merely intelligent guesses.

One business application that can be easily modeled with a formula is simple interest. **Simple interest** is the money paid for the use of money (called the **principal**) over a specific time.

Simple Interest

$$I = Prt$$

where $P =$ principal, $r =$ annual rate of interest, and $t =$ time in years.

If $P = \$1000$ is invested at 8% for 2 years, then the simple interest earned is

$$I = Prt = 1000(0.08)(2) = \$160.$$

The balance (or amount A) at the end of 2 years would be the principal plus the interest,

$$A = P + I = 1000 + 160 = \$1160.$$

However, most people would prefer to have their money earning interest on interest already earned (**compound interest**).

Compound Interest

A formula (model) for the balance with annual compounding at rate r for t years is

$$A = P(1 + r)^t,$$

where r is in decimal form.

Thus, using the previous data with $P = 1000$, $r = 8\% = 0.08$, and $t = 2$ years, we have

$$A = 1000(1.08)^2$$
$$= 1000(1.1664)$$
$$= \$1166.40.$$

Both of these models accurately describe financial situations. There are other formulas for compound interest that we will discuss later.

Unlike the formulas for simple and compound interest, models that describe general economic theories can be ambiguous and inaccurate because they are affected by uncontrollable variables such as politics, interest rates, and international crises. General theories and models such as these will be left to courses in economics.

In this course we will discuss several models related to specific situations involving particular products that we can analyze with algebra and calculus. The following terms and their interrelationships will be used.

Term	Symbol	Description
Items (or units)	x	Number of items produced
Cost	$C(x)$	Total costs
Fixed costs	$C(0)$	Constant costs that do not depend on the number of items produced (rent, light, heat, etc.), and is the y-intercept for $C(x)$
Variable costs	$C(x) - C(0)$	Costs that depend on the number of items produced (labor, material, etc.)
Revenue	$R(x)$	Income that depends on the number of items sold and the selling price per item
Price	p	Selling price per item
Profit	$P(x)$	The difference between the revenue and the cost $[P(x) = R(x) - C(x)]$
Break-even point		Point where revenue equals cost $[R(x) = C(x)$, or where $P(x) = 0]$
Supply function	$p = S(x)$	p is the price per item at which producers are willing to supply x items. (As supply x increases, the price increases.)
Demand function	$p = D(x)$	p is the price per item at which consumers are willing to buy x items. (As demand x increases, the price decreases.)
Equilibrium price	p_E	Price at which supply is equal to the demand $[S(x) = D(x)]$
Equilibrium point	(x_E, p_E)	Point at which the supply is equal to the demand; consumers buy (demand) all x_E items supplied when the price is p_E
Revenue in terms of the unit price	$R(x) = x \cdot D(x)$ $= x \cdot p$	Revenue is the product of the number of items sold times the price per item. This price is set by the demand function $p = D(x)$.

A General Comment about Price

We have stated that the price per item, given in the supply and demand functions, is a function of (or dependent on) the number x of items supplied or demanded. You should be aware, however, that in the field of economics, supply and demand functions are sometimes written as

$$x = S(p) \quad \text{and} \quad x = D(p)$$

in which the number of items supplied or demanded is a function of the price. That is, price is used as the independent variable to indicate that a change in price affects the number of items supplied or demanded.

One reason we have chosen to write $p = S(x)$ and $p = D(x)$ is so that we can use these functions in conjunction with the revenue function

$$R(x) = x \cdot p = x \cdot D(x),$$

which is dependent on the number of items sold. This approach is taken to simplify the study of applications of calculus to economics.

Example 1: Car Rental

Suppose that the total cost of renting a car consists of a fixed cost of \$35 per day plus a variable cost of 15 cents per mile.

a. Write a cost function that represents the cost of driving x miles in one day.

b. Find the cost of driving 500 miles in one day.

Solutions

a. The formula for calculating cost is $C(x) = (\text{variable cost}) + (\text{fixed cost})$.

Let x = number of miles driven in one day, variable cost = $\$0.15x$, and fixed cost = \$35 per day.

Inserting these into the formula gives $C(x) = 0.15x + 35$.

b. $C(500) = 0.15(500) + 35$ — Find the cost by substituting $x = 500$ miles into the formula.

$ = 75 + 35 = \$110$

Note: Observe that if you do no driving, the fixed costs are $C(0) = \$35$.

Example 2: Revenue

A manufacturer determines that the revenue generated by selling x units of a product is a linear function of x. If the revenue from 20 units is \$380 and the revenue from 15 units is \$285, find the revenue function.

Solution

Since the revenue function is linear, treat the given information as two points on a line: (20, 380) and (15, 285).

We use the formula for slope,

$$m = \frac{y_1 - y_2}{x_1 - x_2}$$

and the point-slope form $y - y_1 = m(x - x_1)$, where $x =$ number of units sold and $y =$ revenue in dollars.

$$m = \frac{380 - 285}{20 - 15} = \frac{95}{5} = 19$$

Substitute the values for x_1, x_2, y_1, and y_2 into the formula for m.

$$y - 380 = 19(x - 20)$$
$$y = 19x - 380 + 380$$
$$y = 19x$$

Using the found value for m and the known values for x_1 and y_1, solve for y in the point-slope formula.

The revenue function is $R(x) = 19x$.

This results in the revenue function, $R(x)$.

Example 3: Break-Even Point

The Green-Belt Company determines that the cost of manufacturing men's belts is \$2 each plus \$300 per day in fixed costs. The company sells the belts for \$3 each. What is the break-even point?

Solution

The break-even point occurs where revenue and costs are equal. Let $x =$ the number of belts manufactured in a day.

$$R(x) = 3x$$
$$C(x) = 2x + 300$$

$$R(x) = C(x)$$
$$3x = 2x + 300$$
$$x = 300$$

So 300 belts must be made and sold each day for the company to break even. The company must sell more than 300 belts each day to make a profit.

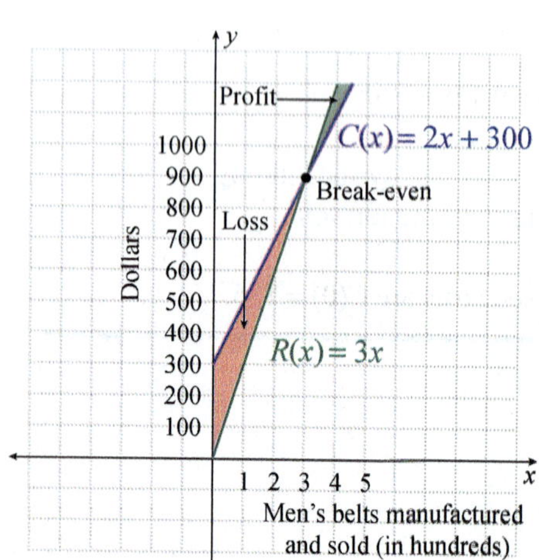

Men's belts manufactured and sold (in hundreds)

In Example 3, revenue and costs are linear functions and there appears to be no limit to the profit that can be made each day. To make more profit, the company need only make and sell more belts. However, in the real world, there are a limited number of customers and the company's plant physically restricts the number of belts that can be produced. Example 4 illustrates a more realistic model in which the revenue is a quadratic function, reflecting the fact that the demand function $p = D(x)$ is a decreasing function. That is, as the number of items sold increases, the price declines, and, eventually, the revenue declines.

Example 4: Modeling in Production

Suppose that a company has determined that the cost of producing x items is $500 + 140x$ and that the price it should charge for one item is $p = 200 - x$.

a. Find the cost function.

b. Find the revenue function.

c. Find the profit function.

d. Find the break-even point.

Solutions

a. The cost function is given as $C(x) = 500 + 140x$.

b. The revenue function is found by multiplying the price for one item by the number of items sold.

$$R(x) = p \cdot x$$
$$= (200 - x)x = 200x - x^2$$

c. Profit is the difference between revenue and cost.

$$P(x) = R(x) - C(x)$$
$$= (200x - x^2) - (500 + 140x)$$
$$= -x^2 + 60x - 500$$

d. To find the break-even point, set the revenue equal to the cost and solve for x.

$$R(x) = C(x)$$
$$200x - x^2 = 500 + 140x$$
$$0 = x^2 - 60x + 500$$
$$0 = (x - 10)(x - 50)$$
$$x = 10 \text{ or } x = 50$$

There are two break-even points.

This model shows that a profit occurs if the company produces between 10 and 50 items. Techniques for maximizing profit are discussed in calculus.

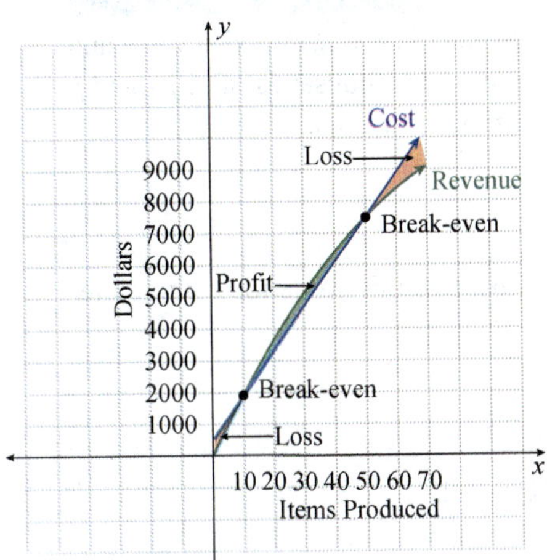

In the marketplace, price is related to both supply and demand. As we stated earlier, these basic relationships can be modeled by the following two functions.

$$p = S(x)$$

Supply function: price per unit at which the supplier will supply x units of an item.

$$p = D(x)$$

Demand function: price per unit at which consumers will buy x units of an item.

The equilibrium point, denoted as (x_E, p_E), is the point where the price is such that the production level (supply) is equal to the purchase level (demand). (See Figure 1.)

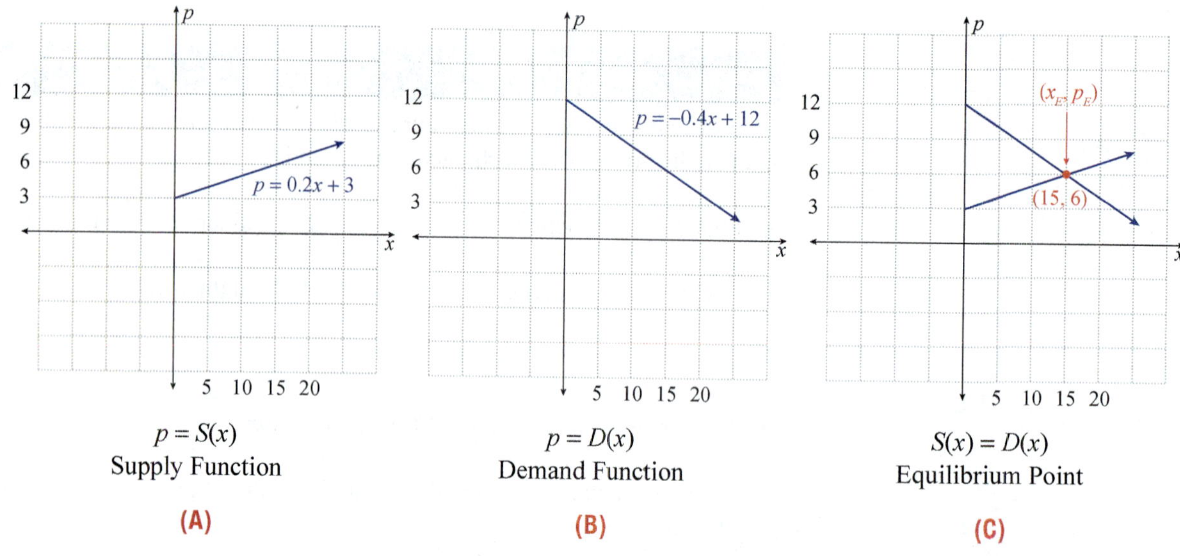

FIGURE 1

In general, supply and demand functions are not linear functions as they are shown to be in Figure 1. However, supply functions are increasing functions, and demand functions are decreasing functions. Suppliers will happily provide more products as prices increase, and consumers will happily buy more products as prices decrease.

Example 5: Equilibrium Point

Suppose that the supply function for a particular product is $p = S(x) = x^2 + x + 3$ and the demand function $p = D(x) = (x - 5)^2$ where x represents thousands of units and p represents thousands of dollars. Find the equilibrium point (x_E, p_E).

Solution

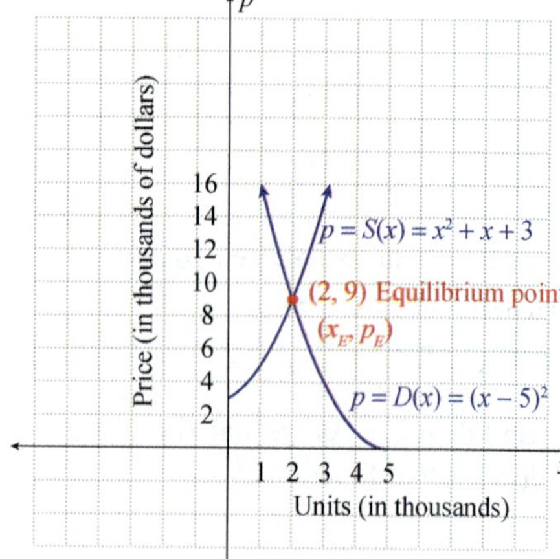

We solve for x_E by setting $S(x) = D(x)$. Then we substitute this value for x into either $S(x)$ or $D(x)$ to find p_E.

$$S(x_E) = D(x_E)$$
$$x_E^2 + x_E + 3 = (x_E - 5)^2$$
$$x_E^2 + x_E + 3 = x_E^2 - 10x_E + 25$$
$$11x_E = 22$$
$$x_E = 2$$
$$p_E = S(x_E) = S(2)$$
$$= 2^2 + 2 + 3 = 9$$

The equilibrium point occurs where $x = 2000$ units and $p = \$9000$.

Other Applications

Whenever one quantity is equal to some constant times another quantity, the first is said to **vary directly with** (or to be **directly proportional to** or just **proportional to**) the second quantity. For example, the circumference C of a circle is directly proportional to the diameter d. We have the general relationship

$$C = kd,$$

where k is called the **constant of proportionality**. For circles, we know that this constant is π. That is,

$$C = \pi d.$$

Other examples are:

$A = \pi r^2$	The area A of a circle varies directly with the square of the radius r.
$d = kw$	Hooke's Law says that the distance d that a hanging spring stretches is proportional to the weight w of the object attached to the spring. (The constant k depends on the characteristics of the spring such as its length and type of material.)

If one quantity increases as the other decreases while their product remains fixed, then the quantities are said to **vary inversely** or to be **inversely proportional**. For example, suppose $y = \dfrac{5}{x}$. As x increases, y is said to "decrease in proportion", and xy stays equal to 5. A famous example from physics is the following: the gravitational force between an object and the earth is inversely proportional to the square of the distance from the object to the center of the earth.

$$F = \frac{m_1 m_2}{G d^2} = \frac{k}{d^2}$$

where F is the gravitational force, d is the distance from the object to the center of the earth, k is the constant of proportionality, m_1 and m_2 are the masses of the objects, G is a constant, and $k = \dfrac{m_1 m_2}{G}$.

Example 6: Physics

A basic law of physics states that, when an object is dropped in a near vacuum, the distance it falls varies directly with the square of the elapsed time. If an object dropped in a near vacuum falls 64 feet in 2 seconds, how far does it fall in 3 seconds? (Assume that the object does not hit the ground before 3 seconds have elapsed.)

Solution

The basic law of physics described above can be modeled by the equation

$$d = kt^2,$$

where d = distance, t = time, and k = constant of proportionality.

$d = kt^2$ First, find k by using $d = 64$ and $t = 2$.

$64 = k(2)^2$

$64 = 4k$

$16 = k$

$d = kt^2$

$d = 16(3)^2 = 144$ ft   Now find d by using $k = 16$ and $t = 3$.

The object will fall 144 feet in 3 seconds.

Examples 7 and 8 feature cost functions described differently from those we have discussed previously and Example 9 presents a function related to a geometric figure.

Example 7: Stock Market

Scott wants to buy a total of 500 shares in the stock market. He will buy x shares at $4 per share and the rest at $6 per share.

a. Write a function for the total cost of the shares.

b. What will be Scott's cost if he buys 200 shares at $4?

Solution

a. Since Scott will buy x shares at $4 per share, he must buy $500 - x$ shares at $6 per share. The total cost function is

$$C(x) = 4x + 6(500 - x) = 3000 - 2x.$$

b. If Scott buys 200 shares at $4, then $x = 200$ and

$$C(200) = 3000 - 2 \cdot 200 = 3000 - 400 = \$2600.$$

Example 8: Phone Call Charges

Suppose that the cost of an overseas call is $9.00 for the first 3 minutes or less plus 95 cents for each additional minute. Write a function for the cost of a call of x minutes. (Assume that a fraction of a minute over 3 minutes is charged the corresponding fraction of 95 cents.)

Solution

Let x = the length of the call in minutes. Then

$$C(x) = \begin{cases} 9.00 & \text{for } 0 < x \le 3 \\ 9.00 + 0.95(x-3) & \text{for } x > 3 \end{cases}$$

The function is defined in pieces because the formula to be used depends on whether or not the call lasts longer than 3 minutes.

Example 9: Perimeter

A rectangular lot is fenced on three sides. An adjacent building forms the fourth side (see diagram). If the total length of the fencing is 48 feet, write a function that represents the area of the rectangle. Determine a good viewing rectangle and graph the function.

Solution

Let x = width of the rectangle. Since the total length of the fence is 48 feet and two sides are each of length x, then the length of the third side must be $48 - 2x$. Since the enclosed area is the product of length times width, we have

$$A(x) = x(48 - 2x).$$

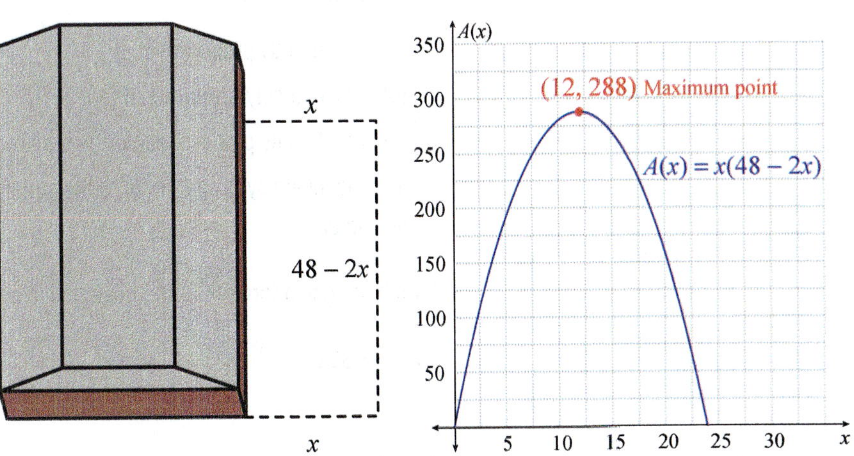

The significance of the highest point (at $x = 12$) is that when the width is 12, the maximum area is enclosed (288 ft²). Any other dimensions for the fence reduce total enclosed area.

3.2 EXERCISES

💡 PRACTICE

Find the break-even point given the revenue and cost functions in Exercises 1–4.

1. $R(x) = 15x$
$C(x) = 5x + 30$

2. $R(x) = 27x$
$C(x) = 14x + 442$

3. $R(x) = 24x - 0.2x^2$
$C(x) = 8x + 300$

4. $R(x) = 9x - 0.7x^2$
$C(x) = 2x + 11.2$

Find the equilibrium point given the supply and demand functions in Exercises 5–8.

5. $S(x) = 2x + 3$
$D(x) = 15 - x$

6. $S(x) = 4x + 7$
$D(x) = 33 - 1.2x$

7. $S(x) = x^2 + x$
$D(x) = 35 - x$

8. $S(x) = x^2 + 3$
$D(x) = 51 - 2x$

In Exercises 9–18, the quantity represented by y is related to the quantity represented by x in one of the following ways:

(a) y is directly proportional to x

(b) y is inversely proportional to x

(c) y is directly proportional to the square of x

(d) y is inversely proportional to the square of x

(e) other

Match the given formula to the best choice of (a) through (e).

9. $y = 32x$

10. $y = \dfrac{32}{x}$

11. $y = 16x^2$

12. $y = \dfrac{-32}{x^2}$

13. $y = -x$

14. $y = 2x + 3$

15. $y = \pi x$

16. $y = (x - 1)^2$

17. $y = \dfrac{x}{10}$

18. $y = \dfrac{wxv}{stu}$

🚀 APPLICATIONS

For Exercises 19–23, use the following situation:

Two students create a downloadable app which connects dots on a grid so that two players can play "Chase the Rabbit." They have developed a website on which to promote and sell the app for $5.00 and they pay an outside vendor $0.50 per purchase to manage the payments. They also pay another student $27/day (5 days a week) to monitor a customer service email box to answer questions and relay orders.

Let x denote the number of copies of the app sold, let $C(x)$ be the weekly total cost function (linear), and let $R(x)$ be the revenue function.

19. a. Write the expression for $C(x)$.
   b. Determine the weekly cost of selling 500 copies of the app.

20. a. Write the revenue function.
   b. How much revenue is produced by the sale of 500 copies of the app.

21. a. Write the profit function.
   b. Determine the profit from the sale of 500 copies of the app.

22. How many copies of the app must be sold in order to break even?

23. A business professor estimates that the campus craze for the game could become national, and, therefore, the game could be marketed nationally. If 100 colleges were to become market sites, find a profit function for all 100 colleges together. Assume total profits and the number of colleges involved are directly proportional.

For Exercises 24–28, use the following information:

An ideal gas satisfies a law which may be stated as $\dfrac{PV}{T} = 0.821n$ where P is the pressure in atmospheres (atm), V is the volume in liters, T is the temperature in degrees Kelvin ($K = 273 + C$, where C is Celsius), and n is the number of moles (gram molecular weights). Thus, for one mole of gas, pressure and volume are indirectly proportional for a constant temperature, pressure and temperature are directly proportional for a fixed volume, and volume and temperature are directly proportional for a fixed pressure.

Assume, if necessary, that there is 1 mole (6.023×10^{23} molecules) present.

24. What volume is occupied by 1 mole of gas at a temperature of 300°K and a pressure of 2 atm?

25. A fixed volume of gas is heated from a temperature of 200°K and 0.5 atm of pressure to 300°K. What is the new pressure?

26. A gas at a pressure of 2 atm expands, at constant temperature, from 10 liters to 15 liters. What is the new pressure?

27. One mole of gas occupies a volume of 2 liters and has a temperature of 136.5°K. What is the pressure in atmospheres?

28. A gas has a volume of 2 liters, a temperature of 30°C and a pressure of 1 atm. When the gas is heated to 60°C and its volume is compressed to a volume of 1.25 liters, what is its new pressure? (**Hint:** For this problem, $n \neq 1$.)

29. Modeling in business: The manager of a pie shop sells his pies for $6.50. The overhead is $378 per day and each pie costs $1.10 to make.
  a. Write the revenue function.
  b. Write the cost function.
  c. Write the profit function.
  d. Find the break-even point.

30. Modeling in business: A certain style of athletic shoe costs $11.80 per pair to produce. The fixed costs are $864 per week. The shoes can be sold for $19.00 per pair.
  a. Write the revenue function.
  b. Write the cost function.
  c. Write the profit function.
  d. Find the break-even point.

31. Modeling in manufacturing: A manufacturer of golf clubs finds that the fixed costs are $5780 per week and the cost of producing each set of clubs is $73.00. Each set of clubs can be sold for $243.00.
  a. Write the revenue function.
  b. Write the cost function.
  c. Write the profit function.
  d. Find the break-even point.

32. Modeling in business: A soft drink company has fixed costs of $4000 per day. The variable costs are $2.75 per case of soda. Each case sells for $5.25.
  a. Write the revenue function.
  b. Write the cost function.
  c. Write the profit function.
  d. Find the break-even point.

33. Modeling in production: The cost of producing 200 pens is $290. Producing 250 pens would cost $297.50.
  a. Find the average cost per pen for additional 50 pens over 200.
  b. Assuming the total cost function is linear, write an equation for the cost of producing x pens.
  c. What are the fixed costs?

34. Modeling in production: The Blue Umbrella Company can produce 500 umbrellas per week at a cost of $1800. It would cost $1950 to produce 600 umbrellas.
  a. Find the average cost of each of the additional 100 umbrellas over 500.
  b. Assuming the total cost is a linear function, write an equation for the cost of producing x umbrellas.
  c. What are the fixed costs?

35. Revenue-profit: It has been determined that the cost of producing x units of a certain item is $11x + 500$. The demand function is given by $p = D(x) = 31 - 0.5x$.
  a. Write the revenue function.
  b. Write the profit function.

36. **Modeling in sales:** The manager of a men's store knows he can sell 60 pairs of a certain style of sock when the price is \$1.20 per pair. If the price is \$1.50, he can sell only 48 pairs of socks. The total cost function for x pairs of socks is $C(x) = 0.70x + 15$ dollars.
    a. Assuming the demand function is linear, write an equation for $D(x)$.
    b. Write the revenue function.
    c. Write the profit function.

37. **Modeling in manufacturing:** A manufacturer of TVs can sell 800 TVs to his dealers at \$384 each. If the price is \$380, he can sell 1000 TVs. The total cost of producing x TVs is $C(x) = 3600 + 250x - 0.01x^2$ dollars.
    a. Assuming the demand function is linear, write an equation for $D(x)$.
    b. Write the revenue function.
    c. Write the profit function.

38. **Revenue:** Suppose the revenue R from the sale of a product is directly proportional to the number of units x of the product that are sold. Suppose also that the revenue from the sale of 65 units of the product is \$1820.
    a. Write a function for R in terms of x.
    b. Find the revenue if 75 units are sold.

39. **Interest:** Suppose the annual interest I earned on an investment is directly proportional to the amount of money invested P. Suppose also that an investment of \$8200 earns an annual interest of \$512.50.
    a. Write a function for I in terms of P.
    b. Find the annual interest earned by \$6000.

40. **Interest:** What will \$6000 accumulate to if it is deposited in a bank for three years and earns 5% a year with annual compounding?

41. **Price:** Suppose that for a certain product, the price per item p is inversely proportional to the number of items sold x. Suppose also that the price per item is \$8.50 when 40 items are sold.
    a. Write a function for p in terms of x.
    b. Find the price if 34 items are sold.

42. **Demand:** Pat has decided to produce a limited number of prints from one of her paintings. She plans to issue x prints, where $0 < x \le 50$. If she wants her revenue to be \$5000, write a function for the demand $D(x)$.

43. **Number of orders:** The owner of a camera shop expects to sell 800 cameras of a particular style during the year. How many orders will the dealer need to place with his distributor if each order is for x cameras?

44. **Salary:** A salesperson's weekly salary depends on the amount of her sales. Her salary is \$250 per week plus a commission of 8% of her weekly sales in excess of \$2500. Write a function for her salary if her sales were x dollars.

45. **International calls:** For an international call, the telephone company charges 65 cents for the first 3 minutes or less, plus 15 cents for each additional minute. Write a cost function for a call x minutes long.

46. Car rental: The rate for renting a car at a local agency is $22.50 per day plus $0.10 for each mile driven in excess of 100. If a car is rented for one day, write a function for the cost in terms of the number of miles driven.

47. Agriculture: A farmer cultivates bananas. They cost him 38 cents per bunch to produce. He is able to sell only 85% of those he produces. If he sells his bananas at 75 cents per bunch, find a function for his profit in terms of the number of bunches he produces.

48. Retail profit: A grocery store bought bags of frozen corn for 59 cents per bag and stored it in two freezers. During the night, one freezer defrosted and ruined 14 bags. If the remaining frozen corn was sold for 98 cents per bag, find a function for the profit in terms of the number of bags bought.

49. Retail profit: It costs Liz $12 to build a picture frame. She estimates that, if she charges x dollars per frame, she can sell $60 - x$ frames per week. Write a function for her weekly profit.

50. Retail profit: A toy retailer pays $3 each for a particular doll. He estimates that, if he charges x dollars for each doll, he will be able to sell $300 - 20x$ dolls. Write a function for his profit.

51. Area: The perimeter of a rectangle is 276 feet. If the rectangle is x feet long, write a function for the area $A(x)$.

52. Perimeter: The area of a rectangle is 426 cm^2. If the length of the rectangle is x centimeters, write a function for the perimeter $P(x)$.

53. Perimeter: The area of a rectangle is 288 ft^2. If the length of the rectangle is x feet, write a function for the perimeter $P(x)$.

54. Area: The perimeter of a rectangle is 197 inches. If the rectangle is x inches wide, write a function for the area $A(x)$.

55. Construction: The maintenance department at the city zoo wants to build a pen and divide it as shown in the diagram. If the department has a total of 720 feet of fencing, write a function for the area in terms of x.

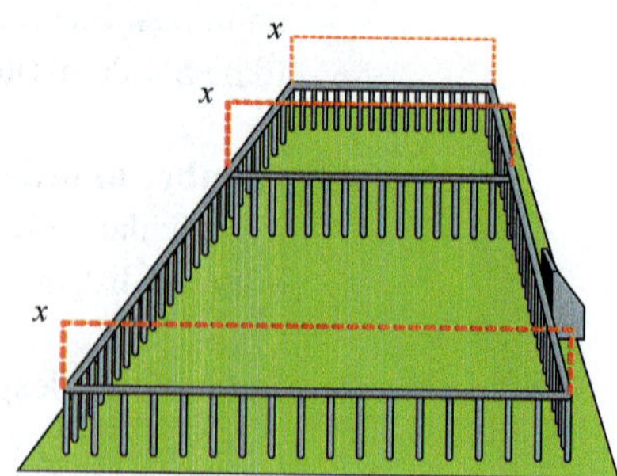

3.3 LINEAR AND QUADRATIC FUNCTIONS

■ TOPICS

- Linear Functions and Their Graphs
- Quadratic Functions and Their Graphs

Linear Functions and Their Graphs

Much of the next several sections of this chapter will be devoted to gaining familiarity with some of the types of functions that commonly arise in mathematics. We will discuss two classes of functions in this section, beginning with linear functions.

Recall that a linear equation in two variables is an equation whose graph consists of a straight line in the Cartesian plane. Similarly, a linear function is a function whose graph is a straight line. We can define such functions algebraically as follows.

Linear Functions

A **linear function** f in the variable x is any function that can be written in the form $f(x) = mx + b$, where m and b are real numbers. If $m \neq 0$, $f(x) = mx + b$ is also called a **first-degree function**.

A function defined by an equation in x and y can be written in function form by solving the equation for y and then replacing y with $f(x)$. This process can be reversed, so the linear function $f(x) = mx + b$ appears in equation form as $y = mx + b$, a linear equation written in slope-intercept form. Thus, the graph of a linear function is a straight line with slope m and y-intercept $(0, b)$.

The graph of a function is a plot of all the ordered pairs that make up the function; that is, the graph of a function f is the plot of all the ordered pairs in the set $\{(x, y)\,|\, f(x) = y\}$. We have experience in plotting such sets if the ordered pairs are defined by an equation in x and y, but we haven't plotted functions that have been defined with functional notation. Any function of x defined with functional notation can be written as an equation in x and y by replacing $f(x)$ with y, so the graph of a function f consists of a plot of the ordered pairs in the set $\{(x, f(x))\,|\, x \in \text{domain of } f\}$.

Consider the function $f(x) = -3x + 5$. Figure 1 contains a table of four ordered pairs defined by the function and a graph of the function with the four ordered pairs noted.

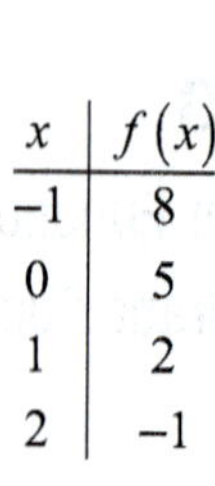

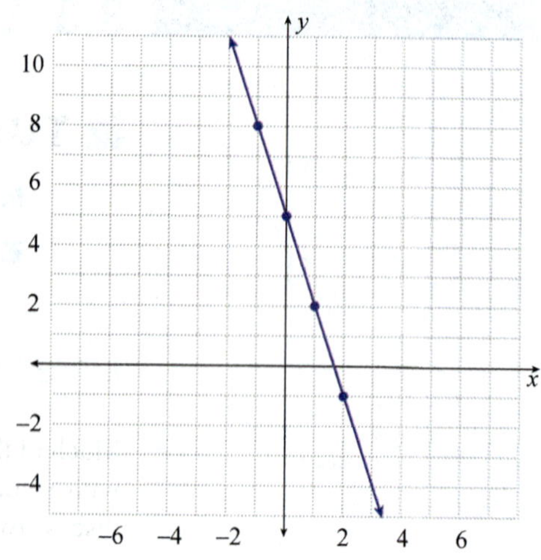

FIGURE 1: Graph of $f(x) = -3x + 5$

Again, note that every point on the graph of the function in Figure 1 is an ordered pair of the form $(x, f(x))$; we have simply highlighted four of them with dots.

We could have graphed the function $f(x) = -3x + 5$ by noting that it is a straight line with a slope of -3 and a y-intercept of 5. We use this approach in the following example.

Graph the following linear functions.

a. $f(x) = 3x + 2$ **b.** $g(x) = 3$

Solution

a. 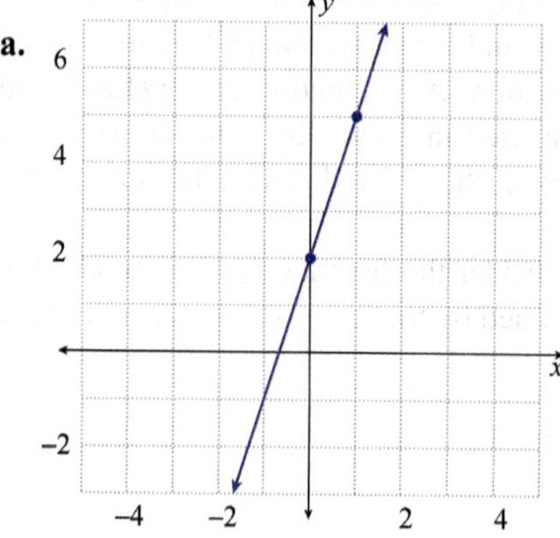

The function f is a line with a slope of 3 and a y-intercept of 2.

To graph the function, plot the ordered pair $(0, 2)$ and locate another point on the line by moving up 3 units and over to the right 1 unit, giving the ordered pair $(1, 5)$. Once these two points have been plotted, connecting them with a straight line completes the process.

b.

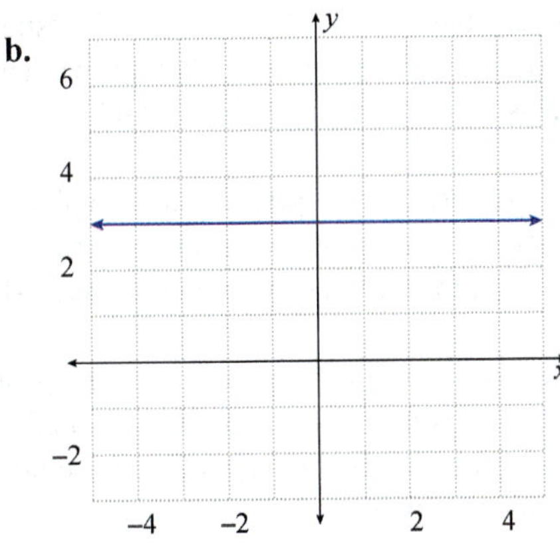

The graph of the function g is a straight line with a slope of 0 and a y-intercept of 3.

A linear function with a slope of 0 is also called a **constant function**, as it turns any input into one fixed constant—in this case the number 3. The graph of a constant function is always a horizontal line.

Quadratic Functions and Their Graphs

Previously, we learned how to solve quadratic equations in one variable. We will now study quadratic *functions* of one variable and relate this new material to what we already know.

Quadratic Functions

A **quadratic function** f in the variable x, also known as a **second-degree function**, is any function that can be written in the form $f(x) = ax^2 + bx + c$, where a, b, and c are real numbers and $a \neq 0$.

The graph of any quadratic function is a roughly U-shaped curve known as a **parabola**. The graph in Figure 2 is the most basic example of a parabola; it is the graph of the quadratic function $f(x) = x^2$. The table contains a few of the ordered pairs on the graph.

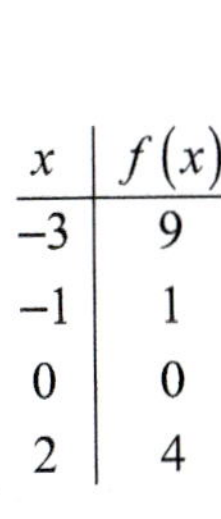

x	$f(x)$
-3	9
-1	1
0	0
2	4

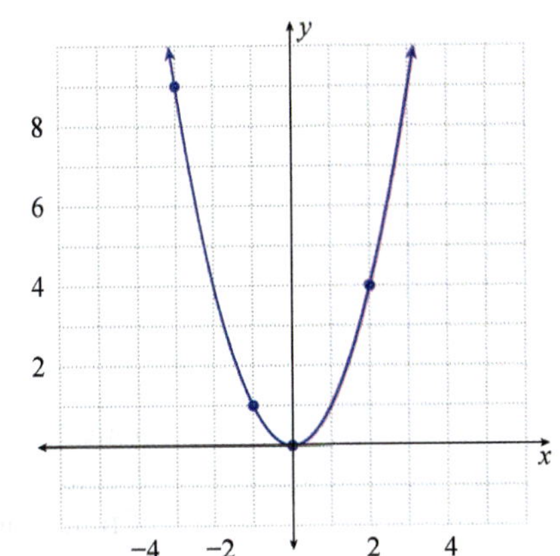

FIGURE 2: Graph of $f(x) = x^2$

Figure 2 demonstrates two key characteristics of parabolas:

- There is one point, known as the **vertex**, where the graph "changes direction". Scanning the graph from left to right, it is the point where the graph stops going down and begins to go up (if the parabola opens upward) or stops going up and begins to go down (if the parabola opens downward).

- Every parabola is symmetric with respect to its **axis**, a line passing through the vertex dividing the parabola into two halves that are mirror images of each other. This line is also called the **axis of symmetry**.

Every parabola that represents the graph of a quadratic function has a vertical axis. Parabolas can be relatively skinny or relatively broad, meaning that the curve of the parabola at the vertex can range from very sharp to very flat.

We will develop the method of graphing quadratic functions by working from the answer backward. We will first see what effects various mathematical operations have on the graphs of parabolas, and then see how this knowledge lets us graph a general quadratic function.

To begin, the graph of the function $f(x) = x^2$, shown in Figure 2, is the basic parabola. We already know its characteristics: its vertex is at the origin, its axis is the y-axis, it opens upward, and the sharpness of the curve at its vertex will serve as a convenient reference when discussing other parabolas.

Now consider the function $g(x) = (x - 3)^2$, obtained by replacing x in the formula for f with $x - 3$. We know x^2 is equal to 0 when $x = 0$. What value of x results in $(x - 3)^2$ equaling 0? The answer is $x = 3$. In other words, the point $(0, 0)$ on the graph of f corresponds to the point $(3, 0)$ on the graph of g. With this in mind, examine the table and graph in Figure 3.

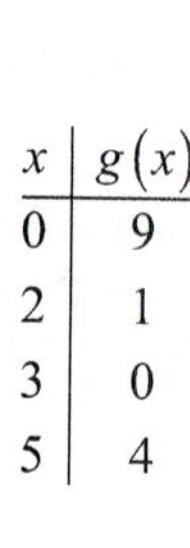

x	$g(x)$
0	9
2	1
3	0
5	4

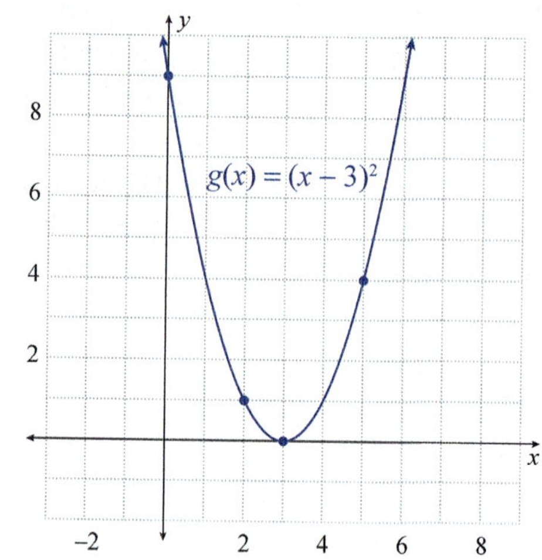

FIGURE 3: Graph of $g(x)$

Notice that the shape of the graph of g is identical to that of f, but it has been shifted over to the right by 3 units. This is our first example of how we can manipulate graphs of functions using transformations.

Now consider the function h obtained by replacing the x in x^2 with $x + 7$. As with the functions f and g, $h(x) = (x + 7)^2$ is nonnegative for all values of x, and only one value for x will return a value of 0: $h(-7) = 0$. Compare the table and graph in Figure 4 with those in Figures 2 and 3.

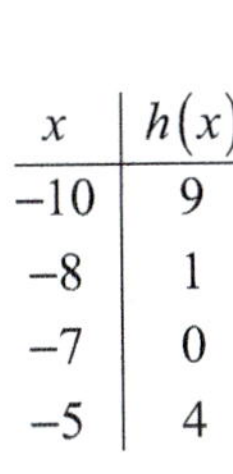

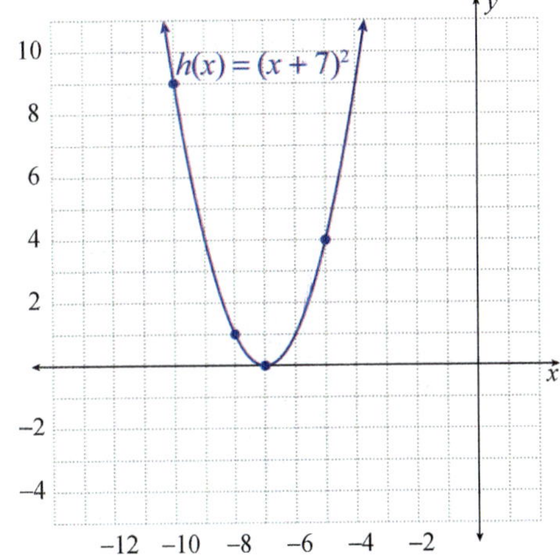

x	$h(x)$
−10	9
−8	1
−7	0
−5	4

FIGURE 4: Graph of $h(x)$

So we have seen how to shift the basic parabola to the left and right: the graph of $g(x) = (x - h)^2$ has the same shape as the graph of $f(x) = x^2$, but it is shifted h units to the right if h is positive and $|h|$ units to the left if h is negative.

How do we shift a parabola up and down? To move the graph of $f(x) = x^2$ up by a fixed number of units, we need to add that number of units to the second coordinate of each ordered pair. Similarly, to move the graph down we subtract the desired number of units from each second coordinate. To see this, consider the table and graphs for the two functions $j(x) = x^2 + 5$ and $k(x) = x^2 - 2$ in Figure 5.

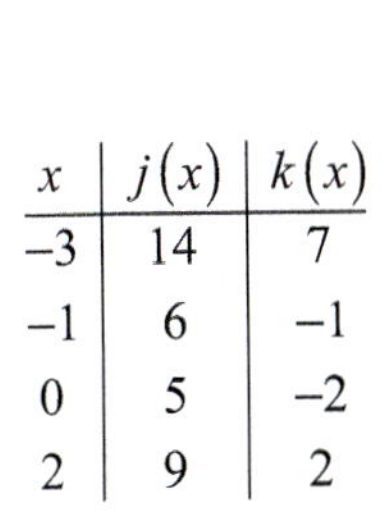

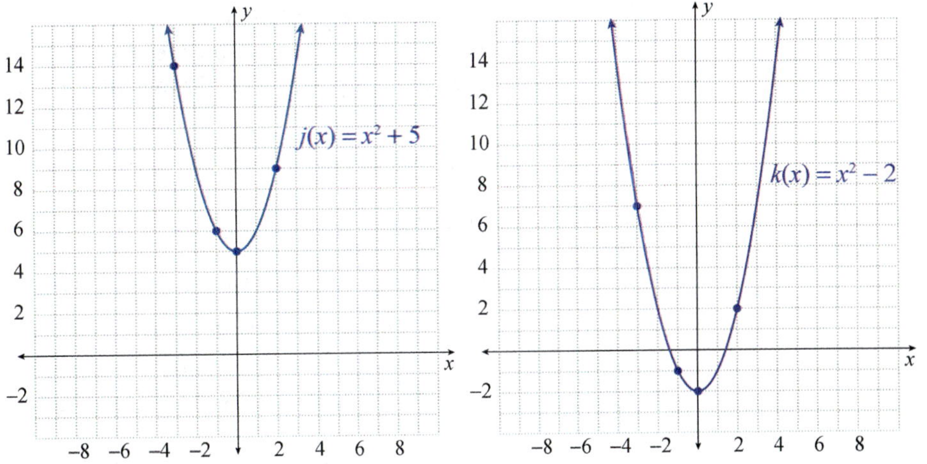

x	$j(x)$	$k(x)$
−3	14	7
−1	6	−1
0	5	−2
2	9	2

FIGURE 5: Graphs of $j(x)$ and $k(x)$

Finally, how do we make a parabola skinnier or broader? To make the basic parabola skinnier (to make the curve at the vertex sharper), we need to stretch the graph vertically. We can do this by multiplying the formula x^2 by a constant a greater than 1 to obtain the formula ax^2. Multiplying the formula x^2 by a constant a that lies between 0 and 1 makes the parabola broader (it makes the curve at the vertex flatter). Finally, multiplying x^2 by a negative constant a turns all of the nonnegative outputs of f into nonpositive outputs, resulting in a parabola that opens downward instead of upward.

Compare the graphs of $l(x) = 6x^2$ and $m(x) = -\dfrac{1}{2}x^2$ in Figure 6 to the basic parabola $f(x) = x^2$.

x	$l(x)$	$m(x)$
-3	54	$-\dfrac{9}{2}$
-1	6	$-\dfrac{1}{2}$
0	0	0
2	24	-2

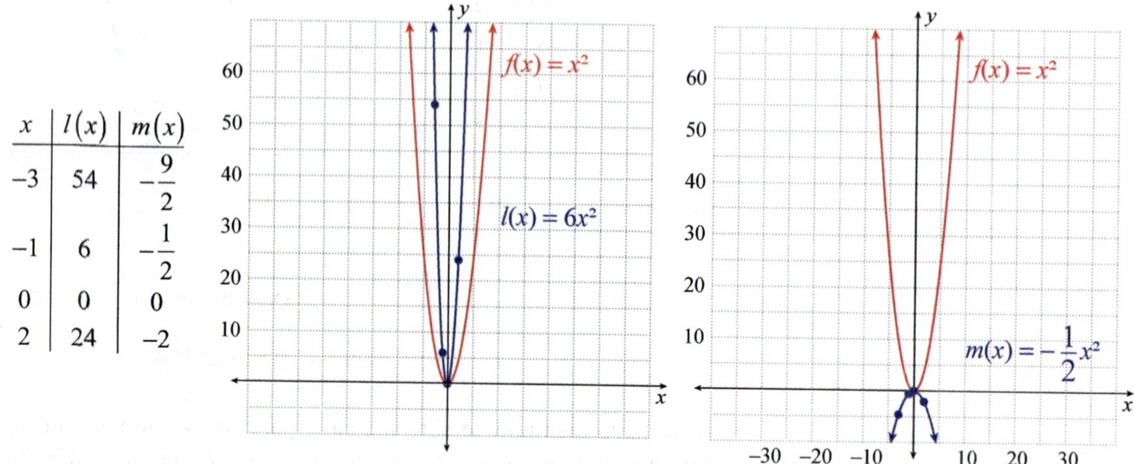

FIGURE 6: Graphs of $l(x)$ and $m(x)$

The following form of a quadratic function brings together all the ways of altering the basic parabola $f(x) = x^2$.

Vertex Form of a Quadratic Function

The graph of the function $g(x) = a(x - h)^2 + k$, where a, h, and k are real numbers and $a \neq 0$, is a parabola whose vertex is at (h, k). The parabola is narrower than $f(x) = x^2$ if $|a| > 1$, and is broader than $f(x) = x^2$ if $0 < |a| < 1$. The parabola opens upward if a is positive and downward if a is negative.

Where does this definition come from? Consider this construction of the vertex form.

$f(x) = x^2$ — Begin with the basic parabola, with vertex $(0, 0)$.

$g(x) = (x - h)^2$ — This represents a horizontal shift of h.

$g(x) = (x - h)^2 + k$ — This step adds a vertical shift of k, making the new vertex (h, k).

$g(x) = a(x - h)^2 + k$ — Finally, apply the stretch/compress factor of a. If a is negative, the parabola opens downward.

The question now is: given a quadratic function $f(x) = ax^2 + bx + c$, how do we determine the location of its vertex, whether it opens upward or downward, and whether it is skinnier or broader than the basic parabola? All of this information is available if the equation is in vertex form, so we need to convert the formula $ax^2 + bx + c$ into the form $a(x - h)^2 + k$. It turns out that we can *always* do this by completing the square on the first two terms of the expression.

Example 2: Graphing Quadratic Functions

Sketch the graph of the function $f(x) = -x^2 - 2x + 3$. Locate the vertex and the x-intercepts.

Solution

Finding and plotting the x-intercepts is a great way to see the shape of the function.

First, identify the vertex of the function by completing the square as follows.

$$f(x) = -x^2 - 2x + 3$$

First, factor out the leading coefficient of -1 from the first two terms.

$$= -\left(x^2 + 2x\right) + 3$$

Complete the square on the x^2 and $2x$ terms.

$$= -\left(x^2 + 2x + 1\right) + 1 + 3$$

$$= -(x+1)^2 + 4$$

Because of the -1 in front of the parentheses, this amounts to adding -1 to the function, so we compensate by adding 1 as well.

Completing the square places the equation in vertex form, and we rewrite the expression $-(x+1)^2 + 4$ as $-(x-(-1))^2 + 4$, so the vertex is $(-1, 4)$.

The instructions also ask us to identify the x-intercepts. An x-intercept of the function f is any point on the x-axis where $f(x) = 0$, so we need to solve the equation $-x^2 - 2x + 3 = 0$. This can be done by factoring.

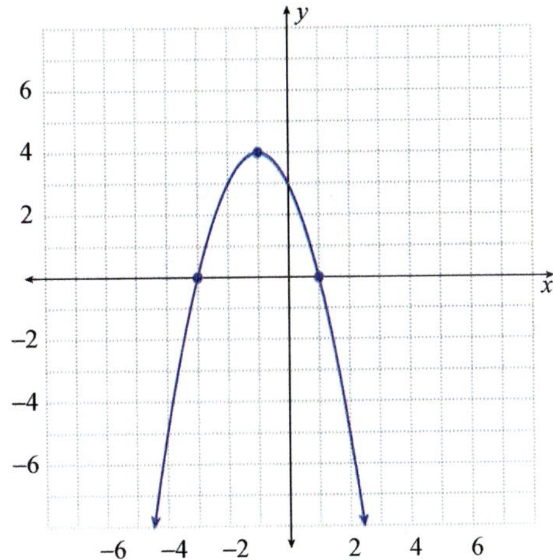

$$-x^2 - 2x + 3 = 0$$

$$x^2 + 2x - 3 = 0$$

First, divide each term by -1.

$$(x+3)(x-1) = 0$$

Factor into two binomials.

$$x = -3, 1$$

The Zero-Factor Property gives us the x-intercepts.

Therefore, the x-intercepts are located at $(-3, 0)$ and $(1, 0)$.

The vertex form of the function, $f(x) = -(x+1)^2 + 4$, tells us that this quadratic opens downward, has its vertex at $(-1, 4)$, and is neither skinnier nor broader than the basic parabola. We now also know that it crosses the x-axis at -3 and 1. Putting this all together, we obtain the graph.

Previously, we completed the square on the generic quadratic equation to develop the quadratic formula. We can use a similar approach to transform the standard form of a quadratic function into vertex form.

$$f(x) = ax^2 + bx + c$$

$$= a\left(x^2 + \frac{b}{a}x\right) + c$$

As always, begin by factoring the leading coefficient a from the first two terms.

$$= a\left(x^2 + \frac{b}{a}x + \frac{b^2}{4a^2}\right) - a\left(\frac{b^2}{4a^2}\right) + c$$

To complete the square, add the square of half of $\dfrac{b}{a}$ inside the parentheses.

$$= a\left(x + \frac{b}{2a}\right)^2 - \frac{b^2}{4a} + c$$

We need to balance the equation by subtracting $a\left(\dfrac{b^2}{4a^2}\right)$ outside the

$$= a\left(x + \frac{b}{2a}\right)^2 + \frac{4ac - b^2}{4a}$$

parentheses, then simplify.

Vertex of a Quadratic Function

Given a quadratic function $f(x) = ax^2 + bx + c$, the graph of f is a parabola with a vertex given by

$$\left(-\frac{b}{2a}, f\left(\frac{-b}{2a}\right)\right) = \left(-\frac{b}{2a}, \frac{4ac - b^2}{4a}\right).$$

Example 3: Using the Vertex Formula

Find the vertex for each of the following quadratic functions using the vertex formula.

a. $f(x) = x^2 - 4x + 8$

b. $g(x) = 3x^2 + 5x - 1$

Solution

NOTE

If the x-coordinate of the vertex is simple, use substitution to find the y-coordinate. If the x-coordinate is complicated, use the explicit formula (the right-hand form in the vertex formula).

a. Begin by using the formula to find the x-coordinate of the vertex.

$1x^2 - 4x + 8$ Note that the value of a is 1.

$-\dfrac{b}{2a} = -\dfrac{(-4)}{2(1)} = 2$ Substitute a and b into the formula and simplify.

At this point, we need to decide how to find the y-coordinate. Since the x-coordinate is an integer, substitute it directly into the original equation, finding $f\left(-\dfrac{b}{2a}\right)$.

$$f(2) = 2^2 - 4(2) + 8$$
$$= 4$$

Thus, the vertex of the graph of $f(x)$ is (2, 4).

b. Again, begin by finding the x-coordinate of the vertex.

$3x^2 + 5x - 1$ Substitute a and b into the formula and simplify.

$-\dfrac{b}{2a} = -\dfrac{(5)}{2(3)} = -\dfrac{5}{6}$

Here, the x-coordinate is a fraction, so substituting it into the original equation leads to messy calculations. Instead, use the explicit formula to find the y-coordinate.

$$\frac{4ac - b^2}{4a} = \frac{4(3)(-1) - (5)^2}{4(3)}$$ Substitute a, b, and c into the formula and simplify.

$$= \frac{-12 - 25}{12}$$

$$= -\frac{37}{12}$$

Thus, the vertex of the graph of $g(x)$ is $\left(-\dfrac{5}{6}, -\dfrac{37}{12}\right)$.

3.3 EXERCISES

💡 PRACTICE

Graph the following linear functions. See Example 1.

1. $f(x) = -5x + 2$

2. $g(x) = \dfrac{3x - 2}{4}$

3. $h(x) = -x + 2$

4. $p(x) = -2$

5. $g(x) = 3 - 2x$

6. $r(x) = 2 - \dfrac{x}{5}$

7. $f(x) = -2(1 - x)$

8. $a(x) = 3\left(1 - \dfrac{1}{3}x\right) + x$

9. $f(x) = 2 - 4x$

10. $g(x) = \dfrac{2x - 8}{4}$

11. $h(x) = 5x - 10$

12. $k(x) = 3x - \dfrac{2 + 6x}{2}$

13. $m(x) = \dfrac{-x + 25}{10}$

14. $q(x) = 1.5x - 1$

15. $w(x) = (x - 2) - (2 + x)$

Match the following functions with their graphs.

16. $f(x) = (8x - 14) - (-17 + 2x)$

17. $f(x) = 3x - \dfrac{7 + 8x}{3}$

18. $f(x) = \dfrac{6}{2} - \dfrac{2}{8}x$

19. $f(x) = 2\left(2 - \dfrac{8}{5}x\right) + x$

a.

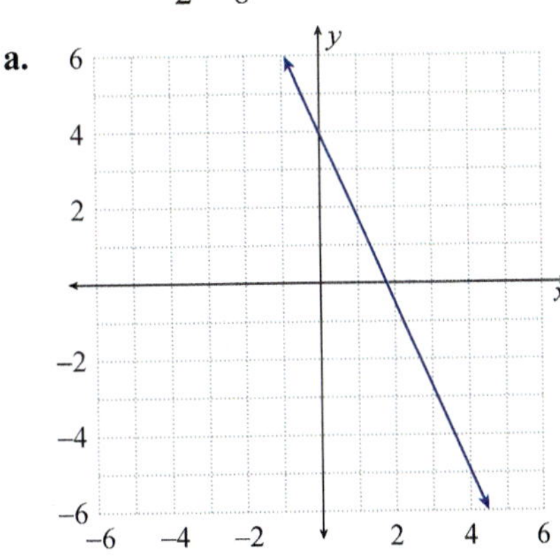

b.

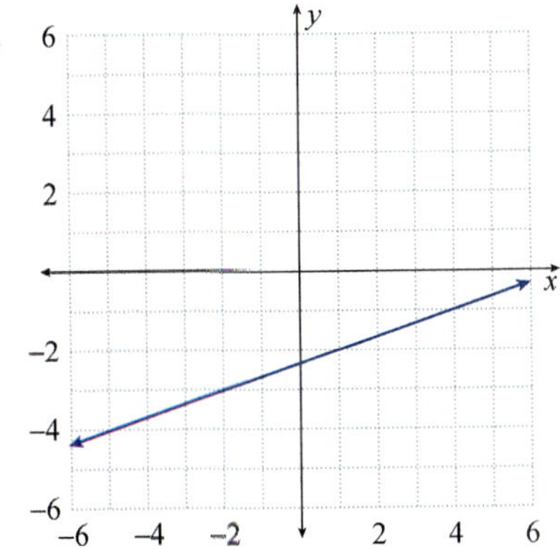

c.

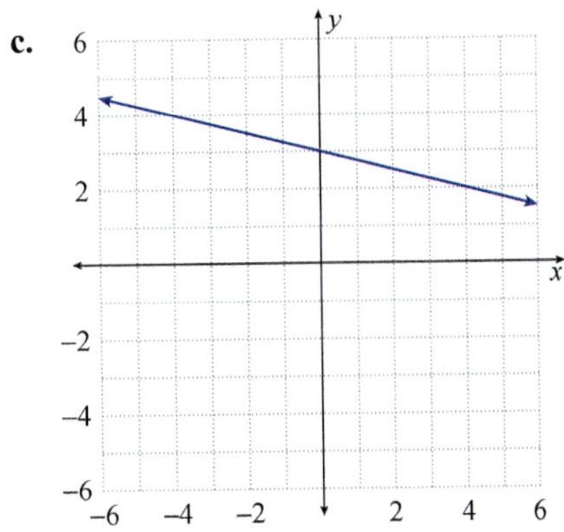

d.

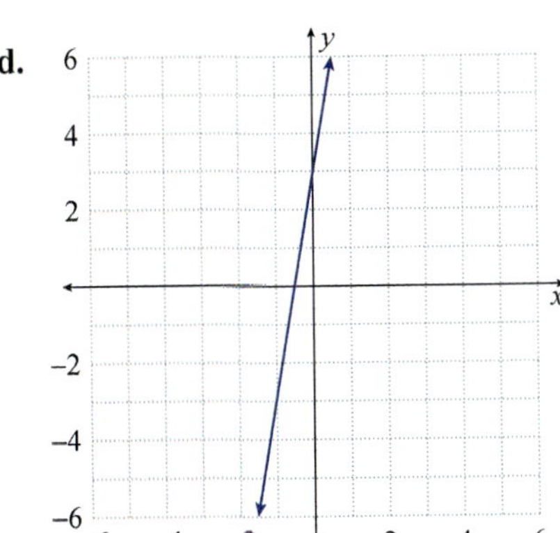

Graph the following quadratic functions, accurately locating the vertices and *x*-intercepts (if any). See Example 2.

20. $f(x) = (x-2)^2 + 3$

21. $g(x) = -(x+2)^2 - 1$

22. $h(x) = x^2 + 6x + 7$

23. $F(x) = 3x^2 + 2$

24. $G(x) = x^2 - x - 6$

25. $p(x) = -2x^2 + 2x + 12$

26. $q(x) = 2x^2 + 4x + 3$

27. $r(x) = -3x^2 - 1$

28. $s(x) = \dfrac{(x-1)^2}{4}$

29. $m(x) = x^2 + 2x + 4$

30. $n(x) = (x+2)(2-x)$

31. $p(x) = -x^2 + 2x - 5$

32. $f(x) = 4x^2 - 6$

33. $k(x) = 2x^2 - 4x$

34. $q(x) = (x+10)(x-2) + 36$

Match the following functions with their graphs.

35. $f(x) = -x^2 + 2x$

36. $f(x) = x^2 + 7x + 6$

37. $f(x) = \dfrac{x^2 - 8x + 16}{2}$

38. $f(x) = (x-5)(x+3) + 16$

a.

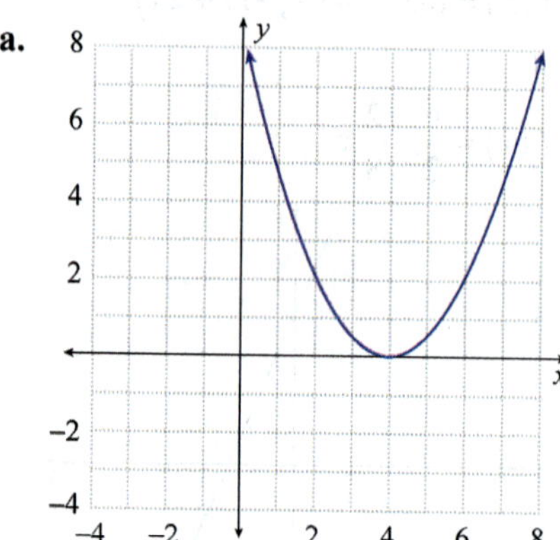

b.

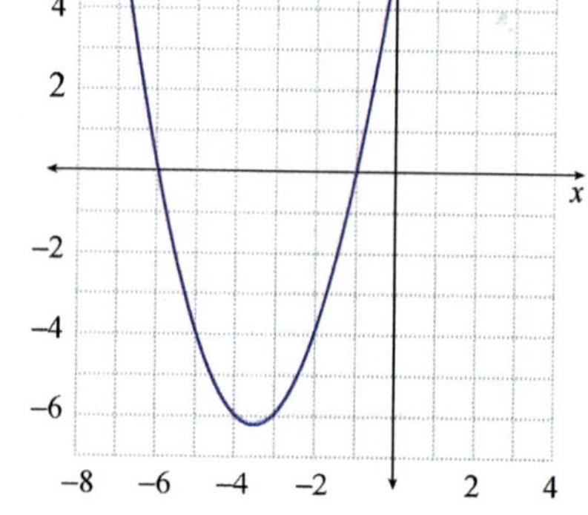

c.

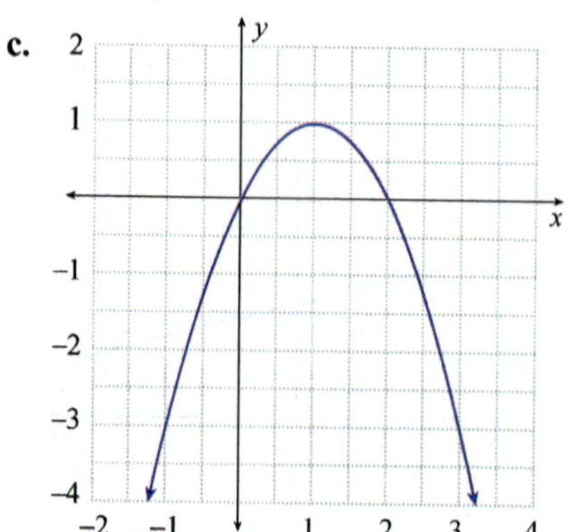

d.

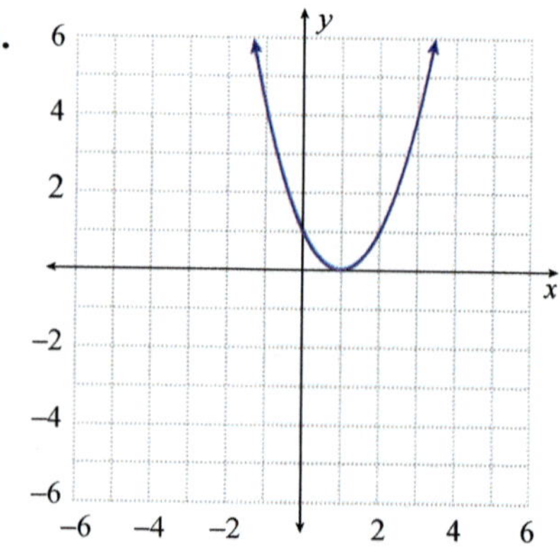

WRITING & THINKING

39. Without graphing, state the number of x-intercepts for each of the following functions and describe the location of the vertex in relation to the x-axis.

a. $y = (x-2)^2$

b. $y = (x-2)(x+2)$

c. $y = -(x-3)(x-1)$

d. $y = -\left(x-\sqrt{3}\right)\left(x+\sqrt{3}\right)$

e. $y = x(x+1)$

f. $y = -\left(x^2+1\right)$

3.4 APPLICATIONS OF QUADRATIC FUNCTIONS

■ TOPICS

- Maximization/Minimization Problems
- Maximum/Minimum of Graphs Using Technology

Maximization/Minimization Problems

Many applications of mathematics involve determining the value (or values) of the variable x that returns either the maximum or minimum possible value of some function $f(x)$. Such problems are called max/min problems for short. Examples from business include minimizing cost functions and maximizing profit functions. Examples from physics include maximizing a function that measures the height of a rocket as a function of time and minimizing a function that measures the energy required by a particle accelerator.

If we have a max/min problem involving a quadratic function, we can solve it by finding the vertex. Recall that the vertex is the only point where the graph of a parabola changes direction. This means it will be the minimum value of a function (if the parabola opens upward) or the maximum value (if the parabola opens downward).

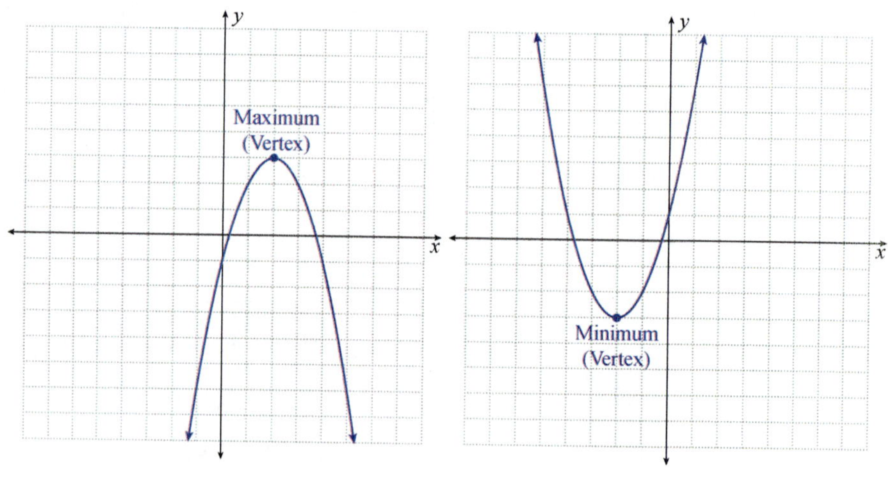

FIGURE 1: Maximum Value of a Quadratic Function

FIGURE 2: Minimum Value of a Quadratic Function

Example 1: Fencing a Garden

A farmer plans to use 100 feet of spare fencing material to form a rectangular garden plot against the side of a long barn, using the barn as one side of the plot. How should he split up the fencing among the other three sides in order to maximize the area of the garden plot?

Solution

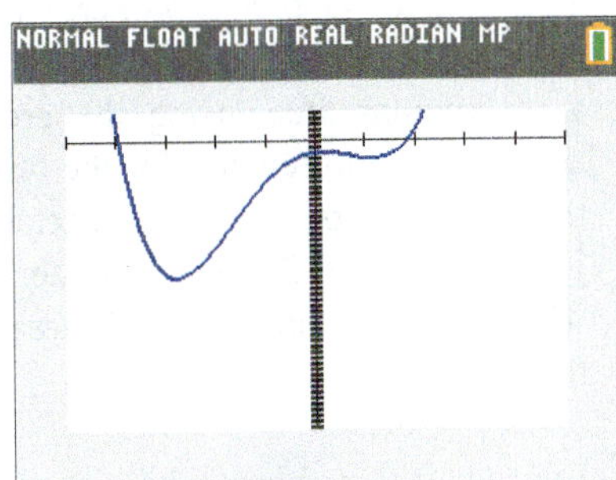

If we let x represent the length of one side of the plot, as shown in the diagram, then the dimensions of the plot are x feet by $100 - 2x$ feet. A function representing the area of the plot is $A(x) = x(100 - 2x)$.

If we multiply out the formula for A, we recognize it as a quadratic function $A(x) = -2x^2 + 100x$. This is a parabola opening downward, so the vertex will be the maximum point on the graph of A.

Using the vertex formula we know that the vertex of A is the ordered pair $\left(-\dfrac{100}{2(-2)}, A\left(-\dfrac{100}{2(-2)}\right)\right)$, or $(25, A(25))$. Thus, to maximize area, we should let $x = 25$, and so $100 - 2x = 50$. The resulting maximum possible area, 25×50, or 1250 square feet, is also the value $A(25)$.

Maximum/Minimum of Graphs Using Technology

As we've seen, finding the maximum or minimum possible values of some function $f(x)$ can be extremely important, and we have a method for doing so when the function is quadratic. But what if we wanted to find the minimum of the function $f(x) = x^4 + 2x^3 - 7x^2 + 2x - 4$? One way is to graph it on a calculator with the following window settings: Xmin = -5, Xmax = 5, Ymin = -100, Ymax = 10.

To find the minimum, press **2nd** **trace** to access the CALC menu and select 3:minimum. (If we were trying to find the maximum, we would select 4:maximum.) The screen should now display the graph with the words "Left Bound?" shown at the bottom. Use the left arrow to move the cursor anywhere to the left of where the minimum appears to be and press ENTER. The screen should now say "Right Bound?" Use the right arrow to move the cursor to the right of where the minimum appears to be and press ENTER again. The text should now read "Guess?"

Press ENTER a third time and the x- and y-values of the minimum will appear at the bottom of the screen.

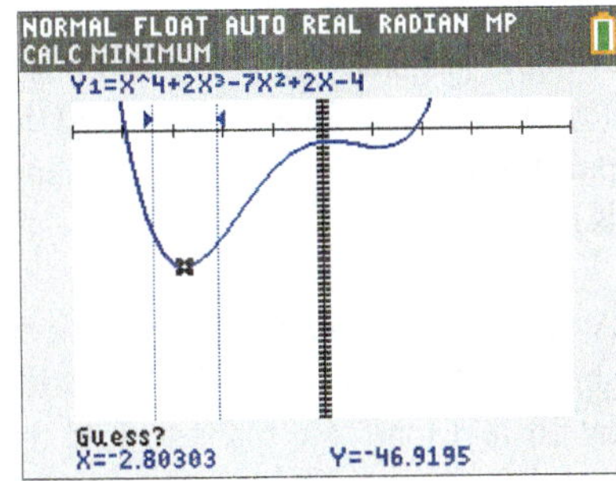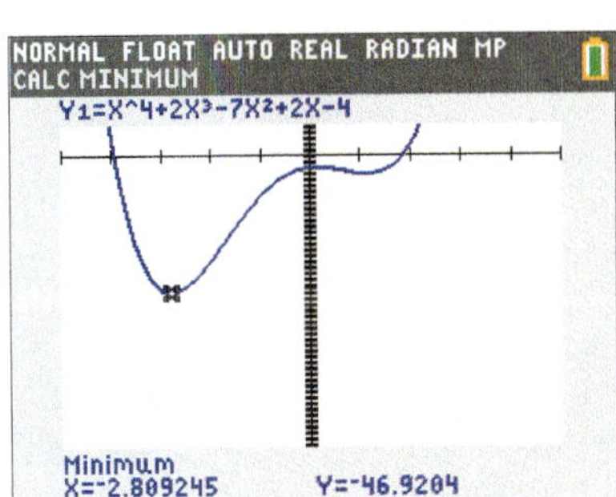

So the minimum is approximately $(-2.809, -46.920)$.

3.4 EXERCISES

APPLICATIONS

1. Cindy wants to construct three rectangular dog-training arenas side by side, as shown, using a total of 400 feet of fencing. What should the overall length and width be in order to maximize the area of the three combined arenas? (**Hint:** Let x represent the width, as shown, and find an expression for the overall length in terms of x.)

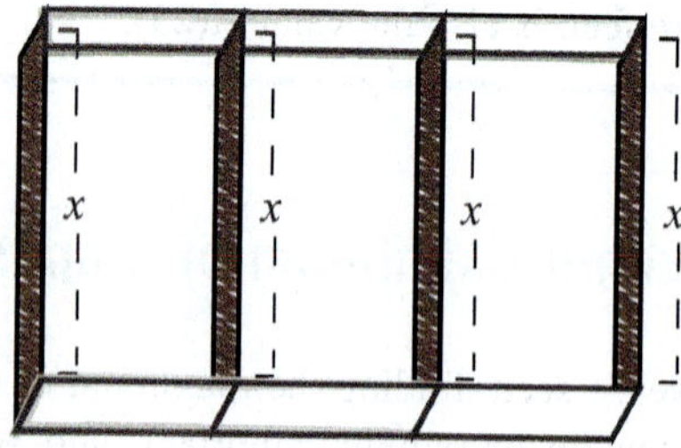

2. Among all the pairs of numbers with a sum of 10, find the pair whose product is maximum.

3. Among all rectangles that have a perimeter of 20, find the dimensions of the one whose area is largest.

4. Find the point on the line $2x + y = 5$ that is closest to the origin. (**Hint:** Instead of trying to minimize the distance between the origin and points on the line, minimize the square of the distance.)

5. Among all the pairs of numbers (x, y) such that $2x + y = 20$, find the pair for which the sum of the squares is minimum.

6. A rancher has a rectangular piece of sheet metal that is 20 inches wide by 10 feet long. He plans to fold the metal to create a narrow three-sided channel and weld two other sheets of metal to the ends to form a watering trough 10 feet long, as shown. How should he fold the metal in order to maximize the volume of the resulting trough?

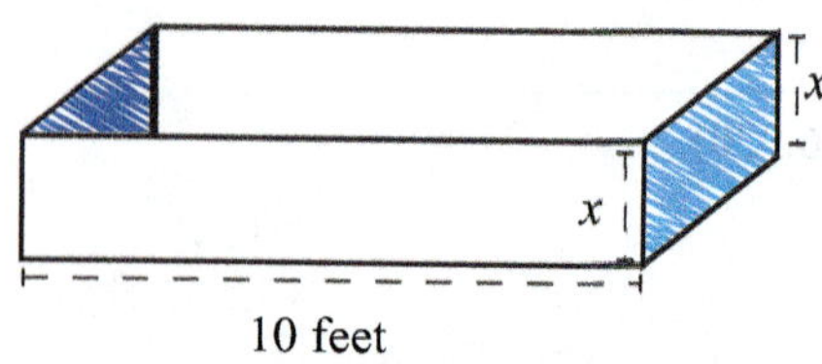

7. Find a pair of numbers whose product is maximum if the pair must have a sum of 16.

8. Search the Seas cruise ship has a conference room onboard that can hold up to 60 people. Companies can reserve the room for groups of 38 or more. If the group contains 38 people, the company pays $60 per person. The cost per person is reduced by $1 for each person in excess of 38. Find the size of the group that maximizes the income for the owners of the ship and find this income.

9. The back of George's property is a creek. George would like to enclose a rectangular area, using the creek as one side and fencing for the other three sides, to create a vegetable garden. If he has 300 feet of material, what is the maximum possible area of the garden?

10. Find a pair of numbers whose product is maximum if two times the first number plus the second number is 48.

11. The total revenue for Thompson's Studio Apartments is given by the function $R(x) = 100x - 0.1x^2$, where x is the number of rooms rented. What number of rooms rented produces the maximum revenue?

12. The total revenue of Tran's Machinery Rental is given by the function $R(x) = 300x - 0.4x^2$, where x is the number of units rented. What number of units rented produces the maximum revenue?

13. The total cost of producing a type of small car is given by $C(x) = 9000 - 135x + 0.045x^2$, where x is the number of cars produced. How many cars should be produced to incur minimum cost?

14. The total cost of manufacturing a set of golf clubs is given by $C(x) = 800 - 10x + 0.20x^2$, where x is the number of sets of golf clubs produced. How many sets of golf clubs should be manufactured to incur minimum cost?

15. The owner of a parking lot is going to enclose a rectangular area with fencing, using an existing fence as one of the sides. The owner has 220 feet of new fencing material (which is much less than the length of the existing fence). What is the maximum possible area that the owner can enclose?

In Exercises 16–18, use the formula $h(t) = -16t^2 + v_0 t + h_0$ for the height at time t of an object thrown vertically upward with velocity v_0 (in feet per second) from an initial height of h_0 (in feet).

16. Sitting in a tree, 48 feet above ground level, Sue shoots a pebble straight up with a velocity of 64 feet per second. What is the maximum height attained by the pebble?

17. A ball is thrown upward with a velocity of 48 feet per second from the top of a 144-foot building. What is the maximum height of the ball?

18. A rock is thrown upward with a velocity of 80 feet per second from the top of a 64-foot-high cliff. What is the maximum height of the rock?

📈 TECHNOLOGY

Use a graphing utility to graph each of the following quadratic functions. Then determine the vertex and x-intercepts.

19. $f(x) = 2x^2 - 16x + 31$

20. $f(x) = -x^2 - 2x + 3$

21. $f(x) = x^2 - 8x - 20$

22. $f(x) = x^2 - 4x$

23. $f(x) = 25 - x^2$

24. $f(x) = 3x^2 + 18x$

25. $f(x) = x^2 + 2x + 1$

26. $f(x) = 3x^2 - 8x + 2$

27. $f(x) = -x^2 + 10x - 4$

28. $f(x) = \frac{1}{2}x^2 + x - 1$

3.5 OTHER COMMON FUNCTIONS

■ TOPICS

- Functions of the Form ax^n
- Functions of the Form $\dfrac{a}{x^n}$
- Functions of the Form $ax^{\frac{1}{n}}$
- The Absolute Value Function
- The Greatest Integer Function
- Piecewise-Defined Functions

In a previous section, we investigated the behavior of linear and quadratic functions, but these are just two types of commonly occurring functions; there are many other functions that arise naturally in solving various problems. In this section, we will explore several other classes of functions, building up a portfolio of functions to be familiar with.

Functions of the Form ax^n

We already know what the graph of any function of the form $f(x) = ax$ or $f(x) = ax^2$ looks like, as these are, respectively, simple linear and quadratic functions. What happens to the graphs as we increase the exponent, and consider functions of the form $f(x) = ax^3$, $f(x) = ax^4$, etc.?

The behavior of a function of the form $f(x) = ax^n$, where a is a real number and n is a natural number, falls into one of two categories. Consider the graphs in Figure 1:

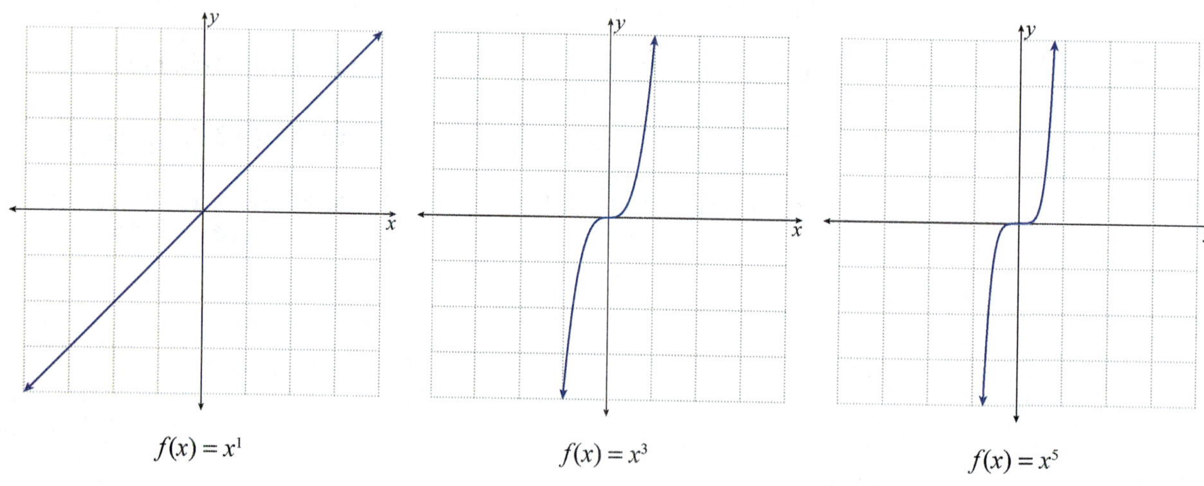

$$f(x) = x^1 \qquad\qquad f(x) = x^3 \qquad\qquad f(x) = x^5$$

FIGURE 1: Odd Exponents

The three graphs in Figure 1 show the behavior of $f(x) = x^n$ for the first three odd exponents. Note that in each case, the domain and the range of the function are both the entire set of real numbers; the same is true for higher odd exponents as well. Now, consider the graphs in Figure 2:

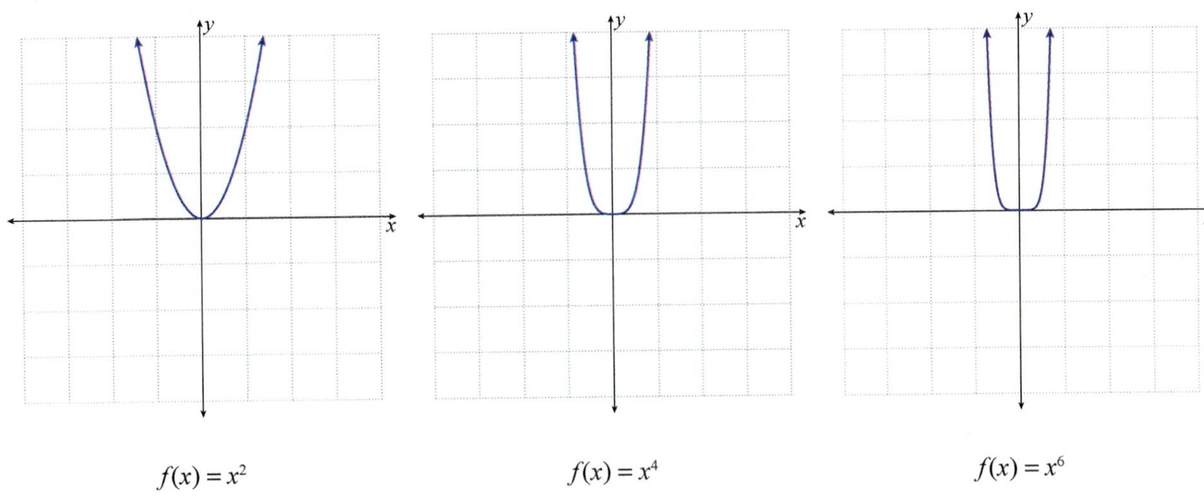

$$f(x) = x^2 \qquad\qquad f(x) = x^4 \qquad\qquad f(x) = x^6$$

FIGURE 2: Even Exponents

These three functions are also similar to one another. The first one is the basic parabola we studied previously. The other two bear some similarity to parabolas, but are flatter near the origin and rise more steeply for $|x| > 1$. For any function of the form $f(x) = x^n$ where n is an even natural number, the domain is the entire set of real numbers and the range is the interval $[0, \infty)$.

Multiplying a function of the form x^n by a constant a has the effect that we noticed previously. If $|a| > 1$, the graph of the function is stretched vertically; if $0 < |a| < 1$, the graph is compressed vertically; and if $a < 0$, the graph is reflected with respect to the x-axis. We can use this knowledge, along with plotting a few specific points, to quickly sketch graphs of any function of the form $f(x) = ax^n$.

Example 1: Functions of the Form ax^n

Sketch the graphs of the following functions.

a. $f(x) = \dfrac{x^4}{5}$

b. $g(x) = -x^3$

Solution

a. The graph of the function f will have the same basic shape as the function x^4, but compressed vertically because of the factor of $\dfrac{1}{5}$. To make the sketch more accurate, calculate the coordinates of a few points on the graph. The graph illustrates that $f(-1) = \dfrac{1}{5}$ and that $f(2) = \dfrac{16}{5}$.

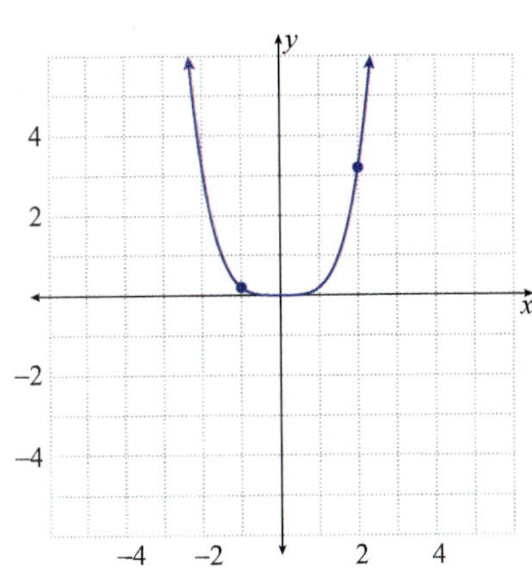

b. We know that the function g will have the same shape as the function x^3, but reflected over the x-axis because of the factor of -1. The graph illustrates this.

We also plot a few points on the graph of g, namely $(-1, 1)$ and $(1, -1)$, as a check.

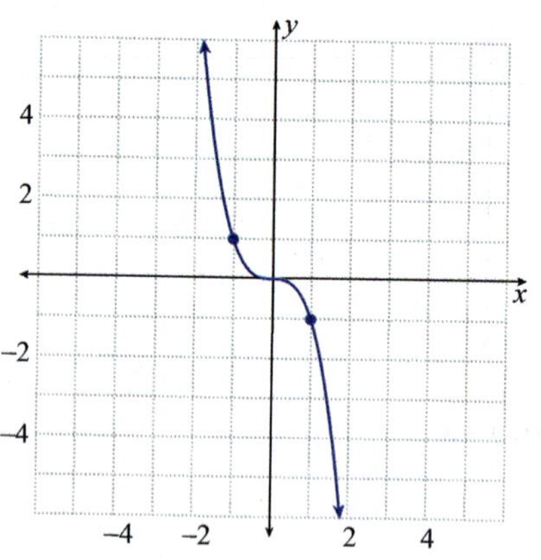

Functions of the Form $\dfrac{a}{x^n}$

We could also describe the following functions as having the form ax^{-n}, where a is a real number and n is a natural number. Once again, the graphs of these functions fall roughly into two categories, as illustrated in Figures 3 and 4.

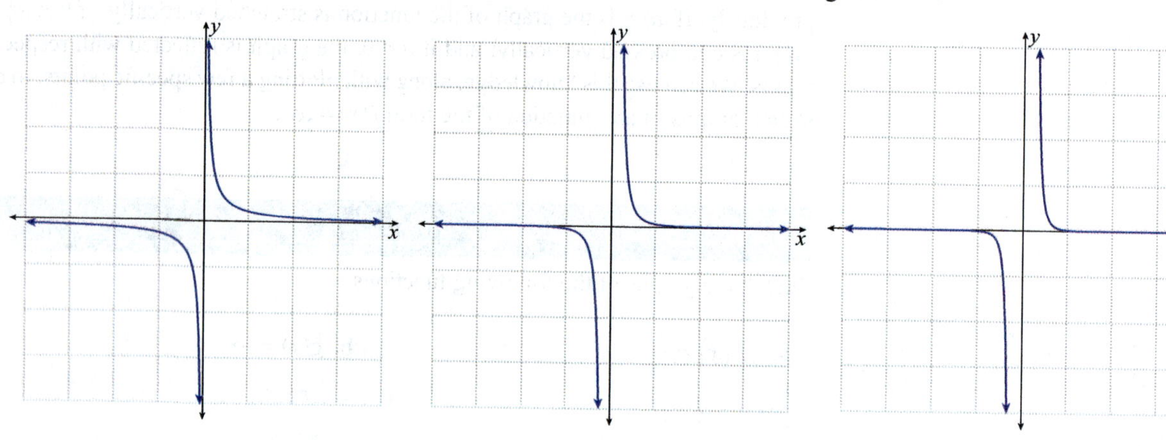

$$f(x) = \frac{1}{x} \qquad\qquad f(x) = \frac{1}{x^3} \qquad\qquad f(x) = \frac{1}{x^5}$$

FIGURE 3: Odd Exponents

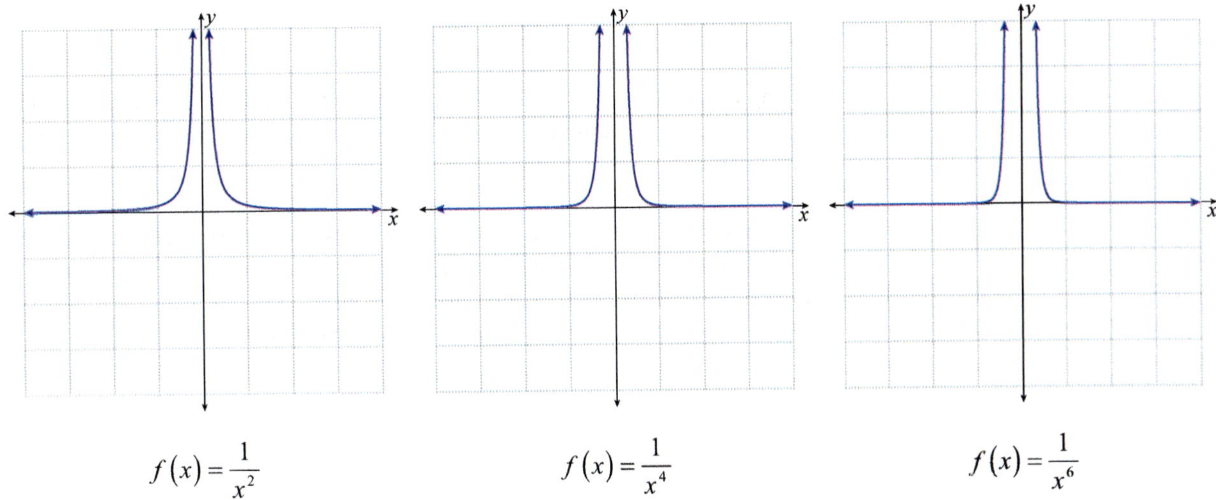

$$f(x) = \frac{1}{x^2} \qquad f(x) = \frac{1}{x^4} \qquad f(x) = \frac{1}{x^6}$$

FIGURE 4: Even Exponents

As with functions of the form ax^n, increasing the exponent on functions of the form $\dfrac{a}{x^n}$ sharpens the curve of the graph near the origin. Note that the domain of any function of the form $f(x) = \dfrac{a}{x^n}$ is $(-\infty, 0) \cup (0, \infty)$, but that the range depends on whether n is even or odd. When n is odd, the range is also $(-\infty, 0) \cup (0, \infty)$, and when n is even the range is $(0, \infty)$.

> ### Example 2: Functions of the Form $\dfrac{a}{x^n}$

Sketch the graph of the function $f(x) = -\dfrac{1}{4x}$.

Solution

The graph of the function f is similar to that of the function $\dfrac{1}{x}$, with two differences. We obtain the formula $-\dfrac{1}{4x}$ by multiplying $\dfrac{1}{x}$ by $-\dfrac{1}{4}$, a negative number between -1 and 1. So one difference is that the graph of f is the reflection of $\dfrac{1}{x}$ with respect to the x-axis. The other difference is that the graph of f is compressed vertically. With this in mind, we can calculate the coordinates of a few points (such as $\left(-\dfrac{1}{4}, 1\right)$ and $\left(1, -\dfrac{1}{4}\right)$) and sketch the graph of f.

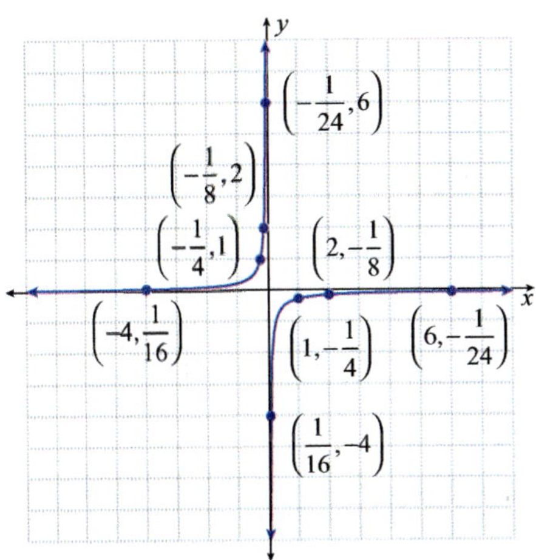

Functions of the Form $ax^{\frac{1}{n}}$

Using radical notation, these are functions of the form $a\sqrt[n]{x}$, where a is again a real number and n is a natural number. Square root and cube root functions, in particular, are commonly seen in mathematics.

Functions of this form again fall into one of two categories, depending on whether n is odd or even. To begin with, note that the domain and range are both the entire set of real numbers when n is odd, and that both are the interval $[0, \infty)$ when n is even. Figures 5 and 6 illustrate the two basic shapes of functions of this form.

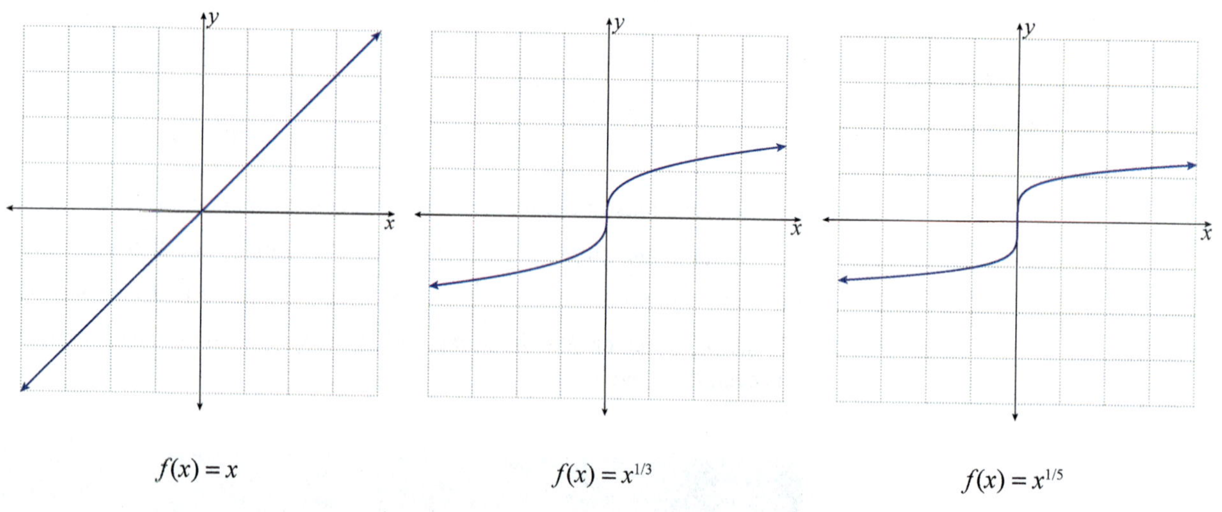

$$f(x) = x \qquad\qquad f(x) = x^{1/3} \qquad\qquad f(x) = x^{1/5}$$

FIGURE 5: Odd Roots

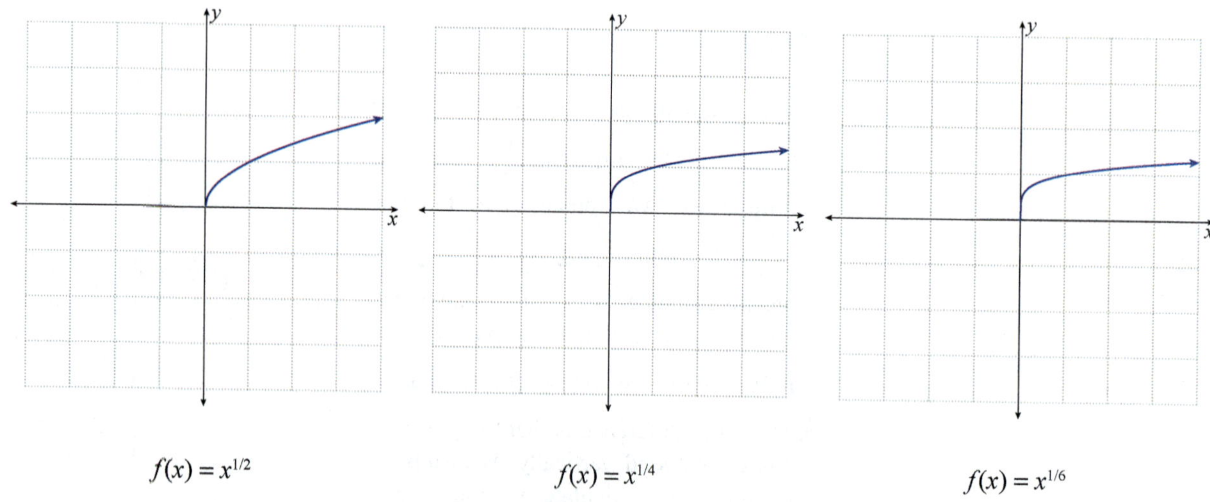

$$f(x) = x^{1/2} \qquad\qquad f(x) = x^{1/4} \qquad\qquad f(x) = x^{1/6}$$

FIGURE 6: Even Roots

At this point, you may be thinking that the graphs in Figures 5 and 6 appear familiar. The shapes in Figure 5 are the same as those seen in Figure 1, but rotated by 90 degrees and reflected with respect to the x-axis. Similarly, the shapes in Figure 6 bear some resemblance to those in Figure 2, except that half of the graphs appear to have been erased. This resemblance is no accident, given that n^{th} roots undo n^{th} powers.

The Absolute Value Function

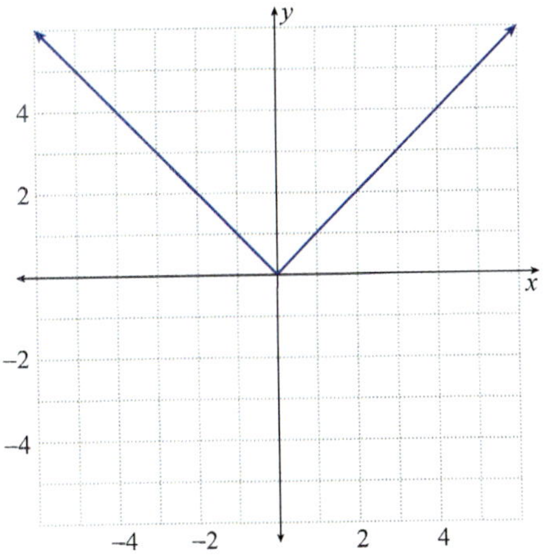

FIGURE 7: The Absolute Value Function

The basic absolute value function is $f(x) = |x|$. Note that for any value of x, $f(x)$ is nonnegative, so the graph of f should lie on or above the x-axis. One way to determine its exact shape is to review the definition of absolute value.

$$|x| = \begin{cases} x & \text{if } x \geq 0 \\ -x & \text{if } x < 0 \end{cases}$$

This means that for nonnegative values of x, $f(x)$ is a linear function with a slope of 1, and for negative values of x, $f(x)$ is a linear function with a slope of -1. Both linear functions have a y-intercept of 0, so the complete graph of f is as shown in Figure 7.

The effect of multiplying $|x|$ by a real number a is what we have come to expect: if $|a| > 1$, the graph is stretched vertically; if $0 < |a| < 1$, yes, the graph is compressed vertically; and if a is negative, the graph is reflected with respect to the x-axis.

Example 3: The Absolute Value Function

Sketch the graph of the function $f(x) = -2|x|$.

Solution

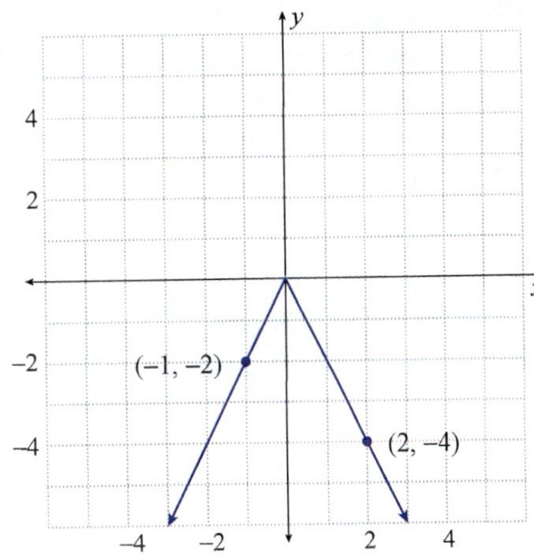

The graph of f will be a vertically stretched version of $|x|$, reflected over the x-axis.

As always, we can plot a few points to verify that our reasoning is correct. We have plotted the values of $f(-1)$ and $f(2)$.

The Greatest Integer Function

The Greatest Integer Function

The greatest integer function, $f(x) = [\![x]\!]$, is a function commonly encountered in computer science applications. It is defined as follows: the **greatest integer of** x is the largest integer less than or equal to x. For instance, $[\![4.3]\!] = 4$ and $[\![-2.9]\!] = -3$ (note that -3 is the largest integer to the left of -2.9 on the real number line).

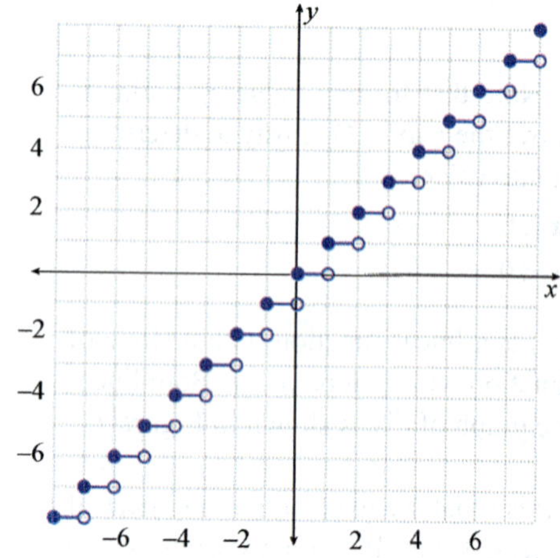

FIGURE 8: The Greatest Integer Function

Careful study of the greatest integer function reveals that its graph must consist of intervals where the function is constant, and that these portions of the graph must be separated by discrete "jumps," or breaks, in the graph. For instance, any value for x chosen from the interval $[1, 2)$ results in $f(x) = 1$, but $f(2) = 2$. Similarly, any value for x chosen from the interval $[-3, -2)$ results in $f(x) = -3$, but $f(-2) = -2$.

Our graph of the greatest integer function must somehow indicate this repeated pattern of jumps. In cases like this, it is conventional to use an open circle on the graph to indicate that the function is either undefined at that point or is defined to be another value. Closed circles are used to emphasize that a certain point really does lie on the graph of the function. With these conventions in mind, the graph of the greatest integer function appears in Figure 8.

Piecewise-Defined Functions

There is no rule stating that a function needs to be defined by a single formula. In fact, we have worked with such a function already; in evaluating the absolute value of x, we use one formula if x is greater than or equal to 0 and a different formula if x is less than 0. Obviously, we can't have two rules govern the same input, but we can have multiple formulas on separate pieces of a function's domain.

Piecewise-Defined Function

A **piecewise-defined function** is a function defined in terms of two or more formulas, each valid for its own unique portion of the real number line. In evaluating a piecewise-defined function f at a certain value for x, it is important to correctly identify which formula is valid for that particular value.

Example 4: Piecewise-Defined Function

Sketch the graph of the following function.

$$f(x) = \begin{cases} -2x - 2 & \text{if } x \le -1 \\ x^2 & \text{if } x > -1 \end{cases}$$

Solution

The function f is a piecewise function with a different formula for two intervals. To graph f, graph each portion separately, making sure that each formula is applied only on the appropriate interval.

The function f is a linear function on the interval $(-\infty, -1]$ and a quadratic function on the interval $(-1, \infty)$.

The complete graph appears with the points $f(-4) = 6$, $f(-1) = 0$, and $f(2) = 4$ noted in particular.

Note the use of a closed circle at $(-1, 0)$ to emphasize that this point is part of the graph, and the use of an open circle at $(-1, 1)$ to indicate that this point is not part of the graph. That is, the value of $f(-1)$ is 0, not 1.

NOTE

Always pay close attention to the boundary points of each interval. Remember that only one rule applies at each point.

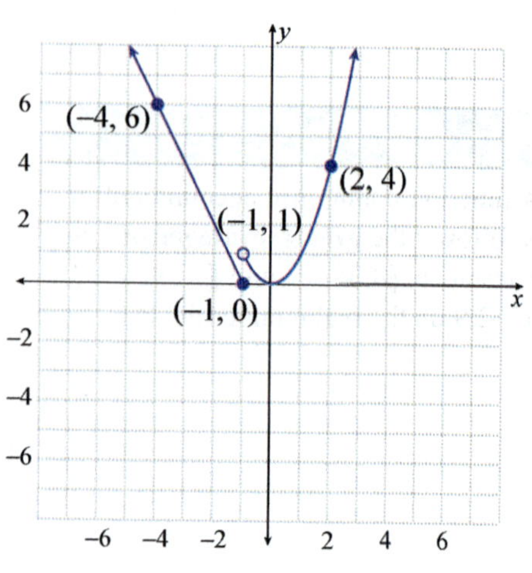

3.5 EXERCISES

PRACTICE

Sketch the graphs of the following functions. Pay particular attention to intercepts, if any, and locate these accurately. See Examples 1 through 4.

1. $f(x) = -\frac{x}{2}$

2. $g(x) = 2x^2$

3. $F(x) = x^{\frac{1}{2}}$

4. $h(x) = x^{-1}$

5. $p(x) = -\frac{2}{x}$

6. $q(x) = -\sqrt[3]{x}$

7. $G(x) = -|x|$

8. $k(x) = \frac{1}{x^3}$

9. $G(x) = \frac{\sqrt{x}}{2}$

10. $H(x) = 0.5x^{\frac{1}{3}}$

11. $r(x) = 3|x|$

12. $p(x) = -\frac{1}{x^2}$

13. $W(x) = \frac{x^4}{16}$

14. $k(x) = \frac{x^3}{9}$

15. $h(x) = 2\sqrt[3]{x}$

16. $d(x) = 2x^5$

17. $S(x) = 4x^{-2}$

18. $f(x) = -x^2$

19. $r(x) = \frac{\sqrt[3]{x}}{3}$

20. $s(x) = \frac{|x|}{3}$

21. $t(x) = \frac{x^6}{4}$

22. $f(x) = 2[\![x]\!]$

23. $P(x) = -[\![x]\!]$

24. $m(x) = \left[\!\left[\frac{x}{2}\right]\!\right]$

25. $f(x) = \begin{cases} 3 - x & \text{if } x < -2 \\ x^{\frac{1}{3}} & \text{if } x \geq -2 \end{cases}$

26. $g(x) = \begin{cases} -x^2 & \text{if } x \leq 1 \\ x^2 & \text{if } x > 1 \end{cases}$

27. $r(x) = \begin{cases} \frac{1}{x} & \text{if } x < 1 \\ -x & \text{if } x > 1 \end{cases}$

28. $p(x) = \begin{cases} x + 1 & \text{if } x < -2 \\ x^3 & \text{if } -2 \leq x < 3 \\ -1 - x & \text{if } x \geq 3 \end{cases}$

29. $q(x) = \begin{cases} -1 & \text{if } x \in \mathbb{Z} \\ 1 & \text{if } x \notin \mathbb{Z} \end{cases}$

30. $s(x) = \begin{cases} \frac{x^2}{3} & \text{if } x < 0 \\ -\frac{x^2}{3} & \text{if } x \geq 0 \end{cases}$

31. $v(x) = \begin{cases} x^2 & \text{if } -1 \leq x \leq 1 \\ |x| & \text{if } x < -1 \text{ or } x > 1 \end{cases}$

32. $M(x) = \begin{cases} x & \text{if } x \in \mathbb{Z} \\ -x & \text{if } x \notin \mathbb{Z} \end{cases}$

33. $t(x) = \begin{cases} x^4 & \text{if } x \leq 1 \\ [\![x]\!] & \text{if } x > 1 \end{cases}$

34. $N(x) = \begin{cases} x^2 & \text{if } x \in \mathbb{Z} \\ [\![x]\!] & \text{if } x \notin \mathbb{Z} \end{cases}$

35. $h(x) = \begin{cases} -|x| & \text{if } x < 2 \\ [\![x]\!] & \text{if } x \geq 2 \end{cases}$ **36.** $u(x) = \begin{cases} [\![x]\!] & \text{if } x \leq 1 \\ 2x - 2 & \text{if } x > 1 \end{cases}$

Match the following functions to their graphs.

37. $f(x) = -2x^4$

38. $f(x) = -\dfrac{7}{9x^4}$

39. $f(x) = -4\left[\!\left[\dfrac{x}{4}\right]\!\right]$

40. $f(x) = -\dfrac{7\sqrt[3]{x}}{3}$

41. $f(x) = -\dfrac{8}{9}|x|$

42. $f(x) = -4\sqrt{x}$

43. $f(x) = \dfrac{3}{7}|x|$

44. $f(x) = \begin{cases} -4x - 12 & \text{if } x \leq -3 \\ \dfrac{5}{10}x^2 & \text{if } x > -3 \end{cases}$

45. $f(x) = \begin{cases} \dfrac{-1}{3}|x| & \text{if } x < 2 \\ \left[\!\left[\dfrac{x}{2}\right]\!\right] & \text{if } x \geq 2 \end{cases}$

46. $f(x) = \begin{cases} -\dfrac{1}{3}|x| & \text{if } x < 2 \\ \dfrac{x}{2} & \text{if } x \geq 2 \end{cases}$

a. 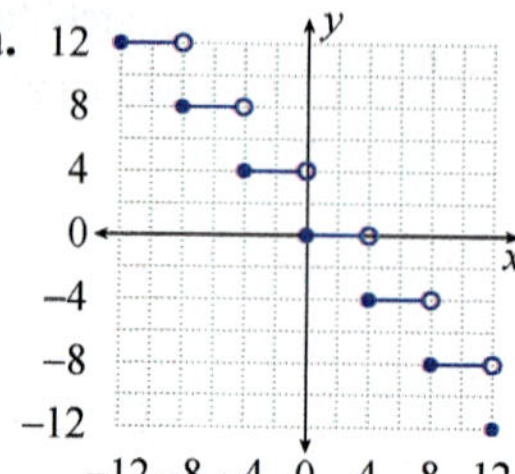

b. 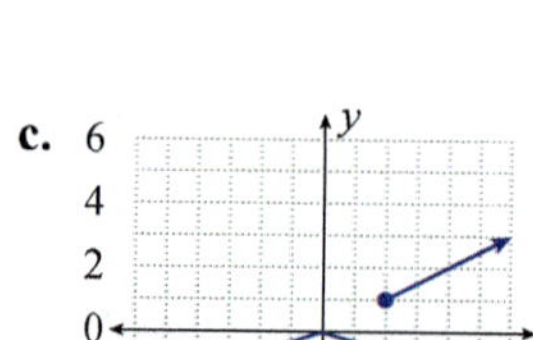

c. 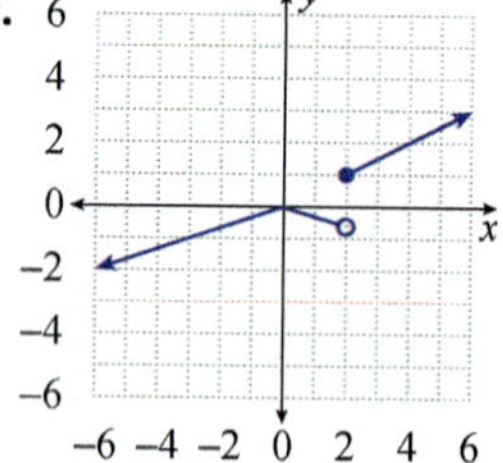

d. 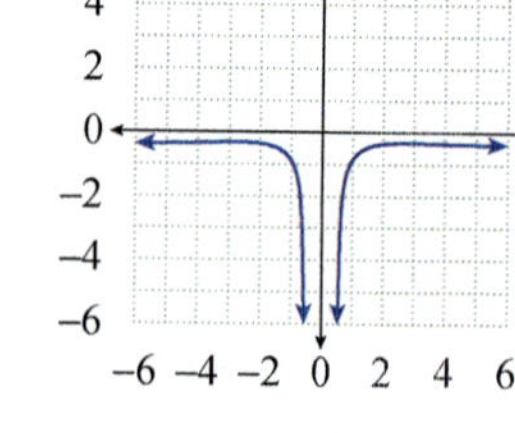

e. 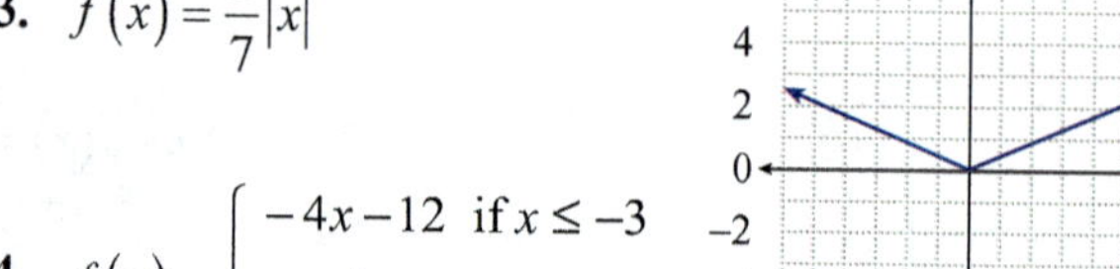

f. 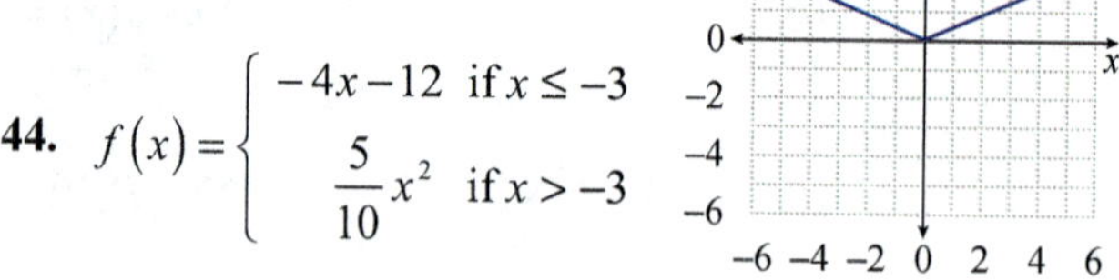

g. 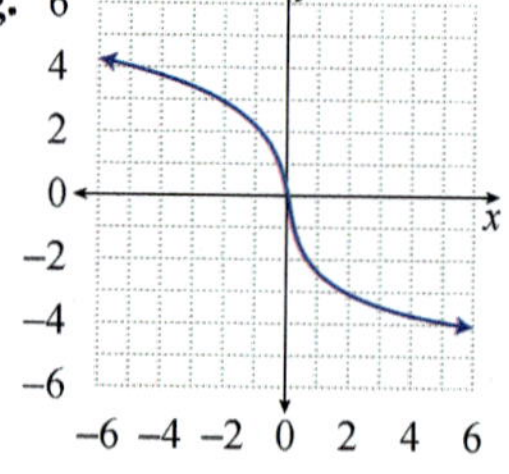

h. 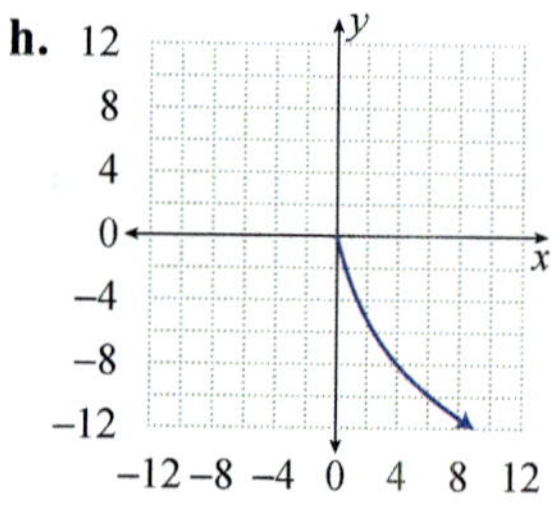

i. 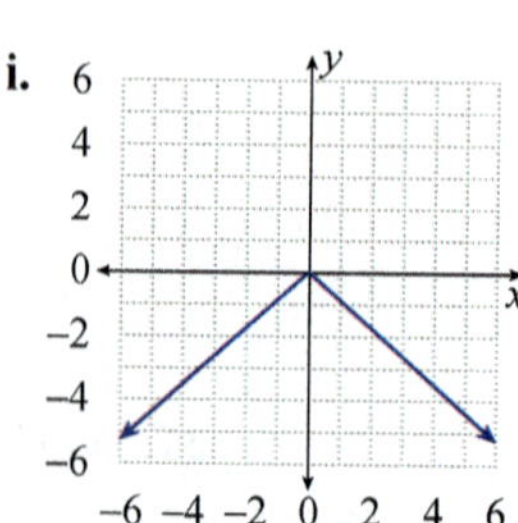

j. 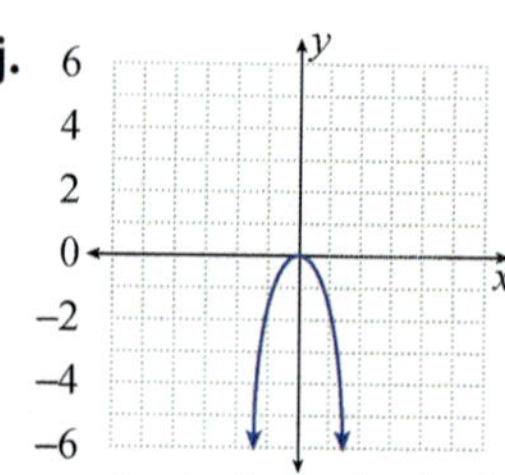

Use a graphing utility to graph the following functions. Experiment with different viewing windows until you obtain a sketch that seems to capture the meaningful parts of the graph.

47. $f(x) = 10x^5 - x^3$

48. $g(x) = x^5 + x^2$

49. $f(x) = x^3 - 5x^2 + x$

50. $g(x) = \sqrt{x} - x^2$

51. $f(x) = \sqrt{x} + 3x - 1$

52. $g(x) = x^4 - 3x^3 + 2$

3.6 TRANSFORMATIONS OF FUNCTIONS

■ TOPICS

- Shifting, Reflecting, and Stretching Graphs
- Symmetry of Functions and Equations
- Intervals of Monotonicity

Shifting, Reflecting, and Stretching Graphs

Much of the material in this section was introduced earlier, in our discussion of quadratic functions. You may want to review the ways in which the basic quadratic function $f(x) = x^2$ can be shifted, stretched, and reflected as you work through the more general ideas here.

Horizontal Shifting/Translation

Let $f(x)$ be a function, and let h be a fixed real number. If we replace x with $x - h$, we obtain a new function $g(x) = f(x - h)$. The graph of g has the same shape as the graph of f, but shifted h units to the right if $h > 0$ and shifted $|h|$ units to the left if $h < 0$.

Example 1: Horizontal Shifting/Translation

Sketch the graphs of the following functions.

a. $f(x) = (x + 2)^3$ **b.** $g(x) = |x - 4|$

Solution

NOTE

Begin by identifying the underlying function that is being shifted.

a.

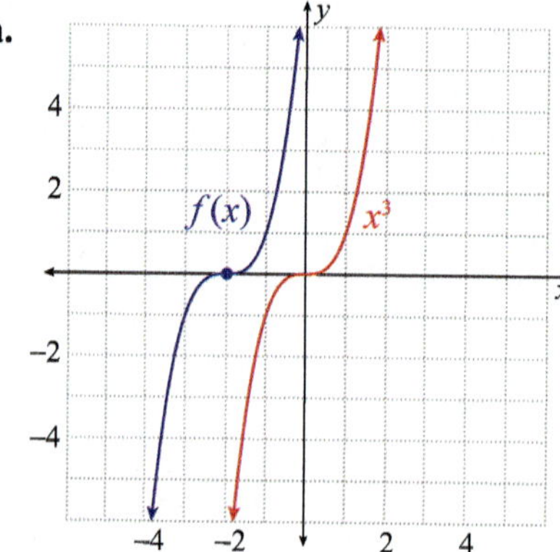

The basic function being shifted is x^3.

Begin by drawing the basic cubic shape (the shape of $y = x^3$).

Since x is replaced with $x + 2$, the graph of $f(x)$ is the graph of x^3 shifted 2 units to the left.

Note, for example, that $(-2, 0)$ is one point on the graph.

The minus sign in the expression $x - h$ is critical. When you see an expression in the form $x + h$ you must think of it as $x - (-h)$.

Consider a specific example: replacing x with $x - 5$ shifts the graph 5 units to the *right*, since 5 is positive. Replacing x with $x + 5$ shifts the graph 5 units to the *left*, since we have actually replaced x with $x - (-5)$.

b.

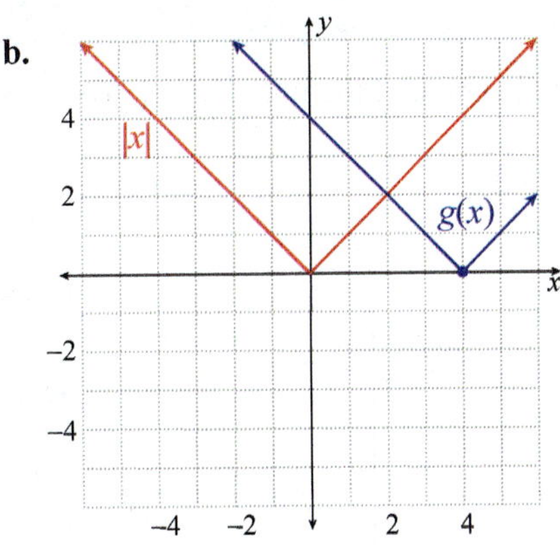

The basic function being shifted is $|x|$.

Start by graphing the basic absolute value function.

The graph of $g(x) = |x - 4|$ has the same shape, but shifted 4 units to the right.

Note, for example, that $(4, 0)$ lies on the graph of g.

Vertical Shifting/Translation

Let $f(x)$ be a function whose graph is known, and let k be a fixed real number. The graph of the function $g(x) = f(x) + k$ is the same shape as the graph of f, but shifted k units up if $k > 0$ and $|k|$ units down if $k < 0$.

Example 2: Vertical Shifting/Translation

Sketch the graphs of the following functions.

a. $f(x) = \dfrac{1}{x} + 3$ **b.** $g(x) = \sqrt[3]{x} - 2$

Solution

> **NOTE**
>
> As before, begin by identifying the basic function being shifted.

a.

The basic function being shifted is $\dfrac{1}{x}$.

The graph of $f(x) = \dfrac{1}{x} + 3$ is the graph of $y = \dfrac{1}{x}$ shifted 3 units up. Note that this doesn't change the domain.

However, the range is affected; the range of f is $(-\infty, 3) \cup (3, \infty)$.

b.

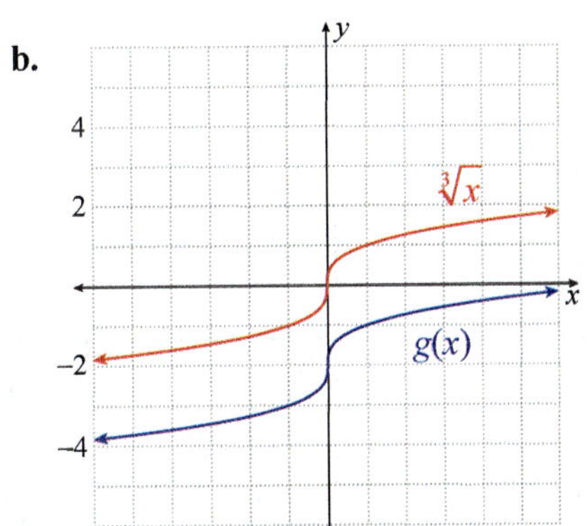

The basic function being shifted is $\sqrt[3]{x}$.

Begin by graphing the basic cube root shape.

To graph $g(x) = \sqrt[3]{x} - 2$, the graph of $y = \sqrt[3]{x}$ is shifted 2 units down.

In this case, it doesn't matter which shift we apply first. However, when functions get more complicated, it is usually best to apply horizontal shifts before vertical shifts.

Example 3: Horizontal and Vertical Shifting

Sketch the graph of the function $f(x) = \sqrt{x+4} + 1$.

Solution

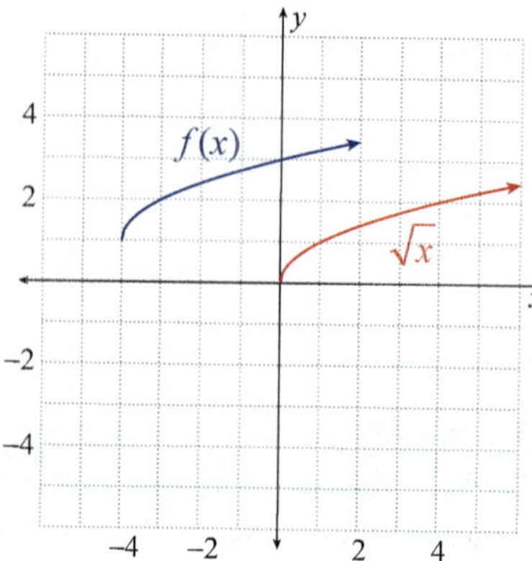

The basic function being shifted is $\sqrt{x}$.

Begin by graphing the basic square root shape.

In $f(x)$ we have replaced x with $x + 4$, so shift the basic function 4 units to the left.

Then shift the resulting function 1 unit up.

Reflecting with Respect to the Axes

Let $f(x)$ be a function.

1. The graph of the function $g(x) = -f(x)$ is the reflection of the graph of f with respect to the x-axis.

2. The graph of the function $g(x) = f(-x)$ is the reflection of the graph of f with respect to the y-axis.

In other words, a function is reflected with respect to the x-axis by multiplying the entire function by -1, and it is reflected with respect to the y-axis by replacing x with $-x$.

Example 4: Reflecting with Respect to the Axes

Sketch the graphs of the following functions.

a. $f(x) = -x^2$ **b.** $g(x) = \sqrt{-x}$

We state that a function is reflected with respect to a particular axis. Visually, this means the function is reflected over (across) that axis.

Solution

a.

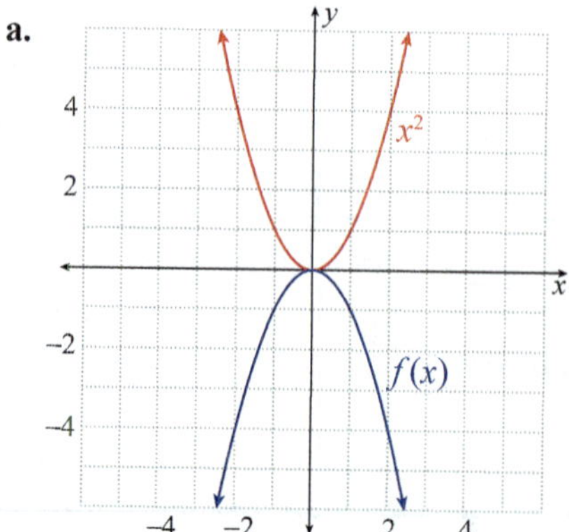

To graph $f(x) = -x^2$, begin with the graph of the basic parabola $y = x^2$.

The entire function is multiplied by -1, so reflect the graph over the x-axis, resulting in the original shape turned upside down.

Note that the domain is still the entire real line, but the range of f is the interval $(-\infty, 0]$.

b.

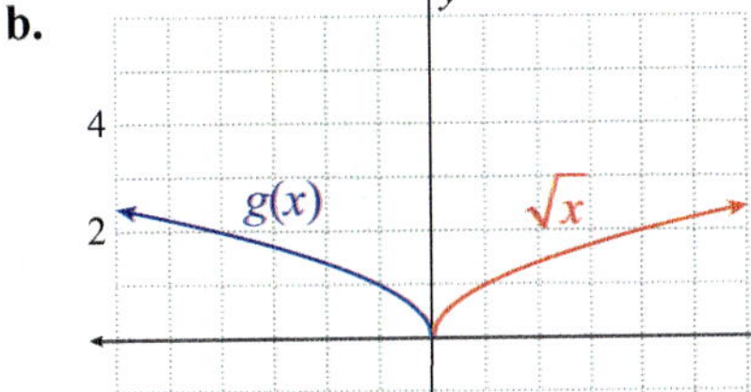

To graph $g(x) = \sqrt{-x}$, begin by graphing $y = \sqrt{x}$, the basic square root function.

In $g(x)$, x has been replaced by $-x$, so reflect the graph with respect to the y-axis.

Note that this changes the domain but not the range. The domain of g is the interval $(-\infty, 0]$ and the range is $[0, \infty)$.

Vertical Stretching and Compressing

Let $f(x)$ be a function and let a be a positive real number.

1. The graph of the function $g(x) = af(x)$ is stretched vertically compared to the graph of f by a factor of a if $a > 1$.

2. The graph of the function $g(x) = af(x)$ is compressed vertically compared to the graph of f by a factor of a if $0 < a < 1$.

Example 5: Vertical Stretching and Compressing

Sketch the graphs of the following functions.

a. $f(x) = \dfrac{\sqrt{x}}{10}$

b. $g(x) = 5|x|$

Solution

NOTE

When graphing stretched or compressed functions, it may help to plot a few points of the new function.

a.

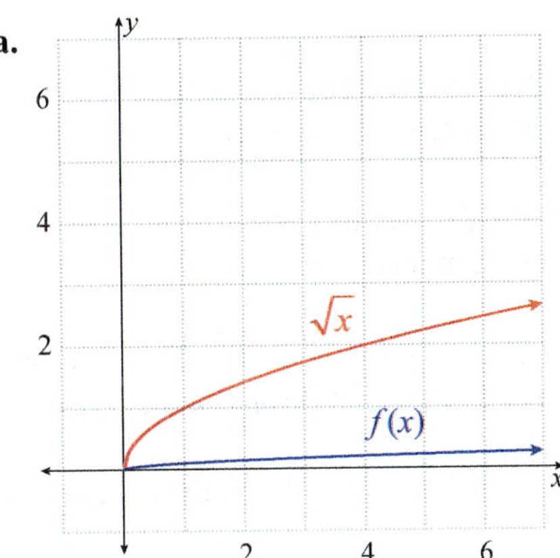

Begin with the graph of $\sqrt{x}$.

The shape of $f(x)$ is similar to the shape of $\sqrt{x}$ but all of the y-coordinates have been multiplied by the factor of $\dfrac{1}{10}$, and are consequently much smaller.

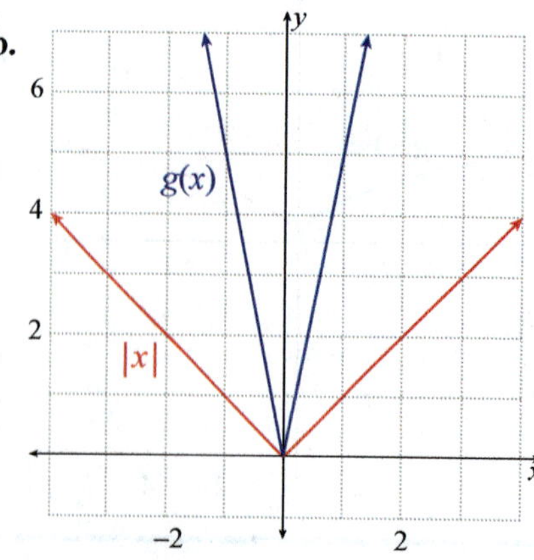

Begin with the graph of the absolute value function.

In contrast to the last example, the graph of $g(x) = 5|x|$ is stretched compared to the standard absolute value function.

Every second coordinate is multiplied by a factor of 5.

If the function g is obtained from the function f by multiplying f by a negative real number, think of the number as the product of -1 and a positive real number (namely, its absolute value). This is a simple example of a function going under multiple transformations. When dealing with more complicated functions, undergoing numerous transformations, we need a procedure for untangling the individual transformations in order to find the correct graph.

Order of Transformations

If a function g has been obtained from a simpler function f through a number of transformations, g can be understood by looking for transformations in the following order.

1. Horizontal shifts

2. Stretching and compressing

3. Reflections

4. Vertical shifts

Consider, for example, the function $g(x) = -2\sqrt{x+1} + 3$, which has been "built up" from the basic square root function through a variety of transformations.

1. First, $\sqrt{x}$ has been transformed into $\sqrt{x+1}$ by replacing x with $x + 1$, and we know that this corresponds graphically to a shift of 1 unit to the left.

2. Next, the function $\sqrt{x+1}$ has been multiplied by 2 to get the function $2\sqrt{x+1}$, and we know that this has the effect of stretching the graph of $\sqrt{x+1}$ vertically.

3. The function $2\sqrt{x+1}$ has then been multiplied by -1, giving us $-2\sqrt{x+1}$, and the graph of this is the reflection of $2\sqrt{x+1}$ with respect to the x-axis.

4. Finally, the constant 3 has been added to $-2\sqrt{x+1}$, shifting the entire graph 3 units up.

These transformations are illustrated, in order, in Figure 1, culminating in the graph of $g(x) = -2\sqrt{x+1} + 3$.

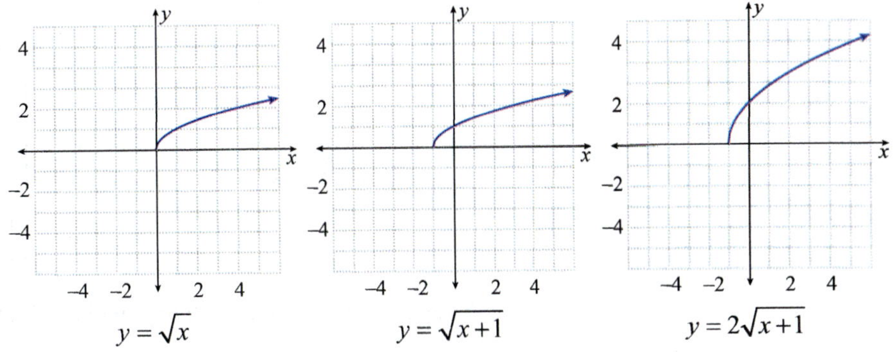

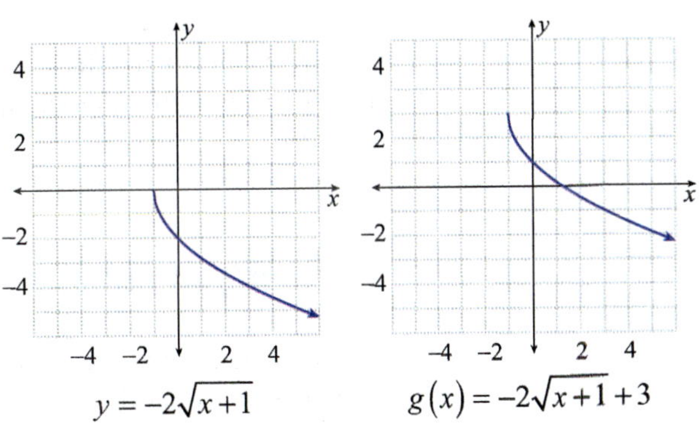

FIGURE 1: Building the Graph of $g(x) = -2\sqrt{x+1} + 3$

Example 6: Order of Transformations

Sketch the graph of the function $f(x) = \dfrac{1}{2-x}$.

Solution

The basic function that f is similar to is $\dfrac{1}{x}$. Following the order of transformations, we determine how to sketch the graph of f.

1. If we replace x by $x+2$ (shifting the graph 2 units to the left), we obtain the function $\dfrac{1}{x+2}$, which is closer to what we want.

2. There does not appear to be any stretching or compressing transformation.

3. If we replace x by $-x$, we have $\dfrac{1}{-x+2} = \dfrac{1}{2-x}$, which is equal to f. This reflects the graph of $\dfrac{1}{x+2}$ with respect to the y-axis.

4. Since we have already found f, we know there is no vertical shift.

The entire sequence of transformations is shown next, ending with the graph of f.

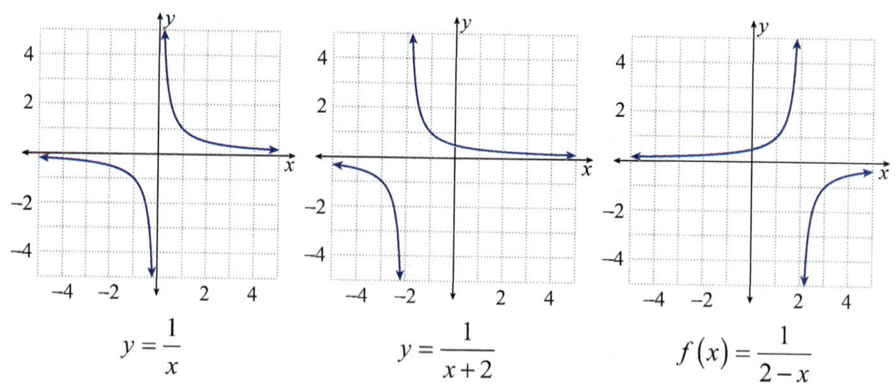

$$y = \frac{1}{x}$$

$$y = \frac{1}{x+2}$$

$$f(x) = \frac{1}{2-x}$$

Note: An alternate approach to graphing $f(x) = \dfrac{1}{2-x}$ is to rewrite the function in the form $f(x) = -\dfrac{1}{x-2}$. In this form, the graph of f is the graph of $\dfrac{1}{x}$ shifted two units to the right, and then reflected with respect to the x-axis. The result is the same, as you should verify. Rewriting an equation in a different, equivalent form never changes its graph.

Symmetry of Functions and Equations

We know that replacing x with $-x$ reflects the graph of a function with respect to the y-axis, but what if $f(-x) = f(x)$? In this case the original graph is the same as the reflection! This means the function f is symmetric with respect to the y-axis.

y-Axis Symmetry

The graph of a function f has **y-axis symmetry**, or is **symmetric with respect to the y-axis**, if $f(-x) = f(x)$ for all x in the domain of f. Such functions are called **even functions**.

Functions whose graphs have y-axis symmetry are called even functions because polynomial functions with only even exponents form one large class of functions with this property. Consider the function $f(x) = 7x^8 - 5x^4 + 2x^2 - 3$. This function is a polynomial of four terms, all of which have even degree. If we replace x with $-x$ and simplify the result, we obtain the function f again.

$$f(-x) = 7(-x)^8 - 5(-x)^4 + 2(-x)^2 - 3$$
$$= 7x^8 - 5x^4 + 2x^2 - 3$$
$$= f(x)$$

Be aware, however, that such polynomial functions are not the only even functions. We will see more examples as we proceed.

There is another class of functions for which replacing x with $-x$ results in the exact negative of the original function. That is, $f(-x) = -f(x)$ for all x in the domain, and this means changing the sign of the x-coordinate of a point on the graph also changes the sign of the y-coordinate.

What does this mean geometrically? Suppose f is such a function and $(x, f(x))$ is a point on the graph of f. If we change the sign of both coordinates, we obtain a new point that is the original point reflected through the origin (we can also think of this as reflected over the y-axis, then the x-axis).

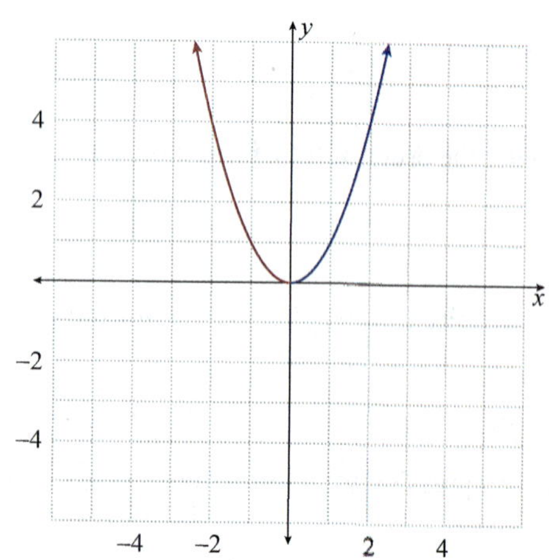

FIGURE 2: A Function with y-Axis Symmetry

For instance, if $(x, f(x))$ lies in the first quadrant, $(-x, -f(x))$ lies in the third, and if $(x, f(x))$ lies in the second quadrant, $(-x, -f(x))$ lies in the fourth. But since $f(-x) = -f(x)$, the point $(-x, -f(x))$ can be rewritten as $(-x, f(-x))$.

Written in this form, we know that $(-x, f(-x))$ is a point on the graph of f, since *any* point of the form $(?, f(?))$ lies on the graph of f. So a function with the property $f(-x) = -f(x)$ has a graph that is symmetric with respect to the origin.

Origin Symmetry

The graph of a function f has **origin symmetry**, or is **symmetric with respect to the origin**, if $f(-x) = -f(x)$ for all x in the domain of f. Such functions are called **odd functions**.

As you might guess, such functions are called odd because polynomial functions with only odd exponents serve as simple examples. For instance, the function $f(x) = -2x^3 + 8x$ is odd.

$$f(-x) = -2(-x)^3 + 8(-x)$$
$$= -2(-x^3) + 8(-x)$$
$$= 2x^3 - 8x$$
$$= -f(x)$$

As far as functions are concerned, y-axis and origin symmetry are the two principal types of symmetry. What about x-axis symmetry? It is certainly possible to draw a graph that displays x-axis symmetry; but unless the graph lies entirely on the x-axis, such a graph cannot represent a function. Why not? Draw a few graphs that are symmetric with respect to the x-axis, then apply the vertical line test to these graphs. In order to have x-axis symmetry, if (x, y) is a point on the graph, then $(x, -y)$ must also be on the graph, and thus the graph can not represent a function.

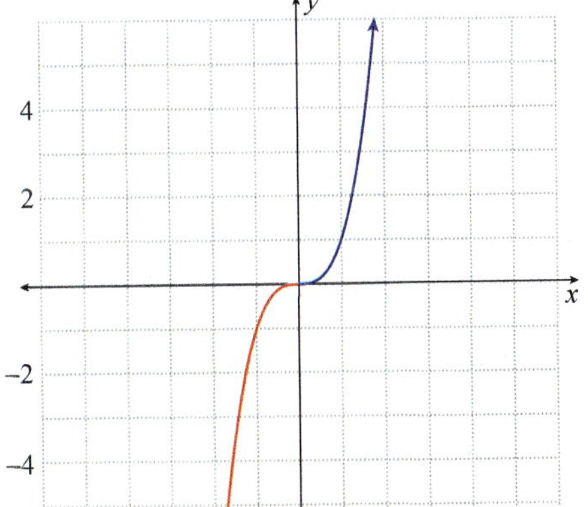

FIGURE 3: A Function with Origin Symmetry

Any equation in x and y defines a relation between the two variables. There are three principal types of symmetry that equations can possess.

Symmetry of Equations

We say that an equation in x and y is **symmetric with respect to**

1. the **y-axis** if replacing x with $-x$ results in an equivalent equation;

2. the **x-axis** if replacing y with $-y$ results in an equivalent equation;

3. the **origin** if replacing x with $-x$ and y with $-y$ results in an equivalent equation.

Knowing the symmetry of a function or an equation can serve as a useful aid in graphing. For instance, when graphing an even function it is only necessary to graph the part to the right of the y-axis, as the left half of the graph is the reflection of the right half with respect to the y-axis. Similarly, if a function is odd, the left half of its graph is the reflection of the right half through the origin.

Example 7: Symmetry of Equations

Sketch the graphs of the following relations, making use of symmetry.

a. $f(x) = \dfrac{1}{x^2}$ **b.** $g(x) = x^3 - x$ **c.** $x = y^2$

Solution

a.

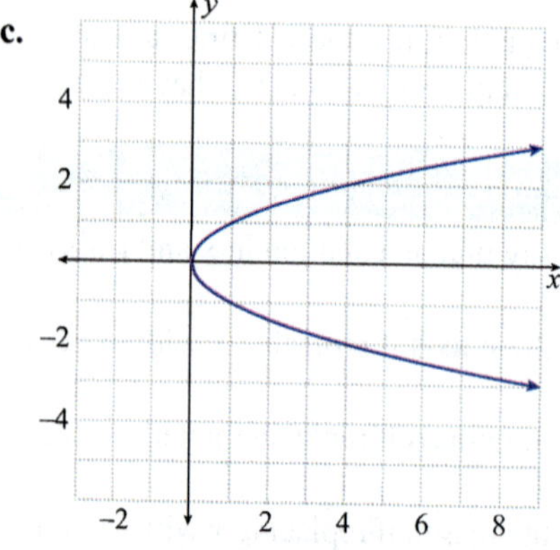

This relation is a function. Note that it is indeed an even function and exhibits y-axis symmetry.

$$f(-x) = \frac{1}{(-x)^2}$$

$$= \frac{1}{x^2}$$

$$= f(x)$$

b.

While we do not yet have the tools to graph general polynomial functions, we can obtain a good sketch of $g(x) = x^3 - x$.

First, g is odd: $g(-x) = -g(x)$ (verify this).

If we calculate a few values, such as

$g(0) = 0,\ g\left(\dfrac{1}{2}\right) = -\dfrac{3}{8},\ g(1) = 0,$ and

$g(2) = 6$, and then reflect these through the origin, we get a good idea of the shape of g.

c.

The equation $x = y^2$ is not a function, but it is a relation in x and y that has x-axis symmetry. If we replace y with $-y$ and simplify the result, we obtain the original equation.

$$x = (-y)^2$$

$$x = y^2$$

The upper half of the graph is the function $y = \sqrt{x}$, so drawing this and its reflection over the x-axis gives us the complete graph of $x = y^2$.

Summary of Symmetry

The first column in the following table summarizes the behavior of a graph in the Cartesian plane if it possesses any of the three types of symmetry we covered. If the graph is of an equation in x and y, the algebraic method in the second column can be used to identify the symmetry. The third column gives the algebraic method used to identify the type of symmetry if the graph is that of a function $f(x)$. Finally, the fourth column contains an example of each type of symmetry.

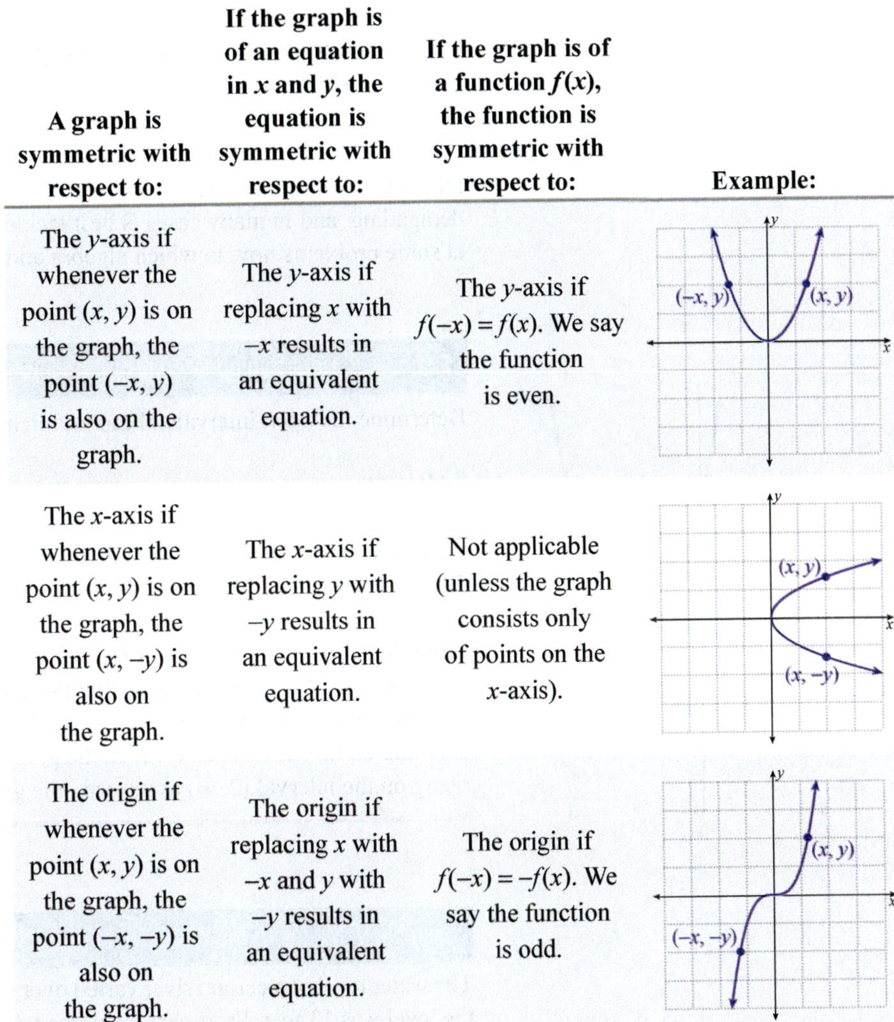

A graph is symmetric with respect to:	If the graph is of an equation in x and y, the equation is symmetric with respect to:	If the graph is of a function $f(x)$, the function is symmetric with respect to:	Example:
The y-axis if whenever the point (x, y) is on the graph, the point $(-x, y)$ is also on the graph.	The y-axis if replacing x with $-x$ results in an equivalent equation.	The y-axis if $f(-x) = f(x)$. We say the function is even.	
The x-axis if whenever the point (x, y) is on the graph, the point $(x, -y)$ is also on the graph.	The x-axis if replacing y with $-y$ results in an equivalent equation.	Not applicable (unless the graph consists only of points on the x-axis).	
The origin if whenever the point (x, y) is on the graph, the point $(-x, -y)$ is also on the graph.	The origin if replacing x with $-x$ and y with $-y$ results in an equivalent equation.	The origin if $f(-x) = -f(x)$. We say the function is odd.	

Intervals of Monotonicity

In many applications, it is useful to identify the intervals of the x-axis for which a function f is increasing in value, decreasing in value, or remaining constant. In this context, we say that these are intervals on which f is **monotone**. In practice, we are usually interested in identifying the largest open intervals of monotonicity (recall that open intervals do not include their endpoints).

Increasing, Decreasing, and Constant

We say that a function f is

1. **increasing on an interval** if, for any x_1 and x_2 in the interval with $x_1 < x_2$, it is the case that $f(x_1) < f(x_2)$;

2. **decreasing on an interval** if, for any x_1 and x_2 in the interval with $x_1 < x_2$, it is the case that $f(x_1) > f(x_2)$;

3. **constant on an interval** if, for any x_1 and x_2 in the interval, it is the case that $f(x_1) = f(x_2)$.

Determining the intervals of monotonicity of a function is a task that can be quite demanding, and in many cases is best tackled with the tools of calculus. We will look at some problems now in which algebra and our intuition will be sufficient.

Example 8: Determining Intervals of Monotonicity

Determine the open intervals of monotonicity of the function $f(x) = (x - 2)^2 - 1$.

Solution

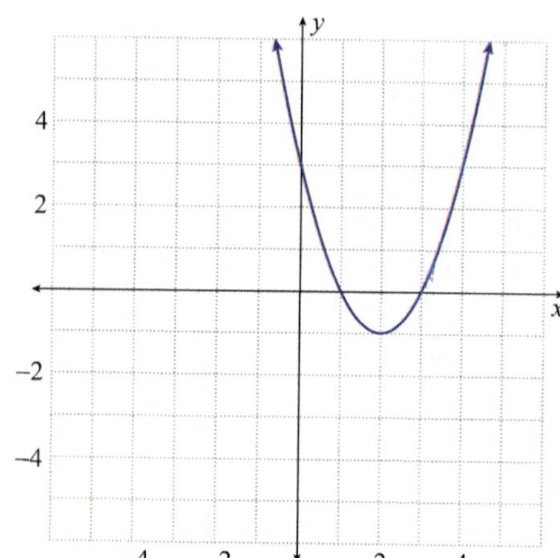

We know that the graph of f is the basic parabola shifted 2 units to the right and 1 unit down, as shown.

From the graph, we can see that f is decreasing on the interval $(-\infty, 2)$ and increasing on the interval $(2, \infty)$. Remember that these are intervals of the x-axis: if x_1 and x_2 are any two points in the interval $(-\infty, 2)$, with $x_1 < x_2$, then $f(x_1) > f(x_2)$. In other words, f is falling on this interval as we scan the graph from left to right. On the other hand, f is rising on the interval $(2, \infty)$ as we scan the graph from left to right.

Example 9: Modeling the Water Level of a River

The water level of a certain river varied over the course of a year as follows. In January, the level was 13 feet. From that level, the water increased linearly to a level of 18 feet in May. The water remained constant at that level until July, at which point it began to decrease linearly to a final level of 11 feet in December. Graph the water level as a function of time and determine the intervals of monotonicity.

Solution

If we let 1 to 12 represent January through December, the intervals of monotonicity are as follows: increasing on $(1, 5)$, constant on $(5, 7)$, and decreasing on $(7, 12)$. The graph of the water level as a function of the month is shown.

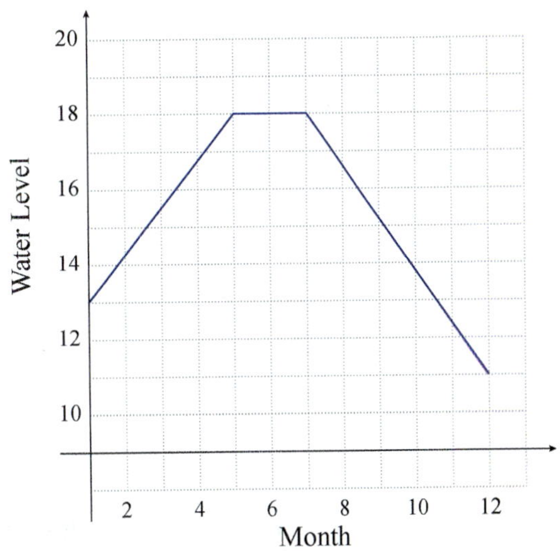

3.6 EXERCISES

💡 PRACTICE

For each function or graph, determine the basic function that has been shifted, reflected, stretched, or compressed.

1. $f(x) = -(1-x)^2 + 2$

2. $f(x) = \dfrac{1}{x-4} + 5$

3. $f(x) = \sqrt[3]{x+6} - 2$

4. $f(x) = -2 + 2|x-3|$

5. $f(x) = \sqrt{x+2} - 5$

6. $f(x) = [\![-2-x]\!]$

7. $f(x) = \dfrac{1}{(x+2)^2} + 1$

8. $f(x) = \dfrac{\sqrt{-x}}{2} + 4$

9. $f(x) = (x+6)^3$

10.

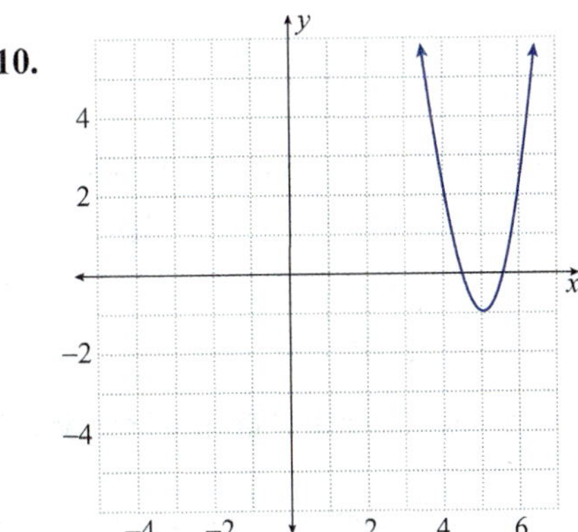

11.

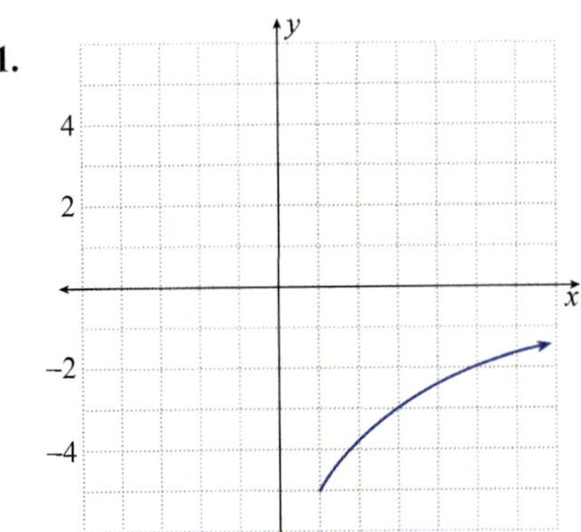

12. 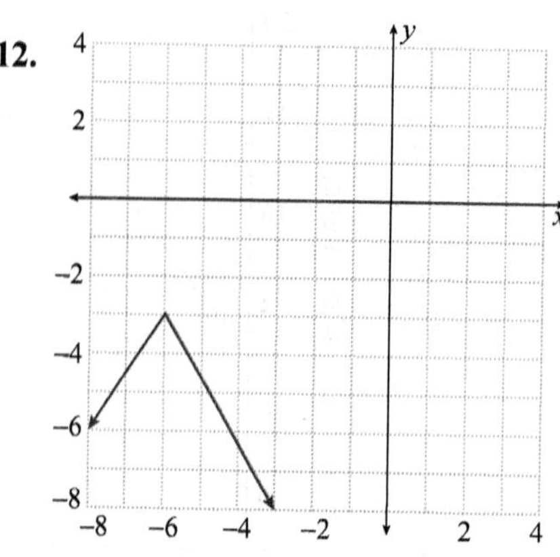

13. 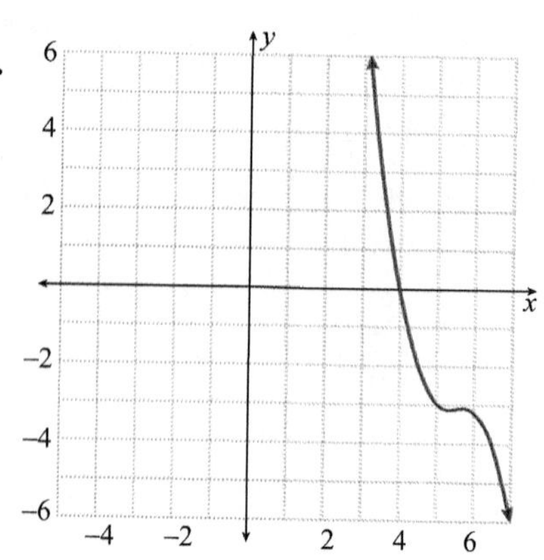

Sketch the graphs of the following functions by first identifying the more basic functions that have been shifted, reflected, stretched, or compressed. Then determine the domain and range of each function. See Examples 1 through 6.

14. $f(x) = (x+2)^3$

15. $G(x) = |x-4|$

16. $p(x) = -(x+1)^2 + 2$

17. $g(x) = \sqrt{x+3} - 1$

18. $q(x) = (1-x)^2$

19. $r(x) = -\sqrt[3]{x}$

20. $s(x) = \sqrt{2-x}$

21. $F(x) = \dfrac{|x+2|}{3} + 3$

22. $w(x) = \dfrac{1}{(x-3)^2}$

23. $v(x) = \dfrac{1}{3x} - 2$

24. $f(x) = \dfrac{1}{2-x}$

25. $k(x) = \sqrt{-x} + 2$

26. $b(x) = \sqrt[3]{x+2} - 5$

27. $b(x) = [\![x-4]\!] + 4$

28. $R(x) = 4 - 2|x|$

29. $S(x) = (3-x)^3$

30. $g(x) = -\dfrac{1}{x+1}$

31. $h(x) = \dfrac{x^2}{2} - 3$

32. $W(x) = 1 - |4-x|$

33. $W(x) = -\dfrac{|x-1|}{4}$

34. $S(x) = \dfrac{1}{x^2} + 3$

35. $V(x) = -3\sqrt{x-1} + 2$

36. $g(x) = x^2 - 6x + 9$ (**Hint:** Find a better way to write the function.)

37. $h(x) = \dfrac{|x|}{x}$ (**Hint:** Evaluate h at a few points to understand its behavior.)

38. $W(x) = \dfrac{x-1}{|x-1|}$

39. $s(x) = [\![x-2]\!]$

Write a formula for each of the functions described.

40. Use the function $g(x) = x^2$. Move the function 3 units to the left and 4 units down.

41. Use the function $g(x) = x^2$. Move the function 4 units to the right and 2 units up.

42. Use the function $g(x) = x^2$. Reflect the function across the x-axis and move it 6 units up.

43. Use the function $g(x) = x^2$. Move the function 2 units to the right and reflect across the y-axis.

44. Use the function $g(x) = x^3$. Move the function 1 unit to the left and reflect across the y-axis.

45. Use the function $g(x) = x^3$. Move the function 10 units to the right and 4 units up.

46. Use the function $g(x) = \sqrt{x}$. Move the function 5 units to the left and reflect across the x-axis.

47. Use the function $g(x) = \sqrt{x}$. Reflect the function across the y-axis and move it 3 units down.

48. Use the function $g(x) = |x|$. Move the function 7 units to the left, reflect across the x-axis, and reflect across the y-axis.

49. Use the function $g(x) = |x|$. Move the function 8 units to the right, 2 units up, and reflect across the x-axis.

Use your knowledge about transformations to find a possible formula for the function $f(x)$ given its graph.

50.

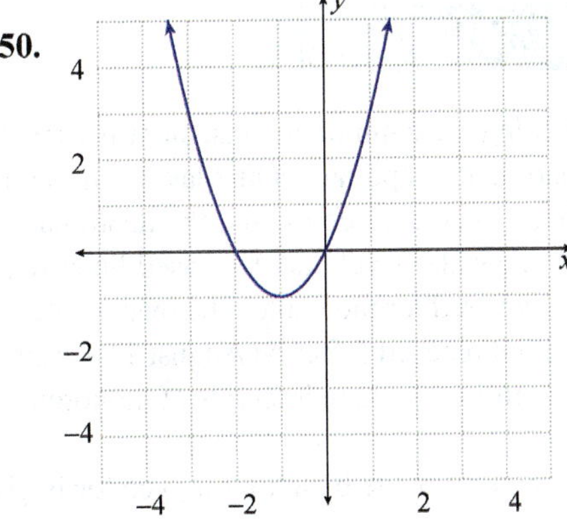

51.

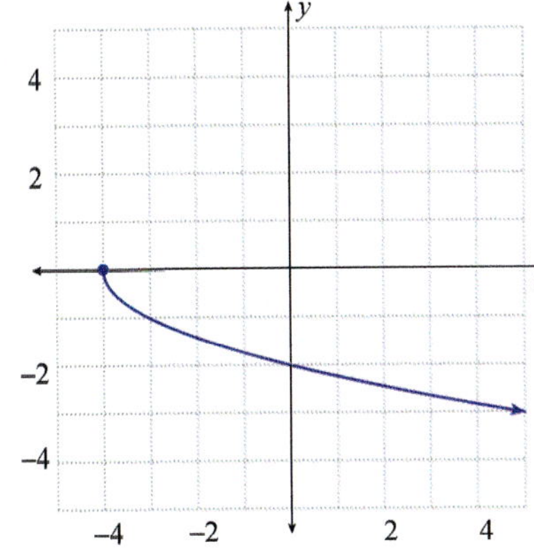

52.

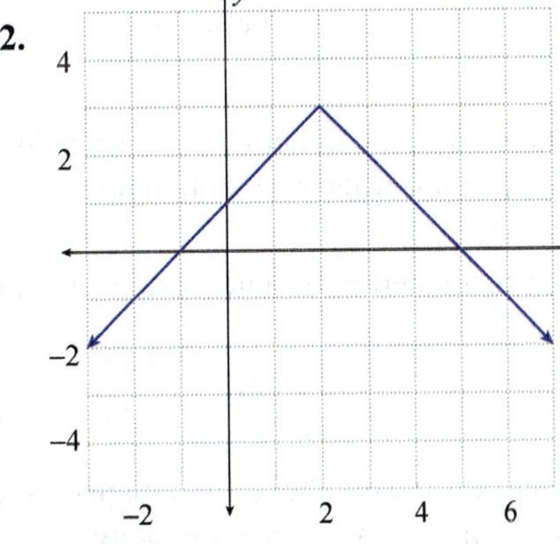

53.

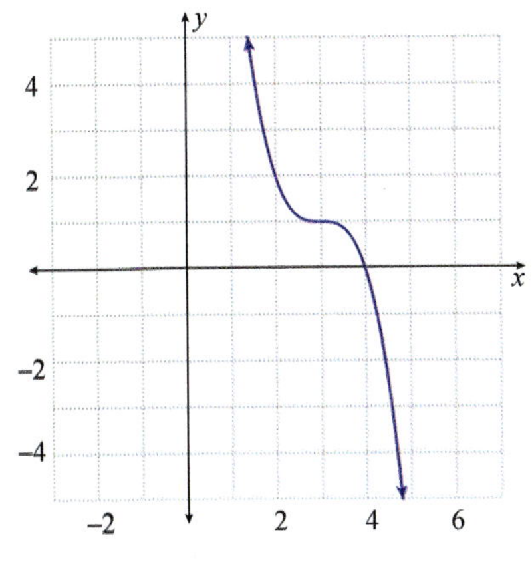

Determine if each of the following relations is a function. If so, determine whether it is even, odd, or neither. Also determine if it has *y*-axis symmetry, *x*-axis symmetry, origin symmetry, or none of these symmetries, and then sketch the graph of the relation. See Example 7.

54. $f(x) = |x| + 3$

55. $g(x) = x^3$

56. $h(x) = x^3 - 1$

57. $w(x) = \sqrt[3]{x}$

58. $x = -y^2$

59. $3y - 2x = 1$

60. $x + y = 1$

61. $F(x) = (x-1)^2$

62. $x = y^2 + 1$

63. $x = 2|y|$

64. $g(x) = \dfrac{x^2}{5} - 5$

65. $s(x) = \left\| x + \dfrac{1}{2} \right\|$

66. $m(x) = \sqrt[3]{x} - 1$

67. $xy = 2$

68. $x + y^2 = 3$

For each of the following functions, find the open intervals of monotonicity where the function is increasing, decreasing, or constant. See Examples 8 and 9.

69. $f(x) = (x+3)^2$

70. $g(x) = -|x-2|$

71. $h(x) = \dfrac{1}{x-1}$

72. $H(x) = \dfrac{1}{(x+3)^2}$

73. $G(x) = \sqrt{x+1}$

74. $F(x) = -2$

75. $p(x) = -30|x-1|$

76. $q(x) = (4-x)^2 + 1$

77. $r(x) = \dfrac{(x-7)^4}{-2} + 4$

78. $P(x) = \begin{cases} (x+3)^2 & \text{if } x < -1 \\ 1 & \text{if } x \geq -1 \end{cases}$

79. $Q(x) = \begin{cases} |x-1| & \text{if } x \leq 3 \\ 5-x & \text{if } x > 3 \end{cases}$

🚀 APPLICATIONS

80. During the summer months, the water level of a garden pool varies as water is added and as it evaporates. On May 1st the pool was 3.4 feet deep. After a steady and linear increase due to rain, the depth had increased to 4.9 feet on June 1st. By July 1st the water level had decreased linearly to 4.2 feet. Knowing that the pool would be covered for the winter, the owner filled the pool (in an essentially linear fashion) until it reached 5 feet on August 1st. Graph the water level as a function of time and determine the open intervals of monotonicity.

81. The profit made by a hot dog vendor is given by the function

$$P(x) = \begin{cases} 2x - 3 & \text{if } x \geq 0 \text{ and } x < 7 \\ \dfrac{1}{4}x^2 & \text{if } x \geq 7 \end{cases}$$

where *x* is the number of hot dogs sold. Graph the profit function and determine the open intervals of monotonicity.

82. The cost incurred by a newspaper stand is given by the function

$$C(x) = \begin{cases} -2\sqrt{x} + 8 & \text{if } x \geq 0 \text{ and } x < 3 \\ -x + 8 & \text{if } x \geq 3 \end{cases}$$

where *x* is the number of newspapers sold. Graph the cost function and determine the open intervals of monotonicity.

📈 TECHNOLOGY

Mentally sketch the graph of the given function by identifying the basic shape that has been shifted, reflected, stretched, or compressed. Then use a graphing utility to graph the function and check your reasoning.

83. $f(x) = -2(3-x)^3 + 5$

84. $f(x) = \dfrac{3}{x+5} - 1$

85. $f(x) = \dfrac{-1}{(x-2)^2} - 3$

86. $f(x) = -3|x+2| - 4$

87. $f(x) = -\sqrt{1-x} + 2$

88. $f(x) = \sqrt[3]{2+x} - 1$

Write a possible equation for the function depicted on the graphing utility. The function is shown in a $[-10,10]$ by $[-10,10]$ viewing window.

89.

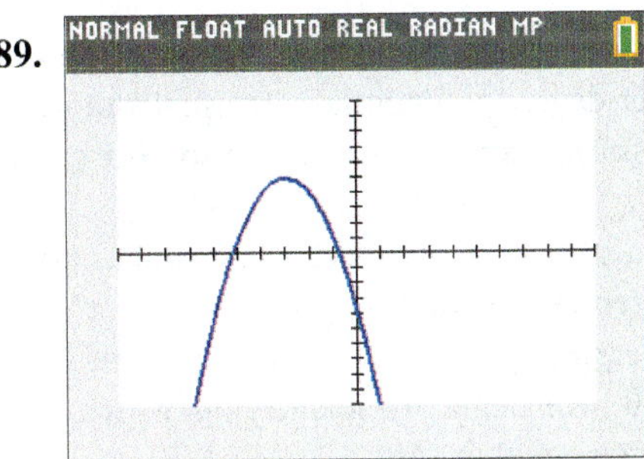

90.

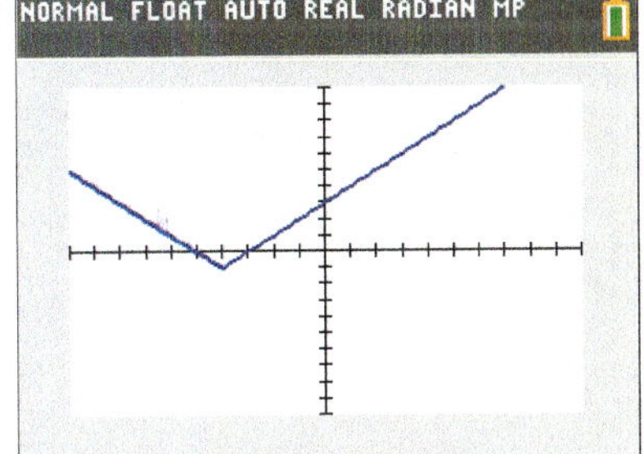

91.

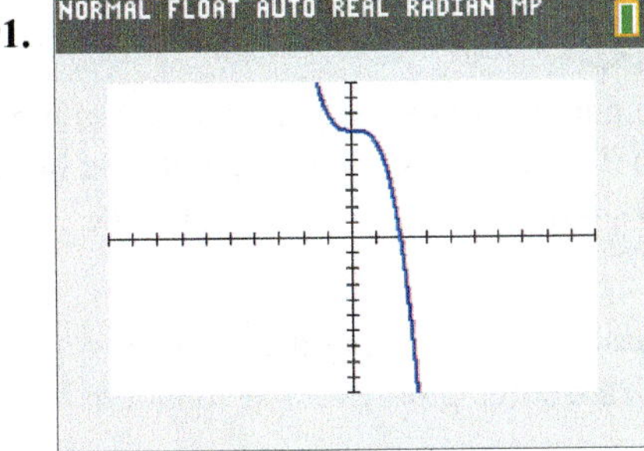

92.

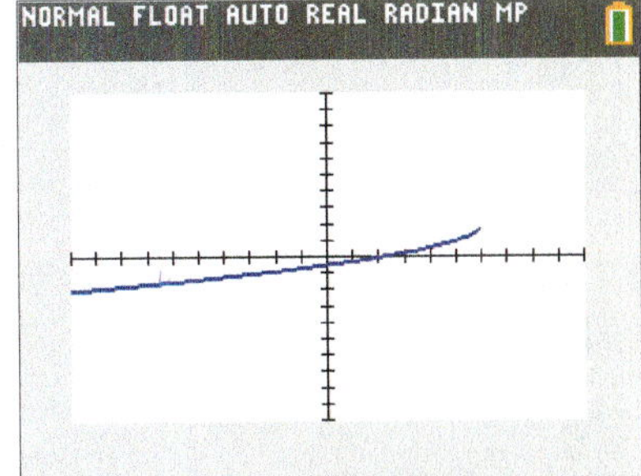

93.

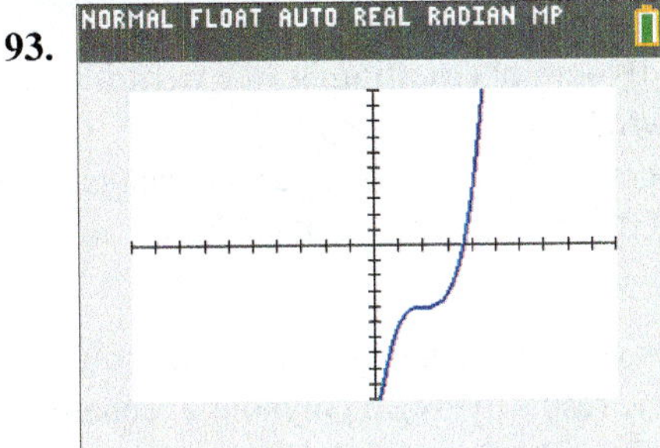

94.

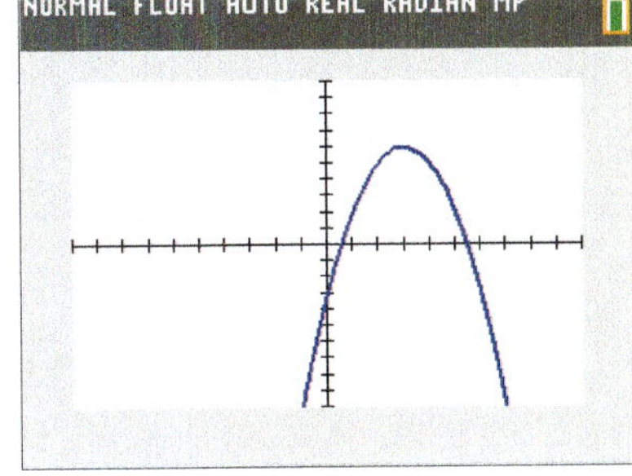

3.7 POLYNOMIAL FUNCTIONS

■ TOPICS

- ■ Zeros of Polynomials and Solutions of Polynomial Equations
- ■ Graphing Factored Polynomials
- ■ Solving Polynomial Inequalities
- ■ Finding Zeros of Polynomials Using Technology

Zeros of Polynomials and Solutions of Polynomial Equations

At this point, we have studied how linear and quadratic polynomial functions behave, and we have tools guaranteed to solve all linear and quadratic equations. We have also studied some elementary higher-degree polynomials (those of the form ax^n). In this section, we begin a more complete exploration of higher-degree polynomials.

Not surprisingly, the complexity of polynomial functions increases with the degree; higher-degree polynomials are usually more difficult to graph accurately, and we cannot necessarily expect to find exact solutions to higher-degree polynomial equations.

In order to make our work as general as possible, we will refer to a generic n^{th}-degree polynomial function $p(x) = a_n x^n + a_{n-1} x^{n-1} + \cdots + a_1 x + a_0$, where n is a nonnegative integer, $a_n, a_{n-1}, \ldots, a_1, a_0$ all represent constants (that may be real or complex) and $a_n \neq 0$. We begin by identifying which values of the variable make a polynomial function equal to zero.

Zeros of a Polynomial

The number k (k may be a complex number) is a **zero** of the polynomial function $p(x)$ if $p(k) = 0$. This is also expressed by saying that k is a **root** of the polynomial or a **solution** of the equation $p(x) = 0$.

The task of determining the zeros of a polynomial arises in many contexts, two of which are solving polynomial equations and graphing polynomials.

Polynomial Equations

A **polynomial equation in one variable**, say the variable x, is an equation that can be written in the form $a_n x^n + a_{n-1} x^{n-1} + \cdots + a_1 x + a_0 = 0$, where n is a nonnegative integer, $a_n, a_{n-1}, \ldots, a_1, a_0$ are constants. Assuming $a_n \neq 0$, we say such an equation is of degree n and call a_n the **leading coefficient**.

Just as with linear and quadratic equations (which are polynomial equations of degree 1 and 2, respectively), a polynomial equation may not appear in the form of the above definition. The first task is often to rewrite the equation so that one side is zero. Then, the zeros of the polynomial on the other side are the solutions of the equation.

Note that, given a polynomial equation in the form $p(x) = 0$, the zeros of p are precisely the solutions of the equation (and vice versa).

Example 1: Solutions of Polynomial Equations

Verify that the given values of x solve the corresponding polynomial equations.

a. $6x^2 - x^3 = 12 + 5x; \quad x = 4$

b. $x^2 = 2x - 5; \quad x = 1 + 2i$

c. $\dfrac{x}{1-i} = 3x^2; \quad x = 0$

NOTE

When verifying a zero, there is no need to rewrite the equation in the form of the definition.

Solution

a.
$$6x^2 - x^3 = 12 + 5x$$
$$6(4)^2 - (4)^3 \overset{?}{=} 12 + 5(4) \qquad \text{Substitute } x = 4 \text{ throughout the equation.}$$
$$96 - 64 \overset{?}{=} 12 + 20 \qquad \text{Simplify both sides.}$$
$$32 = 32 \qquad \text{This is a true statement, so 4 is a solution.}$$

b.
$$x^2 = 2x - 5$$
$$(1+2i)^2 \overset{?}{=} 2(1+2i) - 5 \qquad \text{Substitute } x = 1 + 2i \text{ in the equation.}$$
$$1 + 4i + 4i^2 \overset{?}{=} 2 + 4i - 5 \qquad \text{Expand both sides.}$$
$$1 + 4i - 4 \overset{?}{=} 2 + 4i - 5 \qquad \text{Simplify, using the fact that } i^2 = -1.$$
$$-3 + 4i = -3 + 4i \qquad \text{This is a true statement, so } 1 + 2i \text{ is a solution.}$$

c.
$$\frac{x}{1-i} = 3x^2$$
$$\frac{0}{1-i} \overset{?}{=} 3(0)^2 \qquad \text{Substitute 0 for } x \text{ in the equation.}$$
$$0 = 0 \qquad \begin{array}{l} \text{After simplifying, we arrive at a true statement.} \\ \text{Thus, 0 is a solution to the polynomial equation.} \end{array}$$

Graphing Factored Polynomials

Consider a generic polynomial function $p(x)$ with all real coefficients. Our goal is to be able to sketch the graph of such a function, paying particular attention to the behavior of p as $x \to -\infty$ and as $x \to \infty$, and the x- and y-intercepts of p. We will begin by looking at the behavior of a polynomial function as $x \to \pm\infty$.

The graph of $p(x)$ is similar to the graph of $a_n x^n$ for values of x that are very large in magnitude; the leading term of $p(x)$ dominates the behavior. Take the function $f(x) = x^4 - 3x^3 - 5x^2 + 8$. The graph in Figure 1 shows that as x gets very large, $f(x)$ takes values very similar to x^4 and thus has a very similar graph.

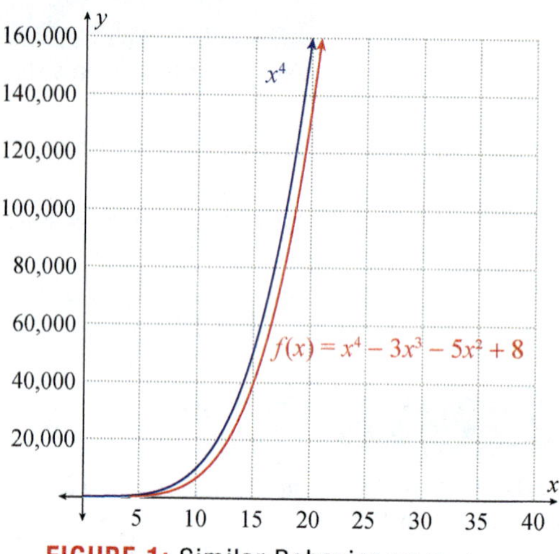

FIGURE 1: Similar Behavior as $x \to \infty$

This means we can understand the behavior of all polynomials as $x \to \pm\infty$ just by understanding how functions of the form $f(x) = a_n x^n$ behave. We know that if n is even, $x^n \to \infty$ as $x \to -\infty$ and as $x \to \infty$, and if n is odd, then $x^n \to -\infty$ as $x \to -\infty$ and $x^n \to \infty$ as $x \to \infty$. We also know that if a_n is negative, the graph undergoes a reflection across the x-axis; this reverses the sign of every y-value, changing the behavior as $x \to \pm\infty$. Figure 2 shows a few examples.

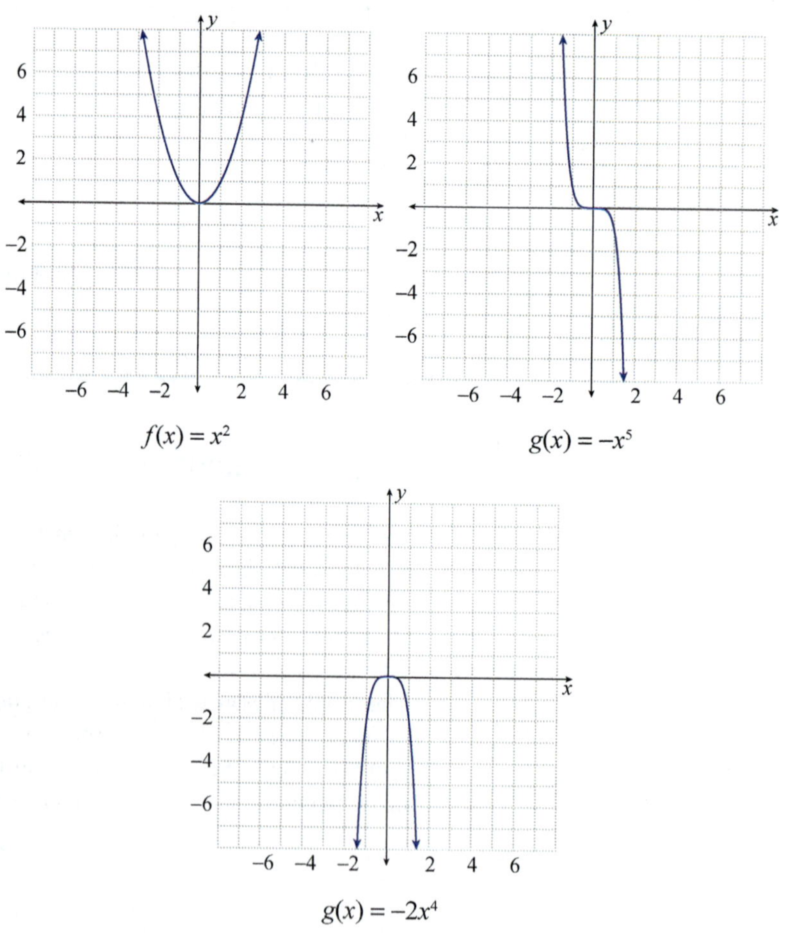

FIGURE 2: Examples of Behavior as $x \to \infty$

Behavior of Polynomials as $x \to \pm\infty$

Given a polynomial function $p(x)$ with degree n, the behavior of $p(x)$ as $x \to \pm\infty$ can be determined from the leading term $a_n x^n$ using the table below.

	n is even	**n is odd**
a_n is positive	as $x \to -\infty$, $p(x) \to +\infty$ as $x \to +\infty$, $p(x) \to +\infty$ The graph rises to the left and rises to the right.	as $x \to -\infty$, $p(x) \to -\infty$ as $x \to +\infty$, $p(x) \to +\infty$ The graph falls to the left and rises to the right.
a_n is negative	as $x \to -\infty$, $p(x) \to -\infty$ as $x \to +\infty$, $p(x) \to -\infty$ The graph falls to the left and falls to the right.	as $x \to -\infty$, $p(x) \to +\infty$ as $x \to +\infty$, $p(x) \to -\infty$ The graph rises to the left and falls to the right.

Near the origin, however, the graph of $p(x)$ is likely to be quite different from the graph of $a_n x^n$. Recall our example from before: x^4 and $f(x) = x^4 - 3x^3 - 5x^2 + 8$ both behave the same as $x \to \pm\infty$. Observe how differently these functions behave near the origin.

Finding the x- and y-intercepts of a polynomial function will help us sketch a more complete graph. Given a polynomial in the form $p(x) = a_n x^n + a_{n-1}x^{n-1} + \cdots + a_1 x + a_0$, finding the y-intercept is not difficult, as it simply requires evaluating $p(0)$.

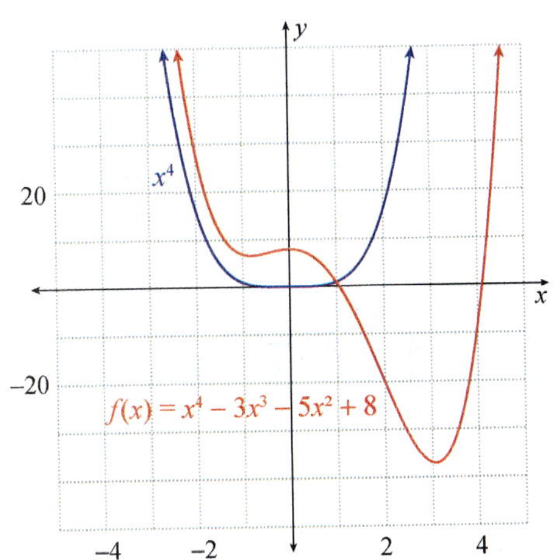

FIGURE 3: Different Behavior near the Origin

$$p(0) = a_n (0)^n + a_{n-1} (0)^{n-1} + \cdots + a_1 (0) + a_0$$
$$= a_0$$

Thus, the y-intercept is $(0, a_0)$. On the other hand, finding the x-intercepts requires solving the polynomial equation $0 = a_n x^n + a_{n-1} x^{n-1} + \cdots + a_1 x + a_0$, which often takes more effort.

Writing the equation in a different form can make our task much easier. If we factor a given polynomial f into a product of linear factors, each linear factor with real coefficients corresponds to an x-intercept of the graph of f. To see why this is true, consider the polynomial function

$$f(x) = (3x - 5)(x + 2)(2x - 6).$$

To determine the x-intercepts of this polynomial we need to solve the equation

$$0 = (3x - 5)(x + 2)(2x - 6).$$

The Zero-Factor Property tells us that the only solutions are those values of x for which $3x - 5 = 0$, $x + 2 = 0$, or $2x - 6 = 0$. Solving these three linear equations gives us the x-coordinates of the three x-intercepts of f: $\left\{ \dfrac{5}{3}, -2, 3 \right\}$.

Working with a polynomial in factored form almost always makes finding the x-intercepts easier, but we do have to adjust how we calculate the y-intercept and the behavior as $x \to \pm\infty$. If a polynomial is in factored form, we cannot simply read off the value of the y-intercept. However, substituting $x = 0$ is not much more work.

$$f(0) = \big(3(0) - 5\big)\big((0) + 2\big)\big(2(0) - 6\big)$$
$$= (-5)(2)(-6)$$
$$= 60$$

Similarly, when a polynomial is in factored form, we can't directly see the term $a_n x^n$ to determine the end behavior. Instead of multiplying out f completely, we just determine how the leading x^3 term arises. The third degree term comes from multiplying together the $3x$ from the first factor, the x from the second factor, and the $2x$ from the third factor. Thus, $a_n x^n = 6x^3$. Since the leading coefficient is positive and the degree is odd, we know that $f(x) \to -\infty$ as $x \to -\infty$ and $f(x) \to \infty$ as $x \to \infty$.

Putting it all together, along with a few computed values of f, we obtain the sketch in Figure 4 (note the difference in the horizontal and vertical scales).

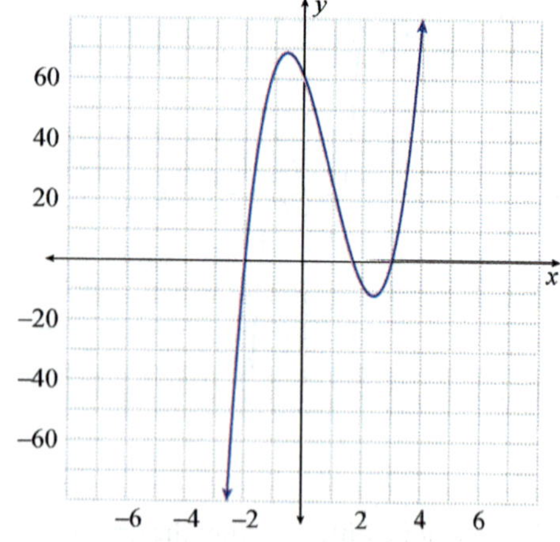

FIGURE 4: Graph of
$f(x) = (3x - 5)(x + 2)(2x - 6)$

As always, plotting additional points will help in sketching an accurate graph.

Example 2: Graphing Polynomial Functions

Sketch the graphs of the following polynomial functions, paying particular attention to the x-intercept(s), the y-intercept, and the behavior as $x \to \pm\infty$.

a. $f(x) = -x(2x + 1)(x - 2)$ **b.** $g(x) = x^2 + 2x - 3$ **c.** $h(x) = x^4 - 1$

Solution

a. Begin with the x-intercepts. Using the Zero-Factor Property, we solve $f(x) = 0$.

$$-x(2x+1)(x-2) = 0$$

$$x = -\frac{1}{2}, 0, 2$$

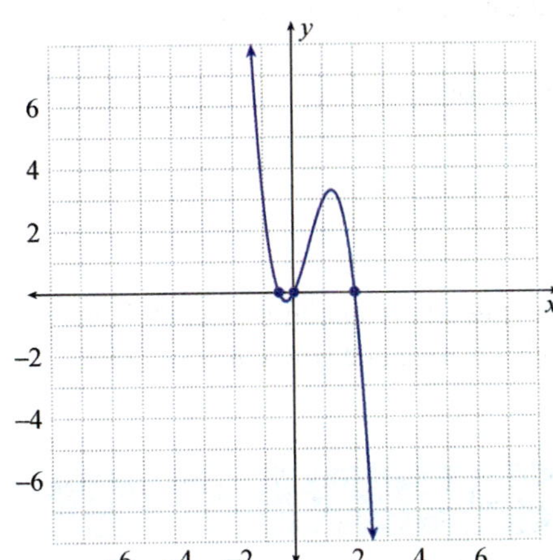

Thus, the x-intercepts are $\left(-\frac{1}{2}, 0\right)$, $(0, 0)$, and $(2, 0)$.

We could plug in $x = 0$ to find the y-intercept, but we already found it when calculating the x-intercepts! The y-intercept is the origin $(0, 0)$.

All that remains is to determine the end behavior. Find the highest degree term by multiplying $(-x)(2x)(x) = -2x^3$. The leading coefficient is negative, and the degree is odd, so $f(x) \to \infty$ as $x \to -\infty$ and $f(x) \to -\infty$ as $x \to \infty$.

Putting all this together, we obtain the graph.

b. This polynomial is not in factored form. Before factoring, we can collect information about the y-intercept and behavior as $x \to \pm\infty$.

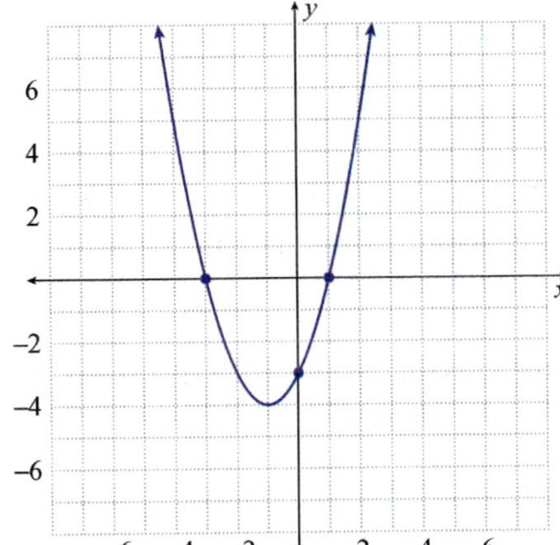

$g(x) = x^2 + 2x - 3$, so $a_0 = -3$ and the leading term is x^2. This means that the y-intercept is $(0, -3)$ and that $g(x) \to \infty$ as $x \to \pm\infty$.

Now, to find the x-intercepts, we factor the polynomial.

$$x^2 + 2x - 3 = 0$$
$$(x+3)(x-1) = 0$$
$$x = -3, 1$$

Thus, the x-intercepts are $(-3, 0)$ and $(1, 0)$.

c. As with the previous example, we determine the y-intercept and end behavior before factoring. $h(x) = x^4 - 1$, so $a_0 = -1$ and the leading term is x^4. This tells us that the y-intercept is $(0, -1)$ and that $h(x) \to \infty$ as $x \to \pm\infty$.

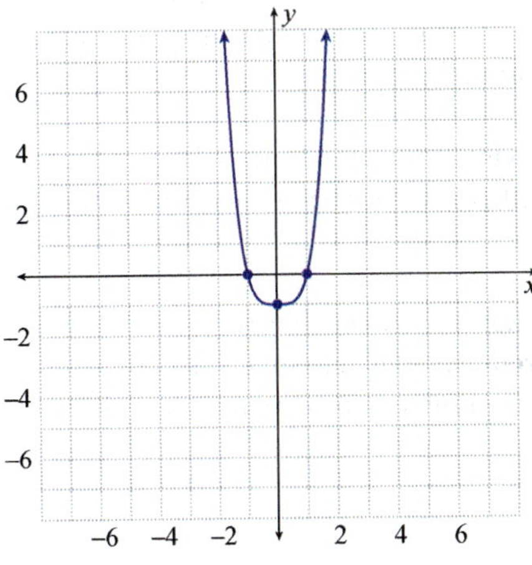

Once again, we factor the polynomial to calculate the x-intercepts.

$$x^4 - 1 = 0 \qquad \text{A difference of squares.}$$
$$(x^2 - 1)(x^2 + 1) = 0 \qquad \text{Another difference of squares.}$$
$$(x-1)(x+1)(x^2 + 1) = 0$$
$$x = -1, 1 \qquad \text{Only the real solutions lead to } x\text{-intercepts. Here,}$$

the x-intercepts are $(-1, 0)$ and $(1, 0)$.

Solving Polynomial Inequalities

Polynomial Inequalities

A **polynomial inequality** is any inequality that can be written in the form

$$p(x) < 0, \quad p(x) \leq 0, \quad p(x) > 0, \quad \text{or} \quad p(x) \geq 0,$$

where $p(x)$ is a polynomial function.

If we have an accurate graph of the function, solving a polynomial inequality is very straightforward; simply read where the function is positive and where it is negative.

Example 3: Solving Polynomial Inequalities Using a Graph

Solve the polynomial inequalities, given the graph of $f(x) = (x + 3)(x + 1)(x - 2)$.

a. $(x + 3)(x + 1)(x - 2) < 0$ 　　　　　　　 **b.** $(x + 3)(x + 1)(x - 2) \geq 0$

Solution

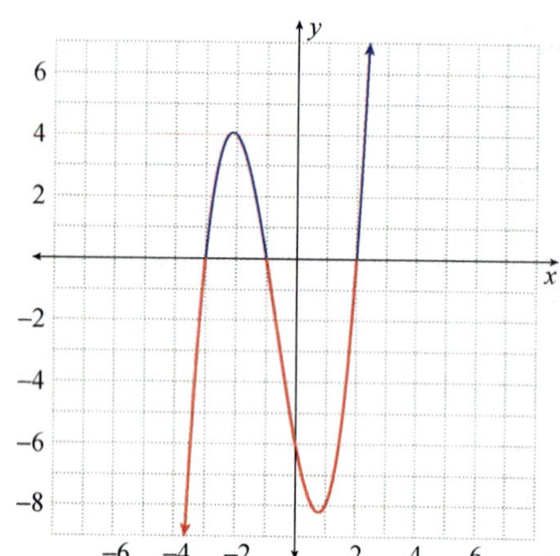

a. From the graph, we see that $f(x) < 0$ on the intervals $(-\infty, -3)$ and $(-1, 2)$. Since the inequality is strict, we do not include the endpoints. Thus, the solution to this inequality is $(-\infty, -3) \cup (-1, 2)$.

b. Similarly, the graph reveals that $f(x) > 0$ on the intervals $(-3, -1)$ and $(2, \infty)$. This inequality is not strict, so we include the x-values where $f(x) = 0$. Therefore, the solution to this inequality is $[-3, -1] \cup [2, \infty)$.

What if we do not have an accurate graph of the function? The graphing methods we have covered so far tell us if a polynomial $p(x)$ is positive or negative as $x \rightarrow \pm\infty$, but give no information about the sign of $p(x)$ near the origin.

As such, we need an algebraic method for solving polynomial inequalities when a detailed graph is not available. Having such a method will be a helpful tool in sketching graphs of polynomials.

Our method depends on a property of polynomials called **continuity**. While a more formal definition of continuity requires tools from calculus, intuitively, a continuous function has no "breaks" in it, in other words its graph can be drawn without lifting the pencil off the paper.

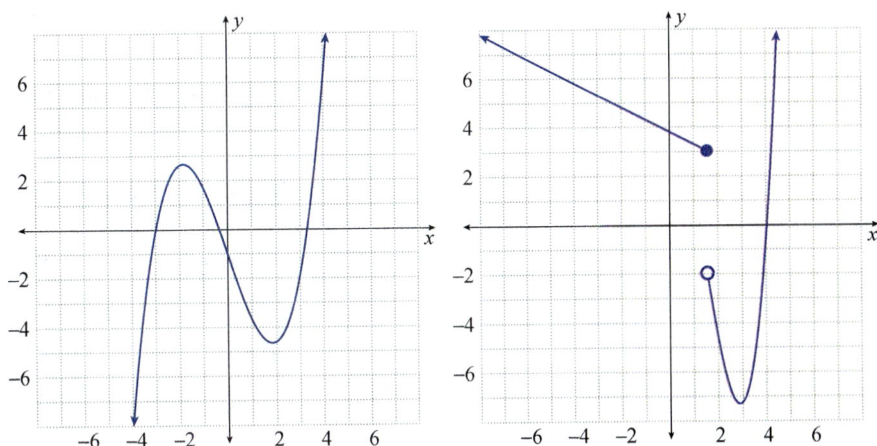

FIGURE 5: Continuous Function   **FIGURE 6:** Discontinuous Function

Why is continuity important? Look at the graphs in Figures 5 and 6. If a graph is not continuous, it can change sign without passing through the x-axis (thus, the value of the function "skips" zero). However, with a polynomial (or any continuous function), the only way for the graph to change sign is to pass through zero! This means that between each pair of zeros, the graph of $p(x)$ is always positive or always negative. Knowing this gives us a method for solving polynomial inequalities.

Solving Polynomial Inequalities: Sign-Test Method

To solve a polynomial inequality $p(x) < 0$, $p(x) \leq 0$, $p(x) > 0$, or $p(x) \geq 0$:

Step 1: Find the real zeros of $p(x)$. Equivalently, find the real solutions of $p(x) = 0$.

Step 2: Place the zeros on a number line, splitting it into intervals.

Step 3: Within each interval, select a **test point** and evaluate p at that number. If the result is positive, then $p(x) > 0$ for all x in the interval. If the result is negative, then $p(x) < 0$ for all x in the interval.

Step 4: Write the solution set, consisting of all of the intervals that satisfy the given inequality. If the inequality is not strict (uses $\leq$ or $\geq$), then the zeros are included in the solution set as well.

Example 4: Solving Polynomial Inequalities: Sign-Test Method

Solve the polynomial inequality $(x - 2)(x + 5)(x - 4) \leq 0$.

Solution

> **NOTE**
>
> When choosing test points, integers (especially 0) are usually easiest to work with.

Follow the steps in the procedure for the Sign-Test Method.

Step 1: Find the zeros of $p(x) = (x - 2)(x + 5)(x - 4)$. By the Zero-Factor Property, we can see that $p(x) = 0$ when $x = -5$, 2, or 4.

Step 2: Place the zeros on a number line.

This splits the number line into the intervals $(-\infty, -5)$, $(-5, 2)$, $(2, 4)$, and $(4, \infty)$.

Step 3: Evaluate $p(x)$ for a test point in each interval.

Interval	Test Point	Evaluate	Result
$(-\infty, -5)$	$x = -6$	$\begin{aligned}p(-6) &= (-6-2)(-6+5)(-6-4)\\ &= (-8)(-1)(-10)\\ &= -80\end{aligned}$	$p(x) < 0$ on $(-\infty, -5)$ **Negative**
$(-5, 2)$	$x = 0$	$\begin{aligned}p(0) &= (0-2)(0+5)(0-4)\\ &= (-2)(5)(-4)\\ &= 40\end{aligned}$	$p(x) > 0$ on $(-5, 2)$ **Positive**
$(2, 4)$	$x = 3$	$\begin{aligned}p(3) &= (3-2)(3+5)(3-4)\\ &= (1)(8)(-1)\\ &= -8\end{aligned}$	$p(x) < 0$ on $(2, 4)$ **Negative**
$(4, \infty)$	$x = 6$	$\begin{aligned}p(6) &= (6-2)(6+5)(6-4)\\ &= (4)(11)(2)\\ &= 88\end{aligned}$	$p(x) > 0$ on $(4, \infty)$ **Positive**

Step 4: Write the solution set to the original inequality, $p(x) \leq 0$. From our table, we see that $p(x)$ is negative on the intervals $(-\infty, -5)$ and $(2, 4)$. Since the inequality is not strict, we need to include the zeros in our solution set. Thus, the solution to the inequality is $(-\infty, -5] \cup [2, 4]$.

Finding Zeros of Polynomials Using Technology

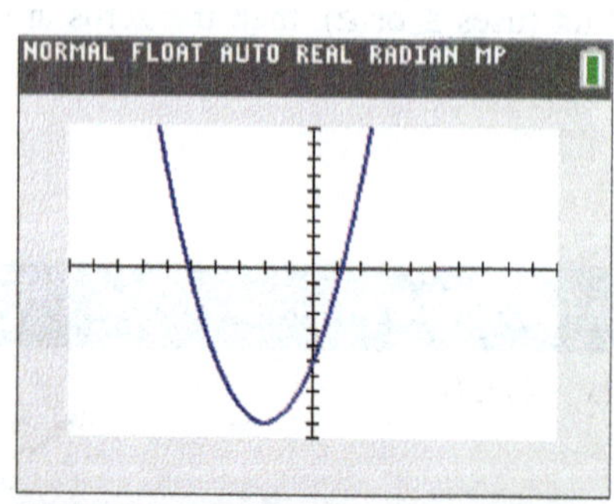

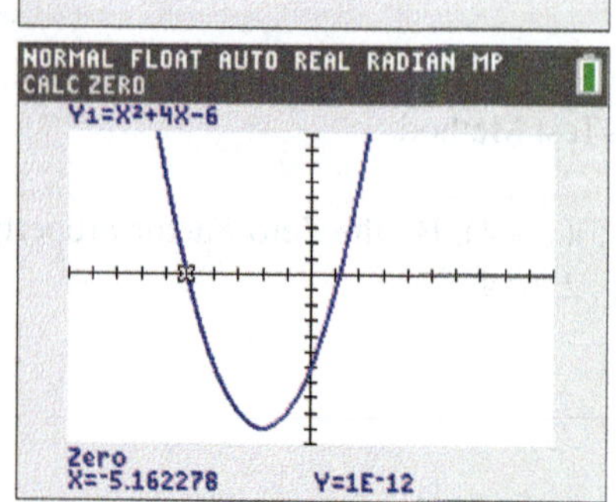

Previously, we saw how to find the x-intercepts of a linear equation on a calculator. The same method can be used to find the x-intercepts, or zeros, of any function graphed on a calculator. The main difference is that with linear functions, there can be no more than one zero, but other functions might have more. Consider the graph of the function $f(x) = x^2 + 4x - 6$.

We can see that there are two zeros that appear to be located near $x = -5$ and $x = 1$. To check more accurately, press **2nd** **trace** to access the **CALC** menu, then select **2:zero**. The screen should now display the graph with the words "**Left Bound?**" shown at the bottom. Choose which zero you want to find and use the arrows to move the cursor anywhere to the left of that intercept and press **ENTER**. The screen should now say "**Right Bound?**" Use the right arrow to move the cursor to the right of that same intercept and press **ENTER** again. (Be sure there is only one x-intercept between what you select as the left bound and what you select as the right bound.) The text should now read "**Guess?**" Press **ENTER** a third time and the x- and y-values of that x-intercept will appear at the bottom of the screen.

The leftmost zero occurs at approximately $x \approx -5.162$. Note that sometimes, as in this example, the display will read a very small number rather than exactly zero. To find the other zero, repeat the process, this time focusing on the rightmost zero. We find that it occurs at $x \approx 1.162$.

3.7 EXERCISES

PRACTICE

Verify that the given values of x solve the corresponding polynomial equations. See Example 1.

1. $9x^2 - 4x = 2x^3 + 15;\ x = -1$

2. $x^2 - 4x = -13;\ x = 2 - 3i$

3. $x^2 + 13 = 4x;\ x = 2 + 3i$

4. $3x^3 + (5 - 3i)x^2 = (2 + 5i)x - 2i;\ x = i$

5. $9x^2 - 4x = 2x^3 + 15;\ x = 3$

6. $9x^2 - 4x = 2x^3 + 15;\ x = \dfrac{5}{2}$

7. $3x^3 + (5 - 3i)x^2 = (2 + 5i)x - 2i;\ x = -2$

8. $x^5 - 10x^4 - 80x^2 = 32 - 80x - 40x^3;\ x = 2$

9. $4x^5 - 8x^4 - 12x^3 = 16x^2 - 25x - 69;\ x = 3$

10. $x^2 - 4x - 12 = 0;\ x = 6$

11. $23x^7 - 12x^5 = 63x^4 - 3x^2;\ x = 0$

12. $x^2 + 74 = 10x;\ x = 5 + 7i$

13. $4x^2 + 32x + (8 + i)x^3 = -8;\ x = 2i$

14. $8x - 17 = x^2;\ x = 4 - i$

15. $(5 - 3i)x - 3x = 4 - 6i;\ x = 2$

16. $x^6 - x^5 + 7x^4 + x^3 - 9x = -1;\ x = 1$

17. $6x^7 - 3x^5 = 3x^4 - 6x^2;\ x = -1$

Determine if the given values of x are solutions of the corresponding polynomial equations. See Example 1.

18. $16x = x^3 + x^2 + 20;\ x = -5$

19. $x^4 - 13x^2 + 12 = -x^3 + x;\ x = -1$

20. $x^4 - 3x^3 - 10x^2 = 0;\ x = 2$

21. $4x^5 - 216x^2 = 36x^3 - 24x^4;\ x = -6$

22. $x^3 - 8ix + 30 = 15x + 2x^2 + 16i;\ x = -i$

23. $x^3 - 7x^2 + 4x - 28 = 0;\ x = 2i$

Solve the following polynomial equations by factoring and/or using the quadratic formula, making sure to identify all the solutions.

24. $x^3 - x^2 - 6x = 0$

25. $x^2 - 2x + 5 = 0$

26. $x^4 + x^2 - 2 = 0$

27. $2x^2 + 5x = 3$

28. $9x^2 = 6x - 1$

29. $x^4 - 8x^2 + 15 = 0$

30. $x^3 - x^2 = 72x$

31. $x^2 + 5x = -\dfrac{25}{4}$

32. $2x^2 + 5 = 11x$

33. $x^4 - 8x^3 + 25x^2 = 0$

34. $x^4 - 13x^2 + 36 = 0$

35. $x^4 + 7x^2 = 8$

For each of the following polynomials, determine the degree and the leading coefficient; then determine the behavior of the graph as $x \to \pm\infty$.

36. $p(x) = 2x^4 - 3x^3 - 6x^2 - x - 23$

37. $j(x) = 4x^7 + 5x^5 + 12$

38. $r(x) = (3x+5)(x-2)(2x-1)(4x-7)$

39. $h(x) = -6x^5 + 2x^3 - 7x$

40. $g(x) = (x-5)^3(2x+1)(-x-1)$

41. $f(x) = -2(x+4)(x-4)(x^2)$

For each of the following polynomial functions, determine the behavior of its graph as $x \to \pm\infty$ and identify the x- and y-intercepts. Use this information to sketch the graph of each polynomial. See Example 2.

42. $f(x) = (x-3)(x+2)(x+4)$

43. $g(x) = (3-x)(x+2)(x+4)$

44. $f(x) = (x-2)^2(x+5)$

45. $h(x) = -(x+2)^3$

46. $r(x) = x^2 - 2x - 3$

47. $s(x) = x^3 + 3x^2 + 2x$

48. $f(x) = -(x-2)(x+1)^2(x+3)$

49. $g(x) = (x-3)^5$

In Exercises 50–55, use the behavior as $x \to \pm\infty$ and the intercepts to match each polynomial with its graph.

50. $g(x) = (x+1)^2(x-3)^2$

51. $h(x) = 1 - (x+2)^2$

52. $f(x) = (x-1)(x+2)(3-x)$

53. $r(x) = x^2 - x - 6$

54. $s(x) = (x-1)^3 - 2$

55. $f(x) = (x-3)^2(4x+1)(x+2)(x-2)$

a.  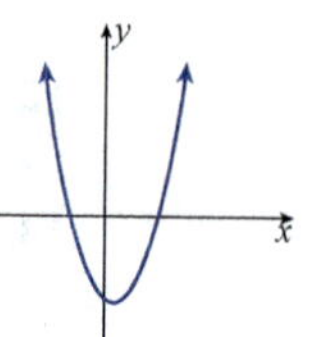

b. 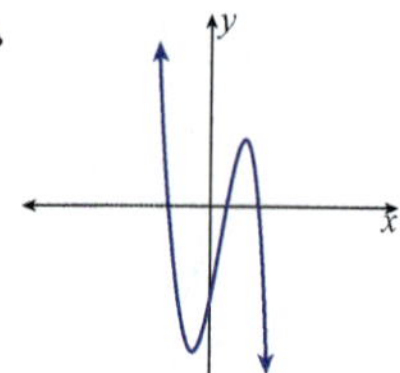

c. 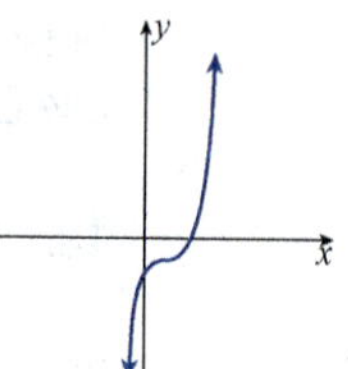

d. 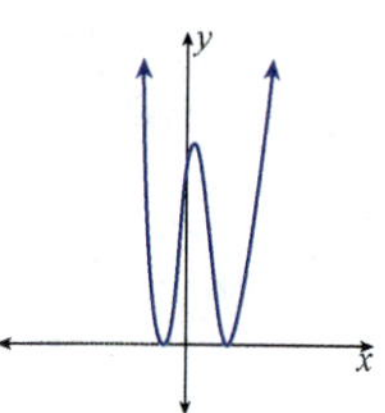

e. 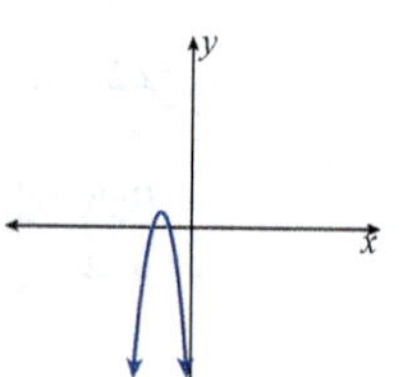

f. 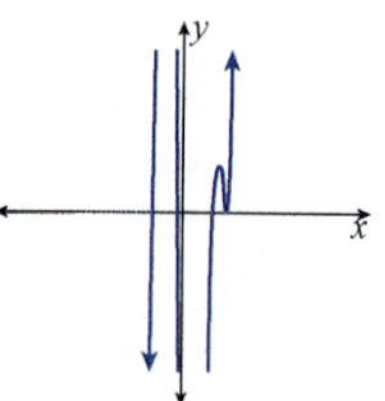

Match each of the following functions to the appropriate description.

56. $z(x) = (x-1)(x+2)(4-x)$

a. cubic curve increasing as $x \to \infty$, has x-intercepts of 0, −1, and −2, and crosses the y-axis at 0

57. $r(x) = x^2 - 6x - 7$

b. parabola that opens up, has x-intercepts at 6 and −1, crosses the y-axis at −6

c. cubic curve increasing as $x \to \infty$, has x-intercepts of 0, −1, and −4, and crosses the y-axis at 0

58. $s(x) = x^3 + 3x^2 + 2x$

d. parabola that opens up, has x-intercepts at 7 and −1, crosses the y-axis at −7

59. $g(x) = (x-1)(x+4)(3-x)$

e. cubic curve decreasing as $x \to \infty$, has x-intercepts at 1, 4, and −2, crosses the y-axis at −8

60. $s(x) = x^3 + 5x^2 + 4x$

f. cubic curve decreasing as $x \to \infty$, has x-intercepts of 1, 3, and −4, and crosses the y-axis at −12

61. $s(x) = x^2 - 5x - 6$

Solve the following polynomial inequalities. See Example 4.

62. $x^2 - x - 6 \leq 0$

63. $x^2 > x + 6$

64. $(x+2)^2 (x-1)^2 > 0$

65. $x^3 + 3x^2 + 2x < 0$

66. $(x-2)(x+1)(x+3) \geq 0$

67. $(x-1)(x+2)(3-x) \leq 0$

68. $-x^3 - x^2 + 30x > 0$

69. $(x^2 - 1)(x-4)(x+5) \leq 0$

70. $x^4 + x^2 > 0$

71. $4x^2 < 6x + 4$

72. $x^2(x+4)(x-3) > 0$

73. $(x-3)(x+4)(2-x) > 0$

🚀 **APPLICATIONS**

For Exercises 74–78, use the fact that profit is equal to revenue minus cost.

74. A small start-up skateboard company projects that the cost per month of manufacturing x skateboards will be $C(x) = 10x + 300$, and the revenue per month from selling x skateboards will be $R(x) = -x^2 + 50x$. For what value(s) of x will the company break even or make a profit?

75. A manufacturer has determined that the revenue from the sale of x cameras is given by $R(x) = -x^2 + 15x$. The cost of producing x cameras is $C(x) = 135 - 17x$. For what value(s) of x will the company break even or make a profit?

76. The revenue from the sale of x fire extinguishers is estimated to be $R(x) = 9 - x^2$. The total cost of producing x fire extinguishers is $C(x) = 209 - 33x$. For what value(s) of x will the company break even or make a profit?

77. A manufacturer has determined that the cost and revenue of producing and selling x telescopes are $C(x) = 253 - 7x$ and $R(x) = 27x - x^2$, respectively. For what value(s) of x will the company break even or make a profit?

78. A company that produces and sells compact refrigerators has found that the revenue from the sale of x compact refrigerators is $R(x) = -x^2 + 30x - 370$. The cost function is given by $C(x) = 6 - 25x$. For what value(s) of x will the company break even or make a profit?

79. An electronics company is deciding whether or not to begin producing phones. The company must determine if a profit can be made on the phones. The profit function is modeled by the equation $P(x) = x + 0.27x^2 - 0.0015x^3 - 300$, where x is the number of phones produced in hundreds. Given this equation, how many phones must the company produce to make a profit?

80. The population of sea lions on an island is represented by the function $L(m) = 110m^2 - 0.35m^4 + 750$, where m is the number of months the sea lions have been observed on the island. Given this information, how many more months will there be sea lions on the island?

81. The population of mosquitoes in a city in Florida is modeled by the function $M(w) = 200w^2 - 0.01w^4 + 1200$, where w is the number of weeks since the town began spraying for mosquitoes. How many weeks will it take for all the mosquitoes to die?

3.8 RATIONAL FUNCTIONS

■ TOPICS

- ■ Definitions and Useful Notation
- ■ Vertical Asymptotes
- ■ Horizontal and Oblique Asymptotes
- ■ Graphing Rational Functions

Definitions and Useful Notation

The study of polynomials leads directly to a study of rational functions, which are ratios of polynomials. Since rational functions can have variables in the denominators of fractions, their behavior can be significantly more complex than that of polynomials.

Rational Functions

A **rational function** is a function that can be written in the form

$$f(x) = \frac{p(x)}{q(x)},$$

where $p(x)$ and $q(x)$ are polynomial functions and $q(x) \neq 0$. Even though q is not allowed to be identically zero, there will often be values of x for which $q(x)$ is zero, and at these values the function is undefined. Consequently, the **domain of** f consists of all real numbers except those for which $q(x) = 0$.

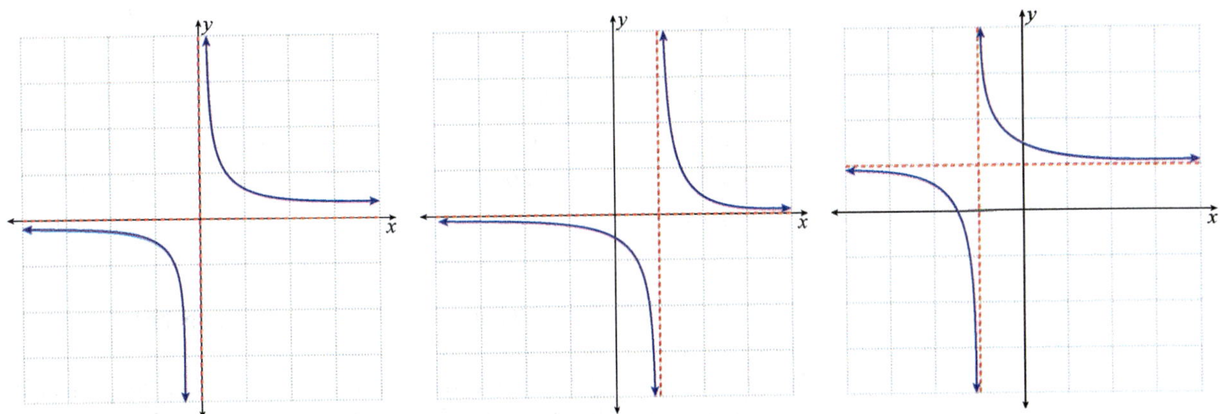

FIGURE 1: Graphs of Three Rational Functions

Each of the three graphs in Figure 1 has a new feature: vertical and horizontal dashed lines. These dashed lines are *not* part of the function, they are examples of *asymptotes*, and they serve as guides to understanding the function. Roughly speaking, an asymptote is a line that the graph of a function approaches, but does not touch. Three kinds of asymptotes will appear in our study of rational functions: vertical, horizontal, and oblique.

Vertical Asymptotes

The vertical line $x = c$ is a **vertical asymptote** of a function f if $f(x)$ increases in magnitude without bound as x approaches c. Examples of vertical asymptotes appear in Figure 2. The graph of a rational function cannot intersect a vertical asymptote.

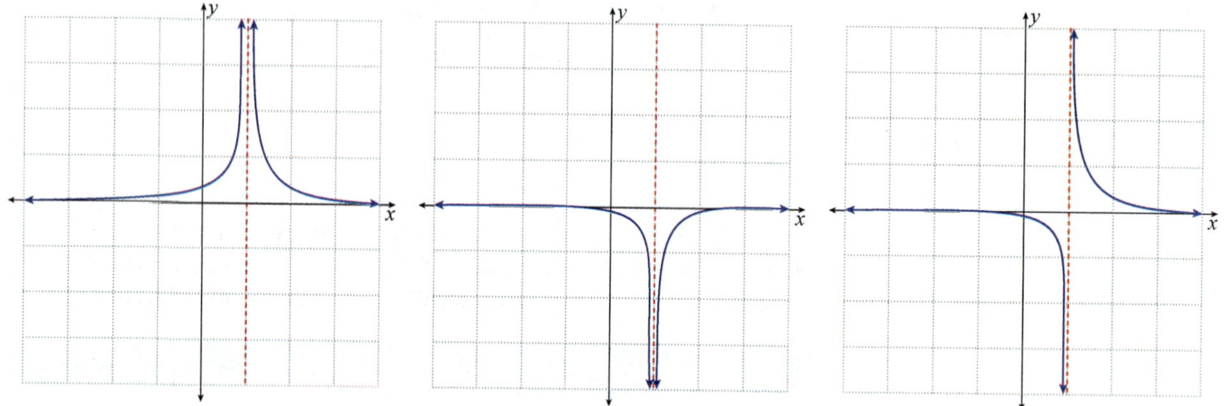

FIGURE 2: Vertical Asymptotes

To understand how vertical asymptotes arise, let's observe what happens to the function $f(x) = \dfrac{1}{x}$ as x gets closer to 0 from the left and from the right.

x	$f(x) = \dfrac{1}{x}$	x	$f(x) = \dfrac{1}{x}$
-1	-1	1	1
-0.1	-10	0.1	10
-0.01	-100	0.01	100
-0.001	-1000	0.001	1000
-0.0001	$-10,000$	0.0001	10,000
-0.00001	$-100,000$	0.00001	100,000

TABLE 1: Values of $f(x) = \dfrac{1}{x}$ as x Approaches 0

We can see that as x gets closer and closer to 0 (where the function is undefined), the value of f increases in magnitude without bound. The graph reflects this, as the curve gets steeper and steeper, never touching the line $x = 0$.

Horizontal Asymptotes

The horizontal line $y = c$ is a **horizontal asymptote** of a function f if $f(x)$ approaches the value c as $x \to -\infty$ or as $x \to \infty$. Examples of horizontal asymptotes appear in Figure 3. The graph of a rational function may intersect a horizontal asymptote near the origin, but will eventually approach the asymptote from one side only as x increases in magnitude.

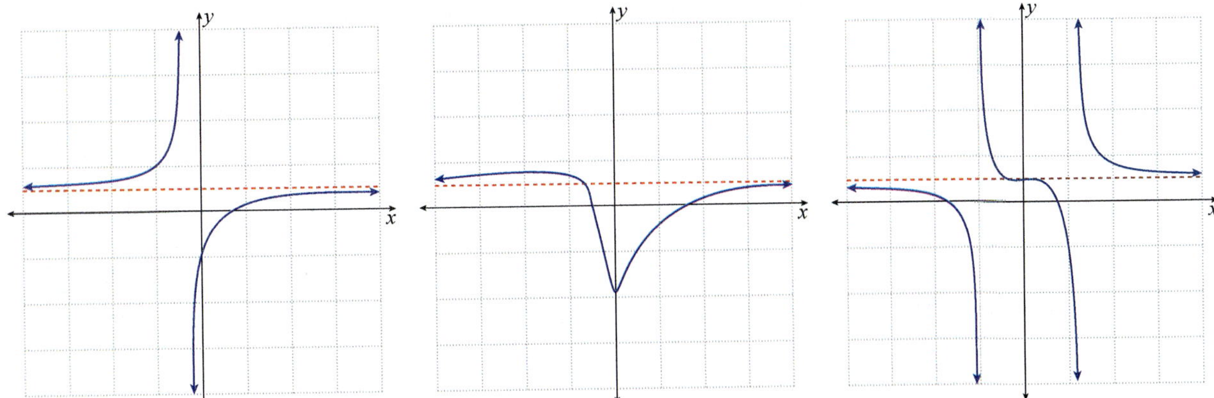

FIGURE 3: Horizontal Asymptotes

Oblique Asymptotes

A nonvertical, nonhorizontal line may also be an asymptote of a function f. Examples of **oblique** (or **slant**) **asymptotes** appear in Figure 4. Again, the graph of a rational function may intersect an oblique asymptote near the origin, but will eventually approach the asymptote from one side only as $x \to \infty$ or $x \to -\infty$.

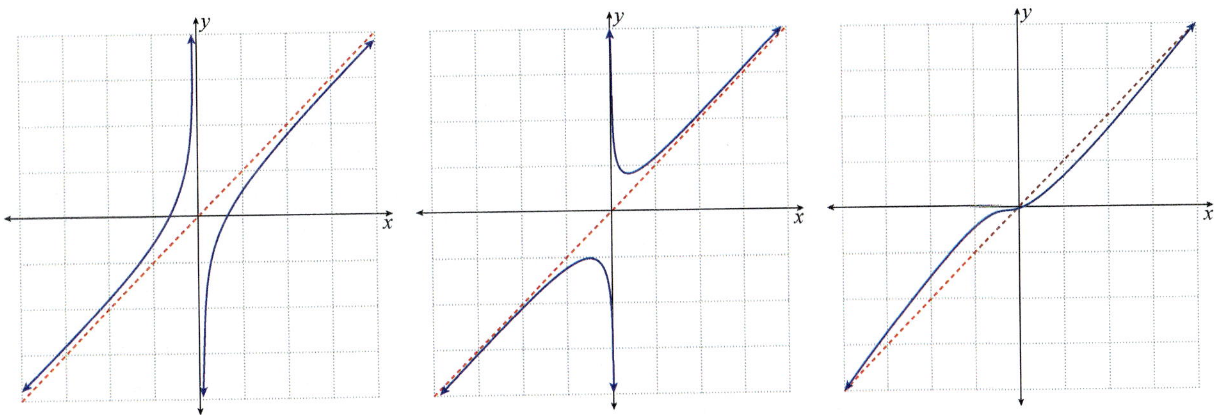

FIGURE 4: Oblique Asymptotes

As Figures 2, 3, and 4 illustrate, the behavior of rational functions with respect to asymptotes can vary considerably. In order to describe the behavior of a given rational function more easily, we have specific asymptote notation.

Asymptote Notation

The notation $x \to c^-$ is used when describing the behavior of a graph as x approaches the value c from the left (the negative side). The notation $x \to c^+$ is used when describing behavior as x approaches c from the right (the positive side). The notation $x \to c$ is used when describing behavior that is the same on both sides of c.

Figure 5 illustrates how the asymptote notation can be used to describe the behavior of functions.

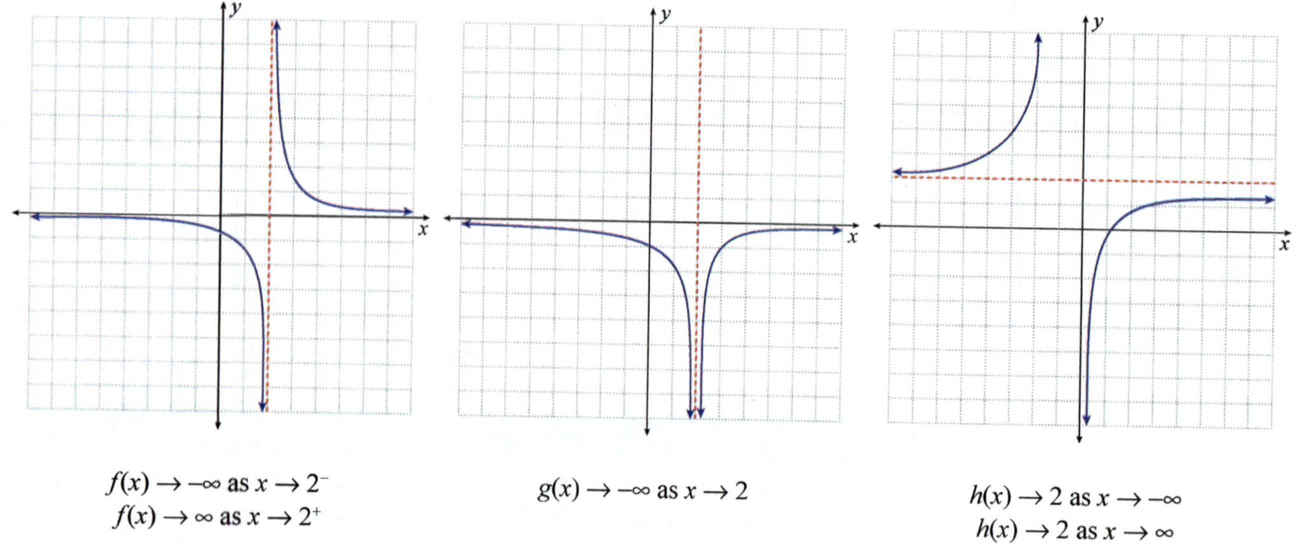

FIGURE 5: Asymptote Notation

Vertical Asymptotes

With the asymptote notation and examples as background, we are ready to delve into the details of identifying asymptotes for rational functions.

Equations for Vertical Asymptotes

If the rational function $f(x) = \dfrac{p(x)}{q(x)}$ has been written in reduced form (so that p and q have no common factors), the vertical line $x = c$ is a **vertical asymptote** of f if and only if c is a zero of the polynomial q. In other words, f has vertical asymptotes at the x-intercepts of q.

Note that the numerator of a rational function is irrelevant in locating the vertical asymptotes, assuming that all common factors in the fraction have been canceled. However, if the numerator and denominator share a common factor of $(x - c)$, the value c will be out of the domain of the function, but the line $x = c$ will not be a vertical asymptote.

Example 1: Vertical Asymptotes

Find the domains and the equations for the vertical asymptotes of the following functions.

a. $f(x) = \dfrac{32}{x+2}$ **b.** $g(x) = \dfrac{x^2+1}{x^2+2x-15}$ **c.** $h(x) = \dfrac{x^2-x}{x-1}$

Solution

a. To answer both questions, we need to calculate the zeros of the denominator (which is already in factored form).

$$x + 2 = 0$$
$$x = -2$$

Since the function f is in reduced form, this zero is the only point excluded from the domain. This means the domain of f is $(-\infty, -2) \cup (-2, \infty)$.

Further, we know that f has a vertical asymptote of $x = -2$.

b. In this case, we need to factor the denominator before calculating the domain and vertical asymptotes.

$$g(x) = \frac{x^2 + 1}{x^2 + 2x - 15} = \frac{x^2 + 1}{(x+5)(x-3)}$$

Since the values -5 and 3 both make the denominator zero, the domain of g is $(-\infty, -5) \cup (-5, 3) \cup (3, \infty)$.

The numerator is a sum of two squares and cannot be factored, so the rational function is already in reduced form. This means that the equations of the two vertical asymptotes are $x = -5$ and $x = 3$.

c. The denominator is already in factored form, so we can see that its only zero is at $x = 1$. Thus, the domain of h is $(-\infty, 1) \cup (1, \infty)$.

Now that we have found the domain, we can look for common factors to cancel.

$$h(x) = \frac{x^2 - x}{x - 1} = \frac{x(x-1)}{x-1}$$
$$= x$$

By canceling the common factor of $(x - 1)$, we have the reduced form $h(x) = x$, which applies only for values in the domain of h (all real numbers except for 1). Since the reduced form of h has no denominator, h has no vertical asymptotes.

Horizontal and Oblique Asymptotes

To determine horizontal and oblique asymptotes, we are interested in the behavior of a function $f(x)$ as $x \to -\infty$ and as $x \to \infty$. If f is a rational function, f is a ratio of two polynomials p and q, so we can begin by considering the effect p and q have on one another.

Consider a rational function $f(x) = \dfrac{p(x)}{q(x)}$ in which the polynomial p has degree n and the polynomial q has degree m. Note that f can be written as a result of the division of polynomial p by polynomial q. That is, f equals a polynomial called a *quotient* of degree $n - m$ plus a *remainder* term. The key fact is that the behavior of rational functions as the magnitude of x gets very large tends to approach the behavior of the quotient. Thus, the horizontal and oblique asymptotes of a rational function depend on the difference in degrees between p and q.

Equations for Horizontal and Oblique Asymptotes

Let $f(x) = \dfrac{p(x)}{q(x)}$ be a rational function, where p is an nth-degree polynomial with leading coefficient a_n, q is an mth-degree polynomial with leading coefficient b_m, and $p(x)$ and $q(x)$ have no common factors other than constants. Then the asymptotes of f are found as follows.

1. If $n < m$, the horizontal line $y = 0$ (the x-axis) is the **horizontal asymptote** for f.

2. If $n = m$, the horizontal line $y = \dfrac{a_n}{b_m}$ is the **horizontal asymptote** for f.

3. If $n = m + 1$, the line $y = g(x)$ is an **oblique asymptote** for f, where g is the quotient polynomial obtained by dividing p by q.

4. If $n > m + 1$, there is **no** straight-line **horizontal** or **oblique asymptote** for f.

Example 2: Horizontal and Oblique Asymptotes

Find the equation for the horizontal or oblique asymptote of each of the following functions.

a. $f(x) = \dfrac{x^2 + 1}{x^2 + 2x - 15}$

b. $g(x) = \dfrac{x^3 + x^2 + 2x + 2}{x^2 + 9}$

c. $h(x) = \dfrac{3x^4 + 10x - 7}{x^6 + x^5 - x^2 - 1}$

d. $j(x) = \dfrac{2x^4 - 3x^2 + 8}{x^2 - 25}$

☑ NOTE

Always begin by comparing the degrees of the numerator and the denominator.

Solution

a. First, note that the degrees of the numerator and denominator of f are both equal to two. This means that the line $y = \dfrac{a_2}{b_2} = 1$ is the horizontal asymptote of f.

b. Here, the numerator and denominator have no common factors and the degree of the numerator is one more than the degree of the denominator. So we know g has an oblique asymptote, equal to the quotient polynomial of the numerator and denominator. To find it, we need to perform polynomial division.

$$
\begin{array}{r}
x + 1 \\
x^2 + 9 \overline{\smash{)}\, x^3 + x^2 + 2x + 2} \\
-\left(x^3 + 0x^2 + 9x\right) \\
\hline
x^2 - 7x + 2 \\
-\left(x^2 + 0x + 9\right) \\
\hline
-7x - 7
\end{array}
$$

This tells us that $g(x) = x + 1 + \dfrac{-7x - 7}{x^2 + 9}$, but we only need the quotient, $x + 1$, to find that the equation for the oblique asymptote is $y = x + 1$.

c. In this case, the degree of the numerator of h is two *less* than the degree of the denominator. This means that the line $y = 0$ is the horizontal asymptote of h.

d. For $j(x)$, the degree of the numerator is two *more* than the degree of the denominator. Thus, j has no horizontal or oblique asymptotes.

Graphing Rational Functions

Much of our experience in graph sketching will be useful as we graph rational functions. In addition to the standard steps of identifying the x-intercepts (if any) and y-intercept (if there is one), we will make use of asymptotes when graphing rational functions. The following is a list of suggested steps.

Graphing Rational Functions

Given a rational function f,

Step 1: Factor the denominator in order to determine the domain of f. Any points excluded from the domain may appear as "holes" in the graph or as vertical asymptotes.

Step 2: Factor the numerator as well and cancel any common factors.

Step 3: Examine the remaining linear factors in the denominator to determine the equations for any vertical asymptotes.

Step 4: Compare the degrees of the numerator and denominator to determine if there is a horizontal or oblique asymptote. If so, find its equation.

Step 5: Determine the y-intercept, if 0 is in the domain of f.

Step 6: Determine the x-intercepts, if there are any, by setting the numerator of the reduced fraction equal to 0.

Step 7: Plot enough points to determine the behavior of f between x-intercepts and between vertical asymptotes.

Example 3: Graphing Rational Functions

Sketch the graphs of the following rational functions.

a. $f(x) = \dfrac{x^2 - x}{x - 1}$

b. $g(x) = \dfrac{x^2 + 1}{x^2 + 2x - 15}$

c. $h(x) = \dfrac{x^3 + x^2 + 2x + 2}{x^2 + 9}$

Solution

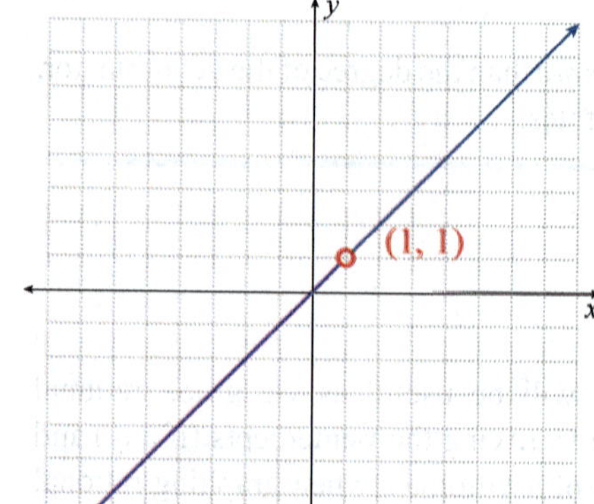

a. The denominator of f is already factored, so we know that the domain of f consists of all real numbers except for $x = 1$.

As in Example 1c, we factor the numerator to see if there are any common factors.

$$f(x) = \frac{x^2 - x}{x - 1} = \frac{x(x-1)}{x-1}$$
$$= x$$

This means that, except for at $x = 1$, where f is undefined, we have $f(x) = x$. We already know how to graph this function, so the remaining steps are unnecessary. The graph of f is the line $y = x$, excluding the point $(1, 1)$, since $x = 1$ is not in the domain of f. The result is that a "hole" appears in the graph of f at $x = 1$.

b. In Example 1b, we factored the denominator as follows.

$$g(x) = \frac{x^2 + 1}{x^2 + 2x - 15} = \frac{x^2 + 1}{(x+5)(x-3)}$$

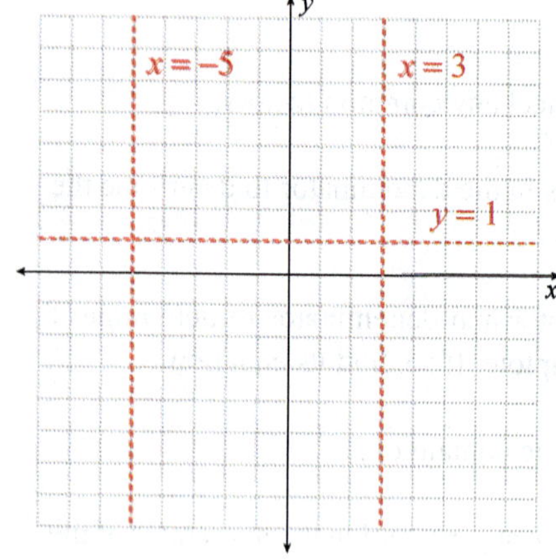

This means the domain of g excludes the values $x = -5$ and $x = 3$. Since the numerator cannot be factored, we also know that the lines $x = -5$ and $x = 3$ are vertical asymptotes of g.

Next, we look at the degrees of the numerator and denominator. As we saw in Example 2a, the degrees are the same, so the line $y = 1$ is the horizontal asymptote. Plotting the asymptotes provides us a framework for graphing $g(x)$.

Setting $x = 0$, we find that the y-intercept lies at $\left(0, -\frac{1}{15}\right)$. Since there is no real solution to the equation $x^2 + 1 = 0$, g has no x-intercepts.

Plotting a few points in each region between asymptotes gives us an idea of the general shape of the graph.

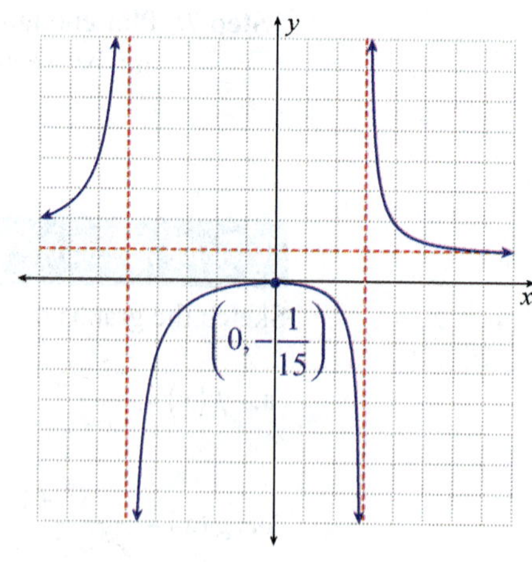

c. The denominator of this function cannot be factored, so there are no restrictions on the domain of h. Further, we saw in Example 2b that this function has an oblique asymptote of $y = x + 1$.

As usual, we calculate the y-intercept by substituting $x = 0$.

$$h(0) = \frac{0^3 + 0^2 + 2(0) + 2}{0^2 + 9} = \frac{2}{9}$$

There are different approaches to finding the x-intercepts. Looking at the numerator, we might guess that -1 is a zero of the numerator. A quick calculation confirms this: $(-1)^3 + (-1)^2 + 2(-1) + 2 = 0$. This means we can factor the numerator. Using division, we find $x^3 + x^2 + 2x + 2 = (x + 1)(x^2 + 2)$. Thus, $(-1, 0)$ is the only x-intercept.

With the intercepts and a few other plotted points, we obtain the graph of h.

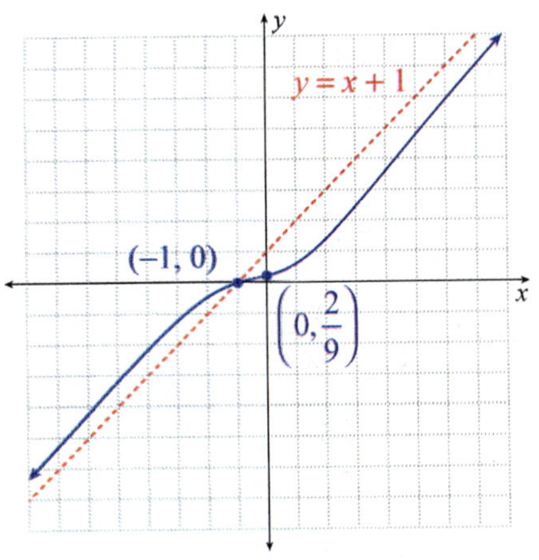

3.8 EXERCISES

💡 PRACTICE

Find equations for the vertical asymptotes, if any, for each of the following rational functions. See Example 1.

1. $f(x) = \dfrac{5}{x-1}$

2. $f(x) = \dfrac{x^2 + 3}{x + 3}$

3. $f(x) = \dfrac{x^2 - 4}{x + 2}$

4. $f(x) = \dfrac{-3x + 5}{x - 2}$

5. $f(x) = \dfrac{3x^2 + 1}{x - 2}$

6. $f(x) = \dfrac{x^2 + 2x}{x + 1}$

7. $f(x) = \dfrac{x^2 - 4}{2x - x^2}$

8. $f(x) = \dfrac{x + 2}{x^2 - 9}$

9. $f(x) = \dfrac{x^2 - 2x - 3}{2x^2 - 5x - 3}$

10. $f(x) = \dfrac{2x^2 + 2x - 4}{x^2 + 2x + 1}$

11. $f(x) = \dfrac{x^3 - 27}{x^2 + 5}$

12. $f(x) = \dfrac{x^2 + 5}{x^3 - 27}$

13. $f(x) = \dfrac{x^2 - 1}{x^2 - 8x + 7}$

14. $f(x) = \dfrac{2x^2 + 7x - 14}{2x^2 + 7x - 15}$

15. $f(x) = \dfrac{x^3 - 6x^2 + 11x - 6}{x^3 + 8}$

16. $f(x) = \dfrac{x^2 - 2x - 15}{x - 5}$

17. $f(x) = \dfrac{x^2 - 16}{x^2 - 4}$

18. $f(x) = \dfrac{x^2 + 4x + 4}{x^2 + x - 2}$

Find equations for the horizontal or oblique asymptotes, if any, for each of the following rational functions. See Example 2.

19. $f(x) = \dfrac{5}{x-1}$

20. $f(x) = \dfrac{x^2+3}{x+3}$

21. $f(x) = \dfrac{x^4-4}{x^2+2}$

22. $f(x) = \dfrac{x^2-4}{2x-x^2}$

23. $f(x) = \dfrac{x+2}{x^2-9}$

24. $f(x) = \dfrac{x^2-2x-3}{2x^2-5x-3}$

25. $f(x) = \dfrac{2x^2+2x-4}{x^2+2x+1}$

26. $f(x) = \dfrac{-3x+5}{x-2}$

27. $f(x) = \dfrac{3x^2+1}{x-2}$

28. $f(x) = \dfrac{x^3-27}{x^2+5}$

29. $f(x) = \dfrac{x^2+5}{x^3-27}$

30. $f(x) = \dfrac{x^2+2x}{x+1}$

31. $f(x) = \dfrac{x^2-81}{x^3+7x-12}$

32. $f(x) = \dfrac{x^3-3x^2+2x}{x-7}$

33. $f(x) = \dfrac{x^2-9x+4}{x+2}$

34. $f(x) = \dfrac{-x^5+2x^2}{5x^5+3x^3-7}$

35. $f(x) = \dfrac{5x^2-x+12}{x-1}$

36. $f(x) = \dfrac{2x^2-5x+6}{x-3}$

Sketch the graphs of the following rational functions, making use of your work in the previous exercises and additional information about intercepts and any other points that may be useful. See Example 3.

37. $f(x) = \dfrac{5}{x-1}$

38. $f(x) = \dfrac{x^2+3}{x+3}$

39. $f(x) = \dfrac{x^2-4}{x+2}$

40. $f(x) = \dfrac{x^2-4}{2x-x^2}$

41. $f(x) = \dfrac{x+2}{x^2-9}$

42. $f(x) = \dfrac{x^2-2x-3}{2x^2-5x-3}$

43. $f(x) = \dfrac{2x^2+2x-4}{x^2+2x+1}$

44. $f(x) = \dfrac{-3x+5}{x-2}$

45. $f(x) = \dfrac{3x^2+1}{x-2}$

46. $f(x) = \dfrac{x^3-27}{x^2+5}$

47. $f(x) = \dfrac{x^2+5}{x^3-27}$

48. $f(x) = \dfrac{x^2+2x}{x+1}$

For each graph, find any **a.** vertical asymptotes, **b.** horizontal asymptotes, **c.** oblique asymptotes, **d.** visible *x*-intercepts, or **e.** visible *y*-intercepts.

49.

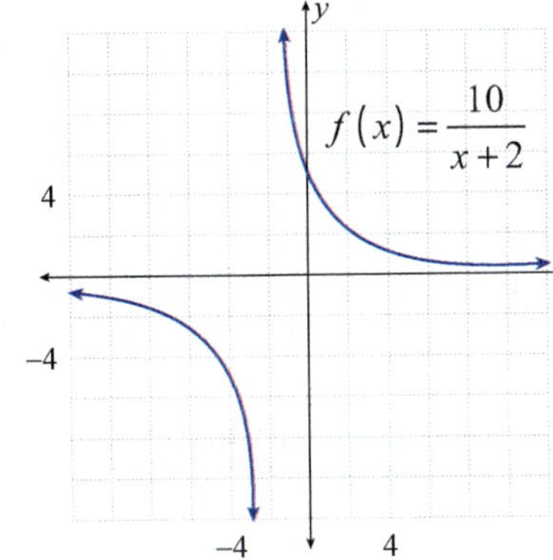

50.

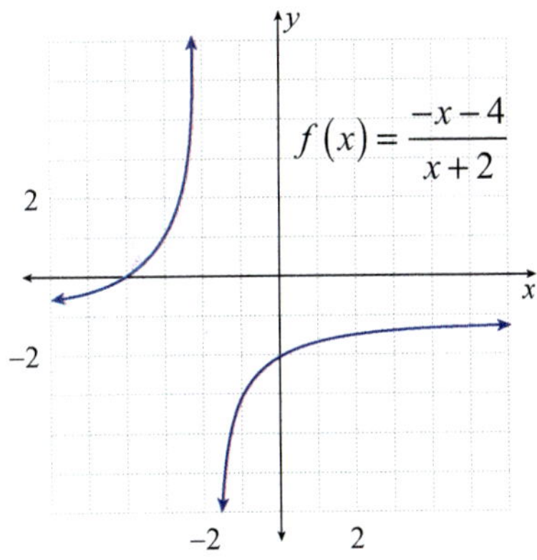

51.

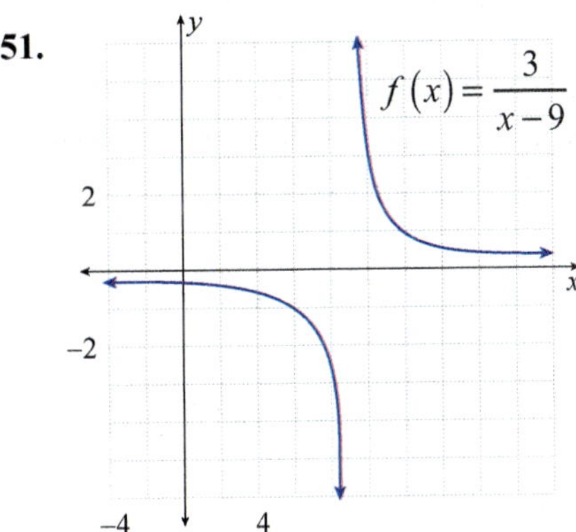

52.

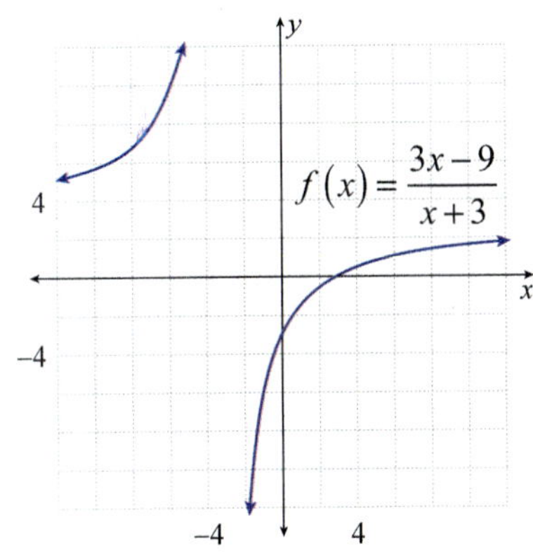

53.

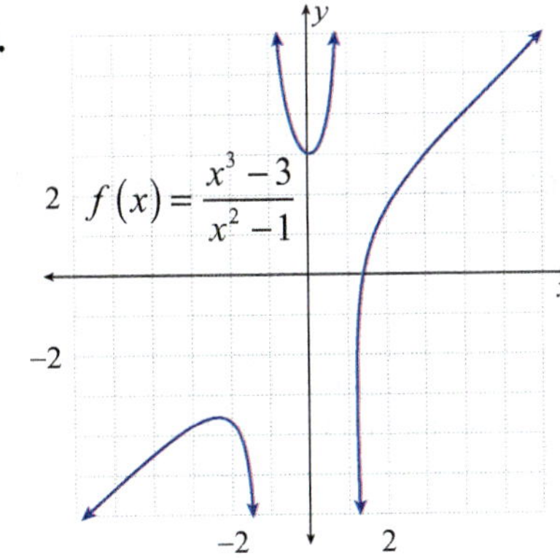

54.

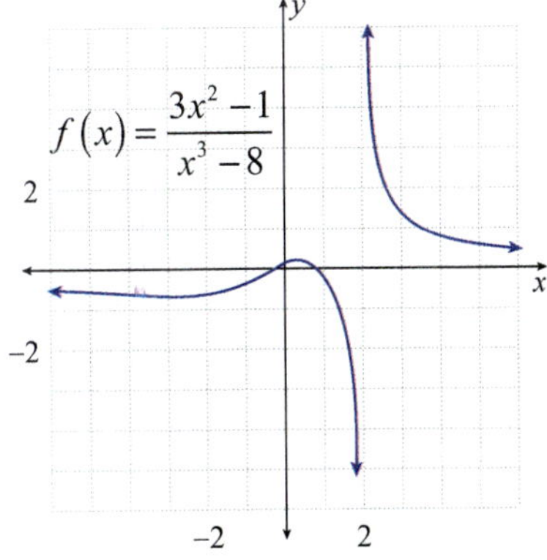

55.

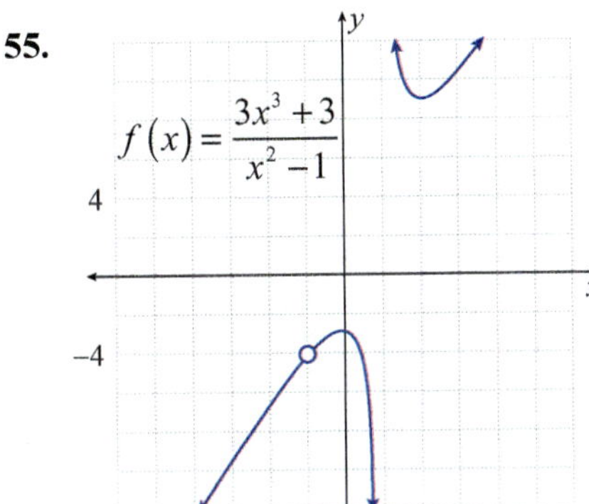

56.

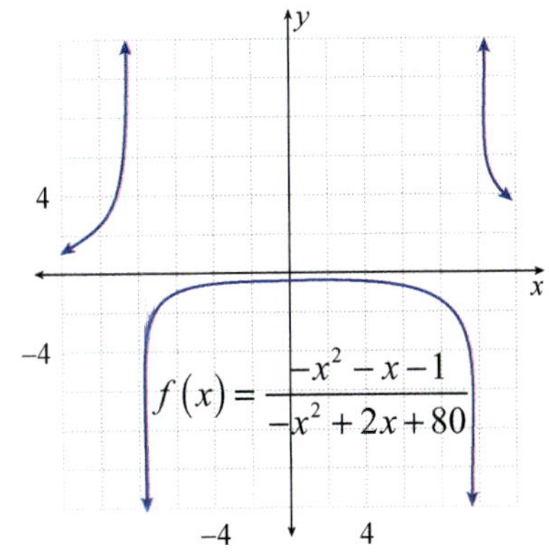

57. April raises a species of aquarium fish, and the total number of fish she has follows the formula

$$p(t) = \frac{200t}{t+1},$$

where $t \geq 0$ represents the number of months since she began.
a. Sketch the graph of $p(t)$ for $t \geq 0$.
b. What happens to April's fish population in the long run?

58. If an object is placed a distance x from a lens with a focal length of f, the image of the object will appear a distance y on the opposite side of the lens, where x, f, and y are related by the equation $\dfrac{1}{x} + \dfrac{1}{y} = \dfrac{1}{f}$.

a. Express y as a function of x and f.
b. Graph your function for a lens with a focal length of 30 mm $(f = 30)$. What happens to y as the distance x increases?

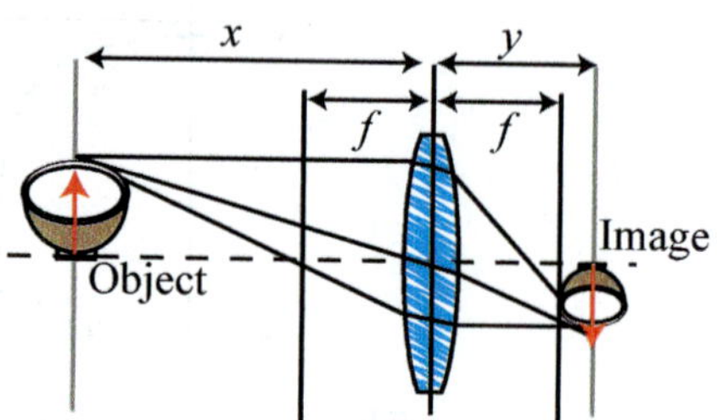

59. At t minutes after injection, the concentration (in mg/L) of a certain drug in the bloodstream of a patient is given by the formula

$$c(t) = \frac{20t}{t^2 + 1}.$$

a. Sketch the graph of $c(t)$ for $t \geq 0$.
b. What happens to the concentration of the drug in the long run?

3.9 RATIONAL INEQUALITIES

■ TOPICS

■ Solving Rational Inequalities

Solving Rational Inequalities

Now that we have discussed rational functions and equations, we can consider rational inequalities. We will see that solving rational inequalities involves similar work as solving polynomial inequalities.

Rational Inequalities

A **rational inequality** is any inequality that can be written in the form

$$f(x) < 0, \quad f(x) \leq 0, \quad f(x) > 0, \quad \text{or} \quad f(x) \geq 0,$$

where $f(x)$ is a rational function.

Just as with polynomial inequalities, solving rational inequalities is relatively simple if we have an accurate graph of the function. All we need to do is identify the intervals on which the function is positive and negative, then determine which intervals satisfy the inequality.

However, since graphing rational functions is even more difficult than graphing polynomial functions, we should depend on an algebraic method.

Recall that to solve a polynomial inequality, we made use of the fact that polynomial functions are continuous. This let us apply the *sign-test method,* in which testing a single point on each interval (between x-intercepts) described the sign behavior of the entire function.

Although rational functions are not always continuous, we know exactly where they are discontinuous: at their vertical asymptotes. This means that we can apply the sign-test method, as long as we account for a possible sign change at each vertical asymptote.

Solving Rational Inequalities: Sign-Test Method

To solve a rational inequality $f(x) < 0$, $f(x) \leq 0$, $f(x) > 0$, or $f(x) \geq 0$, where the rational function $f(x) = \dfrac{p(x)}{q(x)}$ is in reduced form, perform the following steps.

Step 1: Find the real zeros of the numerator $p(x)$. These values are the **zeros** of f.

Step 2: Find the real zeros of the denominator $q(x)$. These values are the locations of the **vertical asymptotes** of f.

Step 3: Place the values from Steps 1 and 2 on a number line, splitting it into intervals.

Step 4: Within each interval, select a **test point** and evaluate f at that number. If the result is positive, then $f(x) > 0$ for all x in the interval. If the result is negative, then $f(x) < 0$ for all x in the interval.

Step 5: Write the **solution set**, consisting of all of the intervals that satisfy the given inequality. If the inequality is not strict (uses $\leq$ or $\geq$), then the zeros of p are included in the solution set as well. The zeros of q are never included in the solution set (as they are not in the domain of f).

Example 1: Solving Rational Inequalities

Solve the rational inequality $\dfrac{x^2 + 1}{x^2 + 2x - 15} > 0$.

Solution

Begin by finding the zeros of the numerator and denominator. The rational expression can be factored as follows.

$$\frac{x^2 + 1}{x^2 + 2x - 15} = \frac{x^2 + 1}{(x+5)(x-3)}$$

Thus, we can see that the numerator has no zeros, while the denominator has zeros at $x = -5$ and $x = 3$.

Therefore, we place the values -5 and 3 on a number line.

This splits the number line into the intervals $(-\infty, -5)$, $(-5, 3)$, $(3, \infty)$. We then evaluate $f(x)$ for a test point in each interval.

Interval	Test Point	Evaluate	Result
$(-\infty, -5)$	$x = -6$	$f(-6) = \dfrac{(-6)^2 + 1}{(-6)^2 + 2(-6) - 15} = \dfrac{37}{9}$	$f(x) > 0$ on $(-\infty, -5)$ **Positive**
$(-5, 3)$	$x = 0$	$f(0) = \dfrac{(0)^2 + 1}{(0)^2 + 2(0) - 15} = -\dfrac{1}{15}$	$f(x) < 0$ on $(-5, 3)$ **Negative**
$(3, \infty)$	$x = 4$	$f(4) = \dfrac{(4)^2 + 1}{(4)^2 + 2(4) - 15} = \dfrac{17}{9}$	$f(x) > 0$ on $(3, \infty)$ **Positive**

Then, the solution to the inequality consists of those intervals where $f(x) > 0$, which is the set $(-\infty, -5) \cup (3, \infty)$.

Example 2: Solving Rational Inequalities

Solve the rational inequalities $\dfrac{x}{x+2} < 3$ and $\dfrac{x}{x+2} \le 3$.

Solution

Most of the work can be done for both inequalities at the same time.

We begin by subtracting 3 from both sides and then write the left-hand side as a single rational function.

$$\frac{x}{x+2} - 3 < 0 \quad \text{and} \quad \frac{x}{x+2} - 3 \le 0$$

$$\frac{x}{x+2} - \frac{3(x+2)}{x+2} < 0 \qquad \frac{x}{x+2} - \frac{3(x+2)}{x+2} \le 0$$

$$\frac{-2x - 6}{x+2} < 0 \qquad \frac{-2x - 6}{x+2} \le 0$$

Then factor -2 from the numerator of the fraction and divide both sides by -2 (reversing the inequality symbol) to obtain the simpler inequalities.

$$\frac{x+3}{x+2} > 0 \quad \text{and} \quad \frac{x+3}{x+2} \ge 0$$

Now that the inequalities are in standard form, we can follow the procedure for solving rational inequalities.

The zeros of the numerator and denominator are -3 and -2, respectively. Placing these values on a number line,

---

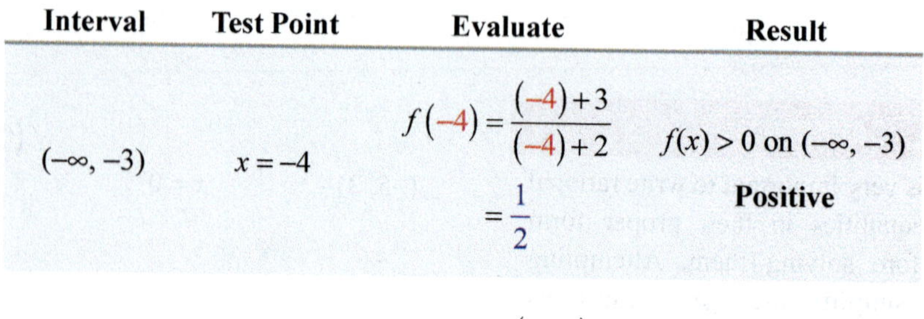

we have the intervals $(-\infty, -3)$, $(-3, -2)$, and $(-2, \infty)$.

Interval	Test Point	Evaluate	Result
$(-\infty, -3)$	$x = -4$	$f(-4) = \dfrac{(-4)+3}{(-4)+2} = \dfrac{1}{2}$	$f(x) > 0$ on $(-\infty, -3)$ **Positive**
$(-3, -2)$	$x = -2.5$	$f(-2.5) = \dfrac{(-2.5)+3}{(-2.5)+2} = -1$	$f(x) < 0$ on $(-3, -2)$ **Negative**
$(-2, \infty)$	$x = 0$	$f(0) = \dfrac{(0)+3}{(0)+2} = \dfrac{3}{2}$	$f(x) > 0$ on $(-2, \infty)$ **Positive**

The final step is to evaluate which intervals satisfy each inequality. The solution to the first inequality is the union of the two intervals where f is positive.

$$(-\infty, -3) \cup (-2, \infty)$$

For the second inequality, we have to decide which endpoints to include. We include $x = -3$, since this is a zero of the rational function, but we do not include $x = -2$, since the value is not in the domain of f. Thus, the solution to the second inequality is

$$(-\infty, -3] \cup (-2, \infty).$$

3.9 EXERCISES

💡 PRACTICE

Solve the following rational inequalities. See Examples 1 and 2.

1. $\dfrac{x+4}{2x} \geq 0$

2. $\dfrac{x}{x-4} \geq 0$

3. $\dfrac{x+6}{x^2} < 0$

4. $\dfrac{3x^2}{x+1} < 0$

5. $\dfrac{x+3}{x+9} > 0$

6. $\dfrac{2x+3}{x-4} < 0$

7. $\dfrac{3x-6}{2x-5} < 0$

8. $\dfrac{4-3x}{2x+4} \leq 0$

9. $\dfrac{x+5}{x-7} \geq 1$

10. $\dfrac{2x+3}{x-1} > 2$

11. $\dfrac{2x+5}{x-4} \leq -3$

12. $\dfrac{3x+2}{4x-1} < 3$

13. $\dfrac{5-2x}{3x+4} < -1$

14. $\dfrac{8-x}{x+5} < -4$

15. $\dfrac{x(x+4)}{x-3} \leq 0$

16. $\dfrac{(x+3)(x-2)}{x+1} > 0$

17. $\dfrac{x-5}{x(x+2)} \geq 0$

18. $\dfrac{-(x-3)^2}{(x-1)(x-4)} < 0$

19. $2x < \dfrac{4}{x+1}$

20. $\dfrac{5}{x-2} \geq \dfrac{3x}{x-2}$

21. $\dfrac{5}{x-2} > \dfrac{3}{x+2}$

22. $\dfrac{x}{x^2-x-6} \leq \dfrac{-1}{x^2-x-6}$

23. $\dfrac{x}{x^2-x-6} \leq \dfrac{-2}{x^2-x-6}$

24. $x > \dfrac{1}{x}$

25. $\dfrac{4}{x-3} \leq \dfrac{4}{x}$

26. $\dfrac{x-7}{x-3} \geq \dfrac{x}{x-1}$

27. $\dfrac{x}{x^2+3x+2} > \dfrac{1}{x^2+3x+2}$

28. $\dfrac{1}{x-4} \geq \dfrac{1}{x+1}$

29. $\dfrac{x}{x+1} \geq \dfrac{x+1}{x}$

30. $\dfrac{x}{x^2-2x-3} > \dfrac{3}{x^2-2x-3}$

📈 TECHNOLOGY

31. Use a graphing utility to graph the rational function $y = \dfrac{x^2+3x-4}{x}$.

   a. Use the graph to find the solution set for $y \geq 0$.

   b. Use the graph to find the solution set for $y < 0$.

   c. Explain the effect of $x = 0$ on the graph and why $x = 0$ is not included in either parts **a.** or **b.**

EXPONENTIAL AND LOGARITHMIC FUNCTIONS

4.1 EXPONENTIAL FUNCTIONS AND THEIR GRAPHS

■ TOPICS

- Definition and Classification of Exponential Functions
- Graphing Exponential Functions
- Solving Elementary Exponential Equations
- Graphing Exponential Functions Using Technology

Definition and Classification of Exponential Functions

We have studied many functions in which a variable is raised to a constant power, including polynomial functions such as $f(x) = x^3 + 2x^2 - 1$, radical functions like $g(x) = x^{\frac{1}{3}}$, and rational functions such as $h(x) = x^{-2}$.

An *exponential function* is a function in which a constant is raised to a variable power. Exponential functions are extremely important because of the large number of natural situations in which they arise. Examples include radioactive decay, population growth, compound interest, spread of epidemics, and rates of temperature change.

Exponential Functions

Let a be a fixed, positive real number not equal to 1. The **exponential function with base a** is the function

$$f(x) = a^x.$$

Why do we have the restrictions $a > 0$, $a \neq 1$? The base of the exponent can't be negative, since a^x would not be real for many values of x. For example, if $a = -1$ and $x = \dfrac{1}{2}$, a^x is not real.

If we let $a = 1$, we don't have a problem producing real numbers. Instead, the "exponential" function turns out to be constant. Recall that for all values of x, $1^x = 1$. For this reason, 1^x is considered a constant function, not an exponential function, and should always be written in its simplified form, 1.

Note that for any positive constant a, a^x is defined for all real numbers x. Consequently, the domain of $f(x) = a^x$ is the set of real numbers. What about the range of f? Since a is positive, we know that a^x must be positive. We will see that the range of all exponential functions is the set of all positive real numbers.

Recall that if a is any nonzero number, a^0 is defined to be 1. This means the y-intercept of any exponential function, regardless of the base, is the point $(0, 1)$. Beyond this, exponential functions fall into two classes, depending on whether a lies between 0 and 1 or if a is larger than 1.

Consider the following calculations for two sample exponential functions:

x	$f(x) = \left(\dfrac{1}{3}\right)^x$	$g(x) = 2^x$
-2	$f(-2) = 9$	$g(-2) = \dfrac{1}{4}$
-1	$f(-1) = 3$	$g(-1) = \dfrac{1}{2}$
0	$f(0) = 1$	$g(0) = 1$
1	$f(1) = \dfrac{1}{3}$	$g(1) = 2$
2	$f(2) = \dfrac{1}{9}$	$g(2) = 4$

Note that the values of f decrease as x increases while the values of g do just the opposite. We say that f is an example of a *decreasing* function and g is an example of an *increasing* function. If we plot these points and then fill in the gaps with a smooth curve, we get the graphs of f and g that appear in Figure 1.

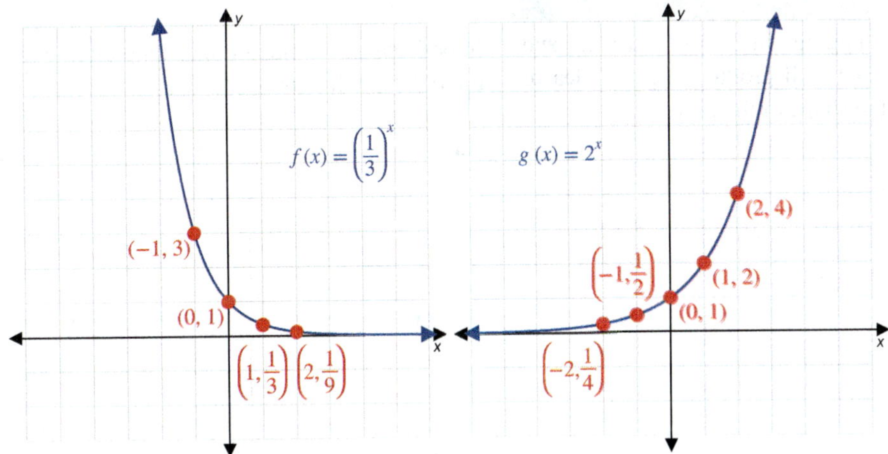

FIGURE 1: Two Exponential Functions

The graphs in Figure 1 suggest that the range of an exponential function is $(0, \infty)$, and that is indeed the case. Note that the base does not matter: the range of a^x is the positive real numbers for any allowable base (that is, for any positive a not equal to 1).

Behavior of Exponential Functions

Given a positive real number a not equal to 1, the function $f(x) = a^x$ is:

- a **decreasing function** if $0 < a < 1$, with $f(x) \to \infty$ as $x \to -\infty$ and $f(x) \to 0$ as $x \to \infty$,

- an **increasing function** if $a > 1$, with $f(x) \to 0$ as $x \to -\infty$ and $f(x) \to \infty$ as $x \to \infty$.

In either case, the point $(0, 1)$ lies on the graph of f, the domain of f is the set of real numbers, and the range of f is the set of positive real numbers.

Graphing Exponential Functions

Given that all exponential functions take one of two basic shapes (depending on whether a is less than 1 or greater than 1), they are relatively easy to graph. Plotting a few points, including the y-intercept of $(0, 1)$, will provide an accurate sketch.

Example 1: Graphing Exponential Functions

Sketch the graph of the following exponential functions.

a. $f(x) = 3^x$

b. $g(x) = \left(\dfrac{1}{2}\right)^x$

Solution

In both cases, we plot 3 points by plugging in $x = -1$, $x = 0$, and $x = 1$. We then connect the points with a smooth curve.

a.

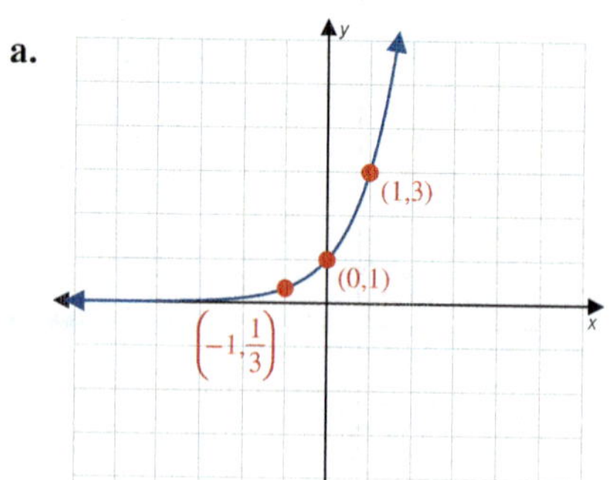

b.

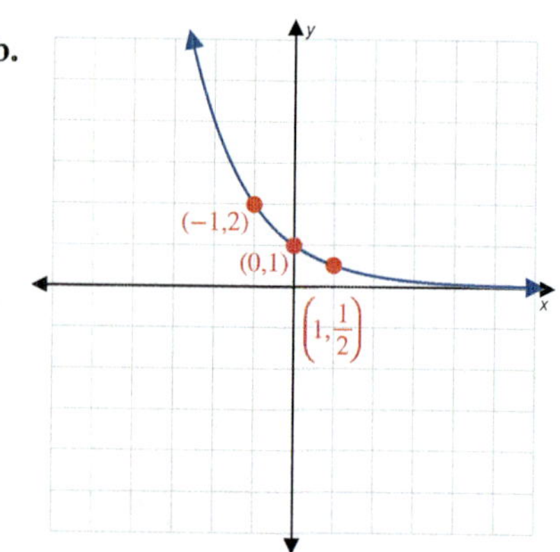

An exponential function, like any function, can be transformed in ways that result in the graph being shifted, reflected, stretched, or compressed. It often helps to graph the base function before trying to graph the transformed one.

📝 **NOTE**

Plugging in $x = -1$, $x = 0$, and $x = 1$ will produce a good idea of the shape of the graph.

Example 2: Graphing Exponential Functions

Sketch the graph of each of the following functions.

a. $f(x) = \left(\dfrac{1}{2}\right)^{x+3}$ **b.** $g(x) = -3^x + 1$ **c.** $h(x) = 2^{-x}$

Solution

a.

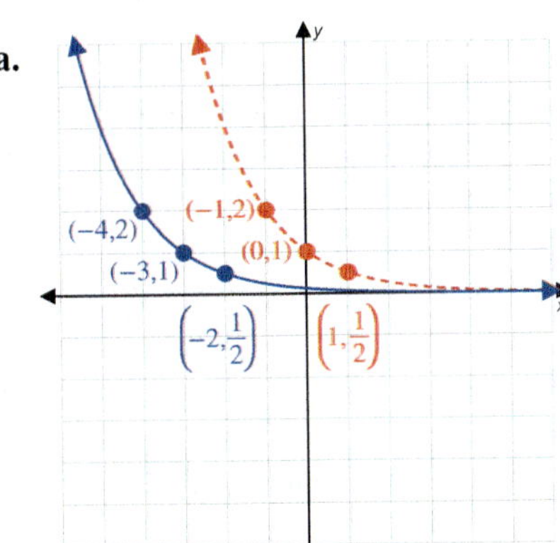

First, draw the graph of the function $\left(\dfrac{1}{2}\right)^{x}$, as in Example 1b.

Then, since x has been replaced by $x + 3$, shift the graph to the left by 3 units.

b.

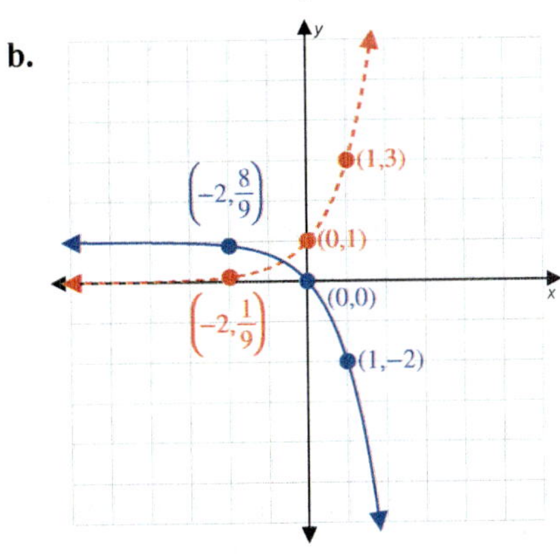

Begin with the graph of 3^x shown as a dashed curve.

The effect of multiplying a function by −1 is to reflect the graph with respect to the x-axis.

Following this, the second transformation of adding 1 to a function causes a vertical shift of the graph. The solid curve is the graph of $g(x) = -3^x + 1$.

c.

Begin by graphing the base function 2^x, shown as a dashed curve.

For h, x has been replaced by $-x$ so we reflect the graph of 2^x across the y-axis to obtain the graph of h, shown as the solid curve.

Note that this graph is also the graph of $\left(\dfrac{1}{2}\right)^{x}$.

Using properties of exponents is another way to think about this problem as

$$2^{-x} = \left(2^{-1}\right)^{x} = \left(\dfrac{1}{2}\right)^{x}.$$

Solving Elementary Exponential Equations

As you might expect, an equation in which the variable appears as an exponent is called an *exponential equation*. We are not yet ready to tackle exponential equations in full generality. Even something as simple as

$$2^x = 5$$

currently stumps us. We know that $2^2 = 4$ and $2^3 = 8$, so the answer must be between 2 and 3, but beyond this we don't have a method to proceed. This must wait until we have discussed a class of functions called *logarithms*.

However, we *are* ready to solve exponential equations that can be written in the form

$$a^x = a^b,$$

where a is an exponential base (positive and not equal to 1) and b is a constant.

You might guess, just from the form of the equation $a^x = a^b$ that the solution is $x = b$. This guess is correct, but we need to investigate why it is true.

The reason that the single value b is the solution of an exponential equation of the form $a^x = a^b$ is that the exponential function $f(x) = a^x$ is one-to-one (its graph passes the horizontal line test). Note that if g is a **one-to-one function**, then the only way for $g(x_1)$ to equal $g(x_2)$ is if $x_1 = x_2$.

In the case of the function $f(x) = a^x$, the equation $a^x = a^b$ is equivalent to the statement $f(x) = f(b)$, and this implies $x = b$, since f is one-to-one.

An exponential equation may not appear in the simple form $a^x = a^b$ initially. The following procedure describes the steps you may need to take to solve an elementary exponential equation.

Solving Elementary Exponential Equations

To solve an elementary exponential equation, complete the following steps.

Step 1: Isolate the exponential. Move the exponential containing x to one side of the equation and any constants or other variables in the expression to the other side. Simplify, if necessary.

Step 2: Find a base that can be used to rewrite both sides of the equation.

Step 3: Equate the powers, and solve the resulting equation.

Example 3: Solving Elementary Exponential Equations

Solve the following exponential equations.

a. $25^x - 125 = 0$ **b.** $8^{y-1} = \dfrac{1}{2}$ **c.** $\left(\dfrac{2}{3}\right)^x = \dfrac{9}{4}$

> **📝 NOTE**
>
> As always, it is good practice to check your solution in the original equation.

Solution

a. $25^x - 125 = 0$ Begin by isolating the term with the variable on one side.

$$25^x = 125$$

$$\left(5^2\right)^x = 5^3$$ We can write both sides with the same base since 25 and 125 are both powers of 5.

$$5^{2x} = 5^3$$ Simplify using properties of exponents.

$$2x = 3$$ We then equate the powers, resulting in a linear equation that we can easily solve.

$$x = \frac{3}{2}$$

b. $8^{y-1} = \dfrac{1}{2}$ Again, we need to rewrite both sides using the same base.

$$\left(2^3\right)^{y-1} = 2^{-1}$$ We see that 8 and $\dfrac{1}{2}$ can both be written as a power of 2.

$$2^{3y-3} = 2^{-1}$$ Simplify using properties of exponents.

$$3y - 3 = -1$$ Set the exponents equal to each other.

$$3y = 2$$ Solve the resulting linear equation.

$$y = \frac{2}{3}$$

c. $\left(\dfrac{2}{3}\right)^x = \dfrac{9}{4}$ Sometimes, the choice of base is not obvious.

$$\left(\frac{2}{3}\right)^x = \left(\frac{3}{2}\right)^2$$ Initially, we write the right-hand side as shown.

$$\left(\frac{2}{3}\right)^x = \left(\frac{2}{3}\right)^{-2}$$ Then, we can make the two bases equal by using properties of exponents.

$$x = -2$$ After making both bases the same, we equate the exponents to find the solution.

Graphing Exponential Functions Using Technology

> **⚠ CAUTION**
>
> Regardless of which method is used, be careful with your negative signs. We know that $-3 \cdot -3 = 9$ and not -9 because a negative times a negative is a positive. However, if we type -3^2 into a calculator, the output is -9. The number that we are squaring is -3, not 3, so we need to type $(-3)^2$ or $(-3)^{\wedge}2$ into the calculator to get the correct answer, 9.

To input exponents into a graphing calculator, we can use one of two methods. If the exponent is a 2, we can type the base and then press $\boxed{x^2}$. Otherwise, we need to use the caret symbol, $\boxed{\wedge}$. For example, to calculate 6^4, we would type the base, 6, then $\boxed{\wedge}$ followed by the exponent, 4.

To graph an exponential function, where the exponent is the variable, we use the technique that incorporates the caret symbol. Consider the graph of the function $f(x) = 3\left(\dfrac{1}{2}\right)^x$. To graph this in a calculator, press $\boxed{Y=}$ and type in the following:

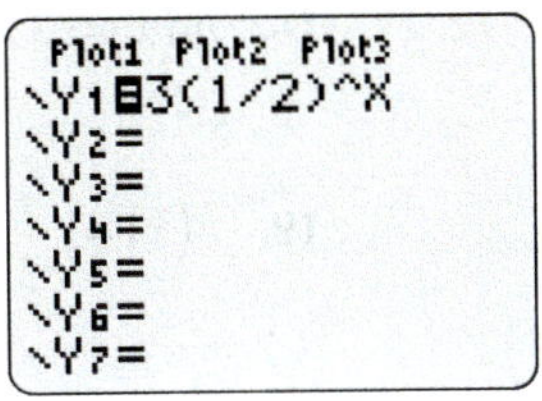

Notice that only the fraction $\dfrac{1}{2}$ is being raised to the exponent, so we put it in parentheses. The graph looks like

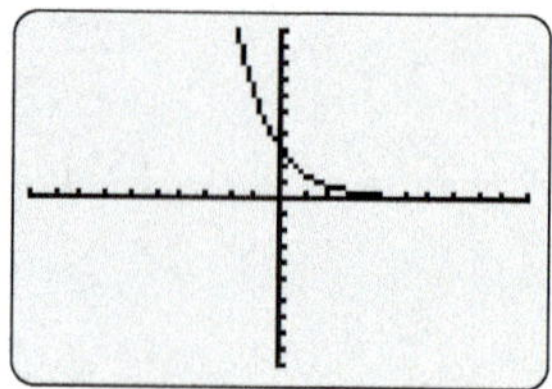

Later, we will learn about the irrational number, e, which is used often in exponential equations. To calculate or input a value such as $e^{0.25}$ into a graphing calculator, type **2nd** **LN**. Then, type in the exponent, 0.25, and close the parentheses by pressing **)**. Then press **ENTER**:

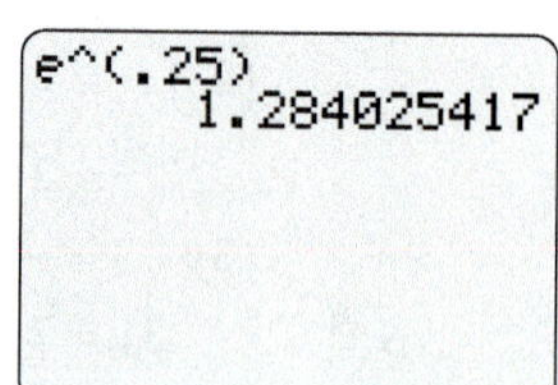

4.1 EXERCISES

💡 PRACTICE

Sketch the graphs of the following functions. State their domain and range. See Examples 1 and 2.

1. $f(x) = 4^x$

2. $g(x) = (0.5)^x$

3. $s(x) = 3^{x-2}$

4. $f(x) = \left(\dfrac{1}{3}\right)^{x+1}$

5. $r(x) = 5^{x-2} + 3$

6. $h(x) = 1 - 2^{x+1}$

7. $f(x) = 2^{-x}$

8. $r(x) = 3^{2-x}$

9. $g(x) = 3\left(2^{-x}\right)$

10. $h(x) = 2^{2x}$

11. $s(x) = (0.2)^{-x}$

12. $f(x) = \dfrac{1}{2^x} + 1$

13. $g(x) = 3 - 2^{-x}$

14. $r(x) = \dfrac{1}{2^{3-x}}$

15. $h(x) = \left(\dfrac{1}{2}\right)^{5-x}$

16. $m(x) = 3^{2x+1}$

17. $p(x) = 2 - 4^{2-x}$

18. $q(x) = 5^{3-2x}$

19. $r(x) = \left(\dfrac{9}{2}\right)^{-x}$

20. $p(x) = \left(\dfrac{1}{3}\right)^{2-x}$

21. $r(x) = 1 - \left(\dfrac{15}{4}\right)^x$

Solve the following exponential equations. See Example 3.

22. $5^x = 125$

23. $3^{2x-1} = 27$

24. $9^{2x-5} = 27^{x-2}$

25. $10^x = 0.01$

26. $4^{-x} = 16$

27. $2^x = \left(\dfrac{1}{2}\right)^{13}$

28. $2^{x+1} = 64^3$

29. $\left(\dfrac{2}{3}\right)^{x+3} = \left(\dfrac{9}{4}\right)^{-x}$

30. $\left(\dfrac{1}{5}\right)^{x-4} = 625^{\frac{1}{2}}$

31. $4^{3x+2} = \left(\dfrac{1}{4}\right)^{-2x}$

32. $5^x = 0.2$

33. $7^{x^2+3x} = \dfrac{1}{49}$

34. $3^{x^2+4x} = 81^{-1}$

35. $\left(\dfrac{1}{2}\right)^{x-3} = \left(\dfrac{1}{4}\right)^{x-5}$

36. $64^{x+\frac{7}{6}} = 2$

37. $6^{2x} = 36^{2x-3}$

38. $4^{2x-5} = 8^{\frac{x}{2}}$

39. $\left(\dfrac{2}{5}\right)^{2x+4} = \left(\dfrac{4}{25}\right)^{11}$

40. $4^{4x-7} = \dfrac{1}{64}$

41. $-10^x = -0.001$

42. $3^x = 27^{x+4}$

43. $1000^{-x} = 10^{x-8}$

44. $1^{3x-7} = 4^{2-x}$

45. $5^{3x-1} = 625^x$

46. $3^{2x-7} = 81^{\frac{x}{2}}$

Match the graphs of the following functions to the appropriate equation.

47. $f(x) = 2^{3x}$

48. $h(x) = 5^x - 1$

49. $g(x) = 2\left(4^{x-1}\right)$

50. $p(x) = 1 - 2^{-x}$

51. $f(x) = 6^{4-x}$

52. $r(x) = \dfrac{1}{3^x}$

53. $m(x) = -2 + 2^{-3x}$

54. $g(x) = \left(\dfrac{1}{4}\right)^{1+x}$

55. $h(x) = 3^{\frac{1}{2}x}$

56. $s(x) = 1^x - 4$

a.

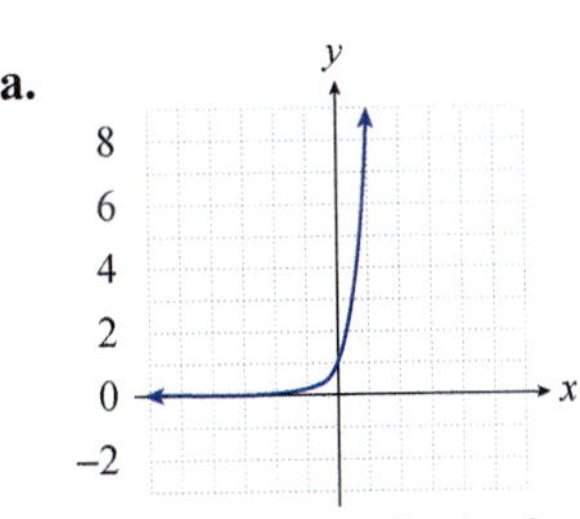

b.

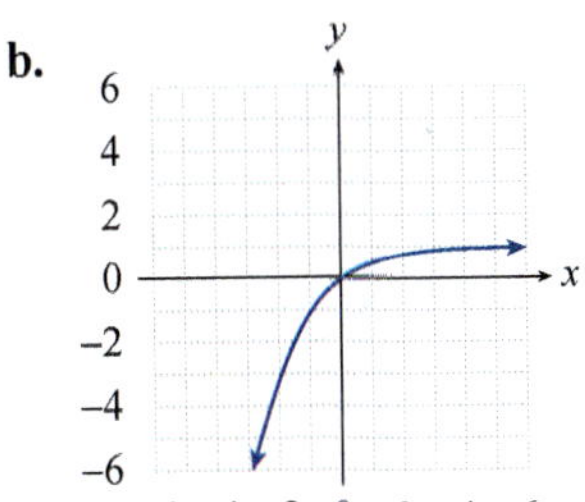

c.

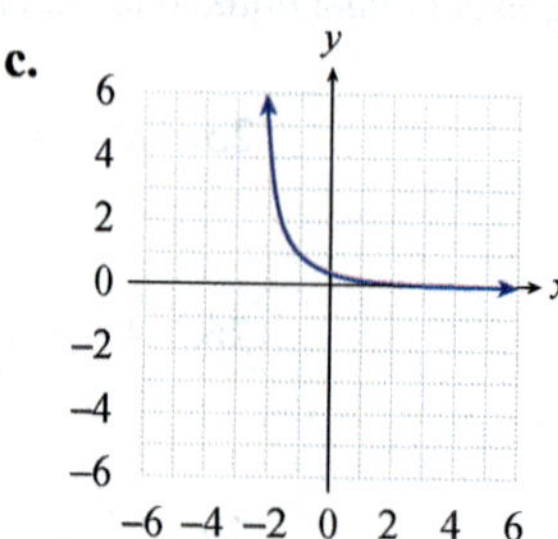

d.

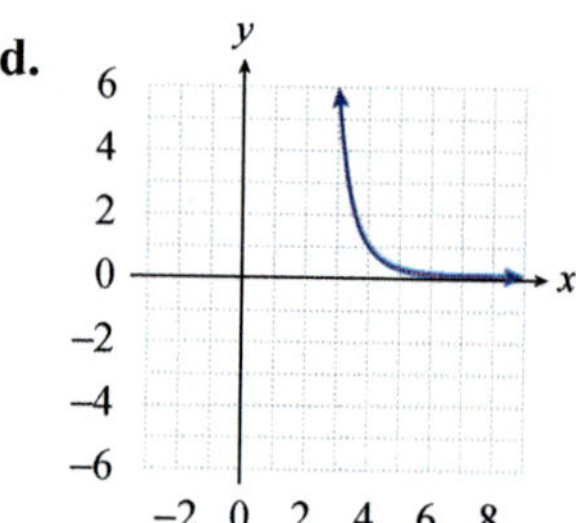

e.

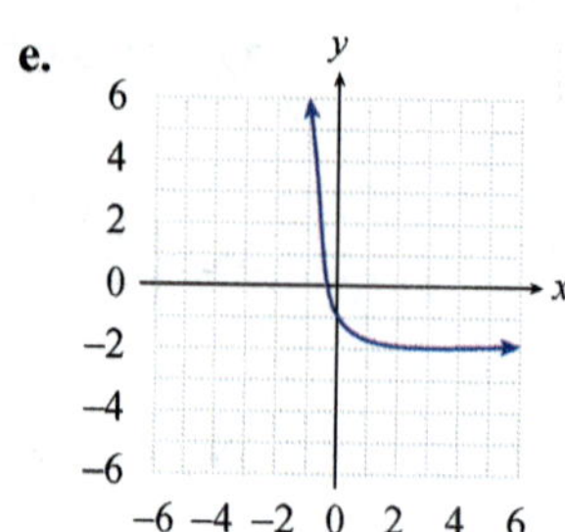

f.

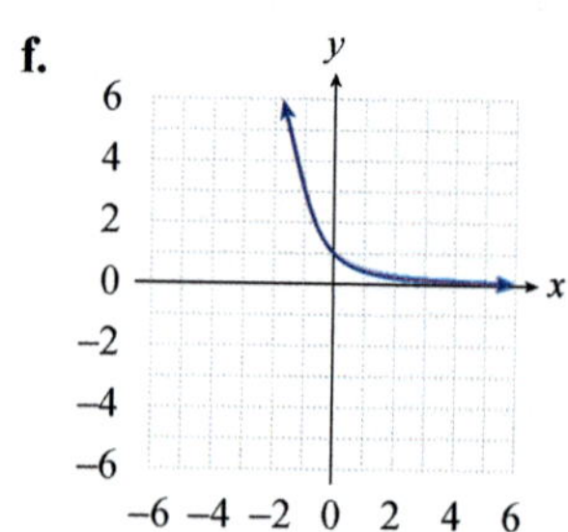

g.

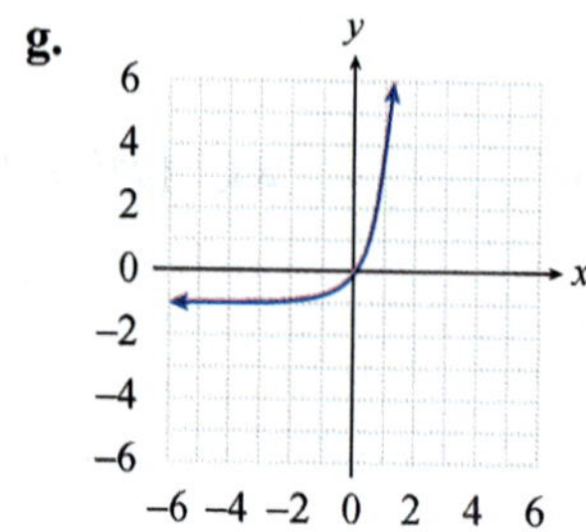

h.

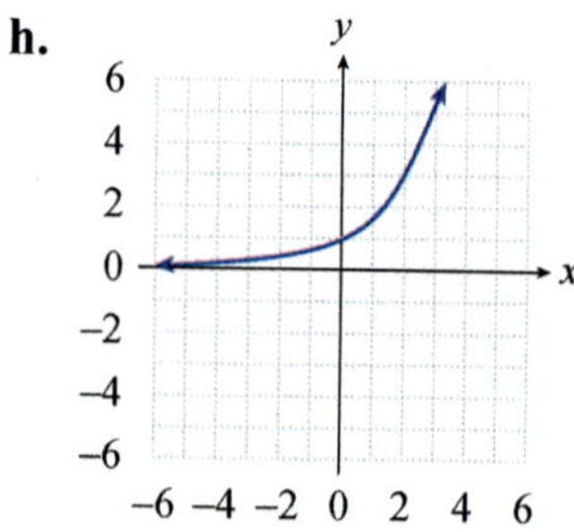

i.

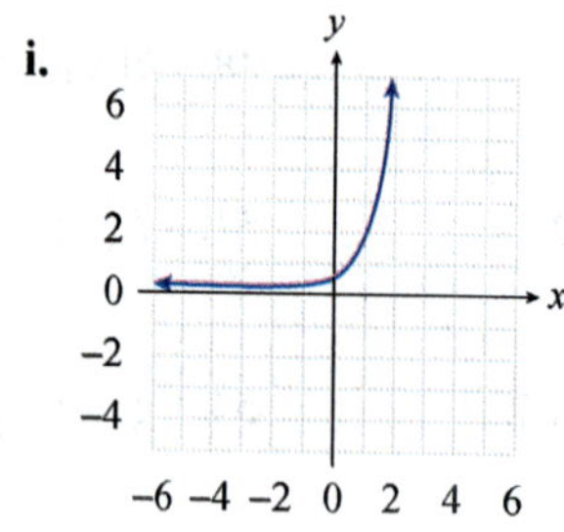

j.

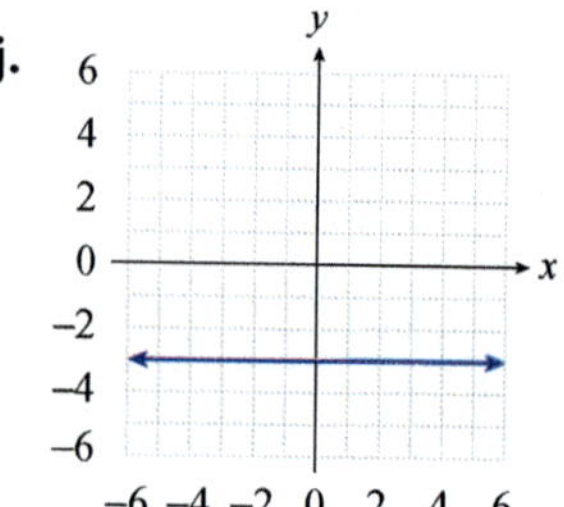

4.2 APPLICATIONS OF EXPONENTIAL FUNCTIONS

■ TOPICS

- Models of Population Growth
- Models of Radioactive Decay
- Compound Interest and the Number e

Models of Population Growth

Exponential functions arise naturally in a wide array of situations. In this section, we will study a few such situations in some detail, beginning with population models.

Many people working in such areas as mathematics, biology, and sociology study mathematical models of population. The models represent many types of populations, such as people in a given city or country, wolves in a wildlife habitat, or number of bacteria in a Petri dish. While such models can be quite complex, depending on factors like availability of food, space constraints, and effects of disease and predation, at their core, many population models assume that population growth displays exponential behavior.

The reason for this is that the growth of a population usually depends to a large extent on the number of members capable of producing more members. This assumes an abundant food supply and no constraints on population from lack of space, but at least initially this is often the case. In any situation where the rate of growth of a population is proportional to the size of the population, the population will grow exponentially, so we can write the function

$$P(t) = P_0 a^t.$$

This function tells us the size of a population $P(t)$ at time t. What do the other variables in the function represent?

Consider what happens if we substitute $t = 0$: $P(0) = P_0 a^0$, and since $a^0 = 1$ for all a, we have $P(0) = P_0$. Thus, P_0 is the *initial population* (population at time 0).

Recall from our work in graphing exponential equations that when $a > 1$, the larger the value of a is, the faster the exponential function grows. In models of population growth, a is greater than 1 (otherwise we would have population decay). For this reason, the value of a is the *growth rate* of the population.

Often, we will need to use information about a population to determine the values of P_0 and a. Once we have the function for population growth, we can answer other questions about the population.

Example 1 illustrates this with a specific model of bacterial population growth.

Example 1: Population Growth

A biologist is culturing bacteria in a large Petri dish. She begins with 1000 bacteria, and supplies sufficient food so that for the first five hours the bacteria population grows exponentially, doubling every hour.

a. Find a function that models the population growth of this bacteria culture.

b. Determine when the population reaches 16,000 bacteria.

c. Calculate the population two and a half hours after the scientist begins.

Solution

a. We know that we seek a function of the form $P(t) = P_0 a^t$. Since the scientist starts with 1000 bacteria, the initial population, $P_0 = 1000$.

To solve for a, we use the fact that the population doubles every hour.

$$2000 = P(1)$$
$$2000 = 1000a^1$$
$$2000 = 1000a$$
$$a = 2$$

Substitute, using the fact that the population after one hour equals 2000 bacteria.

Simplify, then solve for a.

Thus, a function that models the population growth of the bacteria culture is $P(t) = 1000(2)^t$, where t is measured in hours.

b. We wish to find the time t for which $P(t) = 16{,}000$.

$$16{,}000 = 1000(2)^t$$
$$16 = 2^t$$
$$2^4 = 2^t$$
$$t = 4$$

Substitute the desired population value.

Divide both sides by 1000.

Rewrite both sides with the same base, 2.

Equate the exponents and solve.

Thus, the bacteria culture reaches a population of 16,000 in 4 hours.

c. To calculate the population two and half hours after growth begins, we substitute $t = 2.5$ into the population model function.

$$P(2.5) = 1000(2)^{2.5}$$

While we could rewrite this using rational exponents, and try to find an exact value, frequently this will be very tedious or impossible, so we use a calculator to evaluate.

$$P(2.5) \approx 5657 \text{ bacteria}$$

Models of Radioactive Decay

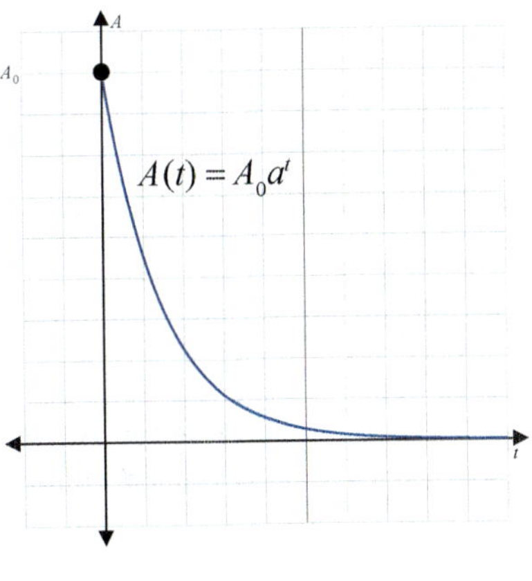

$$A(t) = A_0 a^t$$

FIGURE 1: Radioactive Decay

In contrast to populations (at least healthy populations), radioactive substances diminish with time. To be exact, the mass of a radioactive element decreases over time as the substance decays into other elements. Since exponential functions with a base between 0 and 1 are decreasing functions, this suggests that radioactive decay is modeled by

$$A(t) = A_0 a^t,$$

where $A(t)$ represents the amount of a given substance at time t, A_0 is the amount at time $t = 0$, and a is a number between 0 and 1.

The fact of radioactive decay is important (indirectly) in using radioactivity for power generation, and important again in working out the details of storing spent radioactive fuel rods. Radioactive decay also has other uses, one of which goes by the name of *radiocarbon dating*. This is a technique used by archaeologists, anthropologists, and others to estimate how long ago an organism died. The method depends on the fact that living organisms constantly absorb molecules of the radioactive substance carbon-14 while alive, but the intake of carbon-14 ceases once the organism dies. It is believed that the percentage of carbon-14 on Earth (relative to other isotopes of carbon) has been relatively constant over time, so the first step in the method is to determine the percentage of carbon-14 in the remains of a given organism. By comparing this (smaller) percentage to the percentage found in living tissue, an estimate of when the organism died can then be made.

The mathematics of the age estimation depends on the fact that half of a given mass of carbon-14 decays over a period of 5730 years. This is known as the *half-life* of carbon-14; every radioactive substance has a half-life, and the half-life is usually an important feature when working with such substances. In the case of carbon-14, the half-life of 5730 years means that if, for instance, an organism contained 12 grams of carbon-14 at death, it would contain 6 grams after 5730 years, 3 grams after another 5730 years, and so on. (In exponential growth, there is the related concept of *doubling time*, which is the length of time needed for the function to double in value. The doubling time for the bacteria population in Example 1 is 1 hour.)

Example 2: Radioactive Decay

Determine the base a so that the function $A(t) = A_0 a^t$ accurately describes the decay of carbon-14 as a function of t years.

Solution

Note that $A(0) = A_0 a^0 = A_0$, so A_0 represents the amount of carbon-14 at time $t = 0$. Since we are seeking a general formula, we don't know what A_0 is specifically; that is, the value of A_0 will vary depending on the details of the situation. But we can still determine the base constant a.

What we know is that half of the original amount of carbon-14 decays over a period of 5730 years, so $A(5730)$ will be half of A_0. This gives us the following equation.

$$A_0\left(a^{5730}\right) = \frac{A_0}{2}$$

To solve this for a, we can first divide both sides by A_0; the fact that A_0 then cancels from the equation just emphasizes that its exact value is irrelevant for the task at hand. Now we have the following equation.

$$a^{5730} = \frac{1}{2}$$

At this point a calculator is called for, as we need to take the 5730th root of both sides. This gives us the value for a that we seek.

$$a = \left(\frac{1}{2}\right)^{\frac{1}{5730}} \approx 0.999879$$

The function is thus $A(t) = A_0(0.999879)^t$ (using our approximate value for a). We can now verify that the function behaves as expected by evaluating A at various multiples of the half-life for carbon-14, as shown.

$$A(1 \cdot 5730) = A_0(0.999879)^{5730} \approx 0.5A_0$$

$$A(2 \cdot 5730) = A_0(0.999879)^{11,460} \approx 0.25A_0$$

$$A(3 \cdot 5730) = A_0(0.999879)^{17,190} \approx 0.125A_0$$

Compound Interest and the Number e

One of the most commonly encountered applications of exponential functions is in compounding interest. We run into compound interest when earning money (by interest on a savings account or investment) and when spending money (on car loans, mortgages, and credit cards).

The basic compound interest formula can be understood by considering what happens when money is invested in a savings account. Typically, a savings account is set up to pay interest at an annual rate of r (which we will write in decimal form) compounded n times per year. For instance, a bank may offer an annual interest rate of 5% (meaning $r = 0.05$) compounded monthly (so $n = 12$).

Compounding is the act of calculating the interest earned on an investment and adding that amount to the investment. An investment in a monthly compounded account will have interest added to it twelve times over the course of a year, once each month.

Suppose an amount of P (for principal) dollars is invested in a savings account at an annual rate of r compounded n times per year. We want a formula for the amount of money $A(t)$ in the account after t years.

If we say that a period is the length of time between compoundings, interest is calculated at the rate of $\dfrac{r}{n}$ per period (for instance, if $r = 0.05$ and $n = 12$, interest is earned at a rate of $\dfrac{0.05}{12} \approx 0.00417$ per month). Table 1 illustrates how compounding increases the amount in the account over the course of several periods.

Period	Amount
0	$A = P$
1	$A = P\left(1 + \dfrac{r}{n}\right)$
2	$A = P\left(1 + \dfrac{r}{n}\right)\left(1 + \dfrac{r}{n}\right) = P\left(1 + \dfrac{r}{n}\right)^2$
3	$A = P\left(1 + \dfrac{r}{n}\right)^2\left(1 + \dfrac{r}{n}\right) = P\left(1 + \dfrac{r}{n}\right)^3$
k	$A = P\left(1 + \dfrac{r}{n}\right)^{k-1}\left(1 + \dfrac{r}{n}\right) = P\left(1 + \dfrac{r}{n}\right)^k$

TABLE 1: Effect of Compounding Interest on an Investment of P Dollars

Since there are nt compounding periods in t years, we obtain the following formula.

Compound Interest Formula

An investment of P dollars, compounded n times per year at an annual interest rate of r, has a value after t years of

$$A(t) = P\left(1 + \frac{r}{n}\right)^{nt}.$$

Example 3: Compound Interest Formula

Sandy invests \$10,000 in a savings account earning 4.5% annual interest compounded quarterly. What is the value of her investment after three and a half years?

Solution

We know that $P = 10{,}000$, $r = 0.045$ (remember to express the interest rate in decimal form), $n = 4$ (since the account is compounded four times per year), and $t = 3.5$. Now we substitute and evaluate.

$$A(3.5) = 10{,}000\left(1 + \frac{0.045}{4}\right)^{(4)(3.5)}$$

$$= 10{,}000\left(1.01125\right)^{14}$$

$$\approx \$11{,}695.52$$

Thus, after three and a half years, Sandy's investment grows to \$11,695.52.

The compound interest formula can also be used to determine the interest rate of an existing savings account, as shown in Example 4.

Example 4: Compound Interest Formula

Nine months after depositing $520.00 in a monthly compounded savings account, Frank checks his balance and finds the account has $528.84. Being the forgetful type, he can't remember what the annual interest rate for his account is, and sees the bank is advertising a rate of 2.5% for new accounts. Should he close out his existing account and open a new one?

Solution

As in Example 3, we will begin by identifying the known quantities in the compound interest formula: $P = 520$, $n = 12$ (12 compoundings per year), and $t = 0.75$ (nine months is three-quarters of a year).

Further, the amount in the account, A, at this time is $528.84. This gives us the equation

$$528.84 = 520\left(1 + \frac{r}{12}\right)^{(12)(0.75)}$$

to solve for r, the annual interest rate.

$$528.84 = 520\left(1 + \frac{r}{12}\right)^{9} \qquad \text{Simplify the exponent.}$$

$$1.017 = \left(1 + \frac{r}{12}\right)^{9} \qquad \text{Divide both sides by 520.}$$

$$1.001875 \approx 1 + \frac{r}{12} \qquad \text{Take the ninth root of both sides.}$$

$$0.001875 \approx \frac{r}{12} \qquad \text{Simplify to solve for } r.$$

$$0.0225 \approx r$$

Thus Frank's current savings account is paying an annual interest rate of 2.25%, so he would gain a slight advantage by switching to a new account.

Even though the interest rate is divided by n, the number of periods per year, increasing the frequency of compounding always increases the total interest earned on the investment. This is because interest is always calculated on the current balance. For example, consider how much interest you earn in one year if you invest $1000 in an account with a 5% annual interest rate. If the interest is compounded once per year, you earn $50 in interest. However, if the interest is compounded *twice* per year, you earn 2.5%, or $25, for the first period, then 2.5% *of the new balance* of $1025 in the second period. This brings the total interest earned to $50.63! Table 2 shows the interest earned in one year on an investment of $1000 in an account with a 5% annual interest rate, compounded n times per year.

n	Calculation	Value after 1 Year
1 (annually)	$A = 1000(1 + 0.05)$	$1050.00
2 (biannually)	$A = 1000\left(1 + \dfrac{0.05}{2}\right)^2$	$1050.63
4 (quarterly)	$A = 1000\left(1 + \dfrac{0.05}{4}\right)^4$	$1050.95
12 (monthly)	$A = 1000\left(1 + \dfrac{0.05}{12}\right)^{12}$	$1051.16
52 (weekly)	$A = 1000\left(1 + \dfrac{0.05}{52}\right)^{52}$	$1051.25
365 (daily)	$A = 1000\left(1 + \dfrac{0.05}{365}\right)^{365}$	$1051.27
8760 (hourly)	$A = 1000\left(1 + \dfrac{0.05}{8760}\right)^{8760}$	$1051.27

TABLE 2: Value of an Investment with Interest Compounded n Times per Year

Looking at Table 2, we see that the amount of interest keeps growing as we divide the year into more compounding periods, but that this growth slows dramatically. It looks as if there is a *limit* to how much interest we can earn, and this is in fact the case.

Let's examine the formula as $n \to \infty$. In order to do this we need to perform some algebraic manipulation of the formula.

$$A(t) = P\left(1 + \frac{r}{n}\right)^{nt}$$

$$= P\left(1 + \frac{1}{m}\right)^{rmt} \qquad \text{Substitute } m = \frac{n}{r} \text{, so } \frac{1}{m} = \frac{r}{n}.$$

$$= P\left(\left(1 + \frac{1}{m}\right)^{m}\right)^{rt} \qquad \text{Bring together the instances of the variable } m \text{ using the properties of exponents.}$$

Although the manipulation may appear strange, it has accomplished the important task of isolating the part of the formula that changes as $n \to \infty$. Looking at the change of variables $m = \dfrac{n}{r}$, we see that letting $n \to \infty$ means that $m \to \infty$ as well. Since every other quantity remains fixed, we only have to understand what happens to

$$\left(1 + \frac{1}{m}\right)^{m}$$

as m grows without bound.

This is not a trivial undertaking. We might think that letting m get larger and larger would make the expression grow larger and larger, as m is the exponent in the expression and the base is larger than 1. But at the same time, the base approaches 1 as m increases without bound, and 1 raised to any power is simply 1. It turns out that these two effects balance one another out, as we can see in Table 3.

m	$\left(1+\dfrac{1}{m}\right)^{m}$
10	2.59374
100	2.70481
1000	2.71692
10,000	2.71815
100,000	2.71827

TABLE 3: Values of $\left(1+\dfrac{1}{m}\right)^{m}$ as m Increases

As m gets larger and larger, we find that the expression approaches a fixed number.

$$\left(1+\frac{1}{m}\right)^{m} \rightarrow 2.718281828459\ldots$$

This value is a very important number in mathematics, so important that it gets its own symbol, the letter e (which highlights its connection to exponential functions).

The Number e

The number e is defined as the value of $\left(1+\dfrac{1}{m}\right)^{m}$ as $m \rightarrow \infty$.

$$e \approx 2.71828182846$$

In most applications, the approximation $e \approx 2.7183$ is sufficiently accurate.

Let's look back at the compound interest formula from before.

$$A(t) = P\left(1+\frac{r}{n}\right)^{nt} = P\left(\left(1+\frac{1}{m}\right)^{m}\right)^{rt}$$

This means that if P dollars is invested in an account that is **compounded continuously**, that is, with $n \rightarrow \infty$, then the amount in the account is determined by the following formula, which results from substituting the definition of e.

Continuous Compounding Formula

An investment of P dollars, compounded continuously at an annual interest rate of r, has a value after t years of

$$A(t) = Pe^{rt}.$$

Example 5: Continuous Compounding Formula

If Sandy (last seen in Example 3) has the option of investing her $10,000 in a continuously compounded account earning 4.5% annual interest, what will be the value of her account in three and a half years?

Solution

Again, the solution boils down to substituting the correct values and evaluating the result. Here, $P = 10,000$, $r = 0.045$, and $t = 3.5$.

$$A(3.5) = 10,000e^{(0.045)(3.5)}$$
$$= 10,000e^{0.1575}$$
$$\approx \$11,705.81$$

This account earns $10.29 more than the quarterly compounded account in Example 3.

All exponential functions can be expressed with the base e (or any other base, for that matter). The base e is so commonly used for exponential functions that it is often called the *natural base*. For instance, the formula for the radioactive decay of carbon-14, using the base e, is

$$A(t) = A_0 e^{-0.000121} t.$$

You should verify that this version of the decay formula does indeed give the same values for $A(t)$ as the version derived in Example 2.

4.2 EXERCISES

APPLICATIONS

1. A new virus has broken out in isolated parts of Africa and is spreading exponentially through tribal villages. The growth of this new virus can be mapped using the following formula where V stands for the number of people in the village who are infected with the virus, P stands for the number of people in a village and d stands for the number of days since the virus first appeared. According to this equation, how many people in a village of 300 will be infected after 5 days?

$$V = P\left(1 - e^{-0.18d}\right)$$

2. A prototype for an electric motorcycle uses a battery whose energy capacity $C(d)$, in kilowatt-hours (kWh), is given by the formula $C(d) = 12e^{-0.02d}$, where d represents the number of days since receiving a full charge. What is the battery's energy capacity 30 days after being fully charged?

3. A young economics student has come across a very profitable investment scheme in which his money will accrue interest according to the equation listed below, where C represents the investment value after m months for an initial investment of I dollars. If this student invests $1250 into this lucrative endeavor, how much money will he have after 24 months? I represents the investment and m represents the number of months the money has been invested for.

$$C = Ie^{0.08m}$$

4. A family releases a couple of pet rabbits into the wild. Upon being released the rabbits begin to reproduce at an exponential rate, as shown in the formula below. After 2 years how large is the rabbit population P, where n stands for the initial rabbit population (2) and m stands for the number of months?

$$P = ne^{0.5m}$$

5. Inside a business network, an email worm was downloaded by an employee. This worm goes through the infected computer's address book and sends itself to all the listed email addresses. This worm very rapidly works its way through the network following the equation below, where C is the number of computers in the network and W is the number of computers infected h hours after the worm is initially downloaded. After only 8 hours, how many computers has the worm infected if there are 150 computers in the network?

$$W = C\left(1 - e^{-0.12h}\right)$$

6. A construction crew has been assigned to build an apartment complex. The work of the crew can be modeled using the exponential formula below, where A is the total number of apartments to be built, w is the number of weeks, and F is the number of finished apartments. Out of a total of 100 apartments, how many apartments have been finished after 4 weeks of work?

$$F = A\left(1 - e^{-0.1w}\right)$$

7. The half-life of radium is approximately 1600 years.
   a. Determine a so that $A(t) = A_0 a^t$ describes the amount of radium left after t years, where A_0 is the amount at time $t = 0$.
   b. How much of a 1-gram sample of radium would remain after 100 years?
   c. How much of a 1-gram sample of radium would remain after 1000 years?

8. The radioactive element polonium-210 has a relatively short half-life of 138 days, and one way to model the amount of polonium-210 remaining after t days is with the function $A(t) = A_0 e^{-0.005023t}$, where A_0 is the mass at time $t = 0$ (note that $A(138) = \dfrac{A_0}{2}$.) What percentage of the original mass of a sample of polonium-210 remains after 365 days?

9. A certain species of fish is to be introduced into a new man-made lake, and wildlife experts estimate the population will grow according to $P(t) = (1000)2^{\frac{t}{3}}$, where t represents the number of years from the time of introduction.
   a. What is the doubling time for this population of fish?
   b. How long will it take for the population to reach 8000 fish, according to this model?

10. The population of a certain inner-city area is estimated to be declining according to the model $P(t) = 237,000e^{-0.018t}$, where t is the number of years from the present. What does this model predict the population will be in ten years?

11. In an effort to control vegetation overgrowth, 100 rabbits are released in an isolated area that is free of predators. After one year, it is estimated that the rabbit population has increased to 500. Assuming exponential population growth, what will the population be after another six months?

12. Assuming a current world population of 7.75 billion people, an annual growth rate of 1.9% per year, and a worst-case scenario of exponential growth, what will the world population be in **a.** 10 years? **b.** 50 years?

13. Madiha has \$3500 that she wants to invest in a simple savings account for two and a half years, at which time she plans to close out the account and use the money as a down payment on a car. She finds one local bank offering an annual interest rate of 2.75% compounded monthly, and another bank offering an annual interest rate of 2.7% compounded daily (365 times per year). Which bank should she choose?

14. Madiha, from the last problem, does some more searching and finds an online bank offering an annual rate of 2.75% compounded continuously. How much more money will she earn over two and a half years if she chooses this bank rather than the local bank offering the same rate compounded monthly?

15. Tom hopes to earn \$1000 in interest in three years time from \$10,000 that he has available to invest. To decide if it's feasible to do this by investing in a simple monthly compounded savings account, he needs to determine the annual interest rate such an account would have to offer for him to meet his goal. What would the annual rate of interest have to be?

16. An investment firm claims that its clients usually double their principal in five years time. What annual rate of interest would a savings account, compounded monthly, have to offer in order to match this claim?

17. The function $C(t) = C_0(1 + r)^t$ models the rise in the cost of a product that has a cost of C_0 today, subject to an average yearly inflation rate of r for t years. If the average annual rate of inflation over the next decade is assumed to be 3%, what will the inflation-adjusted cost of a \$100,000 house be in 10 years? Round your answer to the nearest dollar.

18. Given the inflation model $C(t) = C_0(1 + r)^t$ (see Exercise 17), and given that a loaf of bread that currently sells for \$3.60 sold for \$3.10 six years ago, what has the average annual rate of inflation been for the past six years?

19. The function $N(t) = \dfrac{10{,}000}{1 + 999e^{-t}}$ models the number of people in a small town who have caught the flu t weeks after the initial outbreak.

  a. How many people were ill initially?
  b. How many people have caught the flu after eight weeks?
  c. Determine what happens to the function $N(t)$ as $t \to \infty$.

20. The concentration $C(t)$, in milligrams per liter, of a certain drug in the bloodstream after t minutes is given by the formula $C(t) = 0.05\left(1 - e^{-0.2t}\right)$. What is the concentration after 10 minutes?

21. Carbon-11 has a radioactive half-life of approximately 20 minutes, and is useful as a diagnostic tool in certain medical applications. Because of the relatively short half-life, time is a crucial factor when conducting experiments with this element.

  a. Determine a so that $A(t) = A_0 a^t$ describes the amount of carbon-11 left after t minutes, where A_0 is the amount at time $t = 0$.
  b. How much of a 2 kg sample of carbon-11 would be left after 30 minutes?
  c. How many milligrams of a 2 kg sample of carbon-11 would be left after six hours?

22. Charles has recently inherited \$8000 that he wants to deposit into a savings account. He has determined that his two best bets are an account that compounds annually at a rate of 3.20% and an account that compounds continuously at an annual rate of 3.15%. Which account would pay Charles more interest?

23. Marshall invests $1250 in a mutual fund which boasts a 5.7% annual return compounded semiannually (twice a year). After three and a half years, Marshall decides to withdraw his money.

    a. How much is in his account?

    b. How much has he made in interest from his investment?

24. Adam is working in a lab testing bacteria populations. After starting out with a population of 375 bacteria, he observes the change in population and notices that the population doubles every 27 minutes.

    a. Find the equation for the population P in terms of time t in minutes, rounding a to the nearest thousandth.

    b. Find the population after two hours.

25. Your credit union offers a special interest rate of 10% compounded monthly for the first year for a student savings account opened in August if the student deposits $5000 or more. You received a total of $9000 for graduation, and you decide to deposit all of it in this special account. Assuming you open your account in August and make no withdrawals for the first year, how much money will you have in your account at the end of February (after six months)? How much will you have at the end of the following July (after one full year)?

26. You have a savings account of $3000 with an interest rate of 6.8%.

    a. How much interest would be earned in two years if the interest is compounded annually?

    b. How much interest would be earned in two years if the interest is compounded semiannually?

    c. In which case do you make more money on interest? Explain why this is so.

27. If $2500 is invested in a continuously compounded certificate of deposit with an annual interest rate of 4.2%, what would be the account balance at the end of three years?

28. The new furniture store in town boasts a special in which you can buy any set of furniture in their store and make no monthly payments for the first year. However, the fine print says that the interest rate of 7.25% is compounded quarterly beginning when you buy the furniture. You are considering buying a set of living room furniture for $4000 but know you cannot save up more than $4500 in one year's time. Can you fully pay off your furniture on the one year anniversary of having bought the furniture? If so, how much money will you have left over? If not, how much more money will you need?

29. When Nicole was born, her grandmother was so excited about her birth that she opened a certificate of deposit in Nicole's honor to help send her to college. Now at age 18, Nicole's account has $81,262.93. How much did her grandmother originally invest if the interest rate has been 8.1% compounded annually?

30. Inflation is a relative measure of your purchasing power over time. The formula for inflation is the same as the compound interest formula, but with $n = 1$. Given the current values below, what will the values of the following items be 10 years from now if inflation is at 6.4%?

    a. an SUV: $38,000

    b. a loaf of bread: $1.79

    c. a gallon of milk: $3.40

    d. your salary: $34,000

31. Depreciation is the decrease of an item's value and can be determined using a formula similar to that for compound interest: $V = P(1-r)^t$, where V is the new value. If the particular car you buy upon graduation from college costs $17,500 and depreciates at a rate of 16% per year, what will the value of the car be in 5 years when you pay it off?

32. Assume the interest on your credit card is compounded continuously with an APR (annual percentage rate) of 19.8%. If you put your first term bill of $3984 on your credit card, but do not have to make payments until you graduate (4 years later), how much will you owe when you start making payments?

33. Suppose you deposit $5000 in an account for five years at an annual interest rate of 8.5%.

   a. What would be the ending account balance if the interest is continuously compounded?

   b. What would be the ending account balance if the interest is compounded daily?

   c. Are these two answers similar? Why or why not?

⤳ TECHNOLOGY

Use a graphing utility to sketch the graphs of the following functions.

34. $m(x) = 1 - 3e^x$

35. $p(x) = e^{4x} - 2$

36. $b(x) = \dfrac{1}{e^{x-2}}$

37. $m(x) = e^{2x^2 - 3x + 1}$

38. $g(x) = e^{x+3} - 3$

39. $m(x) = 6e^{2x} - 2$

4.3 LOGARITHMIC FUNCTIONS AND THEIR GRAPHS

■ TOPICS

- ■ Definition of Logarithmic Functions
- ■ Graphing Logarithmic Functions
- ■ Evaluating Elementary Logarithmic Expressions
- ■ Solving Elementary Logarithmic Equations
- ■ Common and Natural Logarithms
- ■ Graphing Logarithmic Functions Using Technology

Definition of Logarithmic Functions

Currently, we are only able to solve a small subset of possible exponential equations.

Solvable	Not Easily Solvable Yet
We can solve the equation $$2^x = 8$$ by writing 8 as 2^3 and equating exponents. This is an example of an elementary exponential equation we learned to solve in an earlier section.	We cannot solve the equation $$2^x = 9$$ in the same way, although this equation is only slightly different. All we can say at the moment is that x must be a bit larger than 3.
If we know A, P, n, and t, we can solve the compound interest equation $$A = P\left(1 + \frac{r}{n}\right)^{nt}$$ for the annual interest rate r.	If we know A, P, and t, it is not so easy to solve the continuously compounded interest equation $$A = Pe^{rt}$$ for the annual interest rate r.

In both of the first two equations, the variable x appears in the exponent (making them exponential equations), and we aren't able to rewrite the equation with the variable outside the exponent. We can only solve the first equation because we can rewrite 8 as a power of 2, and even this doesn't remove the variable from the exponent.

In the second pair of equations, we are able to solve for r in the first case, but it is difficult in the continuous-compounding case: once again the variable is inconveniently "stuck" in the exponent. We need a way of undoing exponentiation.

To "undo" functions we actually find their inverses. Therefore, we need to find the inverse of the general exponential function $f(x) = a^x$. Because every horizontal line in the plane intersects the graph of $f(x) = a^x$ no more than once (regardless of whether $0 < a < 1$ or $a > 1$), f has an inverse function, denoted f^{-1}.

The outline of the algorithm for finding the inverse of $f(x) = a^x$ is as follows.

$$f(x) = a^x$$

$\quad y = a^x$ Rewrite the function as an equation by replacing $f(x)$ with y.

$\quad x = a^y$ Then interchange x and y and proceed to solve for y.

$\quad y = ?$ At this point, we are stuck again. What is y?

No concept or notation that we have encountered up to this point allows us to solve the equation $x = a^y$ for the variable y, and this is the reason for introducing a new class of functions called *logarithms*.

Logarithmic Functions

Let a be a fixed positive real number not equal to 1. The **logarithmic function with base a** is defined to be the inverse of the exponential function with base a, and is denoted $\log_a x$. In symbols, if $f(x) = a^x$, then $f^{-1}(x) = \log_a x$.

In equation form, the definition of logarithm means that the equations

$$x = a^y \quad \text{and} \quad y = \log_a x$$

are equivalent. Note that a is the base in both equations: either the base of the exponential function or the base of the logarithmic function.

Example 1: Exponential and Logarithmic Equations

Use the definition of logarithmic functions to rewrite the following exponential equations as logarithmic equations.

a. $8 = 2^3$ **b.** $5^4 = 625$ **c.** $7^x = z$

Then rewrite the following logarithmic equations as exponential equations.

d. $\log_3 9 = 2$ **e.** $3 = \log_8 512$ **f.** $y = \log_2 x$

Solution

Note that in each case, the base of the exponential equation is the base of the logarithmic equation.

a. $3 = \log_2 8$ **b.** $4 = \log_5 625$ **c.** $x = \log_7 z$

d. $3^2 = 9$ **e.** $8^3 = 512$ **f.** $2^y = x$

Graphing Logarithmic Functions

Since logarithmic functions are inverses of exponential functions, we can learn a great deal about the graphs of logarithmic functions by recalling the graphs of exponential functions. For instance, the domain of $\log_a x$ (for any allowable a) is the positive real numbers, because the positive real numbers constitute the range of a^x (the domain of f^{-1} is the range of f). Similarly, the range of $\log_a x$ is the entire set of real numbers, because the real numbers make up the domain of a^x.

The graphs of a function and its inverse are reflections of one another with respect to the line $y = x$. Since exponential functions come in two forms based on the value of a ($0 < a < 1$ and $a > 1$), logarithmic functions also fall into two categories. In both of the graphs in Figure 1, the dashed curve is the graph of an exponential function and the solid curve is the corresponding logarithmic function.

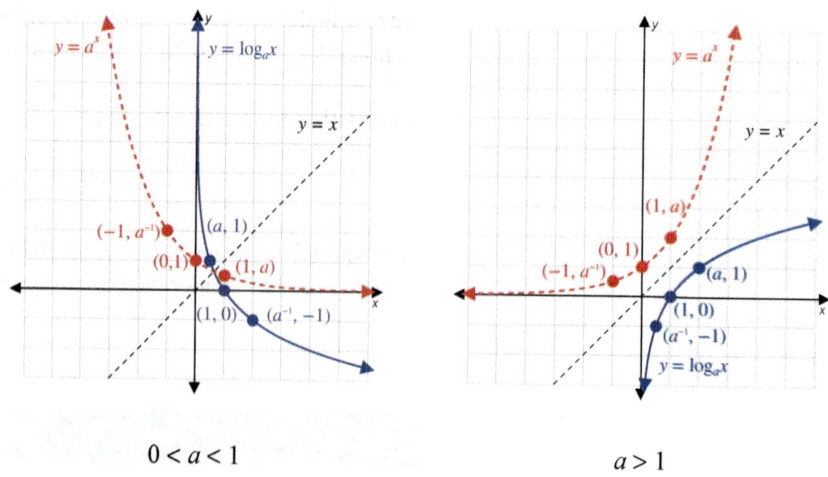

FIGURE 1: The Two Classes of Logarithmic Functions

In all cases, the points $(1, 0)$, $(a, 1)$, and $(a^{-1}, -1)$ lie on the graph of a logarithmic function with base a. Frequently, these points will be enough to get a good idea of the shape of the graph.

Note that the domain of each of the logarithmic functions in Figure 1 is indeed $(0, \infty)$, and that the range in each case is $(-\infty, \infty)$. Also note that the y-axis is a vertical asymptote for both, and that neither has a horizontal asymptote.

> **⚠ CAUTION**
>
> While it may appear so on the graph, logarithmic functions do **not** have a horizontal asymptote. The reason that logarithmic functions often look like they have a horizontal asymptote is because they are among the slowest growing functions in mathematics! Consider the function $f(x) = \log_2 x$. We have $f(1024) = 10$, $f(2048) = 11$, and $f(4096) = 12$. While the function may increase its value very slowly as x increases, it never approaches an asymptote.

> **Example 2: Graphing Logarithmic Functions**

Sketch the graphs of the following logarithmic functions.

a. $f(x) = \log_3 x$

b. $g(x) = \log_{\frac{1}{2}} x$

Solution

a.

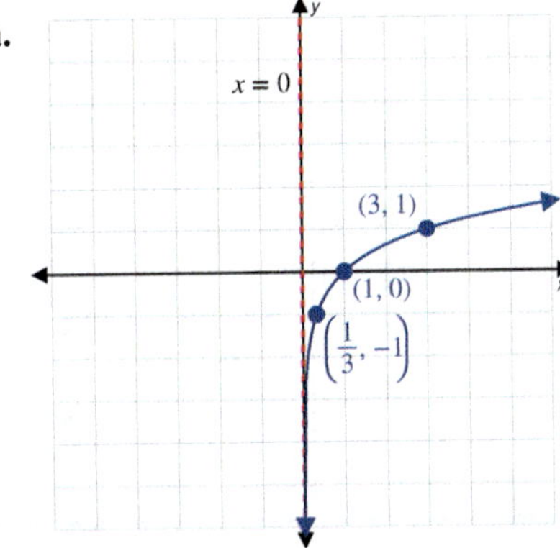

b.

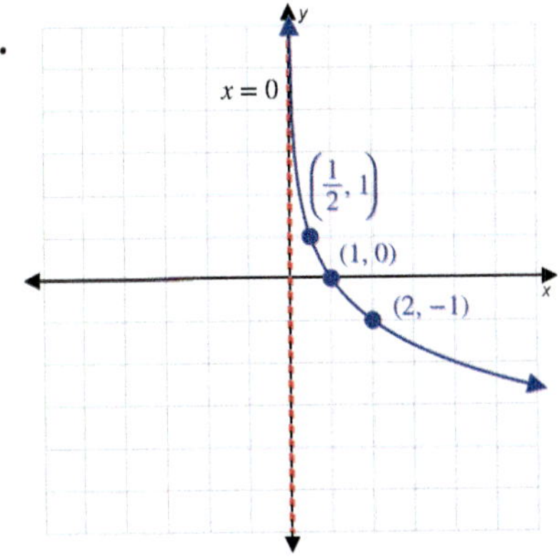

Example 3: Graphing Logarithmic Functions

Sketch the graphs of the following functions.

a. $f(x) = \log_3(x+2) + 1$ **b.** $g(x) = \log_2(-x-1)$ **c.** $h(x) = \log_{\frac{1}{2}}(x) - 2$

Solution

a.

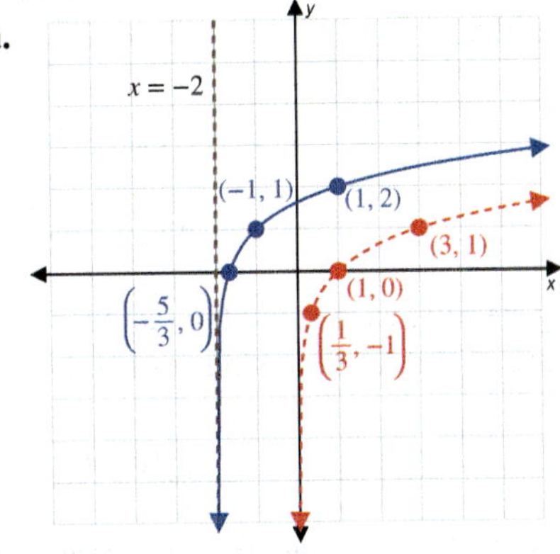

Begin by graphing the base function, which is $\log_3 x$. This is shown as a dashed curve.

Since x has been replaced by $x + 2$, we shift the graph 2 units to the left.

To find the graph of f, we shift the result up 1 unit, since 1 has been added to the function. This graph is the solid curve.

Note that the asymptote has also shifted to the left.

b.

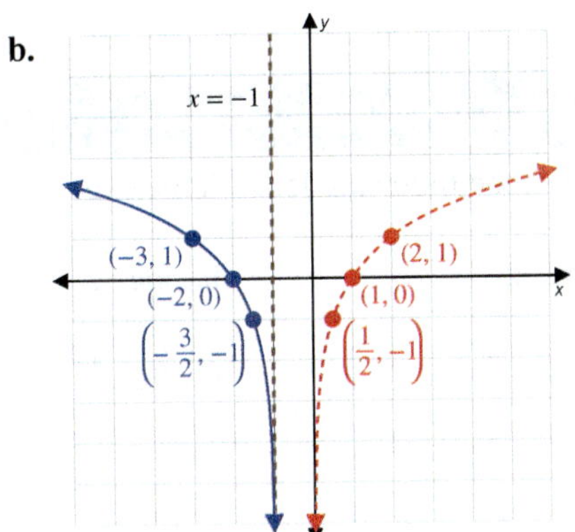

The basic shape of the graph of g is the same as the shape of $y = \log_2 x$, shown as a dashed curve.

To obtain g from $\log_2 x$, the variable x is replaced with $x - 1$, which shifts the graph 1 unit to the right, and then x is replaced by $-x$, which reflects the graph with respect to the y-axis.

c.

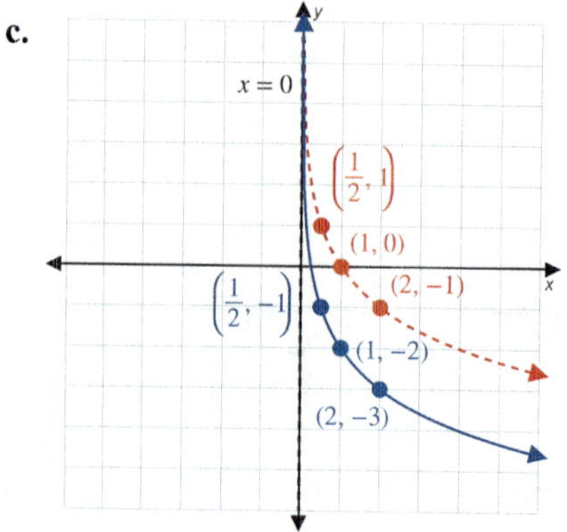

We begin with the dashed curve, which is the graph of $\log_{\frac{1}{2}} x$.

We then shift the graph 2 units down to obtain the graph of the function h.

Evaluating Elementary Logarithmic Expressions

Now that we have graphed logarithmic functions, we can augment our understanding of their behavior with a few algebraic observations. These will enable us to evaluate some logarithmic expressions and solve some elementary logarithmic equations.

First, our work in Example 2 suggests that the point (1, 0) is always on the graph of $\log_a x$ for any allowable base a, and this is indeed the case. A similar observation is that $(a, 1)$ is always on the graph of $\log_a x$. These two facts are actually just restatements of two corresponding facts about exponential functions, a consequence of the definition of logarithms.

$$\log_a 1 = 0, \text{ because } a^0 = 1$$

$$\log_a a = 1, \text{ because } a^1 = a$$

More generally, we can use the fact that the functions $\log_a x$ and a^x are inverses of one another to write the following equations.

$$\log_a (a^x) = x \quad \text{and} \quad a^{\log_a x} = x$$

In Example 4, we use the first statement to evaluate logarithmic expressions. Note the similarity to solving exponential equations; we rewrite the argument of the logarithm as a power of the base, allowing us to simplify the expression.

Example 4: Logarithmic Expressions

Evaluate the following logarithmic expressions.

a. $\log_5 25$

b. $\log_{\frac{1}{2}} 2$

c. $\log_\pi \sqrt{\pi}$

d. $\log_{17} 1$

e. $\log_{16} 4$

f. $\log_{10}\left(\dfrac{1}{100}\right)$

> **✎ NOTE**
>
> We write each of the equivalent exponential equations as a reference.

Solution

a. $\log_5 25 = \log_5 5^2$

$\qquad\quad = 2$

Rewrite 25 as a power of 5.

Equivalent exponential equation: $25 = 5^2$

b. $\log_{\frac{1}{2}} 2 = \log_{\frac{1}{2}} \left(\frac{1}{2}\right)^{-1}$

$\qquad\quad = -1$

Rewrite 2 as a power of $\frac{1}{2}$.

Equivalent exponential equation: $2 = \left(\frac{1}{2}\right)^{-1}$

c. $\log_\pi \sqrt{\pi} = \log_\pi \pi^{\frac{1}{2}}$

$\qquad\quad = \dfrac{1}{2}$

Rewrite the radical as a rational exponent.

Equivalent exponential equation: $\sqrt{\pi} = \pi^{\frac{1}{2}}$

d. $\log_{17} 1 = 0$

Equivalent exponential equation: $1 = 17^0$

e. $\log_{16} 4 = \log_{16} 16^{\frac{1}{2}}$

$\qquad\quad = \dfrac{1}{2}$

Rewrite 4 as a power of 16.

Equivalent exponential equation: $4 = 16^{\frac{1}{2}}$

f. $\log_{10}\left(\dfrac{1}{100}\right) = \log_{10} 10^{-2}$

$\qquad\quad = -2$

Rewrite $\dfrac{1}{100}$ as a power of 10.

Equivalent exponential equation: $\dfrac{1}{100} = 10^{-2}$

Solving Elementary Logarithmic Equations

In Example 5, we use both statements to solve logarithmic and exponential equations. Observe how converting between logarithmic and exponential form can make an equation easier to solve.

Example 5: Solving Logarithmic Equations

Solve the following equations involving logarithms.

a. $\log_6 (2x) = -1$

b. $3^{\log_{3x} 2} = 2$

c. $\log_2 8^x = 5$

Solution

a. $\log_6 (2x) = -1$

$\qquad 2x = 6^{-1}$

$\qquad 2x = \dfrac{1}{6}$

$\qquad\;\; x = \dfrac{1}{12}$

Convert the equation to exponential form.

Simplify the exponent, then solve for x.

b. $3^{\log_{3x} 2} = 2$ This time, we convert from the exponential form to

$\log_{3x} 2 = \log_3 2$ the logarithmic form.

$3x = 3$ We can equate the bases, just like we equate the

$x = 1$ exponents in an elementary exponential equation.

c. $\log_2 8^x = 5$

$8^x = 2^5$ Rewrite the equation in exponential form.

$\left(2^3\right)^x = 2^5$ Rewrite 8 as a power of 2.

$2^{3x} = 2^5$ Simplify using properties of exponents.

$3x = 5$ Set the exponents equal to each other, then solve.

$x = \dfrac{5}{3}$

Common and Natural Logarithms

We have already mentioned the fact that the number e, called the natural base, plays a fundamental role in many important real-world situations and in higher mathematics, so it is not surprising that the logarithmic function with base e is worthy of special attention. For historical reasons, namely the fact that our number system is based on powers of 10, the logarithmic function with base 10 is also singled out.

Common and Natural Logarithms

- The function $\log_{10} x$ is called the **common logarithm**, and is usually written $\log x$.

- The function $\log_e x$ is called the **natural logarithm**, and is usually written $\ln x$.

Another way in which these particular logarithms are special is that most calculators, if they are capable of calculating logarithms at all, are only equipped to evaluate common and natural logarithms. Such calculators normally have a button labeled "LOG" for the common logarithm and a button labeled "LN" for the natural logarithm.

Properties of Natural Logarithms

$$\ln x = y \iff e^y = x$$

Properties	Reasons
1. $\ln 1 = 0$	Raise e to the power 0 to get 1.
2. $\ln e = 1$	Raise e to the power 1 to get e.
3. $\ln e^x = x$	Raise e to the power x to get e^x.
4. $e^{\ln x} = x$	$\ln x$ is the power to which e must be raised to get x.

Example 6: Evaluating Logarithmic Expressions

Evaluate the following logarithmic expressions.

a. $\ln \sqrt[3]{e}$ **b.** $\log 1000$ **c.** $\ln (4.78)$ **d.** $\log (10.5)$

Solution

a. $\ln \sqrt[3]{e} = \ln e^{\frac{1}{3}} = \dfrac{1}{3}$ No calculator is necessary for this problem, just an application of an elementary property of logarithms.

b. $\log 1000 = \log 10^3 = 3$ Again, no calculator is required.

c. $\ln (4.78) \approx 1.564$ This time, a calculator is needed, and only an approximate answer can be given. Be sure to use the correct logarithm.

d. $\log (10.5) \approx 1.021$ Again, we must use a calculator, though we can say beforehand that the answer should be only slightly larger than 1, as $\log 10 = 1$ and 10.5 is only slightly larger than 10.

Graphing Logarithmic Functions Using Technology

Common and natural logarithms can be entered into a graphing calculator using the **LOG** and **LN** buttons, respectively. For instance, to graph the function $f(x) = \log(3 - x)$, press **Y=** and **LOG**. The opening parenthesis appears automatically, so we just have to type in the argument, $3 - x$, and the right-hand parenthesis.

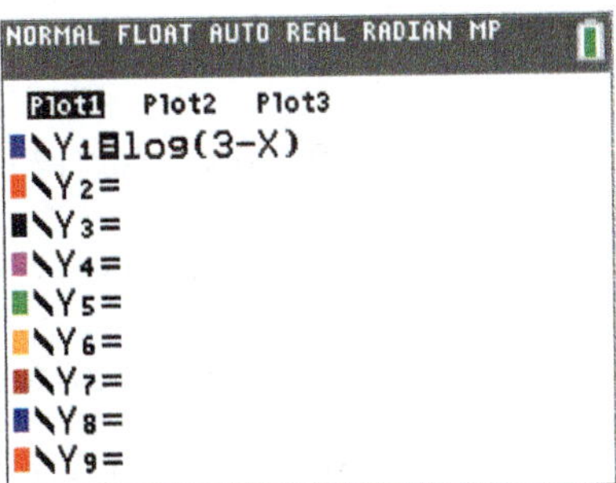

The following graph should appear:

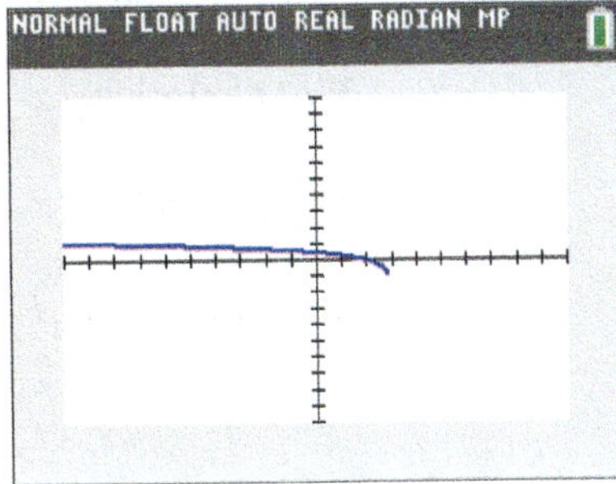

⚠ CAUTION

Notice that the graph appears to cut off around the point where $x = 3$. We know, however, that the graph has a vertical asymptote at $x = 3$, and approaches that asymptote even though it does not show on the graphing calculator.

4.3 EXERCISES

PRACTICE

Write the following equations in logarithmic terms.

1. $625 = 5^4$

2. $216 = 6^3$

3. $x^3 = 27$

4. $b^2 = 3.2$

5. $4.2^3 = C$

6. $1.3^2 = V$

7. $4^x = 31$

8. $16^{2x} = 215$

9. $(4x)^{\sqrt{3}} = 13$

10. $e^x = \pi$

11. $2^{e^x} = 11$

12. $4^e = N$

Write the following logarithmic equations as exponential equations.

13. $\log_3 81 = 4$

14. $\log_2\left(\dfrac{1}{8}\right) = -3$

15. $\log_b 4 = \dfrac{1}{2}$

16. $\log_y 9 = 2$

17. $\log_2 15 = b$

18. $\log_5 8 = d$

19. $\log_5 W = 12$

20. $\log_7 T = 6$

21. $\log_\pi (2x) = 4$

22. $\log_{\sqrt{3}} (2\pi) = x$

23. $\ln 2 = x$

24. $\ln(5x) = 3$

Sketch the graphs of the following functions. State their domain and range. See Examples 2 and 3.

25. $f(x) = \log_3 (x - 1)$

26. $g(x) = \log_5 (x + 2) - 1$

27. $r(x) = \log_{\frac{1}{2}} (x - 3)$

28. $p(x) = 3 - \log_2 (x + 1)$

29. $q(x) = \log_3 (2 - x)$

30. $s(x) = \log_{\frac{1}{3}} (5 - x)$

31. $h(x) = \log_7 (x - 3) + 3$

32. $m(x) = \log_{\frac{1}{2}} (1 - x)$

33. $f(x) = \log_3 (6 - x)$

34. $p(x) = 4 - \log(x + 3)$

35. $s(x) = -\log_{\frac{1}{3}} (-x)$

36. $g(x) = \log_5 (2x) - 1$

Match the graph of the appropriate equation to the logarithmic function.

37. $f(x) = \log_2 (x - 1)$

38. $f(x) = \log_2 (2 - x)$

39. $f(x) = \log_2 (-x)$

40. $f(x) = \log_2 (x - 3)$

41. $f(x) = 1 - \log_2 x$

42. $f(x) = -\log_2 x$

43. $f(x) = -\log_2 (-x)$

44. $f(x) = \log_2 x$

45. $f(x) = \log_2 (x + 3)$

a. 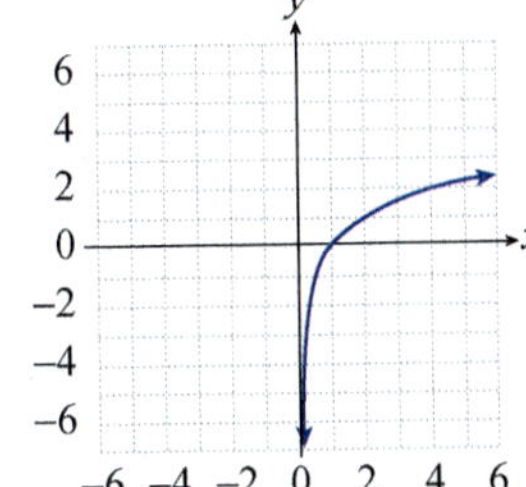

b. 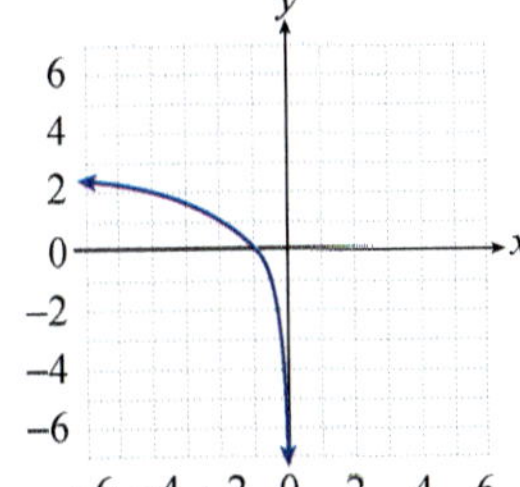

c. 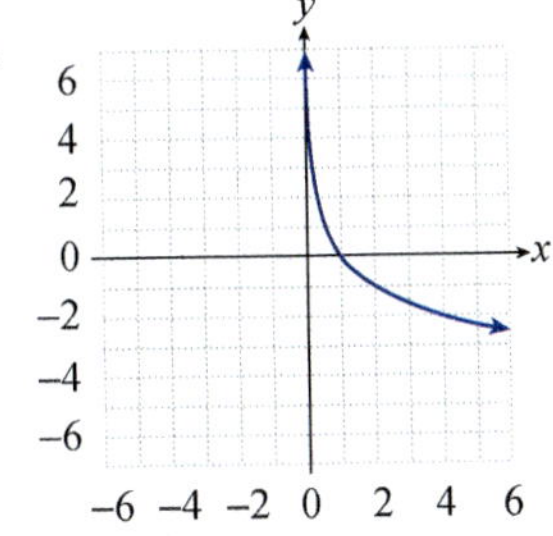

d. 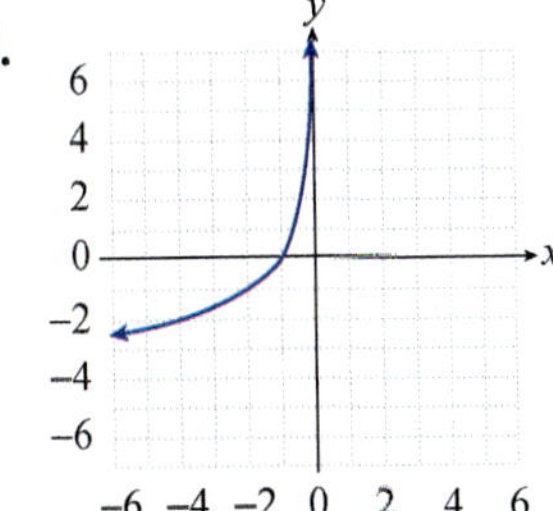

e. 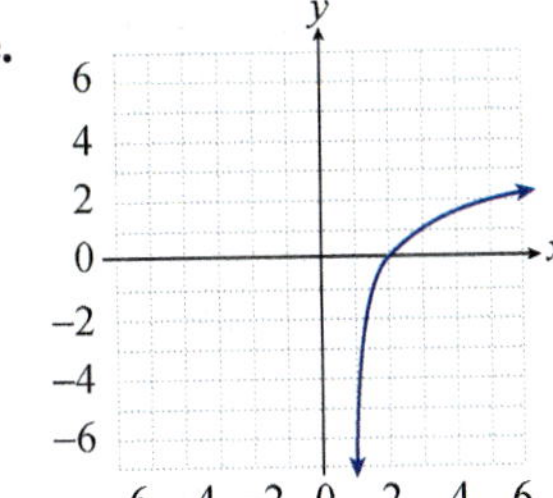

f. 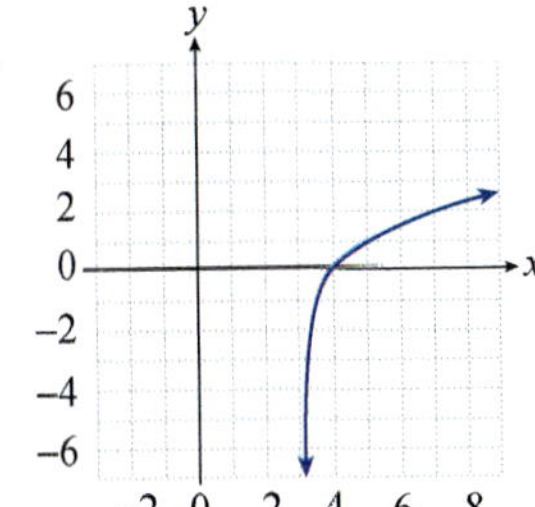

g. 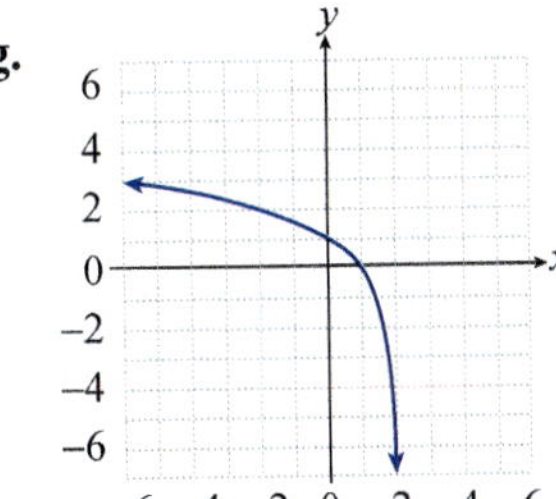

h. 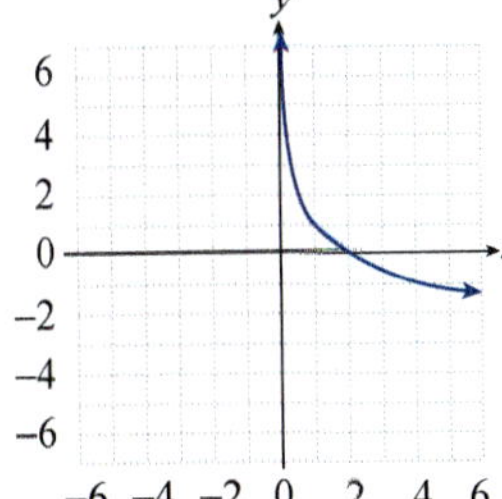

i. 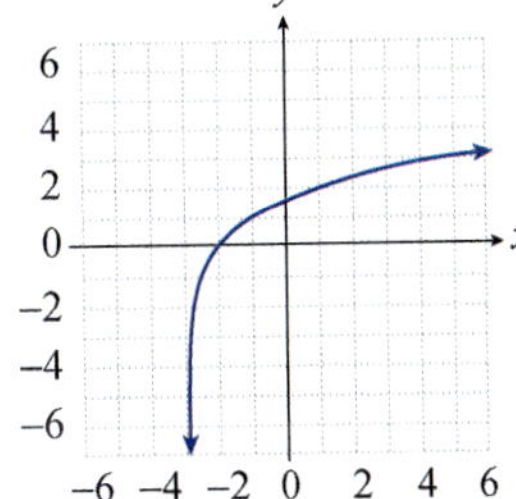

Evaluate the following logarithmic expressions without the use of a calculator. See Examples 4 and 6.

46. $\log_7\left(\sqrt{7}\right)$

47. $\log_{\frac{1}{2}} 4$

48. $\log_9\left(\dfrac{1}{81}\right)$

49. $\log_3 27$

50. $\log_{27} 3$

51. $\log_9\left(\dfrac{1}{3}\right)$

52. $\log_{27} 9$

53. $\log_{\frac{1}{16}}\left(\dfrac{1}{8}\right)$

54. $\log_3\left(\log_{27} 3\right)$

55. $\ln e^{2.89}$

56. $\log(0.0001)$

57. $\log_a\left(a^{\frac{5}{3}}\right)$

58. $\ln\left(\dfrac{1}{e}\right)$

59. $\log\left(\log\left(10^{10}\right)\right)$

60. $\log_3 1$

61. $\ln\left(\sqrt[5]{e}\right)$

62. $\log_{\frac{1}{16}} 4$

63. $\log_8 4^{\log 1000}$

Use the elementary properties of logarithms to solve the following equations. See Example 5.

64. $\log_{16} x = \dfrac{3}{4}$

65. $\log_{16}\left(x^{\frac{1}{2}}\right) = \dfrac{3}{4}$

66. $\log_{16} x = -\dfrac{3}{4}$

67. $\log_5\left(5^{\log_3 x}\right) = 2$

68. $\log_a\left(a^{\log_b x}\right) = 0$

69. $\log_3\left(9^{2x}\right) = -2$

70. $\log_{\frac{1}{3}}\left(3^x\right) = 2$

71. $\log_7\left(3x\right) = -1$

72. $4^{\log_3 x} = 0$

73. $\log\left(x^{10}\right) = 10$

74. $\log_x\left(\log_{\frac{1}{2}}\left(\dfrac{1}{4}\right)\right) = 1$

75. $6^{\log_x\left(e^2\right)} = e$

Hint: Note that $\log_a b = \log_{a^2} b^2$. This follows from the fact that

$$\log_a b = y \Leftrightarrow b = a^y \Leftrightarrow b^2 = a^{2y} = \left(a^2\right)^y \Leftrightarrow \log_{a^2} b^2 = y.$$

Solve the following logarithmic equations, using a calculator if necessary to evaluate the logarithms. See Examples 5 and 6. Express your answer either as a fraction or a decimal rounded to two decimal places.

76. $\log(3x) = 2.1$

77. $\log\left(x^2\right) = -2$

78. $\ln(x + 1) = 3$

79. $\ln(2x) = -1$

80. $\ln\left(e^x\right) = 5.6$

81. $\ln\left(\ln\left(x^2\right)\right) = 0$

82. $\log 19 = 3x$

83. $\log\left(e^x\right) = 5.6$

84. $\log_9\left(2x - 1\right) = 2$

85. $\log\left(\log\left(x - 2\right)\right) = 1$

86. $\log\left(300^{\log x}\right) = 9$

4.4 APPLICATIONS OF LOGARITHMIC FUNCTIONS

■ TOPICS

- Properties of Logarithms
- The Change of Base Formula
- Applications of Logarithmic Functions

Properties of Logarithms

Previously, we introduced logarithmic functions and studied some of their elementary properties. The motivation was our inability, at that time, to solve certain exponential equations. Let us reconsider the two sample problems that initiated our discussion of logarithms and see if we have made progress. We begin with the continuously compounding interest problem.

Example 1: Continuously Compounded Interest

Anne reads an ad in the paper for a new bank in town. The bank is advertising "continuously compounded savings accounts" in an attempt to attract customers, but fails to mention the annual interest rate. Curious, she goes to the bank and is told by an account agent that if she were to invest $10,000 in an account, her money would grow to $10,202.01 in one year's time. But, strangely, the agent also refuses to divulge the yearly interest rate. What rate is the bank offering?

Solution

We need to solve the equation $A = Pe^{rt}$ for r, given that $A = 10{,}202.01$, $P = 10{,}000$, and $t = 1$.

$$10{,}202.01 = 10{,}000e^{r(1)} \qquad \text{Substitute the given values.}$$
$$1.020201 = e^{r} \qquad \text{Divide both sides by 10,000.}$$
$$\ln(1.020201) = r \qquad \text{Convert to logarithmic form.}$$
$$r \approx 0.02 \qquad \text{Evaluate using a calculator.}$$

Note that we use the natural logarithm since the base of the exponential function is e. While we must use a calculator, we can now solve the equation for r.

Example 2: Solving Exponential Equations

Solve the equation $2^{x} = 9$.

Solution

We convert the equation to logarithmic form to obtain the solution $x = \log_2 9$. Unfortunately, this answer still doesn't tell us anything about x in decimal form, other than that it is bound to be slightly more than 3. Further, we can't use a calculator to evaluate $\log_2 9$ since the base is neither 10 nor e.

As Example 2 shows, our ability to work with logarithms is still incomplete. In this section we will derive some important properties of logarithms that allow us to solve more complicated equations, as well as provide a decimal approximation to the solution of $2^x = 9$.

The following properties of logarithmic functions are analogs of corresponding properties of exponential functions, a consequence of how logarithms are defined.

Properties of Logarithms

Let a (the logarithmic base) be a positive real number not equal to 1, let x and y be positive real numbers, and let r be any real number.

1. $\log_a(xy) = \log_a x + \log_a y$ ("the log of a product is the sum of the logs")

2. $\log_a\left(\dfrac{x}{y}\right) = \log_a x - \log_a y$ ("the log of a quotient is the difference of the logs")

3. $\log_a(x^r) = r \log_a x$ ("the log of something raised to a power is the power times the log")

We illustrate the link between these properties and the related properties of exponents by proving the first one. Try proving the second and third as further practice.

Proof:

Let $m = \log_a x$ and $n = \log_a y$. The equivalent exponential forms of these two equations are $x = a^m$ and $y = a^n$.

Since we are interested in the product xy, note that

$$xy = a^m a^n = a^{m+n}.$$

The statement $xy = a^{m+n}$ can then be converted to logarithmic form, giving us

$$\log_a(xy) = m + n.$$

Referring back to the definition of m and n, we have $\log_a(xy) = \log_a x + \log_a y$.

If the properties of logarithms appear strange at first, remember that they are just the properties of exponents restated in logarithmic form.

In some situations, we will find it useful to use properties of logarithms to decompose a complicated expression into a sum or difference of simpler expressions, while in other situations we will do the reverse, combining a sum or a difference of logarithms into one logarithm. Examples 3 and 4 illustrate these processes.

Example 3: Expanding Logarithmic Expressions

Use the properties of logarithms to expand the following expressions as much as possible (that is, decompose the expressions into sums or differences of the simplest possible terms).

a. $\log_4\left(64x^3\sqrt{y}\right)$ **b.** $\log_a\sqrt[3]{\dfrac{xy^2}{z^4}}$ **c.** $\log\left(\dfrac{2.7\times10^4}{x^{-2}}\right)$

Solution

a. $\log_4\left(64x^3\sqrt{y}\right) = \log_4 64 + \log_4 x^3 + \log_4 \sqrt{y}$ Use the first property to rewrite the expression as three terms.

$$= \log_4 4^3 + \log_4 x^3 + \log_4 y^{\frac{1}{2}}$$

$$= 3 + 3\log_4 x + \frac{1}{2}\log_4 y$$ We can evaluate the first term and rewrite the second and third terms using the third property.

b. $\log_a\sqrt[3]{\dfrac{xy^2}{z^4}} = \log_a\left(\dfrac{xy^2}{z^4}\right)^{\frac{1}{3}}$ Rewrite the radical as an exponent.

$$= \frac{1}{3}\log_a\left(\frac{xy^2}{z^4}\right)$$ Bring the exponent in front of the logarithm using the third property.

$$= \frac{1}{3}\left(\log_a x + \log_a y^2 - \log_a z^4\right)$$ Expand the expression using the first two properties.

$$= \frac{1}{3}\left(\log_a x + 2\log_a y - 4\log_a z\right)$$ Apply the third property to the terms that result.

c. Recall that if a base is not explicitly written, it is assumed to be 10.

$$\log\left(\frac{2.7\times10^4}{x^{-2}}\right) = \log(2.7) + \log\left(10^4\right) - \log x^{-2}$$ Expand using the first and second properties.

$$= \log(2.7) + 4 + 2\log x$$ Evaluate the first two terms and use the third property on the last term.

$$\approx 4.43 + 2\log x$$

It is appropriate to either evaluate $\log(2.7)$ or leave it in exact form. Use the context of the problem to decide which form is more convenient.

Example 4: Condensing Logarithmic Expressions

Use the properties of logarithms to condense the following expressions as much as possible (that is, rewrite the expressions as a sum or difference of as few logarithms as possible).

a. $2\log_3\left(\dfrac{x}{3}\right) - \log_3\left(\dfrac{1}{y}\right)$ **b.** $\ln x^2 - \dfrac{1}{2}\ln y + \ln 2$ **c.** $\log_b 5 + 2\log_b x^{-1}$

NOTE

Often, there will be multiple orders in which we can apply the properties to find the final result.

Solution

a. $2\log_3\left(\dfrac{x}{3}\right) - \log_3\left(\dfrac{1}{y}\right) = \log_3\left(\dfrac{x}{3}\right)^2 + \log_3\left(\dfrac{1}{y}\right)^{-1}$

Use the third property to make the coefficients appear as exponents.

$$= \log_3\left(\dfrac{x^2}{9}\right) + \log_3 y$$

Evaluate the exponents.

$$= \log_3\left(\dfrac{x^2 y}{9}\right)$$

Combine terms using the first property.

b. $\ln x^2 - \dfrac{1}{2}\ln y + \ln 2 = \ln x^2 - \ln y^{\frac{1}{2}} + \ln 2$

Rewrite each term to have a coefficient of 1 or −1 using the third property.

$$= \ln\left(\dfrac{x^2}{y^{\frac{1}{2}}}\right) + \ln 2$$

We can then combine the terms using the second property.

$$= \ln\left(\dfrac{2x^2}{y^{\frac{1}{2}}}\right) \text{ or } \ln\left(\dfrac{2x^2}{\sqrt{y}}\right)$$

The final answer can be written in several different ways, two of which are shown.

c. $\log_b 5 + 2\log_b x^{-1} = \log_b 5 + \log_b x^{-2}$

$$= \log_b 5x^{-2} \text{ or } \log_b\left(\dfrac{5}{x^2}\right)$$

Rewrite the coefficient as an exponent, then combine terms.

The Change of Base Formula

The properties we just derived can be used to provide an answer to a question about logarithms that has been left unanswered thus far. A specific illustration of the question arose in Example 2: how do we determine the decimal form of a number like $\log_2 9$?

Surprisingly, to answer this question we will undo our work in Example 2. We assign a variable to the result $\log_2 9$, convert the resulting logarithmic equation into exponential form, take the natural logarithm of both sides, and then solve for the variable.

$x = \log_2 9$ — Let x equal the result from before, $\log_2 9$.

$2^x = 9$ — Convert the equation to exponential form.

$\ln\left(2^x\right) = \ln 9$ — Take the natural logarithm of both sides.

$x\ln 2 = \ln 9$ — Move the variable out of the exponent using the third property of logarithms.

$x = \dfrac{\ln 9}{\ln 2}$ — Simplify.

$x \approx 3.17$ — Evaluate with a calculator.

While using a logarithm with any base will give the correct solution to this problem, if a calculator is to be used to approximate the number $\log_2 9$, there are (for most calculators) only two good choices: the natural log and the common log. If we had done the work with the common logarithm, the final answer would have been the same. That is,

$$\frac{\log 9}{\log 2} \approx 3.17.$$

And even though it would not be easy to evaluate,

$$\frac{\log_a 9}{\log_a 2} \approx 3.17$$

for any allowable logarithmic base a.

More generally, a logarithm with base b can be converted to a logarithm with base a through the same reasoning. This allows us to evaluate all logarithmic expressions.

Change of Base Formula

Let a and b both be positive real numbers, neither of them equal to 1, and let x be a positive real number. Then

$$\log_b x = \frac{\log_a x}{\log_a b}.$$

Example 5: Change of Base Formula

Evaluate the following logarithmic expressions, using the base of your choice.

a. $\log_7 15$ **b.** $\log_{\frac{1}{2}} 3$ **c.** $\log_\pi 5$

Solution

> **NOTE**
>
> Both the common and natural logarithms work in solving these problems.

a. $\log_7 15 = \dfrac{\ln 15}{\ln 7}$ — Apply the change of base formula.

≈ 1.392 — Evaluate using a calculator.

b. $\log_{\frac{1}{2}} 3 = \dfrac{\log 3}{\log\left(\dfrac{1}{2}\right)}$ — Apply the change of base formula. This time we use the common logarithm.

≈ -1.585 — Since the base of the logarithm is a fraction, we should expect a negative answer.

c. $\log_\pi 5 = \dfrac{\log 5}{\log \pi}$ — Once again, we apply the change of base formula, then evaluate using a calculator.

≈ 1.406

Applications of Logarithmic Functions

Logarithms appear in many different contexts and have a wide variety of uses. This is due partly to the fact that logarithmic functions are the inverses of exponential functions, and partly to the logarithmic properties we have discussed. In fact, the mathematician who can be most credited for "inventing" logarithms, John Napier (1550–1617) of Scotland, was inspired in his work by the convenience of what we now call logarithmic properties.

Computationally, logarithms are useful because they relocate exponents as coefficients, thus making them easier to work with. Consider, for example, a very large number such as 3×10^{17} or a very small number such as 6×10^{-9}. The common logarithm (used because it has a base of 10) expresses these numbers on a more comfortable scale:

$$\log\left(3 \times 10^{17}\right) = \log 3 + \log\left(10^{17}\right) = 17 + \log 3 \approx 17.477$$

$$\log\left(6 \times 10^{-9}\right) = \log 6 + \log\left(10^{-9}\right) = -9 + \log 6 \approx -8.222$$

Napier, working long before the advent of electronic calculating devices, devised logarithms in order to take advantage of this property.

In chemistry, the concentration of hydronium ions in a solution determines its acidity. Since concentrations are small numbers that vary over many orders of magnitude, it is convenient to express acidity in terms of the pH scale, as follows.

The pH Scale

The **pH** of a solution is defined to be $-\log[H_3O^+]$, where $[H_3O^+]$ is the concentration of hydronium ions in units of moles/liter. Solutions with a pH less than 7 are said to be *acidic*, while those with a pH greater than 7 are *basic*.

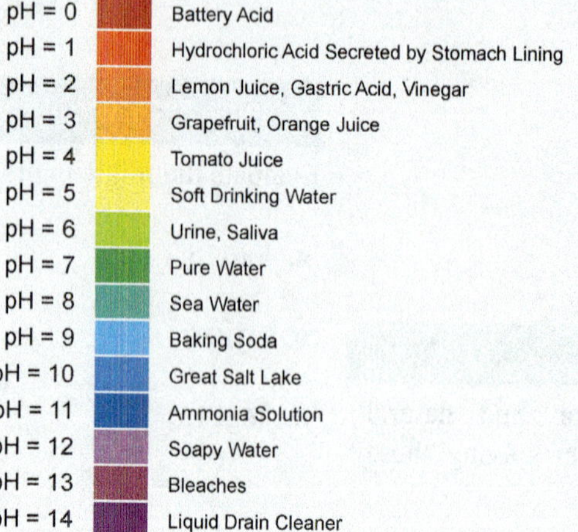

FIGURE 1: pH of Common Substances

Example 6: The pH Scale

If a sample of orange juice is determined to have a $[H_3O^+]$ concentration of 1.58×10^{-4} moles/liter, what is its pH?

Solution

Applying the formula $pH = -\log[H_3O^+]$ we can find the pH of the sample as follows.

$$pH = -\log\left(1.58 \times 10^{-4}\right) \approx -(-3.80) = 3.8$$

After doing this calculation, the reason for the minus sign in the formula is more apparent. By multiplying the log of the concentration by -1, the pH of a solution is positive, which is convenient for comparative purposes.

The energy released during earthquakes can vary greatly, but logarithms provide a convenient way to analyze and compare the intensity of earthquakes.

The Richter Scale

Earthquake intensity is measured on the **Richter scale** (named for the American seismologist Charles Richter, 1900–1985). In the original formula that follows, I_0 is the intensity of a just-discernible earthquake, I is the intensity of an earthquake being analyzed, and R is its ranking on the Richter scale.

$$R = \log\left(\frac{I}{I_0}\right)$$

By this measure, earthquakes range from a classification of minor ($R < 4$), to light ($4 \leq R < 5$), to moderate ($5 \leq R < 6$), to strong ($6 \leq R < 7$), to major ($7 \leq R < 8$), to great ($8 \leq R$).

The base 10 logarithm means that every increase of 1 unit on the Richter scale corresponds to an increase by a factor of 10 in the intensity. This is a characteristic of all logarithmic scales. Also, note that a barely discernible earthquake has a rank of 0, since $\log 1 = 0$.

Example 7: The Richter Scale

The January 2001 earthquake in the state of Gujarat in India was 7,940,000 times as intense as a 0-level earthquake. What was the Richter ranking of this devastating event?

Solution

If we let I denote the intensity of the Gujarat earthquake, then $I = 7,940,000I_0$, so

$$\begin{aligned}
R &= \log\left(\frac{7,940,000I_0}{I_0}\right) \\
&= \log\left(7.94 \times 10^6\right) \\
&= \log\left(7.94\right) + \log\left(10^6\right) \\
&= \log\left(7.94\right) + 6 \\
&\approx 6.9.
\end{aligned}$$

The Gujarat earthquake thus fell in the category of strong on the Richter scale.

Sound intensity is another quantity that varies greatly, and the measure of how the human ear perceives intensity, in units called decibels, is very similar to the measure of earthquake intensity.

The Decibel Scale

In the **decibel scale**, I_0 is the intensity of a just-discernible sound, I is the intensity of the sound being analyzed, and D is its decibel level.

$$D = 10\log\left(\frac{I}{I_0}\right)$$

Decibel levels range from 0 for a barely discernible sound, to 60 for the level of normal conversation, to 80 for heavy traffic, to 120 for a loud rock concert, and finally (as far as humans are concerned) to around 160, at which point the eardrum is likely to rupture.

Example 8: The Decibel Scale

Given that $I_0 = 10^{-12}$ watts/meter^2, what is the decibel level of a jet airliner's engines at a distance of 45 meters, for which the sound intensity is 50 watts/meter^2?

Solution

$$D = 10\log\left(\frac{50}{10^{-12}}\right)$$
$$= 10\log\left(5\times10^{13}\right)$$
$$= 10\left(\log 5 + 13\right)$$
$$\approx 137$$

In other words, the sound level would probably not be literally earsplitting, but it would be very painful.

4.4 EXERCISES

PRACTICE

Use the properties of logarithms to expand the following expressions as much as possible. Simplify any numerical expressions that can be evaluated without a calculator. See Example 3.

1. $\log_5\left(125x^3\right)$

2. $\ln\left(\dfrac{x^2 y}{3}\right)$

3. $\ln\left(\dfrac{e^2 p}{q^3}\right)$

4. $\log(100x)$

5. $\log_9\left(9xy^{-3}\right)$

6. $\log_6\left(\sqrt[3]{\dfrac{p^2}{q}}\right)$

7. $\ln\left(\dfrac{\sqrt{x^3}\,pq^5}{e^7}\right)$

8. $\log_a\sqrt[5]{\dfrac{a^4 b}{c^2}}$

9. $\log\left(\log\left(100x^3\right)\right)$

10. $\log_3\left(9x + 27y\right)$

11. $\log\left(\dfrac{10}{\sqrt{x+y}}\right)$

12. $\ln\left(\ln\left(e^{ex}\right)\right)$

13. $\log_2\left(\dfrac{y^2+z}{16x^4}\right)$ **14.** $\log\left(\log\left(100,000^{2x}\right)\right)$ **15.** $\log_b\left(\sqrt{\dfrac{x^4y}{z^2}}\right)$

16. $\ln\left(7x^2-42x+63\right)$ **17.** $\log_b\left(ab^2c^b\right)$ **18.** $\ln\left(\ln\left(e^{e^x}\right)\right)$

Use the properties of logarithms to condense the following expressions as much as possible, writing each answer as a single term with a coefficient of 1. See Example 4.

19. $\log x-\log y$

20. $\log_5 x-2\log_5 y$

21. $\log_5\left(x^2-25\right)-\log_5\left(x-5\right)$

22. $\ln\left(x^2y\right)-\ln y-\ln x$

23. $\dfrac{1}{3}\log_2 x+\log_2\left(x+3\right)$

24. $\dfrac{1}{5}\left(\log_7\left(x^2\right)-\log_7\left(pq\right)\right)$

25. $\ln 3+\ln p-2\ln q$

26. $2\left(\log_5\left(\sqrt{x}\right)-\log_5 y\right)$

27. $\log\left(x-10\right)-\log x$

28. $2\log a^2b-\log\left(\dfrac{1}{b}\right)+\log\left(\dfrac{1}{a}\right)$

29. $3\left(\ln\left(\sqrt[3]{z^2}\right)-\ln\left(xy\right)\right)$

30. $\log_2\left(4x\right)-\log_2 x$

31. $\log_5 20-\log_5 5$

32. $\log 30-\log 2-\log 5$

33. $\ln 15+\ln 3$

34. $\ln 8-\ln 4+\ln 3$

35. $0.5\log_3 16-\log_3 4$

36. $3\log_7 2-2\log_7 4$

37. $0.25\ln 81+\ln 4$

38. $2\left(\log 4-\log 1+\log 2\right)$

39. $\log 11+0.5\log 9-\log 3$

40. $3\log_4\left(x^2\right)+\log_4\left(x^6\right)$

41. $\log_8\left(2x^2-2y\right)-0.25\log_8 16$

42. $\log_{3x} x^2+\log_{3x} 18-\log_{3x} 6$

Use the properties of logarithms to write each of the following as a single term that does not contain a logarithm.

43. $5^{2\log_5 x}$ **44.** $10^{\log y^2-3\log x}$ **45.** $e^{2-\ln x+\ln p}$

46. $e^{5\left(\ln\sqrt[5]{3}+\ln x\right)}$ **47.** $10^{\log x^3-4\log y}$ **48.** $a^{\log_a b+4\log_a\sqrt{a}}$

49. $10^{2\log x}$ **50.** $10^{4\log x-2\log x}$ **51.** $\log_4 16\cdot\log_x x^2$

52. $e^{\ln x+2+\ln x^2}$ **53.** $4^{\log_4\left(3x\right)+0.5\log_4\left(16x^2\right)}$ **54.** $4^{2\log_2 6-\log_2 9}$

Evaluate the following logarithmic expressions. See Example 5.

55. $\log_4 17$ **56.** $2\log_{\frac{1}{3}} 5$ **57.** $\log_9 8$

58. $\log_2 0.01$ **59.** $\log_{12} 10.5$ **60.** $\log\left(\ln 2\right)$

61. $\log_6 3^4$ **62.** $\log_7 14.3$ **63.** $\log_{\frac{1}{2}} \pi^{-2}$

64. $\log_{\frac{1}{5}} 626$ **65.** $\ln\left(\log 123\right)$ **66.** $\log_{17} 0.041$

67. $\log 16$

68. $\log_3 9$

69. $\log_5 20$

70. $\log_8 26$

71. $\log_4 0.25$

72. $\log_{1.8} 9$

73. $\log_{2.5} 34$

74. $\log_{0.5} 10$

75. $\log_4 2.9$

76. $\log_{0.4} 14$

77. $\log_{0.2} 17$

78. $\log_{0.16} 2.8$

Without using a calculator, evaluate the following expressions.

79. $\log_4 16$

80. $\log_5 25^3$

81. $\ln e^4 + \ln e^3$

82. $\log_4 \dfrac{1}{64}$

83. $\ln e^{1.5} - \log_4 2$

84. $\log_2 8^{(2\log_2 4 - \log_2 4)}$

Find the value of x in each of the following equations. Express your answer in exact form, or rounded to two decimal places.

85. $\log_x 1024 = 4$

86. $\log_6 729 = x$

87. $\log_2 529 = x$

88. $\log_4 625 = x$

89. $\log_x 729 = 9$

90. $\log_4 x = 8$

91. $\log_{12} x = 1$

92. $\log_x 16{,}807 = 7$

93. $\log_4 x = 10$

🚀 APPLICATIONS

94. A certain brand of tomato juice has a $\left[H_3O^+\right]$ concentration of 6.31×10^{-5} moles/liter. What is the pH of this brand?

95. One type of detergent, when added to neutral water with a pH of 7, results in a solution with a $\left[H_3O^+\right]$ concentration that is 5.62×10^{-4} times weaker than that of the water. What is the pH of the solution?

96. What is the concentration of $\left[H_3O^+\right]$ in lemon juice with a pH of 2.1?

97. An earthquake in Chile in 2019 measured 6.7 on the Richter scale. What was the intensity, relative to a 0-level earthquake, of this event?

98. How much stronger was the 2001 Gujarat earthquake (6.9 on the Richter scale) than the 2019 earthquake described in Exercise 97?

99. A construction worker operating a jackhammer would experience noise with an intensity of 20 watts/meter^2 if it weren't for ear protection. Given that $I_0 = 10^{-12}$ watts/meter^2, what is the decibel level for such noise?

100. A microphone picks up the sound of a thunderclap and measures its decibel level as 105. Given that $I_0 = 10^{-12}$ watts/meter^2, with what sound intensity did the thunderclap reach the microphone?

101. Matt, a lifeguard, has to make sure that the pH of the swimming pool stays between 7.2 and 7.6. If the pH is out of this range, he has to add chemicals that alter the pH level of the pool. If Matt measures the $\left[H_3O^+\right]$ concentration in the swimming pool to be 2.40×10^{-8} moles/liter, what is the pH? Does he need to change the pH by adding chemicals to the water?

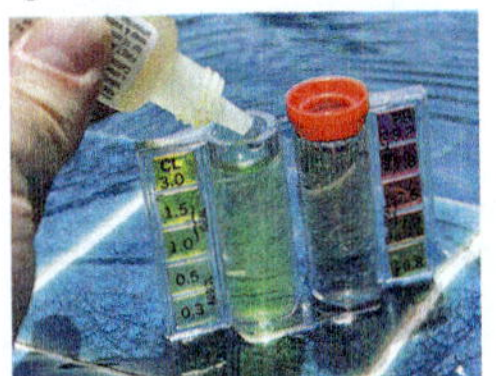

102. The intensity of a cat's soft purring is measured to be 2.19×10^{-11}. Given that $I_0 = 10^{-12}$ watts/meter^2, what is the decibel level of this noise?

103. Newton's Law of Cooling states that the rate at which an object cools is proportional to the difference between the temperature of the object and the surrounding temperature. If C denotes the surrounding temperature and T_0 denotes the temperature at time $t = 0$, the temperature of an object at time t is given by $T(t) = C + (T_0 - C)e^{-kt}$, where k is a constant that depends on the particular object under discussion.

    a. You are having friends over for tea and want to know how long after boiling the water it will be drinkable. If the temperature of your kitchen stays around 74°F and you found online that the constant k for tea is approximately 0.049, how many minutes after boiling the water will the tea be drinkable (you prefer your tea no warmer than 140°F)? Recall that water boils at 212°F.

    b. As you intern for your local crime scene investigation department, you are asked to determine at what time a victim died. If you are told k is approximately 0.1947 for a human body and the body's temperature was 72°F at 1:00 a.m., and the body has been in a storage building at a constant 60°F, approximately what time did the victim die? Recall the average temperature for a human body is 98.6°F. Note in this situation, t is measured in hours.

    c. When helping your father cook a turkey, you were told to remove the turkey when the thickest part had reached 180°F. If you remove the turkey and place it on the table in a room that is 72°F, and it cools to 155°F in 20 minutes, what will the temperature of the turkey be at lunch time (an hour and 15 minutes after the turkey is removed from the oven)? Should you warm the turkey before eating?

Kundenbetreuung: 04106 - 708 25 00
www.com
M. Mustermann
Rechtsverbindliche Unterschrift
comdirect
25449 Qui
Germany
Karten-Nr: 7
BIC: COB

Chapter 5

MATHEMATICS OF FINANCE

5.1 BASICS OF PERSONAL FINANCE

◼ TOPICS

- ◼ Budgeting
- ◼ Understanding Percentages for Consumer Purchases

As adults we all face financial challenges that affect our decisions in daily life. Determining if we have enough money to eat at a certain restaurant and go to a movie is just one example. What about if we wanted to start saving for a home or a car? Learning how to manage the money earned from a job by making a budget and sticking to it is difficult for most people. This section will introduce you to some basic concepts of personal finance that can be used for a lifetime.

Many financial decisions we make, either in college or just after graduating, have long-term effects on our ability to save for retirement, put a down payment on a house, or buy a car. Learning a small number of financial concepts that can help you build good financial habits provides a powerful framework for life. Some of the concepts that will guide us throughout this chapter are listed below.

Sound Financial Concepts

- Learn how to make a budget.

- Determine what your long-term goals are for your money.

- Learn how to research major purchases that may need financing.

- Develop a credit history in a smart, positive manner.

- Always know your bank balance by keeping up with all transactions such as deposits, written checks, debit purchases, and credit card payments.

- Understand how interest works on credit cards.

- Understand how interest works on large purchases such as cars, houses, etc.

These concepts are not meant to be all-encompassing, but rather a foundation for learning to maintain a healthy financial livelihood. Let's begin our study of personal finance with budgeting.

Budgeting

A task that many people find challenging, including the federal government, is setting up a budget based on income. While budgeting may seem overwhelming when you first start out, with practice and dedication it can quickly become an important and useful part of your financial life.

Steps to Create a Budget

1. Calculate monthly and periodical income.

2. Calculate monthly and periodical expenses.

3. Subtract total expenses from income.

4. Allocate any remaining funds to savings, long-term goals, or an emergency fund.

A budget is essentially based on income. One of the biggest hurdles for young workers is to determine the **net income**, which is income after taxes, that they will receive from a given yearly salary. There are many factors that affect net income such as state taxes, federal taxes, social security/medicare taxes—usually listed as FICA (Federal Insurance Contributions Act) on a pay stub—and unemployment taxes. The amount a worker pays for each of these is based on their salary. The 2013 rates for each of these taxes for a worker earning between $18,000 per year and $72,000 per year are: income tax is 15%; social security tax is 6.2%; and 6.0% for unemployment taxes. State taxes vary by state with some states having no state income tax.

Net Income

Net income is equal to total, or gross, income minus all taxes.

Example 1: Determining Federal Income Tax

Jamie graduated from college in 2021 with a degree in accounting. Her first job pays an annual salary of $52,000. How much should Jamie expect to pay in federal income tax for this salary if the federal tax rate is 15%?

Solution

According to the 2021 tax rates, Jamie should expect to pay about 15% of her salary in federal income tax.

$$\$52,000 \cdot 15\% = \$52,000 \cdot 0.15 = \$7800$$

Therefore, Jamie can expect to pay about $7800 in federal income tax at her current salary.

Example 2: Determining Monthly Take Home Pay

Pria just graduated from a liberal arts college and acquired a job as a sociologist. Her yearly salary is $34,500. If the federal tax rate is 15%, the social security tax rate is 6.2%, and the unemployment tax rate is 6%, determine the following.

a. Pria's yearly taxes

b. Pria's monthly take home pay

Solution

a. Pria's salary of $34,500 means that her income tax rate is 15%, social security tax is 6.2%, and unemployment tax is 6%. Therefore, Pria's yearly taxes would be as follows.

$$\begin{aligned}
\text{income tax:} \quad & \$34,500(0.15) = \$5175 \\
\text{social security:} \quad & \$34,500(0.062) = \$2139 \\
\text{unemployment:} \quad & \underline{\$34,500(0.06) = \$2070} \\
\text{total} \quad & \qquad\qquad\quad = \$9384
\end{aligned}$$

So, Pria can expect to pay approximately $9384 in taxes for the year.

b. Pria's monthly take home pay would be equal to her yearly salary minus her yearly taxes divided over the 12 months of the year. So,

$$\frac{\$34,500 - \$9384}{12} = \frac{\$25,116}{12} = \$2093 \text{ per month.}$$

Therefore, Pria's monthly take home pay is $2093.

Example 3: Budgeting on a Given Income

In Example 2, Pria had a monthly net income of $2093. Use the steps to create a budget to allocate Pria's monthly income if her expenses consist of the following.

$$\begin{aligned}
\text{rent:} \quad & \$600 \\
\text{gas:} \quad & \$250 \\
\text{utilities:} \quad & \$150 \\
\text{food/entertainment:} \quad & \$600 \\
\text{student loans:} \quad & \underline{\$210} \\
& \$1810
\end{aligned}$$

Solution

The steps to create a budget indicate that we need to do the following.

1. Determine monthly income.

2. Determine total monthly expenses.

3. Subtract the amount from Step 2 from the amount from Step 1.

4. Allocate additional funds to savings.

In Example 2, we calculated Pria's monthly income to be $2093. Given Pria's total monthly expenses of $1810, we can now find the difference between monthly income and total monthly expenses.

$$\$2093 - \$1810 = \$283$$

This means that Pria has $283 of remaining income each month that she can put toward savings, long-term goals, or an emergency fund. Making a sound financial decision would mean that Pria should understand that she may not always have an extra $283 each month. Therefore, instead of spending the funds, placing the money into a savings account would be a financially responsible decision.

Example 4: Creating a Budget with a Spreadsheet

Gail works for an art gallery with an annual income of $44,000. Her net monthly pay is $2,669.33. Her basic monthly expenses are the following.

rent:	$550
utilities (water, gas, electricity, internet, cable):	$310
cell phone:	$60
insurance:	$40
transportation:	$65
food:	$300
loan repayment:	$290

In addition, Gail receives $30 a week on average from a side consulting job and gives $45 a month to charities. Use a spreadsheet, such as Microsoft Excel, to create a budget for Gail. How much does she have each month for expenses not listed in her budget?

Solution

Creating a budget using a spreadsheet allows us to let the computer do the arithmetic. This means that even though expenses might change month-to-month or new incomes are introduced, we can easily add them to the spreadsheet and the budget will be adjusted accordingly. Recall that to prepare a budget, we need to add both the income and expenses separately. Because spreadsheets can only perform calculations on numbers and not words, we need to list the numerical figures in one cell and their labels in another. Therefore, we'll create a column for expenses and one for incomes. Let's start with expenses. In cell A1 of the spreadsheet, type "Expense", and in the adjacent cell, B1, type "Amount". The spreadsheet should look like the following screenshot.

	A	B
1	Expense	Amount
2		
3		
4		

Under the column labeled "Expense" enter the name of each of Gail's expenses and under the column labeled "Amount" enter the amount associated with each expense. The new spreadsheet should now look like the following.

	A	B
1	**Expense**	**Amount**
2	rent	$550
3	utilities (water, gas, electricity, internet, cable)	$310
4	cell phone	$60
5	insurance	$40
6	transportation	$65
7	food	$300
8	loan repayment	$290
9	charities	$45

<table>
<tr><td colspan="2">🖝 **HELPFUL HINT**</td></tr>
</table>

All formulas in Excel must begin with an equal sign, =.

To obtain the total amount of expenses, we need a place for the total on row 10. Enter "Total" in A10. In B10, we'll put a formula that will sum the expenses. Enter =SUM(B2:B9), which instructs the computer to add together the data from column B, row 2 through column B, row 9.

fx	=SUM(B2:B9)	
	A	B
1	**Expense**	**Amount**
2	rent	$550
3	utilities (water, gas, electricity, internet, cable)	$310
4	cell phone	$60
5	insurance	$40
6	transportation	$65
7	food	$300
8	loan repayment	$290
9	charities	$45
10	Total	$1,660

The spreadsheet calculates that Gail has monthly expenses of $1,660.

Next, we'll follow the same process for Gail's income. Create a set of columns for Income and Income Amount in columns D and E. (We've intentionally left column C blank for clarity.)

	A	B	C	D	E
1	**Expense**	**Amount**		**Income**	**Income Amount**
2	rent	$550			
3	utilities (water, gas, electricity, internet, cable)	$310			
4	cell phone	$60			
5	insurance	$40			
6	transportation	$65			
7	food	$300			
8	loan repayment	$290			
9	charities	$45			
10	Total	$1,660			

Enter Gail's income labels in column D and the amount in column E. Note that when we enter the amount for her consulting income, we'll need to write a formula for the monthly income, since we're given the weekly amount. The formula will multiply the weekly amount by 4; that is, "=30*4." (Note that during some months, Gail will be paid 5 times for her side consulting job.) Doing it this way means that if the amount Gail receives each week changes, she can simply replace the number 30 with the new weekly income, and the spreadsheet will adjust the monthly amount accordingly. To calculate Gail's total income for the month, enter "=sum(E2:E3)" into cell E10. The budget spreadsheet now looks like the following.

fx	=30*4				
	A	**B**	**C**	**D**	**E**
		Amount		**Income**	**Income Amount**
1	**Expense**				
2	rent	$550		Salary	$2,669.33
3	utilities (water, gas, electricity, internet, cable)	$310		Consulting	$120
4	cell phone	$60			
5	insurance	$40			
6	transportation	$65			
7	food	$300			
8	loan repayment	$290			
9	charities	$45			
10	Total	$1,660		Total	$2,789.33

Finally, we need to subtract the Expense Total from the Income Total to find the remaining funds that Gail has available. We'll put this in column G; again, leaving a blank column F for the sake of clarity. We need to tell the spreadsheet to subtract the expense total from the income total by directing it to the cells that contain the total for expenses and incomes. The formula will be in the form income total - expense total, so type the formula "=E10-B10" into cell G10.

fx	=E10-B10						
	A	**B**	**C**	**D**	**E**	**F**	**G**
		Amount		**Income**	**Income Amount**		**Remaining Funds**
1	**Expense**						
2	rent	$550		Salary	$2,669.33		
3	utilities (water, gas, electricity, internet, cable)	$310		Consulting	$120		
4	cell phone	$60					
5	insurance	$40					
6	transportation	$65					
7	food	$300					
8	loan repayment	$290					
9	charities	$45					
10	Total	$1,660		Total	$2,789.33		$1,129.33

Using the spreadsheet from Example 4, find Gail's remaining income if her utilities increase by $50 a month.

We can now see that Gail has $1,129.33 to put towards expenses not in her budget each month. One of the values of using a spreadsheet like this is that if any of the expenses or incomes change, the spreadsheet automatically adjusts the final calculation for us.

Understanding Percentages for Consumer Purchases

In an attempt to be financially responsible, purchasing items on sale whenever possible is quite prudent. We have all been in a department store and noticed items that were on sale for an announced percentage off of the list price, or full-purchase price. Sometimes the reduced prices are indicated on the price tag, but often we are left calculating the discount on our own before reaching the cash register. Either way, knowing the amount that we save can help us stay within our budget. Similarly, having a sense of how much a store marks up an item from their original purchase price in order to determine the retail price often gives us an incentive to wait for those sales!

A few common concepts in consumer purchasing and financial mathematics are necessary for us to continue.

List Price

The price of an item as it is listed for public sale is the **list price**.

Discount

The **discount** is the reduction from the list price. This is usually given as a percentage of the list price.

Sale Price

The **sale price**, sometimes called the **net price**, is the actual cost of an item after a discount of some kind is applied.

$$\text{sale price} = \text{list price} - \text{discount}$$

Example 5: Finding the Sale Price

Let's begin with a simple example regarding the price of a pair of shoes.

Tennis Star Shoes is having a sale. A pair of tennis shoes has a list price of $129.99. The store is offering a discount of 25% off of the list price. Determine the sale price of the shoes before taxes.

Solution

We are to find the sale price of the shoes after a discount of 25% is taken. Since the list price is $129.99, the discount amounts to 25% of $129.99.

$$25\% \text{ of } \$129.99 = 0.25(\$129.99) = \$32.50$$

So, the discount amounts to $32.50. Now, the final price before taxes is determined by subtracting the discount from the list price.

$$\begin{aligned} \text{sale price} &= \text{list price} - \text{discount} \\ &= \$129.99 - \$32.50 \\ &= \$97.49 \end{aligned}$$

The sale price of the pair of shoes is $97.49 before taxes.

Example 6: Determining a Discount

Andrea bought a stereo on sale for $198.45. The list price of the stereo was $330.75. What percentage is the discount?

Solution

We begin by recognizing that a percentage off the list price leads to our sale price. A discount of some percentage is subtracted from the original amount of $330.75. This means that the amount of the discount is $330.75 − $198.45 = $132.30. So, what we really need to know is what percentage $132.30 is of $330.75. Mathematically we can interpret the problem as the following.

$$330.75x = 132.30$$
$$x = \frac{132.30}{330.75}$$
$$= 0.4$$

So, the amount of the discount represents a 40% discount.

Skill Check 2

Max bought a computer on sale for $427. The original price of the computer was $1220. What percentage is the discount?

Percentage Change

The **percentage change** over a period of time can be calculated by the following formula.

$$\text{percentage change} = \frac{\text{new value} - \text{reference value}}{\text{reference value}} \cdot 100\%$$

Example 7: Computing Percentage Decrease

Freddie bought a tablet PC for $1200. One year later, the same tablet PC costs $900. Determine the percentage decrease for the price of the tablet PC.

Solution

To find the percentage change, we need to compare the difference between the values (that is, new price − old price) and divide by the reference value (old price). So, the percentage change is calculated as follows.

$$\text{percentage change} = \frac{\text{new value} - \text{reference value}}{\text{reference value}} \cdot 100\%$$
$$= \frac{900 - 1200}{1200} \cdot 100\%$$
$$= \frac{-300}{1200} \cdot 100\%$$
$$= -0.25 \cdot 100\%$$
$$= -25\%$$

Skill Check 3

Zola bought a camera for $1125. If the price of the camera six months later was $761, what is the percentage decrease?

This means the price of the tablet has decreased 25% over the last year. Note that if the percentage change in price was an increase of the price, then the percentage change would be positive.

The list price of a product in a store is based on two factors.

1. The cost of the product for the store—usually called the wholesale price

2. The percentage the product is marked up from the wholesale price

Skill Check 4

Use the list price of $630 for the bike in Example 8 to find the sale price if the bike was reduced 40%. Compare the 40% reduction in price to the wholesale price in Example 8.

Example 8: Computing Percentage Increase

Pear Bike Company uses a 40% profit margin to determine the list price of their products. If the company buys a bike at a wholesale price of $450, what would be the list price for the bike?

Solution

To determine a percentage increase, we need to recognize that the final price will be equal to the original price of $450 increased by 40%. So, the amount of the increase is

$$\$450(0.40) = \$180.$$

This means the total price for the bike would be

$$\$450 + \$180 = \$630.$$

Skill Check Answers

1. $1079.33

2. 65%

3. −32.36%, or a 32.36% decrease

4. $378; It is less than the wholesale price in Example 8.

5.1 EXERCISES

🚀 APPLICATIONS

1. Jeff was recently hired for a job with an annual income of $48,000. Using the federal income tax rate of 15%, what amount should Jeff expect to pay in federal taxes?

2. Megan has a job where she has a take-home salary each month of $1800. If Megan wants to spend no more than 25% of her income on rent, how much rent can Megan afford?

3. Erica earns $670 weekly at her job at a local newspaper. Calculate the weekly taxes for Erica's salary. If Erica wishes to purchase a car where her payments are no more than 15% of her take home pay for a month, what is the maximum monthly car payment she can afford? Assume that the tax rate is 27.2% and that a month has 4 weeks.

4. Jack rents an apartment for $650 per month, pays his car payment of $470 per month, has utilities that cost $420 per month, and spends $850 per month on food and entertainment. Determine Jack's monthly expenses.

5. Using the information from Exercise 4, if Jack has $1500 remaining after his monthly expenses, what is his take-home pay?

6. The given table represents the estimated cost for an on-campus student to attend the University of California, Los Angeles (UCLA) in the 2013–2014 academic year (10 months). Create a monthly expense budget for attending UCLA and determine a monthly income that would allow a student to attend UCLA without incurring any debt.

Estimated Cost of Attending UCLA for the 2013–2014 Academic Year

Budget Category	On-Campus Student
University Fees	$12,685
Room & Board	$14,454
Books & Supplies	$1536
Transportation	$807
Personal	$1395
Health Insurance	$1323
Loan Fees	$156
Total	$32,356

Source: Financial Aid Office, "2013-2014 Undergraduate and Graduate Budgets", UCLA. Accessed April 2014. http://www.fao.ucla.edu/publications/2013-2014/Budget_Figures.pdf

7. A car with a list price of $18,000 will be discounted 35% at the time of purchase. What is the sale price?

8. The discount on a TV amounts to $240. If the sale price is $1575, what was the list price of the TV?

9. You purchase a pair of jeans with a list price of $175. If the sales tax is 8.5%, what is the total cost of the jeans? Round your answer to the nearest cent.

10. At a restaurant, your total bill is $48.55. You wish to give a tip of 18% of the total bill. What is the amount of the tip? Round your answer to the nearest cent.

11. Jamie found a receipt for a wireless speaker for $127.59, tax included. If the sales tax rate was 9%, what was the selling price of the wireless speaker before taxes? Round your answer to the nearest cent.

12. The average cost of a car in 1990 was $12,500. In 2010, the average price of a car was $21,800. What is the percentage increase in the average price of a car? Round your answer to the nearest tenth of a percent.

13. During the housing price decline of 2009, the value of a house decreased by 30% in one year. If the 2010 value of the house was $115,000, what was the house worth prior to the decline? Round your answer to the nearest cent.

14. The wholesale price of a coat is 50% less than the retail price. If the retail price is $199, what is the wholesale price?

15. A store is having a 60% off sale. The sale price of an item is $125. What is the list price?

16. The local sales tax is 8%. If a pair of shoes sells for $115, what is the total cost after tax?

17. The value of your house is $245,000. If your local property tax rate is 2.5% of the value, how much are your property taxes?

✎ WRITING & THINKING

18. You receive a 10% decrease in your pay. What percentage increase in pay would you have to receive in order to gain your original pay rate again? Round your answer to the nearest whole percent.

5.2 SIMPLE AND COMPOUND INTEREST

■ TOPICS

- Simple Interest
- Compound Interest
- Continuous Compound Interest
- Annual Percentage Yield

Albert Einstein is reported to have once said, "the most powerful force in the universe is compound interest. " Whether Einstein actually said this or not, it's a great financial principle to live by. Without attention, compound interest can rapidly become a force of ill intent and create financial burdens we never planned on. Compound interest can also become the means to our financial dreams through a prudent investment. Why? Because compound interest is an exponential function. As we've seen, exponential functions can quickly create extremely large numbers. This might explain why feelings of disappointment set in after years of paying a mortgage only to discover that most of the money went toward paying the lender for the privilege of borrowing the money, or feelings of elation when discovering that a small nest egg invested at the time of your birth is now available to help you pay for college!

We are often placed in a situation where we need to borrow money in order make a purchase and we must rely on credit to do so. Whether that is in the form of a personal loan, payday loan, or credit card, when we borrow money, we do so with the understanding that we will in turn reimburse the credit holder in the form of **interest**. In other words, we pay the lender a fee for allowing us to borrow money.

Interest

Interest is the amount charged by a lender for borrowing money.

Interest can be a positive thing if you are investing money; that is, if you are the lender. Consider an 18-year-old making a one-time $5000 contribution to an investment fund that had an average 6% annual return. If she never touches the money, $5000 will grow to over $77,000 by the time she retires at age 65. However, if she waits until she's 40 years old to make her investment, that $5000 would only grow to about $21,000. The key factor is in the length of time over which the compounding of interest takes place.

However, if you are the one needing to borrow money, interest is not as gratifying. For instance, if you place $1000 on a credit card with 15% interest rate and pay only the minimum payments required each month, it will take you almost nine years to pay off the debt and cost you $730 in interest.

Simple Interest

In order to calculate the amount of interest to be paid to a lender, you need to know a few things: the amount of the principal, the interest rate being charged, and the amount of time over which the interest will be calculated. The **principal** amount is the sum of money on which the interest is charged. It is also referred to as the initial value of an investment or the starting value of an investment. The **interest rate** is the amount charged expressed as a percentage of the principal. It is usually noted on an annual basis and called the **annual percentage rate (APR)**.

> ### Principal
>
> The **principal** is the sum of money on which interest is charged.

> ### Interest Rate
>
> The **interest rate** is the amount charged to the borrower expressed as a percentage of the principal.

> ### Annual Percentage Rate (APR)
>
> **Annual percentage rate** is the yearly interest rate that is charged for borrowing. APR is normally given as a percentage per year.

The most straightforward way to calculate interest is by using **simple interest**. This method of calculating interest only considers the principal amount.

Since interest is nothing more than a percentage, or rate, of the principal, for any given amount of principal P, we can see that the interest on that principal for a one-year loan can be found by multiplying the principal by the interest rate. When we combine this idea with the fact that we are introducing the variable of time, we can compute the amount of interest that accrues, or accumulates, for a given amount of time.

> ### Simple Interest Formula
>
> The amount of interest on a simple interest loan with principal P, annual interest rate r (written as a decimal), and loan term of t (usually in years) is calculated with the following formula.
>
> $$I = Prt$$

> ### Example 1: Calculating Simple Interest

Determine the interest that is accrued on $5500 for five years at a rate of 8.5%.

Solution

We have that the principal is $P = \$5500$ and the interest rate is 8.5%. Note that we need to change the interest rate to a decimal before substituting it in the formula, so $r = 0.085$. We also have that $t = 5$. Using our formula, we can determine the amount of interest owed on the $5500 as follows.

$$I = Prt$$
$$= (5500)(0.085)(5)$$
$$= \$2337.50$$

So, if we borrowed $5500 at the given rate of 8.5% for five years, we would owe $2337.50 in interest along with the original principal of $5500 for a total purchase of $7837.50.

Retail stores often offer special financing through their store for purchases over a certain amount. They entice consumers by advertising a "no interest, same as cash" loan. But beware, these interest free loans are often short term and have very high interest rates after the free period is over. Example 2 examines such an offer.

Example 2: Calculating Simple Interest on Purchases

Ian is purchasing a new television with a "deal" from Big Screens R Us. The deal offers 90 days same-as-cash to make the purchase. This means that at the end of 90 days, if Ian has paid off the cost of the television, he owes no interest charge. However, if Ian does not pay off the amount, he owes simple interest for the original purchase amount calculated over the entire 90 days. If the price of the television is $2650 (tax included) with an annual interest rate of 21.99%, how much would Ian owe on the 91st day if he made no payments during the first 90 days?

Solution

First, we want to determine the entire amount due after the 90 days has expired assuming that Ian has not paid any money toward the amount he owes. This means the principal of $2650 is still due in addition to the amount of interest that accumulated over the 3 months that he borrowed the money. Since the formula for simple interest requires that time t must be given in years, we need to convert three months into a fraction of a year; that is, $\dfrac{3}{12} = \dfrac{1}{4}$ of a year. Remember that when using the formula, we need the interest rate to be in decimal form. Therefore, we have that $P = 2650$, $r = 0.2199$, and $t = \dfrac{1}{4}$. The amount of interest gained over three months can then be calculated as follows.

$$I = Prt$$
$$= (\$2650)(0.2199)\left(\frac{1}{4}\right)$$
$$\approx \$145.68$$

So, the amount of interest due at the end of 90 days would be $145.68. Therefore, the total amount Ian must pay for the television on the 91st day is the sum of the principal and the interest.

$$\$2650 + \$145.68 = \$2795.68$$

Skill Check 1

Find the total cost of a loan for $4320 that has a simple interest rate of 15% for 18 months.

Compound Interest

Each of the previous examples was based on the use of simple interest. For certain types of loans, interest is calculated and applied more than one time per year. This is known as compound interest.

Compound interest is based on simple interest, except that there is a gain of interest added to the principal at each interval. In each compounding interval, interest is calculated on the principal as well as any interest accumulated during the previous interval. Thus, the principal plus the interest equals the new adjusted principal for the next interval.

Compound Interest

Compound interest is interest that is computed based on both the principal and the accrued interest as additional principal at each interval. The **future value** is the total amount of money A that has been accrued after compounding at an annual percentage rate r based on the initial principal P with n compounding intervals per year for t years. Future value of a compound interest account is calculated with the following formula.

$$A = P\left(1 + \frac{r}{n}\right)^{nt}$$

For example, if we let the principal be $500 and the rate be 4% per year, then the interest after the first year would be 4% of $500, or $0.04 \cdot 500 = \$20$. This amount is then added to the principal of $500 and the new adjusted principal will become $520. The second year's interest will be 4% of $520, or $20.80. In the same manner as before, the interest is added to the principal and the adjusted principal will be $540.80. After just two years, we have accrued $40.80 in interest.

Interest can be compounded more than once a year. Table 1 shows some common compounding intervals along with their terminology for reference.

Compounding	Number per Year
Annually	1
Semiannually	2
Quarterly	4
Monthly	12
Weekly	52
Daily	365

TABLE 1: Compounding Intervals

Example 3: Computing Compound Interest

Lilly deposits $12,000 into an account with an annual interest rate of 4.5% compounded monthly. If she leaves the money in the account for 10 years, what will the future value be at the end of this time period?

Solution

We can use the compound interest formula to compute the future value for Lilly after 10 years. In her case, $P = 12{,}000$, $r = 0.045$, and $t = 10$. Since interest is compounded monthly, $n = 12$. So we have the following.

$$A = P\left(1 + \frac{r}{n}\right)^{nt}$$

$$= 12{,}000\left(1 + \frac{0.045}{12}\right)^{12 \cdot 10}$$

$$= 12{,}000\left(1.00375\right)^{120}$$

$$\approx \$18{,}803.91$$

So, after 10 years, Lilly will have accumulated a total of $18,803.91, which includes her initial investment of $12,000.

⌁ TECH TRAINING

Wolfram|Alpha can be used to calculate compound interest. Go to www.wolframalpha.com and type "compound interest" into the input bar. Then, click the = button. Wolfram|Alpha will return a worksheet with fields to input the data, but it uses the names instead of variables to label the inputs. Note that we need to choose *future value* from the drop-down menu next to *Calculate*. We also need to click on *compounding frequency* at the bottom of the worksheet since the number of compounding intervals is more than one. Enter "$12000" for *present value*, "4.5%" for *interest rate*, and "10" for *interest periods*—this is asking for the number of years. Choose *monthly* from the drop-down menu next to "compounding frequency". Wolfram|Alpha displays the following. (Notice that Wolfram|Alpha rounds the future value to the nearest $10.)

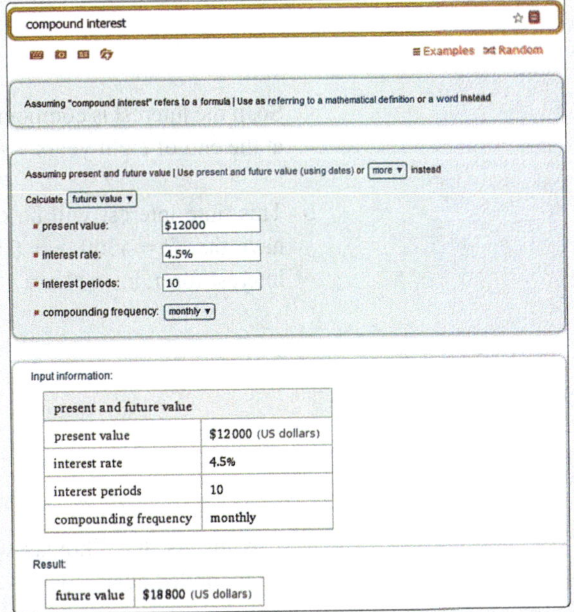

Source: Wolfram Alpha LLC. 2009. Wolfram|Alpha. http://www.wolframalpha.com/input/?i=compound+interest&a=*C.compound+interest-_*Formula. dflt-&a=FSelect_**PresentValueFutureValue-.dflt-&a=*FS-_**PresentValueFutureValue.FV-.*PresentValueFutureValue.PV-.*PresentValueFutureValue. i-.*PresentValueFutureValue.n--&f4=%2412000&f=PresentValueFutureValue.PV_%2412000&f5=4.5%25&f =PresentValueFutureValue.i_4.5&f6=10&f=PresentValueFutureValue.n_10&a=*FP. PresentValueFutureValue. compoundingfreq-_Monthly (accessed July 8, 2014).

Skill Check 2

Use the information in Example 3 to find the future value of Lilly's account after 20 years.

The power of compound interest lies not only in the interest rate, but also in the number of times the interest is compounded per year. The following example compares the growth of an investment given different numbers of yearly compound intervals.

Example 4: Comparing Compound Interest for Different Compounding Intervals

Thomas uses $4500 to open an IRA (Individual Retirement Account) savings account that earns 3.8% APR. If he leaves the money alone, what is the future value after eight years for the following compounding intervals?

a. The interest is compounded yearly.

b. The interest is compounded quarterly.

c. The interest is compounded weekly.

Solution

a. We have that Thomas' principal is $P = 4500$. The interest rate is $r = 0.038$ and time is $t = 8$. If the interest is compounded yearly, then $n = 1$. Therefore, we have

$$A = P\left(1 + \frac{r}{n}\right)^{nt}$$

$$= 4500\left(1 + \frac{0.038}{1}\right)^{1 \cdot 8}$$

$$\approx \$6064.45.$$

So, if the interest is compounded yearly, Thomas will have approximately \$6064.45 at the end of eight years.

b. This time interest will be compounded quarterly. In other words, $n = 4$. We still have that $P = 4500$, $r = 0.038$, and $t = 8$. Substituting these into the compound interest formula, we have

$$A = P\left(1 + \frac{r}{n}\right)^{nt}$$

$$= 4500\left(1 + \frac{0.038}{4}\right)^{4 \cdot 8}$$

$$\approx \$6089.97.$$

This means that Thomas will have \$6089.97 after eight years if the interest is compounded quarterly.

c. Finally, if the interest is compounded weekly, $n = 52$. Substituting this into the formula, we have

$$A = P\left(1 + \frac{r}{n}\right)^{nt}$$

$$= 4500\left(1 + \frac{0.038}{52}\right)^{52 \cdot 8}$$

$$\approx \$6098.03.$$

So, after eight years, Thomas' investment has a future value of \$6098.03 with interest compounded weekly.

You might notice that the more compounding intervals occur in a given year, the larger the total accumulated amount of money. The other factor that has an impact on the accumulated amount is time. The longer an amount is invested, the larger the accumulated amount. Making sound financial decisions requires that we also understand the power of compounding interest over a long period of time.

Continuous Compound Interest

Sometimes there are instances where, instead of compounding interest a certain number of times per year, interest is compounding literally *continuously*—every moment of every day of every week of every year. We call this **continuous compound interest**. Interest that is continuously compounded uses a formula similar to that of interest compounded n times per year, except that the variation among the number of compounding intervals per year is accounted for using the irrational number e.

Continuous Compound Interest Formula

The future value A of a continuous compound interest account after t years at an annual interest rate of r and an initial amount, or principal, P is calculated with the following formula.

$$A = Pe^{rt}$$

Use the following keystrokes on a TI-30XIIS/B or TI-83/84 Plus calculator for the calculation of continuously compounded interest in Example 5.

8900 [×] [2nd] [LN] 0.0205 [×] 15 [)] [=]

Wolfram|Alpha can be used to calculate the result from Example 5. Go to www.wolframalpha.com and type "8900*e^(0.0205*15)" into the input bar. Then click the = button.

In order to calculate the value of Pe^{rt}, notice that we must use the exponential function e^x. In order to enter this function on a calculator, we must press the buttons [2nd] [LN].

If you are using a TI-30XS/B Multiview calculator, the keystrokes will be slightly different for your calculator.

Example 5: Calculating Continuous Compound Interest

Find the future value of $8900 invested at a rate of 2.05% that is compounded continuously over 15 years.

Solution

We know that $P = 8900$, $t = 15$, and $r = 0.0205$. Using the continuous compound interest formula, we have

$$A = Pe^{rt}$$
$$= 8900e^{0.0205 \cdot 15}$$
$$\approx \$12,104.19.$$

So, the future value is $12,104.19 when interest is compounded continuously.

Although it is possible, you very seldom find banks offering interest that is compounded continuously. However, since continuous compounding represents the greatest possible number of compound intervals over a year, it can help us find the maximum amount of interest that is possible to earn in one year.

Example 6: Finding the Maximum Amount of Interest Possible in One Year

Assume you wish to deposit $2500 into an account bearing 6% interest for 10 years.

a. What is the maximum future value possible after 10 years?

b. What is the maximum amount of interest possible after 10 years?

Solution

a. Because we want to know the maximum future value, we need to use the continuous compound interest formula. We are given that the principal is $P = 2500$, $r = 0.06$, and $t = 10$. Substituting the given values into the formula we have the following.

$$A = Pe^{rt}$$

$$= 2500e^{0.06 \cdot 10}$$

$$\approx 4555.30$$

Therefore, the largest possible future value of $2500 invested at 6% over 10 years is $4555.30.

b. We can find the largest amount of interest the principal can earn over the 10-year period by subtracting the principal amount from the future value.

$$\text{interest earned} = \text{future value} - \text{principal}$$

$$I = A - P$$

$$I = 4555.30 - 2500.00$$

$$= 2055.30$$

So, the principal can earn at most $2055.30 in interest over the 10 years.

Annual Percentage Yield

An important component to managing your money and investments is determining the actual amount that an account will earn in a given year. In particular, the effect of the number of compounding intervals per year can make a significant difference in the total amount of money earned in a year. Consider investing $100 at 5% simple interest. At the end of the year, the future value will be $105 since the interest was not compounded during the year. However, if we had $100 invested at 5% compounded quarterly, the future value would be as follows.

$$A = P\left(1 + \frac{r}{n}\right)^{nt}$$

$$= 100\left(1 + \frac{0.05}{4}\right)^{4 \cdot 1}$$

$$\approx \$105.09$$

After we take quarterly compounding into account, the practical interest rate is no longer 5%, it is effectively 5.09% over the year. It should make sense that this rate is higher, since at each compounding period, we earn interest on the original principal plus the interest earned in the previous periods. This practical interest rate is called the **annual percentage yield (APY)**, or **effective interest rate**, for a given period.

Annual Percentage Yield (APY)

The **annual percentage yield** (APY) is the effective annual interest rate earned in a given year that accounts for the effects of compounding. APY is calculated with the formula

$$APY = \left[\left(1 + \frac{r}{n}\right)^{n} - 1\right] \cdot 100\%,$$

where r is the annual percentage rate and n is the number of compounding intervals per year.

Samantha deposits $5000 in an account paying 5% interest per year.

a. Find the APY for Samantha's investment if the interest is compounded monthly.

b. Find the APY if the interest is compounded daily on Samantha's investment.

Solution

a. Notice that to determine the APY, we only need the rate of interest and the number of compound intervals per year. The amount of the principal plays no role in finding the APY. So, we need $r = 0.05$ and $n = 12$ for monthly compounding intervals.

$$\text{APY} = \left[\left(1+\frac{r}{n}\right)^n - 1\right] \cdot 100\%$$

$$= \left[\left(1+\frac{0.05}{12}\right)^{12} - 1\right] \cdot 100\%$$

$$\approx (0.05116) \cdot 100\%$$

$$= 5.116\%$$

This means with monthly compounding, the APY for Samantha's investment is 5.116%.

b. Daily compounding means that $n = 365$. Therefore, our APY calculation is as follows.

$$\text{APY} = \left[\left(1+\frac{r}{n}\right)^n - 1\right] \cdot 100\%$$

$$= \left[\left(1+\frac{0.05}{365}\right)^{365} - 1\right] \cdot 100\%$$

$$\approx (0.05127) \cdot 100\%$$

$$= 5.127\%$$

With daily compounding, Samantha has an annual percentage yield of 5.127%.

Notice that although the advertised APR for Samantha's investment was 5%, when interest is compounded, none of the annual percentage yields were actually 5%.

Since the APY is a larger percentage than the annual percentage rate (APR) when compounding occurs, banks will often publicize the APY for investment accounts instead of the APR to entice investors to deposit money with their institution. Conversely, when advertising interest rates for loans, lenders will use the APR instead of the APY to entice borrowers.

Suppose that the APD Bank of the South advertises the following rates for their personal loans.

Loan Amount	APR*
< $20,000	10.49%
$20,000–$99,999	9.99%
≥ $100,000	7.50%

*interest is compounded quarterly

TABLE 2: APD Bank of the South Personal Loan Rates

Find the APY, or effective interest rate, for each of the loan categories.

Solution

To find the APY for each loan category we need the published APR as well as the number of compounding intervals per year. In this case, each is compounded quarterly, so $n = 4$.

The APY for an APR of 10.49% compounded quarterly is calculated as follows.

$$\text{APY} = \left[\left(1 + \frac{r}{n}\right)^n - 1 \right] \cdot 100\%$$

$$= \left[\left(1 + \frac{0.1049}{4}\right)^4 - 1 \right] \cdot 100\%$$

$$\approx (0.1091) \cdot 100\%$$

$$= 10.91\%$$

Therefore, the APY is 10.91%.

For an APR of 9.99%, we can calculate the APY as follows.

$$\text{APY} = \left[\left(1 + \frac{r}{n}\right)^n - 1 \right] \cdot 100\%$$

$$= \left[\left(1 + \frac{0.0999}{4}\right)^4 - 1 \right] \cdot 100\%$$

$$\approx (0.1037) \cdot 100\%$$

$$= 10.37\%$$

That gives an APY of 10.37%.

Lastly, to calculate the APY for an APR of 7.50%, we have the following.

$$\text{APY} = \left[\left(1 + \frac{r}{n}\right)^n - 1 \right] \cdot 100\%$$

$$= \left[\left(1 + \frac{0.075}{4}\right)^4 - 1 \right] \cdot 100\%$$

$$\approx (0.07714) \cdot 100\%$$

$$= 7.714\%$$

Therefore, the APY is 7.714%.

Table 3 shows the comparison of the publicized APR versus the APY.

Loan Amount	APR	APY
< $20,000	10.49%	10.91%
$20,000–$99,999	9.99%	10.37%
≥ $100,000	7.50%	7.714%

TABLE 3: APR vs. APY

Therefore, when trying to determine which loan is best, look at the APY to get a clearer picture of the actual cost of the loan.

A great example of how the APR can differ substantially from the advertised interest rate is with a payday loan. A payday loan, sometimes called a cash advance, is a short-term loan that requires no collateral, just a history of having a paycheck. A typical payday loan is for two weeks, where the amount of interest owed per $100 borrowed varies from $15 to $25—for just two weeks! These loans became popular during the 1990s and have since been made illegal in many states because of the excessive interest rates charged. So, how expensive is it to get a payday loan?

Example 9: Calculating Interest on Payday Loans

Assume you wish to borrow $300 for two weeks in the form of a payday loan and the amount of interest you must pay is $25 per $100 borrowed. This means that at the end of two weeks, you owe $375. What is the APR?

Solution

Recall that APR is defined as the interest rate over a year. Since the loan was for two weeks, we need to convert this to a yearly rate. We can find the APR by using

$$\text{APR} = \text{short term interest rate} \cdot \frac{52 \text{ weeks}}{\text{length of loan}},$$

where the short term rate of interest is 25% for two weeks, and the loan is for two weeks.

Now, when we fill in the numbers, we obtain the following.

$$\text{APR} = 25\% \cdot \frac{52 \text{ weeks}}{2 \text{ weeks}}$$
$$= 650\%$$

Thus, the APR is 650%. That is a ridiculous rate of interest for one year. Although this rate is never paid because the loans are for a very short amount of time, the APR is the reason these types of loans have become illegal in many states.

Skill Check Answers

1. $5292

2. $29,465.60

5.2 EXERCISES

💡 PRACTICE

Calculate the simple interest for each situation. Round your answer to the nearest cent, if necessary.

1. Determine the interest owed on $800 for 5 years at a rate of 8.5%.

2. Determine the interest owed on $5000 for 2 years at a rate of 10%.

3. Determine the interest owed on $1200 for 30 months at a rate of 19.5%.

4. Determine the interest owed on $550 for 5 years at a rate of 8.5%.

Use the compound interest formula to find the following for each situation.

    a. Calculate the total amount in the account after the given time period.

    b. Determine the amount of interest earned for each.

5. $P = \$2500$, $r = 6.5\%$ compounded weekly, $t = 10$ years

6. $P = \$3500$, $r = 4.5\%$ compounded monthly, $t = 10$ years

7. $P = \$2500$, $r = 6.5\%$ compounded daily, $t = 10$ years

8. $P = \$5650$, $r = 8\%$ compounded biannually, $t = 15$ years

9. $P = \$2500$, $r = 6.5\%$ compounded yearly, $t = 10$ years

10. $P = \$15,000$, $r = 6\%$ compounded semiannually, $t = 25$ years

11. $P = \$12,500$, $r = 8\%$ compounded biweekly, $t = 15$ years

12. $P = \$7300$, $r = 19.9\%$ compounded weekly, $t = 20$ years

An account is compounded monthly for five years with an APR of 9%. For each principal amount, calculate the following.

    a. The total amount paid

    b. The amount of interest paid

13. $15,000

14. $30,000

15. $60,000

16. $120,000

17. You are purchasing a new computer using the store's "90 days same as cash" deal. If the cost of the computer is $1575 (tax included) with an annual interest rate of 19.99%, how much would you owe on the 91st day if you make no payments during the first 90 days?

18. Assume you wish to borrow $500 for two weeks and the amount of interest you must pay is $20 per $100 borrowed. What is the APR at which you are borrowing money? Round your answer to the nearest hundredth.

19. A couple deposits $25,000 into an account earning 6% annual interest for 25 years.
  a. Calculate the future value of the investment if interest is compounded monthly.
  b. Calculate the future value if the interest on the investment is compounded weekly.

20. Suppose your salary in 2016 was $65,000. If the annual inflation rate is 4%, what salary do you need to make in 2024 in order for it to keep up with inflation?

21. Suppose that $15,000 is deposited for eight years at 5% APR.
  a. Calculate the simple interest earned.
  b. Calculate the interest earned if interest is compounded weekly.
  c. Calculate the interest earned if interest is compounded monthly.

22. A payday loan is made for six weeks, where the amount of interest owed per $100 borrowed is $20. If you borrow $500 for six weeks, answer the following questions.
  a. How much do you owe at the end of six weeks?
  b. What is the APR for this transaction?

23. Angela deposits $2500 into an account with an APR of 5.5% for 10 years. Find the amount of interest earned for each of the following situations.
  a. Interest is compounded annually.
  b. Interest is compounded monthly.
  c. Interest is compounded weekly.
  d. Interest is compounded daily.
  e. Interest is compounded continuously.

24. David deposits $4000. Determine the APY for each of the following.
  a. APR of 5% compounded monthly
  b. APR of 5% compounded weekly
  c. APR of 5% compounded daily
  d. APR of 7.5% compounded monthly
  e. APR of 7.5% compounded weekly
  f. APR of 7.5% compounded daily

25. Determine the simple interest earned on $10,000 after 10 years if the APR is each of the following rates.
    a. 3%
    b. 6%
    c. 12%
    d. 24%

26. Suppose the First Bank of Lending offers a CD (Certificate of Deposit) that has a 6.45% interest rate and is compounded quarterly for three years. You decide to invest $5500 into this CD.
    a. Determine how much money you will have at the end of the three years?
    b. Find the APY.

27. The First Bank of Lending lists the following APR for loans. Determine the APY, or effective interest rate for each category.

First Bank of Lending Loan APR

Loan Amount	APR*
< $20,000	11.25%
$20,000–$99,999	8.99%
≥ $100,000	5.75%

*interest is compounded quarterly

5.3 ANNUITIES: PRESENT AND FUTURE VALUE

■ TOPICS

- ■ Saving Money
- ■ Working toward a Goal

Saving Money

Interest is not always a bad thing. It can lend a helping hand when we are saving money. Whether it is for a particular short term goal or a long term retirement plan, we can use interest to our benefit.

Suppose we have a particular monetary goal in mind that we would like to save for. We can rearrange the interest formula we saw previously so that it indicates how much money needs to be invested *now* in order to meet that goal in the *future*. This amount needed to invest now is called the **present value**.

Present Value (*PV*)

Present value (*PV*) is the amount of principal needed now in order to reach a future value amount. Present value is calculated with the formula

$$PV = \frac{A}{\left(1+\dfrac{r}{n}\right)^{nt}}$$

where A is the future value, r is the APR, n is the number of compound intervals per year, and t is time, or term, in years.

For example, many states have college savings plans that allow parents to deposit a lump sum amount of money into their child's college fund based on current college costs. The amount deposited will then earn enough interest to cover the predicted cost of college 18 years in the future. The only catch is you must have a lump sum payment to invest now.

Example 1: Computing Principal Needed for Future Value

State College predicts that in 18 years it will take $150,000 to attend the college for four years. Amber has a substantial amount of cash and wishes to save money for her newborn child's college fund. How much should Amber put aside in an account with an APR of 4%, compounded monthly, in order to have $150,000 in the account in 18 years?

Solution

We know that Amber wishes to have a future value of $150,000 for her child's college fund. Since she has found an investment opportunity that will give her monthly compounding at 4%, we know that $A = 150{,}000$, $r = 0.04$, $n = 12$, and $t = 18$. Substituting these values into the present value formula, we have the following.

$$PV = \frac{150{,}000}{\left(1+\dfrac{0.04}{12}\right)^{12\cdot18}}$$

$$\approx 73{,}100.31$$

This means that if Amber could deposit \$73,100.31 into the account now, then after 18 years there will be \$150,000 in the account.

If calculating large formulas in pieces on a calculator, round each step to at least six decimal places to avoid rounding errors in the subsequent calculations.

How much should you deposit now in an account with an APR of 7% compounded monthly if you wish for the account to have a balance of \$200,000 in 25 years?

To perform the calculation from Example 1 on a TI-30XIIS/B or TI-83/84 Plus calculator, use the following keystrokes.

150000 ÷ ((1 + ((0.04 ÷ 12)) ^ ((12 × 18)) =

To perform the calculation with Wolfram|Alpha, go to www.wolframalpha.com and type "150000/(1+(0.04/12))^(12*18)" into the input bar. Then, click the = button.

Often it is preferable to make regular payments into a savings program rather than depositing a lump sum of money at once. Although these payments might be considerably smaller than a lump sum amount, as we discussed in the previous example, compound interest can work in our favor in just the same manner. The significant point with this type of savings plan is that consistent payments are ongoing. For instance, suppose you deposit \$50 each month into an account. The first month, interest is calculated on the initial \$50 deposit. However, in the next period, after you've made your second deposit of \$50, interest is calculated on both \$50 payments as well as the interest you earned in the first period. So, although you personally are building up your savings by contributing \$50 a month, interest continues to add to your savings as well. In broad terms, whenever fixed regular payments are made we refer to them as **annuity** payments.

Annuity

An **annuity** is a sequence of regular payments made into an account, or taken out of an account, over time.

We can calculate the future value of an annuity payment plan by using the following formula. It takes into account not only the additional compound interest each period, but also the regular payment added each time.

Annuity Formula for Finding Future Value

The future value (FV) of an annuity savings account is calculated with the formula

$$FV = PMT\,\frac{\left[\left(1+\dfrac{r}{n}\right)^{nt}-1\right]}{\left(\dfrac{r}{n}\right)}$$

where PMT is the payment amount that is deposited on a regular basis, r is the APR, n is the number of regular payments made each year, and FV is the future value after t years.

Example 2: Calculating Future Value Using an Annuity Plan

Daigle began placing $75 per month into an annuity savings plan when he received his first official paycheck in August 2020. The savings plan earns 2.5% APR, compounded monthly.

a. Determine the amount of money that Daigle will personally put in the savings plan over 10 years.

b. Calculate the future value of Daigle's savings in August 2030.

Solution

a. To find the amount of money that Daigle will personally add to the savings plan, we need to multiply the monthly payment amount by the number of payments each year and then by the number of years he plans to continue his payments. This gives us the following.

$$\text{payment amount} \cdot \text{payments per year} \cdot \text{number of years} = \$75 \cdot 12 \cdot 10$$
$$= \$9000$$

So, over the 10 years, Daigle will contribute $9000 to his savings plan.

b. In order to find the future value of Daigle's savings, we need several values. First, we are told that he will contribute $75.00 each month. So, we know that $PMT = \$75.00$ and the number of payments made each year gives us $n = 12$. The APR given is 2.5%, or $r = 0.025$. The time period is from August 2020 through August 2030, so $t = 10$ years. Now, we can substitute these values in the annuity formula to find the future value.

$$FV = PMT \cdot \frac{\left[\left(1+\dfrac{r}{n}\right)^{nt} - 1\right]}{\left(\dfrac{r}{n}\right)}$$

$$= 75 \cdot \frac{\left[\left(1+\dfrac{0.025}{12}\right)^{12 \cdot 10} - 1\right]}{\left(\dfrac{0.025}{12}\right)}$$

$$\approx 10{,}212.90$$

Therefore, after 10 years, Daigle's savings account will contain $10,212.90. Since we know that he contributed $9000, over 10 years Daigle's savings earned interest in the amount of: $10,212.90 − $9000 = $1212.90.

Skill Check 2

Find the future value of the annuity plan in Example 2 after 15 years.

We can use Microsoft Excel to calculate the annuity formula by using the built-in formula "=FV(rate,nper,pmt,[pv],[type])". (Note that another spreadsheet program could be used, but the directions may be slightly different.) The built-in function uses the following notation.

rate: interest rate per period (note this is per period, so this value is $\frac{r}{n}$). Excel doesn't mind whether interest is in the form of a decimal or a percentage.

nper: total number of payment periods—because this is the total number of periods, this value is $n \cdot t$.

pmt: payment amount made each period—this must be the same for each period; the value you enter here must be negative if you are making a deposit into an account and positive if you are receiving a payment from an account.

fv: future value, or a cash balance you want to attain after the last payment is made.

[pv]: present value, or the lump-sum amount that a series of future payments is worth right now. If left blank, Excel will assume the value is $0.

[type]: a value of 0 if the payments are made at the beginning of the compounding periods or 1 if the payments are made at the end of the periods. If left blank, Excel will assume that payments are made at the beginning of the compounding periods.

To calculate the future value of an annuity, write the formula in any cell with the values filled in.

Therefore, for part **b.** of Example 2, to calculate the amount in the account after 10 years (or 120 months) using Microsoft Excel, we can type in "=FV(2.5%/12,120, -75)"—note that the payment is negative since we are making a payment into the account. The value given by Excel is $10,212.90.

Use the annuity formula to calculate the future value after six months of monthly $50 payments for an annuity savings plan having an APR of 12%.

Solution

We have monthly payments of $PMT = \$50$, an annual interest rate of $r = 0.12$, $n = 12$ because the payments are made monthly, and $t = \frac{1}{2}$ because six months is a half year. Using the savings formula, we can find the future value after six months. Replacing the variables in the formula with their values we obtain the following.

$$FV = PMT \cdot \frac{\left[\left(1+\dfrac{r}{n}\right)^{nt} - 1\right]}{\left(\dfrac{r}{n}\right)}$$

$$= 50 \cdot \frac{\left[\left(1+\dfrac{0.12}{12}\right)^{12 \cdot 0.5} - 1\right]}{\left(\dfrac{0.12}{12}\right)}$$

$$\approx 307.60$$

Therefore, after six months, the savings plan has accrued $307.60.

HELPFUL HINT

If you are using a TI-30XS/B Multiview calculator, the keystrokes will be slightly different for your calculator.

TECH TRAINING

To perform the calculation from Example 3 on a TI-83/84 Plus calculator, use the following keystrokes.

50 × (((1 + 0.12 ÷ 12) ^ (12 × 0.5) − 1) ÷ (0.12 ÷ 12)) =

To perform the calculation with Excel, input the following to obtain the future value. "= FV(12%/12,6,-50)"

To perform the calculation with Wolfram|Alpha, go to www.wolframalpha.com and type "50*((1+0.12/12)^(12*0.5)-1)/(0.12/12)" into the input bar. Then, click the = button.

Example 4: Calculating Future Value of an IRA Using the Annuity Formula

At the age of 25, Angela starts an IRA (individual retirement account) to save for retirement. She deposits $200 into the account each month. Based on past performance of similar accounts, she expects to obtain an annual interest rate of 8%.

a. How much money will Angela have saved upon retirement at the age of 65?

b. Compare the future value to the total amount Angela deposited over the time period.

Solution

a. Because Angela starts making payments at age 25 and stops at 65, she is making payments over 40 years. Using the savings plan annuity formula, we need the following: payments of $PMT = \$200$, an interest rate of $r = 0.08$, and $n = 12$ for monthly deposits. The balance after $t = 40$ years is as follows.

$$FV = PMT \cdot \frac{\left[\left(1+\dfrac{r}{n}\right)^{nt} - 1\right]}{\dfrac{r}{n}}$$

$$= 200 \cdot \frac{\left[\left(1+\dfrac{0.08}{12}\right)^{12\cdot 40} - 1\right]}{\dfrac{0.08}{12}}$$

$$\approx 698,201.57$$

Therefore, Angela will have saved $698,201.57 upon retirement.

b. Now we wish to calculate the total deposit amount over the 40 years. Keep in mind that over the 40 years, Angela has made 480 monthly deposits of $200. That means she has deposited a total of

$$480 \cdot \$200 = \$96,000.$$

The rest is all interest. Or in other words, even though Angela only deposited $96,000, she made $602,201.57 in interest. Just think if she had started saving that same $200 per month when she was 18!

Repeat Example 4 with Angela saving $200/month from age 18 to age 65.

⬕ TECH TRAINING

To perform the calculation from Example 4 on a TI-83/84 Plus calculator, use the following keystrokes.

200 **×** **(** **(** **(** 1 **+** 0.08 **÷** 12 **)** **^** **(** 12 **×** 40 **)** **−** 1 **)** **÷** **(** 0.08 **÷** 12 **)** **)** **=**

To perform the calculation with Excel, input the following to obtain the future value. "=FV(8%/12,(12*40),-200)"

To perform the calculation with Wolfram|Alpha, go to www.wolframalpha.com. In the input bar, type "200*((1+0.08/12)^(12*40)-1)/(0.08/12)". Then, click the = button.

To further show the power of compound interest, let's consider two different investment strategies. Suppose two friends, Kurt and Johnny, graduate from college on the same day. Each has a different view on saving for the future. Kurt decides to save for the future right away and begins depositing $200.00 per month into an account with an APR of 7.5%. After 15 years, Kurt decides to stop investing money into the account and NEVER make another deposit. Without withdrawing any of the money, the account will still continue to earn the same level of interest.

Johnny, however, waits to start saving. On the same day that Kurt stops adding monthly payments into his savings account, Johnny decides to begin depositing $200.00 per month into a savings account of his own, having the same APR of 7.5%. How many years will it take Johnny to catch up with Kurt's savings balance?

Kurt's savings plan after 15 years has a future value of $66,222.46. We find this just as we have before with the annuity formula, as shown.

$$FV = 200 \cdot \frac{\left[\left(1+\dfrac{0.075}{12}\right)^{12 \cdot 15} - 1\right]}{\dfrac{0.075}{12}}$$

$$= 66,222.46$$

Since Kurt decides not to withdraw any money from the account, the savings will continue to grow at 7.5% interest, compounded monthly even though he has stopped depositing money each month. The following graph shows the growth of Kurt's account since graduation.

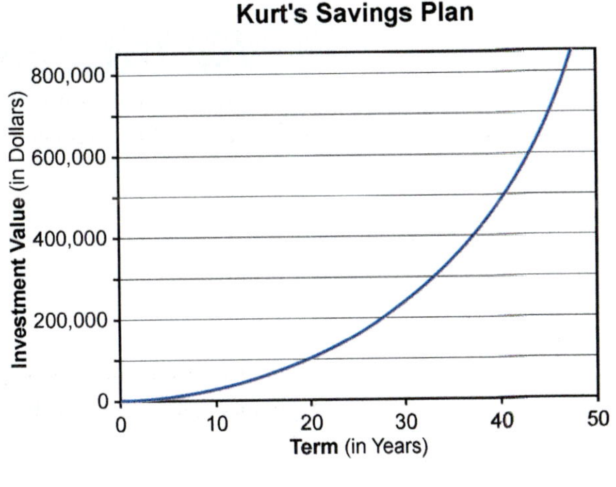

FIGURE 1

Johnny begins depositing money when he turns 37 years of age, 15 years after graduation. At this time, he begins to deposit $200.00 per month, just as Kurt did, into an account paying 7.5% interest. Figure 2 compares the values of Kurt's and Johnny's savings since graduation.

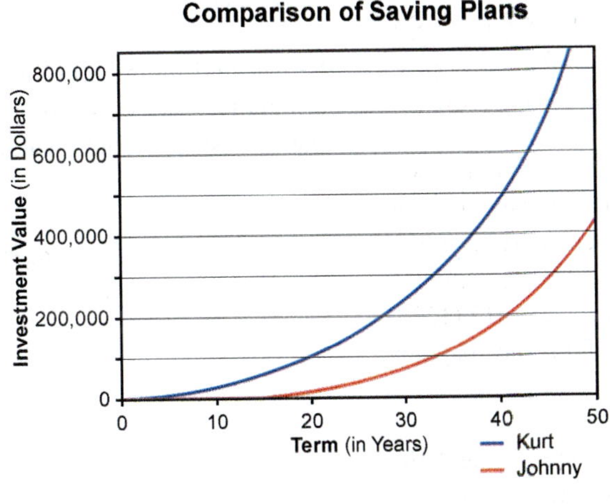

FIGURE 2

Remember that Kurt has a big head start on the amount of money in the account, even though he stopped his monthly payments. Will the amount of money in Johnny's account ever exceed the amount in Kurt's account?

When we consider the graphs of these two functions, we can see that Kurt has such a huge lead in money that, if the money continues to grow, Johnny will never catch up to Kurt even though he continues to deposit money long after the initial 15 years. Therein lies the power of compound interest over long periods of time! Lesson to learn: invest early, even if it's a small amount.

Working toward a Goal

When saving money, there is often a goal amount we wish to reach for things such as college or retirement. In these cases, it is usually easier to make regular payments toward the goal, just as we have been looking at, rather than depositing a lump sum. We can use the same formula, but solve it for the payment amount instead of future value.

Annuity Formula for Finding Payment Amounts

The regular payment amount (*PMT*) needed to reach a future goal with an annuity savings account is calculated with the formula

$$PMT = FV \cdot \frac{\left(\dfrac{r}{n}\right)}{\left[\left(1+\dfrac{r}{n}\right)^{nt} - 1\right]}$$

where *FV* is the future value desired, *r* is the APR, *n* is the number of regular payments made each year for *t* years.

TECH TRAINING

To perform the calculation from Example 5 on a TI-30XIIS/B or a TI-83/84 Plus calculator, use the following keystrokes.

3000 [×] [(] 0.0125 [÷] 12 [)] [÷] [(] [(] 1 [+] 0.0125 [÷] 12 [)] [^] [(] 12 [×] 3 [)] [−] 1 [)] [=]

To perform the calculation with Excel, use the built-in formula "=PMT(rate,nper,pv,[fv],[type])". Using the values from Example 5, input "=PMT(1.25%/12,(12*3), 0,3000,0)" to find the payment amount.

To perform the calculation using Wolfram|Alpha, go to www.wolframalpha.com and type "3000*(0.0125/12)/((1+0.0125/12) ^(12*3)-1)" into the input bar. Then, click the = button.

Example 5: Finding Monthly Payments in Order to Meet a Goal

Suppose you'd like to save enough money to pay cash for your next computer. The goal is to save an extra $3000 over the next three years. What amount of monthly payments must you make into an account that earns 1.25% interest in order to reach your goal?

Solution

We can find the monthly payment required to meet the goal by using the annuity formula for payments. The future value is $FV = 3000$, $r = 0.0125$, $n = 12$, and $t = 3$ years. Substituting these values into the formula, we have the following.

$$PMT = FV \cdot \frac{\left(\dfrac{r}{n}\right)}{\left[\left(1+\dfrac{r}{n}\right)^{nt} - 1\right]}$$

$$= 3000 \cdot \frac{\left(\dfrac{0.0125}{12}\right)}{\left[\left(1+\dfrac{0.0125}{12}\right)^{12\cdot 3} - 1\right]}$$

$$\approx 81.82$$

Therefore, in order to reach $3000 in three years with a 1.25% APR, you need to make a monthly payment of $81.82.

Let's return to the first example at the beginning of the section in which Amber wished to save up enough money for her newborn child to go to State College in 18 years. Recall that State College predicted that the cost of a degree would be $150,000 in 18 years. We calculated that Amber could put a lump sum of $73,100.31 into an account and leave it for 18 years in order to have enough money saved up. The next example determines how much Amber would need to deposit on a monthly basis in order to save the $150,000 for her child's college fund in 18 years. We will use the same APR of 4% as in Example 1.

Example 6: Calculating Monthly Payments

Amber wishes to deposit a fixed monthly amount into an annuity account for her child's college fund. She wishes to accumulate a future value of $150,000 in 18 years.

a. Assuming an APR of 4%, how much money should Amber deposit monthly in order to reach her goal?

b. How much of the $150,000 will Amber ultimately deposit in the account, and how much is interest earned?

Solution

a. Amber's goal is to have accumulated a balance of $150,000 after 18 years. She has an interest rate of 4% and is going to make monthly payments. Therefore, we know that $P = 150{,}000$, $r = 0.04$, $n = 12$, and $t = 18$. Now, substituting these into the formula, we can calculate the monthly payment needed.

$$PMT = FV \cdot \frac{\left(\dfrac{r}{n}\right)}{\left[\left(1+\dfrac{r}{n}\right)^{nt} - 1\right]}$$

$$= 150{,}000 \cdot \frac{\left(\dfrac{0.04}{12}\right)}{\left[\left(1+\dfrac{0.04}{12}\right)^{12 \cdot 18} - 1\right]}$$

$$\approx 475.30$$

So, Amber will need to deposit at least $475.30 per month in order to meet her goal of $150,000.

b. Amber will need to make 18 years worth of monthly payments of $475.30. We can find her contribution to the savings plan by multiplying these together.

$$18 \cdot 12 \cdot \$475.30 = \$102{,}664.80$$

The interest earned is then the difference between the goal and the contributed amount from Amber.

So, we can subtract and have the following.

$$\$150{,}000 - \$102{,}664.80 = \$47{,}335.20$$

So, the savings plan earned $47,335.20 in interest over the 18 years.

Retirement is such a long way off that it may seem silly to start thinking about it before you have even started your first job out of college. However, we know from our study on interest that time is on our side and interest works in our favor when saving money. The longer we can let money accrue interest, the better off we'll be. So, yes, now is the time to start saving for your retirement, even if it's just a small amount each month. Ideally, later on in life, our retirement funds will provide us with a consistent yearly pay without the worry of outliving our savings. In other words, the annuity payments would come solely from interest earned on the principal. That way the principal would not be affected at all and continue to earn us more money! Let's take a look at an example to help you imagine how you can create a nice nest egg for your own retirement.

Example 7: Calculating Monthly Payments for a Retirement Fund

Assume you have just graduated from college and have obtained your first job making $44,250 per year. Currently, the national average retirement income for people with a bachelor's degree is approximately $60,000 per year. Assuming you are 22 years of age at graduation and you will retire at age 67, then you will be working—and investing—for 45 years. At current rates, a mutual fund has an APR of 8%. How much do you need to invest each month in order to live off of $60,000 per year when you retire?

Solution

You want to build an account large enough so that you can have an annual annuity payment of $60,000 without reducing the principal on your retirement fund. In other words, you will live on the interest the account earns. We can phrase this another way by asking, "What balance is needed so that it earns $60,000 in annual interest?" Since we are assuming there is an APR of 8%, the $60,000 must be equal to 8% of the total balance. We can write this as a mathematical equation, letting x equal the principal needed, and then solve for x as shown below.

$$0.08 \cdot x = 60,000$$
$$x = \frac{60,000}{0.08}$$
$$x = 750,000$$

With an 8% APR, a balance of $750,000 allows you to withdraw $60,000 per year without reducing the principal. Now that we know the total amount needed to generate our $60,000 per year income, we can use the monthly annuity formula to compute the monthly payment amount needed to meet this future value goal.

We have that $FV = \$750,000$ from our calculation, $r = 0.08$, $t = 45$ years, and $n = 12$.

$$PMT = FV \cdot \frac{\left(\dfrac{r}{n}\right)}{\left[\left(1+\dfrac{r}{n}\right)^{nt} - 1\right]}$$

$$= 750{,}000 \cdot \frac{\left(\dfrac{0.08}{12}\right)}{\left[\left(1+\dfrac{0.08}{12}\right)^{12 \cdot 45} - 1\right]}$$

$$\approx 142.19$$

So, if you wish to retire and make $60,000 per year without affecting the principal in your retirement fund, you need to start at age 22 depositing approximately $143 per month. With a fixed APR of 8% over the years of your retirement, you will never outlive your retirement savings.

5.3 EXERCISES

PRACTICE

For each situation, use the formula for present value of money to calculate the amount you need to invest now in one lump sum to reach the amount given. Round your answer to the nearest cent, if necessary.

1. $25,000 after 10 years with an APR of 8% compounded monthly

2. $25,000 after 10 years with an APR of 12% compounded monthly

3. $100,000 after 18 years with an APR of 6% compounded quarterly

4. $1,000,000 after 40 years with an APR of 10% compounded monthly

APPLICATIONS

5. Alexis and Will are purchasing a home. They wish to save money for 10 years and purchase a house that has a value of $180,000 with cash. If they deposit money into an account paying 12% interest, how much do they need to deposit each month in order to make the purchase?

6. Marilyn wishes to retire at age 65 with $2,000,000 in the bank. At the age of 21, she decides to begin depositing money into an account with an APR of 11%. What is the monthly payment Marilyn must make in order to make this happen?

7. Repeat Exercise 6 with an APR of 6%.

8. Repeat Exercise 6 with a desired retirement amount of $1,500,000.

9. Suppose you wish to retire at the age of 65 with $80,000 in savings. Determine your monthly payment into an IRA (Individual Retirement Account) if the APR is 7.5% and you begin making payments at
   a. 20 years old.
   b. 30 years old.
   c. 40 years old.

10. Revere College predicts that in 18 years it will take $200,000 to attend the college for four years. Debbie wishes to save money for her child's college fund. How much should Debbie put aside in an account with an APR of 9% compounded monthly in order to have $200,000 in the account in 18 years?

11. Repeat Exercise 10 with an interest rate earned of 5%.

12. Repeat Exercise 10 with an interest rate earned of 3.5%.

13. Suppose you'd like to save enough money to pay cash for your next car. The goal is to save an extra $26,000 over the next 6 years. What amount of quarterly payments must you make into an account that earns 5.5% interest in order to reach your goal?

14. Repeat Exercise 13 if the interest rate earned is 6.5%.

15. Repeat Exercise 13 if the interest rate earned is 3.5%.

16. Willie deposits a fixed monthly amount into an annuity account for his child's college fund. He wishes to accumulate a future value of $75,000 in 15 years.
   a. Assuming an APR of 3.5%, how much money should Willie deposit monthly in order to reach his goal?
   b. How much of the $75,000 will Willie ultimately deposit in the account, and how much is interest earned?

17. Repeat Exercise 16 with an APR of 6%.

18. Repeat Exercise 16 with an accumulated amount of $125,000.

19. Blake starts an IRA (Individual Retirement Account) to save for retirement at the age of 22. He deposits $450 each month. The IRA has an average annual interest rate of 7%.
   a. How much money will he have saved upon retirement at the age of 65?
   b. Determine the amount of money Blake deposited over the length of the investment and how much he made in interest.

20. Jimmie has a job at an advertising agency earning $54,000 per year. Jimmie is currently 26 years old and wishes to retire at age 67 with a retirement income of $75,000. How much money would Jimmie need to invest each month into a growth stock mutual fund with an interest rate of 6.5% in order to withdraw $75,000 per year without reducing the principal?

5.4 BORROWING MONEY

■ TOPICS

- Credit Cards
- Car Loans and Mortgages
- Good Financial Habits

Good financial habits, such as creating a budget and saving for the future, are key to staying financially healthy. However, even with the best budgeting skills, sometimes we must borrow money in order to make large purchases, such as a car or a house, or to help bridge the gap when emergencies arise. In this section, we will look at credit cards, car loans, and mortgages to learn best practices of borrowing money.

Credit Cards

HELPFUL HINT

A creditor is any organization, such as a bank or credit card company, that lends money with the idea that when you pay the money back, you will also pay the interest charged as well.

Just as interest worked in our favor when saving money, borrowing money can make interest work against us. The financial institutions that loan the money are the ones earning interest, not the individual. Most savings accounts offer an APR of 10% or less. However, when borrowing money, institutions often charge a much higher rate of interest, such as 14.99%, 19.99%, or even 29.99%. In some cases the interest rate can soar even higher if payments are missed or are late.

We will begin by looking at credit cards and how they work. Before we do, one important fact we need to point out is that despite the fact that we're very used to seeing most plastic cards with a Visa or MasterCard emblem on them, not all of these cards are actual credit cards. Many are in fact debit cards, also known as *bank cards* or *check cards*. Debit cards are not cards for borrowing money, they are simply an electronic means by which consumers can access the money already in their bank accounts.

Credit cards work on the principle that money is lent out for a month. If the money is not repaid in full at the end of the month, and a balance remains, a charge of interest is incurred upon the balance. The month where no interest is accruing is called a **grace period**. However, if a balance is carried over to the next month, interest continues to accrue until the balance is 0.

Grace Period

A **grace period** is a period of time in which no interest accrues on a debt.

The interest rate on a credit card is often a *variable interest rate*. This means that the interest rate fluctuates over time based on a particular benchmark, often a country's prime rate of interest. Notice that Figure 1 shows the APR for a customer's credit card. The APR on the left-hand side is for purchases only and is a variable rate, which means it is subject to change.

FIGURE 1: Credit Card Interest Rates

The APR on the right-hand side is for cash advances. It also is a variable interest rate, but is quite a bit higher than the rate for purchases. This is standard practice for most credit cards. Unfortunately, most cardholders don't realize that other fees also exist for cash advances. Initially, an up-front fee is charged, usually 2% of the amount. This is to cover the fee that is normally incurred by the merchant when a purchase is made. Along with the higher interest rate and the fee, there is also no grace period. In other words, the cash advance starts accruing interest the minute it is dispensed from the ATM. Lesson to learn: cash advances on credit cards should only be used in emergencies!

Using credit cards wisely means that you only charge what you can afford to pay off in one month. If it is the case that you cannot pay off the balance by the end of the month, then you must make a partial payment. The **minimum payment**, which is the minimum amount the lender requires you to pay each month, is usually equal to a percentage of your average balance due, say 2%. So, if your average daily balance for a certain month is $1000, your minimum payment for that month will be $20. Many people assume that it's a good deal to pay $20 per month in order to pay off a debt of $1000. However, what people tend to forget is that they are accruing interest on the balance every month it goes unpaid. Therefore, this is really not a good deal at all. If the interest rate is as high as 20% on the balance owed, the $20 payment each month goes almost completely towards interest, and has very little impact on reducing the total balance of the credit card. By requiring that you pay only small amounts each month, creditors know that it will take you much longer to pay off your debt, and that you'll end up paying a lot more interest over time. Of course, this is how lenders actually make money and stay in business.

Consider paying off a $1000 credit card balance by paying the minimum payment each month. Most of the payment at the beginning is simply interest, but as more and more payments are made, the balance is reduced. When the balance shrinks, so does the interest accrued. Since the minimum payment is a percentage of the balance, it shrinks as well. As you might imagine, this smaller and smaller payment method looks good to the consumer, but it is actually prolonging the debt.

Figure 2 is a graph showing the payment timeline for a $1000 credit card balance with 14% interest where the minimum payment due is 4% of the balance each month—this is assuming no new additional purchases are made. Notice that after about six years (72 months) the balance is finally less than $120.00. This means that the minimum payment amount is less than $5.00 each month. Sounds good, but the payments will continue to go on and on. At this rate, the balance will never be completely $0. To prevent this from happening, credit card companies have a set minimum payment amount that will eventually force you to pay the balance in full.

Payment Timeline Using Minimum Payments

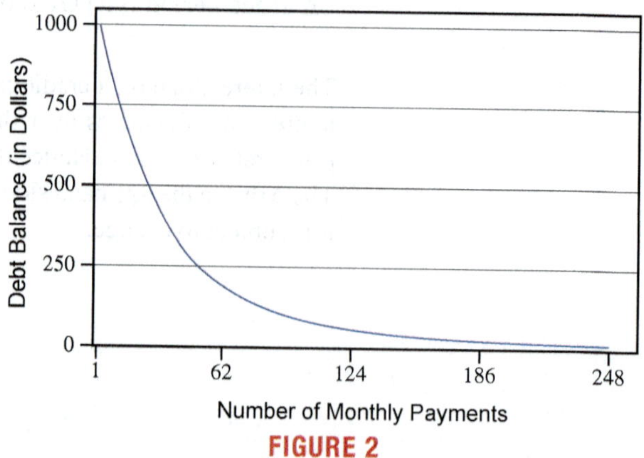

FIGURE 2

Let's contrast that with a different picture of making payments. Suppose the same $1000 debt was paid off monthly with a consistent payment of $40 per month. Even though this is the starting minimum payment, it will soon become more than the required minimum payment and hence bring down the balance more quickly. Figure 3 shows the payment timeline for this regular minimum payment. You can see that after about 30 payments, the debt is paid off.

Payment Timeline Using a Fixed Payment

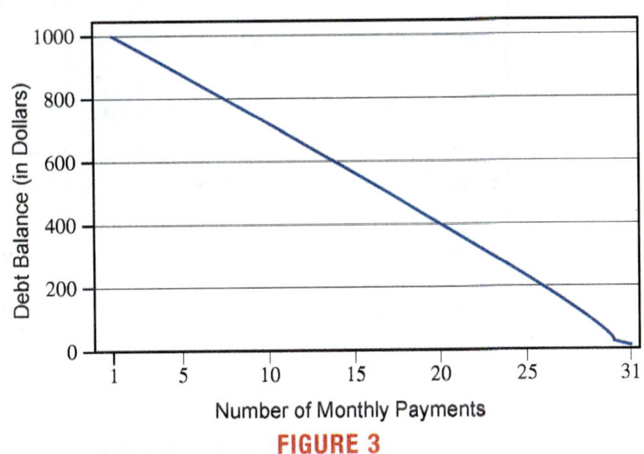

FIGURE 3

Figure 4 shows just how pronounced the difference is by graphing both payoff schemes on one graph.

Minimum Payments vs. Fixed Payments

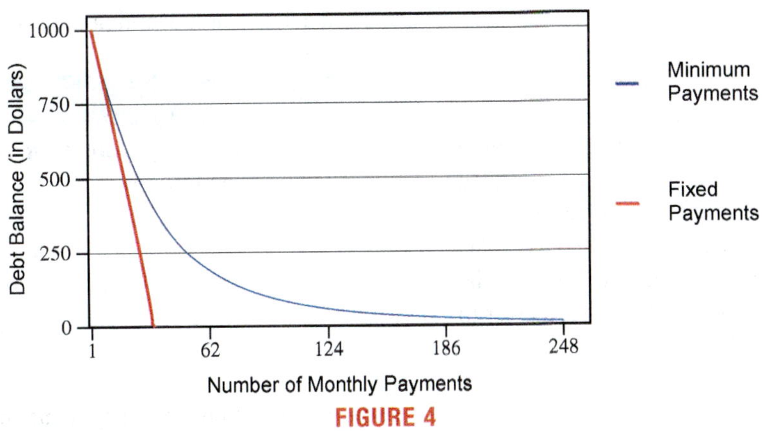

FIGURE 4

In May 2009, Congress concluded that certain practices in the credit card industry were neither fair nor transparent to consumers, and so they signed into law the Credit Card Accountability, Responsibility, and Disclosure Act (Credit CARD Act) with strong support in both the Senate and House of Representatives. Among other things, this Act protects consumers from excessive charges and wildly changing APRs. As a result of the Credit CARD Act, many credit card statements now provide information to their customers about paying off the debt.[1]

Figure 5 shows an example of the information given on a credit card statement. Notice also that there is a help line number for consumer debt. Although this number is fictitious as shown in this course, the Credit CARD Act requires that creditors provide a toll-free telephone number for the purposes of providing information about accessing credit counseling and debt management services.

---

1   **Source:** The Library of Congress THOMAS, s.v. "Bill Summary & Status 111th Congress (2009-2010) H.R.627." Accessed February 2013. http://thomas.loc.gov/cgi-bin/bdquery/z?d111:H.R.627:

<table>
<tr><td colspan="3">Late Payment Warning: If we do not receive your minimum payment due by the payment due date listed, you may have to pay a late fee of up to $35.00 and your purchase APR may be increased to the penalty APR of 27.24%.</td></tr>
<tr><td colspan="3">Minimum Payment Warning: If you make only the minimum payment each period, you will pay more in interest and it will take you longer to pay off your balance. For example:</td></tr>
<tr><td>If you make no additional charges and each month you pay. . .</td><td>You will pay off the balance shown on this statement in about. . .</td><td>You will pay an estimated total of. . .</td></tr>
<tr><td>Only the minimum payment due</td><td>7 years</td><td>$2771</td></tr>
<tr><td>$60</td><td>3 years</td><td>$2152 (Savings = $619)</td></tr>
<tr><td colspan="3">If you would like information about credit counseling services, call 1-800-555-0199.</td></tr>
</table>

FIGURE 5: Credit Card Warnings

Although many consumers may have never taken the time to notice this type of warning on their monthly statement, the creditor is actually showing you how bad it is for you to only pay them the minimum payment. They even go as far as to tell you how much money you could save!

We can use the following formula to calculate the number of fixed payments required to pay down a debt that is accruing interest monthly, just like the credit card company did.

The payment amount must be more than the amount of interest added on each period. Otherwise, the balance will never decrease. The formula will return an error if the payment is less than the interest to be added.

Number of Fixed Payments Required to Pay Off Credit Card Debt

The **number of fixed payments R required to pay off a credit card debt** is calculated with the formula

$$R = \frac{-\log\left[1 - \dfrac{r}{n}\left(\dfrac{A}{PMT}\right)\right]}{\log\left(1 + \dfrac{r}{n}\right)}$$

where n is the number of payments made per year for a loan of amount A, with interest rate r and monthly payment PMT.

Example 1: Paying Off Credit Card Debt with a Fixed Payment

Assume you want to buy a new computer that costs $2200 using a credit card that has an APR of 19.99%.

a. How long will it take you to pay off the computer if you make regular monthly payments of $40?

b. How much will you pay in the long run for the computer if you make monthly payments of $40?

c. How long will it take you to pay off the computer if you make regular monthly payments of $80?

d. How much will you pay in the long run for the computer if you make monthly payments of $80?

Solution

a. We wish to find the number of fixed payments required to pay off a credit card debt. We are told that the debt is $A = 2200$ and the APR $= 19.99\%$, so $r = 0.1999$. The monthly payment is $PMT = 40$ and $n = 12$. Substituting these values into the formula, we have the following.

$$R = \frac{-\log\left[1 - \dfrac{r}{n}\left(\dfrac{A}{PMT}\right)\right]}{\log\left(1 + \dfrac{r}{n}\right)}$$

$$= \frac{-\log\left[1 - \dfrac{0.1999}{12}\left(\dfrac{2200}{40}\right)\right]}{\log\left(1 + \dfrac{0.1999}{12}\right)}$$

$$\approx 150.0760203$$

Therefore, it will take approximately 150 monthly payments, or $12\frac{1}{2}$ years, to pay off the debt.

b. If we make 150 payments at $40 each, then over $12\frac{1}{2}$ years we will have paid a total of

$$150 \cdot \$40 = \$6000.$$

As you might imagine, the computer would be considerably out of date in $12\frac{1}{2}$ years, and you paid nearly triple the original price of the computer. This is not a wise method for purchasing a computer.

c. This time, we will double the monthly fixed payment so the $PMT = 80$. The remaining variables stay the same; that is, $A = 2200$, $r = 0.1999$, and $n = 12$. Using the formula, we have the following.

$$R = \frac{-\log\left[1 - \dfrac{r}{n}\left(\dfrac{A}{PMT}\right)\right]}{\log\left(1 + \dfrac{r}{n}\right)}$$

$$= \frac{-\log\left[1 - \dfrac{0.1999}{12}\left(\dfrac{2200}{80}\right)\right]}{\log\left(1 + \dfrac{0.1999}{12}\right)}$$

$$\approx 37.084776$$

Therefore, it will take approximately 37 monthly payments, or just over three years, to pay off the debt.

d. If we make 37 payments at \$80 each, then over three years we will have paid a total of

$$37 \cdot \$80 = \$2960.$$

This is a much more reasonable method for purchasing the computer. Although the computer still might be outdated in three years, it will probably still be functionally acceptable.

⬆ TECH TRAINING

Use the following keystrokes on a TI-30XIIS/B or TI-83/84 Plus calculator for the calculation in part c. of Example 1.

[(-)] [LOG] 1 [-] 0.1999 [÷] 12 [x] [( ] [( ] 2200 [÷] 80 [ )] [ )] [÷]
[LOG] 1 [+] 0.1999 [÷] 12 [ )]

To perform this calculation with Wolfram|Alpha, go to www.wolframalpha.com. In the input bar, type "-log(1-(0.1999/12)(2200/80))/log(1+0.1999/12)" then click the = button.

Car Loans and Mortgages

Credit cards lend money that is paid back over an unspecified period of time often with a variable interest rate. One of the dangers in using this type of loan is the risk of buying more than you can afford. Loans such as car loans or house mortgages don't use open-ended time frames like credit cards. Instead, they operate as fixed installment loans. For **fixed installment loans**, terms for the amount of the loan and the length of time it will take to pay off the loan are agreed upon up front so that both the lender and the borrower know exactly how much the loan will cost. Because everything is decided at the beginning of the loan period, fixed installment loans use *fixed interest rates*—rates where the APR never changes over the life of the loan. Normally, a **down payment** is required for a fixed installment loan. A down payment is a cash payment made up front towards the total purchase price. It usually equals a certain percentage, often 5% to 25%, of the total value of the purchase. Once the down payment is subtracted from the original price, the amount remaining is the principal that needs to be *financed*, or borrowed with interest.

Fixed Installment Loans

Fixed installment loans have a fixed interest rate and are paid off with monthly payments over a specified amount of time.

Down Payment

A **down payment** is a cash payment made up front towards the total purchase price of the goods or service.

When a fixed installment loan is taken out, several factors go into determining the monthly payment amount. The initial consideration is the size of the down payment. A larger down payment can often mean a better interest rate. For new car purchases, it is a wise rule of thumb to put down at least 20% of the car's price. Because vehicles *depreciate*, or decrease in value, very quickly, the value of your car depreciates by about 20% the minute you drive the new car off the lot. So, if you had an emergency and had to sell your new car, it would now only be worth 80% of its original value, and you don't want to owe more than it's worth!

Since the recession of 2007, down payment requirements have become stricter for the average house loan, or *mortgage*. In general, most mortgages require anywhere from 10% to 25% of the purchase price as a down payment. There are exceptions, however, that may decrease down payments to around 3% for first time home buyers, or increase them to 35% for "credit-challenged" borrowers.

The other consideration is the length of time over which the loan will be paid off. For car loans, the time is usually between three and six years. Although a longer time period means lower payments, it also can mean higher interest rates. We already know from the earlier sections that higher interest rates over longer periods of time mean more interest accrues and hence you pay more for the car! For house mortgages, the length of a loan with a fixed interest rate is usually either 15 years or 30 years. Just as with car loans, a bigger down payment and shorter amount of time can mean a lower interest rate.

The following formula is used to calculate the monthly payment amount for fixed installment loans.

HELPFUL HINT

Notice that the formula contains a negative exponent in the denominator. Be careful not to drop the negative sign during calculations.

Monthly Payment Formula for Fixed Installment Loans

The amount of a **monthly payment** (*PMT*) on a **fixed installment loan** is calculated with the formula

$$PMT = \frac{\left(P \cdot \dfrac{r}{n}\right)}{\left[1 - \left(1 + \dfrac{r}{n}\right)^{-nt}\right]}$$

where P is the principal amount of money borrowed, r is the APR, n is the number of payments per year, and t is the number of years of the loan.

Example 2: Purchasing a New Car

Kelly wishes to purchase a new car. The car she has chosen has a price of $34,000, including taxes and fees. She chooses to make a down payment of 20% of the price and wants to finance the remainder. If Kelly has acquired an APR of 3.99% for a 72-month loan, what is the amount of her monthly payment?

Solution

First we need to know the amount of the down payment Kelly will pay up front. We can find 20% of $34,000 by multiplying the two together. Remember to change the percentage to a decimal in order to multiply.

$$\text{down payment} = 0.20 \cdot \$34{,}000 = \$6800$$

Therefore, Kelly will need to finance the remainder, found by subtracting the down payment from the original price of the car.

$$\text{principal to finance} = \$34{,}000 - \$6800 = \$27{,}200$$

Now we have that Kelly will borrow $P = 27{,}200$ for $t = 6$ years with an APR of 3.99%, so $r = 0.0399$. Substituting into the formula for fixed installment loans, we can calculate what Kelly's monthly payment will be.

$$PMT = \frac{\left(P \cdot \dfrac{r}{n}\right)}{\left[1 - \left(1 + \dfrac{r}{n}\right)^{-nt}\right]}$$

$$= \frac{\left(27{,}200 \cdot \dfrac{0.0399}{12}\right)}{\left[1 - \left(1 + \dfrac{0.0399}{12}\right)^{-12 \cdot 6}\right]}$$

$$\approx 425.425058$$

So Kelly's monthly payment for her new car would be \$425.43.

Jeff wants to buy a new truck that costs \$48,000. After a down payment of \$10,000 he finances \$38,000. If Jeff has an APR of 5% for a 48-month loan, what is Jeff's monthly payment?

📈 TECH TRAINING

To find the payment amount using Excel, type in "=-PMT(rate,nper,pv,fv, type)" where

- $rate = \text{APR}/n$

- $nper$ = number of payment periods for the loan

- pv = the principal of the loan

- fv = the future value that is assumed to be 0 for a loan

- $type = 0$ for payments due at the end of the period and 1 for payments due at the beginning of the period

We type a negative sign in front of the payment formula so that the payment amount appears as a positive amount.

For our example, enter the following.

"=-PMT(3.99%/12,72,27200,0,0)"

The payment amount can be found using Wolfram|Alpha. Go to www.wolframalpha.com and type "(27200*(0.0399/12))/(1-(1+0.0399/12)^(-12*6))" into the input bar. Then, click the = button.

Example 3: Determining Best New Car Incentive

Dean gets to choose from one of the new car incentives when he purchases his car next week. He can either choose 0.9% APR financing for 60 months or $1500 cash back with a 3.75% APR over 48 months. Compare the two incentives that Dean has to choose from if the new car he wishes to buy is $27,465 and he has saved a down payment of $5000.

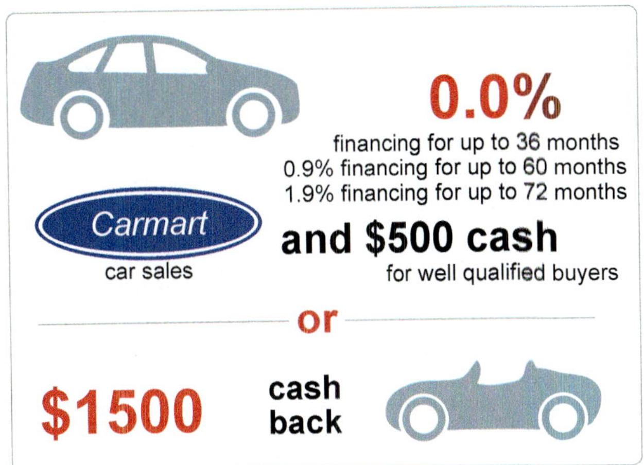

Solution

Let Option A be the 0.9% APR financing for 60 months and Option B will be $1500 cash back with a 3.75% APR over 48 months.

Option A: For this option, the principal to be financed is

$$\$27{,}465 - \$5000 = \$22{,}465.$$

In addition, $r = 0.009$ and $t = 60 \div 12 = 5$. The payments each month are calculated as follows.

$$PMT = \frac{\left(P \cdot \dfrac{r}{n} \right)}{\left[1 - \left(1 + \dfrac{r}{n} \right)^{-nt} \right]}$$

$$= \frac{\left(22{,}465 \cdot \dfrac{0.009}{12} \right)}{\left[1 - \left(1 + \dfrac{0.009}{12} \right)^{-12 \cdot 5} \right]}$$

$$\approx 383.0445874$$

Dean would have a monthly payment of $383.04 for 60 months with Option A. His total cost for the car would be

$$\$383.04 \cdot 60 + \$5000 = \$27{,}982.40.$$

Option B: For this option the APR is considerably more at 3.5%, but Dean can use the $1500 cash back to add to his down payment so that he has less to finance. Therefore, he will only need to finance

$$\$27{,}465 - \$5000 - \$1500 = \$20{,}965.$$

For this option $r = 0.0375$ and the length of time is now $t = 48 \div 12 = 4$. So the payments each month will be the following.

$$PMT = \frac{\left(P \cdot \dfrac{r}{n}\right)}{\left[1 - \left(1 + \dfrac{r}{n}\right)^{-nt}\right]}$$

$$= \frac{\left(20{,}965 \cdot \dfrac{0.0375}{12}\right)}{\left[1 - \left(1 + \dfrac{0.0375}{12}\right)^{-12 \cdot 4}\right]}$$

$$\approx 471.0281082$$

Therefore, Dean's payments for Option B will be \$471.03 for 48 months. His total cost for the car would be

$$\$471.03 \cdot 48 + \$5000 = \$27{,}609.44.$$

Now that the calculations for both options have been done, you can see that Dean will pay less overall if he uses Option B. Note that the difference is about \$400 between the two options. There are other factors that might sway Dean's decision. First, Option B only saves him money if he is disciplined and puts the cash back that he received towards the purchase of the car. However, the car is completely paid off an entire year earlier with this option. Dean will also need to consider which monthly payment option he can afford. Option A has the lower monthly payment. Now that Dean understands the outcomes of total cost based on length of the loan and interest rate, he can use his knowledge of budgeting and financial decision-making to choose the best option for his financial situation.

⌁ TECH TRAINING

Using the Excel payment formula, enter the following to obtain the monthly payment from Example 3.

For 48 months: "=-PMT(0.0375/12,48,20965,0,0)"

For 60 months: "=-PMT(0.009/12,60,22465,0,0)"

Now let's look at longer term fixed installment loans used for purchasing a house.

Example 4: Calculating Monthly Mortgage Payments

William and Helen are preparing to buy a new home in the Midwestern United States. They have saved \$21,300 for a down payment. The total price of the house is \$131,800, including taxes and fees. Find their monthly mortgage payment if the fixed interest rate is 3.52% for a 30-year loan.

Solution

Since William and Helen have a down payment, we can subtract that from the purchase price to find the amount they will need to finance to buy the home.

$$\text{principal amount to finance} = \$131{,}800 - \$21{,}300 = \$110{,}500$$

We are told that the APR $= 3.52\%$, so $r = 0.0352$ and $t = 30$ years. Using the payment formula for fixed installment loans we can calculate their monthly mortgage payment.

$$PMT = \dfrac{\left(P \cdot \dfrac{r}{n}\right)}{\left[1 - \left(1 + \dfrac{r}{n}\right)^{-nt}\right]}$$

$$= \dfrac{\left(110{,}500 \cdot \dfrac{0.0352}{12}\right)}{\left[1 - \left(1 + \dfrac{0.0352}{12}\right)^{-12 \cdot 30}\right]}$$

$$\approx 497.4288455$$

Therefore, William and Helen will have a monthly mortgage of $497.43.

> **⊠ TECH TRAINING**
>
> Using the Excel payment formula, enter the following to obtain the monthly payment from Example 4.
>
> "=-PMT(3.52%/12,12*30,110500,0,0)"

In 2007, the United States had a major financial crisis occur when variable interest rates on home loans escalated, causing many homeowners to lose their homes. As a result, financial education about mortgages became key for consumers. According to financial experts, monthly home mortgage payments should cost less than 25% of your monthly take-home salary to avoid being overburdened with home loan debt. This means that if your monthly take-home pay is $2000.00, your monthly mortgage payment should be no more than $500.00 per month.

If we want to stay within the recommended monthly mortgage payment—that is, 25% of our take-home pay—we can rearrange the payment formula to help us determine the maximum amount we should spend on buying a house. The following formula can be used to calculate the maximum purchase price attainable based on a set monthly payment and a predetermined interest rate and time period.

Maximum Purchase Price Formula

The **maximum purchase price** for a house can be found by

$$\text{maximum purchase price} = PMT \cdot \dfrac{\left[1 - \left(1 + \dfrac{r}{n}\right)^{-nt}\right]}{\left(\dfrac{r}{n}\right)}$$

where PMT is the monthly mortgage payment you wish to make, r is the APR, n is the number of payments per year, and t is the number of years of the loan.

Example 5: How Much House Can You Afford?

Suppose you have recently graduated from college and want to purchase a house. Your take-home pay is $3220 per month and you wish to stay within the recommended guidelines for mortgage amounts by only spending $\frac{1}{4}$ of your take-home pay on a house payment. You have $15,300 saved for a down payment. With your good credit and the down payment you can get an APR from your bank of 3.37% compounded monthly.

a. What is the total cost of a house you could afford with a 15-year mortgage?

b. What is the most that you could afford with a traditional 30-year mortgage instead of a 15-year?

Solution

a. The first thing to do is to calculate the size of the monthly mortgage payment you are willing to spend. Since you have $3220 per month in take-home pay, multiply this by 25% to find your maximum monthly payment.

$$\text{advised monthly payment} = \$3220 \cdot 0.25 = \$805$$

We know that $r = 0.0337$ and that because this is a 15-year mortgage, $n = 12$ and $t = 15$. Substituting these values in the formula, we have the following.

$$\text{maximum purchase price} = PMT \cdot \frac{\left[1-\left(1+\dfrac{r}{n}\right)^{-nt}\right]}{\left(\dfrac{r}{n}\right)}$$

$$= 805 \cdot \frac{\left[1-\left(1+\dfrac{0.0337}{12}\right)^{-12\cdot15}\right]}{\left(\dfrac{0.0337}{12}\right)}$$

$$\approx 113,617.8221$$

So, you could afford a 15-year mortgage of approximately $113,617.82.

Remember that the amount of down payment you have available will add to the maximum amount you can spend on a house. Therefore, with a 15-year mortgage you can afford to buy a house with a maximum price of

$$\$113,617.82 + \$15,300 = \$128,917.82.$$

🖢 HELPFUL HINT

If you are using a TI-30XS/B Multiview calculator, the keystrokes will be slightly different for your calculator.

📈 TECH TRAINING

Use the following keystrokes on a TI-30XIIS/B or TI-83/84 Plus calculator for the calculation of the maximum purchase price for part a. of Example 5.

805 ⊗ ((1 ⊖ ((1 ⊞ 0.0337 ÷ 12)) ^ (((-) 12 ⊗ 15))) ÷ (0.0337 ÷ 12)) ⊜

Wolfram|Alpha can be used to find the maximum purchase price. Go to www.wolframalpha.com and type "805*(1-(1+0.0337/12)^(-12*15))/(0.0337/12)" into the input bar. Then, click the = button.

b. For a 30-year mortgage, the only thing that changes is t. Now $t = 30$. The monthly payment you can afford stays the same, as well as the interest rate and n. Therefore, we have the following.

$$\text{maximum purchase price} = PMT \cdot \frac{\left[1 - \left(1 + \dfrac{r}{n}\right)^{-nt}\right]}{\left(\dfrac{r}{n}\right)}$$

$$= 805 \cdot \frac{\left[1 - \left(1 + \dfrac{0.0337}{12}\right)^{-12 \cdot 30}\right]}{\left(\dfrac{0.0337}{12}\right)}$$

$$\approx 182{,}201.1079$$

By this calculation, you should be able to afford a house that costs approximately $182,200 plus your down payment.

With the down payment added in, your total purchase price with a 30-year mortgage could be as much as

$$\$182{,}200 + \$15{,}300 = \$197{,}500.$$

This amount is considerably more than the 15-year mortgage, but remember that you will incur much more interest over the 30 years as well. In addition, we assumed that you could get the same interest rates for both time periods, but in reality, the 30-year mortgage would probably have a higher interest rate as well.

Good Financial Habits

To end this chapter we want to point out some of the best practices for healthy financial living.

Make a budget.   First and foremost, begin the habit of budget making early. Why not start today? It might surprise you how much money you are spending in certain areas, or how much money you could afford to be putting into an interest earning savings account for later. Always make room for savings. You can't afford not to. Remember to let time be on your side when saving for the big things.

Control impulse spending.   One of the biggest budget breakers is impulse spending, or spending on a whim. Eating out, online purchases, or even picking up a small thing here and there when shopping can add up quickly. Evaluate your needs versus your wants. A budget will help you keep up with this.

Have an emergency fund.   Saving for a goal is great, but don't forget that life unexpectedly throws you financial curve balls. It can happen to any of us.

Be a smart consumer.   Look in the fine print or ask about extra fees or penalties when borrowing money. Some loans, especially mortgages, can carry a penalty for paying the loan off early.

Monitor your accounts and pay your bills on time.   Use the online resources that most banks provide free of charge and check your account balance often. One of the best ways to get good interest rates on loans is to have a healthy credit score. Although we don't go into detail in the course about the components that make up a personal credit score, paying your bills on time, every time, will give you a good start to a good score.

Ask questions.   Finally, don't assume you can't ask for help with financial matters. Ask questions and shop around to look for the best possibilities for you. Often, there isn't just one right answer that fits everyone's needs. There are plenty of resources available to the public, even online resources. Use them!

Skill Check Answers

1. $875.11

5.4 EXERCISES

💡 PRACTICE

Consider a credit card with a balance of $7000. You wish to pay off the credit card in each scenario. Round your answer to the nearest cent, if necessary.

    a. Calculate the amount of a monthly payment within the time frame given.

    b. Calculate the total amount paid over the time period.

1.   APR of 17.99% paid off within 1 year

2.   APR of 12.5% paid off within 2 years

3.   APR of 24% paid off within 3 years

Consider a credit card with a balance of $5560. You wish to pay off the credit card in each scenario. Calculate the following. Round your answer to the nearest cent, if necessary.

    a. Calculate the amount of a monthly payment within the time frame given.

    b. Calculate the total amount paid over the time period.

4.   APR of 14.99% paid off within 1 year

5.   APR of 11.99% paid off within 2 years

6.   APR of 5.9% paid off within 3 years

🚀 APPLICATIONS

Round your answer to the nearest cent, if necessary.

7. Given the chart below, solve the following problems.

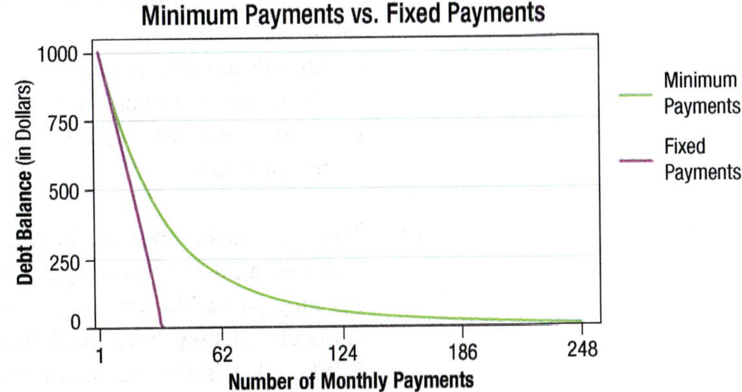

   a. Estimate the total amount paid when a debt balance was paid using a fixed monthly payment of $40.
   b. Estimate the total amount paid when a debt balance was paid using the minimum monthly payment of $18.

8. Rachel is purchasing a new camera that costs $3800 for her photography business. Rachel uses a credit card that has an APR of 16.99%.
   a. How long will it take her to pay off the camera if she makes monthly payments of $75?
   b. How much will she pay in the long run for the camera if she makes monthly payments of $75?
   c. How long will it take her to pay off the camera if she makes monthly payments of $150?
   d. How much will she pay in the long run for the camera if she makes monthly payments of $150?

9. Tommy gets to choose from one of the new car incentives when he purchases his car next week. He can either choose 0.9% APR financing for 48 months or $1000 cash back with a 4.75% APR over 48 months. Compare the two incentives that Tommy has to choose from if the new car he wishes to buy is $32,457 and he has saved a down payment of $3500.

10. Mike bought a new car and financed $25,000 to make the purchase. He financed the car for 60 months with an APR of 6.5%. Determine each of the following.
   a. Mike's monthly payment
   b. Total cost of Mike's car
   c. Total interest Mike pays over the life of the loan

11. Omar wants to purchase three vans for his delivery business. Each van costs $38,000. He wishes to finance the purchase for 48 months and has acquired an APR of 4.5%. Determine each of the following.
   a. Omar's monthly payment
   b. Total cost of Omar's vans
   c. Total interest paid by Omar over the life of the loan

12. Jamal bought a new car for $32,000. He paid a 10% down payment and financed the remaining balance for 36 months with an APR of 4.5%. Determine each of the following.
   a. Jamal's monthly payment
   b. Total cost of Jamal's car
   c. Total interest Jamal pays over the life of the loan

13. Susan wants to buy a new computer from Banana Computers. The company sells a laptop model for $2650. Susan decides to finance the computer for 24 months at an APR of 12.5%. Determine each of the following.
  a. Susan's monthly payment
  b. Total cost of the computer
  c. Total interest paid over the 24 months

14. Amanda and Ferobee are buying a house on a 30-year mortgage. They can only pay $800 per month for a mortgage. If they have an APR of 3.75%, what is the maximum price of a mortgage that they can take out?

15. Brad decides to purchase a $250,000 house. He wants to finance the entire balance. He has received an APR of 4.5% for a 30-year mortgage.
  a. What is Brad's monthly payment?
  b. Over the course of the loan, how much interest will Brad pay?
  c. What is Brad's total cost if he takes all 30 years to pay off the house?
  d. If he changed the term to 15 years instead of 30 years, what would his monthly payment be?
  e. With a 15-year mortgage, how much interest will Brad pay?
  f. With a 15-year mortgage, what is the total cost of the house?

16. The city of Nettleton recently completed a new school building. The entire cost of the project was $19,000,000. The city has put the project on a 20-year loan with an APR of 2.4%. There are 15,000 families that will be responsible for paying the loan.
  a. Determine the amount of the monthly payment for the loan.
  b. Determine the amount that each family should be required to pay each year to cover the cost of the school.
  c. Determine the total cost of the school.

17. You want to buy a car and finance $20,000 to do so. You can afford a payment of up to $450 per month. The bank offers three choices for the loan: a four-year loan with an APR of 7%, a five-year loan with an APR of 7.5%, and a six-year loan with an APR of 8%. Which option best meets your needs, assuming you want to pay the least amount of interest?

18. A credit card has a balance of $5000 at an APR of 9.99%. You plan to pay $500 each month in an effort to clear the debt quickly. How long will it take you to pay off the balance?

19. A credit card has a balance of $11,500 at an APR of 14.99%. You plan to pay $650 each month in an effort to clear the debt quickly. How long will it take you to pay off the balance?

20. Suppose you have a student loan of $80,000 with an APR of 4.5% for 25 years.
  a. What is your monthly payment?
  b. If you decide you want to pay off the loan in 15 years instead of 25, what is your monthly payment?
  c. What is your savings for paying the loan off in 15 years instead of 25?

21. Suppose you have graduated from college and want to purchase a house. Your take-home pay is $4560 per month and you wish to stay within the recommended guidelines for mortgage amounts by only spending $\frac{1}{4}$ of your take-home pay on a house payment. You have $18,500 saved for a down payment. With your good credit and the down payment you can get an APR from your bank of 4.35%, compounded monthly.
  a. What is the total cost of a house you could afford with a 15-year mortgage?
  b. What is the most that you could afford with a traditional 30-year mortgage instead of a 15-year?

22. What if you won the Powerball Lottery with a jackpot of $150 million? Calculate the amount of money you would receive over a 25-year period with each of the following two options. Which option gives you the most money over the 25 years?

    Option 1: Taking all the money at once with a 40% penalty, and pay the income tax of 38% on the lump sum, and investing the remaining amount into an account earning 6% interest for 25 years.

    Option 2: Acquire the money as part of an annuity to be paid out in 25 equal payments over a 25-year period, paying the income tax of 38% on the income from the winnings each year.

# Chapter 6

SYSTEMS OF LINEAR EQUATIONS; MATRICES

6.1 SOLVING SYSTEMS OF LINEAR EQUATIONS BY SUBSTITUTION AND ELIMINATION

■ TOPICS

- Systems of Linear Equations
- Solving Systems of Linear Equations by Substitution
- Solving Systems of Linear Equations by Elimination
- Systems of Linear Equations in Three or More Variables
- Applications of Systems of Linear Equations
- Solving Systems of Equations Using Technology

Systems of Linear Equations

Many problems in mathematics can be described by two or more equations in two or more variables. When the equations are all linear, such a collection of equations is called a *system of linear equations*, or sometimes *simultaneous linear equations*. The word *simultaneous* refers to the goal of identifying the values of the variables (if there are any) that solve all of the equations simultaneously.

In the case of two linear equations in two variables, it turns out that there are only three possible configurations of solutions, which we see in Figure 1.

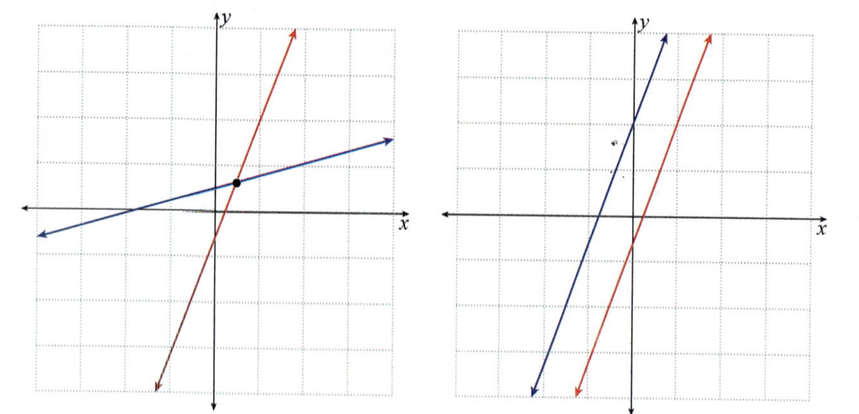

Consistent and independent:
One point common to both equations

Inconsistent:
No points common to both equations

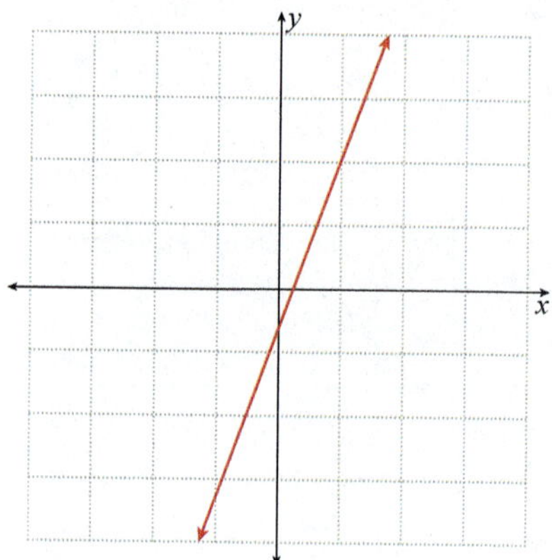

Consistent and dependent:
Two equations that happen to coincide

FIGURE 1: Solutions to Systems of Two Linear Equations

Now, each equation individually has an infinite number of solutions. What we are interested in are the points that solve both equations. These are *solutions to the system of equations*.

In the first graph of Figure 1, the two lines intersect in exactly one point. In the second graph, the two lines are parallel, so the system of equations has no solution since there is no point lying on both lines. In the final graph, the two lines actually coincide and appear as one, meaning that any ordered pair that solves one of the equations in the system solves the other as well, so the system has an infinite number of solutions.

We will encounter systems consisting of more than two equations and/or more than two variables, but larger systems have one important similarity to the two-variable, two-equation systems in Figure 1: every system of linear equations will have exactly one solution, no solution, or an infinite number of solutions.

Solutions to Systems of Linear Equations

- A system of linear equations with no solution is called **inconsistent**.

- A system of linear equations that has at least one solution is called **consistent**.

  1. A consistent system with exactly one solution is called **independent**.

  2. A consistent system with more than one solution must have an infinite number of solutions and is called **dependent**.

Our goal is to develop systematic and effective methods of solving systems of linear equations. Because such systems arise in so many different contexts and are of great importance for both theoretical and practical reasons, many solution methods have been devised. We will learn two of them, the method of substitution and the method of elimination, in this section.

Solving Systems of Linear Equations by Substitution

The solution method of substitution hinges on solving one equation in a system for one of the variables, and substituting the result for that variable in the remaining equations. This can be a time-consuming process if the system is large (meaning more than a few equations and more than a few variables), and in fact the task may have to be repeated many times. But it is a very natural method to use for some small systems, as illustrated in Examples 1 and 2.

Example 1: Solving an Independent System by Substitution

Use the method of substitution to solve the system $\begin{cases} 2x - y = 1 \\ x + y = 5 \end{cases}$.

Solution

NOTE

Solving for a variable with a coefficient of 1 will make the process of substitution a bit easier.

Either equation can be solved for either variable, and the choice doesn't affect the final answer. We will solve the second equation for x, and then substitute the result in the first equation, giving us one equation in the variable y. Once we have solved for y, we can substitute the value of y into either original equation to find x.

$$x = -y + 5 \qquad \text{Solve the second equation for } x.$$
$$2(-y+5) - y = 1 \qquad \text{Substitute the result in the first equation.}$$
$$-2y + 10 - y = 1 \qquad \text{Simplify and solve for } y.$$
$$-3y = -9$$
$$y = 3$$

We then substitute our answer for y into the original second equation.

$$x + 3 = 5 \qquad \text{Substitute } y = 3 \text{ in the second equation.}$$
$$x = 2 \qquad \text{Solve for } x.$$

Thus, the solution to the system of equations is $(2,3)$.

Note how the following graph corresponds to the system and the ordered pair solution that we have found.

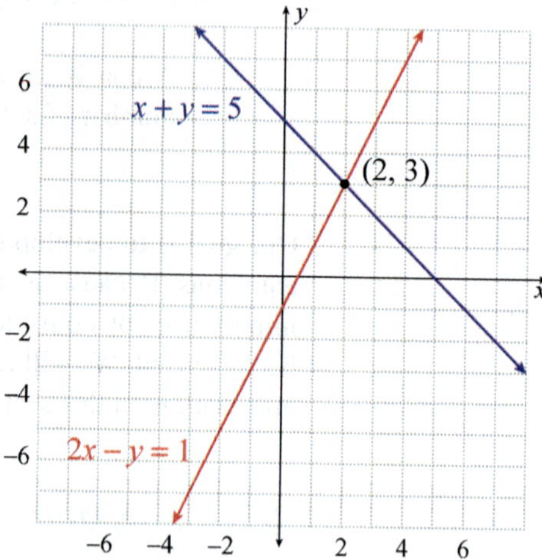

Example 2: Solving a Dependent System by Substitution

Use the method of substitution to solve the system $\begin{cases} -2x + 6y = 6 \\ x + 3 = 3y \end{cases}$.

Solution

$$x = 3y - 3 \qquad \text{Solve the second equation for } x.$$
$$-2(3y - 3) + 6y = 6 \qquad \text{Substitute the result in the first equation.}$$
$$-6y + 6 + 6y = 6 \qquad \text{Simplify.}$$
$$0 = 0$$

The resulting equation is always true. This means that for any value of y, letting $x = 3y - 3$ results in an ordered pair $(x, y) = (3y - 3, y)$ that solves both equations. Since there are an infinite number of solutions, the system is dependent.

Graphically, this means the graphs of the two equations are exactly the same. In fact, the first equation is simply a rearranged multiple of the second equation.

Algebraically, we can describe the solution set as $\{(3y-3, y) \mid y \in \mathbb{R}\}$. If we had solved either equation for y instead of x, we would have obtained the alternative but equivalent

solution $\left\{\left(x, \dfrac{x+3}{3}\right) \mid x \in \mathbb{R}\right\}$.

Solving Systems of Linear Equations by Elimination

In some systems, the expressions obtained by solving one equation for one variable are difficult to work with, no matter which variable we choose. In such cases, the solution method of elimination may be a more efficient choice. The elimination method is also often a better choice for larger systems.

The method of elimination is based on the goal of eliminating one variable in one equation by adding two equations together, and in fact, the elimination method is often called the *addition method*.

The method works by making sure that the resulting equation has fewer variables than either of the original two, simplifying the system. In fact, if the system is of the two-variable, two-equation variety, the new equation is ready to be solved for its remaining variable, and the solution of the system is then straightforward to find.

> ### Example 3: Solving an Independent System by Elimination

Use the method of elimination to solve the system $\begin{cases} 5x + 3y = -7 \\ 7x - 6y = -20 \end{cases}$.

Solution

The coefficient of y in the second equation is -6, while the coefficient of y in the first equation is 3. This means that if we multiply all the terms in the first equation by 2, the coefficients of y will be negatives of one another, so adding the two equations will eliminate the y variable.

In order to keep track of these steps, we annotate our work with labeled arrows. When we modify the system, we are not changing the solutions, we are just writing an equivalent system that is easier to solve. The ultimate goal is to rewrite the system so that we can "read off" the answer.

The notation above the arrow indicates that we have modified the system by multiplying each term in equation 1 by the constant 2.

$$\begin{cases} 5x + 3y = -7 \\ 7x - 6y = -20 \end{cases} \xrightarrow{2E_1} \begin{cases} 10x + 6y = -14 \\ 7x - 6y = -20 \end{cases}$$

$$17x = -34 \qquad \text{Add the equations.}$$

$$x = -2 \qquad \text{Solve for } x.$$

We can then substitute $x = -2$ into either of the original equations to determine y. Here, we substitute into the first equation of the original system.

$$5(-2) + 3y = -7$$
$$3y = 3$$
$$y = 1$$

The ordered pair $(-2, 1)$ is thus the solution of the system. Note that using the second equation gives the same y-value and is a good way to check our work.

Example 4: Solving an Inconsistent System by Elimination

Use the method of elimination to solve the system $\begin{cases} 2x - 3y = 3 \\ 3x - \dfrac{9}{2}y = 5 \end{cases}$.

Solution

Eliminate the variable x by multiplying both equations by a constant:

$$\begin{cases} 2x - 3y = 3 \\ 3x - \dfrac{9}{2}y = 5 \end{cases} \xrightarrow[-2E_2]{3E_1} \begin{cases} 6x - 9y = 9 \\ -6x + 9y = -10 \end{cases}$$
$$\overline{\ 0 = -1}$$

Although the intent was to obtain coefficients of x that were negatives of one another, we have achieved the same thing for y.

When we add the equations, the result is $0 = -1$, a false statement. This means that no ordered pair solves both equations, and the system is inconsistent. Graphically, the two lines defined by the equations are parallel.

Systems of Linear Equations in Three or More Variables

Algebraically, larger systems of equations can be dealt with in the same way as the two-variable, two-equation systems that we have studied (though the number of steps needed to obtain a solution might increase). Geometrically, however, larger systems can mean something quite different.

For example, if an equation contains three variables, say x, y, and z, a given solution of the equation must consist of an *ordered triple* of numbers, not an ordered pair. A graphical representation of the ordered triple requires three coordinate axes, as opposed to two. This leads to the concept of three-dimensional space, and a coordinate system with three axes.

Figure 2 is an illustration of the way in which the positive x, y, and z axes are typically represented on a two-dimensional surface, such as a piece of paper, a computer monitor, or a blackboard. The negative portion of each of the axes is not drawn, and the three axes meet at the origin (the point with $(0, 0, 0)$ as its coordinates) at right angles. As an illustration of how ordered triples appear plotted in Cartesian *space*, the point $(1, 2, 4)$ is plotted (the dotted colored lines are drawn merely for reference and are not part of the plot).

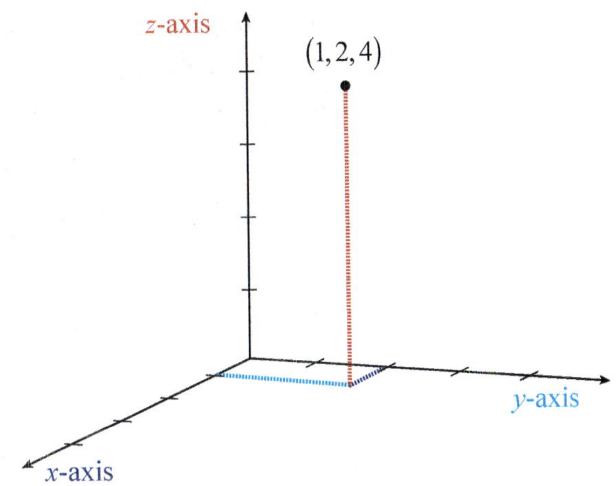

FIGURE 2: Plotting the Point (1, 2, 4)

We are interested in the graph of a linear equation in three variables. It turns out that any equation of the form

$$Ax + By + Cz = D,$$

where not all of A, B, and C are 0, depicts a plane in three-dimensional space. A linear system of equations in three variables will thus describe a collection of planes, one per equation. In a system of three linear equations in three variables, it is possible for the three planes to intersect in exactly one point. It is also possible, however, for two of the planes to intersect in a line while the third plane contains no point of that line, for the three planes to intersect in a common line, for two or three of the planes to be parallel, or for two or three of the planes to coincide. The addition of another variable and another equation leads to many more possibilities than we have seen thus far. However, it is still the case that a linear system has no solution, exactly one solution, or an infinite number of solutions. Figure 3 illustrates some of the possible configurations of a three-variable, three-equation linear system.

FIGURE 3: Three-Variable, Three-Equation Systems

Solve the system $\begin{cases} 2x+y+z=6 \\ 2x+3y-z=-2 \\ -3x+2y-z=5 \end{cases}$.

Solution

We will follow the same type of approach we used in Example 4. If we add the first equation to the second equation or the third equation, we will eliminate z, resulting in a two-equation system in the variables x and y.

$$\begin{array}{ll} \text{Equation 1:} & \begin{cases} 2x+y+z=6 \\ 2x+3y-z=-2 \end{cases} \\ \text{Equation 2:} & \overline{4x+4y=4} \end{array} \qquad \begin{array}{ll} \text{Equation 1:} & \begin{cases} 2x+y+z=6 \\ -3x+2y-z=5 \end{cases} \\ \text{Equation 3:} & \overline{-x+3y=11} \end{array}$$

Putting these two equations together, we have the system.

$$\begin{cases} 4x+4y=4 \\ -x+3y=11 \end{cases}$$

We use elimination once more, multiplying the second equation by 4 to eliminate x.

$$\begin{cases} 4x+4y=4 \\ -x+3y=11 \end{cases} \xrightarrow{\;4E_2\;} \begin{cases} 4x+4y=4 \\ -4x+12y=44 \end{cases}$$
$$\overline{16y=48}$$
$$y=3$$

Now that we have solved for y, we can plug this back into one of the equations from the two-variable system to solve for x.

$$-x+3(3)=11$$
$$-x+9=11$$
$$x=-2$$

Finally, we substitute both x and y into one of the equations from the original system.

$$2(-2)+3+z=6$$
$$-4+3+z=6$$
$$z=7$$

Thus, the solution to the system of equations is the ordered triple $(-2,3,7)$.

Solve the system $\begin{cases} 3x-5y+z=-10 \\ -x+2y-3z=-7 \\ x-y-5z=-24 \end{cases}$.

Solution

In general, a good approach to solving a large system of equations is to try to eliminate a variable and obtain a smaller system.

There are many possible ways to proceed. One option is to use the second equation (or a multiple of it) to eliminate x when we add it to the first and third equations. The result will be a two-equation system in the variables y and z.

$$\text{Equation 1:} \begin{cases} 3x-5y+z=-10 \\ -x+2y-3z=-7 \end{cases} \text{Equation 2:} \quad \xrightarrow{3E_2} \quad \begin{cases} 3x-5y+z=-10 \\ -3x+6y-9z=-21 \end{cases}$$
$$\underline{\phantom{\begin{cases}3x-5y+z=-10\\-3x+6y-9z=-21\end{cases}}}$$
$$y-8z=-31$$

$$\text{Equation 2:} \begin{cases} -x+2y-3z=-7 \\ x-y-5z=-24 \end{cases} \text{Equation 3:}$$
$$\underline{\phantom{\begin{cases}-x+2y-3z=-7\\x-y-5z=-24\end{cases}}}$$
$$y-8z=-31$$

The two resulting equations are identical, and tell us that $y=8z-31$. We can now use any equation that contains x to determine the relation between x and z. For instance, the third equation in the system tells us that $x=y+5z-24$, or

$$x=(8z-31)+5z-24=13z-55.$$

One description of the solution set is thus $\{(13z-55,\ 8z-31,\ z)\,|\,z\in\mathbb{R}\}$. Geometrically, the three planes described by the equations of the system intersect in a line.

Applications of Systems of Linear Equations

Many applications that we have previously analyzed using a single equation are more naturally stated in terms of two or more equations. Consider, for example, the following mixture problem.

Example 7: Mixing Alloys

A foundry needs to produce 75 tons of an alloy that is 34% copper. It has supplies of 9% copper alloy and 84% copper alloy. How many tons of each alloy must be mixed to obtain the desired result?

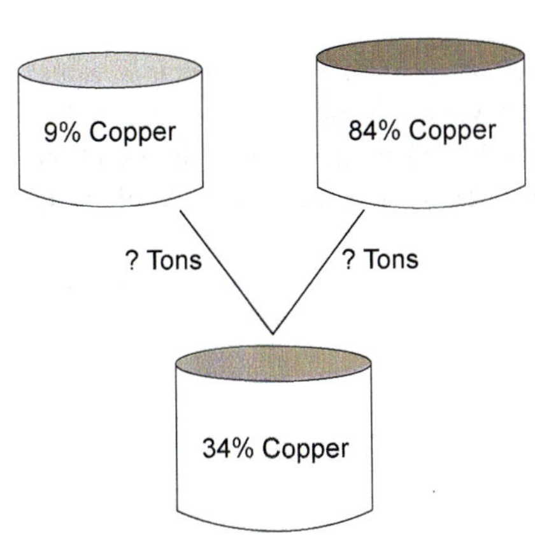

Solution

Let x represent the number of tons of 9% copper alloy needed, and y the number of tons of 84% copper alloy needed. We have two variables, so we will need two equations to find the solution.

Since we need 75 total tons of alloy, one equation is $x+y=75$. We also know that 9% of the x tons and 84% of the y tons represent the total amount of copper, and this amount must equal 34% of 75 tons. The second equation is thus $0.09x+0.84y=0.34(75)$. This gives us a system that can be solved by elimination.

$$\begin{cases} x+y = 75 \\ 0.09x + 0.84y = 0.34(75) \end{cases} \xrightarrow[\;100E_2\;]{-9E_1} \begin{cases} -9x - 9y = -675 \\ 9x + 84y = 2550 \end{cases}$$

$$75y = 1875$$

$$y = 25$$

Substituting back, we have $x + 25 = 75$, so $x = 50$. Thus, 50 tons of 9% alloy and 25 tons of 84% alloy are needed.

Example 8: Determining Ages

If the ages of three girls, Xenia, Yolanda, and Zsa Zsa, are added, the result is 30. The sum of Xenia's and Yolanda's ages is Zsa Zsa's age, while Xenia's age subtracted from Yolanda's is half of Zsa Zsa's age a year ago. How old is each girl?

Solution

Let x, y, and z represent Xenia's, Yolanda's, and Zsa Zsa's ages, respectively. The first sentence tells us that $x + y + z = 30$. The second sentence tells us that $x + y = z$, and that $y - x = \dfrac{z-1}{2}$. To make the work easier, these equations can be rewritten as follows.

$$\begin{cases} x + y + z = 30 \\ x + y = z \\ y - x = \dfrac{z-1}{2} \end{cases} \xrightarrow{\;2E_3\;} \begin{cases} x + y + z = 30 \\ x + y - z = 0 \\ -2x + 2y - z = -1 \end{cases}$$

From this, we see that the sum of the first and second equations results in $2x + 2y = 30$, or $x + y = 15$, and the sum of the first and third equations is $-x + 3y = 29$. The method of elimination is the best choice for solving the new system

$$\begin{cases} x + y = 15 \\ -x + 3y = 29 \end{cases}$$

as the sum of the two equations gives us $4y = 44$, or $y = 11$. We can use this value to determine that $x = 4$, and then use these two values to determine that $z = 15$. Geometrically, it means that the ordered triple $(4,11,15)$ is the point of intersection of the three planes described by these equations. In context, it means that Xenia is 4, Yolanda is 11, and Zsa Zsa is 15.

Solving Systems of Equations Using Technology

The solution to a consistent pair of linear equations is the point common to both equations. Graphically speaking, the solution is the point where the graphs of the two equations intersect. We can use a graphing calculator to find this point. Consider the following system of equations: $\begin{cases} 2x - 3y = -13 \\ \quad\quad x = y - 6 \end{cases}$. One way to solve this system using a calculator is to graph each equation. Remember to solve for y before entering the equation in **Y=** and selecting **graph**.

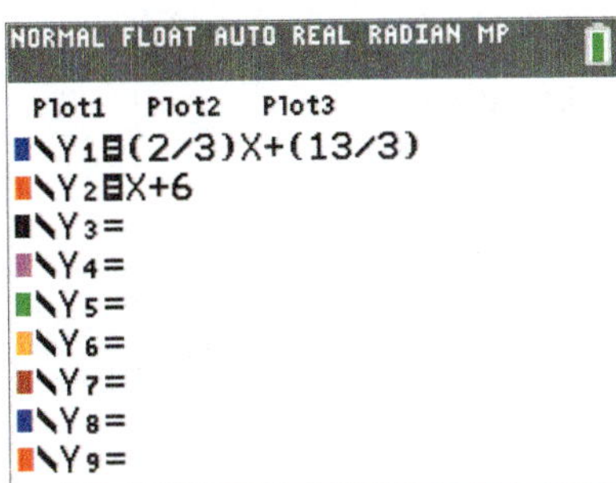
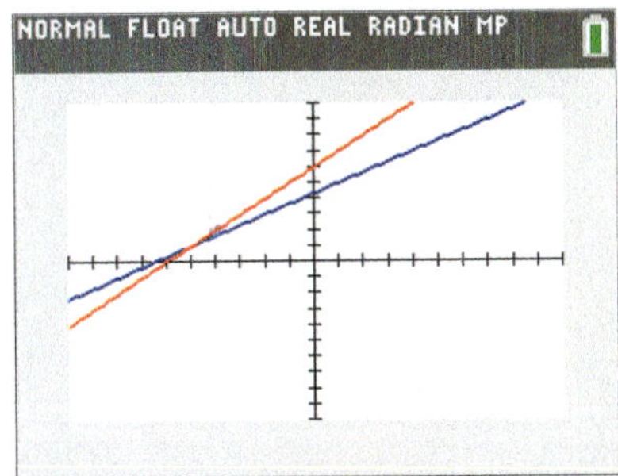

Once the graph of the two lines is displayed, press **2nd** **trace** to access the CALC menu and select `intersect`. The phrase "`First curve?`" should appear. Use the arrows to move the cursor along the first line to where it appears to intersect the other line and press **enter**. When the phrase "`Second curve?`" appears, press **enter** again (as the cursor should now be on the second line, still near the point of intersection). Now the word "`Guess?`" should appear. Press **enter** a final time and the x- and y-values of the point of intersection will appear at the bottom.

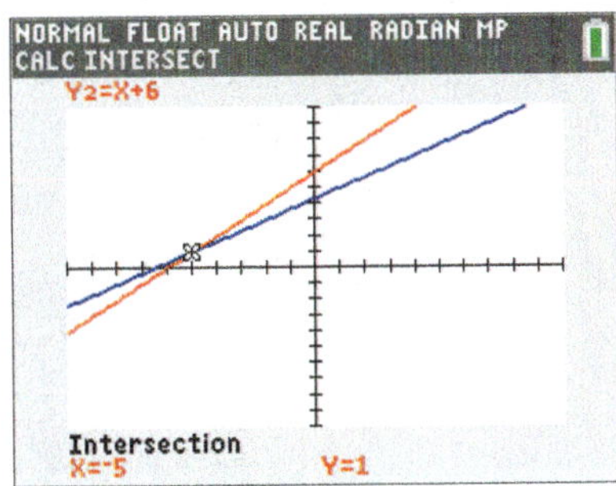

So the point where the lines intersect, and thus the solution to this system of equations, is $(-5, 1)$. This method works with any system of equations that can be graphed on a calculator, not just linear ones.

6.1 EXERCISES

🔆 PRACTICE

Use the method of substitution to solve the following systems of equations. If a system is dependent, express the solution set in terms of one of the variables. See Examples 1 and 2.

1. $\begin{cases} 2x - y = -12 \\ 3x + y = -13 \end{cases}$
    2. $\begin{cases} 2x - 4y = -6 \\ 3x - y = -4 \end{cases}$
    3. $\begin{cases} 3y = 9 \\ x + 2y = 11 \end{cases}$

4. $\begin{cases} -3x - y = 2 \\ 9x + 3y = -6 \end{cases}$
    5. $\begin{cases} 2x + y = -2 \\ -4x - 2y = 5 \end{cases}$
    6. $\begin{cases} 5x - y = -21 \\ 9x + 2y = -34 \end{cases}$

7. $\begin{cases} 2x - y = -3 \\ -4x + 2y = 6 \end{cases}$
    8. $\begin{cases} 3x + 6y = -12 \\ 2x + 4y = -8 \end{cases}$
    9. $\begin{cases} 2x + 5y = 33 \\ 3x = -3 \end{cases}$

10. $\begin{cases} 5x + 2y = 8 \\ 2x + y = 6 \end{cases}$
    11. $\begin{cases} -2x + y = 5 \\ 9x - 2y = 5 \end{cases}$
    12. $\begin{cases} 3x + y = 4 \\ -2x + 3y = 1 \end{cases}$

13. $\begin{cases} 4x - y = -1 \\ -8x + 2y = 2 \end{cases}$
    14. $\begin{cases} 4x - 2y = 3 \\ -2x + y = -7 \end{cases}$
    15. $\begin{cases} 9x - y = -1 \\ 3x + 2y = 44 \end{cases}$

Use the method of elimination to solve the following systems of equations. If a system is dependent, express the solution set in terms of one of the variables. See Examples 3 and 4.

16. $\begin{cases} 2x - 3y = 8 \\ 8x + 5y = -2 \end{cases}$
    17. $\begin{cases} -2x + 3y = 13 \\ 4x + 2y = -18 \end{cases}$
    18. $\begin{cases} 5x + 7y = 1 \\ -2x + 3y = -12 \end{cases}$

19. $\begin{cases} x + 2y = 17 \\ 3x + 4y = 39 \end{cases}$
    20. $\begin{cases} 5x - 10y = 9 \\ -x + 2y = -3 \end{cases}$
    21. $\begin{cases} -2x - 2y = 4 \\ 3x + 3y = -6 \end{cases}$

22. $\begin{cases} 4x + y = 11 \\ 3x - 2y = 0 \end{cases}$
    23. $\begin{cases} 7x + 8y = -3 \\ -5x - 4y = 9 \end{cases}$
    24. $\begin{cases} -2x - y = 9 \\ 4x + 2y = 1 \end{cases}$

25. $\begin{cases} -2x + 4y = 6 \\ 3x - y = -4 \end{cases}$
    26. $\begin{cases} 5x - 6y = -1 \\ -4x + 3y = -10 \end{cases}$
    27. $\begin{cases} \dfrac{2}{3}x + y = -3 \\ 3x + \dfrac{5}{2}y = -\dfrac{7}{2} \end{cases}$

28. $\begin{cases} \dfrac{x}{5} - y = -\dfrac{11}{5} \\ \dfrac{x}{4} + y = 4 \end{cases}$
    29. $\begin{cases} \dfrac{2}{3}x + 2y = 1 \\ x + 3y = 0 \end{cases}$
    30. $\begin{cases} -x - 5y = -6 \\ \dfrac{3}{5}x + 3y = 1 \end{cases}$

Use any convenient method to solve the following systems of equations. If a system is dependent, express the solution set in terms of one or more of the variables, as appropriate. See Examples 5 and 6.

31. $\begin{cases} x - y + 4z = -4 \\ 4x + y - 2z = -1 \\ -y + 2z = -3 \end{cases}$

32. $\begin{cases} x + 2y = -1 \\ y + 3z = 7 \\ 2x + 5z = 21 \end{cases}$

33. $\begin{cases} x + y = 4 \\ y + 3z = -1 \\ 2x - 2y + 5z = -5 \end{cases}$

34. $\begin{cases} 2x - y = 0 \\ 5x - 3y - 3z = 5 \\ 2x + 6z = -10 \end{cases}$

35. $\begin{cases} 3x - y + z = 2 \\ -6x + 2y - 2z = -4 \\ -3x + y - z = -2 \end{cases}$

36. $\begin{cases} 2x - 3y = -2 \\ x - 4y + 3z = 0 \\ -2x + 7y - 5z = 0 \end{cases}$

37. $\begin{cases} 3x - y + z = 2 \\ -6x + 2y - 2z = 1 \\ 5x + 2y - 3z = 2 \end{cases}$

38. $\begin{cases} 4x - y + 5z = 6 \\ 4x - 3y - 5z = -14 \\ -2x - 5z = -8 \end{cases}$

39. $\begin{cases} 3x + 8z = 3 \\ -3x + y - 7z = -2 \\ x + 2y + 3z = 3 \end{cases}$

40. $\begin{cases} x + 2y + z = 8 \\ 2x - 3y - 4z = -16 \\ x - 5y + 5z = 6 \end{cases}$

41. $\begin{cases} 2x - 7y - 4z = 7 \\ -x + 4y + 2z = -3 \\ 3y - 4z = -1 \end{cases}$

42. $\begin{cases} 4x + 4y - 2z = 6 \\ x - 5y + 3z = -2 \\ -2x - 2y + z = 3 \end{cases}$

43. $\begin{cases} 2x + 3y + 4z = 1 \\ 3x - 4y + 5z = -5 \\ 4x + 5y + 6z = 5 \end{cases}$

44. $\begin{cases} x - 4y + 2z = -1 \\ 2x + y - 3z = 10 \\ -3x + 12y - 6z = 3 \end{cases}$

45. $\begin{cases} x + 2y + 3z = 29 \\ 2x - y - z = -2 \\ 3x + 2y - 6z = -8 \end{cases}$

46. $\begin{cases} 5x - 2y + z = 14 \\ 8x + 4y = 12 \\ 9x = 18 \end{cases}$

47. $\begin{cases} 2x + 5y = 6 \\ 3y + 8z = -6 \\ x + 4y = -5 \end{cases}$

48. $\begin{cases} 4x + 3y + 4z = 5 \\ 5x - 6y - 2z = -12 \\ 5z = 20 \end{cases}$

49. $\begin{cases} 9x + 4y - 8z = -4 \\ -6x + 3y - 9z = -9 \\ 8y - 3z = 18 \end{cases}$

50. $\begin{cases} 21x - 7y + 51z = 141 \\ 13x + 9y - 5z = -19 \\ 19x - 8y + 23z = 30 \end{cases}$

APPLICATIONS

51. Karen empties out her purse and finds 45 loose coins, consisting entirely of nickels and pennies. If the total value of the coins is $1.37, how many nickels and how many pennies does she have?

52. What choice of a, b, and c will force the graph of the polynomial $f(x) = ax^2 + bx + c$ to have a y-intercept of 5 and to pass through the points $(1,3)$ and $(2,0)$?

53. A tour organizer is planning on taking a group of 40 people to a musical. Balcony tickets cost $29.95 and regular tickets cost $19.95. The organizer collects a total of $1048.00 from her group to buy the tickets. How many people chose to sit in the balcony?

54. How many ounces each of a 12% alcohol solution and a 30% alcohol solution must be combined to obtain 60 ounces of an 18% solution?

55. Eliza's mother is 20 years older than Eliza, but 3 years younger than Eliza's father. Eliza's father is 7 years younger than three times Eliza's age. How old is Eliza?

56. An investor decides at the beginning of the year to invest some of his cash in an account paying 8% annual interest, and to put the rest in a stock fund that ends up earning 15% over the course of the year. He puts $2000 more in the first account than in the stock fund, and at the end of the year he finds he has earned $1310 in interest. How much money was invested at each of the two rates?

57. Jack and Tyler went shopping for summer clothes. Shirts were $12.47 each, including tax, and shorts were $17.23 per pair, including tax. Jack and Tyler spent a total of $156.21 on 11 items. How many shirts and pairs of shorts did they buy?

58. Three years ago, Bob was twice as old as Marla. Fifteen years ago, Bob was three times as old as Marla. How old is Bob?

59. Deyanira empties her pockets and finds 42 coins consisting of quarters, dimes, and pennies. There are twice as many pennies as dimes and quarters total. If the total value of the coins is $2.13, how many coins of each denomination does she have?

60. If an investor has invested $1000 in stocks and bonds, how much has he invested in stocks if he invested four times more in stocks than in bonds?

61. Twelve years ago, Jim was twice as old as Kristin. Sixteen years ago, Jim was three times older than Kristin. How old is Jim?

62. A movie brought in $740 in ticket sales in one day. Tickets during the day cost $5 and tickets at night cost $7. If 120 tickets were sold, how many were sold during the day?

63. A computer has 24 screws in its case. If there are 7 times more slotted screws than thumb screws, how many thumb screws are in the computer?

64. Jael has $10,000 she would like to invest. She has narrowed her options down to a certificate of deposit paying 5% annually, bonds paying 4% annually, and stocks with an expected annual rate of return of 13.5%. If she wants to invest twice as much in the stocks as in the certificate of deposit and she wants to earn $1000 in interest by the end of the year, how much should she invest in each type of investment?

65. Lea ordered fruit baskets for three of her coworkers. One contained 5 apples, 2 oranges, and 1 mango and cost \$6.81. Another contained 2 mangos, 8 oranges, and 3 apples and cost \$11.88. The third contained 4 apples, 4 oranges, and 4 mangos and cost \$11.04. How much did each type of fruit cost?

📈 TECHNOLOGY

Solve each of the following systems of equations using a graphing utility.

66. $\begin{cases} 98x + 43y - 82z = -784 \\ -65x + 34y = 3032 \\ 28y - 13z = 966 \end{cases}$

67. $\begin{cases} 7.5x + 5.2y - 9.3z = -23.971 \\ -6.8x + 4.4y = 2.708 \\ 0.9x - 1.88y = -2.0194 \end{cases}$

68. $\begin{cases} -5x + 2y - 20z = 14 \\ 2x - 3y + 10z = -19 \\ 7x + 4y - 7z = -7 \end{cases}$

69. $\begin{cases} -5.5x + 2.2y - 5.1z = 11.29 \\ 1.8x + 4.9y - 0.5z = 7.066 \\ 3.9x - 2.6y + 6.3z = -3.698 \end{cases}$

70. $\begin{cases} 5x - 10y + 11z = 19 \\ 27x + 9y + 7z = -44 \\ 2x + 19y - 4z = -3 \end{cases}$

71. $\begin{cases} -23x + 17y - 7z = -51 \\ -13x + 25y - 11z = 45 \\ 51x - 21y - 28z = -58 \end{cases}$

6.2 MATRIX NOTATION AND GAUSS-JORDAN ELIMINATION

■ TOPICS

- Linear Systems, Matrices, and Augmented Matrices
- Gaussian Elimination and Row Echelon Form
- Gauss-Jordan Elimination and Reduced Row Echelon Form
- Matrices and Technology

Linear Systems, Matrices, and Augmented Matrices

Many applications of linear systems involve *thousands* of variables, including the scheduling of planes and passengers at an airport and the modeling of materials at the atomic level. The solution methods of substitution and elimination, learned in the last section, are inadequate for solving such large systems. Further, it can become impractical to even *write down* these systems of equations. In this section, we explore the matrix, a mathematical tool that can be very helpful in understanding linear systems, both large and small.

Matrices and Matrix Notation

A **matrix** is a rectangular array of numbers, called **elements** or **entries** of the matrix. As the numbers are in a rectangular array, they naturally form **rows** and **columns**. It is often important to determine the size of a given matrix; we say that a matrix with m rows and n columns is an $m \times n$ matrix (read "m by n"), or of **order** $m \times n$. By convention, the number of rows is always stated first.

$$
\text{rows}
\begin{bmatrix}
4 & 0 & 1 & -3 \\
12 & -5 & -5 & 8 \\
0 & -7 & 25 & 2
\end{bmatrix}
\quad \text{columns}
$$

This matrix has 3 rows and 4 columns, so it is a 3×4 matrix.

Matrices are often labeled with capital letters. The same letter in lowercase, with a pair of subscripts attached, is usually used to refer to its individual elements. For instance, if A is a matrix, a_{ij} refers to the element in the ith row and the jth column.

If we call the above matrix A, then we have $a_{23} = -5$ and $a_{32} = -7$.

Example 1: Matrices and Matrix Notation

Given the matrix $A = \begin{bmatrix} -27 & 0 & 1 \\ 5 & -\pi & 13 \end{bmatrix}$, determine

a. the order of A,   **b.** the value of a_{13},   **c.** the value of a_{21}.

Solution

a. A has 2 rows and 3 columns, and is thus a 2×3 matrix.

b. The value of the entry in the first row and third column is $a_{13} = 1$.

c. The value of the entry in the second row and first column is $a_{21} = 5$.

Standard Form of a System of Linear Equations

A linear system of equations is in **standard form** when each equation has been simplified with its variables on the left-hand side and its constant term on the right-hand side. Each equation should have its variable terms listed in the same order.

For example, the system $\begin{cases} 3x+4=7y \\ -2x+8y=18 \end{cases}$ is written in standard form as $\begin{cases} 3x-7y=-4 \\ -x+4y=9 \end{cases}$.

Now that we have a standard way to organize the terms in a system of equations, we can put the matrix to use as a way to represent a system of equations concisely.

Augmented Matrices

Given a linear system of equations written in standard form, the **augmented matrix** of that system is a matrix consisting of the coefficients of the variables listed in their relative positions with an adjoined column of the constants of the system. The matrix of coefficients and the column of constants are customarily separated by a vertical bar.

For example, the augmented matrix for the system $\begin{cases} 3x-7y=-4 \\ -x+4y=9 \end{cases}$ is $\left[\begin{array}{cc|c} 3 & -7 & -4 \\ -1 & 4 & 9 \end{array}\right]$.

The augmented matrix will have as many rows as there are equations in the system, and one more column than there are variables.

Example 2: Augmented Matrices

Construct the augmented matrix for the linear system $\begin{cases} \dfrac{2x-6y}{2} = 3-z \\ z-x+5y=12 \\ x+3y-2=2z \end{cases}$.

Solution

Write each equation in standard form, then read off the coefficients and constants to construct the augmented matrix.

Linear System	Standard Form	Augmented Matrix
$\dfrac{2x-6y}{2}=3-z$	$x-3y+z=3$	
$z-x+5y=12$	$-x+5y+z=12$	$\begin{bmatrix} 1 & -3 & 1 & \vert & 3 \\ -1 & 5 & 1 & \vert & 12 \\ 1 & 3 & -2 & \vert & 2 \end{bmatrix}$
$x+3y-2=2z$	$x+3y-2z=2$	

Gaussian Elimination and Row Echelon Form

Consider the following augmented matrix:

$$\begin{bmatrix} 1 & 2 & -2 & \vert & 11 \\ 0 & 1 & -1 & \vert & 3 \\ 0 & 0 & 1 & \vert & -1 \end{bmatrix}$$

We can translate this back into system form to obtain

$$\begin{cases} x+2y-2z=11 \\ \qquad y-z=3 \\ \qquad\qquad z=-1 \end{cases}$$

and we note that this system can be solved when we substitute the last equation $z=-1$ into the second equation to obtain $y-(-1)=3$, or $y=2$, and then substitute again in the first equation to obtain $x+2(2)-2(-1)=11$, or $x=5$.

Typically, systems of equations aren't so straightforward to solve, at least not at first. Recall that the methods of substitution and elimination made systems of equations simpler to solve by reducing the number of equations and/or variables present.

Matrices are a powerful organizational tool that make it easier to transform complicated systems into ones like the example above. The process of *Gaussian elimination* (named after the mathematician Carl Friedrich Gauss) can transform any augmented matrix into a form like the one above, and the solution of the corresponding system then solves the original system as well. The technical name for an augmented matrix of this form is *row echelon form*.

Row Echelon Form

A matrix is in **row echelon form** if each of the following statements is true.

1. The first nonzero entry in each row is 1. We call this a **leading 1**.

2. Every entry below a leading 1 is 0, and each leading 1 appears farther to the right than the leading 1s in the rows above it.

3. All rows consisting entirely of 0s (if there are any) appear at the bottom.

Example 3: Row Echelon Form

Determine if the following matrices are in row echelon form. If not, explain why not.

a. $\begin{bmatrix} 2 & 0 & 13 & | & -1 \\ 0 & 5 & 1 & | & 1 \\ 0 & 0 & 1 & | & -4 \end{bmatrix}$
   b. $\begin{bmatrix} 1 & 2 & -8 & | & 0 \\ 0 & 1 & 1 & | & 3 \\ 0 & 0 & 1 & | & 10 \end{bmatrix}$

c. $\begin{bmatrix} 1 & 0 & 4 & | & -2 \\ 1 & 2 & 0 & | & 7 \\ 1 & -5 & -3 & | & 6 \end{bmatrix}$
   d. $\begin{bmatrix} 1 & -3 & 1 & | & 6 \\ 0 & 0 & 0 & | & 0 \\ 0 & 1 & -8 & | & -10 \end{bmatrix}$

Solution

a. This matrix fails the first condition, as the first nonzero entry is not always 1.

$$\begin{bmatrix} 2 & 0 & 13 & | & -1 \\ 0 & 5 & 1 & | & 1 \\ 0 & 0 & 1 & | & -4 \end{bmatrix}$$

b. This matrix meets each of the three conditions and is in row echelon form.

c. This matrix fails the second condition, as nonzero values are present beneath a leading 1.

$$\begin{bmatrix} 1 & 0 & 4 & | & -2 \\ 1 & 2 & 0 & | & 7 \\ 1 & -5 & -3 & | & 6 \end{bmatrix}$$

d. This matrix fails the third condition, since there is a row of all zeros that lies above a row with nonzero entries.

$$\begin{bmatrix} 1 & -3 & 1 & | & 6 \\ 0 & 0 & 0 & | & 0 \\ 0 & 1 & -8 & | & -10 \end{bmatrix}$$

Since the real goal of Gaussian elimination is to solve systems of equations that are represented by augmented matrices, we need to make sure any changes we apply to the matrix do not change the solution of the system.

These "legal" operations are called *elementary row operations*. There are three such matrix operations, based on the fact that the following actions have no effect on the solutions of a system of equations.

- interchanging the order of two equations in a system

- multiplying all the terms in an equation by a nonzero constant

- adding one equation to another (or a nonzero multiple of an equation)

Elementary Row Operations

Let A be an augmented matrix corresponding to a system of equations. Each of the following operations on A results in the augmented matrix of an equivalent system. In the notation, R_i refers to row i of the matrix A.

1. Rows i and j can be interchanged. (Denoted as $R_i \leftrightarrow R_j$.)
2. Each entry in row i can be multiplied by a nonzero constant c. (Denoted as cR_i.)
3. Row j can be replaced with the sum of itself and a constant multiple of row i. (Denoted as as $cR_i + R_j$.)

Typically, we write that an operation has been performed by connecting the original matrix to the new matrix with an arrow, writing the type of operation above it. For example, the matrix from Example 3d can be placed in row echelon form as follows.

$$\begin{bmatrix} 1 & -3 & 1 & 6 \\ 0 & 0 & 0 & 0 \\ 0 & 1 & -8 & -10 \end{bmatrix} \xrightarrow{R_2 \leftrightarrow R_3} \begin{bmatrix} 1 & -3 & 1 & 6 \\ 0 & 1 & -8 & -10 \\ 0 & 0 & 0 & 0 \end{bmatrix}$$

Example 4: Elementary Row Operations

Perform the indicated elementary row operation.

$$\begin{bmatrix} 1 & 2 & 3 & 0 \\ 5 & 4 & 1 & -7 \\ 6 & 1 & 8 & -9 \end{bmatrix} \xrightarrow{-5R_1 + R_2} ?$$

Solution

The row operation indicates that we should add -5 times Row 1 to Row 2.

$$\begin{bmatrix} 1 & 2 & 3 & 0 \\ 5 & 4 & 1 & -7 \\ 6 & 1 & 8 & -9 \end{bmatrix} \xrightarrow{-5R_1 + R_2} \begin{bmatrix} 1 & 2 & 3 & 0 \\ -5(1)+5 & -5(2)+4 & -5(3)+1 & -5(0)+(-7) \\ 6 & 1 & 8 & -9 \end{bmatrix}$$

$$= \begin{bmatrix} 1 & 2 & 3 & 0 \\ 0 & -6 & -14 & -7 \\ 6 & 1 & 8 & -9 \end{bmatrix}$$

Note that the first row now begins with 1 and the second row now begins with 0. The row operation has begun to change this matrix towards row echelon form.

In summary, Gaussian elimination refines the method of elimination by removing the need to write variables at each step and by providing a framework for systematic application of row operations. Just as with substitution and elimination, there is no one correct method to apply Gaussian elimination. With the upcoming examples, we will see some general guidelines to follow.

Example 5: Gaussian Elimination

Use Gaussian elimination to solve the system $\begin{cases} -2x + y - 5z = -6 \\ x + 2y - z = -8 \\ 3x - y + 2z = 2 \end{cases}$.

Solution

First, we read off the augmented matrix corresponding to this system.

$$\left[\begin{array}{ccc|c} -2 & 1 & -5 & -6 \\ 1 & 2 & -1 & -8 \\ 3 & -1 & 2 & 2 \end{array}\right]$$

Now, we transform it into row echelon form. It is usually easiest to work one column at a time. After getting a leading 1 in the first row, use row operations to obtains 0s below it. Repeat this process with each successive column.

$$\left[\begin{array}{ccc|c} -2 & 1 & -5 & -6 \\ 1 & 2 & -1 & -8 \\ 3 & -1 & 2 & 2 \end{array}\right] \xrightarrow{R_1 \leftrightarrow R_2} \left[\begin{array}{ccc|c} 1 & 2 & -1 & -8 \\ -2 & 1 & -5 & -6 \\ 3 & -1 & 2 & 2 \end{array}\right]$$

Exchange Rows 1 and 2 to make 1 the first entry of the first row.

$$\xrightarrow{2R_1 + R_2} \left[\begin{array}{ccc|c} 1 & 2 & -1 & -8 \\ 0 & 5 & -7 & -22 \\ 3 & -1 & 2 & 2 \end{array}\right]$$

Add 2 times Row 1 to Row 2 to get a 0 as the first entry of Row 2.

$$\xrightarrow{-3R_1 + R_3} \left[\begin{array}{ccc|c} 1 & 2 & -1 & -8 \\ 0 & 5 & -7 & -22 \\ 0 & -7 & 5 & 26 \end{array}\right]$$

Add −3 times Row 1 to Row 3 to get a 0 as the first entry of Row 3.

$$\xrightarrow{\frac{1}{5}R_2} \left[\begin{array}{ccc|c} 1 & 2 & -1 & -8 \\ 0 & 1 & -\frac{7}{5} & -\frac{22}{5} \\ 0 & -7 & 5 & 26 \end{array}\right]$$

Multiply Row 2 by $\frac{1}{5}$ to make its first nonzero entry 1.

$$\xrightarrow{7R_2 + R_3} \left[\begin{array}{ccc|c} 1 & 2 & -1 & -8 \\ 0 & 1 & -\frac{7}{5} & -\frac{22}{5} \\ 0 & 0 & -\frac{24}{5} & -\frac{24}{5} \end{array}\right]$$

Add 7 times Row 2 to Row 3 to get a 0 as the second entry of Row 3.

$$\xrightarrow{-\frac{5}{24}R_3} \left[\begin{array}{ccc|c} 1 & 2 & -1 & -8 \\ 0 & 1 & -\frac{7}{5} & -\frac{22}{5} \\ 0 & 0 & 1 & 1 \end{array}\right]$$

Multiply Row 3 by $-\frac{5}{24}$ to make its first nonzero entry a 1.

The third row in the last matrix tells us that $z = 1$. When we substitute this in the second equation, we obtain

$$y - \frac{7}{5}(1) = -\frac{22}{5},$$

so $y = -3$. From the first row equation, we then obtain

$$x + 2(-3) - 1(1) = -8,$$

or $x = -1$. Thus, the ordered triple $(-1, -3, 1)$ solves the system of equations.

Gauss-Jordan Elimination and Reduced Row Echelon Form

Once a matrix has been put in row echelon form, we can back-substitute, beginning with the last equation and working backward to solve the system of equations. We reached the row echelon form with matrix operations, and it turns out that we can perform the remaining steps in matrix form as well.

Just as Gaussian elimination is a refinement of the method of elimination, Gauss-Jordan elimination, co-named for Wilhelm Jordan (1842–1899), is a refinement of Gaussian elimination. The goal in the method is to put a given matrix into *reduced row echelon form*.

Reduced Row Echelon Form

A matrix is said to be in **reduced row echelon form** if

1. it is in row echelon form and
2. each entry *above* a leading 1 is also 0.

Consider, for instance, the last matrix obtained in Example 5. If we begin with that matrix and eliminate entries (that is, convert to 0) above the leading 1 in the third column, and then do the same in the second column, we obtain a matrix in reduced row echelon form:

$$\begin{bmatrix} 1 & 2 & -1 & -8 \\ 0 & 1 & -\dfrac{7}{5} & -\dfrac{22}{5} \\ 0 & 0 & 1 & 1 \end{bmatrix} \xrightarrow[\ R_3 + R_1\ ]{\ \frac{7}{5}R_3 + R_2\ } \begin{bmatrix} 1 & 2 & 0 & -7 \\ 0 & 1 & 0 & -3 \\ 0 & 0 & 1 & 1 \end{bmatrix} \xrightarrow{\ -2R_2 + R_1\ } \begin{bmatrix} 1 & 0 & 0 & -1 \\ 0 & 1 & 0 & -3 \\ 0 & 0 & 1 & 1 \end{bmatrix}$$

If we now write this in system form, we have

$$\begin{cases} x = -1 \\ y = -3 \\ z = 1 \end{cases}$$

which is equivalent to the original system, but in a form that directly tells us the solution of the system.

Example 6: Reduced Row Echelon Form

Use Gauss-Jordan elimination to solve the system $\begin{cases} x - 2y + 3z = -5 \\ 2x + 3y - z = 1 \\ -x - 5y + 4z = -6 \end{cases}$.

Solution

$$
\begin{bmatrix}
1 & -2 & 3 & | & -5 \\
2 & 3 & -1 & | & 1 \\
-1 & -5 & 4 & | & -6
\end{bmatrix}
\xrightarrow[R_1+R_3]{-2R_1+R_2}
\begin{bmatrix}
1 & -2 & 3 & | & -5 \\
0 & 7 & -7 & | & 11 \\
0 & -7 & 7 & | & -11
\end{bmatrix}
$$

$$
\xrightarrow{R_2+R_3}
\begin{bmatrix}
1 & -2 & 3 & | & -5 \\
0 & 7 & -7 & | & 11 \\
0 & 0 & 0 & | & 0
\end{bmatrix}
$$

Note that the row of 0s at the bottom corresponds to the true statement $0 = 0$, indicating that this system has an infinite number of solutions. In order to describe the solution set algebraically, we continue the row operations until the matrix is in reduced row echelon form.

$$
\begin{bmatrix}
1 & -2 & 3 & | & -5 \\
0 & 7 & -7 & | & 11 \\
0 & 0 & 0 & | & 0
\end{bmatrix}
\xrightarrow{\frac{1}{7}R_2}
\begin{bmatrix}
1 & -2 & 3 & | & -5 \\
0 & 1 & -1 & | & \dfrac{11}{7} \\
0 & 0 & 0 & | & 0
\end{bmatrix}
$$

$$
\xrightarrow{2R_2+R_1}
\begin{bmatrix}
1 & 0 & 1 & | & -\dfrac{13}{7} \\
0 & 1 & -1 & | & \dfrac{11}{7} \\
0 & 0 & 0 & | & 0
\end{bmatrix}
$$

The last matrix is in reduced row echelon form, as every leading 1 has 0s both below and above it. The corresponding system of equations is

$$
\begin{cases} x + z = -\dfrac{13}{7} \\ y - z = \dfrac{11}{7} \end{cases}, \text{ which is equivalent to } \begin{cases} x = -z - \dfrac{13}{7} \\ y = z + \dfrac{11}{7} \end{cases}.
$$

Thus, we can describe the solution set as $\left\{ \left(-z - \dfrac{13}{7}, z + \dfrac{11}{7}, z \right) \,\middle|\, z \in \mathbb{R} \right\}$.

Matrices and Technology

Operations with matrices can also be done on a graphing calculator. Suppose we wanted to find the row reduced form of the matrix $\begin{bmatrix} 8 & -1 & 2 & 30 \\ -3 & 4 & 5 & -20 \\ 1 & 2 & -1 & -6 \end{bmatrix}$. First,

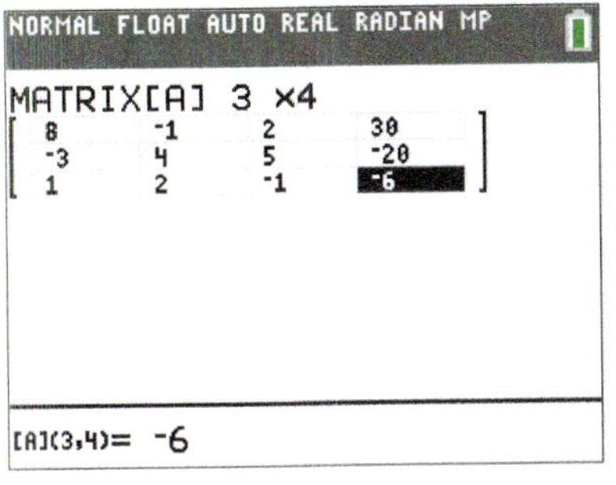

we need to enter in this matrix by pressing **2nd** **x^{-1}** to access the **MATRIX** menu and use the arrows to highlight **EDIT**. Select **1 : [A]** for matrix A and enter the dimensions, 3×4, and the values for each entry. If there are already numbers present on the display, type over them and the calculator will adjust automatically. To enter the values for each entry, move the cursor to the upper left entry position and type the value that goes in that position. Press **enter** and the cursor will move to the next position in the matrix. The double subscripts appear at the bottom of the display as each value is entered.

Return to the main screen by pressing [2nd] [mode]. Now that the matrix is stored in the calculator as matrix A, we can perform operations. Press [2nd] [x^{-1}] to access the MATRIX menu again, but this time select MATH. Scroll down to select A:ref(. When you press [enter], ref(will appear on the main screen.

Now we need to select which matrix we want to put in row echelon form, matrix A. To do this, press [2nd] [x^{-1}] once more. NAMES and 1:[A] should already be highlighted, so press [enter]. Add the right-hand parenthesis and press [enter].

To view this matrix with fractional entries, press [math] and [enter], since 1:Frac is already highlighted. Press [enter] again on the main screen.

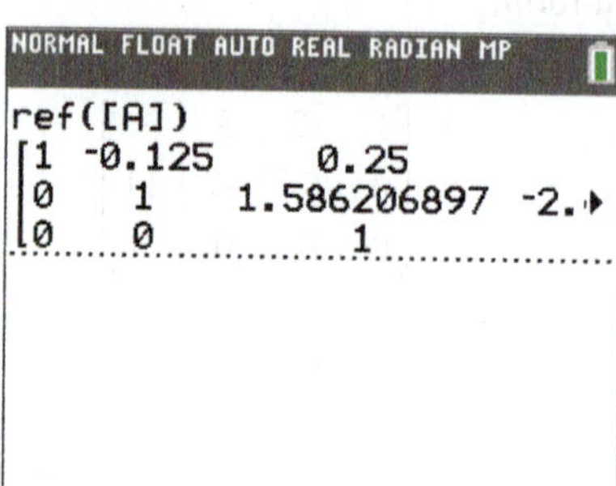

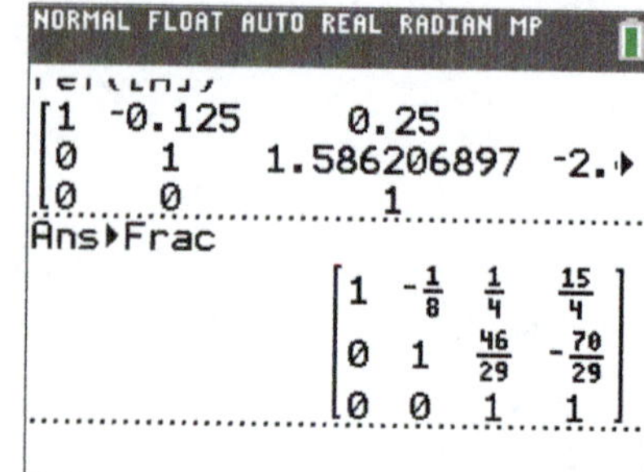

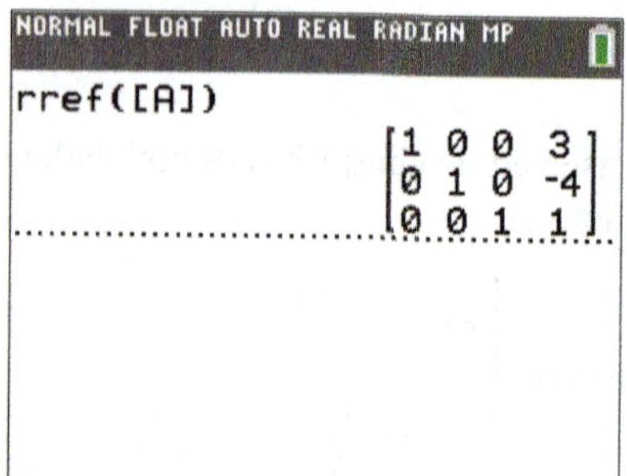

Since the entire matrix does not fit on the screen, use the arrow keys to view the right side of the matrix.

Finding the reduced row echelon form of this matrix is similar. Press [2nd] [x^{-1}] and select MATH but this time arrow down farther to highlight B:rref(and press [enter]. Select matrix A again, add the right-hand parenthesis and press [enter].

6.2 EXERCISES

💡 PRACTICE

1. Let $A = \begin{bmatrix} 4 & -1 \\ 0 & 3 \\ 9 & -5 \end{bmatrix}$. Determine the following, if possible:

    a. The order of A **b.** The value of a_{12} **c.** The value of a_{23}

2. Let $B = \begin{bmatrix} -7 & 2 & 11 \end{bmatrix}$. Determine the following, if possible:

    a. The order of B **b.** The value of b_{12} **c.** The value of b_{31}

3. Let $C = \begin{bmatrix} 1 & 0 \\ 5 & -3 \\ 2 & 9 \\ \pi & e \\ 10 & -7 \end{bmatrix}$. Determine the following, if possible:

    a. The order of C **b.** The value of c_{23} **c.** The value of c_{51}

4. Let $D = \begin{bmatrix} -8 & 13 & -1 \\ 0 & 6 & 3 \\ 0 & -9 & 0 \end{bmatrix}$. Determine the following, if possible:

    a. The order of D **b.** The value of d_{23} **c.** The value of d_{33}

5. Let $E = \begin{bmatrix} -443 & 951 & 165 & 274 \\ 286 & -653 & 812 & -330 \\ 909 & 377 & 429 & -298 \end{bmatrix}$. Determine the following, if possible:

    a. The order of E **b.** The value of e_{42} **c.** The value of e_{21}

6. Let $A = \begin{bmatrix} 9 & 5 & 0 \\ 7 & 4 & 2 \end{bmatrix}$. Determine the following, if possible:

    a. The order of A **b.** The value of a_{22} **c.** The value of a_{13}

7. Let $B = \begin{bmatrix} 8 & 1 \\ 3 & 0 \\ 6 & 7 \end{bmatrix}$. Determine the following, if possible:

    a. The order of B **b.** The value of b_{12} **c.** The value of b_{13}

8. Let $C = \begin{bmatrix} 65 & 32 & 91 & 45 \\ 23 & 18 & 75 & 47 \\ 8 & 63 & 28 & 31 \end{bmatrix}$. Determine the following, if possible:

    a. The order of C **b.** The value of c_{43} **c.** The value of c_{23}

9. Let $D = \begin{bmatrix} 4 & 9 & 7 & 1 & 8 \\ 5 & 3 & 0 & 2 & 6 \end{bmatrix}$. Determine the following, if possible:

    a. The order of D **b.** The value of d_{21} **c.** The value of d_{24}

Construct the augmented matrix that corresponds to each of the following systems of equations. See Example 2. (Answers may appear in slightly different, but equivalent, form.)

10. $\begin{cases} 4x + 5y - 3z = 8 \\ 7x - 2y + 9 = 3 \\ 5x - 6y + 3z = 0 \end{cases}$

11. $\begin{cases} y - 2z + 4 = 3x \\ \dfrac{x}{2} - 4y - 1 = z \\ 3(-y + z) - 1 = 0 \end{cases}$

12. $\begin{cases} 5x + \dfrac{y-z}{2} = 3 \\ 7(z-x) + y - 2 = 0 \\ x - (4-z) = y \end{cases}$

13. $\begin{cases} \dfrac{2-3x}{2} = y \\ 3z + 2(x+y) = 0 \\ 2x - y = 2(x-3z) \end{cases}$

14. $\begin{cases} 2(z+3) - x + y = z \\ -3(x-2y) - 1 = 5z \\ \dfrac{x}{3} - (y-2z) = x \end{cases}$

15. $\begin{cases} \dfrac{12x-1}{5} + \dfrac{y}{2} = \dfrac{3z}{2} \\ y - (x+3z) = -(1-y) \\ 2x - 2 - z - 2y = 7x \end{cases}$

16. $\begin{cases} \dfrac{3x+4y}{2} - 3z = 6 \\ 3(x - 2y + 9z) = 0 \\ 2x + 6y = 3 - z \end{cases}$

17. $\begin{cases} \dfrac{2x-4y}{3} = 2z \\ 8x = 2(y-3z) + 7 \\ 3x = 2y \end{cases}$

18. $\begin{cases} \dfrac{2(2x-y)}{3} + z = 7 \\ 4 = \dfrac{3}{-x+y+3z} \\ 4x - 8y + 4 = 9x \end{cases}$

19. $\begin{cases} 0.5x - 14y = \dfrac{z}{4} - 8 \\ \dfrac{x}{5} - y + \dfrac{z}{4} = \dfrac{y}{6} - 3 \\ \dfrac{2}{3}\left(\dfrac{4}{y-x-1}\right) = \dfrac{5}{z} \end{cases}$

Construct the system of equations that corresponds to each of the following matrices.

20. $\begin{bmatrix} 5 & 3 & | & 9 \\ 1 & 4 & | & 12 \end{bmatrix}$

21. $\begin{bmatrix} 1 & 0 & | & 8 \\ 0 & 1 & | & 3 \end{bmatrix}$

22. $\begin{bmatrix} 14 & 0 & 1 & | & 16 \\ 3 & 6 & 4 & | & 0 \\ 8 & 2 & 5 & | & 21 \end{bmatrix}$

23. $\begin{bmatrix} 1 & 3 & 6 & | & 16 \\ 0 & 1 & 2 & | & 9 \\ 0 & 0 & 1 & | & 4 \end{bmatrix}$

24. $\begin{bmatrix} 2 & 1 & 1 & | & 22 \\ 1 & 3 & 1 & | & 17 \\ 1 & 1 & 4 & | & 8 \end{bmatrix}$

25. $\begin{bmatrix} 0 & 9 & 13 & | & 27 \\ 2 & 0 & 21 & | & 19 \\ 7 & 18 & 0 & | & 32 \end{bmatrix}$

Fill in the blanks by performing the indicated row operations. See Example 4.

26. $\begin{bmatrix} 3 & 2 & | & -7 \\ 1 & 3 & | & 5 \end{bmatrix} \xrightarrow{-3R_2 + R_1} \underline{\ ?\ }$

27. $\begin{bmatrix} 2 & -5 & | & 3 \\ -4 & 3 & | & -1 \end{bmatrix} \xrightarrow{2R_1 + R_2} \underline{\ ?\ }$

28. $\begin{bmatrix} 4 & 2 & | & -8 \\ 3 & -9 & | & 0 \end{bmatrix} \xrightarrow[\frac{1}{3}R_2]{\frac{1}{2}R_1} \underline{\ ?\ }$

29. $\begin{bmatrix} 9 & -2 & | & 7 \\ 1 & 3 & | & -2 \end{bmatrix} \xrightarrow{R_1 \leftrightarrow R_2} \underline{\ ?\ }$

30. $\begin{bmatrix} 4 & 1 & | & 5 \\ 3 & 6 & | & 0 \end{bmatrix} \xrightarrow{2R_1} \underline{\ ?\ }$

31. $\begin{bmatrix} 8 & -2 & | & -4 \\ 3 & -1 & | & 7 \end{bmatrix} \xrightarrow{-2R_2} \underline{\ ?\ }$

32. $\begin{bmatrix} 9 & 12 & | & -6 \\ 15 & -3 & | & 0 \end{bmatrix} \xrightarrow{-\frac{1}{3}R_1} \underline{\ ?\ }$

33. $\begin{bmatrix} 4 & 12 & | & -6 \\ 7 & 3 & | & 9 \end{bmatrix} \xrightarrow{\frac{1}{2}R_1 + R_2} \underline{\ ?\ }$

34. $\begin{bmatrix} 3 & 0 & | & 1 \\ 5 & 7 & | & -2 \end{bmatrix} \xrightarrow{3R_1 + R_2} \underline{\ ?\ }$

35. $\begin{bmatrix} 8 & -2 & | & 10 \\ 9 & -3 & | & 0 \end{bmatrix} \xrightarrow[-\frac{2}{3}R_2]{\frac{1}{2}R_1} \underline{\ ?\ }$

36. $\begin{bmatrix} 5 & 2 & 9 & 7 \\ 1 & 3 & -5 & 0 \\ 2 & -4 & 1 & 8 \end{bmatrix} \xrightarrow[-R_1+R_3]{2R_2} \underline{\ ?\ }$

37. $\begin{bmatrix} 6 & -2 & 5 & 14 \\ -7 & 19 & 2 & 3 \\ -9 & 11 & -4 & 7 \end{bmatrix} \xrightarrow[0.5R_3]{3R_1} \underline{\ ?\ }$

38. $\begin{bmatrix} 5 & 3 & 13 & 15 \\ 17 & 9 & -8 & -14 \\ 4 & -11 & 19 & 8 \end{bmatrix} \xrightarrow{-2R_2+R_3} \underline{\ ?\ }$

39. $\begin{bmatrix} 8 & 11 & 18 & 2 \\ 14 & 33 & -3 & -5 \\ -9 & 21 & 12 & 9 \end{bmatrix} \xrightarrow[-2R_3+R_2]{\frac{1}{3}R_3+R_1} \underline{\ ?\ }$

40. $\begin{bmatrix} 1 & 3 & -2 & 4 \\ 3 & -1 & 8 & 2 \\ -5 & 0 & 2 & 7 \end{bmatrix} \xrightarrow[5R_1+R_3]{-3R_1+R_2} \underline{\ ?\ }$

41. $\begin{bmatrix} 2 & 3 & -3 & 5 \\ 1 & 1 & 3 & 4 \\ 3 & 3 & 9 & 12 \end{bmatrix} \xrightarrow[-3R_2+R_3]{-2R_2+R_1} \underline{\ ?\ }$

42. $\begin{bmatrix} -3 & 2 & 2 \\ 5 & -4 & 1 \end{bmatrix} \xrightarrow{2R_1+R_2} \underline{\ ?\ }$

43. $\begin{bmatrix} -5 & 20 & -15 \\ 2 & -12 & 5 \end{bmatrix} \xrightarrow[\frac{1}{2}R_2]{\frac{1}{5}R_1} \underline{\ ?\ }$

44. $\begin{bmatrix} 2 & 2 & 3 & 7 \\ -3 & 2 & 8 & -2 \\ 1 & 5 & 2 & 6 \end{bmatrix} \xrightarrow[3R_3+R_2]{-2R_3+R_1} \underline{\ ?\ }$

45. $\begin{bmatrix} 1 & 5 & -9 & 11 \\ 1 & 4 & -1 & 4 \\ 4 & 3 & 5 & 45 \end{bmatrix} \xrightarrow[-4R_1+R_3]{-R_1+R_2} \underline{\ ?\ }$

For each matrix, determine if it is in row echelon form, reduced row echelon form, or neither.

46. $\begin{bmatrix} 1 & 5 & 4 \\ 0 & 1 & 3 \end{bmatrix}$

47. $\begin{bmatrix} 1 & 2 & 0 & 9 \\ 0 & 1 & 3 & 4 \\ 0 & 1 & 1 & 12 \end{bmatrix}$

48. $\begin{bmatrix} 1 & 0 & 0 & 4 \\ 0 & 1 & 0 & 1 \\ 0 & 0 & 1 & 8 \end{bmatrix}$

49. $\begin{bmatrix} 1 & 0 & 0 & 7 \\ 5 & 1 & 0 & 14 \\ 3 & 4 & 1 & -16 \end{bmatrix}$

50. $\begin{bmatrix} 1 & 2 & 5 & 0 \\ 0 & 1 & 9 & 3 \\ 0 & 0 & 0 & 1 \end{bmatrix}$

51. $\begin{bmatrix} 0 & 1 & 3 \\ 1 & 0 & 6 \end{bmatrix}$

Use Gaussian elimination and back-substitution to solve the following systems of equations. See Example 5.

52. $\begin{cases} 2x - 4y = -6 \\ 3x - y = -4 \end{cases}$

53. $\begin{cases} 2x - 5y = 11 \\ 3x + 2y = 7 \end{cases}$

54. $\begin{cases} 5x - y = -21 \\ 9x + 2y = -34 \end{cases}$

55. $\begin{cases} x - 4y = -11 \\ 7x - y = 4 \end{cases}$

56. $\begin{cases} x + 2y = 17 \\ 3x + 4y = 39 \end{cases}$

57. $\begin{cases} 2x + 6y = 4 \\ -4x - 7y = 7 \end{cases}$

58. $\begin{cases} 3x - 2y = 5 \\ -5x + 4y = -3 \end{cases}$

59. $\begin{cases} 2x + y = -2 \\ -4x - 2y = 5 \end{cases}$

60. $\begin{cases} 6x - 16y = 10 \\ -3x + 8y = 4 \end{cases}$

61. $\begin{cases} 2x - 3y = 0 \\ 5x + y = 17 \end{cases}$

62. $\begin{cases} 6x + 3y = 3 \\ x + y = 3 \end{cases}$

63. $\begin{cases} 3x + 6y = -12 \\ 2x + 4y = -8 \end{cases}$

64. $\begin{cases} 4x + 5y = 9 \\ 8x + 3y = -17 \end{cases}$

65. $\begin{cases} \dfrac{2}{3}x + 2y = 1 \\ x + 3y = 0 \end{cases}$

66. $\begin{cases} 13x - 17y = -3 \\ -19x + 15y = -35 \end{cases}$

67. $\begin{cases} 3x - 9y - 7z = -9 \\ 5x + 11y - z = 17 \\ -4x - 8y + 7z = 5 \end{cases}$ **68.** $\begin{cases} 8x - y + 5z = -8 \\ 11x - 2y + 9z = -9 \\ 7x - 3y + 13z = 4 \end{cases}$ **69.** $\begin{cases} 17x + 13y + 8z = 46 \\ -12x + 3y + 28z = -19 \\ 14x + 5y - 15z = -15 \end{cases}$

Use Gauss-Jordan elimination to solve the following systems of equations. See Example 6.

70. $\begin{cases} 2x - 3y = 8 \\ 8x + 5y = -2 \end{cases}$ **71.** $\begin{cases} \dfrac{2}{3}x + y = -3 \\ 3x + \dfrac{5}{2}y = -\dfrac{7}{2} \end{cases}$ **72.** $\begin{cases} 3y = 9 \\ x + 2y = 11 \end{cases}$

73. $\begin{cases} 6x + 2y = -4 \\ -9x - 3y = 6 \end{cases}$ **74.** $\begin{cases} 3y = 6 \\ 5x + 2y = 4 \end{cases}$ **75.** $\begin{cases} 3x + 8y = -4 \\ x + 2y = -2 \end{cases}$

76. $\begin{cases} -3x + 2y = 5 \\ 5x - 2y = 1 \end{cases}$ **77.** $\begin{cases} 9x - 11y = 10 \\ -4x + 3y = -12 \end{cases}$ **78.** $\begin{cases} 9x - 15y = -6 \\ -3x + 11y = -10 \end{cases}$

79. $\begin{cases} 3x - 8y = 7 \\ 18x - 35y = -23 \end{cases}$ **80.** $\begin{cases} 4x + y - 3z = -9 \\ 2x - 3z = -19 \\ 7x - y - 4z = -29 \end{cases}$ **81.** $\begin{cases} -5x + 9y + 3z = 1 \\ 3x + 2y - 6z = 9 \\ x + 4y - z = 16 \end{cases}$

82. $\begin{cases} 2x - y = 0 \\ 5x - 3y - 3z = 5 \\ 2x + 6z = -10 \end{cases}$ **83.** $\begin{cases} x + y = 4 \\ y + 3z = -1 \\ 2x - 2y + 5z = -5 \end{cases}$ **84.** $\begin{cases} 2x - 3y = -2 \\ x - 4y + 3z = 0 \\ -2x + 7y - 5z = 0 \end{cases}$

85. $\begin{cases} 3x + 8z = 3 \\ -3x - 7z = -3 \\ x + 3z = 1 \end{cases}$ **86.** $\begin{cases} 3x - y + z = 2 \\ -6x + 2y - 2z = 1 \\ 5x + 2y - 3z = 2 \end{cases}$ **87.** $\begin{cases} x + 2y = -1 \\ y + 3z = 7 \\ 2x + 5z = 21 \end{cases}$

88. $\begin{cases} 2x + 8y - z = -5 \\ -5x + 3y + 4z = -6 \\ x - 4y - 5z = -8 \end{cases}$ **89.** $\begin{cases} 7x - 8y + 2z = -2 \\ 5x - 3y - z = -3 \\ 8x + y - 3z = 7 \end{cases}$

90. $\begin{cases} 8x + 14y - 3z = 3 \\ -6x + 2y + 7z = -13 \\ 8x + 19y + 3z = 11 \end{cases}$ **91.** $\begin{cases} 8x + 5y + 3z = -2 \\ 12x - y - 18z = 1 \\ 7x + 6y + 10z = 19 \end{cases}$

92. $\begin{cases} 4x + 8y + 7z = 27 \\ -2x + 9y - 8z = -15 \\ 9x + 13y + 7z = -33 \end{cases}$ **93.** $\begin{cases} w - x + 2z = 9 \\ 2w + 3y = -1 \\ -2w - 5y - z = 0 \\ x + 2y = -4 \end{cases}$

94. $\begin{cases} 3w - x + 5y + 3z = 2 \\ -4w - 10y - 2z = 10 \\ w - x + 2z = 7 \\ 4w - 2x + 5y + 5z = 9 \end{cases}$

◆ APPLICATIONS

95. The sum of three integers is 155. The first integer is sixteen more than the second. The third integer is seven less than the sum of the first integer and twice the second. What are the three integers?

96. Mario bought a pound of bacon, a dozen eggs, and a loaf of bread to make breakfast for his family. The total cost was $7.42. The bacon cost $0.03 more than twice the price of the bread and the eggs cost $0.03 less than half the price of the bread. Find the price of each item.

97. The Pizza House sells three sizes of pizzas: small, medium, large. The prices of the pizzas are $9.00, $12.00, and $15.00, respectively. In one day, they sold 82 pizzas for a total of $1098.00. If the number of large pizzas sold was twice the number of medium pizzas sold, how many of each size pizza did the Pizza House sell?

6.3 DETERMINANTS AND CRAMER'S RULE

■ TOPICS

- Determinants and Their Evaluation
- Solving Systems of Linear Equations Using Cramer's Rule
- Evaluating Determinants Using Technology

Determinants and Their Evaluation

While Gaussian elimination and Gauss-Jordan elimination both provide significant improvements in ease and speed when solving systems of equations, they are not truly new methods. Each simply uses the matrix as a powerful, agile organizer to make the substitution and elimination methods simpler. Cramer's Rule, named after the Swiss mathematician Gabriel Cramer (1704–1752), is a solution method that truly brings something new to the discussion.

Cramer's Rule relies on the computation of a number, called the *determinant*, associated with every *square* matrix (a matrix with the same number of rows as columns).

> **⚠ CAUTION**
>
> If A is a matrix, $|A|$ stands for the determinant of A, not the absolute value of A. In fact, the concept of absolute value does not apply to matrices.

Determinant of a 2 × 2 Matrix

The **determinant** of the matrix $A = \begin{bmatrix} a_{11} & a_{12} \\ a_{21} & a_{22} \end{bmatrix}$, denoted $|A|$, is $a_{11}a_{22} - a_{21}a_{12}$.

Unfortunately, this only defines determinants for matrices of one size, and the simplicity of the definition does not carry over to larger sizes. The good news, however, is that determinants of 3×3, 4×4, and larger matrices can all be related, ultimately, to determinants of 2×2 matrices. We will do a few calculations before considering larger matrices.

Example 1: Determinant of a 2 × 2 Matrix

Evaluate the determinants of each of the following matrices.

a. $A = \begin{bmatrix} -2 & 3 \\ -1 & 3 \end{bmatrix}$
b. $B = \begin{bmatrix} 2 & -3 \\ -4 & 6 \end{bmatrix}$

Solution

a. $|A| = a_{11}a_{22} - a_{21}a_{12}$

$\quad = (-2)(3) - (-1)(3) = -6 + 3 = -3$

b. $|B| = b_{11}b_{22} - b_{21}b_{12}$

$\quad = (2)(6) - (-4)(-3) = 12 - 12 = 0$

Calculating the determinant of a 3×3 matrix involves calculating three specific 2×2 determinants. Similarly, finding the determinant of a 4×4 matrix requires evaluating four specific 3×3 determinants.

These specific smaller determinants are called the *cofactors* of a matrix.

Minors and Cofactors

Let A be an $n \times n$ matrix, and let i and j be numbers between 1 and n, so that a_{ij} is an element of A.

- The **minor** of the element a_{ij} is the determinant of the $(n-1) \times (n-1)$ matrix formed from A by deleting its ith row and j^{th} column.

- The **cofactor** of the element a_{ij} is $(-1)^{i+j}$ times the minor of a_{ij}. Thus, if $i+j$ is even, the cofactor of a_{ij} is the same as the minor of a_{ij}; if $i+j$ is odd, the cofactor is the negative of the minor.

$$\begin{bmatrix} + & - & + & - & \cdots \\ - & + & - & + & \cdots \\ + & - & + & - & \cdots \\ - & + & - & + & \cdots \\ \vdots & \vdots & \vdots & \vdots & \ddots \end{bmatrix}$$

FIGURE 1: The Cofactor Sign Matrix

The matrix of signs in Figure 1 may help you remember which minors to multiply by -1: if a_{ij} is in a position occupied by a $-$ sign, change the sign of the minor of a_{ij} to obtain the cofactor.

Example 2: Minors and Cofactors

Consider the matrix $A = \begin{bmatrix} -5 & 3 & 2 \\ 1 & 0 & -1 \\ -3 & 1 & 0 \end{bmatrix}$.

a. Evaluate the minor of a_{12}.

b. Evaluate the cofactor of a_{23}.

Solution

a. Finding the minor of a_{12} requires deleting the first row and second column of the matrix A. This gives us the following.

$$\begin{bmatrix} -5 & 3 & 2 \\ 1 & 0 & -1 \\ -3 & 1 & 0 \end{bmatrix}$$

Thus, the minor of $a_{12} = \begin{vmatrix} 1 & -1 \\ -3 & 0 \end{vmatrix} = (1)(0) - (-3)(-1) = 0 - 3 = -3$.

b. Similarly, the cofactor of $a_{23} = (-1)^{2+3} \begin{vmatrix} -5 & 3 \\ -3 & 1 \end{vmatrix}$

$$= (-1)^{5} \left[(-5)(1) - (-3)(3) \right]$$
$$= (-1)(-5+9)$$
$$= (-1)(4)$$
$$= -4 .$$

Determinant of an $n \times n$ Matrix

Evaluation of an $n \times n$ determinant is accomplished by **expansion** along a fixed row or column. The result does not depend on which row or column is chosen.

- To expand along the ith row, each element of that row is multiplied by its cofactor, and the n products are then added.

- To expand along the jth column, each element of that column is multiplied by its cofactor, and the n products are then added.

For example, if we expand along the first column of a 3×3 matrix, we get the following. Note the minus sign in front of a_{21}.

$$\begin{vmatrix} a_{11} & a_{12} & a_{13} \\ a_{21} & a_{22} & a_{23} \\ a_{31} & a_{32} & a_{33} \end{vmatrix} = a_{11}\begin{vmatrix} a_{22} & a_{23} \\ a_{32} & a_{33} \end{vmatrix} - a_{21}\begin{vmatrix} a_{12} & a_{13} \\ a_{32} & a_{33} \end{vmatrix} + a_{31}\begin{vmatrix} a_{12} & a_{13} \\ a_{22} & a_{23} \end{vmatrix}$$

We could expand along a row or a different column in a similar manner.

Example 3: Determinant of an $n \times n$ Matrix

Evaluate the determinant of the matrix $A = \begin{bmatrix} -1 & 3 & 2 \\ -2 & 0 & 0 \\ 4 & 1 & 5 \end{bmatrix}$.

Solution

NOTE

Minimize the number of computations by choosing which row or column to expand along carefully.

First, we decide which row or column to expand along. A row or column with many zeros is generally a good choice, since it makes the multiplication much easier. In this case, Row 2 has the most zeros, so expand along it.

$$\begin{vmatrix} -1 & 3 & 2 \\ -2 & 0 & 0 \\ 4 & 1 & 5 \end{vmatrix} = -(-2)\begin{vmatrix} 3 & 2 \\ 1 & 5 \end{vmatrix} + (0)\begin{vmatrix} -1 & 2 \\ 4 & 5 \end{vmatrix} - (0)\begin{vmatrix} -1 & 3 \\ 4 & 1 \end{vmatrix}$$

$$= -(-2)(13) + 0 - 0$$

$$= 26$$

Thus, $|A| = 26$. We get the same answer if we expand along a different row or column.

$$\begin{vmatrix} -1 & 3 & 2 \\ -2 & 0 & 0 \\ 4 & 1 & 5 \end{vmatrix} = (-1)\begin{vmatrix} 0 & 0 \\ 1 & 5 \end{vmatrix} - (-2)\begin{vmatrix} 3 & 2 \\ 1 & 5 \end{vmatrix} + (4)\begin{vmatrix} 3 & 2 \\ 0 & 0 \end{vmatrix}$$

$$= (-1)(0) - (-2)(13) + (4)(0)$$

$$= 26$$

As we saw in the previous example, even 3×3 determinants can involve a large number of calculations, but by taking advantage of zeros, we are able to reduce the amount of work. We can take this a step further by applying a few properties of determinants.

Properties of Determinants

1. A constant can be factored out of each of the terms in a given row or column when computing determinants.

$$\begin{vmatrix} 2 & -1 \\ 15 & 5 \end{vmatrix} = 5 \begin{vmatrix} 2 & -1 \\ 3 & 1 \end{vmatrix} \quad \text{and} \quad \begin{vmatrix} 4 & 7 \\ 12 & 9 \end{vmatrix} = 4 \begin{vmatrix} 1 & 7 \\ 3 & 9 \end{vmatrix}$$

2. Interchanging two rows or two columns changes the determinant by a factor of -1.

$$\begin{vmatrix} 2 & -1 \\ 15 & 5 \end{vmatrix} = - \begin{vmatrix} 15 & 5 \\ 2 & -1 \end{vmatrix} \quad \text{and} \quad \begin{vmatrix} 3 & -2 \\ 7 & 1 \end{vmatrix} = - \begin{vmatrix} -2 & 3 \\ 1 & 7 \end{vmatrix}$$

3. The determinant is unchanged by adding a multiple of one row (or column) to another row (or column).

$$\begin{vmatrix} 3 & -2 \\ 1 & -1 \end{vmatrix} \overset{-3R_2 + R_1}{=} \begin{vmatrix} 0 & 1 \\ 1 & -1 \end{vmatrix}$$

Example 4: Properties of Determinants

Evaluate the determinant of the matrix $B = \begin{bmatrix} 4 & -2 & 3 & 0 \\ 2 & 1 & -1 & 3 \\ 3 & 0 & 1 & 1 \\ 2 & -2 & 0 & 0 \end{bmatrix}$.

Solution

Use the properties of determinants to try to obtain rows or columns with as many zeros as possible.

$$\begin{vmatrix} 4 & -2 & 3 & 0 \\ 2 & 1 & -1 & 3 \\ 3 & 0 & 1 & 1 \\ 2 & -2 & 0 & 0 \end{vmatrix} \overset{-3R_3 + R_2}{=} \begin{vmatrix} 4 & -2 & 3 & 0 \\ -7 & 1 & -4 & 0 \\ 3 & 0 & 1 & 1 \\ 2 & -2 & 0 & 0 \end{vmatrix}$$

Applying the third property makes the fourth column have only one nonzero entry.

Now expand along the fourth column (remembering that the minus sign is part of the cofactor):

$$\begin{vmatrix} 4 & -2 & 3 & 0 \\ -7 & 1 & -4 & 0 \\ 3 & 0 & 1 & 1 \\ 2 & -2 & 0 & 0 \end{vmatrix} = -(1) \begin{vmatrix} 4 & -2 & 3 \\ -7 & 1 & -4 \\ 2 & -2 & 0 \end{vmatrix}$$

We can continue to apply the third property of determinants to simplify the evaluation of the 3×3 determinant.

$$|B| = -\begin{vmatrix} 4 & -2 & 3 \\ -7 & 1 & -4 \\ 2 & -2 & 0 \end{vmatrix} \overset{C_1+C_2}{=} -\begin{vmatrix} 4 & 2 & 3 \\ -7 & -6 & -4 \\ 2 & 0 & 0 \end{vmatrix}$$

Add the first column to the second.

$$= -(2)\begin{vmatrix} 2 & 3 \\ -6 & -4 \end{vmatrix}$$

Expand along the third row, which now has two zeros.

$$= (-2)(10)$$

$$= -20$$

Notice that we have reduced the work to the evaluation of only one 3×3 determinant, which in turn involved evaluating only one 2×2 determinant.

Solving Systems of Linear Equations Using Cramer's Rule

To understand the form of Cramer's Rule, we will solve the general two-variable, two-equation linear system by elimination. To do this, we note that any such system can be put into the form

$$\begin{cases} ax + by = e \\ cx + dy = f \end{cases}$$

where a, b, c, d, e, and f are all constants. If we can solve this system for x and y, then we will have a formula for the solution of any such system.

Using the method of elimination, we can obtain an equation in x alone by multiplying the first equation by d and the second equation by $-b$.

$$\begin{cases} ax + by = e \\ cx + dy = f \end{cases} \xrightarrow[-bE_2]{dE_1} \begin{cases} adx + bdy = ed \\ -bcx - bdy = -bf \end{cases}$$
$$(ad - bc)x = ed - bf$$

This equation can then be solved for x to obtain

$$x = \frac{ed - bf}{ad - bc}.$$

Similarly, the system can be solved for y to obtain

$$y = \frac{af - ce}{ad - bc}.$$

Note that these formulas only make sense if the denominator is not zero. We will deal with this possibility shortly.

These formulas are worthwhile on their own, but they are a bit complex and would be difficult to memorize. Note that each term in the two fractions appears in the form of a 2×2 determinant. In fact, the above formulas are equivalent to the following equations.

$$x = \frac{\begin{vmatrix} e & b \\ f & d \end{vmatrix}}{\begin{vmatrix} a & b \\ c & d \end{vmatrix}} \quad \text{and} \quad y = \frac{\begin{vmatrix} a & e \\ c & f \end{vmatrix}}{\begin{vmatrix} a & b \\ c & d \end{vmatrix}}$$

The denominator D in both formulas is the determinant of the coefficient matrix, the square matrix consisting of the coefficients of the variables. If we let D_x and D_y represent the numerators of the formulas for x and y, respectively, then D_x is the determinant of the coefficient matrix with the first column (the x-column) replaced by the column of constants, and D_y is the determinant of the coefficient matrix with the second column replaced by the column of constants. Putting these observations together, we obtain Cramer's Rule for the two-variable, two-equation case.

> ⚠ **CAUTION**
>
> Whenever a fraction appears in our work, we need to ask if the expression in the denominator can ever be zero, and what it means if this happens. In Cramer's Rule, the determinant D can equal 0, which prevents us from using the given formulas.
>
> If $D = 0$, the system is either dependent or inconsistent. If both D_x and D_y are also zero, the system is dependent. If at least one of D_x and D_y is nonzero, the system has no solution.

Cramer's Rule for Two-Variable, Two-Equation Linear Systems

The solution of a two-variable, two-equation linear system $\begin{cases} ax + by = e \\ cx + dy = f \end{cases}$ is given by

$$x = \frac{D_x}{D} \quad \text{and} \quad y = \frac{D_y}{D},$$

where D is the determinant of the coefficient matrix $\begin{vmatrix} a & b \\ c & d \end{vmatrix}$, D_x is the determinant of the matrix formed by replacing the column of x-coefficients with the column of constant terms $\begin{vmatrix} e & b \\ f & d \end{vmatrix}$, and D_y is the determinant of the matrix formed by replacing the column of y-coefficients with the column of constant terms $\begin{vmatrix} a & e \\ c & f \end{vmatrix}$.

Example 5: Cramer's Rule

Use Cramer's Rule to solve the following systems.

a. $\begin{cases} 4x - 5y = 3 \\ -3x + 7y = 1 \end{cases}$
b. $\begin{cases} -x + 2y = -1 \\ 3x - 6y = 3 \end{cases}$

Solution

a. $D = \begin{vmatrix} 4 & -5 \\ -3 & 7 \end{vmatrix} = 28 - 15 = 13$ — Calculate D first. Since $D \neq 0$, we know there is a single solution to the system.

$D_x = \begin{vmatrix} 3 & -5 \\ 1 & 7 \end{vmatrix} = 21 - (-5) = 26$ — Calculate D_x and D_y.

$D_y = \begin{vmatrix} 4 & 3 \\ -3 & 1 \end{vmatrix} = 4 - (-9) = 13$

Applying Cramer's Rule, we have $x = \dfrac{D_x}{D} = \dfrac{26}{13} = 2$ and $y = \dfrac{D_y}{D} = \dfrac{13}{13} = 1$, so the solution is $(2, 1)$.

b. $D = \begin{vmatrix} -1 & 2 \\ 3 & -6 \end{vmatrix} = 6 - 6 = 0$ Again we calculate D first. Since $D = 0$ either the system has no solution or it has an infinite number of solutions.

$D_x = \begin{vmatrix} -1 & 2 \\ 3 & -6 \end{vmatrix} = 6 - 6 = 0$

$D_y = \begin{vmatrix} -1 & -1 \\ 3 & 3 \end{vmatrix} = -3 - (-3) = 0$ Since D_x and D_y both equal zero, the system is dependent.

The solution set can be found by solving either equation for either variable: $\{(2y+1, y)\,|\,y \in \mathbb{R}\}$.

Cramer's Rule can be extended to solve any system of n linear equations in n variables. While Cramer's Rule is an extremely powerful method and remarkable for its succinctness, keep in mind that using Cramer's Rule to solve an n-equation, n-variable system entails calculating $(n+1)$ $n \times n$ determinants, so it is important to make use of the laborsaving properties of determinants.

Cramer's Rule

A system of n linear equations in the n variables $x_1, x_2, \ldots, x_n$ can be written in the following form.

$$\begin{cases} a_{11}x_1 + a_{12}x_2 + \cdots + a_{1n}x_n = b_1 \\ a_{21}x_1 + a_{22}x_2 + \cdots + a_{2n}x_n = b_2 \\ \quad\quad\quad\quad \vdots \\ a_{n1}x_1 + a_{n2}x_2 + \cdots + a_{nn}x_n = b_n \end{cases}$$

The solution of the system is given by the formulas $x_1 = \dfrac{D_{x_1}}{D}$, $x_2 = \dfrac{D_{x_2}}{D}, \ldots,$ $x_n = \dfrac{D_{x_n}}{D}$, where D is the determinant of the coefficient matrix and D_{x_i} is the determinant of the same matrix with the ith column replaced by the column of constants $b_1, b_2, \ldots, b_n$.

If $D = 0$ and if each $D_{x_i} = 0$ as well, the system is dependent and has an infinite number of solutions. If $D = 0$ and D_{x_i} is nonzero for at least one value of i, the system has no solution.

Example 6: Cramer's Rule

Use Cramer's Rule to solve the system $\begin{cases} 3x - 2y - 2z = -1 \\ \quad\quad 3y + z = -7. \\ \quad x + y + 2z = 0 \end{cases}$

Solution

Note how the properties of determinants are used to simplify each calculation. In each case, the row or column used for expansion is in red.

$$D = \begin{vmatrix} 3 & -2 & -2 \\ 0 & 3 & 1 \\ 1 & 1 & 2 \end{vmatrix} \overset{-3R_3+R_1}{=} \begin{vmatrix} 0 & -5 & -8 \\ 0 & 3 & 1 \\ 1 & 1 & 2 \end{vmatrix} = (1)\begin{vmatrix} -5 & -8 \\ 3 & 1 \end{vmatrix} = 19$$

Since $D \neq 0$, the system has a unique solution.

$$D_x = \begin{vmatrix} -1 & -2 & -2 \\ -7 & 3 & 1 \\ 0 & 1 & 2 \end{vmatrix} \overset{-2C_2+C_3}{=} \begin{vmatrix} -1 & -2 & 2 \\ -7 & 3 & -5 \\ 0 & 1 & 0 \end{vmatrix} = -(1)\begin{vmatrix} -1 & 2 \\ -7 & -5 \end{vmatrix} = -19$$

$$D_y = \begin{vmatrix} 3 & -1 & -2 \\ 0 & -7 & 1 \\ 1 & 0 & 2 \end{vmatrix} \overset{-2C_1+C_3}{=} \begin{vmatrix} 3 & -1 & -8 \\ 0 & -7 & 1 \\ 1 & 0 & 0 \end{vmatrix} = (1)\begin{vmatrix} -1 & -8 \\ -7 & 1 \end{vmatrix} = -57$$

$$D_z = \begin{vmatrix} 3 & -2 & -1 \\ 0 & 3 & -7 \\ 1 & 1 & 0 \end{vmatrix} \overset{-C_1+C_2}{=} \begin{vmatrix} 3 & -5 & -1 \\ 0 & 3 & -7 \\ 1 & 0 & 0 \end{vmatrix} = (1)\begin{vmatrix} -5 & -1 \\ 3 & -7 \end{vmatrix} = 38$$

After evaluating D_x, D_y, and D_z, we know the solution is the single ordered triple

$$(x, y, z) = \left(\frac{D_x}{D}, \frac{D_y}{D}, \frac{D_z}{D} \right) = \left(\frac{-19}{19}, \frac{-57}{19}, \frac{38}{19} \right) = (-1, -3, 2).$$

Evaluating Determinants Using Technology

To find the determinant of a matrix using the calculator, we must first edit a matrix and enter in the dimensions and elements of the matrix whose determinant we wish

to find, for example, the matrix $\begin{bmatrix} 2 & 4 & -8 \\ 1 & 3 & 6 \\ -7 & 5 & 1 \end{bmatrix}$. With that saved as matrix A, we press

2nd x^{-1} and select MATH. Press enter since $1:\text{det}($ is already

highlighted. We then press 2nd x^{-1} , select $1:[A]$ under NAMES, add the
right-hand parenthesis and press enter .

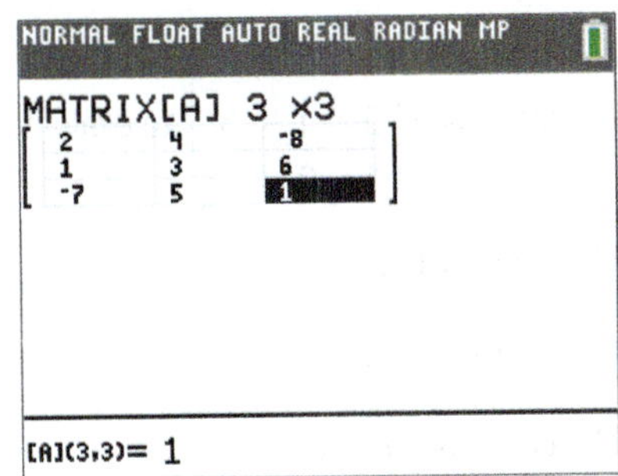

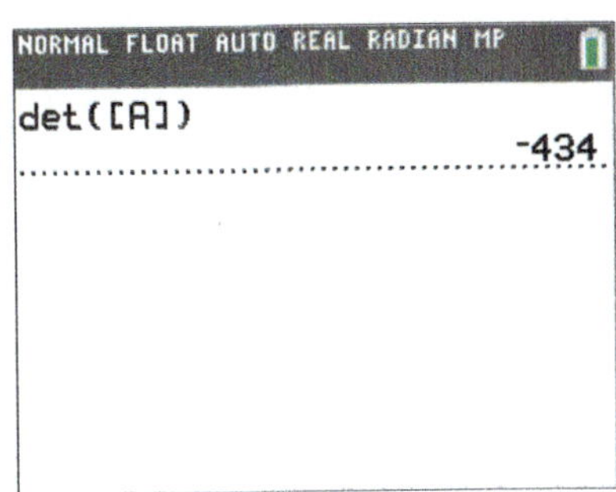

Notice that if we try to find the determinant of a 3×4 matrix, for instance, we get the following error message.

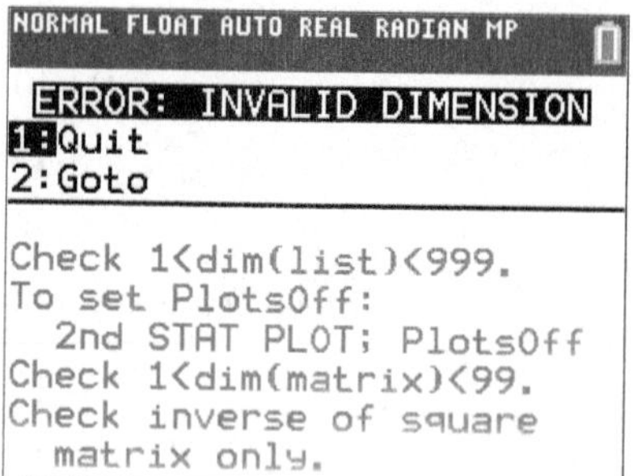

Remember that the definition of determinant only applies to square matrices; matrices that are not square matrices do not have determinants.

6.3 EXERCISES

💡 PRACTICE

Evaluate each of the following determinants. See Example 1.

1. $\begin{vmatrix} 4 & -3 \\ 1 & 2 \end{vmatrix}$
2. $\begin{vmatrix} 5 & -2 \\ 5 & -2 \end{vmatrix}$
3. $\begin{vmatrix} 0 & 3 \\ -5 & 2 \end{vmatrix}$
4. $\begin{vmatrix} 34 & -2 \\ 17 & -1 \end{vmatrix}$

5. $\begin{vmatrix} a & x \\ x & b \end{vmatrix}$
6. $\begin{vmatrix} 5x & 2 \\ -x & 1 \end{vmatrix}$
7. $\begin{vmatrix} -2 & 2 \\ -2 & -2 \end{vmatrix}$
8. $\begin{vmatrix} ac & 2ad \\ bc & db \end{vmatrix}$

9. $\begin{vmatrix} -1 & 2 \\ 3 & 4 \end{vmatrix}$
10. $\begin{vmatrix} w & x \\ y & z \end{vmatrix}$
11. $\begin{vmatrix} -2 & 9 \\ 5 & -3 \end{vmatrix}$
12. $\begin{vmatrix} 2y & 3x \\ y-1 & x^2 \end{vmatrix}$

Solve for *x* by calculating the determinant.

13. $\begin{vmatrix} x-2 & 2 \\ 2 & x+1 \end{vmatrix} = 0$
14. $\begin{vmatrix} x+7 & -2 \\ 9 & x-2 \end{vmatrix} = 0$
15. $\begin{vmatrix} x+1 & 8 \\ 1 & x+3 \end{vmatrix} = 0$

16. $\begin{vmatrix} x-8 & 11 \\ -2 & x+5 \end{vmatrix} = 0$
17. $\begin{vmatrix} x+6 & 2 \\ -1 & x+3 \end{vmatrix} = 0$
18. $\begin{vmatrix} x-4 & -4 \\ 3 & x+9 \end{vmatrix} = 0$

19. $\begin{vmatrix} x+5 & 3 \\ 3 & x-3 \end{vmatrix} = 0$
20. $\begin{vmatrix} x+3 & 6 \\ 5 & x+7 \end{vmatrix} = 0$
21. $\begin{vmatrix} x-3 & 2 \\ 1 & x-4 \end{vmatrix} = 0$

Use the matrix $A = \begin{bmatrix} 2 & -1 & 5 \\ 0 & 1 & 3 \\ 1 & 0 & -2 \end{bmatrix}$ to evaluate the following. See Example 2.

22. The minor of a_{12}
23. The cofactor of a_{12}

24. The minor of a_{22}
25. The cofactor of a_{22}

26. The cofactor of a_{32}
27. The cofactor of a_{33}

28. The minor of a_{13}
29. The cofactor of a_{21}

30. The cofactor of a_{31}

Find the determinant by the method of expansion by cofactors along the given row or column. See Example 3.

31. $\begin{vmatrix} 4 & 5 & 3 \\ -1 & 2 & 7 \\ 11 & 6 & 2 \end{vmatrix}$ Expand along Row 3

32. $\begin{vmatrix} 8 & 2 & 0 \\ 3 & 4 & 7 \\ 1 & 0 & 2 \end{vmatrix}$ Expand along Column 3

33. $\begin{vmatrix} 5 & 8 & 5 \\ 0 & -6 & 3 \\ 2 & 4 & -1 \end{vmatrix}$ Expand along Row 2

34. $\begin{vmatrix} -4 & 2 & 1 \\ 9 & 12 & 8 \\ 0 & 6 & -3 \end{vmatrix}$ Expand along Column 1

35. $\begin{vmatrix} 13 & 0 & -7 \\ 4 & 2 & 3 \\ 1 & 4 & 0 \end{vmatrix}$ Expand along Row 2

36. $\begin{vmatrix} 7 & 0 & 1 \\ 2 & 5 & 3 \\ 8 & 6 & 2 \end{vmatrix}$ Expand along Column 3

37. $\begin{vmatrix} 8 & 0 & -7 & 5 \\ 4 & -2 & 3 & 3 \\ -1 & 1 & 0 & 2 \\ 2 & 0 & 6 & 0 \end{vmatrix}$ Expand along Row 4

38. $\begin{vmatrix} 4 & -2 & 9 & 2 \\ 7 & 0 & 1 & 7 \\ -6 & 3 & 0 & 1 \\ 3 & 1 & 2 & 0 \end{vmatrix}$ Expand along Column 2

Evaluate each of the following determinants. In each case, minimize the required number of computations by carefully choosing a row or column to expand along, and use the properties of determinants to simplify the process. See Examples 3 and 4.

39. $\begin{vmatrix} 2 & 0 & 1 \\ -5 & 1 & 0 \\ 3 & -1 & 1 \end{vmatrix}$

40. $\begin{vmatrix} 12 & 3 & 1 \\ 1 & 1 & -1 \\ 0 & 2 & 0 \end{vmatrix}$

41. $\begin{vmatrix} 12 & 3 & 6 \\ 2 & 2 & -4 \\ 0 & 2 & 0 \end{vmatrix}$

42. $\begin{vmatrix} 1 & 2 & 3 \\ 4 & 5 & 6 \\ 7 & 8 & 9 \end{vmatrix}$

43. $\begin{vmatrix} 2 & 1 & -3 & 0 \\ 1 & -2 & 1 & 0 \\ 0 & 1 & 0 & 1 \\ 2 & 0 & 1 & 1 \end{vmatrix}$

44. $\begin{vmatrix} x & 0 & 0 & 0 \\ 0 & x & 0 & 0 \\ 0 & 0 & x & 0 \\ 0 & 0 & 0 & x \end{vmatrix}$

45. $\begin{vmatrix} x & x & x & x \\ 0 & x & x & x \\ 0 & 0 & x & x \\ 0 & 0 & 0 & x \end{vmatrix}$

46. $\begin{vmatrix} 0 & 2 & 0 & 0 \\ -2 & -4 & 5 & 9 \\ 1 & 3 & -1 & 1 \\ 0 & 7 & 0 & 2 \end{vmatrix}$

47. $\begin{vmatrix} x & x & 0 & 0 \\ yz & x^3 & z & x^4 \\ z & xy & x & 0 \\ x^2 & 0 & 0 & 0 \end{vmatrix}$

Use Cramer's Rule to solve each system of equations. See Examples 5 and 6.

48. $\begin{cases} 2x - 3y = 8 \\ 8x + 5y = -2 \end{cases}$

49. $\begin{cases} 5x + 7y = 9 \\ 2x + 3y = -7 \end{cases}$

50. $\begin{cases} 5x - 10y = 9 \\ -x + 2y = -3 \end{cases}$

51. $\begin{cases} -2x - 2y = 4 \\ 3x + 3y = -6 \end{cases}$

52. $\begin{cases} \dfrac{2}{3}x + y = -3 \\ 3x + \dfrac{5}{2}y = -\dfrac{7}{2} \end{cases}$

53. $\begin{cases} \dfrac{2}{3}x + 2y = 1 \\ x + 3y = 0 \end{cases}$

54. $\begin{cases} x+2y=-1 \\ y+3z=7 \\ 2x+5z=21 \end{cases}$

55. $\begin{cases} 2x-y=0 \\ 5x-3y-3z=5 \\ 2x+6z=-10 \end{cases}$

56. $\begin{cases} 3x+8z=3 \\ -3x-7z=-3 \\ x+3z=1 \end{cases}$

57. $\begin{cases} 3w-x+5y+3z=2 \\ -4w-10y-2z=10 \\ w-x+2z=7 \\ 4w-2x+5y+5z=9 \end{cases}$

58. $\begin{cases} 2w+x-3y=3 \\ w-2x+y=1 \\ x+z=-2 \\ y+z=0 \end{cases}$

59. $\begin{cases} 3w-2x+y-5z=-1 \\ w+x-y+4z=2 \\ 4w-x-z=1 \\ 5w-x=9 \end{cases}$

60. $\begin{cases} -4x+y=1 \\ 7x+2y=407 \end{cases}$

61. $\begin{cases} 5x-4y=-49 \\ 24x-19y=179 \end{cases}$

62. $\begin{cases} 2w-3x+4y-z=21 \\ w+5x=2 \\ -2x+3y+z=12 \\ -3w+4z=-5 \end{cases}$

63. $\begin{cases} -5x+10y=3 \\ \dfrac{7}{2}x-7y=20 \end{cases}$

64. $\begin{cases} 23x+21y=-4 \\ x-3y=-8 \end{cases}$

65. $\begin{cases} w-x+y-z=2 \\ 2w-x+3y=-5 \\ x-2z=7 \\ 3w+4x=-13 \end{cases}$

🚀 APPLICATIONS

66. The three sides of a triangle are related as follows: the perimeter is 43 feet, the second side is 5 feet more than twice the first side, and the third side is 3 feet less than the sum of the other two sides. Find the lengths of the three sides of the triangle.

67. Eric's favorite candy bar and ice cream flavor have fat and calorie contents as follows: each candy bar has 5 grams of fat and 280 calories; each serving of ice cream has 10 grams of fat and 150 calories. How many candy bars and servings of ice cream did he eat during the weekend he consumed 85 grams of fat and 2300 calories from these two treats?

68. A farmer plants soybeans, corn, and wheat and rotates the planting each year on her 500-acre farm. In a particular year, the profits from her crops were $120 per acre of soybeans, $100 per acre of corn, and $80 per acre of wheat. She planted twice as many acres of corn as soybeans. How many acres did she plant with each crop that year if she made a total profit of $51,800?

Using a graphing utility, find the determinant of the matrix.

69. $\begin{vmatrix} 0.1 & 0.4 & -0.7 \\ 0.3 & -0.1 & 0.2 \\ 0.5 & -0.2 & 0.3 \end{vmatrix}$

70. $\begin{vmatrix} 0.1 & 0.3 & 0.1 \\ 0.2 & -0.2 & -0.1 \\ -0.1 & -0.4 & 0.5 \end{vmatrix}$

71. $\begin{vmatrix} 2.2 & 0.3 & -1.7 \\ 0.4 & -0.2 & 0.1 \\ 0.2 & 0.3 & -1.6 \end{vmatrix}$

72. $\begin{vmatrix} 3.1 & 0.6 & -1.1 \\ 1.2 & 5.2 & -7.3 \\ -0.1 & -4.1 & 6.5 \end{vmatrix}$

73. $\begin{vmatrix} 13 & 23 & -21 \\ 17 & -32 & 14 \\ 15 & 12 & -16 \end{vmatrix}$

74. $\begin{vmatrix} 25 & 32 & 17 \\ -13 & 14 & -24 \\ 16 & 26 & 36 \end{vmatrix}$

Use a graphing utility and Cramer's Rule to solve each system of equations.

75. $\begin{cases} x - 2y + 3z = 9 \\ -x + 3y = -4 \\ 2x - 5y + 5z = 17 \end{cases}$

76. $\begin{cases} 2x + 4y + z = 1 \\ x - 2y - 3z = 2 \\ x + y - z = -1 \end{cases}$

77. $\begin{cases} w + x + y + z = 6 \\ -w + 2x + 3y = 0 \\ 2w - 3x + 4y + z = 4 \\ w + x + 2y - z = 0 \end{cases}$

6.4 BASIC MATRIX OPERATIONS

■ TOPICS

- Matrix Addition
- Scalar Multiplication
- Matrix Multiplication
- Transition Matrices
- Basic Matrix Operations Using Technology

Matrix Addition

In this section, we will start exploring how to define the mathematical operations of addition, subtraction, multiplication, and division on matrices.

⚠ CAUTION

It is very important to note the restriction on the order of the two matrices in the definition of matrix addition; a matrix with m rows and n columns can only be added to another matrix with m rows and n columns. In all aspects of matrix algebra, the orders of the matrices involved must be considered.

Matrix Addition

Two matrices A and B can be added to form the new matrix $A + B$ only if A and B are of the same order. The addition is performed by adding corresponding entries of the two matrices together; that is, the element in the ith row and the jth column of $A + B$ is given by $a_{ij} + b_{ij}$.

Matrix Equality

Two matrices A and B are **equal**, denoted $A = B$, if they are of the same order and all corresponding entries of A and B are equal.

Example 1: Matrix Addition

Perform the indicated addition, if possible.

a. $\begin{bmatrix} -3 & 2 \\ 0 & -5 \\ 11 & -9 \end{bmatrix} + \begin{bmatrix} 3 & 17 \\ 5 & 4 \\ -10 & 4 \end{bmatrix}$

b. $\begin{bmatrix} 2 & -5 \\ 1 & 0 \\ 0 & 3 \\ -7 & 10 \end{bmatrix} + \begin{bmatrix} 2 & 1 & 0 & -7 \\ -5 & 0 & 3 & 10 \end{bmatrix}$

Solution

a. Both matrices are 3×2, so the sum is defined and

$$\begin{bmatrix} -3 & 2 \\ 0 & -5 \\ 11 & -9 \end{bmatrix} + \begin{bmatrix} 3 & 17 \\ 5 & 4 \\ -10 & 4 \end{bmatrix} = \begin{bmatrix} 0 & 19 \\ 5 & -1 \\ 1 & -5 \end{bmatrix}$$

Each entry in the first matrix is added to its corresponding entry in the second matrix.

b. The first matrix is 4×2 and the second is 2×4, so the addition cannot be performed.

The entries in matrices do not all have to be constants. In many applications of matrices, it is convenient to represent some of the entries initially as variables, with the intent of eventually solving for the variables. We can solve some examples of such *matrix equations* now, using only matrix addition and matrix equality.

Example 2: Matrix Equations

Determine the values of the variables that will make each of the following statements true.

a. $\begin{bmatrix} -3 & a & b \\ -2 & a+b & 5 \end{bmatrix} = \begin{bmatrix} c & 3 & 7 \\ -2 & d & 5 \end{bmatrix}$
  b. $\begin{bmatrix} 3x \\ 4 \end{bmatrix} + \begin{bmatrix} -y \\ 2x \end{bmatrix} = \begin{bmatrix} 13 \\ 7y \end{bmatrix}$

Solution

a. Solving for the four variables a, b, c, and d is just a matter of comparing the entries one by one.

$\begin{bmatrix} -3 & a & b \\ -2 & a+b & 5 \end{bmatrix} = \begin{bmatrix} c & 3 & 7 \\ -2 & d & 5 \end{bmatrix}$ Comparing the top rows tells us that $a = 3$, $b = 7$, and $c = -3$.

In the bottom rows, note the -2 in the left corner of each matrix and the 5 in each right corner. If these constants were not equal we would have a contradiction, and there would be no way to make the matrix equation true.

The only variable left to solve for is d, which we see is equal to $a+b$. So, $d = 3+7 = 10$.

b. Each matrix consists of only two entries, and after performing the matrix addition on the left and comparing corresponding entries, we arrive at the following system of equations.

$$\begin{cases} 3x - y = 13 \\ 4 + 2x = 7y \end{cases}$$

We know a number of ways of solving such a system. If we choose to use Cramer's Rule, we first rewrite the system as follows.

$$\begin{cases} 3x - y = 13 \\ 2x - 7y = -4 \end{cases}$$

From here we can determine

$$D = \begin{vmatrix} 3 & -1 \\ 2 & -7 \end{vmatrix} = (3)(-7) - (2)(-1) = -19,$$

$$D_x = \begin{vmatrix} 13 & -1 \\ -4 & -7 \end{vmatrix} = (13)(-7) - (-4)(-1) = -95, \text{ and}$$

$$D_y = \begin{vmatrix} 3 & 13 \\ 2 & -4 \end{vmatrix} = (3)(-4) - (2)(13) = -38,$$

so $x = \dfrac{D_x}{D} = \dfrac{-95}{-19} = 5$ and $y = \dfrac{D_y}{D} = \dfrac{-38}{-19} = 2.$

Scalar Multiplication

In the context of matrix algebra, a *scalar* is a real number, and *scalar multiplication* refers to the product of a real number and a matrix. It may seem strange to combine two very different sorts of objects (a scalar and a matrix), but consider the following matrix sum:

$$\begin{bmatrix} -5 & 2 \\ 1 & -3 \\ -2 & 7 \end{bmatrix} + \begin{bmatrix} -5 & 2 \\ 1 & -3 \\ -2 & 7 \end{bmatrix} = \begin{bmatrix} -10 & 4 \\ 2 & -6 \\ -4 & 14 \end{bmatrix}$$

Since the result of adding a matrix to itself has the effect of doubling each entry, it makes sense to write

$$2\begin{bmatrix} -5 & 2 \\ 1 & -3 \\ -2 & 7 \end{bmatrix} = \begin{bmatrix} -10 & 4 \\ 2 & -6 \\ -4 & 14 \end{bmatrix}.$$

Extending the idea to all scalars (real numbers) leads to the following definition.

Scalar Multiplication

If A is an $m \times n$ matrix and c is a scalar, cA stands for the $m \times n$ matrix for which each entry is c times the corresponding entry of A. In other words, the entry in the ith row and j^{th} column of cA is ca_{ij}.

Example 3: Scalar Multiplication

Given the matrices $A = \begin{bmatrix} -1 & 6 & 2 \\ -8 & 0 & 1 \end{bmatrix}$ and $B = \begin{bmatrix} 0 & -3 & 4 \\ 1 & -2 & 6 \end{bmatrix}$, write $-3A + 2B$ as a single matrix.

Solution

Before calculating, note that the operation can be performed since both matrices are of the same order, 2×4.

This problem has two steps. First, multiply each entry of A and B by its corresponding scalar. Second, add the resulting matrices together.

$$-3A + 2B = -3\begin{bmatrix} -1 & 6 & 2 \\ -8 & 0 & 1 \end{bmatrix} + 2\begin{bmatrix} 0 & -3 & 4 \\ 1 & -2 & 6 \end{bmatrix} \qquad \text{Multiply each matrix by its scalar.}$$

$$= \begin{bmatrix} 3 & -18 & -6 \\ 24 & 0 & -3 \end{bmatrix} + \begin{bmatrix} 0 & -6 & 8 \\ 2 & -4 & 12 \end{bmatrix} \qquad \text{Add the resulting matrices.}$$

$$= \begin{bmatrix} 3 & -24 & 2 \\ 26 & -4 & 9 \end{bmatrix}$$

As we try to understand algebraic operations on matrices, we can use real numbers as a model. Subtraction of real numbers is defined in terms of addition, and we define matrix subtraction in the same way.

Matrix Subtraction

Let A and B be two matrices of the same order. The difference $A - B$ is defined by

$$A - B = A + (-B).$$

Example 4: Matrix Subtraction

Perform the indicated subtraction: $\begin{bmatrix} 3 & -5 & 2 \end{bmatrix} - \begin{bmatrix} -2 & -5 & 3 \end{bmatrix}$.

Solution

Since both matrices are of order 1×3, we know the subtraction is possible. Subtract each entry in the second matrix from the corresponding entry in the first.

$$\begin{bmatrix} 3 & -5 & 2 \end{bmatrix} - \begin{bmatrix} -2 & -5 & 3 \end{bmatrix} = \begin{bmatrix} 3-(-2) & -5-(-5) & 2-3 \end{bmatrix}$$
$$= \begin{bmatrix} 5 & 0 & -1 \end{bmatrix}$$

Matrix Multiplication

The definition of matrix multiplication is not as straightforward as that of matrix addition or subtraction. Simply put, matrix multiplication does not refer to multiplying the corresponding entries of two matrices together. To understand the process of matrix multiplication, it helps to first think about matrices as functions.

Note that the system of equations

$$\begin{cases} x' = ax + by \\ y' = cx + dy \end{cases}$$

can be thought of as a function that transforms the ordered pair (x, y) into the ordered pair (x', y'), and that the function is characterized entirely by the matrix

$$A = \begin{bmatrix} a & b \\ c & d \end{bmatrix}.$$

Similarly, the system

$$\begin{cases} x' = ex + fy \\ y' = gx + hy \end{cases}$$

is a function that transforms ordered pairs into ordered pairs, and it is characterized by the matrix

$$B = \begin{bmatrix} e & f \\ g & h \end{bmatrix}.$$

Now we can look at the result of plugging the output of the first function into the second function (*composing* the two functions so that B acts on the result of A).

If we let (x', y') denote the output of the first function, we obtain the new ordered pair (x'', y'') given by

$$\begin{cases} x'' = ex' + fy' \\ y'' = gx' + hy' \end{cases} \quad \text{or} \quad \begin{cases} x'' = e(ax + by) + f(cx + dy) \\ y'' = g(ax + by) + h(cx + dy) \end{cases}.$$

We can change the way the last system is written to obtain

$$\begin{cases} x'' = (ea + fc)x + (eb + fd)y \\ y'' = (ga + hc)x + (gb + hd)y \end{cases}$$

so the composition of the two functions, in the order BA, is characterized by the matrix

$$BA = \begin{bmatrix} ea + fc & eb + fd \\ ga + hc & gb + hd \end{bmatrix}.$$

Note that the entries of this last matrix are the sums of products, where the products in each sum are between elements of the rows of B with elements of the columns of A. This pattern is the basis of our formal definition of matrix multiplication.

⚠ CAUTION

Unlike numerical multiplication, matrix multiplication is not commutative. That is, given two matrices A and B, AB *in general* is not equal to BA. As an illustration of this fact, suppose A is a 3×4 matrix and B is a 4×4 matrix. Then AB is defined (and is of order 3×2), but BA doesn't even exist. Even when both AB and BA are defined, they are generally not equal.

Matrix Multiplication

Two matrices A and B can be multiplied together, resulting in a new matrix denoted AB, only if the number of columns of A (the matrix on the left) is the same as the number of rows of B (the matrix on the right). Thus, if A is of order $m \times n$, the product AB is only defined if B is of order $n \times p$. The order of AB will be $m \times p$.

If we let c_{ij} denote the entry in the ith row and j^{th} column of AB, c_{ij} is obtained from the ith row of A and the jth column of B by the formula

$$c_{ij} = a_{i1}b_{1j} + a_{i2}b_{2j} + \cdots + a_{in}b_{nj}.$$

In words, c_{ij} equals the product of the first element of row i of matrix A and the first element of column j of matrix B, plus the product of the second element of row i and the second element of column j, and so on.

Example 5: Matrix Multiplication

Given the matrices $A = \begin{bmatrix} 2 & 0 \\ -5 & 1 \end{bmatrix}$ and $B = \begin{bmatrix} 7 & -2 \\ 3 & 1 \end{bmatrix}$, find AB.

Solution

Since both matrices are 2×2, they can be multiplied, and the result will also be a 2×2 matrix. For this example, we will follow the procedure one entry at a time.

$$AB = \begin{bmatrix} 2 & 0 \\ -5 & 1 \end{bmatrix} \begin{bmatrix} 7 & -2 \\ 3 & 1 \end{bmatrix} = \begin{bmatrix} 2(7)+0(3) & AB_{12} \\ AB_{21} & AB_{22} \end{bmatrix}$$

Sum the products of entries in the first row of A and first column of B.

$$= \begin{bmatrix} 2 & 0 \\ -5 & 1 \end{bmatrix} \begin{bmatrix} 7 & -2 \\ 3 & 1 \end{bmatrix} = \begin{bmatrix} 14 & 2(-2)+0(1) \\ AB_{21} & AB_{22} \end{bmatrix}$$

Then the first row of A and second column of B

$$= \begin{bmatrix} 2 & 0 \\ -5 & 1 \end{bmatrix} \begin{bmatrix} 7 & -2 \\ 3 & 1 \end{bmatrix} = \begin{bmatrix} 14 & -4 \\ -5(7)+1(3) & AB_{22} \end{bmatrix}$$

Then the second row of A and first column of B

$$= \begin{bmatrix} 2 & 0 \\ -5 & 1 \end{bmatrix} \begin{bmatrix} 7 & -2 \\ 3 & 1 \end{bmatrix} = \begin{bmatrix} 14 & -4 \\ -32 & -5(-2)+1(1) \end{bmatrix}$$

Finally, the second row of A and second column of B

$$= \begin{bmatrix} 14 & -4 \\ -32 & 11 \end{bmatrix}$$

Given the matrices $A = \begin{bmatrix} 2 & -3 \\ 4 & -1 \\ 1 & 0 \end{bmatrix}$ and $B = \begin{bmatrix} 5 & 0 & -2 \\ -4 & 1 & 3 \end{bmatrix}$, find the following products.

a. AB **b.** BA

Solution

a. A is of order 3×2 and B is of order 2×3, so AB is defined and is of order 3×3. Each entry of AB is formed from a row of A and a column of B.

$$AB = \begin{bmatrix} 2 & -3 \\ 4 & -1 \\ 1 & 0 \end{bmatrix} \begin{bmatrix} 5 & 0 & -2 \\ -4 & 1 & 3 \end{bmatrix} = \begin{bmatrix} 2(5)+(-3)(-4) & 2(0)+(-3)(1) & 2(-2)+(-3)(3) \\ 4(5)+(-1)(-4) & 4(0)+(-1)(1) & 4(-2)+(-1)(3) \\ 1(5)+0(-4) & 1(0)+0(1) & 1(-2)+0(3) \end{bmatrix}$$

$$= \begin{bmatrix} 22 & -3 & -13 \\ 24 & -1 & -11 \\ 5 & 0 & -2 \end{bmatrix}$$

b. From the orders of the two matrices, we know BA exists and will be a 2×2 matrix. Note that each entry of BA is a sum of three products.

$$BA = \begin{bmatrix} 5 & 0 & -2 \\ -4 & 1 & 3 \end{bmatrix} \begin{bmatrix} 2 & -3 \\ 4 & -1 \\ 1 & 0 \end{bmatrix} = \begin{bmatrix} 5(2)+0(4)+(-2)(1) & 5(-3)+0(-1)+(-2)(0) \\ (-4)(2)+1(4)+3(1) & (-4)(-3)+1(-1)+3(0) \end{bmatrix}$$

$$= \begin{bmatrix} 8 & -15 \\ -1 & 11 \end{bmatrix}$$

Now that matrix multiplication has been defined, it can be used in another way to illustrate the fact that matrices can be considered as functions. As already mentioned, the matrix

$$A = \begin{bmatrix} a & b \\ c & d \end{bmatrix}$$

characterizes the function that transforms s (x, y) into (x', y') in the following system.

$$\begin{cases} x' = ax + by \\ y' = cx + dy \end{cases}.$$

This is even more clear if we associate the ordered pair (x, y) with the following 2×1 matrix.

$$\begin{bmatrix} x \\ y \end{bmatrix}$$

Then the matrix product

$$\begin{bmatrix} a & b \\ c & d \end{bmatrix}\begin{bmatrix} x \\ y \end{bmatrix},$$

which has the "look" of a function acting on its argument, gives us the expressions in the system. Verify for yourself that

$$\begin{bmatrix} a & b \\ c & d \end{bmatrix}\begin{bmatrix} x \\ y \end{bmatrix} = \begin{bmatrix} ax + by \\ cx + dy \end{bmatrix}.$$

Transition Matrices

In a variety of important applications, matrices can be used to model how the "state" of a situation changes over time. To illustrate the idea, we will consider one such situation in some detail.

Suppose a new grocery store opens in a small town, with the intent of competing with the one existing store. The new store, which we will call store **A**, begins an aggressive advertising campaign, with the result that every month 45% of the customers of the existing store (store **B**) decide to start shopping at store **A**. Store **B**, however, responds with a strong appeal to win back customers and every month 30% of the customers of store **A** return to store **B**. In this highly dynamic situation, a number of questions may arise in the minds of the managers of the two stores, among them:

1. Given the known number of customers one month, how many customers can be expected the following month?

2. Can a certain percentage of the total number of the town's shoppers be expected in the long run? If so, what is that percentage?

To answer the first question, let a denote the number of customers of store **A** in a given month, and let b denote the number of customers of store **B** in the same month. In this simple example, we will assume that $a + b$, the total number of grocery store customers in the town, remains fixed over time.

In the following month, store **A** gains $0.45b$ customers (45% of its competitor's customers) and retains $0.7a$ of its existing customers. At the same time, store **B** gains $0.3a$ new customers (the 30% of **A**'s clientele that switch back to store **B**), and retains $0.55b$ existing customers. In algebraic form, the number of customers of each store the following month are given by the following two expressions.

$$\begin{cases} 0.7a + 0.45b \\ 0.3a + 0.55b \end{cases}$$

Matrix multiplication allows us to write this information as

$$\begin{bmatrix} 0.7 & 0.45 \\ 0.3 & 0.55 \end{bmatrix}\begin{bmatrix} a \\ b \end{bmatrix},$$

and if we let P stand for the 2×2 matrix of percentages, we can think of P as a function that transforms one month's distribution of customers into the next month's. For instance, if store **A** has 360 customers and store **B** has 640 customers one month, then the following month their respective number of customers will be 540 and 460, as

$$\begin{bmatrix} 0.7 & 0.45 \\ 0.3 & 0.55 \end{bmatrix}\begin{bmatrix} 360 \\ 640 \end{bmatrix} = \begin{bmatrix} 540 \\ 460 \end{bmatrix}.$$

In such situations, a matrix like P is called a **transition matrix**, as it characterizes the transition of the system from one state to the next. Transition matrices are characterized by the fact that all of their entries are positive, and the sum of the entries in each column is 1.

To answer the second question, we might continue with this specific example and ask how the 540 and 460 customers of the respective stores realign themselves one month later. Note that

$$\begin{bmatrix} 0.7 & 0.45 \\ 0.3 & 0.55 \end{bmatrix}\begin{bmatrix} 540 \\ 460 \end{bmatrix} = \begin{bmatrix} 585 \\ 415 \end{bmatrix},$$

meaning that at the end of the second month, store **A** has 585 customers and store **B** has 415. But we can obtain the same result by applying the matrix P^2 to the original distribution of 360 and 640 customers, as

$$P^2\begin{bmatrix} 360 \\ 640 \end{bmatrix} = P\left(P\begin{bmatrix} 360 \\ 640 \end{bmatrix}\right) = P\begin{bmatrix} 540 \\ 460 \end{bmatrix}.$$

In other words, we can form the composition of P with itself to obtain a function that transforms one month's distribution of customers into the distribution two months later. Note that

$$P^2 = \begin{bmatrix} 0.625 & 0.5625 \\ 0.375 & 0.4375 \end{bmatrix}$$

and that

$$\begin{bmatrix} 0.625 & 0.5625 \\ 0.375 & 0.4375 \end{bmatrix}\begin{bmatrix} 360 \\ 640 \end{bmatrix} = \begin{bmatrix} 585 \\ 415 \end{bmatrix}.$$

At this point, a calculator or computer software with matrix capability may be useful. We can use either aid to calculate high powers of P to see if there is a long-term trend in the distribution of the customers. For instance, to six decimal places,

$$P^5 = \begin{bmatrix} 0.600391 & 0.599414 \\ 0.399609 & 0.400586 \end{bmatrix},$$

and after a certain point calculators and software will round off the entries and give us a result of

$$P^n = \begin{bmatrix} 0.6 & 0.6 \\ 0.4 & 0.4 \end{bmatrix}$$

for large n (the value for n at which this happens will vary depending on the technology used). This means that, after a few months, store **A** can count on roughly 60% of the town's customers and store **B** can count on roughly 40% (the actual identities of the customers will keep changing from month to month, but the relative proportions will have stabilized). We can verify that the situation is stable by applying the transition matrix to an assumed 1000 customers split 60 : 40.

$$\begin{bmatrix} 0.7 & 0.45 \\ 0.3 & 0.55 \end{bmatrix} \begin{bmatrix} 600 \\ 400 \end{bmatrix} = \begin{bmatrix} 600 \\ 400 \end{bmatrix}$$

Basic Matrix Operations Using Technology

To perform algebra on matrices using the calculator, we must first create the matrices we wish to use by defining them in the MATRIX menu under EDIT. Once the matrices have been defined, we can select them in the MATRIX menu under NAMES and use them in the operations we wish to perform.

For instance, suppose we have two matrices, $A = \begin{bmatrix} -1 & 4 \\ 9 & -2 \end{bmatrix}$ and $B = \begin{bmatrix} 2 & 10 \\ -1 & 4 \end{bmatrix}$,

and we want to find $A + 3B$. After creating matrices A and B, we would enter in the following.

```
[A]+3[B]
            [[5 34]
             [6 10]]
```

Note that if we try to perform an operation that isn't possible due to dimension size, we will get an error.

6.4 EXERCISES

💡 PRACTICE

Given $A = \begin{bmatrix} 3 & -2 \\ 1 & 0 \\ 0 & 5 \end{bmatrix}$, $B = \begin{bmatrix} 4 & -5 \\ 3 & 0 \\ -2 & 2 \end{bmatrix}$, $C = \begin{bmatrix} 2 & -1 \\ 6 & 10 \\ -3 & 7 \end{bmatrix}$, and $D = \begin{bmatrix} 3 & 2 & 5 \\ -2 & -4 & 1 \end{bmatrix}$,

determine the following, if possible. See Examples 1, 3, and 4.

1. $3A - B$ **2.** $B - 2D$ **3.** $3C$ **4.** $\dfrac{1}{2}D$

5. $3D + C$ **6.** $A + B + C$ **7.** $2A + 2B$ **8.** $\dfrac{3}{2}B + \dfrac{1}{2}C$

9. $C - 3A$ **10.** $3C - A$ **11.** $4A - 3D$ **12.** $2(A - 3B)$

Determine the values of the variables that will make each of the following equations true, if possible. See Examples 1–4.

13. $\begin{bmatrix} 2a & b & 3 \\ -5 & 9 & 7 \end{bmatrix} = \begin{bmatrix} 6 & -1 & 3 \\ -5 & 9 & c-3 \end{bmatrix}$

14. $\begin{bmatrix} x \\ -9 \\ -1+z \end{bmatrix} = \begin{bmatrix} 8 \\ 3y \\ 5 \end{bmatrix}$

15. $\begin{bmatrix} a & 2b & c \end{bmatrix} + 3\begin{bmatrix} a & 2 & -c \end{bmatrix} = \begin{bmatrix} 8 & 2 & 2 \end{bmatrix}$

16. $\begin{bmatrix} w & 5x \\ 2y & z \end{bmatrix} - 5\begin{bmatrix} w & x \\ y & -z \end{bmatrix} = \begin{bmatrix} w+5 & 0 \\ 6 & 1 \end{bmatrix}$

17. $\begin{bmatrix} 3x \\ 2y \end{bmatrix} + \begin{bmatrix} x \\ -y \\ z \end{bmatrix} = \begin{bmatrix} 4 \\ 0 \\ 2 \end{bmatrix}$

18. $\begin{bmatrix} 2a & 3b & c \end{bmatrix} = \begin{bmatrix} 4 \\ 3 \\ 0 \end{bmatrix}$

19. $\begin{bmatrix} x \\ 3x \end{bmatrix} - \begin{bmatrix} y \\ 2y \end{bmatrix} = \begin{bmatrix} 5 \\ 20 \end{bmatrix}$

20. $7\begin{bmatrix} -1 \\ y \end{bmatrix} = \begin{bmatrix} 2x \\ 5x \end{bmatrix} + 3\begin{bmatrix} y \\ 1 \end{bmatrix}$

21. $2\begin{bmatrix} x \\ 2y \end{bmatrix} - 3\begin{bmatrix} 5y \\ -3x \end{bmatrix} = \begin{bmatrix} -9 \\ 31 \end{bmatrix}$

22. $2\begin{bmatrix} 3r & s & 2t \end{bmatrix} - \begin{bmatrix} r & s & t \end{bmatrix} = \begin{bmatrix} 15 & 3 & 9 \end{bmatrix}$

23. $2\begin{bmatrix} 2x^2 & x \\ 7x & 4 \end{bmatrix} - \begin{bmatrix} 5x \\ x-2 \end{bmatrix} = \begin{bmatrix} 2x & 0 \\ 6 & x^2 \end{bmatrix}$

24. $\begin{bmatrix} -x \\ 3 \end{bmatrix} - 5\begin{bmatrix} 2 \\ y \end{bmatrix} = \begin{bmatrix} -2y \\ 3x \end{bmatrix}$

25. $3\begin{bmatrix} 2a \\ -a \end{bmatrix} - 3\begin{bmatrix} 3b \\ 2b \end{bmatrix} = \begin{bmatrix} 3 \\ -54 \end{bmatrix}$

26. $2\begin{bmatrix} -s \\ -7 \end{bmatrix} + 2\begin{bmatrix} -2r \\ r \end{bmatrix} = -2\begin{bmatrix} 8 \\ s \end{bmatrix}$

Evaluate the following matrix products, if possible. See Examples 5 and 6.

27. $\begin{bmatrix} 3 & -2 & 1 \end{bmatrix}\begin{bmatrix} 5 & -1 \\ 0 & 3 \\ 9 & 4 \end{bmatrix}$

28. $\begin{bmatrix} 0 & -8 \\ 5 & 6 \end{bmatrix}\begin{bmatrix} 3 & 7 \end{bmatrix}$

29. $\begin{bmatrix} 3 & 7 \end{bmatrix}\begin{bmatrix} 0 & -8 \\ 5 & 6 \end{bmatrix}$

30. $\begin{bmatrix} 5 & 0 & -3 \end{bmatrix}\begin{bmatrix} 4 \\ 2 \\ -6 \end{bmatrix}$

31. $\begin{bmatrix} 3 & 9 & -4 \\ 0 & 0 & 2 \\ 5 & -2 & 7 \end{bmatrix}\begin{bmatrix} 3 & 2 \\ 2 & 1 \end{bmatrix}$

32. $\begin{bmatrix} 4 \\ 2 \\ -6 \end{bmatrix}\begin{bmatrix} 5 & 0 & -3 \end{bmatrix}$

33. $\begin{bmatrix} -3 & -6 & -3 \end{bmatrix}\begin{bmatrix} 6 & 9 \\ 6 & -8 \\ -8 & 8 \end{bmatrix}$

34. $\begin{bmatrix} 4 & -5 \\ 7 & -9 \end{bmatrix}\begin{bmatrix} -8 & 3 \end{bmatrix}$

35. $\begin{bmatrix} -3 \\ -5 \\ -6 \end{bmatrix} \begin{bmatrix} -5 & 1 & 8 \end{bmatrix}$

Given $A = \begin{bmatrix} -3 & 1 \\ 2 & 3 \end{bmatrix}$, $B = \begin{bmatrix} 8 & -5 \end{bmatrix}$, $C = \begin{bmatrix} 4 \\ 7 \\ -2 \end{bmatrix}$, and $D = \begin{bmatrix} -5 & 4 \\ -1 & -1 \end{bmatrix}$, determine the following, if possible. See Examples 5 and 6.

36. AB **37.** BA **38.** $BA + B$ **39.** A^2

40. C^2 **41.** CB **42.** D^2 **43.** $CD + C$

44. DA **45.** AD **46.** DB **47.** $(BD)A$

🚀 APPLICATIONS

48. Suppose that each month 20% of store B's customers switch to store A, and 10% of store A's customers switch back to store B. At the start of January, store A has 300 customers and store B has 700. How many customers can each store expect at the start of February? At the start of March?

49. Given the percentages stated in the last problem, what long-term proportion of the town's customers can each store expect? (A graphing utility may be used to compute high powers of the transition matrix, or you can use the method described in the following exercise.)

✏️ WRITING & THINKING

50. Suppose P is a 2×2 transition matrix, and we want to determine the effect of applying high powers of P to the matrix

$$\begin{bmatrix} x \\ y \end{bmatrix},$$

where $x + y$ is a fixed constant, say c. (In our competing store situation, $x + y = 1000$.) If the long-term behavior approaches a steady state, as in our two-store example, then there is some value for x and some value for y such that $x + y = c$ and

$$P \begin{bmatrix} x \\ y \end{bmatrix} = \begin{bmatrix} x \\ y \end{bmatrix}.$$

In other words, once the steady state has been reached, applying the matrix P to it has no effect on the state.

We can use this fact to actually solve for x and y as follows. Given the matrix

$$P = \begin{bmatrix} 0.7 & 0.45 \\ 0.3 & 0.55 \end{bmatrix},$$

write the equation

$$P\begin{bmatrix} x \\ y \end{bmatrix} = \begin{bmatrix} x \\ y \end{bmatrix}$$

in system form. You should find that the two equations that result are actually identical. But if we now also use the fact that $x + y = 1000$, we can solve for the variables and find that $x = 600$ and $y = 400$. Verify that this is indeed the case.

51. Your friend Jared is having trouble with matrices, so you offer to study with him. Check his solution of the following problem. If the solution is incorrect, explain the error that has been made.

$$2\begin{bmatrix} -8 & -9 & -1 \\ -8 & 1 & 5 \end{bmatrix} - 2\begin{bmatrix} -2 & 3 \\ 5 & -7 \\ -8 & -1 \end{bmatrix}$$

$$= \begin{bmatrix} -16 & -18 & -2 \\ -16 & 2 & 10 \end{bmatrix} + \begin{bmatrix} 4 & -6 \\ -10 & 14 \\ 16 & 2 \end{bmatrix}$$

$$= \begin{bmatrix} -16+4-18-10-2+16 & -16-6-18+14-2+2 \\ -16+4+2-10+10+2 & -16-6+2+14+10+2 \end{bmatrix}$$

$$= \begin{bmatrix} -26 & -26 \\ -8 & 6 \end{bmatrix}$$

⬐ TECHNOLOGY

Given $A = \begin{bmatrix} 3.8 & -1.2 & 4.6 \end{bmatrix}$, $B = \begin{bmatrix} -8.2 & -4.9 \\ 7.4 & -1.3 \\ 3.5 & -2.1 \end{bmatrix}$, $C = \begin{bmatrix} 6.3 \\ 5.7 \end{bmatrix}$, and $D = \begin{bmatrix} 2.8 & -7.1 \\ -5.4 & 6.6 \end{bmatrix}$,

use a graphing utility to determine the following, if possible.

52. BD **53.** CA **54.** D^2

55. AB **56.** DC **57.** BC

6.5 INVERSES OF SQUARE MATRICES

■ TOPICS

- The Matrix Form of a Linear System
- Finding the Inverse of a Matrix
- Solving Linear Systems Using Matrix Inverses
- Finding the Inverse of a Matrix Using Technology

The Matrix Form of a Linear System

As we saw in the last section, if we express the ordered pair (x, y) as a 2×1 matrix, then the linear system

$$\begin{cases} ax + by = e \\ cx + dy = f \end{cases}$$

can be written as

$$\begin{bmatrix} a & b \\ c & d \end{bmatrix} \begin{bmatrix} x \\ y \end{bmatrix} = \begin{bmatrix} e \\ f \end{bmatrix}$$

The fact that the matrix equation is equivalent to the system of equations above it is a great leap in efficiency: it converts a system of any number of equations into a single matrix equation. More importantly, the function interpretation of a matrix allows us to express a *whole system* of equations in a form like that of a *single linear equation* of a single variable.

Since the generic linear equation $ax = b$ can be solved by dividing both sides by a,

$$ax = b \quad \Leftrightarrow \quad x = \frac{b}{a} \ \left(\text{assuming } a \neq 0\right),$$

it is tempting to solve

$$\begin{bmatrix} a & b \\ c & d \end{bmatrix} \begin{bmatrix} x \\ y \end{bmatrix} = \begin{bmatrix} e \\ f \end{bmatrix}$$

by "dividing" both sides by the 2×2 matrix of coefficients. Unfortunately, we don't yet have a way of making sense of "matrix division."

We will return to this thought soon, but first we will see how some specific linear systems appear in matrix form.

Write each linear system as a matrix equation.

a. $\begin{cases} -3x + 5y = 2 \\ x - 4y = -1 \end{cases}$

b. $\begin{cases} 3y - x = -2 \\ 4 - z + y = 5 \\ z - 3x + 3 = y - x \end{cases}$

Solution

a. Since the system is in standard form, we can just read off the coefficients of x and y to form the equation

$$\begin{bmatrix} -3 & 5 \\ 1 & -4 \end{bmatrix} \begin{bmatrix} x \\ y \end{bmatrix} = \begin{bmatrix} 2 \\ -1 \end{bmatrix}.$$

b. First, we write each equation in standard form.

$$\begin{cases} 3y - x = -2 \\ 4 - z + y = 5 \\ z - 3x + 3 = y - x \end{cases} \longrightarrow \begin{cases} -x + 3y = -2 \\ y - z = 1 \\ -2x - y + z = -3 \end{cases}$$

Now we can read off the coefficients to form the matrix equation

$$\begin{bmatrix} -1 & 3 & 0 \\ 0 & 1 & -1 \\ -2 & -1 & 1 \end{bmatrix} \begin{bmatrix} x \\ y \\ z \end{bmatrix} = \begin{bmatrix} -2 \\ 1 \\ -3 \end{bmatrix}.$$

Finding the Inverse of a Matrix

In order to solve matrix equations like the two we obtained in Example 1, we need a way to "undo" the matrix of coefficients that appears in front of the column of variables.

In order to figure out how to "undo" a matrix, we will first need to understand how to do *nothing* to a matrix. Consider the following matrix products:

$$\begin{bmatrix} 1 & 0 \\ 0 & 1 \end{bmatrix} \begin{bmatrix} x \\ y \end{bmatrix} \quad \text{and} \quad \begin{bmatrix} 1 & 0 & 0 \\ 0 & 1 & 0 \\ 0 & 0 & 1 \end{bmatrix} \begin{bmatrix} x \\ y \\ z \end{bmatrix}$$

Evaluating these products, we have

$$\begin{bmatrix} 1 & 0 \\ 0 & 1 \end{bmatrix} \begin{bmatrix} x \\ y \end{bmatrix} = \begin{bmatrix} 1x + 0y \\ 0x + 1y \end{bmatrix} = \begin{bmatrix} x \\ y \end{bmatrix}$$

and

$$\begin{bmatrix} 1 & 0 & 0 \\ 0 & 1 & 0 \\ 0 & 0 & 1 \end{bmatrix} \begin{bmatrix} x \\ y \\ z \end{bmatrix} = \begin{bmatrix} 1x + 0y + 0z \\ 0x + 1y + 0z \\ 0x + 0y + 1z \end{bmatrix} = \begin{bmatrix} x \\ y \\ z \end{bmatrix}.$$

If these matrix products appear as the left-hand sides of matrix equations, the equations would correspond to solutions of linear systems:

$$\begin{bmatrix} 1 & 0 \\ 0 & 1 \end{bmatrix} \begin{bmatrix} x \\ y \end{bmatrix} = \begin{bmatrix} a \\ b \end{bmatrix} \quad \text{corresponds to} \quad \begin{cases} x = a \\ y = b \end{cases}$$

and

$$\begin{bmatrix} 1 & 0 & 0 \\ 0 & 1 & 0 \\ 0 & 0 & 1 \end{bmatrix} \begin{bmatrix} x \\ y \\ z \end{bmatrix} = \begin{bmatrix} a \\ b \\ c \end{bmatrix} \quad \text{corresponds to} \quad \begin{cases} x = a \\ y = b \\ z = c \end{cases}.$$

When we multiply any matrix by one of these matrices, the original matrix is *unchanged.* This fact is very useful in solving matrix equations.

Identity Matrices

The $n \times n$ **identity matrix**, denoted I_n (just I when there is no possibility of confusion), is the $n \times n$ matrix consisting of 1s on the *main diagonal* and 0s everywhere else. The **main diagonal** consists of those entries in the first row–first column, the second row–second column, and so on down to the nth row–n^{th} column. Every identity matrix has the form

$$I = \begin{bmatrix} 1 & 0 & 0 & \cdots & 0 \\ 0 & 1 & 0 & \cdots & 0 \\ 0 & 0 & 1 & \cdots & 0 \\ \vdots & \vdots & \vdots & \ddots & \vdots \\ 0 & 0 & 0 & \cdots & 1 \end{bmatrix}.$$

If the matrices A and B have appropriate order, so that the matrix products are defined, then $AI = A$ and $IB = B$. Thus, the identity matrix serves as the multiplicative identity on the set of appropriately sized matrices. In this sense, I serves the same purpose as the number 1 in the set of real numbers.

We know that a linear system of n equations and n variables can be expressed as a matrix equation $AX = B$, where A is an $n \times n$ matrix of coefficients, X is an $n \times 1$ matrix containing the n variables, and B is an $n \times 1$ matrix of the constants from the right-hand sides of the equations. If we could find a matrix, which we call A^{-1}, with the property that $A^{-1}A = I$, then we could use A^{-1} to "undo" the matrix A. We call the matrix A^{-1}, if it exists, the *inverse* of A. This is analogous to the fact that $\dfrac{1}{a}$, sometimes denoted a^{-1}, is the (multiplicative) inverse of the real number a.

The Inverse of a Matrix

Let A be an $n \times n$ matrix. If there exists an $n \times n$ matrix A^{-1} such that

$$A^{-1}A = I_n \text{ and } AA^{-1} = I_n,$$

we call A^{-1} the **inverse** of A.

Example 2: Finding the Inverse of a Matrix

Find the inverse of the matrix $A = \begin{bmatrix} 2 & -3 \\ -1 & 2 \end{bmatrix}$.

Solution

If we let $A^{-1} = \begin{bmatrix} w & x \\ y & z \end{bmatrix}$, we can use the equation $AA^{-1} = I$ to find $w, x, y,$ and z.

$$\begin{bmatrix} 2 & -3 \\ -1 & 2 \end{bmatrix} \begin{bmatrix} w & x \\ y & z \end{bmatrix} = \begin{bmatrix} 1 & 0 \\ 0 & 1 \end{bmatrix}$$

Multiplying the left-hand side out, we see that we need to solve the equation

$$\begin{bmatrix} 2w-3y & 2x-3z \\ -w+2y & -x+2z \end{bmatrix} = \begin{bmatrix} 1 & 0 \\ 0 & 1 \end{bmatrix}$$

which, if we equate columns on each side, means we need to solve the two linear systems

$$\begin{cases} 2w-3y=1 \\ -w+2y=0 \end{cases} \quad \text{and} \quad \begin{cases} 2x-3z=0 \\ -x+2z=1 \end{cases}.$$

We have covered many methods for solving such systems. If we write the augmented matrix for each system, we get

$$\begin{bmatrix} 2 & -3 & | & 1 \\ -1 & 2 & | & 0 \end{bmatrix} \quad \text{and} \quad \begin{bmatrix} 2 & -3 & | & 0 \\ -1 & 2 & | & 1 \end{bmatrix}.$$

Note that the left-hand sides of these matrices are the same. This allows us to combine them into a new kind of augmented matrix so we can use Gauss-Jordan elimination to solve the systems at the same time. Combining the matrices, we get

$$\begin{bmatrix} 2 & -3 & | & 1 & 0 \\ -1 & 2 & | & 0 & 1 \end{bmatrix}.$$

When we change this new matrix into reduced row echelon form, we will have solved the first system with the numbers in the third column and the second system with the numbers in the fourth column.

$$\begin{bmatrix} 2 & -3 & | & 1 & 0 \\ -1 & 2 & | & 0 & 1 \end{bmatrix} \xrightarrow{R_1 \leftrightarrow R_2} \begin{bmatrix} -1 & 2 & | & 0 & 1 \\ 2 & -3 & | & 1 & 0 \end{bmatrix} \xrightarrow{2R_1+R_2} \begin{bmatrix} -1 & 2 & | & 0 & 1 \\ 0 & 1 & | & 1 & 2 \end{bmatrix}$$

$$\xrightarrow{-R_1} \begin{bmatrix} 1 & -2 & | & 0 & -1 \\ 0 & 1 & | & 1 & 2 \end{bmatrix} \xrightarrow{2R_2+R_1} \begin{bmatrix} 1 & 0 & | & 2 & 3 \\ 0 & 1 & | & 1 & 2 \end{bmatrix}$$

This tells us that $w=2$ and $y=1$ (from the third column) and $x=3$ and $z=2$ (from the fourth column). So

$$A^{-1} = \begin{bmatrix} 2 & 3 \\ 1 & 2 \end{bmatrix}.$$

We can now verify that

$$\begin{bmatrix} 2 & -3 \\ -1 & 2 \end{bmatrix}\begin{bmatrix} 2 & 3 \\ 1 & 2 \end{bmatrix} = \begin{bmatrix} 1 & 0 \\ 0 & 1 \end{bmatrix} \quad \text{and also} \quad \begin{bmatrix} 2 & 3 \\ 1 & 2 \end{bmatrix}\begin{bmatrix} 2 & -3 \\ -1 & 2 \end{bmatrix} = \begin{bmatrix} 1 & 0 \\ 0 & 1 \end{bmatrix},$$

so we have indeed found A^{-1}.

Note how, during the solution process, the identity matrix passed from the right side of the matrix to the left, resulting in reduced row echelon form. With this observation, we can omit the intermediate step of constructing the systems of equations, and skip to the process of putting the appropriate augmented matrix into reduced row echelon form.

Finding the Inverse of a Matrix

Let A be an $n \times n$ matrix. The inverse of A can be found by

Step 1: Forming the augmented matrix $[A \,|\, I]$, where I is the $n \times n$ identity matrix.

Step 2: Using Gauss-Jordan elimination to put $[A \,|\, I]$ into the form $[I \,|\, B]$, if possible.

Step 3: Defining A^{-1} to be B.

If it is not possible to put $[A \,|\, I]$ into reduced row echelon form, then A doesn't have an inverse, and we say A is **not invertible**.

If the coefficient matrix of a system of equations is not invertible, it means that the system either has an infinite number of solutions or has no solution.

There is a shortcut for finding inverses of 2×2 matrices that can save you some time. This shortcut also quickly identifies those 2×2 matrices that are not invertible.

Inverse of a 2 × 2 Matrix

Let $A = \begin{bmatrix} a & b \\ c & d \end{bmatrix}$. Then $A^{-1} = \dfrac{1}{|A|} \begin{bmatrix} d & -b \\ -c & a \end{bmatrix}$, where $|A| = ad - bc$ is the determinant

of A. Since $|A|$ appears in the denominator of a fraction, A^{-1} fails to exist if $|A| = 0$.

Since $|A|$ appears in the denominator of a fraction, A^{-1} fails to exist if $|A| = 0$.

In fact, we can extend this last observation to all square matrices.

Invertible Matrices

A square matrix A is **invertible** if and only if $|A| \neq 0$.

Example 3: Finding the Inverse of a Matrix

Find the inverses of the following matrices, if possible.

a. $A = \begin{bmatrix} 3 & -5 \\ 2 & 1 \end{bmatrix}$
    b. $B = \begin{bmatrix} 2 & 4 & 2 \\ -1 & 5 & -1 \\ 3 & 1 & 3 \end{bmatrix}$

Solution

a. Since A is 2×2, we can use the shortcut,

$$A^{-1} = \frac{1}{3-(-10)}\begin{bmatrix} 1 & 5 \\ -2 & 3 \end{bmatrix} = \begin{bmatrix} \dfrac{1}{13} & \dfrac{5}{13} \\ \dfrac{-2}{13} & \dfrac{3}{13} \end{bmatrix}.$$

We can verify our work as follows:

$$\begin{bmatrix} 3 & -5 \\ 2 & 1 \end{bmatrix}\begin{bmatrix} \dfrac{1}{13} & \dfrac{5}{13} \\ \dfrac{-2}{13} & \dfrac{3}{13} \end{bmatrix} = \begin{bmatrix} 1 & 0 \\ 0 & 1 \end{bmatrix} = \begin{bmatrix} \dfrac{1}{13} & \dfrac{5}{13} \\ \dfrac{-2}{13} & \dfrac{3}{13} \end{bmatrix}\begin{bmatrix} 3 & -5 \\ 2 & 1 \end{bmatrix}$$

b. Since B is a 3×3 matrix, we use the augmented identity matrix process.

$$\left[\begin{array}{ccc|ccc} 2 & 4 & 2 & 1 & 0 & 0 \\ -1 & 5 & -1 & 0 & 1 & 0 \\ 3 & 1 & 3 & 0 & 0 & 1 \end{array}\right] \xrightarrow[3R_2+R_3]{2R_2+R_1} \left[\begin{array}{ccc|ccc} 0 & 14 & 0 & 1 & 2 & 0 \\ -1 & 5 & -1 & 0 & 1 & 0 \\ 0 & 16 & 0 & 0 & 3 & 1 \end{array}\right]$$

$$\xrightarrow{R_1 \leftrightarrow R_2} \left[\begin{array}{ccc|ccc} -1 & 5 & -1 & 0 & 1 & 0 \\ 0 & 14 & 0 & 1 & 2 & 0 \\ 0 & 16 & 0 & 0 & 3 & 1 \end{array}\right]$$

$$\xrightarrow[\frac{1}{14}R_2]{-R_1} \left[\begin{array}{ccc|ccc} 1 & -5 & 1 & 0 & -1 & 0 \\ 0 & 1 & 0 & \dfrac{1}{14} & \dfrac{1}{7} & 0 \\ 0 & 16 & 0 & 0 & 3 & 1 \end{array}\right]$$

$$\xrightarrow{-16R_2+R_3} \left[\begin{array}{ccc|ccc} 1 & -5 & 1 & 0 & -1 & 0 \\ 0 & 1 & 0 & \dfrac{1}{14} & \dfrac{1}{7} & 0 \\ 0 & 0 & 0 & -\dfrac{8}{7} & \dfrac{5}{7} & 1 \end{array}\right]$$

At this point, we can stop. Once the first three entries of any row are 0 in the 3×3 matrix, there is no way to put the matrix into reduced row echelon form. Thus, B has no inverse.

We could have seen this fact earlier by considering the determinant of B, which is 0, meaning B has no inverse. To quickly see that $|B| = 0$, note that B has two identical columns. Recall that switching two columns changes the sign of the determinant, but if we switch identical columns, the determinant must remain the same! The only way this is possible is if $|B| = 0$.

Solving Linear Systems Using Matrix Inverses

We have now assembled all the tools we need to solve linear systems by the inverse matrix method.

The Inverse Matrix Method

To solve a linear system of n equations in n variables:

Step 1: Write the system in matrix form as $AX = B$, where A is the $n \times n$ matrix of coefficients, X is the $n \times 1$ matrix of variables, and B is the $n \times 1$ matrix of constants.

Step 2: Calculate A^{-1}, if it exists. If A^{-1} does not exist, the system either has an infinite number of solutions, or no solution, and a different method is required.

Step 3: Multiply both sides of the equation $AX = B$ by A^{-1}. We obtain

$$A^{-1}AX = A^{-1}B$$
$$IX = A^{-1}B$$
$$X = A^{-1}B$$

Step 4: The entries in the $n \times 1$ matrix $A^{-1}B$ are the solutions for the variables listed in the $n \times 1$ matrix X.

⚠ CAUTION

It is crucial to multiply both sides of the equation $AX = B$ by A^{-1} on the *left-hand side*. Recall that matrix multiplication is not commutative, so failing to do this can result in an incorrect answer, and often the multiplication will not even be defined.

Example 4: Inverse Matrix Method

Solve the following systems by the inverse matrix method.

a. $\begin{cases} 4x - 5y = 3 \\ -3x + 7y = 1 \end{cases}$
    b. $\begin{cases} -x + 2y = 3 \\ 3x - 6y = -5 \end{cases}$

Solution

a. $\begin{cases} 4x - 5y = 3 \\ -3x + 7y = 1 \end{cases}$

$$\begin{bmatrix} 4 & -5 \\ -3 & 7 \end{bmatrix}\begin{bmatrix} x \\ y \end{bmatrix} = \begin{bmatrix} 3 \\ 1 \end{bmatrix}$$

Write the system in matrix form.

$$\begin{bmatrix} 4 & -5 \\ -3 & 7 \end{bmatrix}^{-1} = \frac{1}{28-15}\begin{bmatrix} 7 & 5 \\ 3 & 4 \end{bmatrix}$$
$$= \begin{bmatrix} \dfrac{7}{13} & \dfrac{5}{13} \\ \dfrac{3}{13} & \dfrac{4}{13} \end{bmatrix}$$

Find the inverse of the coefficient matrix. Since the matrix is of order 2×2, we can use the shortcut to obtain the inverse quickly.

It is a good idea to check your work at this stage by making sure that the product of the original matrix and its inverse is the identity matrix.

$$\begin{bmatrix} x \\ y \end{bmatrix} = \begin{bmatrix} \dfrac{7}{13} & \dfrac{5}{13} \\ \dfrac{3}{13} & \dfrac{4}{13} \end{bmatrix} \begin{bmatrix} 3 \\ 1 \end{bmatrix}$$

Rewrite the equation $AX = B$ in the form $X = A^{-1}B$ and carry out the matrix multiplication.

$$= \begin{bmatrix} \dfrac{21}{13} + \dfrac{5}{13} \\ \dfrac{9}{13} + \dfrac{4}{13} \end{bmatrix}$$

$$= \begin{bmatrix} 2 \\ 1 \end{bmatrix}$$

Thus, the solution to the system is $(2,1)$.

b. $\begin{cases} -x + 2y = 3 \\ 3x - 6y = -5 \end{cases}$

$$\begin{bmatrix} -1 & 2 \\ 3 & -6 \end{bmatrix} \begin{bmatrix} x \\ y \end{bmatrix} = \begin{bmatrix} 3 \\ -5 \end{bmatrix}$$

Again, write the system in matrix form.

$$\begin{bmatrix} -1 & 2 \\ 3 & -6 \end{bmatrix}^{-1} = \frac{1}{6-6} \begin{bmatrix} -6 & -2 \\ -3 & -1 \end{bmatrix}$$

The coefficient matrix has no inverse (since its determinant is 0), so the system has no solution or an infinite number of solutions.

$$D_x = \begin{vmatrix} 3 & 2 \\ -5 & -6 \end{vmatrix} = -18 - (-10) \neq 0$$

Since $D_x \neq 0$, Cramer's Rule tells us the system has no solution.

This system has no solution.

We have now reached the end of our list of solution methods for linear systems. One important question has not yet been addressed. How should we choose a method of solution, given a system of equations?

Unfortunately, there is no simple answer. If the system is small, or looks especially simple, then the methods of substitution or elimination may be the quickest route to a solution. Generally, the larger and/or the more complicated the system, the more likely we are to prefer a matrix method of solution, such as Gauss-Jordan elimination, Cramer's Rule, or the inverse matrix method. If you find that your first choice leads to a computational headache, don't be reluctant to stop and try another method.

Finding the Inverse of a Matrix Using Technology

We can also use a graphing calculator to find the inverse of a matrix by first defining the matrix whose inverse we want to find. Then, enter the matrix, press `x⁻¹`, and press `enter`. To show the answer in fraction form, press `math` and select `1:Frac`.

If we defined matrix A to be $\begin{bmatrix} 7 & 4 \\ 1 & 2 \end{bmatrix}$, we would find the following to be its inverse.

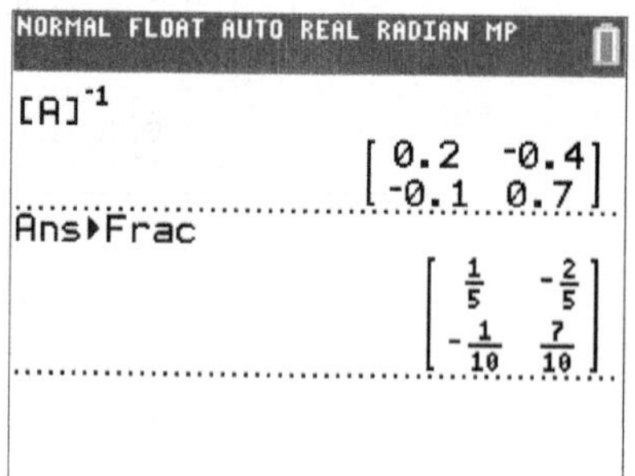

6.5 EXERCISES

PRACTICE

Write each of the following systems of equations as a single matrix equation. See Example 1.

1. $\begin{cases} 14x - 5y = 7 \\ x + 9y = 2 \end{cases}$

2. $\begin{cases} x - 5 = 9y \\ 3y - 2x = 8 \end{cases}$

3. $\begin{cases} -6 - 2y = x \\ 9x + 14 = 3y \end{cases}$

4. $\begin{cases} x - y = 5 \\ 2 - z = x \\ z - 3y = 4 \end{cases}$

5. $\begin{cases} 3x_1 - 7x_2 + x_3 = -4 \\ x_1 - x_2 = 2 \\ 8x_2 + 5x_3 = -3 \end{cases}$

6. $\begin{cases} x_3 = x_2 \\ x_2 = x_1 \\ x_1 = x_3 \end{cases}$

7. $\begin{cases} \dfrac{3x - 8y}{5} = 2 \\ y - 2 = 0 \end{cases}$

8. $\begin{cases} x - 7y = 5 \\ \dfrac{6 + x}{2} = 3y - 2 \end{cases}$

9. $\begin{cases} 4x = 3y - 9 \\ 13 - 2x = -4y \end{cases}$

10. $\begin{cases} -\dfrac{7}{3}y = \dfrac{5 - x}{6} \\ x - 5(y - 3) = -2 \end{cases}$

11. $\begin{cases} 2x - y = -3z \\ y - x = 17 \\ 2 + z + 4x = 5y \end{cases}$

12. $\begin{cases} 2x_1 - 3x_3 = 7 \\ x_2 - 10x_3 = 0 \\ 2x_1 - x_2 + x_3 = 1 \end{cases}$

Find the inverse of each of the following matrices, if possible. See Examples 2 and 3.

13. $\begin{bmatrix} 0 & 4 \\ -5 & -1 \end{bmatrix}$

14. $\begin{bmatrix} -2 & -2 \\ -1 & 2 \end{bmatrix}$

15. $\begin{bmatrix} 3 & 4 \\ -4 & -5 \end{bmatrix}$

16. $\begin{bmatrix} -1 & -1 \\ -\dfrac{1}{4} & -\dfrac{1}{2} \end{bmatrix}$

17. $\begin{bmatrix} -\dfrac{1}{5} & 0 \\ \dfrac{1}{5} & \dfrac{1}{2} \end{bmatrix}$

18. $\begin{bmatrix} -7 & 2 \\ 7 & -2 \end{bmatrix}$

19. $\begin{bmatrix} -2 & -4 & -2 \\ 1 & -4 & 1 \\ 4 & -3 & 4 \end{bmatrix}$

20. $\begin{bmatrix} -3 & 0 & -4 \\ 2 & 5 & 4 \\ 1 & -5 & -2 \end{bmatrix}$

21. $\begin{bmatrix} -\dfrac{5}{11} & -\dfrac{8}{11} & 1 \\ \dfrac{13}{11} & \dfrac{12}{11} & -2 \\ -\dfrac{2}{11} & -\dfrac{1}{11} & 0 \end{bmatrix}$

22. $-\dfrac{1}{31}\begin{bmatrix} 17 & -8 & -2 \\ 1 & 5 & 9 \\ -6 & 1 & 8 \end{bmatrix}$ **23.** $\begin{bmatrix} -1 & 2 & -1 \\ 0 & 3 & -1 \\ 0 & 4 & -1 \end{bmatrix}$ **24.** $\begin{bmatrix} -1 & 0 & -1 \\ \dfrac{3}{2} & 1 & \dfrac{3}{2} \\ -\dfrac{1}{2} & 0 & -\dfrac{1}{4} \end{bmatrix}$

25. $\begin{bmatrix} -\dfrac{6}{5} & -\dfrac{2}{5} & -1 \\ \dfrac{3}{5} & \dfrac{1}{5} & 1 \\ 1 & 0 & 1 \end{bmatrix}$ **26.** $\begin{bmatrix} 2 & -2 & 1 \\ -2 & 2 & -3 \\ 1 & 0 & 2 \end{bmatrix}$ **27.** $\begin{bmatrix} 0 & 1 & 1 \\ 1 & 1 & 0 \\ 0 & 1 & 2 \end{bmatrix}$

28. $\begin{bmatrix} 9 & 8 & 7 \\ 6 & 5 & 4 \\ 3 & 2 & 1 \end{bmatrix}$ **29.** $\begin{bmatrix} \dfrac{2}{3} & \dfrac{8}{9} & \dfrac{1}{9} \\ -\dfrac{1}{3} & \dfrac{2}{9} & -\dfrac{2}{9} \\ -\dfrac{1}{3} & -\dfrac{7}{9} & -\dfrac{2}{9} \end{bmatrix}$ **30.** $\begin{bmatrix} -3 & -3 & -4 \\ 0 & \dfrac{1}{4} & \dfrac{1}{2} \\ 2 & 2 & 3 \end{bmatrix}$

For each pair of matrices, determine if either matrix is the inverse of the other.

31. $\begin{bmatrix} -5 & -2 \\ -7 & 4 \end{bmatrix}, \begin{bmatrix} 10 & 4 \\ 14 & -8 \end{bmatrix}$ **32.** $\begin{bmatrix} 9 & -18 \\ 3 & 12 \end{bmatrix}, \begin{bmatrix} -3 & -6 \\ -1 & -4 \end{bmatrix}$

33. $\begin{bmatrix} -6 & -1 & 1 \\ 4 & -1 & -2 \\ 1 & -1 & -1 \end{bmatrix}, \begin{bmatrix} -1 & -2 & 3 \\ 2 & 5 & -8 \\ -3 & -7 & 10 \end{bmatrix}$ **34.** $\begin{bmatrix} -1 & 4 & 5 \\ 3 & -11 & -17 \\ 4 & -17 & -19 \end{bmatrix}, \begin{bmatrix} -80 & -9 & -13 \\ -11 & -1 & -2 \\ -7 & -1 & -1 \end{bmatrix}$

35. $\begin{bmatrix} 2 & 0 & -1 \\ 3 & 4 & 2 \\ 1 & 1 & -3 \end{bmatrix}, \begin{bmatrix} 4 & 0 & -2 \\ 6 & 8 & 4 \\ 2 & 2 & -6 \end{bmatrix}$ **36.** $\begin{bmatrix} -7 & 0 & -2 \\ -10 & -1 & -2 \\ -7 & -1 & -1 \end{bmatrix}, \begin{bmatrix} -1 & 2 & -2 \\ 4 & -7 & 6 \\ 3 & -7 & 7 \end{bmatrix}$

Solve the following systems by the inverse matrix method, if possible. If the inverse matrix method doesn't apply, use any other method to determine if the system is inconsistent or dependent. See Example 4.

37. $\begin{cases} -2x - 2y = 9 \\ -x + 2y = -3 \end{cases}$ **38.** $\begin{cases} 3x + 4y = -2 \\ -4x - 5y = 9 \end{cases}$ **39.** $\begin{cases} -2x + 3y = 1 \\ 4x - 6y = -2 \end{cases}$

40. $\begin{cases} -2x + 4y = 5 \\ x - 4y = -3 \end{cases}$ **41.** $\begin{cases} -5x = 10 \\ 2x + 2y = -4 \end{cases}$ **42.** $\begin{cases} -3x + y = 2 \\ 9x - 3y = 5 \end{cases}$

43. $\begin{cases} 8x + 2y = 26 \\ -16x - 2y = -90 \end{cases}$ **44.** $\begin{cases} 3x - 7y = -2 \\ -6x + 14y = 4 \end{cases}$ **45.** $\begin{cases} 3y = 15 \\ 8x + 4y = 20 \end{cases}$

46. $\begin{cases} 4y + 3z = -254 \\ 2x - 2y - z = 100 \\ -x + y - 2z = 155 \end{cases}$ **47.** $\begin{cases} 2x - y - 3z = -10 \\ 2y - z = 11 \\ -x + 4z = 0 \end{cases}$ **48.** $\begin{cases} 3y - 4z = 15 \\ x + 2y - 3z = 9 \\ -x - y + 2z = -5 \end{cases}$

✏ WRITING & THINKING

Solve the following sets of systems by the inverse matrix method. **Hint:** The coefficient matrix is the same for the three systems within a set.

49.
$$\begin{cases} x+2y-z=2 \\ 3x+3y-z=-5 \\ 4x+4y-z=1 \end{cases} \qquad \begin{cases} x+2y-z=1 \\ 3x+3y-z=1 \\ 4x+4y-z=1 \end{cases} \qquad \begin{cases} x+2y-z=0 \\ 3x+3y-z=1 \\ 4x+4y-z=1 \end{cases}$$

50.
$$\begin{cases} -x-y-2z=4 \\ x+3y+3z=0 \\ -3y-2z=9 \end{cases} \qquad \begin{cases} -x-y-2z=1 \\ x+3y+3z=0 \\ -3y-2z=0 \end{cases} \qquad \begin{cases} -x-y-2z=-2 \\ x+3y+3z=-3 \\ -3y-2z=1 \end{cases}$$

51.
$$\begin{cases} -x+z=6 \\ -x+3y+2z=-11 \\ 2x-4y-3z=13 \end{cases} \qquad \begin{cases} -x+z=-2 \\ -x+3y+2z=2 \\ 2x-4y-3z=-1 \end{cases} \qquad \begin{cases} -x+z=-4 \\ -x+3y+2z=2 \\ 2x-4y-3z=0 \end{cases}$$

〰 TECHNOLOGY

Using a graphing utility, find the inverse of each of the following matrices, if possible. Round your answers to three decimal places if necessary.

52. $\begin{bmatrix} -7 & 3 \\ -1 & 2 \end{bmatrix}$ **53.** $\begin{bmatrix} -6 & 2 \\ -5 & 5 \end{bmatrix}$ **54.** $\begin{bmatrix} -2 & 0 & 2 \\ 2 & -3 & 1 \\ 1 & -2 & 3 \end{bmatrix}$

55. $\begin{bmatrix} 2.3 & 7.8 \\ -3.4 & 1.6 \end{bmatrix}$ **56.** $\begin{bmatrix} 4.5 & -9.4 & 6.9 \\ 8.6 & -2.8 & 1.2 \\ 3.1 & 0.3 & -7.0 \end{bmatrix}$ **57.** $\begin{bmatrix} 38 & -44 & 72 \\ -93 & 16 & 29 \\ 65 & 23 & -19 \end{bmatrix}$

6.6 LEONTIEF INPUT-OUTPUT ANALYSIS

■ TOPICS

- ■ The Leontief Model
- ■ Closed Systems

Matrix algebra is a useful tool in exploring the interdependent relationships between the inputs and outputs of various industries within an economy. In 1973, Russian-American economist Wassily Leontief was awarded the Nobel Prize in economics for his work using systems of equations and matrices to analyze these relationships. We will now investigate his method of input-output analysis by way of example and then generalize the result so that it can be applied to any appropriate system of equations.

The Leontief Model

Example 1: Leontief Model

Suppose that in a certain fuel and power system we have the gasoline, electricity, and coal industries. To produce one dollar in output, each industry needs the following input.

1. The gasoline industry requires \$0.20 from itself (gas needed to produce more gas), \$0.40 from electricity, and no input from coal.

2. The electricity industry requires \$0.40 from gasoline, \$0.20 from itself, and \$0.20 from coal.

3. The coal industry requires \$0.40 from gasoline, \$0.20 from electricity, and \$0.20 from itself.

This can be summarized in the following table.

	Outputs (Users)		
Inputs (Suppliers)	**Gasoline**	**Electricity**	**Coal**
Gasoline	0.20	0.40	0.40
Electricity	0.40	0.20	0.20
Coal	0.00	0.20	0.20

From this table we form the **input-output matrix** (also known as the **technology matrix**) A.

$$A = \begin{bmatrix} 0.2 & 0.4 & 0.4 \\ 0.4 & 0.2 & 0.2 \\ 0.0 & 0.2 & 0.2 \end{bmatrix}$$

If we let x_1 = gasoline units produced, x_2 = electricity units produced, and x_3 = coal units produced, then the total demand within the system for gasoline can be represented by the following equation.

$$x_1 = 0.2x_1 + 0.4x_2 + 0.4x_3$$

Similarly, we can set up the equations for the demand within the system for electricity and coal. This combined system of equations represents the amount each industry needs to produce in order to keep all three operating. Solving it will yield the **internal demand**.

$$x_1 = 0.2x_1 + 0.4x_2 + 0.4x_3$$
$$x_2 = 0.4x_1 + 0.2x_2 + 0.2x_3$$
$$x_3 = 0.0x_1 + 0.2x_2 + 0.2x_3$$

Now, let's further suppose that in addition to the internal demand, there is a surplus demand from outside of the industries for \$5000 in gasoline, \$2000 in electricity, and \$8000 in coal. This is called an **external demand** and the previous equations can be altered as follows to reflect this additional need in production.

$$x_1 = 0.2x_1 + 0.4x_2 + 0.4x_3 + 5000$$
$$x_2 = 0.4x_1 + 0.2x_2 + 0.2x_3 + 2000$$
$$x_3 = 0.0x_1 + 0.2x_2 + 0.2x_3 + 8000$$

This new system of equations allows us to find the total production necessary to satisfy all demands, internal plus external. We can rewrite the new equations in matrix notation and use matrix operations to solve for the total production.

$$\text{Let } X = \begin{bmatrix} x_1 \\ x_2 \\ x_3 \end{bmatrix} \text{ (a } 3 \times 1 \text{ matrix), } A = \begin{bmatrix} 0.2 & 0.4 & 0.4 \\ 0.4 & 0.2 & 0.2 \\ 0.0 & 0.2 & 0.2 \end{bmatrix} \text{ (a } 3 \times 3 \text{ matrix), and } E = \begin{bmatrix} 5000 \\ 2000 \\ 8000 \end{bmatrix}$$

(a 3×1 matrix). Then we can rewrite the system of equations as $X = AX + E$.

$$\begin{bmatrix} x_1 \\ x_2 \\ x_3 \end{bmatrix} = \begin{bmatrix} 0.2 & 0.4 & 0.4 \\ 0.4 & 0.2 & 0.2 \\ 0.0 & 0.2 & 0.2 \end{bmatrix} \begin{bmatrix} x_1 \\ x_2 \\ x_3 \end{bmatrix} + \begin{bmatrix} 5000 \\ 2000 \\ 8000 \end{bmatrix}$$

We call X the **production matrix** and E the **external demand matrix** (or simply **demand matrix**). Recall that we previously named A the input-output matrix.

In most cases, as in this example, we seek to solve for the production matrix X. In some cases, rather than X we may wish to find the external demand matrix E if we know the total production capability of the industrial system. The flexibility of matrix algebra allows us to do so quickly and efficiently, and we will see an example of this later. It is this process of using matrix algebra to solve for the unknown quantities in an economic system that we call **Leontief input-output analysis**. Now, continuing with our example, we will solve for the production matrix and then develop a general form for its calculation.

Returning to our matrix notation, we solve for X as follows. For this example, I is the 3×3 identity matrix.

$$X = AX + E$$
$$X - AX = E$$
$$(I - A)X = E$$
$$X = (I - A)^{-1} E$$

Therefore, to solve for the production matrix X, we first find $(I - A)^{-1}$, which is the inverse of the matrix obtained by subtracting the input-output matrix A from the identity matrix I of appropriate size. Then we multiply by the external demand matrix E. Returning once more to our example, we complete the solution process.

First, calculate $I - A$.

$$I - A = \begin{bmatrix} 1 & 0 & 0 \\ 0 & 1 & 0 \\ 0 & 0 & 1 \end{bmatrix} - \begin{bmatrix} 0.2 & 0.4 & 0.4 \\ 0.4 & 0.2 & 0.2 \\ 0.0 & 0.2 & 0.2 \end{bmatrix} = \begin{bmatrix} 0.8 & -0.4 & -0.4 \\ -0.4 & 0.8 & -0.2 \\ 0.0 & -0.2 & 0.8 \end{bmatrix}$$

At this point, the simplest way to obtain the solution is to use a calculator or computer algebra system.

$$(I - A)^{-1} E = \begin{bmatrix} 0.8 & -0.4 & -0.4 \\ -0.4 & 0.8 & -0.2 \\ 0.0 & -0.2 & 0.8 \end{bmatrix}^{-1} \begin{bmatrix} 5000 \\ 2000 \\ 8000 \end{bmatrix}$$

You can, however, use Gauss-Jordan elimination to find

$$\begin{bmatrix} 0.8 & -0.4 & -0.4 \\ -0.4 & 0.8 & -0.2 \\ 0.0 & -0.2 & 0.8 \end{bmatrix}^{-1} = \begin{bmatrix} 1.875 & 1.25 & 1.25 \\ 1 & 2 & 1 \\ 0.25 & 0.5 & 1.5 \end{bmatrix}.$$

Regardless of our approach, we can now solve for the total production of the system.

$$X = \begin{bmatrix} x_1 \\ x_2 \\ x_3 \end{bmatrix} = (I - A)^{-1} E = \begin{bmatrix} 0.8 & -0.4 & -0.4 \\ -0.4 & 0.8 & -0.2 \\ 0.0 & -0.2 & 0.8 \end{bmatrix}^{-1} \begin{bmatrix} 5000 \\ 2000 \\ 8000 \end{bmatrix}$$

$$= \begin{bmatrix} 1.875 & 1.25 & 1.25 \\ 1 & 2 & 1 \\ 0.25 & 0.5 & 1.5 \end{bmatrix} \begin{bmatrix} 5000 \\ 2000 \\ 8000 \end{bmatrix}$$

$$= \begin{bmatrix} 21,875 \\ 17,000 \\ 14,250 \end{bmatrix}$$

The solution implies that in order to meet the combined internal and external demands of the system, the following amounts should be produced.

Gasoline: \$21,875

Electricity: \$17,000

Coal: \$14,250

In summary, we generalize the solution process in the previous example.

Solving for a Production Matrix

Given a system of n industries, complete the following steps to solve for the production matrix X.

Step 1: Construct the $n \times 1$ production matrix X.

$$X = \begin{bmatrix} x_1 \\ x_2 \\ \vdots \\ x_n \end{bmatrix}$$

- x_i = the total production of the ith industry
- This step is basically a formality since X is not used in the following constructions and computations. It is, however, a good practice to write the matrix as it assists in keeping the production variables properly ordered.

Step 2: Construct the $n \times n$ input-output matrix A.

$$A = \begin{bmatrix} a_{11} & a_{12} & \cdots & a_{1n} \\ a_{21} & a_{22} & \cdots & a_{2n} \\ \vdots & \vdots & \ddots & \vdots \\ a_{n1} & a_{n2} & \cdots & a_{nn} \end{bmatrix}$$

- a_{ij} = the input needed from the ith industry for the jth industry to produce one unit of output
- The inputs are always between 0 and 1 since each input is the fraction of a unit needed from the ith industry to produce one unit of output for the jth industry.

Step 3: Construct the $n \times 1$ external demand matrix E.

$$E = \begin{bmatrix} e_1 \\ e_2 \\ \vdots \\ e_n \end{bmatrix}$$

- e_i = the external demand of the ith industry

Step 4: Calculate $X = (I - A)^{-1} E$ where I is the $n \times n$ identity matrix.
   Note: If technology is available, this may be calculated in a straightforward manner. Otherwise, completing the following steps may be necessary.
   1. Calculate $I - A$.
   2. Calculate $(I - A)^{-1}$.
   3. Calculate $(I - A)^{-1} E$.

Step 5: Use the results of Step 4 to write the solution for the production matrix X and interpret the results in terms of units of production of each industry that are necessary to satisfy the internal and external demands of the system.

Note that the numbers in each of the columns of the input-output matrix in the previous example sum to less than one. This means the economic model is an **open system** and gasoline, electricity, and coal are not all of the industries in the economic system. For instance, labor is usually omitted in an open system.

$$A = \begin{array}{c} \\ \\ \\ \end{array} \overset{\begin{array}{ccc} G & E & C \end{array}}{\begin{bmatrix} 0.2 & 0.4 & 0.4 \\ 0.4 & 0.2 & 0.2 \\ 0.0 & 0.2 & 0.2 \end{bmatrix}} \begin{array}{l} \text{Gasoline} \\ \text{Electricity} \\ \text{Coal} \end{array}$$

We can inspect the values in the first column of the input-output matrix A and see that their sum is $0.2 + 0.4 + 0.0 = 0.6$. This means that for every dollar's worth of gasoline produced, 60 cents were required for its production and 40 cents' worth is available for external use. However, if each of the columns of an input-output matrix sums to one, this indicates that all of the production is used to keep the system operating and nothing is available for consumption outside the system. In this case, it is called a **closed system** and there is no surplus. As you might imagine, a closed system, or *closed economy*, in practice is difficult to maintain. It is rare that a community and its industries are completely self-sufficient with no need or desire to import or export goods and services.

Closed Systems

When finding the production matrix for a closed system, we will need to adjust the steps developed in Example 1 and outlined in the procedure table. Example 2 illustrates why this is necessary.

Example 2: Closed Economy

Suppose a closed economy consists of three industries—with x_1 equal to the value of shipping output, x_2 equal to the value of manufacturing output, and x_3 equal to the government budget—and has the following input-output matrix.

$$A = \overset{\begin{array}{ccc} S & M & G \end{array}}{\begin{bmatrix} 0.4 & 0.2 & 0.2 \\ 0.2 & 0.3 & 0.3 \\ 0.4 & 0.5 & 0.5 \end{bmatrix}} \begin{array}{l} \text{Shipping} \\ \text{Manufacturing} \\ \text{Government} \end{array}$$

Find each industry's production (that is, find the shipping output x_1, manufacturing output x_2, and government budget x_3).

Solution

The production matrix X is $\begin{bmatrix} x_1 \\ x_2 \\ x_3 \end{bmatrix}$. The input-output matrix A is given as $\begin{bmatrix} 0.4 & 0.2 & 0.2 \\ 0.2 & 0.3 & 0.3 \\ 0.4 & 0.5 & 0.5 \end{bmatrix}$.

The external demand matrix E is $\begin{bmatrix} 0 \\ 0 \\ 0 \end{bmatrix}$ because there is no external demand. The Leontief input-output model is $X = AX + E$ or $(I - A)X = E$. Making the appropriate substitutions, we can now solve for the production matrix X.

$$\left(\begin{bmatrix} 1 & 0 & 0 \\ 0 & 1 & 0 \\ 0 & 0 & 1 \end{bmatrix} - \begin{bmatrix} 0.4 & 0.2 & 0.2 \\ 0.2 & 0.3 & 0.3 \\ 0.4 & 0.5 & 0.5 \end{bmatrix} \right) X = \begin{bmatrix} 0 \\ 0 \\ 0 \end{bmatrix}$$

$$\begin{bmatrix} 0.6 & -0.2 & -0.2 \\ -0.2 & 0.7 & -0.3 \\ -0.4 & -0.5 & 0.5 \end{bmatrix} \begin{bmatrix} x_1 \\ x_2 \\ x_3 \end{bmatrix} = \begin{bmatrix} 0 \\ 0 \\ 0 \end{bmatrix}$$

We can now write the following augmented matrix for this system.

$$\left[\begin{array}{ccc|c} 0.6 & -0.2 & -0.2 & 0 \\ -0.2 & 0.7 & -0.3 & 0 \\ -0.4 & -0.5 & 0.5 & 0 \end{array} \right]$$

Next, we write the system in reduced row echelon form. This step may be done manually but is typically accomplished with the aid of a calculator with matrix functionality.

$$\left[\begin{array}{ccc|c} 1 & 0 & -\dfrac{10}{19} & 0 \\ 0 & 1 & -\dfrac{11}{19} & 0 \\ 0 & 0 & 0 & 0 \end{array} \right]$$

From this matrix, we get the following equations.

$$x_1 - \frac{10}{19}x_3 = 0 \quad \Rightarrow \quad x_1 = \frac{10}{19}x_3$$

$$x_2 - \frac{11}{19}x_3 = 0 \quad \Rightarrow \quad x_2 = \frac{11}{19}x_3$$

Since x_3 is equal to the government budget, this system is satisfied when shipping output is $\dfrac{10}{19}$ of the government budget and manufacturing output is $\dfrac{11}{19}$ of the government budget. In other words, for every \$19 in the government budget, shipping needs to produce \$10 and manufacturing needs to produce \$11.

In Example 2, the government budget was undetermined. It will always be the case that in a closed system, one of the unknown values in the production matrix X will be left unknown. Mathematically, this means that there is no unique solution to the system and the matrix $(I - A)^{-1}$ does not exist and therefore cannot be used to find the solution.

Our next and final example illustrates how to find the external demand that can be met given the total production capacity of a system. We will use fractional internal demands in this example in order to illustrate that we can interchange fractions and decimals depending on which may be more convenient or appropriate. The units may be interpreted as something other than dollars.

Example 3: External Demand

Suppose each of the industries A, B, and C requires the following input in order to produce one unit in output.

A requires nothing from itself, $\dfrac{1}{2}$ unit from B, and $\dfrac{1}{4}$ unit from C.

B requires $\dfrac{1}{4}$ unit from A, none from itself, and $\dfrac{1}{4}$ unit from B.

C requires $\dfrac{1}{3}$ unit from A, $\dfrac{1}{4}$ unit from B, and none from itself.

So, the input-output matrix A is given as follows.

$$A = \begin{bmatrix} 0 & \dfrac{1}{4} & \dfrac{1}{3} \\ \dfrac{1}{2} & 0 & \dfrac{1}{4} \\ \dfrac{1}{4} & \dfrac{1}{4} & 0 \end{bmatrix}$$

Suppose further that we know the total production capacity of each industry is as follows.

A: 60 units
B: 52 units
C: 48 units

This determines the production matrix X.

$$X = \begin{bmatrix} 60 \\ 52 \\ 48 \end{bmatrix}$$

Our goal is to find the external demand matrix E or, equivalently, the surplus units available from each industry.

Since the Leontief input-output model is $(I - A)X = E$, to find the external demand, we should first calculate $I - A$ and then multiply the result by X.

$$I - A = \begin{bmatrix} 1 & 0 & 0 \\ 0 & 1 & 0 \\ 0 & 0 & 1 \end{bmatrix} - \begin{bmatrix} 0 & \dfrac{1}{4} & \dfrac{1}{3} \\ \dfrac{1}{2} & 0 & \dfrac{1}{4} \\ \dfrac{1}{4} & \dfrac{1}{4} & 0 \end{bmatrix} = \begin{bmatrix} 1 & -\dfrac{1}{4} & -\dfrac{1}{3} \\ -\dfrac{1}{2} & 1 & -\dfrac{1}{4} \\ -\dfrac{1}{4} & -\dfrac{1}{4} & 1 \end{bmatrix}$$

$$E = (I - A)X = \begin{bmatrix} 1 & -\dfrac{1}{4} & -\dfrac{1}{3} \\ -\dfrac{1}{2} & 1 & -\dfrac{1}{4} \\ -\dfrac{1}{4} & -\dfrac{1}{4} & 1 \end{bmatrix} \begin{bmatrix} 60 \\ 52 \\ 48 \end{bmatrix} = \begin{bmatrix} 31 \\ 10 \\ 20 \end{bmatrix}$$

We interpret this result to mean that at peak production, A can produce 31 units, B can produce 10 units, and C can produce 20 units for use outside the system.

Our examples had only three industries, but real-world economic systems may consist of 100 or more products and services. For complex systems such as these, we would need the help of a computer algebra system to solve for the production matrix. We will close this lesson with a list of the biggest industries in the United States and some of their products according to *WorldAtlas.com*.

1. Real Estate, Renting, Leasing (apartments, single-family homes, condominiums)

2. State and Local Government (House of Representatives, Senate, mayor, police and fire departments)

3. Finance and Insurance (insurance carriers, credit intermediation, commodity contracts, trusts)

4. Health and Social Care (curative, rehabilitative, preventative)

5. Durable Manufacturing (computers, automobiles, firearms, sports equipment, house appliances, aircraft)

6. Retail Trade (Walmart, Amazon, eBay)

7. Wholesale Trade (Uline, Allied Building Products, Pfizer)

8. Nondurable Manufacturing (gasoline, electricity)

9. Federal Government (Social Security, Internal Revenue)

10. Information (media, data processing, telephone companies)

11. Arts and Entertainment (museums, theatres, television)

12. Construction (dwellings, roads and bridges, public and private businesses)

13. Waste Services (sewage treatment, water filtration and storage, landfills)

(Others listed are Other Services, Utilities, Mining, Corporate Management, Education Services, and Agriculture.)

6.6 EXERCISES

💡 PRACTICE

1. The input-output matrix A and corresponding demand matrix E for a local economy are given.

$$A = \begin{bmatrix} 0.7 & 0.1 \\ 0.2 & 0.6 \end{bmatrix} \begin{matrix} \text{Agriculture} \\ \text{Manufacturing} \end{matrix}, \quad E = \begin{bmatrix} 8000 \\ 5000 \end{bmatrix}$$

   (columns labeled A   M)

   a. Identify the inputs needed from each industry to produce one unit of output from manufacturing.
   b. Identify the inputs needed from each industry to produce 3 units of output from agriculture.
   c. Find $I - A$ and $(I - A)^{-1}$.
   d. Find the production matrix X.

2. The input-output matrix A and corresponding demand matrix E for a local real estate market are given.

$$A = \begin{bmatrix} 0.4 & 0.3 \\ 0.3 & 0.1 \end{bmatrix} \begin{matrix} \text{Apartments} \\ \text{Single-Family Homes} \end{matrix}, \quad E = \begin{bmatrix} 900 \\ 600 \end{bmatrix}$$

   (columns labeled A   S)

   a. Identify the inputs needed from each industry to produce one unit of output from single-family homes.
   b. Identify the inputs needed from each industry to produce 4 units of output from apartments.
   c. Find $I - A$ and $(I - A)^{-1}$.
   d. Find the production matrix X.

3. The input-output matrix A and corresponding demand matrix E for a local economy market are given.

$$A = \begin{bmatrix} 0.1 & 0.2 & 0.3 \\ 0.1 & 0.2 & 0.3 \\ 0.1 & 0.2 & 0.3 \end{bmatrix} \begin{matrix} \text{Agriculture} \\ \text{Mining} \\ \text{Manufacturing} \end{matrix}, \quad E = \begin{bmatrix} 400 \\ 1200 \\ 800 \end{bmatrix}$$

   (columns labeled A   Mi   Ma)

   a. Identify the inputs needed from each industry to produce one unit of output from agriculture.
   b. Identify the inputs needed from each industry to produce 2 units of output from mining.
   c. Find $I - A$ and $(I - A)^{-1}$.
   d. Find the production matrix X.

4. The input-output matrix A and corresponding demand matrix E for a local real estate market are given.

$$A = \begin{bmatrix} 0.2 & 0.3 & 0.3 \\ 0.1 & 0.2 & 0.2 \\ 0.1 & 0.1 & 0.1 \end{bmatrix} \begin{matrix} \text{Single-Family} \\ \text{Multi-Family} \\ \text{Rental} \end{matrix} \quad , \quad E = \begin{bmatrix} 600 \\ 200 \\ 400 \end{bmatrix}$$

with column headings S M R.

a. Identify the inputs needed from each industry to produce one unit of output from rentals.
b. Identify the inputs needed from each industry to produce 3 units of output from single family.
c. Find $I - A$ and $(I - A)^{-1}$.
d. Find the production matrix X.

Given the following input-output matrix A and corresponding production matrix X, find the external demand matrix E.

5. $A = \begin{bmatrix} 0.3 & 0.1 \\ 0.5 & 0.5 \end{bmatrix} \begin{matrix} \text{Hospitality} \\ \text{Health Care} \end{matrix}$, $X = \begin{bmatrix} 1700 \\ 2900 \end{bmatrix}$

with column headings H HC.

6. $A = \begin{bmatrix} 0.4 & 0.2 \\ 0.5 & 0.5 \end{bmatrix} \begin{matrix} \text{Construction} \\ \text{Manufacturing} \end{matrix}$, $X = \begin{bmatrix} 2200 \\ 2600 \end{bmatrix}$

with column headings C M.

7. $A = \begin{bmatrix} 0.4 & 0.2 & 0 \\ 0.3 & 0.3 & 0.3 \\ 0.2 & 0.1 & 0.3 \end{bmatrix} \begin{matrix} \text{Coal} \\ \text{Oil} \\ \text{Natural Gas} \end{matrix}$, $X = \begin{bmatrix} 800 \\ 1200 \\ 600 \end{bmatrix}$

with column headings C O G.

8. $A = \begin{bmatrix} 0.2 & 0.3 & 0.1 \\ 0.1 & 0.4 & 0.2 \\ 0.6 & 0.2 & 0.3 \end{bmatrix} \begin{matrix} \text{Computers} \\ \text{Automobiles} \\ \text{Aircraft} \end{matrix}$, $X = \begin{bmatrix} 800 \\ 600 \\ 1000 \end{bmatrix}$

with column headings C Au Ai.

🚀 APPLICATIONS

9. Suppose that in a certain local economy we have manufacturing and agriculture industries. To produce one dollar in output, each industry needs the following input.
   - The manufacturing industry requires $0.10 from itself and $0.30 from agriculture.

   - The agriculture industry requires $0.30 from manufacturing and $0.40 from itself.

   Suppose further that in addition to the internal demand, there is a surplus demand from outside of the industries for $300 in manufacturing and $600 in agriculture. Solve for the production necessary to meet these internal and surplus demands.

10. Suppose that in a certain local economy we have natural gas and coal industries. To produce one dollar in output, each industry needs the following input.
    - The natural gas industry requires $0.40 from itself and $0.30 from coal.

    - The coal industry requires $0.20 from natural gas and $0.40 from itself.

    Suppose further that in addition to the internal demand, there is a surplus demand from outside of the industries for $600 in natural gas and $900 in coal. Solve for the production necessary to meet these internal and surplus demands.

11. Suppose that in a certain local economy we have computer manufacturing and automobile manufacturing industries. To produce one dollar in output, each industry needs the following input.
    - The computer manufacturing industry requires $0.10 from itself and $0.20 from automobile manufacturing.

    - The automobile manufacturing industry requires $0.20 from computer manufacturing and $0.40 from itself.

    Suppose further that in addition to the internal demand, there is a surplus demand from outside of the industries for $400 in computer manufacturing and $800 in automobile manufacturing. Solve for the production necessary to meet these internal and surplus demands.

12. Suppose that in a certain local economy we have coal, oil, and natural gas industries. To produce one dollar in output, each industry needs the following input.
    - The coal industry requires $0.10 from itself, $0.20 from oil, and $0.30 from natural gas.

    - The oil industry requires no input from coal, $0.40 from itself, and $0.20 from natural gas.

    - The natural gas industry requires $0.10 from coal, $0.20 from oil, and $0.30 from itself.

    Suppose further that in addition to the internal demand, there is a surplus demand from outside of the industries for $10,000 in coal, $10,000 in oil, and $10,000 in natural gas. Solve for the production necessary to meet these internal and surplus demands.

13. Suppose that in a certain local entertainment economy we have museums, theaters, and sporting events industries. To produce one dollar in output, each industry needs the following input.

    • The museum industry requires $0.40 from itself, $0.30 from theater, and $0.20 from sporting events.

    • The theater industry requires $0.20 from museums, $0.40 from itself, and $0.30 from sporting events.

    • The sporting events industry requires $0.20 from museums, $0.40 from theater, and $0.30 from itself.

    Suppose further that in addition to the internal demand, there is a surplus demand from outside of the industries for $500 in museums, $750 in theaters, and $400 in sporting events. Solve for the production necessary to meet these internal and surplus demands.

14. Suppose that in a certain local information economy we have media, data processing, and telephone company industries. To produce one dollar in output, each industry needs the following input.

    • The media industry requires $0.30 from itself, $0.20 from data processing, and no input from telephone companies.

    • The data processing industry requires $0.40 from media, $0.20 from itself, and $0.10 from telephone companies.

    • The telephone company industry requires $0.10 from media, $0.10 from data processing, and $0.20 from itself.

    Suppose further that in addition to the internal demand, there is a surplus demand from outside of the industries for $300 in media, $300 in data processing, and $300 in telephone companies. Solve for the production necessary to meet these internal and surplus demands.

15. Suppose that in a certain local health economy we have curative, rehabilitative, and preventative industries. To produce one dollar in output, each industry needs the following input.

    • The curative industry requires $0.20 from itself, $0.10 from rehabilitative, and $0.20 from preventative.

    • The rehabilitative industry requires $0.20 from curative, $0.30 from itself, and $0.20 from preventative.

    • The preventative industry requires $0.20 from curative, $0.40 from rehabilitative, and $0.20 from itself.

    Suppose further that in addition to the internal demand, there is a surplus demand from outside of the industries for $500 in curative, $1500 in rehabilitative, and $1000 in preventative. Solve for the production necessary to meet these internal and surplus demands.

16. Suppose that in a certain local economy we have coal, gasoline, electric, and natural gas industries. To produce one dollar in output, each industry needs the following input.
    - The coal industry requires $0.10 from itself, $0.20 from gasoline, no input from electric, and $0.20 from natural gas.

    - The gasoline industry requires $0.40 from coal, $0.10 from itself, no input from electric, and $0.10 from natural gas.

    - The electric industry requires no input from coal, $0.10 from gasoline, $0.20 from itself, and $0.30 from natural gas.

    - The natural gas industry requires $0.20 from coal, $0.10 from gasoline, no input from electric, and $0.10 from itself.

    Suppose further that in addition to the internal demand, there is a surplus demand from outside of the industries for $300 in coal, $600 in gasoline, $600 in electric, and $900 in natural gas. Solve for the production necessary to meet these internal and surplus demands.

17. Suppose that in a certain local economy we have manufacturing, agricultural, health, and energy industries. To produce one dollar in output, each industry needs the following input.
    - The manufacturing industry requires $0.20 from itself, $0.20 from agriculture, $0.20 from health, and $0.20 from energy.

    - The agricultural industry requires $0.20 from manufacturing, $0.20 from itself, $0.20 from health, and $0.20 from energy.

    - The health industry requires $0.20 from manufacturing, $0.20 from agriculture, $0.20 from itself, and $0.20 from energy.

    - The energy industry requires $0.10 from manufacturing, no input from agriculture, $0.30 from health, and $0.40 from itself.

    Suppose further that in addition to the internal demand, there is a surplus demand from outside of the industries for $1000 in manufacturing, $500 in agriculture, $500 in health, and $500 in energy. Solve for the production necessary to meet these internal and surplus demands.

18. Suppose that in a certain local economy we have durable manufacturing and nondurable manufacturing industries. To produce one dollar in output, each industry needs the following input.
    - The durable manufacturing industry requires $0.10 from itself and $0.30 from nondurable manufacturing.

    - The nondurable manufacturing industry requires $0.30 from durable manufacturing and $0.60 from itself.

    Suppose further that the total production capacity of durable manufacturing is $500 and of nondurable manufacturing is $800. Find the external demand.

19. Suppose that in a certain local economy we have agriculture, manufacturing, and energy industries. To produce one dollar in output, each industry needs the following input.

- The agriculture industry requires $0.30 from itself, no input from manufacturing, and $0.20 from energy.

- The manufacturing industry requires $0.10 from agriculture, $0.40 from itself, and $0.20 from energy.

- The energy industry requires $0.30 from agriculture, $0.40 from manufacturing, and $0.20 from itself.

Suppose further that the total production capacity of agriculture is $800, of manufacturing is $800, and of energy is $1000. Find the external demand.

20. Suppose that in a certain local economy we have coal, gasoline, electric, and natural gas industries. To produce one dollar in output, each industry needs the following input.

- The coal industry requires $0.30 from itself, $0.20 from gasoline, $0.10 from electric, and $0.20 from natural gas.

- The gasoline industry requires $0.20 from coal, $0.10 from itself, $0.10 from electric, and $0.10 from natural gas.

- The electric industry requires no input from coal, $0.10 from gasoline, $0.20 from itself, and $0.30 from natural gas.

- The natural gas industry requires $0.20 from coal, $0.10 from gasoline, no input from electric, and $0.10 from natural gas.

Suppose further that the total production capacity of coal is $800, of gasoline is $400, of electric is $300, and of natural gas is $1000. Find the external demand.

21. Suppose a closed economy consists of two industries—with x_1 equal to the value of the agriculture output, and x_2 equal to the value of the manufacturing output— and has the following input-output matrix.

$$\begin{array}{cc} \text{A} & \text{M} \end{array}$$

$$A = \begin{bmatrix} 0.2 & 0.3 \\ 0.8 & 0.7 \end{bmatrix} \begin{array}{l} \text{Agriculture} \\ \text{Manufacturing} \end{array}$$

a. Find the productions (outputs of agriculture and manufacturing) for each industry x_1 and x_2.

b. For every $8 produced by the manufacturing industry, how many dollars' worth of production is needed from the agricultural industry?

22. Suppose a closed economy consists of two industries—with x_1 equal to the value of the energy output, and x_2 equal to the value of the real estate output—and has the following input-output matrix.

$$A = \begin{matrix} & E & R \\ & \begin{bmatrix} 0.4 & 0.2 \\ 0.6 & 0.8 \end{bmatrix} & \begin{matrix} \text{Energy} \\ \text{Real Estate} \end{matrix} \end{matrix}$$

a. Find the productions (outputs of energy and real estate) for each industry x_1 and x_2.

b. For every \$3 produced by the real estate industry, how many dollars' worth of production is needed from the energy industry?

23. Suppose a closed economy consists of three industries—with x_1 equal to the value of the agriculture output, x_2 equal to the value of the manufacturing output, and x_3 equal to the value of the energy output—and has the following input-output matrix.

$$A = \begin{matrix} & A & M & E \\ & \begin{bmatrix} 0.3 & 0.1 & 0.3 \\ 0.5 & 0.4 & 0.4 \\ 0.2 & 0.5 & 0.3 \end{bmatrix} & & \begin{matrix} \text{Agriculture} \\ \text{Manufacturing} \\ \text{Energy} \end{matrix} \end{matrix}$$

a. Find the productions (outputs of agriculture, manufacturing, and energy) for each industry x_1, x_2, and x_3.

b. For every \$33 produced by the energy industry, how many dollars' worth of production is needed from the agricultural industry and how much is needed from the manufacturing industry?

✏ WRITING & THINKING

24. Use the given input-output matrix to answer the questions that follow.

$$\begin{matrix} & C & G & E & N \\ & \begin{bmatrix} 0.3 & 0.2 & 0.1 & 0.2 \\ 0.2 & 0.1 & 0.1 & 0.1 \\ 0.1 & 0.1 & 0.2 & 0.1 \\ 0.2 & 0.1 & 0.4 & 0 \end{bmatrix} & & & \begin{matrix} \text{Coal} \\ \text{Gasoline} \\ \text{Electric} \\ \text{Natural Gas} \end{matrix} \end{matrix}$$

a. Which industry is most dependent on its own production for its operation?

b. Which industry is least dependent on its own production?

c. Which industry is most dependent on the natural gas industry?

d. Which industry would be most affected by a rise in the cost of coal?

7

Chapter 7

INEQUALITIES AND LINEAR PROGRAMMING

7.1 LINEAR INEQUALITIES IN TWO VARIABLES

■ TOPICS

- Solving Linear Inequalities in Two Variables
- Solving Linear Inequalities Joined by "And" or "Or"
- Systems of Linear Inequalities
- Graphing Inequalities Using Technology

Solving Linear Inequalities in Two Variables

Just as in one variable, linear inequalities in two variables have much in common with linear equations in two variables. The first similarity is in the definition: if the equality symbol in a linear equation in two variables is replaced with $<$, $>$, $\leq$, or $\geq$, the result is a **linear inequality in two variables**. In other words, a linear inequality in the two variables x and y is an inequality that can be written in the form

$$ax + by < c, \quad ax + by > c,$$

$$ax + by \leq c, \quad ax + by \geq c,$$

where a, b, and c are constants and a and b are not both 0.

Another similarity lies in the solution process. The solution set of a linear inequality in two variables consists of all the ordered pairs in the Cartesian plane that lie on one side of a line in the plane, possibly including those points on the line. The first step in solving such an inequality then is to identify and graph this line. This line is simply the graph of the equation that results from replacing the inequality symbol in the original problem with an equality symbol.

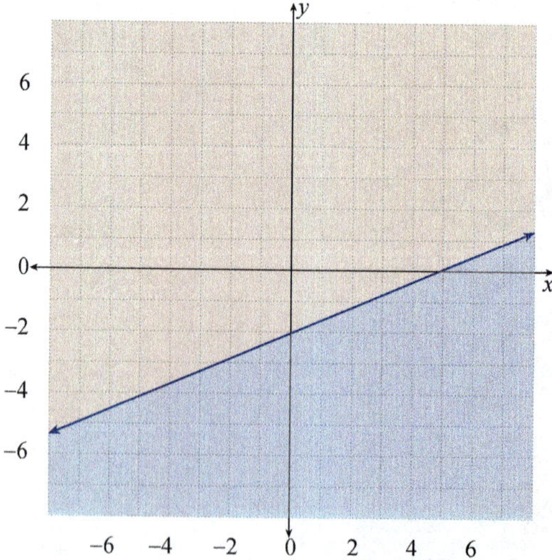

FIGURE 1: Dividing the Plane

Any line divides the plane into two **half-planes**; given a linear inequality, all of the points in one of the two half-planes will solve the inequality. In addition, the points on the **boundary line** also solve the inequality if the inequality symbol is $\leq$ or $\geq$, and this fact must be denoted graphically.

Solving Linear Inequalities in Two Variables

Step 1: Graph the line that results from replacing the inequality symbol with =.

Step 2: Make the line solid if the inequality symbol is $\leq$ or $\geq$ (not strict) and dashed if the symbol is $<$ or $>$ (strict). A solid line indicates that points on the line are included in the solution set while a dashed line indicates that points on the line are excluded from the solution set.

Step 3: Determine which of the half-planes defined by the boundary line solves the inequality by substituting a **test point** from one of the two half-planes into the inequality. If the resulting numerical statement is true, all the points in the same half-plane as the test point solve the inequality. Otherwise, the points in the other half-plane solve the inequality. Shade in the half-plane that solves the inequality.

Example 1: Solving Linear Inequalities

Solve the following linear inequalities by graphing their solution sets.

a. $3x + 2y < 12$ 　　　　**b.** $x - y \leq 0$ 　　　　**c.** $x > 3$

Solution

NOTE

Whenever possible, use the origin as a test point, since it makes for a very simple calculation.

a.
$$3x + 2y = 12$$
$$3(0) + 2y = 12 \qquad 3x + 2(0) = 12$$
$$y = 6 \qquad\qquad x = 4$$

To graph the boundary line, replace the inequality symbol with an equal sign, then find the x- and y-intercepts.

x-intercept of boundary: $(4, 0)$

y-intercept of boundary: $(0, 6)$

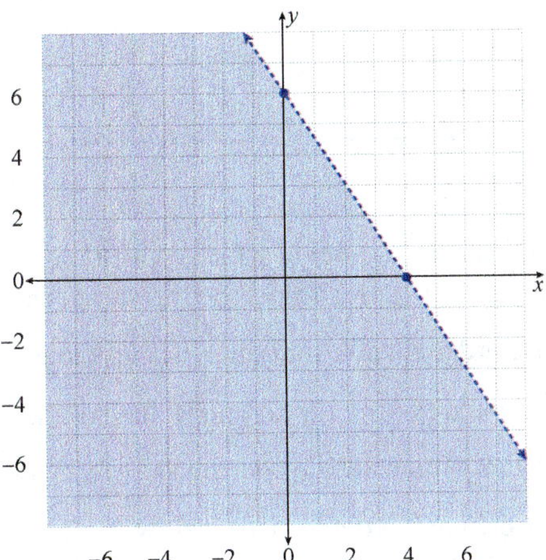

The graph of the equation is drawn with a dashed line since the inequality is strict.

The origin $(0, 0)$ clearly lies on one side of the boundary line, so we can use it as our test point.

When both x and y in the inequality are replaced with 0, we obtain the true statement $0 < 12$. This tells us to shade the half-plane that contains $(0, 0)$.

b. $x - y \le 0$

$x - y = 0$

$y = x$

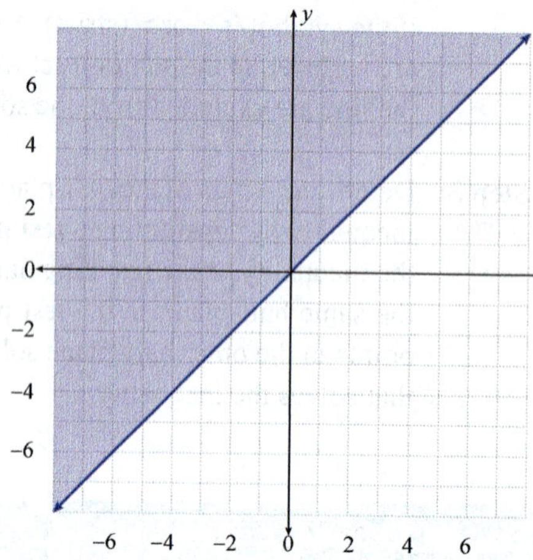

In slope-intercept form, the equation for the boundary is $y = x$, a line with a y-intercept of $(0, 0)$ and a slope of 1.

Graph the boundary with a solid line since the inequality is not strict. Thus, points on the boundary do solve the inequality.

This time, the origin cannot be used as a test point, since it lies directly on the boundary line. The point $(1, -1)$ lies below the boundary, and if we substitute $x = 1$ and $y = -1$ into the inequality $x - y \le 0$, we obtain the false statement $2 \le 0$. Thus, we shade the half-plane that does not contain $(1, -1)$.

c. $x > 3$

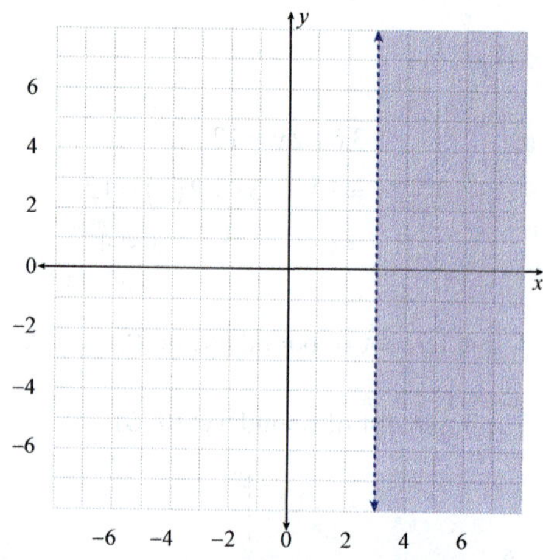

In this example, the boundary line is a vertical line with x-intercept $(3, 0)$.

Since the inequality is strict, we graph the boundary line as a dashed line.

The origin can be used as a convenient test point, and results in the false statement $0 > 3$, leading us to shade the half-plane to the right of the boundary.

Another way to decide which half-plane to shade is to consider each inequality after solving for y (or for x if y is not present). The inequality in Example 1a is satisfied by all those ordered pairs for which y is less than $-\dfrac{3}{2}x + 6$. That is, if the line $y = -\dfrac{3}{2}x + 6$ is graphed (with a dashed line), the solution set of the inequality consists of all those ordered pairs whose y-coordinate is *less than* (lies below) $-\dfrac{3}{2}$ times the x-coordinate, plus 6 (those ordered pairs below the boundary line). Similarly, in Example 1b, we shade in those ordered pairs for which the y-coordinate is *greater than or equal to* the x-coordinate. Finally, in Example 1c, we shade in the ordered pairs for which the x-coordinate is *greater than* (lies to the right of) 3.

Solving Linear Inequalities Joined by "And" or "Or"

Often, we need to identify those points in the Cartesian plane that satisfy more than one inequality. In the problems that we will examine, we will identify the portion of the plane that satisfies two or more linear inequalities joined by the word "and" or the word "or".

Earlier, we defined the union of two sets A and B, denoted $A \cup B$, as the set containing all elements that are in set A **or** set B, and we defined the intersection of two sets A and B, denoted $A \cap B$, as the set containing all elements that are in both A **and** B.

If we let A denote the portion of the plane that solves one inequality and B the portion of the plane that solves a second inequality, then $A \cup B$ represents the solution set of the two inequalities joined by the word "or" and $A \cap B$ represents the solution set of the two inequalities joined by the word "and".

Visually, we can think of the union as the combining of two regions, while we can think of the intersection as the overlaps between two regions. Keep in mind that $A \cup B$ *contains* both of the sets A and B (and so is at least as large as either one individually), while $A \cap B$ is *contained in* both A and B (and so is no larger than either individual set).

To find the solution sets in the following problems, we will solve each linear inequality individually and then form the union or the intersection of the individual solutions, as appropriate.

Example 2: Solving Linear Inequalities

Graph the solution sets that satisfy the following inequalities.

a. $5x - 2y < 10$ and $y \leq x$ **b.** $x + y < 4$ or $x \geq 4$

Solution

a. $5x - 2y < 10$ and $y \leq x$

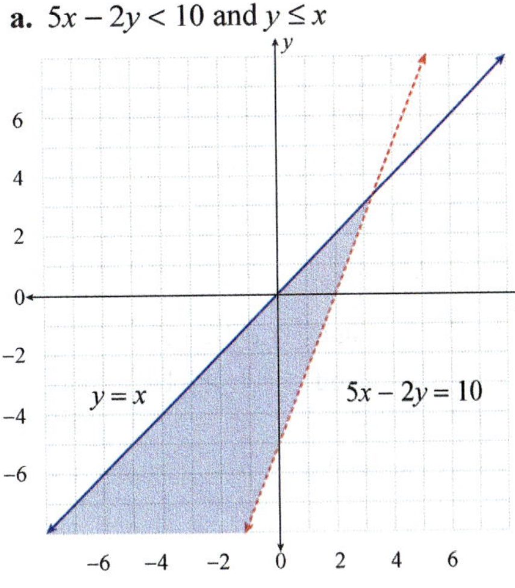

First graph the individual inequalities. We graph the line $5x - 2y = 10$ with a dashed line and the line $y = x$ with a solid line.

We then note that the half-plane lying above $5x - 2y = 10$ solves the first inequality (verify this by using the test point $(0, 0)$) and that the half-plane lying below $y = x$ solves the second inequality (this can be verified using the test point $(1, -1)$).

Since the two inequalities are joined by the word "and", we shade in the intersection of the two half-planes.

b. $x + y < 4$ or $x \geq 4$

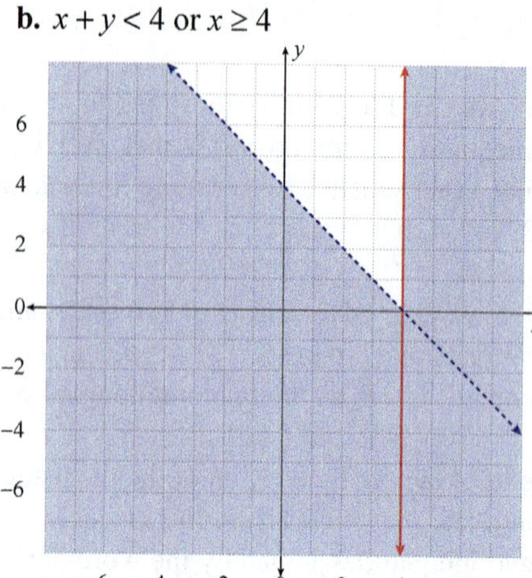

Again, begin by solving each inequality separately. We graph the line $x + y = 4$ with a dashed line, and the line $x = 4$ with a solid line, and note that the half-plane below $x + y = 4$ solves the first inequality and that the half-plane to the right of $x = 4$ solves the second.

Since the two inequalities are joined by "or", we shade the *union* of the two half-planes. Note that this results in a larger shaded region than either of the two individual solutions. Any ordered pair in the shaded region will satisfy one or both of the inequalities.

Unions and intersections of regions of the plane can also arise when solving inequalities involving absolute values. Earlier, we saw that an inequality of the form $|x| < a$ can be rewritten as the joint condition $x > -a$ and $x < a$, which corresponds to the intersection of two sets. Similarly, an inequality of the form $|x| > a$ can be rewritten as $x < -a$ or $x > a$, a union of two sets.

In the Cartesian plane, solutions of such inequalities will be, respectively, intersections and unions of half-planes. The next example demonstrates how such solution sets can be found.

Example 3: Solving Absolute Value Linear Inequalities

Graph the solution set in $\mathbb{R}^2$ that satisfies the joint conditions $|x - 3| > 1$ and $|y - 2| \leq 3$.

Solution

First, we need to find the solution to each individual inequality.

$$|x - 3| > 1$$

$$x - 3 > 1 \quad \text{or} \quad x - 3 < -1$$

$$x > 4 \quad \text{or} \quad x < 2$$

Begin by rewriting the absolute value inequality as two linear inequalities.

Simplify. We have a union of solutions.

$$|y - 2| \leq 3$$

$$y - 2 \leq 3 \quad \text{and} \quad y - 2 \geq -3$$

$$y \leq 5 \quad \text{and} \quad y \geq -1$$

Again, rewrite the inequality without absolute value signs.

This time, we have an intersection of solutions.

Now we can graph the solutions to the individual conditions:

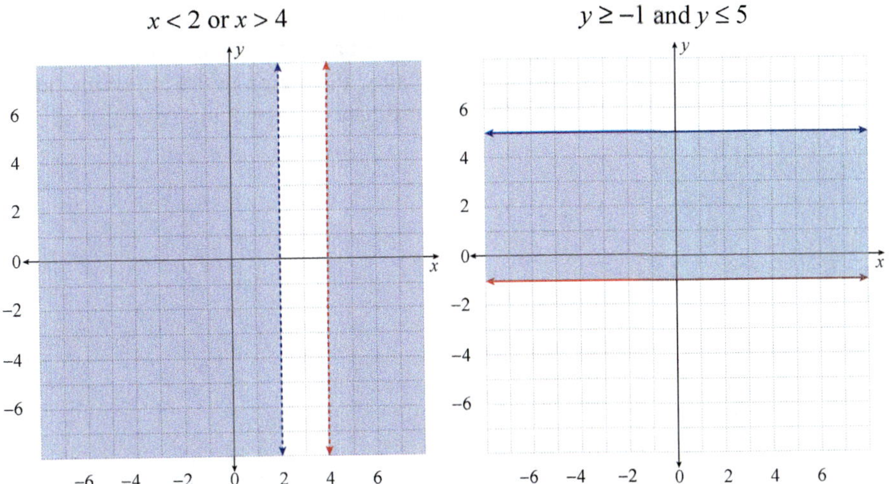

We now intersect the solution sets to obtain the final answer:

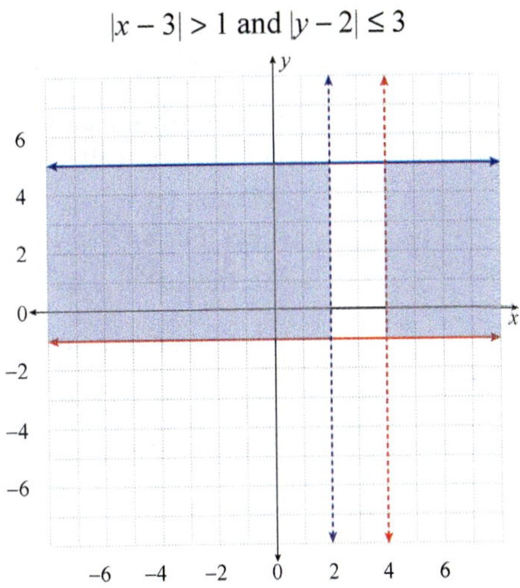

Systems of Linear Inequalities

Just as the solution set of a system of linear equations consists of all points that solve each of the equations simultaneously, the solution set of a system of linear inequalities consists of all points that solve each of the inequalities simultaneously. The process is similar to graphing the solutions of linear inequalities joined by the word "and".

Example 4: Graphing the Solution of a System of Linear Inequalities

Graph the solution set to the following system of linear inequalities.

$$\begin{cases} x < 2 \\ y \leq \dfrac{x}{2} + 1 \\ x + y \geq -3 \end{cases}$$

Solution

The boundary of the solution set of the first inequality is the vertical line $x = 2$, which we draw as a dashed line to reflect the fact that the inequality is strict. For this inequality, the region to the left of the boundary constitutes the solution set, since the x-coordinate of any solution point must be less than 2.

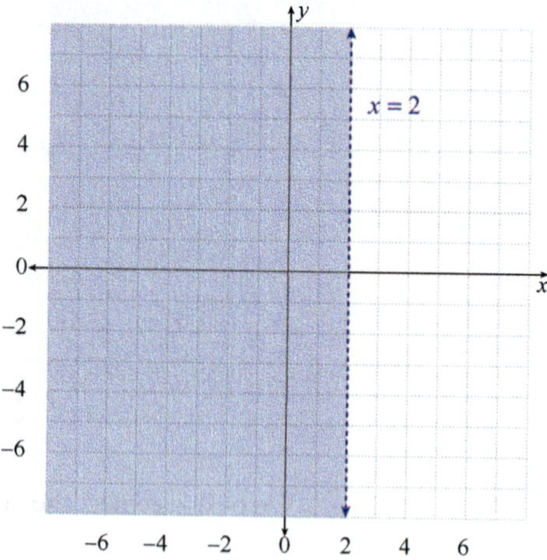

We similarly draw the lines $y = \dfrac{x}{2} + 1$ and $x + y = -3$ for the boundaries of the second and third inequalities, respectively, this time using solid lines as the inequalities are not strict. Recall that one way to determine which side of a boundary constitutes the solution set of a given inequality is to use a convenient test point on one side or the other. For both of these inequalities, the test point $(0, 0)$ is convenient. Alternatively, the inequality $y \le \dfrac{x}{2} + 1$ explicitly states that the y-coordinate of any solution point must lie on or below the line $y = \dfrac{x}{2} + 1$, and the inequality $x + y \ge -3$, when rewritten as $y \ge -x - 3$, says that the y-coordinate of any solution point must lie on or above the boundary line.

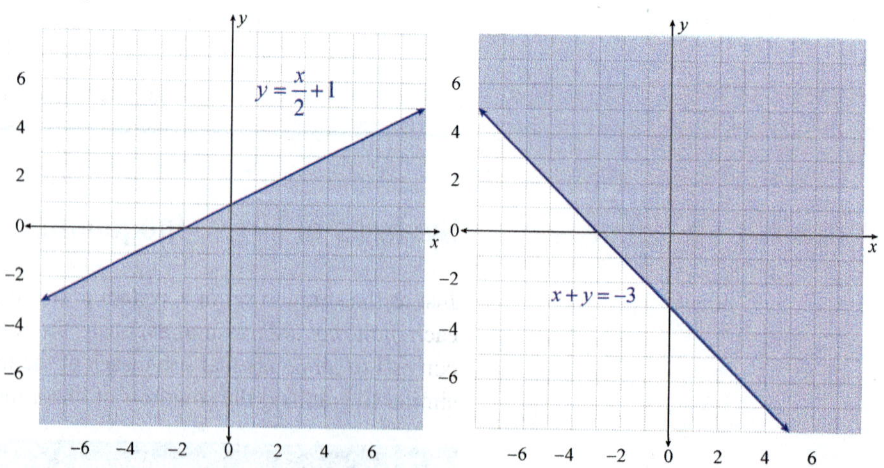

To obtain the region representing the solution set of the system of inequalities, we plot the intersection of the three previously graphed regions.

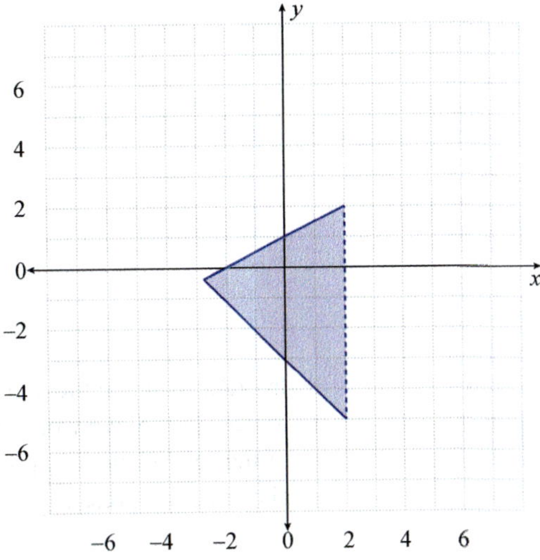

Graphing Inequalities Using Technology

Just as we can use a calculator to graph linear equations, we can also graph linear inequalities like $y < -3x + 5$. Press **Y=** and type in the right-hand side of the inequality. Pressing **graph** would display the line $y = -3x + 5$, which is the boundary line of the solution set. If we test the point $(0, 0)$, we find that it is a solution to the inequality, so we know to shade the half-plane below the boundary line. On the **Y=** screen, use the left arrow to move the cursor to the left of the Y1, where there is a small diagonal line. Press **enter** until that line appears with shading like in the screenshot.

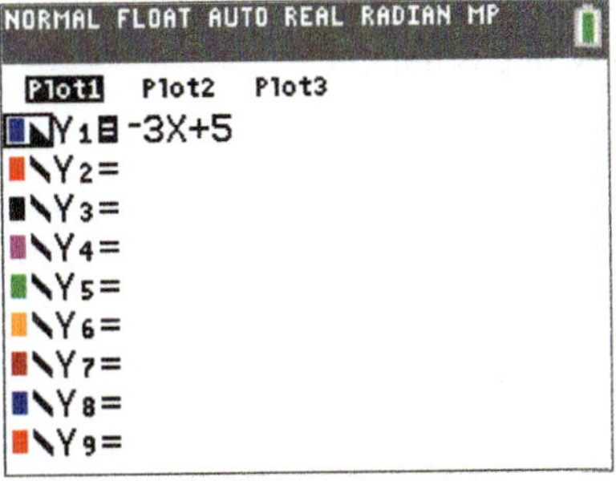

With that selection, pressing **graph** will display the boundary line with the half-plane below it shaded, which is the solution to the inequality.

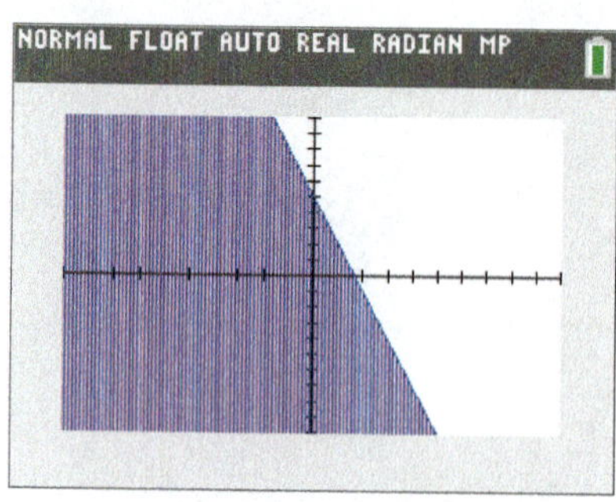

In order to shade the half-plane above the boundary line, which would be the solution to $y > -3x + 5$, we would need to return to the Y= screen, again place the cursor to the left of the Y1, press enter until the diagonal line appeared with shading above it, and press graph.

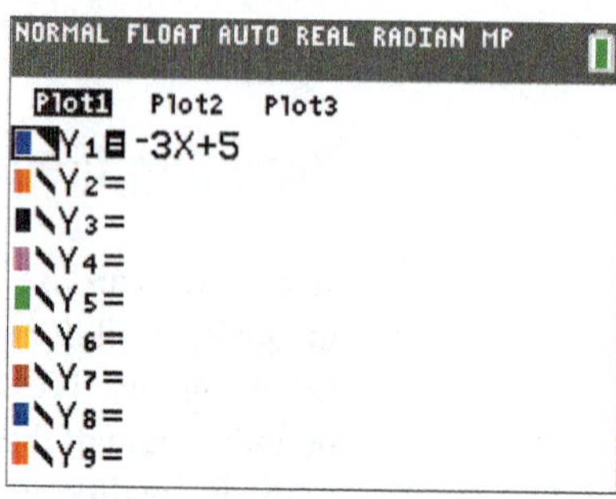

> **⚠ CAUTION**
>
> Notice that the boundary line appears in the calculator as a solid line. However, because we are finding the solution to a strict inequality, the boundary line is not included in the answer.

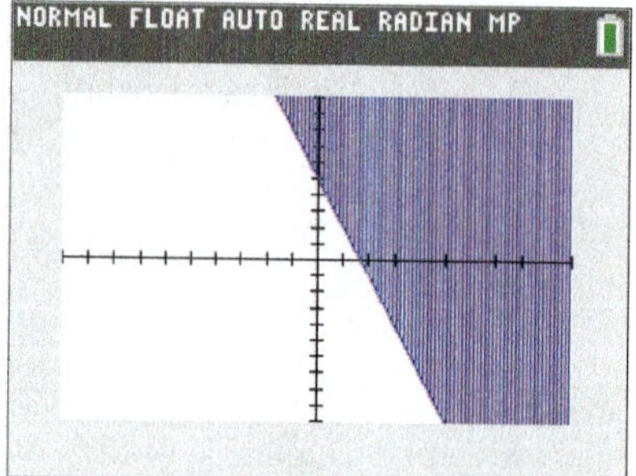

7.1 EXERCISES

PRACTICE

Solve the following linear inequalities by graphing their solution sets. See Example 1.

1. $x - 3y < 6$

2. $y < 2x - 1$

3. $x > \dfrac{3}{4}y$

4. $x - 3y \geq 6$

5. $3x - y \leq 2$

6. $\dfrac{2x - y}{4} > 1$

7. $y < -2$

8. $x + 1 \geq 0$

9. $x + y < 0$

10. $x + y > 0$

11. $-(y - x) > -\dfrac{5}{2} - y$

12. $-2y \leq -x + 4$

13. $5(y + 1) \geq -x$

14. $3x - 7y \geq 7(1 - y) + 2$

15. $x - y < 2y + 3$

Graph the solution sets that satisfy the following inequalities. See Example 2.

16. $y > -3x - 6$ or $y \leq 2x - 7$

17. $y \geq -2$ and $y > 1$

18. $y \geq -2x - 5$ and $y \leq -6x - 9$

19. $y \leq 4x + 4$ and $y > 7x + 7$

20. $x - 3y \geq 6$ and $y > -4$

21. $x - 3y \geq 6$ or $y > -4$

22. $3x - y \leq 2$ and $x + y > 0$

23. $x > 1$ and $y > 2$

24. $x > 1$ or $y > 2$

25. $x + y > -2$ and $x + y < 2$

26. $y > -2$ and $2y > -3x - 4$

27. $3y > x + 2$ or $4y \leq -x - 2$

28. $y \leq -x$ and $2y + 3x > -4$

29. $5x + 6y < -30$ and $x \geq 2$

30. $6y - 2x > -6$ or $y > 6$

31. $x > -3$ or $y \geq 4$

32. $-2y < -3x - 6$ or $-3y \geq -6x - 18$

33. $x < 6$ and $x \geq -5$

Graph the solution sets that satisfy the following linear absolute value inequalities. See Example 3.

34. $|x - 3| < 2$

35. $|x - 3| > 2$

36. $|3y - 1| \leq 2$

37. $|2x - 4| > 2$

38. $1 - |y + 3| < -1$

39. $|x + 1| < 2$ and $|y - 3| \leq 1$

40. $|x - 3| \geq 1$ or $|y - 2| \leq 1$

41. $|x - y| < 1$

42. $|x + y| \geq 1$

43. $|4x - 2y - 3| \leq 5$

44. $|2x - 3| \geq 1$ or $|2y + 3| \geq 1$

45. $|y - 3x| \leq 2$ and $|y| < 2$

Match the following inequalities to the appropriate graph.

46. $-8y + 5x \geq -8y + 5$

47. $x < -2$ and $x \geq -5$

48. $|-7x - 4y + 23| \leq 16$

49. $y \leq 3x - 6$ and $y > 2x - 4$

50. $|3y - 2x| > 17$ and $|y + 6| \geq 1$

51. $4(y + 2) < -x$

52. $-y < 6x + 3$ or $4y \geq 3x - 6$

53. $|7x + 4| \leq 5$ or $|7y + 4| \leq 5$

a.

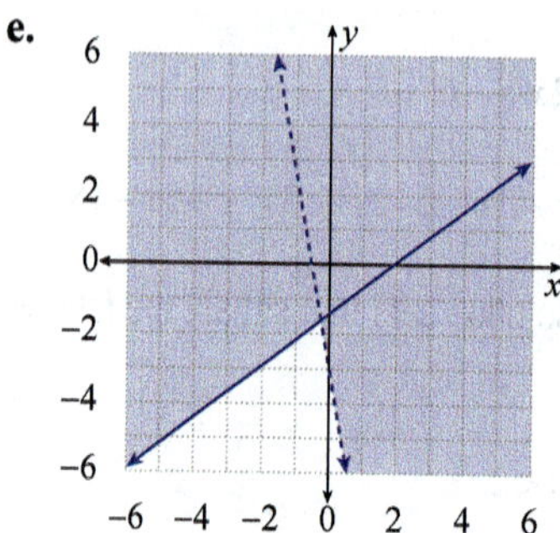

b.

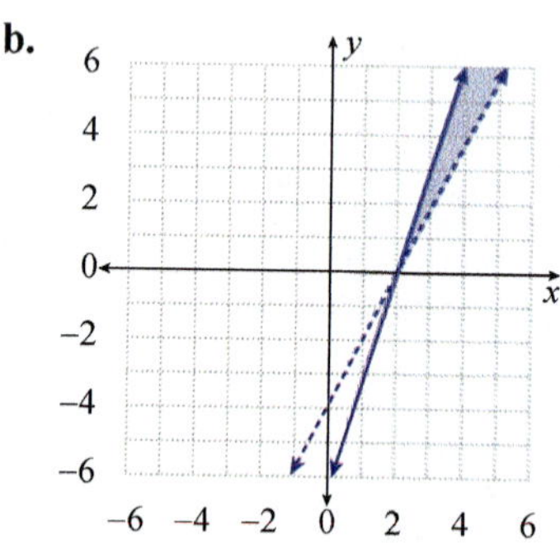

c.

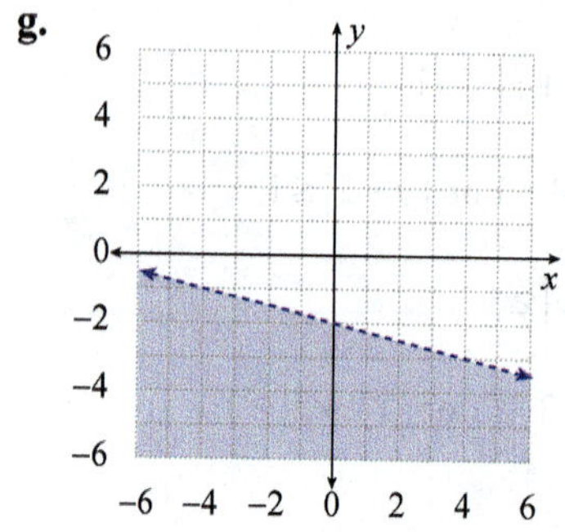

d.

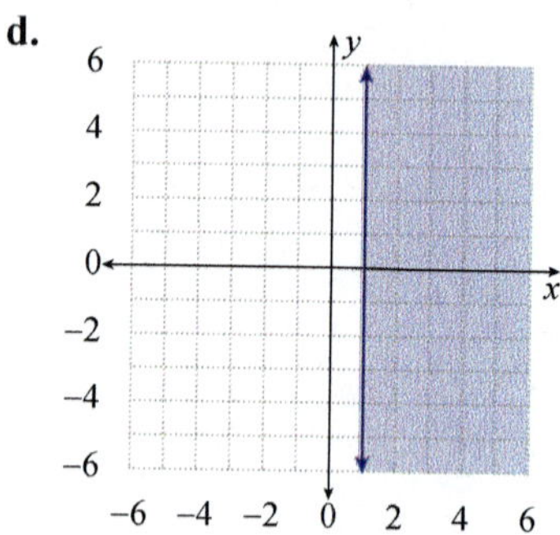

e.

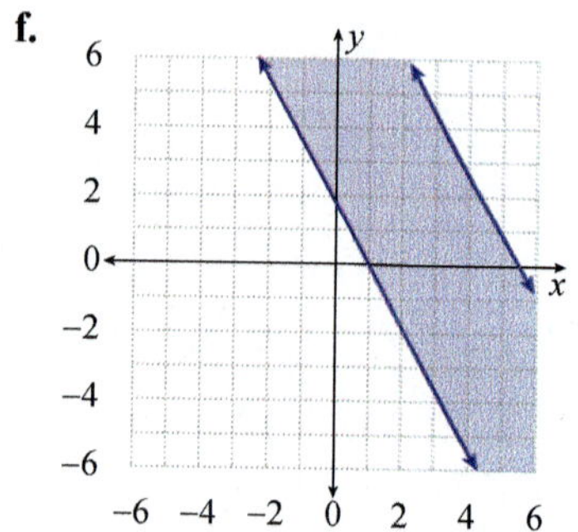

f.

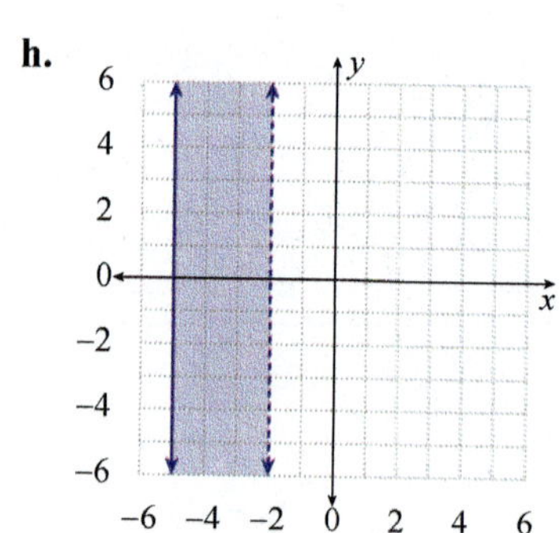

g.

h.

Graph the solution set of each of the following systems of inequalities.

54. $\begin{cases} y \geq -2 \\ y > 1 \end{cases}$

55. $\begin{cases} y \geq -2x - 5 \\ y \leq -6x - 9 \end{cases}$

56. $\begin{cases} y \leq 4x + 4 \\ y > 7x + 7 \end{cases}$

57. $\begin{cases} x - 3y \geq 6 \\ y > -4 \end{cases}$

58. $\begin{cases} 3x - y \leq 2 \\ x + y > 0 \end{cases}$

59. $\begin{cases} x > 1 \\ y > 2 \end{cases}$

60. $\begin{cases} x + y > -2 \\ x + y < 2 \end{cases}$

61. $\begin{cases} y > -2 \\ 2y > -3x - 4 \end{cases}$

62. $\begin{cases} y \leq -x \\ 2y + 3x > -4 \end{cases}$

63. $\begin{cases} 5x + 6y < -30 \\ x \geq 2 \end{cases}$

64. $\begin{cases} x < 6 \\ x \geq -5 \end{cases}$

65. $\begin{cases} |x + 1| < 2 \\ |y - 3| \leq 1 \end{cases}$

66. $\begin{cases} |y - 3x| \leq 2 \\ |y| < 2 \end{cases}$

🚀 APPLICATIONS

67. It costs Happy Land Toys $5.50 in variable costs per doll produced. If total costs must remain less than $200, write a linear inequality describing the relationship between cost and dolls produced.

68. Trish is having a garden party where she wants to have several arrangements of lilies and orchids for decoration. The lily arrangements cost $12 each and the orchids cost $22 each. If Trish wants to spend less than $150 on flowers, write a linear inequality describing the number of each arrangement she can purchase. Graph the inequality.

69. Rob has 300 feet of fencing he can use to enclose a small rectangular area of his yard for a garden. Assuming Rob may or may not use all the fencing, write a linear inequality describing the possible dimensions of his garden. Graph the inequality.

70. Flowertown Canoes produces two types of canoes. The two-person model costs $73 to produce and the one-person model costs $46 to produce. Write a linear inequality describing the number of each canoe the company can produce and keep costs under $1750. Graph the inequality.

7.2 LINEAR PROGRAMMING: THE GRAPHICAL APPROACH

■ TOPICS

- Planar Feasible Regions
- Linear Programming in Two Variables

Planar Feasible Regions

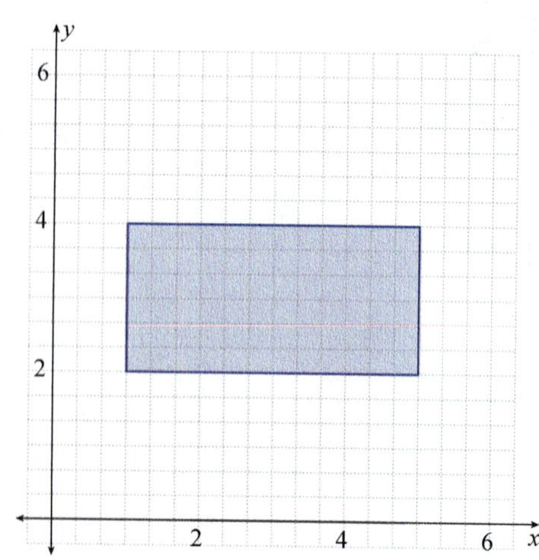

Many applications of mathematics require identifying the values that a set of variables is allowed to take. If an application imposes limitations on the values that two variables can assume, the set of all allowable values forms a portion of the plane called the **feasible region** or **region of constraint**. For instance, the constraints $1 \leq x \leq 5$ and $2 \leq y \leq 4$ define the rectangular feasible region shown in Figure 1.

More often, constraint inequalities involve both variables and the corresponding feasible region is a bit more interesting than a rectangle. The collection of all constraints in a problem comprises a system of linear inequalities, and the techniques learned so far can be used to determine the feasible region described by such systems.

FIGURE 1: A Rectangular Feasible Region

Example 1: Feasible Regions

A family orchard sells peaches and nectarines. To prevent a pest infestation, the number of nectarine trees in the orchard cannot exceed the number of peach trees. Also, because of the space requirements of each type of tree, the number of nectarine trees plus twice the number of peach trees cannot exceed 100 trees. Construct the constraints and graph the feasible region for this situation.

Solution

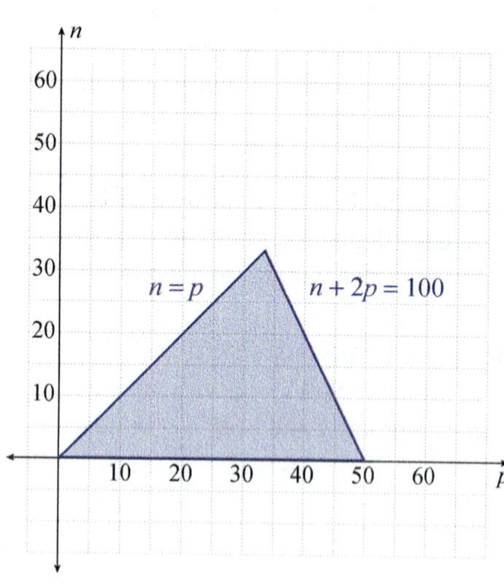

Let p be the number of peach trees and n be the number of nectarine trees. The constraints given in the problem can now be written in terms of p and n.

The second sentence tells us that $n \leq p$, and the third sentence tells us that $n + 2p \leq 100$. Since we can't have a negative number of trees, we also know that $n \geq 0$ and $p \geq 0$.

Letting the horizontal axis represent p and the vertical axis represent n, we can graph the feasible region defined by these constraints.

Any point (with integer coordinates) in the shaded feasible region corresponds to a certain number of peach trees and nectarine trees, and all such points meet the stated conditions.

Linear Programming in Two Variables

Often, given a feasible region, we want to find the point in the region such that the associated values of the variables maximize or minimize some quantity. Two important examples come from business: profit is a quantity that people usually wish to maximize, and cost is a quantity best kept to a minimum. When the quantity to be optimized (that is, either maximized or minimized) is a linear function of the variables concerned, the problem is known as a **linear programming** problem. Linear programming is used in a wide variety of applications, some involving thousands of variables, and a complete study of its techniques is beyond the scope of this course. However, the basic concepts are simple and easily applied, especially when working with applications involving just two variables.

Bounded and Unbounded Regions

A **bounded region** is one in which all the points lie within some finite distance of the origin. An **unbounded region** has points that are arbitrarily far away from the origin.

Linear Programming

Step 1: Identify the variables to be considered in the problem, determine all the constraints on the variables imposed by the problem, and sketch the feasible region described by the constraints.

Step 2: Determine the function that is to be either maximized or minimized. This function is called the **objective function**.

Step 3: Evaluate the function at each of the vertices of the feasible region and compare the values found. If the feasible region is bounded, the optimum value of the function will occur at a vertex. If the feasible region is unbounded, the optimum value of the function *may not exist*, but if it does exist, it will occur at a vertex.

Example 2: Linear Programming

Find the maximum and minimum values of $f(x, y) = 3x + 2y$ subject to the following constraints.

$$\begin{cases} x \geq 0 \\ y \geq 0 \\ y \geq 3 - 3x \\ 2x + y \leq 10 \\ x + y \leq 7 \\ y \geq x - 2 \end{cases}$$

Solution

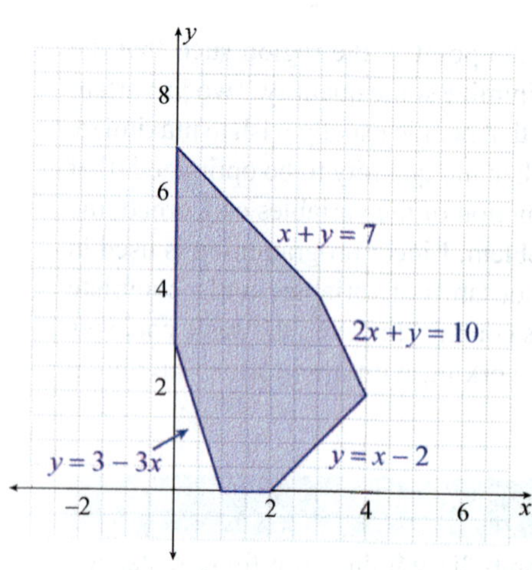

The two variables are given to us, as are the constraints; all we must do to complete the first step is sketch the region defined by the constraints. The graph contains the sketch, with selected boundaries identified by their corresponding equations.

The function to be optimized is also given to us in this problem, so step two is complete. According to the linear programming method, the only remaining task is to evaluate the function at the vertices of the feasible region. Why should we expect the maximum and minimum values of f to occur at vertices, and not in the interior of the region?

Another diagram will answer this question. The function $f(x, y)$ takes on a variety of values at different ordered pairs (x, y). But for any particular value, say k, there will be a line of ordered pairs in the plane such that $f(x, y) = k$. Such lines for different values of k will be parallel. Two particular values for k are labeled in the diagram, and these values are the minimum and maximum values of f.

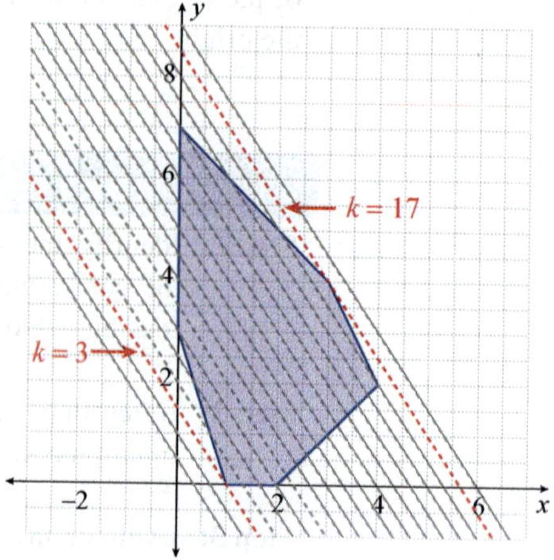

The extreme values will thus occur where a line of constant value *just touches* the region. This will be at a vertex or one of the edges of the region, if the edge is parallel to the line of constant value. Thus, evaluating the function at the vertices of the feasible region will suffice to identify the extreme values.

Each vertex of a feasible region is the intersection of two edges, so the coordinates of each vertex are found by solving a system of two equations. The six vertices of the region in this problem are determined by the following six systems:

$$\begin{cases} x + y = 7 \\ 2x + y = 10 \end{cases} \begin{cases} 2x + y = 10 \\ y = x - 2 \end{cases} \begin{cases} y = x - 2 \\ y = 0 \end{cases} \begin{cases} y = 0 \\ y = 3 - 3x \end{cases} \begin{cases} y = 3 - 3x \\ x = 0 \end{cases} \begin{cases} x = 0 \\ x + y = 7 \end{cases}$$

Solving each of these systems, we find the vertices to be $(3, 4)$, $(4, 2)$, $(2, 0)$, $(1, 0)$, $(0, 3)$, and $(0, 7)$. Substituting each of these ordered pairs into $f(x, y) = 3x + 2y$, we find the maximum value of 17 occurs at the vertex $(3, 4)$, and the minimum value of 3 occurs at the vertex $(1, 0)$.

Example 3: Linear Programming

Find the maximum and minimum values of $f(x, y) = x + 2y$ subject to the following constraints.

$$\begin{cases} x \geq 0 \\ y \geq 3 - x \\ x - y \leq 1 \\ y \leq 2x + 4 \end{cases}$$

Solution

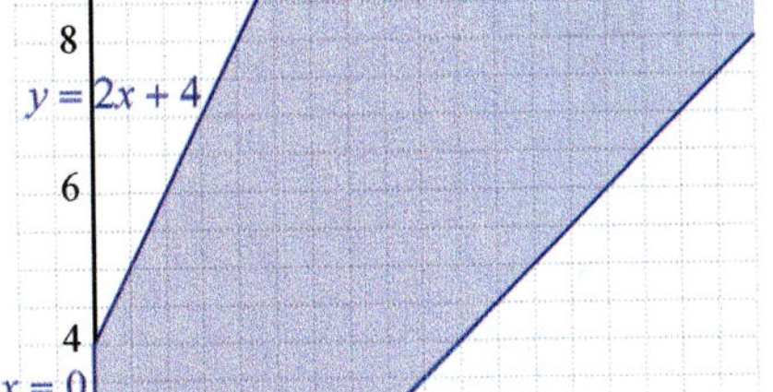

The feasible region defined by the constraints is unbounded.

Because the region is unbounded, the function f may not have a maximum or a minimum value over the region. In this case, f has no maximum value. One way to see this is to note that ordered pairs of the form (x, x) lie in the region for all $x \geq \dfrac{3}{2}$, and $f(x, x) = x + 2x = 3x$ grows without bound as x increases.

On the other hand, f does have a minimum value over the feasible region. The three vertices are $(0, 4)$, $(0, 3)$, and $(2, 1)$. Evaluating the function at these points, we obtain $f(0, 4) = 8$, $f(0, 3) = 6$, and $f(2, 1) = 4$, so the minimum value is 4.

Our last example illustrates how linear programming can be applied in a typical business application.

Example 4: Linear Programming

A manufacturer makes two models of a certain electronic device. Model A requires 3 minutes to assemble, 4 minutes to test, and 1 minute to package. Model B requires 4 minutes to assemble, 3 minutes to test, and 6 minutes to package. The manufacturer can allot 7400 minutes total for assembly, 8000 minutes for testing, and 9000 minutes for packaging. Model A generates a profit of \$7.00 and model B generates a profit of \$8.00. How many of each model should be made in order to maximize profit?

Solution

If we let x denote the number of Model A devices made and y the number of Model B devices made, the constraint inequalities are as follows.

$$\begin{cases} x \geq 0,\, y \geq 0 \\ 3x + 4y \leq 7400 \\ 4x + 3y \leq 8000 \\ x + 6y \leq 9000 \end{cases}$$

Next we sketch the feasible region.

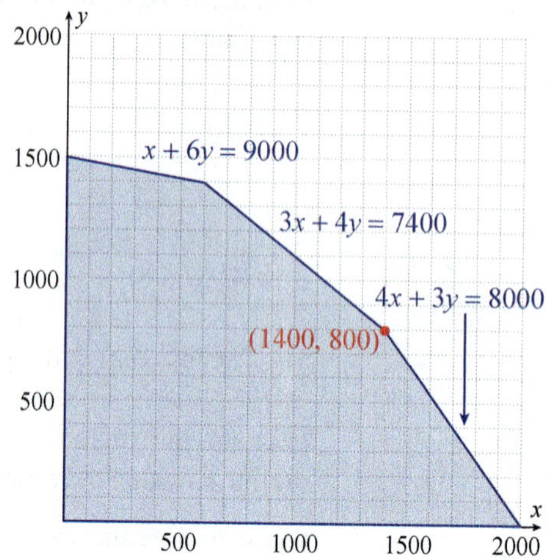

The profit function in this situation is $f(x, y) = 7x + 8y$, and our goal is to maximize this function. The vertices of the feasible region are (0, 1500), (600, 1400), (1400, 800), (2000, 0), and (0, 0). Of course, manufacturing 0 units of Model A and Model B is not going to be profitable, so we really only need to evaluate f at four vertices:

$$f(0,1500) = 12,000$$

$$f(600,1400) = 15,400$$

$$f(1400,800) = 16,200$$

$$f(2000,0) = 14,000$$

From these calculations, we conclude that the maximum profit is generated from making 1400 units of Model A and 800 units of Model B. The maximum profit would be $16,200.

7.2 EXERCISES

💡 PRACTICE

Find the minimum and maximum values of the given functions, subject to the given constraints. See Examples 2 and 3.

1. Objective Function:
$$f(x, y) = 2x + 3y$$
Constraints:
$$\begin{cases} x \geq 0, y \geq 0 \\ x + y \leq 7 \end{cases}$$

2. Objective Function:
$$f(x, y) = 4x + y$$
Constraints:
$$\begin{cases} x \geq 0, y \geq 0 \\ x + y \leq 3 \end{cases}$$

3. Objective Function:
$$f(x, y) = 2x + 5y$$
Constraints:
$$\begin{cases} x \geq 0, y \geq 0 \\ x + y \leq 7 \end{cases}$$

4. Objective Function:
$$f(x, y) = 7x + 4y$$
Constraints:
$$\begin{cases} x \geq 0, y \geq 0 \\ 3x + y \leq 3 \end{cases}$$

5. Objective Function:
$$f(x,y) = 5x + 6y$$
Constraints:
$$\begin{cases} 0 \leq x \leq 7 \\ 0 \leq y \leq 10 \\ 8x + 5y \geq 40 \end{cases}$$

6. Objective Function:
$$f(x,y) = 9x + 7y$$
Constraints:
$$\begin{cases} 0 \leq x \leq 20 \\ 0 \leq y \leq 10 \\ 6x + 12y \geq 140 \end{cases}$$

7. Objective Function:
$$f(x,y) = 6x + 4y$$
Constraints:
$$\begin{cases} 0 \leq x \leq 4 \\ 0 \leq y \leq 5 \\ 4x + 3y \geq 10 \end{cases}$$

8. Objective Function:
$$f(x,y) = 3x + 7y$$
Constraints:
$$\begin{cases} 0 \leq x \leq 8 \\ 0 \leq y \leq 6 \\ 7x + 10y \geq 50 \end{cases}$$

9. Objective Function:
$$f(x,y) = 6x + 8y$$
Constraints:
$$\begin{cases} x \geq 0, y \geq 0 \\ 4x + y \leq 16 \\ x + 3y \leq 15 \end{cases}$$

10. Objective Function:
$$f(x,y) = x + 2y$$
Constraints:
$$\begin{cases} x \geq 0, y \geq 0 \\ 3x + y \leq 45 \\ x + 3y \leq 24 \end{cases}$$

11. Objective Function:
$$f(x,y) = 6x + y$$
Constraints:
$$\begin{cases} x \geq 0, y \geq 0 \\ 3x + 4y \geq 24 \\ 3x + 4y \leq 48 \end{cases}$$

12. Objective Function:
$$f(x,y) = 15x + 30y$$
Constraints:
$$\begin{cases} x \geq 0, y \geq 0 \\ 5x + 7y \geq 70 \\ 5x + 7y \leq 140 \end{cases}$$

13. Objective Function:
$$f(x,y) = 3x + 10y$$
Constraints:
$$\begin{cases} x \geq 0 \\ 2x + 4y \geq 8 \\ 5x - y \leq 10 \\ x + 3y \leq 40 \end{cases}$$

14. Objective Function:
$$f(x,y) = 20x + 30y$$
Constraints:
$$\begin{cases} x \geq 0 \\ 12x + 6y \geq 120 \\ 9x - 6y \leq 144 \\ x + 4y \leq 12 \end{cases}$$

🚀 APPLICATIONS

15. A plane carrying relief food and water can carry a maximum of 50,000 pounds and is limited in space to carrying no more than 6000 cubic feet. Each container of water weighs 60 pounds and takes up 1 cubic foot, and each container of food weighs 50 pounds and takes up 10 cubic feet. What is the region of constraint for the numbers of containers of food and water that the plane can carry?

16. A furniture company makes two kinds of sofas, the Standard model and the Deluxe model. The Standard model requires 40 hours of labor to build, and the Deluxe model requires 60 hours of labor to build. The finish of the Deluxe model uses both teak and fabric, while the Standard uses only fabric, with the result that each Deluxe sofa requires 5 square yards of fabric and each Standard sofa requires 8 square yards of fabric. Given that the company can use 200 hours of labor and 25 square yards of fabric per week building sofas, what is the region of constraint for the numbers of Deluxe and Standard sofas the company can make per week?

17. Sarah is looking through a clothing catalog, and she is willing to spend up to $80 on clothes and $10 for shipping. Shirts cost $12 each plus $2 shipping, and a pair of pants costs $32 plus $3 shipping. What is the region of constraint for the numbers of shirts and pairs of pants Sarah can buy?

18. Suppose you inherit $75,000 from a previously unknown (and highly eccentric) uncle and that the inheritance comes with certain stipulations regarding investments. First, the dollar amount invested in bonds must not exceed the dollar amount invested in stocks. Second, a minimum of $10,000 must be invested in stocks, and a minimum of $5000 must be invested in bonds. Finally, a maximum of $40,000 can be invested in stocks. What is the region of constraint for the dollar amounts that can be invested in the two categories of stocks and bonds?

19. A manufacturer produces two models of computers. The times (in hours) required for assembling, testing, and packaging each model are listed in the following table.

Process	Model X	Model Y
Assemble	2.5	3
Test	2	1
Package	0.75	1.25

The total times available for assembling, testing, and packaging are 4000 hours, 2500 hours, and 1500 hours, respectively. The profits per unit are $50 for Model X and $52 for Model Y. How many of each type should be produced to maximize profit? What is the maximum profit?

20. A manufacturer produces two types of fans. The times (in minutes) required for assembling, packaging, and shipping each type are listed in the following table.

Process	Type X	Type Y
Assemble	20	25
Package	40	10
Ship	10	7.5

The total times available for assembling, packaging, and shipping are 4000 minutes, 4800 minutes, and 1500 minutes, respectively. The profits per unit are $4.50 for Type X and $3.75 for Type Y. How many of each type should be produced to maximize profit? What is the maximum profit?

21. Ashley is making a set of patchwork curtains for her apartment. She needs a minimum of 16 yards of the solid material, at least 5 yards of the striped material, and at least 20 yards of the flowered material. She can choose between two sets of precut bundles. The olive-based bundle costs $10 per bundle and contains 8 yards of the solid material, 1 yard of the striped material, and 2 yards of the flowered material. The cranberry-based bundle costs $20 per bundle and includes 2 yards of the solid material, 1 yard of the striped material, and 7 yards of the flowered material. How many of each bundle should Ashley buy to minimize her cost and yet buy enough material to complete the curtains? What is her minimum cost?

22. A volunteer has been asked to drop off some supplies at a facility housing victims of a hurricane evacuation. The volunteer would like to bring at least 60 bottles of water, 45 first aid kits, and 30 security blankets on his visit. The relief organization has a standing agreement with two companies that provide victim packages. Company A can provide packages of 5 water bottles, 3 first aid kits, and 4 security blankets at a cost of $1.50. Company B can provide packages of 2 water bottles, 2 first aid kits, and 1 security blanket at a cost of $1.00. How many of each package should the volunteer pick up to minimize the cost? What total amount does the relief organization pay?

23. On your birthday your grandmother gave you $25,000, but told you she would like you to invest the money for 10 years before you use any of it. Since you wish to respect your grandmother's wishes, you seek out the advice of a financial adviser. She suggests you invest at least $15,000 in municipal bonds yielding 6% and no more than $5000 in Treasury bills yielding 9%. How much should be placed in each investment so that income is maximized?

24. A boutique cell phone manufacturer produces two models: a retro model flip phone and a smart phone. The manufacturer's quota per day is to produce at least 100 flip phones and 80 smart phones. No more than 200 flip phones and 170 smart phones can be produced per day due to limitations on production. A total of at least 200 phones must be shipped every day.

    a. If the production costs are $5 for a flip phone and $7 for a smart phone, how many of each model should be produced on a daily basis to minimize cost and what would that cost be?

    b. If each flip phone results in a $2 loss but each smart phone results in a $5 gain, how many of each model should be manufactured daily to maximize profit? What is the maximum profit if this number of phones is produced?

7.3 THE SIMPLEX METHOD: MAXIMIZATION

Previously, we used the graphical approach to solve various linear programming problems. The reason why this approach worked so well is because all problems involved only two variables. We were able to graph the feasible region because it was planar, and we used the following fact:

> If the feasible region is bounded, the optimum value of the function will occur at a vertex. If the feasible region is unbounded, the optimum value of the function may not exist, but if it does it will occur at a vertex.

The next question to consider is this: How do we solve a similar linear programming problem involving three or more variables? One can imagine that the three-variable case would involve a three-dimensional coordinate system; the feasible region would be a three-dimensional solid, and the boundaries of the feasible region would be planes but four or more variables would be impossible to represent graphically. Thus, we need a different method for solving a general linear programming problem. This method is called the **simplex method**.

To illustrate this method, we will consider finding the maximum value of $f(x,y) = 5x + 8y$ subject to the constraints $x + 8y \leq 56$ and $2x + 2y \leq 21$. This example has only two unknowns, x and y, which we'll call **decision variables**. In application problems, these are the variables that decision makers would want to determine in order to find an optimal solution. Since there are only two decision variables here, we can compare how the simplex method works along with the graphical approach. Before we illustrate the solution to our first problem, a couple of remarks are in order. First, we always assume that the decision variables are nonnegative. This is because negative values for decision variables won't make sense in the context of the problems we will solve. Secondly, all of the problems we solve in this lesson will be maximization problems. In general, if a maximization problem has n decision variables, each constraint will be of the form $a_1 x_1 + a_2 x_2 + \cdots + a_n x_n \leq b$, where $b > 0$. With this notation, our first example reads as follows: Find the maximum value of $f(x_1, x_2) = 5x_1 + 8x_2$ subject to the constraints $x_1 + 8x_2 \leq 56$ and $2x_1 + 2x_2 \leq 21$. But since we will display the graphical approach, we will denote the decision variables as x and y for this example.

Example 1: Comparing the Simplex Method to the Graphical Approach

Find the maximum value of $f(x, y) = 5x + 8y$ subject to the following constraints.

$$\begin{cases} x + 8y \leq 56 \\ 2x + 2y \leq 21 \end{cases}$$

Solution

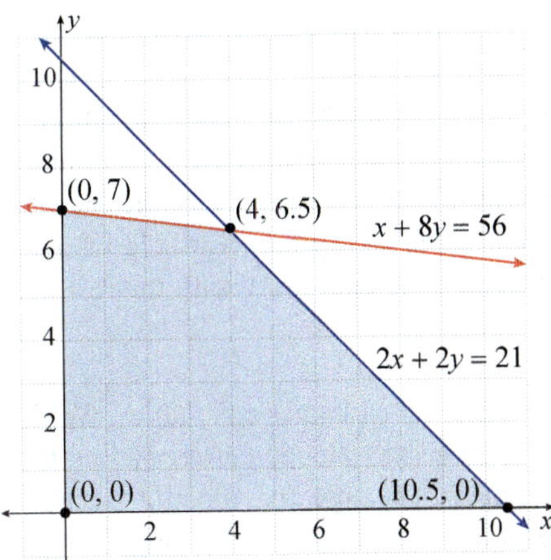

The Graphical Approach: The feasible region, shown in the graph, is in the first quadrant bounded by the lines $x + 8y = 56$ and $2x + 2y = 21$.

Now we find the value of $f(x, y)$ at each vertex.

Vertex	$f(x, y)$
$(0, 0)$	0
$(10.5, 0)$	52.5
$(4, 6.5)$	72
$(0, 7)$	56

Therefore, the maximum value of f is 72, and it is attained at $(4, 6.5)$.

The Simplex Method: First define a new variable, say z, and set $z = f(x, y)$. Then we rewrite the objective function as follows.

$$z = 5x + 8y \quad \Rightarrow \quad -5x - 8y + z = 0$$

In particular, notice that the objective function is written in standard form because all of the terms involving variables appear on the left-hand side of the equation.

Next, we rewrite the constraints as equations. We can do this by introducing nonnegative quantities, say s_1 and s_2, called **slack variables**. The idea is that s_1 takes up the slack in the first inequality and s_2 takes up the slack in the second. This leads to the following.

$$x + 8y + s_1 = 56$$
$$2x + 2y + s_2 = 21$$

Note: Use a different slack variable for each constraint. Reason: We don't want to assume that the quantity $2x + 2y$ is short of 21 by the same amount that $x + 8y$ is short of 56.

We now write the system with the constraint equations in the first two rows and the objective function in the bottom row as follows.

$$\begin{cases} x + 8y + s_1 \quad\quad\quad = 56 \\ 2x + 2y \quad + s_2 \quad\quad = 21 \\ -5x - 8y \quad\quad\quad + z \ = 0 \end{cases}$$

Notice that each equation in this system is written in standard form; all terms involving any variables are on the left-hand side and the constant term is on the right-hand side.

We then form the augmented matrix corresponding to this system. This is called the **initial simplex tableau**. In this example, the initial simplex tableau is as follows.

$$\begin{array}{ccccc} x & y & s_1 & s_2 & z \end{array}$$
$$\left[\begin{array}{ccccc|c} 1 & 8 & 1 & 0 & 0 & 56 \\ 2 & 2 & 0 & 1 & 0 & 21 \\ \hline -5 & -8 & 0 & 0 & 1 & 0 \end{array}\right]$$

At this stage, we can consider a basic feasible solution to the linear programming problem. A basic feasible solution, in general (with n decision variables and m constraints), will have the form $(x_1, x_2, \ldots, x_n, s_1, s_2, \ldots, s_m, z)$ in which at most m of the variables are 0. All variables in the basic feasible solution, with the exception of z, that are nonzero are called **basic variables**, and those that are 0 are the **nonbasic variables**. Typically, the number of basic variables is the same as the number of constraints, which would be m in the general case. Any variable can be a basic variable, but it is most convenient to choose each of the m basic variables to be a variable whose corresponding column in the simplex tableau has a single entry of 1 with the rest of the column entries being 0.

In our example, since we have two constraints, we will have two basic variables. We'll let s_1 and s_2 be our basic variables since the columns in the tableau corresponding to these variables have a 1 and the rest of the entries are 0. The objective function z is not considered a basic variable even though it corresponds to a column with a single entry of 1 and all other entries 0. Thus, the remaining variables, x and y, will be the nonbasic variables. To get an idea of how to read the basic feasible solution, notice that the bottom row of the tableau corresponds to the equation $-5x - 8y + z = 0$, so $z = 5x + 8y$ depends on x and y (the nonbasic variables). If these nonbasic variables equal 0 (namely x and y are both 0), then rows 1 and 2 of the tableau lead to the equations $s_1 = 56$ and $s_2 = 21$. Therefore, the basic feasible solution is $(x, y, s_1, s_2, z) = (0, 0, 56, 21, 0)$. This solution is not desired because $z = 0$ here, and we want to maximize z.

The next question is this: How can we obtain another solution that yields a more desirable value of z? The answer involves performing elementary row operations on the simplex tableau. During this process, we will need to choose an entry in the matrix as a reference and perform row operations that make the entry a 1 and all other entries in the column 0. This process is called **pivoting**, and the entry that gets chosen is called the **pivot element**.

In our example, the value of our objective function $z = 5x + 8y$ will be increased by a greater amount if y is increased by one unit compared to x being increased by one unit. Thus, the column in the tableau corresponding to the y variable will be our **pivot column**. Since the bottom row in the tableau corresponds to the objective function being written as $-5x - 8y + z = 0$, we look for the negative entry in the bottom row with largest magnitude (-8 in our example) to determine our pivot column. Next, we divide each entry in the far-right column by the corresponding entry (in the same row) in the pivot column to determine the **pivot row**. Note that the bottom row cannot be the pivot row. In our example, the second column is the pivot column. Looking at the quotients of the entries in the far-right column divided by those in the second column, we find $\dfrac{56}{8} = 7$ and $\dfrac{21}{2} = 10.5$. The smaller of these two quotients will determine the pivot row; in our case, row 1 will be the pivot row. Therefore, we choose 8 (row 1, column 2 entry) as our pivot element.

$$
\begin{array}{ccccc}
x & y & s_1 & s_2 & z \\
\end{array}
$$

$$
\left[
\begin{array}{ccccc|c}
1 & \boxed{8} & 1 & 0 & 0 & 56 \\
2 & 2 & 0 & 1 & 0 & 21 \\
\hline
-5 & -8 & 0 & 0 & 1 & 0
\end{array}
\right]
$$

Next, we illustrate the pivoting process on column 2. Perform elementary row operations using the pivot row so that a 1 appears in the pivot position and all other entries in the pivot column are 0.

$$\begin{array}{ccccc} x & y & s_1 & s_2 & z \\ \end{array}$$

$$\left[\begin{array}{ccccc|c} 1 & \boxed{8} & 1 & 0 & 0 & 56 \\ 2 & 2 & 0 & 1 & 0 & 21 \\ \hline -5 & -8 & 0 & 0 & 1 & 0 \end{array}\right] \xrightarrow{\;R_1+R_3\;} \left[\begin{array}{ccccc|c} 1 & 8 & 1 & 0 & 0 & 56 \\ 2 & 2 & 0 & 1 & 0 & 21 \\ \hline -4 & 0 & 1 & 0 & 1 & 56 \end{array}\right]$$

$$\xrightarrow{\;\frac{1}{8}R_1\;} \left[\begin{array}{ccccc|c} \frac{1}{8} & 1 & \frac{1}{8} & 0 & 0 & 7 \\ 2 & 2 & 0 & 1 & 0 & 21 \\ \hline -4 & 0 & 1 & 0 & 1 & 56 \end{array}\right]$$

$$\xrightarrow{\;-2R_1+R_2\;} \left[\begin{array}{ccccc|c} \frac{1}{8} & 1 & \frac{1}{8} & 0 & 0 & 7 \\ \frac{7}{4} & 0 & -\frac{1}{4} & 1 & 0 & 7 \\ \hline -4 & 0 & 1 & 0 & 1 & 56 \end{array}\right]$$

In this tableau, y and s_2 are the basic variables, and x and s_1 are nonbasic. Reading the solution from this tableau, we have $(x, y, s_1, s_2, z) = (0, 7, 0, 7, 56)$. At this point, it is natural to ask the following: Is 56 the optimal value for z under our constraints? The answer to that question lies in the bottom row of the most recent tableau. The equation corresponding to this row is $-4x + s_1 + z = 56$, which is equivalent to $z = 4x - s_1 + 56$. If we increase x, we increase the value of z. Since z still has room to grow in this manner, we can conclude that 56 is not the maximum of z. The quickest way to see that 56 is not an optimal value of z is to note that there's still a negative entry in the bottom row of the simplex tableau, namely -4. This means we will need to repeat the pivoting process by choosing a pivot element as we did before.

Since -4 in the only negative entry in the bottom row, column 1 is our pivot column.

To choose the pivot row, look at $\dfrac{7}{\frac{1}{8}} = 56$ and $\dfrac{7}{\frac{7}{4}} = 4$. Since 4 is the smaller quotient,

row 2 is our pivot row, and thus our pivot element is $\dfrac{7}{4}$.

$$\begin{array}{ccccc} x & y & s_1 & s_2 & z \\ \end{array}$$

$$\left[\begin{array}{ccccc|c} \frac{1}{8} & 1 & \frac{1}{8} & 0 & 0 & 7 \\ \boxed{\frac{7}{4}} & 0 & -\frac{1}{4} & 1 & 0 & 7 \\ \hline -4 & 0 & 1 & 0 & 1 & 56 \end{array}\right]$$

We now do the pivoting process on column 1 by performing elementary row operations until a 1 appears in the pivot position and all other entries in the pivot column are 0.

$$
\begin{array}{ccccc}
x & y & s_1 & s_2 & z \\
\end{array}
\left[\begin{array}{ccccc|c}
\dfrac{1}{8} & 1 & \dfrac{1}{8} & 0 & 0 & 7 \\[6pt]
\dfrac{7}{4} & 0 & -\dfrac{1}{4} & 1 & 0 & 7 \\[6pt]
-4 & 0 & 1 & 0 & 1 & 56
\end{array}\right]
\xrightarrow{\;\frac{4}{7}R_2\;}
\begin{array}{ccccc}
x & y & s_1 & s_2 & z \\
\end{array}
\left[\begin{array}{ccccc|c}
\dfrac{1}{8} & 1 & \dfrac{1}{8} & 0 & 0 & 7 \\[6pt]
1 & 0 & -\dfrac{1}{7} & \dfrac{4}{7} & 0 & 4 \\[6pt]
-4 & 0 & 1 & 0 & 1 & 56
\end{array}\right]
$$

$$
\xrightarrow{\;-\frac{1}{8}R_2+R_1\;}
\begin{array}{ccccc}
x & y & s_1 & s_2 & z \\
\end{array}
\left[\begin{array}{ccccc|c}
0 & 1 & \dfrac{1}{7} & -\dfrac{1}{14} & 0 & \dfrac{13}{2} \\[6pt]
1 & 0 & -\dfrac{1}{7} & \dfrac{4}{7} & 0 & 4 \\[6pt]
-4 & 0 & 1 & 0 & 1 & 56
\end{array}\right]
$$

$$
\xrightarrow{\;4R_2+R_3\;}
\begin{array}{ccccc}
x & y & s_1 & s_2 & z \\
\end{array}
\left[\begin{array}{ccccc|c}
0 & 1 & \dfrac{1}{7} & -\dfrac{1}{14} & 0 & \dfrac{13}{2} \\[6pt]
1 & 0 & -\dfrac{1}{7} & \dfrac{4}{7} & 0 & 4 \\[6pt]
0 & 0 & \dfrac{3}{7} & \dfrac{16}{7} & 1 & 72
\end{array}\right]
$$

In this tableau, x and y are the basic variables, and s_1 and s_2 are nonbasic. Reading the solution from this tableau, we have $\left(x, y, s_1, s_2, z\right) = \left(4, \dfrac{13}{2}, 0, 0, 72\right)$. Now we consider the following question: Is 72 the optimal value of z? At this point, the bottom row of the simplex tableau corresponds to $\dfrac{3}{7}s_1 + \dfrac{16}{7}s_2 + z = 72$, which is equivalent to $z = 72 - \dfrac{3}{7}s_1 - \dfrac{16}{7}s_2$.

Substituting any positive values of s_1 and s_2 into this equation will decrease the value of z. Therefore, we can be assured that 72 is indeed the maximum value of z. The easiest way to see we are done is to note that there are no negative entries in the bottom row of the simplex tableau.

We conclude that $f(x, y) = 5x + 8y$ attains a maximum value of 72 when $x = 4$ and $y = \dfrac{13}{2}$.

Note: In reviewing the solution in Example 1, we can see that each of the basic feasible solutions found using the simplex method corresponds to a vertex of the feasible region. Moreover, it is interesting to note that the pivoting process has the effect of moving from one vertex to another until an optimal solution is reached. Figure 1 illustrates this.

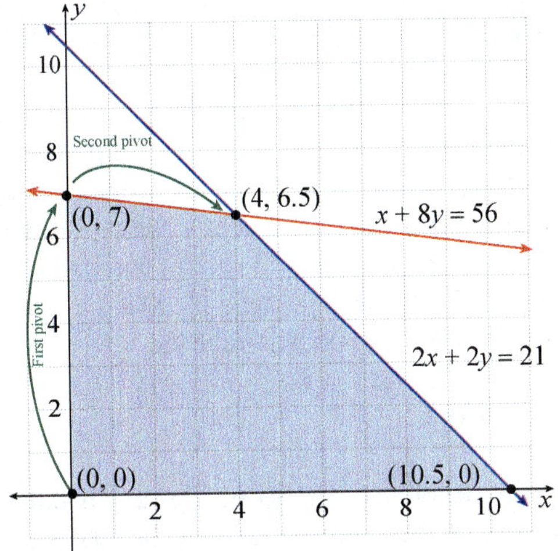

FIGURE 1: Summary of the Pivoting Process

The Simplex Method

1. Rewrite the objective function as an equation in standard form with all variables on the left side and a nonnegative constant on the right side.

2. Convert each constraint inequality into an equation with all variables on the left side and a positive constant on the right side by introducing a different slack variable for each constraint.

3. Form the system to solve with the constraint equations in the first rows and the objective function in the bottom row.

4. Create the initial simplex tableau by forming the augmented matrix for the system of equations with the objective function in the bottom row.

5. Locate the negative entry in the bottom row with largest magnitude to determine the pivot column.

6. In each row except for the bottom one, divide the entry in the far-right column by the one in the pivot column. The row with the smallest quotient will be the pivot row.

7. Perform elementary row operations on the simplex tableau to make the pivot element 1 and the remaining entries in the pivot column 0. Always use the pivot row to do any row operations.

8. If all entries in the bottom row are nonnegative, stop. If not, repeat steps 5 through 7 until all entries in the bottom row are nonnegative.

9. The maximum value of the objective function is in the lower-right corner of the final simplex tableau.

Example 2: An Application of the Simplex Method

An amusement park vendor sells two types of bagged snack mix: Chocolaty Crunch and Pretzel Delight. Each bag of Chocolaty Crunch consists of 2 cups of M&M's, 2 cups of chocolate chips, and 1 cup of pretzels; and each bag of Pretzel Delight consists of 1 cup of M&M's, 2 cups of chocolate chips, and 2 cups of pretzels. Suppose the vendor has 70 cups of M&M's, 90 cups of chocolate chips, and 65 cups of pretzels in his inventory. Suppose also that one bag of Chocolaty Crunch sells for $15 and one bag of Pretzel Delight sells for $10. How many bags of each mix should the vendor sell in order to maximize revenue.

Solution

Let x_1 denote the number of bags of Chocolaty Crunch sold and x_2 denote the number of bags of Pretzel Delight. Since the goal is to maximize revenue, we need to maximize $f(x_1, x_2) = 15x_1 + 10x_2$ (our objective function).

Since the vendor has 70 cups of M&M's at his disposal, we need $2x_1 + x_2 \le 70$. Similarly, we need $2x_1 + 2x_2 \le 90$ for the chocolate chip constraint.

Finally, we need $x_1 + 2x_2 \le 65$ for the pretzel constraint.

Our objective is to find the maximum value of $f(x_1, x_2) = 15x_1 + 10x_2$ subject to the following constraints.

$$\begin{cases} 2x_1 + x_2 \le 70 \\ 2x_1 + 2x_2 \le 90 \\ x_1 + 2x_2 \le 65 \end{cases}$$

We first let $z = 15x_1 + 10x_2$ and rewrite this in standard form as $-15x_1 - 10x_2 + z = 0$. Then we introduce the slack variables s_1, s_2, and s_3 for the three constraints and write the system as follows.

$$\begin{cases} 2x_1 + x_2 + s_1 &= 70 \\ 2x_1 + 2x_2 \quad + s_2 &= 90 \\ x_1 + 2x_2 \quad\quad + s_3 &= 65 \\ -15x_1 - 10x_2 \quad\quad\quad + z = 0 \end{cases}$$

The initial simplex tableau is as follows.

$$\begin{array}{cccccc|c} x_1 & x_2 & s_1 & s_2 & s_3 & z & \\ \hline 2 & 1 & 1 & 0 & 0 & 0 & 70 \\ 2 & 2 & 0 & 1 & 0 & 0 & 90 \\ 1 & 2 & 0 & 0 & 1 & 0 & 65 \\ \hline -15 & -10 & 0 & 0 & 0 & 1 & 0 \end{array}$$

Since -15 is the negative entry in the bottom row with largest absolute value, we choose column 1 as our pivot column. Then in each row (except for the bottom one), take the entry in column 7 divided by the entry in column 1 to obtain the following quotients: $\dfrac{70}{2} = 35$, $\dfrac{90}{2} = 45$, and $\dfrac{65}{1} = 65$. Since 35 is the smallest quotient, row 1 is our pivot row, and our pivot element is the 2 in row 1, column 1.

Next, we do a sequence of row operations using the pivot row so that all entries in the pivot column are 0 except for the 1 in the pivot position.

$$
\begin{array}{cccccc}
x_1 & x_2 & s_1 & s_2 & s_3 & z \\
\end{array}
\left[\begin{array}{cccccc|c}
\boxed{2} & 1 & 1 & 0 & 0 & 0 & 70 \\
2 & 2 & 0 & 1 & 0 & 0 & 90 \\
1 & 2 & 0 & 0 & 1 & 0 & 65 \\
\hline
-15 & -10 & 0 & 0 & 0 & 1 & 0
\end{array}\right]
\xrightarrow{-R_1+R_2}
\begin{array}{cccccc}
x_1 & x_2 & s_1 & s_2 & s_3 & z \\
\end{array}
\left[\begin{array}{cccccc|c}
2 & 1 & 1 & 0 & 0 & 0 & 70 \\
2 & 1 & -1 & 1 & 0 & 0 & 20 \\
1 & 2 & 0 & 0 & 1 & 0 & 65 \\
\hline
-15 & -10 & 0 & 0 & 0 & 1 & 0
\end{array}\right]
$$

$$
\xrightarrow{\frac{1}{2}R_1}
\begin{array}{cccccc}
x_1 & x_2 & s_1 & s_2 & s_3 & z \\
\end{array}
\left[\begin{array}{cccccc|c}
1 & \frac{1}{2} & \frac{1}{2} & 0 & 0 & 0 & 70 \\
0 & 1 & -1 & 1 & 0 & 0 & 20 \\
1 & 2 & 0 & 0 & 1 & 0 & 65 \\
\hline
-15 & -10 & 0 & 0 & 0 & 1 & 0
\end{array}\right]
$$

$$
\xrightarrow{-R_1+R_3}
\begin{array}{cccccc}
x_1 & x_2 & s_1 & s_2 & s_3 & z \\
\end{array}
\left[\begin{array}{cccccc|c}
1 & \frac{1}{2} & \frac{1}{2} & 0 & 0 & 0 & 35 \\
0 & 1 & -1 & 1 & 0 & 0 & 20 \\
0 & \frac{3}{2} & -\frac{1}{2} & 0 & 1 & 0 & 30 \\
\hline
-15 & -10 & 0 & 0 & 0 & 1 & 0
\end{array}\right]
$$

$$
\xrightarrow{15R_1+R_4}
\begin{array}{cccccc}
x_1 & x_2 & s_1 & s_2 & s_3 & z \\
\end{array}
\left[\begin{array}{cccccc|c}
1 & \frac{1}{2} & \frac{1}{2} & 0 & 0 & 0 & 35 \\
0 & 1 & -1 & 1 & 0 & 0 & 20 \\
0 & \frac{3}{2} & -\frac{1}{2} & 0 & 1 & 0 & 30 \\
\hline
0 & -\frac{5}{2} & \frac{15}{2} & 0 & 0 & 1 & 525
\end{array}\right]
$$

Since we still have a negative entry in the bottom row, namely $-\frac{5}{2}$, we need to do the pivoting process again. This time, column 2 will be the pivot column. Moreover, we have the following quotients formed by dividing each entry in column 7 by the entry in column 2: $\dfrac{35}{\frac{1}{2}} = 70$, $\dfrac{20}{1} = 20$, and $\dfrac{30}{\frac{3}{2}} = 20$. This time, we have a tie for the smallest quotient of 20. We can choose either row 2 or row 3 as our pivot row. Since there's already a 1 in the row 2, column 2 entry, it will be convenient to choose this as our pivot element.

Once again, we do a sequence of row operations using the pivot row so that all entries in the pivot column are 0 except for the 1 in the pivot position.

$$
\begin{array}{c}
\begin{array}{cccccc} x_1 & x_2 & s_1 & s_2 & s_3 & z \end{array} \\
\left[\begin{array}{cccccc|c}
1 & \tfrac{1}{2} & \tfrac{1}{2} & 0 & 0 & 0 & 35 \\
0 & \boxed{1} & -1 & 1 & 0 & 0 & 20 \\
0 & \tfrac{3}{2} & -\tfrac{1}{2} & 0 & 1 & 0 & 30 \\ \hline
0 & -\tfrac{5}{2} & \tfrac{15}{2} & 0 & 0 & 1 & 525
\end{array}\right]
\end{array}
\xrightarrow{-\tfrac{1}{2}R_2+R_1}
\begin{array}{c}
\begin{array}{cccccc} x_1 & x_2 & s_1 & s_2 & s_3 & z \end{array} \\
\left[\begin{array}{cccccc|c}
1 & 0 & 1 & -\tfrac{1}{2} & 0 & 0 & 25 \\
0 & 1 & -1 & 1 & 0 & 0 & 20 \\
0 & \tfrac{3}{2} & -\tfrac{1}{2} & 0 & 1 & 0 & 30 \\ \hline
0 & -\tfrac{5}{2} & \tfrac{15}{2} & 0 & 0 & 1 & 525
\end{array}\right]
\end{array}
$$

$$
\xrightarrow{-\tfrac{3}{2}R_2+R_3}
\begin{array}{c}
\begin{array}{cccccc} x_1 & x_2 & s_1 & s_2 & s_3 & z \end{array} \\
\left[\begin{array}{cccccc|c}
1 & 0 & 1 & -\tfrac{1}{2} & 0 & 0 & 25 \\
0 & 1 & -1 & 1 & 0 & 0 & 20 \\
0 & 0 & 1 & -\tfrac{3}{2} & 1 & 0 & 0 \\ \hline
0 & -\tfrac{5}{2} & \tfrac{15}{2} & 0 & 0 & 1 & 525
\end{array}\right]
\end{array}
$$

$$
\xrightarrow{\tfrac{5}{2}R_2+R_4}
\begin{array}{c}
\begin{array}{cccccc} x_1 & x_2 & s_1 & s_2 & s_3 & z \end{array} \\
\left[\begin{array}{cccccc|c}
1 & 0 & 1 & -\tfrac{1}{2} & 0 & 0 & 25 \\
0 & 1 & -1 & 1 & 0 & 0 & 20 \\
0 & 0 & 1 & -\tfrac{3}{2} & 1 & 0 & 0 \\ \hline
0 & 0 & 5 & \tfrac{5}{2} & 0 & 1 & 575
\end{array}\right]
\end{array}
$$

We no longer have any negative entries in the bottom row. From the final simplex tableau, we obtain the following solution.

$$(x_1, x_2, s_1, s_2, s_3, z) = (25, 20, 0, 0, 0, 575)$$

Therefore, the vendor should sell 25 bags of Chocolaty Crunch and 20 bags of Pretzel Delight to make a maximum revenue of $575. Notice that all of the slack variables are 0 ($s_1 = s_2 = s_3 = 0$). This means all of the supplies in the inventory need to be used to maximize revenue.

Example 3: An Application of the Simplex Method Involving Three Decision Variables

Suppose a couple makes pies to supplement their income. During a particular month, they make apple, grape, and blueberry pies. Each apple pie sells for $1.50 to the grocer, grape pies sell for $1.20, and blueberry pies sell for $1.60. Suppose the couple has 1200 cups of sugar and 2100 cups of flour in their pantry for this month. Each apple pie requires $1\tfrac{1}{2}$ cups of sugar and 3 cups of flour, each grape pie requires 2 cups of sugar and 3 cups of flour, and each blueberry pie requires $1\tfrac{3}{4}$ cups of sugar and $2\tfrac{1}{2}$ cups of flour. Working together, the couple takes 6 minutes to make each apple pie, 3 minutes to make each grape pie, and 4 minutes to make each blueberry pie; and they plan to work no more than 62 hours. Determine how many of each kind of pie the couple should produce and sell in order to maximize revenue.

Solution

To start with, let x_1 denote the number of apple pies, x_2 denote the number of grape pies, and x_3 denote the number of blueberry pies. Our objective is to maximize the revenue. Based on the pie prices given in the problem, the revenue is given by $f\left(x_1,x_2,x_3\right)=\frac{3}{2}x_1+\frac{6}{5}x_2+\frac{8}{5}x_3$ (we converted the prices to fractions to make row operations easier when working with the simplex tableau). This is our objective function. We define $z=\frac{3}{2}x_1+\frac{6}{5}x_2+\frac{8}{5}x_3$, which is $-\frac{3}{2}x_1-\frac{6}{5}x_2-\frac{8}{5}x_3+z=0$ in standard form.

As for the constraints, first consider the amounts of sugar needed. Each apple, grape, and blueberry pie requires $1\frac{1}{2}, 2$, and $1\frac{3}{4}$ cups of sugar, respectively; and we have a limit of 1200 cups. Thus, the first constraint is

$$\frac{3}{2}x_1+2x_2+\frac{7}{4}x_3 \leq 1200.$$

We proceed similarly with looking at the amounts of flour needed and obtain the second constraint:

$$3x_1+3x_2+\frac{5}{2}x_3 \leq 2100.$$

Finally, the third constraint is concerned with the amount of time spent making the pies. Since they won't work for more than 62 hours, or 3720 minutes, we have

$$6x_1+3x_2+4x_3 \leq 3720.$$

Hence, our objective is to maximize $f\left(x_1,x_2,x_3\right)=\frac{3}{2}x_1+\frac{6}{5}x_2+\frac{8}{5}x_3$ subject to the following constraints.

$$\begin{cases} \frac{3}{2}x_1+2x_2+\frac{7}{4}x_3 \leq 1200 \\ 3x_1+3x_2+\frac{5}{2}x_3 \leq 2100 \\ 6x_1+3x_2+4x_3 \leq 3720 \end{cases}$$

We introduce the slack variables s_1, s_2, and s_3 for the three constraints to obtain the following system (which includes the equation corresponding to the objective function in the bottom row).

$$\begin{cases} \frac{3}{2}x_1+2x_2+\frac{7}{4}x_3+s_1 &= 1200 \\ 3x_1+3x_2+\frac{5}{2}x_3 \quad +s_2 &= 2100 \\ 6x_1+3x_2+4x_3 \quad\quad +s_3 &= 3720 \\ -\frac{3}{2}x_1-\frac{6}{5}x_2-\frac{8}{5}x_3 \quad\quad\quad +z &= 0 \end{cases}$$

The initial simplex tableau is as follows.

$$
\begin{array}{ccccccc}
x_1 & x_2 & x_3 & s_1 & s_2 & s_3 & z \\
\end{array}
$$

$$
\left[
\begin{array}{ccccccc|c}
\dfrac{3}{2} & 2 & \dfrac{7}{4} & 1 & 0 & 0 & 0 & 1200 \\[2mm]
3 & 3 & \dfrac{5}{2} & 0 & 1 & 0 & 0 & 2100 \\[2mm]
6 & 3 & 4 & 0 & 0 & 1 & 0 & 3720 \\[1mm]
\hline
-\dfrac{3}{2} & -\dfrac{6}{5} & -\dfrac{8}{5} & 0 & 0 & 0 & 1 & 0
\end{array}
\right]
$$

Since $-\dfrac{8}{5}$ is the negative entry in the bottom row with largest absolute value, the third column is our pivot column. To determine the pivot row, we look at the quotients formed by dividing each entry in column 8 by the entry in column 3. We have $\dfrac{1200}{\frac{7}{4}} = \dfrac{4800}{7} \approx 685.714$, $\dfrac{2100}{\frac{5}{2}} = 840$, and $\dfrac{3720}{4} = 930$. Thus our pivot row is row 1, and our pivot element is $\dfrac{7}{4}$.

Now we do a sequence of row operations using the pivot row so that all entries in the pivot column are 0 except for the 1 in the pivot position.

$$
\left[
\begin{array}{ccccccc|c}
\dfrac{3}{2} & 2 & \boxed{\dfrac{7}{4}} & 1 & 0 & 0 & 0 & 1200 \\[2mm]
3 & 3 & \dfrac{5}{2} & 0 & 1 & 0 & 0 & 2100 \\[2mm]
6 & 3 & 4 & 0 & 0 & 1 & 0 & 3720 \\[1mm]
\hline
-\dfrac{3}{2} & -\dfrac{6}{5} & -\dfrac{8}{5} & 0 & 0 & 0 & 1 & 0
\end{array}
\right]
\xrightarrow{\frac{4}{7}R_1}
\left[
\begin{array}{ccccccc|c}
\dfrac{6}{7} & \dfrac{8}{7} & 1 & \dfrac{4}{7} & 0 & 0 & 0 & \dfrac{4800}{7} \\[2mm]
3 & 3 & \dfrac{5}{2} & 0 & 1 & 0 & 0 & 2100 \\[2mm]
6 & 3 & 4 & 0 & 0 & 1 & 0 & 3720 \\[1mm]
\hline
-\dfrac{3}{2} & -\dfrac{6}{5} & -\dfrac{8}{5} & 0 & 0 & 0 & 1 & 0
\end{array}
\right]
$$

$$
\xrightarrow{-\frac{5}{2}R_1+R_2}
\left[
\begin{array}{ccccccc|c}
\dfrac{6}{7} & \dfrac{8}{7} & 1 & \dfrac{4}{7} & 0 & 0 & 0 & \dfrac{4800}{7} \\[2mm]
\dfrac{6}{7} & \dfrac{1}{7} & 0 & -\dfrac{10}{7} & 1 & 0 & 0 & \dfrac{2700}{7} \\[2mm]
6 & 3 & 4 & 0 & 0 & 1 & 0 & 3720 \\[1mm]
\hline
-\dfrac{3}{2} & -\dfrac{6}{5} & -\dfrac{8}{5} & 0 & 0 & 0 & 1 & 0
\end{array}
\right]
$$

$$\xrightarrow{2R_1+R_3}
\begin{array}{ccccccc}
x_1 & x_2 & x_3 & s_1 & s_2 & s_3 & z
\end{array}
\left[
\begin{array}{ccccccc|c}
\dfrac{6}{7} & \dfrac{8}{7} & 1 & \dfrac{4}{7} & 0 & 0 & 0 & \dfrac{4800}{7} \\[2ex]
\dfrac{6}{7} & \dfrac{1}{7} & 0 & -\dfrac{10}{7} & 1 & 0 & 0 & \dfrac{2700}{7} \\[2ex]
\dfrac{18}{7} & -\dfrac{11}{7} & 0 & -\dfrac{16}{7} & 0 & 1 & 0 & \dfrac{6840}{7} \\[2ex]
\hline
-\dfrac{3}{2} & -\dfrac{6}{5} & -\dfrac{8}{5} & 0 & 0 & 0 & 1 & 0
\end{array}
\right]$$

$$\xrightarrow{\frac{8}{5}R_1+R_4}
\begin{array}{ccccccc}
x_1 & x_2 & x_3 & s_1 & s_2 & s_3 & z
\end{array}
\left[
\begin{array}{ccccccc|c}
\dfrac{6}{7} & \dfrac{8}{7} & 1 & \dfrac{4}{7} & 0 & 0 & 0 & \dfrac{4800}{7} \\[2ex]
\dfrac{6}{7} & \dfrac{1}{7} & 0 & -\dfrac{10}{7} & 1 & 0 & 0 & \dfrac{2700}{7} \\[2ex]
\dfrac{18}{7} & -\dfrac{11}{7} & 0 & -\dfrac{16}{7} & 0 & 1 & 0 & \dfrac{6840}{7} \\[2ex]
\hline
-\dfrac{9}{70} & \dfrac{22}{35} & 0 & \dfrac{32}{35} & 0 & 0 & 1 & \dfrac{7680}{7}
\end{array}
\right]$$

The $-\dfrac{9}{70}$ in the bottom row indicates that our next pivot column is column 1. Also,

$$\dfrac{\dfrac{4800}{7}}{\dfrac{6}{7}}=800,\ \dfrac{\dfrac{2700}{7}}{\dfrac{6}{7}}=450,\text{ and }\dfrac{\dfrac{6840}{7}}{\dfrac{18}{7}}=380.\text{ Since 380 is the smallest quotient, our}$$

pivot row is row 3 and our pivot element is $\dfrac{18}{7}$. Once again, we do a sequence of row

operations using the pivot row so that all entries in the pivot column are 0 except for
the 1 in the pivot position.

$$\begin{array}{ccccccc}
x_1 & x_2 & x_3 & s_1 & s_2 & s_3 & z
\end{array}
\left[
\begin{array}{ccccccc|c}
\dfrac{6}{7} & \dfrac{8}{7} & 1 & \dfrac{4}{7} & 0 & 0 & 0 & \dfrac{4800}{7} \\[2ex]
\dfrac{6}{7} & \dfrac{1}{7} & 0 & -\dfrac{10}{7} & 1 & 0 & 0 & \dfrac{2700}{7} \\[2ex]
\boxed{\dfrac{18}{7}} & -\dfrac{11}{7} & 0 & -\dfrac{16}{7} & 0 & 1 & 0 & \dfrac{6840}{7} \\[2ex]
\hline
-\dfrac{9}{70} & \dfrac{22}{35} & 0 & \dfrac{32}{35} & 0 & 0 & 1 & \dfrac{7680}{7}
\end{array}
\right]
\xrightarrow{-\frac{1}{3}R_3+R_1}
\begin{array}{ccccccc}
x_1 & x_2 & x_3 & s_1 & s_2 & s_3 & z
\end{array}
\left[
\begin{array}{ccccccc|c}
0 & \dfrac{5}{3} & 1 & \dfrac{4}{3} & 0 & -\dfrac{1}{3} & 0 & 360 \\[2ex]
\dfrac{6}{7} & \dfrac{1}{7} & 0 & -\dfrac{10}{7} & 1 & 0 & 0 & \dfrac{2700}{7} \\[2ex]
\dfrac{18}{7} & -\dfrac{11}{7} & 0 & -\dfrac{16}{7} & 0 & 1 & 0 & \dfrac{6840}{7} \\[2ex]
\hline
-\dfrac{9}{70} & \dfrac{22}{35} & 0 & \dfrac{32}{35} & 0 & 0 & 1 & \dfrac{7680}{7}
\end{array}
\right]$$

$$\xrightarrow{-\frac{1}{3}R_3+R_2}
\begin{array}{ccccccc}
x_1 & x_2 & x_3 & s_1 & s_2 & s_3 & z
\end{array}
\left[
\begin{array}{ccccccc|c}
0 & \dfrac{5}{3} & 1 & \dfrac{4}{3} & 0 & -\dfrac{1}{3} & 0 & 360 \\[2ex]
0 & \dfrac{2}{3} & 0 & -\dfrac{2}{3} & 1 & -\dfrac{1}{3} & 0 & 60 \\[2ex]
\dfrac{18}{7} & -\dfrac{11}{7} & 0 & -\dfrac{16}{7} & 0 & 1 & 0 & \dfrac{6840}{7} \\[2ex]
\hline
-\dfrac{9}{70} & \dfrac{22}{35} & 0 & \dfrac{32}{35} & 0 & 0 & 1 & \dfrac{7680}{7}
\end{array}
\right]$$

$$\frac{7}{18}R_3 \longrightarrow \quad \begin{array}{c} \begin{array}{ccccccc} x_1 & x_2 & x_3 & s_1 & s_2 & s_3 & z \end{array} \\ \left[\begin{array}{ccccccc|c} 0 & \dfrac{5}{3} & 1 & \dfrac{4}{3} & 0 & -\dfrac{1}{3} & 0 & 360 \\[2mm] 0 & \dfrac{2}{3} & 0 & -\dfrac{2}{3} & 1 & -\dfrac{1}{3} & 0 & 60 \\[2mm] 1 & -\dfrac{11}{8} & 0 & -\dfrac{8}{9} & 0 & \dfrac{7}{18} & 0 & 380 \\[2mm] -\dfrac{9}{70} & \dfrac{22}{35} & 0 & \dfrac{32}{35} & 0 & 0 & 1 & \dfrac{7680}{7} \end{array}\right] \end{array}$$

$$\frac{9}{70}R_3 + R_4 \longrightarrow \quad \begin{array}{c} \begin{array}{ccccccc} x_1 & x_2 & x_3 & s_1 & s_2 & s_3 & z \end{array} \\ \left[\begin{array}{ccccccc|c} 0 & \dfrac{5}{3} & 1 & \dfrac{4}{3} & 0 & -\dfrac{1}{3} & 0 & 360 \\[2mm] 0 & \dfrac{2}{3} & 0 & -\dfrac{2}{3} & 1 & -\dfrac{1}{3} & 0 & 60 \\[2mm] 1 & -\dfrac{11}{8} & 0 & -\dfrac{8}{9} & 0 & \dfrac{7}{18} & 0 & 380 \\[2mm] 0 & \dfrac{11}{20} & 0 & \dfrac{4}{5} & 0 & \dfrac{1}{20} & 1 & 1146 \end{array}\right] \end{array}$$

We no longer have any negative entries in the bottom row. From the final simplex tableau, we obtain the following solution.

$$(x_1, x_2, x_3, s_1, s_2, s_3, z) = (380, 0, 360, 0, 60, 0, 1146)$$

Therefore, the couple should make 380 apple pies, no grape pies, and 360 blueberry pies to make a maximum revenue of $1146.

Note: In our solution, $s_1 = 0$ means all of the sugar is used, $s_2 = 60$ means 60 cups of flour will remain unused, and $s_3 = 0$ means the couple will need to work the full 62 hours.

Example 4: A Maximization Problem with Multiple Solutions

Use the simplex method to maximize $f(x_1, x_2) = 6x_1 + 8x_2$ subject to the following constraints.

$$\begin{cases} 3x_1 + 4x_2 \le 48 \\ 2x_1 + 5x_2 \le 80 \end{cases}$$

Solution

Since we only have two decision variables, we will first consider the graphical approach. The feasible region for this problem is shown in the graph.

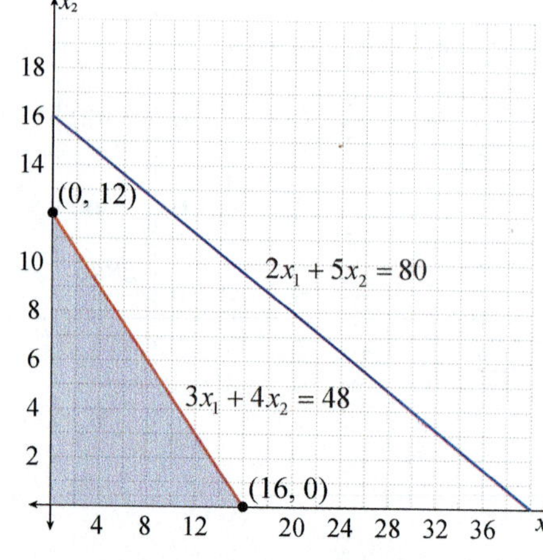

The value of $f(x_1, x_2)$ at each vertex is in the following table.

Vertex	$f(x_1, x_2)$
(0, 0)	0
(16, 0)	96
(0, 12)	96

Therefore, the maximum value of 96 is attained at multiple points. In fact, the maximum value of 96 is attained at all points on the line $3x_1 + 4x_2 = 48$ because for each of these points, $f(x_1, x_2) = 6x_1 + 8x_2 = 2(3x_1 + 4x_2) = 2(48) = 96$.

Next, we'll look at what happens when we apply the simplex method. If we set $z = f(x_1, x_2)$, we obtain the equation $-6x_1 - 8x_2 + z = 0$ in standard form. After introducing the slack variables s_1 and s_2, we write the system and obtain the following initial simplex tableau.

$$\begin{array}{ccccc} x_1 & x_2 & s_1 & s_2 & z \end{array}$$
$$\left[\begin{array}{ccccc|c} 3 & 4 & 1 & 0 & 0 & 48 \\ 2 & 5 & 0 & 1 & 0 & 80 \\ \hline -6 & -8 & 0 & 0 & 1 & 0 \end{array}\right]$$

Our pivot column is column 2 since -8 is the negative entry in the bottom row with largest absolute value. Moreover, row 1 is our pivot row since $\dfrac{48}{4} = 12$ is smaller than $\dfrac{80}{5} = 16$. We apply the pivoting process in column 2 as follows.

$$\begin{array}{ccccc} x_1 & x_2 & s_1 & s_2 & z \end{array}$$
$$\left[\begin{array}{ccccc|c} 3 & \boxed{4} & 1 & 0 & 0 & 48 \\ 2 & 5 & 0 & 1 & 0 & 80 \\ \hline -6 & -8 & 0 & 0 & 1 & 0 \end{array}\right] \xrightarrow{2R_1 + R_3} \left[\begin{array}{ccccc|c} 3 & 4 & 1 & 0 & 0 & 48 \\ 2 & 5 & 0 & 1 & 0 & 80 \\ \hline 0 & 0 & 2 & 0 & 1 & 96 \end{array}\right]$$

$$\xrightarrow{\frac{1}{4}R_1} \begin{array}{ccccc} x_1 & x_2 & s_1 & s_2 & z \end{array}$$
$$\left[\begin{array}{ccccc|c} \frac{3}{4} & 1 & \frac{1}{4} & 0 & 0 & 12 \\ 2 & 5 & 0 & 1 & 0 & 80 \\ \hline 0 & 0 & 2 & 0 & 1 & 96 \end{array}\right]$$

$$\xrightarrow{-5R_1 + R_2} \begin{array}{ccccc} x_1 & x_2 & s_1 & s_2 & z \end{array}$$
$$\left[\begin{array}{ccccc|c} \frac{3}{4} & 1 & \frac{1}{4} & 0 & 0 & 12 \\ -\frac{7}{4} & 0 & -\frac{5}{4} & 1 & 0 & 20 \\ \hline 0 & 0 & 2 & 0 & 1 & 96 \end{array}\right]$$

Notice that x_2 and s_2 are basic variables in this tableau since each of their corresponding columns has a single entry of 1 and all other entries are 0. In addition, x_1 and s_1 are nonbasic. In the previous examples, columns corresponding to nonbasic variables have nonzero entries in the bottom row. This example is different because x_1 is a nonbasic variable but its entry in the bottom row is 0. The bottom row corresponds to the equation $2s_1 + z = 96$, or equivalently $z = 96 - 2s_1$. Therefore, increasing s_1 by any amount will decrease z. Thus, the feasible solution of $(x_1, x_2, s_1, s_2, z) = (0, 12, 0, 20, 96)$ tells us that $f(x_1, x_2) = 6x_1 + 8x_2$ attains a maximum value of 96 at the point $(x_1, x_2) = (0, 12)$.

If we continue the pivoting process in column 1, it won't matter which row (either 1 or 2) we choose to determine our pivot position. This is because the bottom entry of the first column is already 0, and thus we don't need to apply any further row operations to change the bottom row. Moreover the bottom row will still correspond to the equation $z = 96 - 2s_1$. To see if there are other feasible solutions, we use the row 1, column 1 entry (namely $\dfrac{3}{4}$) as the pivot element and obtain following simplex tableau.

$$\begin{array}{ccccc} x_1 & x_2 & s_1 & s_2 & z \end{array}$$
$$\left[\begin{array}{ccccc|c} \boxed{\dfrac{3}{4}} & 1 & \dfrac{1}{4} & 0 & 0 & 12 \\[2mm] -\dfrac{7}{4} & 0 & -\dfrac{5}{4} & 1 & 0 & 20 \\[2mm] \hline 0 & 0 & 2 & 0 & 1 & 96 \end{array}\right] \xrightarrow[\frac{4}{3}R_1]{\frac{7}{3}R_1+R_2} \begin{array}{ccccc} x_1 & x_2 & s_1 & s_2 & z \end{array} \left[\begin{array}{ccccc|c} 1 & \dfrac{4}{3} & \dfrac{1}{3} & 0 & 0 & 16 \\[2mm] 0 & \dfrac{7}{3} & -\dfrac{2}{3} & 1 & 0 & 48 \\[2mm] \hline 0 & 0 & 2 & 0 & 1 & 96 \end{array}\right]$$

In this simplex tableau, we once again have a nonbasic variable (this time x_2) whose entry in the bottom row is 0. The feasible solution $(x_1, x_2, s_1, s_2, z) = (16, 0, 0, 48, 96)$ tells us that $f(x_1, x_2) = 6x_1 + 8x_2$ attains a maximum value of 96 at the point $(x_1, x_2) = (16, 0)$. If we were to continue the pivoting process, we may find other points at which the objective function attains a maximum value of 96.

Thus we conclude that $f(x_1, x_2) = 6x_1 + 8x_2$ attains a maximum value of 96 at multiple points. In addition, we observe that when using the simplex method to solve maximization problems with multiple solutions, there is some point in the process where a column corresponding to a nonbasic variable in a simplex tableau has an entry in the bottom row of 0. Furthermore, we can use this column as a pivot column to discover an additional solution.

Example 5: A Maximization Problem with No Solution

Use the simplex method to maximize $f(x_1, x_2, x_3) = x_1 + x_2 + 2x_3$ subject to the following constraints.

$$\begin{cases} x_1 - 2x_2 + x_3 \le 2 \\ x_1 - 4x_2 + 2x_3 \le 4 \end{cases}$$

Solution

If we set $z = f(x_1, x_2, x_3)$, we obtain the equation $-x_1 - x_2 - 2x_3 + z = 0$ in standard form. After introducing the slack variables s_1 and s_2, we write the system and obtain the following initial simplex tableau.

$$\begin{array}{ccccccc} x_1 & x_2 & x_3 & s_1 & s_2 & z \end{array}$$
$$\left[\begin{array}{cccccc|c} 1 & -2 & 1 & 1 & 0 & 0 & 2 \\ 1 & -4 & 2 & 0 & 1 & 0 & 4 \\ \hline -1 & -1 & -2 & 0 & 0 & 1 & 0 \end{array}\right]$$

Our pivot column is column 3 since -2 is the negative entry in the bottom row with largest absolute value. Since $\dfrac{2}{1} = 2$ and $\dfrac{4}{2} = 2$, we could choose either row 1 or row 2 to be our pivot row. Since there's a 1 in the row 1, column 3 entry, we'll use that as our pivot element. We then obtain the following simplex tableau.

$$\begin{array}{cccccc} x_1 & x_2 & x_3 & s_1 & s_2 & z \end{array}$$
$$\left[\begin{array}{cccccc|c} 1 & -2 & \boxed{1} & 1 & 0 & 0 & 2 \\ 1 & -4 & 2 & 0 & 1 & 0 & 4 \\ \hline -1 & -1 & -2 & 0 & 0 & 1 & 0 \end{array}\right] \xrightarrow[2R_1+R_3]{-2R_1+R_2} \begin{array}{cccccc} x_1 & x_2 & x_3 & s_1 & s_2 & z \end{array} \left[\begin{array}{cccccc|c} 1 & -2 & 1 & 1 & 0 & 0 & 2 \\ -1 & 0 & 0 & -2 & 1 & 0 & 0 \\ \hline 1 & -5 & 0 & 2 & 0 & 1 & 4 \end{array}\right]$$

This time, column 2 has a negative entry in the bottom row. Thus, we choose column 2 to be the pivot column. However, this example is different than previous ones because now there are no positive entries in the second column. Thus, the method we used in earlier examples to identify a pivot row won't work here. What does this mean?

Note that the bottom row corresponds to the equation $x_1 - 5x_2 + 2s_1 + z = 4$, or equivalently $z = 4 - x_1 + 5x_2 - 2s_1$. Since increasing x_1 and s_1 will decrease z (and because x_1 and s_1 are nonbasic variables), z will be optimized if $x_1 = 0$ and $s_1 = 0$. The second row corresponds to the equation $-x_1 - 2s_1 + s_2 = 0$, and if we substitute $x_1 = 0$ and $s_1 = 0$ into this equation, we obtain $s_2 = 0$. The first row corresponds to the equation $x_1 - 2x_2 + x_3 + s_1 = 2$, and if we substitute $x_1 = 0$ and $s_1 = 0$ into this equation, we obtain $-2x_2 + x_3 = 2$; or equivalently $x_3 = 2 + x_2$. Going back to $z = 4 - x_1 + 5x_2 - 2s_1$, we see that increasing x_2 will increase the value of z. However, looking at $x_3 = 2 + x_2$, we see that increasing x_2 will increase the value of x_3.

Therefore, if we increase the size of x_2 without bound, we will obtain a value of x_3 that gets infinitely large, and thus $f(x_1, x_2, x_3) = x_1 + x_2 + 2x_3$ gets infinitely large.

So, we conclude that $f(x_1, x_2, x_3)$ has no maximum value because it is unbounded. In addition, we observe that when using the simplex method to solve maximization problems with no solution, there is some point in the process in which a pivot column in a simplex tableau will have no positive entries.

7.3 EXERCISES

♥ PRACTICE

For each given simplex tableau:
    a. identify the basic and nonbasic variables,
    b. find a basic feasible solution corresponding to this tableau,
    c. find the pivot element, and
    d. perform one pivot operation.

1.
$$\begin{array}{ccccc|c} x_1 & x_2 & s_1 & s_2 & z & \\ 3 & 1 & 1 & 0 & 0 & 6 \\ 1 & 2 & 0 & 1 & 0 & 4 \\ \hline -7 & -5 & 0 & 0 & 1 & 0 \end{array}$$

2.
$$\begin{array}{ccccc|c} x_1 & x_2 & s_1 & s_2 & z & \\ 3 & 0 & 1 & 4 & 0 & 10 \\ 6 & 1 & 0 & 2 & 0 & 15 \\ 9 & 0 & 0 & -8 & 1 & 30 \end{array}$$

3.
$$\begin{array}{cccccc|c} x_1 & x_2 & x_3 & s_1 & s_2 & z & \\ 1 & 1 & 1 & 1 & 0 & 0 & 5 \\ 3 & 0 & 3 & -1 & 1 & 0 & 3 \\ \hline -3 & 0 & -11 & 7 & 0 & 1 & 35 \end{array}$$

4.
$$\begin{array}{ccccccc|c} x_1 & x_2 & x_3 & s_1 & s_2 & s_3 & z & \\ 9 & 0 & 3 & 1 & -1 & 0 & 0 & 54 \\ 1 & 1 & 1 & 0 & 1 & 0 & 0 & 18 \\ 3 & 0 & 7 & 0 & -1 & 1 & 0 & 58 \\ \hline -14 & 0 & -6 & 0 & 6 & 0 & 1 & 108 \end{array}$$

For each given maximization problem, find an optimal solution if one exists. If there are multiple solutions, find at least two. If there are no solutions, explain why. See Examples 1, 4, and 5.

5. Maximize $f(x_1, x_2) = 5x_1 + 6x_2$ subject to the following constraints.

$$\begin{cases} x_1 + 2x_2 \leq 10 \\ 3x_1 + x_2 \leq 15 \end{cases}$$

6. Maximize $f(x_1, x_2) = 3x_1 + x_2$ subject to the following constraints.

$$\begin{cases} x_1 + 4x_2 \leq 5 \\ 2x_1 + x_2 \leq 3 \\ 3x_1 + 2x_2 \leq 5 \end{cases}$$

7. Maximize $f(x_1, x_2) = 4x_1 + 8x_2$ subject to the following constraints.

$$\begin{cases} 3x_1 + x_2 \leq 6 \\ x_1 + 2x_2 \leq 4 \end{cases}$$

8. Maximize
$f(x_1, x_2, x_3) = 5x_1 + 7x_2 + 2x_3$
subject to the following constraints.

$$\begin{cases} x_1 + 2x_2 - x_3 \leq 5 \\ 2x_1 + x_2 - 2x_3 \leq 4 \end{cases}$$

9. Maximize
$f(x_1, x_2, x_3) = 5x_1 + 6x_2 + 3x_3$
subject to the following constraints.

$$\begin{cases} 3x_1 + x_2 + x_3 \leq 18 \\ x_1 + 4x_2 + x_3 \leq 18 \\ x_1 + x_2 + 2x_3 \leq 19 \end{cases}$$

10. Maximize $f(x_1, x_2) = 7x_1 - 3x_2$ subject to the following constraints.

$$\begin{cases} x_1 + x_2 \leq 15 \\ 4x_1 + 2x_2 \leq 21 \end{cases}$$

11. Maximize
$f(x_1, x_2, x_3) = 6x_1 + 5x_2 + 4x_3$
subject to the following constraints.

$$\begin{cases} 2x_1 + x_2 \leq 46 \\ x_1 + 3x_3 \leq 54 \\ x_1 + x_2 + x_3 \leq 60 \end{cases}$$

12. Maximize
$f(x_1, x_2, x_3) = 7x_1 + 4x_2 + x_3$
subject to the following constraints.

$$\begin{cases} 2x_1 + x_2 + x_3 \leq 40 \\ x_1 + x_2 + 2x_3 \leq 30 \end{cases}$$

🚀 APPLICATIONS

Use the simplex method to solve each given application. See Examples 2 and 3.

13. A tool company manufactures drills and table saws. Suppose it costs the company $30 to produce each drill, and the company makes a profit of $6 for each drill sold. Moreover, it costs the company $80 to produce each table saw, and the company makes a profit of $11 for each table saw sold. In addition, suppose the company does not expect the total daily demand for the tools to exceed 100, and the company will halt production for the day once they spend a total of $4500 in inventory. How many drills and table saws should the company produce and sell to optimize daily profit?

14. A company sells two kinds of fruit juices: Tropical Blend and Florida Sunshine. Suppose the Tropical Blend juice is 40% orange juice and 60% pineapple juice, and the Florida Sunshine juice is 60% orange juice and 40% pineapple juice. The company has a daily supply of 100 gallons of orange juice and 120 gallons of pineapple juice. In addition, suppose the company makes a profit of $0.50 per gallon of Tropical Blend sold and $0.40 per gallon of Florida Sunshine sold. How many gallons of Tropical Blend and Florida Sunshine juice should be sold in order for the company to make an optimal daily profit?

15. A company produces and sells two models of roller skates. The following table summarizes how long it takes (in minutes) to assemble and package one pair of each model.

	A	B
Assemble	45	60
Package	10	5

    Suppose the company makes a $3.50 profit for each pair of model A roller skates sold and a $4 profit for each pair of model B roller skates sold. In addition, suppose the time available for assembling and packaging is 1200 hours and 200 hours, respectively. How many pairs of each model of roller skate should the company sell to maximize profit?

16. A candy store sells 2 kinds of trail mix. The standard trail mix consists of 50% peanuts and 50% M&M's, and the Chocolate Lover's trail mix consists of 37.5% peanuts and 62.5% M&M's. Suppose the candy store has a daily inventory of 100 pounds of peanuts and 150 pounds of M&M's. In addition, suppose the store makes $0.25 in profit for each pound of standard trail mix sold and $0.20 in profit for every pound of Chocolate Lover's sold. How many pounds of each kind of trail mix should the store sell to maximize daily profit?

17. A contractor builds two types of homes. Type A requires one lot, $200,000 in capital, and 165 worker-days of labor. Type B requires one lot, $250,000 of capital, and 150 worker-days of labor. The contractor owns 170 lots, has $40,000,000 available in capital, and has 27,600 worker-days of labor. The profit for selling each Type A home is $45,000, and the profit for each Type B home is $56,000. How many of each type of home should the contractor build to maximize profit?

18. A shoe manufacturer makes running shoes and basketball shoes. Each pair of running shoes requires 2 units of leather, 5 units of cotton, and 1 unit of rubber. Each pair of basketball shoes requires 4 units of leather, 4 units of cotton, and 2 units of rubber. Shipments are such that leather is limited to 120 units per day, cotton is limited to 150 units per day, and rubber is limited to 80 units per day. If the profits of the shoe manufacturer are $15 per pair of running shoes and $25 per pair of basketball shoes, how many pairs of each kind of shoe should be produced to maximize profit? What is the maximum profit?

19. A landlord has a brand new, empty apartment complex with 110 units available for rent; 25 of them have one bedroom, 50 have two bedrooms, and the other 35 have three bedrooms. He has set the rent at $550 per month for a one-bedroom unit, $850 per month for two bedrooms, and $1150 per month for three bedrooms. Assume that he must rent to one person per bedroom, and the laws restrict him to at most 200 occupants in the complex. How many of each type of apartment should the landlord rent to maximize the revenue? What is the maximum revenue?

20. A manufacturer produces 3 models of wooden rocking chairs. The time required for assembling, finishing, and packaging for each model are as follows.

	Model A	Model B	Model C
Assembling	2 hr 30 min	2 hr 30 min	3 hr 30 min
Finishing	1 hr 30 min	1 hr 20 min	1 hr 30 min
Packaging	30 min	45 min	1 hr

The total time for assembling, finishing, and packaging is 4500 hours, 2200 hours, and 1200 hours, respectively. The profit for each model is $55 (Model A), $60 (Model B), and $77 (Model C). How many of each type should be produced to maximize profit? What is the maximum profit?

21. A furniture store manufactures bookshelves, bed frames, and recliners. Each bookshelf uses 9 units of wood, each bed frame uses 23 units of wood, and each recliner uses 5 units of wood. Each bookshelf uses 1 unit of steel, each bed frame uses 4.5 units of steel, and each recliner uses 7.5 units of steel. It takes 35 minutes to assemble a bookshelf, 80 minutes to assemble a bed frame, and 75 minutes to assemble a recliner. Suppose the store has 600 units of wood and 200 units of steel in its inventory. Also, suppose the total time available for assembly is 3025 minutes. Finally, the price the store sells each item is $200 (bookshelf), $550 (bed frame), and $600 (recliner). How many of each kind of furniture should the store produce to maximize revenue? What is the maximum revenue?

22. A grower has 60 acres of land available for planting 3 crops. It costs $150 to produce an acre of carrots and the profit is $50 per acre. It costs $120 to produce an acre of potatoes and the profit is $42 per acre. Finally, it costs $180 to produce an acre of beets and the profit is $55 per acre. Assume the grower's cost cannot exceed $8700, and the grower must plant at least 10 acres of beets. How many acres of each crop should the grower plant in order to maximize profit? What is the maximum profit?

23. A shoe company is advertising their latest running shoe. They have a budget of $8000 per month for advertising. Newspaper ads cost $120 each and can occur a maximum of 20 times per month. Radio ads cost $400 each and can occur a maximum of 24 times a month. Suppose each newspaper ad reaches 8000 women under the age of 24 and each radio ad reaches 10,000 women of this age group. If the company wants to maximize its ad exposure to women under 24, how many of each ad should it purchase? What is the maximum number of exposures?

24. A candidate running for president wants to purchase radio and television ads to maximize his exposure. Suppose he has a $1 million budget for ads in his campaign, and election day is 6 months away. Each radio ad costs him $800 and can occur a maximum of 100 times over the next 6 months. Each television ad costs $4000, and can occur a maximum of 300 times over the next 6 months. Suppose each radio ad reaches 20 thousand voters and each television ad reaches 80 thousand voters. How many of each kind of ad should this presidential candidate purchase in order to maximize how many voters he reaches?

25. A researcher is studying how the rabbit population grows in two controlled environments. He wants each male rabbit to be in environment A for 30 minutes and each female rabbit to be in environment A for 20 minutes. He wants each male rabbit to be in environment B for 18 minutes and each female rabbit to be in environment B for 30 minutes. Suppose the researcher has the following time limits: 1040 minutes available for environment A and 750 minutes for environment B. How many male and female rabbits should the researcher use to maximize the total number of rabbits he uses in his study?

26. Biologists at a university have a research pond on their grounds that is stocked with bluegill and bass. Suppose each bluegill consumes 2 units of food per day while each bass consumes 8 units of food per day. In addition, suppose each bluegill produces 1 unit of waste per day while each bass produces 5 units of waste per day. Suppose the total daily supply of food for this pond is no more than 220 units. Moreover, in order to sustain life in the pond, the biologists want no more than a total of 120 units of waste excreted in the pond per day, and the bluegill population should be no more than 50. How many bluegill and bass should the biologists stock in this pond to maximize the number of fish they can study?

27. A botanist is studying how three different types of plant (onion, turnip, and radish) respond to three different controlled environments (A, B, and C). The following table describes how many days each plant is to be tested in each environment.

	Onion	Turnip	Radish
Environment A	4	1	3
Environment B	2	3	2
Environment C	1	2	1.5

There is limited time available in these environments: the researcher has 27 days for Environment A, 23 days for Environment B, and 15 days for Environment C. How many of each kind of plant should the researcher test in order to maximize the number of plants she is able to test?

28. A natural habitat in a zoo contains several feeding areas for lions, tigers, and bears. There are 3 foods (A, B, and C) available at each of these feeding areas. The following table describes how many pounds of each type of food is required to feed one animal.

	Food A	Food B	Food C
Lion	4	1	3
Tiger	5	3	2
Bear	3	6	2

Suppose the zoo has the following amounts of food available: 700 pounds (A), 490 pounds (B), and 455 pounds (C). How many of each type of animal can the zoo support so that the number of lions, tigers, and bears is a maximum?

✏ WRITING & THINKING

29. Look once again at Example 4 in this section. Is the constraint $2x_1 + 5x_2 \leq 80$ necessary? Explain your answer.

30. Consider a store that manufactures and sells two brands of clocks, say brand A and brand B. Suppose x_1 and x_2 denote the number of brand A and brand B clocks sold, respectively. Suppose the profit to be maximized (in dollars) is $f(x_1, x_2) = 5x_1 + 6x_2$. Moreover, suppose we have the following constraints.

$$\begin{cases} x_1 + 2x_2 \leq 15 \\ 3x_1 + 2x_2 \leq 26 \end{cases}$$

a. Use the simplex method to find the exact values of x_1 and x_2 that maximize f with the constraints. Do these values make sense in the context of this problem? Explain.

b. If your answer to part a. is no, can you find a solution that is optimal, feasible, and makes sense in the context of this problem? Explain.

31. Recall that it is important to always use the pivot row when doing row operations in any given iteration of the pivoting process. This exercise explores a possible pitfall of failing to do this. Look back at Example 3 of this section. At one stage in this problem, we arrived at the following simplex tableau.

$$
\begin{array}{ccccccc|c}
x_1 & x_2 & x_3 & s_1 & s_2 & s_3 & z & \\
\dfrac{6}{7} & \dfrac{8}{7} & 1 & \dfrac{4}{7} & 0 & 0 & 0 & \dfrac{4800}{7} \\
\dfrac{6}{7} & \dfrac{1}{7} & 0 & -\dfrac{10}{7} & 1 & 0 & 0 & \dfrac{2700}{7} \\
\dfrac{18}{7} & -\dfrac{11}{7} & 0 & -\dfrac{16}{7} & 0 & 1 & 0 & \dfrac{6840}{7} \\
\hline
-\dfrac{9}{70} & \dfrac{22}{35} & 0 & \dfrac{32}{35} & 0 & 0 & 1 & \dfrac{7680}{7}
\end{array}
$$

a. Explain why $\dfrac{18}{7}$ is the pivot element.

b. Since $\dfrac{6}{7}$ appears in rows 1 and 2 of the first column, the row operation $-R_2 + R_1$ seems like an intuitive first step in introducing a 0 in the top left entry. Also note that it doesn't use the pivot row, R_3. Try the sequence of row operations: $-R_2 + R_1$, $-\dfrac{1}{3}R_3 + R_1$, $\dfrac{1}{20}R_3 + R_4$, then $\dfrac{7}{18}R_3$. Do you obtain the optimal feasible solution of $(x_1, x_2, x_3, s_1, s_2, s_3, z) = (380, 0, 360, 0, 60, 0, 1146)$ with this new tableau?

c. How many basic variables are in the tableau you obtained in part b.? Is the number of basic variables different than 3, which is the number of constraints in this problem?

d. Can you do any additional row operations on the tableau you obtained in part b. to introduce another basic variable and arrive at the optimal feasible solution? Explain your answer.

7.4 THE SIMPLEX METHOD: DUALITY AND MINIMIZATION

■ TOPICS

- Simplex Method for Maximization vs. the Duality Principle for Minimization

Previously, we looked at using the simplex method to solve maximization problems. These types of problems arise naturally; for example, in economics where it is common to optimize quantities such as revenue and profit. Another quantity an economist would be interested in optimizing is cost. Naturally, we would like to minimize the cost and this then begs the question: How does one go about solving minimization problems? In this lesson, we discover that the answer to this question involves solving a corresponding maximization problem. Therefore we can adapt the methods of the simplex method we've used before to solving minimization problems.

In general, a minimization problem is of the following form if we have n decision variables. Find the minimum value of $f(x_1, x_2, ..., x_n)$ subject to several constraints; all of which have the form

$$a_1 x_1 + a_2 x_2 + \cdots + a_n x_n \geq c.$$

This idea of solving a minimization problem by way of solving a corresponding maximization problem is summarized in a theorem called the Theorem of Duality. Before formally stating this theorem, it will be helpful to consider an example and establish a few terms.

Example 1: Minimization and Duality

Find the minimum value of $f(x_1, x_2) = 5x_1 + 4x_2$ subject to the following constraints.

$$\begin{cases} 10x_1 + 2x_2 \geq 15 \\ x_1 + 2x_2 \geq 6 \end{cases}$$

Solution

This minimization problem we are to solve is called the **primal problem**. Our first step is to form an augmented matrix for this primal problem. The first rows correspond to the constraints without any slack variables, and the last row corresponds to the objective function $5x_1 + 4x_2 = f$.

$$A = \begin{bmatrix} 10 & 2 & | & 15 \\ 1 & 2 & | & 6 \\ \hline 5 & 4 & | & f \end{bmatrix}$$

This matrix A is called the **matrix for minimization with constraints**. Next, we take the **transpose** of A. This matrix is denoted A^T. It is formed by taking each of the rows in A and writing them as columns in A^T (or one can take the columns of A and write them as rows in A^T).

$$A^T = \begin{bmatrix} 10 & 1 & | & 5 \\ 2 & 2 & | & 4 \\ \hline 15 & 6 & | & g \end{bmatrix}$$

Notice that we relabel the f in A as g in A^T because we will eventually maximize an objective function that is different than f (it will be g). Using A^T, we can write a new problem called the **dual problem** to the primal problem. This problem will be a maximization problem. Using A^T, we state the corresponding dual problem.

Find the maximum value of $g(y_1, y_2) = 15y_1 + 6y_2$ subject to the following constraints.

$$\begin{cases} 10y_1 + y_2 \le 5 \\ 2y_1 + 2y_2 \le 4 \end{cases}$$

Notice that we changed the names of the variables; instead of x_1, x_2, we have y_1 and y_2. The reason for this is that the objective function in the dual problem attains a maximum value at different coordinates than where the objective function in the primal problem attains a minimum value.

At this stage, we can apply the simplex method to solve this maximization problem. We define slack variables $s_1 \ge 0$ and $s_2 \ge 0$ to convert the inequalities in the constraints as equalities, rewrite the objective function as $-15y_1 - 6y_2 + z = 0$ (relabel g as z), and write down the following simplex tableau.

$$\begin{array}{ccccc} y_1 & y_2 & s_1 & s_2 & z \\ \end{array}$$
$$\left[\begin{array}{ccccc|c} 10 & 1 & 1 & 0 & 0 & 5 \\ 2 & 2 & 0 & 1 & 0 & 4 \\ \hline -15 & -6 & 0 & 0 & 1 & 0 \end{array}\right]$$

Now apply the same methods as employed before in choosing a pivot element and performing row operations to introduce a 1 in the pivot position and 0 elsewhere in the pivot column. We find that 10 in column 1 is the pivot element, and after a few row operations, we arrive at the following tableau.

$$\left[\begin{array}{ccccc|c} \boxed{10} & 1 & 1 & 0 & 0 & 5 \\ 2 & 2 & 0 & 1 & 0 & 4 \\ \hline -15 & -6 & 0 & 0 & 1 & 0 \end{array}\right] \xrightarrow[\substack{\frac{1}{10}R_1 \\ -2R_1 + R_2}]{\frac{3}{2}R_1 + R_3} \left[\begin{array}{ccccc|c} 1 & \dfrac{1}{10} & \dfrac{1}{10} & 0 & 0 & \dfrac{1}{2} \\ 0 & \dfrac{9}{5} & -\dfrac{1}{5} & 1 & 0 & 3 \\ \hline 0 & -\dfrac{9}{2} & \dfrac{3}{2} & 0 & 1 & \dfrac{15}{2} \end{array}\right]$$

Next, $\dfrac{9}{5}$ in column 2 is the pivot element. After a few more row operations, we obtain the following tableau.

$$\left[\begin{array}{ccccc|c} 1 & \dfrac{1}{10} & \dfrac{1}{10} & 0 & 0 & \dfrac{1}{2} \\ 0 & \boxed{\dfrac{9}{5}} & -\dfrac{1}{5} & 1 & 0 & 3 \\ \hline 0 & -\dfrac{9}{2} & \dfrac{3}{2} & 0 & 1 & \dfrac{15}{2} \end{array}\right] \xrightarrow[\substack{-\frac{1}{10}R_2 + R_1 \\ \frac{9}{2}R_2 + R_3}]{\frac{5}{9}R_2} \left[\begin{array}{ccccc|c} 1 & 0 & \dfrac{1}{9} & -\dfrac{1}{18} & 0 & \dfrac{1}{3} \\ 0 & 1 & -\dfrac{1}{9} & \dfrac{5}{9} & 0 & \dfrac{5}{3} \\ \hline 0 & 0 & 1 & \dfrac{5}{2} & 1 & 15 \end{array}\right]$$

Therefore, the maximum value of $g(y_1, y_2) = 15y_1 + 6y_2$ is 15, and it is attained at

$$(y_1, y_2) = \left(\frac{1}{3}, \frac{5}{3}\right).$$

To find the minimum value of $f(x_1, x_2) = 5x_1 + 4x_2$, we use duality to conclude that it is the same value as the maximum value of $g(y_1, y_2) = 15y_1 + 6y_2$. Thus, the minimum value of $f(x_1, x_2) = 5x_1 + 4x_2$ is 15. The values of x_1 and x_2 at which the minimum is attained are found in the bottom entries of the columns corresponding to the slack variables. Therefore, the minimum is attained at $(x_1, x_2) = \left(1, \dfrac{5}{2}\right)$.

Since this example has only 2 decision variables, we can check our result with a graphical approach. The graph shows the planar region determined by the constraints.

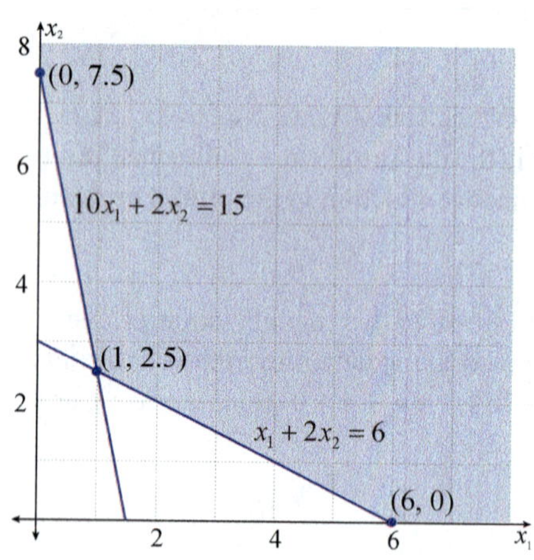

Now we find the value of $f(x_1, x_2)$ at each vertex.

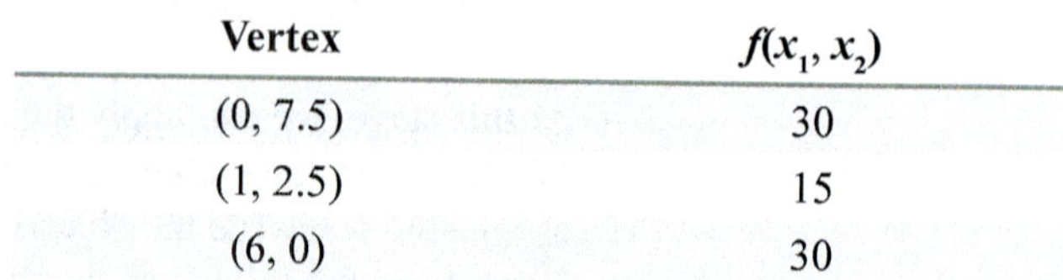

Vertex	$f(x_1, x_2)$
$(0, 7.5)$	30
$(1, 2.5)$	15
$(6, 0)$	30

This confirms that the minimum value of f is 15, and it is attained at $\left(1, \dfrac{5}{2}\right)$.

The Theorem of Duality formalizes the process employed in this example.

Theorem of Duality

If a minimization problem and its dual maximization problem have a solution, the minimum value of the primal problem is the same as the maximum value of the dual problem. Moreover, when the simplex method is used to solve the dual maximization problem, the coordinates at which the minimum value is attained is given by the bottom entry of the columns corresponding to the slack variables of the maximization problem.

Solving Minimization Problems

1. Form the matrix for minimization with constraints, A, for the primal problem.

2. Find A^T, and relabel the objective function in the lower-right.

3. Write down the dual maximization problem. Use variable names different from those in the primal problem.

4. Create the initial simplex tableau by forming the augmented matrix for the system of equations with the objective function in the bottom row.

5. Solve the dual maximization problem using the simplex method.

6. The minimum value of the objective function is in the lower-right corner of the final simplex tableau, and the bottom entry in the columns corresponding to the slack variables give the coordinates at which the minimum is attained.

Example 2: Minimizing Cost

A tool manufacturing company produces drills and saws at two different factories. At factory A, 100 drills and 150 saws can be manufactured in one hour, and the hourly cost is \$800. At factory B, 200 drills and 150 saws are manufactured in one hour, and the hourly cost is \$1000. The company is expecting orders each week for at least 8000 drills and 10,000 saws. How many hours per week should each factory be operated in order to provide the inventory for these orders at the minimum cost?

Solution

First, we let x_1 and x_2 denote the number of operating hours per week for factories A and B, respectively. The objective function to be minimized is $f(x_1, x_2) = 800x_1 + 1000x_2$. We obtain the first constraint by considering the number of drills that can be produced. We obtain $100x_1 + 200x_2 \geq 8000$. Similarly we obtain $150x_1 + 150x_2 \geq 10{,}000$ by looking at the number of saws that can be produced.

Next, we form the matrix for minimization with constraints A as follows.

$$A = \begin{bmatrix} 100 & 200 & 8000 \\ 150 & 150 & 10{,}000 \\ \hline 800 & 1000 & f \end{bmatrix}$$

This matrix corresponds to the primal problem. To find the matrix corresponding to the dual maximization problem, we take A^T and relabel f as g to obtain the following:

$$A^T = \begin{bmatrix} 100 & 150 & 800 \\ 200 & 150 & 100 \\ \hline 8000 & 10{,}000 & g \end{bmatrix}$$

From this, we can state the following dual maximization problem.

Find the maximum value of $g(y_1, y_2) = 8000y_1 + 10{,}000y_2$ subject to the following constraints.

$$\begin{cases} 100y_1 + 150y_2 \leq 800 \\ 200y_1 + 150y_2 \leq 1000 \end{cases}$$

We proceed by applying the simplex method. Letting z denote the function g, the objective function is written as $-8000y_1 - 10{,}000y_2 + z = 0$. If we let s_1 and s_2 denote the slack variables, our initial simplex tableau is as follows.

y_1	y_2	s_1	s_2	z	
100	150	1	0	0	800
200	150	0	0	1	1000
−8000	−10,000	0	0	1	0

Column 2 is our pivot column, and row 1 is our pivot row since $\dfrac{800}{150} \approx 5.33$ is less than $\dfrac{1000}{150} \approx 6.67$. The first set of row operations (with the pivot element boxed) leads to the following simplex tableau.

$$\begin{bmatrix} y_1 & y_2 & s_1 & s_2 & z & \\ 100 & \boxed{150} & 1 & 0 & 0 & 800 \\ 200 & 150 & 0 & 1 & 0 & 1000 \\ -8000 & -10{,}000 & 0 & 0 & 1 & 0 \end{bmatrix} \xrightarrow[\frac{1}{150}R_1]{\substack{-R_1+R_2 \\ \frac{200}{3}R_1+R_3}} \begin{bmatrix} y_1 & y_2 & s_1 & s_2 & z & \\ \frac{2}{3} & 1 & \frac{1}{150} & 0 & 0 & \frac{16}{3} \\ 100 & 0 & -1 & 1 & 0 & 200 \\ -\frac{4000}{3} & 0 & \frac{200}{3} & 0 & 1 & \frac{160{,}000}{3} \end{bmatrix}$$

Column 1 is the next pivot column, and row 2 is the next pivot row since $\dfrac{200}{100} = 2$ is less than $\dfrac{\frac{16}{3}}{\frac{2}{3}} = 8$. A few more row operations lead to the following simplex tableau.

$$\begin{bmatrix} y_1 & y_2 & s_1 & s_2 & z & \\ \frac{2}{3} & 1 & \frac{1}{150} & 0 & 0 & \frac{16}{3} \\ \boxed{100} & 0 & -1 & 1 & 0 & 200 \\ -\frac{4000}{3} & 0 & \frac{200}{3} & 0 & 1 & \frac{160{,}000}{3} \end{bmatrix} \xrightarrow[-\frac{2}{3}R_2+R_1]{\substack{\frac{1}{100}R_2 \\ \frac{4000}{3}R_2+R_3}} \begin{bmatrix} y_1 & y_2 & s_1 & s_2 & z & \\ 0 & 1 & \frac{1}{75} & -\frac{1}{150} & 0 & 4 \\ 1 & 0 & -\frac{1}{100} & \frac{1}{100} & 0 & 2 \\ 0 & 0 & \frac{160}{3} & \frac{40}{3} & 1 & 56{,}000 \end{bmatrix}$$

According the Theorem of Duality, the minimum value of $f(x_1, x_2) = 800x_1 + 1000x_2$ is 56,000. Moreover, this minimum value is attained at $(x_1, x_2) = \left(\dfrac{160}{3}, \dfrac{40}{3} \right)$.

In the context of this problem, we reach the following conclusion: The company minimizes cost when factory A operates at $\dfrac{160}{3}$ hours (53 hours and 20 minutes) per week and factory B operates at $\dfrac{40}{3}$ hours (13 hours and 20 minutes) per week, and the minimum cost to the manufacturing company each week is \$56,000.

Example 3: Low-Carb Diet

Suppose a salad bar has three different options of salads. The table below gives the protein content (in grams), vitamin C content (in milligrams), calorie count, and carbohydrate content (in grams) in a 2.5-ounce serving of each salad.

Salad	Protein	Vitamin C	Calorie Count	Carbs
A	6 g	40 mg	60	16 g
B	4 g	10 mg	30	8 g
C	2 g	30 mg	120	12 g

Suppose that you are on a low-carb diet in which you need a minimum of 12 grams of protein, 70 milligrams of vitamin C, and 240 calories. How many 2.5-ounce servings of each salad should you eat at this salad bar to minimize the carbohydrate content?

Solution

First, we set x_1 to be the number of servings of salad A, x_2 to be the number of servings of salad B, and x_3 to be the number of servings of salad C. Our objective is to minimize the function $f(x_1, x_2, x_3) = 16x_1 + 8x_2 + 12x_3$. The constraints are as follows:

$$\begin{cases} 6x_1 + 4x_2 + 2x_3 \geq 12 & \text{(Protein)} \\ 40x_1 + 10x_2 + 30x_3 \geq 70 & \text{(Vitamin C)} \\ 60x_1 + 30x_2 + 120x_3 \geq 240 & \text{(Calories)} \end{cases}$$

The matrix for minimization with constraints for the primal problem is as follows.

$$A = \left[\begin{array}{ccc|c} 6 & 4 & 2 & 12 \\ 40 & 10 & 30 & 70 \\ 60 & 30 & 120 & 240 \\ \hline 16 & 8 & 12 & f \end{array}\right]$$

To obtain the corresponding matrix for the dual problem, we take the transpose of A and relabel f as g.

$$A^T = \left[\begin{array}{ccc|c} 6 & 40 & 60 & 16 \\ 4 & 10 & 30 & 8 \\ 2 & 30 & 120 & 12 \\ \hline 12 & 70 & 240 & g \end{array}\right]$$

The dual maximization problem is to maximize $g(y_1, y_2, y_3) = 12y_1 + 70y_2 + 240y_3$ subject to the following constraints.

$$\begin{cases} 6y_1 + 40y_2 + 60y_3 \leq 16 \\ 4y_1 + 10y_2 + 30y_3 \leq 8 \\ 2y_1 + 30y_2 + 120y_3 \leq 12 \end{cases}$$

After setting $z = g(x_1, x_2, x_3)$, rewriting the objective function as $-12y_1 - 70y_2 - 240y_3 + z = 0$, and defining the slack variables s_1, s_2, and s_3, we obtain the following initial simplex tableau.

$$\begin{array}{ccccccc} y_1 & y_2 & y_3 & s_1 & s_2 & s_3 & z \\ \end{array}$$
$$\left[\begin{array}{ccccccc|c} 6 & 40 & 60 & 1 & 0 & 0 & 0 & 16 \\ 4 & 10 & 30 & 0 & 1 & 0 & 0 & 8 \\ 2 & 30 & 120 & 0 & 0 & 1 & 0 & 12 \\ \hline -12 & -70 & -240 & 0 & 0 & 0 & 1 & 0 \end{array}\right]$$

Since -240 is the negative entry in the bottom row with largest absolute value, column 3 will be our pivot column. Moreover, $\dfrac{12}{120} = 0.1$ is a smaller quotient than $\dfrac{8}{30}$ and $\dfrac{16}{60}$ (both being approximately 0.2667). Thus 120 is the pivot element, and we obtain the following simplex tableau.

$$\begin{array}{ccccccc} y_1 & y_2 & y_3 & s_1 & s_2 & s_3 & z \\ \end{array}$$
$$\left[\begin{array}{ccccccc|c} 6 & 40 & 60 & 1 & 0 & 0 & 0 & 16 \\ 4 & 10 & 30 & 0 & 1 & 0 & 0 & 8 \\ 2 & 30 & \boxed{120} & 0 & 0 & 1 & 0 & 12 \\ \hline -12 & -70 & -240 & 0 & 0 & 0 & 1 & 0 \end{array}\right]$$

$$\xrightarrow{\;2R_3 + R_4\;}$$
$$\xrightarrow{\;-\frac{1}{2}R_3 + R_1\;}$$
$$\xrightarrow{\;-\frac{1}{4}R_3 + R_2\;}$$
$$\xrightarrow{\;\frac{1}{120}R_3\;}$$

$$\begin{array}{ccccccc} y_1 & y_2 & y_3 & s_1 & s_2 & s_3 & z \\ \end{array}$$
$$\left[\begin{array}{ccccccc|c} 5 & 25 & 0 & 1 & 0 & -\dfrac{1}{2} & 0 & 10 \\[2mm] \dfrac{7}{2} & \dfrac{5}{2} & 0 & 0 & 1 & -\dfrac{1}{4} & 0 & 5 \\[2mm] \dfrac{1}{60} & \dfrac{1}{4} & 1 & 0 & 0 & \dfrac{1}{120} & 0 & \dfrac{1}{10} \\[2mm] \hline -8 & -10 & 0 & 0 & 0 & 2 & 1 & 24 \end{array}\right]$$

Column 2 is our next pivot column. We have a tie for the smallest quotients in rows 1 and 3, since $\dfrac{10}{25} = \dfrac{\frac{1}{10}}{\frac{1}{4}} = 0.4$; if we use 25 as our next pivot element, we obtain the following simplex tableau:

$$
\begin{array}{ccccccc|c}
y_1 & y_2 & y_3 & s_1 & s_2 & s_3 & z & \\
5 & \boxed{25} & 0 & 1 & 0 & -\dfrac{1}{2} & 0 & 10 \\
\dfrac{7}{2} & \dfrac{5}{2} & 0 & 0 & 1 & -\dfrac{1}{4} & 0 & 5 \\
\dfrac{1}{60} & \dfrac{1}{4} & 1 & 0 & 0 & \dfrac{1}{120} & 0 & \dfrac{1}{10} \\
\hline
-8 & -10 & 0 & 0 & 0 & 2 & 1 & 24
\end{array}
$$

$$\xrightarrow[\ \ -\frac{1}{4}R_1+R_3\ \ ]{\ \ \frac{1}{25}R_1,\ \ -\frac{5}{2}R_1+R_2,\ \ 10R_1+R_4\ \ }$$

$$
\begin{array}{ccccccc|c}
y_1 & y_2 & y_3 & s_1 & s_2 & s_3 & z & \\
\dfrac{1}{5} & 1 & 0 & \dfrac{1}{25} & 0 & -\dfrac{1}{50} & 0 & \dfrac{2}{5} \\
3 & 0 & 0 & -\dfrac{1}{10} & 1 & -\dfrac{1}{5} & 0 & 4 \\
-\dfrac{1}{30} & 0 & 1 & -\dfrac{1}{100} & 0 & \dfrac{1}{75} & 0 & 0 \\
\hline
-6 & 0 & 0 & \dfrac{2}{5} & 0 & \dfrac{9}{5} & 1 & 28
\end{array}
$$

Column 1 is our next pivot column, but there is a $-\dfrac{1}{30}$ entry in this column. At this point, we have a basic feasible solution of $(y_1, y_2, y_3, s_1, s_2, s_3, z) = \left(0, \dfrac{2}{5}, 0, 0, 4, 0, 28\right)$, and the basic variables are y_2, y_3, and s_2. It would be helpful to do the pivoting process again using column 1 so that y_1 becomes a basic variable instead of s_2. Since there is a 1 in the second row in the column corresponding to s_2, we'll choose row 2 as the next pivot row (thus 3 is the next pivot element). We obtain the following simplex tableau.

$$
\begin{array}{ccccccc|c}
y_1 & y_2 & y_3 & s_1 & s_2 & s_3 & z & \\
\dfrac{1}{5} & 1 & 0 & \dfrac{1}{25} & 0 & -\dfrac{1}{50} & 0 & \dfrac{2}{5} \\
\boxed{3} & 0 & 0 & -\dfrac{1}{10} & 1 & -\dfrac{1}{5} & 0 & 4 \\
-\dfrac{1}{30} & 0 & 1 & -\dfrac{1}{100} & 0 & \dfrac{1}{75} & 0 & 0 \\
\hline
-6 & 0 & 0 & \dfrac{2}{5} & 0 & \dfrac{9}{5} & 1 & 28
\end{array}
$$

$$\xrightarrow[\ \ \frac{1}{30}R_2+R_3\ \ ]{\ \ \frac{1}{3}R_2,\ \ -\frac{1}{5}R_2+R_1,\ \ 6R_2+R_4\ \ }$$

$$
\begin{array}{ccccccc|c}
y_1 & y_2 & y_3 & s_1 & s_2 & s_3 & z & \\
0 & 1 & 0 & \dfrac{7}{150} & -\dfrac{1}{15} & -\dfrac{1}{150} & 0 & \dfrac{2}{15} \\
1 & 0 & 0 & -\dfrac{1}{30} & \dfrac{1}{3} & -\dfrac{1}{15} & 0 & \dfrac{4}{3} \\
0 & 0 & 1 & -\dfrac{1}{90} & \dfrac{1}{90} & \dfrac{1}{90} & 0 & \dfrac{2}{45} \\
\hline
0 & 0 & 0 & \dfrac{1}{5} & 2 & \dfrac{7}{5} & 1 & 36
\end{array}
$$

We are now done because there are no negative entries in the last row. By the Theorem of Duality, the minimum value of $f(x_1, x_2, x_3) = 16x_1 + 8x_2 + 12x_3$ is 36. Moreover, this minimum value is attained at $(x_1, x_2, x_3) = \left(\dfrac{1}{5}, 2, \dfrac{7}{5}\right)$.

Therefore, we should choose $\dfrac{1}{5}$ servings of salad A, 2 servings of salad B, and $\dfrac{7}{5}$ servings of salad C. Since each serving is 2.5 ounces, we can write the conclusion as follows: We need to consume 0.5 ounces of salad A, 5 ounces of salad B, and 3.5 ounces of salad C to meet the requirements of the diet and minimize the carbohydrate intake to 36 grams.

Simplex Method for Maximization vs. the Duality Principle for Minimization

The following table gives a summary and a comparison of the techniques for maximization and minimization for solving optimization problems.

Simplex Method: Maximization	**Duality Principle: Minimization**
1. Maximize $f(x_1, x_2, \ldots, x_n)$ subject to constraints of the form $$a_1x_1 + a_2x_2 + \cdots + a_nx_n \leq c.$$	**1.** Minimize $f(x_1, x_2, \ldots, x_n)$ subject to constraints of the form $$a_1x_1 + a_2x_2 + \cdots + a_nx_n \geq c.$$
2. Introduce slack variables $s_1, s_2, \ldots, s_m$, and form the initial simplex tableau.	**2.** Write down the matrix for minimization with constraints, A.
3. Repeat the pivoting process until an optimal solution is reached.	**3.** Write down the matrix for the dual maximization problem, A^T.
4. Look at the final simplex tableau. The maximum value of f is in the lower-right corner. The nonzero coordinates at which the maximum is attained are in the last column in the rows corresponding to the 1 entry of each x_i that is basic, and each x_i that is nonbasic is 0.	**4.** Introduce slack variables $s_1, s_2, \ldots, s_m$, and form the initial simplex tableau.
	5. Repeat the pivoting process until an optimal solution to the dual problem is reached.
	6. Look at the final simplex tableau. The minimum value of f is in the lower-right corner. The coordinates at which the minimum is attained are in the last row of the columns corresponding to the slack variables.

7.4 EXERCISES

PRACTICE

For each given matrix A, find the transpose A^T.

1. $A = \begin{bmatrix} 2 & 3 & 7 \\ 4 & 6 & 2 \end{bmatrix}$

2. $A = \begin{bmatrix} 1 & 3 & 2 \\ 3 & 0 & 1 \\ 2 & 1 & 4 \end{bmatrix}$

3. $A = \begin{bmatrix} 1 & 3 \\ 6 & -5 \\ 7 & 2 \\ -1 & 0 \end{bmatrix}$

4. $A = \begin{bmatrix} 5 & -1 & 2 & 4 \\ 6 & 10 & -3 & 8 \\ 4 & -1 & 3 & 7 \end{bmatrix}$

For each given (primal) minimization problem, state the dual maximization problem.

5. Minimize $f(x_1, x_2) = 7x_1 + 2x_2$ subject to the following constraints.

$$\begin{cases} 3x_1 + 5x_2 \geq 18 \\ 2x_1 + 6x_2 \geq 15 \end{cases}$$

6. Minimize $f(x_1, x_2) = 3x_1 + 8x_2$ subject to the following constraints.

$$\begin{cases} x_1 \geq 3 \\ 2x_1 + x_2 \geq 5 \end{cases}$$

7. Minimize $f(x_1, x_2, x_3) = 3x_1 + x_2 + 2x_3$ subject to the following constraints.

$$\begin{cases} 2x_1 + 3x_2 \geq 16 \\ 5x_1 + 4x_3 \geq 20 \end{cases}$$

8. Minimize $f(x_1, x_2, x_3) = 3x_1 + 4x_2 + 5x_3$ subject to the following constraints.

$$\begin{cases} x_1 + x_2 + x_3 \geq 8 \\ 2x_1 + 2x_2 + x_3 \geq 14 \\ x_1 + 3x_2 + x_3 \geq 12 \end{cases}$$

Solve each given minimization problem. See Example 1.

9. Minimize $f(x_1, x_2) = 3x_1 + x_2$ subject to the following constraints.

$$\begin{cases} 4x_1 + 5x_2 \geq 13 \\ 2x_1 + x_2 \geq 5 \end{cases}$$

10. Minimize $f(x_1, x_2) = 2x_1 + 4x_2$ subject to the following constraints.

$$\begin{cases} 3x_1 + x_2 \geq 6 \\ x_1 + x_2 \geq 4 \end{cases}$$

11. Minimize $f(x_1, x_2, x_3) = x_1 + 2x_2 + x_3$ subject to the following constraints.

$$\begin{cases} x_1 + 3x_2 \geq 6 \\ 2x_1 + x_3 \geq 10 \end{cases}$$

12. Minimize $f(x_1, x_2, x_3) = 11x_1 + 8x_2 + 7x_3$ subject to the following constraints.

$$\begin{cases} x_1 + 2x_2 + 2x_3 \geq 8 \\ x_1 + 2x_2 + x_3 \geq 6 \\ 4x_1 + x_2 + x_3 \geq 4 \end{cases}$$

13. Minimize $f(x_1, x_2, x_3) = 7x_1 + 10x_2 + 7x_3$ subject to the following constraints.

$$\begin{cases} x_1 + x_2 + 2x_3 \geq 5 \\ x_1 + 3x_2 \geq 6 \\ 2x_1 + x_2 + x_3 \geq 4 \end{cases}$$

14. Minimize $f(x_1, x_2) = 3x_1 + 4x_2$ subject to the following constraints.

$$\begin{cases} 2x_1 + 2x_2 \geq 7 \\ 2x_1 + 3x_2 \geq 10 \end{cases}$$

15. Minimize $f(x_1, x_2, x_3) = 8x_1 + 3x_2 + 7x_3$ subject to the following constraints.

$$\begin{cases} x_1 + x_2 + x_3 \geq 9 \\ 2x_1 + x_2 + 2x_3 \geq 11 \\ x_1 + 3x_2 + 3x_3 \geq 17 \end{cases}$$

16. Minimize $f(x_1, x_2) = 6x_1 + 8x_2$ subject to the following constraints.

$$\begin{cases} x_1 + x_2 \geq 8 \\ 3x_2 \geq 20 \\ -x_1 + x_2 \geq 1 \end{cases}$$

APPLICATIONS

Use the simplex method to solve each given application. See Examples 2 and 3.

17. A tool company produces wrenches and screwdrivers at factories in two locations (plant A and plant B). It has orders to produce 12,000 wrenches and 10,000 screwdrivers. For each hour, plant A can produce 400 wrenches and 200 screwdrivers, and the operation costs are $1200 per hour. At plant B, 200 wrenches and 500 screwdrivers can be produced in an hour, and the operation costs are $1500 per hour. How many hours should the company run each plant to fill the orders at a minimum cost? What is the minimum cost?

18. An electronics factory produces televisions and DVD players on two assembly lines. Line 1 can produce 50 televisions and 75 DVD players at a cost of $225 per hour. Line 2 can produce 60 televisions and 60 DVD players at a cost of $210 per hour. The factory needs to produce at least 300 televisions and 375 DVD players to fill an order. How many hours should each line be run to fill the order at a minimum cost? What is the minimum cost?

19. An athlete drinks two kinds of dietary drinks to help maintain a healthy diet. Drink I has 3 units of protein, 3 units of carbohydrates, and 2 units of vitamin D per one-quart (32 oz) bottle. Drink II has 2 units of protein, 3 units of carbohydrates, and 5 units of vitamin D per bottle. Suppose each bottle of drink I costs $8 and each bottle of drink II costs $12. Finally, suppose the athlete wants to consume a minimum of 7 units of protein per day, 9 units of carbohydrates per day, and 12 units of vitamin D per day. Find the combination of drinks that the athlete should purchase and consume per day in order to minimize the cost while meeting the dietary requirements.

20. A swine farmer is considering two different types of feed for fattening his pigs. Brand 1 feed costs 40 cents per pound and brand 2 feed costs 35 cents per pound. Brand 1 feed contains 10 units of fiber and 4 units of protein per pound. Brand 2 feed has 12 units of fiber and 3 units of protein per pound. Suppose the farmer would like his pigs to consume at least 160 units of fiber and 55 units of protein for each feeding. How many pounds of each feed should the farmer buy per feeding to satisfy the nutritional requirements at a minimum cost?

21. A petroleum company owns 3 refineries (A, B, and C), and each one produces 3 grades of oil (5W30, 10W30, and 10W40). The following table summarizes how many barrels of each kind of oil per day each refinery produces for the petroleum company.

	5W30	10W30	10W40
Refinery A	300	300	200
Refinery B	200	400	250
Refinery C	250	300	450

Suppose the daily costs to operate the refineries are $25,000 (A), $30,000 (B), and $40,000 (C). Also, suppose the petroleum company has orders totaling 5300 barrels of 5W30, 6900 barrels of 10W30, and 6500 barrels of 10W40. How many days should the petroleum company run each of the three refineries so that the costs of operation are minimized and yet the orders are fulfilled?

22. Three factories dump waste water containing three kinds of pollutants into a lagoon. In order to reduce the pollution levels, the factories must treat the waste water they dump. The following table shows the possible percent reduction of each pollutant at each factory.

	Factory 1	Factory 2	Factory 3
Pollutant 1	40%	10%	30%
Pollutant 2	50%	15%	20%
Pollutant 3	30%	25%	35%

Suppose the costs to dump each ton of waste at each factory is as follows: $100 (factory 1), $40 (factory 2), and $60 (factory 3). Moreover, suppose the state requires a reduction of at least 110 tons per day of pollutant 1, at least 120 tons per day of pollutant 2, and at least 122 tons per day of pollutant 3. Find the number of tons of waste that must be treated each day in order to minimize the cost of treatment. What is the minimum cost?

23. Biologists in a lab feed two different foods (I and II) to mice; each food contains three ingredients (A, B, and C). Food I has 6 units of ingredient A per gram, 2 units of ingredient B per gram, and 1 unit of ingredient C per gram. Food II has 4 units of ingredient A per gram, 8 units of ingredient B per gram, and 3 units of ingredient C per gram. Suppose that the lab requires the mice to have consumed a total of least 60 grams of ingredient A and 50 grams of ingredient B. Also suppose that ingredient C can be harmful to the mice. How many grams of each type of food should be given to the mice to meet the requirements for ingredients A and B, while at the same time minimizing the amount of ingredient C?

24. A nutritionist advises an individual who is suffering from iron and vitamin B deficiency to take at least 3000 milligrams (mg) of iron, 2000 mg of vitamin B1, and 1500 mg of vitamin B2 over a period of time. Two vitamin pills are suitable, brand A and brand B. Each brand A pill costs 10 cents and contains 50 mg of iron, 15 mg of vitamin B1, and 10 mg of vitamin B2. Each brand B pill costs 15 cents and contains 10 mg of iron, 28 mg of vitamin B1, and 22 mg of vitamin B2. What combination of pills should the individual purchase in order to meet the minimum iron and vitamin requirements at the lowest cost?

25. A candidate wishes to use a combination of radio and television advertisements in his campaign. Suppose that every one-minute spot on the radio reaches 10 thousand voters, and every one-minute spot on television reaches 70 thousand voters. The candidate feels that he needs to reach at least 3 million voters, and he must buy at least 90 minutes of advertisements. If a radio ad costs $75 per minute, and a television ad costs $450 per minute, how many minutes of each medium should the candidate use to minimize the cost of advertisements?

26. Biologists at a university have a research pond on their grounds that is stocked with bass and catfish. Suppose each bass consumes 8 units of food per day while each catfish consumes 20 units of food per day. In addition, suppose each bass produces 6 units of waste per day while each catfish produces 14 units of waste per day. Suppose the biologists require that the total amount of food the fish eat in this pond is at least 220 units. In addition, the biologists want to keep at least 20 fish in this pond. How many bass and catfish should the biologists stock in this pond to minimize the amount of waste produced?

27. A dieting company offers three types of packaged meals (I, II, and III), and it groups its customers into three groups (A, B, and C) depending on the diet. The following table gives the percent of the daily nutritional requirements that a serving of each meal provides and the number of ounces of detrimental substances in each serving of food.

	Meal I	Meal II	Meal III
Group A	20% per serving	30% per serving	10% per serving
Group B	30% per serving	10% per serving	10% per serving
Group C	10% per serving	20% per serving	20% per serving
Detrimental Substances	0.4 oz per serving	0.5 oz per serving	0.2 oz per serving

Determine the combination of food types that will provide at least 100% of the daily requirements for each group and will minimize the amount of detrimental substances.

✏ WRITING & THINKING

28. Suppose we want to minimize $f(x_1, x_2) = 4x_1 + 6x_2$ subject to the following constraints.

$$\begin{cases} 2x_1 + 2x_2 \geq 5 \\ x_1 - 3x_2 \leq -6 \end{cases}$$

   a. Notice that this optimization problem is different because one of the constraints has the $\leq$ inequality. Rewrite the second constraint so that it has the $\geq$ inequality.
   b. Use the methods of this section to solve this problem.

29. Suppose we want to minimize $f(x_1, x_2) = 4x_1 + 8x_2$ subject to the following constraints.

$$\begin{cases} -3x_1 + 6x_2 \geq 1 \\ x_1 - 2x_2 \geq 3 \\ x_1 + x_2 \geq 2 \end{cases}$$

   a. Try using the methods of this section to solve this problem. What do you find?
   b. Since there are only two decision variables, draw a graph of the feasible region. Then use the graphical approach to solve this problem.

30. Suppose we want to minimize $f(x_1, x_2) = 6x_1 + 8x_2$ subject to the following constraints.

$$\begin{cases} 3x_1 + 4x_2 \geq 7 \\ 2x_1 + x_2 \geq 3 \end{cases}$$

Find two solutions to this problem. (**Hint:** At some point, you should obtain a tableau in which you can either choose row 1 or row 2 as the pivot row. Try each one.)

7.5 THE SIMPLEX METHOD: MIXED CONSTRAINTS

■ TOPICS

■ Comparison of the Optimization Methods

We have previously applied the simplex method to maximization and minimization problems. In particular, we recall that all constraints for the maximization problems are written in standard form as $a_1 x_1 + a_2 x_2 + \cdots + a_n x_n \leq c$. Similarly, all constraints were equivalent to $a_1 x_1 + a_2 x_2 + \cdots + a_n x_n \geq c$ for the minimization problems. We next consider solving optimization problems involving a combination of these two constraints. In addition, a third type of constraint, namely one of the form $a_1 x_1 + a_2 x_2 + \cdots + a_n x_n = c$, may arise in some of the problems we encounter in this lesson. An optimization problem that involves some combination of these three types of constraints described above are called **mixed constraint problems**. There are two approaches to solving mixed constraint problems. We lay out one procedure for solving such problems with the following example.

Example 1: A Maximization Problem with Mixed Constraints

Maximize the objective function $f(x_1, x_2) = 5x_1 + 7x_2$ subject to the following constraints.

$$\begin{cases} 8x_1 + 4x_2 \leq 28 \\ -3x_1 + 2x_2 \geq 7 \end{cases}$$

Solution

In previous maximization problems, all constraints in standard form involved the $\leq$ inequality. Since the second constraint has the $\geq$ inequality, we multiply the second constraint by -1 and rewrite this as $3x_1 - 2x_2 \leq -7$.

Next, we form a simplex tableau with the usual steps: we introduce $z = f(x_1, x_2)$, rewrite the objective function as $-5x_1 - 7x_2 + z = 0$, introduce slack variables s_1 and s_2, and write the following tableau.

$$\begin{array}{ccccc|c} x_1 & x_2 & s_1 & s_2 & z & \\ 8 & 4 & 1 & 0 & 0 & 28 \\ 3 & -2 & 0 & 1 & 0 & -7 \\ \hline -5 & -7 & 0 & 0 & 1 & 0 \end{array}$$

Note the negative entry -7 in the last column above the bottom row. Because of this, we cannot apply the simplex method to this tableau. We will call the above tableau a **preliminary simplex tableau** because it is in a form such that the simplex method cannot be applied. We need to do at least one iteration of the pivoting process, hoping that this will result in this entry changing sign because this simplex tableau implies that $s_2 = -7$, violating the condition that slack variables are nonnegative. Note that there is another entry in row 2 that is negative, namely -2. We choose this to be our pivot column. Then, since $\dfrac{-7}{-2} = 3.5$ is a smaller positive quotient than $\dfrac{28}{7} = 4$, we let the second row be the pivot row (so -2 is our pivot element).

The pivoting process leads to the following simplex tableau.

$$
\begin{array}{ccccc|c}
x_1 & x_2 & s_1 & s_2 & z & \\
8 & 4 & 1 & 0 & 0 & 28 \\
3 & \boxed{-2} & 0 & 1 & 0 & -7 \\
-5 & -7 & 0 & 0 & 1 & 0
\end{array}
\xrightarrow[\substack{-\frac{1}{2}R_2 \\ 7R_2+R_3}]{2R_2+R_1}
\begin{array}{ccccc|c}
x_1 & x_2 & s_1 & s_2 & z & \\
14 & 0 & 1 & 2 & 0 & 14 \\
-\frac{3}{2} & 1 & 0 & -\frac{1}{2} & 0 & \frac{7}{2} \\
-\frac{31}{2} & 0 & 0 & -\frac{7}{2} & 1 & \frac{49}{2}
\end{array}
$$

Now we can proceed with the simplex method since there are no negative entries in the last column above the bottom row in the updated simplex tableau. We will refer to this tableau as the **initial simplex tableau** because it is in a form on which we can apply the simplex method. Our next pivot column is 1, and since the only positive quotient is $\dfrac{14}{14} = 1$ in the first row, row 1 will be our next pivot row. We obtain the following simplex tableau.

$$
\begin{array}{ccccc|c}
x_1 & x_2 & s_1 & s_2 & z & \\
\boxed{14} & 0 & 1 & 2 & 0 & 14 \\
-\frac{3}{2} & 1 & 0 & -\frac{1}{2} & 0 & \frac{7}{2} \\
-\frac{31}{2} & 0 & 0 & -\frac{7}{2} & 1 & \frac{49}{2}
\end{array}
\xrightarrow[\substack{\frac{3}{2}R_1+R_2 \\ \frac{31}{2}R_1+R_3}]{\frac{1}{14}R_1}
\begin{array}{ccccc|c}
x_1 & x_2 & s_1 & s_2 & z & \\
1 & 0 & \frac{1}{14} & \frac{1}{7} & 0 & 1 \\
0 & 1 & \frac{3}{28} & -\frac{2}{7} & 0 & 5 \\
0 & 0 & \frac{31}{28} & -\frac{9}{7} & 1 & 40
\end{array}
$$

Column 3 is our next pivot column because of the negative entry in the bottom row, and row 1 is our next pivot row because $\dfrac{\frac{1}{14}}{\frac{1}{7}} = 7$ is the only positive quotient in the column. We obtain the following simplex tableau.

$$
\begin{array}{ccccc|c}
x_1 & x_2 & s_1 & s_2 & z & \\
1 & 0 & \frac{1}{14} & \boxed{\frac{1}{7}} & 0 & 1 \\
0 & 1 & \frac{3}{28} & -\frac{2}{7} & 0 & 5 \\
0 & 0 & \frac{31}{28} & -\frac{9}{7} & 1 & 40
\end{array}
\xrightarrow[\substack{9R_1+R_3 \\ 7R_1}]{2R_1+R_2}
\begin{array}{ccccc|c}
x_1 & x_2 & s_1 & s_2 & z & \\
7 & 0 & \frac{1}{2} & 1 & 0 & 7 \\
2 & 1 & \frac{1}{4} & 0 & 0 & 7 \\
9 & 0 & \frac{7}{4} & 0 & 1 & 49
\end{array}
$$

We are done since there are no negative entries in the bottom row. Thus, the objective function $f(x_1, x_2) = 5x_1 + 7x_2$ attains a maximum value of 49 at $(x_1, x_2) = (0, 7)$.

Next, we will consider a minimization problem.

> **Example 2: A Minimization Problem with Mixed Constraints**

Minimize $f(x_1, x_2) = 4x_1 + 3x_2$ subject to the following constraints.

$$\begin{cases} 6x_1 - 5x_2 \leq 4 \\ 2x_1 + 5x_2 \geq 8 \end{cases}$$

Solution

Instead of minimizing f, we will think of this as a maximization problem by multiplying the objective function by -1. Thus, we will maximize $-f(x_1, x_2) = -4x_1 - 3x_2$. Moreover, we will rewrite constraints that contain $\geq$ so that they use $\leq$ instead. We multiply the second constraint by -1 to obtain the updated constraint $-2x_1 - 5x_2 \leq -8$. We let $z = -f(x_1, x_2)$, write the objective function as $4x_1 + 3x_2 + z = 0$, and introduce slack variables s_1 and s_2 to obtain the following preliminary simplex tableau.

$$\begin{array}{ccccc|c} x_1 & x_2 & s_1 & s_2 & z & \\ \hline 6 & -5 & 1 & 0 & 0 & 4 \\ -2 & -5 & 0 & 1 & 0 & -8 \\ \hline 4 & 3 & 0 & 0 & 1 & 0 \end{array}$$

Again, we have a negative entry in the last column above the last row; namely, -8. We look at the other entries in this same row (row 2) as the -8 and note the entries -2 and -5. We can choose either of these columns to be our pivot column. If we choose column 2 to be our pivot column, we look at the quotients $\dfrac{4}{-5} = -0.8$ and $\dfrac{-8}{-5} = 1.6$. Row 2 is our pivot row since the quotient is positive. The pivoting process yields the following tableau.

$$\begin{array}{ccccc|c} x_1 & x_2 & s_1 & s_2 & z & \\ \hline 6 & -5 & 1 & 0 & 0 & 4 \\ -2 & \boxed{-5} & 0 & 1 & 0 & -8 \\ \hline 4 & 3 & 0 & 0 & 1 & 0 \end{array} \quad \begin{array}{c} \xrightarrow{-R_2 + R_1} \\ \xrightarrow{-\frac{1}{5}R_2} \\ \xrightarrow{-3R_2 + R_3} \end{array} \quad \begin{array}{ccccc|c} x_1 & x_2 & s_1 & s_2 & z & \\ \hline 8 & 0 & 1 & -1 & 0 & 12 \\ \frac{2}{5} & 1 & 0 & -\frac{1}{5} & 0 & \frac{8}{5} \\ \hline \frac{14}{5} & 0 & 0 & \frac{3}{5} & 1 & -\frac{24}{5} \end{array}$$

Since there are no negative entries in the last row, we don't need to do any more pivoting. Hence, we have found that $-f(x_1, x_2) = -4x_1 - 3x_2$ attains a maximum value of $-\dfrac{24}{5}$ at $(x_1, x_2) = \left(0, \dfrac{8}{5}\right)$. Therefore, the objective function $f(x_1, x_2) = 4x_1 + 3x_2$ obtains a minimum value of $\dfrac{24}{5}$ at $(x_1, x_2) = \left(0, \dfrac{8}{5}\right)$.

Solving Mixed Constraint Problems Using the Simplex Method

1. If the optimization problem involves minimizing an objective function, f, view it as maximizing $-f$.

2. For each constraint with the $\geq$ inequality (if any), multiply by -1 to change it to a $\leq$ inequality.

3. Introduce slack variables and create the preliminary simplex tableau by forming the augmented matrix for the system of equations with the objective function in the bottom row.

4. If there are any negative entries in the last column above the bottom row, use the following steps.

   a. Choose any other negative value in the same row and use its column as the pivot column.

   b. In each row except for the last one, divide the entry in the far-right column by the one in the pivot column. The row with the smallest positive quotient will be the pivot row.

   c. Complete the iteration of the pivoting process, and repeat **a.** and **b.** until there are no negative entries in the last column above the bottom row.

5. Apply the simplex method to the resulting initial tableau, and arrive at an optimal solution.

Next, we will look at a different method for solving optimization problems with mixed constraints called the **big M** method. Its name comes from a step in the process that introduces a positive constant M that is understood to be very large. We illustrate the big M method in solving the problem in Example 1 again to compare the steps without the big M method and to verify that we arrive at the same solution.

Example 3: Solution to Example 1 Using the Big M Method

Consider once again the problem from Example 1. We want to maximize the objective function $f(x_1, x_2) = 5x_1 + 7x_2$ subject to the following constraints.

$$\begin{cases} 8x_1 + 4x_2 \leq 28 \\ -3x_1 + 2x_2 \geq 7 \end{cases}$$

Solution

For the first constraint, we introduce the slack variable $s_1 \geq 0$ to write $8x_1 + 4x_2 + s_1 = 28$, since it has the $\leq$ inequality. For all constraints with a $\geq$ inequality, we introduce a variable called a **surplus variable**. We will denote this by s_2 and use it in the second constraint. The idea is that the quantity $-3x_1 + 2x_2$ provides a surplus of s_2 beyond 7 in the inequality $-3x_1 + 2x_2 \geq 7$. We write $-3x_1 + 2x_2 - s_2 = 7$. If we form a preliminary simplex tableau now, we would obtain the following:

$$
\begin{array}{ccccc}
x_1 & x_2 & s_1 & s_2 & z \\
\end{array}
$$
$$
\left[\begin{array}{ccccc|c}
8 & 4 & 1 & 0 & 0 & 28 \\
-3 & 2 & 0 & -1 & 0 & 7 \\
\hline
-5 & -7 & 0 & 0 & 1 & 0
\end{array}\right]
$$

This presents two issues. The first one being that the second row implies the equation $-s_2 = 7$ (since x_1 and x_2 are nonbasic variables and thus equal to 0). This is not feasible because s_2 must be nonnegative. Another issue is that only one column besides the z column has a 1 with all other entries being 0 (namely, the one corresponding to s_1); therefore, we only have one basic variable. Recall that the number of basic variables is typically equal to the number of constraints, which is two in our case. To remedy these problems, we introduce new variables called **artificial variables**.

Since we are short one basic variable in our current problem, we need to introduce only one artificial variable. We introduce it in the second constraint since it has the $\geq$ inequality. If we denote it a_1, we can update the equation in the second constraint as $-3x_1 + 2x_2 - s_2 + a_1 = 7$. Notice that if we are to obtain a solution to this problem, we must have $a_1 = 0$. To help us arrive at $a_1 = 0$ (and thus a feasible solution), we subtract a positive constant multiple of a_1, say Ma_1, from the objective function. This is because our objective is to maximize $f(x_1, x_2) = 5x_1 + 7x_2$, and if $a_1 > 0$ the quantity $5x_1 + 7x_2 - Ma_1$ will not be maximized. This positive constant M is understood to be extremely large; so even if a_1 is extremely small, subtracting Ma_1 involves deducting a large amount from the objective function. Hence, we rewrite the objective function as $f(x_1, x_2) = 5x_1 + 7x_2 - Ma_1$. Introducing $z = f(x_1, x_2)$, we write the objective function as the equivalent equation $-5x_1 - 7x_2 + Ma_1 + z = 0$. We can now form the preliminary simplex tableau.

$$
\begin{array}{cccccc}
x_1 & x_2 & s_1 & s_2 & a_1 & z \\
\end{array}
$$
$$
\left[\begin{array}{cccccc|c}
8 & 4 & 1 & 0 & 0 & 0 & 28 \\
-3 & 2 & 0 & -1 & 1 & 0 & 7 \\
\hline
-5 & -7 & 0 & 0 & M & 1 & 0
\end{array}\right]
$$

Even though we have negative entries in the bottom row, we first want to perform any row operations to remove the M in the bottom entry of the column corresponding to a_1. This will yield a tableau in which a_1 becomes a basic variable (and thus giving us the correct number of basic variables). Adding $-M$ times row 2 to row 3 yields the following tableau.

$$
\begin{array}{cccccc}
x_1 & x_2 & s_1 & s_2 & a_1 & z \\
\end{array}
$$
$$
\left[\begin{array}{cccccc|c}
8 & 4 & 1 & 0 & 0 & 0 & 28 \\
-3 & 2 & 0 & -1 & 1 & 0 & 7 \\
\hline
-5 & -7 & 0 & 0 & M & 1 & 0
\end{array}\right]
\xrightarrow{-MR_2 + R_3}
\left[\begin{array}{cccccc|c}
8 & 4 & 1 & 0 & 0 & 0 & 28 \\
-3 & 2 & 0 & -1 & 1 & 0 & 7 \\
\hline
3M-5 & -2M-7 & 0 & 0 & M & 0 & 1 & -7M
\end{array}\right]
$$

We consider the resulting tableau as our initial simplex tableau, and we proceed as usual with the simplex method. Keeping in mind that M is a large positive number, $-2M - 7$ is a negative entry in the last row not in the last column with largest magnitude. Thus, our first pivot column will be column 2. Moreover, the quotient $\dfrac{7}{2} = 3.5$ is smaller than $\dfrac{28}{4} = 7$, and thus row 2 is our pivot row.

The pivoting process results in the following.

$$
\begin{array}{c}
\begin{array}{cccccc} x_1 & x_2 & s_1 & s_2 & a_1 & z \end{array} \\
\left[\begin{array}{cccccc|c}
8 & 4 & 1 & 0 & 0 & 0 & 28 \\
-3 & \boxed{2} & 0 & -1 & 1 & 0 & 7 \\
\hline
3M-5 & -2M-7 & 0 & M & 0 & 1 & -7M
\end{array}\right]
\end{array}
$$

$$
\begin{array}{c}
\xrightarrow{-2R_2+R_1} \\
\xrightarrow{\frac{1}{2}R_2} \\
\xrightarrow{(2M+7)R_2+R_3}
\end{array}
\begin{array}{c}
\begin{array}{cccccc} x_1 & x_2 & s_1 & s_2 & a_1 & z \end{array} \\
\left[\begin{array}{cccccc|c}
14 & 0 & 1 & 2 & -2 & 0 & 14 \\
-\dfrac{3}{2} & 1 & 0 & -\dfrac{1}{2} & \dfrac{1}{2} & 0 & \dfrac{7}{2} \\
-\dfrac{31}{2} & 0 & 0 & -\dfrac{7}{2} & M+\dfrac{7}{2} & 1 & \dfrac{49}{2}
\end{array}\right]
\end{array}
$$

At this point, we have a feasible solution of $\left(x_1,x_2,s_1,s_2,a_1,z\right)=\left(0,\dfrac{7}{2},14,0,0,\dfrac{49}{2}\right)$.

Our next pivot column is column 1, and since $\dfrac{14}{14}=1$ is our only positive quotient, row 1 is our next pivot row. A few more operations yield the following simplex tableau.

$$
\begin{array}{c}
\begin{array}{cccccc} x_1 & x_2 & s_1 & s_2 & a_1 & z \end{array} \\
\left[\begin{array}{cccccc|c}
\boxed{14} & 0 & 1 & 2 & -2 & 0 & 14 \\
-\dfrac{3}{2} & 1 & 0 & -\dfrac{1}{2} & \dfrac{1}{2} & 0 & \dfrac{7}{2} \\
-\dfrac{31}{2} & 0 & 0 & -\dfrac{7}{2} & M+\dfrac{7}{2} & 1 & \dfrac{49}{2}
\end{array}\right]
\end{array}
\begin{array}{c}
\xrightarrow{\frac{1}{14}R_1} \\
\xrightarrow{\frac{3}{2}R_1+R_2} \\
\xrightarrow{\frac{31}{2}R_1+R_3}
\end{array}
\begin{array}{c}
\begin{array}{cccccc} x_1 & x_2 & s_1 & s_2 & a_1 & z \end{array} \\
\left[\begin{array}{cccccc|c}
1 & 0 & \dfrac{1}{14} & \dfrac{1}{7} & -\dfrac{1}{7} & 0 & 1 \\
0 & 1 & \dfrac{3}{28} & -\dfrac{2}{7} & \dfrac{2}{7} & 0 & 5 \\
0 & 0 & \dfrac{31}{28} & -\dfrac{9}{7} & M+\dfrac{9}{7} & 1 & 40
\end{array}\right]
\end{array}
$$

Finally, we find that $\dfrac{1}{7}$ is our pivot element (row 1, column 4), and we obtain the following.

$$
\begin{array}{c}
\begin{array}{cccccc} x_1 & x_2 & s_1 & s_2 & a_1 & z \end{array} \\
\left[\begin{array}{cccccc|c}
1 & 0 & \dfrac{1}{14} & \boxed{\dfrac{1}{7}} & -\dfrac{1}{7} & 0 & 1 \\
0 & 1 & \dfrac{3}{28} & -\dfrac{2}{7} & \dfrac{2}{7} & 0 & 5 \\
0 & 0 & \dfrac{31}{28} & -\dfrac{9}{7} & M+\dfrac{9}{7} & 1 & 40
\end{array}\right]
\end{array}
\begin{array}{c}
\xrightarrow{7R_1} \\
\xrightarrow{\frac{2}{7}R_1+R_2} \\
\xrightarrow{\frac{9}{7}R_1+R_3}
\end{array}
\begin{array}{c}
\begin{array}{cccccc} x_1 & x_2 & s_1 & s_2 & a_1 & z \end{array} \\
\left[\begin{array}{cccccc|c}
7 & 0 & \dfrac{1}{2} & 1 & -1 & 0 & 7 \\
2 & 1 & \dfrac{1}{4} & 0 & 0 & 0 & 7 \\
9 & 0 & \dfrac{7}{4} & 0 & M & 1 & 49
\end{array}\right]
\end{array}
$$

We are done because there are no negative entries in the bottom row, and a_1 is a nonbasic variable (thus making $a_1=0$, as desired). We have reached the same solution as in Example 1; the objective function $f(x_1,x_2)=5x_1+7x_2$ attains a maximum of 49 at $(x_1,x_2)=(0,7)$.

Solving Mixed Constraint Problems Using the Big M Method

1. If the optimization problem involves minimizing an objective function, f, view it as maximizing $-f$.

2. Introduce slack variables $s_i \geq 0$ in each constraint involving the $\leq$ inequality.

3. Introduce surplus variables $s_j \geq 0$ (where $j > i$) in each constraint involving the $\geq$ inequality.

4. Introduce artificial variables $a_k \geq 0$ in each constraint which originally involved a $\geq$ or $=$ (before step 3 was done).

5. For each artificial variable a_k, subtract a term of Ma_k from the function to be maximized, where $M > 0$ is a large number. Use the same coefficient, M, in each of these terms.

6. Set z equal to the function to be maximized and rewrite it in the form where the left side of the equation includes all variables and the right side is 0.

7. Form the preliminary simplex tableau with the constraints in the top rows and the objective function in the bottom row.

8. Perform any row operations to eliminate M at the bottom entry of each column corresponding to an artificial variable, and obtain a second simplex tableau.

9. Apply the simplex method to the resulting initial tableau, and arrive at an optimal solution.

Example 4: An Application Using the Big M Method to Minimize Cost

Suppose a company produces silverware at three different plants. Plant A can manufacture 1500 forks, 2000 spoons, and 2500 knives in an hour, and the hourly cost for this plant is \$600 per hour. Plant B does not manufacture forks, but it can manufacture 1800 spoons and 3000 knives in an hour. The hourly cost for plant B is \$400 per hour. Plant C does not manufacture knives, but it can manufacture 2400 forks and 1800 spoons in an hour. The hourly cost for plant C is \$500 per hour. The company is preparing an order this upcoming week for a large event at which food is being served to thousands of people. The organizers of this event are requesting at least 15,000 forks, exactly 15,000 spoons, and no more than 9000 knives. How many hours this coming week should each plant be operated in order to fulfill the demands of the event organizers at a minimum cost to the company?

Solution

Let x_1, x_2, and x_3 denote the hours of operation this coming week for plants A, B, and C; respectively. Based on the operating costs of each plant, we find that the total cost of operation is $600x_1 + 400x_2 + 500x_3$. Therefore, our objective is to minimize $f(x_1, x_2, x_3) = 600x_1 + 400x_2 + 500x_3$. The constraints are as follows:

$$\begin{cases} 1500x_1 + 2400x_3 \geq 15,000 & \text{(forks)} \\ 2000x_1 + 1800x_2 + 1800x_3 = 15,000 & \text{(spoons)} \\ 2500x_1 + 3000x_2 \leq 9000 & \text{(knives)} \end{cases}$$

Since we are doing a minimization problem, we will consider maximizing $z = -600x_1 - 400x_2 - 500x_3$. The last constraint has $\leq$, so we introduce the slack variable s_1 and write $2500x_1 + 3000x_2 + s_1 = 9000$. The first constraint involves $\geq$, and so we define the surplus variable s_2 and write $1500x_1 + 2400x_3 - s_2 = 15,000$.

Next, we introduce artificial variables in the first and second constraints (since they involve $\geq$ and $=$, respectively). If we introduce a_1 in the first constraint and a_2 in the second constraint, we obtain

$$1500x_1 + 2400x_3 - s_2 + a_1 = 15,000 \quad \text{and} \quad 2000x_1 + 1800x_2 + 1800x_3 + a_2 = 15,000.$$

Next we subtract Ma_1 and Ma_2 from the objective function, thus giving us $z = -600x_1 - 400x_2 - 500x_3 - Ma_1 - Ma_2$. Rewriting this as

$$600x_1 + 400x_2 + 500x_3 + Ma_1 + Ma_2 + z = 0,$$

we are ready to form the preliminary simplex tableau.

$$
\begin{array}{ccccccccc|c}
x_1 & x_2 & x_3 & s_1 & s_2 & a_1 & a_2 & z & \\
1500 & 0 & 2400 & 0 & -1 & 1 & 0 & 0 & 15,000 \\
2000 & 1800 & 1800 & 0 & 0 & 0 & 1 & 0 & 15,000 \\
2500 & 3000 & 0 & 1 & 0 & 0 & 0 & 0 & 9000 \\
\hline
600 & 400 & 500 & 0 & 0 & M & M & 1 & 0
\end{array}
$$

We first perform row operations to eliminate M at the bottom of the columns corresponding to the artificial variables. Two row operations will take care of this and yield the following result.

$$
\begin{array}{ccccccccc|c}
x_1 & x_2 & x_3 & s_1 & s_2 & a_1 & a_2 & z & \\
1500 & 0 & 2400 & 0 & -1 & 1 & 0 & 0 & 15,000 \\
2000 & 1800 & 1800 & 0 & 0 & 0 & 1 & 0 & 15,000 \\
2500 & 3000 & 0 & 1 & 0 & 0 & 0 & 0 & 9000 \\
\hline
600 & 400 & 500 & 0 & 0 & M & M & 1 & 0
\end{array}
$$

$$\xrightarrow{-MR_1 + R_4}$$

$$\xrightarrow{-MR_2 + R_4}$$

$$
\begin{array}{ccccccccc|c}
x_1 & x_2 & x_3 & s_1 & s_2 & a_1 & a_2 & z & \\
1500 & 0 & 2400 & 0 & -1 & 1 & 0 & 0 & 15,000 \\
2000 & 1800 & 1800 & 0 & 0 & 0 & 1 & 0 & 15,000 \\
2500 & 3000 & 0 & 1 & 0 & 0 & 0 & 0 & 9000 \\
\hline
-3500M & -1800M & -4200M & 0 & M & 0 & 0 & 1 & -30,000M \\
+600 & +400 & +500 & & & & &
\end{array}
$$

Now we apply the simplex method to this initial simplex tableau. Column 3 will be our pivot column since $-4200M + 500$ is the negative entry in the last row not in the last column with largest magnitude. Since $\dfrac{15,000}{2400} = 6.25$ is a smaller quotient than $\dfrac{15,000}{1800} \approx 8.33$, row 1 is our pivot row. We obtain the following.

$$\begin{array}{c} \begin{array}{cccccccc} x_1 & x_2 & x_3 & s_1 & s_2 & a_1 & a_2 & z \end{array} \\ \left[\begin{array}{cccccccc|c} 1500 & 0 & \boxed{2400} & 0 & -1 & 1 & 0 & 0 & 15{,}000 \\ 2000 & 1800 & 1800 & 0 & 0 & 0 & 1 & 0 & 15{,}000 \\ 2500 & 3000 & 0 & 1 & 0 & 0 & 0 & 0 & 9000 \\ \hline \begin{array}{c}-3500M\\+600\end{array} & \begin{array}{c}-1800M\\+400\end{array} & \begin{array}{c}-4200M\\+500\end{array} & 0 & M & 0 & 0 & 1 & -30{,}000M \end{array}\right] \end{array}$$

$$\xrightarrow{\;\frac{1}{2400}R_1\;}$$

$$\xrightarrow{\;-1800R_1+R_2\;}$$

$$\xrightarrow{\;(4200M-500)R_1+R_4\;}$$

$$\begin{array}{c} \begin{array}{cccccccc} x_1 & x_2 & x_3 & s_1 & s_2 & a_1 & a_2 & z \end{array} \\ \left[\begin{array}{cccccccc|c} \dfrac{5}{8} & 0 & 1 & 0 & -\dfrac{1}{2400} & \dfrac{1}{2400} & 0 & 0 & \dfrac{25}{4} \\ 875 & 1800 & 0 & 0 & \dfrac{3}{4} & -\dfrac{3}{4} & 1 & 0 & 3750 \\ 2500 & 3000 & 0 & 1 & 0 & 0 & 0 & 0 & 9000 \\ \hline \begin{array}{c}-875M\\+\dfrac{575}{2}\end{array} & \begin{array}{c}-1800M\\+400\end{array} & 0 & 0 & \begin{array}{c}-\dfrac{3}{4}M\\+\dfrac{5}{24}\end{array} & \begin{array}{c}\dfrac{7}{4}M\\-\dfrac{5}{24}\end{array} & 0 & 1 & \begin{array}{c}-3750M\\-3125\end{array} \end{array}\right] \end{array}$$

Column 2 is our next pivot column, and row 2 is the next pivot row since $\dfrac{3750}{1800} \approx 2.08$ is smaller than $\dfrac{9000}{3000} = 3$. Another iteration of the pivoting process yields the following:

$$\begin{array}{c} \begin{array}{cccccccc} x_1 & x_2 & x_3 & s_1 & s_2 & a_1 & a_2 & z \end{array} \\ \left[\begin{array}{cccccccc|c} \dfrac{5}{8} & 0 & 1 & 0 & -\dfrac{1}{2400} & \dfrac{1}{2400} & 0 & 0 & \dfrac{25}{4} \\ 875 & \boxed{1800} & 0 & 0 & \dfrac{3}{4} & -\dfrac{3}{4} & 1 & 0 & 3750 \\ 2500 & 3000 & 0 & 1 & 0 & 0 & 0 & 0 & 9000 \\ \hline \begin{array}{c}-875M\\+\dfrac{575}{2}\end{array} & \begin{array}{c}-1800M\\+400\end{array} & 0 & 0 & \begin{array}{c}-\dfrac{3}{4}M\\+\dfrac{5}{24}\end{array} & \begin{array}{c}\dfrac{7}{4}M\\-\dfrac{5}{24}\end{array} & 0 & 1 & \begin{array}{c}-3750M\\-3125\end{array} \end{array}\right] \end{array}$$

$$\xrightarrow{\;\frac{1}{1800}R_2\;}$$

$$\xrightarrow{\;-3000R_2+R_3\;}$$

$$\xrightarrow{\;(1800M-400)R_2+R_4\;}$$

$$\begin{array}{c} \begin{array}{cccccccc} x_1 & x_2 & x_3 & s_1 & s_2 & a_1 & a_2 & z \end{array} \\ \left[\begin{array}{cccccccc|c} \dfrac{5}{8} & 0 & 1 & 0 & -\dfrac{1}{2400} & \dfrac{1}{2400} & 0 & 0 & \dfrac{25}{4} \\ \dfrac{35}{72} & 1 & 0 & 0 & \dfrac{1}{2400} & -\dfrac{1}{2400} & \dfrac{1}{1800} & 0 & \dfrac{25}{12} \\ \dfrac{3125}{3} & 0 & 0 & 1 & -\dfrac{5}{4} & \dfrac{5}{4} & -\dfrac{5}{3} & 0 & 2750 \\ \hline \dfrac{1675}{18} & 0 & 0 & 0 & \dfrac{1}{24} & M-\dfrac{1}{24} & M-\dfrac{2}{9} & 1 & -\dfrac{11{,}875}{3} \end{array}\right] \end{array}$$

Since there are no negative entries in the last row not in the last column, we have an optimal solution; it is given by

$$\left(x_1, x_2, x_3, s_1, s_2, a_1, a_2, z\right) = \left(0, \frac{25}{12}, \frac{25}{4}, 2750, 0, 0, 0, -\frac{11{,}875}{3}\right)$$

Since $z = -f(x_1, x_2, x_3)$, we have that $f(x_1, x_2, x_3) = 600x_1 + 400x_2 + 500x_3$ attains a minimum of $\dfrac{11{,}875}{3} \approx 3958.33$ at $\left(x_1, x_2, x_3\right) = \left(0, \dfrac{25}{12}, \dfrac{25}{4}\right)$. Therefore, we conclude that the company should not use plant A, run plant B for $\dfrac{25}{12}$ hours (2 hours and 5 minutes), and run plant C for $\dfrac{25}{4}$ hours (6 hours and 15 minutes) this coming week to achieve a minimum cost of \$3958.33 to meet the demands of the event organizers.

Note: The value of the slack variable $s_1 = 2750$ means that only $9000 - 2750 = 6250$ knives need to be manufactured to fulfill the needs of the event organizers.

Comparison of the Optimization Methods

We conclude this lesson by providing two tables that summarize and compare all of the methods we used to solve optimization problems. Table 1 compares the methods covered in this section for solving problems with mixed constraints. Table 2 compares the regular simplex method for maximization with the duality method for minimization.

Mixed Constraints without Big M	**Mixed Constraints with Big M**
1. Optimize objective function f subject to constraints, each involving either $\leq$ or $\geq$.	1. Optimize objective function f subject to constraints, each involving $\leq$, $\geq$, or $=$.
2. If minimizing f, maximize $-f$.	2. If minimizing f, maximize $-f$.
3. Multiply all $\geq$ constraints by -1 to convert them all to $\leq$ constraints.	3. Introduce slack variables in all $\leq$ constraints.
4. Introduce slack variables in all constraints.	4. Introduce surplus variables in all $\geq$ constraints.
5. Form the preliminary simplex tableau.	5. Introduce artificial variables in all constraints which involved $\geq$ or $=$ (before the surplus variables were introduced).
6. For each negative entry in the last column above the bottom row, use a negative value in the same row to determine a pivot column, then choose a pivot element to do the pivoting process. Repeat until there are no negative entries in the last column above the bottom row.	6. Subtract Ma_i from the objective function for each artificial variable a_i introduced.
7. Apply the simplex method to the resulting initial simplex tableau to obtain an optimal solution.	7. Form the preliminary simplex tableau.
	8. Perform any row operations to eliminate M at the bottom entry of each column corresponding to an artificial variable.
	9. Apply the simplex method to the resulting initial simplex tableau to obtain an optimal solution.

TABLE 1

Simplex Method: Maximization	**Duality Principle: Minimization**
1. Maximize objective function f subject to constraints involving $\leq$.	**1.** Minimize objective function f subject to constraints involving $\geq$.
2. Form the initial simplex tableau.	**2.** Write down the matrix for minimization with constraints, A.
3. Apply the simplex method to obtain an optimal solution.	**3.** Write down the matrix for the dual maximization problem, A^T.
	4. Form the initial simplex tableau for the dual problem.
	5. Apply the simplex method to obtain an optimal solution. Use entries in the last row of columns corresponding to slack variables for the coordinates.

TABLE 2

7.5 EXERCISES

💡 PRACTICE

For each given system of inequalities (constraints), rewrite as a system of equations by adding slack or subtracting surplus variables.

1. $\begin{cases} 5x_1 + x_2 \leq 15 \\ 3x_1 + 2x_2 \geq 7 \end{cases}$

2. $\begin{cases} 2x_1 + 3x_2 \leq 7 \\ 4x_1 - x_2 \geq 5 \\ x_1 + x_2 \leq 4 \end{cases}$

3. $\begin{cases} 8x_1 + 3x_2 - x_3 \geq 10 \\ 6x_2 - x_3 \geq 7 \\ x_1 + x_2 + x_3 \leq 8 \end{cases}$

4. $\begin{cases} x_1 + x_2 + 2x_3 \leq 12 \\ 2x_1 + x_2 + 3x_3 \leq 17 \\ x_1 - x_2 + x_3 \geq 7 \\ 2x_2 \leq 5 \end{cases}$

5. $\begin{cases} 4x_1 - x_3 \geq 9 \\ x_2 + 2x_3 \leq 13 \\ 5x_1 - x_2 + 4x_3 \geq 14 \\ 3x_1 + x_2 + x_3 \leq 17 \end{cases}$

6. Suppose you want to minimize $f(x_1, x_2) = x_1 + 3x_2$ subject to the following constraints.

$$\begin{cases} x_1 + x_2 \leq 5 \\ 2x_1 - x_2 \geq 3 \end{cases}$$

Rewrite this as a maximization problem and rewrite the system of constraints using slack variables as in the simplex method.

7. Suppose you want to minimize $f(x_1, x_2, x_3) = 4x_1 + 7x_2 + 9x_3$ subject to the following constraints.

$$\begin{cases} x_1 + x_2 + x_3 \leq 9 \\ -x_1 + 3x_3 \geq 6 \\ 5x_1 - x_2 - x_3 \geq 7 \end{cases}$$

Rewrite this as a maximization problem and rewrite the system of constraints using slack variables as in the simplex method.

8. Consider again the optimization problem in Exercise 6. Rewrite this as a maximization problem, but this time rewrite the system of constraints using slack, surplus, and artificial variables as in the big M method. Also, write the objective function that results after introducing the artificial variables and the constant M.

9. Consider again the optimization problem in Exercise 7. Rewrite this as a maximization problem, but this time rewrite the system of constraints using slack, surplus, and artificial variables as in the big M method. Also, write the objective function that results after introducing the artificial variables and the constant M.

10. Suppose you want to minimize $f(x_1, x_2, x_3) = 3x_1 + 2x_2 + 6x_3$ subject to the following constraints.

$$\begin{cases} 3x_1 + 2x_2 + x_3 \le 8 \\ 2x_1 + x_3 = 5 \\ -x_1 - x_2 + 4x_3 \ge 7 \end{cases}$$

Rewrite this as a maximization problem, and rewrite the system of constraints using slack, surplus, and artificial variables as in the big M method. Also, write the objective function that results after introducing the artificial variables and the constant M.

Use either the simplex method or the big M method to solve each given optimization problem. See Examples 1, 2, and 3.

11. Minimize $f(x_1, x_2) = 4x_1 + 6x_2$ subject to the following constraints.
$$\begin{cases} 2x_1 + 3x_2 \le 10 \\ 4x_1 - x_2 \ge 4 \end{cases}$$

12. Maximize
$f(x_1, x_2, x_3) = 3x_1 + 2x_2 + 4x_3$
subject to the following constraints.
$$\begin{cases} 2x_1 + x_2 + x_3 \le 7 \\ 3x_1 - x_2 + 2x_3 \ge 4 \\ x_1 + x_3 \ge 5 \end{cases}$$

13. Maximize
$f(x_1, x_2, x_3) = 2x_1 - 2x_2 + 6x_3$
subject to the following constraints.
$$\begin{cases} x_1 + x_2 \le 20 \\ x_1 + x_3 = 5 \\ x_2 + x_3 \ge 10 \end{cases}$$

14. Minimize $f(x_1, x_2) = 3x_1 + 4x_2$ subject to the following constraints.
$$\begin{cases} 4x_1 + 5x_2 \ge 18 \\ x_1 + x_2 = 4 \end{cases}$$

15. Minimize
$f(x_1, x_2, x_3) = 7x_1 + 2x_2 + 3x_3$
subject to the following constraints.
$$\begin{cases} x_1 + x_2 + x_3 \ge 6 \\ 3x_1 - x_2 + 2x_3 = 7 \\ -x_1 + x_3 \le 2 \end{cases}$$

16. Minimize $f(x_1, x_2) = 3x_1 + 5x_2$ subject to the following constraints.
$$\begin{cases} x_1 + 3x_2 \le 8 \\ 3x_1 - x_2 = 5 \end{cases}$$

17. Maximize
$f(x_1, x_2, x_3) = 6x_1 + 4x_2 + 5x_3$
subject to the following constraints.
$$\begin{cases} 2x_1 + x_2 + 3x_3 \le 18 \\ x_1 + 2x_2 + 2x_3 \le 16 \\ x_1 + x_2 + x_3 = 10 \end{cases}$$

18. Minimize
$f(x_1, x_2, x_3) = 10x_1 + 8x_2 + 15x_3$
subject to the following constraints.
$$\begin{cases} 2x_1 + x_2 - x_3 \ge 6 \\ x_1 + x_2 + 2x_3 = 6 \\ -x_1 + 4x_2 - x_3 \ge 24 \end{cases}$$

🚀 APPLICATIONS

Use either the simplex method or the big M method to solve each given application. See Example 4.

19. A truck driver is tasked with delivering stereos for an electronics company. Suppose this company has stereos at two warehouses; the warehouse in Charlotte has 500 stereos, and the warehouse in Charleston has 200 of them. A store in Raleigh needs 350 stereos, and a store in Spartanburg needs 250 stereos. Suppose the truck driver earns $5 in profit per stereo delivered to Raleigh from Charlotte and $4 for each stereo delivered to Spartanburg from Charlotte. Moreover, he earns $7 for each stereo delivered to Raleigh from Charleston and $5 for each stereo delivered to Spartanburg from Charleston. How many stereos should the truck driver deliver to each store from each warehouse to maximize his earnings?

20. A Mexican food truck produces and sells beef tacos and beef burritos. Suppose each taco uses 8 ounces of beef and 1 ounce of cheese, and suppose each burrito uses 12 ounces of beef and 5 ounces of cheese. The supply of beef is limited to 1180 ounces, but the supply of cheese is plentiful, and the workers on the truck wish to use up at least 200 ounces of cheese to minimize waste. If each taco sells for $6 and each burrito sells for $7, how many of each kind of food should the truck produce to maximize revenue?

21. A car company has two factories. Factory A has 600 cars (of a particular model) in stock, and factory B has 400 cars (of that same model) in stock. Two dealerships order this car model. Dealership I needs 300 cars and dealership II needs 400 cars. The cost of shipping per car from the two factories to the dealerships is given below.

	Dealership I	Dealership II
Factory A	$40	$30
Factory B	$35	$45

How should the company ship the cars from the two factories to the dealerships in order to minimize the shipping cost? (**Hint:** Use an approach similar to the one in Exercise 19.)

22. An ice cream company sells ice cream bars and ice cream sandwiches. One of its plants makes each treat and packs the bars and sandwiches into 12-count packages. There are enough lines running at this plant so that each package of bars takes 2 seconds to produce, and each package of sandwiches takes 5 seconds. Suppose the plant has at most 15 hours (54,000 seconds) available each day for producing and packaging the ice cream products. Also, limitations and product demand indicate that the number of packages bars plus the number of packages of sandwiches should be at least 18,000, and the number of bars should be at least 1.5 times the number of sandwiches. If the plant's manufacturing costs are $2.50 for each package of bars and $2 for each package of sandwiches, how many packages of each ice cream product should the plant produce to satisfy the constraints at a minimum cost?

23. Three water purification facilities can handle at most 8 million gallons during a certain time period. Plant I leaves 11% of certain impurities and costs $40,000 per million gallons. Plant II leaves 8% of certain impurities and costs $50,000 per million gallons. Plant III leaves 5% of certain impurities and costs $70,000 per million gallons. The desired level of impurities in the water from all plants combined is at most 8%. If plants I and III must handle at least 5 million gallons, find the number of gallons each plant should handle so the water is treated at a minimum cost.

24. A silverware manufacturer produces forks, knives, and spoons. The manager feels that restricting the types of silverware produced could increase revenue. The following table gives the price of each piece of silverware, the raw materials cost for each piece, and the profit made.

	Price	Materials Cost	Profit
Knives	$4	$0.10	$0.60
Forks	$6	$0.10	$0.25
Spoons	$5	$0.15	$1.10

The monthly revenue must be at least $12,000 and the raw material costs should be no more than $280. Also, suppose there should be at least as many forks produced as spoons. How many of each type of silverware should be produced to maximize profit?

25. A candidate for president has budgeted a maximum of $900,000 for political advertisements during her presidential campaign. Suppose each minute of television time costs $30,000 and each one-page newspaper ad costs $12,000. Each television ad is expected to be viewed by 2 million voters, and each newspaper ad is expected to be seen by 200 thousand voters. The candidate is advised to use at least 8 television ads and at least 5 newspaper ads. How should the candidate allocate the advertising budget to reach as many voters as possible?

26. An aquarium contains several feeding areas for dolphins, sharks, and stingrays. There are 3 foods (A, B, and C) available at each of these feeding areas. The following table describes how many pounds of each type of food is required to feed one animal.

	Food A	Food B	Food C
Dolphin	3	1	2
Shark	2	6	6
Stingray	1	2	2

Suppose the aquarium has 300 pounds of food A available and 400 pounds of food B available. Also suppose the aquarium has a surplus of food C and would like to use at least 450 pounds of it. How many of each type of animal can the aquarium support so that the number of dolphins, sharks, and stingrays is a maximum?

27. A family has a side business of selling three kinds of snow cones from their trailer: Tropical Paradise, Arctic Chill, and Paradise Blast. Each Tropical Paradise snow cone uses 3 ounces of mango syrup, 1 ounce of blueberry syrup, and 8 ounces of shaved ice. Each Arctic Chill snow cone uses 3 ounces of blueberry syrup and 8 ounces of shaved ice. Each Paradise Blast snow cone uses 2 ounces of mango syrup, 2 ounces of blueberry syrup, and 8 ounces of shaved ice. Suppose 225 ounces of blueberry syrup and 52.5 pounds (840 ounces) of shaved ice are available, and suppose that all of the ice has to get used up. In addition, suppose there is plenty of mango syrup in stock, and the family would like to use at least 150 ounces of it. If the family makes a profit of $0.40 for each Tropical Paradise snow cone, $0.60 profit for each Arctic Chill snow cone, and $0.50 profit for each Paradise Blast snow cone, how many of each kind of snow cone should the family sell in order to maximize their profit?

28. A company produces three kinds of 3-ring binders depending on their thickness: 1.5-inch, 2-inch, and 3-inch. The number of units of steel, cardboard, and plastic needed to make each kind of 3-ring binder are in the table below.

	1.5-inch	2-inch	3-inch
Steel	2	3	6
Cardboard	10	18	30
Plastic	8	12	16

Suppose the company wants to use no more than 280 units of steel, no less than 1520 units of cardboard, and exactly 1000 units of plastic per day. If it costs the company $2 to produce each 1.5-inch binder, $4 to produce each 2-inch binder, and $6 to produce each 3-inch binder, then how many of each kind of binder should the company produce in order to minimize cost?

29. A plant that makes popsicles has a mixing tank. Currently, it contains 60 gallons of a mixture that is 15% orange syrup, 18% high fructose corn syrup, and 41% water. The foreman at the plant wants to add more mixture to the tank before the lines run out of fluid to make the popsicles. He has 3 different mixes available. Mix I contains 20% orange syrup, 40% high fructose corn syrup, and 35% water. Mix II contains 10% orange syrup, 30% high fructose corn syrup, and 40% water. Mix III contains 25% orange syrup, 40% high fructose corn syrup, and 15% water. The desired final mixture should have at most 18% orange syrup, exactly 30% high fructose corn syrup, and at least 35% water. If the costs of each mix per gallon are $60 (I), $45 (II), and $30 (III), how many gallons of each mix should the foreman use to minimize the cost? **Hint:** If the final mix needs at least 35% water, that means

$$41\%\left(\text{mix in tank}\right)+35\%\left(\text{mix I}\right)+40\%\left(\text{mix II}\right)+15\%\left(\text{mix III}\right)$$
$$\geq 35\%\left(\text{mix in tank}+\text{mix I}+\text{mix II}+\text{mix III}\right).$$

30. An eccentric barbecue restaurant produces and sells three kinds of bratwursts well-known to the locals in the area: Plain Jane, Spicy Samantha, and Cheezy Chelsea. All of these bratwursts consist of pork, spices, and cheese. The following table summarizes how much of each ingredient are in each kind of bratwurst.

	Plain Jane	Spicy Samantha	Cheezy Chelsea
Pork	8 oz	6 oz	7 oz
Spices	1 oz	3 oz	1 oz
Cheese	1 oz	1 oz	2 oz

Suppose the restaurant has 200 pounds (3200 ounces) of pork available, and they want to use at least 800 ounces of spices. Suppose they also have 600 ounces of cheese and must use all of it due to an obscure health code regulation preventing the restaurant from having any traces of unused cheese. Finally, suppose each Plain Jane brings in a profit of $0.70, each Spicy Samantha brings in a profit of $0.50, and each Cheezy Chelsea brings in a profit of $0.60. How many of each kind of sausage should the restaurant produce and sell to maximize its profit?

✎ WRITING & THINKING

31. Use the big M method to try and maximize $f(x_1, x_2) = 3x_1 + 9x_2$ subject to the following constraints.

$$\begin{cases} x_1 + 3x_2 = 7 \\ 2x_1 + x_2 \le 5 \end{cases}$$

What do you find?

32. Use the big M method to try and maximize $f(x_1, x_2) = x_1 + 2x_2$ subject to the following constraints.

$$\begin{cases} x_1 - x_2 \le 5 \\ 2x_1 - x_2 = 20 \end{cases}$$

What do you find?

33. Consider the problem of minimizing $f(x_1, x_2) = 2x_1 + 4x_2$ subject to the following constraints.

$$\begin{cases} 3x_1 + x_2 = 6 \\ x_1 + x_2 \ge 4 \end{cases}$$

a. Use the big M method to solve.
b. According to the first constraint, $3x_1 + x_2 = 6$ holds, and therefore the inequality $3x_1 + x_2 \ge 6$ must also hold. It seems intuitive that we could use the Theorem of Duality (instead of the big M method) to solve this problem if we replace the first constraint with $3x_1 + x_2 \ge 6$. Try it. Do you obtain the same solution as in part a.? If you obtain a different solution, can you explain why?

# Chapter 8

PROBABILITY

8.1 SET NOTATION

Have you ever wondered how a pizza restaurant organizes all of their toppings in a manner that allows them to efficiently create pizzas to order? How is it possible that a supermarket keeps all of the products organized in a manner that allows shoppers to search for and find products? The answer is quite simple: the use of mathematical sets. From the ways in which you select clothes in your closet to how you organize the apps on your smartphone, the use of sets allows us to have a systematic way to analyze groups of items and information.

A flock of sheep, a litter of kittens, and an ensemble of musicians are all sets of things with living members. The alphabet is a set of letters; a line is a set of points. No matter what the set consists of, the mathematical concept of sets is the building block for many areas of mathematics.

Set

A **set** is a collection of objects made up of specified elements, or members.

To begin, we need to explain some notation for describing a set. Mathematical sets are commonly represented using capital letters and the elements of the set are represented by lowercase letters. We can express the member x as being an element of a set G symbolically by $x \in G$ (read "x is an element of G"). If y is not an element of G, we write $y \notin G$ (read "y is not an element of G").

One of the most common ways to describe a set is to simply make a list of all the elements in the set. This notation is called the **roster method**. The common way to write a set using the roster method is for the elements of the set to be surrounded by braces and separated with commas. For example, the following sets A, B, C, and D are described using the roster method.

$$A = \{1, 2, x, y, z\}$$

$$B = \{x, 1, 2, y, z\}$$

$$C = \left\{\frac{1}{2}, \frac{2}{3}, \frac{3}{4}, \frac{4}{5}\right\}$$

$$D = \{\text{Gwen, King Charles spaniel, Zoë, taxi}\}$$

Roster Method

Roster notation is a way to describe a set by listing all of the elements in the set.

Notice that set D does not contain any numerical elements at all. Sets are not limited to numbers. In fact, they can contain anything—real or imaginary. An element of a set may be a number, but it could also be a griffin, a lychee, or a caterpillar. The only thing the elements of a set might have in common is that they are all members of that set. In fact, did you wonder what the elements in D all have in common when you first looked at that set? The only obvious thing that they have in common is that they are all members of the set D.

In set theory, sets themselves can even be elements of other sets. Consider the set G, which is a set consisting of sets of paired numbers.

$$G = \{\{1, 2\}, \{3, 4\}, \{5, 6\}, \{7, 8\}, \{9, 10\}\}$$

The individual numbers 1 through 10 are not members of the set G; however the sets of paired numbers such as $\{1, 2\}$ are.

It's important to point out that there is no concept of order in a set. Thus, the two sets A and B that we introduced previously are really the same set because they contain the same elements. We've listed them again so that you can check for yourself.

$$A = \{1, 2, x, y, z\}$$

$$B = \{x, 1, 2, y, z\}$$

When two sets contain exactly the same elements, the sets are said to be equal. So, we can write $A = B$.

Equal Sets

Two sets are said to be **equal** if they contain exactly the same elements. If sets A and B are equal, we write $A = B$.

Sets come in all shapes and sizes: some have infinitely many elements and some have a finite number of elements. Recall that a set having infinitely many elements means that a list of the elements would go on forever; whereas a finite set has a specific number of elements. The roster notation for an infinite set often includes a series of three dots, called an ellipsis, that indicates that the set continues on without ceasing. For instance, the set of positive integers can be written as $Z = \{1, 2, 3, 4, \ldots\}$. An ellipsis can also be used to indicate when a certain pattern is continued within a list of set members. For instance, when describing the even numbers between 2 and 100, we could write $L = \{4, 6, 8, \ldots, 98\}$.

We will focus most of our discussion on finite sets. Since finite sets have a specific number of elements, we can discuss their cardinal numbers. The *cardinal number* of a finite set is the number of elements contained in the set, or in other words, the size of the set. The cardinal number is denoted by $| \ |$. For example, if $A = \{a, b, c, d, e\}$, then $|A| = 5$.

👉 HELPFUL HINT

Some courses may use $n(\ )$ to indicate the cardinal number of a set.

Cardinal Number

The number of elements contained in a finite set is called the **cardinal number**, or **cardinality**. The cardinal number is denoted by $| \ |$.

Example 1: Using the Roster Method to Represent a Set

Use the roster method to represent S, the set of states in the United States that begin with the letter M. Then, find $|S|$.

Solution

There are eight states in the United States that begin with the letter M.

$S = \{$Maine, Maryland, Massachusetts, Michigan, Minnesota, Mississippi, Missouri, Montana$\}$

Because S contains eight elements, $|S| = 8$.

When two finite sets have the same cardinal number, that is, the same number of elements (no matter if the elements are of the same type or not), the sets are said to be **equivalent**. For instance, sets C and D both contain 4 elements, so C is equivalent to D. We've listed the sets again so that you can check for yourself.

$$C = \left\{\frac{1}{2}, \frac{2}{3}, \frac{3}{4}, \frac{4}{5}\right\}$$

$$D = \{\text{Gwen, King Charles spaniel, Zoë, taxi}\}$$

Symbolically, we write $C \sim D$, which is read "C is equivalent to D".

Equivalent Sets

Sets are **equivalent** if they have the same cardinal number; that is, the same number of elements. If sets C and D are equivalent, we write $C \sim D$.

Example 2: Determining Equal and Equivalent Sets

Determine if the given pairs of sets are equal, equivalent, or neither.

a. $A = \{$Public Health, International Studies, Mechanical Engineering, Music Education, Political Science$\}$

   $B = \{$Tim, Gloria, Alan, Warren, Karen$\}$

b. $X = \{2, 4, 6, 8, 10, 12, 14, 16, 18, 20\}$

   $Y = \{20, 18, 16, 14, 12, 10, 8, 6, 4, 2\}$

Solution

a. $|A| = 5$ and $|B| = 5$, therefore, the sets are equivalent to one another, and we can write $A \sim B$. However, because they do not contain the same elements, they are not equal to one another.

b. $|X| = 10$ and $|Y| = 10$. Even though the order of the elements is different, both contain the even numbers from 2 to 20. Therefore, the sets are equal, and we can write $X = Y$. When two sets are equal, they are by definition also equivalent.

$\mathbb{R}$ represents the set of real numbers, $\mathbb{N}$ represents the set of natural numbers, $\mathbb{Z}$ represents the set of integers, and $\mathbb{W}$ represents the set of whole numbers. All of these sets are infinite sets.

Describing a set using the roster method is very convenient for small sets, however it quickly loses its usefulness for big sets. An alternative when the members of a set all share a certain property is to describe the elements using **set-builder notation**. For instance, the following describes the integers, $\mathbb{Z}$, using set-builder notation.

$$\mathbb{Z} = \{n \mid n \text{ is an integer}\}$$

It is read "$\mathbb{Z}$ is the set of all n such that n is an integer."

Set-Builder Notation

Set-builder notation is used to describe a set when the members all share certain properties.

Set-builder notation may also use a colon instead of a vertical line to represent "such that." For example, $Y = \{x : x \text{ is an even integer}\}$.

Set-builder notation is extremely helpful when representing an infinite set of numbers like the natural numbers. It is impossible to list out every member of an infinite set, but by describing it in this manner, set-builder notation encompasses the entire set.

Example 3: Using Set-Builder Notation to Represent a Set

Use set-builder notation to represent Y, the set of all even integers.

Solution

$$Y = \{x \mid x \text{ is an even integer}\}$$

Note that there are many correct ways to write the solution. Alternative solutions include

$$Y = \{x \mid x \in \mathbb{Z} \text{ and } x \text{ is even}\} \quad \text{and} \quad Y = \{2x \mid x \text{ is an integer}\}.$$

Skill Check 1

Represent the set J of all natural numbers less than 10, using both the roster method and set-builder notation.

There are a couple of very special types of sets that we need to define. The first of these is the **empty set**, also called the **null set**. Think for a moment about the set A, which is the set of states that border Hawaii. Since Hawaii is made up of islands that do not border any other state, A has no elements and is an empty set. If a set is empty, we denote this symbolically by writing $A = \varnothing$, or sometimes $A = \{\ \}$.

The cardinality of the empty set is 0. The empty set should not be confused with the set $\{\varnothing\}$ or the set $\{0\}$, both of which contain a single element. The first is the set whose only element is an empty set and the second is the set containing the number 0, both of which have a cardinality of 1.

Empty Set

The **empty set**, or **null set**, is the set that contains no elements. If set A is empty, we write $A = \varnothing$.

> ### Example 4: Determining Empty Sets

Determine if the following sets are empty sets.

a. The set A of negative numbers less than 100.

b. The set B of any state that contains the letter q in its name.

Solution

a. Since it is the case that all negative numbers are less than 100, the set A is not empty. In fact, it is a set with infinitely many elements.

b. Since there are not any states whose names contain the letter q, this set is empty. That is, $B = \varnothing$.

Give an example of an empty set.

The second special set to mention is the **universal set**. The universal set is the set of all elements that are being considered in any particular situation. For instance, suppose you are buying a car. The universal set could be the set of all new cars available, it could be the set of all used cars available, or it might be both—the set of all new and used cars available. Choosing a universal set "sets the scene" for those elements that will be considered.

> ### Universal Set
>
> The set of all elements being considered for any particular situation is called the **universal set** and is denoted by U.

The universal set is especially useful in that it allows us to express everything that is not in a particular set. For instance, let's go back to our car example. Suppose the universal set is the set of all new and used cars that are available. Assume you can afford a car under \$25,000. Then the set of cars you can afford is $A = \{c \mid c \in U$ and cost $< \$25,000\}$. Since not everyone has the luxury of choosing from every possible make and model of car in the world, not all cars are included in A. For instance, in most situations, a \$4,000,000 Lamborghini would not be up for consideration, and therefore would not be in the set A. It would be in the **complement** of A, denoted A'. In other words, it is an element of the universal set that is not in the set A. In our car example, $A' = \{c \mid c \in U$ and cost $\geq \$25,000\}$.

> ### HELPFUL HINT
>
> The complement of A can be denoted by A', A^c, or $\overline{A}$.

> ### Complement of a Set
>
> The **complement** of set A consists of all the elements in the given universal set that are not contained in A. The complement of A is denoted A'.

The set-builder notation for the complement of set A is $A' = \{x \mid x \in U, x \notin A\}$.

Consider for a moment two special sets: the universal set and the empty set. One contains all of the elements being considered and the other contains no elements. Hence, the universal set and the empty set will always be complements of one another.

$$U' = \varnothing \quad \text{and} \quad \varnothing' = U$$

Example 5: Determining the Complement of a Set

Twitter is a type of online social media where registered users can post "tweets" that are up to 280 characters long. In the world of Twitter, you can keep up with what other users post by "following" them and in turn, other users can "follow" you and your tweets.

Let

$$U = \{x \mid x \text{ is a registered user on Twitter}\}$$

$$A = \{x \mid x \text{ is a registered user on Twitter and } x \text{ follows you on Twitter}\}$$

$$B = \{x \mid x \text{ is a registered user on Twitter and you do not follow } x \text{ on Twitter}\}$$

Determine the complements of A and B.

Solution

Since A is the set of all Twitter users who follow you on Twitter, the complement of A is the set of all registered Twitter users who do not follow you on Twitter. We write the complement of A in the following manner.

$$A' = \{x \mid x \text{ is a registered user on Twitter and } x \text{ does not follow you on Twitter}\}$$

Since Twitter currently has more than half a billion users, it is likely that A' is much larger than A.

B is the set of all registered users who you do not follow on Twitter; so, the complement of B is the set of registered users who you do follow on Twitter. We can write the complement of B as

$$B' = \{x \mid x \text{ is a registered user on Twitter and you follow } x \text{ on Twitter}\}.$$

Skill Check Answers

1. Roster notation: $J = \{1, 2, 3, 4, 5, 6, 7, 8, 9\}$;

   Set-builder notation: $J = \{x \mid x \in \mathbb{N}, x < 10\}$

2. Answers will vary. Examples may include $A = \{x \mid x \in \mathbb{N}, x < 0\}$;

   $B = \{x \mid x \text{ is a US President before Barack Obama and } x \text{ is a woman}\}$.

3. $A' = \{x \mid x \in U, \text{ and you did not read } x \text{ as an e-book}\}$; Yes, but not as an e-book.

8.1 EXERCISES

✔ CONCEPT CHECK

Determine whether each statement is true or false. If the statement is false, explain why.

1. It is always possible to list every element of a set using the roster method.

2. $\varnothing \sim \{0\}$

3. Let $U = \{\text{set of all students enrolled at Xavier University}\}$ and $A = \{x \mid x \in U$ and x is a student with less than 30 earned credit hours$\}$. Then $A' = \{x \mid x \in U$ and x is a student with at least 30 earned credit hours$\}$.

4. $0 \in \varnothing$

5. $x \in \{x, y, z\}$

6. $\{x\} \in \{x, y, z\}$

7. $|\varnothing| = 0$

8. Let $Y = \{\text{Tim, Emilia, Aleesa, Whit}\}$, then $|Y| = 4$.

9. Let $A = \{\text{Sounds, Angels, Ravens, Titans}\}$ and $B = \{201, 38, 46, 23\}$, then $A \sim B$.

💡 PRACTICE

Write each set using the roster method.

10. A is the set of months of the year that have exactly 30 days.

11. B is the set of states whose names begin with the letter N.

12. C is the set of positive numbers smaller than 100 that have 2 digits that are the same.

13. D is the set of last names of people who teach this course at your school.

14. E is the set of planets in our solar system.

15. F is the set of weekdays.

Write each set using set-builder notation.

16. Let G be the set of whole numbers.

17. Let H be the set of natural numbers less than or equal to 50.

18. Let U be the set of all the states in the United States of America and J be the set of all states that border an ocean.

19. Let U be the set of all students enrolled at a school of higher education and K be the set of all collegiate athletes.

Write a description for each set. It is possible for more than one description to be correct.

20. $J = \{-11, -9, -7, -5, -3, -1\}$

21. $K = \{\$1, \$2, \$5, \$10, \$20, \$50, \$100\}$

22. $M = \{\text{Saturday, Sunday}\}$

23. $N = \{\text{January, June, July}\}$

Write each set using the roster method.

24. $A = \{x \mid x \text{ is an odd number and } 20 < x < 30\}$

25. $B = \{x \mid x \leq 15 \text{ and } x \text{ is a positive multiple of } 3\}$

26. $C = \{x \mid 2x = 4\}$

27. $D = \{x \mid x \text{ is a state that shares a common border with Colorado}\}$

Use the given sets to answer each question.

$P = \{\text{lasagna, rotini, orzo, tortellini, penne}\}$

$Q = \{x \mid x \text{ is a pasta shape}\}$

$R = \{\text{penne, tortellini, orzo, rotini, lasagna}\}$

$S = \{\text{marinara, pesto, alfredo, Bolognese, carbonara}\}$

28. Is $P = Q$? Why or why not?

29. Is $P = R$? Why or why not?

30. Is $P = S$? Why or why not?

31. Are any of P, Q, R, and S equivalent? Explain.

Use the given sets to answer each question.

$A = \{\text{Business, Physics, Psychology, Kinesiology, Graphic Design, History}\}$

$B = \{\text{History, Graphic Design, Kinesiology, Physics, Business}\}$

$C = \{\text{Art History, Education, Nursing, Biology, Statistics}\}$

$D = \{x \mid x \text{ is a university major}\}$

32. Is $A = B$? Why or why not?

33. Is $B = C$? Why or why not?

34. Is $A = D$? Why or why not?

35. Are any of A, B, C, and D equivalent? Explain.

Determine the cardinal number of each set.

36. $W = \{0, 3, 4, 5, 6, 7, 8, 9, 10\}$

37. $X = \{x \mid x \in \mathbb{Z}, x \text{ is even, and } |x| < 10\}$

38. The empty set

39. $Y = \{x \mid x$ is a United States president, past or present$\}$

Use the set $A = \{b, a, s, k, e, t\}$ to solve each problem.

40. Find $|A|$.

41. If $U = \{a, b, c, d, \ldots, x, y, z\}$, find A'.

42. If $U = \{a, b, c, d, \ldots, x, y, z\}$, find $|A'|$.

43. If $U = \{a, b, c, d, \ldots, x, y, z, A, B, C, D, \ldots, X, Y, Z\}$, find A'.

44. If $U = \{a, b, c, d, \ldots, x, y, z, A, B, C, D, \ldots, X, Y, Z\}$, find $|A'|$.

Use the set $B = \{1, 2, 3, 4\}$ to solve each problem.

45. Find $|B|$.

46. If $U = \{0, 1, 2, 3, 4, 5, 6, 7, 8, 9\}$, find B'.

47. If $U = \{0, 1, 2, 3, 4, 5, 6, 7, 8, 9\}$, find $|B'|$.

48. If $U = \{-9, -8, -7, -6, \ldots, 6, 7, 8, 9\}$, find B'.

49. If $U = \{-9, -8, -7, -6, \ldots, 6, 7, 8, 9\}$, find $|B'|$.

◆ APPLICATIONS

The table shows the total number of official Olympic medals for the time period from the first modern Olympics in 1896 through the Winter Games of 2014 for the top 10 countries, some of which are no longer countries. Let the universal set consist of the 10 countries listed. Solve each problem.

Top 10 All-Time Olympic Medal Winning Countries

Team	Gold	Silver	Bronze	Combined Total
United States (USA)	1072	860	749	2681
Soviet Union (URS)	473	376	355	1204
Great Britain (GBR)	246	276	284	806
Germany (GER)	252	260	270	782
France (FRA)	233	254	293	780
Italy (ITA)	235	200	228	663
Sweden (SWE)	193	204	230	627
China (CHN)	213	166	147	526
Russia (RUS)	182	162	177	521
East Germany (GDR)	192	165	162	519

Source: Wikipedia, s.v. "All-time Olympic Games medal table," accessed July 2014, http://en.wikipedia.org/wiki/All-time_Olympic_Games_medal_table

50. Let X equal the set of countries who have won more than 1000 medals overall. Write the set X using the roster method.

51. Let Y equal the set of countries who have won between 500 and 1000 medals overall. Write the set Y using the roster method.

52. Let Z equal the set of countries who have won less than 500 medals overall. Write the set Z using the roster method.

53. Let G equal the set of countries who have won more than 200 Gold medals. Write the set G using the roster method.

54. Is $X = G$? Explain.

55. Is $X \sim G$? Explain.

✎ WRITING & THINKING

56. Give an example of a set that cannot be represented using the roster method.

57. Describe two sets of which you are a member.

58. Let A be a set in which one of the elements is President Barack Obama. List at least three different universal sets for A.

59. Let X be a set in which one of the elements is π. List at least three different universal sets for X.

8.2 OPERATIONS WITH SETS

■ TOPICS

- ■ Intersection and Union of Sets
- ■ Disjoint Sets
- ■ De Morgan's Laws
- ■ Inclusion-Exclusion Principle

Intersection and Union of Sets

The introduction to sets and set notation provided us with the foundation for discussing operations with sets. In this section, we will use set operations to compare the manner in which two separate sets are related.

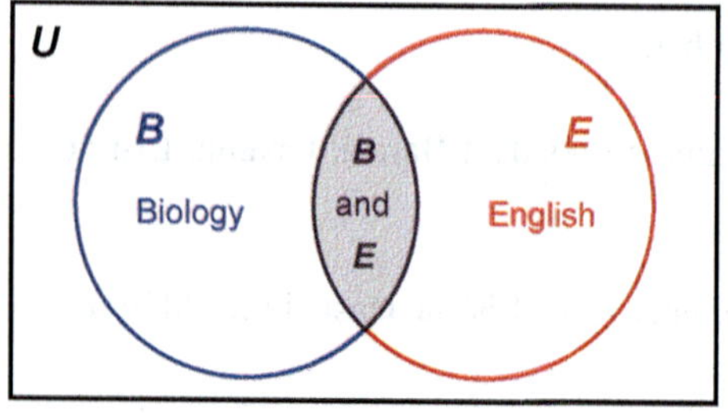

FIGURE 1

Consider the members of a mathematics class as the universal set. A teacher wants to conduct a survey of her students who are also taking English and biology courses. Suppose the teacher defines the set E of students taking English and the set B of students taking biology. We can use a Venn diagram, as shown in Figure 1, to represent how the two sets of students are related.

The teacher is interested in surveying students that are common to both sets B and E. We call this commonality between the two sets the *intersection* of the sets B and E. When some, but not all, of the elements of one set are contained in the other, the sets are represented as overlapping circles in a Venn diagram. In Figure 1 the overlapping region represents the intersection of the two sets B and E.

Intersection

The **intersection** of two sets A and B is the set of all elements common to both A and B. We denote the intersection of A and B as $A \cap B = \{x \mid x \in A \text{ and } x \in B\}$.

Example 1: Determining the Intersection of Sets

Find the intersection of the sets $A = \{n, u, m, b, e, r, s\}$ and $B = \{r, u, l, e\}$.

Solution

Since the intersection of two sets consists of all of the elements that appear in *both* sets, we can see that the intersection of A and B consists of the elements r, u, and e.

$$A \cap B = \{n, u, m, b, e, r, s\} \cap \{r, u, l, e\}$$
$$= \{r, u, e\}$$

We can also use a Venn diagram to find the intersection. Notice that in the Venn diagram, the elements in the intersection are only listed once.

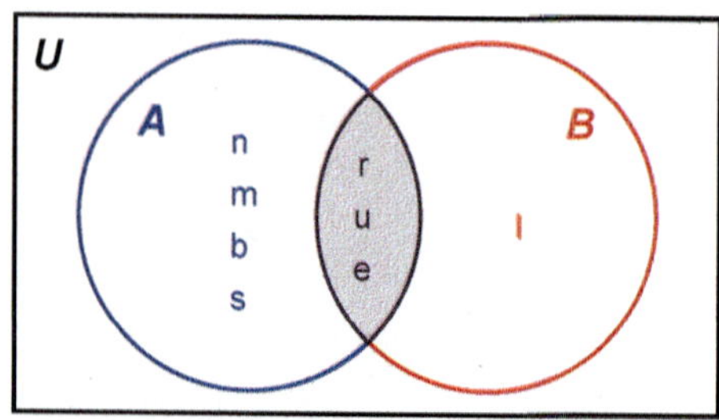

Some diagrams display the number of elements in each region instead of listing the elements themselves. The following example illustrates this.

Example 2: Using a Venn Diagram to Find the Intersection

The given Venn diagram represents the number of students that participated in certain activities while on a spring break trip. Determine the number of students that went both hiking and skiing over spring break.

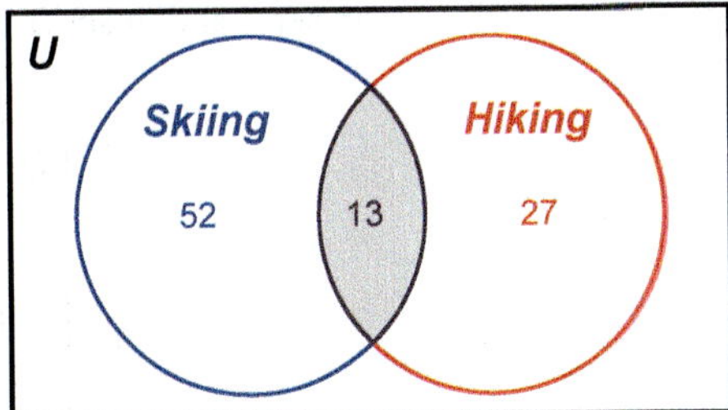

Solution

Using the Venn diagram, we can see that 52 students went skiing exclusively on their break, 27 students went hiking exclusively, and 13 students went both skiing *and* hiking.

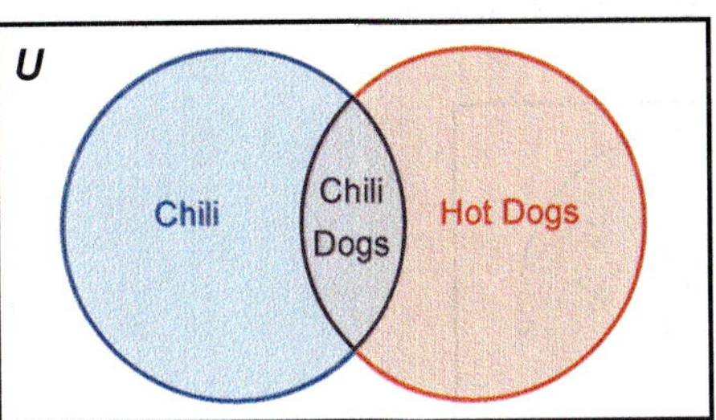

FIGURE 2

Big's Diner is famous for their hot dogs and chili. They have hot chili, mild chili, chili with beans, chili without beans, vegan chili, and their specialty, Texas chili. Their hot dog selection consists of jumbo dogs, turkey dogs, bratwurst, Vienna, and Cumberland. You can even order a chili dog, if you like. If we draw a Venn diagram of all of the offerings at Big's Diner, we get the Venn diagram in Figure 2.

We can form the complete menu offered at Big's Diner from these two sets of dishes. We do so by combining the elements of the two sets into one set. When we combine the elements of two or more sets together, we call this the *union*.

Union

The **union** of two sets A and B is the set of all elements in A or in B. We denote the union of A and B as $A \cup B = \{\, x \mid x \in A \text{ or } x \in B \}$.

Example 3: Determining the Union of Sets

Let $A = \{1, 2, 3, 4, 5\}$, $B = \{2, 4, 6, 8\}$, $C = \{1, 3, 5, 7, 9\}$. Find the following unions.

a. $A \cup B$ **b.** $A \cup C$ **c.** $B \cup C$

Solution

a. To find the union of sets A and B, we need to find the elements that appear in either set A or set B. The elements of set A are 1, 2, 3, 4, 5 and the elements of set B are 2, 4, 6, 8. We simply combine the items together to create a new set to represent the union.

$$A \cup B = \{1,2,3,4,5\} \cup \{2,4,6,8\}$$
$$= \{1,2,3,4,5,6,8\}$$

The Venn diagram shown gives us a visual illustration of the union of the elements in set A and in set B.

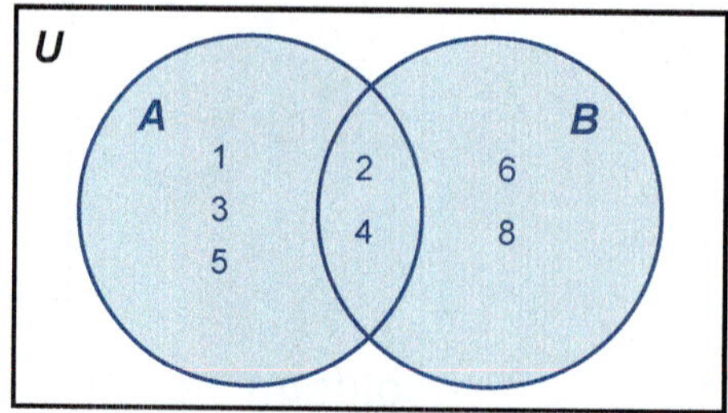

Note that $A \cap B = \{2, 4\}$.

b. Finding the union of sets A and C is done in a similar manner.

$$A \cup C = \{1, 2, 3, 4, 5\} \cup \{1, 3, 5, 7, 9\} = \{1, 2, 3, 4, 5, 7, 9\}$$

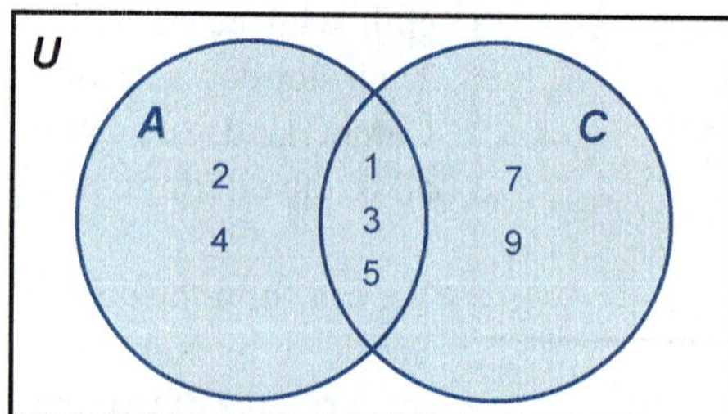

Note that $A \cap C = \{1, 3, 5\}$.

c. Finding the union of sets B and C is also done in a similar manner.

$$B \cup C = \{2, 4, 6, 8\} \cup \{1, 3, 5, 7, 9\} = \{1, 2, 3, 4, 5, 6, 7, 8, 9\}$$

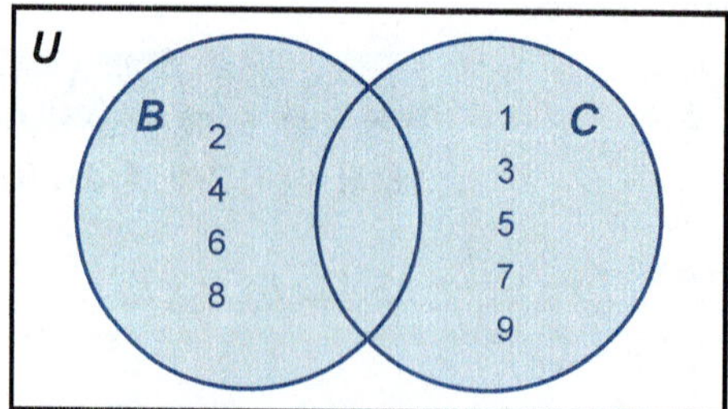

Note that the intersection of these two sets is empty. Therefore, $B \cap C = \varnothing$.

In addition to finding the intersection and the union of two sets, we can also combine the operations of intersection and union. Let's try an example of combining the operations.

> **Example 4: Combining Intersection and Union**
>
> Let $U = \{a, b, c, d, e, \ldots, z\}$, $M = \{m, a, t, h\}$, $N = \{m, o, n, e, y\}$, and $K = \{i, n, v, e, s, t, o, r\}$. Find
>
> **a.** $M \cup (N \cap K)$ **b.** $M \cap (N \cup K)$

Solution

a. In order to find the solution when combining the operations of intersections and unions of sets, it might be best to describe the set first. $M \cup (N \cap K)$ is the set of all elements that are in set M or are in N and in K. Let's do each part and see what we get.

$$M \cup (N \cap K) = \{m, a, t, h\} \cup (\{m, o, n, e, y\} \cap \{i, n, v, e, s, t, o, r\})$$

Just as in order of operations with numbers, we need to perform the operation in parentheses first.

$$N \cap K = (\{m, o, n, e, y\} \cap \{i, n, v, e, s, t, o, r\}) = \{o, n, e\} \text{ and}$$

$$M \cup (N \cap K) = \{m, a, t, h\} \cup \{o, n, e\}$$
$$= \{m, a, t, h, o, n, e\}$$

b. Similarly, we can find $M \cap (N \cup K)$.

$$M \cap (N \cup K) = \{m, a, t, h\} \cap (\{m, o, n, e, y\} \cup \{i, n, v, e, s, t, o, r\})$$
$$= \{m, a, t, h\} \cap \{m, o, n, e, y, i, v, s, t, r\}$$
$$= \{m, t\}$$

> **Skill Check 1**
>
> Let $M = \{m, a, t, h\}$, $N = \{m, o, n, e, y\}$, and $K = \{i, n, v, e, s, t, o, r\}$.
>
> Find $K \cap (M \cup N)$.

Disjoint Sets

Each of the examples involving the intersection and union of sets had elements that were common to both sets except Example 3c, where the intersection was the empty set. Just as in Example 3c, it is possible that two sets could have no common elements. For instance, if you were to draw a single card out of a standard deck of playing cards, the card will be either a red card or a black card. It is impossible for the card to be both red and black at the same time. In this case, we say the set of red cards and the set of black cards are *disjoint*.

> **Disjoint**
>
> Two sets A and B are **disjoint** if there are no elements in set A that are also contained in set B. In the case where two sets are disjoint, the intersection of those two sets is the empty set, or null set. Therefore, $A \cap B = \emptyset$ when sets A and B are disjoint.

It might be easier to visualize two sets that are disjoint by creating a Venn diagram. Using our example of playing cards (all playing cards are either red or black, but not both), we can see that two sets with no elements in common are disjoint. See Figure 3.

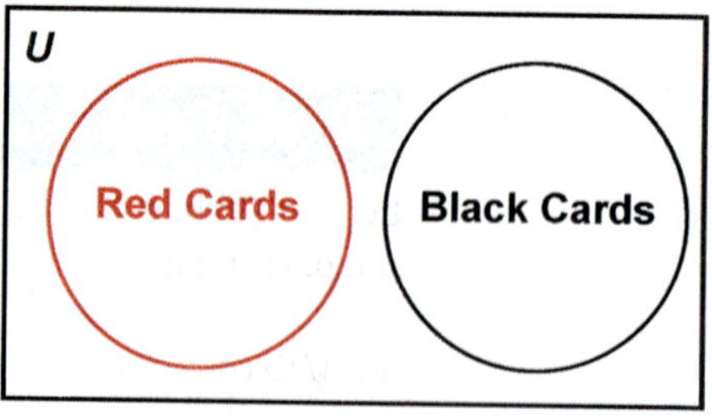

FIGURE 3

Example 5: Identifying Disjoint Sets

Let U = {all students}, A = {students with GPA < 2.5}, and B = {students with GPA > 3.0}. Determine if sets A and B are disjoint and draw a Venn diagram to illustrate the relationship between sets A and B.

Solution

Since it would be impossible for a student to have a GPA that is less than 2.5 and a GPA greater than 3.0 at the same time, sets A and B are disjoint. We can illustrate the relationship between sets A and B using a Venn diagram. The universal set is represented as all students' grade point averages of 0.0 to 4.0. Therefore, when sets A and B are described as students with GPAs less than 2.5 and greater than 3.0, respectively, there are no common GPAs. In addition, any student with a GPA between 2.5 and 3.0 is not represented in either set A or set B.

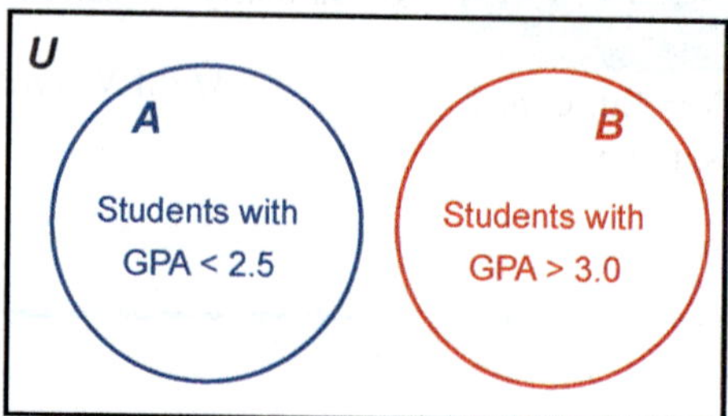

De Morgan's Laws

Up to this point, we have defined the intersection and union of sets as well as what it means for two sets to be disjoint. It is also helpful to discuss the items that are *not* part of a given set.

Previously, we defined the complement of a set A as the set that consists of all the elements in the given universal set that are not contained in A. Recall that the complement of A is denoted A'. We can use Venn diagrams to give a visual illustration of the complement.

Figures 4 and 5 represent the complements of sets A and B, respectively. Notice that everything in the universal set U, **except** what is in set A, is in the complement of A and everything in the universal set U, except what is in set B, is in the complement of B.

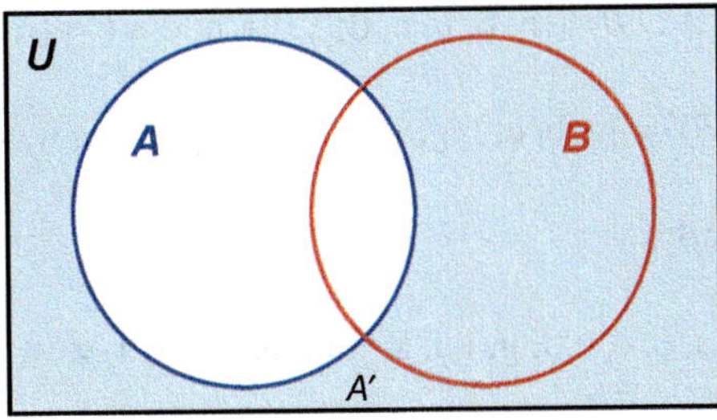
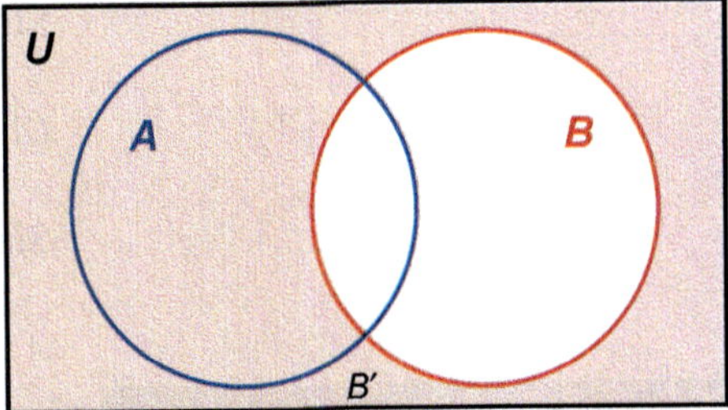

FIGURE 4: Complement of Set A **FIGURE 5:** Complement of Set B

Now that we have visual representations of the complements of sets, we can begin to analyze the complements of the union and intersection of the sets. To make understanding each of these a little easier, we can use De Morgan's Laws. De Morgan's Laws allow us to relate the three operations of sets—intersection, union, and complement—in an effort to analyze sets.

De Morgan's Laws

Let A and B be sets. Then,

$$(A \cup B)' = A' \cap B'$$

and

$$(A \cap B)' = A' \cup B'$$

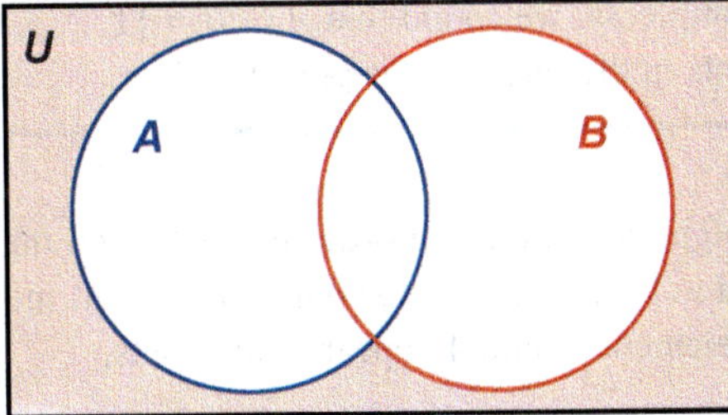
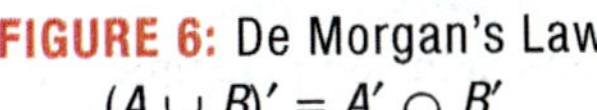
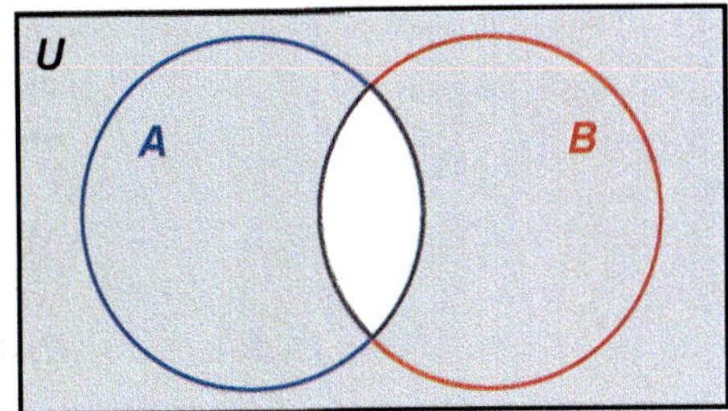

FIGURE 6: De Morgan's Law
$(A \cup B)' = A' \cap B'$

FIGURE 7: De Morgan's Law
$(A \cap B)' = A' \cup B'$

Example 6: Using De Morgan's Laws

Given sets $U = \{a,\ b,\ c,\ d,\ ...,\ z\}$, $A = \{h, o, u, n, d\}$, $B = \{r, o, c, k\}$, verify that $(A \cup B)' = A' \cap B'$.

Solution

The universal set U consists of all of the letters of the alphabet.

Since $A \cup B = \{h, o, u, n, d\} \cup \{r, o, c, k\} = \{r, o, c, k, h, u, n, d\}$, we know that

$(A \cup B)' = \{a, b, e, f, g, i, j, l, m, p, q, s, t, v, w, x, y, z\}$.

Similarly,

$A' = \{a, b, c, e, f, g, i, j, k, l, m, p, q, r, s, t, v, w, x, y, z\}$ and

$B' = \{a, b, d, e, f, g, h, i, j, l, m, n, p, q, s, t, u, v, w, x, y, z\}$,

so $A' \cap B' = \{a, b, e, f, g, i, j, l, m, p, q, s, t, v, w, x, y, z\}$.

Notice that $(A \cup B)' = A' \cap B'$. Hence we have verified De Morgan's Law for A and B.

Skill Check 2

Let $U = \{a, b, c, \ldots, z\}$, $A = \{h, o, u, n, d\}$, and $B = \{r, o, c, k\}$.

Verify $(A \cap B)' = A' \cup B'$.

Inclusion-Exclusion Principle

Recall that a set's cardinal number is the number of elements in the given set. For example, the set $A = \{a, b, c\}$ has 3 elements, and we denote this by $|A| = 3$.

Example 7: Determining the Cardinal Number of a Union

Let $A = \{1, 2, 3, 4, 5\}$ and $B = \{2, 4, 6, 8\}$. Find $|A \cup B|$.

Solution

In Example 3a, we found that $A \cup B = \{1, 2, 3, 4, 5, 6, 8\}$. Therefore, the number of elements in the set $A \cup B$ is $|A \cup B| = 7$.

You might notice that, in order for us to find the cardinality of the union of two sets, we had to be careful not to count an element more than once. In the case of Example 7, the elements 2 and 4 appear in both sets.

Inclusion-Exclusion Principle

The **inclusion-exclusion principle** states that the number of elements in the union of two sets A and B is calculated by adding the number of elements in set A to the number of elements in set B, less the number of elements that appear in both sets. We denote this by $|A \cup B| = |A| + |B| - |A \cap B|$.

The definition of the inclusion-exclusion principle means that when we take $|A| + |B|$, we are counting the total elements in both sets, which means that certain elements are counted twice. To correct this, we must subtract off the number of elements that appear in the intersection of the sets.

Example 8: Applying the Inclusion-Exclusion Principle

A standard deck of playing cards has 52 cards (26 of which are red and 26 of which are black) divided into 4 suits (clubs, spades, diamonds, and hearts), where there are 13 of each suit (ace, 2, 3, 4, 5, 6, 7, 8, 9, 10, jack, queen, king). Of these cards, 12 are considered face cards (4 kings, 4 queens, and 4 jacks). Find the number of cards in a standard deck that are either clubs or face cards.

Solution

We start the solution by writing what we are looking for using set notation.

$$|\text{clubs} \cup \text{face cards}| = |\text{clubs}| + |\text{face cards}| - |\text{clubs} \cap \text{face cards}|$$

If the set A consists of clubs and the set B consists of face cards, then this is equivalent to

$$|A \cup B| = |A| + |B| - |A \cap B|.$$

There are 13 clubs in the deck and there are 12 face cards. However, there are 3 face cards that are also clubs (king of clubs, queen of clubs, and jack of clubs). Therefore,

$$|A \cup B| = 13 + 12 - 3 = 22.$$

So, the number of cards that are either clubs or face cards is **22**.

Skill Check 3

Find the number of playing cards that are either even (2, 4, 6, 8, 10) or are diamonds.

Skill Check Answers

1. {n, e, t, o}

2. $A \cap B = \{o\}$, so $(A \cap B)' = \{a, b, c, d, e, f, g, h, i, j, k, l, m, n, p, q, r, s, t, u, v, w, x, y, z\}$.
   $A' = \{a, b, c, e, f, g, i, j, k, l, m, p, q, r, s, t, v, w, x, y, z\}$ and
   $B' = \{a, b, d, e, f, g, h, i, j, l, m, n, p, q, s, t, u, v, w, x, y, z\}$, so
   $A' \cup B' = \{a, b, c, d, e, f, g, h, i, j, k, l, m, n, p, q, r, s, t, u, v, w, x, y, z\}$.
   Thus, $(A \cap B)' = A' \cup B'$.

3. 28

8.2 EXERCISES

💡 PRACTICE

Use the given sets to solve each problem.

$U = \{1, 2, 3, \ldots, 20\}$ $\qquad$ $A = \{1, 2, 3, 4, 5, 6, 7, 8\}$

$B = \{2, 4, 6, 8, 10, 12\}$ $\qquad$ $C = \{5, 7, 9, 11, 13, 15\}$

1. Find $A \cup B$. $\qquad\qquad$ **2.** Find $A \cup C$.

3. Find $B \cap C$.

4. Find $A \cap C$.

5. Verify $(A \cup B)' = A' \cap B'$.

6. Verify $(A \cap B)' = A' \cup B'$.

Use the given sets to solve each problem.

$$U = \{a, b, c, d, \ldots, z\} \qquad A = \{n, u, m, b, e, r, s\} \qquad B = \{r, u, l, e\}$$

7. Find $A \cup B$.

8. Find $A \cap B$.

9. Find $|A \cap B|$.

10. Verify $(A \cup B)' = A' \cap B'$.

11. Verify $(A \cap B)' = A' \cup B'$.

Use the given sets to solve each problem.

$$U = \{A, B, C, D, \ldots, Z\} \qquad A = \{I, C, E\} \qquad B = \{C, U, B, E\}$$

12. Find $A \cup B$.

13. Find $A \cap B$.

14. Find $|A \cap B|$.

15. Verify $(A \cup B)' = A' \cap B'$.

16. Verify $(A \cap B)' = A' \cup B'$.

Use the given sets to solve each problem.

$$U = \{A, B, C, D, \ldots, Z\} \qquad A = \{F, A, C, T, O, R\} \qquad B = \{P, R, O, D, U, C, T\}$$

17. Find $A \cup B$.

18. Find $A \cap B$.

19. Find $|A \cap B|$.

20. Verify $(A \cup B)' = A' \cap B'$.

21. Verify $(A \cap B)' = A' \cup B'$.

Use the given sets to solve each problem.

$$U = \{a, b, c, d, \ldots, z\} \qquad M = \{b, r, i, d, g, e\}$$

$$N = \{g, a, t, o, r\} \qquad K = \{b, a, l, i, s, t, e, r\}$$

22. Find $N \cap (M \cup K)$.

23. Find $N \cup (M \cap K)$.

24. Find $K \cup (N \cap M)$.

25. Verify $(M \cup K)' = M' \cap K'$.

26. Verify $(M \cap N)' = M' \cup N'$.

27. A grocery store found that 275 of its customers use push carts to shop, 185 used a carry basket to shop, and that 145 used both a push cart and a carry basket. How many customers use only a push cart or a carry basket? Draw the Venn diagram.

Use the Venn diagram to solve each problem.

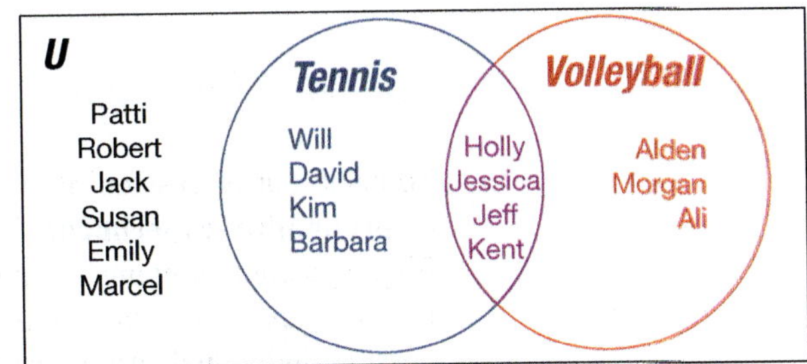

28. Which students played only tennis?

29. Determine which students played tennis or volleyball.

30. Determine which students played tennis and volleyball.

31. Find the number of students that play tennis or volleyball.

Solve each problem.

32. Determine the number of playing cards in a standard deck that are red cards or face cards.

33. Determine the number of playing cards in a standard deck that are odd numbered cards or black cards.

Show that each pair of sets is equal by drawing a Venn diagram of each set.

34. $A \cap B$ and $B \cap A$

35. $A \cup B$ and $B \cup A$

36. $(A \cap B) \cap C$ and $A \cap (B \cap C)$

37. $(A \cup B) \cup C$ and $A \cup (B \cup C)$

38. $A \cup \varnothing$ and A

Use set notation to represent each shaded region.

39. 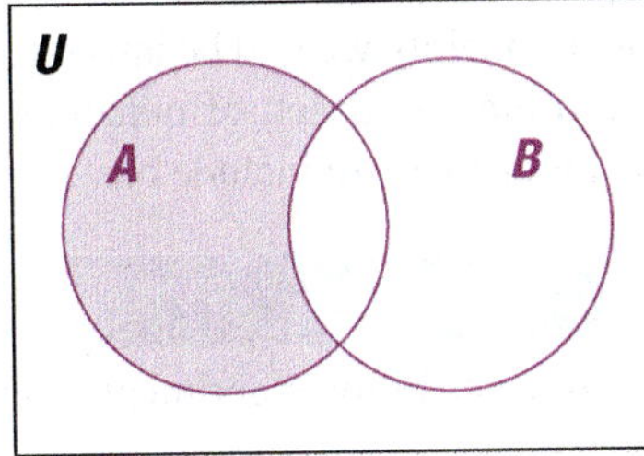

40. 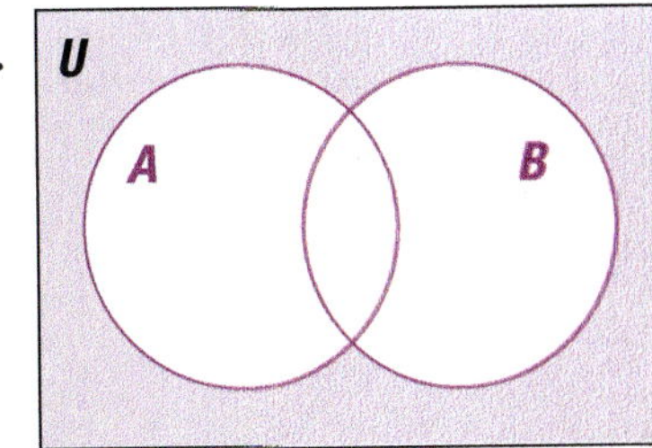

8.3 INTRODUCTION TO PROBABILITY

■ TOPICS

- Classical Probability
- Empirical Probability

Classical Probability

To begin our discussion of probability, we need to define a few terms. First, a **trial**, or **probability experiment**, is any process that produces a random result, such as flipping a coin, drawing a number from one to ten out of a hat, or having a computer randomly generate a three-digit number. Each of these examples easily illustrates the possible individual results, called **outcomes**, in a trial. For instance, when you flip a coin, the outcomes possible are either a head or a tail. When you draw a number from the hat described above, four is a possible outcome, as is nine. In fact, all the numbers from one to ten are outcomes. The set of all possible outcomes from a given probability experiment is called the **sample space** of that experiment. For example, we've already seen that the sample space for a coin flip is {head, tail}. The sample space for the computer example is all possible three-digit numbers.

Example 1: Determining Sample Spaces

Identify the sample space for each of the following experiments. Note that the outcomes in a sample space are listed between brackets and separated by commas.

a. Rolling a single die.

b. Birth order gender for two children in a single family.

Solution

a. The sample space consists of all of the possible outcomes of rolling a die. It can land on any of the six sides of the die. Therefore, the sample space is the following.

b. The sample space consists of all possible gender outcomes for a family with two children. Let B = boy and G = girl. The sample space is

$$\{BB, BG, GB, GG\}.$$

Often, in probability we will be interested in grouping together outcomes in the sample space. A group, or subset, of outcomes in the sample space is called an *event*. It is possible for an event to include one, some, or all the members of the sample space.

Trial

A **trial**, or probability experiment, is any process that produces a random result.

Outcome

The **outcomes** of a trial are the possible individual results.

Sample Space

The **sample space** S of a trial is the set of all possible outcomes.

Event

An **event** E is a group, or subset, of outcomes in the sample space.

Informally, when we talk of the likelihood of an event, we are often referring to either a precise theoretical probability, called classical probability, or more of an experimental approach, called empirical probability. First let's consider classical probability.

The scale used for the **classical probability** of an event is a real number between 0 and 1, where 0 means it will never happen and 1 means it is certain to happen. The notation $P(x)$ refers to the probability that x will happen. It is sometimes easier to think about classical probabilities in terms of fractions.

An event having a probability of $\frac{1}{2}$ that is, $P(x) = \frac{1}{2}$ means that there is an equal chance that the event will happen or will not happen. In other words, the event is one of two equally likely possibilities. Examples of events with probabilities of $\frac{1}{2}$ are a coin coming up heads or rolling an even number on a die. Similarly, $P(x) = \frac{1}{6}$ means that the event is one of six equally likely possibilities, such as rolling a 3 on a standard die. Whenever we can divide the situation into a known number of **equally likely** outcomes, the probability follows immediately. Be careful, the term "equally likely" here is important. There is an obvious flaw in the statement, *There's a 50% chance I'll win the lottery tomorrow.* Even though the sample space contains the two outcomes {win the lottery, don't win the lottery}, the probability of winning the lottery is *not* $\frac{1}{2}$ because the outcomes are definitely not equally likely!

Recall that real numbers are not limited to fractions and whole numbers. Numbers like $\frac{1}{\sqrt{2}}$ and $\frac{3}{\pi}$ are viable values for probabilities since they are numbers between 0 and 1.

When an event includes the entire sample space, the probability of the event occurring is equal to 1.

Classical Probability

If all outcomes are equally likely, **classical probability** is calculated with the formula

$$P(\text{event}) = \frac{\text{the number of possible outcomes in the event}}{\text{the number of outcomes in the sample space}}.$$

P(event) will always be a real number between 0 and 1, inclusive.

Example 2: Calculating Classical Probability

Suppose you were asked to draw a card from a standard deck of 52 cards. A standard deck of cards contains the following cards.

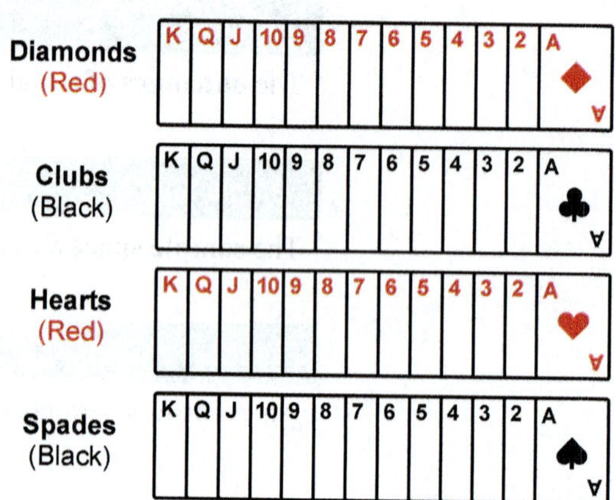

a. What is the probability that the card you draw is red?

b. What is the probability that the card you draw is a diamond?

c. What is the probability that the card you draw is a face card (king, queen, or jack)?

d. What is the probability of drawing a red spade?

Solution

a. Because this is a standard deck of cards we are drawing from, each card has the same probability of being chosen. We know that the sample space contains 52 cards. We also know that since there are two red suits (and two black suits) each with 13 cards, there are 26 possible red cards to choose. So, the probability that the card you draw is red is

$$P(\text{red card}) = \frac{\text{number of red cards}}{\text{total number of cards in deck}} = \frac{26}{52} = \frac{1}{2} = 0.5.$$

b. There are 13 cards in the diamond suit. So the probability that your card is a diamond is

$$P(\text{diamond}) = \frac{13}{52} = \frac{1}{4} = 0.25.$$

c. Each of the four suits contains three face cards (king, queen, and jack), so there are $4 \cdot 3 = 12$ face cards to choose from. So the probability that your card is a face card is

$$P(\text{face card}) = \frac{12}{52} = \frac{3}{13} \approx 0.230769.$$

d. Because all spades are black, it is impossible to draw a red spade. Therefore,

$$P(\text{red spade}) = 0.$$

Example 3: Calculating Classical Probability

Suppose that you grab a snack from a bag of chocolates that contains 4 caramel with milk chocolate, 4 peppermint with white chocolate, 6 dark chocolate with mint, and 2 raspberry with dark chocolate. What is the probability that you randomly grab a raspberry with dark chocolate for your snack?

Solution

Remember, to find the probability, we first need to know the number of outcomes in the sample space. We can add together the numbers of all the different types of chocolate in the bag to find out the number of possible outcomes.

$$4 + 4 + 6 + 2 = 16 \text{ different chocolates in the bag}$$

Next, we know that there are 2 possible outcomes for the event of choosing a raspberry with dark chocolate. Therefore, the probability of the event is

$$P(\text{raspberry with dark chocolate}) = \frac{2}{16} = \frac{1}{8} = 0.125.$$

Empirical Probability

We have seen that the probability of rolling a 3 on a standard die is $\frac{1}{6}$, since it is 1 of 6 equally likely possibilities. Unfortunately, not every situation is so easily analyzed. For instance, what if we suspected that a die is loaded (meaning it is unfair). How would we know? Suppose we were to roll the die 600 times. If it were a fair die, we would expect to roll a 3 about 100 times, maybe not precisely 100, but close. In fact, doing an experiment like this is a way to estimate probability, called **empirical probability**. Empirical probability is built around the **law of large numbers**, which says that the greater the number of trials, the closer the experimental probability will be to the *true* probability.

Empirical Probability

If all outcomes are based on an experiment, **empirical probability** is calculated with the formula

$$P(\text{event}) = \frac{\text{the number of times the event occurs}}{\text{total number of times the experiment is performed}}.$$

$P(\text{event})$ will always be a real number between 0 and 1, inclusive.

Example 4: Empirical Probability with Multiple Trials

For her elementary school science fair project, Libby is conducting research on the accuracy of the weather prediction from her local news channel. She recorded the forecast and the actual weather for two weeks. Table 1 shows her results.

Forecast	Actual Weather
Rain	Rain
Chance of snow	Rain
Snow	Snow
Cloudy	Clear
Cloudy	Cloudy
Rain	Rain
Clear	Drizzling rain
Clear	Clear
Cloudy	Clear
Chance of snow	Cloudy
Clear	Clear
Clear	Clear
Rain	Rain
Chance of rain	Cloudy

TABLE 1: Accuracy of Weather Prediction

Using empirical probability, what is the probability that the news channel accurately predicts the next day's weather?

Solution

For Libby to calculate the probability, she needs to count the number of days the weatherman correctly predicted the weather and divide it by 14 (the total number of days she did the experiment).

$$P(\text{correct prediction}) = \frac{\text{number of times the forecast was correct}}{\text{total number of times the weather was recorded}}$$

$$= \frac{8}{14} = \frac{4}{7} \approx 0.571429.$$

Because Libby can now estimate that the news channel correctly predicts the weather 57% of the time, she knows that this is also the probability that the prediction for the following day's weather will be correct.

Example 5: Classical vs. Empirical Probability

Determine if the scenarios given are examples of classical or empirical probability techniques.

a. Katie is curious about her chances of winning an e-reader from the student government association. She polled her friends to find out how many of them filled out the survey to be entered in the contest.

b. Tristan is interested in his chances of winning at the blackjack table. He determines the probability of what his next card will be by knowing the cards that have already been played.

c. Based on the recent United States Census, the local government estimates the amount of growth the community will experience in the coming years.

Solution

a. Because Katie is conducting an informal survey and not all students are included, the probability is empirical.

b. This is an example of classical probability since all cards have an equal chance of being dealt at the beginning, and Tristan adjusts his chances by accounting for those cards that have already been drawn.

c. Since the United States Census is actually an incomplete count, any probability calculated from it would be empirical.

As you might imagine, carrying out trials and surveys to estimate probabilities using an empirical approach can be very time consuming and potentially costly. Consequently, although using empirical probability may sometimes be unavoidable, it is certainly preferable to use classical probability whenever it can be obtained.

Here's a recap of this section.

Probability Concepts

- A **trial**, or probability experiment, is any process in which the result is random in nature.

- An **outcome** is an individual result that is possible from a probability experiment.

- The **sample space** S is the set of all possible outcomes from a given probability experiment.

- An **event** E is a subset of outcomes from the sample space.

- If all outcomes are equally likely, **classical probability** is calculated with the formula

$$P(\text{event}) = \frac{\text{the number of possible outcomes in the event}}{\text{the number of outcomes in the sample space}}.$$

- If all outcomes are based on an experiment, **empirical probability** is calculated with the formula

$$P(\text{event}) = \frac{\text{the number of times the event occurs}}{\text{the total number of times the experiment is performed}}.$$

- $P(\text{event})$ will always be a real number between 0 and 1, inclusive.

8.3 EXERCISES

PRACTICE

Identify the sample space for each experiment.

1. A coin is flipped and then a single die is rolled.

2. Choosing a number from all positive two-digit integers where the digits are repeated (for example, 11).

3. Five marbles are in a bag, one of each of the following colors: blue (B), clear (C), green (G), yellow (Y), and red (R). Two marbles are drawn consecutively. Assume that the first marble is not put back in the bag before the second marble is drawn. Order of the selection matters. In other words, BG and GB are two different selections.

4. When choosing an outfit, you have a choice of three shirts: white (W), black (B), or patterned (P); a choice of two types of jeans: faded (F) or dark wash (D); and a choice of four pairs of shoes: sandals (S), running shoes (R), climbing boots (C), or mules (M). List the outcomes in the sample space in regard to the outfits (combination of shirt, jeans, and shoes) you could pick from.

Determine whether each probability is empirical or classical.

5. In order to find the percentage of bass that Troy had in his pond this spring, he decided to spend three days catching a total of 15 fish each day and counting how many of those were bass.

6. Jason wants to know how likely he is to win a raffle if he bought 3 of the 1000 tickets that were sold.

7. Virginia wants to know how likely it is for her to win a backgammon game if she only needs to roll a double six to win.

8. Emre wants to see how many students drive a hybrid car. He surveys 100 college freshmen at a Winter Welcome event and asks what kind of car they drive.

🚀 APPLICATIONS

Calculate each empirical probability. Round your answer to the nearest millionth when necessary.

9. A news organization asked a selection of voters exiting a polling place their age bracket in order to paint a picture of turnout on election day. Here's the record of the results collected so far.

Voting Age			
17–29	**30–44**	**45–64**	**65 and Older**
9	8	32	15

   a. What is the probability that the next voter to exit will be between 30 and 44?
   b. What is the probability that the next voter to exit will be in either of the youngest two age groups?
   c. What is the probability that the next voter to exit will be under 65?

10. The blood types of 200 people are collected at a doctor's office. The table shows the breakdown of patients per blood type. Based on the data in the table, what is the probability that a new patient will have type O blood?

Blood Type Survey Results	
Blood Type	**Number of Patients**
A	50
B	65
O	70
AB	15

11. As students were exiting the student center on campus, Vicki took note whether they were listening to headphones. The table shows results that she collected. Based on Vicki's data, what is the probability that a randomly selected student will have headphones on when exiting the student center?

Data for Students Exiting the Student Center	
	Number of Students
Headphones	33
No Headphones	51

12. A sample of 500 active-duty military showed that 82 of them suffered from post-traumatic stress disorder (PTSD). However, 95 of those studied reported that they would be too embarrassed to seek mental health services.
   a. Based on this sample, if a soldier is chosen at random, what is the probability that he/she suffers from PTSD?
   b. Based on this sample, if a soldier is chosen at random, what is the probability that he/she would be willing to seek mental health services?

Use classical probability to calculate each probability. Assume individual outcomes are equally likely. Round your answer to the nearest millionth when necessary.

13. What is the probability that, out of 235 attendees (including yourself) at a conference, you are selected to win the door prize at the opening session?

14. A standard die is rolled.
    a. Find the probability that the roll produces a number less than 3.
    b. Find the probability that the number rolled is an even number.
    c. Find the probability that the number rolled is greater than 0.

15. On an American roulette wheel, there are 18 red pockets, 18 black pockets, and 2 green pockets. What is the probability of landing on a red pocket?

16. The table shows a breakdown for all employees on nonfarm payrolls in the United States during March 2014 (the values are not seasonally adjusted).

Employees on Nonfarm Payrolls (in Thousands), March 2014		
	Area of Employment	**Number of Employees (in Thousands)**
Private Sector	Goods-Producing	18,558.2
	Wholesale Trade	5803.7
	Retail Trade	15,004.0
	Transportation and Warehousing	4524.8
	Utilities	550.3
	Information	2653.0
	Financial Activities	7870.0
	Professional and Business Services	18,832.0
	Education and Health Services	21,481.0
	Leisure and Hospitality	14,143.0
	Other Private Service-Providing Services	5464.0
Public Sector	Federal Government	2705.0
	State Government	5217.0
	Local Government	14,341.0
	Total Nonfarm Employees	**137,147.0**
	Source: Bureau of Labor Statistics. "Table B-1. Employees on nonfarm payrolls by industry sector and selected industry detail." Accessed June 2014. http://www.bls.gov/news.release/empsit.t17.htm	

    a. Find the probability that a random nonfarm employee was employed in the leisure and hospitality sector during March 2014.
    b. Find the probability that a random nonfarm employee was in the public sector during March 2014.

17. What is the probability that a card drawn randomly from a standard deck of cards will be an ace?

18. A book contains 321 pages numbered 1, 2, 3, …, 321. If a student randomly opens the book, what is the probability that the page has a two- or three-digit number and all of its digits are the same?

19. Consider parents with four biological children. Find the probability that all four siblings are boys.

20. Mason has 213 songs on his iPod. He's categorized them in the following manner: 20 from sound tracks, 8 spiritual, 31 jazz, 16 Latin American, 27 R&B, 47 rock, and 64 pop. If Mason puts his iPod on shuffle, what is the probability that the first song played is a pop song?

21. Landon decided to play a joke on his friends at work. When he stocked the vending machine with sodas, he randomly put the drinks into the different slots. If he put in 15 Diet Coke, 15 Coke, 12 Sprite, 17 Dr. Pepper, and 10 Fanta cans, what is the probability that the next person will get a Fanta drink when they put their money into the machine?

8.4 COUNTING PRINCIPLES: COMBINATIONS AND PERMUTATIONS

■ TOPICS

- The Fundamental Counting Principle
- Factorials
- Permutations and Combinations

Recall that classical probability is calculated as follows.

$$P(\text{event}) = \frac{\text{the number of possible outcomes in the event}}{\text{the number of outcomes in the sample space}}$$

You can see that it is important to know how many events are in the sample space. That sounds easy enough, and in many cases it is.

For instance, in how many ways can you roll two dice such that the outcome of the second die is less than the outcome of the first die? An easy way to count these outcomes is by listing out all of the outcomes in an orderly way, as shown in Figure 1.

6, 1	**5, 1**	**4, 1**	**3, 1**	**2, 1**	1, 1
6, 2	**5, 2**	**4, 2**	**3, 2**	2, 2	1, 2
6, 3	**5, 3**	**4, 3**	3, 3	2, 3	1, 3
6, 4	**5, 4**	4, 4	3, 4	2, 4	1, 4
6, 5	5, 5	4, 5	3, 5	2, 5	1, 5
6, 6	5, 6	4, 6	3, 6	2, 6	1, 6

FIGURE 1: Rolling Two Dice (1st Die, 2nd Die)

You can see that only the pairs of dice rolls in bold fit the criteria that the outcome of the second die is less than the outcome of the first die. That gives us 15 possible ways in which to roll two dice this way.

As another example, suppose you won a new iPod in 2011. All you had to do was go to the store and pick it out. You had a number of decisions to make before you could claim your prize. How many different iPods could you choose from? Let's use a **tree diagram** to help us visualize all the possibilities. Just as its name suggests, a tree diagram uses branches to indicate possible choices at the next stage of outcomes.

> ### Tree Diagram
>
> A **tree diagram** uses branches to indicate possible choices at the next stage of outcomes.

Figure 2 shows what your choices would have looked like for an iPod in 2011. As you can see, there were many choices. In fact, if you count the iPods along the ends of all the branches, you can see that there were 48 different iPods for you to choose from, which is the number of possible outcomes in our sample space.

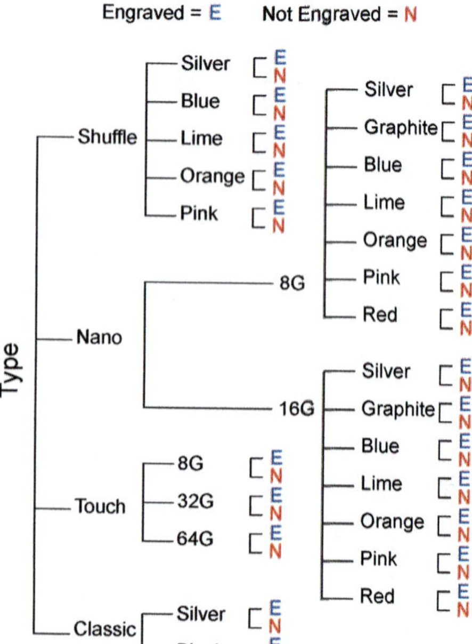

FIGURE 2

The Fundamental Counting Principle

As we illustrated with the iPod example, even with just 48 outcomes a tree diagram can quickly become unmanageable. This is where we'll turn to a counting technique called the **Fundamental Counting Principle**. When there are several outcomes at each stage of an experiment, the Fundamental Counting Principle states that you can multiply together the number of possible outcomes at each stage of the experiment in order to obtain the total number of outcomes for the experiment.

> **✆ HELPFUL HINT**
>
> Although the formal definition for the Fundamental Counting Principle speaks of a sequence of experiments, you can also think of it as different stages of a single experiment.

Fundamental Counting Principle

For a sequence of n experiments where the first experiment has k_1 outcomes, the second experiment has k_2 outcomes, the third experiment has k_3 outcomes, and so forth, the total number of possible outcomes for the sequence of experiments is $(k_1)(k_2)(k_3)\cdots(k_n)$.

When counting outcomes in multistep experiments, we need to determine if repetition of outcomes is permitted (also known as an experiment with replacement) or if each outcome can only be used once (also known as an experiment without replacement).

Replacement

With replacement: When counting possible outcomes with replacement, objects are placed back into consideration for the following choice.

Without replacement: When counting possible outcomes without replacement, objects are *not* placed back into consideration for the following choice.

Example 1: Using the Fundamental Counting Principle with Replacement

In order to log in to your new e-mail account, you must create a password. The requirements are that the password needs to be 8 characters long consisting of 5 lowercase letters followed by 3 numbers. If you are allowed to use a character more than once, that is, **with replacement**, how many different possibilities are there for passwords?

Solution

If we think about each character in the password as a slot to fill, then we have 8 slots that need filling. The first 5 can be filled with letters and the last 3 with digits as the following figure shows.

Slot 1	Slot 2	Slot 3	Slot 4	Slot 5	Slot 6	Slot 7	Slot 8
a b c … x y z	a b c … x y z	a b c … x y z	a b c … x y z	a b c … x y z	0 1 2 … 8 9	0 1 2 … 8 9	0 1 2 … 8 9

Number of Choices: $26 \cdot 26 \cdot 26 \cdot 26 \cdot 26 \cdot 10 \cdot 10 \cdot 10$

The first 5 slots contain 26 possibilities each, one for each letter of the alphabet. The last 3 slots have 10 possible possibilities each, one for each digit 0 through 9. Using the Fundamental Counting Principle, we multiply each of the possibilities together to get $(26)(26)(26)(26)(26)(10)(10)(10) = 11{,}881{,}376{,}000$ possible passwords for the new e-mail account.

Example 2: Using the Fundamental Counting Principle without Replacement

Let's change the previous example slightly. The password still needs to be 8 characters long consisting of 5 lowercase letters followed by 3 numbers. However, now the characters may not be duplicated in the password, that is, we say we're counting **without replacement, or without repetition**.

Solution

We still have the first 5 slots being filled with letters and the last 3 with numbers. This time our picture changes slightly. The first slot still has a possibility of 26 letters, but the second slot now only has 25 choices since we used one letter for the first slot. Similarly, the third slot has 24 choices, and so forth. The same thing happens with the digits in the last 3 spaces.

$a\,b\,c\,\dots x\,y\,z$	$a\,b\,c\,\dots x\,y\,z$	$a\,b\,c\,\dots x\,y\,z$	$a\,b\,c\,\dots x\,y\,z$	$a\,b\,c\,\dots x\,y\,z$	$0\,1\,2\,\dots 8\,9$	$0\,1\,2\,\dots 8\,9$	$0\,1\,2\,\dots 8\,9$
Slot 1	Slot 2	Slot 3	Slot 4	Slot 5	Slot 6	Slot 7	Slot 8

Number of Choices: $26 \cdot 25 \cdot 24 \cdot 23 \cdot 22 \cdot 10 \cdot 9 \cdot 8$

So now we have $(26)(25)(24)(23)(22)(10)(9)(8) = 5{,}683{,}392{,}000$ possible passwords. That is almost half of the original amount of passwords possible if we allowed replacement!

Factorials

Notice that in Example 2, we multiplied successive decreasing numbers together to get our solution for the number of possible passwords. This is a common occurrence when solving this type of problem. Mathematically, we can represent something similar to this by using **factorials**.

$$5! = (5) \cdot (4) \cdot (3) \cdot (2) \cdot (1) = 120$$

However, note in Example 2 we did not multiply all integers less than 26 because $26!$ was not the answer.

n Factorial

In general, $n!$ (read "**n factorial**") is the product of all the positive integers less than or equal to n, where n is a positive integer.

$$n! = n(n-1)(n-2)(n-3)\cdots(2)(1)$$

Note that $0!$ is defined to be 1.

Example 3: Calculating Factorials

Calculate the values of the following factorial expressions.

a. $8!$

b. $\dfrac{3!}{0!}$

c. $\dfrac{89!}{87!}$

d. $\dfrac{7!}{(5-1)!}$

e. $\dfrac{5!}{3!(4-2)!}$

Solution

a. Multiply together all the positive integers less than or equal to 8.

$$8! = (8)(7)(6)(5)(4)(3)(2)(1) = 40{,}320$$

b. Calculate each factorial and then divide.

$$\frac{3!}{0!} = \frac{(3)(2)(1)}{1} = \frac{6}{1} = 6$$

c. Because the numbers are so large here, let's first look at taking a shortcut.

$$\frac{89!}{87!} = \frac{(89)(88)(87)(86)\cdots(2)(1)}{(87)(86)\cdots(2)(1)}$$

Many of the numbers being multiplied in the numerator and denominator will cancel, so let's do that first.

$$\frac{89!}{87!} = \frac{(89)(88)\cancel{(87)}\,\cancel{(86)}\cdots\cancel{(2)}\,\cancel{(1)}}{\cancel{(87)}\,\cancel{(86)}\cdots\cancel{(2)}\,\cancel{(1)}} = (89)(88) = 7832$$

d. Before we can start multiplying numbers, we need to do the subtraction in the denominator. Then we can cancel and multiply.

$$\frac{7!}{(5-1)!} = \frac{7!}{4!} = \frac{(7)(6)(5)\cancel{(4)}\,\cancel{(3)}\,\cancel{(2)}\,\cancel{(1)}}{\cancel{(4)}\,\cancel{(3)}\,\cancel{(2)}\,\cancel{(1)}} = (7)(6)(5) = 210$$

e. Before we can start multiplying numbers, we once again need to perform the subtraction in the denominator. Then, notice that $5! = (5)(4)3!$ allows us to cancel $3!$ in the numerator and the denominator before multiplying the remaining values.

$$\frac{5!}{3!(4-2)!} = \frac{5!}{3!2!} = \frac{(5)(4)\cancel{3!}}{\cancel{3!}(2)(1)} = \frac{(5)(4)}{(2)(1)} = \frac{20}{2} = 10$$

To calculate $n!$ using a TI-83/84 Plus calculator, enter the number for n, then press `math`; scroll over to select PRB, then choose option `4:!` and press `enter`. The following screenshots illustrate 8!.

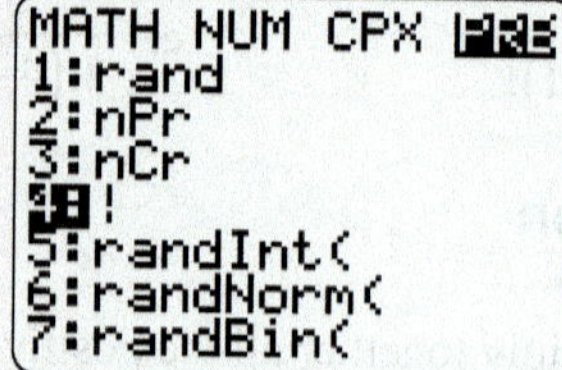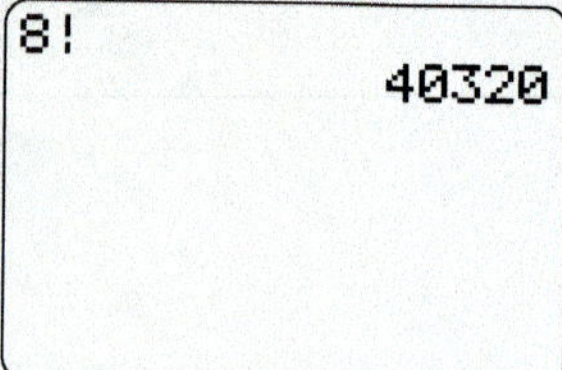

To calculate $n!$ using a TI-30XIIS/B calculator, enter the number for n, then press `PRB`, scroll to the !, and press `enter`.

Wolfram|Alpha can be used to compute factorials as well. Go to www.wolframalpha.com and type "8!" into the input line. Then, click the = button. Wolfram|Alpha will return the following.

Source: Wolfram Alpha LLC. 2009. Wolfram|Alpha. http://www.wolframalpha.com/input/?i=8%21 (accessed July 2, 2014).

Permutations and Combinations

We often want to be able to count the number of ways that we can choose members from a group of objects. For instance, how many different sandwiches can be made with the ingredient choices at a sandwich shop? Or how many ways can the top three spots be filled at the end of a race of 140 people? Both of these are scenarios that can be calculated by using either a **permutation** or a **combination**. Let's define them both so you can see the difference.

The following are all alternate notations for combinations and permutations.

$$_nC_r = C(n,r) = \binom{n}{r} = {}^nC_r = C_{n,r}$$

$$_nP_r = P(n,r) = {}^nP_r = P_{n,r}$$

Combinations

A **combination** involves choosing a specific number of objects from a particular group of objects, using each only once, when the order in which they are chosen is *not* important.

Permutations

A **permutation** involves choosing a specific number of objects from a particular group of objects, using each only once, when the order in which they are chosen *is* important.

Decide whether you would use a permutation or combination to count the number of outcomes for each of the following scenarios.

a. In how many ways can 1st, 2nd, and 3rd place prizes be handed out to science fair winners if there are 30 students participating?

b. If each department needs two student representatives from each major on a campus committee, how many ways can the biology department choose the representatives from the 45 students who are majoring in biology?

You can see, just from their definitions, that the only thing that differentiates the two is whether the order of the objects is important. For the sandwich example, we'll contend that order is not important when deciding if one sandwich with turkey, mayonnaise, lettuce, and tomato is the same as another sandwich with the same ingredients. You might be particular in which order the tomato should be added, but it doesn't make a different sandwich. So, here we'll use a combination technique to count the number of possible sandwiches. However, if there are 140 people in a race of which Chloe, Blake, and Mary finish in the top three, then the order in which they finish makes a difference. Finishing first and winning the blue ribbon is certainly different than finishing third, so we need to use a permutation to count the possibilities.

Now that we know when to use a combination and when to use a permutation, how do we actually calculate the number of outcomes produced by each? Here are the formulas for combinations and permutations.

Combinations and Permutations of n Objects Taken r at a Time

The number of ways to select r objects from a total of n objects is found by the following two formulas. (Note that $r \leq n$.)

When order is not important, use the following formula for a **combination**.

$$_nC_r = \frac{n!}{r!(n-r)!}$$

When order is important, use the following formula for a **permutation**.

$$_nP_r = \frac{n!}{(n-r)!}$$

Notice the difference in the formulas for combinations and permutations. When counting the number of possible permutations your result accounts for ALL possible arrangements. However, when counting the number of combinations you do not want to count groupings of the same r things more than once. So one has to divide by the number of arrangements for each group of r, which is the extra $r!$ in the denominator of $_nC_r$.

Example 4: Using Combinations

Let's calculate the number of possibilities for our sandwich example. Suppose there are 18 toppings to choose from once you've decided on bread, meat, and cheese. How many different possible sandwiches are there if you choose 4 different toppings?

Solution

The order of sandwich toppings does not change the type of sandwich that is made. Therefore, this is a combination problem where we are choosing 4 toppings from a list of 18. Fill in the combination formula using $n = 18$ and $r = 4$.

$$_{18}C_4 = \frac{18!}{4!(18-4)!} = \frac{18!}{4!14!} = \frac{(18)(17)(16)(15)(14)(13)\cdots(2)(1)}{(4)(3)(2)(1)(14)(13)\cdots(2)(1)}$$

$$= \frac{(18)(17)(16)(15)}{(4)(3)(2)(1)} = 3060$$

Therefore, there are 3060 different sandwich possibilities—far too many for you to say to a friend, "Just pick me up a turkey sandwich. It doesn't matter what kind. They're all alike!"

〰 TECH TRAINING

A TI-83/84 Plus can be used to find solutions involving combinations. To calculate $_{18}C_4$, type **1 8**, and then press **math**. Scroll over to PRB and choose option **3:nCr**. Then, type **4** and press **enter**. The screenshot below illustrates this.

A TI-30XIIS/B can also be used to find solutions involving combinations. To calculate $_{18}C_4$, type **1 8**, and then press **PRB**. Scroll to nCr and press **enter**. Then, type **4** and press **enter**.

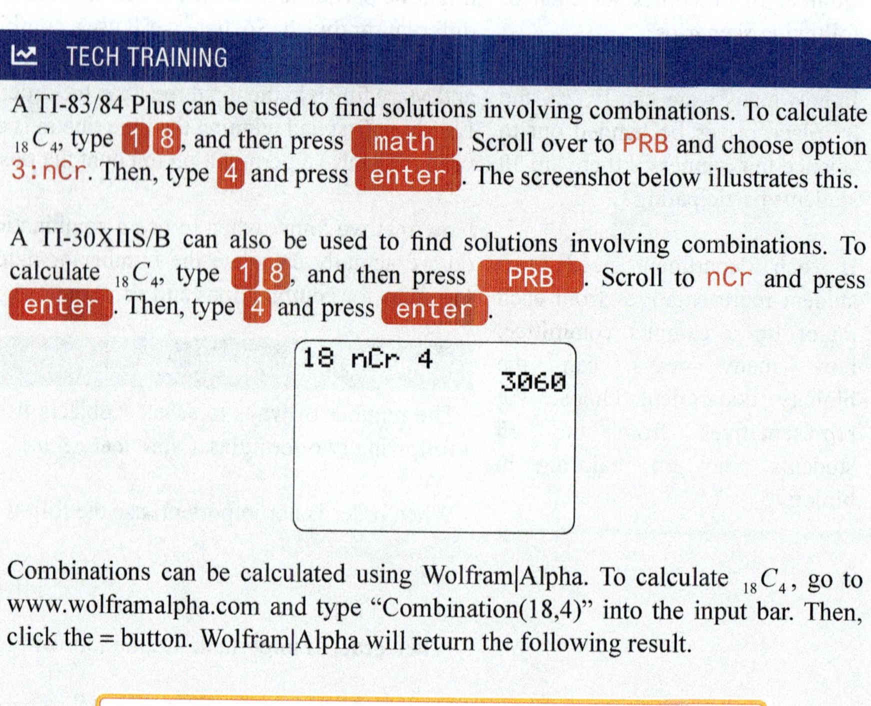

Combinations can be calculated using Wolfram|Alpha. To calculate $_{18}C_4$, go to www.wolframalpha.com and type "Combination(18,4)" into the input bar. Then, click the = button. Wolfram|Alpha will return the following result.

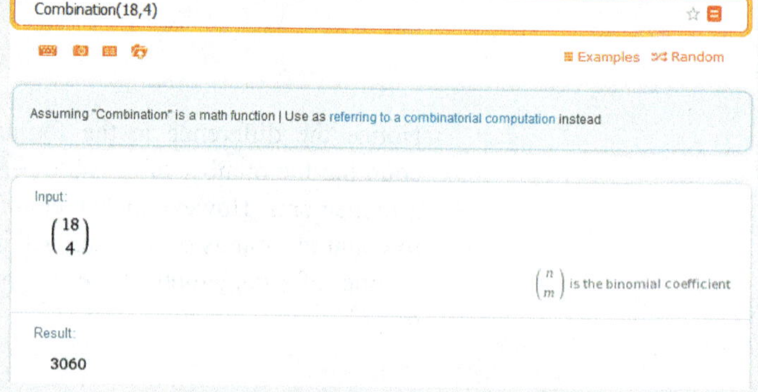

Source: Wolfram Alpha LLC. 2009. Wolfram|Alpha. http://www.wolframalpha.com/input/?i=Combination%2818%2C4%29 (accessed July 2, 2014).

Example 5: Using Permutations

Consider a race with 140 participants. How many possible outcomes are there for the top three positions of gold, silver, and bronze?

Solution

Since the order of the winners matters in this example, we use a permutation to count the possibilities. We are choosing three runners from the original 140 that ran. Therefore, $n = 140$ and $r = 3$. Filling in the permutation formula with these values gives us the following equation.

$$_{140}P_3 = \frac{140!}{(140-3)!} = \frac{140!}{137!} = \frac{(140)(139)(138)\,\cancel{(137)}\,\cancel{(136)}\cdots\cancel{(2)}\,\cancel{(1)}}{\cancel{(137)}\,\cancel{(136)}\cdots\cancel{(2)}\,\cancel{(1)}}$$

$$= (140)(139)(138) = 2{,}685{,}480$$

So, there are 2,685,480 possible ways the top three spots could be awarded.

ⓘ TECH TRAINING

A TI-83/84 Plus can be used to find solutions involving permutations. To calculate $_{140}P_3$, type `1` `4` `0` and then press `math`. Scroll over to **PRB** and choose option `2:nPr`. Then, type `3` and press `enter`.

A TI-30XIIS/B can be used to find solutions involving permutations. To calculate $_{140}P_3$, type `1` `4` `0` and then press `PRB`. Scroll to **nPr** and press `enter`. Then, type `3` and press `enter`.

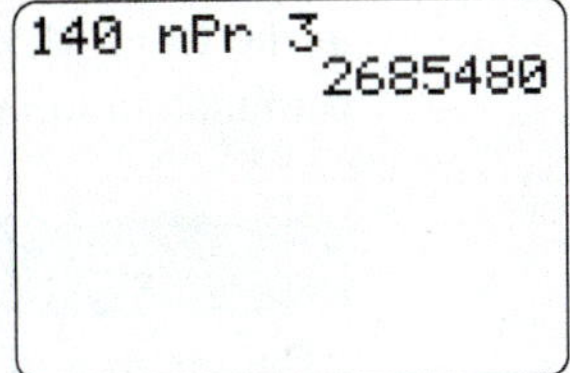

Permutations can be calculated using Wolfram|Alpha. To calculate $_{140}P_3$, go to www.wolframalpha.com and type "Permutation(140,3)" into the input bar. Then, click the = button. Wolfram|Alpha will return the following result.

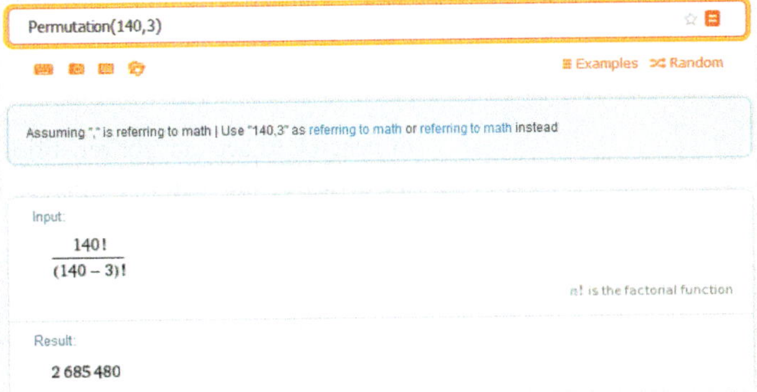

Source: Wolfram Alpha LLC. 2009. Wolfram|Alpha. http://www.wolframalpha.com/input/?i=Permutation%28140%2C3%29 (accessed July 2, 2014).

Example 6: Using Permutations

How many possible ways are there to arrange the order of appearance for the contestants in the local talent show, if there are 15 contestants altogether?

Solution

Again, order is important here because being the first to perform is certainly not the same as performing last, or even second for that matter. So, this is a permutation situation with $n = 15$. However, for this problem, r is also 15 since all of the contestants are to be chosen for the talent show. Using these values in the permutation formula, we have the following.

$$_{15}P_{15} = \frac{15!}{(15-15)!} = \frac{15!}{0!} = \frac{(15)(14)(13)\cdots(2)(1)}{1} = 1{,}307{,}674{,}368{,}000$$

Having 1,307,674,368,000 possible choices for the contestant lineup means that they will probably never choose the order by listing out all the possibilities and then randomly drawing one from a hat! Also, note that the result of $_{15}P_{15}$ is the same as 15!.

So far, using permutations and combinations required us to have distinct objects from which to choose. In other words, none of the objects in the group were the same—no two runners, no two toppings, and no two contestants. However, suppose the objects we have to choose from contain some identical members, that is, objects are repeated within the group, and we'd like to count how many distinct ways we can arrange, or permute, the objects. For instance, how many ways are there to arrange the letters in the word MISSISSIPPI? In order to count this, we must use a slightly different permutation formula.

Permutations with Repeated Objects

The number of **distinguishable permutations** of n objects, of which k_1 are all alike, k_2 are all alike, and so forth is given by

$$\frac{n!}{(k_1!)(k_2!)(k_3!)\cdots(k_p!)},$$

where $k_1 + k_2 + \cdots + k_p = n$.

In general, to account for repetitions of objects when counting distinct permutations, we divide by the factorial representing the number of times each object is duplicated.

Example 7: Using Permutations with Repeated Objects

How many different ways can you arrange the letters in the word MISSISSIPPI?

Solution

Because there are repeated letters in the word, and no real distinction is made between each duplicated letter, we need to count the duplicate letters for our formula.

$$M = 1$$
$$I = 4$$
$$S = 4$$
$$P = 2$$

Note that since the letter M is not duplicated, $M = 1$ and its factorial is $1! = 1$, which will not change our fraction when we include it. This is always the case for unduplicated objects.

There are 11 letters in MISSISSIPPI, so $n = 11$. Substituting these values into the formula, we have

$$\frac{11!}{1!4!4!2!} = \frac{(11)(10)(9)(8)(7)(6)(5)\,\cancel{4!}}{\cancel{4!}4!2!} = \frac{(11)(10)(9)(8)(7)(6)(5)}{(4)(3)(2)(1)(2)(1)} = 34{,}650$$

Thus, there are 34,650 ways to arrange the letters in the word MISSISSIPPI.

📈 TECH TRAINING

Permutations with repetitions can be calculated using Wolfram|Alpha. To calculate the number of rearrangements for the word MISSISSIPPI, we need to input the number of times each letter appears. Go to www.wolframalpha.com and type "Multinomial(1,4,4,2)" into the input bar. Then, click on the = button. Wolfram|Alpha will return the following result.

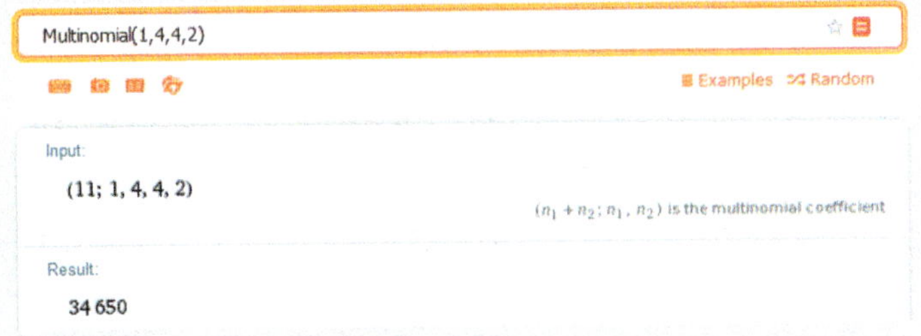

Source: Wolfram Alpha LLC. 2009. Wolfram|Alpha. http://www.wolframalpha.com/input/?i=Multi
nomial%281%2C4%2C4%2C2%29 (accessed July 2, 2014).

Skill Check Answers

1. a. Permutation   **b.** Combination

8.4 EXERCISES

PRACTICE

Create a tree diagram to list the outcomes in each sample space.

1. Find the sample space for the gender of each child in regard to birth order for a family with three children.

2. In picking out a new car, there are several choices to make. The color can be red, white, or silver. The seats can be cloth or leather. Finally, it can have a sunroof, a moonroof, or neither. Find the sample space for the possible new car combinations using a tree diagram.

3. Use a tree diagram to find the sample space for tossing a coin three times.

4. Four students are randomly selected from an algebra class and asked whether they suffer from math anxiety. Find the sample space for the possible outcomes of the survey using a tree diagram.

Evaluate each factorial expression.

5. $5!$

6. $9!$

7. $\dfrac{8!}{6!}$

8. $1!$

9. $0!$

10. $\dfrac{10!}{2!4!}$

11. $\dfrac{5!}{3!2!}$

12. $\dfrac{12!}{8!(3-1)!}$

13. $\dfrac{23!}{11(25-4)!}$

Evaluate each permutation or combination.

14. $_8P_2$

15. $_7P_4$

16. $_5P_1$

17. $_4P_4$

18. $_3C_2$

19. $_{30}C_1$

20. $_5C_5$

21. $\dfrac{_3C_2}{_3P_2}$

22. $\dfrac{_5P_3}{_5C_3}$

23. $_7C_4 + {_7C_3} + {_7C_2} + {_7C_1}$

24. $_7P_4 + {_7P_3} + {_7P_2} + {_7P_1}$

APPLICATIONS

25. How many three-digit area codes can be made from the digits 0 through 9? Assume that the digits may repeat and that area codes beginning with 0 are allowed.

26. How many three-digit area codes can be made from the digits 0 through 9 if the first digit is not allowed to be a 0 and the digits are allowed to repeat?

27. How many six-character password codes are possible if you are allowed both lowercase letters and the digits 0 through 9, and repeating characters are allowed?

28. When ordering a new e-reader, you have several choices to make. You can choose from five price ranges, decide between Wi-Fi and 5G, choose to have a one-year or three-year warranty, and pick a black, white, or silver casing. How many possible e-readers are there for you to choose from?

29. The Youngs are planning their next family night. They always have dinner out somewhere and then do something fun together. There are two adults and four boys in the family. Each family member is allowed two meal suggestions, and each boy is allowed three activity suggestions. Assuming no family members choose the same thing, how many different family night possibilities are there?

30. If there are seven children lining up for recess, how many different ways can they line up?

31. The college's soccer team will play 11 games next fall. Each game can result in one of three outcomes: a win, a loss, or a tie. Find the total number of possible outcomes for the season record.

32. Harper is deciding on her schedule for next semester. She must take each of the following classes: English 102, College Algebra, History 102, and Biology 101. If there are 16 sections of English 102, 14 sections of College Algebra, 7 sections of History 102, and 12 sections of Biology 101, how many different possible schedules are there for Harper to choose from? Assume there are no time conflicts between the different classes.

33. If there were no restrictions on the use of the number zero, in theory, how many seven-digit telephone numbers are possible?

34. How many four-digit even whole numbers exist?

35. In the new ice creamery, several choices need to be made before tasting "a little bit of heaven," as the advertisement suggests. First, there are three cup sizes to choose from, then 39 different ice cream flavors to decide from, and finally several mix-ins to choose from: 10 candy bars, 8 fruits, 8 nuts, and 10 cookies or cakes. If you want a medium cup with one flavor of ice cream and two mix-ins, one candy bar and one nut, how many possible choices are there for you to have your "taste of heaven"?

Determine whether to use a permutation or combination to answer each question, and then determine the total number of outcomes.

36. There are 10 board members on the Community Arts Council. In how many ways can a president and treasurer be chosen? Assume that no member can hold both positions at the same time.

37. In how many ways can a committee of five people be chosen from a pool of 120 employees?

38. If there are 84 runners in a race, in how many ways can 1st, 2nd, and 3rd place ribbons be given out?

39. Elliot has to submit three photographs for the school art show. This semester he has taken 29 photographs that he thinks are show-worthy. In how many ways can he choose the photographs to submit?

40. Avery was born on 10/15/1995. How many eight-digit codes could she make using the digits in her birthday?

41. There are 18 tenured faculty in the biology department on campus. The department needs one tenured faculty member to facilitate undergraduate research, one member to supervise graduate advising, and one to coordinate grant proposals. In how many ways can these tasks be assigned, if a member may be appointed to only one duty?

42. A service organization on campus needs a group of six students from their organization's membership of 123 students to serve as program attendants at graduation. In how many ways can the attendants be chosen?

43. In how many ways can the letters in the word STATISTICS be arranged?

44. In how many ways can the letters in the word TENNESSEE be arranged?

✎ WRITING & THINKING

45. Without calculating the permutations, decide which of the following words would produce the greatest number of four-letter arrangements.
a. PASS
b. TEST
c. FAIR
d. FREE

46. Determine how many ways a group of three students from your class could be selected.

47. Count the number of possible outfits you have in your closet based on the number of pants, shirts, and pairs of shoes you have. Assume that all match well together or that you would be making your own fashion statement some days.

48. Calculate the number of ways to arrange the letters in
a. your first name,
b. your last name,
c. both your first and last names.

8.5 COUNTING PRINCIPLES AND PROBABILITY

■ TOPICS

- ■ Complements

Our goal is to be able to calculate classical probability for certain events. Now that we've looked at several methods of counting the outcomes in a sample space, we can begin to look at calculating probabilities. Recall that the definition of classical probability is

$$P(\text{event}) = \frac{\text{the number of possible outcomes in the event}}{\text{the number of outcomes in the sample space}}, \text{ and }$$

$P(\text{event})$ will always be a real number between 0 and 1, inclusive.

Let's reconsider the statement that *the probability will always be a number between 0 and 1, inclusive.* In other words, for any event E, $0 \leq P(E) \leq 1$. To put this in context, if an event will **not** occur, then its probability is 0 (or $P(E) = 0$). However, if an event is **certain** to happen, its probability is 1 (or $P(E) = 1$). All other probabilities fall somewhere between those two possibilities. The closer a probability is to 0, the less likely the event is to happen, and the closer the probability is to 1, the more likely the event is to happen. A probability of 0.5, or $\frac{1}{2}$ means an event is just as equally likely to happen as to not happen.

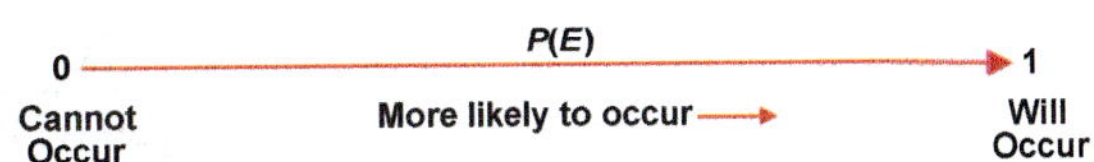

FIGURE 1: Range of Probability

Example 1: Calculating Classical Probability Using Combinations

Suppose that as one of the 20 graduate students in the physics department, you have a chance of being selected for one of the three student spots for a conference trip to Cancun. If the names of all the graduate students were put in a hat and three were drawn, what is the probability that you and your two friends, Leonard and Sheldon, end up being chosen?

Solution

The first thing we need to do is count the number of ways that the three student spots on the trip can be filled. Because the order in which the students are chosen is not important, we can count the outcomes using the combination formula. We have 20 students to choose from, so $n = 20$ and $r = 3$. That means there are

$$_{20}C_3 = \frac{20!}{3!(20-3)!} = \frac{(20)(19)(18)\,\cancel{17!}}{3!\,\cancel{17!}} = \frac{(20)(19)(18)}{(3)(2)(1)} = 1140$$

possible ways to choose 3 students from 20 for the trip. There is only one way in which to choose you and your two friends, so the probability of this event happening is

$$P\left(\text{You, Leonard, and Sheldon in Cancun}\right) = \frac{1}{1140} \approx 0.000877.$$

In other words, it is very unlikely that the three of you would randomly be chosen to go on the conference trip to Cancun together.

Example 2: Calculating Classical Probability Using Permutations

Let's change the previous example slightly. Suppose that as one of the 20 graduate students in the physics department, you have a chance of being selected for one of the three student spots for a conference in Cancun. The names of all the graduate students are put in a hat and three are drawn. However, if your name is drawn first, you get all expenses paid. If you are chosen second, everything is paid for except meals, and if you're chosen third, you must pay for your own meals and hotel. (This means that the department still picks up the tab for the flight and conference fees of the three lucky students, so it's not a bad deal!) What is the probability that you and your two friends, Leonard and Sheldon, all end up being chosen, and that your name is drawn first?

Solution

This time, the order in which the three students are chosen does make a difference when we are counting, so we'll use a permutation. Note that n is still 20 and r is still 3. Now there are

$$_{20}P_3 = \frac{20!}{(20-3)!} = \frac{(20)(19)(18)\,17!}{17!} = (20)(19)(18) = 6840$$

possible ways to choose the three lucky students for the trip. However, let's consider how many outcomes are in the event that you, Sheldon, and Leonard are chosen, and that your name is drawn first; we'll call this event E.

Let's list all the ways that the three of you could be chosen.

Possibility 1	Possibility 2	Possibility 3	Possibility 4	Possibility 5	Possibility 6
1st pick: **You**	1st pick: You	1st pick: Sheldon	1st pick: Sheldon	1st pick: Leonard	1st pick: Leonard
2nd pick: **Leonard**	2nd pick: Sheldon	2nd pick: You	2nd pick: Leonard	2nd pick: Sheldon	2nd pick: You
3rd pick: **Sheldon**	3rd pick: Leonard	3rd pick: Leonard	3rd pick: You	3rd pick: You	3rd pick: Sheldon

TABLE 1: Permutations of You, Leonard, and Sheldon

We can see that there are only two ways in which you are first in the list, and therefore get all expenses paid. So the probability that the event described occurs in this way is

$$P(E) = \frac{2}{6840} \approx 0.000292.$$

It seems that there is an even smaller chance of this happening, so it's better not to be greedy and wish for the top spot!

Complements

As we calculate probabilities, it will often be the case that the type of scenario we want to look at requires more than the basic classical probability formula. Let's look at some other vocabulary that will be helpful when solving probability questions. Recall that the probability of choosing a diamond from a standard deck of cards is $P(\text{diamond}) = 0.25$.

<table>
<tr><td>

🖙 HELPFUL HINT

The complement of an event E can be denoted in several different ways, such as E^c, $\overline{E}$, or E'.

</td><td>

Complement of an Event

The **complement** of event E, denoted by E^c, consists of all outcomes in the sample space that are *not* in event E.

</td></tr>
</table>

Because of the definition of complement, we can determine the probability of not choosing a diamond. The probability of not choosing a diamond would be the probability of choosing any of the other cards; that is,

$$P(\text{not a diamond}) = \frac{39}{52} = \frac{3}{4} = 0.75.$$

Example 3: Finding the Complement of an Event

Describe the complement for each of the following events.

a. Rolling an even number on a die.

b. Choosing a number that doesn't end in 1, from all positive two-digit whole numbers.

c. From a class of 52 students, choosing a student who is over 21 years old.

Solution

a. The complement contains all the odd numbers on a die (that is, 1, 3, and 5).

b. The set of all positive two-digit whole numbers includes the numbers 10 through 99. The complement of our event would be all two-digit numbers that do end in 1 (that is, 11, 21, 31, 41, 51, 61, 71, 81, and 91).

c. The complement consists of the students in the class who are 21 years old or younger.

By now, you're beginning to get the idea that between an event and its complement, the entire sample space is accounted for. It naturally follows that if you add the probability of event E to the probability of its complement E^c, you get 1. This gives us the following rules for complements.

Complement Rules of Probability

1. $P(E) + P(E^c) = 1$

2. $P(E) = 1 - P(E^c)$

3. $P(E^c) = 1 - P(E)$

Sometimes it is easier to find the probability of an event by calculating the probability of its complement rather than the probability of the event itself.

Example 4: Finding Probability Using Complements

Using the data in Table 2, find the following probabilities involving nuts imported into the United States from 2006 to 2011.

	India	4373
	Mexico	1268
Walnuts	Spain	5239
	China	1533
	Austria	1938
	Other countries	4157
	Iran	2012
	Turkey	2030
Pistachios	Hong Kong	262
	Switzerland	64
	Italy	115
	Other countries	323

Source: USDA. "Fruit and Tree Nut Data." http://www.ers.usda.gov/data-products/fruit-and-tree-nut-data/data-by-commodity.aspx

TABLE 2: US Import Sources by Weight (Annual Averages in Thousands of Pounds) for Fresh or Dried Walnuts and Pistachios, 2006–2011

a. Find the probability that imported walnuts purchased in the United States during this period were from Austria.

b. Find the probability that imported walnuts purchased in the United States during this period were from somewhere other than Austria.

c. Assume that you purchased imported pistachios in the United States during this time period. What is the probability that the pistachios came from somewhere other than Italy and Switzerland?

Solution

a. The probability that imported walnuts purchased in the United States during this period were from Austria is found by dividing the weight of walnuts imported from Austria by the total weight of walnuts imported during this time period. The first number is given in the table as 1938 pounds. To find the total weight of walnuts imported, we need to add together all of the weights of walnuts imported.

$$\text{total walnut weight} = 4373 + 1268 + 5239 + 1533 + 1938 + 4157 = 18{,}508$$

The probability that the walnuts were from Austria is then found by

$$P(\text{walnuts from Austria}) = \frac{1938}{18{,}508} \approx 0.104711.$$

b. We could find the probability that the walnuts came from somewhere other than Austria by combining all the remaining places together. However, given that we just calculated the probability that the walnuts were from Austria, it is easier for us to just use the complement.

$$P(\text{not from Austria}) = 1 - P(\text{walnuts from Austria}) \approx 1 - 0.104711 = 0.895289$$

c. Again, it will be easier for us to use the complement here. First calculate the probability that the pistachios you purchased came from either Italy or Switzerland.

$$\text{total pistachio weight} = 2012 + 2030 + 262 + 64 + 115 + 323 = 4806$$

$$P(\text{from Italy or Switzerland}) = \frac{\text{Italy} + \text{Switzerland}}{\text{total pistachio imports}} = \frac{115 + 64}{4806} \approx 0.037245$$

Now, to calculate the probability that the pistachios came from somewhere other than Switzerland or Italy, we'll find the probability of the comple1ment.

$$P(\text{not from Italy or Switzerland}) = 1 - P(\text{from Italy or Switzerland})$$
$$\approx 1 - 0.037245$$
$$= 0.962755$$

8.5 EXERCISES

PRACTICE

1. Describe the complement of the set of odd numbers greater than 0 within the set of positive integers.

2. Let the event E be the sum of a pair of dice that is divisible by 3. List the events in E^c.

3. The following is a table of the ages of boys on a soccer team.

   Let $A = \{$soccer players older than 9$\}$. How many players are in the complement of A?

Ages of Boys on Soccer Team	
Age	Number of Boys
8	3
9	6
10	7
11	2

4. Describe the complement of the set of face cards in a standard deck of cards.

5. In a company, all employees who have worked there for more than five years receive a gift. Describe the complement of this group of employees.

APPLICATIONS

Find the classical probability for each scenario. Round your answer to the nearest millionth when necessary.

6. Find the probability of obtaining exactly one head when flipping four coins.

7. There are two sets of balls numbered 1 through 5 placed in a bowl. If two balls are randomly chosen without replacement, find the probability that the balls have the same number.

8. William and Gavin are going to play video games after work. Together they have 48 games. If they decide to randomly choose two games to play, what is the probability that the two games they choose consist of William's favorite game and Gavin's favorite game? Assume they have different favorites.

9. A local pizza parlor has the following list of toppings available for selection. The parlor is running a special to encourage patrons to try new combinations of toppings. They list all possible three-topping pizzas (three distinct toppings) on individual cards and give away a free pizza every hour to a lucky winner.

Pizza Toppings			
Green Peppers	Onions	Pepperoni	Sausage
Baby Portabello Mushrooms	Black Olives	Ham	Spicy Italian Sausage
Roma Tomatoes	Pineapple	Beef	Grilled Chicken
Jalapeño Peppers	Banana Peppers	Bacon	Extra Cheese

   a. How many three-topping pizza cards are there?
   b. Find the probability that the first winner randomly selects the card with the pizza containing green peppers, ham, and bacon on it.

10. A combination padlock is a lock in which a sequence of numbers is used as the "key" to open the lock. Suppose a combination padlock has 10 digits to choose from for each of the four sections of the lock.
    a. Does the "key" for a combination padlock involve permutations or combinations?
    b. How many possible "keys" are there for the combination padlock?
    c. What is the probability that you randomly buy one of these locks whose "key" is made of 4 of the same digit?

11. Four students, three girls and a boy, have arranged to meet on the first day of class and sit in the front row. Suppose they agree to sit in the first four seats in the order that they arrive.
    a. How many possible seating arrangements are there for the four friends?
    b. What is the probability that all three girls end up sitting next to one another?

12. A committee of four is being formed randomly from the employees at a school: 5 administrators, 37 teachers, and 4 staff.
    a. How many ways can the committee be formed?
    b. What is the probability that all four members are staff?
    c. What is the probability that no member is an administrator?

13. A hand of poker is made up of five cards from a standard deck of cards.
    a. How many possible hands of poker are there in a standard deck of 52 cards?
    b. A royal flush consists of the cards ace, king, queen, jack, and ten, all in the same suit. What is the probability of being dealt a royal flush?

14. Find the probability that three people randomly line up to buy tickets in order of their height (tallest, middle, shortest). Assume that no two people in the line are of the exact same height.

15. Matthew needs to set the passcode on his smartphone. It must be a four-digit number and repeated digits are allowed.
    a. How many possible passcodes are there for Matthew to choose from?
    b. How many possible passcodes are there for Matthew if he decides to choose four distinct numbers?
    c. A spy sneaks a look at Matthew's phone and sees his fingerprints on the screen over four numbers. What is the probability that the spy is able to unlock the phone on his first try?
    d. The spy knows the fingerprint trick and so on his phone he uses a repeated digit in his code. If you could see the three fingerprints on the spy's phone, what is the probability that you could unlock the phone on your first attempt?
    e. Based on parts **c.** and **d.**, is it better to repeat a digit or have four distinct digits in the code on your phone for security purposes?

16. A hand of blackjack consists of two cards. The dealer deals you a hand from a fresh deck.
    a. What is the probability that the two cards have the same face value, for instance, both cards are kings or both cards are 5s?
    b. If aces count 1 or 11, picture cards count 10, and card numbers 2 through 10 are equal to their face value, what is the probability that the two cards sum to 21?

17. One option to play the lottery is called "3-way any order." In order to play this method, you select three digits, from 0 to 9, such that precisely two of the digits are the same (for example, 1, 1, 2). You're a winner if your three digits show up in any order in the lottery's three randomly chosen digits. Digits may be repeated when the lottery chooses the winning number. Find the probability of winning with the "3-way any order" method.

18. Another option of playing the lottery is to choose three numbers (allowing repetition) in the exact order they will appear. Find the probability of winning the lottery with one ticket.

19. Ian is playing Scrabble. What is the probability that the next three letters he draws from the bag spell out his name in the order that he draws them? Assume there is one of each letter in the alphabet left in the bag.

20. For a pickup game of basketball, jerseys are in a box and people start grabbing them. The box contains three extra-large, seven large, and four medium jerseys. If you are first to the box and grab two jerseys, what is the probability that you randomly grab two extra-large jerseys?

21. A junk drawer at home contains a half dozen pens, two of which work. What is the probability that you randomly grab two pens from the drawer and don't end up with a pen that works?

Find each probability using complements.

22. A bag contains each letter of the alphabet. Find the probability that a randomly selected letter from the bag will not be one of the five vowels.

23. Find the probability of randomly choosing a letter other than the letter O from a bag that contains the eighteen letters of the Italian city GUIDONIA MONTECELIO.

24. Using the table containing the breakdown of all employees on nonfarm payrolls in the United States during March 2014, find the probability that a randomly selected US nonfarm worker was not in either retail trade or wholesale trade.

Employees on Nonfarm Payrolls (in Thousands), March 2014		
	Area of Employment	**Number of Employees (in Thousands)**
Private Sector	Goods-Producing	18,558.2
	Wholesale Trade	5803.7
	Retail Trade	15,004.0
	Transportation and Warehousing	4524.8
	Utilities	550.3
	Information	2653.0
	Financial Activities	7870.0
	Professional and Business Services	18,832.0
	Education and Health Services	21,481.0
	Leisure and Hospitality	14,143.0
	Other Private Service-Providing Services	5464.0
Public Sector	Federal Government	2705.0
	State Government	5217.0
	Local Government	14,341.0
	Total Nonfarm Employees	**137,147.0**
	Source: Bureau of Labor Statistics. "Table B-1. Employees on nonfarm payrolls by industry sector and selected industry detail." Accessed June 2014. http://www.bls.gov/news.release/empsit.t17.htm	

25. In June 2011, the week of the final mission of the US space shuttle program, a Pew Research poll asked 1502 US adults whether the United States must continue to be a world leader in space exploration. The following table gives a breakdown of their opinions.

The United States Continuing to be a World Leader in Space Exploration Is...		
Essential	**Not Essential**	**Don't Know**
871	571	60
Source: Pew Research Center. "Majority Sees U.S. Leadership in Space Essential." July 5, 2011. http://www.people-press.org/2011/07/05/majority-sees-u-s-leadership-in-space-as-essential/		

    a. Find the probability that someone responded "essential."

    b. Find the probability that someone did not respond "essential."

26. Find the probability of rolling two dice and not getting the same number on both dice.

27. Suppose a family has five pets. Find the probability that at least one of the pets is male.

8.6 PROBABILITY RULES AND BAYES' THEOREM

■ TOPICS

- ■ Event *A* Happening OR Event *B* Happening
- ■ Event *A* Happening AND Event *B* Happening

We now turn our attention to those probabilities that involve two events, rather than just a singular event. There are really only two possibilities.

1. Event *A* happening ***or*** Event *B* happening

2. Event *A* happening ***and*** Event *B* happening

Of course, there are some subtleties of distinction that we will need to take note of as we go along.

Event *A* Happening OR Event *B* Happening

Let's start with the ***or*** events. Think about the probability of selecting a king *or* a spade from a standard deck of cards. The probability of selecting a king is $\frac{4}{52}$ and the probability of selecting a spade is $\frac{13}{52}$. It's tempting to want to just add the two probabilities together. However, think about the card that is both a king *and* a spade. The king of spades is in both events. We've counted it twice, so we need to take that into account. Therefore, the probability of choosing a king *or* a spade from a standard deck of cards is as follows.

$$P(\text{king or spade}) = P(\text{king}) + P(\text{spade}) - P(\text{king and spade})$$

$$= \frac{4}{52} + \frac{13}{52} - \frac{1}{52}$$

$$= \frac{16}{52}$$

$$\approx 0.307692$$

$$P(\text{king or spade}) = P(\text{king}) + P(\text{spade}) - P(\text{king and spade})$$

FIGURE 1: Probability of a King or a Spade

If *A* and *B* are events that have some outcomes in common, then the probability that *A* ***or*** *B* will happen is calculated by adding the individual probability of each and then subtracting the probability that both events occur simultaneously.

Addition Rule for Probability

The probability of Event A happening *or* Event B happening is

$$P(A \text{ or } B) = P(A) + P(B) - P(A \text{ and } B).$$

Example 1: Applying the Addition Rule for Probability

Suppose that a student is chosen at random to receive a gift card for filling out a survey. The following table shows a breakdown of who filled out the survey.

Class	Student Government Member	Not a Student Government Member	Total
Freshman	3	15	18
Sophomore	1	11	12
Junior	2	7	9
Senior	4	3	7
Total	10	36	46

TABLE 1: Breakdown of Survey Takers

What is the probability that the winner is either a freshman or a member of student government?

Solution

We begin by finding the probability of choosing each of the individual criteria. We can see from the table that there were a total of 46 students who filled out the survey. The number of freshmen who filled it out was 18. So, the probability of choosing a freshman is

$$P(\text{freshman}) = \frac{18}{46}.$$

In previous sections, we converted our probability answers from fractions to decimals. However, since we are going to use the Addition Rule for Probability that requires us to add probabilities together, it's better to leave them as nonreduced fractions so that we avoid any error in rounding.

There were 10 members of the student government who filled out the survey, so the probability of choosing a member of the student governing board is

$$P(\text{student government}) = \frac{10}{46}.$$

There are 3 students who are both freshmen and members of the student government, so the probability of choosing a student who is in both groups is

$$P(\text{freshman and student government}) = \frac{3}{46}.$$

Using the Addition Rule for Probability, we have the following equation.

$$P(\text{freshman or student government}) = \frac{18}{46} + \frac{10}{46} - \frac{3}{46}$$

$$= \frac{25}{46}$$

$$\approx 0.543478$$

Example 2: Applying the Addition Rule for Probability

Recall the example from a previous section where you won a new iPod in 2011. You could have any iPod you want; you just had to go to the store and pick it out. We created a tree diagram to list the possible iPods you could choose from.

Tree diagram of iPod choices

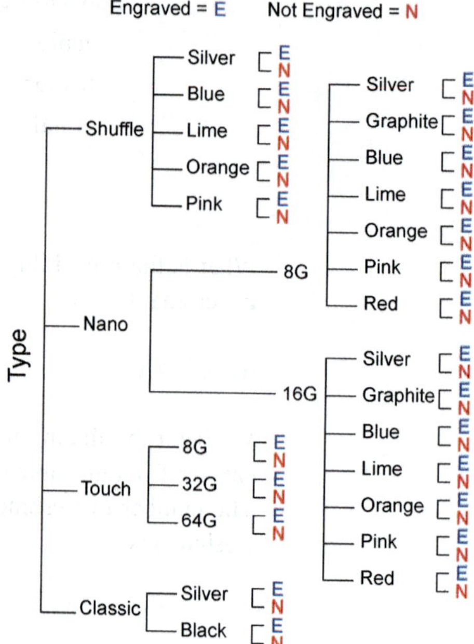

Assuming that you were equally likely to choose any of the 48 iPods, what is the probability that the iPod you chose was orange or not engraved?

Solution

We begin again by finding the probabilities of both criteria individually. The tree diagram shows that of the 48 iPods, there are 6 orange possibilities, so the probability is

$$P(\text{orange}) = \frac{6}{48}.$$

There are 24 iPods that are not engraved, so

$$P(\text{not engraved}) = \frac{24}{48}.$$

There are three orange iPods that are also not engraved, so

$$P(\text{orange and not engraved}) = \frac{3}{48}.$$

Using the Addition Rule for Probability, we have the following equation.

$$P(\text{orange or not engraved}) = \frac{6}{48} + \frac{24}{48} - \frac{3}{48}$$
$$= \frac{27}{48}$$
$$= 0.5625$$

For each pair of events, decide whether they are mutually exclusive or not.

a. Let event A consist of randomly selecting an adult from the mall who has shopped online at least once in the past 6 months. Let event B consist of randomly selecting an adult from the mall who has never shopped online.

b. Let event A consist of selecting an odd number and event B consist of selecting a prime number.

What about the case when the two events do not have any outcomes in common? Think about the following choices.

• Rolling a 1 or a 6 on a single roll of a die

• Living in the city or the country

• Buying a red car or a black truck as your first vehicle

• Going to Hawaii or Sweden for your one week of vacation

• Building a 1500-square-foot home or a 2100-square-foot home on your new single-home property

If we think about the formula for the Addition Rule for Probability, the formula adjusts for overcounting duplicates by subtracting off outcomes that are in both events. So, if two events have no outcomes in common, called **mutually exclusive events**, we end up subtracting 0. We will single out this type of probability by calling it the **Addition Rule for Mutually Exclusive Events**, but it is important to realize that this is simply a special case of the previous formula. There is just no overcounting to subtract.

Addition Rule for Mutually Exclusive Events

The probability of Event A happening *or* Event B happening when A and B have no outcomes in common is

$$P(A \text{ or } B) = P(A) + P(B).$$

Example 3: Applying the Addition Rule for Mutually Exclusive Events

Suppose that you have decided it's time to get a pet. Your apartment complex allows you to have only one pet and you decide to go to the local animal shelter to adopt one of the available pets. Because you can't decide between a dog and a cat, you've left the choice up to chance. You're going to run your finger down the list of available animals without looking and let the lucky pet be the one you stop on. The following graph shows the available animals on the list at the shelter.

Available Animals for Adoption

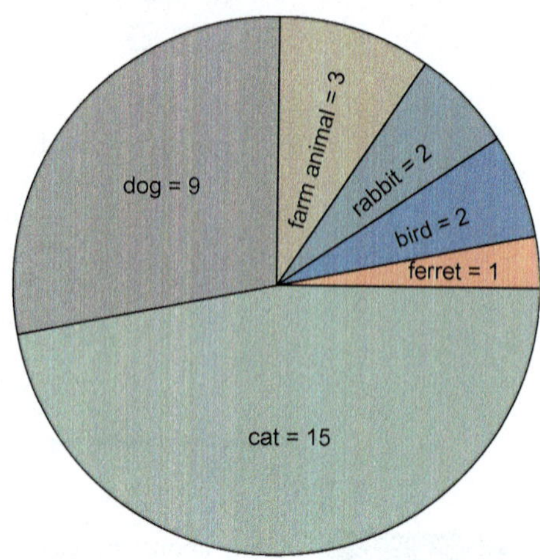

Unfortunately, you didn't consider that other kinds of animals might be on the list. What is the probability that you choose either a cat or a dog to take home with you?

Solution

Once again, we'll begin by finding the probability of choosing a cat and the probability of choosing a dog individually. We can add the totals of each animal category to find out that there are currently 32 animals available at the shelter.

The probability of choosing a cat is $P(\text{cat}) = \dfrac{15}{32}$.

The probability of choosing a dog is $P(\text{dog}) = \dfrac{9}{32}$.

Because these are mutually exclusive events, that is, you cannot choose an animal that is both a cat and dog at the same time, we do not need to worry about duplicating the count of any animal. So, using the formula, we have the following result.

$$P(\text{cat or dog}) = \frac{15}{32} + \frac{9}{32}$$
$$= \frac{24}{32}$$
$$= 0.75$$

Even though you forgot about the fact that other animals might be at the shelter, you still have a 75% chance of randomly choosing a cat or a dog. However, that means there is also a 25% chance you will select one of the other animals, some of which your landlord might not approve!

Choosing a college can be an exciting and nerve-racking experience in the life of a high school student. Emma has finally narrowed down her choices to the top 4. She's also given each school a probability based on certain characteristics.

University	Characteristic	Probability
A	Closest to home	$P(A) = 0.25$
B	Best sports	$P(B) = 0.10$
C	Her best friend's choice	$P(C) = 0.30$
D	Best academic program of her choice	$P(D) = 0.35$

TABLE 2: University Probabilities

What is the probability that Emma ends up at University B or University D?

Solution

Since Emma will choose one or the other, but not both at the same time, these events are mutually exclusive. So we just need to add the probability of her choosing University B to the probability of choosing University D.

$$P(\text{University B or University D}) = 0.10 + 0.35 = 0.45$$

Event *A* Happening AND Event *B* Happening

Now we'll turn our attention to the possibility of two events both happening. Consider the following scenarios.

- Rolling a 6 on a die *and* drawing an ace from a deck of cards

- Rolling six 6s in a row

- You're dealt the queen of hearts *and* then a red face card in a blackjack hand

- Catching a cold *and* breaking your leg on the same day

- Being treated by an emergency room doctor who is female *and* over 35

- Choosing two boys from a group of eight girls and ten boys

- Winning the lottery *and* finding a ten dollar bill on the ground by your car

- Pulling three red Skittles in a row from the same bag

The key to calculating the probabilities for *and* scenarios is determining if one event influences the probability of the other event. Sometimes it does and sometimes it doesn't. Let's take the first scenario in the list. Does rolling a six on a die affect the probability of drawing an ace from a deck of cards? No, but how about being dealt the queen of hearts and then a red face card from a deck of cards? Here, the first event does influence the probability of the second event. If you were dealt the queen of hearts first, the probability of getting a red face card decreases for the second card because there is one less red face card in the deck to choose from.

Let's stop here and make a distinction between the two situations. We say two events are **independent** when the occurrence of one event *does not* influence the probability of the other event happening. If the result of one event *does* influence the probability of the second, we say that the two events are **dependent**.

Independent Events

Independent events are events where the result of one event does not influence the probability of the other.

Dependent Events

Dependent events are events where the result of one event influences the probability of the other.

Example 5: Independent vs. Dependent Events

Determine if the following pairs of events are independent.

a. Event A: Eating a red candy from a new bag of Skittles.

Event B: Pulling a second Skittle from the same bag that is also red.

b. Event A: A woman giving birth to a daughter.

Event B: The same woman's second child is also a girl.

c. Event A: Tina is the first woman to finish the 2020 Boston Marathon.

Event B: Tina is the first woman to finish the 2021 New York Marathon.

Solution

a. These events are dependent. The chances of drawing a second red Skittle from the bag decreases after the first one is drawn, because it obviously was not replaced in the bag. It was eaten.

b. Although these events might appear dependent on one another, the probability that a child is a girl is the same for any given pregnancy. It is not affected by the gender of any previous pregnancies. Therefore, these events are independent.

c. At first glance these events might seem independent of one another. You might think that winning one race has no effect on winning a second race. In fact, the best starting positions for runners in large marathons are given to winners of previous races. Also, running a second marathon in consecutive years will have an effect on your body when training for and running the second one (whether that effect is positive or negative). Therefore, these events are dependent.

This example highlights the fact that the dependence of events isn't always obvious. Sometimes our own personal knowledge or experiences are not enough to rely on and we need to look to experts in other fields to help us determine whether events are dependent or not.

Let's focus on independent events first. If two events are independent, we can find the probability of both events occurring by multiplying the individual probabilities together. This is referred to as the **Multiplication Rule of Probability**. In fact, satisfying the multiplication rule defines independent events.

Multiplication Rule for Independent Events

Events A and B are **independent events** when the probability of Event A happening *and* Event B happening is given by

$$P(A \text{ and } B) = P(A) \cdot P(B).$$

Example 6: Multiplication Rule for Independent Events

Given a fair die and a standard deck of 52 cards, find the probability of rolling a 6 *and* drawing an ace.

Solution

Because the number rolled on the die does not affect the card drawn from the deck and vice versa, the events here are independent. Using the Multiplication Rule for Independent Events, we have

$$P\left(6 \text{ } and \text{ ace}\right) = P\left(6\right) \cdot P\left(\text{ace}\right) = \frac{1}{6} \cdot \frac{4}{52} = \frac{4}{312} \approx 0.012821.$$

Example 7: Multiplication Rule for Independent Events

Suppose we know the following breakdown for internal medicine hospitalists who work at Madison Regional Hospital and the ages of their patients on a given day.

		Number
Hospitalist Gender	Male	6
	Female	7
Patient Age	< 35	16
	35–55	35
	> 55	21

TABLE 3: Hospitalists at Madison Regional Hospital

What is the probability that the first patient treated is over 55 years old and treated by a male hospitalist at Madison Regional Hospital?

Solution

Because the age of the patient and the gender of the hospitalist have no effect on one another, these events are independent. So, we'll have to use the Multiplication Rule of Probability for Independent Events and multiply the individual probabilities together. Let's find the individual probabilities first.

$$P\left(\text{patient age} > 55\right) = \frac{21}{72}$$

$$P\left(\text{male hospitalist}\right) = \frac{6}{13}$$

Using the Multiplication Rule for Independent Events, we have the following.

$$P\left(\text{patient age} > 55 \ and \ \text{male hospitalist}\right) = \frac{21}{72} \cdot \frac{6}{13}$$
$$= \frac{126}{936}$$
$$\approx 0.134615$$

Now, let's consider the case when the two events are dependent. Remember that dependence requires that one event influences the probability of the other event. A measure of that influence is **conditional probability**.

Conditional Probability

The **conditional probability** of Event B happening, given Event A, is the probability of Event B assuming that Event A has already, or will at some point, occur. The conditional probability is written $P(B \mid A)$, and read *the probability of B, given A.*

Once we know the conditional probability of an event, calculating the probability of two dependent events happening together is straightforward.

Multiplication Rule for Dependent Events

If A and B are **dependent events**, the probability of Event A happening *and* Event B happening is

$$P(A \text{ and } B) = P(A) \cdot P(B \mid A).$$

Example 8: Multiplication Rule for Dependent Events

Find the probability that, from a standard deck of cards, you're dealt two cards: the queen of hearts and then a face card.

Solution

As we've already seen, because being dealt the queen of hearts for the first card reduces the number of face cards left in the deck for the second card, these events are dependent. We begin by calculating the probability of first being dealt the queen of hearts. Since all cards are available the probability is

$$P(\text{queen of hearts}) = \frac{1}{52}.$$

When the second card is dealt, there are no longer 12 face cards in the deck. Only 11 face cards remain in a deck of 51 cards (remember that there is one less card). So, the probability is

$$P(\text{face card} \mid \text{queen of hearts}) = \frac{11}{51}.$$

We then use the Multiplication Rule for Dependent Events to get

$$P(\text{queen of hearts } and \text{ face card}) = P(\text{queen of hearts}) \cdot P(\text{face card} \mid \text{queen of hearts})$$

$$= \frac{1}{52} \cdot \frac{11}{51}$$

$$= \frac{11}{2652}$$

$$\approx 0.004148.$$

Example 9: Conditional Probability

Suppose 290 students were asked about their satisfaction with their interactions with the financial aid office on campus. Their responses are given in the following table.

Class	Satisfied	Dissatisfied	Did Not Use
Freshman	55	21	13
Sophomore	15	33	24
Junior	48	6	8
Senior	22	18	3
Graduate	4	1	19

TABLE 4: Financial Aid Office Satisfaction

If one response from the 290 students was selected at random, find the probability that the following occurred.

a. The student was satisfied with their experience.

b. The student was satisfied given that they were a senior.

c. The student was dissatisfied given that they were a freshman or sophomore.

d. The student was a graduate student given that they did not use the financial aid office.

Solution

a. To find the total number of students who were satisfied, we can find the sum of the first column.

$$\text{Students satisfied} = 55 + 15 + 48 + 22 + 4 = 144$$

To find the probability that the student whose response we randomly selected is one of these, we divide the number of satisfied students by the total number of students:

$$P(\text{satisfied}) = \frac{144}{290}$$

$$\approx 0.496552.$$

b. To find the probability that the student was satisfied given that they were a senior, we need to limit our satisfied responses to those of senior students only. First, find the sum of the row of senior responses to find out how many of the 290 students were seniors.

$$\text{Senior students} = 22 + 18 + 3 = 43$$

Now, divide the number of seniors who responded satisfied by the total number of seniors.

$$P(\text{satisfied} \mid \text{senior}) = \frac{22}{43}$$

$$\approx 0.511628$$

c. Again, because we are looking at a conditional probability, we'll need to limit our student responses to only freshmen and sophomores. We can find the sum of their rows to calculate the number of students in these two classes.

$$\text{Freshman and sophomore students} = 55 + 21 + 13 + 15 + 33 + 24 = 161$$

This time we're interested in the number of dissatisfied students. Looking in the dissatisfied column, we see that there were $21 + 33 = 54$ dissatisfied freshmen and sophomores. We can then divide to find the probability that we randomly chose one of these students' responses.

$$P(\text{dissatisfied} \mid \text{freshman or sophomore}) = \frac{54}{161}$$

$$\approx 0.335404$$

d. This conditional probability requires us to consider only the column of students who did not use the financial aid office.

$$\text{Students who did not use the financial aid office} = 13 + 24 + 8 + 3 + 19 = 67$$

Of these 67 students, 19 of them were graduate students. Thus, the probability that we randomly chose a response from a graduate student given that the student did not use the financial aid office is as follows.

$$P(\text{graduate student} \mid \text{did not use financial aid office}) = \frac{19}{67}$$

$$\approx 0.283582$$

Example 10: Putting It All Together

Studies show that men have a 1 in 6 (or about 17%) chance of developing prostate cancer. A PSA test is used to detect prostate cancer and can give either a positive or negative result. Studies also show that of those men who have developed prostate cancer, their PSA test is negative 15% of the time. On the other hand, studies show that 58.5% of all men receive a positive PSA test.

Sources: National Cancer Institute. "Prostate-Specific Antigen (PSA) Test." http://www.cancer.gov/cancertopics/factsheet/Detection/PSA; Mayo Clinic. "Prostate cancer screening: Should you get a PSA test?" http://www.mayoclinic.com/health/prostate-cancer/HQ01273

a. What is the probability that a man develops prostate cancer?

b. What is the probability that a man with cancer has a negative PSA test?

c. What is the probability that a man with cancer has a positive PSA test?

d. What is the probability that a man has cancer and a positive PSA test?

e. What is the probability that a man with a positive PSA test has cancer?

Solution

a. We are told this probability in the information given.

$$P(\text{cancer}) = \frac{1}{6} = 0.1\overline{6}$$

b. Again, we are told this information.

$$P(\text{negative} \mid \text{cancer}) = 15\% = \frac{15}{100} = 0.15$$

c. Here, we want to know the probability of a positive test, given that a man has cancer. Since the information we were given tells us that men with cancer have a negative PSA test 15% of the time, we can use the complement to find the probability of men with cancer having a positive test.

$$P(\text{positive} \mid \text{cancer}) = 1 - \frac{15}{100} = \frac{85}{100} = 0.85$$

d. Use the Multiplication Rule for Dependent Events to find the probability of having cancer and a positive PSA test using the probabilities we previously calculated.

$$P(\text{cancer } and \text{ positive}) = P(\text{cancer}) \cdot P(\text{positive} \mid \text{cancer})$$
$$= \frac{1}{6} \cdot \frac{85}{100}$$
$$= \frac{85}{600}$$
$$\approx 0.141667$$

e. Be careful how you read this question. It's slightly different than the previous one we just answered. We want to know the probability of a man actually having cancer, given that he has a positive test. Since we weren't given this information, we'll need to rearrange the Multiplication Rule for Dependent Events to help us. Recall that the formula is

$$P(A \text{ and } B) = P(A) \cdot P(B \mid A).$$

Using algebra, we can divide both sides by $P(A)$ and get the following equation:

$$\frac{P(A \text{ and } B)}{P(A)} = P(B \mid A).$$

So,

$$P(\text{cancer} \mid \text{positive}) = \frac{P(\text{positive } and \text{ cancer})}{P(\text{positive})}.$$

From part d., we know that $P(\text{positive } and \text{ cancer})$ is $\dfrac{85}{600}$. From the problem statement, we also know that 58.5% of all men receive a positive PSA test, which gives $P(\text{postive}) = \dfrac{58.5}{100}$. Therefore, we calculate the probability as follows.

$$
\begin{aligned}
P(\text{cancer} \mid \text{positive}) &= \frac{P(\text{positive } and \text{ cancer})}{P(\text{positive})} \\[2mm]
&= \frac{\dfrac{85}{600}}{\dfrac{58.5}{100}} \\[2mm]
&= \frac{85}{600} \cdot \frac{100}{58.5} \\[2mm]
&= \frac{8500}{35,100} \\[2mm]
&= \frac{85}{351} \\[2mm]
&\approx 0.242165
\end{aligned}
$$

This means that, given that a man has a positive PSA test, the probability that he actually has cancer is approximately 0.242, or 24.2%.

Parts c. and d. in Example 10 illustrate a common practice of probability. Very often we have the ability to find out $P(A \mid B)$, but what we really want to know is $P(B \mid A)$. A mathematician named Thomas Bayes is credited with a theorem simplifying the relationship between the two conditional probabilities. His theorem, known as Bayes' Theorem, shows how one relates to the other.

Bayes' Theorem

$$P(A \mid B) = \frac{P(B \mid A) \cdot P(A)}{P(B)},$$

when $P(B) > 0$.

Skill Check Answers

1. a. Mutually exclusive **b.** Not mutually exclusive

8.6 EXERCISES

PRACTICE

Determine whether each situation contains independent events.

1. The color of car driven by three randomly chosen classmates.

2. A password must be six characters long with no repeated characters. Are the choices of consecutive characters independent?

3. There are 15 board members, of which seven are men and eight are women. Two randomly chosen members will serve on the United Way campaign committee. If you wish to find the probability that both members chosen are the same sex, do you treat these selections as independent events?

4. Are receiving a bill in Monday's mail and receiving a letter from your grandparents in Monday's mail independent events?

5. Naomi and Amelia both put two business cards into the basket at a coffee shop. The shop owner selects three cards from the basket. Are the two events that Naomi's card is chosen and Amelia's card is chosen independent?

APPLICATIONS

Calculate the probability of each set of mutually exclusive events. Round your answer to the nearest millionth when necessary.

6. Suppose that the probability of obtaining zero defective items in a sample of 50 items off the assembly line is 0.34 while the probability of obtaining 1 defective item in the sample is 0.46. What is the probability of the following?
   a. Obtaining no more than one defective item in a sample.
   b. Obtaining more than one defective item in a sample.

7. A pair of dice is rolled. What is the probability that the sum of the numbers is either 7 or 11?

8. A single letter from the word MISSISSIPPI is chosen. What is the probability of choosing an S or an I?

9. What is the probability that a card selected from a deck will be either an ace or a queen?

10. A reporter for an international newspaper is given an assignment that is randomly chosen from the following destinations worldwide: 13 continental United States assignments, 7 South American assignments, 21 European Union assignments, and 5 Asian assignments. Find the probability that he gets an assignment in Asia or South America.

11. The following table shows the breakdown of opinions for both faculty and students in a recent survey about the new restructuring of the campus to be a walking campus.

Survey Results on Restructuring Campus to a Walking Campus			
	Favor	Oppose	Neutral	Total
Faculty	12	4	3	19
Student	33	57	28	118
Total	**45**	**61**	**31**	**137**

   a. Find the probability that a randomly selected person is either a faculty member in favor of the change or a student who has an opinion either for or against.
   b. Find the probability that a randomly selected person is either neutral or in favor of the restructuring.

12. The probability of the stoplight being green at the intersection of Meeting Street and Main Street is 0.55, while the probability of it being yellow is 0.15. Find the probability that the light is red when you get to the intersection of Meeting Street and Main Street. Assume that the light will be working and will be a solid color: red, yellow, or green.

13. In a box of pens and pencils, the probability of randomly choosing a sharpened pencil is 0.54 and the probability of randomly choosing a pen from the box is 0.39. Find the probability of randomly selecting either an unsharpened pencil or a pen from the box.

Calculate the probability of each set of events that are not mutually exclusive. Round your answer to the nearest millionth when necessary.

14. A pair of dice is rolled. What is the probability that the sum of the numbers is an even number or a multiple of 3?

15. A bag of eleven marbles contains five marbles with red on them, three with green on them, seven with black on them, and four with black and red on them. What is the probability that a randomly chosen marble has either black or red on it?

16. What is the probability that a card selected from a deck will be either an ace or a spade?

17. The following is a table showing the results of a poll taken on campus.

Will You Vote in the Upcoming Election?		
	Male	**Female**
Yes	16	24
No	19	11
Not decided	21	22

   a. What is the probability that a randomly selected student from this poll would be a male who has not decided whether he will vote in the upcoming election?

   b. What is the probability that a randomly selected student from this poll is female or will not vote in the upcoming election?

   c. What is the probability that a randomly selected student from this poll has decided to vote in the upcoming election?

18. Out of a class of 30 students, there are 16 students who study Latin, 21 who study German, and 7 who study both. What is the probability that a randomly selected student from the class will study only Latin?

19. Of the 11 instructors in the English department, four are new to the department and three are female. However, there is only one who fits all of the descriptions. Find the probability that if you randomly choose a course taught by these instructors, you get either a new instructor or a female instructor.

20. The following is a table representing the students who are on the Student Government Board.

Students on the Student Government Board		
	On-Campus Housing	**Off-Campus Housing**
Freshman	3	1
Sophomore	3	2
Junior	2	3
Senior	0	3
Graduate Student	0	2

Find the probability that a randomly chosen member of the Student Government Board is either a sophomore or lives in on-campus housing.

Calculate the probability of each set of independent events. Round your answer to the nearest millionth when necessary.

21. Suppose the probability that my pet will be alive in five years is 0.65 and the probability that my cousin's pet will be alive in five years is 0.48. Find the probability that both of these pets will be alive in five years assuming that they are independent events.

22. Two dice are thrown. Find the probability of getting an even number on the first die and an odd number on the second die.

23. Find the probability of choosing a heart and then an ace from a standard deck of cards with replacement.

24. On any given day at the beach, there is a 49% chance of precipitation. What is the probability that you will get precipitation for three days in a row on your beach vacation? Assume that the weather on a particular day at the beach is independent of the weather the day before.

Calculate each conditional probability. Round your answer to the nearest millionth when necessary.

25. A swim team consists of four boys and three girls. A relay team of four swimmers is chosen at random from the team members. What is the probability that there are two boys on the relay team given that there are two girls on the relay team?

26. Emma is playing Monopoly, a game played with two dice. What is the probability that the sum of the two dice she rolls is less than 4 given that she rolls an odd number?

27. Hunter bets his friend that he can draw two aces in a row from a standard deck of cards. What is the probability that Hunter draws a second ace given that his first card was an ace?

28. The probability that a student passes Intermediate Algebra is 0.55. The probability that a student passes College Algebra given that they pass Intermediate Algebra is 0.70. What is the probability that a student passes both College Algebra and Intermediate Algebra?

29. On each point in racquetball, a player is allowed two serves. Suppose while playing racquetball, Tim gets his first serve in about 75% of the time. He gets his first serve in and wins the point about 50% of the time. What is the probability that he wins the point, given that he gets his first serve in?

Calculate each probability. Round your answer to the nearest millionth when necessary.

30. The following table shows the student demographics for a sociology class.

Sociology 101 Student Demographics		
	Male	Female
Freshman	3	11
Sophomore	4	9
Junior	0	3
Senior	1	0

   a. Find the probability that a randomly selected student from the class is a male.
   b. Find the probability that if two students are randomly selected, without replacement, the first is a female junior and the second is a male sophomore.

31. Arianna likes chicken and apple sausage, but not chicken and asiago cheese sausage. There are 18 pieces of each kind of sausage on a sausage and cheese plate. What is the probability that Arianna randomly skewers three pieces of sausage that she likes given that the first two are to her liking?

32. A swim team consists of four boys and three girls. A relay team of four swimmers is chosen at random.
   a. What is the probability that two boys and two girls are chosen for the relay team?
   b. What is the probability that Jim is one of the two boys and Jane is one of the two girls?

33. James has 20 applications on the home screen of his smartphone. His nephew accidently deletes five of the apps on his home screen. What is the probability that the app originally in the top right corner and the app originally in the bottom left corner have not been deleted?

34. The probability that an e-mail is spam is 0.05, the probability that the word "offer" is in an e-mail is 0.02, and the probability that the word "bank" is in an e-mail is 0.1. The probability that the word "offer" appears given that the e-mail is spam is 0.2, and the probability that the word "bank" appears given that the e-mail is spam is 0.4.
   a. Find the probability that an e-mail contains the word "bank" and is spam.
   b. If the words are assumed to appear independently, find the probability that an e-mail that contains "offer" and "bank" is spam.

8.7 EXPECTED VALUE

Now that we've explored ways to calculate probabilities, let's look at how we can use probabilities to predict the average outcome when an event takes on numerical values. **Expected value** is just that—what you would anticipate getting if you took the average of a sample of events.

For instance, suppose you throw a fair die. What would you "expect" to roll? Each of the numbers on the die has an equal chance of occurring, so you really shouldn't expect one number over another. However, if you threw it 60 times, what would you expect? Well, using the fact that each number has a $\frac{1}{6}$ chance of being rolled, you should expect each number to come up about 10 times, not exactly 10, but close. Suppose we took the average of all of those 60 rolls; in other words, we add them all up and divided by the number of rolls. Since the order of the rolls is irrelevant, we'll list them here in an organized manner so you can have the image in your head.

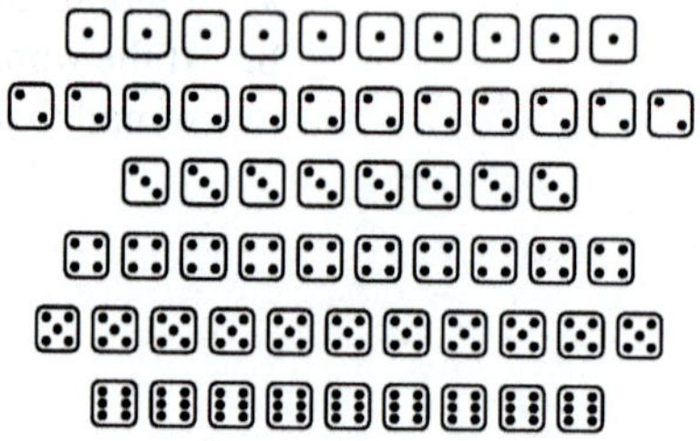

FIGURE 1: Sample of 60 Rolls of a Die

Adding these numbers up, we get 207. Now, divide by 60, and we get

$$\frac{207}{60} = 3.45.$$

So the average of these particular rolls of the die was 3.45. Of course if we did the experiment again, we would get a slightly different set of rolls and probably a slightly different average. If we perform the experiment multiple times, we would begin to build a picture of the *average* value. For instance, we might have the numbers 3.45, 3.52, 3.47, 3.55, etc., all as average rolls of the die. We can see that the average of the rolls is hovering somewhere around 3.5 each time. Mathematically, the average is a number that describes the center of the data set. Although no member of the data set is actually 3.45, it does describe the central tendency of the data.

So how does finding averages link up with probabilities? Expected values are a way to find the anticipated value of a random event without having to go through this elaborate process of repeating a probability experiment many times, listing out all the observed outcomes, and manually calculating their average. To find the expected value, you can simply multiply each outcome by its probability and add the products together.

Expected Value

The formula for **expected value** of an event X is

$$E(X) = x_1 P(x_1) + x_2 P(x_2) + x_3 P(x_3) + \cdots + x_n P(x_n),$$

where x_i is the ith outcome and $P(x_i)$ is the probability of x_i.

Let's use this formula to calculate the expected value for rolling a die. Remember that each number on the die has an equal probability of $\frac{1}{6}$. Constructing a table will make it easier to keep track of the calculations.

Outcome x_i	Probability $P(x_i)$	$x_i P(x_i)$
1	$\frac{1}{6}$	$\frac{1}{6}$
2	$\frac{1}{6}$	$\frac{2}{6}$
3	$\frac{1}{6}$	$\frac{3}{6}$
4	$\frac{1}{6}$	$\frac{4}{6}$
5	$\frac{1}{6}$	$\frac{5}{6}$
6	$\frac{1}{6}$	$\frac{6}{6}$

TABLE 1: Outcomes and Probabilities of Rolling a Die

Now, let's add the products together.

$$E(X) = \frac{1}{6} + \frac{2}{6} + \frac{3}{6} + \frac{4}{6} + \frac{5}{6} + \frac{6}{6}$$
$$= \frac{21}{6}$$
$$= 3.5$$

So, the expected value for a roll of a single die is 3.5. Of course, this isn't the same value that we got when doing the 60 trials before. But, as we observed, our average could vary for each experiment consisting of 60 trials, and these different averages would all cluster around some central value, which is the expected value of 3.5. It is important to recognize that the expected value allows you to quickly calculate the average of a large number of trials without having to carry out any experiments. This is similar to the techniques used in classical probability as opposed to those used in empirical probability. Remember, the expected value gives us a *long-term average*.

One of the nice properties of expectation is that it is *linear*. This means that, since we know that the average roll of one die is 3.5, then we immediately know that the average roll of two dice is 3.5 + 3.5 = 7.0. In fact, the average roll of 20 dice is 20 · 3.5 = 70. This property is not just limited to dice.

Given the following table of outcomes and their probabilities, complete the table to help you calculate the expected value of the event.

Outcomes and Probabilities

Outcome, x_i	$P(x_i)$	$x_i P(x_i)$
3	0.45	
6	0.30	
9	0.25	

The Sum of Expected Values

The expected value of the union of two mutually exclusive events, X and Y, is equal to the sum of the individual expected values. Formally, we can write this as

$$E(X \cup Y) = E(X) + E(Y).$$

You might wonder from this dice example whether this method is actually any quicker than just adding the trials up and dividing. To calculate the average of one die, there are six elements in the sample space to average. For two dice, there are 36 elements to average. For 20 dice, we would have 3,656,158,440,062,976 elements to average. Even if you have all the time in the world, you won't always be able to carry out the experiments you're interested in ahead of time. For instance, suppose you're considering investing in the stock market. You'd like to know the expected value of the return on your investment. Obviously, you'd like to calculate this *before* investing any money. Or, suppose you are a manager who needs to predict sales for the coming year to make hiring decisions. Knowing the average profit from each customer would be critical information.

Example 1: Calculating Expected Value

The campus dining service at State University is preparing for the upcoming semester. Based on past semesters, they have observed the following data on the probability of selling different meal plans. Each plan consists of a set number of meals along with Plus Dollars, which can be used anywhere on campus but are restricted to food purchases. The university predicts an enrollment of 8421 students in the coming semester.

Plan	Description	Price	Probability
1	19 meals per week and $100 Plus Dollars for the semester	$1245	$\frac{1}{10}$
2	14 meals per week and $250 Plus Dollars for the semester	$1245	$\frac{1}{15}$
3	10 meals per week and $350 Plus Dollars for the semester	$1245	$\frac{1}{30}$
4	$815 Plus Dollars to be used anytime during the semester (no meals)	$815	$\frac{1}{10}$
5	90 meals to be used anytime during the semester and $300 Plus Dollars	$870	$\frac{1}{12}$

Note: Students have the option of choosing not to buy a meal plan.

TABLE 2: Meal Plans

a. Suppose a random student from the university is chosen. What is the expected value for the amount that they will spend on their meal plan for the upcoming semester?

b. Suppose that, for each meal plan sold, the university makes $510 from Plan 1, $550 from Plan 2, $600 from Plan 3, $220 from Plan 4, and $270 from Plan 5. What is the expected profit per student in the upcoming semester?

c. What overall profit can dining services expect from student meal plans in the upcoming semester?

Solution

a. To find the expected value, we need to multiply each meal plan cost by the probability of purchasing that meal plan and add these products together. Remember that although they are not listed in the chart, many students will choose no meal plan at all. We've included them in the following calculation although they spend $0 on a meal plan.

$$E(\text{meal plan cost}) = \frac{1}{10}(1245) + \frac{1}{15}(1245) + \frac{1}{30}(1245) + \frac{1}{10}(815) + \frac{1}{12}(870) + \frac{37}{60}(0)$$
$$= 124.50 + 83.00 + 41.50 + 81.50 + 72.50$$
$$= 403$$

Therefore, a random student is expected to pay $403 for their meal plan. Of course, no single student pays $403 for a meal plan. This expected value gives us an average per student across all students whether they buy a plan or not.

b. The profits per meal plan are summarized in Table 3.

Plan	Profit per Meal Plan ($)
1	510
2	550
3	600
4	220
5	270

TABLE 3: Meal Plan Profits

In order to calculate the expected profit per student, multiply the estimated profit for each meal plan by the probability of that meal plan being chosen.

$$E(\text{profit per student}) = \frac{1}{10}(510) + \frac{1}{15}(550) + \frac{1}{30}(600) + \frac{1}{10}(220) + \frac{1}{12}(270) + \frac{37}{60}(0)$$
$$\approx 51 + 36.67 + 20 + 22 + 22.50$$
$$= 152.17$$

Therefore, dining services can expect an average profit of $152.17 per student next semester.

c. Because expectation is linear, we can add together the expected profits for each of the 8421 students. Therefore, the expected profit for dining services is $8421 \cdot \$152.17 = \$1,281,423.57$.

Casinos also rely heavily on expected values in order to turn a profit. Their entire business depends on creating games where the experience is appealing enough for you to play the game, but the probabilities end up in the casino's favor in the long run. Let's look at a simple example.

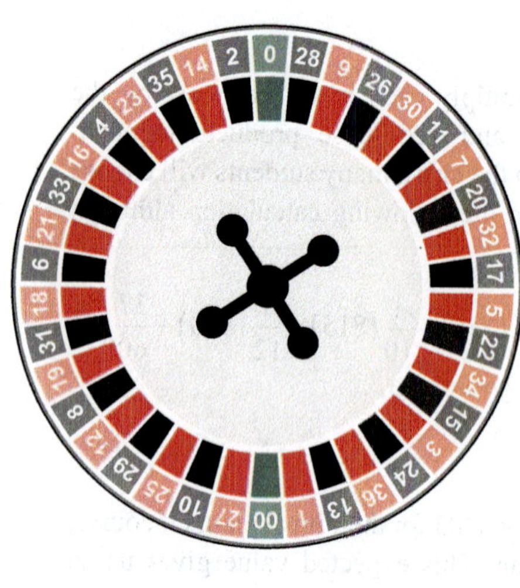

Example 2: Expected Winnings

In American roulette, the wheel contains the numbers 1 through 36, alternating between black and red. There are two green spaces numbered 0 and 00.

a. Calculate the probability of the roulette ball landing on a red pocket (or black for that matter, since the number of red and black pockets is the same).

b. Calculate the probability of the ball not landing on a red pocket.

c. A player bets $1.00 on red to play the game. Calculate his expected winnings.

Solution

a. In calculating the probability for landing on a red, note that there are 18 red pockets out of 38 possible pockets on the wheel. So the probability is

$$P(\text{red}) = \frac{18}{38}.$$

b. The probability of not landing on a red pocket is one minus the probability of landing on a red pocket, so

$$P(\text{not red}) = 1 - \frac{18}{38} = \frac{20}{38}.$$

Note: 20 is the sum of the 18 black pockets and the two green pockets.

c. If a player bets $1.00 on red, his chance of winning $1.00 is $\frac{18}{38}$ and his chance of losing $1.00 (or winning −$1.00) is $\frac{20}{38}$. So, the player's expected value is

$$E(X) = \left(\frac{18}{38}\right)(1) + \left(\frac{20}{38}\right)(-1)$$

$$= \frac{18}{38} - \frac{20}{38} = -\frac{2}{38}$$

$$\approx -0.05.$$

What this means is that the player can expect to lose about 5 cents on average for every dollar he bets throughout the night on red roulette pockets. Of course, he doesn't actually ever lose 5 cents on a single bet, but remember this is his expected average outcome for multiple trials. Suppose he played 10 games betting $1, he could expect to lose

$$(10)(0.05) = \$0.50 \text{ overall.}$$

That doesn't sound very profitable for the casino, but consider if they had 1000 people play that same bet throughout the day. Then their profits would be

$$(1000)(0.05) = \$50.00.$$

Now consider that the casino has millions of roulette players per year. Because the casino has a very slight edge on the roulette wheel, the casino is expected to make money over time. Think about how this advantage changes if you place the same bet on a European roulette wheel; it is exactly the same as the American wheel except that it only has one green 0 pocket. Determining the difference in advantage is left as an exercise.

The last thing that we will mention is the concept of "odds," as you might hear at a casino or racetrack. Although it's a common mistake, the words odds and probability are not interchangeable. However, expected value helps us tie the two together nicely.

Let's take a look at the phrase "odds of 3 : 1 against." (Recall that we read 3 : 1 as "three to one.") Suppose you placed a $1.00 bet on a horse that had odds of 3 : 1. These odds do not mean that you have a $\frac{1}{3}$ chance of winning money. Instead, as we will see, you have a $\frac{1}{4}$ chance (or 25%) of winning. In real terms, it means that for every $1.00 you place on the bet, you win $3.00 if your horse wins. Let's use expected value to see why the probability works this way. Suppose the odds of your horse winning are 3 : 1 against, and you place a bet of $1.00 on the horse. If your horse wins, you gain $3.00, but if it does not win, you lose $1.00 (or win −$1.00). Let the probability of your horse winning be p, which then means that the probability of your horse not winning is $1 - p$ as shown in the following table.

Outcome	Gain	Probability
Winning	$3.00	p
Not Winning	−$1.00	$1 - p$

TABLE 4: Odds of 3 : 1 Against

The following is the expected value formula with this information.

$$E(X) = (\$3.00)(p) + (-\$1.00)(1 - p)$$

Both the bookie and yourself would think that the wager is a fair one if the expected gain and loss for both of you were $0.00. In other words, no money changes hand. So, we can say that the expected value should equal 0 for a fair bet, or

$$E(X) = (\$3.00)(p) + (-\$1.00)(1 - p) = \$0.$$

Using a bit of algebra, we can solve the equation for p.

$$(3)(p) + (-1)(1 - p) = 0$$
$$3p - 1 + p = 0$$
$$4p - 1 = 0$$
$$4p = 1$$
$$p = \frac{1}{4}$$

So, the probability of winning when the odds are 3 : 1 against is $\frac{1}{4}$, and the probability of not winning is $1 - \frac{1}{4} = \frac{3}{4}$.

In fact, rather than being an expression for the probability of winning, the term "odds against" represents the ratio of the chance of losing to the chance of winning. In our example, the phrase "odds against of 3 : 1" means that 3 out of 4 times you're likely to lose. Since odds are ratios, you can also express odds as fractions.

$$\text{Odds against} = \frac{P(\text{losing})}{P(\text{winning})}$$

In this particular case, the fraction is

$$\text{Odds against winning} = \frac{\left(\dfrac{3}{4}\right)}{\left(\dfrac{1}{4}\right)} = \frac{3}{1} = 3.$$

We should note that in the betting world, "odds against" are most often quoted because it is the most convenient way to understand the payout if the bet is a successful one for the player. So odds of 20 : 1 mean that for every \$1.00 you bet, the bookie will pay you \$20.00 if you win. The bookie is out to make money; in reality, the odds represent how much the bookie is willing to pay out rather than the true ratio.

Finally, note that odds may also be expressed as "odds for" (also called "odds in favor"), in which case the ratio is

$$\text{Odds for} = \frac{P(\text{winning})}{P(\text{losing})}.$$

8.7 EXERCISES

💡 PRACTICE

Calculate the expected value of each scenario. Round your answer to the nearest hundredth when necessary.

1.

x_i	$P(x_i)$
1	0.21
2	0.58
3	0.06
4	0.15
5	0

2.

x_i	$P(x_i)$
−\$1.50	0.3
\$0.00	0.5
\$2.75	0.1
\$5.00	0.1

3.

x_i	$P(x_i)$
25	$\dfrac{1}{3}$
15	$\dfrac{2}{5}$
10	$\dfrac{1}{15}$
5	$\dfrac{1}{5}$

4. Let x_i represent the number of even numbers showing when a pair of standard six-sided dice are rolled.

x_i	$P(x_i)$
0	$\dfrac{9}{36} = \dfrac{1}{4} = 0.25$
1	$\dfrac{18}{36} = \dfrac{1}{2} = 0.5$
2	$\dfrac{9}{36} = \dfrac{1}{4} = 0.25$

🚀 APPLICATIONS

5. Suppose Piper eats out twice a week 15% of the time, she eats out once a week 35% of the time, and she doesn't eat out anytime during the week 50% of the time. What is the expected value for the number of times Piper eats out during a week?

6. Suppose that you and a friend are playing cards and decide to make a bet. If you draw two aces in succession from a standard deck of cards without replacing the first card, you win $50.00. Otherwise, you pay your friend $10.00.
   a. What is the expected value of your bet?
   b. If the same bet was made 25 times, how much would you expect to win or lose?

7. A European roulette wheel has only one green slot instead of two. Using Example 2 from this section as a guide, calculate the expected winnings on a European roulette wheel if a player bets $1.00 on red to play the game.

8. Jim likes to day-trade on the Internet. On a good day, he averages a $1100 gain. On a bad day, he averages a $900 loss. Suppose that he has good days 25% of the time, bad days 35% of the time, and the rest of the time he breaks even.
   a. What is the expected value for one day of Jim's day-trading hobby?
   b. If Jim day-trades every weekday for three weeks, how much money should he expect to win or lose?

9. A university in town is raffling off $20,000 for student scholarships. You can buy one ticket for $10, three tickets for $25, or five tickets for $40. Assume that the university sells 10,000 tickets.
   a. Find the expected value for each of the three ticket options: purchasing just one ticket, purchasing three tickets, or purchasing five tickets.
   b. Should you buy one, three, or five tickets in order to maximize the money you expect to have at the end of the raffle?

10. You need to borrow money from your sister. She's feeling quirky on the day you ask and says she wants you to flip a coin. Heads, you get $15, tails you get $5. Thinking this is weird, you ask your mother for money instead. She says she'll let you roll a die and she'll give you $2 times the number that appears on the die. Before agreeing to either of these unique offers from the "mathy" folk in your family, you decide to see which is the better offer by calculating the expected value for each method (realizing that you too fit the bill of a "mathy" member of your family). Which offer should you take? Explain your reasoning.

11. Assume that stock in Degree Compass, a predictive analytics company in higher education, returns the percentages shown in the table.

Degree Compass Stock Returns	
Annual Return Rate	**Probability**
15%	0.17
30%	0.51
45%	0.32

Calculate the expected value of the return rate for stock in Degree Compass.

12. During the NCAA basketball tournament season, affectionately called *March Madness*, part of one team's strategy is to always foul their opponent's tall forward. Because he is so tall, he makes 57% of shots he takes close to the basket. However, when he is fouled, his free throw shooting percentage is only 51.5%. The shots he makes close to the basket are worth two points and each of the two free throw shots after being fouled are worth one point.
    a. Calculate the expected value of the number of points the forward makes when he takes a shot close to the basket.
    b. Calculate the expected value of the number of points the forward makes when he shoots two foul shots.
    c. Based on these expected values, is fouling the tall forward a good strategy? Explain your answer.

13. On your next multiple-choice test, each question has four incorrect answers and one correct answer to choose from. Your professor tells you that each correct answer you make, you receive 1 point, but you lose $\frac{1}{4}$ point for each incorrect answer.
    a. What is your expected gain or loss on a question if you have no idea of the correct answer and end up simply guessing?
    b. What is your expected gain or loss if you guess on all 25 questions?

14. The stock prices of Web Movies on the 1st of the month for the first 6 months of 2014 were \$177.41, \$207.90, \$214.63, \$239.09, \$242.19, and \$259.99 respectively.
    a. Calculate the average change in the stock prices of Web Movies in a month.
    b. Use linearity of expectation to estimate the stock price on January 1st, 2015.

15. If the odds on a bet are $6:1$ against, what is the probability of winning?

16. Suppose the probability of a football team winning a playoff game is 0.25. What are the odds of winning?

17. Suppose the odds of a teenage male having an accident are $2:3$. What is the probability of a teenage male having an accident?

18. An insurance company claims the probability of surviving a certain type of cancer is 95%. What are the odds of surviving?

19. The UVest investment company publishes that the odds of increasing your wealth with their company is $5:2$. What is the probability of UVest increasing your investment?

20. Odds against being struck by lightning in one year are 1,000,000 to 1.[1]
    a. If you live to be 80, what are the odds against being struck by lightning over your lifetime? Assume each year has the same probability.
    b. The National Weather Service gives the odds against being struck by lightning over an 80-year lifetime as 10,000 to 1. Why do you think this is different from the answer you got in part **a.**?

21. Overall odds in favor of winning in a state lottery game are $4.63:1$.
    a. Find the probability of winning in the lottery game.
    b. The prize for this lottery game is $100. If the cost to play the game is $2.00, what is the expected value for playing this game?

22. Suppose the odds on a bet are $10:1$ against. Your friend tells you he thinks the odds are too generous. Odds are considered less generous if the probability of losing is greater. Write down some less generous odds.

---

[1] NWS Lightning Safety, http://www.lightningsafety.noaa.gov

9

Chapter 9

STATISTICS

9.1 COLLECTING DATA

■ TOPICS

- Populations vs. Samples
- Choosing a Sample
- Random Sampling
- Stratified Sampling
- Cluster Sampling
- Systematic Sampling
- Convenience Sampling

Populations vs. Samples

Most statistical studies are brought about by the desire to know the answer to a particular question. For instance, you might want to know which car gets the best gas mileage, which university is the best value for the money, or how often you might expect to be called for jury duty. The search for the answers to many questions just like these is helped along by collecting data and then drawing conclusions.

When we begin a statistical study, we must first define the **population**, or the particular group in which we are interested. Considering our desire to know which car gets the best gas mileage, we might first assume that the desired population for the study is "all cars." However, stop for a moment and think about the implications of trying to study all makes and models of all cars, in all parts of the world! This might take a while and might not be very useful. What if the car with the best gas mileage was only available in Romania 20 years ago? By being more precise about how we define the population of interest, we narrow our potential population to just those in which we are truly interested. In most cases, this also creates a more manageable population with which to work. When defining the statistical group of interest, it is impossible to overclarify the population you wish to study. For our gas mileage example, let's narrow our attention to all passenger cars manufactured in the United States in 2013.

> **Population**
>
> A **population** is the particular group of interest in a study.

Collecting **data**, or information, from each member of the population always results in the best analysis we can have of a particular group. When we are able to achieve this, it is called a **census**. A census gives us the ability to speak with 100% confidence about the data we gathered from the group as a whole. Data gathered from a census allows us to identify population parameters. A **parameter** is the numerical description of a particular population characteristic. For instance, suppose the average miles per gallon (mpg) for all passenger cars manufactured in the United States in 2013 was 23.6. Then 23.6 mpg is a population parameter. All populations have fixed parameters that numerically describe the structure of the population. The goal of a statistical study is to identify as best we can the population parameter we are interested in.

The singular form of data is datum.

Data

Data is information collected for a study.

Census

A **census** involves collecting data from every member of the population.

Parameter

A **parameter** is the numerical description of a particular population characteristic.

Although conducting a census is the ideal way to identify population parameters, as you might imagine, this is often an unrealistic goal. Maybe the population is too big and hard to get to—think gas mileage from all cars around the globe. Even with a small population, data from every member may not be available to you—think obtaining the actual weight of all people in your class right now. When a census cannot be obtained, we gather information from a subset of the population, or a **sample**. The numeric descriptions of particular samples are called **sample statistics**. For instance, suppose you collected data on the gas mileage of the cars driven by half of the students in your class and found that the average mpg was 19.1. Then, 19.1 mpg is a sample statistic.

Sample

A **sample** is a subset of a population.

Sample Statistic

A **sample statistic** is the numerical description of a characteristic of a sample.

Population	Sample
Whole group	Part of a group
Group I want to know about	Group I do know about
Numerical descriptions of characteristics are called parameters	Numerical descriptions of characteristics are called statistics
Parameters are generally unknown	Statistics are always known
Parameters are fixed	Statistics vary with the sample chosen

TABLE 1: Population vs. Sample

Example 1: Identifying Parts of a Survey

Two shortened survey reports are given. In each report, identify the following: the population, the sample, the results, and whether the results represent a sample statistic or a population parameter.

a. A headline about the rising obesity among young people led a school board to survey local high school students. Out of 231 students surveyed, 58% reported eating a "high fat" snack at least 4 times a week.

b. A nonprofit organization interviewed 618 adult shoppers at malls across Louisiana about their views on obesity in youths. The resulting report stated that an estimated 48% of Louisiana adults are in favor of government regulation of "high fat" fast food options.

Solution

a. Population: local high school students

Sample: the 231 students who were surveyed

Results: 58% of students surveyed eat a "high fat" snack at least 4 times a week.

The result refers to only those students who were surveyed, thus the result is a sample statistic.

b. Population: Louisiana adults

Sample: the 618 adult Louisiana mall shoppers who were surveyed

Results: 48% of Louisiana adults are in favor of government regulation of "high fat" fast food options.

The results refer to all Louisiana adults, thus this is a population parameter. This population parameter is an estimate based on the sample statistics, which were not reported.

Since we've already noted that obtaining the values of population parameters by conducting a census is often unrealistic, we will focus our discussion on samples and how their characteristics help us know more about the population. It's important to stop here and note that the way we choose a sample from the population is important. Care must be taken in choosing your sample so that the population is well-represented and the results of the study are meaningful. For instance, consider our earlier example of a study wishing to know the average gas mileage for cars in the United States. If the only data you gathered were from half the students in your class, would this be an accurate picture of the population of cars driven in the United States? What about if we only chose cars of people who lived within the city limits of Nashville? How about in the city limits of San Diego? Let's look more closely at how to choose a sample that is representative of the entire population you are studying.

Choosing a Sample

When choosing a sample from the population, it's important that the sample is **representative** of the entire population.

Representative Sample

A **representative sample** is one that has the same relevant characteristics as the population and does not favor one group of the population over another.

Without a representative sample, it might not be possible to accurately generalize the results to the population as a whole. If the sample is poorly chosen, a form of bias, or favoring of a certain outcome, might occur. Consider the sample of cars driven by residents of the city of San Diego. Because of its dense population, hilly terrain, and popularity with a young affluent crowd, cars driven there tend to be either compact cars, which are more fuel efficient, or luxury vehicles. This sample would certainly favor a narrow sector of cars for our study on average miles per gallon.

There are several standard statistical methods for choosing samples that allow for the best possible representative samples. As you might imagine, some methods are better suited to particular situations than others. Let's look at some of the methods.

Random Sampling

Random Sample

A **random sample** is one in which every member of the population has an equal chance of being selected.

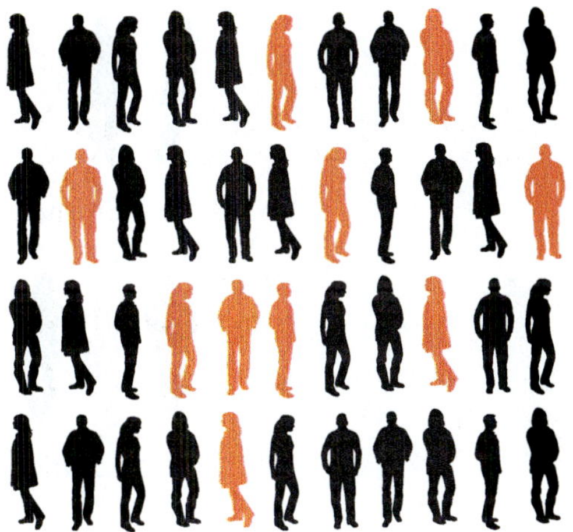

FIGURE 1: Random Sample

Drawing for door prizes by placing all of the ticket stubs in a basket and mixing them up is a very common example of random sampling. However, you should always be cautious when letting a human choose a "random" sample. In reality, it is against our human nature to choose members of the population entirely at random. Take a moment to choose 3 random digits between 0 and 9. Try to make sure they are truly random. Now, consider these questions.

- Are the numbers spaced out or grouped closely together—probably spaced out, right?

- Are they all even, odd, or some of both?

- Did you have a reason for choosing any of the numbers—maybe your favorite?

- Did you alter any of your original responses, and if so, why?

- Did you repeat any digits? Why not? It wasn't against the rules.

- Are your digits consecutive, such as 1, 2, 3?

If any reasons influenced your choice of digits, then your sample is not random. Because it is hard for humans to completely eliminate influences, it's best to take the human element out of it and allow technology to choose a random sample for us. For our fuel efficiency study we mentioned earlier, a random sample would occur if we allowed a computer to choose cars from a list of all possible passenger vehicles manufactured in the United States in 2013.

Stratified Sampling

Stratified Sample

A **stratified sample** is one in which members of the population are divided into two or more subgroups, called strata, that share similar characteristics like age, gender, or ethnicity. A random sample from *each* stratum is then drawn.

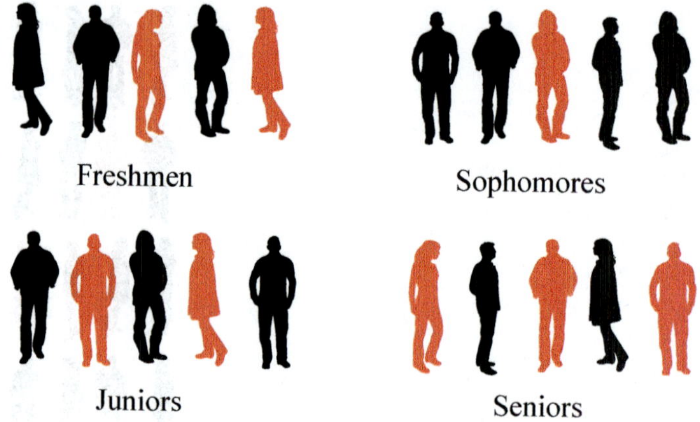

FIGURE 2: Stratified Sample

Skill Check 1

Can you think of some other ways to divide the cars into strata that might represent a broader scope of vehicles on the market?

Stratified sampling is used when we want to be confident that certain subgroups of a population are represented in the sample. For instance, first we could divide our vehicles into strata based on seating capacity, and then randomly choose 10 vehicles from each group. This would produce a stratified sample of cars for us to collect data from while making certain that we include two-seater vehicles as well as those that carry as many as 15 people.

Cluster Sampling

Cluster Sample

A **cluster sample** is one chosen by dividing the population into groups, called clusters, that are each similar to the entire population. The researcher then randomly selects some of the clusters. The sample consists of the data collected from *every* member of each cluster selected.

English 101: T, Th; 10:15 a.m.　　　English 101: M, W, F; 8 a.m.

English 101: M, W, F; 12 p.m.　　　English 101: M, W, F; 2:30 p.m.

FIGURE 3: Cluster Sample

This type of sampling resembles stratified sampling in that the population is first divided into groups. However, whereas stratified sampling chooses some members from each group, cluster sampling chooses all members from some of the groups. Often populations naturally lend themselves to subgroups from which each type of sampling might be done. For instance, if you wanted to know the average grade on the first test of all students in English 101, you could divide the students up by the section they are in. For cluster sampling, you would then randomly choose several sections and collect the grades from all students in each class.

Systematic Sampling

Systematic Sample

A **systematic sample** is one chosen by selecting every n^{th} member of the population.

Every 10th bottle

FIGURE 4: Systematic Sample

Assembly lines in a factory are often tested for quality control by systematic sampling. If every 10th product is checked for defects, then the manufacturer can feel confident that the plant is producing quality products. As a manager of the production line, you would want to ensure that there was no obvious bias in choosing every 10th product as opposed to every 9th one. If there was a unique characteristic about every 10th product, such as they came only from Machine A, you would not want to choose 10 as your counting interval.

Convenience Sampling

Convenience Sample

A **convenience sample** is one in which the sample is "convenient" to select. It is so named because it is convenient for the *researcher*.

Earlier we mentioned gathering data from half the students in your class for our comparison of fuel-efficient cars. Do you think this example of convenience sampling would be an accurate picture of the population of cars driven in the United States? Why or why not?

Sociology Survey

FIGURE 5: Convenience Sample

Although convenience sampling is the least reliable for obtaining a representative sample, in some instances it is all the data we can obtain, and the "some is better than none" factor comes into play. For instance, suppose you need 20 students to fill out a survey for your sociology project. Because of time constraints, you ask students in your English 101 class to participate. This sample of convenience is better than no sample at all. On the whole, however, caution should be used when results are obtained merely from convenience sampling.

Example 2: Identifying Sampling Techniques

Identify the sampling technique used to obtain a sample in each of the following situations.

a. To conduct a survey on collegiate social life, you knock on every 5th dorm room door on campus.

b. Student ID numbers are randomly selected from a computer printout for free tickets to the championship game.

c. Fourth grade reading levels across the county were analyzed by the school board by randomly selecting 25 fourth graders from each school in the county district.

d. In order to determine what ice cream flavors would sell best, a grocery store polls shoppers that are in the frozen foods section.

e. To determine the average number of cars per household, each household in 4 of the 20 local counties were sent a survey regarding car ownership.

Solution

a. Because the sample is obtained by choosing every n^{th} dorm room, this is systematic sampling. This is a representative sample, as long as students were randomly assigned to dorm rooms and there are no hidden potential biases, like only males may live in every n^{th} room.

b. Since every member has an equal chance of being selected, this is random sampling.

c. The students were divided into strata based on their schools and then a random sample from each school was chosen. This is stratified sampling.

d. Because of the ease of choosing shoppers right in their own store, this is convenience sampling. In this case, convenience sampling is a viable method for gaining a representative sample since the store would be interested in knowing the thoughts of their customers.

e. Cluster sampling was used here because the counties are the natural clusters and all of the households in some of the counties received the surveys.

Skill Check Answers

1. Answers will vary. For example: size of engine, manufacturer, make, safety rating, number of doors.

2. No, since cluster sampling is an "all from one group" method, comparing mpg from cars in only certain price ranges would not produce a representative sample.

3. The location of the entrance ramp might lend itself to having cars only on one end of the price scale depending on the businesses located in the area.

4. No, although students in a given class represent a variety of different backgrounds, it is unlikely that everyone would drive a wide range of different types of cars, which would be representative of the population of cars driven in the United States.

9.1 EXERCISES

✔ CONCEPT CHECK

Fill in each blank with the correct term.

1. A ______ involves gathering data from every member of a population.

2. A subset of a population is called a ______. _____________ are numerical descriptors of characteristics of this subset.

3. A ________ is a particular group of interest in a statistical study. ________ ________ describe numerical characteristics of this entire group.

4. A _______________ does not favor one subgroup over another from a population.

Decide if each numerical value from a statistical study is a population parameter or sample statistic.

5. After the presentation by the United Way representative, 67 of the night's 300+ attendees were interviewed about their reaction. Of the 67 people interviewed, 79% said they were motivated to donate more of their time and/or money to helping nonprofit organizations.

6. The Nashville Country Music Marathon reported an average finish time of 4:46:28 for the 4082 finishers.

7. A medical study showed that the heart-rate profile during exercise and recovery is a predictor of sudden death. Of the 5713 men studied, the mean maximum heart rate during exercise (expressed as the percentage of the predicted maximum heart rate) was found to be 96% in subjects who died suddenly from cardiac causes during the 23 years of the follow-up period of the study.[1]

8. An estimated 65% of drivers in rural or town areas view traffic conditions in a good light versus 39% of drivers in the city or suburbs based on a recent news report that aired nationally.

In each scenario, identify the population, the sample, and any population parameters or sample statistics that are given.

9. A medical study followed 675 cancer patients who received either the new drug *Lafent* or a standard chemotherapy drug. 76% of those who received *Lafent* were still living after eight months, compared to only 63% of those who received the standard chemotherapy drug.

10. US realtors across the country are encouraged by the latest reports on the real estate market. Taking into account seasonal factors, sales rose by nearly 9% in the West, 3.5% in the South, 3.4% in the Northeast, and 1% in the Midwest.

11. The online version of *Health* magazine, Health.com, reported the following: "Data from the Women's Health Study . . . examined the medical records of more than 36,000 women who had no history of depression at the start of the study. Roughly 18% of the women were experiencing some form of migraine or had suffered from the headaches in the past. Over the next 14 years, 11% of the study participants received a depression diagnosis." One of the conclusions of the study was that middle-aged women who experience migraines are 40% more likely to become depressed.[2]

12. In a recent study, US graduates of for-profit higher education institutions were between 4.8 and 6.7 percentage points more likely to be unemployed than graduates of nonprofit institutions and community colleges.[3]

---

1 Xavier Jouven et al., "Heart-Rate Profile During Exercise as a Predictor of Sudden Death."
2 Matt McMillen, *Health*, http://www.health.com
3 Dan Berrett, *The Chronicle of Higher Education*, http://www.chronicle.com

Identify the sampling technique used in each scenario.

13. To ensure the quality of its product, a company tests every 15th item off the assembly line.

14. A student committee in the biology department was formed by randomly choosing three biology majors from every level of student, that is, 1st year, 2nd year, etc.

15. In order to choose the winners of free tickets to the campus concert, student housing printed out all 60 pages of student IDs and then chose three of the pages randomly.

16. A candidate for the local school board surveyed parents picking up their children from the public school on three school days in the last month.

17. Surveying moviegoers as they exited the late movie, researchers determined that only young adults under the age of 22 see movies in the theater anymore.

18. In order to sample the production yield of milk from the farm's herd of cows, random samples are taken from each of the five different types of cows.

19. Using Excel to generate an arbitrary list of customers, the marketing department sent promotional materials to the top 250 names on the list.

🚀 APPLICATIONS

Determine an appropriate sampling method for each scenario. Give your reasons for choosing the method. Describe any potential biases that the researcher might need to consider.

20. A state senator wishes to survey his constituents on issues in the upcoming legislature session.

21. A pharmaceutical company wishing to test the outcomes of a new drug for a particular skin disease finds 1200 patients willing to participate in the study. They can only choose 100 of them for the actual study.

22. Medical scientists wish to evaluate the effects of vitamin E and low-dose aspirin in primary prevention of cardiovascular disease and cancer in apparently healthy women.

23. Researchers wish to study the effects of watching too much television at a young age for children diagnosed with autism.

✏️ WRITING & THINKING

24. Explain how cluster sampling is different from stratified sampling.

25. *Self-selected* samples are a type of convenience sampling in which the participants volunteer to participate in the study rather than being chosen by the researcher. One issue with this type of sampling is that people who "self-select" to be in a survey often either have very strong opinions about an issue or desire monetary rewards for their participation. For instance, consider online polls regarding political views that are conducted by popular news outlets. Describe some of the potential responses from those who take the time to log on and respond. Is it reasonable to generalize the results from a study like this to include all of the American public? Why?

26. Consider how you would collect data for a study of illegal immigrants working in the United States. How would you go about collecting the data? Is it possible to conduct a census on this population of people? How reliable do you think your results would be? Discuss any issues you might encounter.

27. Often words like "best" and "worst" are used in reporting data. However, descriptive words like these are hard to measure in a concrete way. Consider how you could measure small towns to compile a list of "America's 50 Best Small Towns to Live In." Name at least five distinct measurements.

28. Researchers try to eliminate as many biases in studies as they can in order to get a true picture of the population they are studying. Describe as many potential sources of bias as you can if you were asked to study the effects of alcohol on college campuses.

9.2 DISPLAYING DATA

■ TOPICS

- Frequency Distributions
- Pie Charts
- Bar Graphs
- Histograms
- Line Graphs
- The Power of Graphs
- Misleading Graphs

Once we have gathered data from the sample, our attention turns to how to communicate the results. As a society that has been inundated with many forms of media trying to get our attention, simply presenting the raw data in the format that it was gathered would be a fruitless act—it would neither elicit attention nor prompt those who do look at it to fully digest the information. As a result, it's important to display data so that it is organized clearly and it effectively conveys the intended message.

Frequency Distributions

Organizing the data into a table format is an easy and natural way to begin. A **distribution** is a way to describe the structure of a particular data set or population.

Frequency Distribution

A **frequency distribution** is a count of every member of the data set and how often each value occurs, that is, the frequency of the data value.

In the simplest form of a frequency distribution, each observed data value is listed along with its corresponding frequency in a table. Sometimes other numbers are also included in a frequency distribution. One such number is that of the **relative frequency** of a data value. The relative frequency of a class is the percentage of all data that fall into that particular class. It can be displayed as a percentage or a fraction. Let's look at an example of this type of frequency distribution.

Example 1: Interpreting a Frequency Distribution

The following frequency distribution shows the number of crashes with roadside objects along State Route 3 in the state of Washington. After looking over the table, answer the questions that follow.

Roadside Object	Number of Crashes (% of total)
Guardrail	57 (15.36)
Earth Bank	55 (14.82)
Ditch	42 (11.32)
Tree	42 (11.32)
Concrete Barrier	38 (10.24)
Over Embankment	31 (8.36)
Utility Pole	20 (5.39)
Wood Sign Support	19 (5.12)
Bridge Rail	17 (4.58)
Culvert	7 (1.89)
Boulder	6 (1.62)
Luminaire	6 (1.62)
Mailbox	5 (1.35)
Fence	5 (1.35)
Building	5 (1.35)
Other Object	16 (4.31)

Source: Jinsun Lee and Fred Manning. "Analysis of Roadside Accident Frequency and Severity and Roadside Safety Management." Washington State Transportation Commission. December 1999. http://www.wsdot.wa.gov/research/reports/fullreports/475.1.pdf

TABLE 1: Crashes with Roadside Objects along State Route 3, Washington

a. What do the numbers in the parentheses represent?

b. Which roadside object was involved in the most crashes?

c. How many total crashes did the survey cover?

Solution

a. The numbers in parentheses represent the relative frequency of each class; that is, the number of crashes in a particular category as a percentage of the total number of crashes. For example, there were 57 crashes involving a guardrail. Crashes involving a guardrail account for 15.36% of the total crashes along State Route 3.

b. Because the "Guardrail" category has the highest frequency listed, guardrails were involved in the most crashes.

c. By adding up the frequencies of all the different categories, we can see that there were 371 crashes involving roadside objects.

Often data will not be in categories that are names or labels, but categories that are numerical. Consider the data in the next example. People were asked to report the number of pets they had in their household. When making the frequency distribution, we'll still list each count of pets as a different category.

The number of household pets for 52 families is recorded. Construct a frequency distribution of the data collected.

0 1 0 2 0 0 3 1 1 1 2 3 1 0 0 2 0 3 4 2 0 1 2 5 1 3
1 3 2 1 1 2 2 2 0 0 0 0 1 6 1 2 2 1 0 0 0 1 0 1 2 2

Solution

List each possible value for the number of household pets in the "Number of Pets per Household" column and the number of times that particular data value occurs in the "Number of Households" column.

Number of Pets per Household	Number of Households
0	16
1	15
2	13
3	5
4	1
5	1
6	1

TABLE 2: Number of Household Pets

Based on the data in Table 2, what is the most popular number of household pets?

You can imagine that if you had 500 unique possible data values, this frequency distribution would be tedious and not much more help than the original list. Consequently, sometimes it is more helpful to group the data into different categories, or **classes**. A frequency distribution of this type is called a **grouped frequency distribution**. When we group data into classes, we have to be mindful of certain important features of the classes.

First of all, the **class width** of a grouped frequency distribution is the number of distinct data points each class may contain. The class widths must be the same for all classes in a grouped frequency distribution. Secondly, the classes should never overlap. In other words, a particular piece of data must have only one possible category into which it could fall.

Suppose that the ages (in years) of children who attended church last week were recorded. Let's walk through creating a grouped frequency distribution of the ages, if they ranged from 0 to 15. We will arbitrarily let the first class contain the ages 0 through 5. This means that there are six possible distinct data points in the first class: 0, 1, 2, 3, 4, and 5. So that our classes do not overlap, but have the same width, the second class should contain the next six distinct points: 6, 7, 8, 9, 10, and 11. Then the final class contains the remaining possible ages: 12, 13, 14, and 15. Since there are only four ages to include in this last class, we must extend the class to include the values 16 and 17 as well. Although our original data didn't contain these last two ages, we must include them so that all of the class widths of the frequency distribution will all be the same.

Age	Number of Children
0–5	10
6–11	14
12–17	7

TABLE 3: Ages of Children Who Attended Church Last Week

The numbers 0, 6, and 12 (found on the left-hand side of the classes) are referred to as the **lower class limits** and the numbers 5, 11, and 17 are the **upper class limits**. The **class width** will always equal the difference between either consecutive lower class limits or consecutive upper class limits. For instance, the difference between consecutive lower class limits is the following.

Lower Class Limits

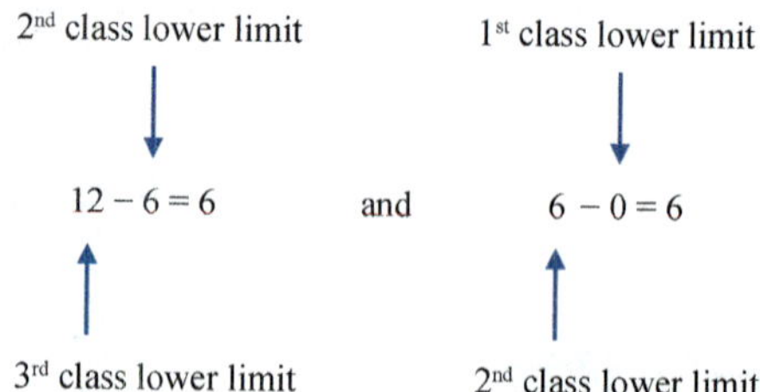

The same is true for the upper class limits.

Upper Class Limits

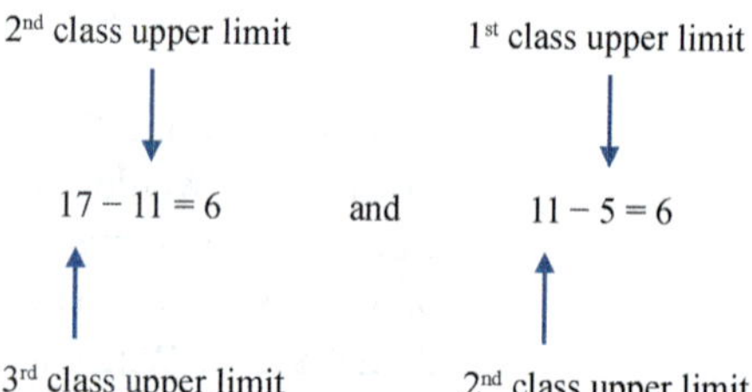

Notice that the class width is 6 for each calculation.

The arbitrary choice of our first class width in the example about the children's ages dictated the layout for the classes that followed. However, the width and the number of classes for any grouped frequency distribution can be adjusted as long as we adhere to the guidelines so the classes maintain equal widths and do not overlap with one another.

Let's try constructing another grouped frequency distribution. Consider the data given and think about how you would divide up the classes so that they make the data easier to digest and analyze. There is really no one correct answer, as long as you follow the guidelines we mentioned. Of course, there will always be cases where the data divides into classes more naturally than others.

Example 3: Constructing a Grouped Frequency Distribution

The number of text messages used per month for 52 individuals was recorded and is listed. Construct a grouped frequency distribution for the data.

357	239	379	381	298	377	130	312	333	287	389	357	328
345	301	278	222	388	391	295	342	327	276	289	367	399
265	298	326	301	333	344	313	389	372	369	328	337	317
352	189	355	318	272	361	340	322	371	358	266	345	207

> **⟲ HELPFUL HINT**
>
> Although choosing the smallest data value as the first lower class limit is easy, it's not necessary to start there. Classes can begin with any number at or below the smallest data value.

Solution

The first thing we need to do is identify the smallest and largest pieces of data to ensure that our distribution will include all of the data values. The smallest number of text messages used was 130 and the largest was 399. When choosing the number of classes, we don't want too many or too few class divisions. Remember, if we have too many divisions, we might as well have listed out each individual piece of data, which is overwhelming. However, if we have too few classes, then the data are all lumped together and don't tell us much. A good rule of thumb is to aim for somewhere between 5 and 20 classes. Since our data set covers a spread of almost 270, a class width of 50 is a good place to start. If our classes each have a width of 50, where should the first class start? Let's try with the smallest data value. By adding the width to 130, we can find each of the lower class limits. Keep adding classes until all data values can be tallied in the distribution. We then have the following start to our table.

Number of Text Messages Used per Month	Frequency
130–	
180–	
230–	
280–	
330–	
380–	
430–	

TABLE 4: Number of Text Messages Used per Month

Since our largest data point is 399, it will safely fall into the class that has a lower limit of 380, since we can now see that the following class would start at 430. Therefore, we do not actually need to add the class beginning with 430, and we can stop with 6 classes. Next, in order to find upper class limits, we can use the lower limits. For the first upper limit, simply count backwards one unit from the lower limit of the second class, making sure the classes don't overlap with one another. For example, the first upper class limit will be $180 - 1 = 179$. We could continue this same process of subtraction to obtain the remaining upper limits. Alternatively, we can simply add the class width of 50 to the upper limit we just found. Either method produces identical upper limits. We then have the following table.

Number of Text Messages Used per Month	Frequency
130–179	
180–229	
230–279	
280–329	
330–379	
380–429	

TABLE 5: Number of Text Messages Used per Month

The only remaining thing to do is to tally the frequencies of each class as shown in Table 6.

Number of Text Messages Used per Month	Frequency
130–179	1
180–229	3
230–279	6
280–329	16
330–379	20
380–429	6

TABLE 6: Number of Text Messages Used per Month

When you complete your grouped frequency distribution, the sum of the frequencies should add up to the total number of data values. Check for yourself that it does in Example 3.

As consumers of information, we're likely to be asked to digest information in the form of a table more often than to construct one. Interpreting data correctly is a crucial step in making informed decisions. Let's look at reading and interpreting a frequency distribution.

Example 4: Interpreting a Frequency Distribution

The San Francisco Bay Area surveyed their residents about the nearly 17 million intraregional daily trips the population makes. Along with other data that they collected, they asked people to identify where their trips originated and how long they traveled. Table 7 organizes the travel data by trip purpose and travel time. The trip purposes recorded in the survey were work, shop, social/recreational, and school. Home-based trips were those that either started or ended at the residence of the trip maker. Non-home based trips were those that neither started nor ended at the residence.

Travel Time (in Minutes)	Home-Based Work	Home-Based Shop (Other)	Home-Based Social/Rec.	Home-Based School	Non-Home Based	Total Purposes
0–5.0	269,350	807,952	272,973	242,906	953,920	2,547,101
5.1–10.0	416,075	881,203	357,196	282,305	917,191	2,853,970
10.1–15.0	886,859	1,217,440	491,636	434,810	1,206,126	4,236,871
15.1–20.0	440,776	316,201	137,957	139,623	331,661	1,366,218
20.1–25.0	283,777	200,142	84,876	103,114	213,205	885,114
25.1–30.0	807,327	391,393	228,241	211,851	455,129	2,093,941
30.1–35.0	155,706	53,270	27,501	40,179	71,154	347,810
35.1–40.0	169,675	50,564	34,060	38,579	79,675	372,553
40.1–45.0	327,028	101,392	60,827	69,478	150,929	709,654
45.1–50.0	88,325	25,253	16,074	11,509	30,240	171,401
50.1–55.0	56,228	15,686	10,876	13,804	25,282	121,876
55.1–60.0	227,844	76,162	48,138	38,221	94,638	485,003
60.1–65.0	45,851	10,352	5287	3284	16,144	80,918
65.1–70.0	42,824	11,974	7697	3952	14,434	80,881
70.1–75.0	81,350	20,182	14,642	10,800	35,675	162,649
75.1–80.0	22,123	5876	3311	3321	9344	43,975
80.1–85.0	13,594	3852	3121	2495	4950	28,012
85.1–90.0	50,220	13,375	12,996	4947	27,684	109,222
90+	68,457	29,496	26,402	9446	47,242	181,043
Total	**4,453,389**	**4,231,765**	**1,843,811**	**1,664,624**	**4,684,623**	**16,878,212**

Source: National Transportation Library. "San Francisco Bay Area 1990 Regional Travel Characteristics - WP #4 - MTC Travel Survey." http://ntl.bts.gov/DOCS/SF.html

TABLE 7: Number of Intraregional Trips

Answer the following questions about the frequency chart and its data.

a. How many classes are there in the frequency distribution?

b. Find the class width. (Exclude the first and last classes.)

c. Which guideline is ignored in this published frequency distribution? What sorts of questions arise from this inconsistency?

d. Which class contains the largest number of school-related trips, and what is its frequency?

e. Is the statement "Most people take 10–15 minutes to travel to school in the San Francisco Bay Area" a true statement?

Solution

a. There are 19 classes in this frequency distribution. This can be found by counting the number of data rows in the frequency chart.

b. The class width is 5 minutes and can be found by finding the difference between the lower class limits (or the upper class limits) of two consecutive classes, excluding the first and last class.

c. The classes are not all of equal width. In fact, the last class might go on forever! Since this class contains more than 181,000 pieces of data (more than some of the other classes) it cannot be ignored. However, we don't know if the data lie within 5 minutes, 10 minutes, or 2 hours of the 90-minute mark. From the table, there is no way to know.

d. The class with the largest number of trips that are school related is the 10.1–15.0 minute class with a frequency of 434,810.

e. Although it is the largest class in that category, in order to say "most people" the class would need to contain more than half of all the people who travel to school. Since we can see that a total of 1,664,624 people travel to school each day, 434,810 is not more than half of these. Therefore, the statement is not true. Would you fall into the trap of saying that statement yourself?

Skill Check 2

Use the frequency distribution from Table 7 to determine how many people take between 10.1 and 25 minutes to travel.

Although a frequency distribution is a reasonable place to start for displaying data, sometimes even a table can contain an overwhelming amount of information. That's when graphical displays such as pie charts, bar graphs, histograms, and line graphs help to convey the data more clearly. Because technology helps us make all of these types of graphs with relative ease, we'll focus our attention on interpreting the graphs.

Pie Charts

When we want to show a comparison between part of the data and the whole, we use a **pie chart**, also called a **circle graph**. A pie chart shows how large each category is in relation to the whole; that is, it uses the relative frequency distribution to divide the "pie" into different-sized wedges. The size of each wedge in the pie chart is determined by the measure of the central angle of the wedge. This central angle is calculated by multiplying the relative frequency of each class by 360° and rounding to the nearest whole degree.

Pie Chart

A **pie chart** shows how large each category is in relation to the whole.

Consider the following pie chart of religious affiliations of adults in the United States made from data gathered by the Pew Research Center. Notice that, although there is quite a bit of data being displayed, the information is visually clear and easy to read.

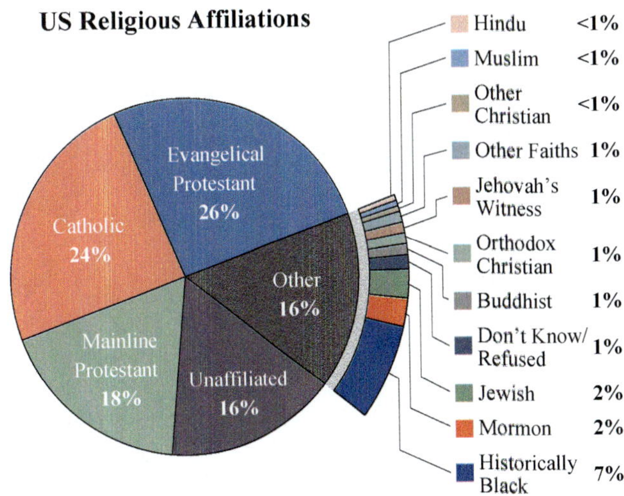

Source: Pew Research Center's U.S. Religious Landscape Survey. "Religious Affiliation: Diverse and Dynamic." Accessed February 2008. http://religions.pewforum.org/affiliations

FIGURE 1

Using the graph, we can easily answer questions along the lines of which religion has the largest US adult following. Obviously it's between the two largest pie slices, Catholics and Evangelical Protestants. By looking at the percentages, we know that Evangelical Protestants have a larger group with 26% of the adult population. However, as we critically look at the graph, one question you might consider is how large of a study was done? Luckily for us, the Pew Research Center gives an exact description of their survey methodology on their website.

"The US Religious Landscape Survey completed telephone interviews with a nationally representative sample of 35,556 adults living in continental United States telephone households. The survey was conducted by Princeton Survey Research Associates International (PSRAI). Interviews were done in English and Spanish by Princeton Data Source, LLC (PDS), and Schulman, Ronca, and Bucuvalas, Inc. (SRBI), from May 8 to Aug. 13, 2007. Statistical results are weighted to correct known demographic discrepancies."[1]

Together, the pie chart and survey methodology give us a better picture of the study that was done. We could have conveyed some of this information on the pie chart by stating how many adults were surveyed when creating the chart. There is a fine line between keeping the information displayed clearly and giving too little information. As a general rule, more information is always better than less.

Let's look at another graph.

---

1 Pew Forum on Religion & Public Life. "U.S. Religious Landscape Survey. Appendix 4: Survey Methodology." Accessed January 2012. http://religions.pewforum.org/pdf/report-religious-landscape-study-appendix4.pdf

Entry-Level Salaries

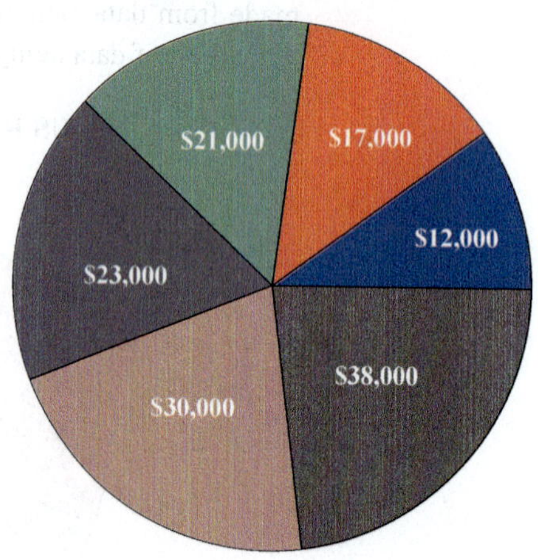

Total Number of Professions Surveyed = 541

FIGURE 2

In this pie chart, the percentages are not given, but instead, each wedge is labeled with the salary level. At a glance, is it possible to tell which salary level occurred most? Certainly we know that it's either the $30,000 or $38,000 wedge that represents the most common salary. In order to determine the underlying frequencies of the original data, we either need the percentages of each category or the angles of the wedges. Figure 3 shows the same graph as Figure 2, but with more information. This enables us to answer more questions about the graph.

Entry-Level Salaries

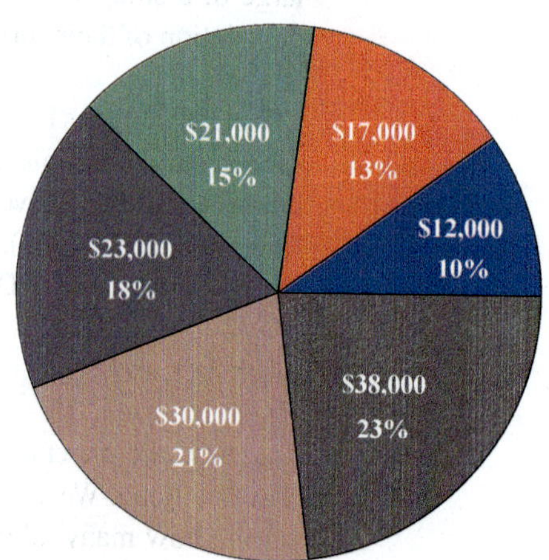

Total Number of Professions Surveyed = 541

FIGURE 3

With this new information, it is clear that the $38,000 salary occurred the most.

Let's calculate how many of the entry-level salaries were $30,000. Since we know that there were 541 salaries in the study, we can multiply 541 by the appropriate percentage for $30,000, which is 21%. Don't forget to change the percentage to a decimal in your calculation.

$$541(21\%) = 541(0.21) = 113.61$$

Consider this answer. We know that there couldn't have been a decimal number of salaries in this category, so we can conclude that the percentages were rounded off when constructing the pie chart. We'll do the same with our answer and estimate that there were 114 salaries that fell into the $30,000 category.

What percentage of salaries were above $25,000 in the study? To answer this question, we can simply add together the percentages for all categories that are above $25,000.

$$23\% + 21\% = 44\%$$

Therefore, 44% of salaries were above the $25,000-level in the study.

Is it then correct to say that most jobs in the study had entry-level salaries that were more than $25,000? Because the percentage is not above 50%, it is incorrect to say that "most salaries" were more than $25,000. However, it would be correct to say that "almost half" of the salaries were more than $25,000.

Bar Graphs

One of the most common ways to display categorical data is to use a **bar graph**. Like a pie chart, a bar graph represents the amount of data in each category. However, it uses either vertical or horizontal bars instead of parts of a circle.

> **Bar Graph**
>
> A **bar graph** is a chart of categorical data in which the height of the bar represents the amount of data in each category.

One advantage of bar graphs over pie charts is the ability to see small differences between individual categories. Thus, we might choose to use a bar graph instead of a pie chart if we wish to compare different categories that have similar amounts of data in them. It's also a useful tool if we are more interested in the frequencies in individual categories rather than how the categories compare to the whole.

When creating a bar graph, one axis displays the categories of data and the other axis displays the frequencies. Traditionally, for vertical bar graphs, the horizontal axis contains the categories and the vertical axis represents the frequencies. Because each bar represents a category, the width of the bar along the horizontal axis does not relay anything about the data itself. So that we don't cause misunderstandings about the data, the bars should be of uniform width and not touching one another. If the bars were different widths, readers might infer that one category held more significance than another. Similarly, bars touching or overlapping in some way might suggest that the categories were not completely separate from one another. The height of the bar should equal the frequency (or relative frequency) of the category.

The following horizontal bar graph shows the percentage of adults in each of four generational categories who think the United States is the greatest country in the world. According to the bar graph, which generation has the largest percentage of approval amongst its age group? Which generation has the second largest approval?

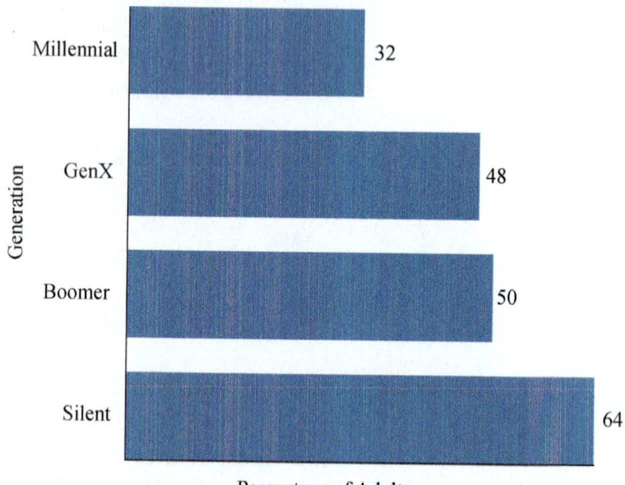

Generational Divide Over American Exceptionalism
% saying the United States is "the greatest country in the world"

Source: Pew Research Center. "The Generation Gap and the 2012 Election, Section 4: Views of the Nation." Accessed January 2012. http://www.people-press.org/2011/11/03/section-4-views-of-the-nation/

Solution

We can easily scan the graph to see that the Silent generation has the largest percentage of approval amongst its age group. 64% of people in the Silent generation feel that the United States is the greatest country in the world.

Which generation has the second largest percentage of approval? To answer this question, we'll need to look a little closer at the labels on the ends of the bars to know the precise percentage of each category. The Boomer generation was second with 50% of its generation believing that the United States is the greatest country in the world.

Bar graphs are also commonly used to show a comparison of categories between multiple populations. We can either use a **side-by-side bar graph** or a **stacked bar graph** to show these comparisons. A side-by-side bar graph is just as the title suggests. The bars are placed next to one another to show the similarities and/or differences between populations. A stacked bar graph places the bars from each population on top of one another. This allows the reader to see a combined total for each category, yet at the same time, see the individual frequencies for each population.

The Surveillance, Epidemiology, and End Results (SEER) Program of the National Cancer Institute works to provide information on cancer statistics in an effort to reduce the burden of cancer among the US population. In order for the results of their studies to have the best impact, SEER researchers need representative samples of the US population. The following side-by-side bar graph shows the demographic breakdown of participants in a SEER study versus the same demographic breakdown in the overall US population. Which two subpopulations are more represented in the SEER group?

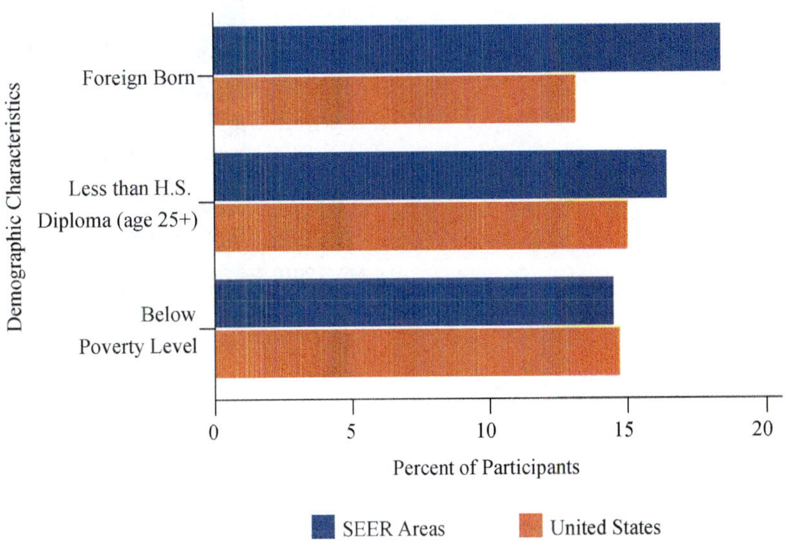

Demographic Characteristics of the SEER Population vs. the Total United States Population

Source: Surveillance, Epidemiology, and End Results Program. "Population Characteristics." Accessed April 2012. http://seer.cancer.gov/registries/characteristics.html

Solution

The bars are obviously longer for SEER than for the United States in both the Foreign Born category and the High School Diploma category, meaning that these have more representation in the SEER group than the general US population. However, if we look carefully at the Poverty Level category, although the percentages are very close, the SEER group slightly underrepresents this demographic.

Is it possible to determine how many people are in the SEER group using the figure from Example 6? Unfortunately, it is not possible. However, by looking at their website we know that there were over 86,000,000 participants in SEER. When producing any type of graph, it's always a good idea to make sure that pertinent information is included for the reader; for example, the size of the sample.

A **stacked bar graph** not only compares different categories, but how the whole of each category is broken down. For instance, a college administrator might want to know how many spaces are available for the most popular classes on campus in the coming spring semester. The graph in Figure 4 shows not only how many total seats will be available for each class, but also how many are already filled. Notice the other categories in which each class is broken down.

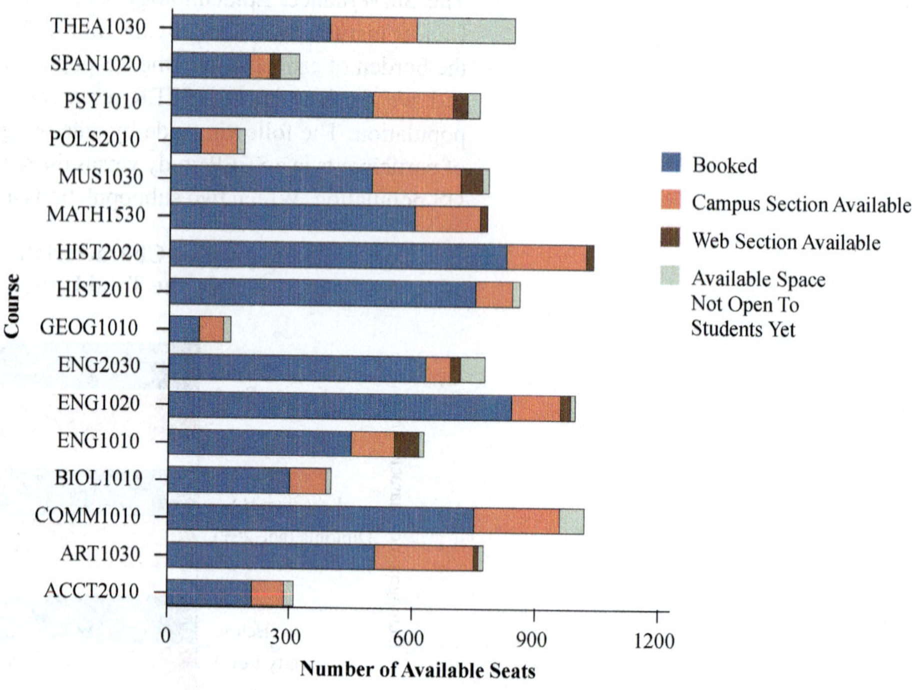

FIGURE 4

Let's take a closer look at the course ENG1010. This is a freshman-level English course required by all liberal arts majors at the college. We can see that there are approximately 625 seats that could be made available for this class for the spring semester. Of those 625 seats, the first section of the bar indicates that a little more than 450 are already booked. The second section tells us the available seats in physical classes that are offered on the campus; there are about 100 of these left. The third section shows the number of spaces available in web classes, which is approximately 50. Finally, we can see by the last section that there are very few spaces to which students do not already have access.

A graph that contains this much information for each course would be good for an administrator who is trying to get an overall picture of the enrollment at his institution. However, if he required the exact numbers of spaces in the courses, this may not be the best way to convey the information.

Histograms

A special type of bar graph that displays the frequency distribution of numerical classes is called a **histogram**. There is a subtle difference between a standard bar graph and a histogram. In a histogram, the height of each bar represents the frequency of the corresponding class. Because the bars represent classes, and class limits are consecutive numbers, the bars touch in a histogram, as shown in Figure 5.

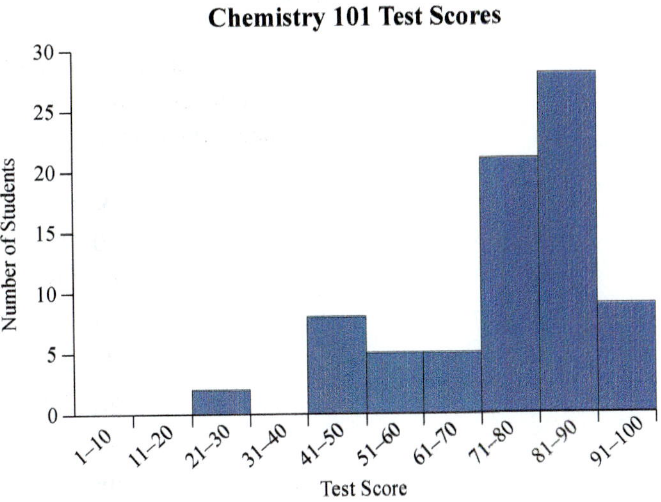

FIGURE 5

Example 7: Interpreting a Histogram

The following histogram displays data collected from delivery truck drivers. They were asked to record their wait times when an address had a closed entrance gate. Use the histogram to answer the questions that follow.

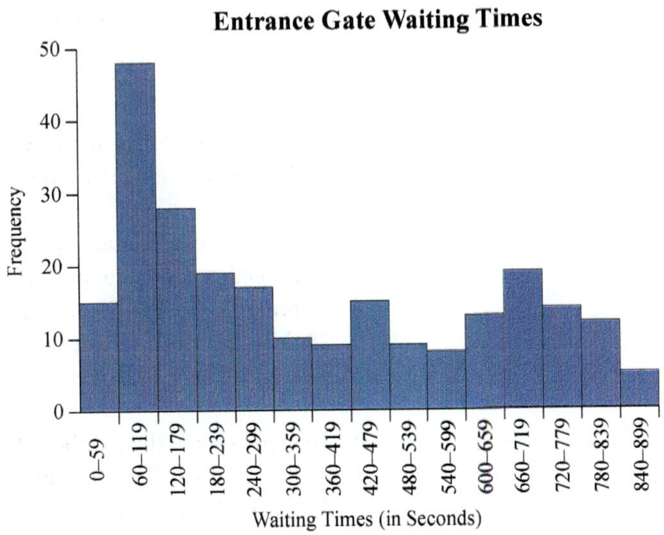

a. Determine approximately how many truck drivers were in the survey?

b. Approximately how many drivers waited between 60 seconds and 180 seconds?

c. Approximately how many drivers waited more than 7 minutes?

d. Is it accurate to say that about half the drivers waited more than 420 seconds? Why?

Solution

a. To determine the number of drivers surveyed, add together each of the frequencies for the different classes. We'll approximate the frequency of each class first and then add them together.

Class	Frequency
0–59	15
60–119	48
120–179	28
180–239	19
240–299	17
300–359	10
360–419	9
420–479	15
480–539	9
540–599	8
600–659	13
660–719	19
720–779	14
780–839	12
840–899	5
Total	**241**

TABLE 8: Entrance Gate Waiting Times

So, we estimate that there were approximately 241 truck drivers in the survey. We'll use these estimates in the following questions.

b. To determine how many drivers waited between 60 and 180 seconds, we need to add together the estimated frequencies of each of the bars between those times. Approximately 48 drivers waited between 60 and 120 seconds, and approximately 28 drivers waited between 120 and 180 seconds. Therefore, we can estimate that about $48 + 28 = 76$ drivers waited between 60 seconds and 180 seconds.

c. To approximate the number of drivers who waited more than 7 minutes, we need to add together all of the bars to the right of 7 minutes. We know that 7 minutes is 420 seconds, so we'll add the frequencies of the last eight bars together.

$$15 + 9 + 8 + 13 + 19 + 14 + 12 + 5 = 95$$

Therefore, there were an estimated 95 drivers who waited more than 7 minutes at the entrance gates.

d. Although 420 seconds represents the middle waiting time for the classes, it is inaccurate to say that half of the drivers recorded times less than this time and half recorded longer times. Instead, we need to think about the numbers of drivers. If there were approximately 241 drivers, then half would be around 120 drivers. We can add together the class frequencies until we reach approximately 120 and then note the waiting time.

The first two classes added together are $15 + 48 = 63$.

Adding the third class gives us $63 + 28 = 91$.

Continuing with the fourth class, we get $91 + 19 = 110$.

This is not quite the 120 that we're looking for, so we can include the fifth class as well: $110 + 17 = 127$. The fifth class has a lower limit of 240 seconds, so it is more accurate to say that about half of the drivers waited more than 240 seconds at entrance gates.

Line Graphs

When your data consists of measurements over time, it is best to display these data using a **line graph**. In a line graph, the horizontal axis represents time. The vertical axis represents the variable being measured. Each point on a line graph represents a data value and the appropriate time period it was observed. By joining the points together with line segments, changes over time are more easily observed. It is these line segments that give the graph its name.

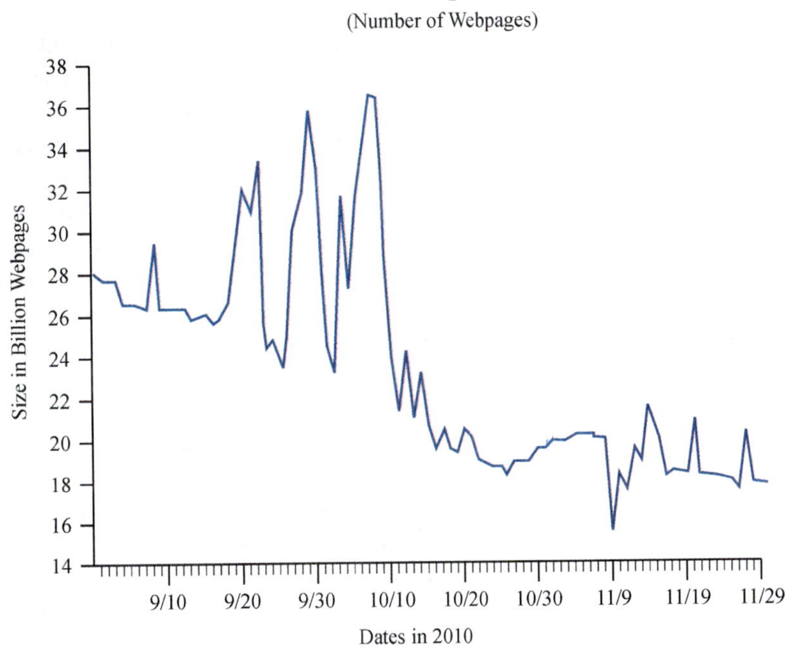

Source: WorldWideWebSize.com. "The size of the World Wide Web (The Internet)." Accessed December, 2011. http://www.worldwidewebsize.com

FIGURE 6

From this line graph of the size of Google, we can see that, overall, the number of web pages decreased over the time period shown from September 1, 2010 to November 29, 2010. Approximately when did the number of pages peak for Google? Determining this date is not as easy as you might think. Although it's easy enough to see the peak, the way the horizontal tick marks are displayed on the graph makes it difficult to easily identify the date of the peak. A good guess would be somewhere around October 8, 2010. Making every 7th tick mark (a week) bold or identifying the first day of the new month on the tick marks would help tremendously with reading and interpreting the graph. All of these small details are helpful to remember when you're the one constructing the graph.

The next line graph shows how you can have data from two different sources with different scales on the same graph.

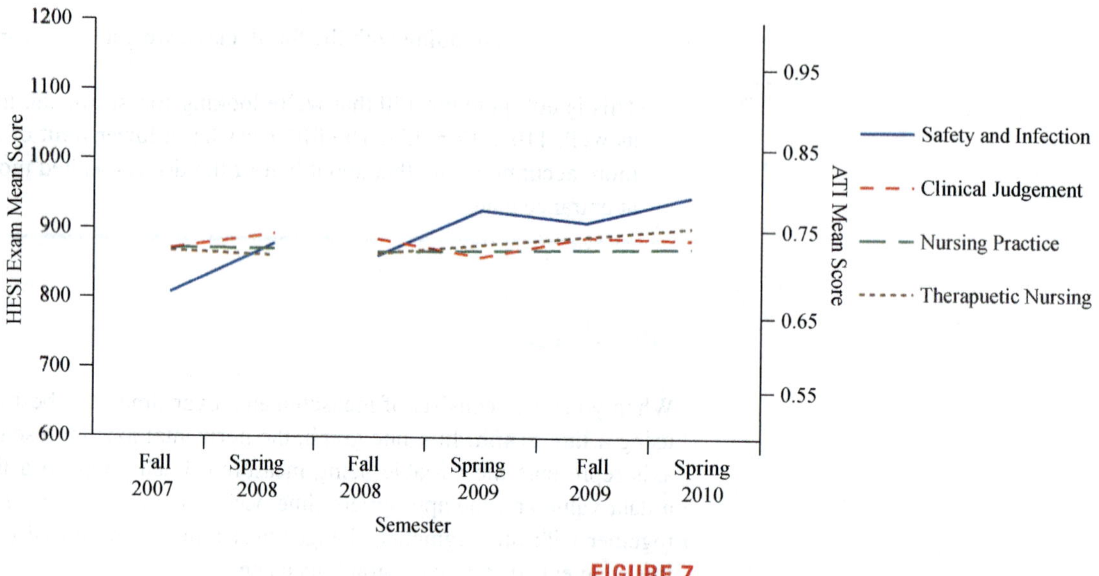

FIGURE 7

Each of the colored lines represents a content strand of knowledge that the nursing department is assessing in their students. However, in the summer of 2008, the test they used for the assessment was changed from the HESI to the ATI. The *y*-axis on the *left* shows the scale out of 1200 points on the HESI test that was given until Spring 2008. The *y*-axis on the *right* shows that the new test, ATI, was based on percentages. By carefully arranging the two scales so that points on the HESI test correspond to percentages on the ATI test, the sequences of outcomes before and after the change of tests can still be appreciated. By placing both test results on a single graph, the nursing department was able to visually show the progress of their students over time, even though the mode of assessment changed. Leaving a break in the lines helps to emphasize the fact that a change in tests was made.

The Power of Graphs

To show how powerful the visual impact of data can be, take a look at the 1861 map of the distribution of the slave population of the Southern United States made from a census taken in 1860, shown in Figure 8. It was one of the first maps to present statistical data, and not just topography and geography. It showed President Lincoln the layout of the Southern States by slave population and played a role in shaping his views about slavery and the framing of the Emancipation Proclamation. The map is even included in a portrait of President Lincoln with his cabinet.

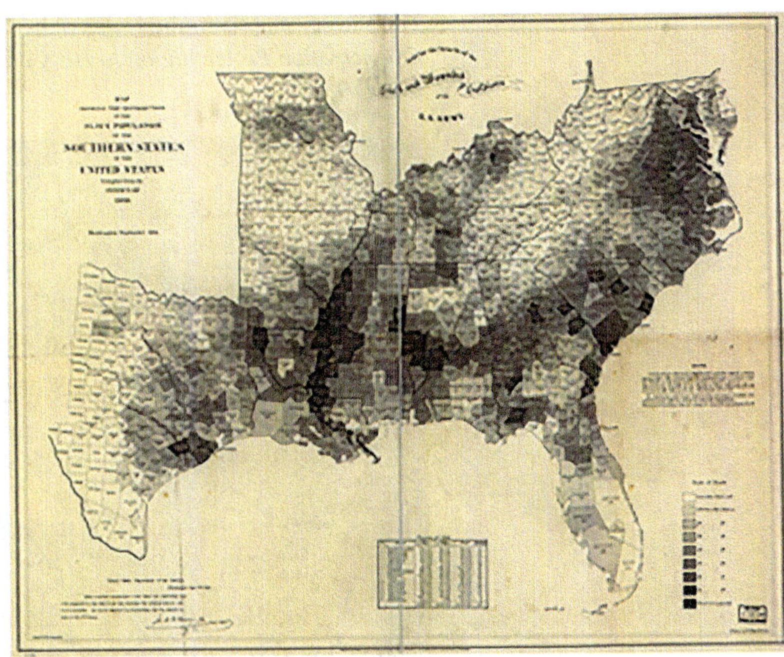

FIGURE 8: Distribution of the Slave Population of the United States, 1861

The data in the table on the map is difficult to read, so we include the data presented there as Table 9.

No.	States	Free Population	Slave Population	Total Population	% of Slaves
1	South Carolina	301,271	402,541	703,812	57.2
2	Mississippi	354,700	436,696	791,396	55.1
3	Louisiana	376,280	333,010	709,290	47.0
4	Alabama	529,164	435,132	964,296	45.1
5	Florida	78,686	61,753	140,439	43.9
6	Georgia	595,097	462,232	1,057,329	43.7
7	North Carolina	661,586	331,081	992,667	33.4
8	Virginia	1,105,192	490,887	1,596,079	30.7
9	Texas	421,750	180,682	602,432	30.0
10	Arkansas	324,323	111,104	435,427	25.5
11	Tennessee	834,063	275,784	1,109,847	24.8
12	Kentucky	930,223	225,490	1,155,713	19.5
13	Maryland	599,846	87,188	687,034	12.7
14	Missouri	1,067,352	114,965	1,182,317	9.7
15	Delaware	110,420	1798	112,218	1.6
		8,289,953	**3,950,343**	**12,240,296**	**32.2**

Source: NOAA Office of Coast Survey, "Map Showing the Distribution of the Slave Population of the Southern States of the United States 1860," http://historicalcharts.noaa.gov/historicals/preview/image/ CWSLAVE

TABLE 9: Census of 1860

The slavery map is in the lower right of the painting *President Lincoln Reading the Emancipation Proclamation to His Cabinet* by Francis Bicknell Carpenter, shown in Figure 9.

Source: *President Lincoln Reading the Emancipation Proclamation to His Cabinet* by Francis Bicknell Carpenter. http://www.noaanews.noaa.gov/stories2011/images/cgs05195.jpg

FIGURE 9: *President Lincoln Reading the Emancipation Proclamation to His Cabinet* by Francis Bicknell Carpenter

Example 8: Reading Graphs

Considering the map of the distribution of the slave population of the Southern United States in Figure 8 and the data in Table 9, answer the following questions. The darker the shading is on the map, the higher the slave population was in that area.

a. Which states had a higher slave population than free population in the 1860 census?

b. Which state had the highest number of slaves in 1860?

c. One of the reasons that made the map so popular was the visual account of slavery by shading. Which areas had the heaviest concentration of slaves?

Solution

a. Using the frequency distribution of the 1860 census found on the bottom part of the map, we can see that South Carolina and Mississippi had a higher slave population than free population.

b. Although only 30.7% of its population were slaves, Virginia had the highest slave population at 490,887.

c. The darkest shadings occur along the Mississippi River, along the coast of South Carolina, in central Alabama, and in the middle of Georgia.

Misleading Graphs

Spirit Week Participation

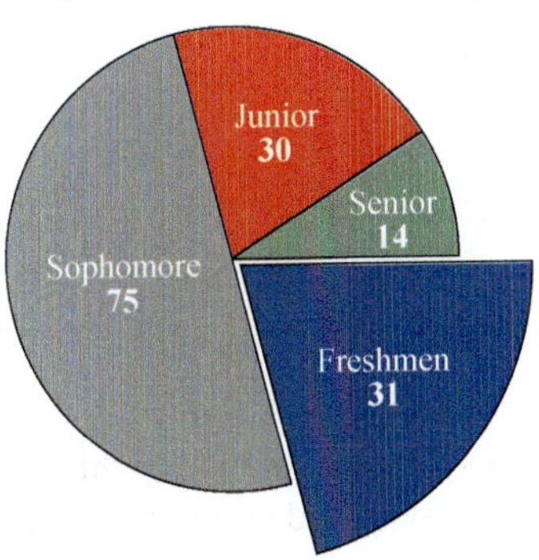

FIGURE 10

As we said at the beginning of the section, it's important to display data so that it is organized clearly and it effectively conveys the intended message. Ideally, graphs should be able to stand alone without the need for additional information in order to be understood. However, sometimes graphs either intentionally or unintentionally convey the wrong message about data or are not quite clear enough to get their message across. It's important to be aware that there are visually misleading and/or ambiguous graphs out there. For example, making the bars different widths on a bar chart might imply that one category is somehow larger than another. Similarly, a distorted piece of a pie graph might inaccurately lead the reader to assume that one section of data is larger than another. An example of this is the graph in Figure 10.

In their graph, the freshmen wanted to emphasize the fact that they came in second place for spirit week—although they just grabbed second place by a tiny margin. If the exact numbers were not on the graph, the emphasis on the freshmen piece of pie might visually imply that the freshman wedge is considerably larger than the junior wedge. Watch out for these visual manipulations when interpreting graphs.

Skill Check Answers

1. 0 pets   **2.** 6,488,203 people

9.2 EXERCISES

✔ CONCEPT CHECK

Fill in each blank with the correct term.

1. A _________ is a graph that represents a frequency distribution.

2. When comparing parts of data to the whole, a ________ visually shows this using sections of a circle.

3. A ___ graph is best to use when showing data over a time period.

4. When comparing multiple categories from different populations, a ____________ ___ graph or a _________ graph can be used.

5. A __________________ is a literal count of each member of a data set and how often it occurs.

APPLICATIONS

6. The grades on the first statistics test for Ms. Seago's class are listed in the following table. Construct a frequency distribution for the grades.

Grades on Statistics Test 1

A	C	F	C	C	D	F
B	D	F	B	A	A	F
B	C	C	A	B	F	D

7. The following table gives the grouped frequency distribution of weights of 194 babies in kilograms. Answer the questions that follow based on the distribution.

Weights of Babies

Birth Weight (kg)	0.00–0.99	1.00–1.99	2.00–2.99	3.00–3.99	4.00–4.99	5.00–5.99
Frequency	2	17	39	89	46	1

   a. How many classes are in the grouped frequency distribution?
   b. What is the class width?
   c. What is the value of the lower class limit of the 3rd class?
   d. What is the value of the upper class limit of the 5th class?
   e. What is the relative frequency of the 4th class? Give your answer as a percentage rounded to the nearest tenth.

8. The following table represents a grouped frequency distribution of the number of hours spent on the computer per week for 55 students.

Hours	Number of Students
0.0–4.4	9
4.5–8.9	14
9.0–13.4	21
13.5–17.9	11

   a. Calculate the relative frequencies (as percentages rounded to the nearest tenth) for each class.
   b. What percentage of the students used the computer between 9 and 13.4 hours per week?
   c. What percentage of the students used the computer less than 9 hours per week?

9. Sisscon is a phone answering service. The following data are the numbers of calls per day reported by the company for the last month.

10	72	64	32	78	62	11
37	45	32	52	38	70	66
13	21	14	13	39	73	62
41	63	44	23	27	22	21
55	24	53	43	20	16	22

   a. Create a grouped frequency distribution for the data using 8 classes and then use it to answer the following questions. Let the first lower class limit be 0 and the class width equal 10.
   b. Calculate the relative frequencies for each class. Give your answers as percentages.
   c. For what percentage of the days is the number of calls between 40 and 49?
   d. For what percentage of the days is the number of calls in the single digits?
   e. What is the most common range for the number of calls per day?

10. Consider the bar graph, constructed from a 2012 study, of the predicted fastest growing occupations between 2010 and 2020.[1]

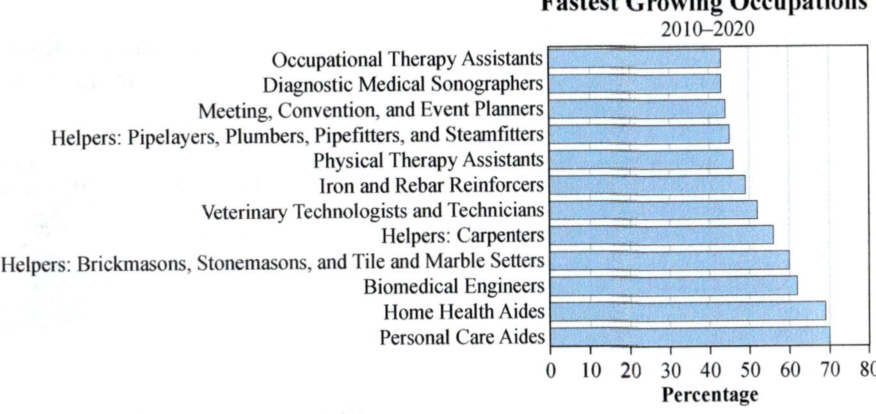

   a. At a quick glance, which occupation was predicted to grow the most between 2010 and 2020? What was the predicted amount of growth?
   b. How many new jobs were predicted to be available in the fastest-growing occupation in 2020?

---

1 BLS Occupational Outlook Handbook, http://www.bls.gov/ooh

11. Answer the following questions about the Job Growth graph.[2]

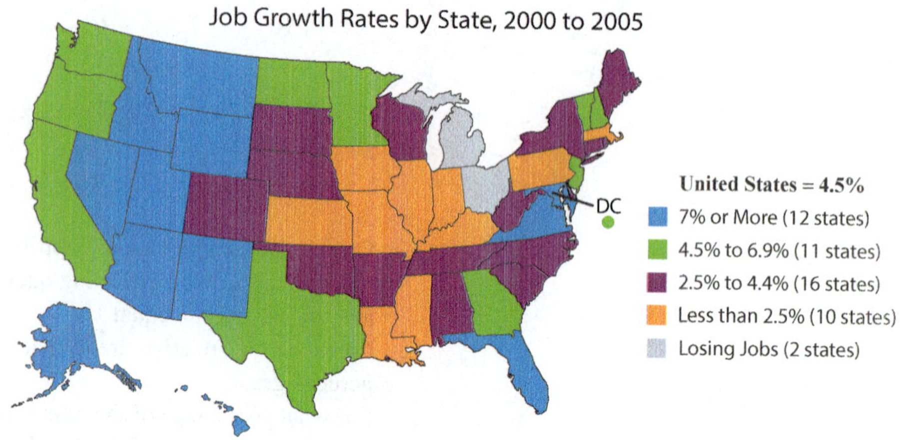

   a.   How many states lost jobs between 2000 and 2005?

   b.   In what part of the country were jobs growing the most in this time period?

12. The following pie chart shows destinations of recent University of Kent mathematics graduates, including business, financial math, and statistics.[3]

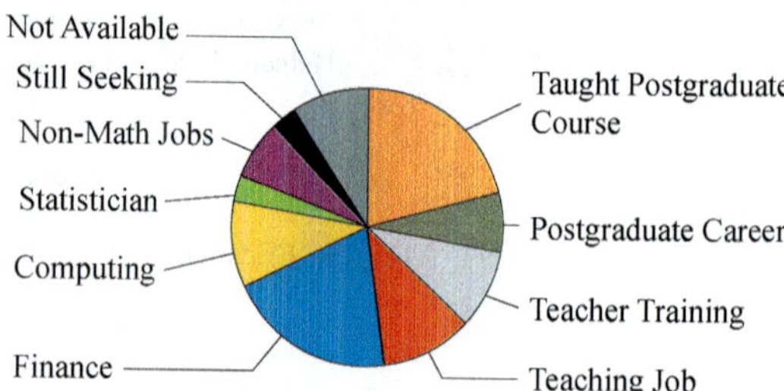

   a.   Does it appear that any destination category accounts for more than 25% of the graduates?

   b.   Which pairs of destinations appear to have similar percentages of graduates?

   c.   How many mathematics graduates from the University of Kent were surveyed?

   d.   Is the graph misleading in any way?

---

2 InContext, http://www.incontext.indiana.edu
3 University of Kent, http://www.kent.ac.uk

13. The following graph shows the US unemployment rate from February 2011 to January 2013.[4]

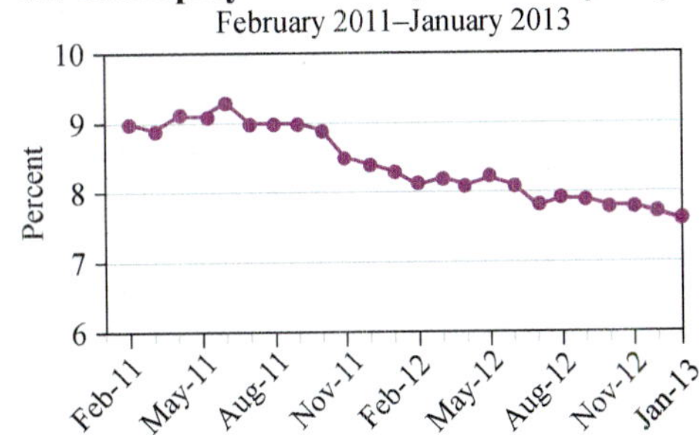

   a. Describe the trend of the unemployment percentage from February 2011 to January 2013.

   b. Approximate the month and rate of the highest unemployment during this time period.

   c. Approximate the month and rate of the lowest unemployment during this time period.

   d. Is the graph misleading in any way?

14. The following is a portion of a table about state health facts. It lists 8 of the 50 states along with the percentage of women age 50 and older who report having had a mammogram between 2008 and 2010.

Percentage of Women Age 50 and Older Who Had a Mammogram between 2008 and 2010

Massachusetts	87.50%
Connecticut	83.80%
North Carolina	81.20%
Virginia	79.10%
Alabama	77.60%
Mississippi	70.90%
Nevada	69.90%
Idaho	68.30%

Source: Centers for Disease Control and Prevention (CDC). Behavioral Risk Factor Surveillance System Survey Data. Atlanta, Georgia: U.S. Department of Health and Human Services, Centers for Disease Control and Prevention, 2010, available at http://apps.nccd.cdc.gov/brfss/list.asp?cat=WH&yr=2010&qkey=4427&state=All

Is the following pie chart a good way to display this data? Explain why or why not.

Percentage of Women Age 50 and Older Who Had a Mammogram between 2008 and 2010

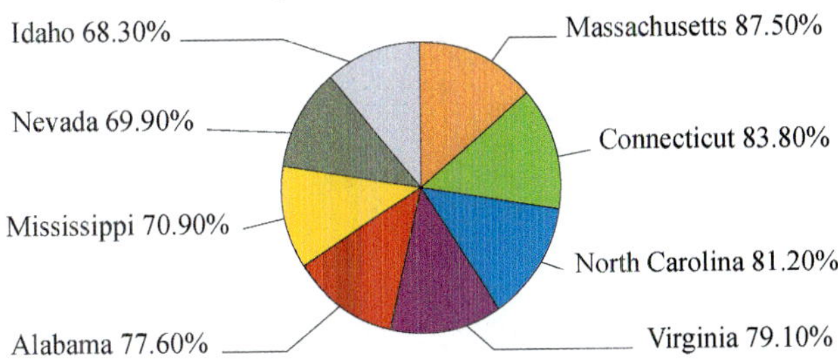

---

4 Bureau of Labor Statistics, http://www.bls.gov

15. The stacked bar graph shows the average number of hours that married people in Japan spend each day doing various activities.[5]

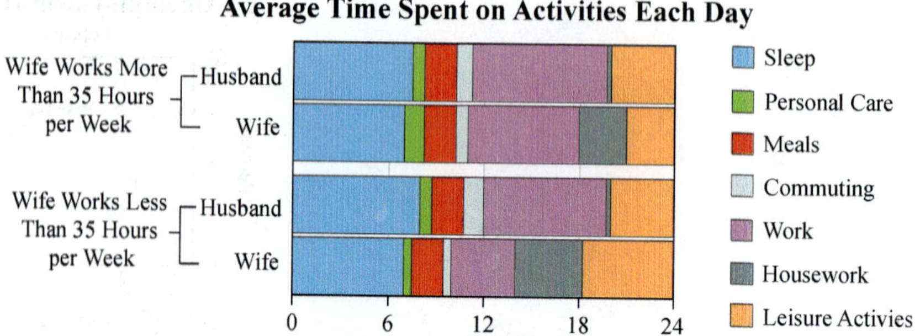

a. For wives who spend less than 35 hours per week working, how many hours on average are spent each day for leisure activities?

b. For husbands whose wives work more than 35 hours per week, approximately how many hours on average are spent on sleep?

c. Compare the number of hours spent sleeping for wives in each category.

d. What type of graph could be used to represent these data in a clearer fashion?

---

5 Statistics Bureau (Japan), http://www.stat.go.jp

9.3 DESCRIBING AND ANALYZING DATA

■ TOPICS

- ■ Measures of Central Tendency: Mean, Median, and Mode
- ■ Measures of Dispersion
- ■ The Empirical Rule
- ■ Measures of Relative Position

As you can tell from the last section, displaying data in a clear and informative way is certainly an important and necessary step in research. Just as important is describing the data numerically, so that they can be compared and analyzed. Imagine if the only way we had of comparing data sets that contain thousands of data points was to list each individual point; we would all throw our hands in the air and scream in frustration. The data would be cumbersome and useless. However, because of statistical tools like measures of central tendency and measures of dispersion (all of which we'll cover in this section) we can begin to get a better picture of what the data are able to tell us by using summary numbers.

Measures of Central Tendency: Mean, Median, and Mode

A number that best describes a typical value in a data set is referred to as a **measure of central tendency**. We'll begin our discussion of measures of central tendency with the most common ways to describe the "center" or "average" of the data: the mean, median, and mode.

Mean

Often we hear people refer to the "average" of a set of data, as in the average age of first-time parents or the average score on a test. Using the term average can refer to any of three measures of central tendency: the mean, the median, or the mode. Most commonly, the use of the word average is in reference to the **arithmetic mean**.

HELPFUL HINT

When calculating the mean, round to one more decimal place than the largest number of decimal places given in the data. Occasional exceptions to this rule can be made when the type of data lends itself to a more natural rounding method such as rounding values of currency to two decimal places.

Arithmetic Mean

The **mean** is the sum of all of the data values divided by the number of data points. Formally, the formula for the **population mean** is

$$\mu = \frac{x_1 + x_2 + \cdots + x_N}{N}.$$

The formula for the **sample mean** is

$$\bar{x} = \frac{x_1 + x_2 + \cdots + x_n}{n}.$$

x_i is the ith data value, N is the number of data values in the population, and n is the number of data values in the sample.

Example 1: Finding the Mean

A sample of the number of sick days employees at Witt's Insurance Agency took during last year is listed below. Calculate the mean of the sample data.

$$14, 5, 7, 11, 9, 7, 12, 6$$

Solution

There are 8 pieces of sample data, so in order to find the sample mean, add all the values together and divide by 8.

$$\bar{x} = \frac{14+5+7+11+9+7+12+6}{8} = \frac{71}{8} \approx 8.9$$

Therefore, the mean of this sample is 8.9.

📈 TECH TRAINING

When calculating the mean using a TI-30XIIS/B calculator, always remember to clear the data list first.

After clearing the data list, enter the data from Example 1 into the calculator using the following commands.

1. Press [2nd] [data].

2. Choose 1–VAR and press [enter].

3. Press [data]; (X1= should appear.

4. Enter a data value and press the down arrow key twice, since the frequency is 1 for each data point.

5. After the last data point is entered, press [enter].

To calculate the mean, press [statvar]. The top of the screen will display a list of values that the calculator computed using the data entered. Use the left arrow key to move the cursor to $\bar{x}$. The mean of the data will be displayed as 8.875.

To calculate the mean using the list function on a TI-83/84 Plus calculator, clear the data list and then perform the following instructions.

1. Press [stat] then choose 1:Edit..., and enter your data in L1.

2. Press [stat] again and now scroll to the right to CALC.

3. Choose option 1:1-Var Stats and press [enter]. If your data are in L1, press [enter] again since L1 is the default list. If you did not type your data in L1, enter the list where your data are located, such as L2 or L3. (These list names are in blue, above the numeric keys.)

🖐 HELPFUL HINT

To clear the data list on the TI-30XIIS/B, press [2nd] [data] and scroll over to CLRDATA and press [enter].

For the TI-83/84 Plus, press the up arrow to highlight the list name, press [clear], and then [enter].

🖐 HELPFUL HINT

If you are using a TI-30XS/B Multiview calculator, the keystrokes will be slightly different for your calculator.

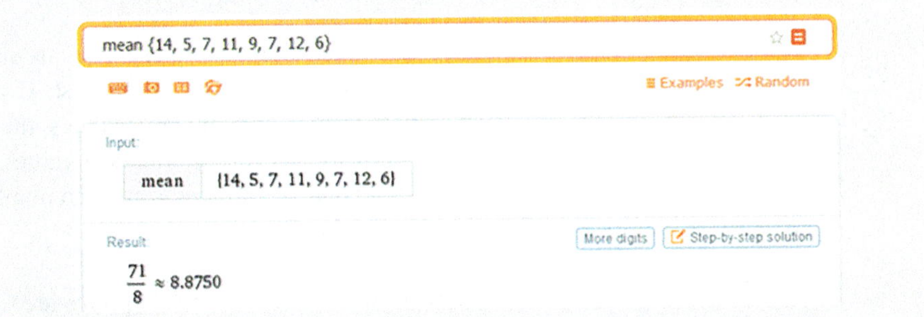

The mean is labeled as $\overline{x}$ in the calculator output, as seen in the screenshot. Therefore, $\overline{x} = 8.875$.

Wolfram|Alpha can be used to calculate the mean of a set of numbers. Go to www. wolframalpha.com and type "mean {14, 5, 7, 11, 9, 7, 12, 6}" into the input line. Then, click the = button. Wolfram|Alpha will return the following.

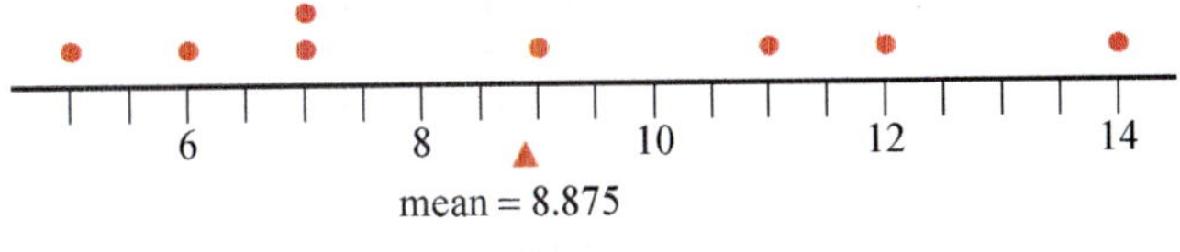

Source: Wolfram Alpha LLC. 2009. Wolfram|Alpha. http://www.wolframalpha.com/input/?i=mean+%7B14%2C+5%2C+7%2c+11%2C+9%2C+7%2c+12%2C+6%7D (accessed July 3, 2014).

Notice that the mean number of sick days in Example 1 isn't actually a member of the sample set. No employee took 8.875 sick days last year. However, it is a description of all the data points.

A nice image to have of the mean is that of a seesaw or teeter-totter. If all the data points were placed on the seesaw with even weights and the pivot point was at the mean, the seesaw would balance evenly. We've illustrated this in Figure 1 with the data from the previous example.

mean = 8.875

FIGURE 1

It's important to note that it is meaningless to find the mean of data that have no measurable values, such as the ratings of a hotel: very satisfactory, satisfactory, needs improvement, or don't plan to visit again. Because there is not a measurable difference between each of the ratings, even assigning a number value to each and calculating the mean, carries little meaning. However, you might often see such calculations being exhibited.

When we calculate the mean, it is often the case that the value we get is not actually a member of the data set, as shown in Example 1. The mean is simply a descriptive value of the entire set of data. It is an especially useful summary statistic when data sets are large and you may not be able to examine all the pieces of data individually.

Median

Another measure of the center of a data set is the **median**.

Median

The **median** of a data set is the middle value in an ordered array of the data.

In other words, after listing the data in either ascending or descending order, the median is the middle value of the list. If there is no single middle value, such as when there is an even number of values, then the median is the piece of data that lies exactly between the two middle data values, that is, the mean of the two middle data values. Note that when there is an even number of data points, the median may not be a member of the data set.

In circumstances where the data points are tightly grouped values, except for one or a few values, the median is a better choice for describing the "average" member of the data set. This is because extreme values do not affect the median in the way that they affect the mean. For example, the highest mean salary earned by University of North Carolina graduates is not earned by Accountancy, Law, or Medicine graduates, but by Geography majors. This is because Michael Jordan was a Geography major at the University of North Carolina. His salary alone is enough to outweigh the influence of all of the other salary points.

Example 2: Finding the Median

A VO_2 max score is the maximum amount of oxygen that one's body can transport and use during exercise. It is measured in liters of oxygen per minute (L/min). Given the following VO_2 max scores for 12 women, find the median score.

28.3, 27.7, 23.0, 25.5, 27.1, 26.94, 27.0, 27.52, 26.8, 27.2, 26.97, 27.53

Solution

First, put the data in ascending numerical order.

23.0, 25.5, 26.8, 26.94, 26.97, 27.0, 27.1, 27.2, 27.52, 27.53, 27.7, 28.3

Since there are 12 pieces of data, the median will be the value between the middle two data points, 27.0 and 27.1. To find this, add the two together and divide by two.

$$\text{Median} = \frac{27.0 + 27.1}{2} = 27.05$$

Once again, the value of this "average" is not a member of the data set. However, it is a typical value in the sense that it is located in the middle of the data set when it is arranged numerically.

Wolfram|Alpha can be used to calculate the median of a set of numbers. Go to www.wolframalpha.com and type "median {28.3, 27.7, 23.0, 25.5, 27.1, 26.94, 27.0, 27.52, 26.8, 27.2, 26.97, 27.53}" into the input bar. Then, click the = button. Wolfram|Alpha will return the following.

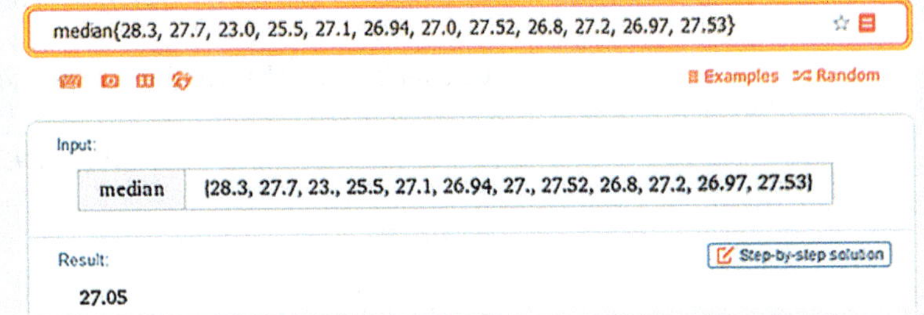

Mode

Sometimes a data set will not lend itself to numerical calculations. For instance, if you survey students on the color cell phone case they prefer, you would have a list of colors, which could not be added, subtracted, or put in ascending or descending order. This is also the case with the hotel ratings we mentioned earlier. It is impossible to add or subtract *very satisfactory*, *satisfactory*, *needs improvement*, or *don't plan to visit again*. However, having a reasonable description for this type of data is still a valuable thing. Even if data are numerical values, and not simply descriptions, sometimes we'd like to know the most common data value.

Mode

The **mode** is the value in the data set that occurs most frequently.

If all the data values occur only once, or they each occur an equal number of times, we say that there is **no mode**. If only one value occurs the most, then the data set is said to be **unimodal**. If exactly two values occur equally often and more than any other data value, the data set is said to be **bimodal**. If more than two values occur equally often and more than any other data value, the data set is **multimodal**. Note that, unlike the mean and the median, if there is a mode, it will always be a value in the data set.

Example 3: Finding the Mode

Find the mode of each of the following sets of data. State if the data set is unimodal, bimodal, multimodal, or has no mode.

a. Preferred color of cell phone cases among students

lemon, gunmetal, violet, turquoise, lime, violet, lemon, orange, red, lemon, pink, violet, lime, violet, lemon, pink, gunmetal, red, turquoise, violet, violet, gunmetal, turquoise, red, violet, turquoise, orange, pink, violet, violet, turquoise, violet, pink

b. Favorite football jersey number

$$32, 18, 99, 12, 7, 10, 28, 56, 13, 16, 19, 51, 23, 78$$

c. Ages of children at the community playground one afternoon

$$12, 4, 2, 7, 8, 4, 10, 6, 5, 7, 7, 4, 3$$

d. Number of ATM withdrawals per hour at the downtown branch of University Bank

$$10, 13, 9, 13, 9, 14, 10, 14$$

Solution

a. The color violet occurs more than any other color, so the mode is violet. This data set is unimodal.

b. Each value occurs only once, so there is no mode.

c. The values 4 and 7 both occur an equal number of times, which is more than any other value. Thus, the set is bimodal with the modes 4 and 7.

d. Be careful here. Since each value occurs the same number of times, there is no mode in this data set.

HELPFUL HINT

SYMMETRIC

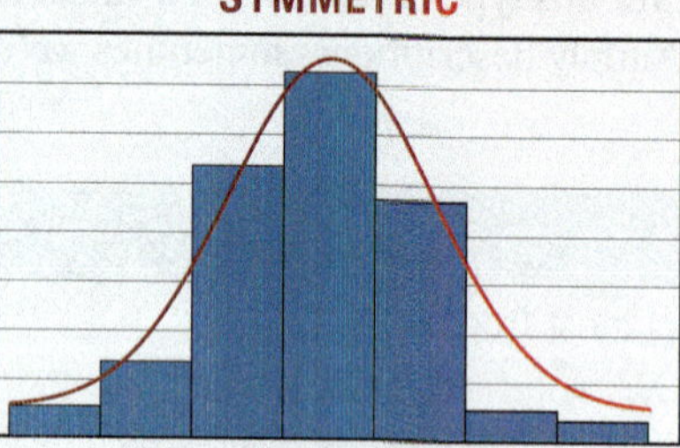

SKEWED TO THE LEFT

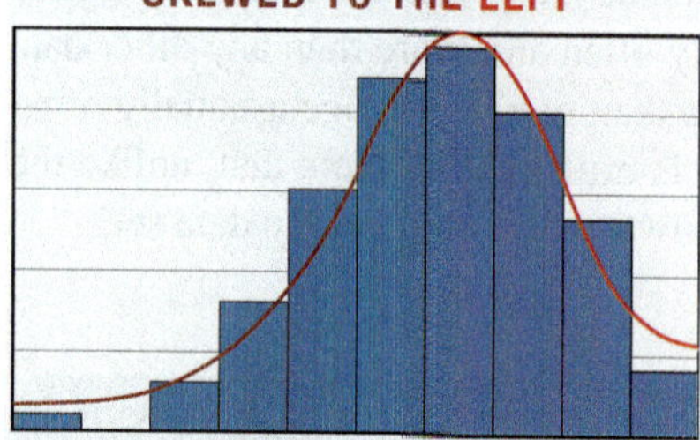

SKEWED TO THE RIGHT

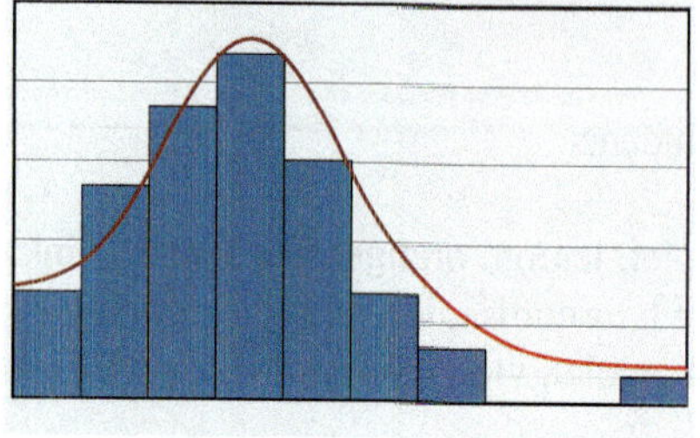

Of the three "averages" (mean, median, and mode), the mean should be used with data consisting of counts and measurements when the data set doesn't include any **outliers**.

Outlier

An **outlier** is a data value that is extreme compared with the rest of the data values in the set.

An outlier will influence the value of the mean of a data set, but will not affect the median or mode. Because it is an extreme value, the outlier drags the value of the mean toward itself. When this happens, the data are said to be skewed, or pulled in the direction of the outlier. The outliers cause the graph of the distribution to be skewed and not symmetrical.

Example 4: Finding the Mean, Median, and Mode

Given the following data set, find the mean, median, and mode, and decide which measure of center you think best describes the data set.

$$16, 44, 15, 48, 14, 77, 11, 84, 26, 61, 15$$

Solution

To find the mean, add up all of the data values and divide by 11 (the number of data values).

$$\text{Mean} = \frac{16+44+15+48+14+77+11+84+26+61+15}{11} = \frac{411}{11} \approx 37.4$$

To find the median, arrange the values in ascending order and find the middle value.

$$11, 14, 15, 15, 16, 26, 44, 48, 61, 77, 84$$

$$\text{Median} = 26$$

The mode is the most commonly occurring value. Notice that 15 occurs twice, while all other values occur only once. Therefore, the mode is 15.

Although there is a mode, because it only occurs twice while all the other data points occur once, this is not the best descriptor of the "average" piece of data. A mode of 15 does not accurately reflect the middle of the data set since the data ranges from 11 to 84. That leaves the mean and the median. Since there are not any outliers it's appropriate to use the mean of the data as the measure of center for this data.

Measures of Dispersion

Measures of dispersion, like the range and standard deviation, describe the "spread" of the data. In other words, they tell us whether the data values are all very similar or if they cover a wide section of the number line. One of the simplest measures of dispersion is the **range** of the data.

Range

The **range** is the difference between the largest and smallest values in the data set, which tells you the distance covered on the number line between the two extremes.

$$\text{range} = \text{maximum data value} - \text{minimum data value}$$

Example 5: Finding the Range

Find the range of the following sets of data.

a. The number of students enrolled as computer science majors over the past 12 semesters

$$5, 21, 54, 33, 12, 14, 36, 40, 27, 29, 37, 22$$

b. The number of shoppers at a gas station downtown Monday through Sunday one week

$$1007, 1010, 1006, 1005, 1054, 1021, 1005$$

Solution

a. The maximum value is 54 and the minimum value is 5, so the range is

$$54 - 5 = 49.$$

b. The maximum value for the data set is 1054 and the minimum value is 1005, so the range is also

$$1054 - 1005 = 49.$$

Calculating the range is very easy. However, the range is not as descriptive as other measures of dispersion. Consider the two data sets in the previous example. Notice that both data sets have the same range. However, almost all of the values in the second data set are similar while the values in the first data set are more spread out. To distinguish between these two situations, we must use another measure of dispersion.

The **standard deviation** is a measure of how much we might expect a member of the data set to differ from the mean. The greater the standard deviation, the more the data values are spread out. Similarly, a smaller standard deviation indicates that the data values lie closer together. The standard deviation is always a number greater than or equal to 0. It is precisely the case when all of the data points are the same value that the standard deviation is equal to 0.

Recall that we presented two formulas for the mean: the population mean and the sample mean. We also have two formulas for the standard deviation: the population standard deviation and the sample standard deviation.

<table>
<tr><td>

⚓ HELPFUL HINT

When calculating the standard deviation, round to one more decimal place than the largest number of decimal places given in the data. Occasional exceptions to this rule can be made when the type of data lends itself to a more natural rounding method, such as rounding values of currency to two decimal places.

</td><td>

Standard Deviation

The **standard deviation** is a measure of how much we might expect a member of the data set to differ from the mean.

The formula for finding the **population standard deviation** is

$$\sigma = \sqrt{\frac{\sum\left(x_i - \mu\right)^2}{N}}$$

where x_i is the ith data value, μ is the population mean, and N is the size of the population.

The **sample standard deviation** is

$$s = \sqrt{\frac{\sum\left(x_i - \bar{x}\right)^2}{n-1}}$$

where x_i is the ith data value, $\bar{x}$ is the sample mean, and n is the sample size.

</td></tr>
</table>

Example 6: Calculating Standard Deviation by Hand

Calculate the sample standard deviation for a sample of nine ages of students working with a university theater production of *Macbeth*.

$$17, 21, 18, 18, 24, 19, 21, 20, 28$$

Solution

When calculating the standard deviation by hand, we need to first note the sample size n and find the sample mean $\bar{x}$. With $n = 9$, the mean is

$$\bar{x} = \frac{17 + 21 + 18 + 18 + 24 + 19 + 21 + 20 + 28}{9}$$

$$\approx 20.67.$$

Note that we will round the mean to the nearest hundredth in an effort to minimize any error introduced from rounding.

When calculating standard deviation by hand, it's helpful to use a table like Table 1 and build up to the formula.

x_i	$x_i - \bar{x}$	$\left(x_i - \bar{x}\right)^2$
17	−3.67	13.47
21	0.33	0.11
18	−2.67	7.13
18	−2.67	7.13
24	3.33	11.09
19	−1.67	2.79
21	0.33	0.11
20	−0.67	0.45
28	7.33	53.73
		$\Sigma = 96.01$

TABLE 1: Sample Standard Deviation

We are now ready to substitute the values into the formula for the sample standard deviation.

$$s = \sqrt{\frac{\Sigma\left(x_i - \bar{x}\right)^2}{n-1}}$$

$$= \sqrt{\frac{96.01}{9-1}}$$

$$= \sqrt{12.00125}$$

$$\approx 3.5$$

So, the sample standard deviation of ages is approximately 3.5. In other words, the age of the average student in the sample is about 3.5 years different (either younger or older) from the mean age of 20.67.

Since technology is so often used to find the standard deviation rather than the pencil and paper method, especially with large data sets, we will focus on showing you how to use technology for the calculations and how to interpret the results of what you find. The following example shows how to calculate standard deviation using a calculator.

Example 7: Calculating Standard Deviation Using a Calculator

Use your TI-30XIIS/B or TI-83/84 Plus calculator to find the standard deviation of the following data sets.

a. The following data represent the average number of Tweets per day posted on Twitter for a sample of 24 college students.

0.8	42.2	20.6	2.8
36.7	12.1	18.6	6.3
5.5	11.3	3.7	0.5
1.2	3.7	14.9	9.4
7.3	9.5	16.0	11.1
4.7	5.6	8.9	10.2

TABLE 2: Tweets per Day

b. The SAT Critical Reading scores for the senior class at Richmond Prep High are given in Table 3.

520	640	750	620	470	520
630	600	590	660	700	580
460	600	640	690	530	490
500	560	630	760	650	760
580	610	710	610	590	570
590	550	610	490	630	550
590	620	610	600	570	690

TABLE 3: SAT Critical Reading Scores

Solution

HELPFUL HINT

If you are using a TI-30XS/B Multiview calculator, the keystrokes will be slightly different for your calculator.

a. In order to calculate the standard deviation using a TI-30XIIS/B calculator, begin by clearing the data lists in the calculator. Then enter the data points by using the following commands.

1. Press `2nd` `data`.

2. Choose `1-VAR` and press `enter`.

3. Press `data`; (X= should appear.

4. Enter a data value and press the down arrow key twice, since the frequency is one for each data point.

5. After the last data point is entered, press `enter`.

Because the values given are only a sample of students, we want the *sample standard deviation*. To calculate the sample standard deviation, press `statvar`. Scroll over to the sample standard deviation, which is denoted by sx in the list of calculated values. From the list we see that $s \approx 10.3$.

To calculate the standard deviation using the list function on a TI-83/84 Plus calculator,

1. Press `stat`, then choose `1:Edit...`, and enter your data in L1.

2. Press `stat` again and now scroll to the right to CALC.

3. Choose option `1:1-Var Stats` and press `enter`. If your data are in L1, press `enter` again since L1 is the default list. If you did not type your data in L1, enter the list where your data are located, such as L2 or L3. (These list names are in blue, above the numeric keys.)

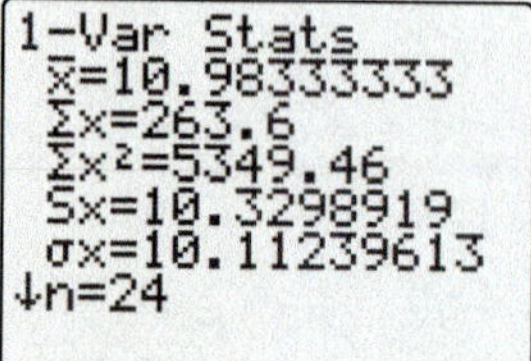

A list of numerical statistics will be generated for the data. The beginning of the list is shown in the margin. Because the values given are only a sample of college students, we want the *sample standard deviation*, which is denoted by s (on the calculator, this is displayed as Sx). From the list we see that $s \approx 10.3$.

Since the standard deviation tells us about the average distance away from the mean, we can conclude that student tweeting behavior usually varies from the mean by tens rather than hundreds of tweets.

```
1-Var Stats
x̄=602.8571429
Σx=25320
Σx²=15487200
Sx=73.72611525
σx=72.84313591
↓n=42
```

b. Begin by clearing the data lists in the calculator. Now enter the data as you did in part **a.** Since we are told that the values given represent an entire senior class, we want the *population standard deviation*, which is denoted by σx. From the list we see that $\sigma \approx 72.8$.

Therefore, we know that SAT Critical Reading scores differ from the mean on average by 72.8 points. While we've got the calculator handy, we can see that the mean is actually approximately 602.9. Although there is not an actual score of 602.9, you can see that many of the students scores fall within about 70 points (or 1 standard deviation) of that mean, either larger or smaller.

📈 TECH TRAINING

Wolfram|Alpha can be used to find the standard deviation of a set of numbers. To find the standard deviation of the data in Example 7a., go to www.wolframalpha.com and type "standard deviation {0.8, 42.2, 20.6, 2.8, 36.7, 12.1, 18.6, 6.3, 5.5, 11.3, 3.7, 0.5, 1.2, 3.7, 14.9, 9.4, 7.3, 9.5, 16.0, 11.1, 4.7, 5.6, 8.9, 10.2}" into the input bar. Then, press the = button. Wolfram|Alpha will return the following.

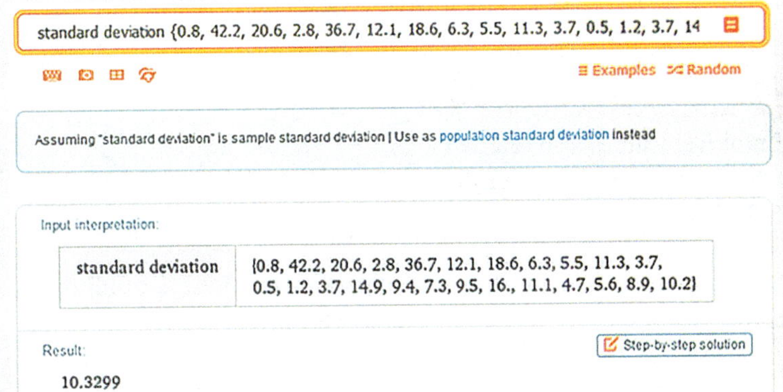

Source: Wolfram Alpha LLC. 2009. Wolfram|Alpha. http://www.wolframalpha.com/input/?i=standard+deviation+%7B0.8%2C+42.2%2C+20.6%2C+2.8%2C+36.7%2C+12.1%2C+18.6%2C+6.3%2C+5.5%2C+11.3%2C+3.7%2C+0.5%2C+1.2%2C+3.7%2C+14.9%2C+9.4%2C+7.3%2C+9.5%2C+16.0%2C+11.1%2C+4.7%2C+5.6%2C+8.9%2C+10.2%7D (accessed July 22, 2014).

Skill Check 2

Find the population standard deviation for the following data.

8, 12, 10, 11, 13, 12, 15, 9, 11, 16

The standard deviation allows us to interpret the variation of the data with a sense of scale. Two standard deviations that are exactly the same number don't necessarily represent the same variation for different populations. For instance, consider Example 6. A standard deviation of 3.50 when referring to ages is not a large variation. However, if the standard deviation for gas prices at local gas stations had a value of $3.50, we would say that the data had a huge variation!

The standard deviation helps us to put into context one of the most useful estimation rules in statistics.

The Empirical Rule

When the distribution of a set of data is approximately bell-shaped, we can estimate the percentage of data values that fall within a few standard deviations of the mean in the following way:

Approximately 68% of all data points lie within 1 standard deviation above and below the mean.

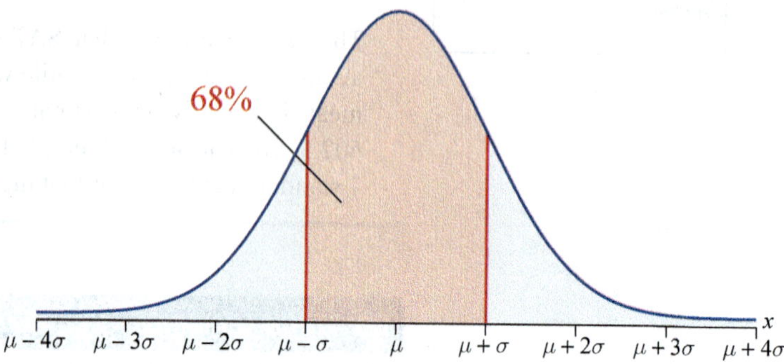

Approximately 95% of all data points lie within 2 standard deviations above and below the mean.

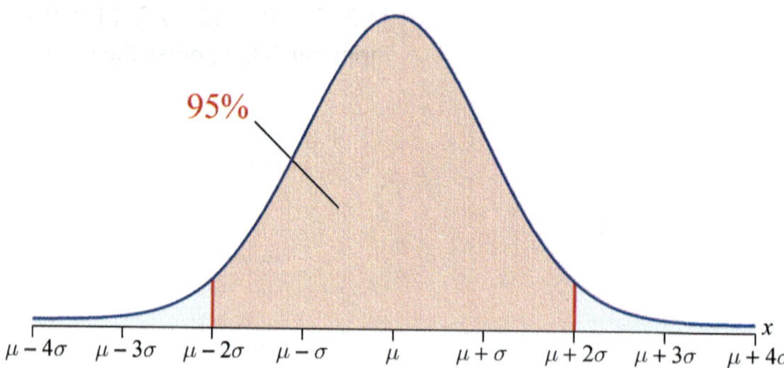

> **⟳ HELPFUL HINT**
>
> The Empirical Rule is also referred to as the Three Sigma Rule, or the 68-95-99.7 Rule.

Approximately 99.7% of all data points lie within 3 standard deviations above and below the mean.

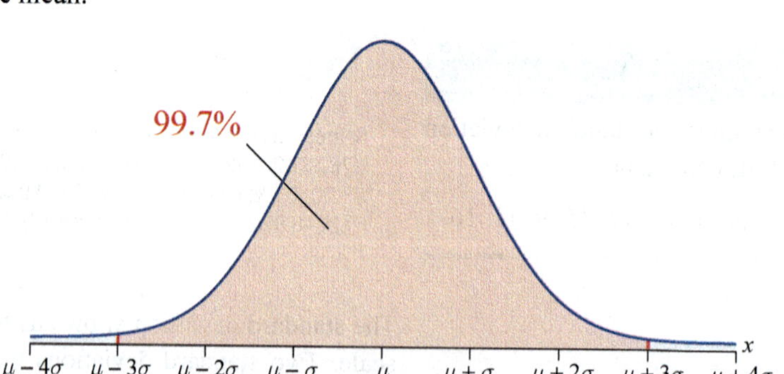

Example 8: Using the Empirical Rule

Suppose you suspect that emergency room waiting times for a local hospital have a bell-shaped distribution. The data have a reported mean of 116.9 minutes and a standard deviation of 56.1 minutes.

a. Identify the range of waiting times that 95% of patients are likely to experience.

b. Estimate the percentage of patients that will wait less than 2 hours and 53 minutes.

Solution

a. By the Empirical Rule, we know that 95% of the data set, which is approximately bell-shaped, will fall within two standard deviations of the mean. Therefore, we can double the standard deviation and then subtract it from and add it to the mean to find the range of times that 95% of patients are likely to experience.

$$(56.1)(2) = 112.2$$

$$116.9 - 112.2 = 4.7$$

$$116.9 + 112.2 = 229.1$$

So, we can estimate that 95% of patients will wait somewhere between 4.7 minutes and 229.1 minutes.

b. To answer this question, we'll need to first convert 2 hours and 53 minutes into minutes only, since our mean and standard deviation are both in minutes. Since 2 hours is 120 minutes, add 53 minutes to get a total time of 173 minutes. We know that this time falls above the mean, but by how much? If we subtract the mean, we have

$$173 - 116.9 = 56.1 \text{ minutes.}$$

In other words, the time we're interested in is one standard deviation above the mean. Don't jump to conclusions here and assume that 68% of the data falls below that. Let's draw a picture first.

Emergency Waiting Room Times

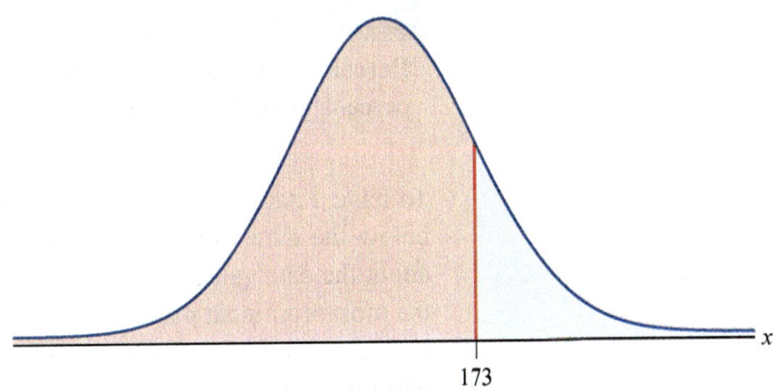

Remember, we want to know what percentage of patients will wait less than 173 minutes. From the Empirical Rule, we know that 68% of data is within one standard deviation, so half of that, 34%, must be between the mean and one standard deviation.

We also know that because of the symmetry of a bell-shaped curve, since the mean is in the middle, half of the data lies below it.

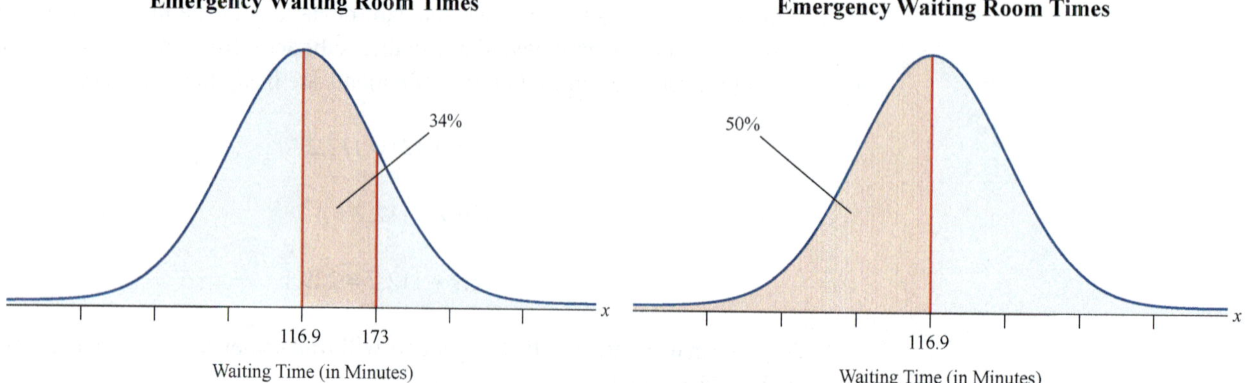

So, putting these two facts together, we have that 50% + 34% = 84% of the data lies below one standard deviation above the mean. In other words, approximately 84% of patients will wait less than 2 hours and 53 minutes.

Measures of Relative Position

When we want to describe the position of particular pieces of data in a set compared to the rest of the data, we use **measures of relative position** like percentiles and quartiles. They are each based on dividing the data into sections and then referencing the section in which the data value falls.

Percentile

Percentiles divide the data into 100 equal parts and tell you approximately what percentage of the data lies at or below a given value.

To have a data point at the 81st percentile means that 81% of the population is at or below the data value. It's important to note that percentiles don't convey anything about the data value itself, only its relative position to the data set as a whole. One of the most prominent places you might have heard of percentiles would be in connection with any standardized tests you've taken, such as the ACT or SAT college entry exams. Birth measurements are also commonly reported in terms of percentiles.

Example 9: Interpreting Percentiles

Sierra received her scores from taking a mathematics placement test for her chosen university. Choose the best explanation for what it means for her to be in the 61st percentile.

a. She correctly answered 61% of the answers on the test.

b. 61% of people taking the test scored the same as Sierra.

c. Sierra's score was at least as good as 61% of the people taking the test.

d. Sierra missed 39% of the test questions.

Solution

The correct interpretation of her score is **c.**: "Sierra's score was at least as good as 61% of the people taking the test." Both **a.** and **d.** are incorrect because they refer to how many questions she answered correctly on the test and not how she did in comparison to others taking the test. Choice **b.** is not quite correct because percentiles tell you the percentage that scored at or below you. They are not all necessarily the same score as Sierra's.

The percentiles that divide the data into four even parts are called **quartiles**. Figure 2 illustrates that to divide something into 4 parts, you only need 3 dividers. We call these "dividers" the first quartile Q_1, the second quartile Q_2, and the third quartile Q_3.

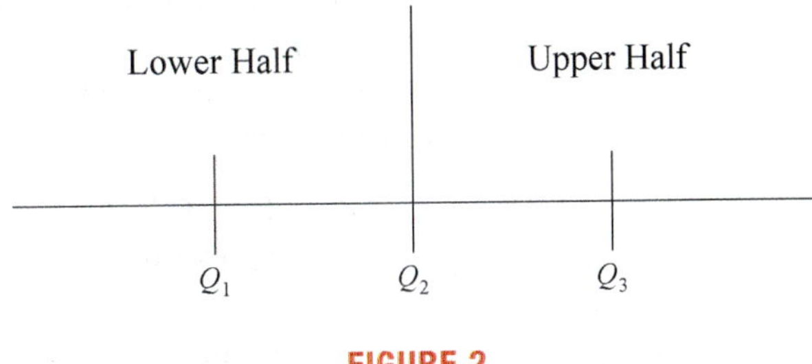

FIGURE 2

Quartiles

Q_1 = **First Quartile** = 25th percentile, that is, 25% of the data is less than or equal to this value.

Q_2 = **Second Quartile** = 50th percentile, that is, 50% of the data is less than or equal to this value.

Q_3 = **Third Quartile** = 75th percentile, that is, 75% of the data is less than or equal to this value.

By definition, Q_2 will be the same as the median.

Example 10: Interpreting Quartiles

On Karl's recent standardized test results, the picture graph of his score showed he was above the third quartile in language arts. His classmate, Asher, said his score was at the 70th percentile, while Rylie said hers was at the 79th percentile. Which of the three had the best language arts test score?

Solution

We know the percentile ranks of both Asher and Rylie are the 70th and 79th respectively. What we know about Karl's score is that it was above the third quartile. Since the third quartile is the same as the 75th percentile, we know that his score was somewhere at or above the 75th percentile. We can conclude that he did better than Asher, whose score was at the 70th percentile, but can make no definite comparison with Rylie, whose score was at the 79th percentile, because we do not know for sure which one had the best language arts score.

As we've already seen, a calculator can be very helpful when calculating the various data descriptors that have been covered in this section. Often it is the case that several descriptors are desired when considering a set of data. Certainly, the more characteristics about the data we have, the better "picture" we can form of the data. One way to easily calculate several descriptors at once is to use Microsoft Excel or similar spreadsheet packages. The following example illustrates the commands for using Excel.

Example 11: Using Excel to Find Data Descriptors

To receive federal financial aid for higher education, students must first complete a Free Application for Federal Student Aid form, commonly called FAFSA. Given the following data gathered from FAFSA on the number of applications by state, use Excel to find the minimum and maximum data values, mean, median, mode, range, and standard deviation for the data.

State	Number of FAFSA Applications	State	Number of FAFSA Applications
Alberta	31	Nebraska	7742
American Samoa	177	Nevada	15,014
Arizona	39,134	New Brunswick	2
Arkansas	16,122	Newfoundland	3
Blank	28,459	New Hampshire	4511
British Columbia	36	New Jersey	34,906
California	197,594	New Mexico	11,335
Canada	104	New York	82,152
Colorado	28,373	North Carolina	57,614
Connecticut	14,757	North Dakota	2045
Delaware	4238	Northern Mariana Islands	238
District of Columbia	2958	Northwest Territories	2
Federated States of Micronesia	877	Nova Scotia	3
Florida	124,874	Nunavut	0
Foreign Country	3042	Ohio	59,066
Georgia	68,669	Oklahoma	18,377
Guam	695	Ontario	89
Hawaii	5211	Oregon	20,937
Idaho	9747	Palau	75
Illinois	59,081	Pennsylvania	45,316
Indiana	30,127	Prince Edward Island	6
Iowa	12,434	Puerto Rico	21,299
Kansas	13,366	Quebec	35
Kentucky	22,078	Rhode Island	4027
Labrador	3	Saskatchewan	7
Louisiana	23,576	South Carolina	27,203

State	Number of FAFSA Applications	State	Number of FAFSA Applications
Maine	4953	South Dakota	2941
Manitoba	5	Tennessee	32,411
Marshall Islands	125	Texas	134,463
Maryland	27,318	Utah	18,628
Massachusetts	23,675	Vermont	1748
Mexico	522	Virginia	41,507
Michigan	60,553	Virgin Islands	558
Minnesota	26,841	Washington	34,980
Mississippi	20,741	West Virginia	7091
Montana	4245	Wisconsin	25,411
Missouri	31,919	Wyoming	2452
		Yukon	7

Source: Federal Student Aid, "Application Volume Reports," http://federalstudentaid.ed.gov/datacenter/application.html

TABLE 4: 2011–2012 Application Cycle for Quarter 4

Solution

You may need to load the Analysis ToolPak in Microsoft Excel. If you do not have the Data Analysis option in the Data tab, follow these steps to load the Analysis ToolPak on a PC.

1. Under the File menu, click on Options.

2. Click Add-ins (listed on the left side), and then in the Manage box, select Excel Add-ins.

3. Click Go.

4. In the Add-ins available box, select the Analysis ToolPak check box, and then click OK.

   If Analysis ToolPak is not listed in the Add-ins available box, click Browse to locate it.

   If you get prompted that the Analysis ToolPak is not currently installed on your computer, click Yes to install it.

5. After you load the Analysis ToolPak, the Data Analysis command is available in the Analysis group on the Data tab.

Begin by typing the data into Columns A and B. Go to the **Data** tab, then **Data Analysis**, then choose **Descriptive Statistics**, and click OK. In the **Input Range** box, enter the cells where your Number of FAFSA Applications data are located. Select **New Worksheet Ply** and type "Descriptive Statistics" in the box. Click on the box in front of **Summary Statistics**. The Descriptive Statistics menu should look similar to the following.

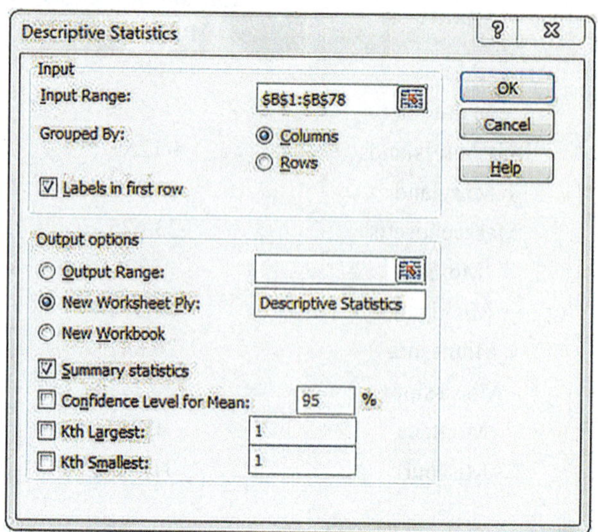

Clicking **OK** should produce the following list of descriptive statistics.

	A	B
1	*Number of FAFSA Applications*	
2		
3	Mean	21611.48
4	Standard Error	3877.837471
5	Median	9747
6	Mode	3
7	Standard Deviation	33583.05761
8	Sample Variance	1127821759
9	Kurtosis	11.48669404
10	Skewness	3.024608257
11	Range	197594
12	Minimum	0
13	Maximum	197594
14	Sum	1620861
15	Count	75

From the output, we can see that the values we are looking for are as follows.

Minimum value = 0

Maximum value = 197,594

Mean = 21,611.5

Median = 9747

Mode = 3

Range = 197,594

Standard deviation = 33,583.1

9.3 EXERCISES

⚡ PRACTICE

Find the mean, median, mode, range, and standard deviation for each data set. When applicable state whether the data set is unimodal, bimodal, or multimodal. Round answers to one more decimal place than the largest number of decimal places given in the data. All data sets are samples unless stated otherwise.

1. 19, 32, 15, 21, 25, 22, 22, 28, 27, 27, 26

2. $11.40, $32.00, $22.50, $12.01, $10.08, $18.30, $18.40, $32.00

3. 45, 21, 26, 26, 45, 37, 22, 33, 26, 21, 42, 37, 41, 43, 46, 35, 31, 29, 46

4. 310, 310, 310, 310, 310, 310

5. 9, 3, −5, −3, −7, 3, 0, 6, −9, −7, −3, −8

6. The following data represent sample ACT scores from students at a local high school.

ACT Scores	
13	26
10	20
24	30
25	31
6	24
35	35
26	15

7. The following are lengths of each movie in the complete Harry Potter film series. Note that because these include all of the films in the series, this is a population.

Time Lengths for Harry Potter Film Series

Movie Title	Time (in Minutes)
Harry Potter and the Philosopher's Stone (2001)	152 minutes
Harry Potter and the Chamber of Secrets (2002)	161 minutes
Harry Potter and the Prisoner of Azkaban (2004)	141 minutes
Harry Potter and the Goblet of Fire (2005)	157 minutes
Harry Potter and the Order of the Phoenix (2007)	138 minutes
Harry Potter and the Half-Blood Prince (2009)	153 minutes
Harry Potter and the Deathly Hallows Part 1 (2010)	146 minutes
Harry Potter and the Deathly Hallows Part 2 (2011)	130 minutes

8. The following table shows the top 15 busiest airports from January to November 2011.

Top 15 Busiest Airports from January to November 2011

Airport	Total Passengers
Amsterdam Schiphol Airport	46,213,944
Beijing Capital International Airport	71,284,796
Dallas/Fort Worth International Airport	53,126,399
Denver International Airport	48,402,802
Dubai International Airport	46,287,234
Frankfurt Airport	52,191,355
Hartsfield-Jackson Atlanta International Airport	85,165,259
Hong Kong International Airport	48,587,000
John F. Kennedy International Airport	44,045,938
London Heathrow Airport	63,912,107
Los Angeles International Airport	56,819,805
Madrid Barajas Airport	46,019,110
O'Hare International Airport	61,370,268
Paris Charles de Gaulle Airport	56,254,938
Soekarno-Hatta International Airport	47,513,248
Tokyo International Airport	56,969,971

Source: Wikipedia, s.v. "World's busiest airports by passenger traffic," http://en.wikipedia.org/wiki/World%27s_busiest_airports_by_passenger_traffic

For each data set, determine the most appropriate measure of center.

9. Styles of houses in a suburb: ranch, colonial, bungalow, etc.

10. Grades on the final in Biology 210 at State University.

11. The ratings on a customer satisfaction survey: strongly disagree, disagree, neither agree nor disagree, agree, and strongly agree.

12. Salaries for janitorial staff at the state governmental buildings that include the Director of Sanitation's salary.

🚀 APPLICATIONS

Use the formula for the mean, $\bar{x} = \dfrac{x_1 + x_2 + \cdots + x_n}{n}$, to find the missing piece of data.

13. John knows that his first 4 tests grades were 84, 79, 82, and 88. Find John's grade on the fifth test if his average was 83.8.

14. A small boat that ferries visitors to a resort island has strict guidelines on the weight allowed for passenger luggage. Consequently the five vacationers are limited to a maximum average luggage weight of 40 pounds (lb). The following are the weights of three out of five pieces of luggage: 39 lb, 32 lb, and 43 lb. The two pieces of luggage that haven't been weighed will have to split the remaining weight allowance. Determine the maximum average possible weight allowance for each remaining bag.

Use the Empirical Rule to answer each question.

15. Although there is some controversy around the precise average body temperature of adults, new data suggest that the mean is 98.2° with a standard deviation of 0.6° and has a bell-shaped distribution.
    a. According to this distribution, approximately what percentage of body temperatures are between 97° and 99.4°?
    b. Approximately what percentage of temperatures are greater than 98.8°?
    c. Approximately what percentage of temperatures are no more than 98.2°?

16. In 2011, high school seniors had the following mean and standard deviation on the mathematics portion of the SAT exam: $\mu = 514$ and $\sigma = 117$.
    a. Approximately what percentage of scores were greater than 397 but less than 631?
    b. What two scores have approximately 95% of the data between them?
    c. Approximately what percentage of high school seniors had mathematics scores between the first and second quartiles?
    d. Describe where the best and worst 0.3% of scores lie.

Solve each problem.

17. Marcel scored in the 91st percentile on the MCAT (Medical College Admissions Test). The medical school he is applying to only accepts students who score in the top 10% on the MCAT. Did Marcel score well enough to be considered for his school of choice?

18. The five-number summary is a numerical description of data that includes the minimum data point, the maximum data point, and the data points representing quartiles Q_1, Q_2, and Q_3. The following are house prices in one neighborhood.

    $181,865 $119,442 $152,750 $100,960 $159,635

    $150,963 $133,702 $149,788 $145,495 $182,500

    $112,021 $120,900 $145,850 $164,590 $144,413

    a. Find the five-number summary of the house prices.
    b. What percentage of house prices is at or below $159,635?
    c. What is the range of house prices for this neighborhood?

19. The following graph contains a box plot. A box plot is a graphic display of a five-number summary, which was introduced in Exercise 18. The endpoints represent the minimum and maximum data values, while the lines sectioning off the box in the middle represent each of the quartiles as shown in the graph.

Birth Weeks

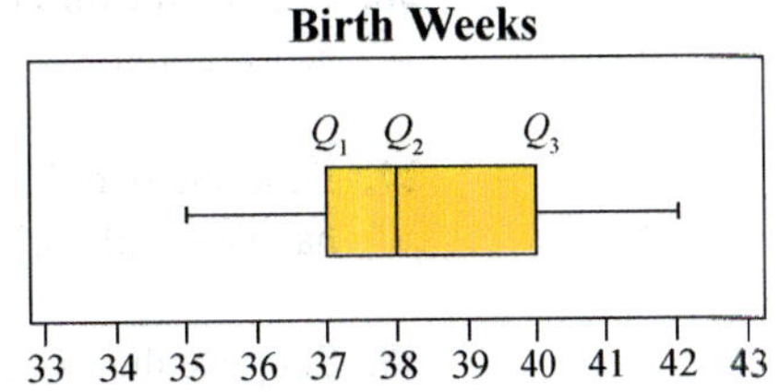

    a. Based on the box plot, estimate each of the values in the five-number summary.
    b. The midrange is the average of the minimum and maximum data values. Estimate the midrange from the box plot.

20. Using the information given in Exercise 19, calculate the values needed to construct a box plot for the following data as described in Exercise 19. Sketch a graph of the box plot.

$$310 \quad 320 \quad 450 \quad 460 \quad 470 \quad 500 \quad 520 \quad 540$$

$$580 \quad 600 \quad 650 \quad 700 \quad 710 \quad 840 \quad 870 \quad 900$$

$$1000 \quad 1200 \quad 1250 \quad 1300 \quad 1400 \quad 1720 \quad 2500 \quad 3700$$

21. Given the five-number summary for three data sets, sketch a box plot for each, side-by-side on the same graph. Then answer the following questions based on your box plots.

Committees and the Ages of Members

	Membership	Finance	Publicity
Min	23	26	25
Q_1	27	32	26
Q_2	29	38	27
Q_3	33	44	29
Max	35	46	33

a. Which committee has the largest range of ages?
b. Which committee has the least variation in the ages?
c. Which committee has the smallest median?

WRITING & THINKING

22. Given the following measures of center, decide the likely shape of the distribution: mean = 22.5, median = 17.0, mode = 17.0.

23. Accounting 101 has five class sections. All five classes took the same final. The mean scores on the final for each class were 72, 78, 76, 74, and 79. Can the mean final score for all students in Accounting 101 be found by averaging the mean scores in each class? Explain your answer.

24. Does the standard deviation of a data set equaling zero imply that all entries in the data set equal zero?

25. Is it possible for a data set to have a standard deviation of −2.5?

26. Explain the difference between Amelia making an 82 on her precalculus exam and scoring in the 82nd percentile in mathematics on the ACT test.

27. Describe two data sets, one that might have a large variation and one that might have a small variation.

28. Suppose that, in a list of data, 37% of the data are greater than 45. True or False: Q_1 must be greater than 45.

29. Lucas received an e-mail containing the five-number summary for the company sales data that he asked for. Unfortunately, the e-mail cut off the summary labels and scrambled their order. Can you still determine which number is the first quartile, Q_1? Explain your answer.

five-number summary: 11, 17.5, 9, 13.5, 19

30. Suppose that 110 male students are surveyed and that 52% have a height less than 1.776 m.
   a. True or False: Of those surveyed, the mean height must be under 1.776 m.
   b. True or False: Of those surveyed, the median height must be under 1.776 m.

31. If we know that a salary of $65,300 was in the 67th percentile in a company survey, can we determine how many employees were in the sample? Why or why not?

9.4 THE BINOMIAL DISTRIBUTION

■ TOPICS

- ■ The Shape of a Binomial Distribution
- ■ Expected Value and Variance

The binomial distribution arises from experiments with repeated two-outcome trials, where only one of the outcomes is counted. Experiments of this kind are rather common in the business world. In market research, a survey respondent (a trial) either will or will not recognize a company's brand. The number that recognize the brand is a count that may be modeled as a **binomial random variable**.

Some examples of binomial random variables include

- the number of left-handed persons in a sample of 200 unrelated people,

- the number of correct guesses out of 20 True or False questions when you randomly choose an answer for each question,

- the number of winning scratch-off lottery tickets when you purchase 15 of the same type.

An experiment is required to meet several conditions in order to qualify as a binomial experiment.

> ### Binomial Experiment
>
> A **binomial experiment** is a random experiment that satisfies all of the following conditions.
>
> 1. There are only two outcomes in each trial of the experiment. (One of the outcomes is usually referred to as a *success*, and the other as a *failure*.)
>
> 2. The experiment consists of n identical trials as described in Condition 1.
>
> 3. The probability of success on any one trial is denoted by p and does not change from trial to trial. (Note that the probability of a failure is $(1 - p)$ and also does not change from trial to trial.)
>
> 4. The trials are independent.
>
> 5. The binomial random variable is the count of the number of successes in n trials.

A binomial random variable is formed by counting the number of successes in n trials of an experiment with two outcomes. One of the simplest of binomial random variables is produced by tossing a coin.

Toss a coin 4 times and record the number of heads. Is the number of heads in 4 tosses a binomial random variable?

Solution

1. There are only two outcomes, heads or tails.

2. The experiment will consist of 4 tosses of a coin. (Hence, $n = 4$.)

3. The probability of getting a head (success) is $\dfrac{1}{2}$ and does not change from trial to trial. (Hence, $p = \dfrac{1}{2}$.)

4. The outcome of one toss will not affect other tosses.

5. The variable of interest is the count of the number of heads in 4 tosses.

All the conditions of a binomial experiment are met, so the number of heads in 4 tosses is a binomial random variable. The probability distribution for this experiment is given in Table 1.

Events	Number of Heads	Probability
TTTT	0	$\dfrac{1}{16}$
HTTT, THTT, TTHT, TTTH	1	$\dfrac{4}{16}$
HHTT, HTHT, HTTH, THHT, THTH, TTHH	2	$\dfrac{6}{16}$
THHH, HTHH, HHTH, HHHT	3	$\dfrac{4}{16}$
HHHH	4	$\dfrac{1}{16}$

TABLE 1: Tossing a Coin

To derive the binomial probability distribution by listing the events in the sample space is unnecessarily tedious. Instead of tossing the coin 4 times, suppose the coin is tossed 10 times in the experiment. The number of events for an experiment with 10 tosses would be $2^{10} = 1024$, and an experiment with 20 tosses would require listing a staggering $2^{20} = 1,048,576$ events. Fortunately, there is a far simpler method of obtaining the probability distribution. The **binomial probability distribution function** provides a relatively simple method of calculating binomial probabilities.

Binomial Probability Distribution Function

The **binomial probability distribution function** is

$$P(X = x) = {}_nC_x\, p^x (1-p)^{n-x}$$

where ${}_nC_x$ represents the number of possible combinations of n objects taken x at a time (without replacement) and is given by

$${}_nC_x = \frac{n!}{x!(n-x)!} \quad \text{where } n! = n(n-1)(n-2)\cdots(2)(1) \text{ and } 0! = 1;$$

$n =$ the number of trials,

$p =$ the probability of a success, and

$x =$ the number of successes in n trials.

To calculate a binomial probability, the parameters of the distribution (x and p) as well as the value of the random variable must be specified. For example, to determine the probability of 3 heads in 4 tosses of a coin, you would substitute the values $x = 3$, $n = 4$, and $p = \dfrac{1}{2}$ into the binomial probability distribution function as follows.

$$P(X = 3) = {}_4C_3 \left(\frac{1}{2}\right)^3 \left(1 - \frac{1}{2}\right)^{4-3}$$

Since ${}_4C_3 = \dfrac{4!}{3!(4-3)!} = \dfrac{(4)\,(3!)}{(3!)\,(1!)} = 4$, then

$$P(X = 3) = 4\left(\frac{1}{2}\right)^3 \left(\frac{1}{2}\right) = \frac{4}{16} = \frac{1}{4} = 0.25.$$

The probability that 2 heads would be tossed can be computed in a similar manner.

$$P(X = 2) = {}_4C_2 \left(\frac{1}{2}\right)^2 \left(1 - \frac{1}{2}\right)^{4-2}$$

$$= \frac{4!}{(2!)(2!)} \left(\frac{1}{2}\right)^2 \left(\frac{1}{2}\right)^2 = \frac{6}{16} = \frac{3}{8} = 0.375$$

The complete distribution shown in Table 2 can be computed by substituting the remaining values of the random variable into the probability distribution function.

The coin toss is a classical binomial experiment easily related to the rules of a binomial experiment. In many instances the relationship of an experiment to the binomial definition is not as clear.

Number of Heads	Probability
0	$\dfrac{1}{16}$
1	$\dfrac{4}{16}$
2	$\dfrac{6}{16}$
3	$\dfrac{4}{16}$
4	$\dfrac{1}{16}$

TABLE 2: Tossing a Coin

Example 2: Determining Whether a Random Variable Is Binomial and Constructing a Probability Distribution

Roll a single six-sided die 4 times and record the number of sixes observed. Does the number of sixes rolled in 4 tosses of a die meet the conditions required of a binomial random variable? Construct the probability distribution for this experiment.

Solution

1. The experiment either produces a six or does not, and thus satisfies the two outcome requirement. (At first glance, there appears to be a problem. There are six sides to a die and there would appear to be six possible outcomes rather than the two required for a single trial of a binomial experiment.)

2. The experiment is repeated 4 times. Hence, the number of trials equals four ($n = 4$).

3. The probability of getting a six, a success, is $\dfrac{1}{6}$. (This probability assumes that the die is fair.) Thus, $p = \dfrac{1}{6}$.

4. The probability remains constant from trial to trial. (One roll of the die does not affect other rolls.)

5. If X is the number of sixes in 4 rolls, then it has a binomial distribution.

To obtain the probability distribution for X, use the binomial probability distribution function with parameters $n = 4$ and $p = \dfrac{1}{6}$.

$$P(X = 0) = {}_4C_0 \left(\frac{1}{6}\right)^0 \left(1 - \frac{1}{6}\right)^{4-0} = (1)(1)\left(\frac{5}{6}\right)^4 \approx 0.4823$$

$$P(X = 1) = {}_4C_1 \left(\frac{1}{6}\right)^1 \left(1 - \frac{1}{6}\right)^{4-1} = (4)\left(\frac{1}{6}\right)\left(\frac{5}{6}\right)^3 \approx 0.3858$$

$$P(X = 2) = {}_4C_2 \left(\frac{1}{6}\right)^2 \left(1 - \frac{1}{6}\right)^{4-2} = (6)\left(\frac{1}{6}\right)^2 \left(\frac{5}{6}\right)^2 \approx 0.1157$$

$$P(X = 3) = {}_4C_3 \left(\frac{1}{6}\right)^3 \left(1 - \frac{1}{6}\right)^{4-3} = (4)\left(\frac{1}{6}\right)^3 \left(\frac{5}{6}\right)^1 \approx 0.0154$$

$$P(X = 4) = {}_4C_4 \left(\frac{1}{6}\right)^4 \left(1 - \frac{1}{6}\right)^{4-4} = (1)\left(\frac{1}{6}\right)^4 \left(\frac{5}{6}\right)^0 \approx 0.0008$$

Number of Sixes, x	Probability
0	0.4823
1	0.3858
2	0.1157
3	0.0154
4	0.0008

TABLE 3: Rolling a Die

The calculations required to produce the distribution were reasonably simple in Example 2. However, if there had been forty rolls rather than four, the determination of the probabilities would have been rather burdensome, at best. Instead of using the binomial probability distribution function to calculate binomial probabilities, we can use a calculator or other technology.

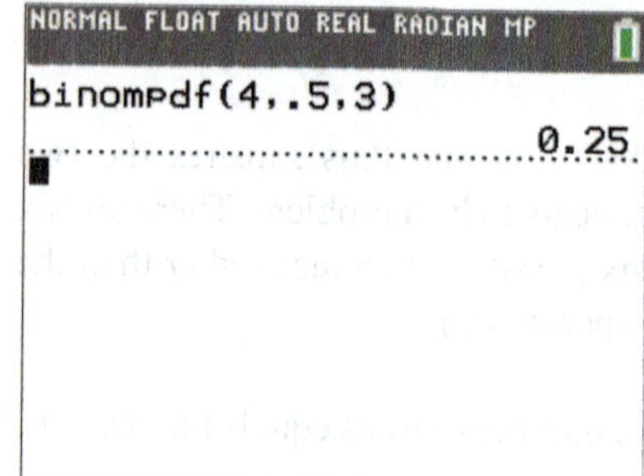

To calculate an individual binomial probability of the form $P(X = x)$ using a TI-83/84 Plus calculator, first press **2nd** and then **vars** to open the **DISTR** menu, scroll down, and choose the `binompdf(` option. Then input the number of trials n, the probability of a success p, and the number of successes x, separated by commas. Finally, press **)** and then **enter** to calculate the probability. For example, to determine the probability of 3 heads in 4 tosses of a coin, you would enter `binompdf(4,.5,3)`.

Example 3: Finding a Binomial Probability (Land Leases)

The US Land Management Office regularly holds a lottery for the lease of government lands. Your company has won the rights to 12 leases. Historically, about 10% of these lands possess sufficient oil reserves for profitable operation. Construct the distribution for the number of leases that will be profitable. What is the probability that at least one of the leases will be profitable?

Solution

Let the binomial random variable X represent the number of profitable leases. The parameters of the binomial distribution are $n = 12$ and $p = 0.1$. Instead of laboriously calculating binomial probabilities using the formula, use the `binompdf(` function on a TI-83/84 Plus calculator. Table 4 shows the binomial distribution for X, with the probabilities rounded to four decimal places.

x	$P(X = x)$
0	0.2824
1	0.3766
2	0.2301
3	0.0852
4	0.0213
5	0.0038
6	0.0005
7	0.0000
8	0.0000
9	0.0000
10	0.0000
11	0.0000
12	0.0000

TABLE 4: Binomial Distribution $n = 12$, $p = 0.1$

In Table 4, notice that the probabilities for 7 through 12 are given as 0.0000. This does not mean that the probability is 0, because it is possible that 7 or more of the leases could be profitable, albeit a rather small probability. (The probability, however, is so small that it is not significant in the fourth decimal place.)

The probability that none of the leases will be profitable is equal to the probability that $X = 0$, which is 0.2824. Hence, the probability that at least one of the leases will be profitable is one minus the probability that none of the leases will be profitable or $1 - 0.2824 = 0.7176$.

Example 4: Finding a Binomial Probability (Belief in Ghosts)

According to a recent national poll, about 40% of Americans believe in ghosts. Assuming this percentage is accurate, if 20 people were randomly selected and asked if they believed in ghosts, what is the probability that 12 or more would say they do?

Solution

For this problem $n = 20$, $p = 0.4$, and x ranges from 12 to 20. So, nine probabilities must be calculated and added together to answer the question. This could be done using the binomial probability distribution function, but it would be cumbersome. Instead, use the `binompdf(` function on a TI-83/84 Plus calculator.

To determine the answer, find the following individual probabilities and add them together.

$$P(X \geq 12) = P(X = 12) + P(X = 13) + \cdots + P(X = 20)$$
$$= 0.0355 + 0.0146 + \cdots + 0.0000$$
$$= 0.0565$$

Alternately, we can use the binomial cumulative distribution function on a TI-83/84 Plus calculator, which computes a cumulative binomial probability of the form $P(X \leq x)$. To calculate the probability of getting no more than x successes using a TI-83/84 Plus calculator, first press `2nd` and then `vars` to open the DISTR menu, scroll down, and choose the `binomcdf(` option. Then input the number of trials n, the probability of a success p, and the number of successes x, separated by commas. Finally, press `)` and then `enter`. For this example, you would enter `binomcdf(20,.4,11)` to calculate the probability that 11 or fewer people said they believe in ghosts, $P(X \leq 11) \approx 0.9435$, and then subtract that from 1 to find the probability of the complement.

$$P(X \geq 12) = 1 - P(X \leq 11) = 1 - 0.9435 = 0.0565$$

The Shape of a Binomial Distribution

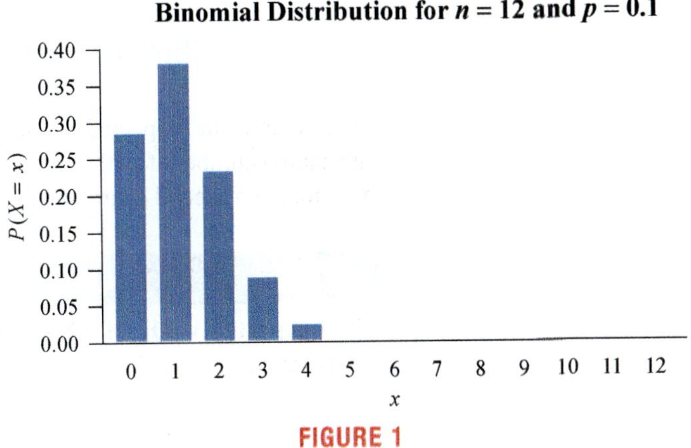

FIGURE 1

Binomial distributions have taken various shapes in the previous problems. The shape of the distribution depends upon the parameters n and p. If p is small the distribution tends to be skewed with a tail on the right (i.e., there are more successes when x is smaller). (See Figure 1.)

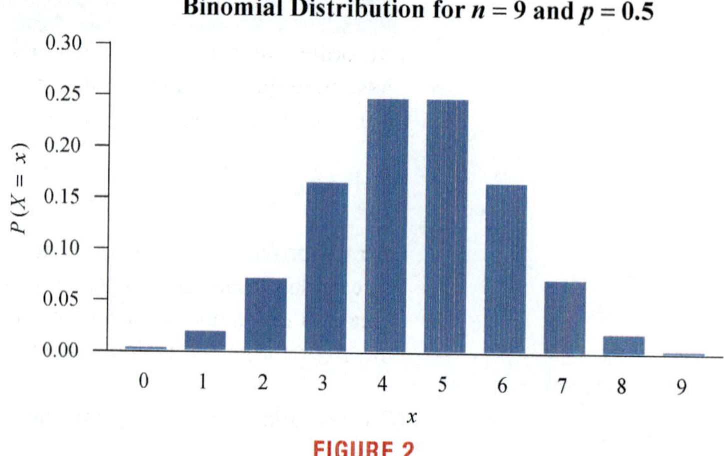

Binomial Distribution for $n = 9$ and $p = 0.5$

FIGURE 2

If p is near 0.5 the distribution is symmetrical. The graph in Figure 2 displays the probabilities for $n = 9$ and $p = 0.5$.

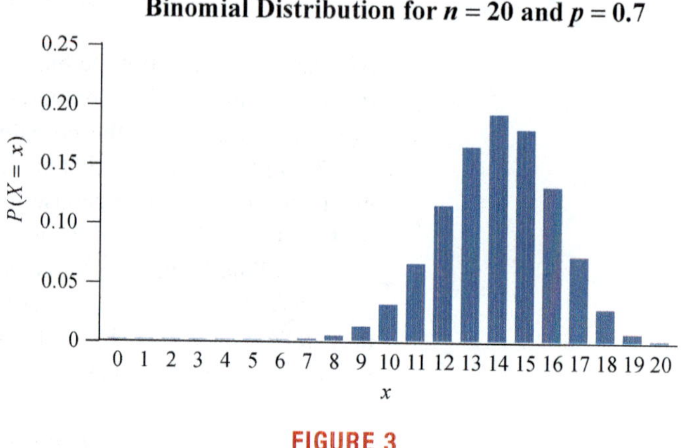

Binomial Distribution for $n = 20$ and $p = 0.7$

FIGURE 3

If p is large the distribution tends to be skewed with a long tail on the left (i.e., there are more successes when x is larger). The graph in Figure 3 displays the binomial probabilities for $n = 20$ and $p = 0.7$.

Expected Value and Variance

To calculate the expected value of a binomial random variable would require a substantial number of arithmetic operations. Fortunately, you can avoid this calculation by utilizing a special characteristic of the binomial distribution.

Expected Value of a Binomial Random Variable

The **expected value** of a binomial random variable can be computed using the expression

$$\mu = E(X) = np,$$

where n and p are the parameters of the binomial distribution.

It is important to remember that this formula is only valid for binomial random variables. Also, there is a simple method for obtaining the variance for a binomial random variable.

Variance and Standard Deviation of a Binomial Random Variable

To find the **variance** of a binomial random variable, use the expression

$$\sigma^2 = V(X) = np(1-p).$$

Thus, the **standard deviation** of a binomial random variable is given by

$$\sigma = \sqrt{V(X)} = \sqrt{np(1-p)}.$$

Example 5: Finding the Expected Value and Standard Deviation of a Binomial Random Variable

Compute the expected value and the standard deviation of the number of profitable leases in Example 3.

Solution

Since the random variable is binomial, we can use the formulas given just before this example. Since $n = 12$ and $p = 0.1$, the expected value is given by the following equation.

$$\mu = E(X) = np$$
$$= 12(0.1)$$
$$= 1.2$$

The variance is

$$\sigma^2 = V(X) = np(1-p)$$
$$= 12(0.1)(0.9)$$
$$= 1.08$$

which implies that the standard deviation is $\sqrt{1.08} \approx 1.039$.

Thus, if groups of 12 oil leases were purchased with the same probability of success (0.10 probability of a profitable lease), then the average number of profitable leases per group of 12 would be 1.2 and the standard deviation would be 1.039 leases.

9.4 EXERCISES

✔ CONCEPT CHECK

1. Describe the characteristics of a binomial experiment.

2. What are the parameters of a binomial probability model?

3. Give an example of a binomial experiment, other than the one used in the section.

4. What is the formula for the binomial probability distribution function?

5. What influences the shape of the binomial probability distribution?

6. How do you calculate the expected value of a binomial random variable? The variance? The standard deviation?

💡 PRACTICE

7. Calculate $_nC_x$ for each of the following combinations of x and n.
   a. $n = 5, x = 4$
   b. $n = 10, x = 8$
   c. $n = 15, x = 1$
   d. $n = 20, x = 0$

8. Calculate $_nC_x$ for each of the following combinations of x and n.
   a. $n = 4, x = 2$
   b. $n = 12, x = 8$
   c. $n = 18, x = 15$
   d. $n = 23, x = 20$

9. The random variable X is a binomial random variable with $n = 9$ and $p = 0.1$.
   a. Find the expected value of X.
   b. Find the standard deviation of X.
   c. Find the probability that X equals 2. (Use the formula for $P(X = x)$.)
   d. Find the probability that X is at most 3.
   e. Find the probability that X is at least 2.
   f. Find the probability that X is less than 5.

10. The random variable X is a binomial random variable with $n = 12$ and $p = 0.8$.
   a. Find the expected value of X.
   b. Find the standard deviation of X.
   c. Find the probability that X equals 7. (Use the formula for $P(X = x)$.)
   d. Find the probability that X is at most 4.
   e. Find the probability that X is at least 1.
   f. Find the probability that X is more than 10.

11. A real estate agent has ten properties that she shows. She feels that there is a ten percent chance of selling any one property during a week. The chance of selling any one property is independent of selling another property.
- **a.** What probability model would be appropriate for describing the number of properties sold each week?
- **b.** Compute the expected number of properties to be sold in a week.
- **c.** Compute the standard deviation of the number of properties sold each week.
- **d.** Compute the probability of selling one property in one week.
- **e.** Compute the probability of selling five properties in one week.
- **f.** Compute the probability of selling at least three properties in one week.

12. A small commuter airline is concerned about reservation no-shows and, correspondingly, how much they should overbook flights to compensate. Assume their commuter planes will hold 15 people. Industry research indicates that 20% of the people making a reservation will not show up for a flight. Whether or not one person takes the flight is considered to be independent of other persons holding reservations.
- **a.** What probability model would be appropriate for the number of passengers that actually take the flight?
- **b.** If the airline decides to book 18 people for each flight, how often will there be at least one person who will not get a seat?
- **c.** If they book 17 people, how often will there be at least one person who will not get a seat?
- **d.** If they book 16 people, how often will there be at least one person who will not get a seat?
- **e.** If they book 18 people for each flight, how often will there be one or more empty seats?
- **f.** If they book 17 people, how often will there be one or more empty seats?
- **g.** If they book 16 people, how often will there be one or more empty seats?
- **h.** Based on the results from parts **b.** to **g.** above, which booking policy do you prefer? Explain your answer.

13. Seven plants are operated by a garment manufacturer. They feel there is a ten percent chance for a strike at any one plant and the risk of a strike at one plant is independent of the risk of a strike at another plant. Let $X =$ the number of plants of the garment manufacturer that strike.
- **a.** Determine the probability distribution for X.
- **b.** Interpret the results for $P(X = 0)$, $P(X = 4)$, and $P(X = 7)$.
- **c.** Compute the expected value of X.
- **d.** Compute the standard deviation for X. Is this value large in relation to the expected value? In what units is the standard deviation expressed?

14. A company that makes traffic signal lights buys switches from a supplier. Out of each shipment of 1000 switches, the company will take a random sample of 10 switches. Let X equal the number of defective switches in the sample.
- **a.** The company has a policy of rejecting a lot if they find any defective switches in the sample. What is the probability that the shipment will be accepted if, in fact, 2% of the switches are actually defective?
- **b.** What is the probability that the shipment will be accepted if the percentage of defective switches is actually 5%?
- **c.** The company decides to change their policy and will accept the lot if they find no more than one defective switch. Repeat parts **a.** and **b.** for this new policy.

15. Parents have always wondered about the sex of a child before it is born. Suppose that the probability of having a male child was 0.5, and that the sex of one child is independent of the sex of other children.

    a.   Determine the probability of having exactly two girls out of four children.

    b.   What is the probability of having four boys out of four children?

16. A certain aspirin is advertised as being preferred by 4 out of 5 doctors. If the advertisement is assumed to be true, answer the following questions.

    a.   What is the probability that at least half of ten doctors chosen at random will prefer this brand of aspirin?

    b.   What is the probability that 9 out of 10 of the doctors will prefer this brand?

17. In manufacturing integrated circuits, the yield of the manufacturing process is the percentage of good chips produced by the process. The probability that an integrated circuit manufactured by the Ace Electronics Company will be defective is $p = 0.05$. If a random sample of 15 circuits is selected for testing, answer the following questions.

    a.   What is the probability that no more than one integrated circuit will be defective in the sample?

    b.   What is the expected number of defective integrated circuits in the sample?

18. The Alvin Secretarial Service procures temporary office personnel for major corporations. They have found that 90% of their invoices are paid within 10 working days. If a random sample of 12 invoices is checked, answer the following questions.

    a.   What is the probability that all of the invoices will be paid within 10 working days?

    b.   What is the probability that six or more of the invoices will be paid within 10 working days?

19. An experiment consists of rolling a pair of dice 10 times. On each roll the sum of the dots on the two dice is noted.

    a.   Find the probability that on any roll of the two dice the sum of the dots is either 7 or 11.

    b.   Find the probability that in the 10 rolls of the pair of dice, a 7 or 11 occurs 5 times.

    c.   Find the probability that in the 10 rolls of the pair of dice, a 7 or 11 does not occur at all.

    d.   Find the mean and variance of the number of times we see a 7 or 11 in the 10 rolls of the dice.

20. "Would you say you eat to live or live to eat?" was asked to each person in a sample of 1001 adults in a Gallup Poll taken in April 1996. Seventy-four percent of the respondents answered eat to live, 23% answered live to eat, and 3% had no opinion. Assuming these percents are accurate, find the probability, in 12 randomly chosen adults, that the number who would answer "eat to live" is

    a.   exactly 7,

    b.   no more than 10,

    c.   at most 11,

    d.   at least 3.

9.5 THE NORMAL DISTRIBUTION

■ TOPICS

- ■ From *z*-Scores to Percentages
- ■ Summary for Using the TI-83/84 Plus Calculator

As we begin to create a better picture of the way data are dispersed, we realize that, in many instances, the "shape" of the data falls into one of a few recognizable forms. Suppose we looked at the histogram of data collected for the heights of 150 random men. It might look something like the following.

Class	Frequency
4.5′–4.9′	10
5.0′–5.4′	44
5.5′–5.9′	49
6.0′–6.4′	35
6.5′–6.9′	12

TABLE 1: Heights of 150 Random Men

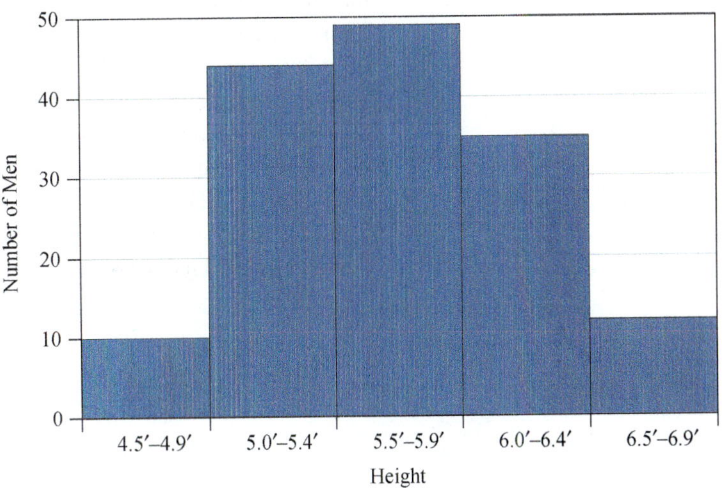

FIGURE 1

You can see that it's what we might describe as approximately bell-shaped and is roughly symmetrical about the middle of the graph. If we were to take bigger and bigger samples of men's heights, our histogram would begin to look even more symmetrical until, eventually, we wouldn't be able to notice where it wasn't exactly symmetrical anymore. It would become almost perfectly bell-shaped. If we were able to sample the heights of all men, we'd find that our histogram would become what we call the **normal distribution**, which is shown in Figure 2.

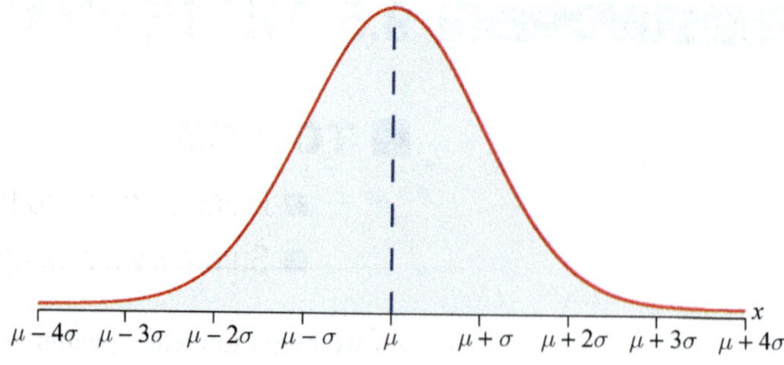

FIGURE 2

Characteristics of the Normal Distribution

1. It is bell-shaped, meaning it has only one mode that is at the center, and symmetrical.
2. The mean, median, and mode are all the same.
3. The total area under the curve is equal to 1.
4. It is completely defined by its mean and standard deviation.

The last characteristic of the normal distribution tells us that if we know a normal distribution's mean and standard deviation, we know how the distribution is dispersed, or spread out. Recall the Empirical Rule we looked at in an earlier section. By knowing the mean and the standard deviation of a given data set, we also know where the data lie in reference to the mean. Recall from the Empirical Rule that 68% of the data will lie within one standard deviation of the mean, 95% of the data will lie within two standard deviations, and 99.7% of the data will lie within three standard deviations of the mean. Remember, the standard deviation tells us how far on average a piece of data is away from the mean. So, the smaller the standard deviation, the closer the data are to one another. The larger the standard deviation, the more dispersed the data. Figure 3 is a picture of several normal distributions with equal means, but different standard deviations. Which distribution do you think has the largest standard deviation?

Various Normal Distributions

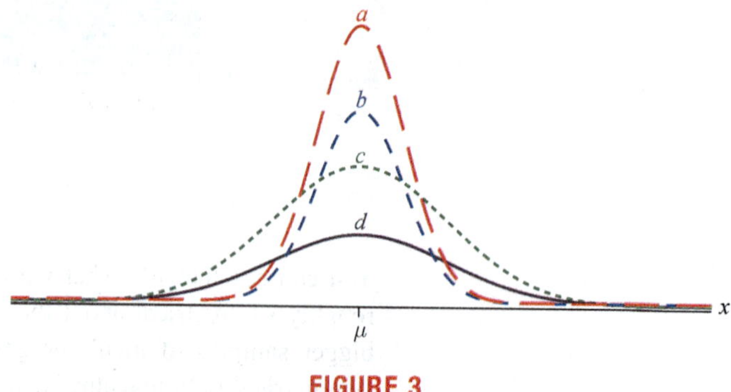

FIGURE 3

Curve *d* has the largest standard deviation, because it is the curve that is most spread out. Note that it's not the one with the highest peak.

Because so many natural data sets follow a normal distribution, we are able to use its features to help estimate and analyze entire populations of data. The **z-score** tells us how many standard deviations a particular piece of data lies away from the mean in a normal distribution. So, if *z* is a positive number, the piece of data is greater than the mean μ; if z is a negative number, the piece of data is less than the mean μ; and if z is equal to 0, the data value equals the mean μ.

z-Score

The **z-score** tells how many standard deviations a particular piece of data lies away from the mean in a normal distribution. The formula for finding a z-score is

$$z = \frac{\text{data value} - \text{mean}}{\text{standard deviation}},$$

or more formally with symbols,

$$z = \frac{x - \mu}{\sigma} \text{ for populations and } z = \frac{x - \bar{x}}{s} \text{ for samples.}$$

Let's first try our hand at calculating some z-scores.

Example 1: Calculating z-Scores

Given that the heights of Canadian women are normally distributed with a mean of 159.5 cm and a standard deviation of 7.1 cm, calculate the z-scores for the following pieces of data.

a. A height of 148.2 cm

b. A height of 160.3 cm

c. A height of 1.7 m

Solution

a. Using a height of 148.2 cm in the z-score formula, we have

$$z = \frac{\text{data value} - \text{mean}}{\text{standard deviation}} = \frac{148.2 - 159.5}{7.1} = \frac{-11.3}{7.1} \approx -1.59.$$

Because the height given is smaller than the mean, we expect the z-score to be negative, which it is. Data points below the mean will always have a negative z-score. What this tells us is that a height of 148.2 cm is 1.59 standard deviations below the mean.

b. The z-score for a height of 160.3 cm is calculated by

$$z = \frac{\text{data value} - \text{mean}}{\text{standard deviation}} = \frac{160.3 - 159.5}{7.1} = \frac{0.8}{7.1} \approx 0.11.$$

This tells us that a height of 160.3 cm, which we know to be only slightly larger than the mean, is 0.11 standard deviations above the mean. Remember, a data point greater than the mean will always have a positive z-score.

c. Before we can use the formula to compute the z-score, we must first convert the data into the same unit of measurement. That is, both should be in either meters or centimeters. Since the previous measurements were in centimeters, we'll change 1.7 m to centimeters. Note that you would get exactly the same z-score if you changed the mean and the standard deviation to meters instead. To change meters to centimeters, simply multiply by 100. So, we have 1.7 m = 170 cm. Substituting into the z-score formula we have

$$z = \frac{\text{data value} - \text{mean}}{\text{standard deviation}} = \frac{170 - 159.5}{7.1} = \frac{10.5}{7.1} \approx 1.48.$$

So, a height of 1.7 m is 1.48 standard deviations above the mean.

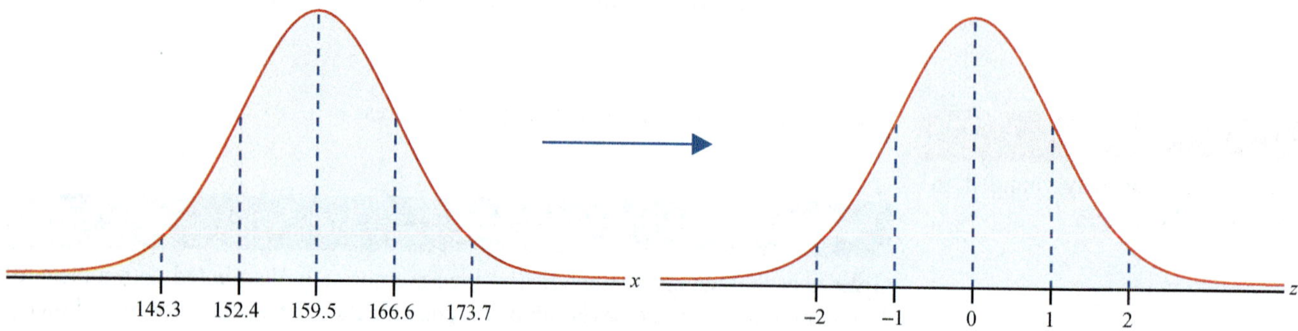

Recall from Example 10 in Section 9.3 that we used percentiles to compare language arts scores from a standardized test amongst three classmates. Another way to compare their results would be to look at their z-scores. But to do that, we'd need to know their actual scores. Let's change it up a bit and suppose that they all took different standardized tests, but we wanted a way to compare their language arts scores. Suppose Karl's score was 18 on his test, where the mean was 15 and the standard deviation was 1.4, Asher scored 18 on his test, where the mean was 17 and the standard deviation was 1.9, and Rylie scored 16 on her test, where the mean was 15.46 and the standard deviation was 0.8. To compare the scores, we need to calculate the z-score for each.

$$\text{Karl: } z = \frac{\text{data value} - \text{mean}}{\text{standard deviation}} = \frac{18 - 15}{1.4} = \frac{3}{1.4} \approx 2.14$$

$$\text{Asher: } z = \frac{\text{data value} - \text{mean}}{\text{standard deviation}} = \frac{18 - 17}{1.9} = \frac{1}{1.9} \approx 0.53$$

$$\text{Rylie: } z = \frac{\text{data value} - \text{mean}}{\text{standard deviation}} = \frac{16 - 15.46}{0.8} = \frac{0.54}{0.8} \approx 0.68$$

Now, knowing each of the respective z-scores, we can see that Karl did the best on his test, even though his raw score was the same as Asher's. Karl's score of 18 was 2.14 standard deviations above the mean. However, Asher and Rylie's scores were fewer standard deviations above their respective means. The larger the z-score, the further above the mean a data point lies.

From *z*-Scores to Percentages

Let's consider Karl's score again. He didn't just do a little better than Asher and Rylie, he actually did considerably better on his test. Initially, we were told that his score was above the third quartile, which means above the 75th percentile. Now that we know his *z*-score, we can actually find out exactly what percentage of students he did better than.

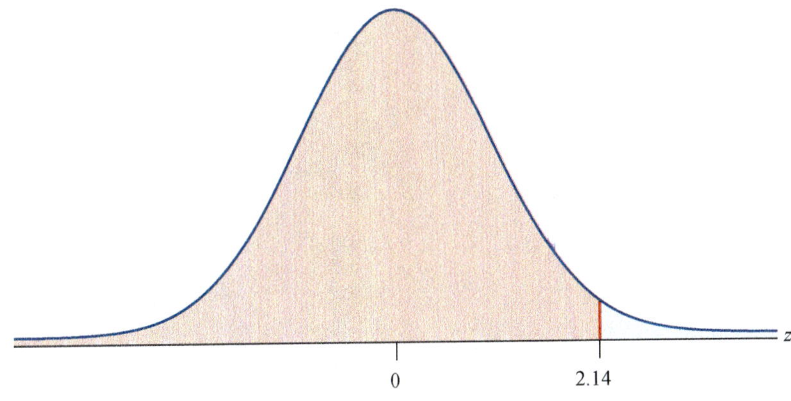

FIGURE 4

We can use a calculator to find what *percentage* of scores is below a *z*-score of 2.14. With a TI-83/84 Plus calculator, first find the correct distribution function by pressing 2nd and then vars. Choose option 2:normalcdf(, which stands for the Normal Cumulative Density Function. The format for entering the statistics is normalcdf(lower bound, upper bound). The terms *"lower bound"* and *"upper bound"* define the boundaries of the area under the curve that we wish to find. In Figure 4, we only have an upper boundary of 2.14. There is essentially no lower boundary since we are interested in all data that fall below 2.14.

Since we want to know the percentage *below z*, we can think of the lower bound as $-\infty$. We cannot enter $-\infty$ into the calculator, so we will enter a very small value for the lower endpoint, such as -10^{99}. This number appears as -1E99 when entered correctly into the calculator. To enter -1E99, use the following keystrokes: (-) 1 2nd , 9 9. Enter normalcdf(-1E99,2.14). The percentage is 0.9838. In other words, 98.38% of the data falls to the left of a *z*-score of 2.14. We now know that Karl's score was better than 98.38% of students taking that test.

It is also possible to use Wolfram|Alpha to find the percentage of scores below a certain value. To find the percentage of scores below $z = 2.14$, go to www.wolframalpha.com and type "z-score calculator" into the input field and then click the = button. Wolfram|Alpha will return a worksheet with a field to enter the endpoint value. Type "2.14" into the endpoint field and then click the Compute button. Wolfram|Alpha will return the following.

Input information:

probabilities for the normal distribution	
endpoint	2.14

Probabilities:

$z < 2.14$ (left-tailed p-value)	0.9838
$z > 2.14$ (right-tailed p-value)	0.01618
$|z| > 2.14$ (two-tailed p-value)	0.03235
$|z| < 2.14$ (confidence level)	0.9676

$|z|$ is the absolute value of z

Plot:

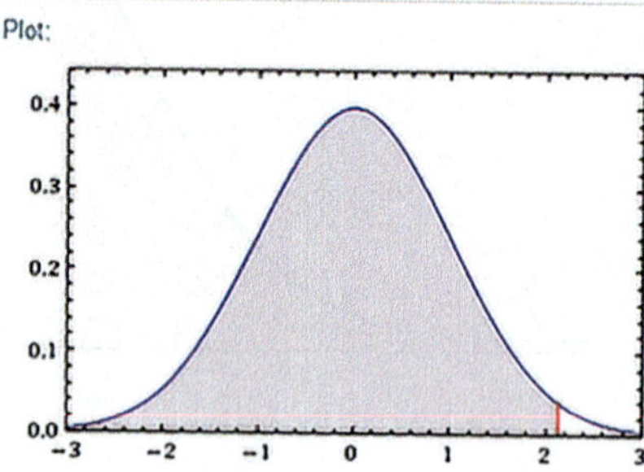

Source: Wolfram Alpha LLC. 2009. Wolfram|Alpha. http://www.wolframalpha.com/input/?i=zscore+calculator&a=FSe lect_**NormalProbabilities-.dflt-&f2=2.14&f=NormalProbabilities.z_2.14&a=*FVarOpt.1_***NormalProbabilities.z.*** NormalProbabilities.pr.**NormalProbabilities.l-.*NormalProbabilities.r---.*--&a=*FVarOpt.2-_**-.***NormalProbabilities. mu--.**NormalProbabilities.sigma---.**NormalProbabilities.z--- (accessed July 14, 2014).

Notice that the left-tailed *p*-value, right-tailed *p*-value, and two-tailed *p*-value are provided. We are interested in finding the percentage of scores below $z = 2.14$, so we use the left-tailed *p*-value answer of 0.9838 or 98.38%.

Since one of the properties of the normal distribution is that the total area under the curve is equal to one, the percentage of data points that lie below a *z*-score together with the percentage of data points that lie above it always add up to 100%. Therefore, with Karl's *z*-score, we also know that $100\% - 98.38\% = 1.62\%$ of students did better than Karl.

```
normalcdf(1.34,1
E99)
            .0901227339
```

Let's look at an example where we find the percentage of data points above a *z*-score. Suppose the score we are interested in is $z = 1.34$. We know the lower bound of the interval is 1.34 and the upper bound is ∞. Once again, we cannot enter ∞ into the calculator, so we'll just use a very large number, say 10^{99}. This is the same entry in the calculator as before (`-1E99`) but without the negative sign. So, we have `normalcdf(1.34,1E99)`. The percentage of data points above $z = 1.34$ is 0.0901 or 9.01%. When using Wolfram|Alpha to find the percentage above a *z*-score, use the right-tailed *p*-value answer.

If we'd like to know the percentage of data points between two *z*-scores, we simply use both *z*-scores as our lower and upper bounds, making sure to enter them in the correct order. For example, to find the percentage of data between a *z*-score of −0.98 and a *z*-score of 2.01, we enter `normalcdf(-0.98,2.01)` and find that 0.8142 or 81.42% of the data are between these *z*-scores.

Using Wolfram|Alpha, go to www.wolframalpha.com and type "z-score calculator" into the input field and then click the = button. Wolfram|Alpha will return a worksheet. Click "left endpoint and right endpoint" under the Compute button. This will return another worksheet with an input field for the left endpoint and the right endpoint. Type "-0.98" into the left endpoint input field and "2.01" into the right endpoint input field, then click the Compute button. Wolfram|Alpha will return the following.

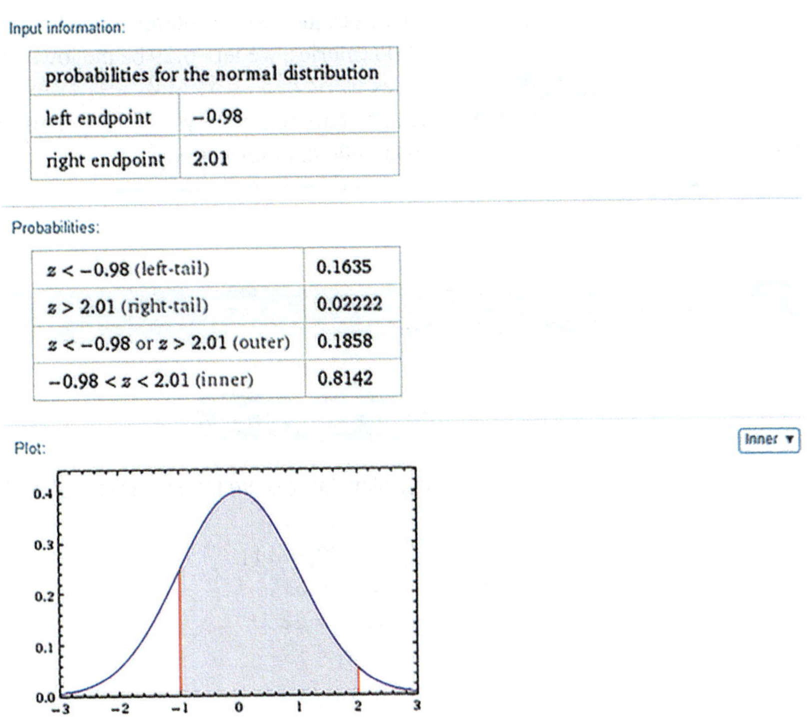

We are interested in the percentage between the two values, so we use the inner probability answer of 0.8142 or 81.42%.

Summary for Using the TI-83/84 Plus Calculator

Percentage of data points **below** z: `normalcdf(-1E99,z)`

Percentage of data points **above** z: `normalcdf(z,1E99)`

Percentage of data points **between** two z-scores: `normalcdf(z1,z2)`

Example 2: Putting It All Together

Suppose that the average caloric intake for women is 2050 calories per day, with a standard deviation of 175 calories. If we assume that caloric intake follows a normal distribution, find the percentage of females in your class that consume more than 2000 calories per day.

Solution

The first thing we need to do is find a z-score for the data point we are interested in of 2000 calories. Substituting into the formula, we have the following.

$$z = \frac{2000 - 2050}{175} \approx -0.29$$

Now, because we're interested in knowing the percentage that consumes more than 2000 calories, we let -0.29 be the lower bound. Our upper bound in this case is ∞. That gives us `normalcdf(-Ø.29,1E99)`. Calculating this using a TI-84 Plus gives a result of approximately `Ø.6141`. This means that 61.41% of females consume more than 2000 calories per day.

Skill Check Answers

1. $z \approx 0.68$

9.5 EXERCISES

💡 PRACTICE

Calculate the z-score for each given value. Round your answer to the nearest hundredth.

1. $\mu = 57, \sigma = 11$
  a. $x_1 = 63$
  b. $x_2 = 38$
  c. $x_3 = 58$

2. $\bar{x} = 1123, s = 241$
  a. $x_1 = 1284$
  b. $x_2 = 900$
  c. $x_3 = 1364$
  d. $x_4 = 1123$

3. $\bar{x} = 3.19, s = 0.06$
  a. $x_1 = 3.13$
  b. $x_2 = 3.22$
  c. $x_3 = 3.00$

4. $\mu = 178.15, \sigma = 49.3$
  a. $x_1 = 73.9$
  b. $x_2 = 267.3$
  c. $x_3 = 199.5$

5. Scores on a test have a mean of 73 and a standard deviation of 11. Steve has a score of 68. Convert Steve's score to a z-score.

Answer each question thoughtfully.

6. The annual rainfall in a town has a mean of 47.22 inches and a standard deviation of 10 inches. Last year there was 51 inches of rain. How many standard deviations from the mean is that?

7. Mason's weekly poker winnings have a mean of $144 and a standard deviation of $51. Last week he won $165. How many standard deviations from the mean is that?

Use the z-score formula to complete each table.

8. Find the missing value in each row of the table.

	z	x	μ	σ
a.		82.1	74.0	6.3
b.	1.05	162.3		8.9
c.	3.04		34.5	5.02
d.	−2.73	379	634	

9. Find the missing value in each row of the table.

	z	x	μ	σ
a.		4.33	6.10	2.04
b.	−2.39	−57		139.8
c.	0.58		118	21.2
d.	2.78	68	43	

Find the percentage of data points that lie below each z-score.

10. $z = -0.19$

11. $z = 1.46$

12. $z = 3.07$

13. $z = -2.22$

14. $z = 0$

Find the percentage of data points that lie above each z-score.

15. $z = 1.03$

16. $z = -1.87$

17. $z = -3.10$

18. $z = 2.84$

19. $z = 0$

Find the percentage of data points that lie between each pair of z-scores.

20. $z_1 = -1.00$
$z_2 = 1.00$

21. $z_1 = -2.40$
$z_2 = 1.73$

22. $z_1 = 2.00$
$z_2 = 3.00$

23. $z_1 = -3.01$
$z_2 = -0.56$

24. $z_1 = 0$
$z_2 = 2.61$

Find the percentage of data points that lie below z_1 and above z_2.

25. $z_1 = -1.10$
$z_2 = 1.10$

26. $z_1 = -2.84$
$z_2 = 2.84$

27. $z_1 = -1.75$
$z_2 = 0.53$

28. $z_1 = 1.09$
$z_2 = 2.88$

29. $z_1 = -0.01$
$z_2 = 0.02$

🚀 **APPLICATIONS**

30. Ava scored a 92 on a test with a mean of 71 and a standard deviation of 15. Charlotte had a score of 688 on a test with a mean of 493 and a standard deviation of 150. Which score was better with respect to their test? Assume the distributions of scores are approximately normal for both tests.

31. Avery started training to run a 5K. Her first race was a 5K for charity. She finished in 37.3 minutes. The average race time for the charity run was 36.42 with a standard deviation of 1.73 minutes. In her second race, Avery finished in 36.5 minutes. The race had a mean time of 33.02 minutes with a standard deviation of 2.45 minutes. In which race did Avery place higher in the list of finishers? Assume the distributions of finishing times are approximately normal for both races.

32. The average IQ score for adults is 100 with a standard deviation of 15. Assume that the distribution of IQ scores is approximately normal.
   a. Find the percentage of adults who have an IQ score less than 90.
   b. Find the percentage of adults who have an IQ score which exceeds the mean by at least 15 points.
   c. Find the percentage of adults who have an IQ score between 100 and 120.
   d. Find the percentage of adults who have an IQ score less than 55 or more than 145.

33. Assume the weights of offensive linemen in the NFL follow a normal distribution with a mean of 300 pounds and a standard deviation of 12.3 pounds.
   a. Find the percentage of linemen in the NFL who weigh more than 320 pounds.
   b. Find the percentage of linemen in the NFL who weigh between 275 and 325 pounds.
   c. Find the percentage of NFL linemen who weigh at least 260 pounds.
   d. Find the percentage of NFL linemen who weigh at most 315 pounds.

✎ WRITING & THINKING

34. The mean score for a set of data is marked by the dotted line on the following graph. Which value is a likely z-score for the indicated value? Choose from **a.** −2.1, **b.** 0, or **c.** 2.7.

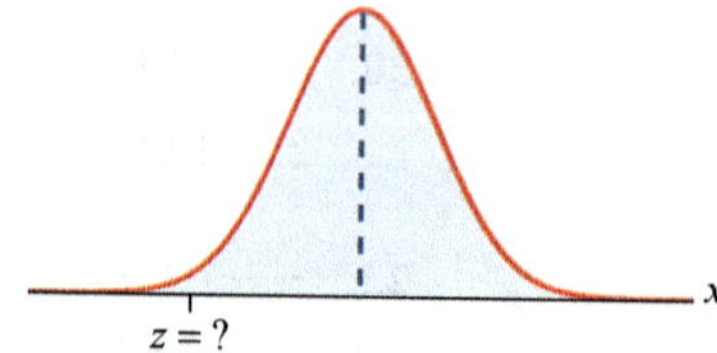

35. What is the minimum z-score that a piece of data would need to have in order to be in the top 10% of a normally distributed set of data?

36. What is an "average" z-score? Explain your answer.

37. What z-score represents the 1st quartile? 2nd quartile? 3rd quartile?

9.6 NORMAL APPROXIMATION TO THE BINOMIAL DISTRIBUTION

■ TOPICS

- ■ The Binomial Distribution

To approximate other distributions, the normal distribution can be very useful. Although it is a continuous distribution, it is used to approximate discrete distributions, specifically the binomial.

The Binomial Distribution

Calculating binomial probabilities can be quite time consuming if n is large. For example, suppose that you intend to sample 2000 subjects for a marketing research survey. If 50 percent of the population believes your product is superior to the competition's, what is the probability of obtaining 600 or fewer subjects who believe your company's product is superior?

$$P(X \leq 600) = P(X = 0) + P(X = 1) + P(X = 2) + \cdots + P(X = 599) + P(X = 600)$$

Determining the appropriate probability using the binomial distribution would require the calculation of 601 individual probabilities, many of which would have extremely large combinations such as the following.

$$_{2000}C_{400}\, 0.5^{400}\left(1 - 0.5\right)^{1600}$$

Computing this and the other 600 similar calculations would be a formidable task. The normal distribution is useful in approximating binomial probabilities. The larger the binomial parameter, n, the more accurate the approximation. Determining the probability described above using the normal approximation is trivial in comparison to calculating the exact probability using the binomial.

Recall that the normal distribution is a function of two parameters, the mean and the standard deviation. Thus, if the normal distribution is used to approximate the binomial distribution, it seems reasonable that the mean and standard deviation of the normal should be the same as the mean and standard deviation of the binomial that is being approximated. Specifically, let

$$\mu = E(X) = np, \text{ and}$$

$$\sigma = \sqrt{V(X)} = \sqrt{np(1-p)}.$$

To approximate a binomial with $n = 20$ and $p = 0.5$ would require a normal distribution with

$$\mu = (20)(0.5) = 10, \text{ and}$$

$$\sigma = \sqrt{(20)(0.5)(1-0.5)} = \sqrt{5} \approx 2.2361.$$

In this example, the shapes of the distributions are quite similar and consequently the approximation will be good.

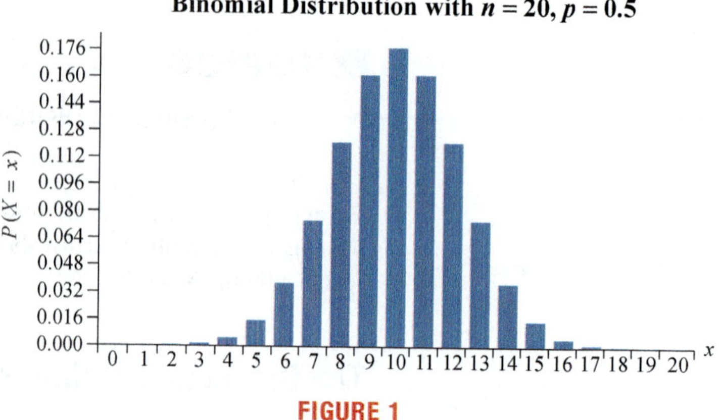

Binomial Distribution with $n = 20$, $p = 0.5$

FIGURE 1

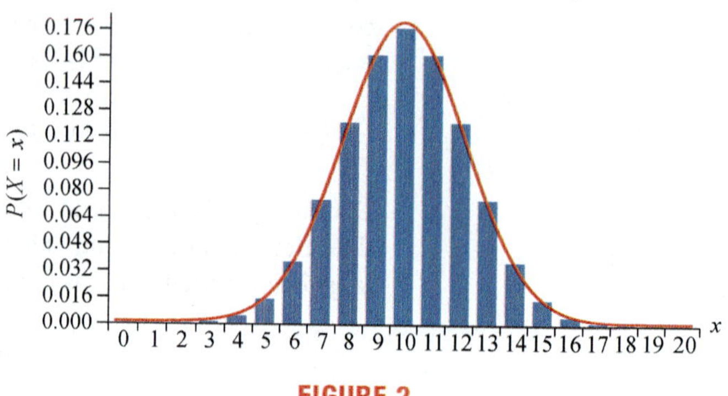

Normal Approximation to the Binomial Distribution with $n = 20$, $p = 0.5$

FIGURE 2

So when should the normal distribution be used to approximate the binomial distribution? Generally, the approximation is reasonable when the mean of the binomial, np, is greater than or equal to 5 and $n(1 - p)$ is greater than or equal to 5. The approximation becomes quite good when np is greater than or equal to 10 and $n(1 - p)$ is greater than or equal to 10.

The normal approximation to the binomial can be improved by using **continuity correction**.

Continuity Correction

Continuity correction is used when a discrete distribution is approximated using a continuous distribution. To apply continuity correction, subtract or add 0.5 (depending on the question at hand) to a selected value in order to find the desired probability.

Suppose that you wished to determine the probability that a binomial random variable ($n = 20$ and $p = 0.5$) is equal to 5. For continuous random variable X, the probability that X is equal to some specific value is equal to zero since there is no area under the curve for a single point, say $X = 5$. Therefore, approximating the probability using the normal would be equivalent to approximating the area of the shaded region given in Figure 3. Approximating the area of the region using the normal would require finding the area under the curve between 4.5 and 5.5.

Normal Approximation to the Binomial, $n = 20, p = 0.5$

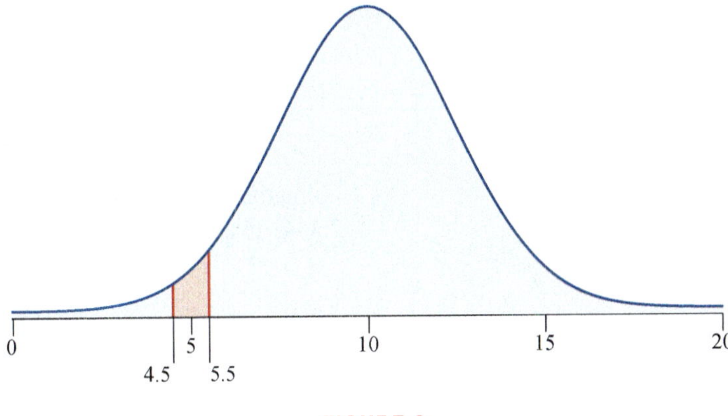

FIGURE 3

The continuity correction should be used whenever the normal distribution is used to approximate the binomial distribution. The following examples will illustrate how to approximate the binomial distribution using the normal distribution with continuity correction.

Example 1: Using the Normal Distribution to Approximate a Probability for a Binomial Random Variable

a. Assuming $n = 20$ and $p = 0.5$, use a normal random variable (Y) to approximate the probability that a binomial random variable (X) is 5 or less.

b. Using a normal distribution to approximate, find the probability that X is greater than 4.

Solution

a. This implies finding the area of the rectangles for 0, 1, 2, 3, 4, and 5.

Binomial Distribution with $n = 20, p = 0.5$

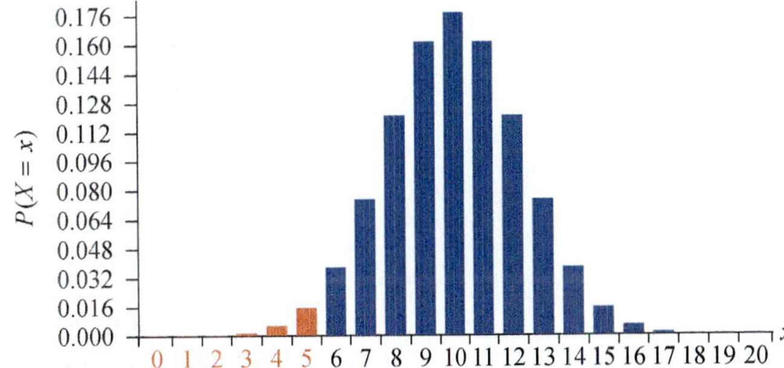

Instead of using the normal approximation $P(Y \leq 5)$, use the continuity correction $P(Y \leq 5.5)$ in order to accumulate all of the probabilities under the normal curve that correspond to the region $X \leq 5$.

To use the normal approximation, the mean and standard deviation of the binomial must be calculated.

$$\mu = E(X) = np = (20)(0.5) = 10$$

$$\sigma = \sqrt{np(1-p)} = \sqrt{(20)(0.5)(1-0.5)} = \sqrt{5} \approx 2.2361$$

Using the normal random variable Y with a mean of 10 and a standard deviation of 2.2361 to approximate the binomial using continuity correction,

$$P(Y \le 5.5) = P\left(z \le \frac{5.5-10}{2.2361}\right)$$

$$= P(z \le -2.01)$$

$$= 0.0222.$$

Normal Approximation to the Binomial, $n = 20$, $p = 0.5$

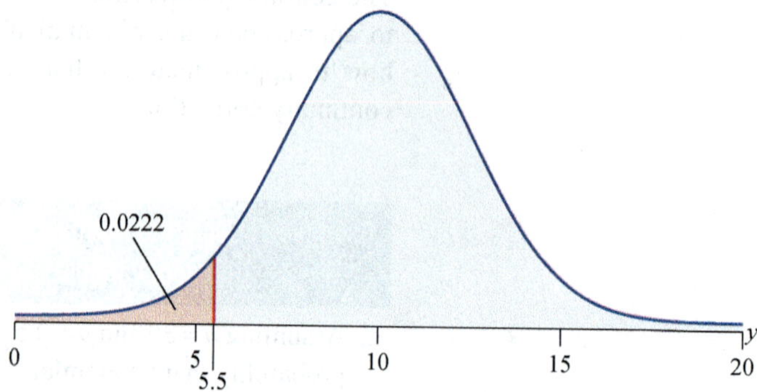

Thus, the probability that the random variable is 5 or less is 0.0222.

b.   **Binomial Distribution with $n = 20$, $p = 0.5$**

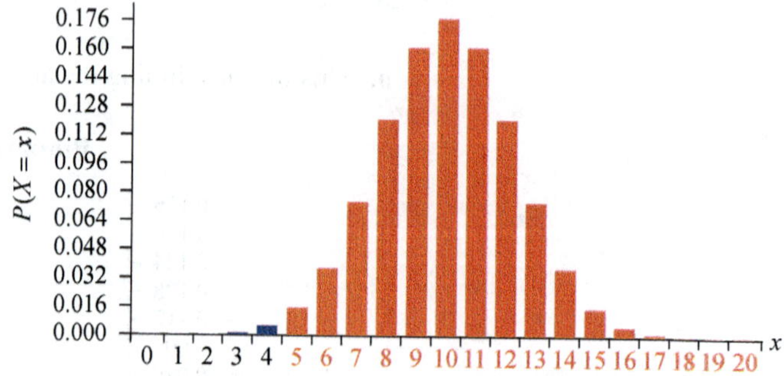

We are interested in the probability that a binomial random variable, X, is greater than 4. Since this is a discrete distribution, the probability that X is greater than 4 is equal to the probability that X is greater than or equal to 5. Thus, when using the normal approximation, we need to apply continuity correction and consider the probability that the normal random variable is greater than or equal to 4.5.

Using the normal random variable Y with a mean of 10 and a standard deviation of 2.2361 to approximate the binomial using continuity correction,

$$P(Y \geq 4.5) = P\left(z \geq \frac{4.5 - 10}{2.2361}\right)$$
$$= P(z \geq -2.46)$$
$$= 1 - P(z < -2.46)$$
$$= 1 - 0.0069$$
$$= 0.9931.$$

Normal Approximation to the Binomial, $n = 20$, $p = 0.5$

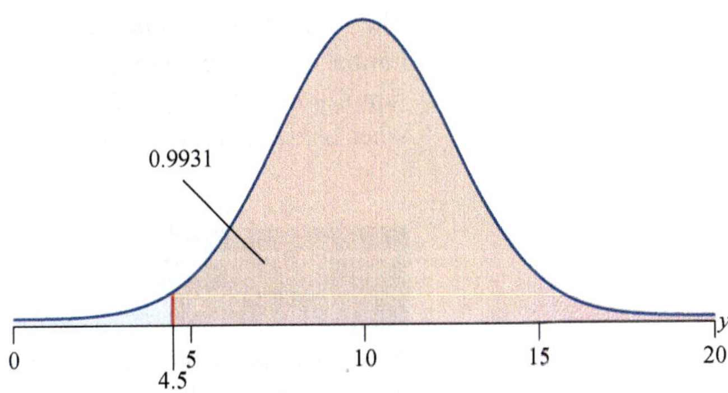

Thus, the probability that the random variable is greater than 4 is 0.9931.

Example 2: Testing a Claim Using the Normal Distribution to Approximate a Binomial Distribution

An advertising agency hired on behalf of Tech's development office conducted an ad campaign aimed at making alumni aware of their new capital campaign. Upon completion of the new campaign, the agency claimed that 20% of alumni in the state of Virginia were aware of the new campaign. To validate the claim of the agency, the development office surveyed 1000 alumni in the state and found that 150 were aware of the campaign. Assuming that the ad agency's claim is true, what is the probability that no more than 150 of the alumni in the random sample were aware of the new campaign?

Solution

Let $X =$ the number of alumni that were aware of the campaign.

X is a binomial random variable with $n = 1000$ and $p = 0.20$.

So, $np = 200$ and $n(1 - p) = 800$. Therefore the normal distribution is appropriate to use as an approximation to the binomial distribution.

The mean is $\mu = np = 200$ and the standard deviation is

$$\sigma = \sqrt{np(1 - p)} = \sqrt{160} \approx 12.6491.$$

We are interested in the probability that no more than 150 of the alumni in the sample were aware of the campaign, or $P(X \leq 150)$. However, since we are using the normal distribution to approximate the binomial, continuity correction must be applied.

Let Y be a normally distributed random variable with a mean of 200 and a standard deviation of 12.6491. Applying continuity correction, we are interested in the following probability.

$$P(Y \leq 150.5) = P\left(z \leq \frac{150.5 - 200}{12.6491}\right) = P(z \leq -3.91) \approx 0.0000$$

Thus, if the marketing agency's claim is true, the probability that 150 or fewer alumni are aware of the campaign is practically zero. This would lead the development office to believe that the agency's claim is false.

The smaller the sample size, the more the binomial distribution deviates from the normal distribution. For this reason, continuity correction is especially useful for small sample sizes. Example 3 illustrates the difference that continuity correction makes when using the normal distribution to approximate the binomial.

Example 3: Applying the Use of the Normal Distribution to Approximate a Binomial Distribution

A popular restaurant near Tech's campus accepts 200 reservations on Saturdays, the day of a Tech football game. Given that many of the reservations are made weeks in advance of game day, the restaurant expects that about eight percent will be no-shows. What is the probability that the restaurant will have no more than 20 no-shows on the next Saturday of a football weekend?

Solution

Let X = the number of no-shows.

Then X is a binomial random variable with $n = 200$ and $p = 0.08$.

Since $np = 16$ and $n(1 - p) = 184$ are both greater than 10, the normal distribution can be used to approximate the binomial probability.

For the binomial, $\mu = np = 16$ and

$$\sigma = \sqrt{np(1-p)} = \sqrt{(200)(0.08)(1-0.08)} = \sqrt{14.72} \approx 3.8367.$$

Using the normal distribution, Y, with a mean of 16 and a standard deviation of 3.8367, to approximate the binomial without continuity correction results in

$$P(Y \leq 20) = P\left(z \leq \frac{20 - 16}{3.8367}\right) = P(z \leq 1.04) = 0.8508.$$

Using continuity correction,

$$P(Y \leq 20.5) = P\left(z \leq \frac{20.5 - 16}{3.8367}\right) = P(z \leq 1.17) = 0.8790.$$

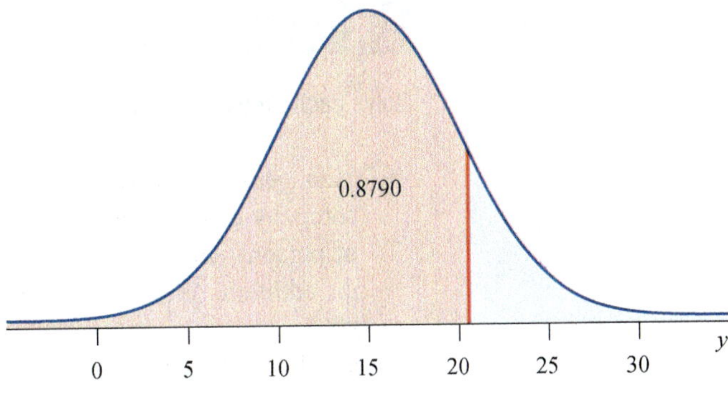

Thus, using the normal approximation and continuity correction, the probability that the restaurant will have no more than 20 no-shows is 0.8790. Notice that the continuity correction has a significant impact on the accuracy of the approximation. Using the binomial distribution, the exact probability is 0.8775.

9.6 EXERCISES

✔ CONCEPT CHECK

1. Why would you want to use the normal distribution to approximate a binomial distribution?

2. What are the parameters of a normal distribution used to approximate a binomial distribution?

3. What is continuity correction? How does it improve the normal approximation to the binomial?

💡 PRACTICE

4. Consider the probability that fewer than 15 out of the 123 people watching a movie have already read the book. Assume that the probability of a given person having read the book is 40%. Verify that a normal distribution can be used to approximate the binomial probability, or show how the conditions have not been met.

5. Consider the probability that at most 2 out of 30 television sets on an assembly line are defective. Assume that the probability of a given television set being defective is 5%. Verify that a normal distribution can be used to approximate the binomial probability, or show how the conditions have not been met.

Solve each problem. Use a normal distribution to approximate each probability.

6. Management at a small engineering company is considering the addition of a company cafeteria area. A random sample of 50 persons out of the total number of persons employed by the firm will be surveyed to see if they are in favor of the addition. Assume that the true percentage of persons that favor the addition is 90%.

   a. Find the expected number of employees in the sample who will favor the addition of the cafeteria area.

   b. Find the standard deviation of the number of employees in the sample who will favor the addition of the cafeteria area.

   c. What is the probability that between 35 and 37 employees (inclusive) in the sample will favor the cafeteria?

   d. What is the probability that more than 40 of the employees in the sample will favor the cafeteria?

   e. What is the probability that at most 38 of the employees in the sample will favor the cafeteria?

7. The accounting department of a large corporation checks the addition of expense reports submitted by executives before paying them. Historically, they have found that 15% of the reports contain addition errors. An auditor randomly selects 60 expense reports and audits them for addition errors.

   a. Find the expected number of reports in the sample that will have addition errors.

   b. Find the standard deviation of the number of reports sampled that will have addition errors.

   c. Find the probability that fewer than 10 of the sampled expense reports will have addition errors.

   d. Find the probability that at least 30 of the sampled expense reports will have addition errors.

   e. Find the probability that between 5 and 15 (inclusive) of the sampled expense reports will have addition errors.

8. A local electronics store purchased a market research study which suggests that 60 percent of all homes have gaming systems. A sample of 200 homes is selected to confirm the study's findings. If the marketing study is correct, answer the following questions.

   a. Find the expected number of homes sampled which will have gaming systems.

   b. Find the standard deviation of the number of homes in the sample which will have gaming systems.

   c. What is the probability that at most 80 of the sampled homes will have gaming systems?

   d. What is the probability that between 100 and 120 (inclusive) homes sampled will have gaming systems?

   e. What is the probability that at least 130 of the sampled homes will have gaming systems?

9. Suppose a virus is believed to infect two percent of the population. If a sample of 3000 randomly selected subjects are tested, answer the following questions.
   a. Find the expected number of subjects sampled that will be infected.
   b. Find the standard deviation of the number of subjects sampled that will be infected.
   c. What is the probability that fewer than 30 of the subjects in the sample will be infected?
   d. What is the probability that between 40 and 80 (inclusive) of the subjects in the sample will be infected?
   e. Find the probability that at least 70 of the subjects in the sample will be infected.

10

Chapter 10

LIMITS AND THE DERIVATIVE

10.1 ONE-SIDED LIMITS

■ TOPICS

- One-Sided Limits of Polynomial Functions
- Unbounded One-Sided Limits

This discussion will be on an intuitive basis, with a table of values and informal definitions.

For example, consider the function $f(x) = \dfrac{x^2 + x - 6}{x - 2}$.

Using a graphing utility with a window $[-6, 3]$ by $[-4, 8]$, we obtain a graph like the one in Figure 1. Using the $\boxed{\texttt{trace}}$ function, the student should make a table of (x, y)-values using the x-values

$$1, 1.5, 1.9, 1.99, 1.999$$

without having to retype the expression.

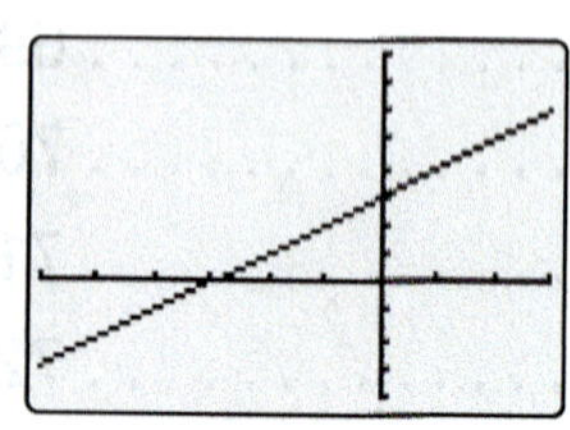

FIGURE 1

The results should be like those shown in the following table. The formula does not define a y-value for the x-input of $x = 2$ or, in other words, $f(2)$ is undefined. However, the table shows what happens to $f(x)$ for x-values less than 2 but close to 2.

x	1	1.5	1.9	1.99	1.999	$\rightarrow$	2
$f(x) = \dfrac{x^2 + x - 6}{x - 2}$	4	4.5	4.9	4.99	4.999	$\rightarrow$	5

Reading the table from left to right, we see that as the values of x become closer and closer to 2, the values for $f(x)$ become closer and closer to 5. We say that, as x **approaches 2 from the left, $f(x)$ approaches 5**. The number 5 is called the **left-hand limit**. Symbolically, we write

The function whose limit is to be found.

$$\lim_{x \to 2^-} \left(\frac{x^2 + x - 6}{x - 2} \right) = 5. \quad \leftarrow \text{The left-hand limit.}$$

The arrow signifies "approaches". The minus sign indicates x-values approaching 2 from the left.

Note that $x = 2$ is not in the domain of f and 5 is not a y-value for f. However, 5 is the limiting value of f as x nears from the left.

The values of the function $f(x) = \dfrac{x^2 - 1}{x - 1}$ in the next table are calculated as x approaches 1 from the right.

x	2	1.5	1.3	1.1	1.01	1.001	$\rightarrow$	1
$f(x) = \dfrac{x^2 - 1}{x - 1}$	3	2.5	2.3	2.1	2.01	2.001	$\rightarrow$	2

We say that, **as x approaches 1 from the right, $f(x)$ approaches 2**. In this case 2 is a
right-hand limit. We write

$$\lim_{x \to 1^+}\left(\frac{x^2-1}{x-1}\right)=2.$$

Read "The limit of $\dfrac{x^2-1}{x-1}$, as x
approaches 1 from the right, is 2."

Similarly, the values of $f(x)=\dfrac{x^2-1}{x-1}$ are calculated as x approaches 1 from the left.

x	0	0.5	0.9	0.99	0.999	$\to$	1
$f(x)=\dfrac{x^2-1}{x-1}$	1	1.5	1.9	1.99	1.999	$\to$	2

For the function $f(x)=\dfrac{x^2-1}{x-1}$, the tables have shown that the left-hand limit
(as $x \to 1^-$) and the right-hand limit (as $x \to 1^+$) are the same number, 2. The graph of
the function is given in Figure 2.

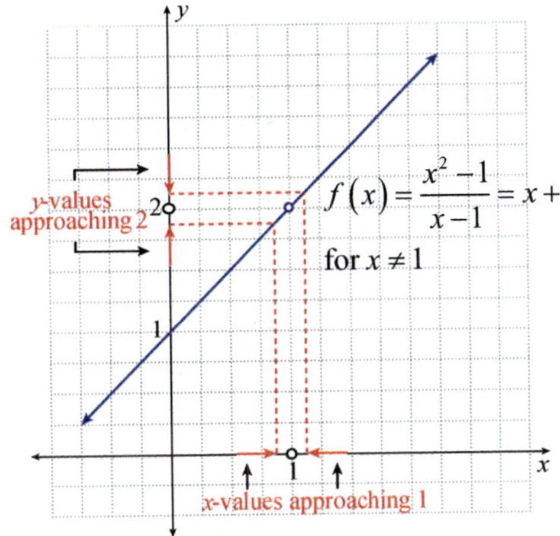

The open circle at the point
(1, 2) indicates that this point
is not included in the graph and
$x = 1$ is not in the domain of this
function.

FIGURE 2

Left-hand limits and right-hand limits need not be the same. For example, consider
another function defined as follows.

$$g(x)=\begin{cases}\dfrac{1}{2}x & \text{for } 0 \le x \le 4 \\ x-1 & \text{for } x > 4\end{cases}$$

This function is defined in pieces, and 4 is a key value for x since the function is
represented by a different expression for values of x on each side of 4. The following
tables show what happens to $g(x)$ as x approaches 4 first from the left and then from
the right.

		x **Approaches 4 from the Left**					
x	3	3.5	3.8	3.9	3.99	$\to$	4
$g(x)=\dfrac{1}{2}x$	1.5	1.75	1.9	1.95	1.995	$\to$	2

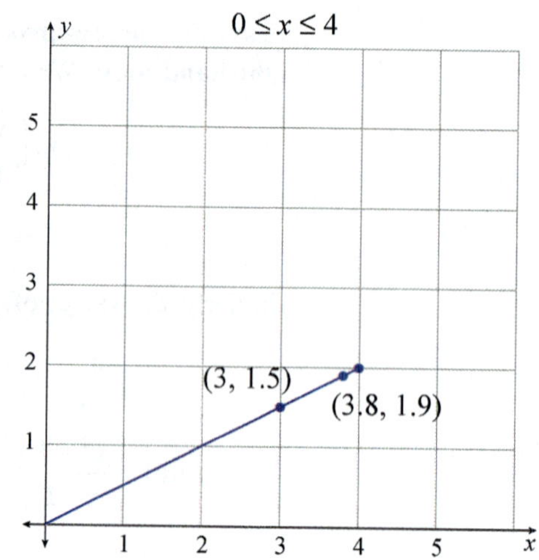

x Approaches 4 from the Right

4	←	4.01	4.1	4.2	4.5	5	x
3	←	3.01	3.1	3.2	3.5	4	$g(x) = x - 1$

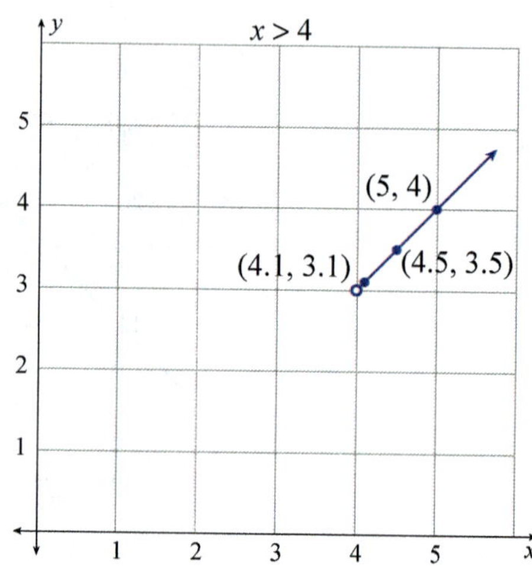

Thus $\lim\limits_{x \to 4^-} g(x) = \lim\limits_{x \to 4^-}\left(\dfrac{1}{2}x\right) = 2$ and $\lim\limits_{x \to 4^+} g(x) = \lim\limits_{x \to 4^+}(x - 1) = 3$.

If we look at the graph of $y = g(x)$ in Figure 3, we can see what happens to the y-values for x-values near 4.

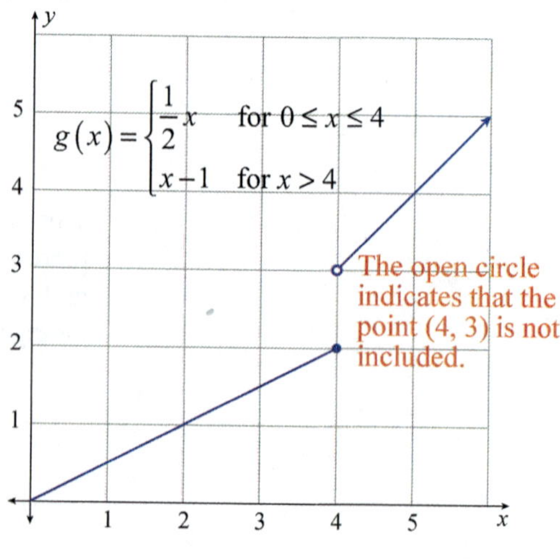

FIGURE 3

An intuitive definition of left- and right-hand limits, also called **one-sided limits**, follows.

One-Sided Limits Defined Informally

1. **Left-Hand Limits**
   If the values of a function $y = f(x)$ get closer and closer to some number K as values of x, that are smaller than some number a, get closer and closer to a, then we say that K is the **limit of $f(x)$ as x approaches a from the left**. We write

$$\lim_{x \to a^-} f(x) = K.$$

2. **Right-Hand Limits**
   If the values of a function $y = f(x)$ get closer and closer to some number M as values of x, that are larger than some number a, get closer and closer to a, then we say that M is the **limit of $f(x)$ as x approaches a from the right**. We write

$$\lim_{x \to a^+} f(x) = M.$$

To help remember one-sided limit notation, think of a^- as representing a **minus** small amounts, that is, numbers near a but slightly less than a. Similarly, think of a^+ as representing a **plus** small amounts, that is, numbers near a but slightly more than a.

In Example 1, the graph of the function is used to find one-sided limits.

Example 1: Finding One-Sided Limits

The graph of the function

$$f(x) = \begin{cases} x^2 & \text{if } 0 \le x \le 2 \\ 2 & \text{if } 2 < x < 4 \\ x & \text{if } 4 \le x \le 6 \end{cases}$$

is shown. Find the following one-sided limits by inspecting the graph.

a. $\displaystyle \lim_{x \to 2^-} f(x)$ **b.** $\displaystyle \lim_{x \to 2^+} f(x)$ **c.** $\displaystyle \lim_{x \to 4^-} f(x)$

d. $\displaystyle \lim_{x \to 4^+} f(x)$ **e.** $\displaystyle \lim_{x \to 0^+} f(x)$

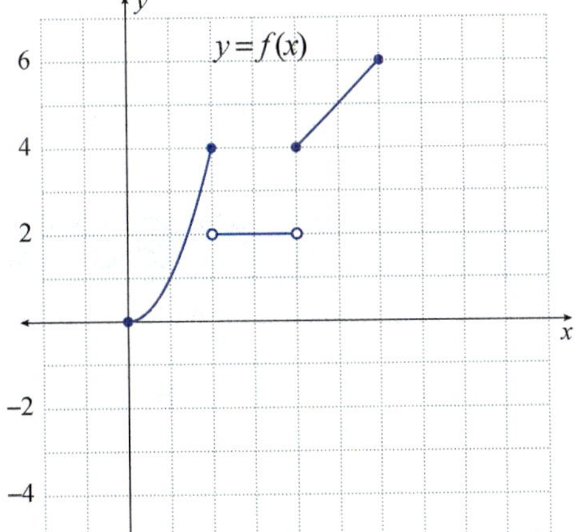

Remarks concerning the graph:

1. The value of the left and right-hand limits as x approaches 2 is independent of the actual y-value at $x = 2$, if there is one.

2. There is no $\displaystyle \lim_{x \to 0^-} f(x)$ and $\displaystyle \lim_{x \to 6^+} f(x)$ since the function is not defined for x-values approaching these numbers from the indicated directions.

Solution

a. $\lim\limits_{x \to 2^-} f(x) = 4$ **b.** $\lim\limits_{x \to 2^+} f(x) = 2$ **c.** $\lim\limits_{x \to 4^-} f(x) = 2$

d. $\lim\limits_{x \to 4^+} f(x) = 4$ **e.** $\lim\limits_{x \to 0^+} f(x) = 0$

One-Sided Limits of Polynomial Functions

In the discussion of one-sided limits thus far, we have dealt with functions whose graphs are missing single points or "jump" from one piece of the curve to another. That is, if you tried to trace any of these curves with a pencil, you would need to lift the pencil in some places. Now we will discuss limits of **polynomial functions**. Since the graphs of polynomial functions behave "nicely" (i.e., the graphs are smooth curves with no jumps or missing points), we will see that finding one-sided limits for these functions is particularly easy.

Polynomial Function

A **polynomial function** is a function of the form

$$p(x) = a_n x^n + a_{n-1} x^{n-1} + \cdots + a_2 x^2 + a_1 x + a_0$$

where each exponent is a positive integer and $a_n, a_{n-1}, \ldots, a_0$ are real numbers.

Two familiar forms of polynomial functions are $p(x) = a_1 x + a_0$ (called **linear functions**) and $p(x) = a_2 x^2 + a_1 x + a_0$ (called **quadratic functions**).

The values for one-sided limits of polynomial functions as $x \to a^-$ or $x \to a^+$ can be found by simply substituting a for x in the function. The following remark emphasizes this.

One-Sided Limits of Polynomial Functions

If p is a polynomial function and a is a real number, then

$$\lim\limits_{x \to a^-} p(x) = p(a) \quad \text{and} \quad \lim\limits_{x \to a^+} p(x) = p(a).$$

This important fact, actually a theorem, makes polynomial functions particularly easy to deal with, and it will be discussed in more detail later.

Example 2: Finding One-Sided Limits

Find the values of each of the following one-sided limits.

a. $\lim\limits_{x \to 3^-} \left(x^2 + 3x - 7\right)$ **b.** $\lim\limits_{x \to 2^+} \left(x^3 - 5x\right)$ **c.** $\lim\limits_{x \to 0^-} \left(4x^2 - 8x + 3\right)$

Solution

Since each function is a polynomial function, the one-sided limits can be found by substituting a for x.

a. $\lim\limits_{x\to 3^-}\left(x^2+3x-7\right)=3^2+3\cdot 3-7=11$ ⠀⠀⠀Substitute $a=3$ for x.

b. $\lim\limits_{x\to 2^+}\left(x^3-5x\right)=2^3-5\cdot 2=-2$ ⠀⠀⠀Substitute $a=2$ for x.

c. $\lim\limits_{x\to 0^-}\left(4x^2-8x+3\right)=4\cdot 0^2-8\cdot 0+3=3$ ⠀⠀⠀Substitute $a=0$ for x.

Unbounded One-Sided Limits

Consider the function $f(x)=\dfrac{1}{x-2}$. This function is undefined at $x=2$. However, in discussing limits, we are more interested in what happens to $f(x)$ as x gets close to 2, rather than at $x=2$. The following table shows what happens to $f(x)$ as $x\to 2^-$.

x	1.5	1.8	1.9	1.99	1.999	$\to$	2
$f(x)=\dfrac{1}{x-2}$	-2	-5	-10	-100	-1000	$\to$	?

The table shows that, as $x\to 2^-$ the values of $f(x)$ are negative and become smaller and smaller (negative, but large in absolute value). In fact, $f(x)$ decreases without bound. We say that $f(x)$ approaches negative infinity ($-\infty$). (**Note:** The symbol, $-\infty$, is not a number. It is used to indicate unboundedness in the negative y direction.) We write

$$\lim_{x\to 2^-}\left(\frac{1}{x-2}\right)=-\infty.$$

The next table shows what happens to $f(x)$ as $x\to 2^+$.

x	2.5	2.2	2.1	2.01	2.001	$\to$	2
$f(x)=\dfrac{1}{x-2}$	2	5	10	100	1000	$\to$	?

The table shows that, as $x\to 2^+$, the values of $f(x)$ are positive and become larger and larger. In fact, $f(x)$ increases without bound. We say that $f(x)$ approaches positive infinity ($+\infty$). (**Note:** The symbol, $+\infty$, is not a number. It is used to indicate unboundedness in the positive y direction.) We write

$$\lim_{x\to 2^+}\left(\frac{1}{x-2}\right)=+\infty.$$

The graph of the function $f(x)=\dfrac{1}{x-2}$ is shown in Figure 4. The vertical line $x=2$ is called a **vertical asymptote** since, as the values of x approach 2, the curve approaches this line but never touches it.

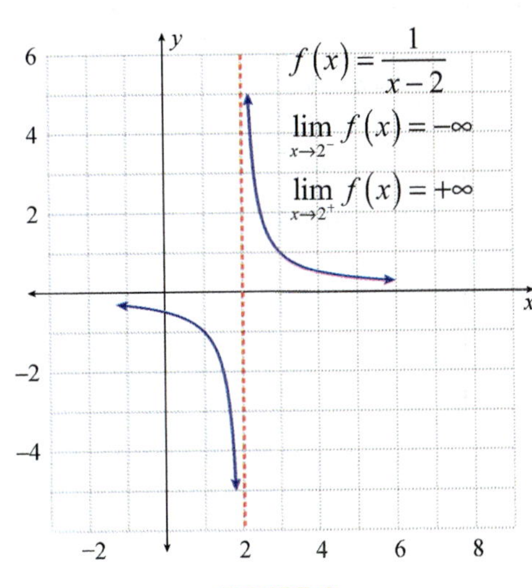

FIGURE 4

NOTE

If a limit is indicated to be $+\infty$ or $-\infty$, then it does not exist. These symbols are used only to indicate the direction of unboundedness.

Infinite One-Sided Limits (Unbounded One-Sided Limits)

1. $+\infty$ Limits

   If $f(x)$ increases without bound as x approaches a from the left (or from the right), then we say that $f(x)$ approaches positive infinity, $+\infty$. (See Figure 5(A) and (B).) We write

   $$\lim_{x \to a^-} f(x) = +\infty \quad \text{or} \quad \lim_{x \to a^+} f(x) = +\infty .$$

2. $-\infty$ Limits

   If $f(x)$ decreases without bound as x approaches a from the left (or from the right), then we say that $f(x)$ approaches negative infinity, $-\infty$. (See Figure 5(C) and (D).) We write

   $$\lim_{x \to a^-} f(x) = -\infty \quad \text{or} \quad \lim_{x \to a^+} f(x) = -\infty .$$

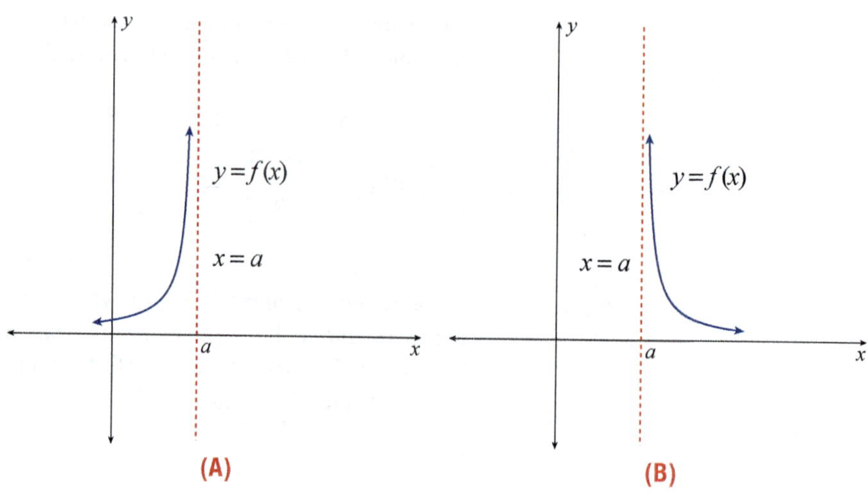

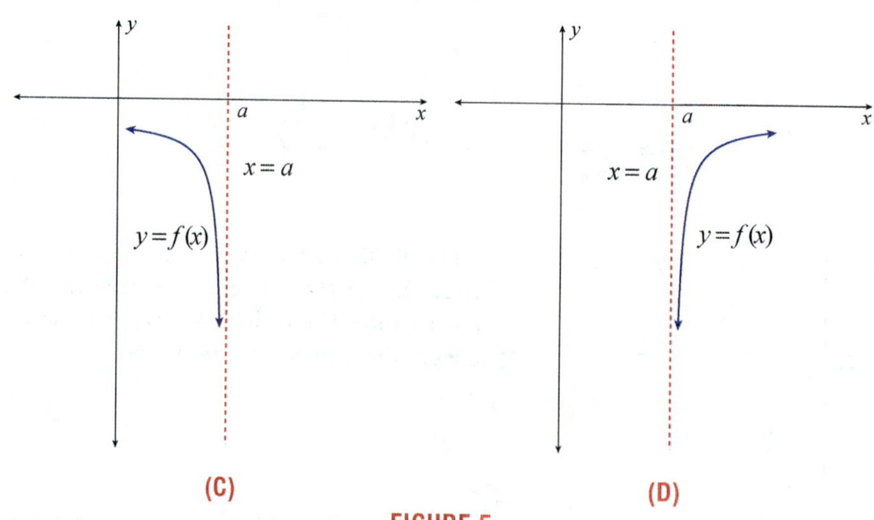

FIGURE 5

Example 3: Finding Infinite One-Sided Limits

a. Find $\lim\limits_{x \to -3^-} \left(\dfrac{x+5}{x+3} \right)$.

b. Find $\lim\limits_{x \to -3^+} \left(\dfrac{x+5}{x+3} \right)$.

Solution

a. As x approaches -3 from the left, the denominator will always be negative but will approach 0; the absolute value of the denominator will get smaller and smaller. For example, $-3.1 + 3 = -0.1$, $-3.01 + 3 = -0.01$, and so on. Meanwhile, the numerator will approach $-3 + 5 = 2$. Thus the fraction $\dfrac{x+5}{x+3}$ will become very large in the negative sense (or unbounded in the negative direction). So $\lim\limits_{x \to -3^-} \left(\dfrac{x+5}{x+3} \right) = -\infty$.

b. Here, as x approaches -3 from the right, $x + 3$ will always be positive. For example, $-2.9 + 3 = +0.1$, $-2.99 + 3 = +0.01$, and so on. Thus the denominator will approach 0 through positive values, and the numerator will approach $-3 + 5 = 2$. Therefore, the fraction $\dfrac{x+5}{x+3}$ will become unbounded in the positive direction, and we have

$$\lim\limits_{x \to -3^+} \left(\frac{x+5}{x+3} \right) = +\infty.$$

Example 4: Finding One-Sided Limits Using a Graph

Study the graph shown for $y = f(x)$ and find the following one-sided limits.

a. $\lim\limits_{x \to 2^-} f(x)$
b. $\lim\limits_{x \to 2^+} f(x)$
c. $\lim\limits_{x \to 4^-} f(x)$
d. $\lim\limits_{x \to 4^+} f(x)$

Solution

a. $\lim\limits_{x \to 2^-} f(x) = +\infty$

b. $\lim\limits_{x \to 2^+} f(x) = 3$

Note: $f(2) = 1$ according to the graph, but this fact does not affect either the left- or right-hand limits in parts **a.** and **b.**

c. $\lim\limits_{x \to 4^-} f(x) = 3$

d. $\lim\limits_{x \to 4^+} f(x) = 3$

10.1 EXERCISES

💡 PRACTICE

In Exercises 1–6, determine the limits. In each case, make a suitable table, with four values, to support your answer. Choose the fourth value ±0.001 from the indicated a-value.

1. $\lim\limits_{x \to 7^-} \left(\dfrac{x^2 - 49}{x - 7} \right)$

2. $\lim\limits_{x \to 7^+} \left(\dfrac{x^2 + 49}{x - 7} \right)$

3. $\lim\limits_{x\to3^+}\left(\dfrac{x^3-9x^2+27x-27}{x-3}\right)$

4. $\lim\limits_{h\to0^+}\left(\dfrac{\sqrt{4+h}}{h}\right)$

5. $\lim\limits_{a\to1^+}\left(\dfrac{a^{10}-1}{a-1}\right)$

6. $\lim\limits_{n\to\sqrt{2}^-}\left(\dfrac{n^2-2}{n-\sqrt{2}}\right)$

Given the table for $\lim\limits_{x\to a^+} f(x)$ in Exercise 7 and $\lim\limits_{x\to a^-} f(x)$ in Exercises 8–9, **a.** give the value for *a* and **b.** determine the limit, if there is one.

7.

x	y
2.500	0.2222
2.100	0.2439
2.010	0.2494
2.001	0.2499

8.

x	y
3.800	15.60
3.900	15.80
3.990	15.98
3.999	15.998

9.

x	y
3.000	3.43
3.100	11.99
3.140	313.90
3.141	843.60

10. a. Complete the table.

x	$f(x)=3x-1$
1	
1.4	
1.8	
1.9	
1.99	
1.999	

b. Find $\lim\limits_{x\to2^-}(3x-1)$.

11. a. Complete the table.

x	$f(x)=x^2-2$
0	
−0.4	
−0.8	
−0.9	
−0.99	
−0.999	

b. Find $\lim\limits_{x\to-1^+}\left(x^2-2\right)$.

12. a. Complete the table.

x	$f(x)=\dfrac{x^2-1}{x+1}$
2	
1.6	
1.2	
1.1	
1.01	
1.001	

b. Find $\lim\limits_{x\to1^+}\left(\dfrac{x^2-1}{x+1}\right)$.

13. a. Complete the table.

x	$f(x)=x^2+3$
2	
2.4	
2.8	
2.9	
2.99	
2.999	

b. Find $\lim\limits_{x\to3^-}\left(x^2+3\right)$.

14. a. Complete the table.

x	$f(x)=\dfrac{1}{x-4}$
3	
3.4	
3.8	
3.9	
3.99	
3.999	

b. Find $\lim\limits_{x\to 4^{-}}\left(\dfrac{1}{x-4}\right)$.

15. a. Complete the table.

x	$f(x)=\dfrac{x}{x+2}$
-3	
-2.6	
-2.2	
-2.1	
-2.01	
-2.001	

b. Find $\lim\limits_{x\to -2^{-}}\left(\dfrac{x}{x+2}\right)$.

16. a. Complete the table.

x	$f(x)=\dfrac{x^{2}-4}{x+2}$
-1	
-1.4	
-1.8	
-1.9	
-1.99	
-1.999	

b. Find $\lim\limits_{x\to -2^{+}}\left(\dfrac{x^{2}-4}{x+2}\right)$.

17. a. Complete the table.

x	$f(x)=\dfrac{x-3}{x^{2}-2x-3}$
4	
3.6	
3.2	
3.1	
3.01	
3.001	

b. Find $\lim\limits_{x\to 3^{+}}\left(\dfrac{x-3}{x^{2}-2x-3}\right)$.

Find In Exercises 18–23, use the graph of $y=f(x)$ to find the limits.

18. $\lim\limits_{x\to -1^{-}} f(x)$

19. $\lim\limits_{x\to -1^{+}} f(x)$

20. $\lim\limits_{x\to 2^{-}} f(x)$

21. $\lim\limits_{x\to 2^{+}} f(x)$

22. $\lim\limits_{x\to 3^{-}} f(x)$

23. $\lim\limits_{x\to 3^{+}} f(x)$

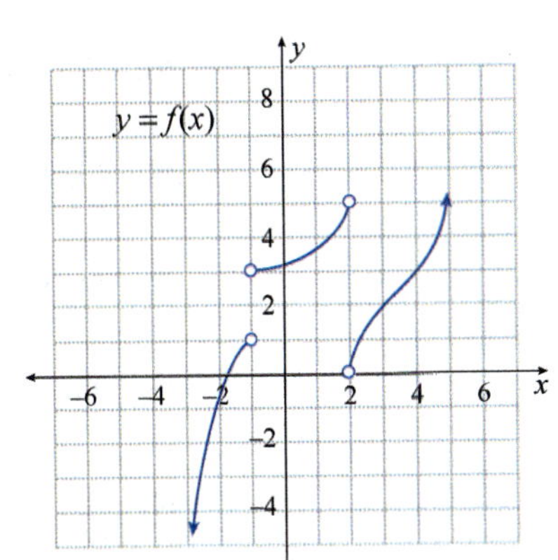

In Exercises 24–29, use the graph of $y = f(x)$ to find the limits.

24. $\lim\limits_{x \to -1^-} f(x)$

25. $\lim\limits_{x \to -1^+} f(x)$

26. $\lim\limits_{x \to 0^-} f(x)$

27. $\lim\limits_{x \to 0^+} f(x)$

28. $\lim\limits_{x \to 3^-} f(x)$

29. $\lim\limits_{x \to 3^+} f(x)$

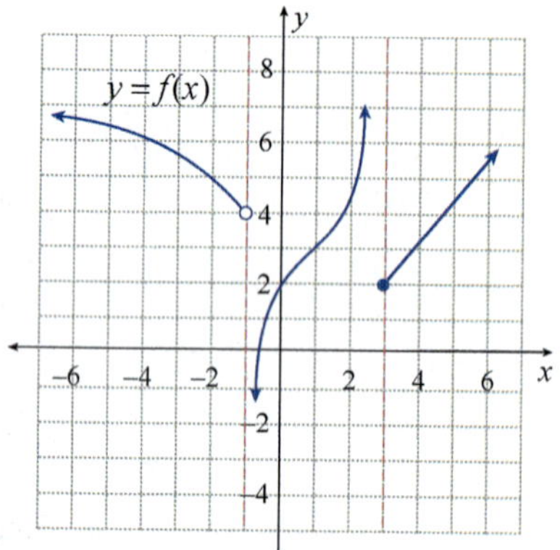

In Exercises 30–35, use the graph of $y = f(x)$ to find the limits.

30. $\lim\limits_{x \to 0^-} f(x)$

31. $\lim\limits_{x \to 0^+} f(x)$

32. $\lim\limits_{x \to 4^-} f(x)$

33. $\lim\limits_{x \to 4^+} f(x)$

34. $\lim\limits_{x \to 2^-} f(x)$

35. $\lim\limits_{x \to 2^+} f(x)$

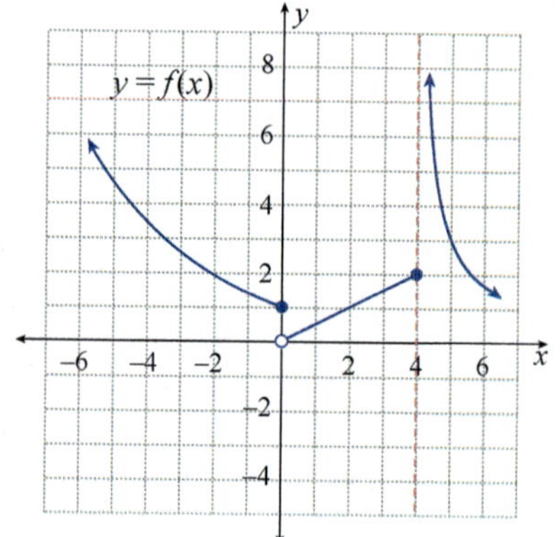

Find the one-sided limits indicated in Exercises 36–59.

36. $\lim\limits_{x \to 2^+} (5x - 3)$

37. $\lim\limits_{x \to -1^+} (2x + 7)$

38. $\lim\limits_{x \to 0^-} (4 - 3x)$

39. $\lim\limits_{x \to 3^-} (1 - 6x)$

40. $\lim\limits_{x \to 2^-} (x^2 - 3x + 1)$

41. $\lim\limits_{x \to -5^+} (x^2 + 4x - 2)$

42. $\lim\limits_{x \to -4^+} (x^2 - x + 3)$

43. $\lim\limits_{x \to -3^-} (x^2 + 2x - 3)$

44. $\lim\limits_{x \to 10^-} (0.01x^2 + 7x - 30)$

45. $\lim\limits_{x \to 10^+} (0.2x^2 - 5x + 6)$

46. $\lim\limits_{x \to 0^+} \left(\dfrac{x - 3}{x} \right)$

47. $\lim\limits_{x \to 0^-} \left(\dfrac{2x + 1}{x} \right)$

48. $\lim\limits_{x \to 1^+} \left(\dfrac{x - 2}{x - 1} \right)$

49. $\lim\limits_{x \to 1^-} \left(\dfrac{x - 2}{x - 1} \right)$

50. $\lim\limits_{x \to 2^-} \left(\dfrac{1}{x + 2} \right)$

51. $\lim\limits_{x \to 2^+} \left(\dfrac{1}{x + 2} \right)$

52. $\lim\limits_{x \to 3^+} \left(\dfrac{1}{x + 1} \right)$

53. $\lim\limits_{x \to 1^+} \left(\dfrac{1}{x - 5} \right)$

54. $f(x) = \begin{cases} 2 - 3x & \text{if } x < 2 \\ x - 1 & \text{if } x \ge 2 \end{cases}$

   a. $\lim\limits_{x \to 2^-} f(x)$

   b. $\lim\limits_{x \to 2^+} f(x)$

55. $f(x) = \begin{cases} x^2 + 2 & \text{if } 0 \le x \le 3 \\ 2x + 5 & \text{if } x > 3 \end{cases}$

   a. $\lim\limits_{x \to 3^-} f(x)$

   b. $\lim\limits_{x \to 3^+} f(x)$

56. $f(x) = \begin{cases} 3x + 1 & \text{if } 0 \le x \le 4 \\ x^2 - 3 & \text{if } x > 4 \end{cases}$

   a. $\lim\limits_{x \to 4^-} f(x)$

   b. $\lim\limits_{x \to 4^+} f(x)$

57. $f(x) = \begin{cases} x^3 & \text{if } x < 2 \\ x^2 + 5 & \text{if } x \ge 2 \end{cases}$

   a. $\lim\limits_{x \to 2^-} f(x)$

   b. $\lim\limits_{x \to 2^+} f(x)$

58. $f(x) = \begin{cases} 3 - 2x & \text{if } x < 1 \\ x & \text{if } 1 \le x \le 4 \\ \dfrac{1}{x - 4} & \text{if } x > 4 \end{cases}$

   a. $\lim\limits_{x \to 1^-} f(x)$ **b.** $\lim\limits_{x \to 1^+} f(x)$

   c. $\lim\limits_{x \to 4^-} f(x)$ **d.** $\lim\limits_{x \to 4^+} f(x)$

59. $f(x) = \begin{cases} x^2 - 1 & \text{if } 0 \le x < 2 \\ 3 & \text{if } 2 \le x \le 5 \\ \dfrac{1}{x - 5} & \text{if } x > 5 \end{cases}$

   a. $\lim\limits_{x \to 2^-} f(x)$ **b.** $\lim\limits_{x \to 2^+} f(x)$

   c. $\lim\limits_{x \to 5^-} f(x)$ **d.** $\lim\limits_{x \to 5^+} f(x)$

🚀 APPLICATIONS

60. Parking rates: The rate for parking in the short-term lot (maximum of 24 hours) at the airport is $1.00 for the first hour plus $0.75 for each additional hour or part thereof, with a maximum cost of $7.00. The function for the cost of parking on this lot for t hours (up to 24 hours) is as follows.

$$C(t) = \begin{cases} 1.00 & \text{for } 0 < t \le 1 \\ 1.75 & \text{for } 1 < t \le 2 \\ 2.50 & \text{for } 2 < t \le 3 \\ 3.25 & \text{for } 3 < t \le 4 \\ 4.00 & \text{for } 4 < t \le 5 \\ 4.75 & \text{for } 5 < t \le 6 \\ 5.50 & \text{for } 6 < t \le 7 \\ 6.25 & \text{for } 7 < t \le 8 \\ 7.00 & \text{for } 8 < t \le 24 \end{cases}$$

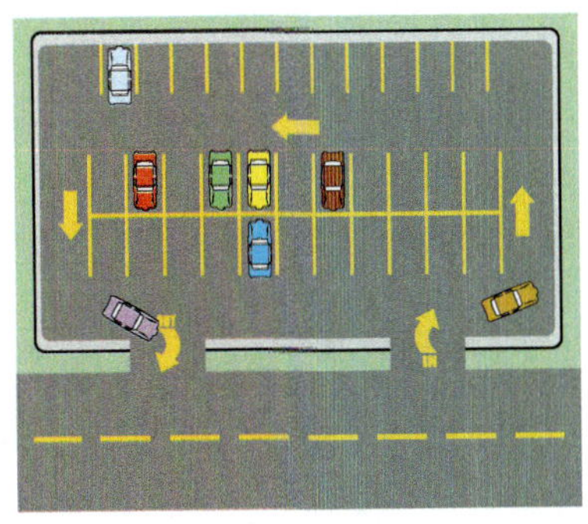

   a. Graph the function for $0 < t \le 24\,\text{hr}$.

   b. Find $\lim\limits_{t \to 3^-} C(t)$.

   c. Find $\lim\limits_{t \to 3^+} C(t)$.

   d. Find $\lim\limits_{t \to 8^-} C(t)$.

   e. Find $\lim\limits_{t \to 8^+} C(t)$.

✏️ WRITING & THINKING

61. Suppose $f(x)$ and $g(x)$ are polynomials and $f(t) = 0 = g(t) = 0$ for some t. If $\lim\limits_{x \to t^-} \dfrac{f(x)}{g(x)} = L$, must $\lim\limits_{x \to t^+} \dfrac{f(x)}{g(x)}$ also be L?

10.2 LIMITS

■ TOPICS

■ Indeterminate Forms and Algebraic Solutions

We have discussed one-sided limits (i.e., left-hand limits and right-hand limits). With the concepts of one-sided limits as a base, we now develop the concept of a **limit** in general by considering left-hand and right-hand limits simultaneously. That is, for the limit of a function of x to exist as x approaches some number a (from both sides of a), the left-hand limit as $x \to a^-$ and the right-hand limit as $x \to a^+$ must be equal. Thus we say $\lim_{x \to a} f(x) = L$ if and only if $\lim_{x \to a^-} f(x) = L$ and $\lim_{x \to a^+} f(x) = L$. For example, consider the function

$$f(x) = \frac{x^2 - x - 6}{x - 3}.$$

The following table will help in analyzing the function for values of x close to 3 on both sides of 3.

	x Approaches 3 from the Left							x Approaches 3 from the Right			
x	2	2.5	2.9	2.99	→	3	←	3.01	3.1	3.5	4
$f(x) = \dfrac{x^2 - x - 6}{x - 3}$	4	4.5	4.9	4.99	→	5	←	5.01	5.1	5.5	6

The table shows that, as $x \to 3^-$ and as $x \to 3^+$, the value of $f(x)$ gets closer and closer to 5. We write

$$\lim_{x \to 3} \left(\frac{x^2 - x - 6}{x - 3} \right) = 5.$$

The graph of the function is shown in Figure 1. There is an open circle on the line at $x = 3$ because the function is not defined at this value.

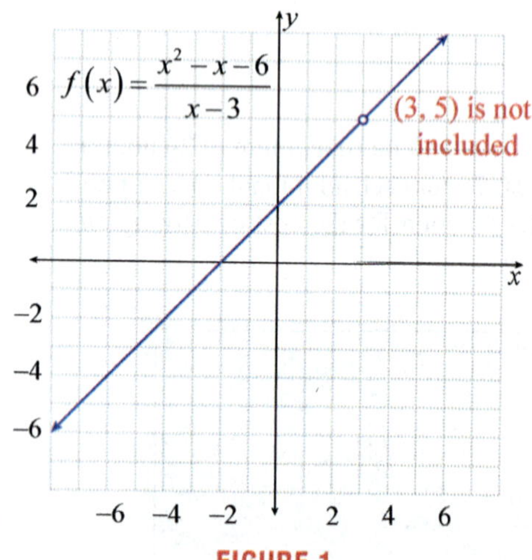

FIGURE 1

These ideas lead to the following definitions of a **limit**.

Limit Defined Informally

If the values of $f(x)$ get closer and closer to some number L as values of x that are smaller than some number t and values of x that are larger than t get closer and closer to t (but not equal to t), then L is the limit of $f(x)$ as x approaches t.

Limit Defined Symbolically

For a function f and a real number t, if $\lim_{x \to t^-} f(x) = L$ and $\lim_{x \to t^+} f(x) = L$, where L is a real number, then $\lim_{x \to t} f(x) = L$.

<table>
<tr><td>

✍ NOTE

Remember, even though the limit does not exist if it is $+\infty$ (or $-\infty$), we still write $+\infty$ (or $-\infty$) for the limit to indicate that the function is unbounded.

</td><td>

Existence of a Limit

1. We say that the limit of a function f as x approaches a real number t **exists** if and only if $\lim_{x \to t} f(x) = L$, where L is a real number.

2. If $\lim_{x \to t^-} f(x) \neq \lim_{x \to t^+} f(x)$, then $\lim_{x \to t} f(x)$ **does not exist**.

3. If $\lim_{x \to t} f(x) = +\infty$ or $\lim_{x \to t} f(x) = -\infty$, then $\lim_{x \to t} f(x)$ **does not exist**.

</td></tr>
</table>

The graphs shown in Figure 2 illustrate a variety of possible situations involving limits. Vertical asymptotes are shown in parts (D), (E), and (F).

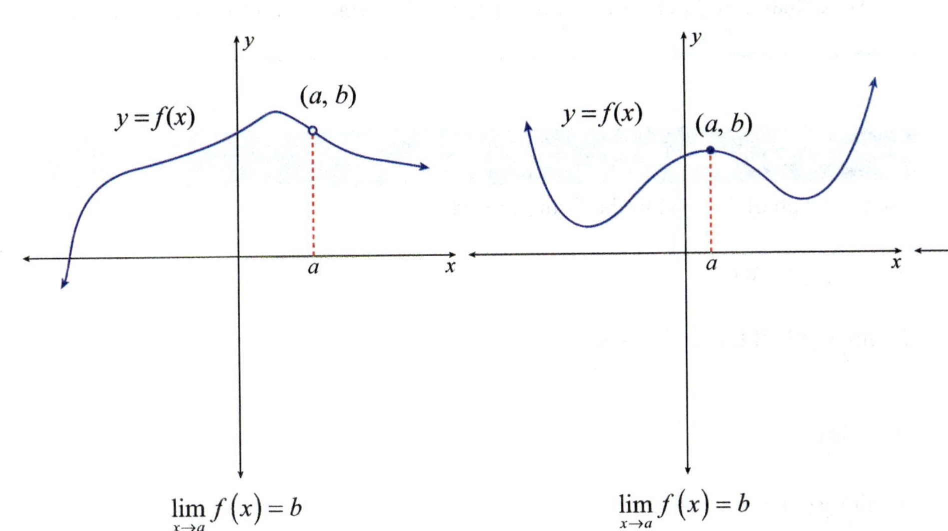

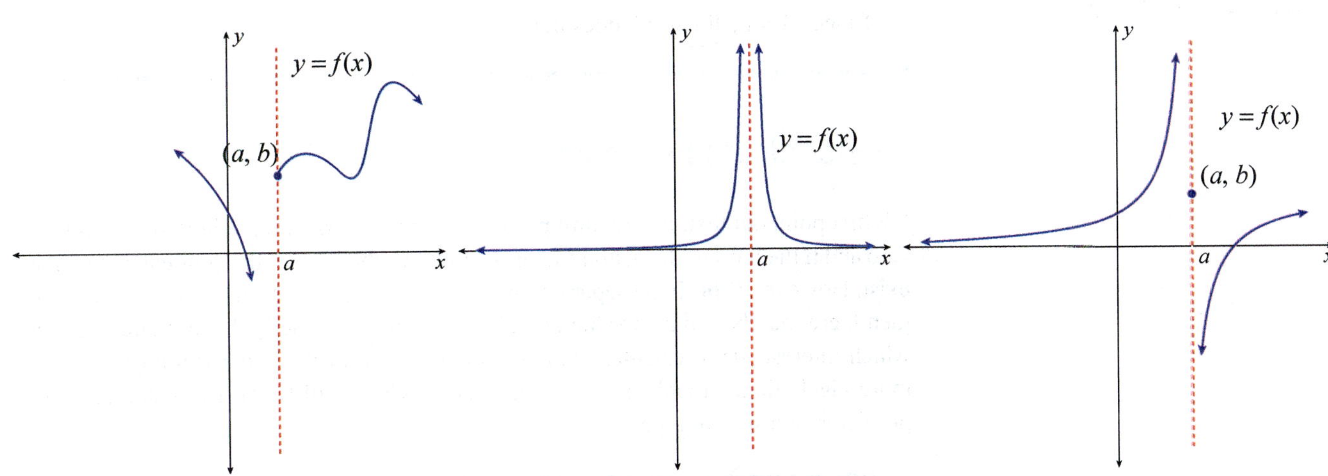

FIGURE 2

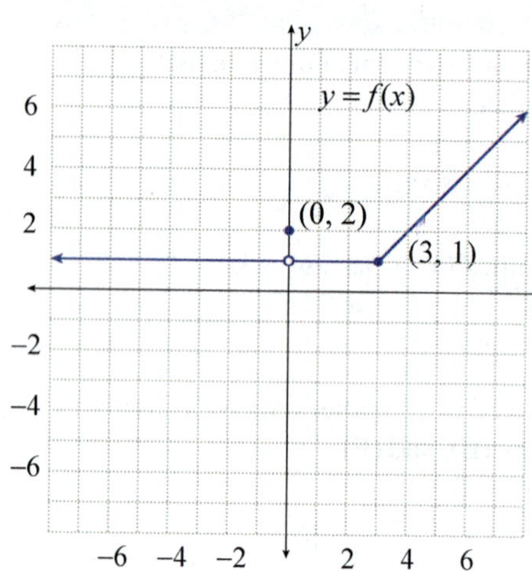

Example 1: Finding a Limit Using a Graph

Use the graph of $y = f(x)$ in the figure to find

a. $\lim\limits_{x \to 3} f(x)$, and

b. $\lim\limits_{x \to 0} f(x)$.

Solution

a. We see that $\lim\limits_{x \to 3^-} f(x) = 1$ and $\lim\limits_{x \to 3^+} f(x) = 1$. Therefore, $\lim\limits_{x \to 3} f(x) = 1$.

b. We see that $\lim\limits_{x \to 0^-} f(x) = 1$ and $\lim\limits_{x \to 0^+} f(x) = 1$. So, even though $f(0) = 2$, $\lim\limits_{x \to 0} f(x) = 1$.

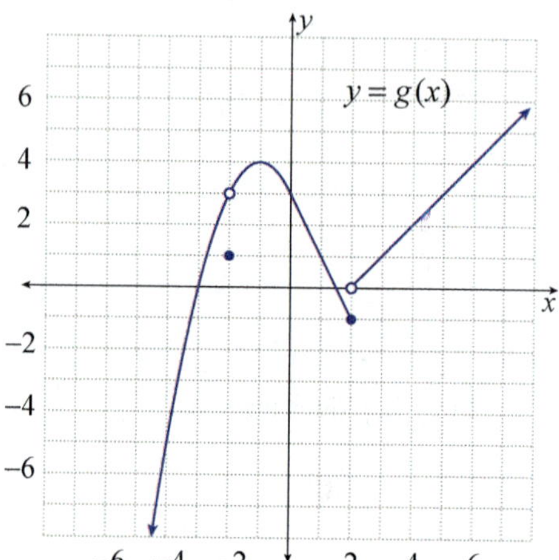

Example 2: Finding a Limit Using a Graph

Use the graph of $y = g(x)$ in the figure to find

a. $\lim\limits_{x \to -2} g(x)$, and

b. $\lim\limits_{x \to 2} g(x)$, if the limits exist.

Solution

a. $\lim\limits_{x \to -2} g(x) = 3$

b. $\lim\limits_{x \to 2^-} g(x) = -1$ and $\lim\limits_{x \to 2^+} g(x) = 0$.

Since $-1 \neq 0$, $\lim\limits_{x \to 2} g(x)$ does not exist.

Indeterminate Forms and Algebraic Solutions

We first point out that, in any limit problem, if the numerator has a limit different from zero and if the limit of the denominator is zero, then the limit of the quotient must fail to exist. However, if **both** the separate limits of a numerator and a denominator are zero, then there may be a limit for the quotient. It is, in fact, exactly this difficult situation which interests us in calculus, and we now move from the numerical approach to a more algebraic approach. The familiar factorizations will be of use in solving limit problems of a special type.

Example 3: Finding Limits Algebraically

a. Determine $\lim\limits_{x \to 2} \left(\dfrac{10(x^2 - 4)}{x - 2} \right)$ by using an algebraically equivalent expression.

Solution

We first factor the numerator and simplify the fraction.

$$\frac{10\left(x^2-4\right)}{x-2} = \frac{10(x+2)\,(x-2)}{x-2} = 10(x+2)$$

The numerical approach, with a table of x-values converging to 2, suggests a limit exists.

Since $\dfrac{10\left(x^2-4\right)}{x-2} = 10(x+2)$ for any $x \neq 2$, we may replace the given expression with the simplified one. Thus

$$\lim_{x\to 2}\left(\frac{10\left(x^2-4\right)}{x-2}\right) = \lim_{x\to 2}\left(\frac{10(x+2)\,(x-2)}{x-2}\right)$$

$$= \lim_{x\to 2}\left(10(x+2)\right)$$

$$= 10(2+2)$$

$$= 40.$$

b. Determine $\displaystyle\lim_{x\to 5}\left(\frac{x^3-125}{\left(x-5\right)^3}\right)$ algebraically.

Solution

We factor the numerator:

$$\frac{x^3-125}{\left(x-5\right)^3} = \frac{x^3-5^3}{\left(x-5\right)^3} = \frac{(x-5)\left(x^2+5x+25\right)}{\left(x-5\right)^{3}}$$

$$= \frac{x^2+5x+25}{\left(x-5\right)^2}.$$

The numerator is a difference of two cubes: $A^3 - B^3$ that can be factored into $(A-B)(A^2+AB+B^2)$.

Then the given limit, if it exists, is equivalent to $\displaystyle\lim_{x\to 5}\left(\frac{x^2+5x+25}{\left(x-5\right)^2}\right)$. However, the numerator does not have a limit of zero, but the limit of the denominator is zero. Thus the limit fails to exist.

c. Determine $\displaystyle\lim_{x\to -3}\left(\frac{2x^2+5x-3}{x+3}\right)$ algebraically.

Solution

We factor first:

$$\frac{2x^2+5x-3}{x+3} = \frac{(2x-1)\,(x+3)}{x+3} = 2x-1.$$

Thus $\displaystyle\lim_{x\to -3}\left(\frac{2x^2+5x-3}{x+3}\right) = \lim_{x\to -3}\left(2x-1\right) = 2(-3)-1 = -7.$

Indeterminate Form

A limit expression of the type $\lim\limits_{x \to a}\left(\dfrac{g(x)}{f(x)}\right)$ is called an **indeterminate form** of type

$\dfrac{0}{0}$ if $\lim\limits_{x \to a} g(x) = 0$ and $\lim\limits_{x \to a} f(x) = 0$.

A strategy of solving such a problem is given by the two-step method illustrated in the previous example.

1. Replace the quotient with a simplified expression after factoring.

2. Evaluate the new limit problem by substitution if the denominator does not have a limit of 0 as $x \to a$.

10.2 EXERCISES

◉ PRACTICE

In Exercises 1–10, use the graph to find the indicated limits, if they exist.

1.

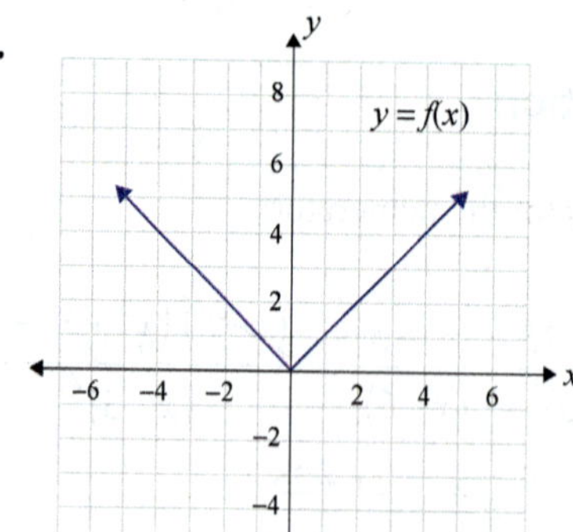

a. $\lim\limits_{x \to 0^-} f(x)$ b. $\lim\limits_{x \to 0^+} f(x)$

c. $\lim\limits_{x \to 0} f(x)$ d. $\lim\limits_{x \to 2} f(x)$

2.

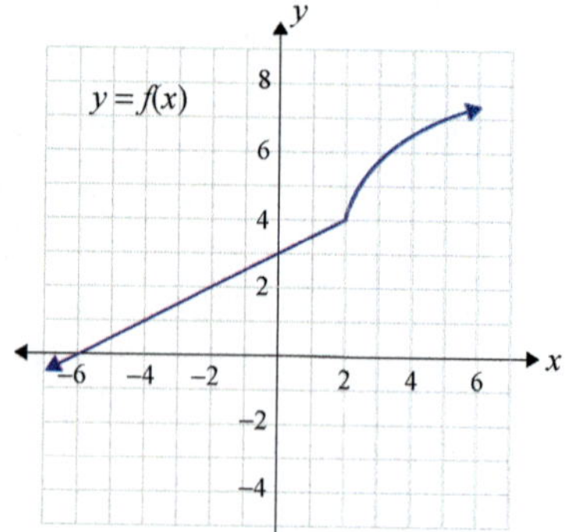

a. $\lim\limits_{x \to 2^-} f(x)$ b. $\lim\limits_{x \to 2^+} f(x)$

c. $\lim\limits_{x \to 2} f(x)$ d. $\lim\limits_{x \to 0} f(x)$

3.

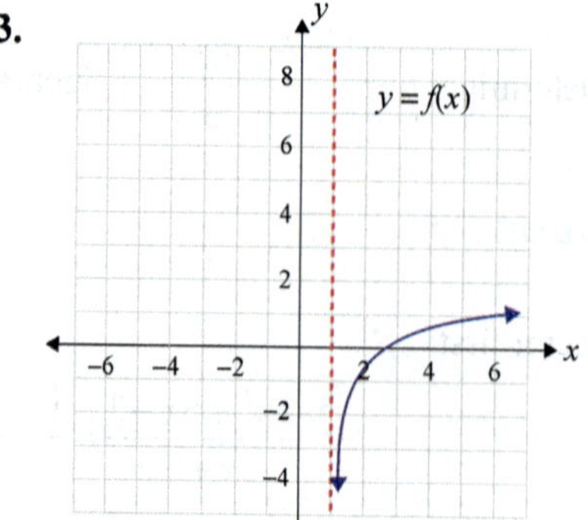

a. $\lim\limits_{x \to 1^+} f(x)$ b. $\lim\limits_{x \to 6^-} f(x)$

c. $\lim\limits_{x \to 6^+} f(x)$ d. $\lim\limits_{x \to 6} f(x)$

4.

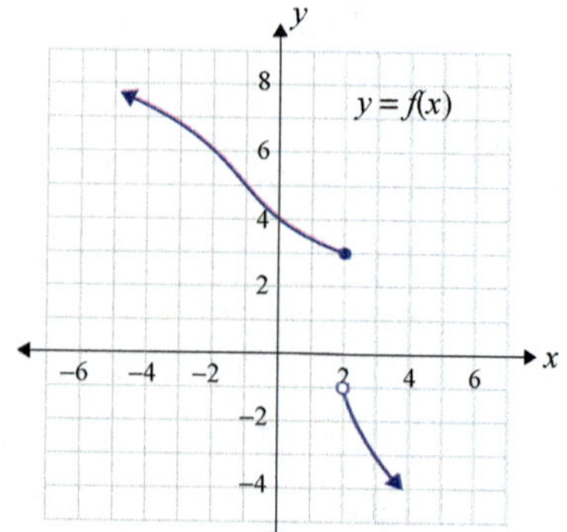

a. $\lim\limits_{x \to 2^-} f(x)$ b. $\lim\limits_{x \to 2^+} f(x)$

c. $\lim\limits_{x \to 2} f(x)$ d. $\lim\limits_{x \to 0} f(x)$

5.

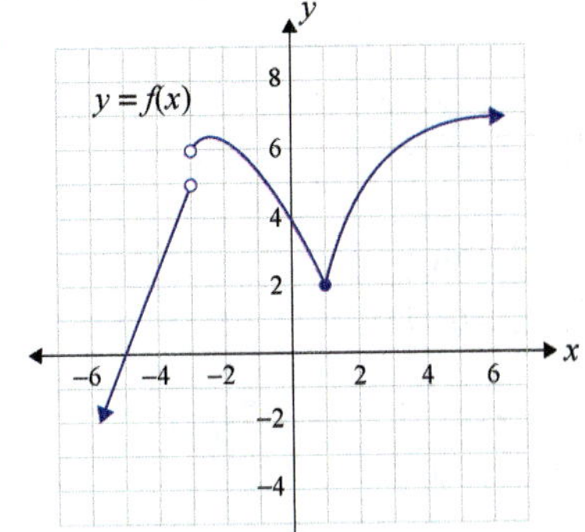

a. $\lim\limits_{x \to -3^-} f(x)$ **b.** $\lim\limits_{x \to -3^+} f(x)$

c. $\lim\limits_{x \to -3} f(x)$ **d.** $\lim\limits_{x \to 1^-} f(x)$

e. $\lim\limits_{x \to 1^+} f(x)$ **f.** $\lim\limits_{x \to 1} f(x)$

6.

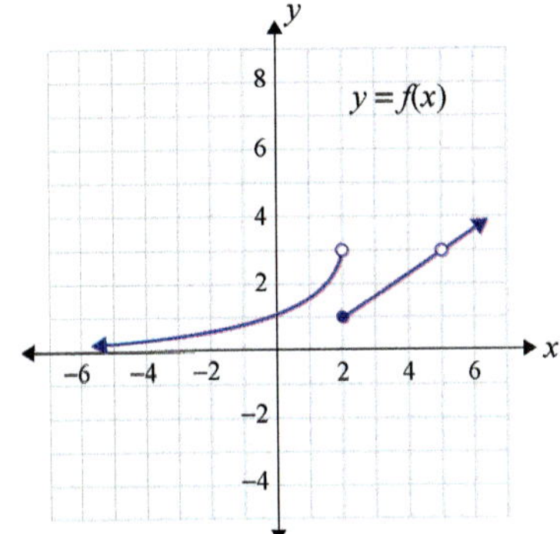

a. $\lim\limits_{x \to 2^-} f(x)$ **b.** $\lim\limits_{x \to 2^+} f(x)$

c. $\lim\limits_{x \to 2} f(x)$ **d.** $\lim\limits_{x \to 5^+} f(x)$

e. $\lim\limits_{x \to 5^-} f(x)$ **f.** $\lim\limits_{x \to 5} f(x)$

7.

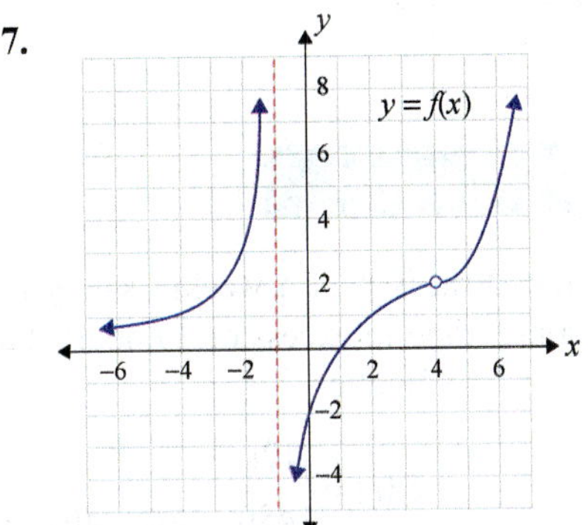

a. $\lim\limits_{x \to -1^-} f(x)$ **b.** $\lim\limits_{x \to -1^+} f(x)$

c. $\lim\limits_{x \to 4} f(x)$ **d.** $\lim\limits_{x \to 1} f(x)$

8.

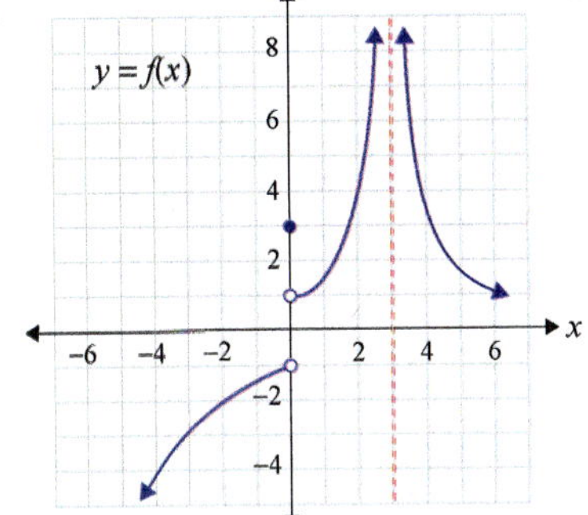

a. $\lim\limits_{x \to 0^-} f(x)$ **b.** $\lim\limits_{x \to 0^+} f(x)$

c. $\lim\limits_{x \to 3^+} f(x)$ **d.** $\lim\limits_{x \to 3^-} f(x)$

9.

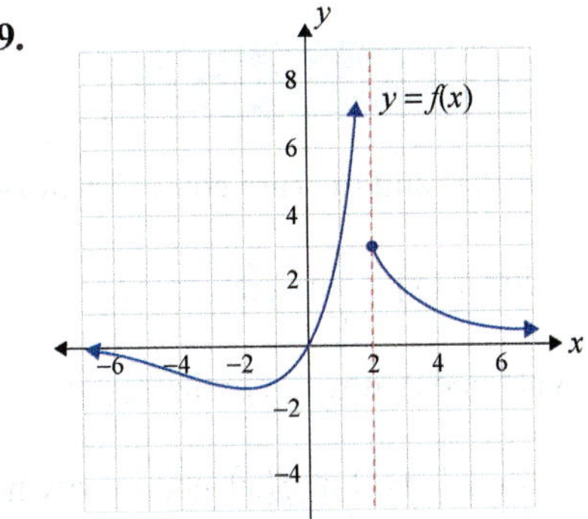

a. $\lim\limits_{x \to 2^-} f(x)$ **b.** $\lim\limits_{x \to 2^+} f(x)$

c. $\lim\limits_{x \to 2} f(x)$ **d.** $\lim\limits_{x \to 0} f(x)$

10.

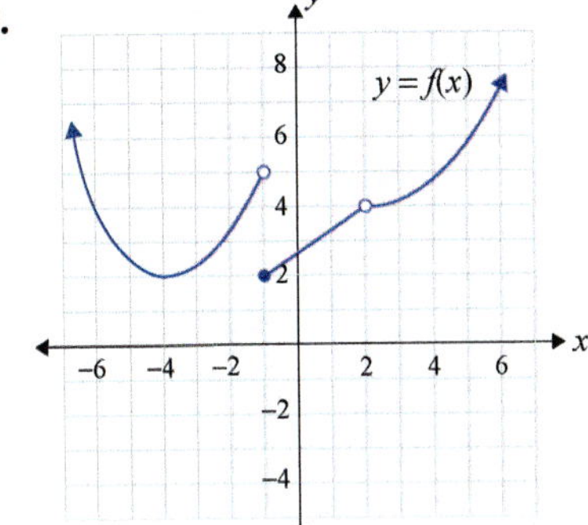

a. $\lim\limits_{x \to -1^-} f(x)$ **b.** $\lim\limits_{x \to -1^+} f(x)$

c. $\lim\limits_{x \to -1} f(x)$ **d.** $\lim\limits_{x \to 2} f(x)$

In Exercises 11–16, determine the limit by first simplifying the expression algebraically.

11. $\lim\limits_{x \to 3}\left(\dfrac{3 - 13x + 4x^2}{x - 3}\right)$

12. $\lim\limits_{x \to 6}\left(\dfrac{x^2 - 36}{x - 6}\right)$

13. $\lim\limits_{x \to -7}\left(\dfrac{x - 7}{x^2 - 49}\right)$

14. $\lim\limits_{h \to 0}\left(\dfrac{f(3 + h) - f(3)}{h}\right), f(x) = x^2 - 2$

15. $\lim\limits_{h \to 0}\left(\dfrac{f(2 - h) - f(2)}{h}\right), f(x) = 1 - x + x^2$

16. $\lim\limits_{x \to 4}\left(\dfrac{x^4 - 256}{x^2 - 16}\right)$

🚀 APPLICATIONS

17. **Salary:** Erin is paid a weekly salary of \$12 per hour plus time-and-a-half for overtime (time in excess of 40 hours, but no more than 60 hours). Her salary is given by the function

$$S(t) = \begin{cases} 12t & \text{if } 0 < t \le 40 \\ 480 + 18(t - 40) & \text{if } 40 < t \le 60 \end{cases}$$

where t is the time in hours, $0 < t \le 60$.

   a. Find $\lim\limits_{t \to 40^-} S(t).$ **b.** Find $\lim\limits_{t \to 40^+} S(t).$ **c.** Find $\lim\limits_{t \to 40} S(t).$

✏️ WRITING & THINKING

18. Suppose $f(x)$ and $g(x)$ are equal for all x-values except $x = t$.

   a. Is $\lim\limits_{x \to t^-} f(x) = \lim\limits_{x \to t^-} g(x)$ true?

   b. What about $\lim\limits_{x \to t^+} f(x) = \lim\limits_{x \to t^+} g(x)$?

   c. Is $\lim\limits_{x \to t} f(x) = \lim\limits_{x \to t} g(x)$ necessarily true?

10.3 MORE ABOUT LIMITS

■ TOPICS

- ■ Properties of Limits
- ■ Limits of Rational Functions as $x \to +\infty$ or $x \to -\infty$

Properties of Limits

Because graphs provide a visual tool that is helpful in an informal approach to limits, we have used them extensively in the discussion thus far. However, as graphs are not always accurate or easy to find, we need a more formal approach to evaluating limits in our development of calculus. The following list of properties of limits, stated here without proof, allows us to study limits on a more rigorous level. Applications of the properties are illustrated in the examples that follow.

Limits as $x \to a$

Suppose that c is any constant, n is a positive integer, a is a real number, $\lim\limits_{x \to a} f(x) = L_1$, and $\lim\limits_{x \to a} g(x) = L_2$, where L_1 and L_2 are real numbers.

Property	Comments
1. $\lim\limits_{x \to a} c = c$	The limit of a constant is the constant itself.
2. $\lim\limits_{x \to a} x = a$	
3. $\lim\limits_{x \to a} x^n = a^n$	
4. $\lim\limits_{x \to a}\left[f(x) \pm g(x) \right]$ $= \lim\limits_{x \to a} f(x) \pm \lim\limits_{x \to a} g(x) = L_1 \pm L_2$	The limit of a sum (or difference) of two functions is the sum (or difference) of the limits.
5. $\lim\limits_{x \to a}\left[f(x) \cdot g(x) \right]$ $= \left[\lim\limits_{x \to a} f(x) \right] \cdot \left[\lim\limits_{x \to a} g(x) \right] = L_1 \cdot L_2$	The limit of a product is the product of the limits provided these limits exist.
6. $\lim\limits_{x \to a}\left[\dfrac{f(x)}{g(x)} \right] = \dfrac{\lim\limits_{x \to a} f(x)}{\lim\limits_{x \to a} g(x)} = \dfrac{L_1}{L_2}$ where $L_2 \neq 0$	The limit of a quotient is the quotient of the limits as long as the limit of the denominator is not 0.
7. $\lim\limits_{x \to a}\left[c \cdot f(x) \right] = c \cdot \lim\limits_{x \to a} f(x) = c \cdot L_1$	The limit of a constant times a function can be found by first finding the limit of the function and then multiplying this limit by the constant.

Property	Comments
8. $\lim\limits_{x \to a} \sqrt[n]{f(x)} = \sqrt[n]{\lim\limits_{x \to a} f(x)}$ $= \sqrt[n]{L_1} = (L_1)^{\frac{1}{n}}$ where $(L_1)^{\frac{1}{n}}$ is defined.	The limit of the nth root of a function can be found by calculating the limit of the function first and then taking the nth root of the limit. (**Note:** $f(x)$ must be defined on intervals to the left of a and to the right of a.)
9. If p is a polynomial function, then $\lim\limits_{x \to a} p(x) = p(a)$.	The limit of a polynomial can be found by substituting a for x. This result is particularly useful and follows directly from Properties 1 through 5.

Example 1: Properties of Limits

Use the properties of limits to find each of the following limits.

a. $\lim\limits_{x \to 4} 10$

b. $\lim\limits_{x \to 3} \left(x^2 + 5x + 7\right)$

c. $\lim\limits_{x \to 2} \sqrt[3]{\dfrac{x-3}{x+6}}$

Solution

a. $\lim\limits_{x \to 4} 10 = 10$ $\qquad\qquad$ Property 1

b. $\lim\limits_{x \to 3} \left(x^2 + 5x + 7\right) = 3^2 + 5(3) + 7$ $\qquad$ Property 9; Substitute $x = 3$.

$\qquad\qquad\qquad\qquad = 9 + 15 + 7$

$\qquad\qquad\qquad\qquad = 31$

c. $\lim\limits_{x \to 2} \sqrt[3]{\dfrac{x-3}{x+6}} = \sqrt[3]{\lim\limits_{x \to 2} \left(\dfrac{x-3}{x+6}\right)}$ $\qquad$ Property 8

$\qquad\qquad = \sqrt[3]{\dfrac{\lim\limits_{x \to 2} (x-3)}{\lim\limits_{x \to 2} (x+6)}}$ $\qquad$ Property 6

$\qquad\qquad = \sqrt[3]{\dfrac{(2-3)}{(2+6)}}$ $\qquad$ Property 9; Substitute $x = 2$.

$\qquad\qquad = \sqrt[3]{\dfrac{-1}{8}} = -\dfrac{1}{2}$

In the special case when the numerator and denominator of a fractional expression both have value 0 at $x = a$, we say that the expression $\dfrac{0}{0}$ is an **indeterminate form**. In this section we will consider such cases only when $x - a$ is a factor of both the numerator and denominator and the expression for $f(x)$ can be simplified before the limit is found. This is called a *removable discontinuity*. This procedure is allowed because we are concerned with what happens to the function as x approaches a and not at $x = a$. For example, if

$$f(x) = \frac{x^2 - 9}{x - 3}, \text{ then } f(3) = \frac{3^2 - 9}{3 - 3} = \frac{0}{0},$$

which is undefined and an indeterminate form. However, by factoring and reducing, we find

$$\lim_{x\to 3} f(x) = \lim_{x\to 3}\left(\frac{x^2-9}{x-3}\right) = \lim_{x\to 3}\frac{(x+3)(x-3)}{x-3} = \lim_{x\to 3}(x+3) = 6.$$

Example 2: Indeterminate Form

If $f(x) = \dfrac{x^3-1}{x-1}$, find $\lim_{x\to 1} f(x)$.

Solution

Since $f(1) = \dfrac{1^3-1}{1-1} = \dfrac{0}{0}$, we simplify $f(x)$ first and then find the limit.

$$\lim_{x\to 1} f(x) = \lim_{x\to 1}\left(\frac{x^3-1}{x-1}\right) = \lim_{x\to 1}\left[\frac{(x-1)(x^2+x+1)}{x-1}\right]$$

$$= \lim_{x\to 1}(x^2+x+1) = 1^2+1+1 = 3$$

In Example 2 we were able to calculate the limit because we found a common factor. This is no coincidence. The Remainder Theorem in college algebra tells us that, for any polynomial $p(x)$, if $p(a) = 0$, then $x - a$ divides evenly into $p(x)$ or, equivalently, that $x - a$ is a factor of $p(x)$. If, for example, $p(x) = x^3 - 1$, we observe $p(1) = 0$. Thus $x - 1$ is a factor of $x^3 - 1$.

From the other point of view, suppose $q(x)$ satisfies $q(a) = 0$ and $p(a) \neq 0$, for two polynomials $p(x)$ and $q(x)$. Then $\lim_{x\to a}\dfrac{p(x)}{q(x)}$ will not exist since, as $x \to a$, the denominator approaches 0 but the numerator does not. There is no factor $x - a$ which will cancel.

Example 3: Undefined Limit

For $f(x) = \dfrac{1}{x-5}$, find $\lim_{x\to 5} f(x)$.

Solution

For $f(x) = \dfrac{1}{x-5}$, $f(5) = \dfrac{1}{0}$, which is not defined, and further, is not of the form $\dfrac{0}{0}$.

The expression $\dfrac{1}{x-5}$ cannot be simplified as the expression in Example 2 was. When the limit of the denominator is 0 and the limit of the numerator is some number other than 0, then the limit of the function does not exist. To show this and to determine the nature of the graph of the function near $x = 5$, we analyze the left- and right-hand limits:

$$\lim_{x\to 5^-}\left(\frac{1}{x-5}\right) = -\infty \quad \text{and} \quad \lim_{x\to 5^+}\left(\frac{1}{x-5}\right) = +\infty.$$

Thus we see that $\lim\limits_{x \to 5} f(x)$ does not exist and that the graph of the function is unbounded in the negative direction as $x \to 5^-$ and unbounded in the positive direction as $x \to 5^+$.

The following example illustrates the technique for finding a limit of a function defined piecewise.

Example 4: Piecewise Function

The function $f(x)$ is defined piecewise as follows.

$$f(x) = \begin{cases} 2x & \text{if } x \le 3 \\ x+3 & \text{if } x > 3 \end{cases}$$

Find $\lim\limits_{x \to 3} f(x)$ if it exists.

Solution

$$\lim\limits_{x \to 3^-} f(x) = \lim\limits_{x \to 3^-} 2x = 2(3) = 6 \qquad \text{Use } f(x) = 2x \text{ since } x \le 3 \text{ when } x \to 3^-.$$

$$\lim\limits_{x \to 3^+} f(x) = \lim\limits_{x \to 3^+} (x+3) = 3+3 = 6 \qquad \text{Use } f(x) = x+3 \text{ since } x > 3 \text{ when } x \to 3^+.$$

Since $\lim\limits_{x \to 3^-} f(x) = \lim\limits_{x \to 3^+} f(x) = 6$, we have $\lim\limits_{x \to 3} f(x) = 6$.

Limits of Rational Functions as $x \to +\infty$ or $x \to -\infty$

In this section we will discuss rational functions and their behavior as x increases (or decreases) without bound (that is, as $x \to +\infty$ or $x \to -\infty$).

Rational Function

A **rational function** is a function of the form

$$f(x) = \frac{p(x)}{q(x)},$$

where $p(x)$ and $q(x)$ are polynomials.

We begin by analyzing $f(x) = \dfrac{1}{x}$ as x increases without bound. If we let $x = 10$, 100, 1000, 10,000, 50,000, 1,000,000, and so on, then $\dfrac{1}{x} = \dfrac{1}{10}, \dfrac{1}{100}, \dfrac{1}{1000}, \dfrac{1}{10,000}, \dfrac{1}{50,000}, \dfrac{1}{1,000,000}$, and so on.

Since $\dfrac{1}{10} > \dfrac{1}{100} > \dfrac{1}{1000} > \dfrac{1}{10,000} > \dfrac{1}{50,000} > \dfrac{1}{1,000,000} > \cdots > 0$, the value of $\dfrac{1}{x}$ becomes smaller and smaller and approaches 0 as x becomes larger and larger. With this intuitive notation as background, we make the following statement: $\lim\limits_{x \to +\infty} \dfrac{1}{x} = 0$.

A similar analysis will show that

$$\lim_{x \to -\infty} \frac{1}{x} = 0.$$

The graph of the function $y = f(x) = \dfrac{1}{x}$ is shown in Figure 1. The line $y = 0$ (the x-axis) is a **horizontal asymptote**.

x	$\dfrac{1}{x}$
10	0.1
100	0.01
1000	0.001
$\downarrow$	$\downarrow$
$+\infty$	0

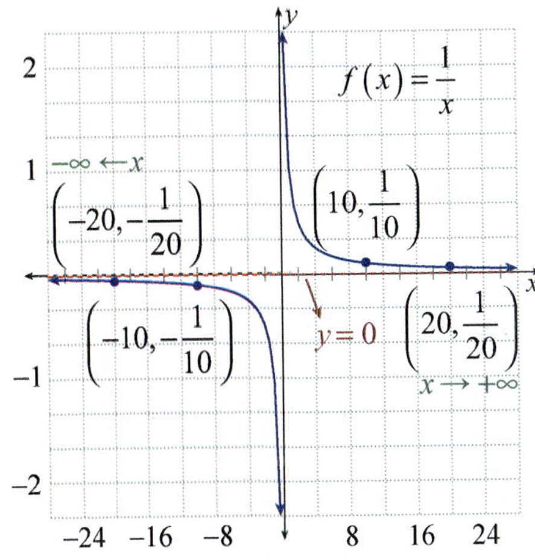

FIGURE 1

Earlier in this section we stated the properties of limits as $x \to a$ where a is a real number. These same properties are valid if $x \to +\infty$ or $x \to -\infty$, and we apply them in the following examples.

Example 5: Properties of Limits

Find $\displaystyle\lim_{x \to +\infty}\left(\frac{3x+5}{2x-1}\right)$, if it exists.

Solution

Use a graphing utility and graph the function $f(x) = \dfrac{(3x+5)}{(2x-1)}$. On the TI-84 Plus, use a window of $[-10, 10]$ by $[-3, 3]$. As x gets larger we see the y-values level off above the x-axis (see Figure 2).

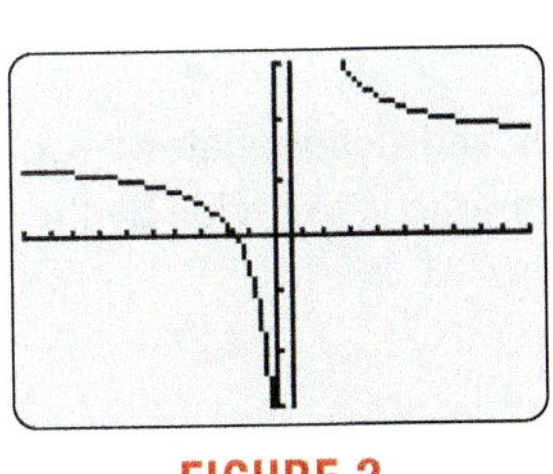

FIGURE 2

There is a natural way to investigate this algebraically.

Individually, both the numerator and denominator get very large. That is, they both approach $+\infty$.

$$\lim_{x \to +\infty}(3x+5) = +\infty \quad \text{and} \quad \lim_{x \to +\infty}(2x-1) = +\infty.$$

Thus this approach leads to an expression of the form $\dfrac{+\infty}{+\infty}$, which is called an indeterminate form. We still do not know what the limit is or even if the limit exists.

Now, if we divide each term in the numerator and denominator by x, we will have an expression with fractions whose limits we can find.

$$\lim_{x \to +\infty}\left(\frac{3x+5}{2x-1}\right) = \lim_{x \to +\infty}\left(\frac{\dfrac{3x}{x}+\dfrac{5}{x}}{\dfrac{2x}{x}-\dfrac{1}{x}}\right) \qquad \text{Divide each term by } x.$$

$$= \lim_{x \to +\infty}\left(\frac{3+\dfrac{5}{x}}{2-\dfrac{1}{x}}\right) \qquad \text{Simplify.}$$

$$= \frac{\displaystyle\lim_{x \to +\infty}\left(3+\dfrac{5}{x}\right)}{\displaystyle\lim_{x \to +\infty}\left(2-\dfrac{1}{x}\right)} \qquad \text{Use limit Property 6.}$$

$$\qquad\qquad\qquad\qquad\qquad \text{Use limit Properties 4 and 7.}$$

$$= \frac{3+0}{2-0} = \frac{3}{2} \qquad \text{Also, use } \lim_{x \to +\infty}\left(\frac{1}{x}\right)=0.$$

$$\textbf{Note: } \lim_{x \to +\infty}\left(\frac{5}{x}\right) = 5 \cdot \lim_{x \to +\infty}\left(\frac{1}{x}\right).$$

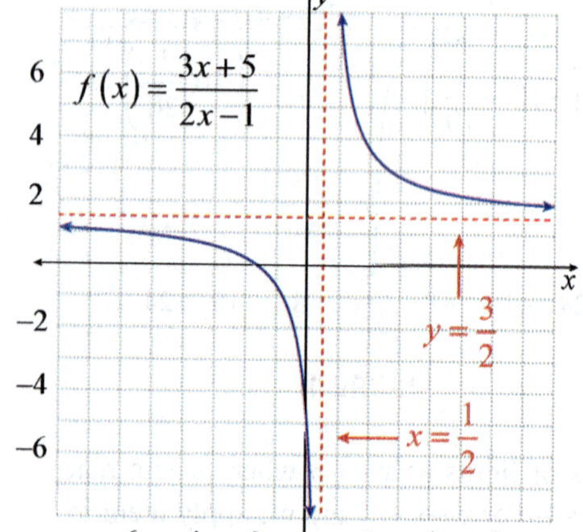

Since the constants $+5$ and -1 will have little effect on the fraction for very large values of x (for example, in the millions), then $\dfrac{3x+5}{2x-1}$ is approximately the same as $\dfrac{3x}{2x}=\dfrac{3}{2}$. Thus the limit of $\dfrac{3}{2}$ seems quite reasonable from this point of view. The graph shows that the line $y=\dfrac{3}{2}$ is indeed a horizontal asymptote.

Find $\displaystyle\lim_{x \to +\infty}\left(\frac{2x+3}{x^3-8}\right)$, if it exists.

Solution

In Example 5 we divided each term in the numerator and denominator by x. In this example, we will divide each term by x^3, since the technique is to divide by the highest power of x present in the function.

$$\lim_{x \to +\infty}\left(\frac{2x+3}{x^3-8}\right) = \lim_{x \to +\infty}\left(\frac{\dfrac{2x}{x^3}+\dfrac{3}{x^3}}{\dfrac{x^3}{x^3}-\dfrac{8}{x^3}}\right) \qquad \text{Divide each term by } x^3.$$

$$= \lim_{x \to +\infty}\left(\frac{\dfrac{2}{x^2}+\dfrac{3}{x^3}}{1-\dfrac{8}{x^3}}\right) \qquad \text{Simplify.}$$

$$= \frac{0+0}{1-0} = \frac{0}{1} = 0 \qquad \begin{array}{l}\text{Since the numerator is 0 and the} \\ \text{denominator is not 0, the fraction has} \\ \text{the value 0.}\end{array}$$

Example 7: Properties of Limits

Find $\lim\limits_{x \to -\infty}\left(\dfrac{x^3 + x^2 - x + 1}{x^2 - 4}\right)$, if it exists.

Solution

$$\lim_{x \to -\infty}\left(\frac{x^3 + x^2 - x + 1}{x^2 - 4}\right) = \lim_{x \to -\infty}\left(\frac{\dfrac{x^3}{x^3} + \dfrac{x^2}{x^3} - \dfrac{x}{x^3} + \dfrac{1}{x^3}}{\dfrac{x^2}{x^3} - \dfrac{4}{x^3}}\right)$$

Divide each term by the highest power of x present, which in this function is x^3.

$$= \lim_{x \to -\infty}\left(\frac{1 + \dfrac{1}{x} - \dfrac{1}{x^2} + \dfrac{1}{x^3}}{\dfrac{1}{x} - \dfrac{4}{x^3}}\right)$$

Simplify.

$$= \frac{1}{0}$$

Since $\dfrac{1}{0}$ is undefined, the limit is either $+\infty$ or $-\infty$. (See Example 3.)

Investigating the expression shows that the highest power is x^3. This term will dominate for very large values of x and will be negative for negative values of x. Thus

$$\lim_{x \to -\infty}\left(\frac{x^3 + x^2 - x + 1}{x^2 - 4}\right) = -\infty.$$

Summary of Limits for Rational Functions as $x \to +\infty$ (or $x \to -\infty$)

Consider the function

$$f(x) = \frac{a_n x^n + a_{n-1}x^{n-1} + \cdots + a_0}{b_m x^m + b_{m-1}x^{m-1} + \cdots + b_0},$$

where $a_n \neq 0$ and $b_m \neq 0$.

Case 1: For $m = n$, $\lim\limits_{x \to +\infty} f(x) = \dfrac{a_n}{b_m}$.

Case 2: For $m > n$, $\lim\limits_{x \to +\infty} f(x) = 0$.

Case 3: For $m < n$, $\lim\limits_{x \to +\infty} f(x) = +\infty$. (Or $-\infty$ depending on the signs of a_n and b_m.)

10.3 EXERCISES

PRACTICE

In Exercises 1–28, find the indicated limit, if it exists.

1. $\lim\limits_{x \to -2} 6$

2. $\lim\limits_{x \to 4} 2x$

3. $\lim\limits_{x \to 3^-}\left(x^2 + 1\right)$

4. $\lim\limits_{x \to -3^+} \left(5 - 2x^2\right)$

5. $\lim\limits_{x \to 1^+} \left(\dfrac{x+2}{x-1}\right)$

6. $\lim\limits_{x \to 1^-} \left(\dfrac{x+2}{x-1}\right)$

7. $\lim\limits_{x \to \frac{1}{3}} \left(\dfrac{3x+1}{x+2}\right)$

8. $\lim\limits_{x \to 0^+} \left(\dfrac{x+4}{x-4}\right)$

9. $\lim\limits_{x \to 0^-} \left(\dfrac{x}{x^2+2x}\right)$

10. $\lim\limits_{x \to 0^-} \left(\dfrac{2x^2+x}{x}\right)$

11. $\lim\limits_{x \to +\infty} \left(\dfrac{x}{x^2+3}\right)$

12. $\lim\limits_{x \to +\infty} \left(\dfrac{2x^2+7}{3x^2-2}\right)$

13. $\lim\limits_{x \to -\infty} \left(\dfrac{x^3+64}{x^2-2x+1}\right)$

14. $\lim\limits_{x \to -\infty} \left(\dfrac{4x-x^3}{x^2+2x-7}\right)$

15. $\lim\limits_{x \to 2} \left(\dfrac{x^2-x-2}{x^2-4}\right)$

16. $\lim\limits_{x \to -3} \left(\dfrac{x^2-9}{x^2+2x-3}\right)$

17. $\lim\limits_{x \to 0^+} \left(4 - \dfrac{3}{x}\right)$

18. $\lim\limits_{x \to 1^-} \left(2x + \dfrac{5}{x-1}\right)$

19. $\lim\limits_{x \to +\infty} \left(8 + \dfrac{1}{x}\right)$

20. $\lim\limits_{x \to -\infty} \left(11 - \dfrac{2}{x^2}\right)$

21. $\lim\limits_{x \to 4} \sqrt{x+5}$

22. $\lim\limits_{x \to 2} \sqrt{3x+10}$

23. $\lim\limits_{x \to 1} \left(\sqrt{x} - 3\right)$

24. $\lim\limits_{x \to 4} \left(\sqrt{x} + 6\right)$

25. a. $\lim\limits_{x \to 4} \left(\dfrac{\sqrt{x}-2}{x-4}\right)$ [**Hint:** $x-4 = \left(\sqrt{x}+2\right)\left(\sqrt{x}-2\right)$]

    b. $\lim\limits_{x \to 9} \left(\dfrac{x-9}{\sqrt{x}-3}\right)$ [**Hint:** $x-9 = \left(\sqrt{x}+3\right)\left(\sqrt{x}-3\right)$]

26. a. $\lim\limits_{h \to 0} \dfrac{\sqrt{2+h}-\sqrt{2}}{h}$ [**Hint:** Multiply by $\dfrac{\sqrt{2+h}+\sqrt{2}}{\sqrt{2+h}+\sqrt{2}}$, simplify the numerator, and calculate the limit.]

    b. $\lim\limits_{h \to 0} \dfrac{\sqrt{5x+h}-\sqrt{5x}}{h}$

27. $\lim\limits_{x \to +\infty} \left(\dfrac{x^2+3x-4}{x^2+5x-9}\right)$

28. $\lim\limits_{x \to -\infty} \left(\dfrac{x^2+x+1}{2x^3+3x^2+x-2}\right)$

29. For the function $f(x) = \begin{cases} -3 & \text{if } x \le 1 \\ x-4 & \text{if } x > 1 \end{cases}$, find the following:

    a. $\lim\limits_{x \to 1^-} f(x)$ **b.** $\lim\limits_{x \to 1^+} f(x)$ **c.** $\lim\limits_{x \to 1} f(x)$ **d.** $\lim\limits_{x \to 2} f(x)$

30. For the function $f(x) = \begin{cases} 2x+1 & \text{if } x < 0 \\ x^2+1 & \text{if } x \ge 0 \end{cases}$, find the following:

    a. $\lim\limits_{x \to 0^-} f(x)$ **b.** $\lim\limits_{x \to 0^+} f(x)$ **c.** $\lim\limits_{x \to 0} f(x)$ **d.** $\lim\limits_{x \to -2} f(x)$

31. For the function $f(x) = \begin{cases} 5-x & \text{if } x \le 2 \\ x^2 - 1 & \text{if } x > 2 \end{cases}$, find the following:

a. $\lim\limits_{x \to 2^-} f(x)$ **b.** $\lim\limits_{x \to 2^+} f(x)$ **c.** $\lim\limits_{x \to 2} f(x)$ **d.** $\lim\limits_{x \to 0} f(x)$

32. For the function $f(x) = \begin{cases} x^3 + 4 & \text{if } x \le -2 \\ \sqrt{x^2 + 5} & \text{if } x > -2 \end{cases}$, find the following:

a. $\lim\limits_{x \to -2^-} f(x)$ **b.** $\lim\limits_{x \to -2^+} f(x)$ **c.** $\lim\limits_{x \to -2} f(x)$ **d.** $\lim\limits_{x \to -1} f(x)$

🚀 APPLICATIONS

33. Utility costs: The Municipal Gas Company uses the following function for computing their customers' monthly gas bills:

$$C(x) = \begin{cases} 0.37x + 3.00 & \text{if } 0 < x \le 24 \\ 0.78x - 6.84 & \text{if } x > 24 \end{cases},$$

where x is the number of therms (thermal units) used and $C(x)$ is the cost in dollars.

a. Find $\lim\limits_{x \to 24^-} C(x)$. **b.** Find $\lim\limits_{x \to 24^+} C(x)$. **c.** Find $\lim\limits_{x \to 24} C(x)$.

34. Income tax: A federal income tax schedule can be given by the function

$$T(x) = \begin{cases} 0.15x & \text{if } 0 < x \le 23{,}900 \\ 0.28x - 3107 & \text{if } 23{,}900 < x \le 61{,}650 \\ 0.33x - 6189.50 & \text{if } 61{,}650 < x \le 123{,}790 \end{cases},$$

where x is the taxable income in dollars, $0 < x \le 123{,}790$, and $T(x)$ is in dollars.

a. Find $\lim\limits_{x \to 23{,}900^-} T(x)$. **b.** Find $\lim\limits_{x \to 23{,}900^+} T(x)$.

c. Find $\lim\limits_{x \to 23{,}900} T(x)$. **d.** Find $\lim\limits_{x \to 61{,}650} T(x)$.

35. Average cost: A manufacturer of golf clubs estimates that if x sets of golf clubs are produced, then the average cost of producing each set is $A(x) = 73 + \dfrac{5780}{x}$ dollars. What will be the average cost of producing each set in the long run $\left(\lim\limits_{x \to +\infty} A(x) \right)$?

36. Dictation rate: It has been determined that after t weeks of class, a certain student in an intermediate shorthand class can take dictation at a rate of $W(t) = 60 + \dfrac{70t^2}{t^2 + 15}$ words per minute. What will be this student's rate of taking dictation in the long run $\left(\lim\limits_{t \to +\infty} W(t) \right)$?

10.4 CONTINUITY

The concept of a continuous function can be explained on an intuitive basis. If you can trace the graph of a function without lifting your pencil from the paper (i.e., there are no "holes" or "jumps" or vertical asymptotes), then the function is *continuous*. (See Figure 1.)

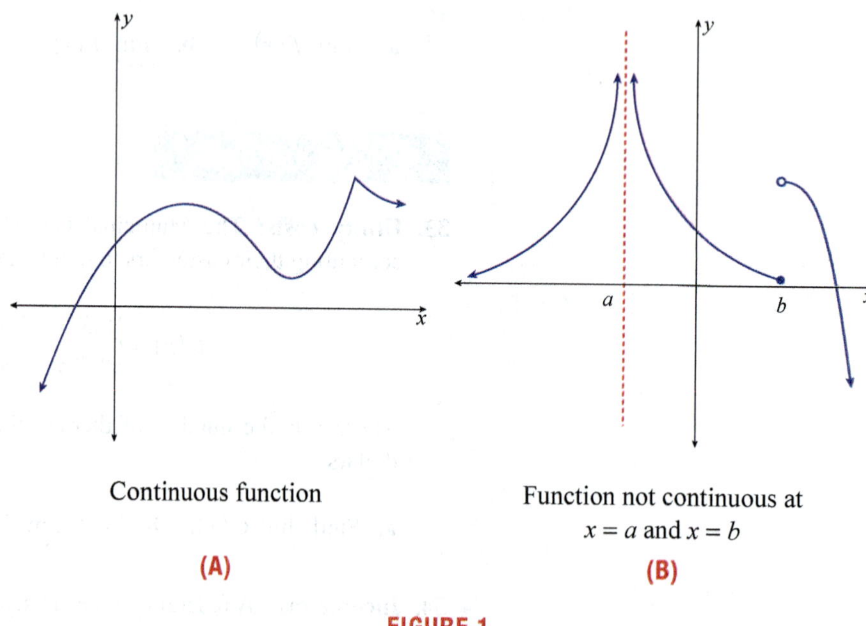

Continuous function

(A)

Function not continuous at
$x = a$ and $x = b$

(B)

FIGURE 1

This approach helps to get a mental picture of a continuous function, but it does not take the place of a formal definition that can be applied algebraically.

The concept of continuity is closely related to limits. In our discussion of limits as $x \to a$, we were concerned with the function values for x close to a but not for $x = a$. In some cases, $f(a)$ did **exist** (was **defined**), and in other cases, $f(a)$ did **not exist** (was **not defined**). In no case did the value of $f(a)$ have any effect on the limit. We will find, however, that the value of $f(a)$ is an integral part of the definition of continuity at $x = a$. That is, we discuss continuity in terms of continuity at a point.

Continuity

A function $y = f(x)$ is **continuous** at $x = a$ if all of the following conditions are true:

1. $f(a)$ exists. ($f(a)$ is a real number.)

2. $\lim\limits_{x \to a} f(x)$ exists. (The limit is a real number.)

3. $\lim\limits_{x \to a} f(x) = f(a)$. (The limit is equal to the value of the function at $x = a$.)

If a function fails to satisfy any one of the three conditions of the definition, then it is not continuous at $x = a$ and is said to be **discontinuous** at $x = a$.

The formal definition not only tells us the technical meaning of "continuous at $x = a$," but it also says something important about computing limits. If a function is known to be continuous at $x = a$, then the limit problem $\lim\limits_{x \to a} f(x)$ is solved merely by substitution of $x = a$ into the formula! For continuous functions, such as polynomial functions, limit problems are trivial. Unfortunately, the important limit problems that occur most often in calculus involve $\dfrac{\Delta y}{\Delta x}$, which is not continuous at $\Delta x = 0$. It is useful to consider the ways in which continuity can fail. A variety of possible situations for $y = f(x)$ at $x = a$ are shown in Figures 2(A)–(E). Only in Figure 2(A) is the function continuous at $x = a$.

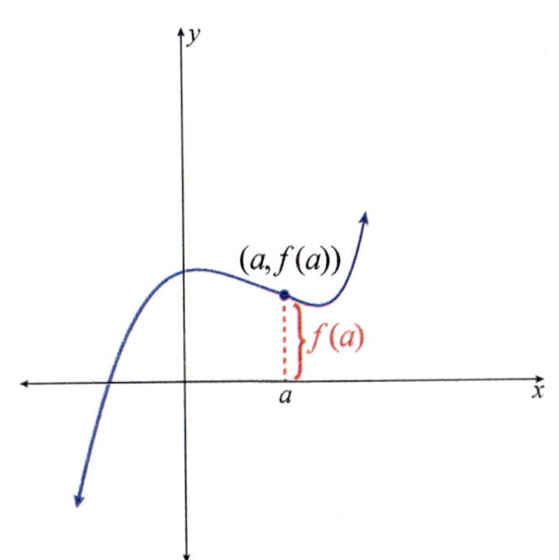

$$\lim_{x \to a} f(x) = f(a)$$

f is continuous at $x = a$.

(A)

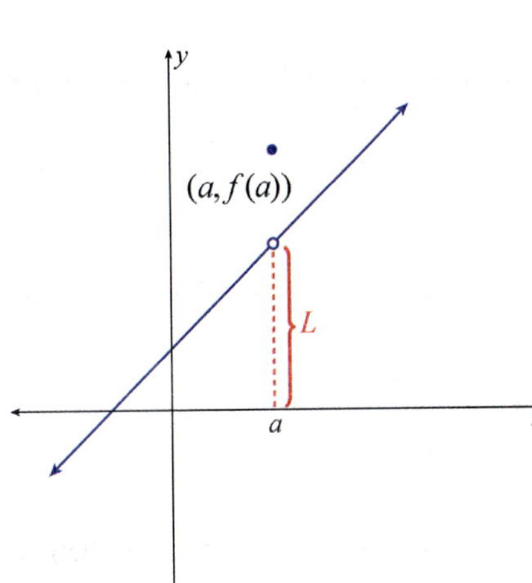

$$\lim_{x \to a} f(x) = L \neq f(a)$$

There is a "hole" in the curve. This is called a **removable discontinuity**.

(B)

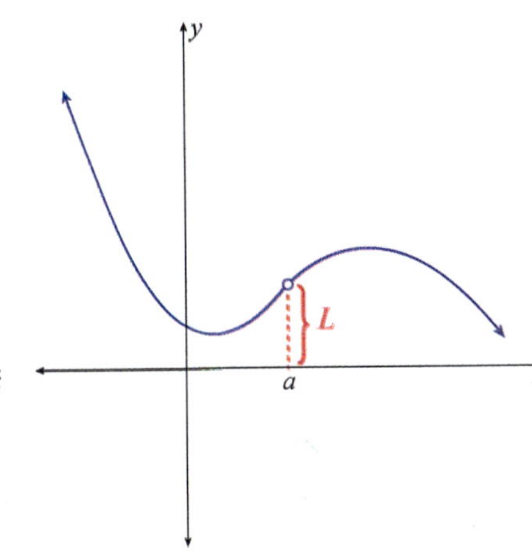

$$\lim_{x \to a} f(x) = L$$

$f(a)$ does not exist. This is a **removable discontinuity**.

(C)

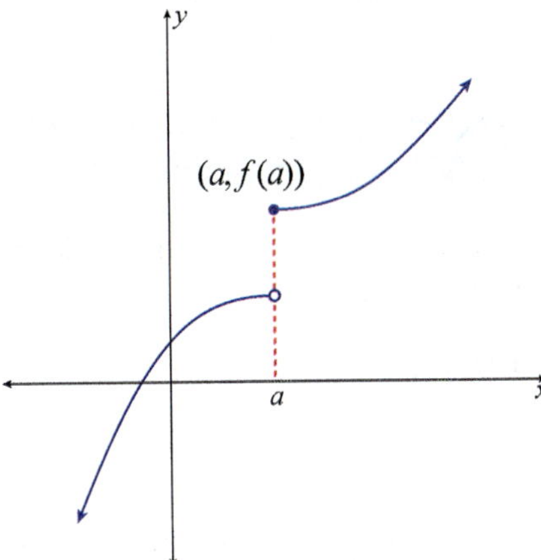

$$\lim_{x \to a^-} f(x) \neq \lim_{x \to a^+} f(x)$$

$\lim\limits_{x \to a} f(x)$ does not exist. This is called a **jump discontinuity**.

(D)

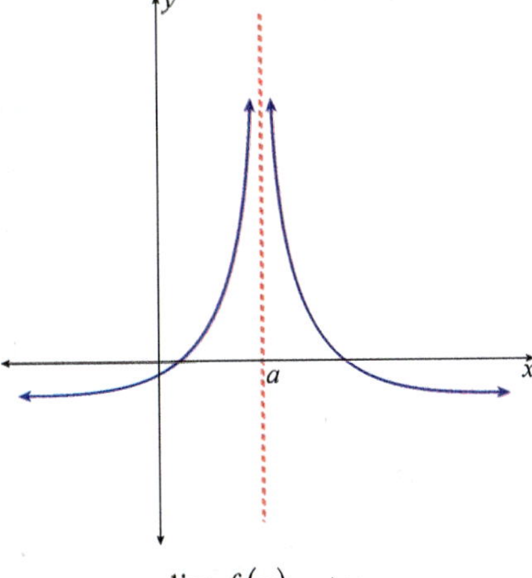

$$\lim_{x \to a} f(x) = +\infty$$

f is discontinuous at $x = a$ since the limit does not exist. Also, $f(a)$ does not exist. This is called an **infinite discontinuity** or a **nonremovable discontinuity**.

(E)

FIGURE 2

Example 1: Continuity

Use the definition of continuity to determine whether the function $f(x) = \dfrac{x^2 - x - 6}{x - 3}$ is continuous at

a. $x = 2$, and   **b.** $x = 3$.

Solution

a. $f(2) = \dfrac{2^2 - 2 - 6}{2 - 3} = \dfrac{-4}{-1} = 4$

Check Condition 1 by substituting $x = 2$.

Thus $f(2)$ exists, and Condition 1 of the definition is met.

$$\lim_{x \to 2} f(x) = \lim_{x \to 2} \frac{(x-3)(x+2)}{x-3} = \lim_{x \to 2}(x+2) = 4$$

Check Condition 2 by finding a limit at $x = 2$.

The limit exists, so Condition 2 of the definition is met.

$$\lim_{x \to 2} f(x) = 4 = f(2)$$

Check Condition 3 by seeing if Conditions 1 and 2 equal one another.

Since all three conditions are met, $f(x)$ is continuous at $x = 2$.

b. $f(3) = \dfrac{3^2 - 3 - 6}{3 - 3} = \dfrac{0}{0}$

Check Condition 1 by substituting $x = 3$.

Since $f(3) = \dfrac{0}{0}$ is undefined, $f(x)$ is not continuous at $x = 3$.

The graph of $f(x)$ shows a removable discontinuity at $x = 3$.

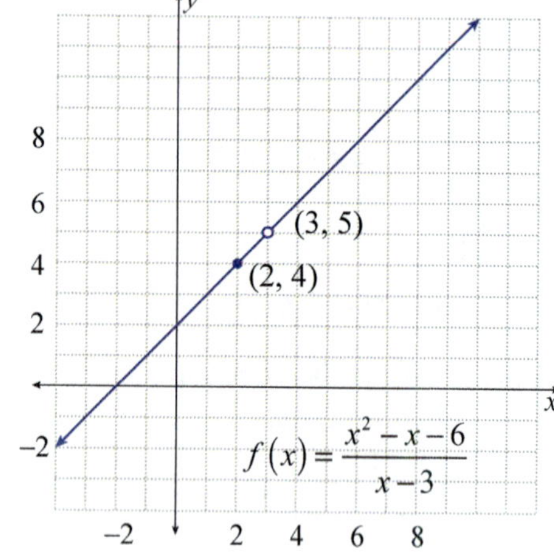

Example 2: Graphing a Discontinuous Function

a. Use the definition of continuity to determine whether the function $g(x) = \dfrac{|x|}{x}$ is continuous at $x = 0$.

b. Draw the graph of $g(x)$.

Solution

a. Since $g(0) = \dfrac{|0|}{0} = \dfrac{0}{0}$ is undefined, g is not continuous at $x = 0$.

b. The graph can be found by using the definition of $|x|$ and rewriting the function as follows.

$$g(x) = \frac{|x|}{x} = \begin{cases} \dfrac{x}{x} = 1 & \text{if } x > 0 \\[2ex] \dfrac{-x}{x} = -1 & \text{if } x < 0 \end{cases}$$

The graph shows that $g(x)$ has a jump discontinuity at $x = 0$.

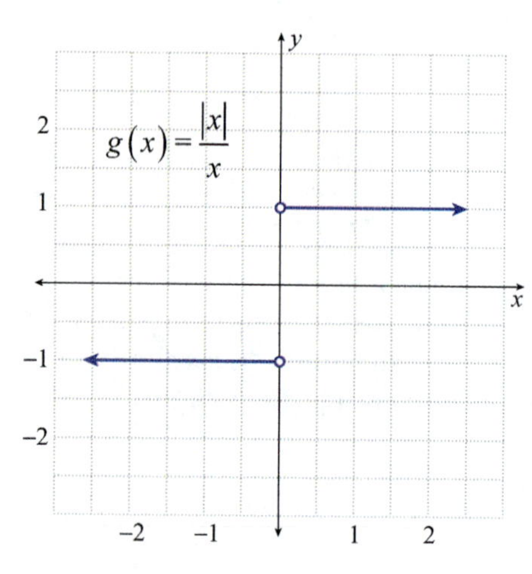

Analysis shows, polynomial functions have no points of discontinuity. That is, for a polynomial function p, $\lim_{x \to a} p(x) = p(a)$ for every real number a.

A function is **continuous** if it has no point of discontinuity. If a function is discontinuous at any one point, then it is a **discontinuous** function.

An interesting function that has several points of discontinuity is the *post office function*, which relates the weight of an item to the price of delivery by the Postal Service. It is an example of a **step function**, and its graph is shown in Figure 3.

U.S. Post Office Fees for First-Class Delivery (2005)

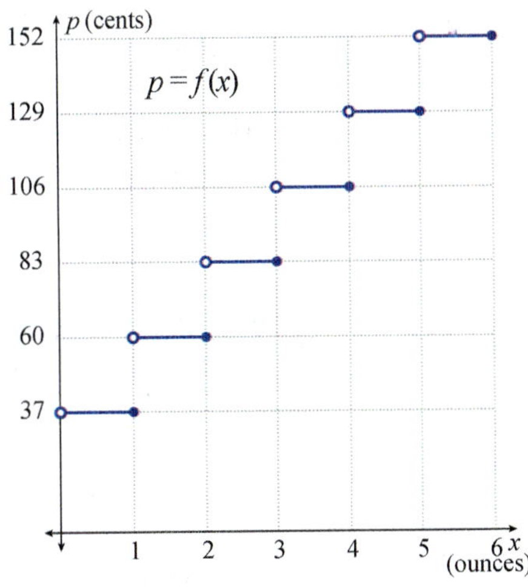

FIGURE 3

Another step function is the **greatest integer function**. It is a particularly interesting example of a function with an infinite number of discontinuities. Each of these discontinuities is a jump discontinuity and occurs at an integer value of x.

Greatest Integer Function

The function $h(x) = [\![x]\!]$ is called the greatest integer function and is defined as follows:

$$h(x) = [\![x]\!] = \text{the greatest integer} \leq x.$$

Informally, $h(x) = [\![x]\!]$ is the integer closest to x but still less than or equal to x.

Thus, for any x, the value of $h(x)$ is an integer. As examples, we show the following:

$$h(3.1) = [\![3.1]\!] = 3$$

$$h(4.6) = [\![4.6]\!] = 4$$

$$h(0.8) = [\![0.8]\!] = 0, \text{ and}$$

$$h(-1.7) = [\![-1.7]\!] = -2.$$

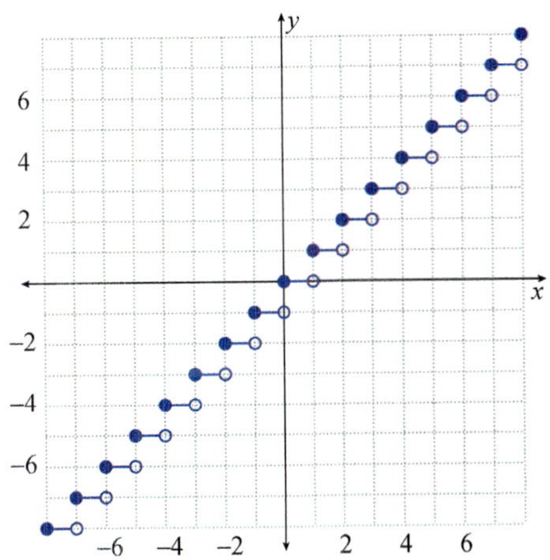

FIGURE 4: The Greatest Integer Function

The graph of $h(x) = [\![x]\!]$ is shown in Figure 4.

Another famous function is the **Dirichlet function**. This function is defined as

$$f(x) = \begin{cases} 1 & \text{if } x \text{ is irrational} \\ 0 & \text{if } x \text{ is rational} \end{cases}$$

and is discontinuous at every point. The reasoning is as follows:

If a is a real number, then $f(a) = 1$ or $f(a) = 0$. Every interval about $x = a$ contains an infinite number of irrational numbers and an infinite number of rational numbers. Thus, as $x \to a$, there are an infinite number of points where $f(x) = 1$ and an infinite number of points where $f(x) = 0$. Neither 1 nor 0 can be $\lim\limits_{x \to a} f(x)$, so $\lim\limits_{x \to a} f(x)$ does not exist.

For a look at continuity from a different viewpoint, consider the problem of determining the form or the value of a function at a particular point so that the function will be continuous at that point.

Example 3: Making a Function Continuous

Find a value for k so that the function

$$f(x) = \begin{cases} x^2 & \text{if } x < 2 \\ -3x + k & \text{if } x \geq 2 \end{cases}$$

will be continuous at $x = 2$.

Solution

The function f is defined in pieces, and each piece is a polynomial function. We must determine a value for k so that both pieces will have the same value when $x = 2$.

Set $x^2 = -3x + k$ and substitute $x = 2$:

$$2^2 = -3(2) + k$$
$$4 = -6 + k$$
$$10 = k$$

Thus, the function f is continuous at $x = 2$ if $k = 10$.

Graphically, we see that the two pieces "fit" together at the point $(2, 4)$.

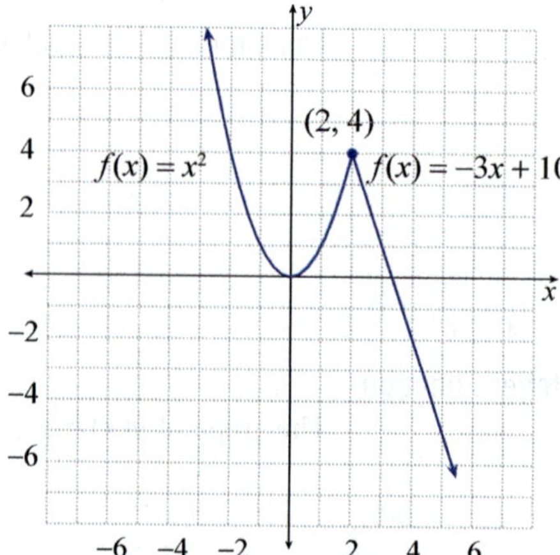

10.4 EXERCISES

◉ PRACTICE

In Exercises 1–8, use the graph of $y = f(x)$ to answer the questions regarding the function.

1.

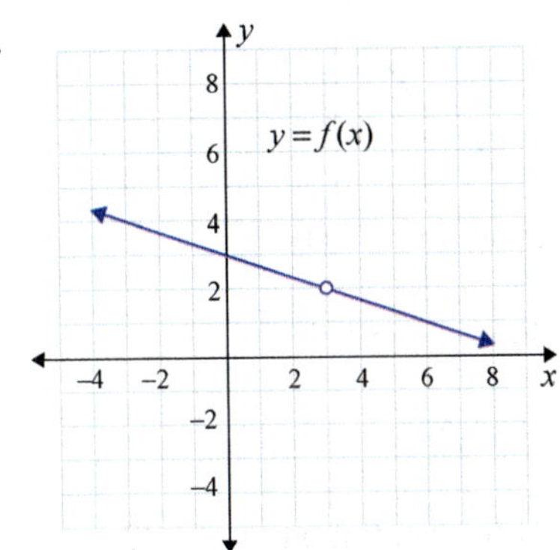

a. Find $\lim_{x \to 3^-} f(x)$.

b. Find $\lim_{x \to 3^+} f(x)$.

c. Find $f(3)$.

d. Is $f(x)$ continuous at $x = 3$? Explain.

2.

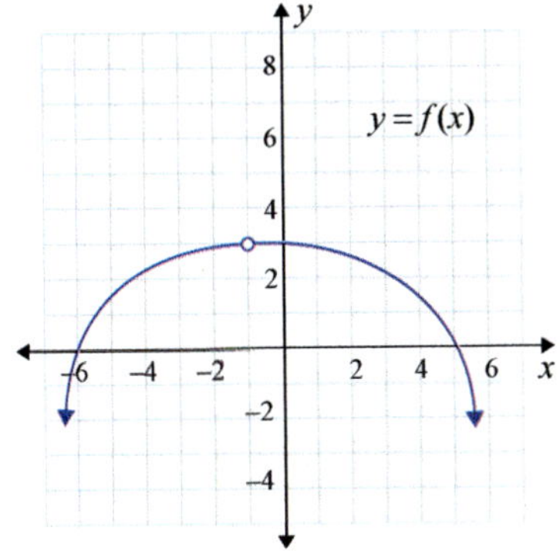

a. Find $\lim_{x \to -1^-} f(x)$.

b. Find $\lim_{x \to -1^+} f(x)$.

c. Find $f(-1)$.

d. Is $f(x)$ continuous at $x = -1$? Explain.

3.

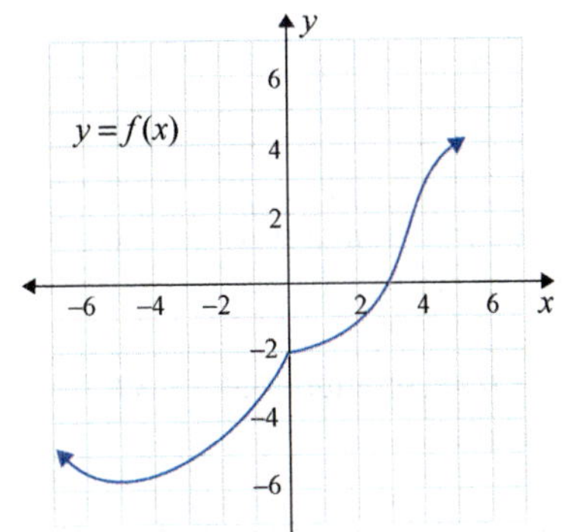

a. Find $\lim_{x \to 0^-} f(x)$.

b. Find $\lim_{x \to 0^+} f(x)$.

c. Find $f(0)$.

d. Is $f(x)$ continuous at $x = 0$? Explain.

4.

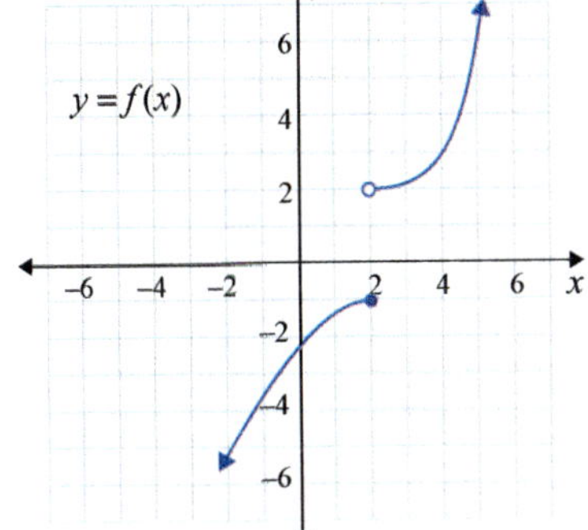

a. Find $\lim_{x \to 2^-} f(x)$.

b. Find $\lim_{x \to 2^+} f(x)$.

c. Find $f(2)$.

d. Is $f(x)$ continuous at $x = 2$? Explain.

5.

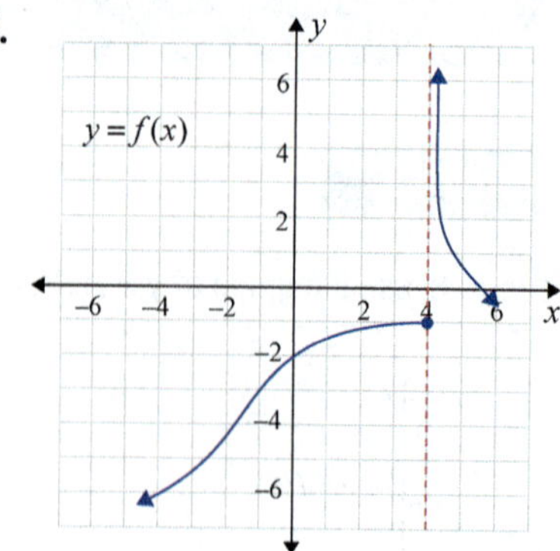

a. Find $\lim\limits_{x \to 4^-} f(x)$.

b. Find $\lim\limits_{x \to 4^+} f(x)$.

c. Find $f(4)$.

d. Is $f(x)$ continuous at $x = 4$? Explain.

6.

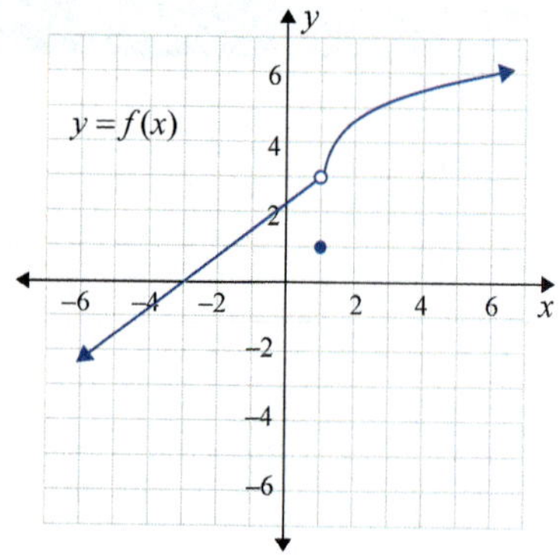

a. Find $\lim\limits_{x \to 1^-} f(x)$.

b. Find $\lim\limits_{x \to 1^+} f(x)$.

c. Find $f(1)$.

d. Is $f(x)$ continuous at $x = 1$? Explain.

7.

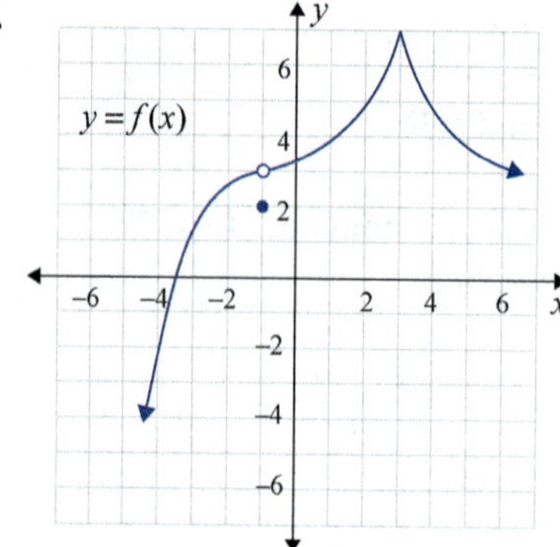

a. Is $f(x)$ continuous at $x = -1$? Explain.

b. Is $f(x)$ continuous at $x = 3$? Explain.

8.

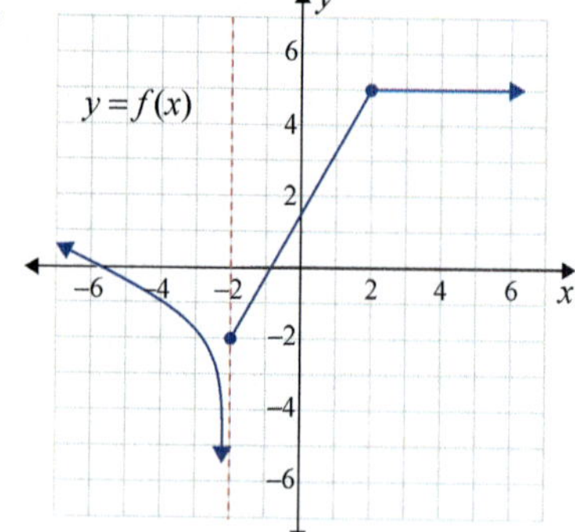

a. Is $f(x)$ continuous at $x = 2$? Explain.

b. Is $f(x)$ continuous at $x = -2$? Explain.

In Exercises 9–12, use the graph of $y = f(x)$ to find the points of discontinuity, if any exist. Determine what type of discontinuity each is.

9.

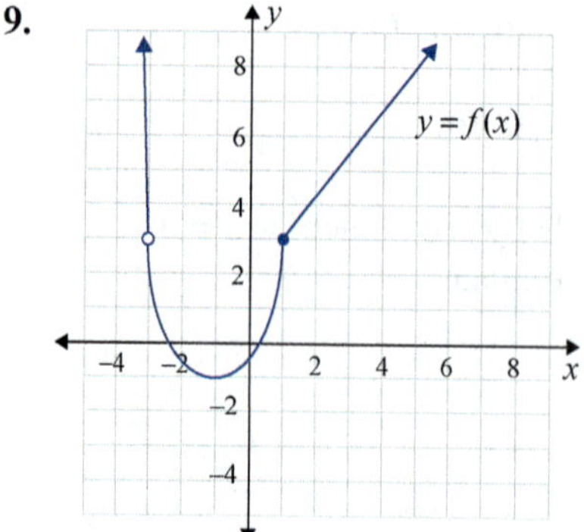

10.

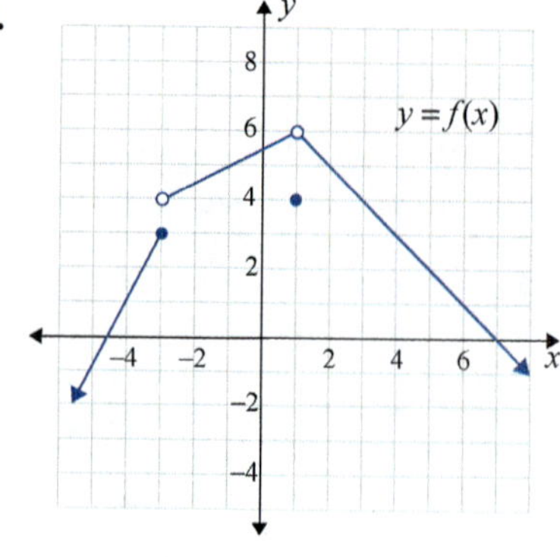

11. 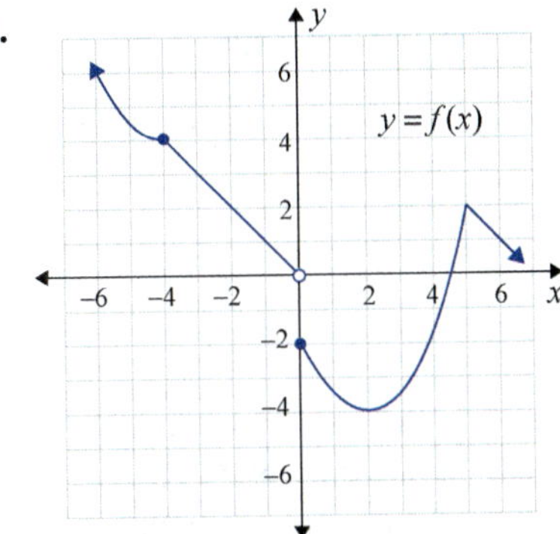

12. 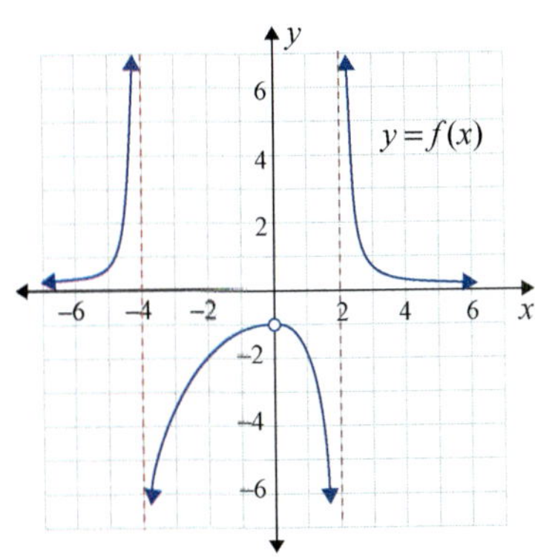

In Exercises 13–20, use the definition of continuity to determine whether or not the function is continuous at the given value of *x*.

13. $f(x) = 3 - 2x; \; x = 1$

14. $f(x) = 5x - x^2; \; x = 0$

15. $f(x) = \dfrac{x^2 - x - 2}{x - 2}; \; x = 2$

16. $f(x) = \dfrac{x + 1}{x^2 - 1}; \; x = 0$

17. $f(x) = \begin{cases} 2x + 1 & \text{if } x \le 1 \\ 3 & \text{if } x > 1 \end{cases}; \; x = 1$

18. $f(x) = \begin{cases} x^2 & \text{if } x \le 2 \\ 2x & \text{if } x > 2 \end{cases}; \; x = 2$

19. $f(x) = \begin{cases} 1 - 3x & \text{if } x < 0 \\ 4x & \text{if } x \ge 0 \end{cases}; \; x = 0$

20. $f(x) = \begin{cases} \dfrac{x}{x - 3} & \text{if } x \ne 3 \\ 2 & \text{if } x = 3 \end{cases}; \; x = 3$

In Exercises 21–28, find the points of discontinuity for each function, if any exist. Determine what type of discontinuity each is.

21. $f(x) = 2x^2 + 3x - 1$

22. $f(x) = 3x^2 - x + 7$

23. $f(x) = \dfrac{5}{x + 3}$

24. $f(x) = \dfrac{x + 8}{x}$

25. $f(x) = \dfrac{x}{x^2 - 9}$

26. $f(x) = \dfrac{2}{x^2 - 4x}$

27. $f(x) = \begin{cases} 2 + 3x & \text{if } x \le 1 \\ x^2 + 4 & \text{if } x > 1 \end{cases}$

28. $f(x) = \begin{cases} x^2 + 1 & \text{if } x \le 2 \\ 2x - 1 & \text{if } x > 2 \end{cases}$

In Exercises 29–32, find a value for *k* so that the given function will be continuous at the indicated value for *x*.

29. $f(x) = \begin{cases} 3x & \text{if } x \le 2 \\ x^2 + k & \text{if } x > 2 \end{cases}; \; x = 2$

30. $f(x) = \begin{cases} 7 & \text{if } x < -3 \\ k - 2x & \text{if } x \ge -3 \end{cases}; \; x = -3$

31. $f(x) = \begin{cases} 3x - k & \text{if } x \le 1 \\ \dfrac{x^2 - 3x + 2}{x - 1} & \text{if } x > 1 \end{cases}; \; x = 1$

32. $f(x) = \begin{cases} \dfrac{x^2 + 3x}{x} & \text{if } x < 0 \\ 2x^2 - k & \text{if } x \ge 0 \end{cases}; \; x = 0$

APPLICATIONS

33. Pricing: A leather craft store has the following pricing policy for a belt buckle:

$$C(x) = \begin{cases} 0.79x & \text{if } 0 < x < 12 \\ 0.71x & \text{if } 12 \le x < 50 \\ 0.67x & \text{if } x \ge 50 \end{cases}$$

where x is the number of buckles and $C(x)$ is in dollars.
   a. Graph the function $C(x)$.
   b. Is $C(x)$ a continuous function? Explain your answer.

34. Salary: A salesperson's weekly salary is determined by the function

$$S(x) = \begin{cases} 550 & \text{if } 0 < x < 10,000 \\ 0.06x & \text{if } x \ge 10,000 \end{cases}$$

where x is the weekly sales.
   a. Graph the function $S(x)$.
   b. Is $S(x)$ a continuous function? Explain your answer.

35. Cost of telephone call: The cost of an overseas call is given by the function

$$C(x) = \begin{cases} 9.00 & \text{if } 0 < x \le 3 \\ 0.95x + 6.15 & \text{if } x > 3 \end{cases}$$

where $C(x)$ is in dollars and x is in minutes.
   a. Graph the function $C(x)$.
   b. Is $C(x)$ a continuous function? Explain your answer.

36. Revenue: The revenue from the sale of a particular model of wireless speaker is given by the function

$$R(p) = \begin{cases} 84p & \text{if } 0 < p \le 20 \\ 144p - 3p^2 & \text{if } p > 20 \end{cases}$$

where p is the price in dollars.
   a. Graph the function $R(p)$.
   b. Is $R(p)$ a continuous function? Explain your answer.

10.5 AVERAGE RATE OF CHANGE

Suppose that you rode your bicycle 24 miles in 2 hours. Then your average speed **(average rate of change of distance with respect to time)** was $\dfrac{24\ \text{miles}}{2\ \text{hours}} = 12$ mph. This does not mean that you were moving at 12 mph all the time you were riding. At some times you were going slower and at some times faster. For example, if you happened to be riding downhill at a particular instant, then your **instantaneous rate of change** at that moment might have been 18 mph.

These two ideas, average rate of change and instantaneous rate of change, form the basis for our development of the derivative of a function.

The Greek capital letter delta (Δ) is used in calculus to indicate change. Thus Δx is read "delta x" (or "the change in x"), and Δy is read "delta y" (or "the change in y"). For a function $y = f(x)$, as the values of x change from x_1 to x_2, the corresponding values of y change from y_1 to y_2 or from $f(x_1)$ to $f(x_2)$. We write

$$\Delta x = x_2 - x_1$$
$$\text{and}$$
$$\Delta y = y_2 - y_1 = f(x_2) - f(x_1).$$

Average Rate of Change

If $y = f(x)$ then the ratio

$$\frac{\Delta y}{\Delta x} = \frac{y_2 - y_1}{x_2 - x_1} = \frac{f(x_2) - f(x_1)}{x_2 - x_1}$$

is called the **average rate of change of y with respect to x** as x changes from x_1 to x_2.

Example 1: Finding the Average Rate of Change

For the function $y = 2x + 1$, find the average rate of change of y with respect to x, $\dfrac{\Delta y}{\Delta x}$, as x changes from 0 to 3.

Solution

For $x_1 = 0$ and $x_2 = 3$ the corresponding values of y are

$$y_1 = 2 \cdot x_1 + 1 = 2 \cdot 0 + 1 = 1 \qquad \text{Substitute } x_1 = 0 \text{ into } y \text{ and solve.}$$

and

$$y_2 = 2 \cdot x_2 + 1 = 2 \cdot 3 + 1 = 7. \qquad \text{Substitute } x_2 = 3 \text{ into } y \text{ and solve.}$$

Thus,

$$\Delta y = y_2 - y_1 = 7 - 1 = 6,$$

$$\Delta x = x_2 - x_1 = 3 - 0 = 3,$$

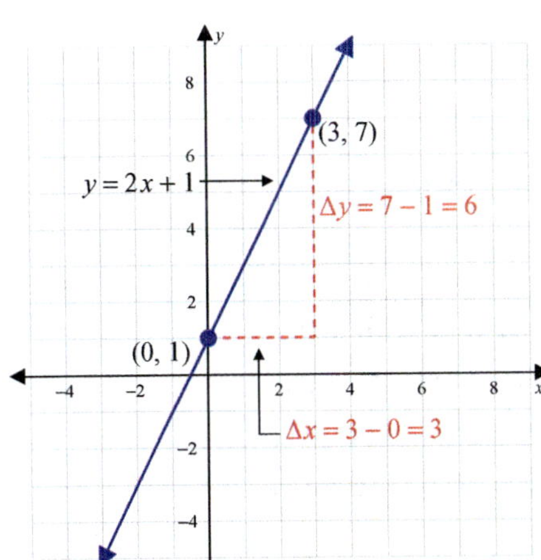

and the average rate of change is $\dfrac{\Delta y}{\Delta x} = \dfrac{6}{3} = 2.$

Any other distinct choices for x_1 and x_2 will show that $\dfrac{\Delta y}{\Delta x} = 2$, the slope of the line.

Example 2: Average Rate of Change

Let $f(x) = \dfrac{2}{x}$. Find and simplify the expression that represents the average rate of change of f between x and $x + h$.

Solution

Here $x_1 = x$ and $x_2 = x + h$. Therefore, $\Delta x = x_2 - x_1 = x + h - x = h$.

So the average rate of change is calculated as follows.

$$\frac{\Delta y}{\Delta x} = \frac{f(x_2) - f(x_1)}{x_2 - x_1} = \frac{f(x + h) - f(x)}{h}$$

$$= \frac{\dfrac{2}{x+h} - \dfrac{2}{x}}{h} = \frac{\dfrac{2 \cdot x}{(x+h) \cdot x} - \dfrac{2 \cdot (x+h)}{x \cdot (x+h)}}{h}$$

$$= \frac{\dfrac{2x - 2x - 2h}{x(x+h)}}{\dfrac{h}{1}}$$

$$= \frac{-2h}{x(x+h)} \cdot \frac{1}{h}$$

$$= -\frac{2}{x(x+h)}$$

Example 3: A Falling Object

A ball is dropped from an airplane. The distance d that it falls in t seconds is given by the function $d = 16t^2$. Table 1 shows the distance the ball falls and the change in distance Δd for 1-second time intervals during the first 5 seconds of its descent.

a. How far does the ball fall during the first 3 seconds?

b. What is the average velocity of the ball during the first 3 seconds? (Velocity is the rate of change of distance with respect to time.)

c. How fast is the ball traveling when $t = 3$?

t (sec)	$d = 16t^2$	Δd (ft)
0	0	—
1	16	$d_1 - d_0 = 16 - 0 = 16$
2	64	$d_2 - d_1 = 64 - 16 = 48$
3	144	$d_3 - d_2 = 144 - 64 = 80$
4	256	$d_4 - d_3 = 256 - 144 = 112$
5	400	$d_5 - d_4 = 400 - 256 = 144$

TABLE 1

Solution

a. $\Delta d = d_3 - d_0 = 144 - 0 = 144$ ft   Using data from Table 1.

b. $\dfrac{\Delta d}{\Delta t} = \dfrac{144 \text{ ft}}{3 \text{ sec}} = 48 \dfrac{\text{ft}}{\text{sec}}$ Using $\Delta t = t_3 - t_0 = 3 - 0 = 3$ sec.

c. We do not know how fast the ball is traveling when $t = 3$ seconds. This is a question of instantaneous velocity, which we will discuss later.

10.5 EXERCISES

💡 PRACTICE

Use the graph to solve Exercises 1–2. The variable t is the number of hours since midnight and $f(t)$ is the temperature at time t.

1. What is the average rate of change of $f(t)$ from $t = 0$ to $t = 2$?

2. What is the average rate of change of $f(t)$ from $t = 6$ to $t = 12$?

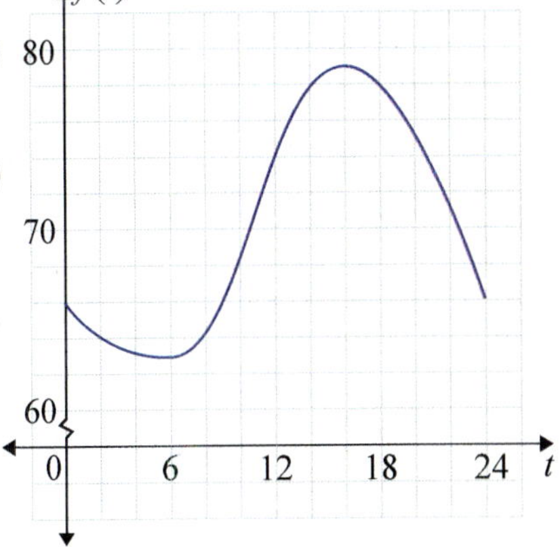

In Exercises 3–10, find the average rate of change of the given functions between the given values of x_1 and x_2.

3. $f(x) = 5x + 3;\ x_1 = 1,\ x_2 = 3$

4. $f(x) = 3x + 8;\ x_1 = -2,\ x_2 = 1$

5. $f(x) = 2x^2 - x - 3;\ x_1 = 2,\ x_2 = 2.5$

6. $f(x) = 3x^2 - 2x - 1;\ x_1 = 1,\ x_2 = 1.5$

7. $f(x) = \dfrac{-2}{2x - 1};\ x_1 = 3,\ x_2 = 3.5$

8. $f(x) = \dfrac{2}{3x + 2};\ x_1 = 0.5,\ x_2 = 1$

9. $f(x) = \sqrt{x};\ x_1 = 1,\ x_2 = 2.5$

10. $f(x) = \sqrt{x - 3};\ x_1 = 4,\ x_2 = 4.44$

10.6 INSTANTANEOUS RATE OF CHANGE

■ TOPICS

- Slopes and Rates of Change
- Slopes and Calculators

In this section, through the use of tables, we will discuss the concept of instantaneous rate of change of a function at one specific value of the variable. Later we will take a more formal algebraic approach.

Consider the problem where a ball is dropped from an airplane. The distance the ball falls is given by the function $d = 16t^2$ where t is time in seconds and d is distance in feet. (See Figure 1.)

To determine how fast the ball is moving at the instant when $t = 3$ seconds, we analyze the average rate of change $\dfrac{\Delta d}{\Delta t}$ for smaller and smaller values of Δt where $t = 3$. Table 1 shows that t_1 is fixed at 3 and t_2 approaches 3. The various values of the function can be found with a calculator.

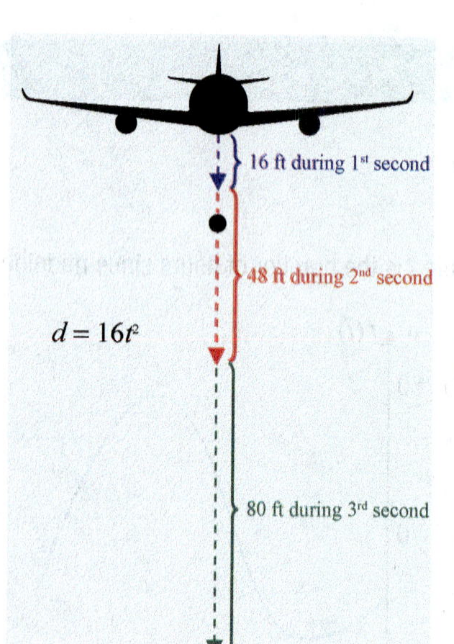

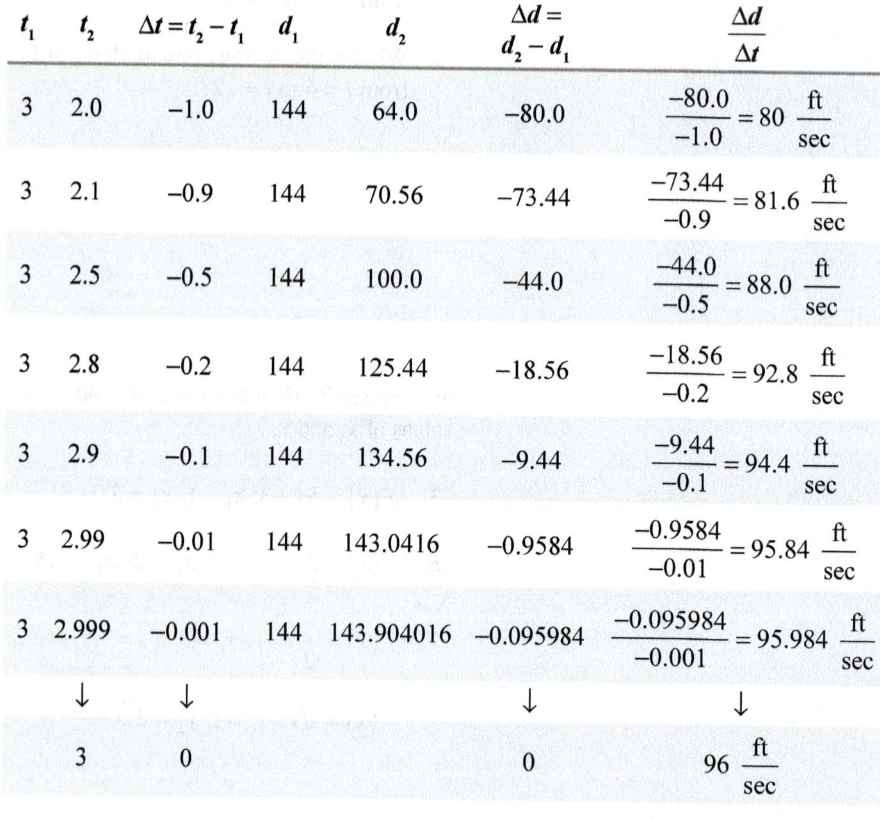

t_1	t_2	$\Delta t = t_2 - t_1$	d_1	d_2	$\Delta d = d_2 - d_1$	$\dfrac{\Delta d}{\Delta t}$
3	2.0	−1.0	144	64.0	−80.0	$\dfrac{-80.0}{-1.0} = 80 \dfrac{\text{ft}}{\text{sec}}$
3	2.1	−0.9	144	70.56	−73.44	$\dfrac{-73.44}{-0.9} = 81.6 \dfrac{\text{ft}}{\text{sec}}$
3	2.5	−0.5	144	100.0	−44.0	$\dfrac{-44.0}{-0.5} = 88.0 \dfrac{\text{ft}}{\text{sec}}$
3	2.8	−0.2	144	125.44	−18.56	$\dfrac{-18.56}{-0.2} = 92.8 \dfrac{\text{ft}}{\text{sec}}$
3	2.9	−0.1	144	134.56	−9.44	$\dfrac{-9.44}{-0.1} = 94.4 \dfrac{\text{ft}}{\text{sec}}$
3	2.99	−0.01	144	143.0416	−0.9584	$\dfrac{-0.9584}{-0.01} = 95.84 \dfrac{\text{ft}}{\text{sec}}$
3	2.999	−0.001	144	143.904016	−0.095984	$\dfrac{-0.095984}{-0.001} = 95.984 \dfrac{\text{ft}}{\text{sec}}$
	↓	↓			↓	↓
	3	0			0	$96 \dfrac{\text{ft}}{\text{sec}}$

TABLE 1: $d = 16t^2$

Notes about Table 1

a. The values of $t_1 = 3$ and $d_1 = 144$ are fixed.

b. Since t_2 is approaching 3, the difference Δt is approaching 0.

c. The difference Δd is also approaching 0.

d. The average rate of change $\dfrac{\Delta d}{\Delta t}$ is approaching the **instantaneous rate of change** of d with respect to t when $t = 3$. We can write

$$\lim_{\Delta t \to 0} \frac{\Delta d}{\Delta t} = 96 \ \frac{\text{ft}}{\text{sec}}.$$

The instantaneous rate of change of distance with respect to time, as illustrated in Table 1, is called **instantaneous velocity**.

Slopes and Rates of Change

Rate of change is a numerical interpretation of $\dfrac{\Delta y}{\Delta x}$ for some function $y = f(x)$. This is a natural geometric interpretation using the familiar concept of **slope of a line**: for any two points (x_1, y_1) and (x_2, y_2) on the graph of $f(x)$, the number (rate of change)

$$\frac{\Delta y}{\Delta x} = \frac{y_2 - y_1}{x_2 - x_1}$$ is also the slope of the line through the two points.

If we use a graphing utility to graph $y = x^2$ in a window $[-5, 5]$ by $[-10, 20]$, we see a parabola.

Let us consider the points $(x_1, y_1) = (3, 9)$ and $(x_2, y_2) = (4, 16)$.

The slope of the segment containing these points is

$$\frac{\Delta y}{\Delta x} = \frac{16 - 9}{4 - 3} = 7.$$

The line $y = mx + b$ through these two points with a y-intercept of b is given by

$$9 = 7(3) + b. \qquad \text{Using point } (x_1, y_1) = (3, 9) \text{ and slope} = 7$$

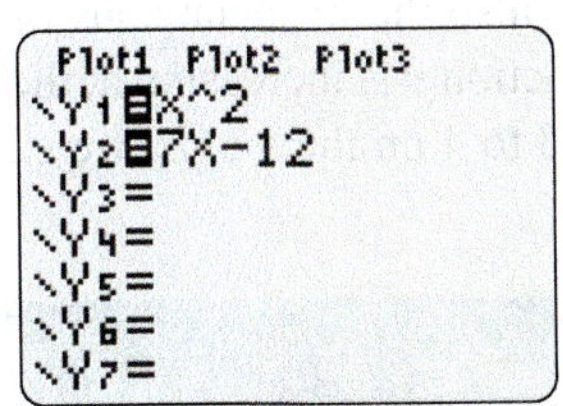

FIGURE 2

Then, solving for b, we find that

$$b = -12. \qquad \text{Adding } -21 \text{ to both sides}$$

Therefore, the equation of the line through both points is

$$y = 7x - 12.$$

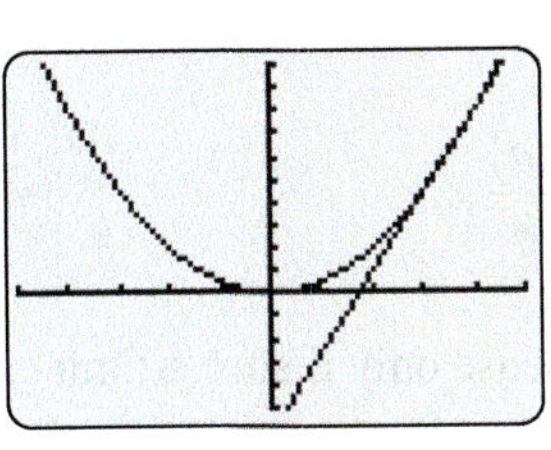

FIGURE 3

This line is also called the **secant line** through the two points. Display this line on the same calculator screen as $y = x^2$. To do so, type the equation of this secant line into the position **Y2** on the graphing calculator, as shown in Figure 2. Figure 3 is what you should see graphed.

Now we want to draw a new view with a smaller Δx. We choose a line through points $(x_1, y_1) = (3, 9)$ and $(x_2, y_2) = (3.1, 9.61)$. The slope of this line is

$$\frac{\Delta y}{\Delta x} = \frac{9.61 - 9}{3.1 - 3} = \frac{0.61}{0.1} = 6.1$$

and the y-intercept is

$$9 = 6.1(3) + b$$
$$9 = 18.3 + b$$
$$-9.3 = b.$$

Therefore, the line through the new points is $y = 6.1x - 9.3$.

Using the window [2.8, 3.3] by [7.5, 9.9], both secant lines are shown graphed in Figure 4. Even under high magnification, the two secant lines are difficult to discern.

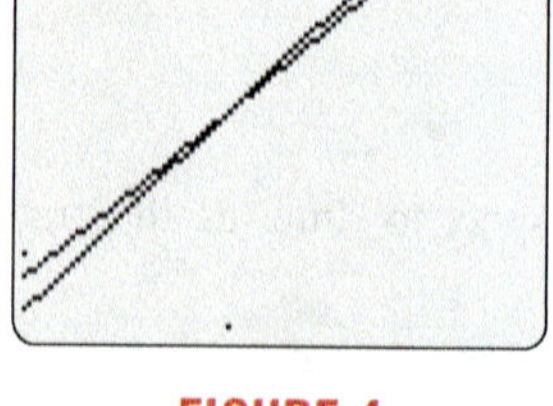

FIGURE 4

We repeat this process a third time with $(x_1, y_1) = (3, 9)$, using an even smaller Δx.

Let $(x_2, y_2) = (3.01, 9.0601)$.

$$\frac{\Delta y}{\Delta x} = \frac{9.0601 - 9}{3.01 - 3} = \frac{0.0601}{0.01} = 6.01$$

As $\Delta x \to 0$, the slope calculated seems to be nearing 6.

That is,

$$\lim_{\Delta x \to 0} \frac{\Delta y}{\Delta x} = 6.$$

One can continue this process, and as $\Delta x \to 0$ the secant line approaches a limiting position. This position is occupied by the line through $(3, 9)$ having slope 6, which has the equation $y = 6x + 9$.

This line touches $y = x^2$ exactly at the point $(3, 9)$ and at no other point. It is called the **tangent line** to $y = x^2$ at the point $(3, 9)$.

As we zoom closer and closer to the point $(3, 9)$, the graph of the function and the graph of the tangent line cannot be separated on the graphing screen (see Figure 5).

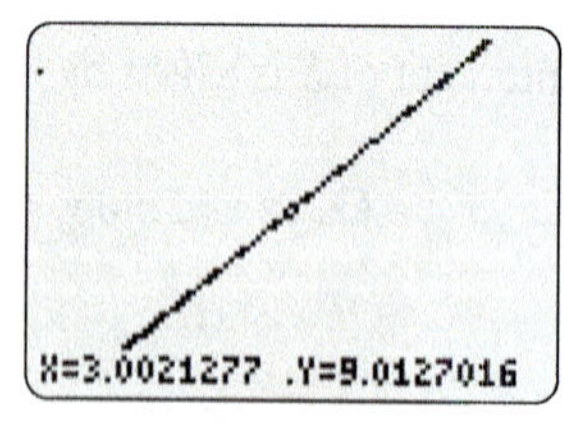

FIGURE 5

The number 6 is the slope of the tangent line, and it is also the instantaneous rate of change of y at the point $(3, 9)$. At the point $(3, 9)$ the function y is increasing at the rate of 6 units (per 1 unit of change in x). Therefore, from 3 to 4 on the x-axis, we expect y to increase by about 6.

Slope at a Point on a Curve

The slope of a curve $f(x)$ at the point (a, b) is exactly the slope of the tangent line to the curve at that point. This is also called the **instantaneous rate of change**.

Numerically, this slope is

$$\lim_{\Delta x \to 0} \frac{\Delta y}{\Delta x} = \lim_{h \to 0} \frac{f(a+h) - f(a)}{h},$$

where $h = \Delta x$. Technically, the definition given makes sense only if such a limit exists.

We will use the notation $f'(a)$, read "f prime of a", to denote the slope at $(a, f(a))$.

Difference Quotient

If $(x, f(x))$ is any fixed point on a curve and $(x + h, f(x + h))$ is another point on the curve, then the **difference quotient** is the slope of the secant line through these two points.

$$m_{\text{sec}} = \frac{f(x+h) - f(x)}{(x+h) - x} = \frac{f(x+h) - f(x)}{h}$$

The **derivative** $f'(x)$, if it exists, can be interpreted as the slope of the tangent line at the point $(x, f(x))$. Figure 6 illustrates this geometric relationship between the derivative and the secant line.

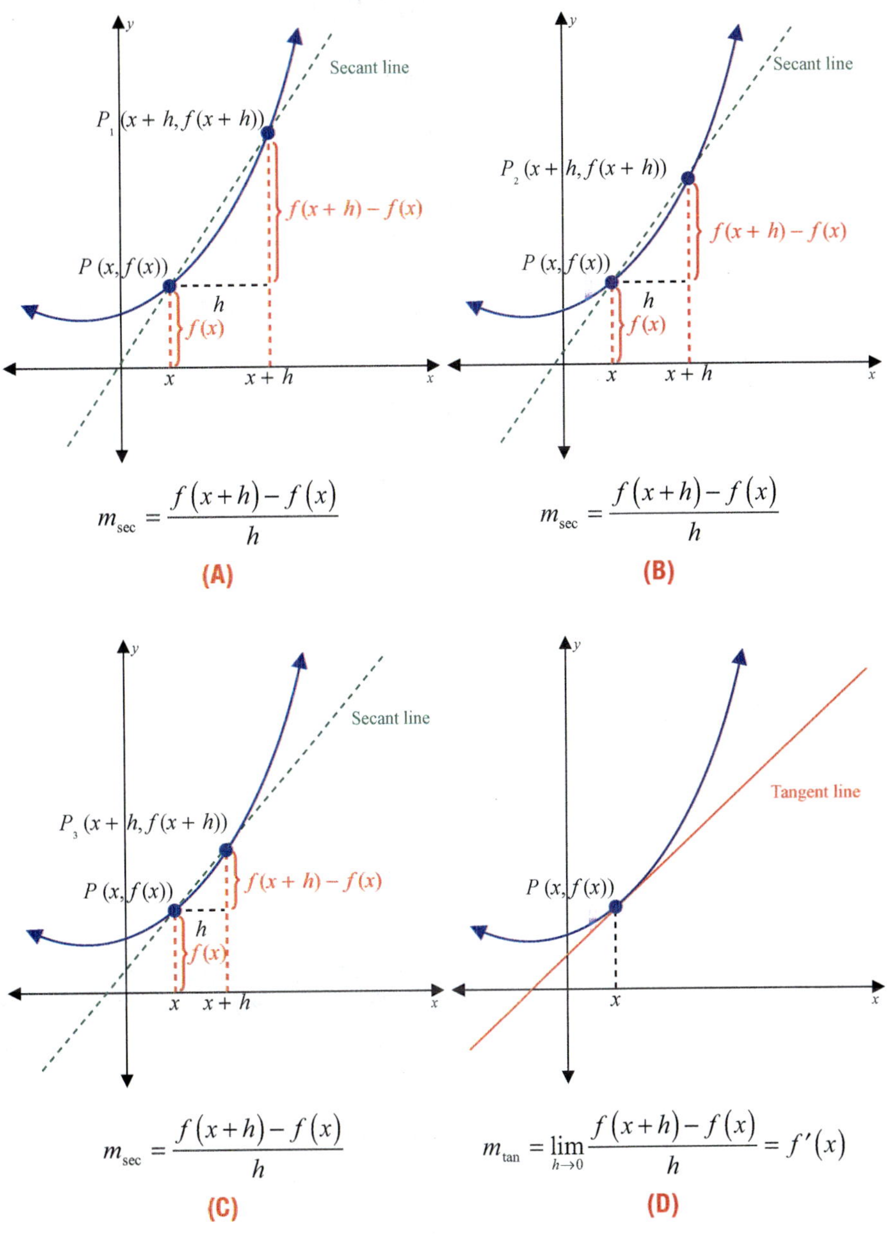

FIGURE 6

The derivative $f'(x)$ is the slope of the tangent line at $(x, f(x))$.

In general, given a function $y = f(x)$, we will be interested in two "output" numbers corresponding to the input $x = a$, namely the corresponding y-value, $f(a)$, and the slope, $f'(a)$.

Example 1: Temperature

Let $f(t)$ denote the temperature in South Carolina on a typical summer day at time t, where t is the number of hours since midnight. For instance, $t = 1$ corresponds to 1:00 a.m. and $t = 13$ corresponds to 1:00 p.m.

a. Is $f'(8)$ positive, negative, or zero?

b. Is $f'(21)$ positive, negative, or zero?

c. Sketch an approximate graph for this $f(t)$.

Solution

a. Since the graph of $f(t)$ will increase from sunup to about midafternoon, the slope at $(8, f(8))$ is positive. That is, $f'(8)$ is positive.

b. At 9:00 p.m. ($t = 21$), the temperature has begun to decline and so the slope will be negative. That is, $f'(21)$ is negative.

c.

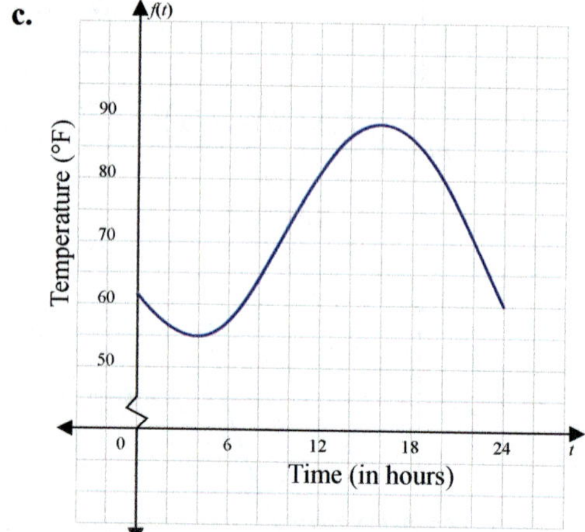

The graph needs to increase from about 6:00 a.m. to 4:00 p.m. and then decline, as described in parts **a.** and **b.** Drawing such a graph will also support the answers found in parts **a.** and **b.**

Example 2: Manufacturing

Let $C(x)$ denote the total cost of manufacturing x units of a certain style of metal office bookcase.

a. Given $C(300) = 31{,}000$ and $C'(300) = 90$, sketch a tangent line to $C(x)$ at the point $(300, 31{,}000)$.

b. Estimate the cost of manufacturing 303 bookcases.

Solution

a. Using the definition of the slope of a curve at a point, $C'(300) = 90$ is the slope of the tangent line at the point $(300, 31{,}000)$. Knowing this, we can draw a line with slope 90 at the point $(300, 31{,}000)$.

b. First, let's further understand the information given in the example. $C(300) = 31{,}000$ means it costs \$31,000 to manufacture 300 bookcases. Using this, we can determine the average cost of producing the first 300 bookcases.

$$\frac{C(300)}{300} = \frac{31{,}000}{300} \approx \$103.33$$

$C'(300) = 90$ is the rate of change of y at the point $(300, 31{,}000)$. This means that y is changing at the rate of \$90 per bookcase at this point. Another way of stating that is \$90 is the cost of an additional bookcase if 301 were produced instead of 300. The average cost is decreasing—a useful fact to be able to calculate for a manufacturing company.

Using these meanings, the cost of producing 3 more bookcases, compared to the cost of producing 300, is about $90 \cdot 3 = 270$ dollars more. So it will cost $\$31{,}000 + \$270 = \$31{,}270$ to manufacture 303 bookcases.

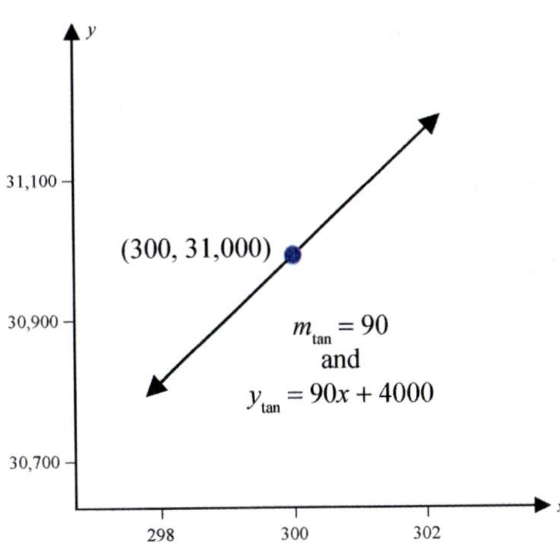

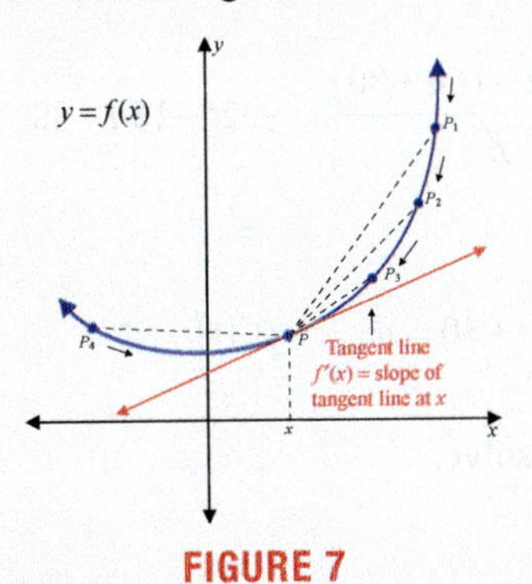

NOTE

We know that $\dfrac{\Delta y}{\Delta x}$ can be interpreted as the slope of a secant line through two points on a curve. Now, if one point is fixed, then the **tangent line at that point is the limiting position of secant lines**, as illustrated in Figure 7.

FIGURE 7

As the points $P_1,\ P_2,\ P_3,\ ...,\ P_n$ approach point P, the secant lines $PP_1,\ PP_2,\ PP_3,\ ...,\ PP_n$ approach the tangent at P.

Example 3: Average Rate of Change

Consider the function $f(x) = \dfrac{1}{3}x^3$. Find the average rate of change of f as x changes from $x_1 = 0$ to $x_2 = 3$.

Solution

$f(x_1) = f(0) = \dfrac{1}{3} \cdot 0^3 = 0$

$f(x_2) = f(3) = \dfrac{1}{3} \cdot 3^3 = 9$

$\dfrac{\Delta y}{\Delta x} = \dfrac{f(x_2) - f(x_1)}{x_2 - x_1}$

$= \dfrac{9 - 0}{3 - 0}$

$= \dfrac{9}{3}$

$= 3$

The average rate of change is the slope of a line secant to $f(x)$ passing through both the start and stop points.

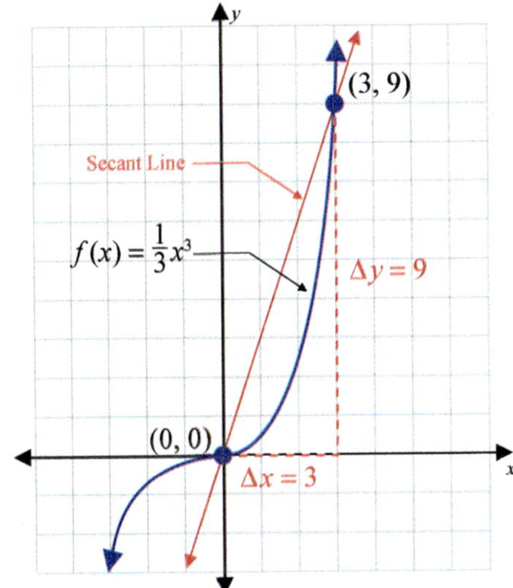

$$\frac{\Delta y}{\Delta x} = m_{\sec} = \text{average rate of change}$$

In Example 4 we show how the derivative can be interpreted as both the slope of a tangent line to a curve and the instantaneous velocity of an object.

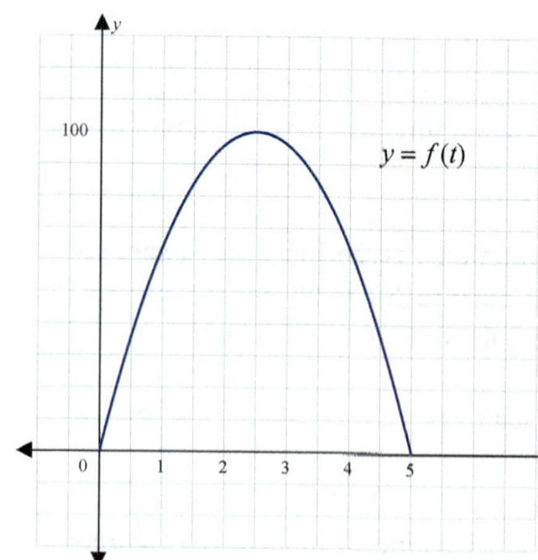

Example 4: Instantaneous Velocity

An arrow is shot into the air and its height in feet after t seconds is given by the function $f(t) = -16t^2 + 80t$. The graph of the curve $y = f(t)$ is the parabola shown.

a. Find the height of the arrow when $t = 3$.

b. Find the instantaneous velocity of the arrow when $t = 3$.

c. Find the slope of the tangent line to the curve at $t = 3$.

d. Find the time it takes the arrow to reach its peak.

Solution

a. $f(3) = -16(3)^2 + 80(3)$ Substitute $t = 3$ into the function $f(t)$.

$$= -144 + 240 = 96$$

So, when $t = 3$, the height is 96 feet.

b. To find the instantaneous velocity when $t = 3$, find the derivative $f'(t)$, which is also the slope of the tangent line at $(t, f(t))$, and then evaluate at $t = 3$.

Step 1: Form the difference quotient.

$$\frac{f(t+h) - f(t)}{h} = \frac{\left[-16(t+h)^2 + 80(t+h)\right] - \left[-16t^2 + 80t\right]}{h}$$

Step 2: Simplify the difference quotient.

$$\frac{\left[-16(t+h)^2 + 80(t+h)\right] - \left[-16t^2 + 80t\right]}{h} = \frac{-16(t^2 + 2th + h^2) + 80t + 80h + 16t^2 - 80t}{h}$$

$$= \frac{-16t^2 - 32th - 16h^2 + 80t + 80h + 16t^2 - 80t}{h}$$

$$= \frac{-32ht - 16h^2 + 80h}{h}$$

$$= \frac{h(-32t - 16h + 80)}{h} = -32t - 16h + 80$$

Step 3: Find the limit.

$$\lim_{h \to 0}(-32t - 16h + 80) = -32t + 80 = m_{\text{tan}} = f'(t)$$

Now, using this derivative, substitute $t = 3$ and solve.

$$f'(t) = -32t + 80$$
$$f'(3) = -32 \cdot 3 + 80 = -16$$

Therefore, when $t = 3$ the arrow is heading towards the ground (this is what the negative sign means) at a rate of 16 $\dfrac{\text{ft}}{\text{sec}}$.

c. From part **b.** we know that $f'(3) = -16$. So, the slope of the tangent line to the curve $f(t) = -16t^2 + 80t$ at $t = 3$ is -16.

Therefore, we see that the derivative can be interpreted in two ways:

1. as an instantaneous rate of change and

2. as the slope of a tangent line.

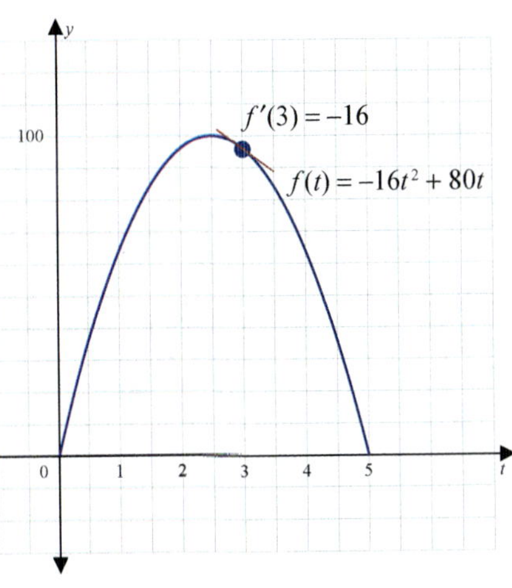

d. At the instant the arrow hits its peak, its velocity will be 0. That is, the arrow will actually be stopped in midair (it will no longer be traveling upwards, and will not yet be going downwards). **Note:** This also corresponds to the fact that the tangent line at the peak will be horizontal and thus have slope 0 at the highest point of the parabola.

Using this information, we set $f'(t) = 0$ and solve for t.

$$-32t + 80 = 0$$
$$-32t = -80$$
$$t = \frac{80}{32} = \frac{\cancel{16} \cdot 5}{\cancel{16} \cdot 2}$$
$$= \frac{5}{2} = 2.5$$

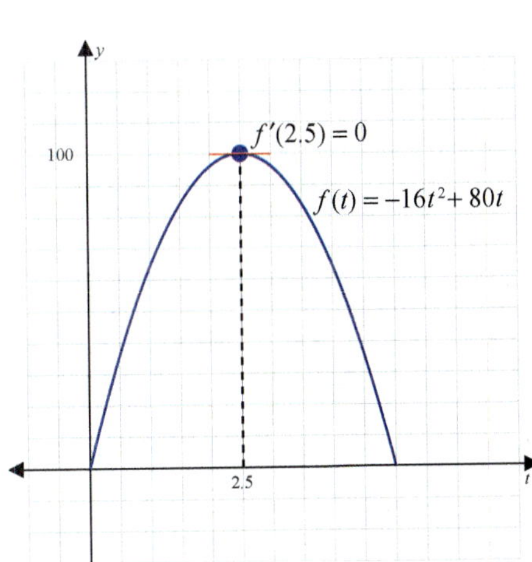

The arrow hits its peak in 2.5 sec.

Example 5: Interpreting a Graph

Estimate the slopes at the marked points as 1, −1, 0, undefined, positive and greater than 1, positive and less than 1, negative and greater than −1, or negative and less than −1. Keep in mind that a tangent line has slope 1 when it makes a 45° angle with the x-axis, has a slope 0 when it is horizontal, and has an undefined slope when it is vertical.

Solution

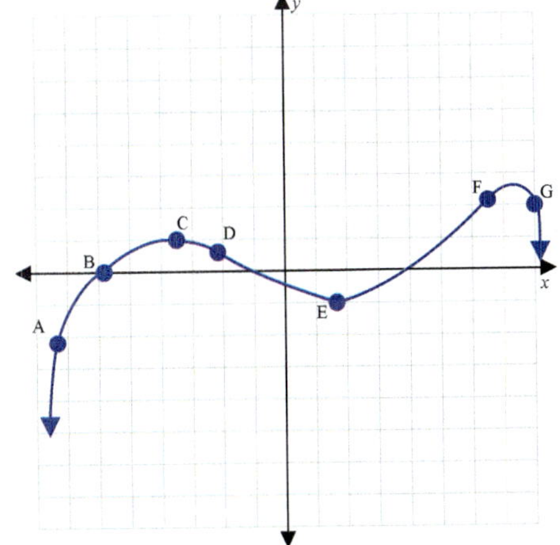

Point	$f'(x)$
A	positive, greater than 1
B	1
C	0
D	negative, greater than −1
E	0
F	positive, less than 1
G	negative, less than −1

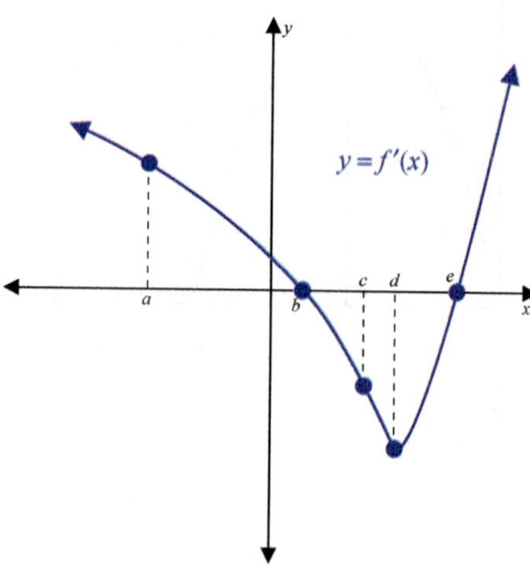

Example 6: Describing and Graphing Rates of Change

a. Describe $f'(x)$ at the points marked on the graph.

b. Sketch a possible graph of $f'(x)$.

Solution

a.

Point	$f'(x)$
A	positive
B	0
C	negative
D	negative
E	0

> **✏ NOTE**
>
> We emphasize that $f'(a)$ denotes the slope of a curve $y = f(x)$ at the point $(a, f(a))$. Also, $f'(a)$ is the instantaneous rate of change of $f(x)$ at $(a, f(a))$. The number $f'(a)$ is called the derivative of $f(x)$ at $x = a$.

b. Using part **a.**, we can draw a possible graph of $f'(x)$. At A, the slope is positive, and at C, the slope is negative. We know that at point B (and E), $f'(x) = 0$ since the tangent line is horizontal. Therefore, the graph of $y = f'(x)$ is above the x-axis at $x = a$ (positive), passes through the x-axis at $x = b$ (0), and lies below the x-axis at $x = c$ (negative). From point C to point D the slopes continue to decrease, since $f'(x)$ is negative, but from point D to point E the slopes increase because we know that at point E the slope is equal to 0.

Slopes and Calculators

One can use the TI-84 Plus, or other graphing calculator, to obtain slopes. The TI-84 Plus has several simple mechanisms for obtaining slopes or the equation of a line tangent to $y = f(x)$ at a specified point.

To illustrate the methods, let us create and complete Table 2 for $f(x) = x^2$.

x	$f(x)$	$f'(x)$
−1		
0		
1		
2		
3		
4		

TABLE 2

Enter x^2 in the Y1 position and graph in a window $[-5, 5]$ by $[-10, 30]$. When the graph appears on screen, press trace and then type the first x-value and press enter. Note that the values $x = -1$ and $y = 1$ are listed at the bottom of the screen. Type the next x-value and press enter. Continue to type the x-values in the first column, one after the other, without having to retype the trace command. Without pressing trace again, the corresponding y-values can be obtained and written down as shown in Table 3.

x	$f(x)$	$f'(x)$
−1	1	
0	0	
1	1	
2	4	
3	9	
4	16	

TABLE 3

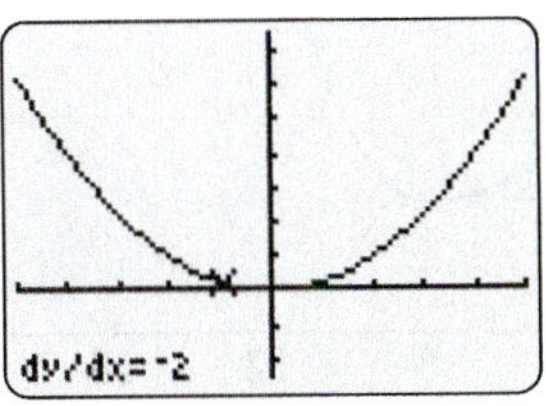

FIGURE 8

Next, with the graph displayed on the screen, type 2nd trace and select item six (type 6). Now type the desired x-value (start with −1). dy/dx=-2 appears at the bottom of the screen (see Figure 8). The symbol dy/dx is another notation for *derivative*.

Retype 2nd trace, select item 6 again, and input the next x-value. Continue until all of the x-values in the table have been used and you have written down the appropriate slopes.

When the table is complete, you should be able to guess a formula for the slope of the function $y = x^2$. Table 4 has been completed for the integers between −1 and 4.

x	$f(x)$	$f'(x)$
−1	1	−2
0	0	0
1	1	2
2	4	4
3	9	6
4	16	8

TABLE 4

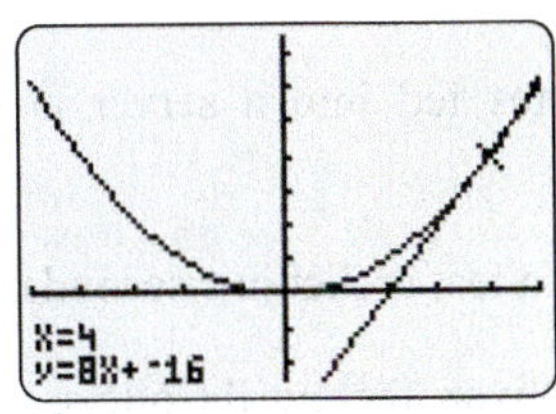

FIGURE 9

The calculator also has the ability to draw a tangent line and calculate the line's equation. With the graph of $y = x^2$ on the screen, press 2nd prgm (this will activate the DRAW function) and select 5:Tangent((see Figure 9).

Next, type in the desired x-value, for instance $x = 4$, and press enter. The calculator draws a tangent line to the graph at $x = 4$ and displays pertinent information at the bottom of the screen as shown in Figure 10. The information printed at the bottom is

$$x=4$$
$$y=8x+-16.$$

The equation of the tangent line is $y = 8x - 16$, with a slope of 8 and a y-intercept of −16. To solve for the remaining slopes using this method, you need only to repeat the commands 2nd, prgm, 5, and type in the next x-value.

FIGURE 10

10.6 EXERCISES

💡 PRACTICE

Use the graph to solve Exercises 1–3. The variable t is the number of hours since midnight and $f(t)$ is the temperature at time t.

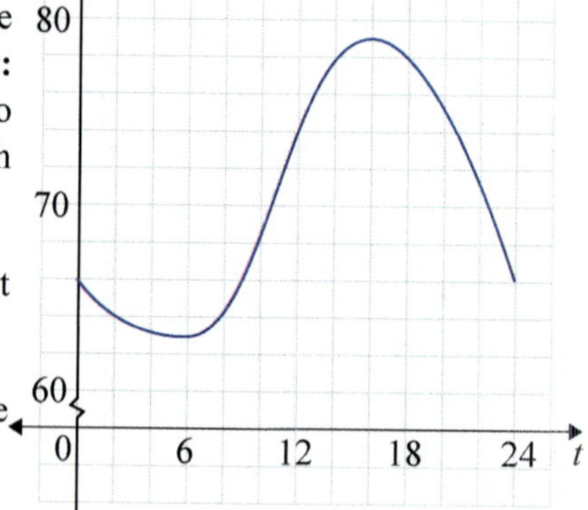

1. Roughly estimate the instantaneous rate of change of $f(t)$ at 3:00 p.m. (**Hint:** Extend an imaginary tangent line so as to come close to or to intersect points with integer coordinates.)

2. Estimate the time t for the lowest temperature $f(t)$.

3. Estimate the time t for the fastest increase in temperature $f(t)$.

4. Sketch one graph so that all of the following statements are true.

    (a) $f'(x)$ is positive for $-2 \leq x \leq 6$.   (b) $f'(x) < 0$ for $x > 6$

    (c) $f'(6) = 0$ (d) $f(6) = 10$ and $f(0) = 1$

5. Interpret the meaning of $f(3) = 14$ and $f'(3) = 7$ for the function $f(x) = 2 + x + x^2$.

🚀 APPLICATIONS

In Exercises 6–15, determine what the slope f' represents in terms of the subject in the problem.

6. $f(t)$ is the distance in feet traveled by a car in t minutes.

7. $f(s)$ is the total money spent in a department store by s customers.

8. $f(n)$ is the number of birds nesting in woods with n trees per acre.

9. $f(x)$ is the total cost of manufacturing x toasters.

10. $f(u)$ is the total revenue from the sale of u car radios.

11. $f(t)$ is the speed of a race car after t seconds.

12. $f(v)$ is the total amount of information in bytes fed into a server at Castle Manufacturing Company in v seconds.

13. $f(x)$ is the vertical distance in meters traveled by a test rocket in x seconds.

14. $f(s)$ is the grade point average of freshmen at Sullivan Technical College where s is the average SAT score of the freshman class.

15. $f(t)$ is the cost of calculus books at college bookstores in the United States where t is the time in years since 1980.

16. Suppose $f(x)$ is the number of gallons of gas used by a car after it has traveled x miles.
    a. Suppose the car gets 20 miles/gallon. What is $f(100)$?
    b. Is $f'(100)$ positive or negative?
    c. Would $f'(100)$ be greater for a subcompact car or for an SUV?

17. Suppose $f(x)$ denotes the production units for input x in labor units (man-hours). Suppose $f(500) = 2000$ and $f'(500) = 3$.
    a. Interpret $f(500) = 2000$ and $f'(500) = 3$.
    b. Estimate the increased production if x is increased from 500 to 501.

18. Suppose $f(x)$ is the total number of students on a college campus that have the flu and x is the number of days after the first case is reported. Interpret $f(8) = 9$ and $f'(8) = 3$.

19. Suppose $f(x)$ is the cost of a Ford Taurus and x is the age of the car.
    a. Is $f'(x)$ positive or negative?
    b. Interpret the meaning of $f'(3) = -2500$.

20. Average prices for one-bedroom condominiums have steadily risen in Charleston, SC since 2000, according to local reports. Suppose $f(x) = 3000x + 72{,}000$ is the cost of a one-bedroom condo and x is the number of years since 2000.
    a. Interpret $f(0) = 72{,}000$.
    b. Interpret $f'(x) = 3000$.
    c. Interpret $f(3) = 81{,}000$.

21. Suppose $f(x)$ denotes the weight of a cancerous tumor x weeks after discovery. Interpret $f(3) = 4$ grams and $f'(3) = 0.4$ grams/week.

22. Water boils at $212°\,$F and at $100°\,$C. Water freezes at $32°\,$F and $0°\,$C. Let F denote temperature in degrees Fahrenheit and let x be temperature in degrees Celsius.
    a. Write a formula $F(x) = mx + b$, which can convert Celsius input x into Fahrenheit output F.
    b. Use the value of m to write a formula for $F'(x)$.

23. **Birth rate:** The fertility decline in many countries can be modeled by an appropriate equation. In Bangladesh, from 1970 to 2000, patterns of fertility changed according to the equation $y = -0.11x + 6.45$, where x is the time in years beginning in 1970 and y is the average number of children per woman. (**Source:** Lori Ashford, "World Population Highlights 2004," BRIDGE Population Reference Bureau, August 2004.)
    a. What number is $f(20)$ and what does it represent?
    b. What number is $f'(20)$ and what does it represent?

24. **Birth rate:** Patterns of fertility changed in India from 1970 to 2000 according to the equation $y = -0.068x + 5.22$, where x is the number of years after 1970 and y is the average number of children per woman. (**Source:** Lori Ashford, "World Population Highlights 2004," BRIDGE Population Reference Bureau, August 2004.)
    a. What number is $f(30)$ and what does it represent?
    b. What number is $f'(30)$ and what does it represent?

25. Birth rate: In China, from 1964 to the present, the death rate has remained nearly constant at about 8 deaths per 1000 persons. However, the yearly birth rate, in births per 1000 people, has declined in most years according to the formula $f(x) = -0.641x + 35.8$, where x is the number of years since 1964. (**Source:** Nancy E. Riley, "China's Population: New Trends and Challenges," *Population Bulletin*, Vol. 20, No. 2, June 2004.)

  a. What number is $f(30)$ and what does it represent?

  b. What number is $f'(30)$ and what does it represent?

  (**Note:** See Exercise 36 in Section 10.8 for a similar problem using a more accurate model than the linear model given here. Both models are based on the same data.)

📈 TECHNOLOGY

Use a graphing utility in Exercises 26–29 to find the slope of $f(x)$ at the given point. Sketch the graph of $f(x)$ and tangent line at the given point on your paper.

26. $f(x) = \dfrac{4 + 2x}{\sqrt{x}}; \ (16, 9)$ **27.** $f(x) = x^3; \ (3, 27)$

28. $f(x) = 2 - 3x + x^2; \ (1, 0)$ **29.** $f(x) = 10^x; \ (2, 100)$

30. For the function $f(x) = 4 - x - 2x^2$, find a window including the point $(0, 4)$ so that the graph of the function and the tangent at $(0, 4)$ are indistinguishable.

31. For the function $f(x) = x^3$, make a table with headings a, $f(a)$, and $f'(a)$. Then substitute numbers using $a = -1, 0, 1, 2, 3, 4$. Give a formula for $f'(x)$.

32. Locate (with a graphing utility) the x- and y-coordinates of the lowest point on the graph $f(x) = x^2 - 6x + 11$. What is the slope at the lowest point?

33. For the function $y = x^2$ add a column to Table 3 to include the y-intercepts of the tangent line. What curiosity do you observe in the table?

34. Find the **LN** button on your graphing calculator and sketch a graph of $y = \ln x$ on your calculator using the window $[-2, 8]$ by $[-0.5, 2]$. What is the slope at the point $(1, 0)$? Sketch the graph and tangent on your paper.

35. On $f(x) = \sqrt{x}$, locate the x- and y-coordinates of the point at which the slope is exactly 1. (**Hint:** Find $\left(a, \sqrt{a}\right)$ so that $f'(a) = 1$.)

36. A bacteria culture in a lab grows according to the formula $y = 1600\left(2^t\right)$ where t is time in hours and y is the quantity of bacteria.

  a. Interpret the meaning of $f'(1)$.

  b. Determine $f'(1)$ using a graphing utility.

37. Sketch $f(x) = (x + 10)(x - 5)(x - 10)$ on a graphing utility. Give the window used.

  a. Locate the points $(a, f(a))$ for which $f'(a) = 0$.

  b. What is the value of $f'(2)$?

38. Sketch $y = \dfrac{1}{x}$ on a graphing utility. Find the x- and y-coordinates of any point with slope -25.

39. Suppose $f(x) = 2x^2$. Create a table of values for the slope of $f'(x)$ (see Table 4). Guess a formula for $f'(x)$.

10.7 DEFINITION OF THE DERIVATIVE AND THE POWER RULE

■ TOPICS

- Definition of the Derivative
- Existence of Derivatives
- The Power Rule
- Polynomial and Polynomial-Like Functions

We have considered numerical calculation of slope and the interpretation of the number as slopes of tangent lines or as ratio of change. In this section we want to begin to concentrate on the derivative as a function, important in its own right, and on the algebraic connections between $f(x)$ and $f'(x)$.

Definition of the Derivative

We begin with the formal definition.

Derivative

For any given function $y = f(x)$, the **derivative** of f at x is defined to be

$$f'(x) = \lim_{h \to 0} \left(\frac{f(x+h) - f(x)}{h} \right)$$

provided this limit exists. (**Note:** $\Delta x = (x + h) - x = h$.)

If $f'(x)$ exists, then f is said to be **differentiable at x**. Finding the formula for $f'(x)$ given $f(x)$ is called **differentiating**, and the process itself is called **differentiation**. We can also use $f'(x)$, $\dfrac{dy}{dx}$, y', or just f' to denote the derivative of $f(x)$.

It is customary for students of calculus to be able to determine the formula for $f'(x)$ given the formula for f. This is done in one of two ways. The first is by application of some rule, or theorem. For instance, there is already one rule that was suggested numerically in the previous section:

$$\text{If } f(x) = x^2, \text{ then } f'(x) = 2x.$$

In the absence of such rules, we determine the formula for a derivative by directly applying the definition of the derivative. We can do this by using an algorithm we will call the Three-Step Method.

> **The Three-Step Method for Finding a Derivative**
>
> 1. Form the ratio $\dfrac{f(x+h)-f(x)}{h}$, called the **difference quotient**.
> 2. Simplify the difference quotient algebraically.
> 3. Calculate $f'(x)=\lim\limits_{h\to 0}\left(\dfrac{f(x+h)-f(x)}{h}\right)$, if it exists.

> **Example 1: Finding a Derivative**

Use the Three-Step Method to find $f'(x)$ for $f(x)=x^3-5x$.

Solution

Step 1: Form the difference quotient.

$$\frac{f(x+h)-f(x)}{h}=\frac{\left[(x+h)^3-5(x+h)\right]-\left(x^3-5x\right)}{h}$$

Step 2: Simplify the difference quotient.

$$\frac{\left[(x+h)^3-5(x+h)\right]-\left(x^3-5x\right)}{h}=\frac{x^3+3x^2h+3xh^2+h^3-5x-5h-x^3+5x}{h}$$

$$=\frac{3x^2h+3xh^2+h^3-5h}{h}=\frac{\cancel{h}\left(3x^2+3xh+h^2-5\right)}{\cancel{h}}$$

$$=3x^2+3xh+h^2-5$$

Step 3: Find the limit.

$$\lim_{h\to 0}\left(3x^2+3xh+h^2-5\right)=3x^2+3x\cdot 0+0^2-5$$

$$=3x^2-5$$

Therefore, $f'(x)=3x^2-5$.

Existence of Derivatives

Since derivatives are limits and limits do not always exist, we can infer that derivatives do not always exist. That is, if $y=f(x)$ and $x=a$, then

$$f'(a)=\lim_{h\to 0}\frac{f(a+h)-f(a)}{h},$$

and if this limit exists, we say that $f'(a)$ exists. If the limit fails to exist, then $f(x)$ is not differentiable at $x=a$. We cannot possibly list all the ways in which a function can fail to be differentiable. However, the following three conditions are commonly discussed.

Conditions of a Nondifferentiable Function

A function $f(x)$ is not differentiable at $x = a$ if any of the following conditions are true:

1. $f(x)$ is discontinuous at $x = a$.

2. The graph of $f(x)$ has a sharp corner at $x = a$.

3. The tangent line at $x = a$ is a vertical line.

Example 2 shows three different functions, each of which is not differentiable due to one of the conditions defined above.

Example 2: Nondifferentiable Functions

a. The function $f(x) = \dfrac{1}{x}$ is not defined at $x = 0$. Thus, there cannot be a tangent line at $x = 0$ and the limit does not exist.

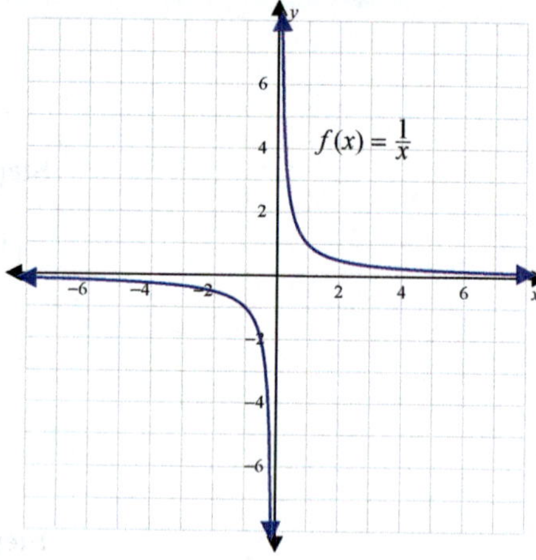

b. The function $f(x) = (x-3)x^{\frac{2}{3}}$ has a sharp point at $x = 0$, so it is not differentiable at $x = 0$.

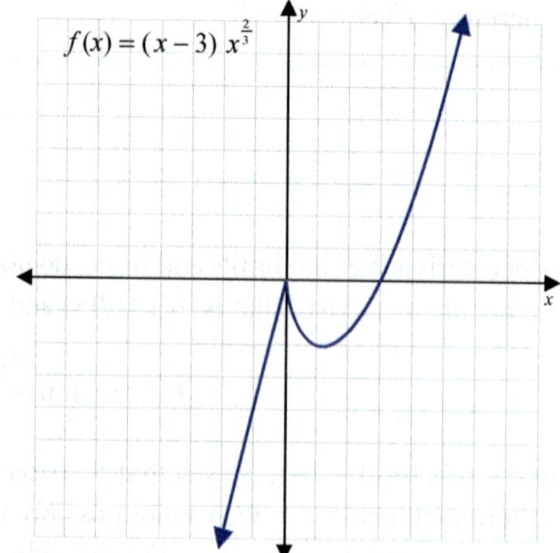

Note: The limit of a set of secant lines from the right of $x = 0$ will have negative slope, but from the left, the limit will have positive slope.

c. The function $f(x) = \sqrt[3]{x}$ has a vertical tangent at $x = 0$. Since vertical lines have no slope, $f'(0)$ is not defined.

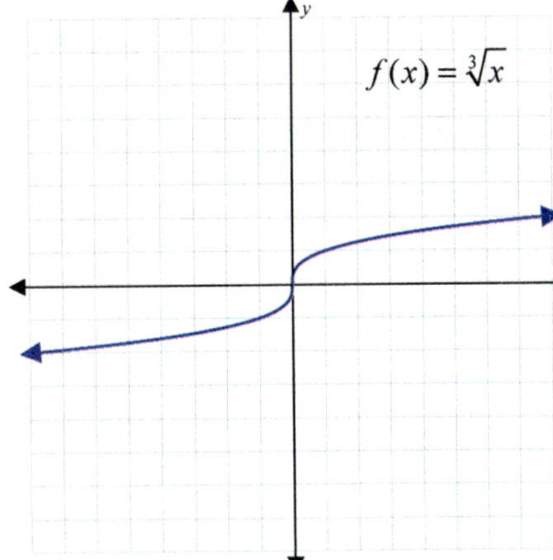

For this function, $f'(x)$ exists at every point except for $x = 0$. In general, the functions usually studied in introductory and intermediate algebra or college algebra are "well-behaved" from the calculus point of view. They usually are differentiable except at obvious points.

The Power Rule

The idea that a physical process, such as movement from one place to another, is continuous was a difficult idea to accept since it involved an infinite process. Infinity is a tricky concept mathematically. Zeno's Paradox is an ancient paradox that states an arrow shot into the air cannot land! It must go halfway, and then half of the remaining distance, and then half that, and so on. Zeno of Elea (490 BC–425 BC) claimed that since no infinite process can ever end, no physical motion could be continuous. It can then be argued (by Zeno, not us) that the mathematical functions we use are not continuous either, as they describe processes with infinitely many steps. However, it is the concept of numerical limit that sweeps away contradictions and allows us to resolve this ancient dilemma.

For this discussion we note that a function $f(x)$ is said to be continuous at $x = a$ provided that the left- and right-hand limits agree:

Continuity at $x = a$

The function f is **continuous at $x = a$** if and only if

$$\lim_{x \to a^-}\left(f(x)\right) = f(a) = \lim_{x \to a^+}\left(f(x)\right).$$

The limit concept is viewed today as the perfect tool for giving a formal structure to the ideas of calculus. From the student's point of view, the converse idea is even more important: if a function $f(x)$ is continuous at $x = a$, then the limit problem at $x = a$ is solved by substitution! (No calculation or table of values is necessary.)

Continuous Function on an Interval

A function f is **continuous on an interval** (x_1, x_2) provided the function is continuous at $x = a$ for every number a with $x_1 < a < x_2$.

Recall the fact that $(a + b)^3 = a^3 + 3a^2b + 3ab^2 + b^3$. Now we can get a formula for y' given $y = x^3$. For this function we consider the point $(x, f(x)) = (x, x^3)$ and a nearby point $(x + h, (x + h)^3)$. Note that $h = \Delta x$.

✑ NOTE

Informally, continuous functions can be drawn without lifting one's pencil from the paper. Every differentiable function is continuous, but not every continuous function is differentiable (see Example 2b). Put another way, differentiability is a more restrictive condition—it requires that a function be continuous, smooth, and without vertical tangents.

Step 1: Form the difference quotient.

$$\frac{\Delta y}{\Delta x} = \frac{(x+h)^3 - x^3}{(x+h) - x}$$

Step 2: Simplify the difference quotient.

$$\frac{\Delta y}{\Delta x} = \frac{(x+h)^3 - x^3}{x+h-x}$$

$$= \frac{x^3 + 3x^2h + 3xh^2 + h^3 - x^3}{h}$$

$$= \frac{3x^2h + 3xh^2 + h^3}{h}$$

$$= \frac{\cancel{h}\left(3x^2 + 3xh + h^2\right)}{\cancel{h}}$$

$$= 3x^2 + 3xh + h^2$$

Step 3: Find the limit.

$$\lim_{h \to 0}\left(3x^2 + 3xh + h^2\right) = 3x^2 + 0 + 0 = 3x^2$$

Example 3: Using the Definition of Derivative

Use the definition of derivative to find $f'(x)$ for $f(x) = x^2$.

Solution

Step 1: Form the difference quotient, $\dfrac{\Delta y}{\Delta x}$. Here we begin with the two points $(x, f(x))$ and $(x + h, f(x + h))$.

So $\Delta y = f(x + h) - f(x)$ and $\Delta x = (x + h) - x = h$.

Inserting these expressions into the difference quotient, we get

$$\frac{\Delta y}{\Delta x} = \frac{f(x+h) - f(x)}{h} = \frac{(x+h)^2 - x^2}{h}.$$

Step 2: Simplify the difference quotient.

$$\frac{\Delta y}{\Delta x} = \frac{(x+h)^2 - x^2}{h}$$

$$= \frac{x^2 + 2xh + h^2 - x^2}{h}$$

$$= \frac{2xh + h^2}{h}$$

$$= \frac{\cancel{h}(2x + h)}{\cancel{h}} = 2x + h$$

Step 3: Compute or determine the limit as $h \to 0$.

$$\lim_{h \to 0}\left(\frac{\Delta y}{\Delta x}\right) = \lim_{h \to 0}(2x + h) = 2x + 0 = 2x$$

y	y'
x	1
x^2	$2x$
x^3	$3x^2$

TABLE 1

It is easy to see that, at any point on the line $y = x$, the slope is 1. We now put several results in Table 1 and we see a pattern.

The pattern suggests that if $f(x) = x^4$, then $f'(x) = 4x^3$. This is exactly what happens for $y = x^4$. Similarly, if $y = x^5$, then $y' = 5x^4$. We are now able to state one of the most useful and striking formulas in mathematics.

The Power Rule

For the function $f(x) = x^r$, where r is any real number, then

$$f'(x) = rx^{r-1}.$$

Especially noteworthy is the fact that the exponent r does not have to be a whole number, or even positive! Also, since 0^0 is undefined, the expression 0^0 is the one case not allowed in the Power Rule.

The second method of obtaining derivatives is by direct application of the definition of derivative itself. Usually, this method is cumbersome and is therefore only used to establish a logical basis for formulae like the Power Rule. In fact, we wish to illustrate our Three-Step Method to justify this very formula.

Our goal for the rest of this section is to provide a way to determine the formula for $f'(x)$ for any polynomial or polynomial-like function $f(x)$. Most of our results here will be proved using the Three-Step Method.

Basic Rule of the Derivative of a Linear Function

For $y = mx + b$, $y' = m$.

This means that we have defined slope for points on a curve consistently since, if the "curve" is a linear function, slope still means slope! The proof is an easy application of the definition. If $(x_1, mx_1 + b)$ and $(x_2, mx_2 + b)$ are any two points on the line, then we calculate the derivative as follows.

Step 1: Form the difference quotient.

$$\frac{\Delta y}{\Delta x} = \frac{(mx_2 + b) - (mx_1 + b)}{x_2 - x_1}$$

Step 2: Simplify the difference quotient.

$$\frac{\Delta y}{\Delta x} = \frac{(mx_2 + b) - (mx_1 + b)}{x_2 - x_1} = \frac{mx_2 + b - mx_1 - b}{x_2 - x_1}$$

$$= \frac{mx_2 - mx_1}{x_2 - x_1} = \frac{m(x_2 - x_1)}{x_2 - x_1} = m$$

Step 3: Determine the limit.

$$\lim_{\Delta x \to 0} \left(\frac{\Delta y}{\Delta x} \right) = m$$

> **Two Special Cases of the Basic Rule**
>
> **1.** If $f(x) = c$, then $f'(x) = 0$. Has slope 0 (constant functions are horizontal lines)
>
> **2.** If $f(x) = x$, then $f'(x) = 1$. Has slope 1

Polynomial and Polynomial-Like Functions

> **Polynomial Functions**
>
> A **polynomial function** is a function of the form
>
> $$f(x) = a_0 + a_1 x + a_2 x^2 + \cdots + a_n x^n,$$
>
> where $a_0, a_1, a_2, \ldots, a_n$ are real numbers and the exponents are positive integers.
>
> Also, for any real number a, $\lim_{x \to a} f(x) = f(a)$, and therefore, $f(x)$ is continuous everywhere.

Consider the following functions:

a. $f(x) = 3 + x^{\frac{1}{2}}$,

b. $f(x) = 3 + x^2 - x^3 + x^{\frac{2}{3}}$, and

c. $f(x) = \dfrac{5}{x^2} = 5x^{-2}$.

Each has the form of a polynomial except that the exponents are not always positive integers. In function **c.**, the expression must be revised using a law of exponents. Each function is a sum or difference of terms of the form bx^r where b is some real number and r is some real number. Such functions are called **polynomial-like**. The derivatives of these, and standard polynomial functions, are the easiest type to determine.

> **Power Rule (General Case)**
>
> For the function $f(x) = c \cdot x^r$ where c and r are any real numbers,
>
> $$f'(x) = c \cdot r \cdot x^{r-1}.$$

> **Example 4: Finding the Derivatives of Functions**

Find the derivative of each of the following functions.

a. $f(x) = 5$ **b.** $f(x) = 5x^3$ **c.** $f(x) = x^{\frac{1}{2}}$

Solution

a. $f(x) = 5$ is a constant function, so $f'(x) = 0$. This is the first special case of the Basic Rule.

b. $f'(x) = 5 \cdot 3 \cdot x^{3-1} = 15x^2$

Here we use the general Power Rule, where $c = 5$ and $r = 3$.

c. $f'(x) = \dfrac{1}{2}x^{\frac{1}{2}-1} = \dfrac{1}{2}x^{-\frac{1}{2}}$

We can use the Power Rule with this polynomial-like function because $r = \dfrac{1}{2}$ is a real number.

Example 5: Rewriting a Fraction to Find the Derivative

Find the derivative of $g(x)$ given that $g(x) = \dfrac{1}{x}$.

Solution

$$g(x) = \frac{1}{x} = x^{-1}$$

$$g'(x) = -1x^{-1-1} = -x^{-2} = -\frac{1}{x^2}$$

We first rewrite the fraction in a form with a negative exponent and then apply the Power Rule.

At the beginning of this section we listed a few other ways to denote the derivative. A more complete list of commonly used notations is as follows:

$$f'(x), \ y', \ \frac{dy}{dx}, \ \frac{d}{dx}f(x), \ f_x, \ D_x(f(x)), \ \text{and} \ f'.$$

Example 6: Using Derivative Notation

Find $D_x(h(x))$ where $h(x) = \dfrac{18}{x^{\frac{2}{3}}}$.

Solution

$$h(x) = \frac{18}{x^{\frac{2}{3}}} = 18x^{-\frac{2}{3}}$$

Rewrite the expression to be a polynomial-like function.

$$D_x(h(x)) = D_x\left(18x^{-\frac{2}{3}}\right) = h'(x) = 18 \cdot \left(-\frac{2}{3} \cdot x^{-\frac{2}{3}-1}\right)$$

$$= \frac{-18(2)}{3}x^{-\frac{5}{3}}$$

$$= -12x^{-\frac{5}{3}}$$

Use the Power Rule with $c = 18$ and $r = -\dfrac{2}{3}$ to find the derivative.

10.7 EXERCISES

💡 PRACTICE

Use the various rules of differentiation to find the derivative for each of the functions in Exercises 1–20.

1. $f(x) = 4$

2. $f(x) = 3x$

3. $f(x) = 7x - 2$

4. $y = 12$

5. $y = 4x^2$

6. $y = 8x^2$

7. $y = \dfrac{7}{x}$

8. $y = \dfrac{4}{x^5}$

9. $y = \dfrac{1}{2x^3}$

10. $g(x) = \dfrac{4}{3x^2}$

11. $g(x) = 3\sqrt{x}$

12. $h(x) = 2\sqrt[3]{x}$

13. $h(t) = t^{2.3}$

14. $h(t) = t^{-1.4}$

15. $f(x) = 3x^{0.8}$

16. $f(u) = 2u^{0.1}$

17. $f(u) = \dfrac{1}{\sqrt{u}}$

18. $f(x) = \dfrac{2}{\sqrt[4]{x}}$

19. $f(x) = -5x^{\frac{3}{4}}$

20. $f(x) = 6x^{-\frac{2}{3}}$

10.8 TECHNIQUES FOR FINDING DERIVATIVES

■ TOPICS

- The Sum and Difference Rule
- Applications of Tangent Lines and Velocity

The Sum and Difference Rule

Now we list two more general rules and then show how these rules, along with the Power Rule, can be implemented to find the derivatives of a variety of functions.

Constant Times a Function Rule

If $f(x)$ is a differentiable function, c is a real constant, and $y = c \cdot f(x)$, then

$$\frac{dy}{dx} = c \cdot f'(x).$$

In words, the derivative of a constant times a function is the constant times the derivative of the function. (The Power Rule is a special case of this rule.)

The Sum and Difference Rule

If $f(x)$ and $g(x)$ are differentiable functions and $y = f(x) \pm g(x)$, then

$$\frac{dy}{dx} = f'(x) \pm g'(x).$$

In words, the derivative of a sum (or difference) of two functions is the sum (or difference) of their derivatives.

Example 1: Using the Rules

Find the derivative of each of the following functions.

a. $y = 5x^2$

Solution

$$\frac{dy}{dx} = \frac{d}{dx}\left(5x^2\right)$$

Rewrite the equation of the derivative by pulling the constant, 5, out in front. By doing so, we have used the Constant Times a Function Rule.

$$= 5 \cdot \frac{d}{dx}\left(x^2\right)$$

$$= 5\left(2x^{2-1}\right)$$

Apply the Power Rule to x^2.

$$= 10x$$

b. $y = -4x$

Solution

$$\frac{dy}{dx} = \frac{d}{dx}(-4x)$$

Rewrite the equation of the derivative by pulling the constant, -4, out in front. By doing so, we have used the Constant Times a Function Rule.

$$= -4 \cdot \frac{d}{dx}(x^1)$$

$$= -4(1x^{1-1})$$

Apply the Power Rule to x^1.

$$= -4x^0 = -4$$

c. $y = 5x^2 - 4x$

Solution

Notice that this function is actually the sum of the functions in parts **a.** and **b.** Therefore, to find the derivative of this function we will add the results from parts **a.** and **b.**

$$\frac{dy}{dx} = \frac{d}{dx}(5x^2 - 4x)$$

Rewrite the equation of the derivative as a sum of the two differentiable functions so we can use the rule.

$$= \frac{d}{dx}(5x^2) + \frac{d}{dx}(-4x)$$

$$= 10x - 4$$

Using the Sum and Difference Rule, we can take the sum of the derivatives of the functions (found in parts **a.** and **b.**).

d. $y = x^3 + 8\sqrt{x} + \dfrac{2}{x}$

Solution

$$y = x^3 + 8x^{\frac{1}{2}} + 2x^{-1}$$

First, rewrite using exponents.

$$\frac{dy}{dx} = \frac{d}{dx}(x^3) + \frac{d}{dx}\left(8x^{\frac{1}{2}}\right) + \frac{d}{dx}(2x^{-1})$$

Next, use the Sum and Difference Rule to rewrite the equation of the derivative.

$$= \frac{d}{dx}(x^3) + 8 \cdot \frac{d}{dx}\left(x^{\frac{1}{2}}\right) + 2 \cdot \frac{d}{dx}(x^{-1})$$

Then apply the Constant Times a Function Rule to extract the constants.

$$= 3x^2 + \overset{4}{\cancel{8}}\left(\frac{1}{\cancel{2}}x^{\frac{1}{2}-1}\right) + 2\left(-1x^{-1-1}\right)$$

Lastly, use the Power Rule.

$$= 3x^2 + 4x^{-\frac{1}{2}} - 2x^{-2} = 3x^2 + \frac{4}{\sqrt{x}} - \frac{2}{x^2}$$

As illustrated in Example 1d, there can be more than one correct algebraic form for a derivative. In Example 2 we show algebraic manipulations before the derivative is found.

Example 2: Algebraic Manipulations

a. Find $f'(u)$ if $f(u) = u^{\frac{1}{2}}\left(u^2 + 2u\right)$.

Solution

$$f(u) = u^{\frac{1}{2}}\left(u^2 + 2u\right) = u^{2+\frac{1}{2}} + 2u^{1+\frac{1}{2}} = u^{\frac{5}{2}} + 2u^{\frac{3}{2}}$$

Multiply and simplify.

$$f'(u) = \frac{d}{du}\left(u^{\frac{5}{2}}\right) + 2 \cdot \frac{d}{du}\left(u^{\frac{3}{2}}\right)$$

Use both the Sum and Difference Rule and the Constant Times a Function Rule.

$$= \frac{5}{2}u^{\frac{5}{2}-1} + 2\left(\frac{3}{2}u^{\frac{3}{2}-1}\right)$$

Apply the Power Rule.

$$= \frac{5}{2}u^{\frac{3}{2}} + 3u^{\frac{1}{2}}$$

b. If $F(v) = \dfrac{v^2 + 1}{\sqrt{v}}$, find $F'(v)$.

Solution

$$F(v) = \frac{v^2 + 1}{\sqrt{v}} = \frac{v^2 + 1}{v^{\frac{1}{2}}}$$

$$= \frac{v^2}{v^{\frac{1}{2}}} + \frac{1}{v^{\frac{1}{2}}} = v^{\frac{3}{2}} + v^{-\frac{1}{2}}$$

Simplify by first dividing each term in the numerator by $v^{\frac{1}{2}}$.

$$F'(v) = \frac{d}{dv}\left(v^{\frac{3}{2}}\right) + \frac{d}{dv}\left(v^{-\frac{1}{2}}\right)$$

Use the Sum and Difference Rule.

$$= \frac{3}{2}v^{\frac{1}{2}} - \frac{1}{2}v^{-\frac{3}{2}}$$

Apply the Power Rule.

Or, we can factor and write the answer in fraction form with a single denominator.

$$F'(v) = \frac{1}{2}v^{-\frac{3}{2}}\left(3v^2 - 1\right) = \frac{3v^2 - 1}{2v^{\frac{3}{2}}}$$

Applications of Tangent Lines and Velocity

We know that a derivative of a function can be interpreted as the slope of a line tangent to the graph of the function. In the following example we show how to find the equation of a tangent line at a particular point.

Example 3: Tangent Lines

For the function $f(x) = 8x - x^2$,

a. find the slopes of the tangent lines at $x = 2$, $x = 4$, and $x = 5$;

b. find the equations of the tangent lines at $x = 2$, $x = 4$, and $x = 5$;

c. sketch the curve and the three tangent lines.

Solution

a. First, find the derivative of the function:

$$f'(x) = 8 \cdot 1 \cdot x^{1-1} - 2x^{2-1} = 8 - 2x.$$

Then evaluate the derivative at each point to find the slope of the tangent line at that point.

$$\text{At } x = 2 \rightarrow f'(2) = 8 - 4 = 4.$$
$$\text{At } x = 4 \rightarrow f'(4) = 8 - 8 = 0.$$
$$\text{At } x = 5 \rightarrow f'(5) = 8 - 10 = -2.$$

b. To find the equation of the tangent line at each of the points, use the point-slope formula for the equation of a line, $y - y_1 = m(x - x_1)$. Insert each value of x into the function to find its corresponding y-value.

At $x = 2 \rightarrow x_1 = 2$, so $y_1 = f(2) = 8 \cdot 2 - 2^2 = 12$. From part **a.**, we know that $f'(2) = 4 = m$.

Putting this information into the point-slope formula, we get

$$y - 12 = 4(x - 2)$$
$$y = 4x + 4.$$

At $x = 4 \rightarrow x_1 = 4$, so $y_1 = f(4) = 8 \cdot 4 - 4^2 = 16$. From part **a.**, we know that $f'(4) = 0 = m$.

Putting this information into the point-slope formula, we get

$$y - 16 = 0(x - 4)$$
$$y = 16.$$

At $x = 5 \rightarrow x_1 = 5$, so $y_1 = f(5) = 8 \cdot 5 - 5^2 = 15$. From part **a.**, we know that $f'(5) = -2 = m$.

Putting this information into the point-slope formula, we get

$$y - 15 = -2(x - 5)$$
$$y = -2x + 25.$$

c. The graph is a parabola that opens downward. The vertex of the parabola occurs at $(4, 16)$, where the tangent line has slope 0.

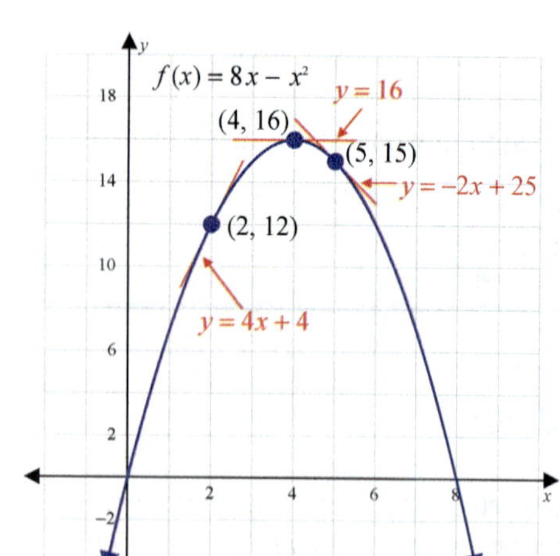

As we learned in an earlier section, velocity is the instantaneous rate of change of distance with respect to time. Mathematicians generally use the letter s to represent distance. Thus a typical distance function might look like

$$s(t) = t^2 - 2t \quad \text{where } t \geq 0.$$

Notations for velocity are

$$v(t) = s'(t) = \frac{ds}{dt}.$$

Example 4: Velocity

Suppose that a sailboat is observed, over a period of 5 minutes, to travel a distance from a starting point according to the function $s(t) = t^3 + 60t$, where t is time in minutes and s is the distance traveled in meters.

a. How fast is the boat moving at the starting point?

b. How fast is the boat moving at the end of 3 minutes?

Solution

a. The velocity at time t is $v(t) = s'(t) = 3t^2 + 60$. The boat is at the starting point when $t = 0$.

$$v(0) = s'(0) = 3 \cdot (0)^2 + 60 = 60 \ \frac{\text{m}}{\text{min}}$$

b. When $t = 3$,

$$v(3) = s'(3) = 3 \cdot (3)^2 + 60 = 27 + 60 = 87 \ \frac{\text{m}}{\text{min}}.$$

Example 5: Identifying Derivatives in Real Life

Determine which of the following represent (or might represent) the value of a derivative at a point, and if so, give a suitable description of a function f.

a. 30 miles per hour

b. 6 fish per day

c. 23 ears of corn

d. 19 neutrons per millisecond

Solution

a. 30 miles per hour is a rate of change; therefore, it is a possible derivative. A suitable function is $f(x)$ where f is the total distance, in miles, from an origin and x is the time spent traveling, in hours.

b. 6 fish per day is also a rate of change, and therefore is a possible derivative. A suitable function $f(x)$ is the total number of fish sold by a pet store during June and x is the number of days since June 1.

c. 23 ears of corn is a quantity rather than a rate. Thus it is not a derivative.

d. 19 neutrons per millisecond is a rate of change; therefore, it is a possible derivative. A suitable function $f(x)$ is the total number of neutrons emitted by a nuclear sample, and x is the time, in milliseconds, since the sample arrived at the college physics laboratory.

10.8 EXERCISES

PRACTICE

Use the various rules of differentiation to find the derivative for each of the functions in Exercises 1–10.

1. $y = x^3 - 7x$

2. $y = 4x^2 - 9x + 2$

3. $y = 0.3x^2 - 4x + 6$

4. $y = 120 + 8x - 0.2x^2$

5. $y = x^3 - 6x^2 + 5x + 2$

6. $y = \dfrac{1}{3}x^3 - \dfrac{1}{2}x^2 - 3x + 4$

7. $y = 2x^{\frac{3}{2}} + 4x^{\frac{1}{2}} - 5$

8. $y = 2x^{-\frac{2}{3}} + 3x^{\frac{1}{3}} + 7$

9. $f(t) = 2t^{-\frac{1}{2}} + t^{\frac{1}{2}} + t$

10. $f(x) = 3x^{-\frac{1}{3}} - 2x^{-\frac{1}{2}} + 1$

In Exercises 11–20, use algebraic techniques to rewrite each function as a sum or difference; then find the derivative.

11. $y = (x+1)(2x-3)$

12. $y = \sqrt{x}\left(2x^2 + x - 3\right)$

13. $f(v) = v^{\frac{3}{2}}\left(v^2 + 2v - 1\right)$

14. $f(v) = v^{\frac{1}{3}}\left(6 - 4v + v^2\right)$

15. $f(x) = \dfrac{x^4 + 5x^3}{x^2}$

16. $f(x) = \dfrac{6x^2 + 1}{x^3}$

17. $g(t) = \dfrac{t-2}{\sqrt{t}}$

18. $g(t) = \dfrac{t^2 + 3}{\sqrt[3]{t}}$

19. $g(x) = \dfrac{4x + 5x^{\frac{1}{2}} - 1}{\sqrt{x}}$

20. $g(x) = \dfrac{3\sqrt{x} + 4x - 2}{x^2}$

In Exercises 21–26, confirm your results with a graphing utility.

21. Let $f(x) = x^3 + 2x - 4$.
   a. Find the slope of the tangent line at $x = -1$.
   b. Find the equation of the tangent line at $x = -1$.

22. Let $f(x) = 2x^3 - x^2 - 3x$.
   a. Find the slope of the tangent line at $x = 2$.
   b. Find the equation of the tangent line at $x = 2$.

23. Let $g(x) = x^2 + 6x + 5$.
   a. Find the slopes of the tangent lines at $x = -1$, $x = -3$, and $x = -4$.
   b. Find the equations of the tangent lines at $x = -1$, $x = -3$, and $x = -4$.
   c. Sketch the graphs of the curve and the three tangent lines.

24. Let $g(x) = x^2 - 8x + 12$.
  a. Find the slopes of the tangent lines at $x = 2$, $x = 4$, and $x = 5$.
  b. Find the equations of the tangent lines at $x = 2$, $x = 4$, and $x = 5$.
  c. Sketch the graphs of the curve and the three tangent lines.

25. Let $f(x) = 8 - x^2$.
  a. Find the slopes of the tangent lines at $x = -2$, $x = 0$, and $x = 1$.
  b. Find the equations of the tangent lines at $x = -2$, $x = 0$, and $x = 1$.
  c. Sketch the graphs of the curve and the three tangent lines.

26. Let $f(x) = 10 - 3x - x^2$.
  a. Find the slopes of the tangent lines at $x = -3$, $x = -2$, and $x = 0$.
  b. Find the equations of the tangent lines at $x = -3$, $x = -2$, and $x = 0$.
  c. Sketch the graphs of the curve and the three tangent lines.

🚀 APPLICATIONS

27. Velocity of a rocket: A model rocket is fired vertically upward. The height after t seconds is $s(t) = 192t - 16t^2$ feet.
  a. Find the velocity at $t = 0$ seconds.
  b. Find the velocity at $t = 4$ seconds.
  c. When will the velocity be zero?

28. Velocity of a particle: A particle moving in a straight line is at a distance of $s(t) = 2.5t^2 + 18t$ feet from its starting point after t seconds, where $0 \le t \le 12$.
  a. Find the velocity at $t = 6$.
  b. Find the velocity at $t = 9$.

29. Population: A city's population t years from now can be estimated from the formula $P(t) = 9000 + 500t - 72\sqrt{t}$.
  a. Find the rate at which the city is growing after 4 years.
  b. Find the rate at which the city is growing after 9 years.

30. Cost: The total cost of producing x units of a product is given by $C(x) = 4000 + 25x - 0.2x^2$ dollars, where $0 \le x \le 50$.
  a. Find the rate of change in the cost when $x = 10$.
  b. Find the rate of change in the cost when $x = 30$.

31. Fuel consumption: When a factory operates from 6:00 a.m. to 6:00 p.m., its total fuel consumption varies according to the formula $f(t) = 0.9t^2 - 0.3t^{0.5} + 20$, where t is the time in hours after 6:00 a.m. and $f(t)$ is number of barrels of fuel oil.
  a. How much fuel oil is consumed by noon?
  b. What is the rate of consumption of fuel at 10:00 a.m.?
  c. What is the average rate of consumption from 6:00 a.m. to 2:00 p.m.?

32. Population: The population of bacteria in a lab experiment for BIOL 403 at Nevada Tech is given by $f(x) = 2.2x^{1.5} - 0.7x + 2$, where x is the time in hours after 2:00 p.m. and $f(x)$ is population in suitable units.
  a. What is the population at 2:00 p.m.?
  b. The lab is over at 4:00 p.m. What is the new population of bacteria?
  c. What is the average rate of change of bacteria from 2:00 p.m. to 4:00 p.m.?
  d. What is the instantaneous rate of change of bacteria at 3:00 p.m.?

33. **Electrical charge:** The electrical charge on a new cell phone declines according to the formula $C(t) = 15 - 0.1t^2 - 0.5t$, where t is the time in hours following a full charge and $C(t)$ is a measure of the charge.
   a. To the nearest hour, how long does one have until the charge is fully depleted?
   b. What is the instantaneous rate of change, in charge units per hour, at $t = 4$?
   c. What is the average rate of change from $t = 0$ to $t = 4$?

34. **Spreading a rumor:** The number of college students at Salis Technical College who have not heard a new rumor is approximated by the formula $N(x) = 300\left(1 - 0.004x^2\right)$, where x is the number of days following the start of a new rumor.
   a. How many days does it take for 90 percent of the students to hear the rumor?
   b. What is the instantaneous rate of change in students after one day? Interpret the meaning of this number.
   c. Why is the slope negative in this exercise?

35. **Birth rate:** The fertility decline in many countries can be modeled by a quadratic equation. In China, from the late 1960s to the present, the number of births per woman has declined according to the formula $f(x) = 0.00675x^2 - 0.3215x + 5.585$, where x is the number of years after 1969. (**Source:** Nancy E. Riley, "China's Population: New Trends and Challenges," *Population Bulletin*, Vol. 20, No. 2, June 2004.)
   a. What was the number of births per woman in 1969?
   b. What was the number of births per woman in 1999?
   c. What was the rate of change of this fertility rate in 1979?

36. **Birth rate:** In China, from 1964 to the present, the death rate has remained nearly constant at approximately 8 deaths per 1000 people. The yearly birth rate, in births per 1000 people, has declined in most years according to the formula $f(x) = -0.00191x^3 + 0.134x^2 - 3.16x + 44.5$, where x is the number of years since 1964. (**Source:** Nancy E. Riley, "China's Population: New Trends and Challenges," *Population Bulletin*, Vol. 20, No. 2, June 2004.)
   a. What number is $f(30)$ and what does it represent?
   b. What number is $f'(30)$ and what does it represent?
   c. In what year, according to the model, did the number of new births equal the number of deaths?

37. **Population growth:** The percentage of older persons in China (age 60 and over) has grown since the 1950s according to the formula $f(x) = 0.003727x^2 - 0.105x + 7.063$, where x is the number of years since 1953. (**Source:** Nancy E. Riley, "China's Population: New Trends and Challenges," *Population Bulletin*, Vol. 20, No. 2, June 2004.)
   a. What was the percentage of older persons in China in 1953?
   b. What is the percentage of older persons projected to be in the year 2025?
   c. At what rate was the percentage changing in 2000?

38. **Population growth:** The population in billions of people in the lesser developed countries has varied according to the formula $P(x) = \dfrac{4.953}{10^4}x^2 - 0.007352x + 1.7748$, where x is the number of years since 1900. (**Source:** Population Reference Bureau, "Transition's in World Population," *Population Bulletin*, Vol. 59, No. 1, 5, March 2004.)
    a. What was the population in 1950?
    b. At what rate was the population changing in 1950?
    c. What population was projected for 2020 in these countries?

39. **Death rate:** The death rate in Mexico has varied since 1920 according to the formula $M(x) = -\dfrac{2.076}{10^4}x^3 + 0.033x^2 - 1.785x + 43.07$, where x is the number of years since 1920. (**Source:** Population Reference Bureau, "Transition's in World Population," *Population Bulletin*, Vol. 59, No. 1, 5, March 2004.)
    a. What was the death rate in 1950?
    b. At what rate was the death rate changing in 1950?
    c. What was the death rate in 2000?
    d. In what year did the formula predict that no one would die?

40. **Birth rate:** The function $f(x) = 0.00375x^2 - 0.2355x + 5.595$ gives the average number of births per woman in Thailand where x denotes the number of years since 1970. (**Source:** Lori Ashford, "World Population Highlights 2004," BRIDGE Population Reference Bureau, August 2004.)
    a. What number is $f(10)$ and what does it represent?
    b. What number is $f'(10)$ and what does it represent?

41. **Birth rate:** In Argentina, from 1970 to 2000, the average number of births per woman was given by the function $f(x) = -0.001x^2 + 0.014x + 3.14$, where x is the number of years after 1970. (**Source:** Lori Ashford, "World Population Highlights 2004," BRIDGE Population Reference Bureau, August 2004.)
    a. What number is $f(30)$ and what does it represent?
    b. What number is $f'(30)$ and what does it represent?

10.9 APPLICATIONS: MARGINAL ANALYSIS

■ TOPICS

- ■ Marginal Cost
- ■ Marginal Revenue
- ■ Marginal Profit
- ■ Marginal Average Cost
- ■ Marginal Propensity to Consume

Recall from our previous discussions about topics related to business that

$C(x)$ represents the total cost of producing x items,
$R(x)$ represents the total revenue when x items are sold, and
$P(x) = R(x) - C(x)$ represents the profit from selling all x items produced.

Each of these functions depends on x and changes as x changes. The term marginal is used in business and economics to indicate a rate of change and, therefore, a derivative. In this section we are going to investigate the use of calculus in marginal analysis and these three derivatives:

$C'(x)$, called **marginal cost**;
$R'(x)$, called **marginal revenue**; and
$P'(x)$, called **marginal profit**.

We emphasize that $C'(x)$ is an estimate for the increase in $C(x)$ if x is increased by 1.

Marginal Cost

If the total cost of producing x items is $C(x)$, then the total cost of producing $x + 1$ items is $C(x + 1)$.

For example, suppose $C(x) = 225 + 2x^2$. Then

$C(10) = 225 + 2 \cdot 10^2 = 225 + 200 = \$425$ is the cost of producing 10 items, and

$C(11) = 225 + 2 \cdot 11^2 = 225 + 242 = \$467$ is the cost of producing 11 items.

The difference $C(11) - C(10) = 467 - 425 = \$42$ is the cost of producing the eleventh item.

If we differentiate $C(x) = 225 + 2x^2$, we get $C'(x) = 4x$ and $C'(10) = 4 \cdot 10 = \$40$, a very close approximation to the cost of producing the eleventh item. This is exactly what marginal cost at x is—an approximation of the cost of producing the $(x + 1)^{th}$ item.

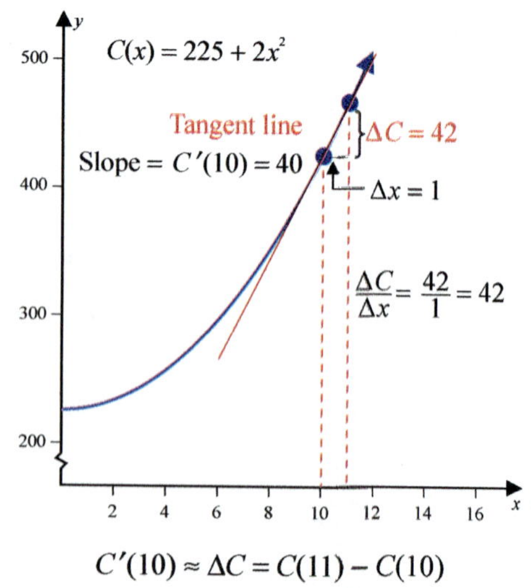

FIGURE 1: Marginal Cost

Marginal Cost

Marginal cost is the rate of change of total cost per unit change in quantity x. This concept is illustrated geometrically in Figure 1.

Example 1: Marginal Cost

A manufacturer of specialty items determines that the cost of producing x ballpoint pens is $C(x) = 500 + 3x$ in dollars.

a. Find $C(101) - C(100)$, the cost of producing the 101st pen.

b. Find $C'(100)$, the marginal cost at $x = 100$ pens.

Solution

a. $C(101) = 500 + 3(101) = 500 + 303 = 803$
$C(100) = 500 + 3(100) = 500 + 300 = 800$
$C(101) - C(100) = 803 - 800 = 3$

So, when the production level is increased from 100 to 101 pens, the total cost is increased by \$3. Therefore, the cost of producing the 101st pen is \$3.

b. $C(x) = 500 + 3x$ The given cost function

$C'(x) = 3$ Marginal cost is the derivative of the cost function.

$C'(100) = \$3$ The marginal cost at $x = 100$

In Example 1 the cost function is linear and the graph is a straight line with slope 3. Therefore, increasing output by one item always leads to an increase in total cost of \$3. In fact, the marginal cost is \$3 per pen, regardless of how many are produced.

Marginal Revenue

Marginal Revenue

Marginal revenue is the rate of change of the total revenue per unit change in sales when the level of sales is x items.

As with marginal cost, marginal revenue is an approximation of the growth or decline in revenue per unit sold. (Note that revenue per unit can decline even if more items are sold because of possible changes in the price per item.)

Remember that if $p = D(x)$ is the **demand function**, then the revenue function is formed as the product of the demand function (price per item) and x, the number of items sold:

$$R(x) = x \cdot p \quad \text{where} \quad p = D(x).$$

Example 2: Marginal Revenue

Suppose that a manufacturer has determined that the price of certain custom-made tables he produces can be determined by the demand function $p = D(x) = 117 - \dfrac{x}{4}$, where x is the number of tables produced and sold.

a. Find the revenue function.

b. Determine the marginal revenue when $x = 16$ tables.

Solution

a. $R(x) = x \cdot p = x\left(117 - \dfrac{x}{4}\right) = 117x - \dfrac{x^2}{4}$

b. $R'(x) = 117 \cdot 1 - \dfrac{1}{4} \cdot 2x = 117 - \dfrac{x}{2}$

So $R'(16) = 117 - \dfrac{16}{2} = 117 - 8 = \$109.$

Thus, the revenue is increasing at a rate of \$109 per table as the sixteenth table is sold.

Marginal Profit

Marginal Profit

Marginal profit is the rate of change of the profit per unit change in sales when x items are produced and sold.

As long as revenue is greater than cost, there is a profit. When revenue is equal to cost (a break-even point), the profit is 0. As with revenue and cost, when items are produced and sold, profit changes (increases or decreases). A basic question for any person in business is, "What production and sales level will yield maximum profit?" Calculus techniques for answering such a question will be developed later.

Example 3: Marginal Profit

If the table manufacturer in Example 2 has a cost function of $C(x) = 225 + 2x^2$ along with his revenue function of $R(x) = 117x - \dfrac{x^2}{4}$, find

a. all break-even points;

b. the marginal profit when $x = 10$, $x = 20$, $x = 30$.

Solution

a. Break-even points occur where $C(x) = R(x)$.

$$225 + 2x^2 = 117x - \dfrac{x^2}{4}$$

$$900 + 8x^2 = 468x - x^2$$

$$9x^2 - 468x + 900 = 0$$

$$9\left(x^2 - 52x + 100\right) = 0$$

$$9(x - 2)(x - 50) = 0$$

$$x - 2 = 0 \quad x - 50 = 0$$

$$x = 2 \qquad x = 50$$

Break-even points occur when $x = 2$ and $x = 50$.

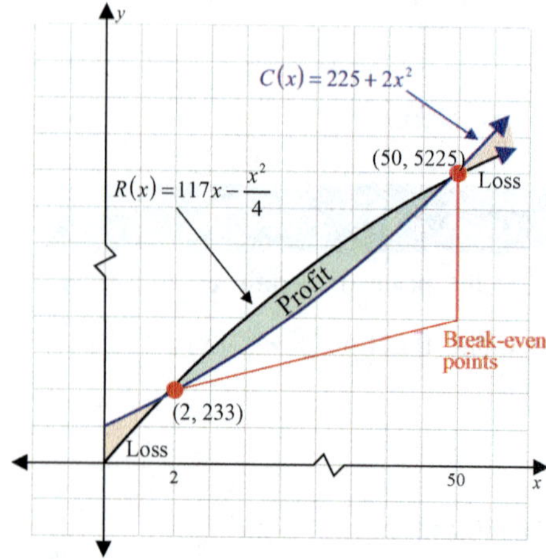

b. $P(x) = R(x) - C(x)$

$\qquad = \left(117x - \dfrac{x^2}{4}\right) - \left(225 + 2x^2\right)$

$\qquad = 117x - \dfrac{x^2}{4} - 225 - 2x^2$

$\qquad = -225 + 117x - \dfrac{9}{4}x^2$

Marginal profit is rate of change of profit, so first we must find the profit function $P(x)$.

The profit function $P(x)$

$P'(x) = 0 + 117 \cdot 1 - \dfrac{9}{4} \cdot 2x$

$\qquad\ \ = 117 - \dfrac{9}{2}x$

Now we can find the derivative of $P(x)$ to calculate the various marginal profits.

$P'(10) = 117 - \dfrac{9}{2}(10) = 117 - 45 = \$72$

$P'(20) = 117 - \dfrac{9}{2}(20) = 117 - 90 = \$27$

$P'(30) = 117 - \dfrac{9}{2}(30) = 117 - 135 = -\$18$

Note that $P'(x)$ is actually getting smaller as x gets larger. Marginal analysis shows that the rate of growth of profit per table is actually decreasing at a rate of $18 per table when 30 tables are manufactured and sold. This does not necessarily mean there is a loss, but as production increases, the profit per table is growing less because costs are increasing faster than revenue.

Marginal Average Cost

Suppose we let $C(x)$ represent the total cost of producing x items. If we divide the total cost by the number of items, we have the **average cost per item** $\overline{C}(x)$. That is,

$$\overline{C}(x) = \frac{C(x)}{x}.$$

For example, if $C(x) = 0.02x^2 + 100x + 2000$ represents the cost in dollars of producing x lawn mowers, then

$$\overline{C}(x) = \frac{C(x)}{x} = \frac{0.02x^2 + 100x + 2000}{x} = 0.02x + 100 + \frac{2000}{x}$$

represents the average cost per lawn mower.

Marginal Average Cost

Marginal average cost $\overline{C}'(x)$ is the rate of change of the average cost per unit change in production. This is an approximation of the change in average cost when one more item is produced.

> ### Example 4: Marginal Average Cost
>
> Suppose $C(x) = 0.02x^2 + 100x + 2000$ represents the cost in dollars of producing x lawn mowers.
>
> **a.** Find the average cost function.
>
> **b.** Find the average cost per lawn mower if 1000 lawn mowers are produced.
>
> **c.** Find the marginal average cost if 1000 lawn mowers are produced.
>
> #### Solution
>
> **a.** $\bar{C}(x) = \dfrac{C(x)}{x} = \dfrac{0.02x^2 + 100x + 2000}{x} = 0.02x + 100 + \dfrac{2000}{x}$
>
> **b.** $\bar{C}(1000) = 0.02(1000) + 100 + \dfrac{2000}{1000} = 20 + 100 + 2 = \$122$ Use $x = 1000$.
>
> **c.** $\bar{C}(x) = 0.02x + 100 + 2000x^{-1}$ Rewrite the equation so that there are no fractions.
>
> $\bar{C}'(x) = 0.02 + 0 + 2000(-1)x^{-1-1}$
>
> $= 0.02 - 2000x^{-2}$ Find the rate of change of the average cost to find the marginal average cost.
>
> $= 0.02 - \dfrac{2000}{x^2}$
>
> $\bar{C}'(1000) = 0.02 - \dfrac{2000}{1000^2}$ Use $x = 1000$ in the function for marginal average cost.
>
> $= 0.02 - 0.002$
>
> $= \$0.018$
>
> The average cost per lawn mower is increasing at a rate of 1.8 cents per lawn mower when 1000 lawn mowers are produced.

Marginal Propensity to Consume

Economists study consumption functions for countries in attempts to relate different parts of a nation's economy. A consumption function gives an estimate of a country's total annual consumption of consumer goods. The rate of change of such a function is called the marginal propensity to consume.

Suppose x represents total national income. If $C(x)$ is the amount of consumption and the national savings is $S(x)$, then

$$x = C(x) + S(x)$$

and

$$1 = C'(x) + S'(x).$$

$C'(x)$ is the marginal propensity to consume and $S'(x)$ is the marginal propensity to save.

It follows that

$$C'(x) = 1 - S'(x).$$

Example 5: Marginal Propensity

If the marginal propensity to save is $S'(x) = 0.02x$, what is the marginal propensity to consume?

Solution

$$C'(x) = 1 - S'(x) \qquad \text{We replace } S'(x) \text{ with } 0.02x.$$
$$= 1 - 0.02x$$

10.9 EXERCISES

PRACTICE

In Exercises 1–10, create an appropriate function of the type indicated.

1. The cost of producing x leather belts is given by $C(x) = 220 + 0.4x$. Determine the average cost function.

2. The cost of manufacturing a certain class of screws for wall hangers is given by $C(x) = 800 + 0.0005x + 0.00002x^2$. Determine the average cost function.

3. A vendor charges $3 for a hot dog. What is the demand function? What is the revenue function?

4. The Knoll Industrial Supply Company charges for 55-gallon drums using the demand function $p(x) = 22.5 - 0.5x$. What is the revenue function?

5. A supplier of souvenir T-shirts charges street vendors for each order based on a setup fee of $50 and an item charge of $1.15 per T-shirt. What is the cost function for the vendor?

6. A supplier of souvenir T-shirts has setup costs of $25 and shirts cost $0.40 each. What is the cost function?

7. The answer to Exercise 5 is also a revenue function for the supplier of T-shirts. Using the answers to Exercises 5 and 6, determine a profit function for the supplier of T-shirts.

8. For the supplier in Exercise 6, what is the average cost function?

9. A department store's cost estimate for a line of rocker-recliner chairs is given by $C(x) = 100 + 150x$. The store sells them for $450 each.
   a. What is the average cost function?
   b. What is the demand function?
   c. What is the revenue function?
   d. What is the profit function?

10. A shoe store estimates a certain line of dress shoes has costs given by $C(x) = 1500 + 30x + 0.05x^2$. The store charges \$75 per pair.
   a. What is the average cost function?
   b. What is the demand function?
   c. What is the revenue function?
   d. What is the profit function?

11. Suppose that total national consumption is given by a function $C(x) = 200 - 0.6x - 0.05x^{0.6}$, where x is the total national income.
   a. Determine the marginal propensity to consume.
   b. Determine the marginal propensity to save.

12. If the marginal propensity to save of a certain country is given by $S'(x) = 0.4x + 0.3$, determine the marginal propensity to consume.

13. The average cost $\overline{C}(x)$ of a product is $\dfrac{C(x)}{x}$, where $C(x)$ is the total cost function.
   a. What is the average cost function if its total cost function is $C(x) = 30 + 2x + 0.003x^2$?
   b. What is the rate of change of average cost?
   c. What value of x results in a minimum average cost?

14. The average cost of a product is given by $A(x) = 20x^{-1} + 3$.
   a. Determine the cost function for the product.
   b. Determine the marginal cost function.

🚀 APPLICATIONS

15. The weekly cost of producing x electric drills is given by the function $C(x) = 2400 + 28x + 0.25x^2$.
   a. Find $C(10)$, $C(20)$, and $C(30)$.
   b. Find the marginal cost function.
   c. Find $C'(10)$, $C'(20)$, and $C'(30)$.
   d. Find the average cost function and the marginal average cost function.
   e. Find the marginal average cost when $x = 10$, $x = 20$, and $x = 30$.

16. The total cost function for producing x units of a product is given by $C(x) = \dfrac{1}{3}x^3 - \dfrac{1}{2}x^2 + 7x + 18$.
   a. Find $C(3)$, $C(4)$, and $C(6)$.
   b. Find the marginal cost function.
   c. Find $C'(3)$, $C'(4)$, and $C'(6)$.
   d. Find the average cost function and the marginal average cost function.
   e. Find the marginal average cost when $x = 3$, $x = 4$, and $x = 6$.

17. The total cost of producing x units of a commodity is given by $C(x) = 60 + 10x - 0.5x^2$.
   a. Find $C(4)$, $C(6)$, and $C(9)$.
   b. Find the marginal cost function.
   c. Find $C'(4)$, $C'(6)$, and $C'(9)$.
   d. Find the average cost function.
   e. Find the marginal average cost when $x = 4$, $x = 6$, and $x = 9$.

18. The total cost of producing x wireless speakers is given by the function $C(x) = 300 + 24x - 0.4x^2 + 0.1x^3$.
    a. Find $C(2)$, $C(3)$, and $C(5)$.
    b. Find the marginal cost function.
    c. Find $C'(2)$, $C'(3)$, and $C'(5)$.
    d. Find the average cost function and the marginal average cost function.
    e. Find the marginal average cost if $x = 4$.

19. A manufacturer has determined that the revenue from the sale of x cell phones is given by $R(x) = 94x - 0.03x^2$ dollars. The cost of producing x cell phones is $C(x) = 10,800 + 34x$ dollars.
    a. Find the profit function $P(x)$.
    b. Find $P(200)$, $P(400)$, and $P(600)$.
    c. Find the marginal profit function $P'(x)$.
    d. Find $P'(200)$, $P'(400)$, and $P'(600)$.
    e. Find any break-even points.

20. The revenue from the sale of x fire extinguishers is estimated to be $R(x) = 54x - 0.4x^2$ dollars. The total cost of producing x fire extinguishers is $C(x) = 400 + 30x - 0.2x^2$ dollars.
    a. Find the profit function $P(x)$.
    b. Find $P(20)$, $P(40)$, and $P(60)$.
    c. Find the marginal profit function $P'(x)$.
    d. Find $P'(20)$, $P'(40)$, and $P'(60)$.
    e. Find any break-even points.

21. A company that produces and sells compact refrigerators has found that the revenue from the sale of x refrigerators is $R(x) = 100x - 0.1x^2$ dollars. The cost function is given by $C(x) = 2070 + 25x + 0.1x^2$ dollars.
    a. Find the profit function $P(x)$.
    b. Find $P(60)$, $P(80)$, and $P(100)$.
    c. Find the marginal profit function $P'(x)$.
    d. Find $P'(60)$, $P'(80)$, and $P'(100)$.
    e. Find any break-even points.

22. A manufacturer has determined that the cost and the revenue of producing and selling x telescopes are $C(x) = x^2 + 20x + 1050$ dollars and $R(x) = 140x - 0.5x^2$ dollars, respectively.
    a. Find the profit function $P(x)$.
    b. Find $P(30)$, $P(35)$, and $P(40)$.
    c. Find the marginal profit function $P'(x)$.
    d. Find $P'(30)$, $P'(35)$, and $P'(40)$.
    e. Find any break-even points.

23. The owner of a leather craft shop has determined that he can sell x attaché cases if the price is $p = D(x) = 46 + 0.25x$ dollars. The total cost for these cases is $C(x) = 0.15x^2 + 6x + 190$ dollars.
    a. Find the revenue function $R(x)$.
    b. Find the profit function $P(x)$.
    c. Find $P(25)$, $P(30)$, and $P(40)$.
    d. Find the marginal profit function $P'(x)$.
    e. Find $P'(25)$, $P'(30)$, and $P'(40)$.

24. A firm can sell x items of a product when the price is $p = D(x) = 3.00 - 0.001x$ dollars. The total production costs are $C(x) = 0.002x^2 + 0.72x + 260$ dollars.
    a. Find the revenue function $R(x)$.
    b. Find the profit function $P(x)$.
    c. Find $P(300)$, $P(375)$, and $P(400)$.
    d. Find the marginal profit function $P'(x)$.
    e. Find $P'(300)$, $P'(375)$, and $P'(400)$.

25. A local publishing company prints a special magazine each month. It has been determined that x magazines can be sold monthly when the price is $p = D(x) = 5.50 - 0.0004x$. The total cost of producing the magazine is $C(x) = 0.0002x^2 + x + 4650$ dollars.
    a. Find the revenue function $R(x)$.
    b. Find the profit function $P(x)$.
    c. Find $P(3000)$, $P(3500)$, and $P(4000)$.
    d. Find the marginal profit function $P'(x)$.
    e. Find $P'(3000)$, $P'(3500)$, and $P'(4000)$.

26. A sales representative for a company that produces skateboards can sell x units of their deluxe model if the price is $p = D(x) = 79.9 - 0.03x$ dollars. The total cost for these skateboards is given by $C(x) = 0.08x^2 + 5.1x + 5800$ dollars.
    a. Find the revenue function $R(x)$.
    b. Find the profit function $P(x)$.
    c. Find $P(320)$, $P(340)$, and $P(350)$.
    d. Find the marginal profit function $P'(x)$.
    e. Find $P'(320)$, $P'(340)$, and $P'(350)$.

27. A certain model car has a valuation in dollars given by the formula $f(x) = 12{,}519.3 - 1391.1x$, for $0 \le x \le 7$, where x is the age of the car in years. $x = 0$ corresponds to this calendar year.
    a. What is $f(0)$? Interpret this number.
    b. What is the marginal valuation? Interpret this number.

28. Based on averaging results at a certain state college, a relationship between grades and SAT scores was found to be $f(s) = 1.36 + 0.00141s$, where s is a student's SAT score and $f(s)$ is the student's graduating GPA (GPA based on 4.0 maximum score).
    a. What is the expected GPA for a student with an SAT score of 1000?
    b. What is the marginal GPA? Interpret this number.

Exercises 29–31 deal with projections of world population based on estimates of fertility around the world. Use the following background information:

The Total Fertility Rate (TFR) is the average number of children a woman will have. The United Nations projects world population according to assumptions about TFR values. In general, lower values promote economic well-being and lower world population. (**Source:** Population Reference Bureau, "Transitions in World Population," *Population Bulletin*, Vol. 59, No. 1, 36, March 2004.)

29. Total fertility rate: Using a TFR of 1.5, the UN projects total world population, in billions, will be modeled by $F(t)$ where t is the number of years after 2000 and F is the function $F(t) = -\dfrac{2.1}{10^6}t^4 + \dfrac{1.6}{10^4}t^3 - 0.00457t^2 + 0.0994t + 6.03$.

a. Give the marginal population function $F'(t)$.
b. Determine the estimate of world population in 2030.
c. What is the marginal population in 2030? Interpret this number.

30. Total fertility rate: Using a TFR of 2.0, the UN projects total world population, in billions, will be modeled by $G(t)$ where t is the number of years after 2000 and G is the function $G(t) = -\dfrac{2.1}{10^6}t^4 + \dfrac{1.66}{10^4}t^3 - 0.0045t^2 + 0.113t + 6.03$.

a. Give the marginal population function $G'(t)$.
b. Determine the estimate of world population in 2030.
c. What is the marginal population in 2030? Interpret this number.

31. Total fertility rate: Using a TFR of 2.5, the UN projects total world population, in billions, will be modeled by $H(t)$ where t is the number of years after 2000 and H is the function $H(t) = -\dfrac{2.6}{10^6}t^4 + \dfrac{1.87}{10^4}t^3 - 0.00396t^2 + 0.116t + 6.05$.

a. Give the marginal population function $H'(t)$.
b. Determine the estimate of world population in 2030.
c. What is the marginal population in 2030? Interpret this number.

✏ WRITING & THINKING

32. An economics professor claimed that the average cost was a minimum if the average cost equaled the marginal cost. Do you agree? Explain why or why not.

33. "The average revenue $\dfrac{R(x)}{x}$ is not usually studied in the context of business economics." Argue for or against this statement.

11

MORE ABOUT THE DERIVATIVE

11.1 THE PRODUCT AND QUOTIENT RULES

■ TOPICS

- Product Rule
- Quotient Rule

Product Rule

Suppose that $f(x) = x^2 + 5$ and $g(x) = 2x - 5$. Now consider the function formed by the product of f and g.

$$y = f(x) \cdot g(x) = (x^2 + 5)(2x - 5)$$

Is the derivative of y equal to the product of the derivatives of f and g? Symbolically, is $\dfrac{dy}{dx} = f'(x) \cdot g'(x)$ a true statement? To answer this question, we first note $f'(x) \cdot g'(x) = 2x \cdot 2 = 4x$. Next, we multiply f and g to express y more simply.

$$y = \left(x^2 + 5\right)\left(2x - 5\right)$$
$$= 2x^3 - 5x^2 + 10x - 25$$

Then we differentiate term by term to obtain

$$\frac{dy}{dx} = 6x^2 - 10x + 10,$$

which is a result considerably different from $4x$. Thus we see that the derivative of a product is **not** the product of the derivatives. Nevertheless, simply multiplying the factors of a function and then differentiating the result term by term is not always possible. Therefore, we need a method to deal with derivatives of products. Hence we introduce the **Product Rule**.

Product Rule

If $f(x)$ and $g(x)$ are differentiable functions and $y = f(x) \cdot g(x)$, then

$$\frac{dy}{dx} = f(x) \cdot g'(x) + g(x) \cdot f'(x).$$

In words, the derivative of the product of two functions is equal to the first function times the derivative of the second plus the second function times the derivative of the first.

Applying the Product Rule to $y = (x^2 + 5)(2x - 5)$ with $f(x) = x^2 + 5$ and $g(x) = 2x - 5$, we get

$$\frac{dy}{dx} = f(x) \quad \cdot \quad g'(x) \quad + \quad g(x) \quad \cdot \quad f'(x)$$

| | First Function | Derivative of Second | Second Function | Derivative of First |

$$\frac{dy}{dx} = (x^2 + 5) \quad \cdot \quad (2) \quad + \quad (2x - 5) \quad \cdot \quad 2x$$

$$\frac{dy}{dx} = 2(x^2 + 5) + 2x(2x - 5) = 2x^2 + 10 + 4x^2 - 10x$$

$$= 6x^2 - 10x + 10.$$

This is the same result we obtained earlier by multiplying $f(x) \cdot g(x)$ and then differentiating. The Product Rule can be written in several forms by incorporating the various notations for derivatives.

Other Forms of the Product Rule

If $f(x)$ and $g(x)$ are differentiable functions and $y = f(x) \cdot g(x)$, then

1. $\dfrac{dy}{dx} = f(x) \cdot \dfrac{d}{dx}\big[g(x)\big] + g(x) \cdot \dfrac{d}{dx}\big[f(x)\big],$

2. $\dfrac{d}{dx}\big[f(x) \cdot g(x)\big] = f(x) \cdot \dfrac{d}{dx}\big[g(x)\big] + g(x) \cdot \dfrac{d}{dx}\big[f(x)\big],$

3. $y' = f(x) \cdot g'(x) + g(x) \cdot f'(x),$

4. $D_x[f(x) \cdot g(x)] = f(x) \cdot D_x[g(x)] + g(x) \cdot D_x[f(x)],$ and

5. $y' = f \cdot g' + g \cdot f'.$

Example 1: Using the Product Rule

Use the Product Rule to find $\dfrac{dy}{dx}$ given $y = (x^2 + 3x - 1)(x^3 - 8x)$.

Solution

$$\frac{dy}{dx} = \left(x^2 + 3x - 1\right) \cdot \frac{d}{dx}\left[x^3 - 8x\right] + \left(x^3 - 8x\right) \cdot \frac{d}{dx}\left[x^2 + 3x - 1\right] \qquad \text{Apply the Product Rule.}$$

$$= \left(x^2 + 3x - 1\right)\left(3x^2 - 8\right) + \left(x^3 - 8x\right)\left(2x + 3\right)$$

$$= 3x^4 - 8x^2 + 9x^3 - 24x - 3x^2 + 8 + 2x^4 + 3x^3 - 16x^2 - 24x$$

$$= 5x^4 + 12x^3 - 27x^2 - 48x + 8$$

In Example 1, $\dfrac{dy}{dx} = \left(x^2 + 3x - 1\right)\left(3x^2 - 8\right) + \left(x^3 - 8x\right)\left(2x + 3\right)$ is a perfectly satisfactory answer. Removing parentheses in an algebraic expression does not necessarily improve appearance or understanding. A rule of thumb we like is to remove parentheses if the end result will be a polynomial of degree 2 or less.

Example 2: Using the Product Rule

Use the Product Rule to find the derivative of the function $f(x) = \left(\sqrt{x} + x^{-1}\right)\left(x^2 + 1\right)$.

Solution

In this case we have used $f(x)$ to represent the entire function rather than one of the functions in the product. Before differentiating, convert radicals to exponential form.

$$f'(x) = \left(x^{\frac{1}{2}} + x^{-1}\right) \cdot \frac{d}{dx}\left[x^2 + 1\right] + \left(x^2 + 1\right) \cdot \frac{d}{dx}\left[x^{\frac{1}{2}} + x^{-1}\right] \quad \text{Apply the Product Rule.}$$

$$= \left(x^{\frac{1}{2}} + x^{-1}\right)(2x) + \left(x^2 + 1\right)\left(\frac{1}{2}x^{-\frac{1}{2}} - x^{-2}\right)$$

You may choose to use radicals and fractions in your answer.

$$f'(x) = \left(\sqrt{x} + \frac{1}{x}\right)(2x) + \left(x^2 + 1\right)\left(\frac{1}{2\sqrt{x}} - \frac{1}{x^2}\right)$$

Remember, however, that the fractional and negative exponents make differentiating easier.

Quotient Rule

Suppose that $f(x) = x^2$ and $g(x) = 5x + 1$. Then the function formed by the quotient of these two functions is

$$y = \frac{f(x)}{g(x)} = \frac{x^2}{5x + 1}.$$

We can use the following **Quotient Rule** to find $\dfrac{dy}{dx}$.

Quotient Rule

If $f(x)$ and $g(x)$ are differentiable functions and $y = \dfrac{f(x)}{g(x)}$, then

$$\frac{dy}{dx} = \frac{g(x) \cdot f'(x) - f(x) \cdot g'(x)}{\left(g(x)\right)^2}.$$

In words, the derivative of a quotient is the denominator times the derivative of the numerator minus the numerator times the derivative of the denominator, all divided by the square of the denominator.

Example 3: Using the Quotient Rule

Use the Quotient Rule to find the derivative of the function $y = \dfrac{x^2}{5x + 1}$.

Solution

Here $f(x) = x^2$ and $g(x) = 5x + 1$. So, $f'(x) = 2x$ and $g'(x) = 5$. Substituting these into the Quotient Rule, we get

$$\frac{dy}{dx} = \frac{(5x+1)\cdot(2x) - (x^2)\cdot(5)}{(5x+1)^2}$$

$$= \frac{10x^2 + 2x - 5x^2}{(5x+1)^2} = \frac{5x^2 + 2x}{(5x+1)^2}.$$

⚠ CAUTION

Because of the minus sign in the Quotient Rule, the order of terms in the numerator is critical.

In Example 3, since the numerator will be a second-degree polynomial, removing parentheses and simplifying is appropriate. The student will never find it necessary to square a binomial expression in the denominator.

In Example 4 we show two techniques for finding the derivative of the same function. Many times the choice of technique is based on what our experience and practice tell us is the easiest to use in a particular situation.

Example 4: Using the Quotient Rule

Let $f(x) = \dfrac{x^2 - x + 1}{x^2}$.

a. Find $f'(x)$ by applying the Quotient Rule.

b. Find $f'(x)$ by first simplifying $f(x)$ algebraically and then using negative exponents.

Solution

a. Here $f(x)$ represents the entire quotient. In this case, we can think of the statement of the Quotient Rule in terms of numerator and denominator rather than in terms of $f(x)$ and $g(x)$.

$$f'(x) = \frac{(x^2)\cdot(2x-1) - (x^2 - x + 1)\cdot(2x)}{(x^2)^2}$$

where the arrows indicate: Denominator · Numerator′ − Numerator · Denominator′, over Denominator Squared.

Numerator $= x^2 - x + 1$
Denominator $= x^2$
Numerator $' = 2x - 1$
Denominator $' = 2x$

$$f'(x) = \frac{2x^3 - x^2 - 2x^3 + 2x^2 - 2x}{x^4}$$

$$= \frac{x^2 - 2x}{x^4} = \frac{x(x-2)}{x^3} = \frac{x-2}{x^3}$$

Note:

$$-(x^2 - x + 1)(2x)$$
$$= -(2x^3 - 2x^2 + 2x)$$
$$= -2x^3 + 2x^2 - 2x$$

b. Since the denominator is a single term, we can simplify algebraically as follows.

$$f(x) = \frac{x^2 - x + 1}{x^2} = \frac{x^2}{x^2} - \frac{x}{x^2} + \frac{1}{x^2}$$

$$= 1 - \frac{1}{x} + \frac{1}{x^2} = 1 - x^{-1} + x^{-2}$$

Therefore,

$$f'(x) = 0 - (-1)x^{-1-1} + (-2)x^{-2-1}$$

$$= x^{-2} - 2x^{-3}$$

Use the Power Rule on each of the terms.

Note that the answers in parts **a.** and **b.** are different forms of the same answer.

$$x^{-2} - 2x^{-3} = \frac{1}{x^2} - \frac{2}{x^3} = \frac{x-2}{x^3}$$

In Example 4, without experience, the student might not realize the enormous simplification which will result. The first expression for f' in solution **a.** is correct; but, by inspection, one can notice two things. First, the cubic terms will subtract out; and second, each term has x as a factor so there will be a cancellation.

Example 5: Growth of Bacteria

Several years of study have shown that t hours after a bacterium is introduced to a particular culture, the number of bacteria is given by $N(t) = \dfrac{t^2 + t}{\sqrt{t} + 1}$. Find the rate of growth of the bacteria after 4 hours.

Solution

$$N(t) = \frac{t^2 + t}{\sqrt{t} + 1} = \frac{t^2 + t}{t^{\frac{1}{2}} + 1}$$

Rewrite to remove the square root sign.

$$N'(t) = \frac{\left(t^{\frac{1}{2}} + 1\right)\dfrac{d}{dt}\left[t^2 + t\right] - \left(t^2 + t\right)\dfrac{d}{dt}\left[t^{\frac{1}{2}} + 1\right]}{\left(t^{\frac{1}{2}} + 1\right)^2}$$

Use the Quotient Rule.

$$= \frac{\left(t^{\frac{1}{2}} + 1\right)(2t + 1) - \left(t^2 + t\right)\left(\dfrac{1}{2}t^{-\frac{1}{2}}\right)}{\left(t^{\frac{1}{2}} + 1\right)^2}$$

$$N'(4) = \frac{\left(4^{\frac{1}{2}} + 1\right)(2 \cdot 4 + 1) - \left(4^2 + 4\right)\left(\dfrac{1}{2} \cdot 4^{-\frac{1}{2}}\right)}{\left(4^{\frac{1}{2}} + 1\right)^2}$$

Substitute $t = 4$ into the formula for the derivative.

$$= \frac{(2 + 1)(9) - (20)\left(\dfrac{1}{2} \cdot \dfrac{1}{2}\right)}{(2 + 1)^2} = \frac{(3)(9) - (20)\left(\dfrac{1}{4}\right)}{(3)^2}$$

$$= \frac{27 - 5}{9} = \frac{22}{9}$$

The rate of growth when $t = 4$ is $\dfrac{22}{9}$ bacteria per hour.

Note that we did not bother to simplify the derivative $N'(t)$ because we were interested in its value at only one time, $t = 4$, rather than a simplified formula. Also, no cancellation is apparent and the rule of thumb suggests this form is as good as any other.

Example 6: Elementary Operations on Functions

You are given that f and g are differentiable functions and that $f(5) = 10$, $g(5) = -3$, $f'(5) = 1$, $g'(5) = 8$. Determine $h'(5)$ given that:

a. $h(x) = \dfrac{f(x)}{g(x)}$

b. $h(x) = (f(x) + 3x - 1)g(x)$

Solution

a. $h'(x) = \dfrac{g(x)f'(x) - f(x)g'(x)}{\left(g(x)\right)^2}$

$$h'(5) = \frac{(-3)(1) - 10(8)}{(-3)^2} = \frac{-3 - 80}{9} = \frac{-83}{9}$$

b. $h'(x) = (f(x) + 3x - 1)g'(x) + g(x)(f'(x) + 3)$

$$h'(5) = (10 + 15 - 1)(8) + (-3)(1 + 3)$$
$$= (24)(8) + (-3)(4) = 192 - 12 = 180$$

We end the section with the following remarks about the possible algebraic forms of answers. The appropriateness of these remarks will be apparent in the homework as the student gains mastery of the techniques of differentiation and as the related algebraic expressions become more complicated.

⚠ CAUTION

The algebraic form of the derivative of a function depends on which rule of differentiation is applied, and there may be several correct forms.

For example, if $f(x) = \dfrac{x^3 + 3x + 1}{2x + 5}$, then we will find (in a later section) that the following two forms for $f'(x)$ are correct.

1. $f'(x) = \dfrac{(2x+5)(3x^2+3) - (x^3+3x+1)(2)}{(2x+5)^2}$

2. $f'(x) = (x^3 + 3x + 1)(-1)(2x + 5)^{-2}(2) + (2x + 5)^{-1}(3x^2 + 3)$

Both of these forms can be simplified to the form

3. $f'(x) = \dfrac{4x^3 + 15x^2 + 13}{(2x+5)^2}.$

> **⚠ CAUTION**
>
> If the rules of differentiation have been followed, then an answer may be correct even though it does not "look like" the answer given. Be sure to check with your instructor to see how much algebraic simplification is expected and whether or not one form is preferred over another.

11.1 EXERCISES

💡 PRACTICE

In Exercises 1–5, find $f'(x)$ two ways: (1) multiply the factors first, then find the derivative, and (2) use the Product Rule.

1. $f(x) = x^2\left(1 + 3x - 2x^2\right)$

2. $f(x) = (x+3)(x-1)$

3. $f(x) = x^{\frac{1}{2}}\left(1 + 3x^2\right)$

4. $f(x) = x^{\frac{1}{2}}\left(1 + x^{\frac{1}{2}} - x^{\frac{3}{2}}\right)$

5. $f(x) = (2x+3)(2x-3)$

In Exercises 6–10, find $g'(x)$ two ways: (1) divide the factors first, then find the derivative, and (2) use the Quotient Rule and simplify the answer.

6. $g(x) = \dfrac{1 + 5x + x^2}{x}$

7. $g(x) = \dfrac{2 + \sqrt{x}}{\sqrt{x}}$

8. $g(x) = \dfrac{x^2 + 1}{x^5}$

9. $g(x) = \dfrac{30x^2 - 10x^6}{5x}$

10. $g(x) = \dfrac{3x^{\frac{1}{2}} - 5x^{\frac{3}{2}} + 7x^{\frac{5}{2}} - 9x^{\frac{7}{2}}}{x^{\frac{1}{2}}}$

In Exercises 11–34, use the Product Rule or Quotient Rule to find the derivative of each of the functions. Simplify your answers.

11. $f(x) = x^3\left(x^2 + 5\right)$

12. $f(x) = x^5\left(2x - x^3\right)$

13. $f(t) = t^{\frac{1}{2}}\left(4t + 3\right)$

14. $f(t) = t^{\frac{2}{3}}\left(4t^2 + 1\right)$

15. $y = x^2\left(\sqrt{x} + \dfrac{1}{\sqrt{x}}\right)$

16. $y = x^{-2}\left(3x + x^{\frac{1}{3}}\right)$

17. $g(u) = \left(2u^2 + 3\right)(5 - 3u)$

18. $g(u) = \left(3u^2 - 8\right)\left(u^2 + u\right)$

19. $g(t) = \left(5 + \dfrac{1}{t}\right)\left(t^2 + \dfrac{1}{5}\right)$

20. $f(t) = \left(1 - \dfrac{3}{t^2}\right)\left(2t^2 + t - 1\right)$

21. $f(x) = \dfrac{3x}{x+6}$

22. $f(x) = \dfrac{7x^2}{2x-1}$

23. $f(x) = \dfrac{x+8}{x-7}$

24. $f(x) = \dfrac{x^2 + 2x - 3}{x+2}$

25. $y = \dfrac{x^3 - 5}{x^2 + 1}$

26. $y = \dfrac{2x^2 + 3x}{x^3 + 6}$

27. $g(x) = \dfrac{\sqrt{x}}{x+9}$ **28.** $g(x) = \dfrac{6\sqrt{x}}{3x-4}$ **29.** $f(u) = \dfrac{u^2}{\sqrt{u}+1}$

30. $f(u) = \dfrac{7}{1-\sqrt[3]{u}}$ **31.** $f(t) = \dfrac{4-\sqrt{t}}{t^2+3}$ **32.** $f(t) = \dfrac{3-t}{4-5\sqrt{t}}$

33. $f(x) = \dfrac{x^2-5x}{1+2\sqrt[3]{x}}$ **34.** $f(x) = \dfrac{x\left(1+3\sqrt{x}\right)}{\sqrt{x}+6}$

In Exercises 35–44, you are given that $f(x)$ and $g(x)$ are differentiable functions and that $f(2) = 3$, $f'(2) = -1$, $g(2) = -11$, and $g'(2) = 6$. In each exercise, find the value of $h'(2)$.

35. $h(x) = x \cdot f(x)$ **36.** $h(x) = \dfrac{f(x)}{2x+1}$

37. $h(x) = \dfrac{f(x)+3x}{f(x)-3x}$ **38.** $h(x) = \dfrac{g(x)}{f(x)}$

39. $h(x) = \dfrac{g(x)}{3x+10}$ **40.** $h(x) = (3x+5)\cdot f(x)$

41. $h(x) = \dfrac{16x+1}{f(x)-11x+1}$ **42.** $h(x) = f(x)\cdot g(x)$

43. $h(x) = \dfrac{f(x)}{g(x)}$ **44.** $h(x) = g(x)\cdot(1+3x)$

In Exercises 45–50, find the equation of the line tangent to the graph of $f(x)$ at the given point.

45. $f(x) = \left(x+5x^{\frac{1}{2}}\right)\left(6x^2-12x+2\right)$; $(4,700)$

46. $f(x) = \dfrac{\left(11x^2-3x+2\right)}{x^2+1}$; $(1,5)$

47. $f(x) = \dfrac{2-3x}{5+2x}$; $(0,0.4)$

48. $f(x) = \left(x^5-5\right)\left(x^3-x-1\right)$; $(0,5)$

49. $f(x) = \dfrac{20}{17x+3}$; $(1,1)$

50. $f(x) = \dfrac{\sqrt{x}+2}{x^2-1}$; $\left(9,\dfrac{1}{16}\right)$

51. Given $f(x) = (1-x)\left(16-x^2\right)$, find the (x, y)-coordinates on the graph where the tangent line is horizontal.

52. Given $g(x) = (x-10)\left(x^2+2x+1\right)$, find any (x, y)-coordinates on $g(x)$ for which the tangent line is horizontal.

53. Find any point or points on the graph of $y = (x-5)(x+10)$ so that the slope equals 25. Sketch a graph of y and the tangent line or lines.

54. Find any point or points on the graph of $G(x) = (2x+1)(x-3)$ so that the slope is -20. Sketch a graph of G and the tangent line or lines.

55. Sketch a graph of $F(x) = \dfrac{30x}{2x^2 + 5}$ on the x-interval $[-5, 10]$. Determine the (x, y)-coordinates of any point with a horizontal tangent line, and sketch this (or these) horizontal tangent(s). Round to the nearest hundredth.

🚀 APPLICATIONS

56. Bacterial growth: It is estimated that the population of a bacterial culture after t hours is approximately $N(t) = \dfrac{t^2 - 2t}{3\sqrt{t} + 2}$, where $N(t)$ is in thousands and $2 \le t \le 10$. Find the rate of growth after 4 hours.

57. Marginal revenue: The demand function for a particular item is given by $D(x) = \dfrac{115}{3x+1}$. Find the marginal revenue when $x = 3$.

58. Marginal profit: The profit from the sale of x items is given by $P(x) = (2 - 0.5x)(0.5x - 5)$, where $P(x)$ is in hundreds of dollars and $2 \le x \le 10$. Find the marginal profit when $x = 5$.

59. Marginal cost: The cost of producing x items of a product is given by $C(x) = (0.1x + 100)(0.1x + 20) - 600$. Find the marginal cost when $x = 60$.

60. Velocity of a particle: A particle is moving slowly along a line. Its position after t seconds is $S(t) = \dfrac{t}{t^2 + 4}$ feet. Find the velocity when the particle has been moving for 3 seconds.

61. Population growth: It is estimated that t years from now the population of a city will be $P(t) = (0.6t - 7)(0.5t + 6) + 85$ in thousands. How fast will the population be growing in 10 years?

11.2 THE CHAIN RULE AND THE GENERAL POWER RULE

■ TOPICS

- The Chain Rule
- The General Power Rule
- Using the Rules in Combination with Each Other

The Chain Rule

Suppose that at a certain production level, production costs are \$18 per unit and 40 units are produced each hour. How fast is the total cost changing per hour? The answer is found by multiplying:

$$\$18 \text{ per unit} \cdot 40 \text{ units per hour} = \$720 \text{ per hour.}$$

The total cost is changing at a rate of \$720 per hour: each hour 40 more units are produced and since each unit costs \$18 to produce, the total cost per hour is \$720. Notice that, in order to get this rate of change of cost, we multiplied the two other rates.

In this example the total cost is a function of the number of units produced, $y = C(x)$, and the number of units produced is a function of time, $x = g(t)$. We find that the rates are related as follows.

$$\frac{dy}{dx} \cdot \frac{dx}{dt} = \frac{dy}{dt}$$

$$\begin{pmatrix} \text{rate of change} \\ \text{of cost with respect} \\ \text{to unit production} \end{pmatrix} \cdot \begin{pmatrix} \text{rate of change} \\ \text{of unit production} \\ \text{with respect to time} \end{pmatrix} = \begin{pmatrix} \text{rate of change} \\ \text{of cost with} \\ \text{respect to time} \end{pmatrix}$$

$$\begin{pmatrix} \$18 \text{ increase} \\ \text{per additional} \\ \text{unit produced} \end{pmatrix} \cdot \begin{pmatrix} 40 \text{ additional} \\ \text{units produced} \\ \text{per hour} \end{pmatrix} = \begin{pmatrix} \$720 \text{ increase} \\ \text{per hour} \end{pmatrix}$$

These ideas lead to the **Chain Rule**.

The Chain Rule

If $y = f(u)$ and $u = g(x)$, then

$$\frac{dy}{dx} = \frac{dy}{du} \cdot \frac{du}{dx},$$

provided that $\dfrac{dy}{du}$ and $\dfrac{du}{dx}$ both exist.

Example 1: Using the Chain Rule

Suppose that $y = u + \sqrt{u}$ and $u = x^3 + 17$. Use the Chain Rule to find $\dfrac{dy}{dx}$. Then evaluate $\dfrac{dy}{dx}$ at $x = 2$. This evaluation is denoted more briefly by $\left.\dfrac{dy}{dx}\right|_{x=2}$.

Solution

$$y = u + \sqrt{u} \qquad u = x^3 + 17 \qquad \text{Rewrite } y \text{ using exponents.}$$

$$= u + u^{\frac{1}{2}}$$

$$\frac{dy}{du} = 1 + \frac{1}{2}u^{-\frac{1}{2}} \qquad \frac{du}{dx} = 3x^2 \qquad \text{Find the derivatives of each function by}$$
$$\text{using the Power Rule.}$$

$$= 1 + \frac{1}{2\sqrt{u}}$$

So, by the Chain Rule,

$$\frac{dy}{dx} = \frac{dy}{du} \cdot \frac{du}{dx} = \left(1 + \frac{1}{2\sqrt{u}}\right) \cdot 3x^2.$$

To evaluate this derivative at $x = 2$, we must first substitute it into the equation for u, and then substitute u and x into the equation of the derivative.

$$u = (2)^3 + 17 = 25 \qquad \text{Substitute } x = 2 \text{ into } u.$$

$$\left.\frac{dy}{dx}\right|_{x=2} = \left(1 + \frac{1}{2\sqrt{25}}\right) \cdot 3(2)^2 = \left(1 + \frac{1}{10}\right) \cdot 12 \qquad \begin{array}{l}\text{Substitute } u = 25 \text{ and } x = 2 \text{ into the} \\ \text{formula for the derivative.}\end{array}$$

$$= \frac{11}{10} \cdot \frac{12}{1} = \frac{66}{5}$$

Example 2: Ripples on a Pond

A pebble is tossed into a still pond, and concentric circles are formed on the pond's surface. The area of each circle depends on the radius of the circle, and each radius depends on the elapsed time. If the radius is changing at the rate of $\dfrac{1}{2}\dfrac{\text{ft}}{\text{sec}}$, how fast is the area of the circle growing when the radius is 3 feet?

Solution

Recall that the area of a circle is given by the formula $A = \pi r^2$, where $\pi \approx 3.14$. Differentiating A with respect to r gives

$$\frac{dA}{dr} = 2\pi r.$$

Since the radius is changing at the rate of $\dfrac{1}{2}\dfrac{\text{ft}}{\text{sec}}$, we have

$$\frac{dr}{dt} = \frac{1}{2}\frac{\text{ft}}{\text{sec}}.$$

Thus, by the Chain Rule,

$$\frac{dA}{dt} = \frac{dA}{dr} \cdot \frac{dr}{dt} = 2\pi\, r \cdot \frac{1}{2} = \pi\, r.$$

Therefore, when $r = 3$,

$$\left.\frac{dA}{dt}\right|_{r=3} = \pi \cdot 3 = 3\pi \ \frac{\text{ft}^2}{\text{sec}}.$$

The area of the circle is growing at the rate of $3\pi \ \dfrac{\text{ft}^2}{\text{sec}} \approx 9.42\, \dfrac{\text{ft}^2}{\text{sec}}$.

In Example 2, as with so many mathematical models of real life situations, there are certain realistic limitations on the validity of the model. In this example, we know (even though it is not explicitly stated) that r is restricted by the size of the pond. Furthermore, if the pebble is small, the circle may dissipate even before the radius becomes 3 feet. In business, such practical considerations can affect a model for profit, revenue, or cost on a yearly, monthly, or weekly basis.

Now we will look at the Chain Rule from the point of view of the composition of functions. For example, if

$$y = f(u) = \sqrt{u} \quad \text{and} \quad u = g(x) = x^2 + 9,$$

then

$$y = f\big(g(x)\big) = \sqrt{g(x)} = \sqrt{x^2 + 9}.$$

By the Chain Rule,

$$\frac{dy}{dx} = \frac{dy}{du} \cdot \frac{du}{dx} = f'(u) \cdot g'(x) = \frac{1}{2} u^{-\frac{1}{2}} \cdot 2x.$$

Substituting $u = g(x)$, we have

$$\frac{dy}{dx} = f'\big(g(x)\big) \cdot g'(x) = \frac{1}{2}\big(x^2 + 9\big)^{-\frac{1}{2}} \cdot 2x.$$

This leads to a second form of the Chain Rule.

The Chain Rule (Second Form)

If $y = f(g(x))$, then

$$\frac{dy}{dx} = f'\big(g(x)\big) \cdot g'(x)$$

provided that $f'(g(x))$ and $g'(x)$ both exist.

Example 3: Second Form of the Chain Rule

Use the second form of the Chain Rule to find $\dfrac{dy}{dx}$ if $y = \big(x^2 + 3x - 7\big)^{\frac{5}{2}}$.

Solution

Let $g(x) = x^2 + 3x - 7$. Then we can write

$$y = f\left(g(x)\right) = \left[g(x)\right]^{\frac{5}{2}}.$$

Using the second form of the Chain Rule, we get

$$\frac{dy}{dx} = f'\left(g(x)\right) \cdot g'(x)$$

Substitute $f'\left(g(x)\right) = \frac{5}{2}\left[g(x)\right]^{\frac{3}{2}}$ and $g'(x) = 2x + 3$ into the formula for the second form of the Chain Rule.

$$= \frac{5}{2}\left[g(x)\right]^{\frac{3}{2}}(2x+3)$$

$$= \frac{5}{2}\left(x^2 + 3x - 7\right)^{\frac{3}{2}}(2x+3).$$

Substitute $g(x) = x^2 + 3x - 7$.

Example 4: Second Form of the Chain Rule

Determine $h'(5)$ if $h(x) = (25 - 2x + x^2)^3$.

Solution

Set $h(x) = f(g(x))$ where $g(x) = 25 - 2x + x^2$. Then

$$h'(x) = f'\left(g(x)\right) \cdot g'(x)$$

$$= 3\left(g(x)\right)^2 \cdot g'(x) = 3\left(25 - 2x + x^2\right)^2 (-2 + 2x).$$

Now $x = 5$, so

$$h'(5) = 3\left(25 - 2(5) + 5^2\right)^2 (-2 + 2(5))$$

Note: $h'(5) = f'(g(5)) \cdot g'(5)$.

$$= 3(40)^2 (8) = 38,400.$$

The General Power Rule

In Example 3 and Example 4, the Chain Rule is applied to a function in the form $f(g(x)) = [g(x)]^n$. That is, the function f is a power function, and $g(x)$ is raised to a power. At this time, we state the **General Power Rule**, which is the special case of the Chain Rule where the function f is a power function.

The General Power Rule

If $y = [g(x)]^n$, then

$$\frac{dy}{dx} = n\left[g(x)\right]^{n-1} \cdot g'(x)$$

provided that $g'(x)$ exists.

Another form of the General Power Rule is

$$\frac{dy}{dx} = n\big[g(x)\big]^{n-1} \cdot \frac{d}{dx}\big(g(x)\big).$$

Example 5: The General Power Rule

Using the General Power Rule, find the derivative of each of the following functions.

a. $y = (x^2 + 8x)^{10}$

Solution

Substitute the following values into the General Power Rule: $n = 10$, $n - 1 = 9$, $g(x) = x^2 + 8x$, $g'(x) = 2x + 8$.

$$\frac{dy}{dx} = n\big[g(x)\big]^{n-1} \cdot g'(x)$$

$$= 10\big(x^2 + 8x\big)^9 \cdot (2x + 8)$$

b. $y = 6\sqrt[3]{2x+5}$

Solution

Rewrite the radical with a fractional exponent; then apply the General Power Rule.

$$y = 6\sqrt[3]{2x+5} = 6(2x+5)^{\frac{1}{3}}$$

Substitute the following values into the General Power Rule: $c = 6$, $n = \frac{1}{3}$, $n - 1 = -\frac{2}{3}$, $g(x) = 2x + 5$, $g'(x) = 2$.

$$\frac{dy}{dx} = n\big[g(x)\big]^{n-1} \cdot g'(x)$$

$$= 6 \cdot \frac{1}{3}(2x+5)^{-\frac{2}{3}} \cdot 2$$

$$= 4(2x+5)^{-\frac{2}{3}} = \frac{4}{(2x+5)^{\frac{2}{3}}}$$

Example 6: The General Power Rule

Use the General Power Rule to find $f'(u)$ if $f(u) = \dfrac{1}{\sqrt{7-u}}$.

Solution

$$f(u) = \frac{1}{\sqrt{7-u}} = (7-u)^{-\frac{1}{2}} \qquad \text{Rewrite } f(u) \text{ with exponents.}$$

Substitute the following values into the General Power Rule: $n = -\frac{1}{2}$, $n - 1 = -\frac{3}{2}$, $g(u) = 7 - u$, $g'(u) = -1$.

$$f'(u) = n \left[g(u) \right]^{n-1} \cdot g'(u)$$

$$= -\frac{1}{2}(7-u)^{\frac{-3}{2}} \cdot (-1)$$

$$= \frac{1}{2}(7-u)^{\frac{-3}{2}} = \frac{1}{2(7-u)^{\frac{3}{2}}}$$

Using the Rules in Combination with Each Other

The following examples use a variety of notations and illustrate how the differentiation rules (Sum and Difference Rule, Product Rule, Quotient Rule, General Power Rule, and so on) can be used in combination in one problem. Try to follow each example through, step by step. By careful analysis of these examples, you will be able to work similar problems.

Example 7: Using Multiple Rules

Find $\dfrac{dy}{dx}$ if $y = (x^2 + 1)\sqrt{2x - 3}$.

Solution

Treat y as the product of two functions, $f(x) = (x^2 + 1)$ and $g(x) = (2x - 3)^{\frac{1}{2}}$, and

use the Product Rule. In using the Product Rule, differentiate $(2x - 3)^{\frac{1}{2}}$ by using the General Power Rule.

$$\frac{dy}{dx} = \overset{f(x)}{(x^2 + 1)} \cdot \frac{d}{dx}\left[\overset{g'(x)}{(2x-3)^{\frac{1}{2}}} \right] + \overset{g(x)}{(2x-3)^{\frac{1}{2}}} \cdot \frac{d}{dx}\left[\overset{f'(x)}{x^2 + 1} \right]$$

Product Rule

$$\frac{dy}{dx} = (x^2 + 1) \cdot \frac{1}{2}(2x-3)^{-\frac{1}{2}}(2) + (2x-3)^{\frac{1}{2}}(2x)$$

To find $g'(x)$, substitute the following values into the General Power Rule:

$$= (x^2 + 1)(2x-3)^{-\frac{1}{2}} + (2x-3)^{\frac{1}{2}}(2x)$$

$$g(x) = h(x)^{\frac{1}{2}},$$

$$= (2x-3)^{-\frac{1}{2}}\left[(x^2 + 1) + (2x-3)(2x) \right]$$

$$n = \frac{1}{2},\ h(x) = 2x - 3,$$

$$= (2x-3)^{-\frac{1}{2}}\left(x^2 + 1 + 4x^2 - 6x \right)$$

$$n - 1 = -\frac{1}{2},\ h'(x) = 2.$$

$$= (2x-3)^{-\frac{1}{2}}\left(5x^2 - 6x + 1 \right)$$

Also substitute $f'(x) = 2x$.

Factor out $(2x-3)^{-\frac{1}{2}}$. Simplify.

Using radical notation, we have

$$\frac{dy}{dx} = \frac{5x^2 - 6x + 1}{\sqrt{2x - 3}}.$$

In Example 7, the useful factoring of the radical term illustrates the simplification of any derivative expression when y is a product of a polynomial and a radical. If the expression in the radical is linear, and the other polynomial is of low degree, then the simplification should be made.

Example 8: Using Multiple Rules

Find $D_x\left[\sqrt[3]{\dfrac{2x-7}{3x+4}}\right]$.

Solution

Apply the General Power Rule and the Quotient Rule.

$$D_x\left[\sqrt[3]{\frac{2x-7}{3x+4}}\right] = D_x\left[\left(\frac{2x-7}{3x+4}\right)^{\frac{1}{3}}\right] \qquad \text{Rewrite using exponents.}$$

$$= \frac{1}{3}\left(\frac{2x-7}{3x+4}\right)^{-\frac{2}{3}}\cdot D_x\left[\frac{2x-7}{3x+4}\right] \qquad \text{Apply the General Power Rule.}$$

$$= \frac{1}{3}\left(\frac{2x-7}{3x+4}\right)^{-\frac{2}{3}}\left(\frac{(3x+4)\cdot 2 - (2x-7)\cdot 3}{(3x+4)^2}\right) \qquad \text{Use the Quotient Rule to find } D_x\left[\frac{2x-7}{3x+4}\right].$$

$$= \frac{1}{3}\left(\frac{2x-7}{3x+4}\right)^{-\frac{2}{3}}\left(\frac{6x+8-6x+21}{(3x+4)^2}\right)$$

$$= \frac{1}{3}\left(\frac{2x-7}{3x+4}\right)^{-\frac{2}{3}}\left(\frac{29}{(3x+4)^2}\right) \qquad \text{Simplify.}$$

Example 9: Air Pollution

Suppose that the average measure of air pollution in Sootville is given by the formula

$$P(t) = \frac{t^2}{\sqrt{t^2+36}},$$

where t is time in years since 1980. How fast was this measure changing in 1988?

Solution

Find the derivative of $P(t)$ by using the Quotient Rule and the General Power Rule.

$$P(t) = \frac{t^2}{\left(t^2+36\right)^{\frac{1}{2}}} \qquad \text{Rewrite using exponents.}$$

$$P'(t) = \frac{\left(t^2+36\right)^{\frac{1}{2}}\cdot 2t - t^2\cdot\frac{1}{2}\left(t^2+36\right)^{-\frac{1}{2}}\cdot 2t}{\left(t^2+36\right)} \qquad \text{Here the Quotient Rule and the General Power Rule are used in combination.}$$

$$= \frac{\left(t^2+36\right)^{-\frac{1}{2}}\left[\left(t^2+36\right)\cdot 2t - t^3\right]}{\left(t^2+36\right)} \qquad \text{Factor out } \left(t^2+36\right)^{-\frac{1}{2}}.$$

$$= \frac{t^3+72t}{\left(t^2+36\right)^{\frac{3}{2}}} \qquad \text{Simplify.}$$

To find how fast this measure was changing in 1988 (8 years since 1980), evaluate the derivative for $t = 8$.

$$P'(8) = \frac{8^3 + 72 \cdot 8}{\left(8^2 + 36\right)^{\frac{3}{2}}} = \frac{512 + 576}{\left(100\right)^{\frac{3}{2}}} = \frac{1088}{1000} = 1.088$$

In 1988, the air pollution measure was increasing at a rate of 1.088 units per year.

11.2 EXERCISES

♀ PRACTICE

Find the derivative for each function given in Exercises 1–40 and simplify your answer.

1. $f(x) = (2x - 5)^4$

2. $f(x) = (7x + 2)^3$

3. $f(x) = (1 - 4x)^3$

4. $f(x) = (3 - 5x)^5$

5. $g(x) = (x^2 + 4)^{-2}$

6. $g(x) = (x^2 - 8)^{-1}$

7. $h(t) = (2t^2 + 3t)^{-3}$

8. $h(t) = (4t^2 - t)^{-2}$

9. $y = (2x^2 + 5x - 7)^2$

10. $y = (4x^2 + 9x - 3)^3$

11. $y = (x^3 + 1)^{\frac{1}{2}}$

12. $y = (2x^3 - 5)^{\frac{1}{3}}$

13. $y = \sqrt[3]{4x^2 + 1}$

14. $y = \sqrt{7 + 4x^2}$

15. $y = \sqrt[4]{1 - 2x^3}$

16. $y = \sqrt[3]{5x^3 - 4}$

17. $f(t) = 5t(t^3 + 3)^4$

18. $f(x) = -7x(x^4 - 2)^3$

19. $f(x) = 2x^3(x^2 - 8)^3$

20. $f(t) = t(4 - 3t^2)^2$

21. $g(x) = \dfrac{1}{\sqrt{x^2 - 6}}$

22. $g(x) = \dfrac{5}{\sqrt{x^3 + 4}}$

23. $g(t) = \dfrac{t}{\sqrt{t^2 + 8}}$

24. $g(x) = \dfrac{x^2}{\sqrt[3]{x^2 + 6}}$

25. $h(x) = \dfrac{\sqrt[3]{2x + 3}}{x^2}$

26. $h(x) = \dfrac{\sqrt{5x - 2}}{x^3}$

27. $y = (2x + 1)\sqrt{3x - 4}$

28. $y = (4x + 3)\sqrt{x^2 + 3}$

29. $y = (3x - 2)^2(5x + 1)^{-2}$

30. $y = (2x + 7)^3(3x + 1)^{-4}$

31. $f(t) = \dfrac{5t + 1}{(t - 1)^{\frac{2}{3}}}$

32. $f(t) = \dfrac{t^2 + t + 1}{\sqrt{t^4 - 1}}$

33. $g(t) = \left(\dfrac{2t + 5}{t + 1}\right)^3$

34. $g(t) = \left(\dfrac{5t + 4}{t^2 - 3}\right)^4$

35. $y = \left(\dfrac{x + 3}{4 - 2x}\right)^{\frac{1}{2}}$

36. $y = \left(\dfrac{x^2}{4x + 1}\right)^{\frac{1}{3}}$

37. $y = \sqrt{\dfrac{x + 2}{3x - 1}}$

38. $y = \sqrt{\dfrac{x^2 + 6}{x^3}}$

39. $y = \dfrac{x^2 + x}{\sqrt{7 - 2x}}$

40. $y = \dfrac{\left(x^2 + 2\right)^2}{\sqrt{5x - 3}}$

In Exercises 41–50, find $\dfrac{dy}{du}, \dfrac{du}{dx}$, and $\dfrac{dy}{dx}$. Then evaluate $\dfrac{dy}{dx}$ for the given value of x.

41. $y = u^2 + 2, \ u = 3x^2 + 1; \ x = -1$

42. $y = \sqrt{u + 4}, \ u = x^2 + x - 1; \ x = 2$

43. $y = \dfrac{1}{u^2}, \ u = 2x^3 - 3x + 3; \ x = 1$

44. $y = \sqrt[3]{u}, \ u = 2x^3 - 4x; \ x = 2$

45. $y = u^{\frac{3}{2}}, \ u = x^3 - 2x^2; \ x = 3$

46. $y = \dfrac{1}{u^3}, \ u = 3x + 1; \ x = 1$

47. $y = \sqrt[3]{u}, \ u = 7x^2 + 1; \ x = -3$

48. $y = u^2 + 3u + 4, \ u = x^3 - 5x - 2; \ x = -2$

49. $y = 2u^2 - 5u + 3, \ u = 5x + 6; \ x = 2$

50. $y = 2u^3 - 3u + 1, \ u = x^3 + 8; \ x = 1$

In Exercises 51–56, use the given information to find $h'(2)$: $f(2) = 3$, $f'(2) = -1$, $g(2) = 4$, $g'(2) = 10$, $g(3) = 8$, and $g'(3) = 7$.

51. $h(x) = \left(f(x) + 1\right)^3$

52. $h(x) = \left(\dfrac{f(x)}{g(x)}\right)^2$

53. $h(x) = g\left(f(x)\right)$

54. $h(x) = \sqrt{f(x) + g(x)}$

55. $h(x) = \left(g(x)\right)^3$

56. $h(x) = \left(2 + 3 \cdot f(x)\right)\left(g(x)\right)^3$

Determine the equation of the tangent line for $f(x)$ at the x-value indicated in Exercises 57–60.

57. $f(x) = (3x - 1)^3; \ x = 1$

58. $f(x) = \left(\dfrac{x^2 + 1}{x + 1}\right)^3; \ x = 2$

59. $f(x) = \left(3x^2 + 2x + 8\right)^{\frac{1}{2}}; \ x = 4$

60. $f(x) = \sqrt{10x + 1}; \ x = 8$

APPLICATIONS

61. Marginal revenue: A dealer of microwave ovens estimates that he can sell x ovens per month when the demand function (price) is $p = D(x) = 20\sqrt{280 - 4x}$ dollars.

   a. Find the revenue function $R(x)$.

   b. Find $R(21)$.

   c. Find the marginal revenue function $R'(x)$.

   d. Find $R'(21)$.

62. **Marginal revenue:** The demand function (price) for a particular product is given by $D(x) = 8\sqrt{25 - 5x + 0.25x^2}$ dollars, where x is the number of units (in hundreds) sold.
    a. Find the revenue function $R(x)$.
    b. Find $R(2)$.
    c. Find the marginal revenue function $R'(x)$.
    d. Find $R'(2)$.

63. **Population growth:** It is estimated that t years from now the population of Castle City will be $P(t) = 10(40 + 2t)^2 - 1600t$.
    a. What will the population be in 8 years?
    b. Find the rate of change in population in 8 years.

64. **Air pollution:** It is estimated that t years from now the level of air pollution in Bohrberg will be $P(t) = \dfrac{0.6\sqrt{8t^2 + 11t + 60}}{(t+1)^2}$ parts per million. Find the rate of change in the pollution level in 7 years.

65. **Pollution:** After a sewage spill, the level of pollution in San Remo Bay is estimated by $P(t) = \dfrac{200t^2}{\sqrt{t^2 + 11}}$, where t is the time in days since the spill occurred.
    How fast is the level changing after 5 days? Round to the nearest whole number.

66. **Bacterial growth:** It is estimated that in t hours the population of bacteria in a culture will be $P(t) = \dfrac{8000}{\sqrt{8 - 0.5t}}$.
    a. What will be the population in 8 hours?
    b. Find the rate of change in the population in 8 hours.

67. **Rate of change of cost:** A manufacturer of vacuum cleaners estimates that the total cost of producing x vacuum cleaners is given by $C(x) = -0.5x^2 + 56x + 800$ dollars. Records show that after t hours on a typical day, the number of units produced is given by $x = 5\sqrt{t^2 + 5t}$. Find the rate of change of total cost with respect to time at the end of
    a. 4 hours.
    b. 5 hours.

68. **Rate of change of profit:** A manufacturer has determined that the weekly profit from the sale of x items is given by $P(x) = -x^2 + 280x - 4000$ dollars. It is estimated that after t days in any week, $x = 0.5t^2 + 5t$ items will have been produced. Find the rate of change of profit with respect to time at the end of
    a. 4 days.
    b. 5 days.

69. Pollution: Studies show that the average level of certain pollutants in the air is given by $L = 1 + 0.2x + 0.001x^2$ parts per million when the population is x thousand people. It is estimated that t years from now the population will be $x = \dfrac{200}{\sqrt{7 - 0.5t}}$ in thousands. Find the rate of change of the level of pollutants after

   a. 6 years.

   b. 12 years.

70. Security costs: The annual cost for campus security is given by $C(x) = 3x^2 - 32x + 16$ in thousands of dollars. It is estimated that the enrollment in t years will be $x = 16 + 0.5t + 0.02t^2$ in thousands. Find the rate of change in security costs after

   a. 3 years.

   b. 4 years.

71. Baseball attendance: The average home attendance per week at a Class AA baseball park varied according to the formula $N(t) = (3 + 0.2t)^{\frac{1}{2}}$, where t is the number of weeks into the season ($0 \le t \le 12$) and N is in thousands of persons.

   a. What was the attendance during the first week into the season?

   b. Determine the number $N'(5)$.

   c. Interpret the meaning of $N'(5)$.

72. Weekly attendance: The semester after its student team won an intercollegiate Duplicate Bridge championship, the average weekly attendance at the University Union Building varied according to the formula $B(t) = 100 - 50\left(1 - \dfrac{t}{16}\right)^{\frac{3}{2}}$, where t is the number of weeks after the championship ($0 \le t \le 16$) and B is the number of persons.

   a. What is $B(0)$ and what does it represent?

   b. Determine $B'(t)$.

   c. What is $B'(7)$? Interpret the meaning of this number.

73. Class registration: The annual registration in university calculus classes varies according to the formula $P(t) = 1 + \left(1 - \dfrac{t}{30}\right)^{2.5}$, where t is the number of years since 1995 and P is in millions of students.

   a. Determine $P(0)$ and explain its meaning.

   b. Determine $P'(t)$.

   c. Calculate $P'(10)$ and explain its meaning.

74. Travel: The number of passengers traveling from California to Central America and back on cruise ships is given by $C(t) = (10t + 50)^{1.5}$, where t is the number of years since 1990 and C is passenger count in thousands.

   a. When did the number of passengers hit 1,000,000 people ($C = 1000$)?

   b. Compute $C'(t)$.

   c. Calculate $C'(12)$ and intrepret its meaning.

11.3 IMPLICIT DIFFERENTIATION AND RELATED RATES

■ TOPICS

- ■ Implicit Differentiation
- ■ Related Rates

Implicit Differentiation

When a function is expressed in the form $y = f(x)$, we say that y is an **explicit function** of x or that the function is in **explicit form**. For example, the functions

$$y = 2x + 3 \quad \text{and} \quad y = \frac{1}{x+1}$$

are in explicit form.

If we want to find $\dfrac{dy}{dx}$ in the equation $xy + y = 1$, where the variable y is **implied** to be a function of x, we can first represent this function in explicit form by solving for y.

$$xy + y = 1 \qquad \text{Implicit form.}$$
$$y(x+1) = 1 \qquad \text{Factor out } y.$$
$$y = \frac{1}{x+1} \qquad \text{Explicit form (solved for } y\text{).}$$

Now that we have the function in explicit form, we can differentiate as follows.

$$y = (x+1)^{-1} \qquad \text{Rewrite } y \text{ using exponents.}$$
$$\frac{dy}{dx} = -(x+1)^{-2} = -\frac{1}{(x+1)^2} \qquad \text{Differentiate.}$$

However, in the equation

$$x^2 + y^2 = 4,$$

we cannot solve for y explicitly in terms of x. In fact, the equation represents a circle with radius 2 and center at the origin, and does not represent a function. (See Figure 1.)

To find $\dfrac{dy}{dx}$ in the equation $x^2 + y^2 = 4$, we use the method of **implicit differentiation** where y is implied to be a function of x. For example, the upper semicircle $\left(y = \sqrt{4-x^2}\right)$ might be the implied function, or the lower semicircle $\left(y = -\sqrt{4-x^2}\right)$ might be the implied function. However, we do not need to know which function can be implied.

In the procedure of **implicit differentiation**, we use the General Power Rule (or, in more general cases, the Chain Rule). Remember that we are differentiating with respect to x.

For example, suppose that we want to find

$$\frac{d}{dx}\left(y^2\right),$$

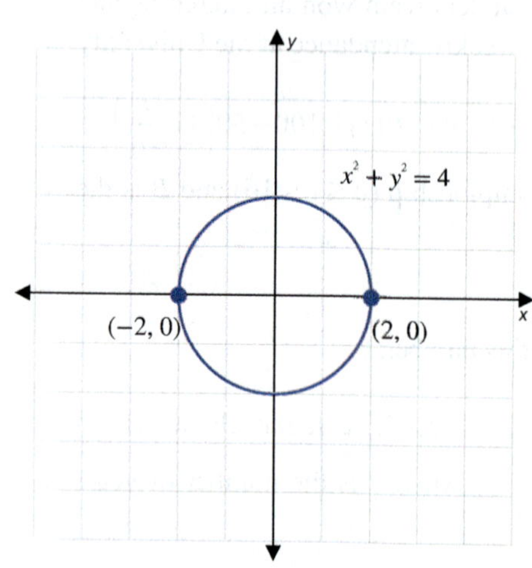

FIGURE 1

■ **NOTE**

There are some cases, such as $x^2 + y^2 = -1$, where there are no real solutions and no function can be implied. However, the process of implicit differentiation may lead to apparently meaningful results. Such cases are not discussed in this course.

where y is implied to be a differentiable function of x. By denoting $y = g(x)$ and $\dfrac{dy}{dx} = g'(x)$, we can write

$$\frac{d}{dx}(y^2) = \frac{d}{dx}\left[g(x)^2\right] \qquad \text{By the General Power Rule}$$

$$= 2 \cdot g(x) \cdot g'(x) \qquad \text{Substitute } y = g(x) \text{ and } \frac{dy}{dx} = g'(x).$$

Or, $\dfrac{d}{dx}(y^2) = (y^2)' = 2y \cdot y'$. (Note that $\dfrac{dy}{dx} = y'$.)

Similarly, by using implicit differentiation, we have

$$\frac{d}{dx}(y^3) = 3y^2 \frac{dy}{dx} \quad \text{and} \quad \frac{d}{dx}(y^4) = 4y^3 \frac{dy}{dx}.$$

Or, $(y^3)' = 3y^2 \cdot y'$ and $(y^4)' = 4y^3 \cdot y'$.

Now, to find $\dfrac{dy}{dx}$ in the equation $x^2 + y^2 = 4$, we differentiate both sides of the equation with respect to x and then solve for $\dfrac{dy}{dx}$.

$$x^2 + y^2 = 4 \qquad \text{Use the Sum and Difference Rule.}$$

$$\frac{d}{dx}(x^2 + y^2) = \frac{d}{dx}(4)$$

$$\frac{d}{dx}(x^2) + \frac{d}{dx}(y^2) = \frac{d}{dx}(4) \qquad \begin{array}{l}\text{Differentiate. Use the Power Rule on}\\ \text{the first term, implicit differentiation on}\\ \text{the second term, and the rule that the}\\ \text{derivative of a constant is 0 on the last}\end{array}$$

$$2x + 2y\frac{dy}{dx} = 0 \qquad \text{term.}$$

$$2y\frac{dy}{dx} = -2x \qquad \text{Subtract } 2x \text{ from both sides.}$$

$$\frac{dy}{dx} = \frac{-\cancel{2}x}{\cancel{2}y} = -\frac{x}{y} \qquad \text{Solve for } \frac{dy}{dx} \text{ and simplify.}$$

<table>
<tr><td>

✐ NOTE

In general, we do not need to know the explicit representation for y in order to obtain $\dfrac{dy}{dx}$.

</td></tr>
</table>

Note that the derivative has both x and y in it. To evaluate such a derivative, we need to know both x and y at a point (x, y) on the curve.

Example 1: Derivative at a Point

Given the equation $x^2 + y^2 = 25$, find $\dfrac{dy}{dx}$ and then evaluate the derivative at $(4, 3)$ and $(0, 5)$.

Solution

$$x^2 + y^2 = 25 \qquad \text{Use the Sum and Difference Rule.}$$

$$\frac{d}{dx}(x^2 + y^2) = \frac{d}{dx}(25) \qquad \begin{array}{l}\text{Differentiate. Use the Power Rule on the first}\\ \text{term, implicit differentiation on the second term,}\end{array}$$

$$\frac{d}{dx}(x^2) + \frac{d}{dx}(y^2) = \frac{d}{dx}(25) \qquad \begin{array}{l}\text{and the rule that the derivative of a constant is 0}\\ \text{on the last term.}\end{array}$$

$$2x + 2y\frac{dy}{dx} = 0$$

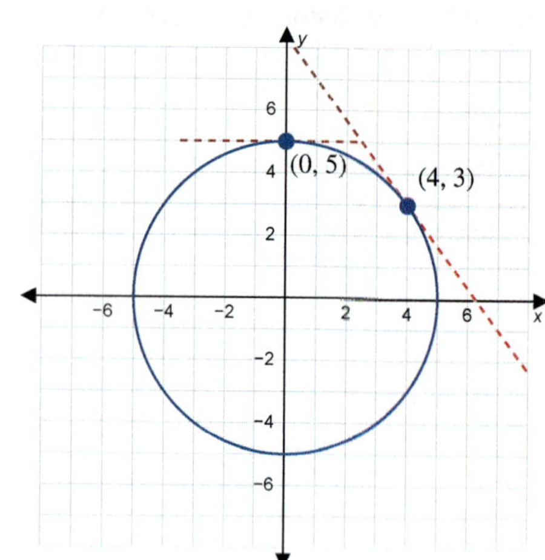

Next, solve for $\dfrac{dy}{dx}$ and simplify.

$$2y\frac{dy}{dx} = -2x$$

$$\frac{dy}{dx} = -\frac{x}{y}$$

Finally, substitute the given x- and y-values of each point to evaluate the derivative at that point.

$$\left.\frac{dy}{dx}\right|_{(4,3)} = -\frac{4}{3} \quad \text{and} \quad \left.\frac{dy}{dx}\right|_{(0,5)} = -\frac{0}{5} = 0$$

Note that $-\dfrac{4}{3}$ is the slope of the tangent line to the circle at $(4, 3)$.

Example 2: Implicit Differentiation

If $xy + x^2y^3 - 1 = 0$, find $\dfrac{dy}{dx}$.

Solution

Note that both xy and x^2y^3 are products and that the Product Rule must be used.

$$xy + x^2y^3 - 1 = 0$$

$$\frac{d}{dx}\left(xy + x^2y^3 - 1\right) = \frac{d}{dx}(0)$$

$$\frac{d}{dx}(xy) + \frac{d}{dx}\left(x^2y^3\right) + \frac{d}{dx}(-1) = \frac{d}{dx}(0)$$

Use the Sum and Difference Rule.

$$\left(x\cdot\frac{dy}{dx} + y\cdot 1\right) + \left(x^2\cdot 3y^2\cdot\frac{dy}{dx} + y^3\cdot 2x\right) + 0 = 0$$

Differentiate. Use the Product Rule in combination with the General Power Rule on the first and second terms and the rule that the derivative of a constant is 0 on the remaining terms.

$$x\frac{dy}{dx} + y + 3x^2y^2\frac{dy}{dx} + 2xy^3 = 0$$

$$x\frac{dy}{dx} + 3x^2y^2\frac{dy}{dx} = -y - 2xy^3$$

$$\left(x + 3x^2y^2\right)\frac{dy}{dx} = -y - 2xy^3$$

$$\frac{dy}{dx} = \frac{-y - 2xy^3}{x + 3x^2y^2} \quad \text{Solve for } \frac{dy}{dx}.$$

Example 3: Implicit Differentiation

Find $\dfrac{dy}{dx}$ if $\sqrt{x} - \sqrt{y} = 16$.

Solution

$$\sqrt{x} - \sqrt{y} = 16$$

$$x^{\frac{1}{2}} - y^{\frac{1}{2}} = 16 \qquad \text{Rewrite using exponents.}$$

$$\frac{d}{dx}\left(x^{\frac{1}{2}} - y^{\frac{1}{2}} \right) = \frac{d}{dx}(16) \qquad \text{Use the Sum and Difference Rule.}$$

$$\frac{d}{dx}\left(x^{\frac{1}{2}} \right) - \frac{d}{dx}\left(y^{\frac{1}{2}} \right) = \frac{d}{dx}(16) \qquad \begin{array}{l}\text{Differentiate using the General Power} \\ \text{Rule and the rule that the derivative of a} \\ \text{constant is 0.}\end{array}$$

$$\frac{1}{2}x^{\frac{-1}{2}} - \frac{1}{2}y^{\frac{-1}{2}} \cdot \frac{dy}{dx} = 0$$

$$-\frac{1}{2}y^{\frac{-1}{2}} \frac{dy}{dx} = -\frac{1}{2}x^{\frac{-1}{2}}$$

$$\frac{dy}{dx} = \frac{x^{\frac{-1}{2}}}{y^{\frac{-1}{2}}} = \frac{y^{\frac{1}{2}}}{x^{\frac{1}{2}}} \qquad \text{Solve for } \frac{dy}{dx}.$$

$$= \frac{\sqrt{y}}{\sqrt{x}} = \sqrt{\frac{y}{x}} \qquad \text{Rewrite using radicals.}$$

We point out that the following notation suggests that x and y are treated differently:

$$(x^n)' = nx^{n-1} \quad \text{but} \quad (y^n)' = ny^{n-1} \cdot y'.$$

When the notation $\dfrac{d}{dx}$ is used for differentiation with respect to x, the rule is seen to be "equal" since $\dfrac{d}{dx}\left(x^n\right) = nx^{n-1} \cdot \dfrac{dx}{dx}$ and $\dfrac{d}{dx}\left(y^n\right) = ny^{n-1} \cdot \dfrac{dy}{dx}$. We actually use, implicitly, the result that $\dfrac{d}{dx}\left(x\right) = \dfrac{dx}{dx} = 1$.

Related Rates

Now we will consider problems in which two or more variables are dependent on time, and we will discuss the use of implicit differentiation to determine their rates of change with respect to time. For example, if x and y are both dependent on time t and

$$y = 5x + 1,$$

then differentiating both sides of the equation with respect to t gives

$$\frac{d}{dt}(y) = \frac{d}{dt}(5x+1)$$

$$\frac{dy}{dt} = 5\frac{dx}{dt}.$$

Thus we see that the rates $\dfrac{dy}{dt}$ and $\dfrac{dx}{dt}$ are **related**, and y is changing 5 times as fast as x.

Example 4: Balloon Inflation

Suppose that a spherical balloon is inflated by a continuous flow of air from an air compressor at a rate of 2 in.³ / sec. How fast is the radius of the balloon growing when the radius is 3 in.?

Solution

We know that $\dfrac{dV}{dt} = 2$ in.³ / sec, and we want to find $\dfrac{dr}{dt}$.

The formula for the volume of a sphere is $V = \dfrac{4}{3}\pi r^3$.

$$V = \frac{4}{3}\pi r^3 \qquad \text{Write an equation.}$$

$$\frac{d}{dt}(V) = \frac{d}{dt}\left(\frac{4}{3}\pi r^3\right) \qquad \text{Differentiate both sides of the equation with respect to } t.$$

$$\frac{dV}{dt} = \frac{4}{\cancel{3}}\pi \cdot \cancel{3}r^2 \cdot \frac{dr}{dt} \qquad \text{Use implicit differentiation.}$$

$$\frac{dV}{dt} = 4\pi r^2 \frac{dr}{dt}$$

Now substitute $\dfrac{dV}{dt} = 2$ and $r = 3$.

$$2 = 4\pi (3)^2 \frac{dr}{dt} \qquad \text{Solve for } \frac{dr}{dt}.$$

$$\frac{\overset{1}{\cancel{2}}}{\underset{18}{\cancel{36}}\pi} = \frac{dr}{dt}$$

Therefore,

$$\frac{dr}{dt} = \frac{1}{18\pi} \text{ in./ sec} \approx 0.018 \text{ in./ sec}.$$

Example 5: Airplane Holding Pattern

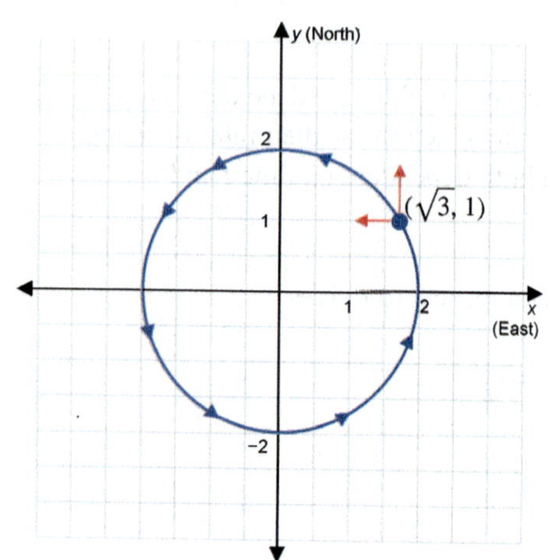

An airplane is flying in a circular "holding" pattern with a radius of 2 miles. The path of the plane can be described by the equation $x^2 + y^2 = 4$. When the plane is at the point $x = \sqrt{3}$ miles and $y = 1$ mile, as shown in the figure, it is traveling north at 300 mph. How fast is it traveling west?

Solution

First, differentiate both sides of the equation $x^2 + y^2 = 4$ with respect to t.

$$\frac{d}{dt}\left(x^2 + y^2\right) = \frac{d}{dt}(4) \qquad \text{Use the Sum and Difference Rule.}$$

$$\frac{d}{dt}\left(x^2\right) + \frac{d}{dt}\left(y^2\right) = \frac{d}{dt}(4) \qquad \text{Apply implicit differentiation.}$$

$$2x\frac{dx}{dt} + 2y\frac{dy}{dt} = 0$$

Now, substituting $x = \sqrt{3}$, $y = 1$, and $\dfrac{dy}{dt} = 300$, we can solve for $\dfrac{dx}{dt}$.

$$2\sqrt{3}\,\frac{dx}{dt} + 2 \cdot 1 \cdot 300 = 0$$

$$\frac{dx}{dt} = -\frac{\overset{300}{\cancel{600}}}{\underset{1}{\cancel{2}}\sqrt{3}} = -\frac{300}{\sqrt{3}} \cdot \frac{\sqrt{3}}{\sqrt{3}} = -100\sqrt{3}\ \text{mph}$$

The plane is traveling west (in the negative x direction) at a rate of $100\sqrt{3}\ \text{mph} \approx 173.2\ \text{mph}$.

Example 6: Moving Cars

Two cars, A and B, are traveling toward the same intersection, and neither driver plans to stop. Car A is 0.5 miles from the intersection moving east at 40 mph, and Car B is 0.75 miles from the intersection moving south at 60 mph. How fast is the distance between the two cars changing?

Solution

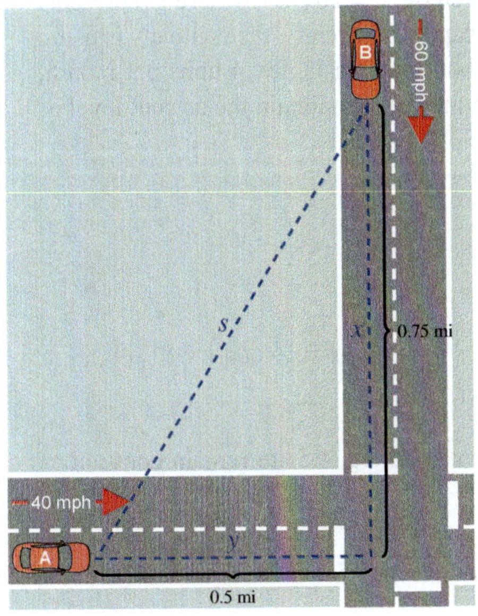

From the figure, $s^2 = x^2 + y^2$.

Differentiating with respect to t, we obtain

$$\frac{d}{dt}\left(s^2\right) = \frac{d}{dt}\left(x^2 + y^2\right)$$

$$2s\,\frac{ds}{dt} = 2x\,\frac{dx}{dt} + 2y\,\frac{dy}{dt}$$

$$\frac{ds}{dt} = \frac{x\,\dfrac{dx}{dt} + y\,\dfrac{dy}{dt}}{s}.$$

Note that as x and y are decreasing, their velocities are negative. Now, substituting $x = 0.75$, $\dfrac{dx}{dt} = -60$, $y = 0.5$, $\dfrac{dy}{dt} = -40$, and

$$s = \sqrt{x^2 + y^2} = \sqrt{(0.75)^2 + (0.5)^2} = \sqrt{0.8125} \approx 0.9,$$

we find

$$\frac{ds}{dt} = \frac{x\,\dfrac{dx}{dt} + y\,\dfrac{dy}{dt}}{s} \approx \frac{0.75(-60) + (0.5)(-40)}{0.9} \approx -72.2\ \text{mph}.$$

The distance between the cars is decreasing at a rate of approximately 72.2 mph.

The costs of a production process can be related to two categories: labor and capital. The cost of labor is essentially the payroll, and the cost of capital is the cost of the physical items such as buildings and machines used in the production process. One relationship frequently used by economists, involving units produced, labor, and capital, is the **Cobb-Douglas Production Formula**.

$$P = Cx^a y^{1-a}$$

where

P = number of units produced,
x = units of labor, and
y = units of capital.
C and a are constants, $0 < a < 1$.

Example 7: Production

Suppose that a firm's level of production is given by $P = 20x^{\frac{1}{4}}y^{\frac{3}{4}}$, where x represents the units of labor and y represents the units of capital. Currently, the company is using 16 units of labor and 81 units of capital. If labor is increasing by 4 units per month, what must the change in units of capital per month be to maintain the current level of production?

Solution

The current level of production is

$$P = 20(16)^{\frac{1}{4}}(81)^{\frac{3}{4}} \qquad\qquad \text{Use } x = 16 \text{ and } y = 81.$$
$$= 20(2)(27) = 1080 \text{ units.}$$

Since the current level of production is to be maintained, P is to remain constant at 1080. Thus we have

$$1080 = 20x^{\frac{1}{4}}y^{\frac{3}{4}}, \text{ which gives } 54 = x^{\frac{1}{4}}y^{\frac{3}{4}}.$$

We want to find $\dfrac{dy}{dt}$. Use implicit differentiation and differentiate both sides of the simplified equation with respect to t.

$$\frac{d}{dt}(54) = x^{\frac{1}{4}}\cdot\frac{d}{dt}\left(y^{\frac{3}{4}}\right) + y^{\frac{3}{4}}\cdot\frac{d}{dt}\left(x^{\frac{1}{4}}\right) \qquad \text{By the Product Rule}$$

Thus

$$0 = x^{\frac{1}{4}}\cdot\frac{d}{dt}\left(y^{\frac{3}{4}}\right) + y^{\frac{3}{4}}\cdot\frac{d}{dt}\left(x^{\frac{1}{4}}\right).$$

The Chain Rule gives $\dfrac{d}{dt}\left(x^{\frac{1}{4}}\right) = \dfrac{1}{4}x^{-\frac{3}{4}}\cdot\dfrac{dx}{dt}$, and $\dfrac{d}{dt}\left(y^{\frac{3}{4}}\right) = \dfrac{3}{4}y^{-\frac{1}{4}}\cdot\dfrac{dy}{dt}$; so

$$0 = x^{\frac{1}{4}}\cdot\left(\frac{3}{4}y^{-\frac{1}{4}}\frac{dy}{dt}\right) + y^{\frac{3}{4}}\cdot\left(\frac{1}{4}x^{-\frac{3}{4}}\frac{dx}{dt}\right).$$

We are given that $x = 16$, $y = 81$, and $\dfrac{dx}{dt} = 4$. After substituting in these values, we can solve for $\dfrac{dy}{dt}$.

$$0 = (16)^{\frac{1}{4}} \cdot \frac{3}{4}(81)^{-\frac{1}{4}} \frac{dy}{dt} + (81)^{\frac{3}{4}} \cdot \frac{1}{4}(16)^{-\frac{3}{4}}(4)$$

$$0 = \cancel{2} \cdot \frac{\cancel{3}}{_2\cancel{4}} \cdot \frac{1}{\cancel{3}} \cdot \frac{dy}{dt} + 27 \cdot \frac{1}{\cancel{4}} \cdot \frac{1}{8} \cdot \cancel{4}$$

$$0 = \frac{1}{2} \cdot \frac{dy}{dt} + \frac{27}{8}$$

$$-\left(\frac{1}{2} \cdot \frac{dy}{dt}\right) = \frac{27}{8}$$

$$\frac{dy}{dt} = \left(-\cancel{2}\right)\frac{27}{\cancel{8}_4}$$

$$\frac{dy}{dt} = -\frac{27}{4} = -6.75$$

Because the derivative is negative, the capital must **decrease** at a rate of $\dfrac{27}{4} = 6.75$ units per month to keep the current level of production.

11.3 EXERCISES

⚲ PRACTICE

Use implicit differentiation to find $\dfrac{dy}{dx}$ for each of the equations in Exercises 1–20.

1. $2x^2 + y^2 = 4$

2. $x^3 + y^3 = 5$

3. $2x^3 + y^3 = 8$

4. $x^2 - y^2 = 16$

5. $\sqrt{x} + \sqrt{y} = 1$

6. $x - \sqrt{y} = 2$

7. $x^2 y = 2$

8. $xy^2 = -1$

9. $x^2 + xy + y^2 = -1$

10. $x^3 + 2xy - y^2 = 3$

11. $4x^2 + 3xy + y^2 = 2x$

12. $x^3 + y^3 = 3xy$

13. $\dfrac{1}{x} + \dfrac{x}{y} = 2x$

14. $x^2 + \dfrac{2x}{y} = \dfrac{1}{x^2}$

15. $x^2 + \sqrt{xy} = 2y^2$

16. $2y + \sqrt{xy} = 5x^2$

17. $x^2 y^2 + xy^3 = x^4$

18. $x^3 y + xy^3 = 3x^3$

19. $x^2 + (y-2)^2 = 16$

20. $x^2 + 4(y+3)^2 = 9$

In Exercises 21–30, use implicit differentiation to find $\dfrac{dy}{dx}$ for the given equations; then find the slope of the tangent line at the given point.

21. $4x^2 - 8y^3 = 24;\ (2, -1)$

22. $3x^3 + 5y^2 + x = 1;\ (-1, 1)$

23. $x^2y - y^2 + 4x + 8 = 0;\ (1, 4)$

24. $5x^2 + xy^2 + 2x = 8;\ (-2, 2)$

25. $x^3 + 2xy - y^2 = 0;\ (3, -3)$

26. $4x^2 - 3xy + y^2 = 7;\ (2, 3)$

27. $\dfrac{1}{x} + \dfrac{x}{y^2} = 2x;\ (1, -1)$

28. $3x - \dfrac{2x}{y^2} + x^2 = 12;\ (3, 1)$

29. $2y + \sqrt{xy} = 5x^2;\ (2, 8)$

30. $x^2 - \sqrt{xy} = 2y^2 + 3x;\ (4, 1)$

In Exercises 31–40, x and y are functions of a third variable, t. Use implicit differentiation to find an expression for $\dfrac{dy}{dt}$.

31. $x^2 - 4y^2 = 16$

32. $3x^2 + y^4 = 4$

33. $x^3 + 5y^2 = 2x$

34. $6x^2 + 5x = 2y^2$

35. $\sqrt{xy} = 4$

36. $\sqrt{x} + \sqrt{y} = 3$

37. $x^2 + xy + y^2 = 3$

38. $x + 2xy - y^2 = 6$

39. $2xy + x^2 = y^3$

40. $x^2y^2 - y^3 + 4x = 0$

🚀 APPLICATIONS

41. Retail sales: The manager of an audio electronics store has determined that the number of stereo receivers and the number of speaker systems sold weekly are related by the equation $0.9y^2 = 10x + xy$, where x is the number of receivers and y is the number of speaker systems. Find $\dfrac{dy}{dx}$ if $x = 12$ and $y = 20$, and interpret your answer.

42. Retail sales: The number of pairs of trousers x and the number of shirts y sold at a department store are related by the equation $36x = 11y + 0.01x^2y$. Find $\dfrac{dy}{dx}$ when $x = 10$ and $y = 30$, and interpret your answer.

43. Cobb-Douglas production: The level of production of a company is given by $P = 30x^{\frac{1}{3}}y^{\frac{2}{3}}$ units monthly, where x is the units of labor and y is the units of capital. The company is currently utilizing 64 units of labor and 27 units of capital. If labor is increased by 2 units per week, what will be the change in units of capital per week to maintain the current level of production?

44. Cobb-Douglas production: The level of production of a company is given by $P = 18x^{0.3}y^{0.7}$ units monthly, where x is the units of labor and y is the units of capital. The company is currently utilizing 35 units of labor and 24 units of capital. If capital is increased by 3 units per week, what will be the change in units of labor per week to maintain the current level of production?

45. Rate of increase in cost: The cost of producing x units of a product is given by the function $C(x) = 0.02x^3 - x^2 + 8x + 200$. The factory is currently producing 60 units per week but plans to increase production at a rate of 3 units per week. What will be the rate of increase in the total cost?

46. Rate of decrease in cost: The cost of producing x units of a commodity is given by the function $C(x) = x^2 - 2x^{\frac{3}{2}} + 7x + 180$. Currently, the production level is 36 units per day. The company plans to decrease production at a rate of 2 units per day. What will be the rate of decrease in the total cost?

47. Sliding ladder: A 17 ft ladder is leaning against a wall. The bottom of the ladder is pulled away from the wall at a rate of 3 ft/sec. How fast is the top of the ladder moving down the wall when the top is 8 feet above the ground?

48. Velocity: Marijean is standing on a boat dock pulling in her boat by means of a rope attached to a boat at water level. Her hands are 6 feet above the water and she is pulling in the rope at a rate of 1.5 ft/sec. How fast is the boat approaching the dock if there are 10 feet of rope still out?

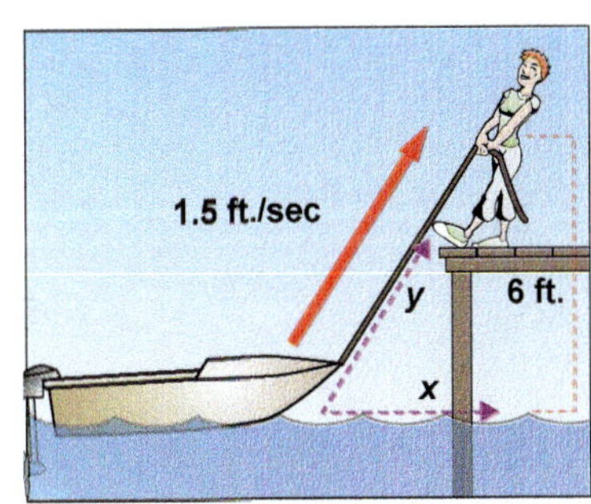

49. Driving: A car traveling south at 30 ft/sec crosses an intersection. When the car is 90 feet past the intersection, a bicyclist crosses the intersection traveling east at a rate of 20 ft/sec. How fast is the distance between the car and the bicycle increasing 5 seconds after the bicycle crosses the intersection?

50. Distance to an airplane: An airplane traveling at a height of 3000 feet crosses directly over an observer. The speed of the plane is 400 ft/sec. How fast is the distance between the observer and the plane changing after 10 seconds?

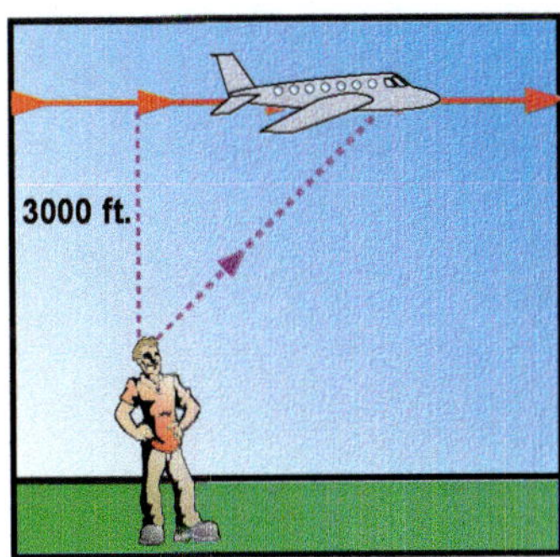

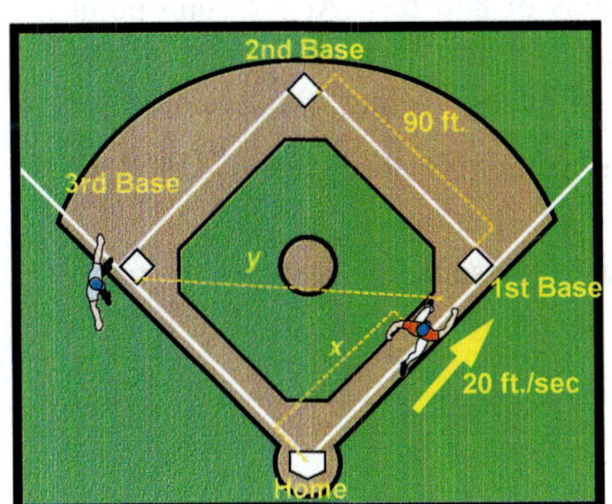

51. Baseball: A base runner heads towards first base with a speed of 20 ft/sec. A baseball diamond is a square, 90 feet on each side.
   a. How long will it take for the runner to reach first base?
   b. What is the runner's rate of change of distance from the umpire standing on third base (we will call this $\dfrac{dy}{dt}$)? (**Hint:** Use the diagram to get a relation between x and y.)
   c. What is the runner's speed when he arrives at first base?

52. Sailing: A Coast Guard radar monitoring station on shore observed a sailboat on its radar grid. It was determined that the boat's east-west distance along the shoreline changed by the formula $x = 12 + 0.1t$ and its seaward distance (north-south) changed by the formula $y = 20 - 0.3t$. Here, x and y are in miles and t is in minutes.

  a. What was the sailboat's position at $t = 0$ minutes? Sketch this situation.

  b. Determine $\dfrac{dx}{dt}$ and $\dfrac{dy}{dt}$.

  c. What is $\dfrac{dy}{dx}$ at $t = 10$? Interpret this number.

  d. Where and when will the sailboat hit the shore (assuming it keeps its present heading)?

53. Bags of oranges: Suppose the wholesale price in Miami of bags of oranges satisfies a demand equation of $xp + 20p = 1040$, where x is the number of bags supplied and p is the demand (unit price) in dollars.

  a. If 500 bags are available today, what is the unit price?

  b. If the supply is increasing at the rate of 100 bags per day, at what rate is the price changing?

54. Dripping water: The radius of a pan of water is 7 cm and water from a tap drips in at the rate of 10 cubic centimeters per minute.

  a. Determine $\dfrac{dh}{dt}$, where h is the height of water in the pan.

  b. How long will it take to fill the pan to a height of 7 cm?

55. Demand: The Arrow Marketing Group sells teddy bears with college logos to college and university gift shops. Their demand equation is $px - 2800 = x$, where x is the quantity of bears and p is the price in dollars that each sells for.

  a. How many bears can be sold (or "demanded") at \$4.50 apiece?

  b. Determine a formula $\dfrac{dx}{dp}$ and evaluate for p and x as in part a.

56. Electricity costs: Electricity costs per semester at Mount State University are calculated by the formula $C = 0.05x + 0.03y - 0.08xy$, where x is a measure of student size and activity, y is dependent on the usage of various buildings and their efficiencies, and C is in millions of dollars.

  a. Determine C if $x = 7$ and $y = 9$.

  b. Determine a general formula for $\dfrac{dy}{dx}$ and evaluate for $x = 7$ and $y = 9$.

57. Race track: A circular race track has a radius of 840 feet. At a certain point in time, t_0, an observer in a maintenance pit 504 feet from the center of the track clocks a car (when it is directly north of him) traveling counterclockwise along the track at a speed of 50 miles per hour from his right to left $\left(\dfrac{dx}{dt}\right)$. (Use 1 mile = 5280 feet.)

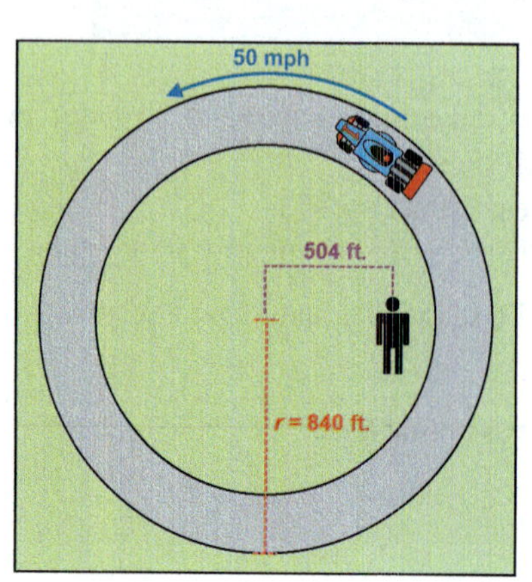

  a. Determine the location of the car on the track. Assume the center of the track is at the origin.

  b. Determine a general equation for $\dfrac{dy}{dx}$.

  c. What is $\left.\dfrac{dy}{dx}\right|_{t=t_0}$? Interpret its meaning.

  d. What is $\dfrac{dy}{dt}$ at $t = t_0$?

58. Chlorination costs: Chlorination costs for the swimming pool at a spa are given by $C = 0.27x + 2y - 0.001xy^2$, where x is the number of weekly swimmers, y is the number of special functions, and C is in dollars.

a. Determine C if $x = 500$ and $y = 4$.

b. Determine a formula for $\dfrac{dy}{dx}$ and evaluate at $x = 500$ and $y = 4$.

59. Probability measurement: The standard deviation, S, of a binomial random variable is given by $S = \sqrt{np(1-p)}$, where n is the number of trials or repetitions of an experiment and p is the probability of success on one outcome. S is a measure of how the data tends to vary from the center (or mean).

a. For $n = 768$ and $p = 0.25$, determine S.

b. Suppose we consider S as a fixed quantity. Determine $\dfrac{dp}{dn}$ for n and p as in part a.

11.4 INCREASING AND DECREASING INTERVALS

■ TOPICS

■ Increasing and Decreasing Intervals of a Function

Increasing and Decreasing Intervals of a Function

Suppose $y = f(x)$ is a function defined on the interval (a, b). We are interested in what happens to the y-values (functional values) as the values of x increase (i.e., move from left to right on the x-axis). Four basic cases are illustrated in Figure 1.

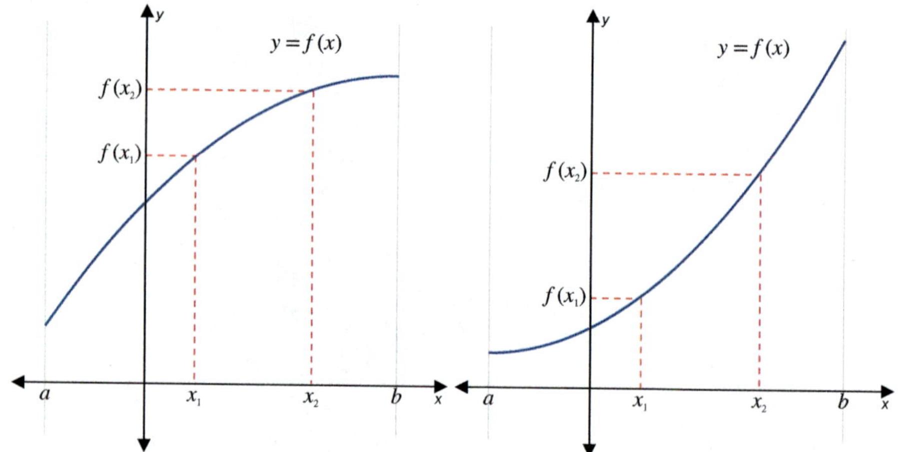

For $x_1 < x_2$, $f(x_1) < f(x_2)$. The function $y = f(x)$ is **increasing**.

(A)

For $x_1 < x_2$, $f(x_1) < f(x_2)$. The function $y = f(x)$ is **increasing**.

(B)

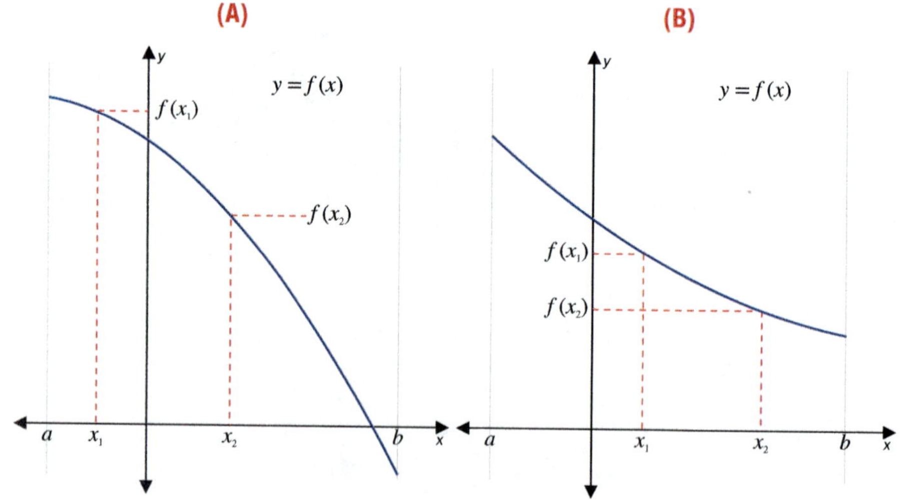

For $x_1 < x_2$, $f(x_1) > f(x_2)$. The function $y = f(x)$ is **decreasing**.

(C)

For $x_1 < x_2$, $f(x_1) > f(x_2)$. The function $y = f(x)$ is **decreasing**.

(D)

FIGURE 1

Increasing and Decreasing Intervals of a Function

Suppose that a function f is defined on the interval (a, b).

1. If $x_1 < x_2$ implies $f(x_1) < f(x_2)$ for every x_1 and x_2 in (a, b), then $f(x)$ is increasing on (a, b).

2. If $x_1 < x_2$ implies $f(x_1) > f(x_2)$ for every x_1 and x_2 in (a, b), then $f(x)$ is decreasing on (a, b).

Example 1: Increasing and Decreasing Intervals

The graph of $y = f(x)$ is given. Find the intervals on which **a.** f is increasing, and **b.** f is decreasing.

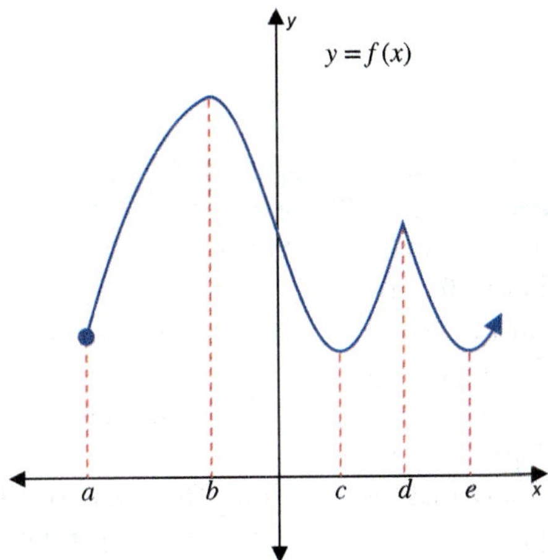

Solution

a. The function f is increasing on the intervals (a, b), (c, d), and $(e, +\infty)$.

b. The function f is decreasing on the intervals (b, c) and (d, e).

The derivative can be used to determine whether a function is increasing or decreasing on an interval. In Figure 2 we illustrate how the slope of a tangent line (the value of the derivative) at a point can indicate whether the graph of the function is rising or falling at that point. Essentially, we are relating three ideas: the derivative of a function, the slopes of lines tangent to the graph of the function, and the graph of the function.

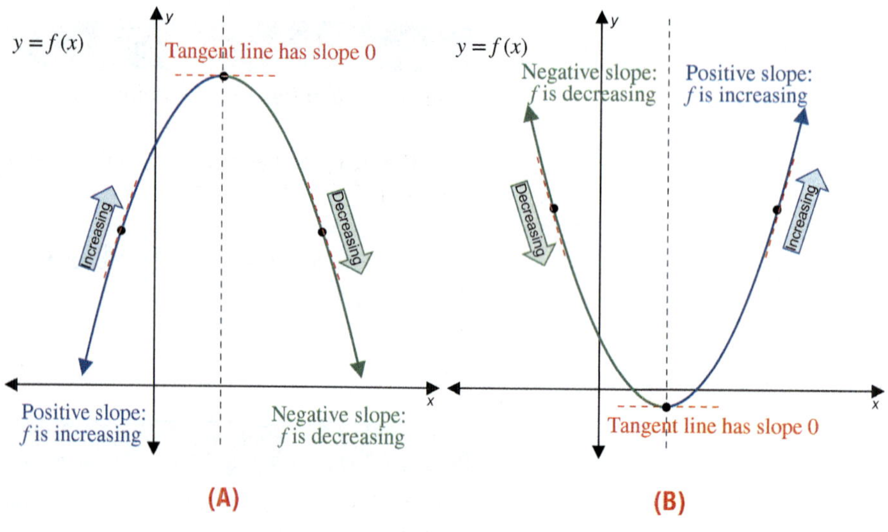

FIGURE 2

The following theorem states these ideas more formally.

<table>
<tr><td>

NOTE

Positive values for the derivative indicate positive slopes for the tangent lines which in turn imply that the function is increasing. Negative values for the derivative indicate negative slopes for the tangent lines which imply that the function is decreasing.

</td><td>

Increasing and Decreasing Intervals of a Differentiable Function

Suppose that $f(x)$ is differentiable on the interval (a, b).

If $f'(x)$ is positive ($f'(x) > 0$) for all x in (a, b), then f is increasing on (a, b).

If $f'(x)$ is negative ($f'(x) < 0$) for all x in (a, b), then f is decreasing on (a, b).

Note: The values of x for which a function is increasing (or decreasing) are always presented as open intervals. These intervals are part (or all) of the domain of the function.

</td></tr>
</table>

The following examples illustrate how the derivative can be used to determine the intervals on which the function is increasing or decreasing and how this information can be applied to sketching the graph.

Example 2: Graphing a Function

Consider the function $f(x) = x^3 - 12x + 1$.

a. Find all values of x that correspond to horizontal tangent lines. These values occur where the derivative is 0.

b. Use the values from part **a.** to find the open intervals on which the function is increasing and the open intervals on which the function is decreasing.

c. Sketch the graph of the function.

Solution

a. $f'(x) = 3x^2 - 12$ Find the derivative of $f(x)$.

We set $f'(x) = 0$ and solve for x.

$$3x^2 - 12 = 0$$
$$3\left(x^2 - 4\right) = 0$$
$$3(x+2)(x-2) = 0$$
$$x = -2, 2$$

Horizontal tangent lines occur at $x = -2$ and $x = 2$.

b. We can use a number line to help determine the intervals on which $f'(x) > 0$ (and so $f(x)$ is increasing) and those on which $f'(x) < 0$ (and so $f(x)$ is decreasing).

Mark the values for x where $f'(x) = 0$ on a number line. In this case, they are $x = 2$ and $x = -2$ (this was determined in part **a.**). Then select any one test point in each interval and find the sign of $f'(x)$ at these test points by substituting the value into $f'(x)$.

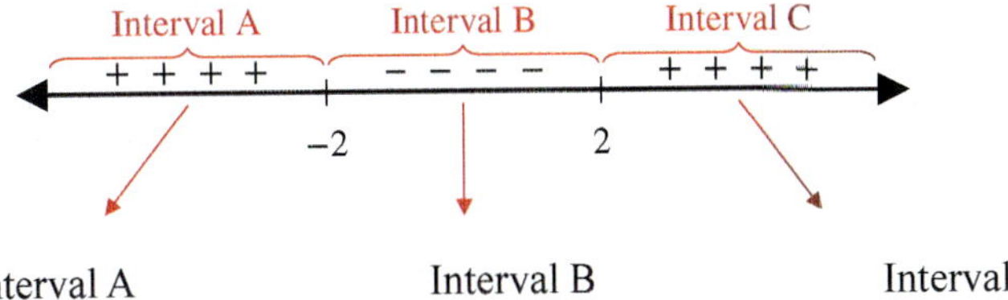

Interval A	Interval B	Interval C

We will use the test point $x = -5$.

We will use the test point $x = 0$.

We will use the test point $x = 3$.

$$f'(-5) = 3(-5)^2 - 12$$
$$= 75 - 12$$
$$= 63$$
$$63 > 0$$

$$f'(0) = 3(0)^2 - 12$$
$$= 0 - 12$$
$$= -12$$
$$-12 < 0$$

$$f'(3) = 3(3)^2 - 12$$
$$= 27 - 12$$
$$= 15$$
$$15 > 0$$

Since $f'(-5)$ is positive, mark interval A with + signs. Since $f'(0)$ is negative, mark interval B with – signs. Since $f'(3)$ is positive, mark interval C with + signs. Note that positive values for the derivative indicate that the function is increasing, and negative values for the derivative indicate that the function is decreasing. Thus, f is increasing on the intervals $(-\infty, -2)$ and $(2, +\infty)$, and f is decreasing on the interval $(-2, 2)$.

c. The goal for the student is to use the information from parts **a.** and **b.**, to sketch the curve without using a calculator. A secondary goal is to sketch the graph using as few (x, y)-coordinates as possible. We do need the important y-values corresponding to the maximum and to the minimum points (which have the horizontal tangents). We have $f(-2) = 17$ and $f(2) = -15$. We recommend in general that the student note the value of the y-intercept, $f(0)$. Here $f(0) = 1$. With only these three coordinates, a smooth curve can be drawn (on one's paper).

The analysis shows that f increases left-to-right until $x = -2$. At this value f reaches a local maximum of 17. Then f declines until, for $x = 2$, the value -15 is reached. Then f increases without bound.

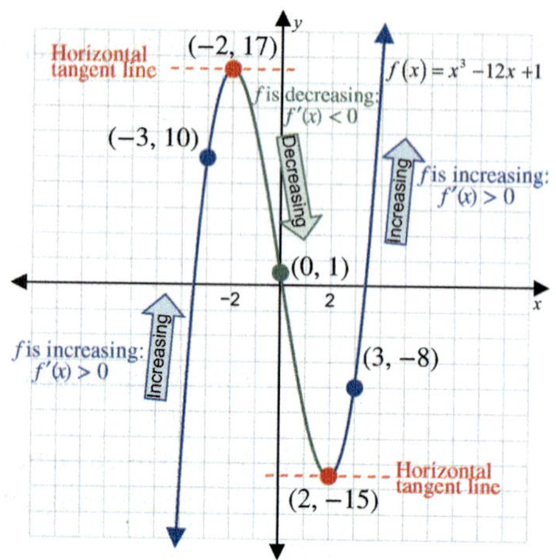

Consider the function $y = x^3 + 1$.

a. Find all values of x where the derivative is 0.

b. Determine where the function is increasing and where it is decreasing.

c. Sketch its graph.

Solution

a. $\dfrac{dy}{dx} = 3x^2$ Find the derivative of y.

Set $\dfrac{dy}{dx} = 0$ and solve for x.

$$3x^2 = 0$$

$$x = 0$$

A horizontal tangent line occurs at $x = 0$.

b. Mark the values for x where $\dfrac{dy}{dx} = 0$ on a number line. In this case, there is only one such point: $x = 0$ (this was determined in part **a.**). Select a test point in each of the two intervals and find the sign of $\dfrac{dy}{dx}$ at these test points by substituting the value into $\dfrac{dy}{dx}$.

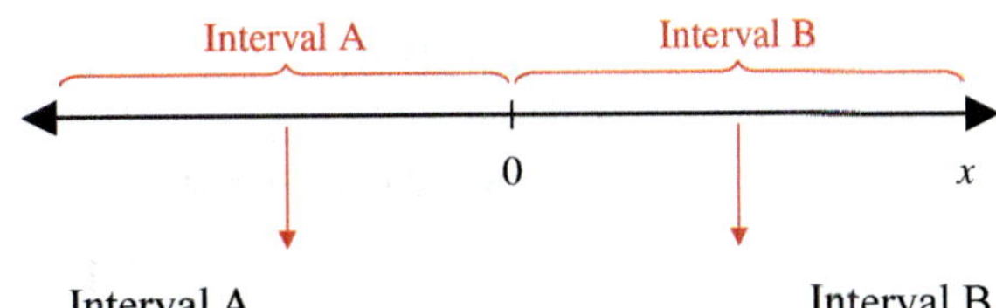

Interval A

We will use the test point $x = -1$.

$$\left.\frac{dy}{dx}\right|_{x=-1} = 3(-1)^2$$

$$= 3$$

$$3 > 0$$

Interval B

We will use the test point $x = 1$.

$$\left.\frac{dy}{dx}\right|_{x=1} = 3(1)^2$$

$$= 3$$

$$3 > 0$$

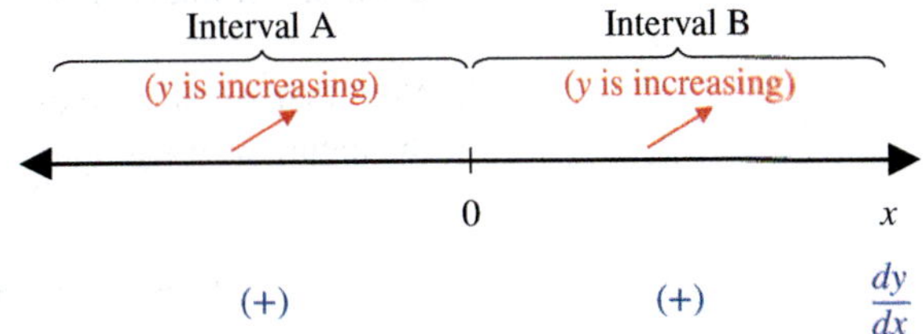

c. Since $\dfrac{dy}{dx} = 3x^2 \geq 0$ for all x, y is increasing on $(-\infty, +\infty)$. Part **a.** determined a horizontal tangent at $x = 0$, and part **b.** showed that y is increasing on the interval $(-\infty, +\infty)$. (**Note:** Since y is increasing on either side of 0, it is also increasing on any interval that contains 0.)

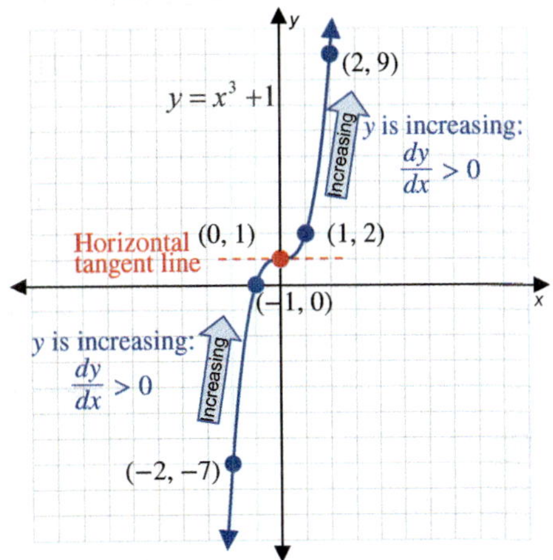

Example 4: Graphing a Function

Consider the function $f(x) = \dfrac{1}{x}$.

a. Find all values of x where the derivative is 0.

b. Determine where the function is increasing and where it is decreasing.

c. Sketch its graph.

Solution

a. $f(x) = \dfrac{1}{x} = x^{-1}$ Rewrite $f(x)$ using exponents.

$f'(x) = -x^{-2} = -\dfrac{1}{x^2} \neq 0$ Find the derivative of $f(x)$.

There are no horizontal tangent lines. There are no x-values which will make the slope equal to 0 in the formula for f'. However, we note there is no slope defined for $x = 0$. It turns out that we need to put on our slope analysis line not only x-values which make $f'(x) = 0$ but also any x-values for which f' is not defined. In this problem, $x = 0$ is the only point in either category, and we put $x = 0$ as the only point on the analysis line.

b. Note that we still investigate the derivative on either side of 0 to help determine the nature of the graph.

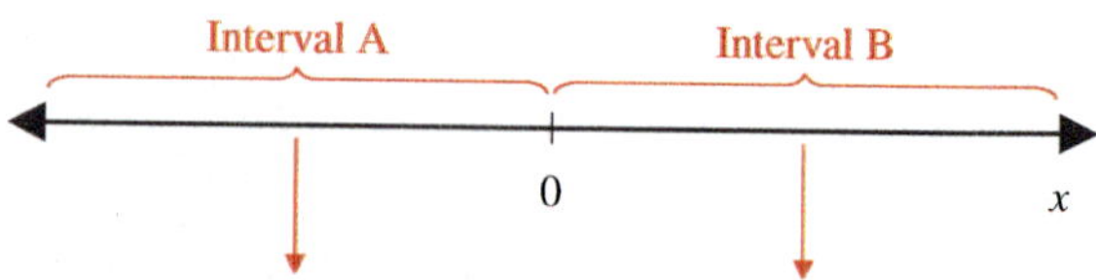

Interval A

We will use the test point $x = -1$.

$$f'(-1) = -\dfrac{1}{(-1)^2}$$

$$= -1$$

$$-1 < 0$$

Interval B

We will use the test point $x = 1$.

$$f'(1) = -\dfrac{1}{(1)^2}$$

$$= -1$$

$$-1 < 0$$

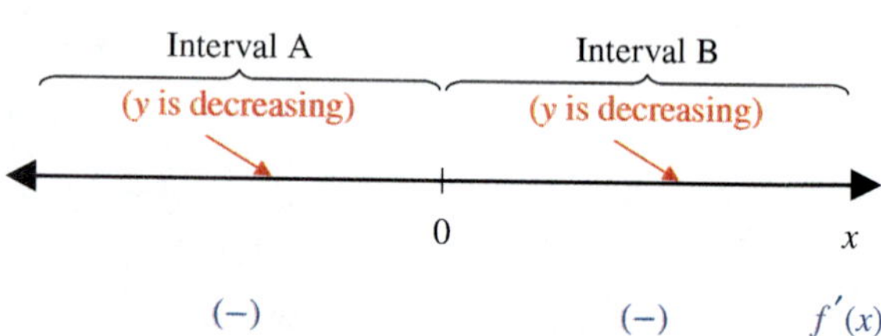

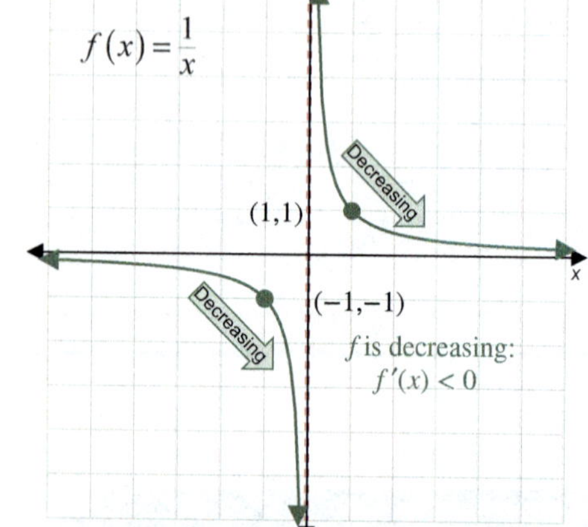

Since $f'(x) = -\dfrac{1}{x^2} < 0$ for all $x \neq 0$, the function f is decreasing on the intervals $(-\infty, 0)$ and $(0, +\infty)$.

c. The line $x = 0$ is a *vertical asymptote*. The curve approaches this line without touching it. There is also a horizontal asymptote at $y = 0$.

In Example 4, $f(0)$ is not defined and so $f'(0)$ is not defined. It can also happen that $f(a)$ is defined, but $f'(a)$ is not defined.

Example 5: Graphing $f'(x)$

The graph for $y = f(x)$ is given below. Identify the portions which show where $f'(x)$ is negative, positive, or zero. Draw a possible graph of $f'(x)$ by estimating the absolute values of the slopes on $f(x)$.

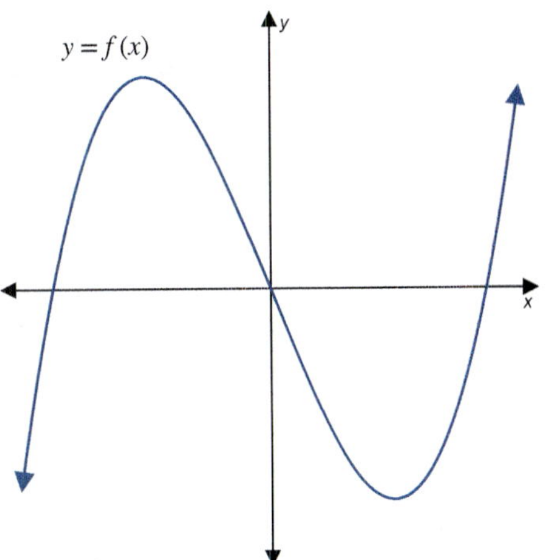

Solution

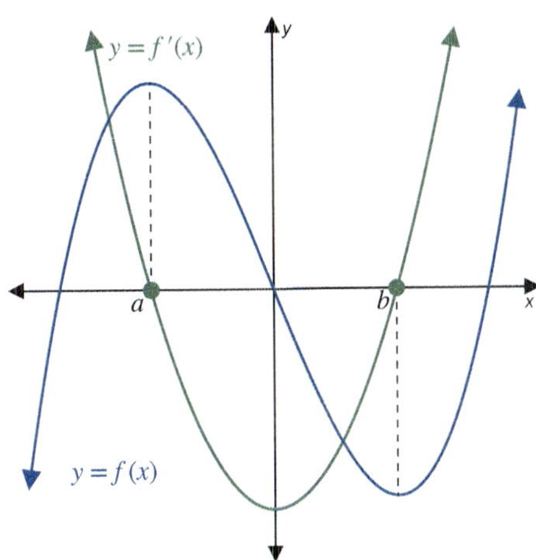

Let $x = a$ and $x = b$ (say $a < b$) denote the two x-values corresponding to the two local extremes. Then $f'(a) = f'(b) = 0$. In between a and b, the y-values decrease on the interval (a, b); therefore y' is negative. At $x = b$, the y-values start to increase (and y' becomes positive). Thus the graph of y' includes points $(a, 0)$ and $(b, 0)$ and lies below the x-axis (as y' is negative) from $x = a$ to $x = b$. For $x > b$ and $x < a$, the graph of y' lies above the x-axis. We draw a smooth curve from $(a, 0)$ to $(b, 0)$ and extend it in both directions above the x-axis. The low point for y' seems to occur near $x = 0$ because the graph of y is steepest near $(0, 0)$.

11.4 EXERCISES

💡 PRACTICE

In Exercises 1–10, find the open intervals on which **a.** f is increasing, and **b.** f is decreasing.

1.

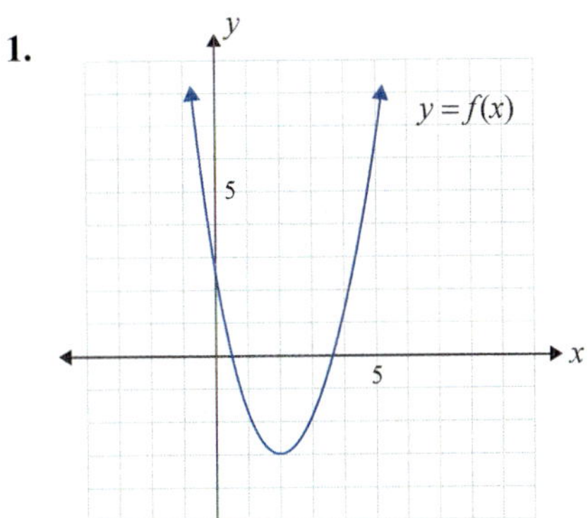

2.

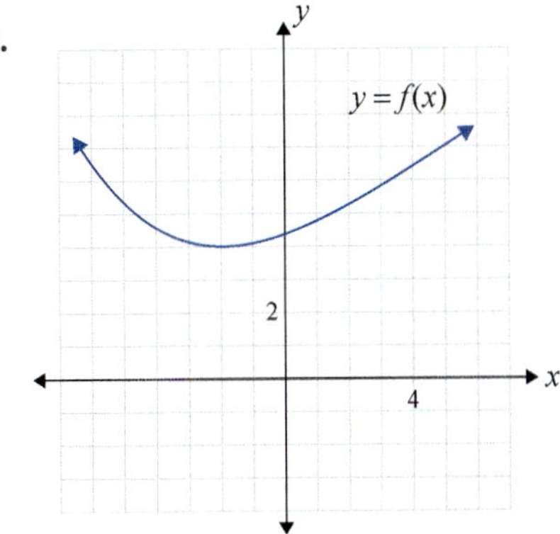

3.

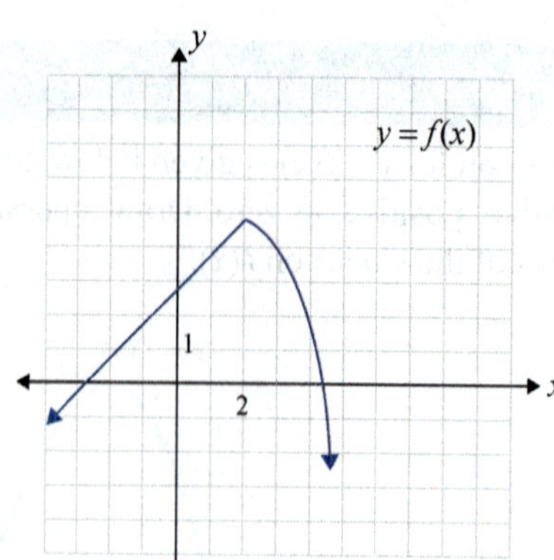

y
x
y = f(x)
1
2

y
x
y = f(x)
5
5

y
x
y = f(x)
1
5

y
x
y = f(x)
5
4

y
x
y = f(x)
y = f(x)
1
5

y
x
y = f(x)
5
5

y
x
y = f(x)
4
4

y
x
y = f(x)
2
2

4.

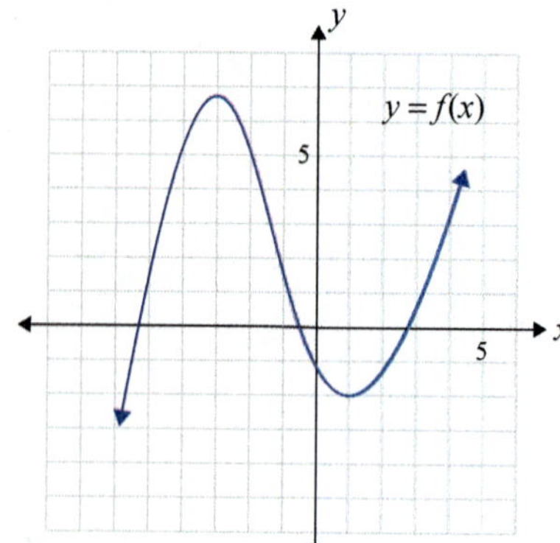

5.

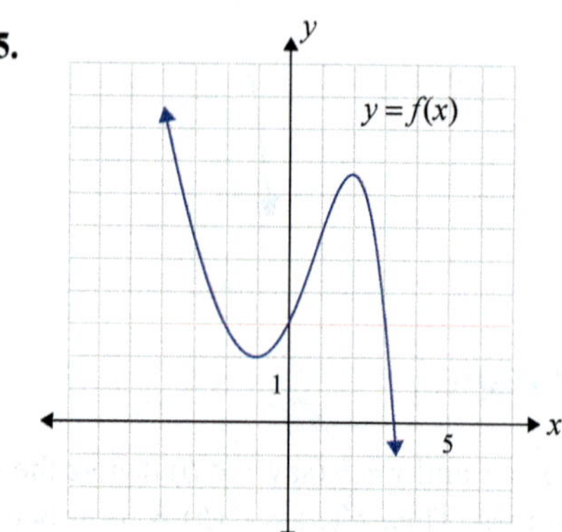

6.

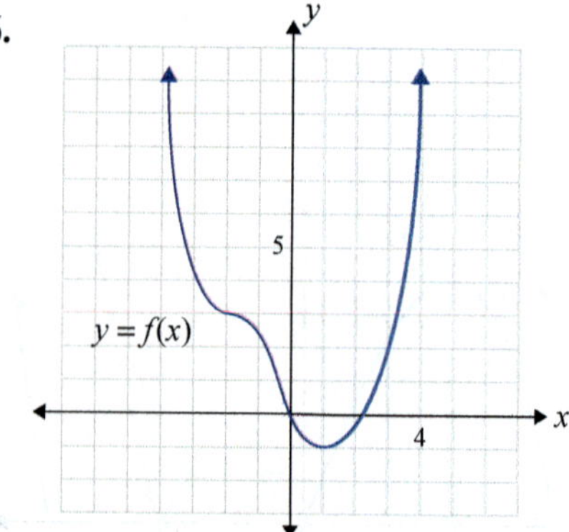

7.

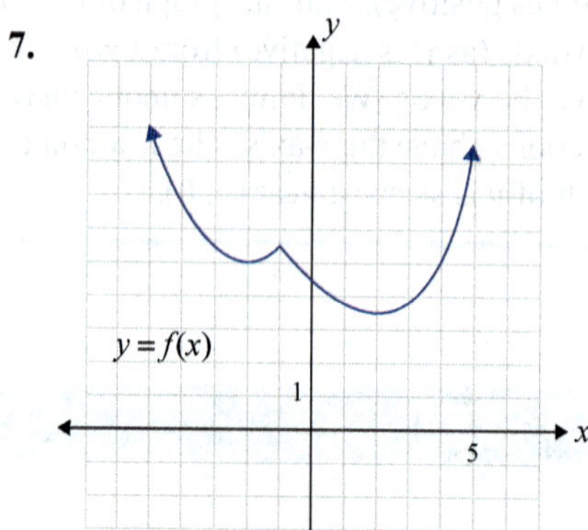

8.

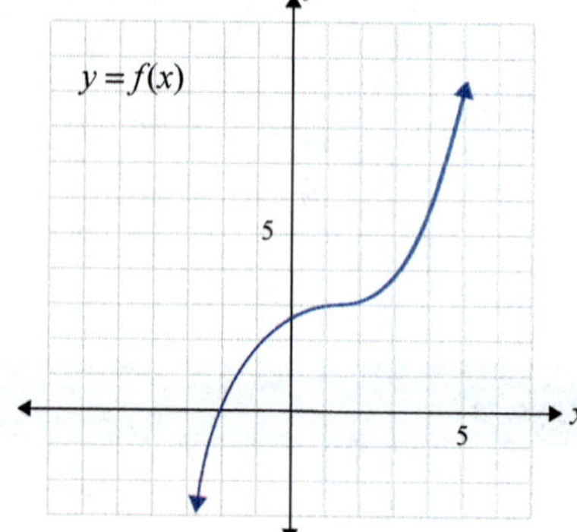

9.

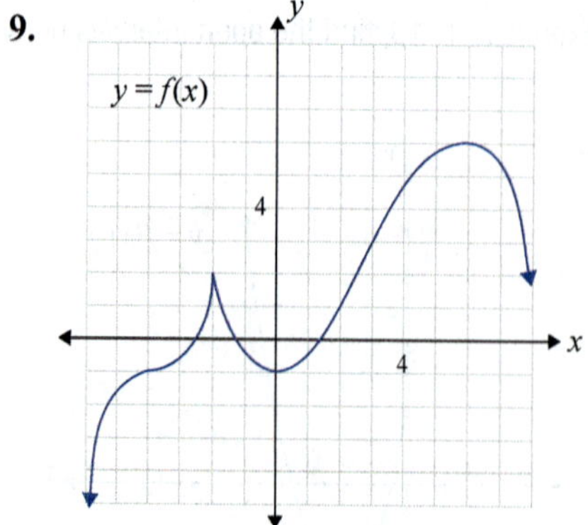

10.

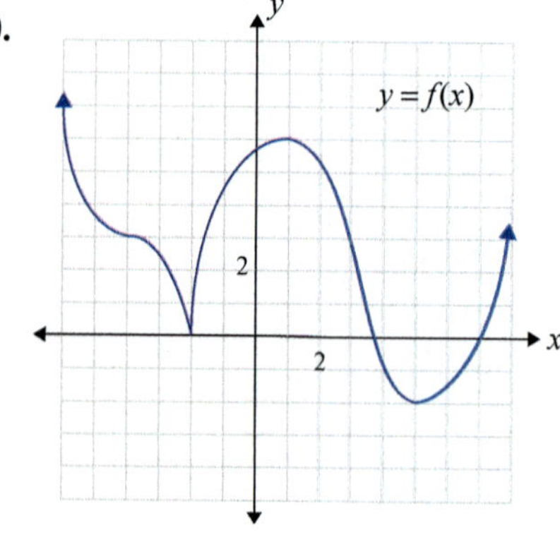

For each of the graphs of $f(x)$ in Exercises 11–13, **a.** determine the open intervals on which the function is increasing and the open intervals on which it is decreasing, and **b.** sketch a possible graph of $f'(x)$.

11.

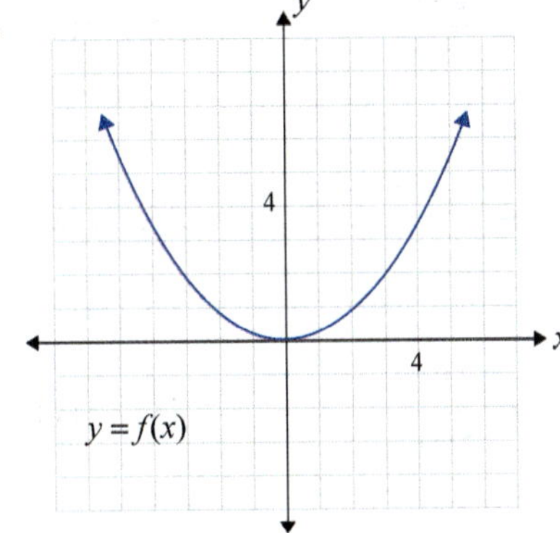

12.

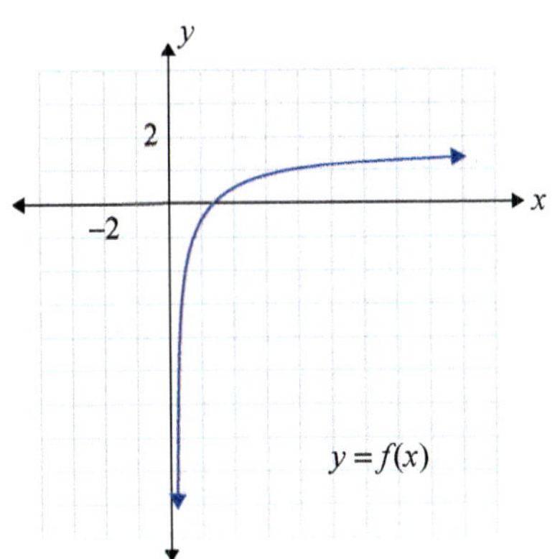

13.

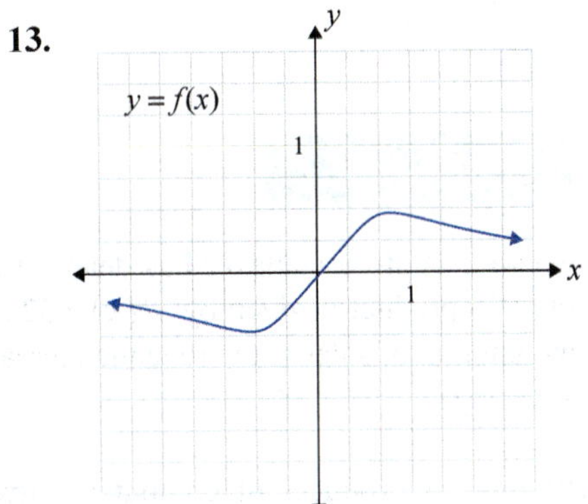

For each of the functions in Exercises 14–33, **a.** find all values of x that correspond to horizontal tangent lines, **b.** find the open intervals on which the function is increasing and the open intervals on which it is decreasing, and **c.** graph the function.

14. $f(x) = x^2 - 8x + 3$

15. $f(x) = 2x^2 + 12x - 1$

16. $f(x) = 5 - 3x - x^2$

17. $f(x) = 7x - 2x^2$

18. $f(x) = 2 - 4x - 2x^2$

19. $f(x) = 3x^2 - 4x + 2$

20. $f(x) = (2x + 3)^2$

21. $f(x) = (3x - 2)^2$

22. $f(x) = 2x^3 - 5$

23. $f(x) = 3x^3 + 4$

24. $f(x) = x^3 - 3x^2 + 7$

25. $f(x) = x^3 - 6x^2 - 4$

26. $f(x) = x^3 - 3x^2 - 9x + 12$

27. $f(x) = x^3 - x^2 - x$

28. $f(x) = x^3 + \dfrac{1}{2}x^2 - 2x + 3$

29. $f(x) = x^3 - x^2 - 5x + 2$

30. $f(x) = \dfrac{1}{3}x^3 - x^2 - 8x + 10$

31. $f(x) = \dfrac{1}{3}x^3 - 2x^2 + 3x - 6$

32. $f(x) = \dfrac{1}{3}x^3 - \dfrac{3}{2}x^2 + 2x - 6$

33. $f(x) = \dfrac{1}{3}x^3 + \dfrac{5}{2}x^2 + 4x + 11$

For each of the functions in Exercises 34–43, **a.** find all values of x that correspond to horizontal tangent lines, and **b.** find the open intervals on which the function is increasing and the open intervals on which it is decreasing.

34. $f(x) = \dfrac{x-1}{x}$

35. $f(x) = \dfrac{x}{x-2}$

36. $f(x) = \dfrac{x^2-4}{x}$

37. $f(x) = \dfrac{x^2-9}{x}$

38. $f(x) = \dfrac{x^3+16}{x}$

39. $f(x) = \dfrac{2x^3-27}{x^2}$

40. $f(x) = \dfrac{x+2}{x-1}$

41. $f(x) = \dfrac{x-5}{x+3}$

42. $f(x) = 2x - \dfrac{125}{x^2}$

43. $f(x) = x^2 + \dfrac{128}{x}$

🚀 APPLICATIONS

44. Revenue: A store manager has determined that the revenue from the sale of x units of a product is given by $R(x) = 32x - 0.4x^2$ dollars, where $0 \le x \le 80$. On what interval of sales is the revenue increasing, and on what interval of sales is it decreasing?

45. Revenue: A producer of computer software has determined that the revenue from the production and sale of x units is given by $R(x) = 48x - 0.003x^2$ dollars, where $0 \le x \le 10,000$. For what interval of production is the revenue increasing, and for what interval is it decreasing?

46. Profit: The revenue from the sale of x coffee makers is given by $R(x) = 40x - 0.4x^2$ dollars. The total cost is given by $C(x) = 370 + 16x - 0.2x^2$ dollars, where $0 \le x \le 100$. Determine the interval(s) where the profit is increasing and where it is decreasing.

47. Profit: The revenue from the sale of x 50-gallon aquariums is given by $R(x) = 54x - 0.3x^2$ dollars. The total cost function is given by $C(x) = 0.1x^2 + 4x + 200$ dollars, where $0 \le x \le 100$. Determine the interval of sales for which the profit is increasing and the interval for which it is decreasing.

48. Population: The population of the inner-city district of a city is given in thousands by $P(t) = 24 - 0.3t + 0.01t^2$, where t is the number of months after the implementation of an urban renewal project. How long will it be before the population starts to increase?

49. Wildlife management: In an attempt to naturally control the elk population in a national park, the U.S. Fish and Game Department has reintroduced the wolf into the area. It is estimated that the population of the elk herd will be $P(t) = 600 + 12t - 4t^{\frac{3}{2}}$, where t is the number of years after the reintroduction of the wolf. How long will it be before the elk population begins to decrease?

50. **Average cost:** The cost of producing x wireless speakers is given in dollars by $C(x) = 320 + 30x + 0.2x^2$, where $x \geq 0$. Determine the interval of production for which the average cost function is increasing.

51. **Average cost:** The cost of producing x units of a product is given in dollars by $C(x) = 250 + 45x - 0.2x^2$, where $x \geq 0$. Show that the average cost function is always decreasing. (This case corresponds to situations in which increased production distributes the cost so that the cost per unit decreases.)

11.5 CRITICAL POINTS AND THE FIRST DERIVATIVE TEST

■ TOPICS

- ■ Critical Values
- ■ The First Derivative Test

Critical Values

We have discussed the concept of a function increasing or decreasing on an interval. Now, we are interested in locating the points where a function changes from increasing to decreasing, or vice versa. These points are called **local extrema** (singular, **local extremum**). Our examples and discussion include continuous functions and discontinuous functions which have a jump discontinuity (say for a vertical asymptote). Figure 1 illustrates the basic terms **local maximum** and **local minimum**.

There is no local extremum at $x = x_4$ since y is increasing on each side of $x = x_4$. However, a tangent line at $x = x_4$ would be horizontal and would cross the graph at $(x_4, f(x_4))$.

There is a local minimum at $(x_3, f(x_3))$ even though there is no tangent line at that point. Note that immediately to the left of x_3, slopes are large negative numbers, but just to the right of x_3, slopes are large positive numbers.

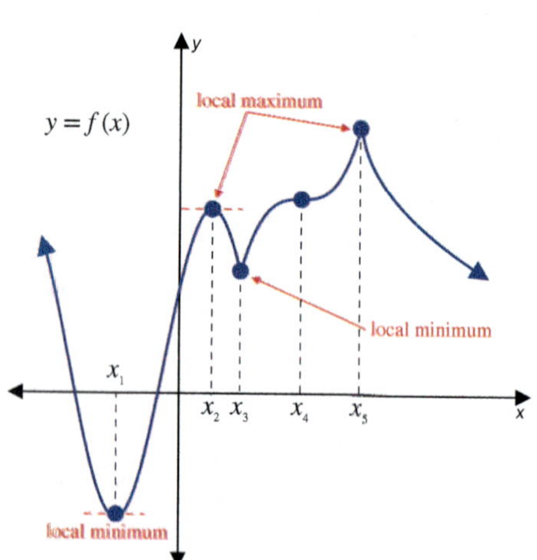

Local extrema occur for x_1, x_2, x_3, and x_5. There is no local extremum for x_4.

FIGURE 1

Local Maximum and Local Minimum

Let f be a function defined at $x = c$.

1. $f(c)$ is called a **local maximum** (or **relative maximum**) if there exists some interval (a, b) that contains c such that

$$f(x) \leq f(c)$$

for all x in (a, b). (We say a local maximum occurs at $x = c$.)

2. $f(c)$ is called a **local minimum** (or **relative minimum**) if there exists some interval (a, b) that contains c such that

$$f(x) \geq f(c)$$

for all x in (a, b). (We say a local minimum occurs at $x = c$.)

Note: We also say that a local extremum occurs at the point $(c, f(c))$ or that the point $(c, f(c))$ is a local extremum, with the understanding that the y-value, $f(c)$, is actually the local extremum.

Finding local extrema without having a graph to refer to can be done with the help of the derivative. We must be careful, however, because there can be several possibilities to consider.

Theorem of Local Extrema

If a function f is continuous on the interval (a, b) and c is in (a, b) and $f(c)$ is either a local maximum or a local minimum, then

1. $f'(c) = 0$, or

2. $f'(c)$ does not exist.

Critical Values

Critical values of x are those values c in the domain of f where $f'(c) = 0$ (indicating horizontal tangent lines) or $f'(c)$ does not exist (indicating vertical tangent lines or sharp points).

For $f(x) = \dfrac{1}{x}$, $x = 0$ is not in the domain for f; however, it is treated as a critical point—that is, it needs to be one of the points marked on an analysis line.

As the theorem implies, **local extrema occur only at critical values; however, a critical value does not guarantee a local extremum**. Figures 2–7 illustrate some of the possibilities.

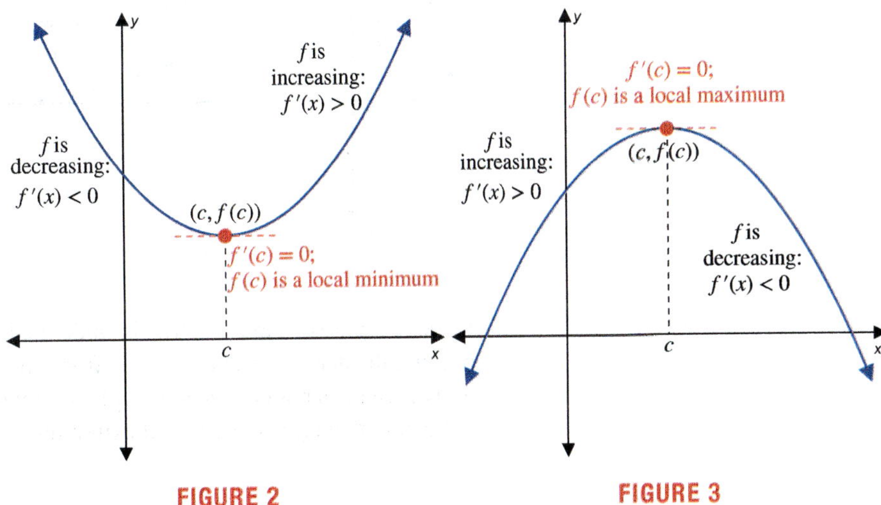

FIGURE 2　　　　　　　　　**FIGURE 3**

In both Figures 2 and 3, $f'(c) = 0$ and the derivative changes sign from one side of $x = c$ to the other. Under these conditions, $f(c)$ is a local extremum.

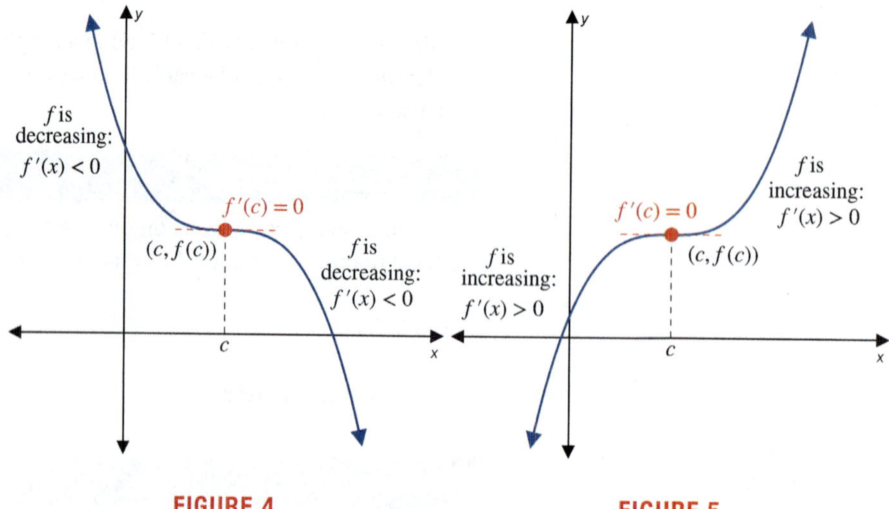

FIGURE 4   **FIGURE 5**

In both Figures 4 and 5, $f'(c) = 0$. However, the derivative does not change sign, so the function is increasing (or decreasing) on both sides of $x = c$. Thus $f(c)$ is not a local extremum.

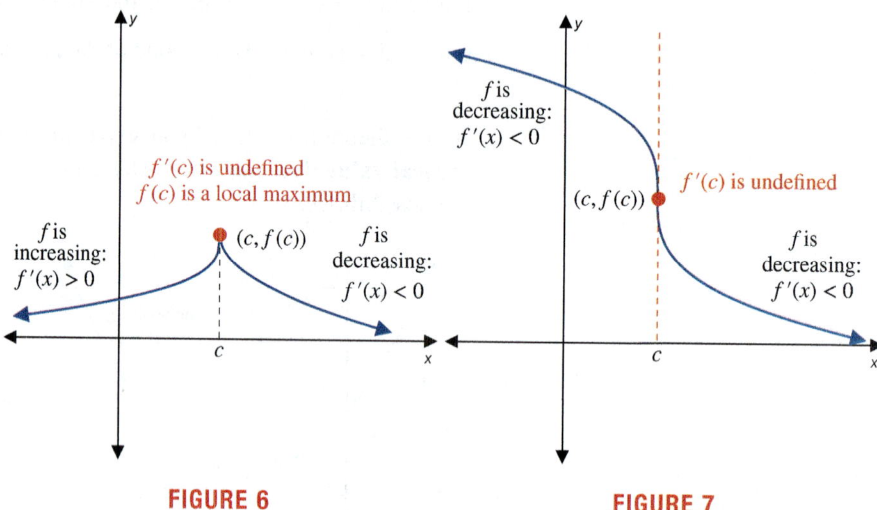

FIGURE 6   **FIGURE 7**

In both Figures 6 and 7, $f'(c)$ is undefined. In Figure 6, $f(c)$ is a local extremum because the derivative changes sign from one side of $x = c$ to the other. In Figure 7, the derivative does not change sign and the function is decreasing on both sides of $x = c$, indicating that $f(c)$ is not a local extremum.

The First Derivative Test

The First Derivative Test is developed as a numerical test to determine whether a particular point is a local maximum (a high point), a local minimum (a low point), or neither.

The First Derivative Test for Local Extrema

To determine the local extrema for a nonconstant function f continuous on the interval (a, b), perform the following steps:

1. Find the critical values, x-values c in (a, b) such that $f'(c) = 0$ or $f'(c)$ is undefined.

2. Draw an analysis line, and mark all critical x-values on it. Check the sign of $f'(x)$ on each side of $x = c$. Moving from left to right, if the sign changes

   i. from $+$ to $-$, then $f(c)$ is a local maximum;

   ii. from $-$ to $+$, then $f(c)$ is a local minimum.

3. If there is no sign change for $f'(x)$ from the left side of $x = c$ to the right side, then $f(c)$ is not a local extremum.

The following examples show how to use the First Derivative Test to find local extrema and sketch the graph of a function.

Example 1: Using the First Derivative Test

Let $f(x) = -\dfrac{1}{3}x^3 + x^2 + 3x + 1$.

a. Find the critical values of f.

b. Use the First Derivative Test to find any local extrema.

c. Sketch the graph of f.

Solution

a. $f'(x) = -\dfrac{1}{\cancel{3}} \cdot \cancel{3}x^2 + 2x + 3$ Find the derivative of $f(x)$.

$\qquad = -x^2 + 2x + 3$

Note that $f'(x)$ is defined for all values of x. Set $f'(x) = 0$ to find the critical values.

$-x^2 + 2x + 3 = 0$ Multiply both sides by -1.

$x^2 - 2x - 3 = 0$

$(x + 1)(x - 3) = 0$ Factor.

$x = -1, 3$

Thus the critical values are $x = -1$ and $x = 3$.

b. Step 1 of the First Derivative Test was completed in part **a.** Now we need to check the sign of $f'(x)$ on either side of each critical value (Step 2).

| Interval A | Interval B | Interval C |

We will use the test point $x = -2$.

We will use the test point $x = 0$.

We will use the test point $x = 5$.

$$f'(-2) = -(-2)^2 + 2(-2) + 3 \qquad f'(0) = -(0)^2 + 2(0) + 3 \qquad f'(5) = -(5)^2 + 2(5) + 3$$

$$= -4 - 4 + 3 \qquad\qquad\qquad = 0 + 0 + 3 \qquad\qquad\qquad = -25 + 10 + 3$$

$$= -5 \qquad\qquad\qquad\qquad\quad = 3 \qquad\qquad\qquad\qquad\quad = -12$$

$$-5 < 0 \qquad\qquad\qquad\qquad 3 > 0 \qquad\qquad\qquad\qquad -12 < 0$$

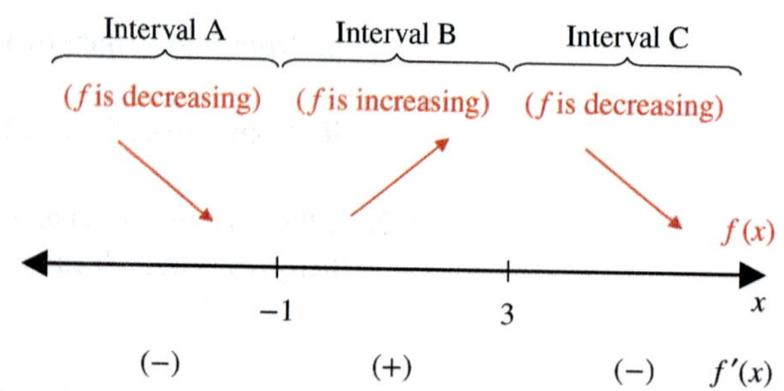

For both of the critical values, there is a sign change from one side of the value to the other. Therefore, local extrema occur at both $x = -1$ and $x = 3$ (Step 3).

We compute the y-values as follows.

$$f(-1) = -\frac{1}{3}(-1)^3 + (-1)^2 + 3(-1) + 1 \qquad f(3) = -\frac{1}{3}(3)^{3/2} + (3)^2 + 3(3) + 1$$

$$= -\frac{2}{3} \qquad\qquad\qquad\qquad\qquad\qquad = 10$$

$$\left(-1, -\frac{2}{3}\right) \text{ is a local minimum.} \qquad\qquad (3, 10) \text{ is a local maximum.}$$

c. Use the following summary of information to sketch the curve.

1. f is decreasing on $(-\infty, -1)$.

2. f is increasing on $(-1, 3)$.

3. f is decreasing on $(3, +\infty)$.

4. A local minimum occurs at the point $\left(-1, -\frac{2}{3}\right)$.

5. A local maximum occurs at the point $(3, 10)$.

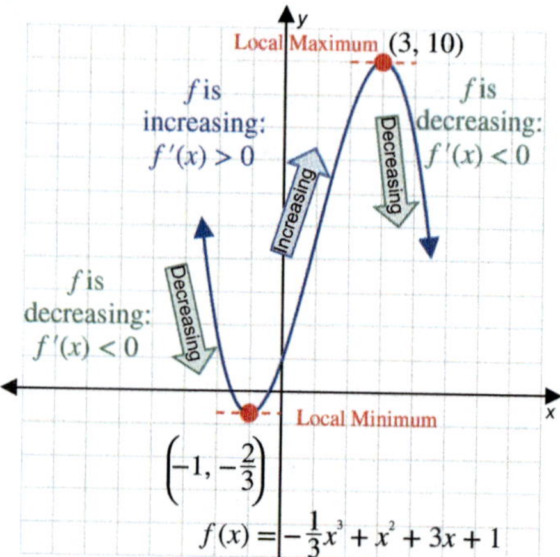

Example 2: Using the First Derivative Test

Let $f(x) = (x - 2)^3 + 1$.

a. Find the critical values of f.

b. Use the First Derivative Test to find any local extrema.

c. Sketch the graph of f.

Solution

a. $f'(x) = 3(x - 2)^2$ Find the derivative of $f(x)$.

Note that $f'(x)$ is defined for all values of x. Set $f'(x) = 0$ to find the critical values.

$$3(x - 2)^2 = 0$$
$$x = 2$$

The only critical value is $x = 2$.

b. Step 1 of the First Derivative Test was completed in part **a**. Now we need to check the sign of $f'(x)$ on either side of the critical value (Step 2).

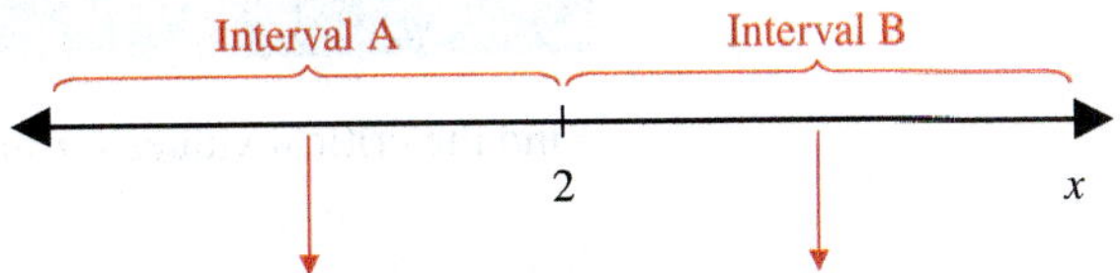

Interval A	Interval B
We will use the test point $x = 1$.	We will use the test point $x = 3$.
$f'(1) = 3(1 - 2)^2$	$f'(3) = 3(3 - 2)^2$
$= 3$	$= 3$
$3 > 0$	$3 > 0$

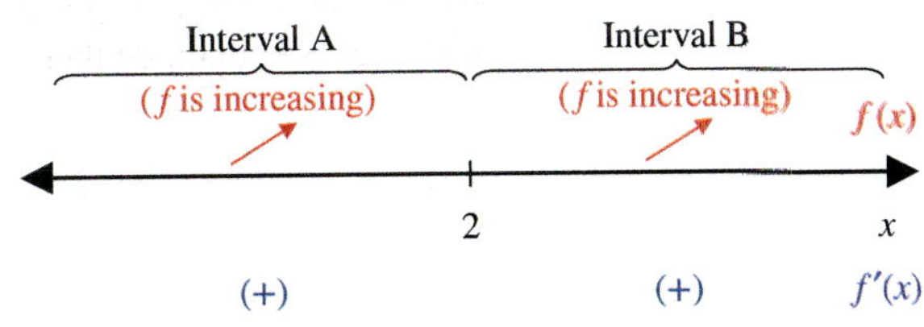

Now we have $f'(x) = 3(x - 2)^2 \geq 0$ for all x. Since there is no sign change from one side of the critical value to the other, $f(2)$ is not a local extremum (Step 3).

c. Use the following summary of information to sketch the curve.

1. f is increasing on $(-\infty, +\infty)$.

2. $f'(2) = 0$, but the point $(2, 1)$ is not a local extremum.

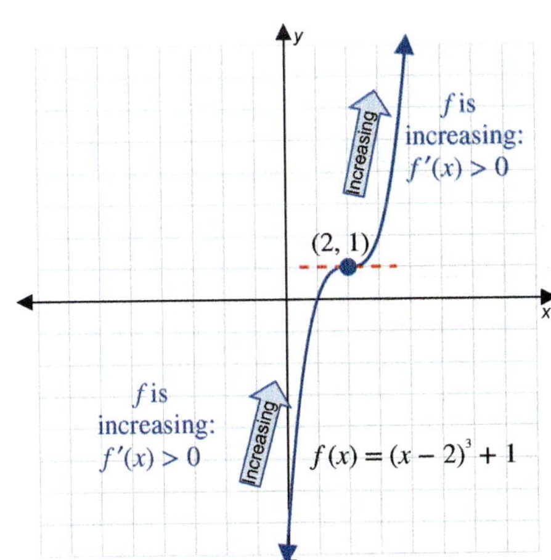

Find the critical values for the function $y = \dfrac{1}{x-2}$.

Solution

$$y = \frac{1}{x-2} = (x-2)^{-1}$$

Rewrite y using exponents.

$$\frac{dy}{dx} = -(x-2)^{-2}$$

Find the derivative of y.

$$= -\frac{1}{(x-2)^2} \neq 0$$

y' cannot equal 0. Recall, an algebraic fraction can be zero if and only if its numerator is zero.

Note that $\dfrac{dy}{dx}$ is undefined at $x = 2$ (and defined for all other values of x). The original function, $y = \dfrac{1}{x-2}$, is also not defined at $x = 2$. Therefore, $x = 2$ is **not** in the domain of the function and is **not** a critical value. There are no critical values.

Find the critical values for the function $y = x^{\frac{2}{3}}$.

Solution

$$y' = \frac{2}{3}x^{-\frac{1}{3}} = \frac{2}{3x^{\frac{1}{3}}} = \frac{2}{3\sqrt[3]{x}} \neq 0$$

Find the derivative of y. Note that y' cannot equal 0 since the numerator cannot equal 0.

In this case, y' is not defined at $x = 0$. However, since $x = 0$ **is** in the domain of the original function, $x = 0$ **is** a critical value. Also, because y' is never equal to 0, there are no other critical values, only $x = 0$. To the left of $x = 0$, y' is negative (y decreases). To the right of $x = 0$, y' is positive (y increases). Near $x = 0$, the absolute values of slope are huge. We conclude there is a sharp point at $x = 0$, where $x = 0$ is a critical value.

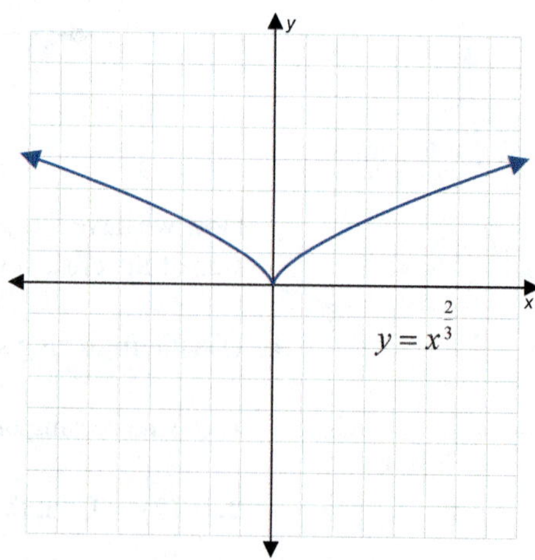

11.5 EXERCISES

PRACTICE

For each of the functions in Exercises 1–20, **a.** find the critical values, and **b.** use the First Derivative Test to find any local extrema.

1. $f(x) = 4x - x^2$

2. $f(x) = 9x - x^2$

3. $f(x) = x^2 + 6x - 2$

4. $f(x) = x^2 - 10x + 12$

5. $f(x) = \dfrac{1}{2}x^2 - 4x + 3$

6. $f(x) = -\dfrac{1}{2}x^2 + 3x - 2$

7. $f(x) = x^3 + x^2 - x + 3$

8. $f(x) = x^3 + 2x^2 + x - 2$

9. $f(x) = -x^3 - \dfrac{3}{2}x^2 + 18x + 6$

10. $f(x) = 2x^3 + x^2 - 4x$

11. $f(x) = x^3 - 3x + 6$

12. $f(x) = x^3 + 3x^2 - 4$

13. $f(x) = x + \dfrac{9}{x}$

14. $f(x) = x - \dfrac{4}{x}$

15. $f(x) = \dfrac{x^2 - 16}{x}$

16. $f(x) = \dfrac{4x^2 - 9}{x}$

17. $f(x) = 9x + x^{-1}$

18. $f(x) = 25x + x^{-1}$

19. $f(x) = 16x + x^{-2}$

20. $f(x) = 54x - x^{-2}$

APPLICATIONS

21. **Rate of dictation:** It has been determined that after t weeks of class, the average students in an intermediate shorthand class can take dictation at a rate of $W(t) = 60 + \dfrac{70t^2}{t^2 + 15}$ words per minute. Show that the rate of dictation is an increasing function which approches an upper bound.

22. **Court reporting:** A typical student in an intermediate court reporting class can reach a level of recording $W(t) = 40 + \dfrac{35t^2}{t^2 + 20}$ words per minute after t hours of instruction and practice. Show that the number of words recorded per minute increases up to a certain level.

23. **Marathon running speed:** The speed at which a marathon runner travels varies over time. The function $F(t) = 5 + \dfrac{5t^2}{t^2 + 200}$ describes the velocity of a particular runner at time t ($F(t)$ is in miles/hour). Determine the intervals over which the function is increasing or decreasing. Is this particular runner an experienced runner? (Experienced runners run faster near the end of the race.)

24. **Heating:** A frozen pizza is placed in the oven at $t = 0$. The function $F(t) = 30 + \dfrac{320t^2}{t^2 + 100}$ approximates the temperature of the pizza at time t. Show that the temperature approaches an upper bound. (The pizza will approach oven temperature over time.)

25. Cooling: A cup of hot coffee is placed in a room. The temperature in degrees Fahrenheit of the coffee is approximated by the function $F(t) = 180 - \dfrac{100t^2}{t^2 + 40}$ where t is the number of minutes the coffee has been in the room. Show that the temperature function is a decreasing function, and find the temperature of the room. (The coffee approaches room temperature over time.)

26. Velocity and acceleration: The velocity of a car varies according to the function $F(t) = \dfrac{t^3}{1875} - \dfrac{119t^2}{1500} + \dfrac{53t}{15} + 5,$ where t is time ($0 < t < 100$). Determine any local maximum and minimum velocities, and the times at which they occur. (Be sure to specify whether each is a maximum or mininum.)

27. Skydiving: The velocity of a skydiver after parachute deployment is given by the formula $F(t) = 180 - \dfrac{165t^2}{t^2 + 10},$ where t is the time after deployment. Show that the function is decreasing, and determine the terminal velocity for the diver-parachute system. (The velocity approaches terminal velocity as t goes to infinity.)

28. Pressure: The pressure in a pressure cooker is given by the function $F(T) = 1 + \dfrac{3T^2}{T^2 + 180},$ where T is the temperature inside the kettle. Show that the function is increasing, and determine the upper bound for pressure.

29. Space probe closing speed: In order to minimize travel time, a fictional probe on its way to a distant star accelerates for part of the voyage, and then decelerates to enter orbit safely. The closing speed of the probe with the star is given by the function $F(t) = \dfrac{t^2}{25} - 4t + 0.5,$ where t is the number of years after launch and $F(t)$ is measured in percentage of light speed. At what time does the probe begin decelerating? What is the probe's maximum closing speed? (**Hint:** Closing speed is negative.)

30. Computation speed: An experimental supercomputer is undergoing testing to determine whether it will meet the necessary specifications. The number of computations it can perform per second is found to be modeled by the function $F(t) = \dfrac{89t^3}{441,000} - \dfrac{757t^2}{22,050} + \dfrac{2227t}{1470},$ where $0 < t < 100$ is the number of minutes after startup and $F(t)$ is in quadrillions of computations/sec. Find all relative maximums and minimums for the function over the interval. Round to the nearest tenth. If the number of computations is zero at any time after startup, the computer crashes. Does the computer crash during testing?

31. Fuel economy: The fuel consumption of an automobile is not constant. Fuel economy depends largely on the speed of the vehicle. The function $F(v) = -\dfrac{8v^3}{19,125} + \dfrac{28v^2}{85} - \dfrac{288v}{85} + 80$ ($F(v)$ is in miles per gallon) describes the fuel consumption of a new hybrid vehicle, where $0 < v < 85$ is the velocity of the vehicle. On what intervals is the consumption increasing? Which velocities yield maximum efficiencies? (Remember, high efficiency means low consumption. Round to the nearest tenth.)

11.6 ABSOLUTE MAXIMUM AND MINIMUM

An engineer wants to design a carburetor that will allow a car to attain maximum acceleration. Another engineer wants to design a carburetor that will use minimum amount of gasoline yet maintain a respectable acceleration capability. Manufacturers want to minimize costs and maximize profits. These situations illustrate how the concepts of maximum and minimum affect activities in business and manufacturing. Whenever the ideas under consideration can be modeled (or represented) with functions, calculus is a useful tool for determining maximum and/or minimum quantities.

Previously, we used the derivative to determine where a function is increasing or decreasing and to find local extrema. As it was illustrated, a function may have several local extrema. However, a function can have only one largest value or **absolute maximum** and only one smallest value or **absolute minimum**.

Absolute Extrema

If c is in the domain of f and, for all x in the domain of f, we have

1. $f(x) \le f(c)$, then $f(c)$ is called the **absolute maximum** of f.

2. $f(x) \ge f(c)$, then $f(c)$ is called the **absolute minimum** of f.

Figure 1 shows some possible situations involving absolute extrema.

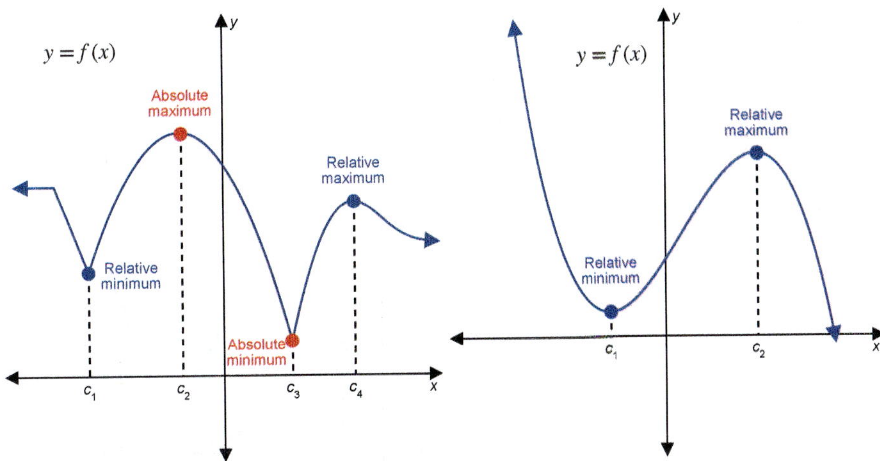

$f(c_2)$ is the absolute maximum of f.
$f(c_3)$ is the absolute minimum of f.

$f(x)$ has no absolute maximum and no absolute minimum.

FIGURE 1

Generally, we are concerned with absolute extrema on a closed interval $[a, b]$ (i.e., an interval in which the endpoints a and b are included). The following theorem guarantees that a continuous function will indeed have an absolute maximum and an absolute minimum on a closed interval. The proof of this theorem is given in more advanced courses.

Theorem of Absolute Extrema of a Continuous Function

If a function f is continuous on a closed interval $[a, b]$, then f will have an absolute maximum value and an absolute minimum value on $[a, b]$.

Now that we are assured of the existence of absolute extrema for a function continuous on a closed interval $[a, b]$, we want to know how to find these values. It can be proved that absolute extrema occur at the endpoints (a or b) or at the critical values c, where $f'(c) = 0$ or $f'(c)$ does not exist. Figure 2 illustrates some possible cases.

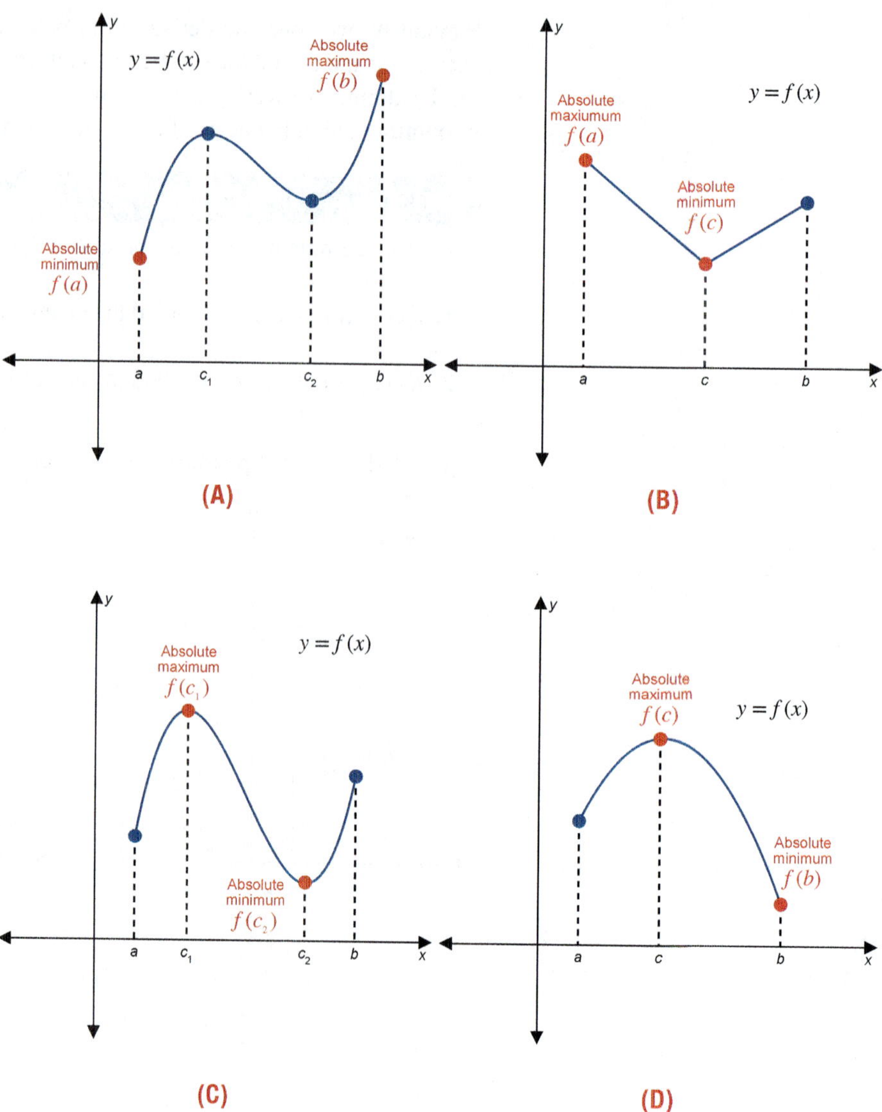

FIGURE 2

Finding Absolute Extrema

Assume that f is a continuous function on the interval $[a, b]$. To find the absolute extrema of f on $[a, b]$, perform the following steps.

1. Find all the critical values for f in $[a, b]$. That is, find all c in $[a, b]$ where

   a. $f'(c) = 0$ or

   b. $f'(c)$ is undefined.

2. Evaluate $f(a), f(b)$, and $f(c)$ for all critical values c.

3. The largest value found in Step 2 is the absolute maximum. The smallest value found in Step 2 is the absolute minimum.

Example 1: Finding Absolute Extrema

Find the absolute extrema for $f(x) = 2x^3 - 15x^2 + 24x + 6$ on the interval $[0, 5]$.

Solution

Step 1: Find all the critical values for $f(x)$ in $[0, 5]$.

$$f'(x) = 6x^2 - 30x + 24$$
$$0 = 6x^2 - 30x + 24$$
$$0 = 6(x^2 - 5x + 4)$$
$$0 = 6(x-1)(x-4)$$
$$x = 1, 4$$

Find the critical values of f by setting $f'(x) = 0$ and solving for x. Note that f' is defined for all x in $[0, 5]$.

Make sure both critical numbers $x = 1, 4$ are in the interval $[0, 5]$.

Step 2: Evaluate $f(a)$, $f(b)$, and $f(c)$ for all c in the given interval. Both the critical values, $x = 1$ and $x = 4$, are in the interval $[0, 5]$. Therefore, evaluate $f(x)$ at $x = 0$, $x = 1$, $x = 4$, and $x = 5$.

$$x = 0 \qquad f(0) = 2(0)^3 - 15(0)^2 + 24 \cdot 0 + 6$$
$$= 6$$

$$x = 1 \qquad f(1) = 2(1)^3 - 15(1)^2 + 24 \cdot 1 + 6$$
$$= 17 \quad \text{Absolute max}$$

$$x = 4 \qquad f(4) = 2(4)^3 - 15(4)^2 + 24 \cdot 4 + 6$$
$$= -10 \quad \text{Absolute min}$$

$$x = 5 \qquad f(5) = 2(5)^3 - 15(5)^2 + 24 \cdot 5 + 6$$
$$= 1$$

Step 3: The largest value found in Step 2 is the absolute maximum and the smallest value is the absolute minimum. So the absolute maximum is $f(1) = 17$ and occurs at $x = 1$. The absolute minimum is $f(4) = -10$ and occurs at $x = 4$.

The full graph of f on $[0, 5]$ is shown to help your understanding of the problem.

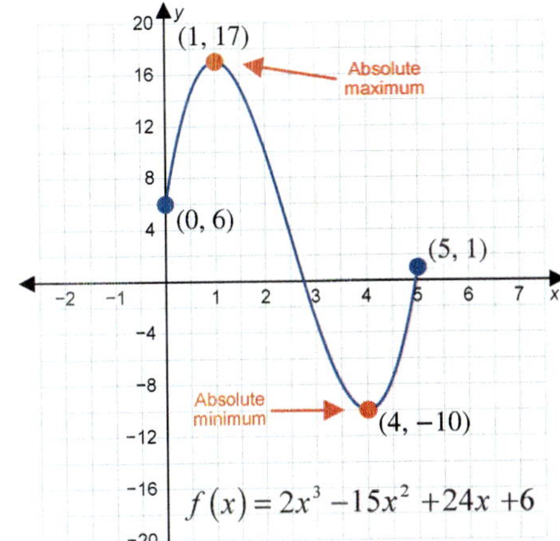

Find the absolute extrema for $f(x) = -2x + 10$ on the interval $[1, 3]$.

Solution

Step 1: Find all the critical values for $f(x)$ in $[1, 3]$.

$$f'(x) = -2$$
$$0 \neq -2$$

Find the critical values of f by setting $f'(x) = 0$ and solving for x. Note that f' is defined for all x in $[1, 3]$.

There are no critical values. Furthermore, since $f'(x) = -2 < 0$, f is decreasing for all x.

Step 2: Evaluate $f(a), f(b)$, and $f(c)$ for all c in the given interval.

Again, there are no critical values, c. Therefore, only the endpoints of the interval, $x = 1$ and $x = 3$, need to be evaluated.

$$x = 1 \qquad f(1) = -2(1) + 10 = 8 \quad \text{Absolute max}$$

$$x = 3 \qquad f(3) = -2(3) + 10 = 4 \quad \text{Absolute min}$$

Step 3: The largest value found in Step 2 is the absolute maximum and the smallest value is the absolute minimum.

So the absolute maximum is $f(1) = 8$ and the absolute minimum is $f(3) = 4$.

The graph of f on $[1, 3]$ is a straight line.

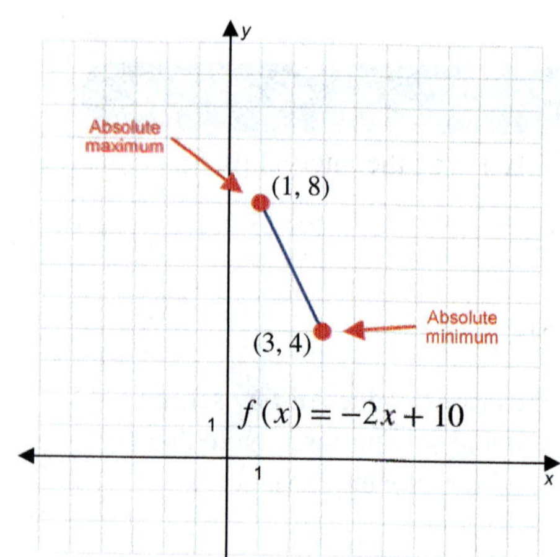

Find the absolute extrema for $f(x) = x^{\frac{2}{3}} + 1$ on the interval $[-8, 8]$.

Solution

Step 1: Find all the critical values for $f(x)$ in $[-8, 8]$.

$$f'(x) = \frac{2}{3} x^{-\frac{1}{3}}$$

Find the critical values of f by setting $f'(x) = 0$ and solving for x.

$$= \frac{2}{3x^{\frac{1}{3}}}$$

Since $f'(0)$ is undefined, $x = 0$ is a critical value. In this case, $x = 0$ is the only critical value.

Step 2: Evaluate $f(a)$, $f(b)$, and $f(c)$ for all c in the given interval. Now evaluate $f(x)$ at $x = -8$, $x = 0$, and $x = 8$.

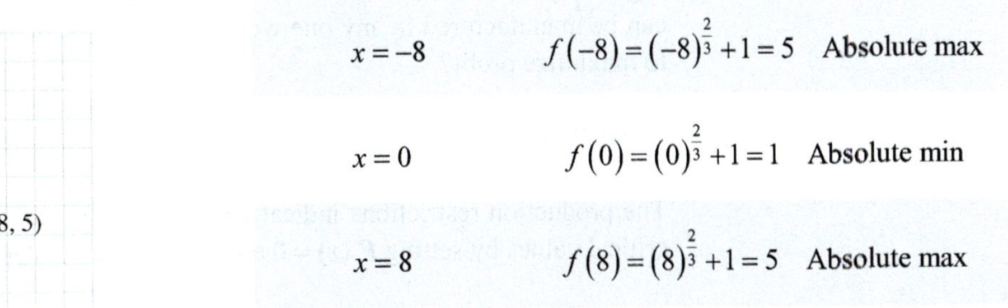

$$x = -8 \qquad f(-8) = (-8)^{\frac{2}{3}} + 1 = 5 \quad \text{Absolute max}$$

$$x = 0 \qquad f(0) = (0)^{\frac{2}{3}} + 1 = 1 \quad \text{Absolute min}$$

$$x = 8 \qquad f(8) = (8)^{\frac{2}{3}} + 1 = 5 \quad \text{Absolute max}$$

Step 3: The largest value found in Step 2 is the absolute maximum and the smallest value is the absolute minimum. In this case, there are two x-values that result in the absolute maximum. These are $x = -8$ and $x = 8$ (note that these are the endpoints) and the value of the absolute maximum is 5. The absolute minimum is $f(0) = 1$.

Example 4: Finding Absolute Extrema

Find the absolute extrema for $f(x) = \dfrac{1}{3}x^3 - 9x$ on the interval $[0, 4]$.

Solution

Step 1: Find all the critical values for $f(x)$ in $[0, 4]$.

$$f'(x) = x^2 - 9$$
$$0 = x^2 - 9$$
$$0 = (x+3)(x-3)$$
$$x = -3, 3$$

Find the critical values of f by setting $f'(x) = 0$ and solving for x. Note that f' is defined for all x in $[0, 4]$.

Although $x = -3$ and $x = 3$ are critical values for the function, $x = -3$ is not in the interval $[0, 4]$, and must be disregarded.

Step 2: Evaluate $f(a)$, $f(b)$, and $f(c)$ for all c in the given interval. Since the critical value $x = -3$ is not in the interval, evaluate $f(x)$ only at $x = 0$, $x = 3$, and $x = 4$.

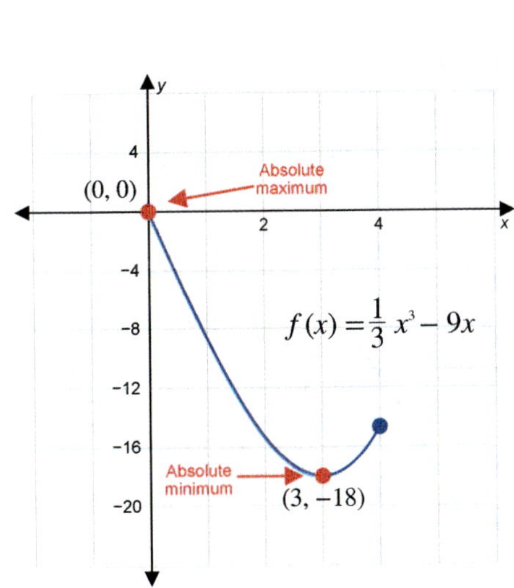

$$x = 0 \qquad f(0) = \frac{1}{3}(0)^3 - 9(0) = 0 \quad \text{Absolute max}$$

$$x = 3 \qquad f(3) = \frac{1}{3}(3)^3 - 9(3) = -18 \quad \text{Absolute min}$$

$$x = 4 \qquad f(4) = \frac{1}{3}(4)^3 - 9(4) = -\frac{44}{3} \approx -14.67$$

Step 3: The largest value found in Step 2 is the absolute maximum and the smallest value is the absolute minimum. So the absolute maximum is $f(0) = 0$ and the absolute minimum is $f(3) = -18$.

Example 5: Maximizing Profits

A company finds that its profit in dollars for producing x units of a product in one week is given by $P(x) = -2x^2 + 1600x$. If the company is set up so that no more than 500 units can be manufactured in any one week, how many units should the company produce to maximize profit?

Solution

The production restrictions indicate that x is in the closed interval $[0, 500]$. Find the critical values by setting $P'(x) = 0$ and solving for x (Step 1).

$$P'(x) = -4x + 1600 \qquad \text{Note that } P' \text{ is defined for all } x \text{ in } [0, 500].$$
$$0 = -4x + 1600$$
$$4x = 1600$$
$$x = 400$$

Now evaluate $P(x)$ for $x = 0$, $x = 400$, and $x = 500$ (Step 2).

$$x = 0 \qquad\qquad P(0) = -2(0)^2 + 1600 \cdot 0 = 0$$

$$x = 400 \qquad P(400) = -2(400)^2 + 1600 \cdot 400 = 320,000 \quad \text{Absolute max}$$

$$x = 500 \qquad\qquad P(500) = -2(500)^2 + 1600 \cdot 500 = 300,000$$

The profit is maximized at \$320,000 when 400 units are produced (Step 3).

11.6 EXERCISES

💡 PRACTICE

In Exercises 1–8, find the absolute extrema for each graph of $f(x)$ on the given interval.

1. $[3, 5]$

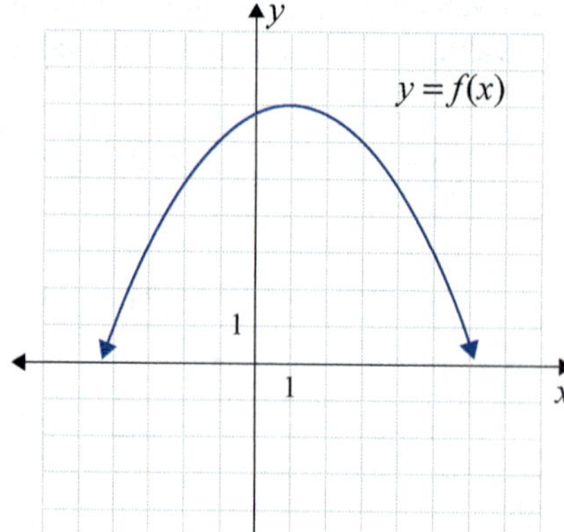

2. $[-1, 4]$

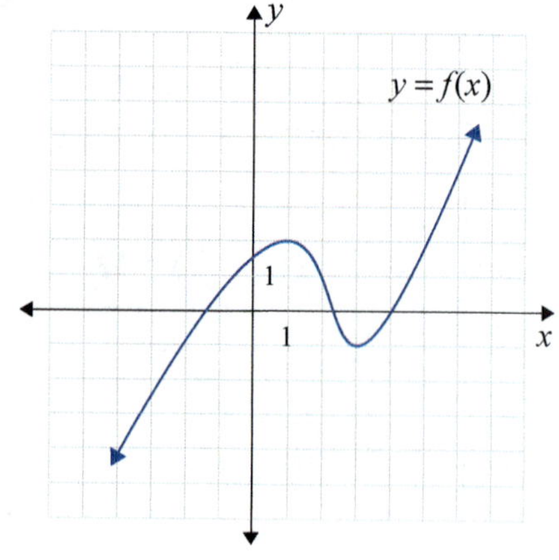

3. $[-5, 3]$

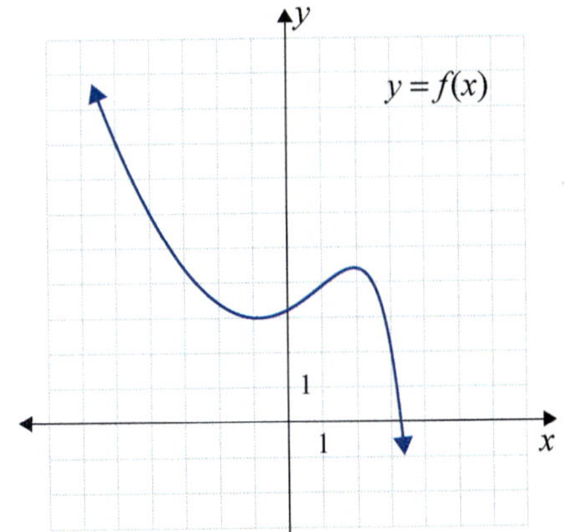

4. $[-4, 4]$

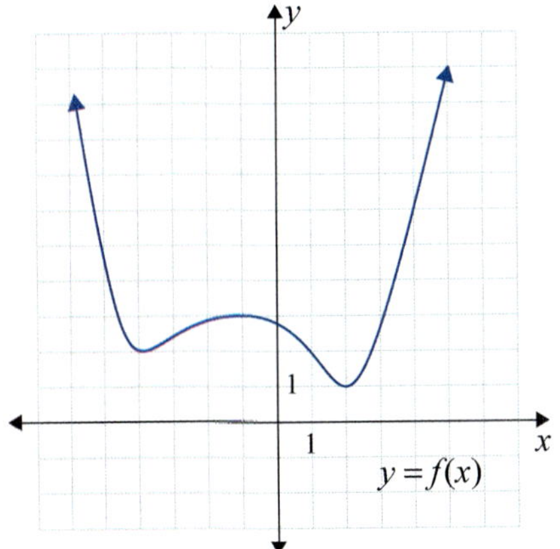

5. $[1, 4]$

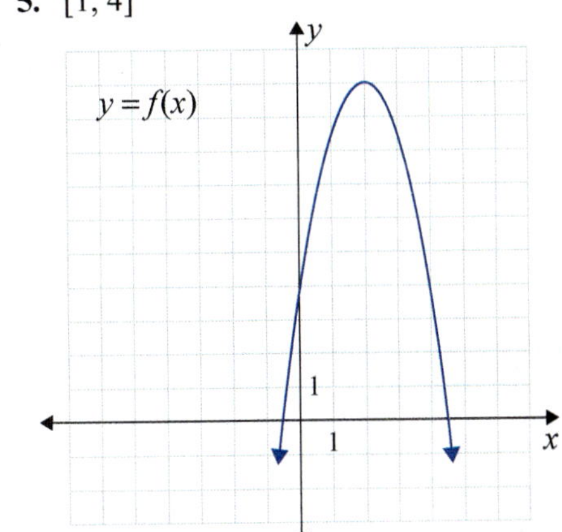

6. $[-2, 2]$

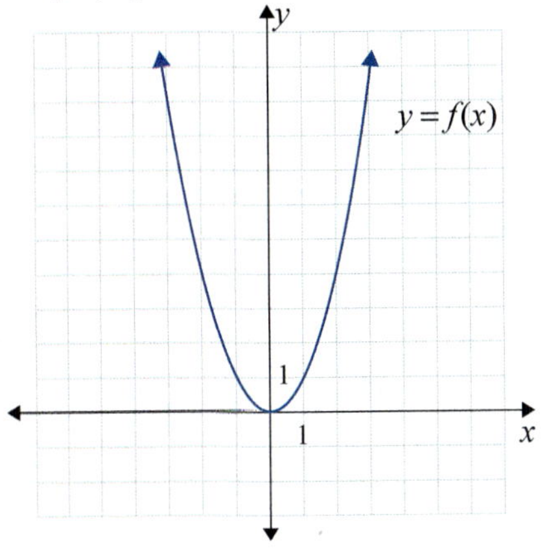

7. $[-3, 5]$

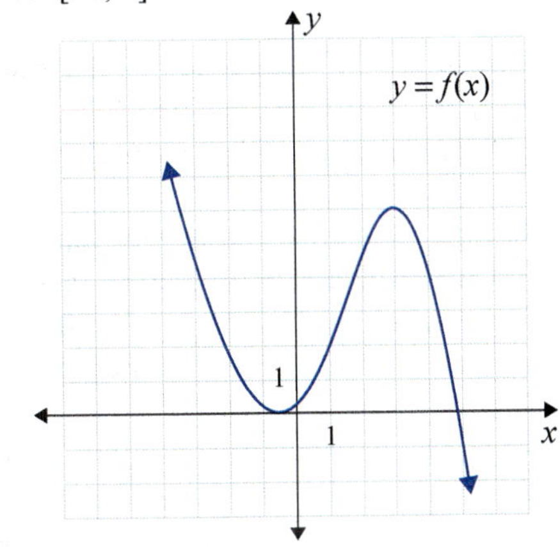

8. $[-4, 5]$

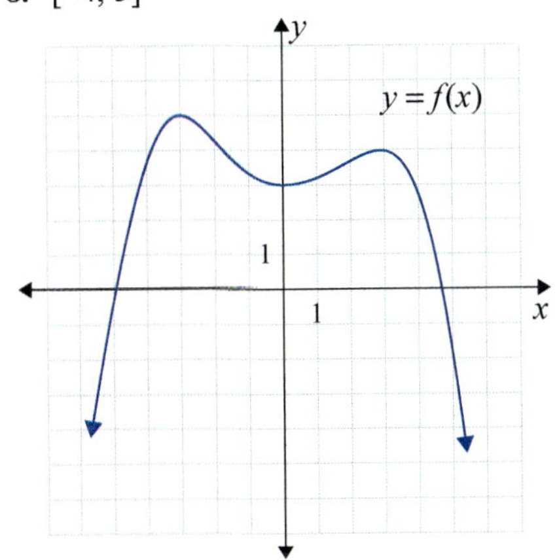

In Exercises 9–38, find the absolute extrema for each function on the given interval.

9. $f(x) = x^2 - 8x;\ [0, 5]$

10. $f(x) = 3x^2 - 12x;\ [0, 4]$

11. $f(x) = 6 + 10x - x^2;\ [3, 6]$

12. $f(x) = 11 - 4x - x^2;\ [-3, 0]$

13. $f(x) = 14 - 3x;\ [0, 4]$

14. $f(x) = 7 + \frac{1}{2}x;\ [-2, 4]$

15. $f(x) = 8 - x^3;\ [-1, 3]$

16. $f(x) = x^3 + 4;\ [-2, 2]$

17. $f(x) = x^3 - 12x$; $[-3, 4]$

18. $f(x) = x^3 - 3x$; $[-2, 3]$

19. $f(x) = 2x^3 - x^2$; $[0, 2]$

20. $f(x) = x^3 + 2x^2$; $[0, 3]$

21. $f(x) = 9x - 3x^2 - x^3$; $[-4, 2]$

22. $f(x) = x^3 - 3x^2 - 24x$; $[-3, 3]$

23. $f(x) = 2x^3 - 3x^2 - 12x - 10$; $[0, 4]$

24. $f(x) = x^3 + 3x^2 - 24x$; $[0, 3]$

25. $f(x) = x^{\frac{2}{3}} - 4$; $[-1, 8]$

26. $f(x) = 3x^{\frac{2}{3}} + 2$; $[-1, 4]$

27. $f(x) = 3x^{\frac{1}{3}} - 4x$; $[0, 1]$

28. $f(x) = 3x^{\frac{2}{3}} + x$; $[-9, 1]$

29. $f(x) = \sqrt{x^2 + 4}$; $[-1, 2]$

30. $f(x) = \sqrt{9 - x^2}$; $[-1, 2]$

31. $f(x) = \sqrt[3]{x^2 - 1}$; $[-2, 2]$

32. $f(x) = (x^2 - 1)^{\frac{2}{3}}$; $[-2, 2]$

33. $f(x) = x + \dfrac{4}{x}$; $\left[\dfrac{1}{2}, 3\right]$

34. $f(x) = 2x + \dfrac{18}{x}$; $[1, 4]$

35. $f(x) = 4x + \dfrac{9}{x}$; $[1, 3]$

36. $f(x) = 9x + \dfrac{16}{x}$; $[1, 2]$

37. $f(x) = x^2 + \dfrac{16}{x}$; $\left[\dfrac{1}{2}, 4\right]$

38. $f(x) = x^2 + \dfrac{2}{x}$; $[1, 4]$

🚀 APPLICATIONS

39. **Revenue:** The weekly revenue from the sale of x units of a product is given by $R(x) = 24x - 0.5x^2$ dollars. If the company is set up so that it can produce no more than 40 units per week, how many units should the company produce to maximize revenue?

40. **Revenue:** The revenue from the sale of x units of a product is given by $R(x) = 12x - 0.04x^2$ thousand dollars, where $0 \le x \le 250$. How many units should be sold to maximize the revenue?

41. **Profit:** A manufacturer of telescopes has determined that the revenue from the production and sale of x telescopes is $R(x) = 140x - 0.5x^2$ dollars. The cost function is given by $C(x) = x^2 + 20x + 1050$ dollars. Find the level of production and sales that will maximize the profit if $0 \le x \le 70$.

42. **Profit:** A marketing analyst for a company that produces skateboards has determined that if the company sells x units of the deluxe model, the revenue function is $R(x) = 79.9x - 0.03x^2$ dollars and the cost function is $C(x) = 0.08x^2 + 5.1x + 5800$ dollars. Find the sales level that will yield maximum profit if $0 \le x \le 380$.

43. **Average cost:** The cost of producing x compact refrigerators is given by $C(x) = 2880 + 35x + 0.2x^2$ dollars. Find the value of x that minimizes the average cost function if $0 \le x \le 150$.

44. **Average cost:** The cost of producing x electronic games is given by $C(x) = 1080 + 42x + 0.3x^2$ dollars. Find the value of x that minimizes the average cost function if $0 \le x \le 90$.

45. **Air quality:** The Air Quality Management District monitors the level of pollution in the air. On a good day, the level is approximately $P(t) = 35 + \dfrac{126t}{0.5t^2 + 18}$ PSI (Pollution Standard Index), where t is the number of hours after 7:00 a.m. and $0 \le t \le 11$. At what time will the pollution level be a maximum?

46. **Air quality:** On a moderately smoggy day, the level of nitrogen dioxide in the air is approximately $N(t) = 0.126 + \dfrac{0.36t}{2t^2 + 40.5}$ ppm (parts per million), where t is the number of hours after 8:00 a.m. and $0 \le t \le 10$. At what time will the level of nitrogen dioxide reach its maximum?

47. **Bacteria:** It is estimated that t hours after a particular bacterium is introduced into a culture, the population of bacteria in the culture will be $P(t) = \dfrac{4800}{\sqrt{12 - 0.5t}}$, where $0 \le t \le 6$. What will be the absolute maximum population? At what time t will this occur?

48. **Population:** It is estimated that t years from now the population of a small community will be $P(t) = \dfrac{5000}{\sqrt{25 + 0.4t}}$ people, where $0 \le t \le 10$. What will be the maximum population?

49. **Altitude:** The altitude of an airplane following a certain flight path is given by the function $F(t) = \dfrac{13t}{20} - \dfrac{3t^2}{200}$, where t is in minutes, and $F(t)$ is in thousands of feet. Find the absolute maximum of the function over the interval $[0, 43]$.

50. **Velocity of a car:** The velocity of a car in miles per hour varies according to the function $F(x) = \dfrac{x^2}{100} - \dfrac{19x}{25} + \dfrac{5349}{100}$, where x denotes time in seconds. Find the absolute maximum and minimum velocities over the interval $[0, 100]$. Find the car's velocity at each endpoint.

51. **Tire distortion:** When a car accelerates, its tires are distorted by the force exerted on them by the motor. For a certain model car, the distortion after the driver floors the accelerator is modeled by the function $F(t) = 0.00026t^3 - 0.0533t^2 + 2.74014t$, where $0 < t < 100$ is the number of milliseconds after acceleration begins, and $F(t)$ is a percentage of the tire's maximum flexibility (at $F(t) = 100$, the tire tears into pieces). Find the time when the maximum distortion occurs, and find the percentage that the tires are distorted. Do the tires survive the acceleration?

52. **Pollution:** The amount of pollution (measured in parts per million) in a small river is given by the function $F(t) = \dfrac{450t^3}{169} - \dfrac{18,225t^2}{169} + \dfrac{23,625t}{23} + \dfrac{260,675}{169}$, where t is the number of years after 1980. Sometime after 1980, an environmental protection law was enacted to reduce the amount of pollution in the river, and in the same year the pollution levels began to decrease. Find the maximum pollution for $0 \le t \le 20$, and determine the year the law was enacted. (The largest integer less than the t-value specifies the year of enactment.)

53. **Wind resistance:** An aeronautics company is testing a new airframe. The company wishes to determine an ideal cruising speed, and one step in the process is to find the speed at which the airframe experiences the least wind resistance (other than when it is not moving). After performing a number of tests, the engineers determine that the wind resistance can be modeled by a function $G(v) = \dfrac{v^3}{3{,}645{,}000} - \dfrac{8v^2}{6075} + \dfrac{350v}{243} + \dfrac{614{,}500}{729}$, where $300 \leq v \leq 4000$ is the wind velocity in feet per second and $G(v)$ is in newtons. Determine the ideal cruising speed for the airframe based on this model.

54. **Bacteria:** A bacteriologist doing research on antibiotics has discovered that a certain type of disease-causing bacteria can be effectively treated using a cocktail of two different antibiotics administered at specific intervals. The population changes in response to the antibiotics according to the function $F(t) = -0.0224t^3 + 4.5676t^2 - 252.4610t + 5000$, where t is in the interval $[0, 200]$. The maximum values of the function correspond to the times when the antibiotics were administered. At what times, t, are the antibiotics administered and what is the population of bacteria at those times?

55. **Gravitational pull:** A rocket traveling to the moon is affected by the gravity of Earth and of the moon. The total gravitational force exerted on the rocket is approximated by the function $F(h) = \dfrac{43{,}750h^2}{3} - \dfrac{70{,}000h}{3} + 10{,}000$, where $0 < h < 1$ is the height of the object, given in percentage of the distance between Earth and the moon, and $F(h)$ is measured in newtons. Find the height at which the minimum occurs and the gravitational force at that altitude. Additionally, there is an onboard experiment which can only be performed if the gravitational force falls below 1000 newtons. Can this experiment be performed?

56. **Thrown object:** A cell phone is thrown into the air. The position of the phone is given by the function $F(t) = -\dfrac{49t^2}{10} + \dfrac{297t}{10}$, where $0 < t < 10$ is the number of seconds after the phone is thrown, and $F(t)$ is measured in feet. Find the maximum height the phone attains. Also find the time when the phone hits the ground. The phone will shatter if it hits the ground with a speed greater than 32 fps (feet per second). Does the phone shatter?

57. **Acceleration:** Due to many variable factors in car engines, acceleration is never constant. The acceleration of a particular car in a particular test is approximated by the function $F(t) = -\dfrac{260t^3}{137} + \dfrac{2418t^2}{137} - \dfrac{7371t}{137} + 75$, where $0 \leq t \leq 5$ is the number of seconds after acceleration begins. What is the maximum acceleration of the car during this test? When does it occur? What is the minimum acceleration, and when does it occur?

58. **Power consumption:** A certain piece of equipment in a chemistry lab draws power according to the function

$$G(m) = \left(-6.087 \times 10^{-6}\right)m^3 + \left(8.641 \times 10^{-3}\right)m^2 - 1.751m + 1000,$$

where $0 < m < 1000$ is the mass of the sample to be analyzed in grams and G is power consumption in watts. Find the sample mass which causes the equipment to draw the most power. How much power does the equipment draw for a sample of this mass?

59. **Chemistry:** A certain chemical procedure requires the addition of reactants and catalysts at precise times to maintain reaction rates. The rate of reaction is given by the function

$$H(t) = (9.423 \times 10^{-7})t^3 - (1.230 \times 10^{-4})t^2 + (4.384 \times 10^{-3})t + 0.001,$$

where $0 < t < 100$ is the number of seconds after beginning the reaction, and $H(t)$ is measured in number of moles formed per second. Find the time at which a relative maximum reaction rate is reached and the number of moles per second being formed at that time. At what time was the second reactant/catalyst mixture added? (The reactant/catalyst mixture is added at a relative minimum.) Round to two decimal places.

60. **Phone traffic:** Phone traffic varies greatly over the course of a day. A phone provider estimates that the number of international calls active per minute on a certain holiday is given by the function $P(t) = -0.001t^3 + t^2 + 50t + 150,000$, where $0 < t < 1000$ is the number of minutes after 6.00 a.m. and $P(t)$ is the number of active international calls. For what t does maximum phone traffic occur? How many calls are active at this time?

61. **Stored energy:** While a rubber ball is moving, it has kinetic energy. When the ball impacts a hard surface, the kinetic energy is converted to potential energy, and then back to kinetic energy. This happens quite rapidly. The amount of potential energy after impact is approximated by the function $E(t) = -\dfrac{3t^2}{80} + \dfrac{3t}{2}$, where t is the number of nanoseconds after impact and t is in $[0, 40]$. When does the ball have maximum potential energy? How much potential energy does it have at its maximum?

62. **Photosynthesis:** Plants absorb different amounts of light, depending on the wavelength of the light. A study of Rhododendron bushes show that they absorb light according to the formula $L(w) = \dfrac{-2.22w^3}{10^{10}} + \dfrac{4.44w^2}{10^6} - \dfrac{2w}{10^3} + 0.444$, where $100 < w < 650$ is the wavelength of light in nanometers, and $L(w)$ is the proportion of the light shone on the bush. What wavelength does Rhododendron absorb the best? What proportion of light of this wavelength does Rhododendron absorb?

63. **Package delivery:** A study says that the package flow in the Southeast USA during the month of August follows the function $D(t) = \dfrac{7t^3}{9300} - \dfrac{7t^2}{248} + \dfrac{7t}{31} + 1$, where $1 \le t \le 31$ is the day of the month, and $D(t)$ is given in millions of packages. On which day are the most packages delivered? How many packages are delivered on this day?

12

Chapter 12

APPLICATIONS OF THE DERIVATIVE

12.1 CONCAVITY AND POINTS OF INFLECTION

■ TOPICS

- Higher-Order Derivatives
- Concavity
- Points of Inflection
- Using a Calculator to Find Inflection Points

Higher-Order Derivatives

Given a function f, we have defined and discussed the derivative f', called the **first derivative**. We have seen that there may be x-values for which f is defined but for which the derivative does not exist. However, in general, the functions defined and applied in this course have slope defined at most points for which the function is defined. This means that f' is itself a function which has its own derivative defined at most points in its domain, and it too can be differentiated. The **second derivative**, f'', if it exists, is the derivative of the first derivative. The **third derivative**, f''' (or $f^{(3)}$), if it exists, is the derivative of the second derivative, and so on.

For example, if

$$f(x) = \frac{1}{x} = x^{-1} \quad (\text{where } x \neq 0)$$

$$f'(x) = -1 \cdot x^{-2},$$

then

$$f''(x) = -1(-2)x^{-3} = 2x^{-3},$$

$$f'''(x) = 2(-3)x^{-4} = -6x^{-4},$$

$$f^{(4)}(x) = -6(-4)x^{-5} = 24x^{-5},$$

and so on.

Although there are important applications that involve an infinite number of derivatives, we will study only first and second derivatives and related applications in this chapter.

Recall, the first derivative of a function $y = f(x)$ can be represented by any of the following symbols.

$$y', \quad f'(x), \quad \frac{dy}{dx}, \quad f_x, \quad \text{and} \quad D_x[y]$$

Similar notation is used for the second derivative.

> **✓ NOTE**
>
> The notation $\dfrac{d^2y}{dx^2}$ deserves special attention. The expression $\dfrac{d}{dx}$ is called an **operator** and is meaningless by itself. When applied to a function, it means to find the derivative of that function with respect to x. Thus
>
> $$\frac{d}{dx}\left[\frac{dy}{dx}\right]$$
>
> indicates the derivative of $\dfrac{dy}{dx}$. This leads to the following simplified notation.
>
> $$\frac{d}{dx}\left[\frac{dy}{dx}\right] = \frac{d^2y}{dx^2}$$

Notation for the Second Derivative

The second derivative of the function $y = f(x)$ can be denoted by any of the following symbols.

$$y'', \quad f''(x), \quad f^{(2)}, \quad \frac{d^2y}{dx^2}, \quad f_{xx}, \quad \text{and} \quad D_x^2[y]$$

Example 1: First and Second Derivatives

Find both the first and second derivatives of the function $f(x) = x^3 - 12x + 1$.

Solution

$$f'(x) = 3x^2 - 12 \quad \text{Differentiate } f(x).$$
$$f''(x) = 6x \quad \text{Differentiate } f'(x).$$

Example 2: First and Second Derivatives

Find $\dfrac{dy}{dx}$ and $\dfrac{d^2y}{dx^2}$ if $y = \sqrt{x^2 + 1}$.

Solution

$$y = \sqrt{x^2 + 1} = \left(x^2 + 1\right)^{\frac{1}{2}} \qquad \text{Rewrite } y \text{ using exponents.}$$

$$\frac{dy}{dx} = \frac{1}{2}\left(x^2 + 1\right)^{\frac{-1}{2}}(2x) \qquad \text{Use the General Power Rule to find the first derivative of } y.$$

$$= x\left(x^2 + 1\right)^{\frac{-1}{2}} \qquad \text{Simplify.}$$

In order to find $\dfrac{d^2y}{dx^2}$ we must take the derivative of $\dfrac{dy}{dx}$. Since $\dfrac{dy}{dx}$ is a product of x and $\left(x^2 + 1\right)^{-\frac{1}{2}}$, we use the Product Rule to find $\dfrac{d^2y}{dx^2}$. When applying the Product Rule, differentiate $\left(x^2 + 1\right)^{\frac{1}{2}}$ using the General Power Rule. If $f(x) = x$, $g(x) = \left(x^2 + 1\right)^{-\frac{1}{2}}$, and $\dfrac{dy}{dx} = f(x) \cdot g(x)$, we have

$$\frac{d^2y}{dx^2} = \underset{f(x)}{x} \cdot \underset{g'(x)}{\frac{d}{dx}\left[\left(x^2 + 1\right)^{\frac{-1}{2}}\right]} + \underset{g(x)}{\left(x^2 + 1\right)^{\frac{-1}{2}}} \cdot \underset{f'(x)}{\frac{d}{dx}[x]} \qquad \text{Use the Product Rule.}$$

$$= x \cdot \left(-\frac{1}{2}\right)\left(x^2 + 1\right)^{\frac{-3}{2}}(2x) + \left(x^2 + 1\right)^{\frac{-1}{2}} \cdot 1 \qquad \text{Use the General Power Rule to find } g'(x).$$

$$= -x^2\left(x^2 + 1\right)^{\frac{-3}{2}} + \left(x^2 + 1\right)^{\frac{-1}{2}}$$

$$= \left(x^2 + 1\right)^{\frac{-3}{2}}\left[-x^2 + \left(x^2 + 1\right)\right] \qquad \text{Factor out } \left(x^2 + 1\right)^{\frac{-3}{2}}.$$

$$= \left(x^2 + 1\right)^{\frac{-3}{2}}(1) = \frac{1}{\left(x^2 + 1\right)^{\frac{3}{2}}} \qquad \text{Simplify.}$$

Example 3: First and Second Derivatives

For the function $f(u) = \dfrac{u^2}{u^2 - 9}$, find $f'(u)$ and $f''(u)$.

Solution

In order to find $f'(u)$, we must use the Quotient Rule.

Numerator $= u^2$
Denominator $= u^2 - 9$

Numerator$' = 2u$
Denominator$' = 2u$

$$f'(u) = \frac{(u^2 - 9)(2u) - (u^2)(2u)}{(u^2 - 9)^2}$$

$$f'(u) = \frac{2u^3 - 18u - 2u^3}{(u^2 - 9)^2}$$

$$= \frac{-18u}{(u^2 - 9)^2} \qquad \text{Simplify.}$$

To find $f''(u)$, we must again use the Quotient Rule. When applying the Quotient Rule, differentiate $(u^2 - 9)^2$ using the General Power Rule.

Numerator $= -18u$
Denominator $= (u^2 - 9)^2$

Numerator$' = -18$
Denominator$' = 2(u^2 - 9)(2u)$

$$f''(u) = \frac{(u^2 - 9)^2(-18) - (-18u)[2(u^2 - 9)(2u)]}{(u^2 - 9)^4}$$

$$f''(u) = \frac{\left[-18(u^2 - 9)\right]\left[(u^2 - 9) - 4u^2\right]}{(u^2 - 9)^{\cancel{4}\,3}} \qquad \begin{array}{l}\text{Factor out } -18(u^2 - 9) \text{ and cancel the} \\ \text{common factor.}\end{array}$$

$$= \frac{-18(-3u^2 - 9)}{(u^2 - 9)^3} = \frac{54(u^2 + 3)}{(u^2 - 9)^3} \qquad \text{Simplify.}$$

Concavity

Earlier, we discussed how the first derivative can be used to indicate the intervals on which a function is increasing or decreasing. When $f'(a)$ is positive, we know that f is increasing on an interval containing the point $(a, f(a))$. The meaning for $f''(a)$ is analogous: if $f''(a)$ is positive, then f' is increasing on an interval containing $x = a$. Thus, it is important, when considering information about f'', to be able to visualize what it means to say that the slope of f is increasing at $x = a$.

There are visually (and numerically) two ways in which the increasing slope of a function can be recognized. One of these occurs in Figure 1 for the graph of $f(x) = x^2$. For example, at $x = 1$, the slope is increasing. To the left of $x = 1$, at $x = 0.75$ say, the slope is 1.5. But $f'(1) = 2$, and $f'(1.5) = 3$. Thus the slope is increasing from small

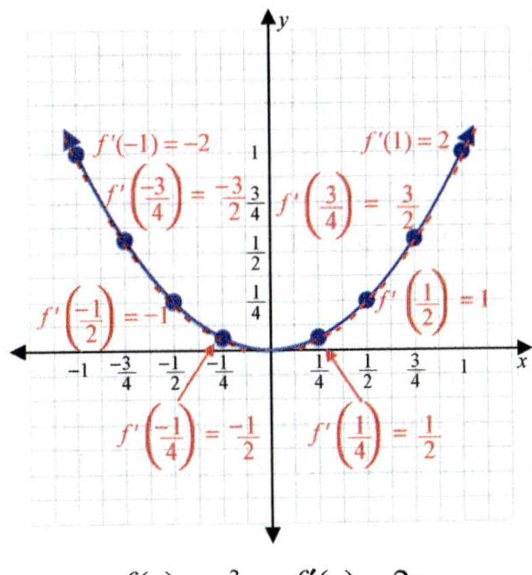

$$f(x) = x^2 \qquad f'(x) = 2x$$

FIGURE 1

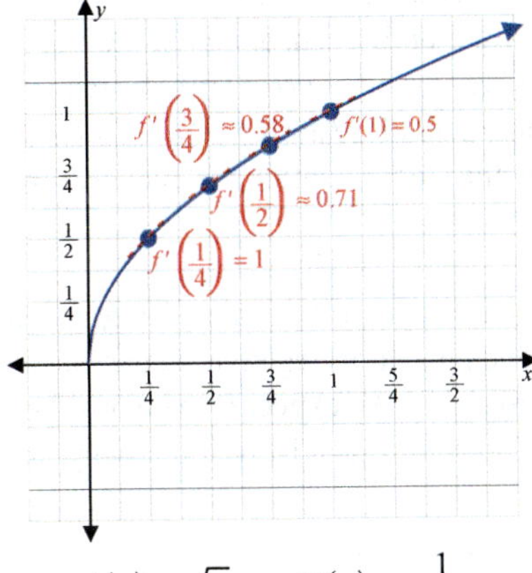

$$f(x) = \sqrt{x} \qquad f'(x) = \frac{1}{2\sqrt{x}}$$

FIGURE 2

✎ NOTE

If the graph is concave upward, we also say that f is concave upward. If the graph is concave downward, we say that f is concave downward. Alternatively, we can think of a function being "cupped up" where it is concave up and we may say a function is "cupped down" where it is concave down.

positive numbers to bigger positive numbers. But if slope increased, then its derivative is positive—that is, $f''(1)$ is positive. Visually, we observe the curve is "steeper" from left to right.

The graph of $y = x^2$ at $x = -1$ is also significant. Notice that the slopes increase here too. At the point $(-1, 1)$, the slope is -2. But just to the left, at $x = -1.5$, the slope is -3, and, just to the right at $x = -0.5$, we have $f'(-0.5) = -1$. The slope increased from the negative value -3 to -1. Thus we expect $f''(-1)$ to be positive as well. Although "steeper" does not seem to describe the graph here, say between $(-1, 1)$ and $(0, 0)$, the second derivative $f''(x)$ is positive and the slope is increasing in this region.

The graph of $f(x) = x^2$ at all of its points can also be characterized by the fact that the tangent lines lie below the curve.

On the other hand, the graph of $f(x) = \sqrt{x}$ pictured in Figure 2 has the property that tangent lines lie above the curve. Notice, as well, that the slopes for $f(x) = \sqrt{x}$ are decreasing.

This property of curves which we are discussing is called **concavity**. As illustrated in Figure 1, a curve that is **concave upward** lies above the lines tangent to the curve. Conversely, as in Figure 2, a curve that is **concave downward** lies below the lines tangent to the curve.

The following definition tells us how the first derivative can be used to determine concavity.

Concavity

Suppose that f is differentiable on the interval (a, b).

1. If f' is increasing on (a, b), then the graph of f is **concave upward** on (a, b).
2. If f' is decreasing on (a, b), then the graph of f is **concave downward** on (a, b).

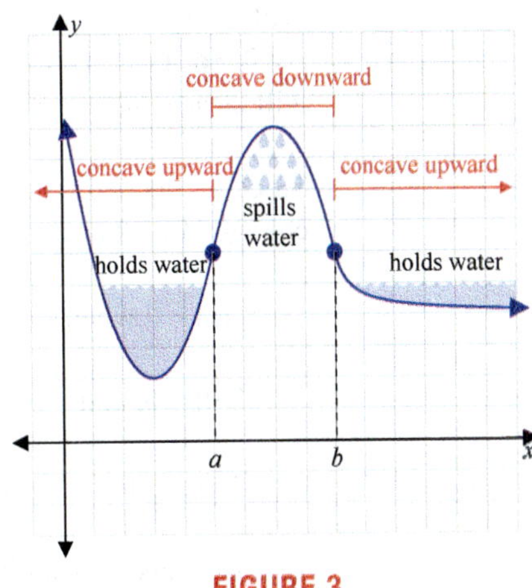

FIGURE 3

Intuitively, we say that the curve will "hold water" on an interval where the curve is concave upward and will "spill water" on an interval where the curve is concave downward. (See Figure 3.)

To determine the x-axis intervals for which a given graph is concave up or concave down, as in Figure 3, one uses the second derivative because it is precisely the tool for determining whether f' is increasing or decreasing.

In Figure 3, one observes that, at the minimum point, say (x_1, y_1), to the left of $x = a$, the graph is certainly concave up. Thus, at x_1, we have $f'(x_1) = 0$ and $f''(x_1) > 0$. At the local maximum, say (x_2, y_2), the graph is certainly concave down, and $f'(x_2) = 0$ and $f''(x_2) < 0$. Somewhere between x_1 and x_2, the concavity changed from concave up to concave down, and somewhere in between x_1 and x_2, the values of $f''(x)$ changed from positive to negative. These changes occur at the same point, namely $(a, f(a))$.

In similar fashion, to the right of the local maximum at (x_2, y_2), there is another point where the concavity changes, namely $(b, f(b))$. The student will be asked both to estimate (visually) the location of such points where the concavity changes and also to determine the exact x-values where such changes occur.

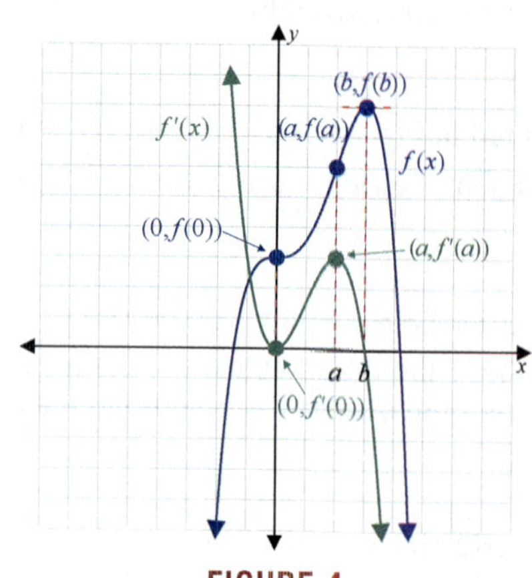

FIGURE 4

In Figure 4, we have marked the point $(a, f(a))$. Slopes decrease for all x-values less than $x = 0$. For the negative x-values not shown, the slopes are large positive numbers which decrease to small positive numbers, and then, at about $x = 0$, the slope is zero. The slopes then increase, becoming positive and larger until $x = a$, and then the slopes start to decrease. They decrease to zero at $x = b$. Then, exactly at $(b, f(b))$, the tangent to $y = f(x)$ is horizontal, and $f'(b) = 0$. The maximum point for f at $(b, f(b))$ lies directly above the point $(b, f'(b)) = (b, 0)$. Following $x = b$, the slopes continue to decrease and do so for all other positive x-values.

Let us examine more closely the graph of $y = f(x)$ in Figure 4. We have also graphed $f'(x)$ on the same axes. These graphs are very revealing. For negative values of the input x, the output values of y are increasing (left to right). We expect f' to be positive. This is shown by the fact that the graph of f' lies above the x-axis. Algebraically, we can write

$$f'(x) > 0 \text{ when } x < 0, \text{ and thus } y \text{ is increasing.}$$

As x approaches zero from the left, the graph of y levels off and the derivative appears to be zero at the point $(0, f(0))$. We observe that the derivative is in fact zero; the graph of y' hits the x-axis at $x = 0$. Since y levels off and increases at the same time, its rate of increase declines. Therefore, we reason y' to be decreasing for negative x, and this is also visible in the graph of y'. Further, tangents to $y = f(x)$ lie above the graph when $x < 0$. So $f(x)$ is concave down to the left of $x = 0$. But tangents just to the right of $x = 0$ will clearly lie below the graph. So y is concave up just to the right of $x = 0$.

Since y' decreases (until $x = 0$) and then increases, we can predict y'' (the rate of change of y') to be zero at the minimum value of y'. That is, for Figure 4, we expect

$$f''(x) < 0 \text{ when } x \text{ is negative,}$$

$$f''(0) = 0, \text{ and}$$

$$f''(x) > 0 \text{ for } x \text{ bigger than zero but less than } a, \text{ stated algebraically } 0 < x < a.$$

This is precisely what it means for $(0, f(0))$ to be a point on the graph of y at which the concavity changes.

Does y have yet another point at which concavity changes? In fact, the rate of change of y (that is, y') increases to a maximum from $x = 0$ to $x = a$ and then decreases. Thus, for Figure 4, we expect y'' is positive to the left of $x = a$ and is negative to the right of $x = a$. This means the point $(a, f(a))$ is another point of change in the concavity of f.

We emphasize that such points occur when y' has a local extrema. Here, for the graph of f', $(0, f'(0)) = (0, 0)$ is a local minimum and $(a, f'(a))$ is a local maximum.

The following test for concavity gives us an analytical method for determining the intervals on which the derivative is increasing or decreasing. This method involves the second derivative, f''. Just as f' tells us where f is increasing or decreasing, f'' tells us where f' is increasing or decreasing.

Interpreting $f''(x)$ to Determine Concavity

Suppose that f is a function and f' and f'' both exist on the interval (a, b).

1. If $f''(x) > 0$ for all x in (a, b), then f' is increasing and f is **concave upward** on (a, b).

2. If $f''(x) < 0$ for all x in (a, b), then f' is decreasing and f is **concave downward** on (a, b).

Example 4: Determining Concavity

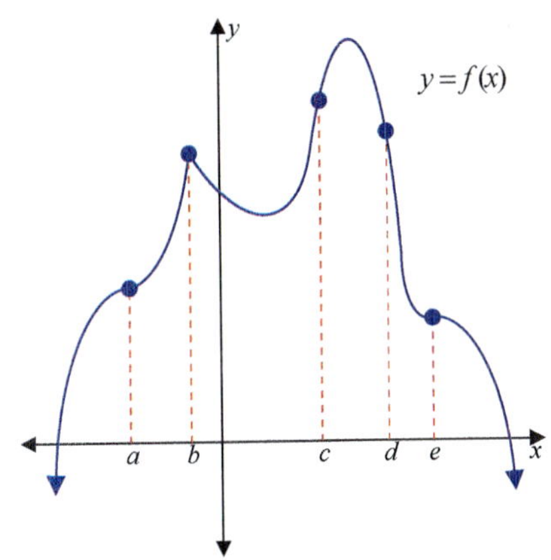

In the figure shown, the graph of $y = f(x)$ is given.

a. List the intervals on which f is concave upward.

b. List the intervals on which f is concave downward.

Solution

a. f is concave upward on the intervals (a, b), (b, c), and (d, e).

b. f is concave downward on the intervals $(-\infty, a)$, (c, d), and $(e, +\infty)$.

Example 5: Sketching a Curve Using Concavity

Let $f(x) = x^4 - 4x^3$.

a. Determine the intervals on which the function is concave upward and the intervals on which it is concave downward.

b. Use the information from part **a.** to draw a rough sketch of the graph of $f(x)$.

Solution

a. To determine concavity, we must interpret $f''(x)$. First, we find the second derivative and then determine the intervals where it is positive and the intervals where it is negative.

$$f'(x) = 4x^3 - 12x^2 \qquad \text{To find the second derivative, find the first derivative of}$$
$$f''(x) = 12x^2 - 24x \qquad f(x) \text{ and then find its derivative.}$$

Now that we have determined $f''(x)$, set $f''(x) = 0$ and solve for x to find the endpoints of the intervals to be tested for concavity.

$$12x^2 - 24x = 0$$
$$12x(x-2) = 0$$
$$x = 0, 2$$

So there are three intervals to be considered:

$$(-\infty, 0), \quad (0, 2), \quad \text{and} \quad (2, +\infty).$$

The values of y'' in the interval $(-\infty, 0)$ are all positive or all negative. (If some were positive and some were negative, then there would have to be some other x-value, say b, such that $f''(b) = 0$. But there is no such x-value in that interval.) Similarly, in each interval, all the values of y'' are negative or all are positive. Therefore, for each interval, we choose one test value in the selected interval which seems convenient for calculation. We use this x-value, say $x = a$, to determine whether $f''(a)$ is positive or negative. If $f''(a)$ is positive, then $f''(x)$ is positive for all x in that interval, and the graph of y is concave up for that interval. Likewise, if $f''(a)$ is negative, then $f''(x)$ is negative for all x in that interval, and the graph of y is concave down for that interval.

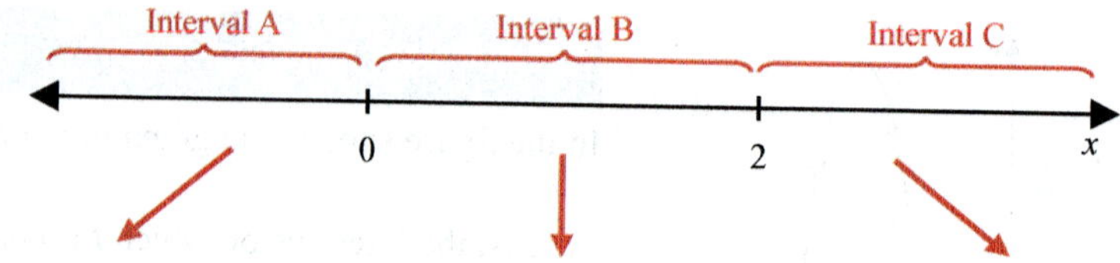

Interval A	Interval B	Interval C
We will use the test point $x = -1$.	We will use the test point $x = 1$.	We will use the test point $x = 3$.

$$f''(-1) = 12(-1)^2 - 24(-1)$$
$$= 12 + 24$$
$$= 36$$
$$36 > 0$$

$$f''(1) = 12(1)^2 - 24(1)$$
$$= 12 - 24$$
$$= -12$$
$$-12 < 0$$

$$f''(3) = 12(3)^2 - 24(3)$$
$$= 108 - 72$$
$$= 36$$
$$36 > 0$$

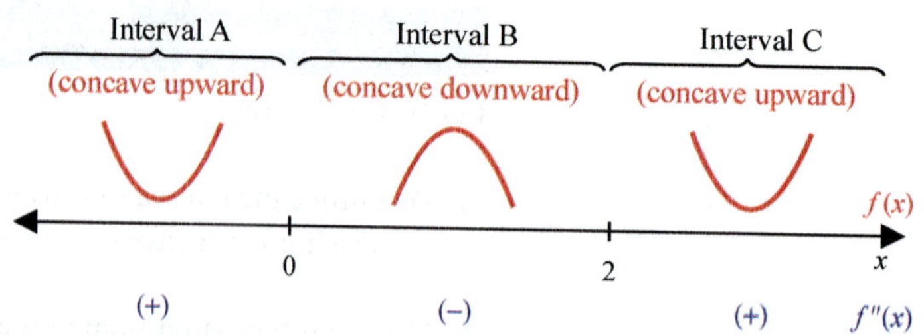

Therefore, f is concave upward on the intervals $(-\infty, 0)$ and $(2, +\infty)$, and it is concave downward on the interval $(0, 2)$.

b. In making a rough sketch of a function using concavity, one wishes to calculate as few (x, y) points as possible. However, the y-intercept, $(0, f(0))$, is usually easy to calculate at a glance (without a calculator). Here $f(0) = 0$, so the origin is a point on the graph. Also, at $x = 0$, the graph changes concavity from concave up to concave down. Near the origin, the graph of $f(x)$ must look like the sketch provided. We have calculated $f(2) = -16$, where the concavity changes from down to up. From the form of the polynomial $f(x) = x^4 - 4x^3$, we can say that eventually all the y-values are positive (since the term of largest degree is positive). This means that there is a global minimum to the right of $(2, -16)$; after this point, y increases without further concavity changes. The x-intercept to the right of $(2, -16)$ is easily found since there is no constant term for y. We set $f(x) = 0$ and solve by factoring.

$$x^4 - 4x^3 = 0$$
$$x^3(x - 4) = 0$$
$$x = 0, \quad x = 4$$

So $(4, 0)$ is a convenient reference point. We have calculated only three (x, y) values.

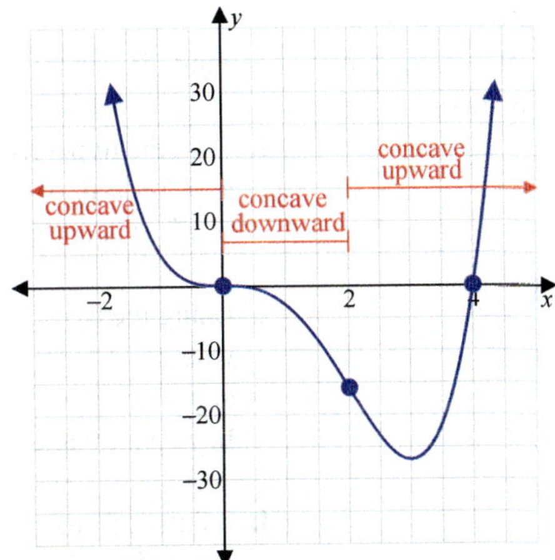

Points of Inflection

The graph in Example 5 changes concavity at $x = 0$ and at $x = 2$. In most applications, such dramatic changes in a function's behavior signify an important event. We concentrate now on locating such points algebraically. The graphs of three continuous functions are shown in Figure 5. In each case, the graph changes concavity at a point marked with the letter I. In Figure 5(A), the tangent lines lie above the graph to the left of I but below the graph to the right of I. At that point I, the tangent line is said to cross through the graph. Such points are called **points of inflection**. Similar situations are presented in parts (B) and (C).

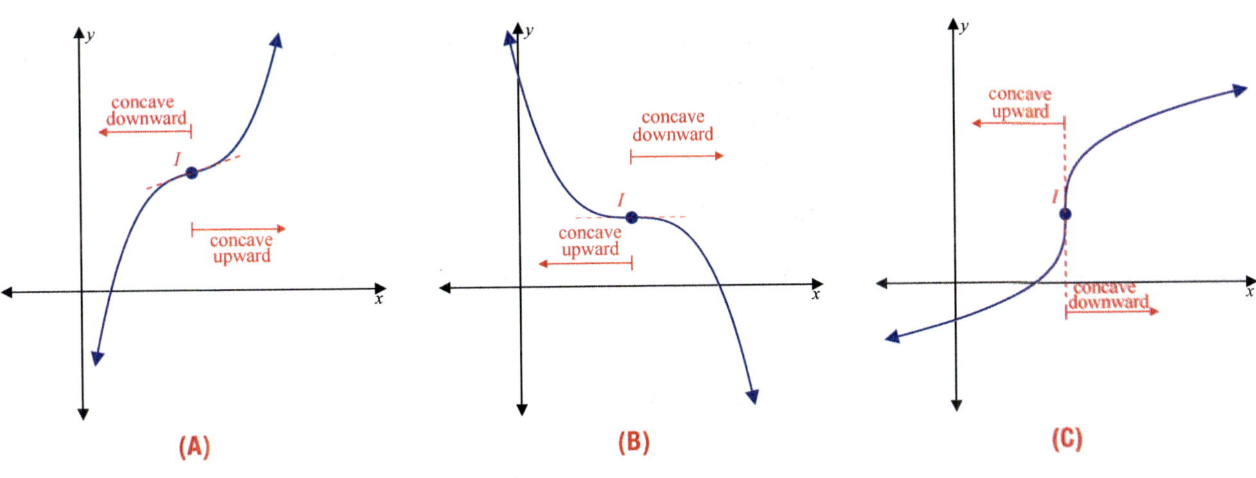

FIGURE 5

✎ NOTE

Hypercritical values of x are those values $x = c$ in the domain of f where $f''(c) = 0$ or $f''(c)$ does not exist. As we will see, points of inflection occur only at hypercritical values. It turns out, however, that hypercritical values do not guarantee points of inflection. Remember that at a point of inflection, the function must change concavity from up to down, or vice versa.

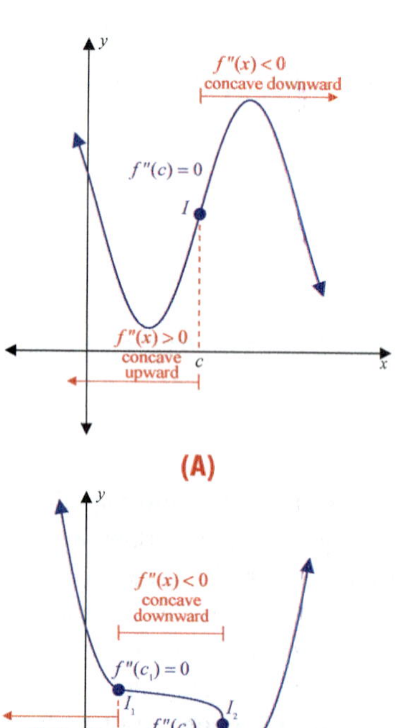

FIGURE 6

Point of Inflection

If the graph of a continuous function has a tangent line at a point (possibly a vertical line) and the graph changes concavity at that point, then the point is called a **point of inflection**.

To Determine Points of Inflection

Suppose that f is a continuous function.

1. Find $f''(x)$.

2. Find the hypercritical values. That is, find the values $x = c$ where
   a. $f''(c) = 0$, or
   b. $f''(c)$ is undefined.

3. Using the hypercritical values as endpoints of intervals, determine the intervals where
   a. $f''(x) > 0$ and f is concave upward, and
   b. $f''(x) < 0$ and f is concave downward.

4. Points of inflection occur at those hypercritical values where f changes concavity. (See Figure 6.)

Example 6: Finding Points of Inflection

Let $f(x) = x^3 - \dfrac{3}{2}x^2 + 5$.

a. Determine the intervals on which f is concave upward and the intervals on which it is concave downward.

b. Locate any points of inflection.

Solution

a. We find the hypercritical values and then determine the concavity on the related intervals.

$$f'(x) = 3x^2 - 3x$$ To find the second derivative, find the first derivative of $f(x)$

$$f''(x) = 6x - 3$$ and then find its derivative.

Now set $f''(x) = 0$.

$$6x - 3 = 0$$ Note that there are no values where $f''(x)$ is undefined.

$$x = \frac{1}{2}$$

There are two intervals to be considered:

$$\left(-\infty, \frac{1}{2}\right) \quad \text{and} \quad \left(\frac{1}{2}, +\infty\right).$$

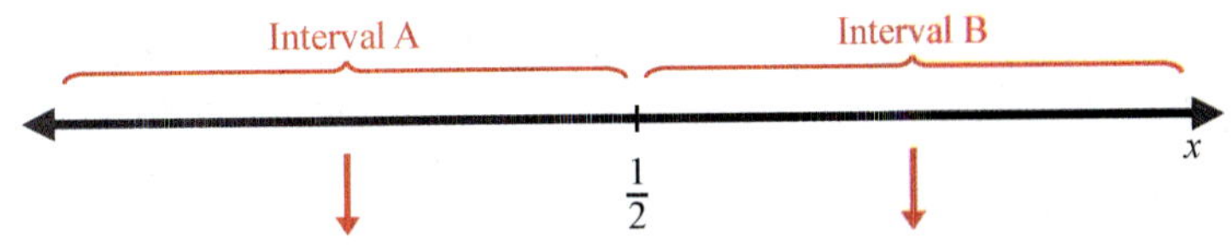

Interval A
We will use the test point $x = -1$.

$$f''(-1) = 6(-1) - 3$$
$$= -6 - 3$$
$$= -9$$
$$-9 < 0$$

Interval B
We will use the test point $x = 1$.

$$f''(1) = 6(1) - 3$$
$$= 6 - 3$$
$$= 3$$
$$3 > 0$$

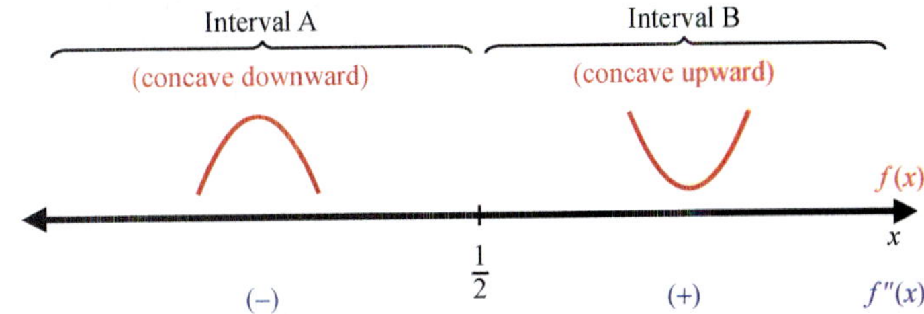

Therefore, f is concave downward on $\left(-\infty, \dfrac{1}{2}\right)$ and is concave upward on $\left(\dfrac{1}{2}, +\infty\right)$.

b. Since we found in part **a.** that f changes concavity at $x = \dfrac{1}{2}$, there must be a point of inflection there.

$$f(x) = x^3 - \frac{3}{2}x^2 + 5$$

Substitute $x = \dfrac{1}{2}$ into $f(x)$ to find the point of the inflection.

$$f\left(\frac{1}{2}\right) = \left(\frac{1}{2}\right)^3 - \frac{3}{2}\left(\frac{1}{2}\right)^2 + 5$$

$$= \frac{1}{8} - \frac{3}{8} + 5 = \frac{19}{4}$$

So $\left(\dfrac{1}{2}, \dfrac{19}{4}\right)$ is a point of inflection of $f(x)$.

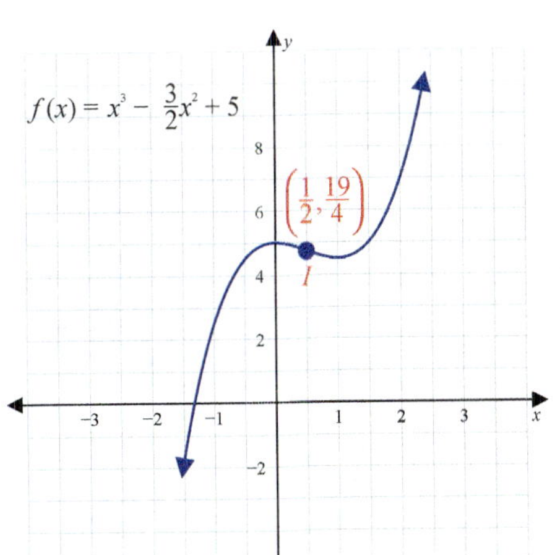

Using a Calculator to Find Inflection Points

Powerful hand-held calculators can sketch y, y', and y'', all on the same axes. They can locate coordinates of maximum points, minimum points, and points of inflection.

A TI-84 Plus can be used to graph y' and from this graph of y', one can find inflection points easily. To illustrate this procedure, we use the function $y = \dfrac{8}{3}x^3 - x^4 + 5$. (This function was actually used to graph Figure 4. Refer to this figure and its accompanying text as necessary.)

Step 1: Type $y = \dfrac{8}{3}x^3 - x^4 + 5$ into the **Y1** position (in the **Y=** menu).

FIGURE 7

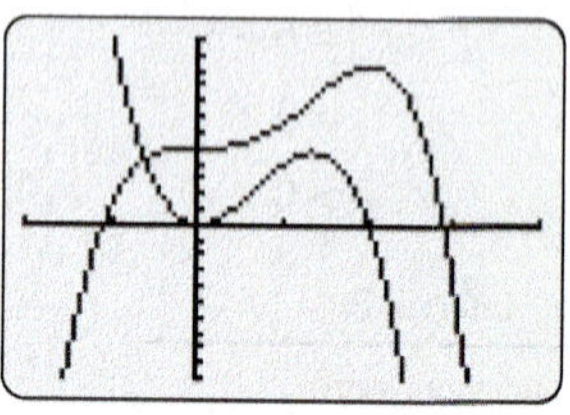

FIGURE 8

Step 2: Put the cursor in the **Y2** position. Next, press the [math] button. Arrow down the menu to item **8:nDeriv(** and press [enter]. This inserts "**nDeriv(**" into the **Y2** slot. Next press [vars], and arrow right one position to **Y-VARS**. The menu cursor will be on item **1:Function**. Press [enter]. When you see the menu of functions, press [enter] again. This selects **Y1** and puts it into the **nDeriv(** function. Next, input "**,X,X)**". The full entry in the **Y2** position should look like **nDeriv(Y1,X,X)**. (See Figure 7.)

Step 3: Using a window of [−2, 3] by [−6, 12], press [graph]. Both graphs should be displayed on the screen as shown in Figure 8.

Note: Typically, you will not want the graph of y' displayed in every window. You can "turn off" the graph of y' by moving the calculator's cursor, using the arrow keys, to the equal sign just after **Y2**. With the cursor on the equal sign, press [enter]. This deselects the **Y2** graph, but leaves the typed equation or commands in place, so that the graph of y' can be reselected when desired.

Step 4: Type [2nd] [trace] and select item **4:Maximum**. If the cursor is blinking on the graph of $f(x)$, press the down arrow once. (This switches graphs so that the graph of $f'(x)$ is now selected.)

Step 5: You are now being prompted to select a left bound for the maximum point on the graph of $f'(x)$. Using the arrow keys, move the blinking cursor to the left of the maximum point, at or to the left of $x = 1$. Press [enter].

Step 6: Now, using the arrow keys, select a right bound by moving the blinking cursor to the right of the maximum point, say to $x = 1.7$. Press [enter].

Step 7: Finally, you are prompted "Guess". You should press the left arrow key once and then press [enter]. The (x, y)-values of the maximum point will be displayed at the bottom of the screen. In this case $x = 1.3333304$ and $y = 4.7407381$ will be given. The y-value here means $f''(1.33) = 4.74$.

As might be expected, the actual value of x is $\dfrac{4}{3}$, but the answer given by the calculator is quite accurate. The computation shows $f''\left(\dfrac{4}{3}\right) = 0$. So $\left(\dfrac{4}{3}, f'\left(\dfrac{4}{3}\right)\right)$ is the high point on the graph of f' and thus $\left(\dfrac{4}{3}, f\left(\dfrac{4}{3}\right)\right)$ is the inflection point on $f(x)$. We can confirm the calculator results.

$$f(x) = \frac{8x^3}{3} - x^4 + 5,$$
$$f'(x) = 8x^2 - 4x^3, \text{ and}$$
$$f''(x) = 16x - 12x^2$$

We set $f''(x) = 0$ and obtain

$$16x - 12x^2 = 0, \text{ so}$$
$$4x(4 - 3x) = 0.$$

Thus $x = 0$ and $x = \dfrac{4}{3}$ will make $f''(x)$ equal to zero. Both values give inflection points for $f(x)$.

We can use $f'(x) = 0$ to locate the maximum and minimum points for $f(x)$.

$$f'(x) = 0$$
$$8x^2 - 4x^3 = 0$$
$$4x^2(2 - x) = 0$$
$$x = 0, 2$$

In Figure 8, we observe $x = 0$ corresponds to an inflection point (neither a maximum nor a minimum) and $x = 2$ corresponds to the maximum point $\left(2, \dfrac{31}{3}\right)$.

In this problem, the algebraic solutions of $f'(x) = 0$ and $f''(x) = 0$ are of a familiar type. When these equations are too difficult for an algebraic solution, a graphing utility is invaluable.

12.1 EXERCISES

PRACTICE

1. At each point marked on the graph of f, determine if f' is positive, negative, or zero. Determine if f'' is positive, negative or zero.

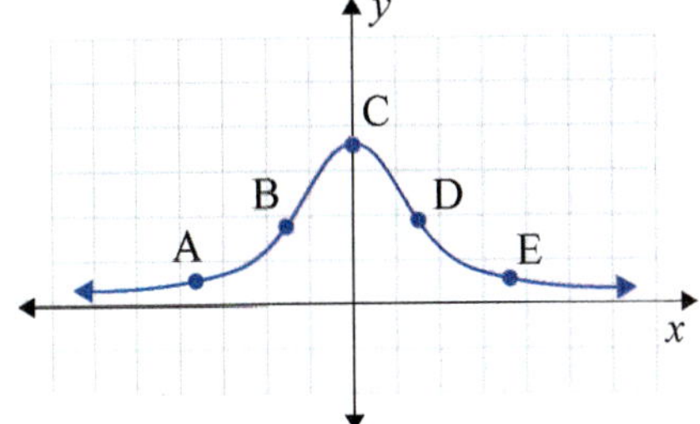

Draw a graph that satisfies the given conditions in Exercises 2–5.

2. $f(5) = 9, \quad f'(5) = 2, \quad f''(5) = -2$

3. $f(-5) = -9, \quad f'(-5) = 2, \quad f''(-5) = 2$

4. $f(5) = -9, \quad f'(5) = 0, \quad f''(5) = 3$

5. $f(0) = 12, \quad f'(0) = 0, \quad f''(0) = -3$

For Exercises 6–13, find $f''(x)$. Then evaluate $f''(0)$, $f''(1)$, and $f''(4)$, if they exist.

6. $f(x) = x^3 + x^2 + 3$

7. $f(x) = x^3 - x^2 + 7$

8. $f(x) = x^2 - 5\sqrt{x} + 1$

9. $f(x) = x^2 + 2\sqrt{x} - 3$

10. $f(x) = \sqrt{x - 4}$

11. $f(x) = \sqrt{2x + 1}$

12. $f(x) = \dfrac{x}{x + 5}$

13. $f(x) = \dfrac{x - 2}{x + 4}$

In Exercises 14–17, sketch a possible graph for $f'(x)$ on the same coordinate axes as $f(x)$. Then locate all inflection points on the graph of $f(x)$.

14.

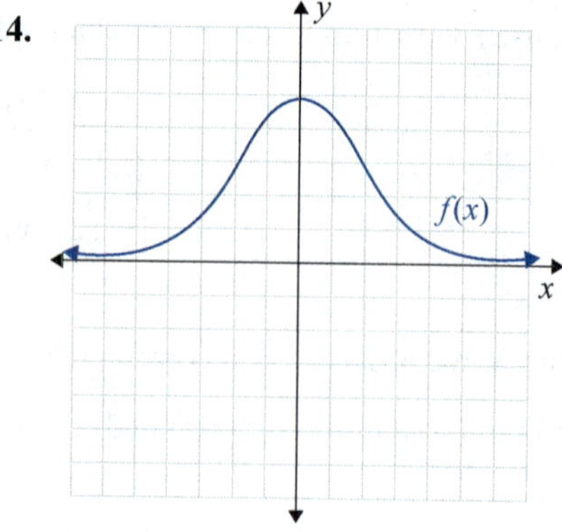

15.

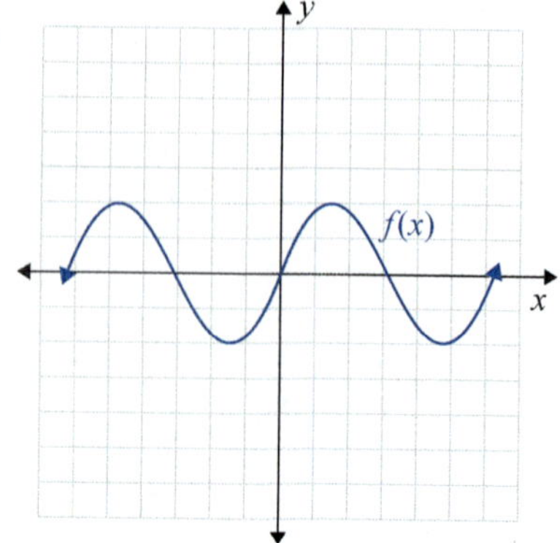

16.

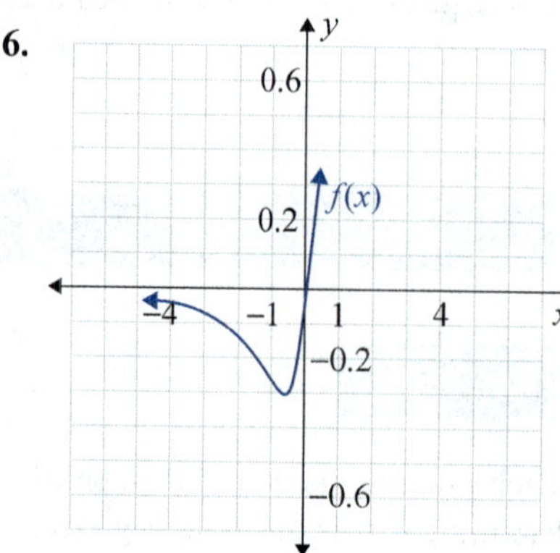

17.

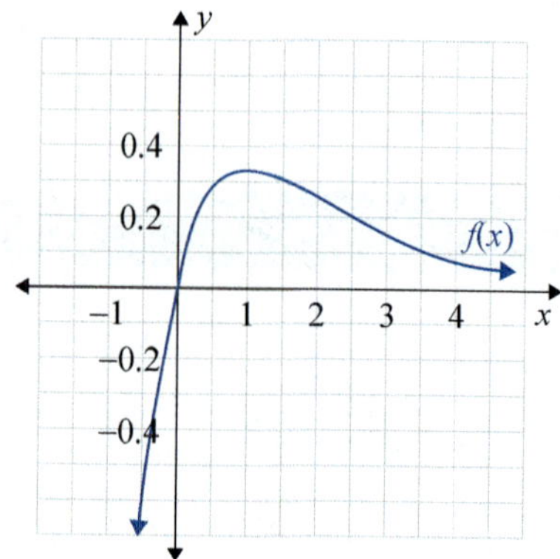

For each of the graphs in Exercises 18–21, list the interval(s) **a.** on which f is concave upward and **b.** on which f is concave downward; then **c.** locate all points of inflection.

18.

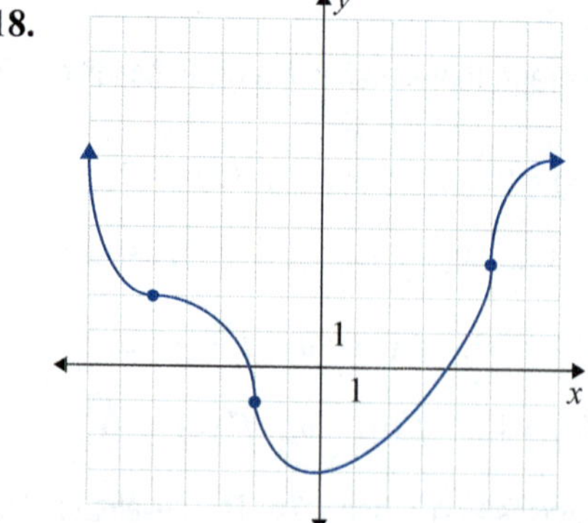

19.

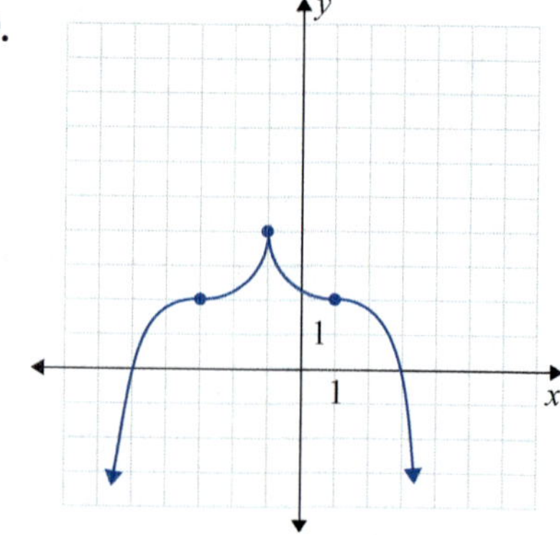

20. 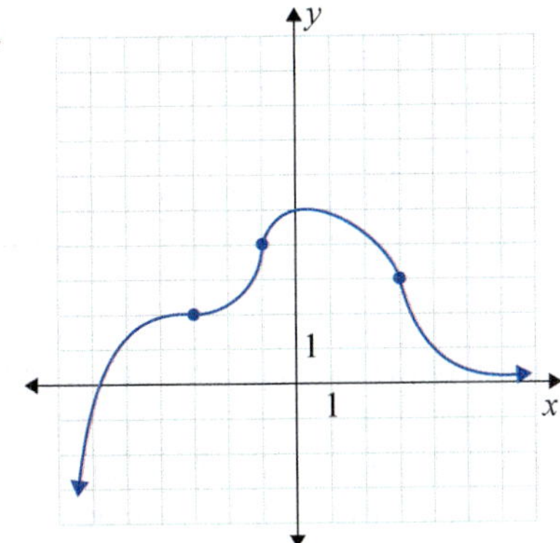

21. 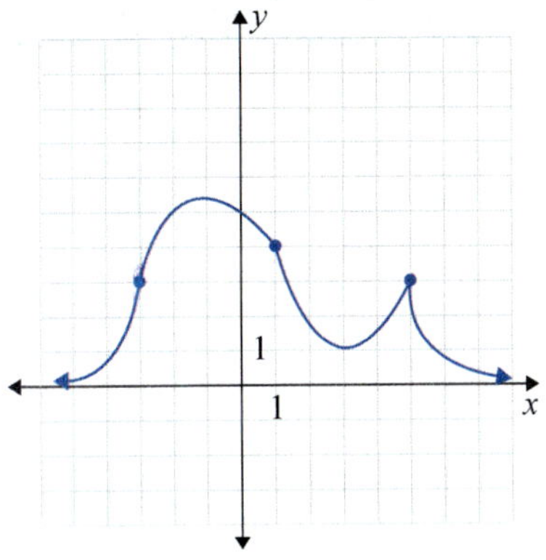

In Exercises 22–33, determine the intervals on which each function is **a.** concave upward and **b.** concave downward; then **c.** locate all points of inflection. Use the information gathered to sketch the function. Confirm the details with a graphing calculator.

22. $f(x) = 2x^2 + 5x - 9$

23. $f(x) = 5x^2 + 8x - 1$

24. $f(x) = x^3 - 3x^2 + 7$

25. $f(x) = x^3 + 6x^2 - 10$

26. $f(x) = x^3 + 11x - 4$

27. $f(x) = 5x^3 + 7x + 2$

28. $f(x) = \dfrac{1}{3}x^3 - 2x^2 + x - 3$

29. $f(x) = \dfrac{1}{3}x^3 + 3x^2 + 2x - 5$

30. $f(x) = \sqrt[3]{2x + 3}$

31. $f(x) = \sqrt[3]{5x - 3}$

32. $f(x) = \dfrac{x}{x^2 - 4}$

33. $f(x) = \dfrac{4x}{x^2 - 5}$

✏ WRITING & THINKING

In Exercises 34–37, give an example of a polynomial function that satisfies the conditions.

34. $F(5) = 15$; $F'(x)$ is nonzero, but $F''(x) = 0$ for all x.

35. $G(0) = 0$, $G'(0) = 0$, and $G''(0) = 0$; $G(x)$ is concave upward everywhere and has no inflection points.

36. $H(4) = 0$; $H'(x)$ is positive for $x > 4$ and negative for $x < 4$. $H(x)$ has no inflection points.

37. $J(4) = 0$; $J'(4)$ is zero but $J'(x)$ is positive if $x \neq 4$; $J''(4) = 0$.

🚀 APPLICATIONS

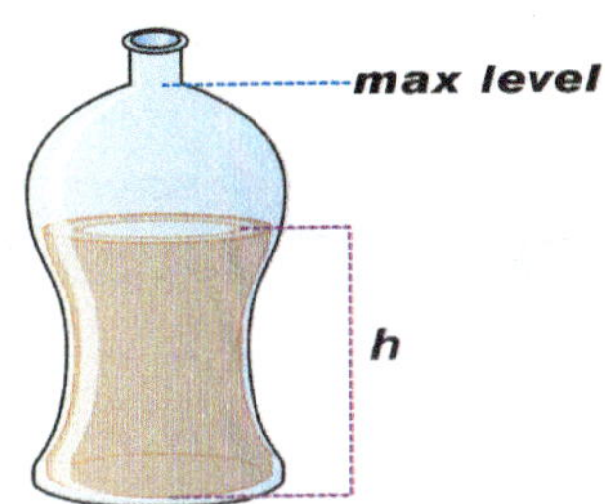

38. Filtrate: In a chemistry lab a filtrate drips slowly but continuously at a constant rate into a glass container shaped like the one shown. The container eventually fills to the base of the neck. Let t denote the passage of time and h be the height of the liquid.

a. Describe at what points on the bottle $\dfrac{dh}{dt}$ will be a maximum and a minimum.

b. Sketch a graph of $\dfrac{dh}{dt}$. Are there any inflection points on a graph of $y = h(t)$?

c. Add a sketch of $y = h(t)$ on the same coordinate axes as in part **b.**

12.2 THE SECOND DERIVATIVE TEST

■ TOPICS

- ■ Application: Point of Diminishing Returns

We have seen how the first derivative of a function can be used to locate local extrema and how the second derivative can be used to analyze concavity and locate points of inflection. Now we will show how the second derivative, if it exists, provides a relatively simple test for local extrema.

Second Derivative Test for Local Extrema

Suppose that f is a function, f' and f'' exist on the interval (a, b), c is in (a, b), and $f'(c) = 0$.

1. If $f''(c) > 0$, then $f(c)$ is a local minimum. (See Figure 1(A).)

2. If $f''(c) < 0$, then $f(c)$ is a local maximum. (See Figure 1(B).)

3. If $f''(c) = 0$, then the test fails to give any information about local extrema.

■ NOTE

The principle being applied in the Second Derivative Test is simple. If $f''(c)$ is defined at a local maximum $(c, f(c))$, a curve is concave down. If $f''(c)$ is defined at a local minimum $(c, f(c))$, a curve is concave up.

Of course, the primary value of the Second Derivative Test occurs only when the algebra formula is available for analysis.

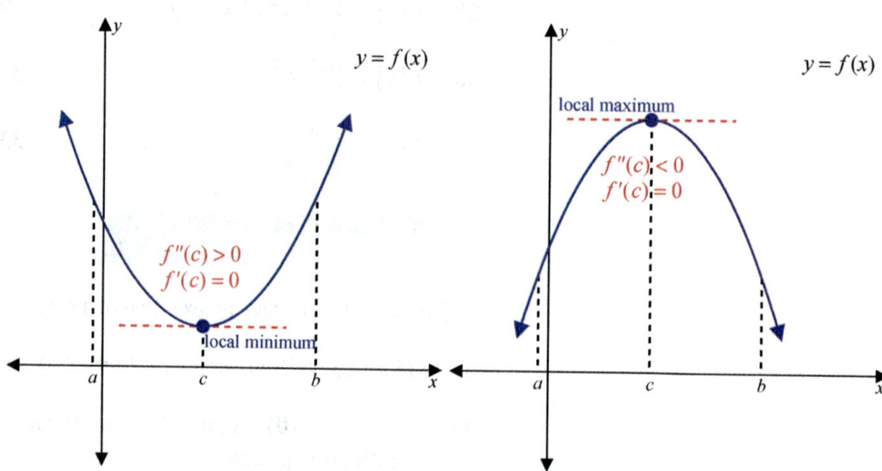

At a local minimum, $x = c$, $f(x)$ must be concave upward ($f''(c) > 0$).

(A)

At a local maximum, $x = c$, $f(x)$ must be concave downward ($f''(c) < 0$).

(B)

FIGURE 1

In the pursuit of fully understanding calculus, it is very important that the student is capable of graphing a few functions successfully by using calculus, without the aid of a calculator. The next example will demonstrate this.

Example 1: Using the Second Derivative Test

Let $f(x) = x^4 - 18x^2$.

a. Find the local extrema.

b. Find the points of inflection.

Solution

a. We will use the Second Derivative Test to find all local extrema of $f(x)$. First, find all values $x = c$ so that $f'(c) = 0$.

$$f'(x) = 4x^3 - 36x \qquad \text{Find the derivative of } f(x).$$
$$0 = 4x^3 - 36x \qquad \text{Set } f'(x) = 0.$$
$$0 = 4x(x^2 - 9)$$
$$0 = 4x(x+3)(x-3)$$
$$x = -3, 0, 3$$

Substituting these x-values into $f(x)$, the corresponding y-values are $f(-3) = -81$, $f(0) = 0$, and $f(3) = -81$.

Now we need to determine how $f''(x)$ responds at these x-values.

$$f''(x) = 12x^2 - 36 \qquad \text{Find } f''(x).$$
$$f''(-3) = 12(-3)^2 - 36 \qquad \text{Substitute each of the values found for } c: -3, 0, \text{ and } 3.$$
$$= 72$$
$$f''(0) = 12(0)^2 - 36$$
$$= -36$$
$$f''(3) = 12(3)^2 - 36$$
$$= 72$$

Use the method for determining concavity in addition to the Second Derivative Test to interpret these results:

$x = -3$	$x = 0$	$x = 3$
$f''(-3) > 0$	$f''(0) < 0$	$f''(3) > 0$
Therefore, f is concave up and $(-3, -81)$ is a local minimum.	Therefore, f is concave down and $(0, 0)$ is a local maximum.	Therefore, f is concave up and $(3, -81)$ is a local minimum.

b. The points of inflection come from setting $f''(x) = 0$ and solving for x.

$$f''(x) = 12x^2 - 36$$
$$0 = 12x^2 - 36 \qquad \text{Set } f''(x) = 0.$$
$$0 = 12(x^2 - 3)$$
$$0 = 12(x - \sqrt{3})(x + \sqrt{3})$$
$$x = \pm\sqrt{3}$$

The corresponding y-values are

$$f(-\sqrt{3}) = (-\sqrt{3})^4 - 18(-\sqrt{3})^2$$
$$= 9 - 18(3) = -45$$
$$f(\sqrt{3}) = (\sqrt{3})^4 - 18(\sqrt{3})^2$$
$$= 9 - 18(3) = -45$$

To verify your graph with a TI-83/84 Plus calculator, perform the following steps:

1. Enter the given function into Y1.

2. Check that your `window` is appropriate.

3. Press `graph`.

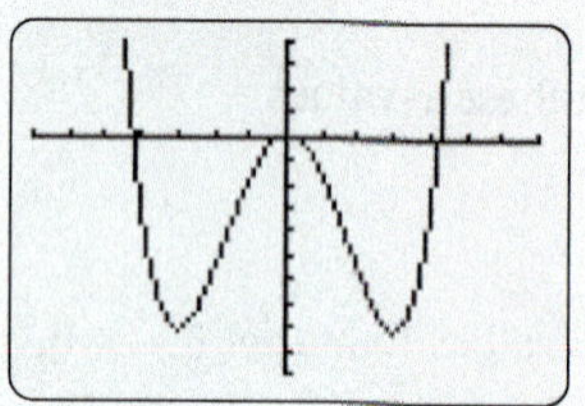

So the points of inflection are $\left(-\sqrt{3}, -45\right)$ and $\left(\sqrt{3}, -45\right)$.

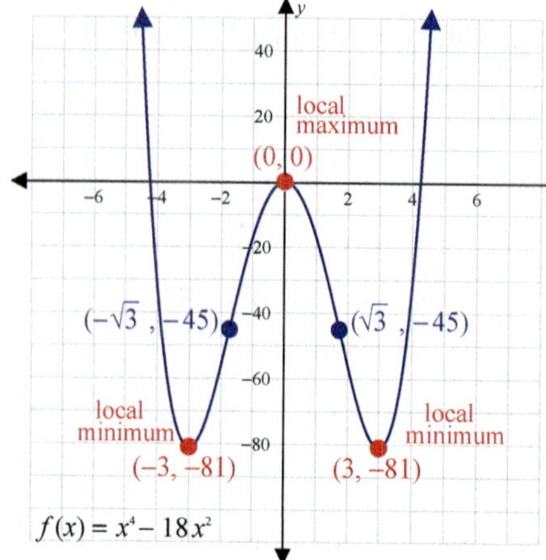

Note that it is easy (without a graphing calculator) to sketch $y = x^4 - 18x^2$. We need only the coordinates of the three extreme points, and we may sketch a smooth curve only knowing which is a maximum and/or a minimum. In this example, the potential inflection points occur between a maximum and a minimum point and so we "know" the concavity changes at these points.

In the next example, we confirm intervals on which the graph of a function is concave up and on which it is concave down.

Locate the local extrema for the function $f(x) = \dfrac{8}{3}x^3 - x^4$ and sketch a graph of the function.

Solution

$$f'(x) = 8x^2 - 4x^3$$
$$0 = 8x^2 - 4x^3$$
$$0 = 4x^2(2 - x)$$
$$x = 0, 2$$

Find $f'(x)$ and all values $x = c$ where $f'(c) = 0$.

Determine the corresponding y-values of $x = 0$ and $x = 2$.

$$f(x) = \frac{8}{3}x^3 - x^4$$
$$f(0) = \frac{8}{3}(0)^3 - (0)^4 = 0$$
$$f(2) = \frac{8}{3}(2)^3 - (2)^4 = \frac{16}{3}$$

The critical points are $(0, 0)$ and $\left(2, \dfrac{16}{3}\right)$.

Now find $f''(0)$ and $f''(2)$ and apply the Second Derivative Test to interpret these points.

$$f''(x) = 16x - 12x^2$$
$$f''(0) = 16(0) - 12(0)^2$$
$$= 0$$
$$f''(2) = 16(2) - 12(2)^2$$
$$= -16$$

Substitute $x = 0$ and $x = 2$ into $f''(x)$.

Since $f''(0) = 0$, the Second Derivative Test fails at this point.

$$x = 0$$
$$f''(0) = 0$$

The Second Derivative Test cannot provide information about the point $(0, 0)$ as a local extremum.

$$x = 2$$
$$f''(2) < 0$$

Therefore, f is concave down at $x = 2$ and $\left(2, \dfrac{16}{3}\right)$ is a maximum point.

A local maximum occurs at the point $\left(2, \dfrac{16}{3}\right)$.

Because the Second Derivative Test provides no information about the point $(0, 0)$ as a local extremum, we must use the First Derivative Test. Let us check the sign of $f'(x)$ on either side of 0. Using $x = -1$ and $x = 1$,

$$f'(-1) = 12 \text{ (positive) and } f'(1) = 4 \text{ (also positive).}$$

This shows that f increases to the left of $x = 0$ as well as to the right. Since there is no sign change from one side of $f'(0)$ to the other, $x = 0$ does not give a local minimum or a local maximum.

In fact, we have shown that slope is positive to the left of the origin, decreases to zero at the origin, and then increases to the right of the origin. This shows that the concavity changed at $(0, 0)$ which must be a point of inflection.

> **📝 NOTE**
>
> The Second Derivative Test requires only one number to be checked. The First Derivative Test requires two numbers to be computed. For polynomial functions, the formula for f'' is one degree lower than that for f', so numbers are easier to calculate. For these reasons, the Second Derivative Test is usually preferred. However, sometimes, as in Example 2, it provides no information, and one must then use the First Derivative Test.

Since the graph is concave up to the right of $(0, 0)$, and concave down at the maximum $\left(2, \dfrac{16}{3}\right)$, there must be an inflection point in between these points. Set $y'' = 0$ and solve for x.

$$f''(x) = 16x - 12x^2$$
$$0 = 16x - 12x^2$$
$$0 = 4x(4 - 3x)$$
$$x = 0, \frac{4}{3}$$

So $\left(\dfrac{4}{3}, f\left(\dfrac{4}{3}\right)\right)$ is the other inflection point. Note as a check that $x = \dfrac{4}{3}$ is between $x = 0$ and $x = 2$, which is already known.

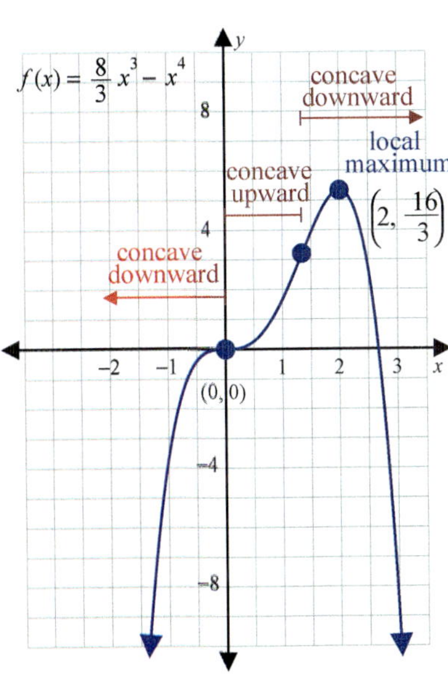

Application: Point of Diminishing Returns

A company expects that as it spends more dollars on an advertising campaign for a product, the sales of the product will increase. A typical sales curve (sales as a function of advertising dollars) is shown in Figure 2.

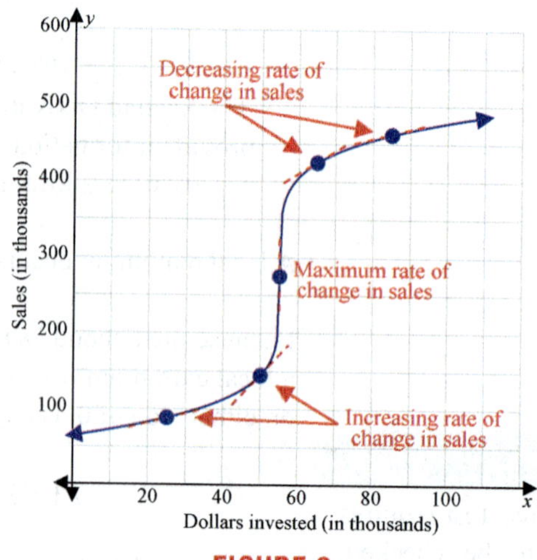

FIGURE 2

At the beginning of a successful advertising campaign, not only do the sales increase, but the rate at which the sales grow also increases (the sales curve is concave upward). As the campaign continues, usually there is some point at which the rate of growth is a maximum. As more money is spent beyond this point, sales continue to increase, but at a slower rate. In economics, this point of rate change is called the **point of diminishing returns**.

Point of Diminishing Returns

The **point of diminishing returns** occurs at a point of inflection where the sales curve changes from concave upward to concave downward.

Example 3: Point of Diminishing Returns

Find the point of diminishing returns for the sales function

$$S(x) = -0.02x^3 + 3x^2 + 100,$$

where x represents thousands of dollars spent on advertising, $0 \le x \le 80$, and S is sales in thousands of dollars for automobile tires.

Solution

To find the point at which concavity changes from concave upward to concave downward, we must first find the hypercritical values of x between 0 and 80. Then we determine whether these points are points of inflection.

$$S(x) = -0.02x^3 + 3x^2 + 100$$
$$S'(x) = -0.06x^2 + 6x \qquad \text{Find } S'(x).$$
$$S''(x) = -0.12x + 6 \qquad \text{Find } S''(x).$$
$$-0.12x + 6 = 0 \qquad \text{Set } S''(x) \text{ equal to 0 to find the hypercritical}$$
$$x = 50 \qquad \text{values.}$$

Note that $S''(10) = -0.12(10) + 6 = 4.8$ and $S''(60) = -1.2.$

Testing shows that

$$S''(x) > 0 \quad \text{for} \quad 0 < x < 50$$

and

$$S''(x) < 0 \quad \text{for} \quad 50 < x < 80.$$

Concavity changes from upward on the left side of $S(50)$ to downward on the other, signifying it is a point of diminishing returns.

Thus the point of diminishing returns is at $(50, S(50)) = (50, 5100)$. At the point of diminishing returns, $50,000 are spent on advertising, and sales in tires are $5,100,000.

12.2 EXERCISES

PRACTICE

Find both the first and second derivatives for each of the functions in Exercises 1–12. Locate any relative maximum or minimum points and any points of inflection. Determine the intervals on which the function is concave upward or concave downward.

1. $f(x) = 7x^2 - 28x + 8$

2. $f(x) = 5x^2 - 9x + 2$

3. $f(x) = 2x^3 + 5x - 1$

4. $f(x) = 3x^3 + 6x - 8$

5. $f(x) = x^3 + 2\sqrt{x} + 5$

6. $f(x) = x^4 - 3\sqrt{x} + 2$

7. $f(x) = (x^2 + 7)^2$

8. $f(x) = (2x^2 - 5)^2$

9. $f(x) = \sqrt{x^2 + 3}$

10. $f(x) = \sqrt[3]{x^2 + 9}$

11. $f(x) = \dfrac{3x}{x^2 + 1}$

12. $f(x) = \dfrac{2x + 1}{x^2 - 4}$

In Exercises 13–16, find all inflection points. Apply the Second Derivative Test at possible maximum/minimum points. Make a sketch of the graph and confirm your results with a graphing calculator.

13. $f(x) = (x + 5)\sqrt[3]{x}$

14. $f(x) = (x^2 + 1)\sqrt[3]{x}$

15. $f(x) = 2x\sqrt[3]{x + 1}$

16. $f(x) = (x + 10)\sqrt[3]{x^2 + 10}$

In Exercises 17–30, use the Second Derivative Test to find all local extrema, if the test applies. Otherwise, use the First Derivative Test.

17. $f(x) = x^2 - 3x + 5$

18. $f(x) = 8 + 7x - 2x^2$

19. $f(x) = x^3 - 3x^2 + 8$

20. $f(x) = x^3 + 6x^2 - 10$

21. $f(x) = x^3 - 12x + 3$

22. $f(x) = x^3 - 3x + 4$

23. $f(x) = \dfrac{2}{3}x^3 - x^2 - 4x - 2$

24. $f(x) = \dfrac{1}{3}x^3 + x^2 - 3x - 1$

25. $f(x) = x^4 - 8x^2 + 7$

26. $f(x) = x^4 - 2x^2 + 3$

27. $f(x) = x^4 + 2x^3 - 4$

28. $f(x) = x^4 - 6x^3 + 8$

29. $f(x) = 2x + \dfrac{8}{x}$

30. $f(x) = \dfrac{x^2 + 9}{x}$

🚀 APPLICATIONS

31. Point of diminishing returns: Find the point of diminishing returns for the sales function $S(x) = 112 + 1.8x^2 - 0.1x^3$, where x represents thousands of dollars spent on advertising, $0 \le x \le 10$, and S is sales in thousands of dollars.

32. Point of diminishing returns: The sales function for a product is given by $S(x) = 204 + 6.3x^2 - 0.25x^3$, where x represents thousands of dollars spent on advertising, $0 \le x \le 12$, and S is sales in thousands of dollars. Find the point of diminishing returns.

33. Marginal cost: The cost function for a particular product is given by $C(x) = 0.1x^3 - 2.4x^2 + 24x + 190$ dollars, where $0 \le x \le 12$. Find the minimum marginal cost.

34. Marginal cost: Find the minimum marginal cost of a product if the cost function is given by $C(x) = 0.0001x^3 - 0.036x^2 + 16.8x + 1900$ dollars, where $0 \le x \le 150$.

35. Law enforcement: Due to the rapid increase in major crimes, the mayor of a large city plans to organize a major crime task force. It is estimated that for every 1000 persons in the city, the numbers of major crimes will be $N(t) = 56 + 3t^2 - 0.8t^{\frac{5}{2}}$, where t is the number of months after the task force is organized and $0 \le t \le 12$.
  a. Find the maximum $N(t)$.
  b. Find the maximum rate of increase in $N(t)$.

36. Meteorology: Meteorology records for a certain city suggest that for the month of June, the temperature between midnight and 6:00 p.m. can be approximated by $T(t) = -0.04t^3 + 1.14t^2 - 7.2t + 66$ degrees, where t is the number of hours after midnight and $0 \le t \le 18$.
  a. Find the maximum and minimum temperatures.
  b. Find the maximum rate of increase in the temperature.

✏️ WRITING & THINKING

37. Given $f(x) = px^3 + bx + 10$, answer the following questions.

  a. Suppose p and b are positive numbers. What can be said about maximum/minimum points and points of inflection?
  b. Suppose p and b have opposite signs (one is positive and the other negative). What can be said about maximum/minimum points and points of inflection?
  c. If the constant term 10 is changed to some other value, do your responses to parts **a.** and **b.** change?
  d. Put your answers to parts **a.**, **b.**, and **c.** together in a "Lab Report" which discusses the coefficients in the given polynomial $y = px^3 + bx + c$.

12.3 CURVE SKETCHING: POLYNOMIAL FUNCTIONS

■ TOPICS

- ■ Curve Sketching
- ■ Polynomial Functions

Curve Sketching

We begin by studying the basic characteristics that graphs must have if they are to satisfy a particular list of conditions. The more information we have (or the more conditions to be satisfied), the more accurate the corresponding graph can be. In each case, we assume that the function is continuous.

List of Given Conditions
1. $f'(x) < 0$ on $(-\infty, 2)$
2. $f'(x) > 0$ on $(2, +\infty)$
3. $f'(2) = 0$

What we know from the given conditions is that

1. f is decreasing on $(-\infty, 2)$,
2. f is increasing on $(2, +\infty)$, and
3. f has a horizontal tangent at $x = 2$.

That is, at $x = 2$, f has a local (and absolute) minimum. We have no idea what the minimum is and have no idea about the concavity of the graph.

Figure 1 shows three possible graphs that satisfy all the conditions given.

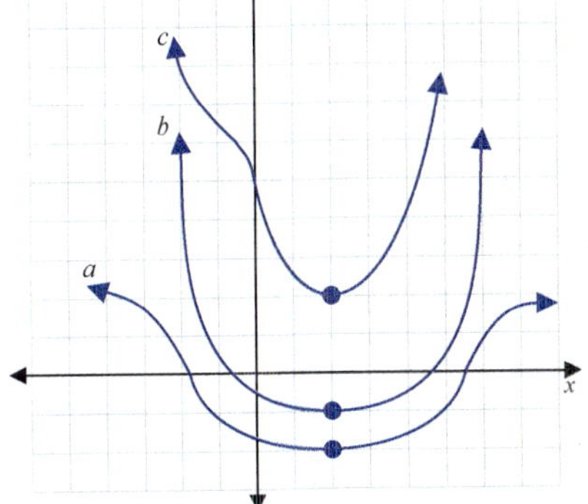

FIGURE 1: Possible Graphs of $f(x)$

Let us additionally suppose

4. $f''(x) > 0$ for all x.

This new condition eliminates graphs a and c since it declares that $f(x)$ is concave up for all x.

Let us add one more condition.

5. $f(2) = -1$

We know that curves a and c in Figure 1 will not satisfy Condition 4 because they are not concave up for all x. Graph b is a likely candidate since the point $(2, -1)$ is on the graph of b; however, there are still an infinite number of possible graphs that satisfy all the given conditions (see Figure 2). Thus we can sketch only a **general curve** that satisfies all the conditions. We cannot expect to graph a specific curve unless we know more conditions that the function must satisfy, or, even better, the equation that represents the function.

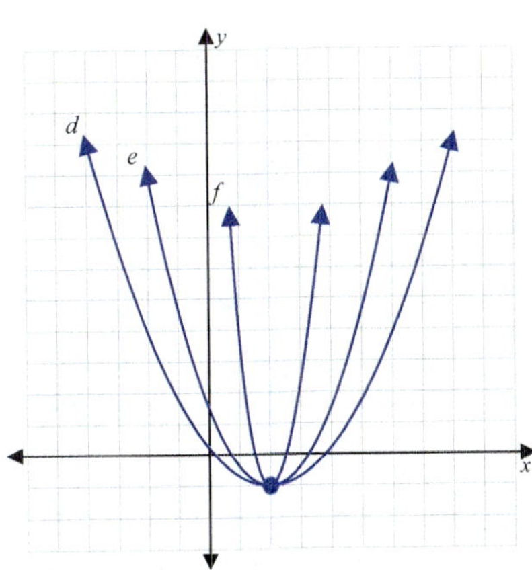

FIGURE 2: Possible General Graphs of $f(x)$

Example 1: Curve Sketching

Sketch a continuous function that satisfies all the given conditions.

1. $f'(x) > 0$ for all x
2. $f''(3) = 0$ and $f(3) = 4$
3. $f''(x) < 0$ on $(-\infty, 3)$
4. $f''(x) > 0$ on $(3, +\infty)$

Solution

Condition 1 means that f is increasing for all x. Conditions 2, 3, and 4 indicate that $(3, 4)$ is a point of inflection and f is concave downward on $(-\infty, 3)$ and concave upward on $(3, +\infty)$.

One possible curve that satisfies all the given conditions is shown in the graph.

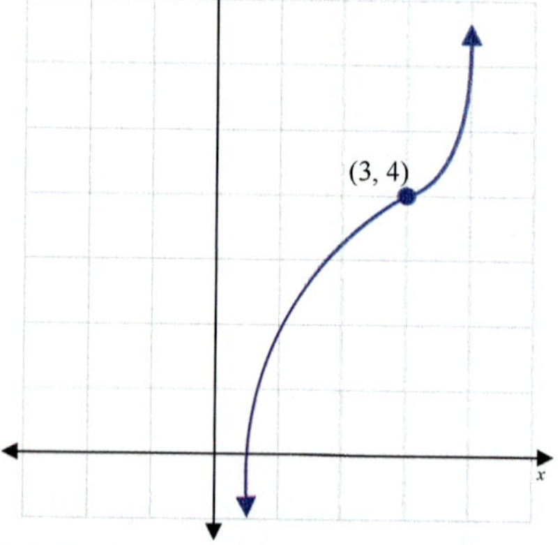

Example 2: Curve Sketching

Sketch a continuous function that satisfies all the given conditions.

1. $g(-1) = 2$, $g(0) = 1$, $g(1) = -2$
2. $g'(0) = 0$, $g'(1) = 0$
3. $g''(0) = 0$, $g(0.5) = 0$
4. $g''(x) < 0$ if $0 < x < 0.5$
5. $g''(x) > 0$ if $x < 0$ or $x > 0.5$

Solution

In Condition 1, three points are given: $(-1, 2)$, $(0, 1)$, and $(1, -2)$. There are horizontal tangent lines at $x = 0$ and $x = 1$ (Condition 2). Due to changes in concavity inferred from Conditions 3 to 5, points $(0, 1)$ and $(0.5, 0)$ are points of inflection. We also know that a local minimum occurs at the point $(1, -2)$ because the function is concave up by Condition 5 but has zero slope by Condition 2.

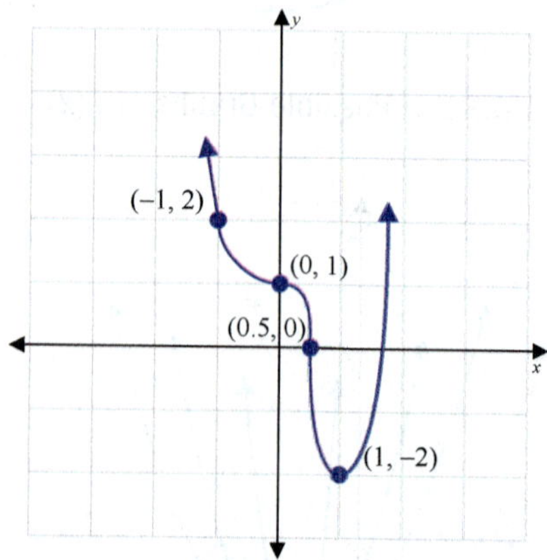

We have previously seen how f' and f'' can be used to describe characteristics of f (whether it is increasing, decreasing, concave up, or concave down) and to locate critical values and hypercritical values. The following systematic plan or strategy shows how to collect all this information for a given function in order to obtain an accurate graph of that function.

Strategy for Curve Sketching Given a Function f

To sketch the graph of a function f, perform the following steps.

1. Find $f'(x)$ and $f''(x)$.

2. Find the critical values. That is, find the values $x = c$ where
   a. $f'(c) = 0$, or
   b. $f'(c)$ is undefined.

3. Using the critical values as endpoints, find intervals
   a. where $f'(x) > 0$ (f is increasing), and
   b. where $f'(x) < 0$ (f is decreasing).

4. Locate the local extrema using the First Derivative Test or the Second Derivative Test.

5. Find the hypercritical values. That is, find the values $x = c$ where
   a. $f''(c) = 0$, or
   b. $f''(c)$ is undefined.

6. Using the hypercritical values as endpoints, find intervals
   a. where $f''(x) > 0$ (f is concave upward), and
   b. where $f''(x) < 0$ (f is concave downward).

7. Locate all points of inflection. These occur at hypercritical values where the curve changes concavity.

8. Using the combined information from Steps 1–7 and any other specific points that might be helpful, sketch the graph. (A helpful point would usually be the y-intercept $(0, f(0))$ or an x-intercept: $(a, f(a))$ if $f(a) = 0$.)

Polynomial Functions

In the following examples we use the plan for curve sketching to sketch the graphs of polynomial functions. Remember that polynomial functions are functions of the form

$$f(x) = a_n x^n + a_{n-1} x^{n-1} + \cdots + a_1 x + a_0,$$

where the coefficients $a_n, a_{n-1}, \ldots, a_0$ are real numbers and the exponents are positive integers.

Example 3: Using the Strategy for Curve Sketching

Sketch the graph of the polynomial function $f(x) = x^2 - 2x - 3$.

Solution

In this example several steps are listed together because the first and second derivatives are easily found.

Step 1: Find $f'(x)$ and $f''(x)$.

$$f'(x) = 2x - 2$$
$$f''(x) = 2$$

Steps 2 and 3: Find the critical values and the intervals where $f'(x) > 0$ and where $f'(x) < 0$.

$$f'(x) = 2x - 2 = 2(x - 1)$$ Find the critical values by setting $f'(x) = 0$ and
$$0 = 2(x - 1)$$ solving for x.
$$x = 1$$

The critical value is $x = 1$.

Now we can determine the following:

If $x < 1$, then $f'(x) = 2(x - 1) < 0$ and f is decreasing.

If $x > 1$, then $f'(x) = 2(x - 1) > 0$ and f is increasing.

Step 4: Find all local extrema.

$$f''(1) = 2$$ Evaluate $f''(x)$ for the critical value to find the local extrema and
        use the Second Derivative Test.
$$2 > 0$$ Local minimum

Find the corresponding y-value of the critical value.

$$f(1) = (1)^2 - 2(1) - 3 = -4$$

The point $(1, -4)$ is the local minimum.

Steps 5, 6, and 7: Find the hypercritical values, the intervals where $f''(x) > 0$ and where $f''(x) < 0$, and all points of inflection.

For all x,
$$f''(x) = 2 > 0.$$

Thus f is concave up for all x, and there are no points of inflection.

Step 8: Sketch the graph.

Interval or Value	Derivative	Nature of Graph
$(-\infty, 1)$	$f'(x) < 0$	Decreasing
$(1, +\infty)$	$f'(x) > 0$	Increasing
$x = 1$	$f'(1) = 0$	Local minimum
$(-\infty, +\infty)$	$f''(x) = 2 > 0$	Concave up

Setting $y = x^2 - 2x - 3 = 0$ gives $(x + 1)(x - 3) = 0$. So the x-intercepts are at $x = -1$ and $x = 3$. We can also see from the graph that -4 is the absolute minimum value of f.

The graph is a parabola, which agrees with the fact that the graph of every second-degree polynomial function is a parabola.

Example 4: Using the Strategy for Curve Sketching

Use the curve-sketching strategy to sketch the graph of the function

$$f(x) = \frac{1}{3}x^3 - 2x^2 + 3x + 1.$$

Solution

Step 1: Find $f'(x)$ and $f''(x)$.

$$f'(x) = x^2 - 4x + 3$$
$$f''(x) = 2x - 4$$

Step 2: Find the critical values.

$$f'(x) = x^2 - 4x + 3 \quad \text{Find the critical values by setting } f'(x) = 0 \text{ and}$$
$$x^2 - 4x + 3 = 0 \quad\quad \text{solving for } x.$$
$$(x-3)(x-1) = 0$$
$$x = 1, 3$$

The critical values are $x = 1$ and $x = 3$.

Step 3: Use the critical values $x = 1$ and $x = 3$ as endpoints of intervals where $f'(x) > 0$ and where $f'(x) < 0$.

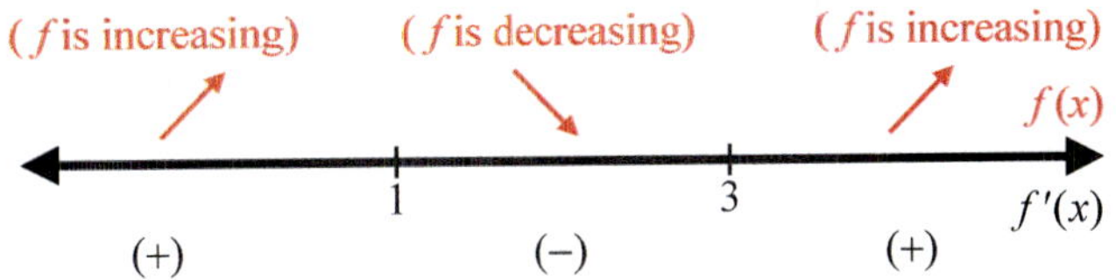

Step 4: Find all local extrema.

$$f''(1) = 2(1) - 4 \quad \text{Evaluate } f''(x) \text{ for the critical values to find the local}$$
$$= -2 \quad\quad\quad \text{extrema and use the Second Derivative Test.}$$
$$-2 < 0 \quad\quad \text{Local maximum}$$
$$f''(3) = 2(3) - 4$$
$$= 2$$
$$2 > 0 \quad\quad \text{Local minimum}$$

Find the corresponding y-value of each of the critical values.

$$f(1) = \frac{1}{3}(1)^3 - 2(1)^2 + 3(1) + 1 = \frac{7}{3}$$
$$f(3) = \frac{1}{3}(3)^3 - 2(3)^2 + 3(3) + 1 = 1$$

Therefore, $\left(1, \frac{7}{3}\right)$ is the local maximum and $(3, 1)$ is the local minimum.

Step 5: Find all hypercritical values.

$$f''(x) = 2x - 4 \quad \text{Find the hypercritical values by setting } f''(x) = 0 \text{ and}$$
$$2x - 4 = 0 \quad\quad \text{solving for } x.$$
$$2(x-2) = 0$$
$$x = 2$$

There is one hypercritical value, $x = 2$.

Step 6: Using the hypercritical value $x = 2$ as an endpoint of intervals, find the intervals where $f''(x) > 0$ and where $f''(x) < 0$.

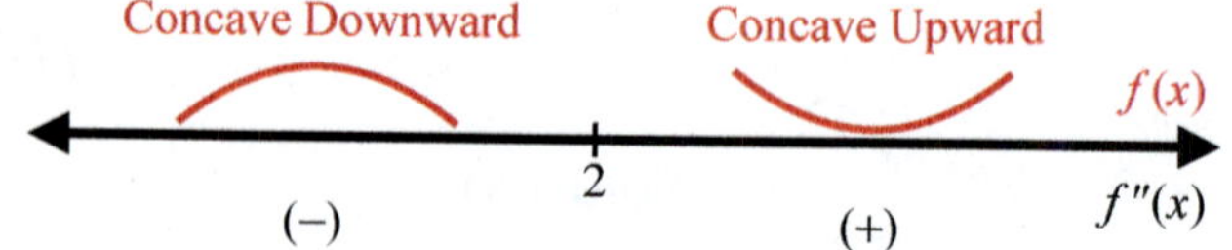

f is concave down on the interval $(-\infty, 2)$ and concave up on the interval $(2, +\infty)$.

Step 7: Find all points of inflection.

$$f(2) = \frac{1}{3}(2)^3 - 2(2)^2 + 3(2) + 1 \quad \text{Find the corresponding } y\text{-value of the hypercritical value.}$$

$$= \frac{5}{3}$$

Since f changes concavity at $x = 2$, the point $\left(2, \frac{5}{3}\right)$ is a point of inflection.

Step 8: Sketch the graph. The following table summarizes the information found in Steps 1–7.

Interval or Value	Derivative	Nature of Graph
$(-\infty, 1)$ or $(3, +\infty)$	$f'(x) > 0$	Increasing
$(1, 3)$	$f'(x) < 0$	Decreasing
$x = 1$	$f'(1) = 0$	Local maximum
$x = 3$	$f'(3) = 0$	Local minimum
$(-\infty, 2)$	$f''(x) < 0$	Concave down
$(2, +\infty)$	$f''(x) > 0$	Concave up
$x = 2$	$f''(2) = 0$	Point of inflection

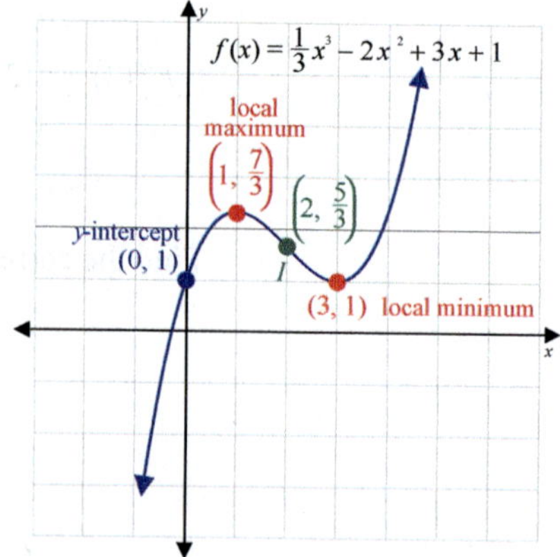

Example 5: Graphing the Derivative

Consider the given graph of a function.

a. Identify the local extrema and locate the point(s) of inflection.

b. Determine the intervals on which $f(x)$ is increasing and on which $f(x)$ is decreasing, and identify the intervals on which $f(x)$ is concave upward and concave downward.

c. Sketch on the same coordinate plane a possible graph of $f'(x)$.

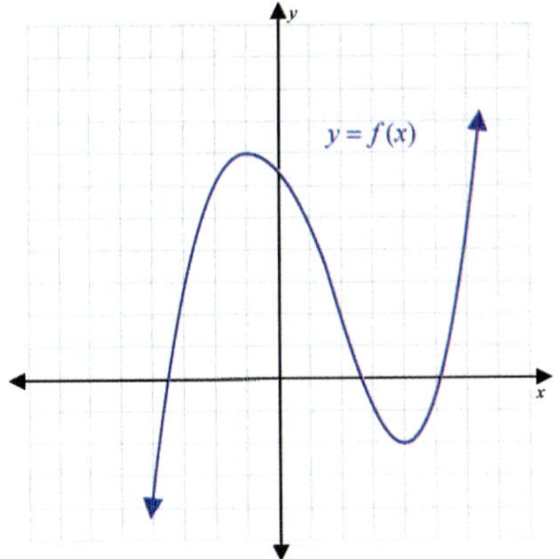

Solution

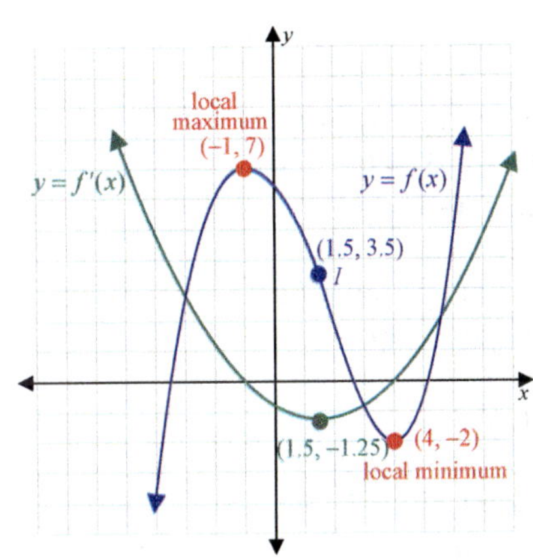

a. A local maximum is located at $(-1, 7)$, and a local minimum is located at $(4, -2)$. There is a point of inflection at about $(1.5, 3.5)$.

b. The function is increasing on the intervals $(-\infty, -1)$ and $(4, +\infty)$. It is decreasing on the interval $(-1, 4)$. The function is concave downward on the interval $(-\infty, 1.5)$ and concave upward on $(1.5, +\infty)$.

c. The slope of $f(x)$ is positive but decreasing from $-\infty < x < -1$, and is 0 at $x = -1$. The slope then becomes negative and continues decreasing until $x = 1.5$ at which point slope is a minimum. Then the slope starts to increase from some negative value (which we estimate to be -1.25) to 0 (at $x = 4$). It then becomes positive and continues increasing.

12.3 EXERCISES

💡 PRACTICE

In Exercises 1–16, sketch the graph of a continuous function that satisfies all the given conditions.

1.   **a.** $f(-1) = 2$
      b. $f'(-1) = 0$
      c. $f'(x) < 0$ if $x < -1$
      d. $f'(x) > 0$ if $x > -1$
      e. $f''(x) > 0$ for all x

2.   **a.** $f(3) = 4$
      b. $f'(3) = 0$
      c. $f'(x) < 0$ if $x > 3$
      d. $f'(x) > 0$ if $x < 3$
      e. $f''(x) < 0$ for all x

3.   **a.** $f(-2) = 4, f(-1) = 1, f(1) = -1$
      b. $f'(-2) = 0, f'(1) = 0$
      c. $f'(x) < 0$ if $-2 < x < 1$
      d. $f'(x) > 0$ if $x < -2$ or $x > 1$
      e. $f''(-1) = 0$
      f. $f''(x) < 0$ if $x < -1$
      g. $f''(x) > 0$ if $x > -1$

4.   **a.** $f(0) = -2, f(2) = 0, f(3) = 3$
      b. $f'(0) = 0, f'(3) = 0$
      c. $f'(x) < 0$ if $x < 0$ or $x > 3$
      d. $f'(x) > 0$ if $0 < x < 3$
      e. $f''(2) = 0$
      f. $f''(x) < 0$ if $x > 2$
      g. $f''(x) > 0$ if $x < 2$

5. **a.** $f(-3) = 5, f(-1) = 2, f(0) = -1$
    b. $f'(-3) = 0, f'(0) = 0$
    c. $f'(x) < 0$ if $x < 0$ and $x \neq -3$
    d. $f'(x) > 0$ if $x > 0$
    e. $f''(-3) = 0, f''(-1) = 0$
    f. $f''(x) < 0$ if $-3 < x < -1$
    g. $f''(x) > 0$ if $x < -3$ or $x > -1$

6. **a.** $f(1) = 2, f(2) = 3, f(4) = 4,$
         $f(6) = 2$
    b. $f'(1) = 0, f'(4) = 0$
    c. $f'(x) < 0$ if $x > 4, x < 1$
    d. $f'(x) > 0$ if $1 < x < 4$

7. **a.** $f(x) = ax^3 + bx + c$
    b. $f(0) = 0$
    c. $f(1) = 15$
    d. $f'(-1) = 0$ and $x = -1$ is a local max
    e. $f''(x) > 0$ if $x < 10$

8. **a.** $f(10) = 5$
    b. $f'(5) = 0$
    c. $f''(5) = 10$
    d. $f''(x) < 0$ if $x > 10$

9. **a.** $f''(x) > 0$ if $x < 5$
    b. $f''(5) = 0$
    c. $f''(x) < 0$ if $x > 5$
    d. $f''(x) > 0$ for all x

10. **a.** $f(4) = 8$
    b. $f'(4) = 0$
    c. $f''(4) = 8$

11. **a.** $f(-5) = 4$
    b. $f'(-5) = 0$
    c. $f''(-5) = -2$

12. **a.** $f(x) = ax^2 + bx + c$
    b. $f'(-3) = 0$
    c. $f''(-3) = 2$

13. **a.** $f(x) = ax^3 + bx^2 + cx + d$
    b. $f(0) = 25$
    c. $f'(4) = 0, f'(-4) = 0$
    d. $f''(4) = 48, f''(-4) = -48$

14. **a.** $f(x) = ax^2 + bx + c$
    b. $f(0) = 79$
    c. $f'(5) = 0$
    d. $f''(x) = 6$

15. **a.** $f(x) = ax^3 + bx^2 + cx + d$
    b. $f(0) = 2$
    c. $f'(0) = 5$
    d. $f''(0) = 4$
    e. $f''(1) = 12$

16. **a.** $f(x) = ax^3 + bx$
    b. $f'(0) = -12$
    c. $f'(2) = 0$

For each of the functions in Exercises 17–36, determine $f'(x)$ and $f''(x)$. Then complete a summary table like those in Examples 3 and 4. Use this table to sketch the graph of the function. (If available, use a graphing utility or calculator to obtain a suitable window and confirm the accuracy of your calculations.)

17. $f(x) = x^2 - 4x + 7$

18. $f(x) = x^2 + 6x - 8$

19. $f(x) = 6 + 5x - x^2$

20. $f(x) = 2 + 3x - 2x^2$

21. $f(x) = x^3 + 3x^2 - 6$

22. $f(x) = \dfrac{1}{3}x^3 - 4x + 3$

23. $f(x) = \dfrac{1}{3}x^3 + x^2 - 3x + 5$

24. $f(x) = 2x^3 - 3x^2 - 12x + 5$

25. $f(x) = x^4 - 2x^2 + 4$

26. $f(x) = x^4 - 8x^2 - 3$

27. $f(x) = \dfrac{1}{4}x^4 - x^3 + 5$

28. $f(x) = x^4 + 4x^3 + 12$

29. $f(x) = x^4 - 4x + 7$

30. $f(x) = 3x^4 - 4x^3 + 3$

31. $f(x) = (x+5)(x-3)^2$

32. $f(x) = (x+1)^2 (x-10)^2$

33. $f(x) = (2x+1)(x-8)^3$

34. $f(x) = (x-5)(x-10)(x+3)$

35. $f(x) = 2x(5x+8)^3$

36. $f(x) = 16x(21+x)^3$

37. Suppose that $f(x) = mx^2 + 6x + 4$. Determine a value of m so that $f(x)$ has a minimum at $x = -1$.

38. Given $y = 4x^2 + nx + 8$, determine a value for n so that y has a minimum at $x = 2$.

39. Determine a value for m such that at $x = 1$ the tangent to the function $f(x) = mx^2 + 6x + 1$ has an equation of $y = 12x - 2$.

40. Determine a value for m so that $y = 4x^3 + mx^2$ has an inflection point at $x = -10$.

🚀 APPLICATIONS

41. In an action movie, the hero is seen fighting the villain inside a plane, which has a large hole in its side. The hero (actually a movie stunt man) is then thrown from the plane. He falls quickly and soon reaches a constant velocity. The hero opens his parachute but it deploys slowly, as if he is having trouble, but finally, in triumph, all is well and he drifts steadily and slowly to the ground.
   a. Draw a graph of the hero's vertical distance to the ground, represented by y, versus time t (in seconds).
   b. Describe any interesting points on the graph with points in the movie narrative.

42. The effectiveness of a certain medical injection is modeled by $E(t) = 0.01\,t(100 - t)$, where t is time in minutes and E is a measure of concentration in the bloodstream. Effectiveness readings above 9.0 are satisfactory and readings above 30 are dangerous.
   a. If an injection is given at midnight, when are the readings satisfactory? When do they become too low?
   b. How high do the effectiveness readings get?
   c. The supervising nurse and the resident pharmacologist must assign a schedule for injections. For the next week, assuming injected dosages are additive, give a reasonable schedule for injections so that the patient's E-reading stays at or above 9 but never exceeds 30.

43. The productivity rating of an individual worker at the Cruz Corporation assembly line is based on the number of tasks accomplished, mistakes made, and responsiveness to difficulties encountered. The average of all scores allows the company to use a simple model based on time on the floor given by $PR = -0.4x^3 + 2x^2 + 10x + 5$, where x is in hours at work. A PR score of 20 is acceptable and a score of 40 is highly unusual.
   a. When are workers' scores the highest?
   b. Design an 8-hour day where workers do the most demanding jobs for about 6 hours and have 2 hours for less stressful work. Explain your reasoning.

44. Is it possible for a polynomial $y = ax^2 + bx + c$ to have an inflection point?

12.4 CURVE SKETCHING: RATIONAL FUNCTIONS

In this section we will apply the graphing strategies to **rational functions**. Examples of rational functions are

$$f(x) = \frac{x-1}{x}, \quad g(x) = \frac{x+2}{x-2}, \quad \text{and} \quad h(x) = \frac{2x}{x^2+1}.$$

Rational Function

A **rational function** is a function of the form

$$f(x) = \frac{P(x)}{Q(x)},$$

where $P(x)$ and $Q(x)$ are polynomials and $Q(x) \neq 0$.

If, in the definition of a rational function, $Q(x)$ is a constant, then the rational function is just a polynomial function. The analysis here will focus on rational functions that are not polynomial functions. Also, we restrict the discussion to rational functions that are completely reduced with no common factors in the numerator and denominator.

A distinguishing characteristic of the graphs of the rational functions we will investigate is that they have one or more asymptotes. Intuitively, an **asymptote** is a line that a curve "approaches" or "gets close to" in a limiting sense. Figure 1 illustrates three kinds of asymptotes: **horizontal**, **vertical**, and **oblique**.

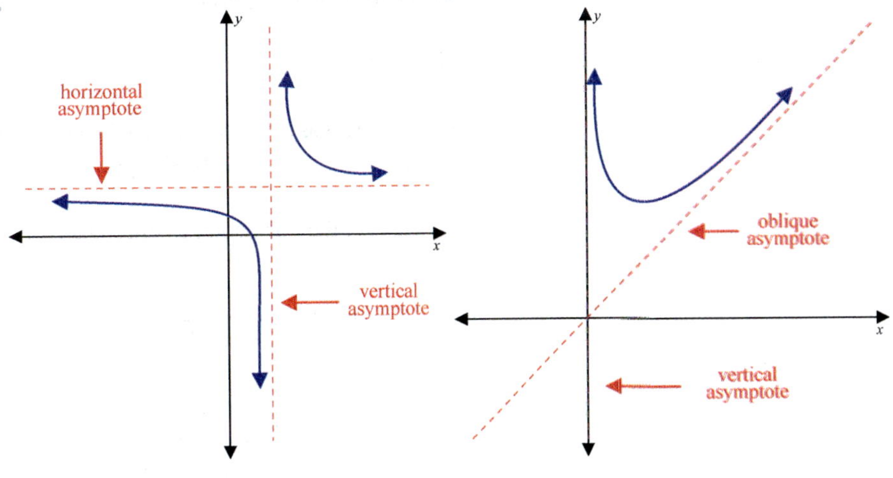

FIGURE 1

Asymptotes can be found by taking limits, as outlined in the following list.

Asymptotes for Rational Functions $f(x) = \dfrac{P(x)}{Q(x)}$

1. **Vertical asymptotes** are of the form $x = a$ and occur where the denominator is 0 and the numerator is not 0 (that is, $Q(a) = 0$ and $P(a) \neq 0$).

2. **Horizontal asymptotes** are of the form $y = b$ and occur where

$$\lim_{x \to +\infty} f(x) = b \quad \text{or} \quad \lim_{x \to -\infty} f(x) = b.$$

3. **Oblique asymptotes** are of the linear form $y = mx + b$ and occur when the numerator is one degree larger than the denominator and $f(x)$ can be written as

$$f(x) = mx + b + \frac{R(x)}{Q(x)},$$

where

$$\lim_{x \to +\infty} \frac{R(x)}{Q(x)} = 0 \quad \text{or} \quad \lim_{x \to -\infty} \frac{R(x)}{Q(x)} = 0.$$

Example 1: Locating Vertical and Horizontal Asymptotes

For the function $f(x) = \dfrac{2x}{x - 5}$, find

a. the vertical asymptotes and

b. the horizontal asymptotes.

Solution

a. A vertical asymptote occurs when the denominator is 0 (i.e., where $x - 5 = 0$). Thus the line $x = 5$ is a vertical asymptote.

b. We find the horizontal asymptotes by finding $\lim\limits_{x \to +\infty} f(x)$ and $\lim\limits_{x \to -\infty} f(x)$.

$$\lim_{x \to +\infty} f(x) = \lim_{x \to +\infty} \frac{2x}{x - 5}$$

$$= \lim_{x \to +\infty} \frac{\dfrac{2x}{x}}{\dfrac{x}{x} - \dfrac{5}{x}} \qquad \text{Divide both numerator and denominator by } x.$$

$$= \lim_{x \to +\infty} \frac{2}{1 - \dfrac{5}{x}} \qquad \lim_{x \to +\infty} \frac{a}{x} = 0$$

$$= \frac{2}{1 - 0} = 2$$

The line $y = 2$ is a horizontal asymptote. Since $\lim\limits_{x \to -\infty} \left(\dfrac{5}{x} \right) = 0$, we see that $\lim\limits_{x \to -\infty} f(x) = 2$ also.

Example 2: Locating Vertical and Horizontal Asymptotes

For the function $f(x) = \dfrac{1}{x^2 + 3}$, find

a. the vertical asymptotes and

b. the horizontal asymptotes.

Solution

a. A vertical asymptote occurs when the denominator is 0. Since the denominator $x^2 + 3$ is never 0, there are no vertical asymptotes.

b. We find the horizontal asymptotes by finding $\lim\limits_{x \to +\infty} f(x)$ and $\lim\limits_{x \to -\infty} f(x)$.

$$\lim_{x \to +\infty} \frac{1}{x^2 + 3} = 0 \quad \text{and} \quad \lim_{x \to -\infty} \frac{1}{x^2 + 3} = 0$$

Therefore, the line $y = 0$ is a horizontal asymptote.

Example 3: Locating Oblique Asymptotes

Find the oblique asymptote for the function $f(x) = \dfrac{2x^2 + 7}{3x}$.

Solution

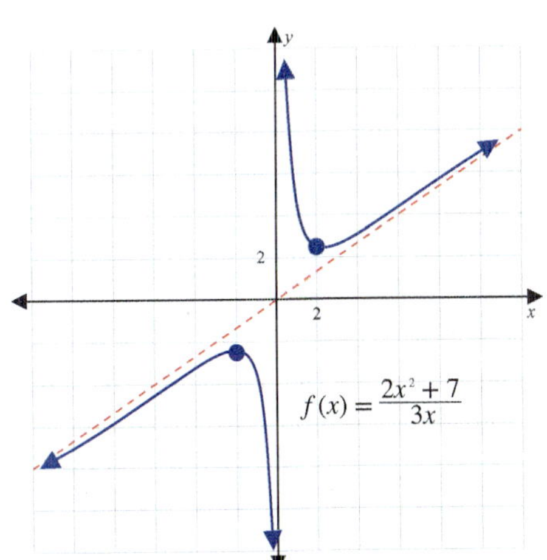

Since the numerator is one degree larger than the denominator, we can write $f(x)$ in the following form:

$$f(x) = \frac{2x^2 + 7}{3x} = \frac{2x^2}{3x} + \frac{7}{3x} = \frac{2}{3}x + \frac{7}{3x}.$$

Now we find that $\lim\limits_{x \to +\infty} \dfrac{7}{3x} = 0.$

So the remaining term gives the oblique asymptote:

$$y = \frac{2}{3}x.$$

In Example 3, $x = 0$ is not in the domain of the function f. Thus $x = 0$ is technically neither a critical value nor a hypercritical value of f. Nevertheless, in the sketch, we notice that f is concave down to the immediate left of $x = 0$. To the immediate right of $x = 0$, f is concave up. Although x-values that determine vertical asymptotes are not critical points, they are "critical" in the sense of having relevance to the analysis.

In Example 4, we apply the structured curve-sketching strategy, along with our knowledge of asymptotes, to sketch the graphs of some rational functions. One useful fact (that will not be proven) is that **a rational function can have at most one horizontal asymptote.**

Example 4: Sketching the Graph of a Rational Function

Sketch the graph of the rational function $f(x) = \dfrac{x - 1}{x - 2}$. Include any asymptotes.

Solution

Step 1: Find $f'(x)$ and $f''(x)$.

$$f'(x) = \frac{(x-2)\frac{d}{dx}(x-1) - (x-1)\frac{d}{dx}(x-2)}{(x-2)^2} = -\frac{1}{(x-2)^2} = -1(x-2)^{-2}$$

$$f''(x) = -1(-2)(x-2)^{-3}\frac{d}{dx}(x-2) = \frac{2}{(x-2)^3}$$

Steps 2, 3, and 4: Find the critical values, the intervals where $f'(x) > 0$ and where $f'(x) < 0$, and the local extrema.

The first derivative, $f'(x) = -\dfrac{1}{(x-2)^2}$, is never 0. Note that f' is undefined at $x = 2$, but $x = 2$ is not a critical value since 2 is not in the domain of f. Thus there are no critical values of f, and f has no local maxima or local minima.

Also, $f'(x) = -\dfrac{1}{(x-2)^2} < 0$ for all $x \neq 2$ and f is decreasing for all x in its domain.

Steps 5, 6, and 7: Find all hypercritical values, intervals where $f''(x) > 0$ and where $f''(x) < 0$, and points of inflection.

Since $f''(x) = \dfrac{2}{(x-2)^3} \neq 0$ and is undefined only for $x = 2$, there are no hypercritical values and no points of inflection.

However, if $x < 2$, then $f''(x) = \dfrac{2}{(x-2)^3} < 0$, and f is concave down, and if $x > 2$, then $f''(x) = \dfrac{2}{(x-2)^3} > 0$, and f is concave up.

Asymptotes: Find any vertical, horizontal, or oblique asymptotes.

The line $x = 2$ is a vertical asymptote because $x - 2$ is not a factor of the numerator and the value of 2 for x will make the denominator 0.

We now find the horizontal asymptote by finding $\lim\limits_{x \to +\infty} f(x)$ and $\lim\limits_{x \to -\infty} f(x)$.

$$\lim_{x \to +\infty} f(x) = \lim_{x \to +\infty} \frac{x-1}{x-2}$$

$$= \lim_{x \to +\infty} \frac{\dfrac{x}{x} - \dfrac{1}{x}}{\dfrac{x}{x} - \dfrac{2}{x}}$$ Divide both the numerator and the denominator by the highest power of x present in the function.

$$= \lim_{x \to +\infty} \frac{1 - \dfrac{1}{x}}{1 - \dfrac{2}{x}} = \frac{1-0}{1-0} = 1$$ We know that $\lim\limits_{x \to +\infty} \dfrac{a}{x} = 0$.

Proceeding as above, we can also find that $\lim\limits_{x \to -\infty} f(x) = 1$. Therefore, the line $y = 1$ is a horizontal asymptote.

We also know that this function does not have an oblique asymptote because the degree of the numerator is not one degree higher than the denominator.

To verify your graph with a TI-83/84 Plus calculator, perform the following steps:

1. Enter the given function into Y1.

2. Check that your [window] is appropriate.

3. Press [graph].

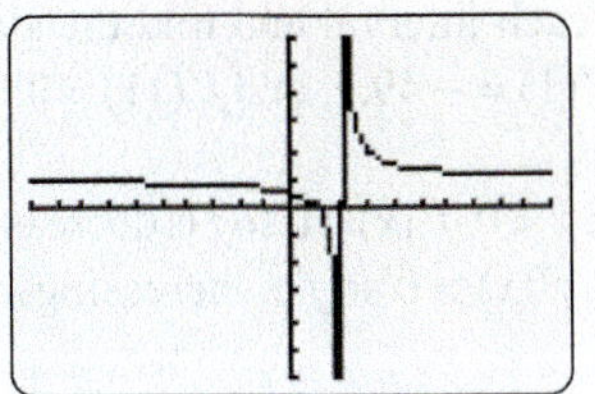

Step 8: Sketch the graph.

Interval or Value	Derivative	Nature of Graph
$(-\infty, 2)$ or $(2, +\infty)$	$f'(x) < 0$	Decreasing
$(-\infty, 2)$	$f''(x) < 0$	Concave down
$(2, +\infty)$	$f''(x) > 0$	Concave up
$x = 2$	NA	Vertical asymp.
$y = 1$	NA	Horizontal asymp.

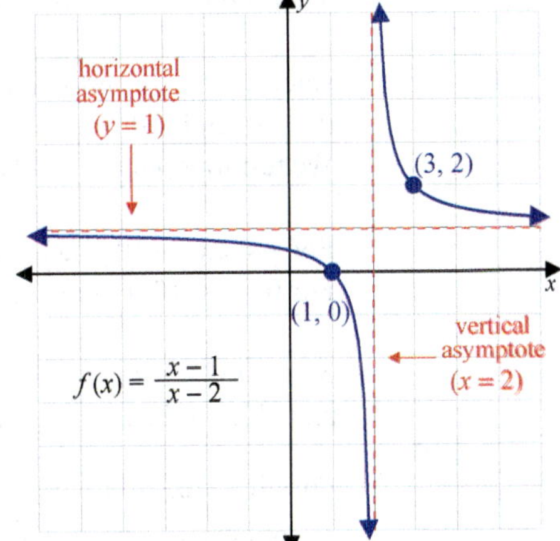

In Example 5, we show all the analysis tools available to the student.

Example 5: Sketching the Graph of a Rational Function

Sketch the graph of the rational function $f(x) = \dfrac{100 + x^2}{2x}$.

Solution

Here we will first identify the function's asymptotes. First, it is easy in this case to see by inspection of the formula that $x = 0$ makes the denominator 0 but not the numerator. Thus $x = 0$ is a vertical asymptote.

$$f(x) = \frac{100 + x^2}{2x}$$
$$= \frac{x^2}{2x} + \frac{100}{2x} = \frac{x}{2} + \frac{50}{x}$$

Since $\lim\limits_{x \to +\infty} \dfrac{50}{x} = 0$, we observe $y = \dfrac{x}{2}$ is an oblique asymptote.

Unlike the function in Example 4, this function's numerator is one degree larger than the denominator, so we know immediately that this function will have an oblique asymptote. Using the above work (equivalent to division of polynomials), the line $y = \dfrac{x}{2}$ is seen to be an oblique asymptote.

Now we will analyze the first and second derivatives.

$$f'(x) = \frac{2x(2x) - (100 + x^2)(2)}{4x^2} = \frac{x^2 - 100}{2x^2}$$

$$f''(x) = \frac{2x^2(2x) - (x^2 - 100)4x}{4x^4} = \frac{100}{x^3}$$

We can factor $f'(x)$ as follows: $f'(x) = \dfrac{x^2 - 100}{2x^2} = \dfrac{(x+10)(x-10)}{2x^2}$.

This shows $x = 10$ and $x = -10$ are critical values, and we must treat $x = 0$ as critical

since $x = 0$ is a vertical asymptote. The intervals to consider are $(-\infty, -10)$, $(-10, 0)$, $(0, 10)$, and $(10, \infty)$. We select a convenient point in each interval and find the slope. We test four points: $f'(-11) \approx 0.087, f'(-1) = -49.5, f'(1) = -49.5$, and $f'(11) \approx 0.087$.

For $-\infty < x < -10, f'(x) > 0$ so f is increasing. For $-10 < x < 0, f'(x) < 0$ so f is decreasing. For $0 < x < 10, f'(x) < 0$ so f is decreasing. For $x > 10, f'(x) > 0$ so f is increasing.

Note: $f(10) = 10$, and $f(-10) = -10$. The analysis shows $(-10, -10)$ is a local maximum and $(10, 10)$ is a local minimum.

For all $x < 0$, $f''(x) = \dfrac{100}{x^3} < 0$, so f is concave down on $(-\infty, 0)$. Since $f'' > 0$ for all $x > 0$, we know f is concave up on $(0, +\infty)$.

Since $f'(10) = 0$ and $f''(10) > 0$, by the Second Derivative Test, we confirm that the point $(10, f(10)) = (10, 10)$ is a local minimum. Similarly $f''(-10) < 0$ and $(-10, -10)$ is a local maximum.

Also, because $f''(x) \neq 0$ and there are no points in the domain of f where $f''(x)$ is undefined, there are no hypercritical values and no points of inflection.

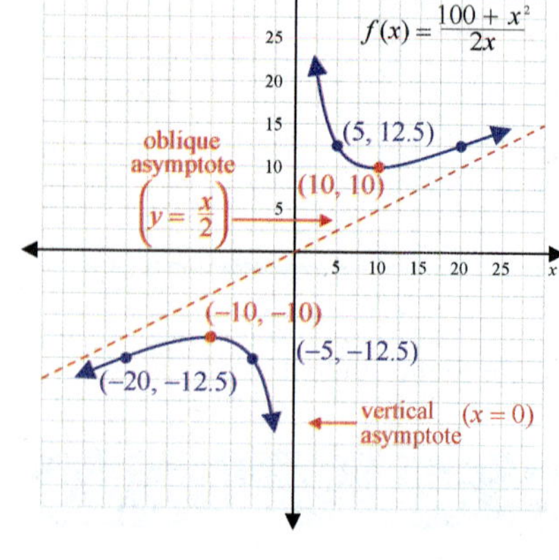

The oblique asymptotes that we have studied thus far have been linear. It is possible for a rational function to have a nonlinear asymptote. See the graph of $f(x) = \dfrac{x^3 + 1}{x} = x^2 + \dfrac{1}{x}$ in Figure 2. This function has an asymptote of $y = x^2$. However, for analysis purposes, we will only consider linear asymptotes.

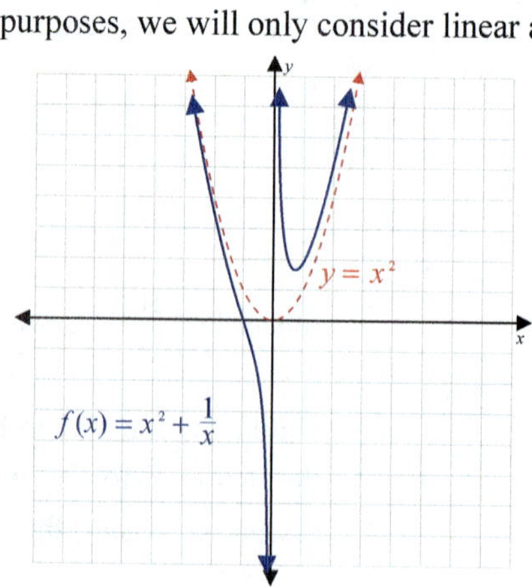

FIGURE 2

12.4 EXERCISES

💡 PRACTICE

For each of the rational functions in Exercises 1–12, find **a.** any vertical asymptotes, **b.** any horizontal asymptotes, and **c.** any oblique asymptotes.

1. $f(x) = \dfrac{1}{x-4}$

2. $f(x) = -\dfrac{3}{x+6}$

3. $f(x) = \dfrac{2x}{x+8}$

4. $f(x) = \dfrac{5x}{2x+1}$

5. $f(x) = \dfrac{x+2}{x^2+1}$

6. $f(x) = \dfrac{x-7}{x^2+3}$

7. $f(x) = \dfrac{5x^2}{3x^2-2x-1}$

8. $f(x) = \dfrac{2x^2}{x^2+3x}$

9. $f(x) = \dfrac{x^2-4}{x}$

10. $f(x) = \dfrac{3x^2+2}{x}$

11. $f(x) = \dfrac{x^2+1}{x+1}$

12. $f(x) = \dfrac{x^2-5}{x-2}$

In Exercises 13–22, sketch the graph of each rational function. Show any asymptotes on each graph.

13. $f(x) = -\dfrac{2}{x+5}$

14. $f(x) = \dfrac{4}{x-3}$

15. $f(x) = \dfrac{2x}{x+1}$

16. $f(x) = \dfrac{3x}{x-2}$

17. $f(x) = \dfrac{x-2}{x-1}$

18. $f(x) = \dfrac{x+4}{2x+1}$

19. $f(x) = 2x + \dfrac{2}{x}$

20. $f(x) = 3x + \dfrac{12}{x}$

21. $f(x) = \dfrac{3x^2+6}{x}$

22. $f(x) = \dfrac{2x^2+1}{3x}$

🚀 APPLICATIONS

23. Junker Renovation completely overhauls junked or abandoned cars. Data shows their 1970s models hold their value quite well. The value $F(x)$ of one of these cars is given by $F(x) = 70 - \dfrac{15x}{x+1}$ where x is the number of years since repurchase and F is in hundreds of dollars.
   a. What is the initial resale price of a car?
   b. Find all asymptotes.
   c. Sketch the function.
   d. What is the long term value of one of these cars?

24. The average cost $A(x)$ is the total cost $C(x)$ divided by the quantity x. Thus $A(x) = \dfrac{C(x)}{x}$. If the total cost function for a product is $C(x) = 3x + 12$, graph the average cost $A(x)$. If there are any asymptotes, locate them and interpret their meaning.

25. A product's total costs are given by $C(x) = 0.03x^2 + 24x + 10$.
   a. Graph the average cost function, locating any asymptotes.
   b. What is the meaning of the asymptotes for average cost?

26. The Polar Pollution Control Company removes debris from old motors. Suppose the cost $F(x)$ of removing x percent of the pollutants is given by $F(x) = \dfrac{100,000}{100 - x}$, where x is a percentage, $0 \le x < 100$, and F is in dollars.
   a. Determine $\lim\limits_{x \to 0^+} F(x)$ and $\lim\limits_{x \to 100^-} F(x)$ and interpret their meanings.
   b. What percentage can be removed at a cost of \$3000?
   c. Show that $F(x)$ is always increasing. Does this make sense in the context of the problem?

27. The cost of camel rides in Tunisia is modeled by the function $F(x) = 40 - \dfrac{20x}{x + 3}$, where x is the number of years since 2000 and F is a national average cost in dinars.

   a. What was the cost of a camel ride in 2002?
   b. What are the asymptotes for F and which is significant in the problem?
   c. When was the average cost 26 dinars?

28. The sugar level concentration in the bloodstream of a certain diabetes patient is modeled by $S(t) = 1 + \dfrac{0.2t}{t^2 + 2}$, where S is in suitable units and t is the time in hours following a meal of allowed carbohydrate content.
   a. Which asymptotes play a role here?
   b. For $0 \le t \le 6$, what is the highest S-value and when does it occur? (If this level exceeds 4, the patient will become ill.)
   c. Are there any inflection points? What is the meaning in the context of the problem of an inflection point?

29. Data suggests a professional football team will win $F(x)$ games (out of 16) if the salary of the superstar players increases. For one team, the function F is given by $F(x) = 8 + \dfrac{6x}{0.125x^2 + 2}$, where x is the average salary (in millions) of the superstars (players earning at least one million dollars).
   a. Are there any asymptotes of consequence in the problem?
   b. What average salary gives the biggest return on games won, according to this model? (Here return is total games won.)

30. If administrative assistants at Bookworm Publications make phone follow-ups after textbook reviews, more colleges and universities will adopt a new statistics book. The publisher noticed that new book sales varied in the second year according to $S(x) = A_0 - \dfrac{(x + 200)}{x^2}$, where S is total sales and x is the number of phone calls made to colleges which have adopted the book. A_0 denotes the sales from the previous year.
   a. Assuming $A_0 = 2500$, what sales can Bookworm Publications expect if they make 100 follow-up phone calls?
   b. What is the horizontal asymptote and what is its significance?

12.5 BUSINESS APPLICATIONS

■ TOPICS

- ■ Minimizing Inventory Costs
- ■ Maximizing Revenue
- ■ Linear Demand Functions

Minimizing Inventory Costs

Retailers' costs of maintaining an inventory of goods can be considered to consist of two categories:

1. Storage costs (warehouse fees, insurance, etc.) and
2. Ordering costs (paperwork and shipping).

A large inventory can tie up valuable space in a store or warehouse. On the other hand, frequent ordering can run up a large shipping bill. We want to find a way to minimize the costs related to storage and shipping. These costs are called **inventory costs**.

For the mathematical model (function) that represents inventory costs to be manageable, the following "ideal" conditions are to be assumed:

1. The total sales for the year are known.
2. The same numbers of items are sold each day until all items in the inventory are sold. That is, the inventory is depleted linearly.
3. Orders are for the same number of items at regular time periods throughout the year.

Figure 1 illustrates these conditions graphically.

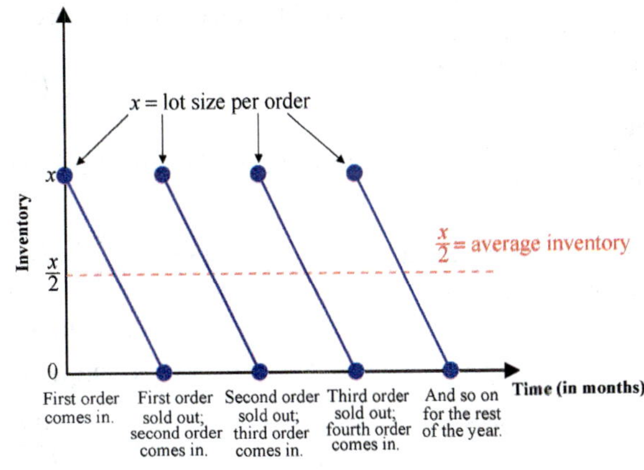

FIGURE 1

The mathematical model has the following components:

$$x = \text{lot size};$$

$$\frac{x}{2} = \text{average number of items in the inventory};$$

$$\frac{\text{total ordered}}{x} = \text{number of orders per year; and}$$

$$C(x) = (\text{storage cost per item}) \cdot \left(\frac{x}{2}\right) + (\text{cost per order}) \cdot \left(\frac{\text{total ordered}}{x}\right)$$

$$\left(\begin{array}{c}\text{inventory} \\ \text{costs}\end{array}\right) = \left(\begin{array}{c}\text{storage} \\ \text{costs}\end{array}\right) + \left(\begin{array}{c}\text{ordering} \\ \text{costs}\end{array}\right)$$

Example 1: Minimizing Inventory Costs

A furniture dealer sells 500 desks per year. The desks take up floor space and warehouse space, and the dealer estimates his storage costs at \$6 per desk. The distributor charges the dealer a \$60 fee for each order. How many times per year and in what lot size should the dealer order to minimize inventory costs?

Solution

Using the given information, we can determine a function for the inventory costs, $C(x)$. Let $x =$ lot size. Then $\dfrac{500}{x}$ is the number of orders per year, and $\dfrac{x}{2}$ is the average inventory.

$$C(x) = (\text{storage cost per item}) \cdot \left(\frac{x}{2}\right) + (\text{cost per order}) \cdot \left(\frac{500}{x}\right)$$

$$C(x) = 6\left(\frac{x}{2}\right) + 60\left(\frac{500}{x}\right)$$

The number of desks ordered is between 1 and 500. At the extremes, one order for 500 desks would cost

$$C(500) = 6\left(\frac{500}{2}\right) + 60\left(\frac{500}{500}\right) = \$1560,$$

and 500 orders for one desk at a time would cost

$$C(1) = 6\left(\frac{1}{2}\right) + 60\left(\frac{500}{1}\right) = \$30,003.$$

Now we need to differentiate $C(x)$ so we can determine the local minima.

$$C(x) = 6\left(\frac{x}{2}\right) + 60\left(\frac{500}{x}\right)$$

$$= 3x + 30,000x^{-1} \qquad \text{Rewrite } C(x) \text{ using exponents.}$$

$$C'(x) = 3 - 30,000x^{-2}$$

$$= 3 - \frac{30,000}{x^2}$$

We set $C'(x) = 0$ and solve for x.

$$3 - \frac{30,000}{x^2} = 0$$

$$3x^2 = 30,000$$

$$x^2 = 10,000$$

$$x = \pm 100$$

To evaluate the function $C(x)$ in Example 1 with a TI-83/84 Plus calculator, perform the following steps:

1. Enter the function $C(x)$ that was found into Y1.

2. Now, on the main screen, press VARS, scroll right to Y-VARS, select 1:Function, and then 1:Y1 to display Y1 on the main screen.

3. With your cursor after Y1, type an opening parenthesis and then the x-value we are evaluating, 500. Type a closing parenthesis.

4. Press **enter** to calculate the corresponding y-value.

5. This process can be repeated for the next x-value, 1.

```
Y₁(500)
              1560
Y₁(1)
              30003
```

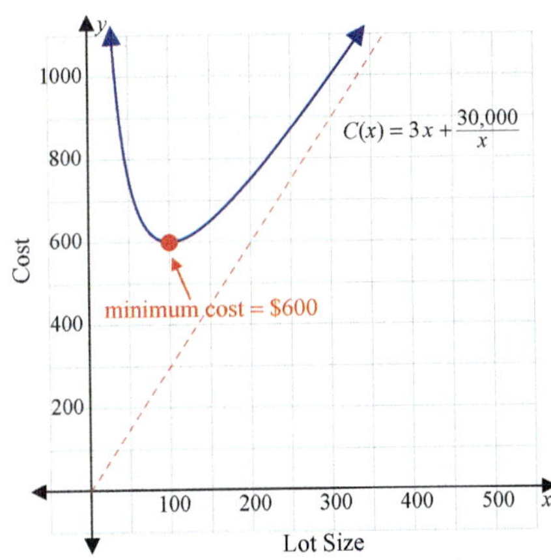

Note that $x = -100$ is not in the interval $[1, 500]$, so it must be disregarded. Substituting $x = 100$ into the equation for the number of orders per year, we get

$$\left(\frac{500}{x}\right) = \left(\frac{500}{100}\right) = 5 \text{ orders per year.}$$

To minimize inventory costs, the dealer should order 100 desks at a time, 5 times per year. The minimum inventory costs are

$$C(100) = 6\left(\frac{100}{2}\right) + 60\left(\frac{500}{100}\right) \qquad \text{Substitute } x = 100 \text{ into } C(x) \text{ to find the minimum inventory costs.}$$

$$= 6(50) + 60(5)$$

$$= 300 + 300 = \$600.$$

Since we are dealing with the closed interval $[1, 500]$ and $C(1) = \$30{,}003$ and $C(500) = \$1560$, $C(100) = \$600$ is also the absolute minimum.

Maximizing Revenue

Consider the following situation. The owner of a business can afford to lower (or raise) prices somewhat. This will have a direct effect on revenue. Generally, raising prices lowers sales and lowering prices increases sales. Sometimes past experience can be a guide in creating appropriate functions, and then calculus can be applied.

Example 2: Maximizing Revenue

Susan knows that she can sell 440 donuts a day at 50 cents per donut. What price should she charge to maximize her revenue if, for every increase of 10 cents in price, she sells 40 fewer donuts? What will this revenue be?

Solution

Let x = number of 10-cent increases in price. (The choice of x here does not represent the unknown price but the number of price increases. This choice for x makes the equation relatively easy to set up.) Then

$$0.50 + 0.10x = \text{new price, and}$$

$$440 - 40x = \text{new sales.}$$

Therefore,

$$R(x) = (\text{price}) \cdot (\text{sales})$$

$$= (0.50 + 0.10x)(440 - 40x)$$

$$= 220 - 20x + 44x - 4x^2$$

$$= 220 + 24x - 4x^2.$$

Differentiating $R(x)$, we get

$$R'(x) = 24 - 8x.$$

Setting $R'(x) = 0$ gives

$$24 - 8x = 0$$

$$-8x = -24$$

$$x = 3.$$

We can verify that $x = 3$ is a maximum by the Second Derivative Test: $R''(x) = -8 < 0$ for all x. Therefore, $x = 3$ does indeed give a maximum revenue.

Thus there should be three 10-cent increases in price. The new price should be

$$0.50 + 0.10(3) = \$0.80 \text{ per donut.}$$

The sales will then be

$$440 - 40(3) = 320 \text{ donuts,}$$

and the maximum revenue will be

$$R(3) = (0.80)(320) = \$256.$$

Note that no mention of profit is made in Example 2. Do you think that the profit will be likely to increase or decrease with an increase in price? What factors should be taken into account?

Linear Demand Functions

Suppose that two points relating sales and price are given. If the demand function is assumed to be linear, then the demand function can be found by using these two points and the slope.

Once the demand function has been determined, then the revenue function ($R(x) = x \cdot D(x)$) can be set up and differentiated to find what sales and price will yield maximum revenue.

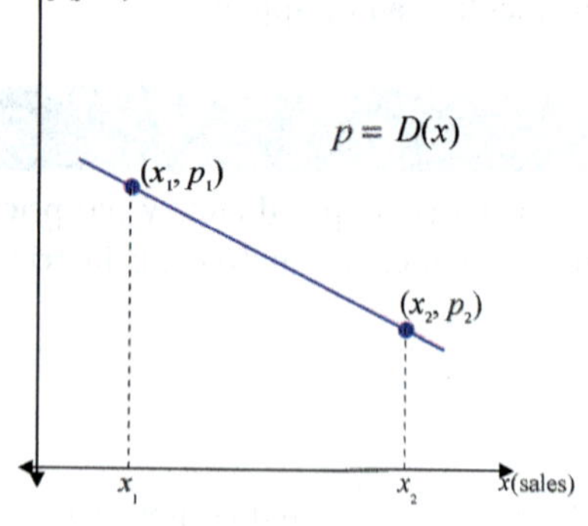

FIGURE 2

Example 3: Finding Maximum Revenue with a Linear Demand Function

A company sells 3000 calculators per month when the price is $7 per calculator. When the price is lowered to $4 per calculator, 6000 are sold. The maximum number of calculators the company can manufacture is 7000 per month. Assuming that the demand function is linear, determine how many calculators the company should produce and sell per month in order to maximize revenue.

Solution

The two points $(3000, 7)$ and $(6000, 4)$ can be used to find the linear demand function.

First, we will use these points in the formula for slope, $m = \dfrac{y_2 - y_1}{x_2 - x_1}$, and then we will use the point-slope form for the equation of a line, $y - y_1 = m(x - x_1)$. For this example we use the points in the form (x, p) instead of (x, y).

$$m = \frac{4 - 7}{6000 - 3000} = \frac{-3}{3000} = -\frac{1}{1000}$$

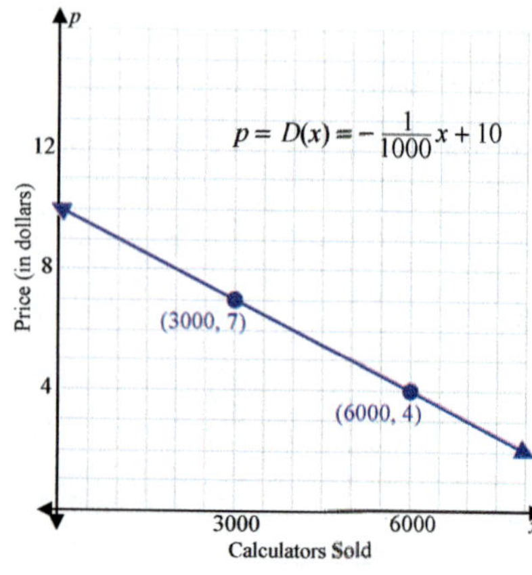

$$p - 7 = -\frac{1}{1000}(x - 3000)$$

$$p - 7 = -\frac{1}{1000}x + 3$$

$$p = -\frac{1}{1000}x + 10 = D(x)$$

$D(x)$ is the demand function where $x =$ number of calculators sold, and $p =$ price per calculator.

The revenue function is then

$$R(x) = x \cdot D(x)$$

$$= x \cdot \left(-\frac{1}{1000}x + 10\right)$$

$$= -\frac{1}{1000}x^2 + 10x.$$

Differentiating $R(x)$ gives

$$R'(x) = -\frac{1}{500}x + 10.$$

Setting $R'(x) = 0$ and solving for x gives

$$-\frac{1}{500}x + 10 = 0$$

$$-x = -5000$$

$$x = 5000.$$

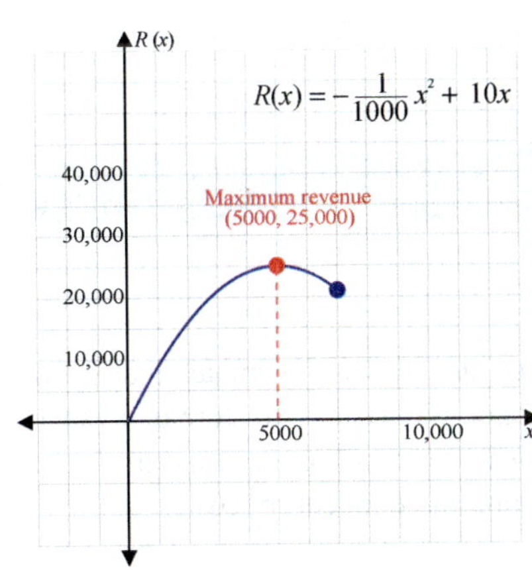

The company should produce 5000 calculators and sell them at a price of

$$p = -\frac{1}{1000}(5000) + 10 = \$5 \text{ each}$$

for the maximum revenue of

$$R(5000) = 5000 \cdot 5 = \$25,000 \text{ per month.}$$

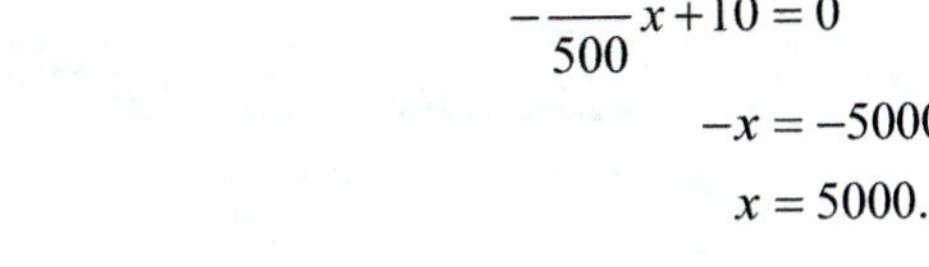

Example 4: Finding Maximum Profit with a Linear Demand Function

Suppose that the company in Example 3 has a cost function $C(x) = 5000 + 3x$. How many calculators should the company produce and sell to maximize profit?

Solution

Find the profit function.

$$P(x) = R(x) - C(x)$$

$$= -\frac{1}{1000}x^2 + 10x - (5000 + 3x)$$

$$= -\frac{1}{1000}x^2 + 10x - 5000 - 3x$$

$$= -\frac{1}{1000}x^2 + 7x - 5000$$

Differentiate $P(x)$.

$$P'(x) = -\frac{1}{500}x + 7$$

Set $P'(x)$ equal to 0, and solve for x.

$$-\frac{1}{500}x + 7 = 0$$

$$-x = -3500$$

$$x = 3500$$

Thus, a maximum profit occurs if the company produces and sells 3500 calculators, which is only half its production capabilities. The price for each calculator would be

$$p = -\frac{1}{1000}(3500) + 10 = \$6.50.$$

The graph shown illustrates the relationship between profit, revenue, and cost. Note that maximum revenue and maximum profit do not necessarily occur at the same level of production and sales.

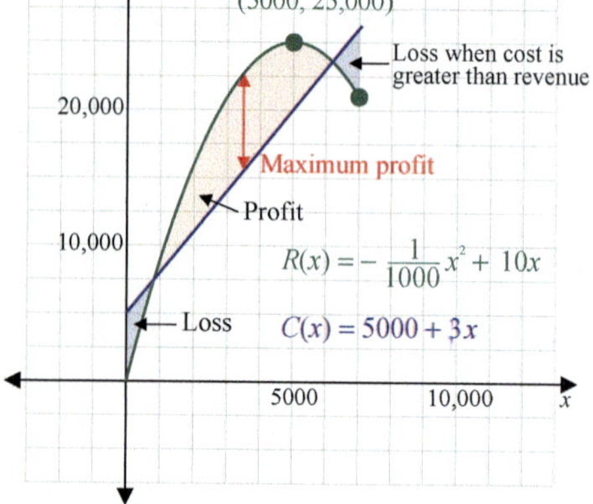

12.5 EXERCISES

🚀 APPLICATIONS

1. **Minimizing inventory costs:** An appliance store owner estimates that he will sell 125 vacuum cleaners of a particular model. It costs \$12 to store one vacuum cleaner for one year. There is a fixed cost of \$30 for each order. Find the lot size and the number of orders per year that will minimize inventory costs.

2. **Minimizing inventory costs:** A hardware store sells 96 chainsaws per year. It costs \$5 to store one chainsaw for one year. There is a fixed reordering cost of \$15. Find the lot size and the number of orders per year that will minimize inventory costs.

3. **Minimizing inventory costs:** An art gallery owner expects to sell 90 copies of a limited-edition print during the next year. It costs \$1.50 to store one copy for one year. For each order she places, there is a fixed cost of \$7.50, plus \$0.50 for each copy. Find the lot size and the number of times the gallery owner should order per year to minimize her inventory costs.

4. **Minimizing inventory costs:** The owner of Lamps-4-U expects to sell 180 brass lamps during the year. For each order he places, there is a fixed cost of \$18, plus \$2 for each lamp ordered. It costs \$5 to store one lamp for one year. In what lot size and how many times per year should he reorder to minimize the inventory costs?

5. **Minimizing inventory costs:** A T-shirt company sells 4000 sweatshirts per year. To reorder, there is a fixed cost of \$6 plus \$0.80 for each sweatshirt. It costs \$1.20 to store one sweatshirt for one year. In what lot size and how many times per year should an order be placed to minimize inventory costs?

6. **Minimizing inventory costs:** An office supply store sells 7500 pink highlighters per year. It costs \$0.15 to store one pink highlighter for one year. To reorder these highlighters, there is a fixed cost of \$22.50, plus \$0.10 for each highlighter. In what lot size and how many times per year should an order be placed to minimize inventory costs?

7. **Minimizing inventory costs:** A snowmobile dealer in Minnesota expects to sell 960 snowmobiles during the next year. It costs \$9 to store one snowmobile for one year. To reorder, there is a fixed cost of \$67.50, plus \$7.50 for each snowmobile. In what lot size and how many times per year should an order be placed to minimize inventory costs?

8. **Minimizing inventory costs:** A car dealer expects to sell 1320 cars during the next year. It costs \$660 to store one car for one year. To reorder, there is a fixed cost of \$225, plus \$304 for each car. Find the lot size and the number of orders that should be placed so inventory costs will be minimized.

9. **Maximizing revenue:** A chain of discount stores sells 84 weather radios per month at \$20 each. The owners estimate that for each \$1 increase in price, they will sell 3 fewer radios per month. How much should they charge for their weather radios to maximize their revenue?

10. **Maximizing revenue:** A farmer estimates that if he plants 30 grapefruit trees per acre, the average yield per tree will be 480 pounds. For each additional tree planted per acre, the yield per tree will be reduced by 12 pounds. How many trees should be planted per acre to maximize the yield?

11. **Maximizing revenue:** Sam operates a chain of convenience stores. He estimates that he can sell 600 small packs of gum per day if he charges 75 cents each. Sam determines that for each 10-cent reduction in price, he will sell an additional 80 packs per day. How much should he charge for the small packs of gum to maximize his revenue?

12. **Maximizing revenue:** A sporting goods store sells 200 baseball gloves per month at \$36 each. The owner estimates that for each \$2 increase in price, he will sell 5 fewer gloves. Find the price that will maximize revenue.

13. **Maximizing revenue:** A sports arena has 40 roaming soda salespeople, each of whom sells 200 sodas per event. Management estimates that for each additional salesperson, the yield per salesperson decreases by 4. How many additional salespeople should management hire to maximize the number of sodas sold?

14. **Maximizing revenue:** Ms. Wills owns a 16-rack dry stack boat storage facility. The unit rent per rack is currently \$400 per month, and all racks are rented. Each time rent is increased by \$20, one boat owner will move out. Find the rental price that will maximize Ms. Wills' revenue.

15. **Linear demand function:** A local amusement park found that if the admission was \$7, about 1000 customers per day were admitted. When the admission was dropped to \$6, the park had about 1200 customers per day. Assuming a linear demand function, determine the admission price that will yield maximum revenue.

16. Linear demand function: A department store manager has determined that when the price of a tank top was $12, she sold 100 tank tops per month. However, only 80 tank tops were sold per month when the price was raised to $14. Assuming a linear demand function, determine the price that would maximize the revenue.

17. Linear demand function: The cost of producing x units of an item is $C(x) = 10x + 20$. When the selling price is $20, twenty-one items are sold. However, when the price is $16, twenty-three items are sold. Assuming the demand function is linear, determine the price per unit and the number of units sold that will maximize the profit.

18. Linear demand function: The manager of a bakery knows he can sell 60 small bags of donut holes when the price is $1.20 each. If the price is $1.50, only 48 bags are sold. The total cost function for x bags is $C(x) = 0.70x + 15$ dollars. Assuming a linear demand function, determine the price per bag and the number of bags sold that will maximize the profit.

19. Linear demand function: The manufacturer of microwave ovens can sell 800 to his dealers at $392 each. If the price is $380, he can sell 1000. The total cost of producing x microwaves is $C(x) = 3600 + 250x - 0.01x^2$ dollars. Assuming the demand function is linear, find the price per microwave and the number of microwaves sold that will maximize profit.

20. Linear demand function: A candy store can sell 180 lollipops at 62 cents each. The store can sell 220 lollipops if the price is 54 cents each. The total cost of producing x lollipops is $C(x) = 3050 - 10x + 0.04x^2$ cents. Find the number of lollipops that should be produced to maximize profit.

21. Profit: Suppose $P(x)$ represents profit on the sales of x cell phones. Suppose $P(25,000) = 12,000$, $P'(25,000) = 2$, and $P''(25,000) = -3$.
  a. Is the company making money or losing it? How much?
  b. If sales are increased, will the profits rise or fall? By how much?
  c. What is the meaning of $P''(25,000) = -3$?

22. Profit: Suppose the monthly marginal profit from John's online tutoring service is $P'(x) = 3$, where x is the number of subscribers.
  a. The profit function is (choose one): linear, quadratic, or a polynomial of degree 3 or higher.
  b. Suppose monthly costs are $C(x) = 9x + 20$. Assuming initial revenue is 0, what is the revenue function?

23. Profit: Suppose the demand for a product is $12 and the total costs are $C(x) = 0.3x^2 + 2x + 5$.
  a. What is the revenue function?
  b. What is the profit function?
  c. What is the maximum value of the profit?

24. Average cost: $A(x) = \dfrac{C(x)}{x}$ gives average cost.

   a. Calculate a formula for $A'(x)$.

   b. Set $A'(x) = 0$ and solve for $C'(x)$.

   c. If average costs are minimal, describe a relationship between average cost and marginal cost. That is, interpret the result of part **b.**

25. Average cost: Suppose that a company's average cost is $A(x) = 0.2x + 3$.

   a. Determine the cost function.

   b. Determine the marginal average cost.

   c. Determine the marginal cost.

26. Earnings: Suppose a company's earnings are given by $E(x) = P(x) + I(x)$, where x is the number of years since 2018, $P(x)$ is the annual profit function, and $I(x)$ is the intangible growth (the growth in value of the company's intangible assets such as its good name). If $P(x) = 1.3x + 2$ and $I(x) = 0.25x + 1$ for a certain company, determine the following.

   a. The marginal earnings for year x

   b. The actual earnings for 2020

   c. The average earnings formula (earnings per year since 2018)

27. Value: The value of a new business franchise grows according to the formula $V(x) = 10 + \dfrac{10x}{1 + 0.5x}$. Here V is in thousands of dollars and x is the number of years after 2012.

   a. What is the expected value in 2022?

   b. Is the value V increasing or decreasing in 2022?

   c. Is the rate of increase in value changing? What has this to do with V' or V''?

28. Value: The timber value of a small stand of pine trees is given by $V(x) = 50\left(1 - \dfrac{1}{x + 2}\right)$, where x is the number of years after 2000 and V is in dollars.

   a. What was the value in 2000?

   b. At what rate was the value changing in 2010?

   c. What asymptotes are present and what is their significance in the problem?

✎ WRITING & THINKING

29. Sales function: A sales function $S(t)$ for a new product is shown in the figure. $S(t)$ is total sales (quantity of items) and t is time in months since the product's release. Copy this graph onto your paper and add a graph of a possible $S'(t)$. Locate approximately the point of inflection on your curve for S.

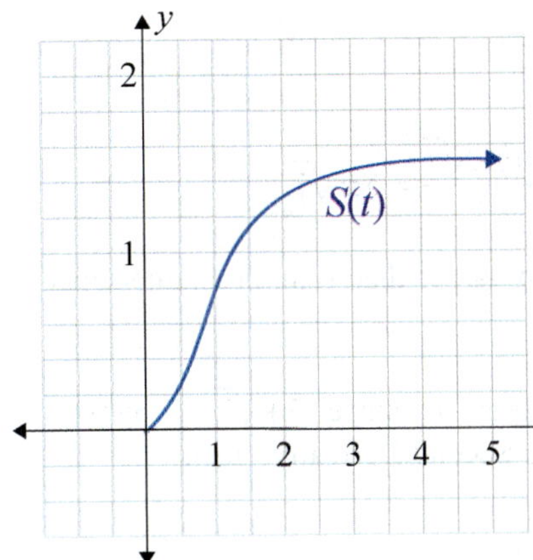

30. Cost function: A certain cost function $C(x)$ satisfies $C(10) = 20$, $C(20) = 40$, and $C(30) = 60$. Suppose $C''(10) = -2$, $C''(20) = 0$, and $C''(30) = 2$. Draw a suitable function.

12.6 OTHER APPLICATIONS: OPTIMIZATION, DISTANCE, AND VELOCITY

■ TOPICS

- ■ Geometry
- ■ Distance and Velocity

Geometry

Example 1: Volume of a Box

A rectangular box with no top is to be made from a rectangular piece of cardboard that is 18 in. long by 12 in. wide. Small squares (all the same size) are to be cut from each corner and the sides folded to form the box. What size should the squares be to give a maximum volume to the box?

Solution

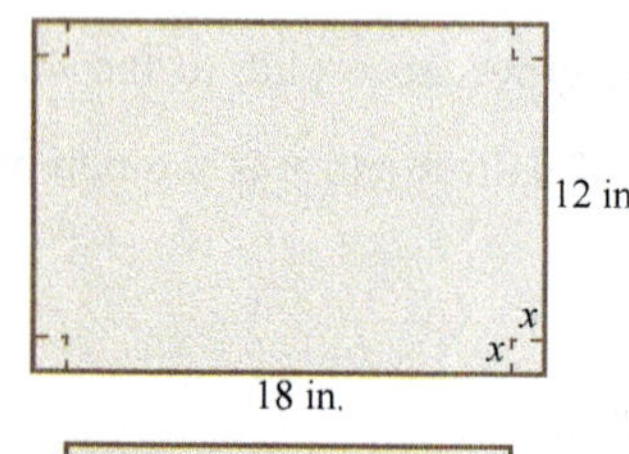

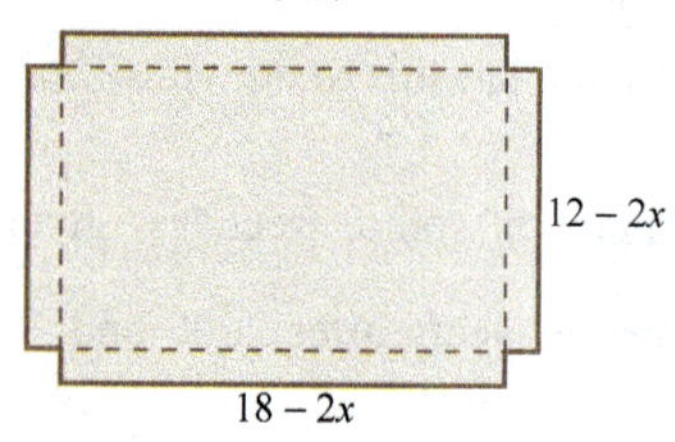

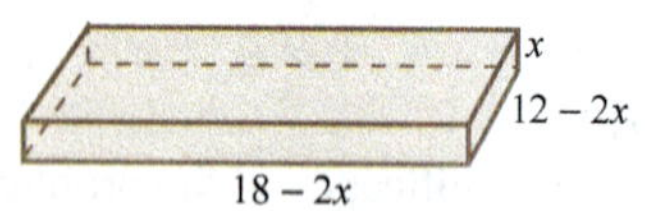

Let x = length of one side of the square to be cut.

$$\text{Volume} = \text{length} \cdot \text{width} \cdot \text{height}$$
$$V = (18 - 2x) \cdot (12 - 2x) \cdot x$$
$$= 216x - 60x^2 + 4x^3$$

Differentiating V, we obtain

$$\frac{dV}{dx} = 216 - 120x + 12x^2$$
$$= 12(x^2 - 10x + 18).$$

Now we set $\dfrac{dV}{dx} = 0$ and solve for x.

$$12(x^2 - 10x + 18) = 0$$
$$x^2 - 10x + 18 = 0$$
$$x = \frac{10 \pm \sqrt{100 - 72}}{2}$$
$$x = \frac{10 \pm \sqrt{28}}{2} = \frac{10 \pm 2\sqrt{7}}{2} = 5 \pm \sqrt{7}$$

We know that the size of the square cannot be any larger than 6 in. or smaller than 0 in.; that is, x must be in the closed interval [0, 6].

Note, $x = 5 + \sqrt{7}$ is too large to apply in the problem since $5 + \sqrt{7} \approx 7.65 > 6$. Therefore, we check $x = 5 - \sqrt{7} \approx 2.35$ in. with the Second Derivative Test.

$$V''(x) = 12(2x - 10)$$
$$V''(2.35) = 12(4.7 - 10) < 0$$

So V is concave down at $x = 5 - \sqrt{7}$ which means this x gives a maximum volume.

Example 2: Surface Area of a Box

A rectangular box is to be made so that the top and bottom are squares. The volume is to be 250 cm³. Material for the top and bottom costs $2 per square centimeter, but the material for the sides costs only $1 per square centimeter. What dimensions will give a minimum cost for the materials? What is the minimum cost?

Solution

Let x = length of an edge along the square bottom.

Let y = length of a vertical edge.

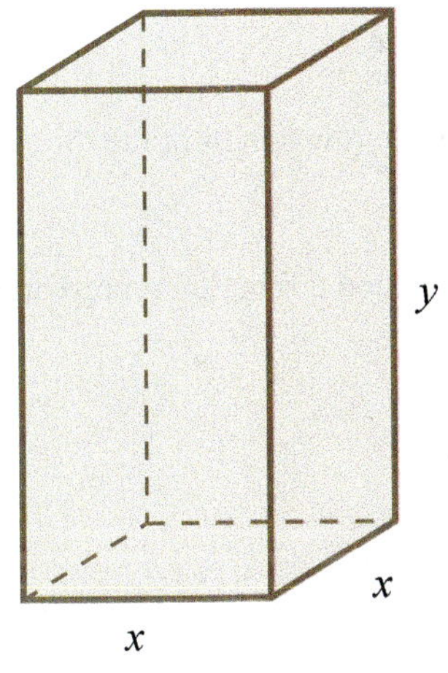

Volume $= x^2 y = 250$ cm³

We are assuming that cost only depends on surface area. The cost, $C(x)$, of the materials is found by multiplying the areas of the faces of the box by $2 and $1 as follows.

$$C(x) = \$2x^2 + \$2x^2 + \$1(4xy) \quad \text{Cost of the top, bottom, and 4 sides}$$
$$= 4x^2 + 4xy$$

Now, using the volume, solve for y and substitute into the cost function.

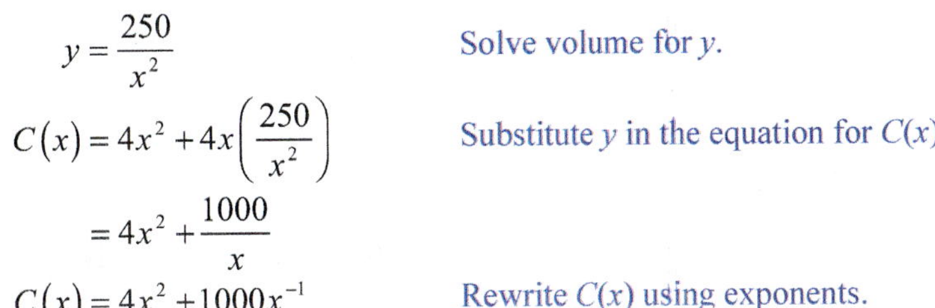

$$y = \frac{250}{x^2} \qquad \text{Solve volume for } y.$$

$$C(x) = 4x^2 + 4x\left(\frac{250}{x^2}\right) \qquad \text{Substitute } y \text{ in the equation for } C(x).$$

$$= 4x^2 + \frac{1000}{x}$$

$$C(x) = 4x^2 + 1000x^{-1} \qquad \text{Rewrite } C(x) \text{ using exponents.}$$

Differentiating $C(x)$, we have

$$C'(x) = 8x - 1000x^{-2}.$$

Setting $C'(x) = 0$ and solving for x gives

$$8x - \frac{1000}{x^2} = 0$$
$$8x = \frac{1000}{x^2}$$
$$8x^3 = 1000$$
$$x^3 = 125$$
$$x = 5 \text{ cm.}$$
$$\text{So} \quad y = \frac{250}{5^2} = 10 \text{ cm.}$$

Note that $C''(x) = 8 + \frac{2000}{x^3}$. Since $C''(x)$ is positive for all positive x, $C(x)$ is concave up everywhere for $x > 0$. Thus, by the Second Derivative Test, the critical point $x = 5$ yields a minimum value.

Therefore, the dimensions of the box are 5 cm by 5 cm by 10 cm, and the minimum cost is

$$C(5) = 4(5)^2 + \frac{1000}{5} = 100 + 200 = \$300.$$

Example 3: Fencing

A farmer wants to enclose a rectangular area next to his barn. If he has 200 feet of fencing to use, what dimensions of the rectangle will maximize the area? What is the maximum area? (Note that the fence is only three sides of the rectangle, since the barn serves as the fourth side of the enclosure.)

Solution

Let x = length of one of the two equal sides of the fence.

Then, since there are only 200 feet of fencing available, the length of the third side is $200 - 2x$.

The area of the rectangle is to be maximized; so we need a function representing the area.

$$A(x) = x(200 - 2x)$$
$$= 200x - 2x^2$$

Then

$$A'(x) = 200 - 4x.$$

Setting $A'(x) = 0$ and solving for x, we have

$$200 - 4x = 0$$
$$4x = 200$$
$$x = 50.$$

We now apply the Second Derivative Test.

Note that $A''(x) = -4$. So $A''(50) = -4$ and A is concave down at $x = 50$. Thus the critical value gives a maximum.

The rectangle should be 50 ft by 100 ft, and the maximum area is

$$A(50) = 50(200 - 2 \cdot 50) = 50(100) = 5000 \text{ ft}^2.$$

Example 4: Oil Pipeline

An oil company wants to lay a pipeline from its offshore drilling rig to a storage tank on shore. The rig (point R) is three miles offshore (point A), and the storage tank (point B) is 8 miles down the shoreline. The cost of laying pipe underwater is $800 per mile and along the shoreline is $400 per mile. Point P is the point on shore where the underwater pipe connects with the shoreline pipe. Where should point P be located so as to minimize the cost of laying pipe?

Solution

Let x be the distance from A to P.

Since $\overline{AB} = 8$, $\overline{PB} = 8 - x$.

By the Pythagorean Theorem,

$$\overline{RP} = \sqrt{x^2 + 9}.$$

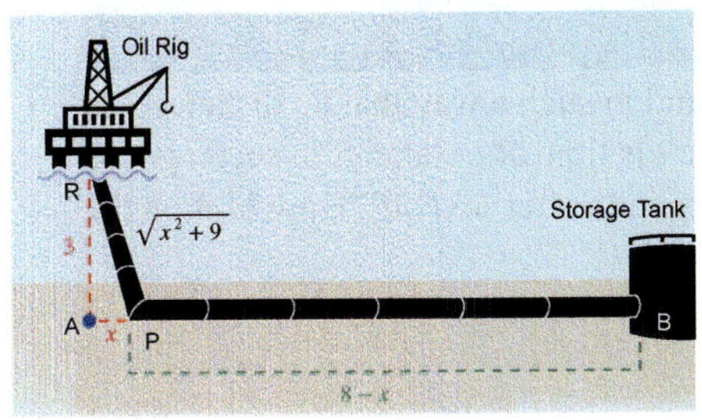

The total cost of laying pipe is

$$C(x) = 800\sqrt{x^2 + 9} + 400(8 - x)$$

$$= 800(x^2 + 9)^{\frac{1}{2}} + 3200 - 400x.$$

Differentiating $C(x)$ gives

$$\frac{dC}{dx} = 800 \cdot \frac{1}{2}(x^2 + 9)^{-\frac{1}{2}}(2x) - 400$$

$$= \frac{800x}{\sqrt{x^2 + 9}} - 400.$$

Note: $\dfrac{d^2C}{dx^2} = C''(x) = 7200(x^2 + 9)^{-\frac{3}{2}}$

Setting $\dfrac{dC}{dx} = 0$ and solving for x, we have

$$\frac{800x}{\sqrt{x^2 + 9}} - 400 = 0$$

$$\frac{800x}{\sqrt{x^2 + 9}} = 400$$

$$2x = \sqrt{x^2 + 9}$$

$$4x^2 = x^2 + 9$$

$$3x^2 = 9$$

$$x^2 = 3$$

$$x = \pm\sqrt{3}.$$

Note that $x = -\sqrt{3}$ is not a reasonable solution because distance cannot be negative. We have assumed x to be in the interval [0, 8]. Now, $C''(\sqrt{3}) = 173.2 > 0$, so $x = \sqrt{3}$ gives a minimum.

So the minimum cost is incurred if point P is located $\sqrt{3}$ miles (approximately 1.732 miles) along the shoreline from point A (or $8 - \sqrt{3} \approx 6.268$ miles from the storage tank at point B).

We should check the endpoints $x = 0$ and $x = 8$ to be sure that they do not give an absolute minimum.

$$C(0) = 800\sqrt{0^2 + 9} + 400(8 - 0) = \$5600$$
$$C(8) = 800\sqrt{8^2 + 9} + 400(8 - 8) \approx \$6835$$

Thus $C(\sqrt{3}) = 800\sqrt{(\sqrt{3})^2 + 9} + 400(8 - \sqrt{3}) \approx \$5278$ is indeed the minimum cost.

Distance and Velocity

Velocity is the rate of change of distance with respect to time. Customary notation in science and mathematics is $s(t)$ for distance and $v(t)$ for velocity, since both distance and velocity depend on time t. We also have

$$\frac{ds}{dt} = s'(t) = v(t).$$

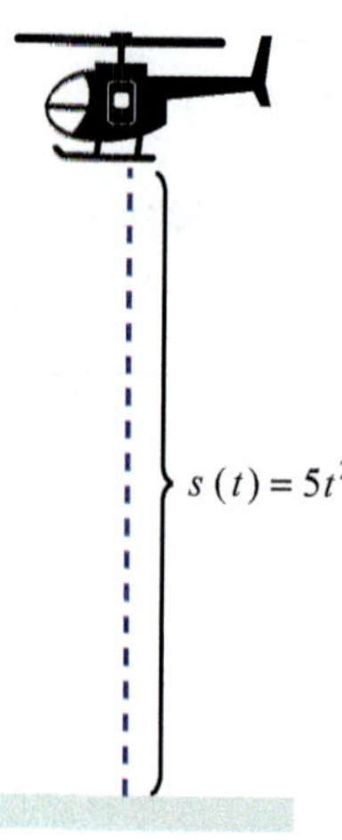

Example 5: Helicopter Takeoff

Suppose that a helicopter is rising from the ground in such a way that its distance s in feet from the ground is given by $s(t) = 5t^2$, where t is time after takeoff in seconds and $0 \le t \le 6$. How fast is the helicopter rising 3 seconds after takeoff? How high is the helicopter at that time?

Solution

Differentiating the distance function with respect to time gives the velocity function.

$$s'(t) = v(t) = 10t$$
$$v(3) = 10(3) = 30 \text{ ft/sec} \quad \text{Substitute } t = 3 \text{ into } v(t) \text{ to find the velocity.}$$

Three seconds after takeoff, the helicopter is rising at 30 ft/sec.

To find the height of the helicopter after 3 seconds, substitute $t = 3$ into $s(t)$.

$$s(3) = 5(3)^2 = 45 \text{ ft}$$

In 3 seconds the helicopter has risen to an altitude of 45 feet.

Example 6: Vertical Projectile

A projectile is fired vertically (straight up) from the top corner of a 176-foot tall building. The height of the projectile at time t (in seconds) is given by the function $s(t) = -16t^2 + 160t + 176$.

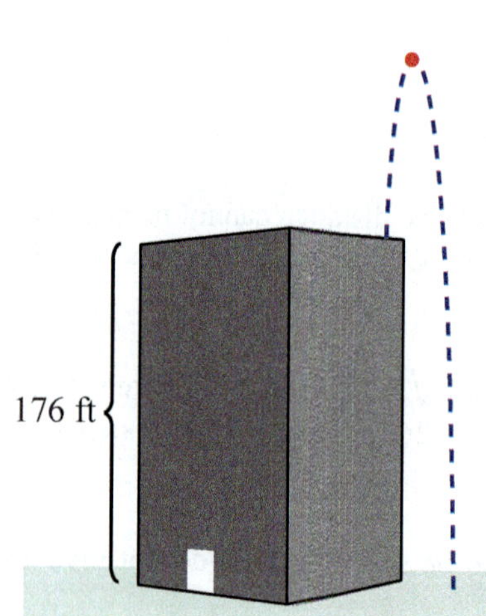

a. What is the maximum height of the projectile?

b. When does the projectile hit the ground?

c. How fast is the projectile moving when it hits the ground?

Solution

a. The maximum height of the projectile occurs at the point where its velocity is 0.

$$s(t) = -16t^2 + 160t + 176$$
$$v(t) = s'(t) = -32t + 160 \quad \text{Differentiate } s(t) \text{ to determine } v(t).$$
$$-32t + 160 = 0 \qquad\qquad\quad \text{Then set } v(t) \text{ equal to 0 and solve for } t.$$
$$-32t = -160$$
$$t = 5$$

Since $s''(t) = -32$ (a constant), the curve $s(t)$ is concave down everywhere. By the Second Derivative Test, the maximum height occurs when $t = 5$ seconds. Substitute this t-value into $s(t)$ to calculate the maximum height.

$$s(5) = -16(5)^2 + 160 \cdot 5 + 176$$
$$= -400 + 800 + 176$$
$$= 576 \text{ ft}$$

b. The projectile hits the ground when $s(t) = 0$.

$$-16t^2 + 160t + 176 = 0 \qquad \text{Solve } s(t) = 0 \text{ for } t.$$
$$-16\left(t^2 - 10t - 11\right) = 0$$
$$-16(t-11)(t+1) = 0$$
$$t = 11 \text{ or } \cancel{t=-1} \qquad \text{Time cannot be negative, so we discard the } t = -1.$$

Thus the projectile hits the ground 11 seconds after it is fired.

c. The projectile's velocity when it hits the ground is $v(11)$. From part **a.** we know

$$v(t) = -32t + 160.$$

So

$$v(11) = -32(11) + 160$$
$$= -352 + 160$$
$$= -192 \text{ ft/sec.} \qquad \text{The negative sign indicates the projectile is falling.}$$

The projectile will be falling at a rate of 192 ft/sec when it hits the ground.

12.6 EXERCISES

🚀 APPLICATIONS

1. **Volume of a box:** A rectangular box with no top is to be made from a piece of cardboard that is 24 in. by 24 in. Equal squares are to be cut from each corner and the sides folded to form the box. What size should the squares be to maximize the volume of the box?

2. **Volume of a box:** A rectangular box with no top is to be made from a piece of cardboard that is 20 in. by 20 in. Equal squares are to be cut from each corner and the sides folded to form the box. What size should the squares be to maximize the volume of the box?

3. **Volume of a box:** Equal squares are to be cut from each corner of a rectangular piece of thin sheet metal, and the sides are to be folded to form a box. If the piece of metal is 8 in. by 15 in., find the dimensions of the box having maximum volume.

4. **Volume of a box:** Equal squares are to be cut from each corner of a rectangular piece of thin sheet metal, and the sides are to be folded to form a box. If the piece of metal is 10 in. by 16 in., find the dimensions of the box having maximum volume.

5. **Surface area:** A rectangular box is designed to have a square base and an open top. The volume is to be 864 in.[3]
   a. What dimensions will give a minimum surface area?
   b. What is the minimum surface area?

6. **Surface area:** A rectangular box is designed to have a square base and an open top. The volume is to be 256 in.³
   a. What dimensions will give a minimum surface area?
   b. What is the minimum surface area?

7. **Container design:** A container manufacturer is asked to design a closed rectangular shipping crate with a square base. The volume is 10 ft³. The material for the top and sides costs \$2 per square foot and the material for the bottom costs \$3 per square foot. Find the dimensions of the box that will minimize the total cost.

8. **Container design:** A container manufacturer is asked to design a closed rectangular shipping crate with a square base. The volume is 36 ft³. The material for the top costs \$1 per square foot, the material for the sides costs \$0.90 per square foot, and the material for the bottom costs \$1.40 per square foot. Find the dimensions of the box that will minimize the total cost of material.

9. **Container design:** An investor plans to manufacture rectangular box containers whose bottom and top measure x by $3x$. The box must contain 18 cubic feet. The top and bottom will cost \$2 per square foot, and the four sides will cost \$3 per square foot. What should the height h be so as to minimize costs?

10. **Area:** A rectangular plot is to be enclosed with an existing block wall as one side. If there are 680 ft of fencing available for the other three sides, find the dimensions that will maximize the area.

11. **Area:** A rectangular play area is to be enclosed with the side of a house as one of the sides. If there are 74 ft of fencing available for the other three sides, find the dimensions that will maximize the area.

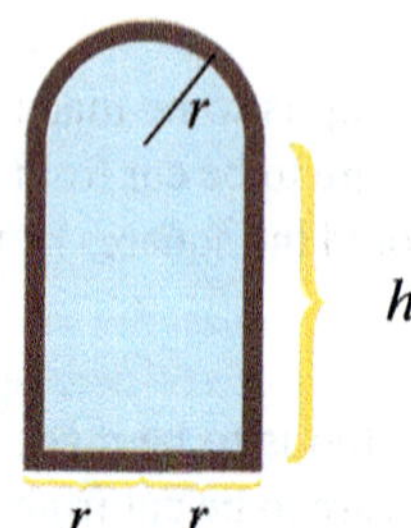

12. **Window area:** A front window on a new home is designed as a rectangle with a semicircle on the top. If the window is designed to let in a maximum amount of light, and the architect fixes the perimeter of the entire window at 600 inches, determine the radius r and rectangular height h so as to maximize the area.

13. **Wall construction:** An old stone wall makes two legs of a right angle, one 40 feet long and the other 20 feet long. A constructor is told to add 220 feet of new stone fence to complete a rectangular fence. How should he complete the fence so as to maximize the enclosed area? Determine the maximum enclosed area he may obtain.

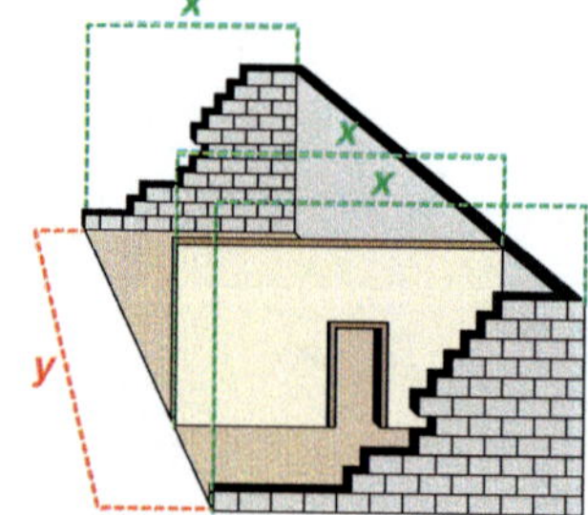

14. **Construction:** A warehouse is being constructed with a total floor area of 2200 ft². A single partition is built to divide the building into storage area and office space. The exterior walls cost \$160 per foot, and the interior wall costs \$120 per foot. Find the dimensions of the warehouse that will minimize the cost.

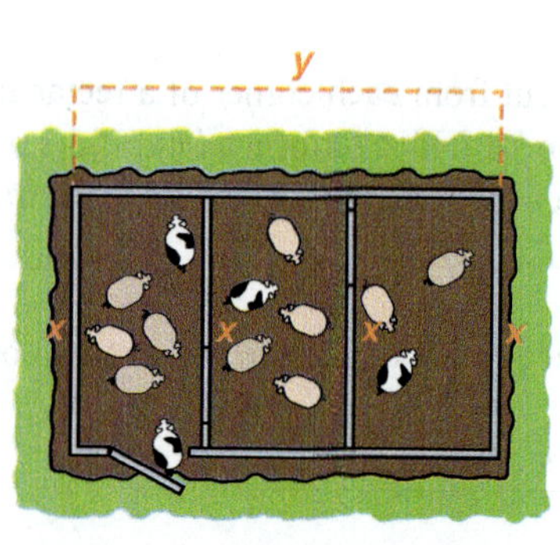

15. **Construction:** A farmer wants to build a rectangular pen and then divide it with two interior fences. The total area is to be 2484 ft². The exterior fence costs \$18 per foot, and the interior fence costs \$16.50 per foot. Find the dimensions of the pen that will minimize the cost.

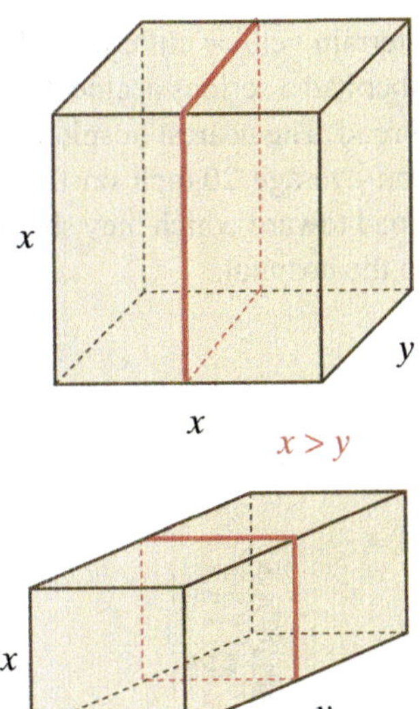

16. Shipping: The Postal Service has a limit of 108 in. on the combined length and girth of a rectangular package to be sent by parcel post. Find the dimensions of the package of maximum volume that has a square cross section. (**Hint:** There are two different answers, depending on the shape of the box. The two shapes are shown here. The girth is defined to be the smallest perimeter of a rectangular cross section of the box.)

17. Shipping: An independent parcel service has a limit of 130 in. on the combined length and girth of a rectangular package it will ship. Find the dimensions of the package of maximum volume that has a square cross section. (**Hint:** There are two different answers. The girth is defined to be the smallest perimeter of a rectangular cross section of the box.)

18. Pipeline construction: An oil company wishes to run a pipeline from a drilling platform located 5 miles offshore to a shipping terminal 16 miles down the coast. The costs are $130,000 per mile to lay the pipeline underwater and $120,000 per mile to lay the pipeline over land. Find the location of point P (as illustrated in the diagram) so that the total cost of laying pipe will be minimized.

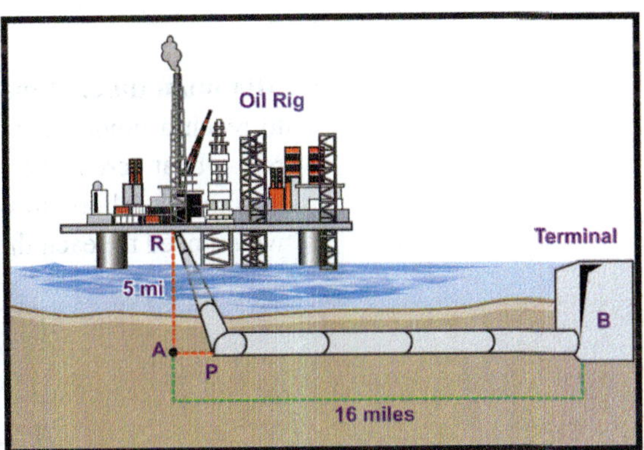

19. Power line construction: The U.S. Forest Service wishes to run a power line to a fire lookout tower located in a wooded area. The tower is 4 miles from the nearest road and the power source is 7 miles down that road. It costs $5000 per mile to run the line through the forest and $3000 per mile to run the line along the road. Find the location of point P (as illustrated in the diagram) so that the total cost of running the power line will be minimized.

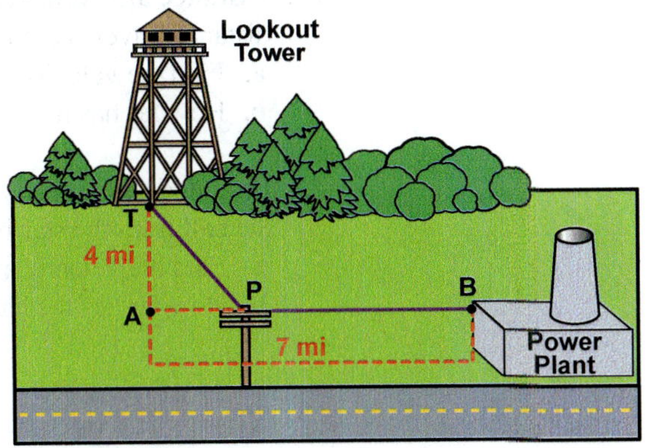

20. Minimum time: The Off-Roaders, an all-terrain vehicle club, were driving their four-wheelers in the desert when one member had a serious accident. At the time, they were 12 miles from the nearest paved road. The nearest hospital was located 11 miles down the paved road. If they can average 20 mph on the desert and 52 mph on the road, locate point P on the road toward which they should drive in order to minimize the time needed to get to the hospital.

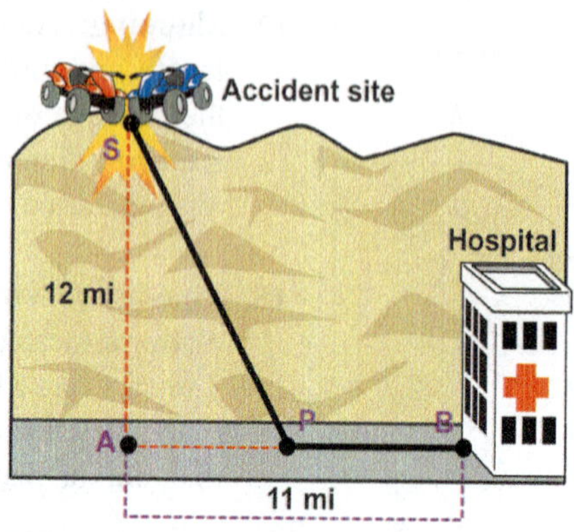

21. Minimum time: A man is on the bank of a river that is 0.6 miles wide. He wants to reach a point on the opposite shore that is 1.5 miles downstream. If he can row a boat across the river at 4 mph and walk at 5 mph, find the location P, on the opposite shore, to which he should row in order to minimize the total time he would need to reach the point downstream.

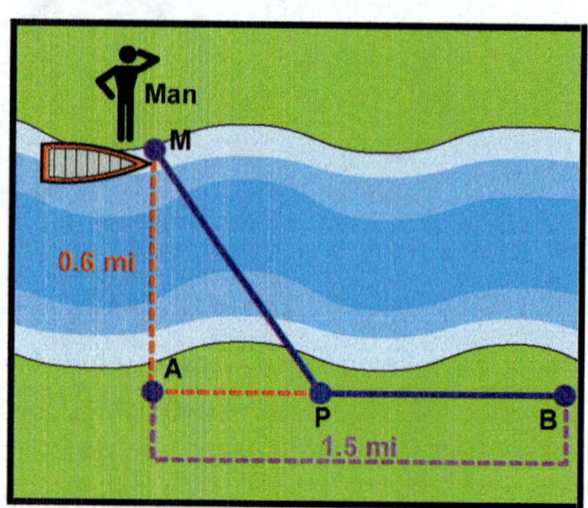

22. Distance and velocity: A particle is moving along a straight line such that the distance traveled at the end of t seconds is given by $s(t) = 7t^2 + 30t$ feet.
a. Find the velocity if $t = 2$ seconds.
b. How far has the particle traveled?

23. Distance and velocity: A ball is rolled down an incline. The distance (in feet) of the ball from the starting point after t seconds is given by $s(t) = 19t + 8t^2$.
a. Find the velocity after 3 seconds.
b. How far has the ball traveled in 3 seconds?

24. **Distance and velocity:** A projectile is fired vertically, and its height (in feet) after t seconds is given by $s(t) = 104t - 16t^2$.
    a. Find the maximum height of the projectile.
    b. When does the projectile hit the ground?
    c. How fast is the projectile moving when it hits the ground?

25. **Distance and velocity:** A stone is projected vertically. The height (in feet) of the stone at time t (in seconds) is given by $s(t) = -16t^2 + 112t + 128$.
    a. What is the maximum height of the stone?
    b. When will the stone hit the ground?
    c. What is the speed of the stone when it hits the ground?

26. **Distance and velocity:** A child rolls a hoop down a hilly street. The distance traveled (in feet) after t seconds is given by $s(t) = 4t + t^2$.
    a. How far has the hoop traveled in 3 seconds?
    b. What was the speed at 3 seconds?
    c. At what rate was the speed changing at $t = 3$?

13

Chapter 13

ADDITIONAL APPLICATIONS OF THE DERIVATIVE

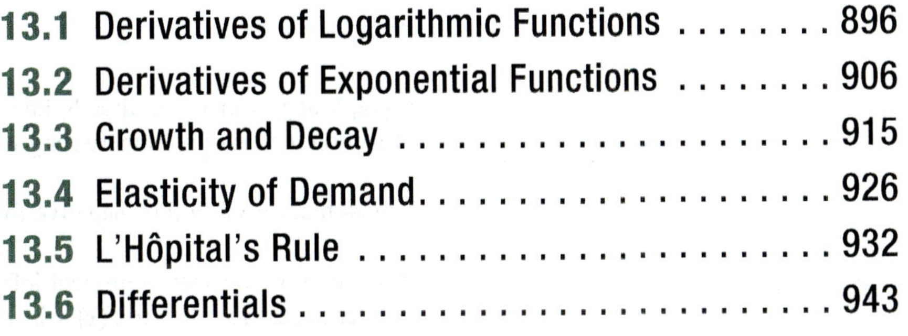

13.1 DERIVATIVES OF LOGARITHMIC FUNCTIONS

■ TOPICS

- The Derivative of $y = \ln x$
- The Derivative of $y = \ln(g(x))$
- Logarithmic Differentiation

The Derivative of $y = \ln x$

From the graph of $y = \ln x$, we already know that $\dfrac{d}{dx}\ln x = (\ln x)'$ will be positive since the function increases with increasing x.

Also, we know that $(\ln x)''$ will be negative for all x since the graph is concave down.

It is always useful to consider numerical information which helps in understanding. Let us consider the tangent line to $f(x) = \ln x$ at $x = 9$. The slope of this line is $f'(9)$. In order to calculate this slope approximately, let point $A = (9, \ln 9)$ and point $B = (9.00001, \ln 9.00001)$. The slope of the segment AB is close to that for the tangent line. This slope is

$$\frac{\Delta y}{\Delta x} = \frac{\ln(9.00001) - \ln(9)}{0.00001} = 0.11111105.$$

Evidently, the slope of the tangent line is quite close to $\dfrac{1}{9} = 0.11111\ldots$. Table 1 can be constructed in the same way or by using a calculator. From Table 1, we will be able to guess the formula for $\dfrac{d}{dx}(\ln x)$.

x	$f(x) = \ln x$	Approximation of $f'(x)$
2	$\ln 2 \approx 0.6931$	$0.499999 \approx \dfrac{1}{2}$
3	$\ln 3 \approx 1.098$	$0.3333327 \approx \dfrac{1}{3}$
4	$\ln 4 \approx 1.386$	$0.249999 \approx \dfrac{1}{4}$
9	$\ln 9 \approx 2.197$	$0.111111 \approx \dfrac{1}{9}$

TABLE 1

⌁ TECH TRAINING

To create a table of x- and y-values for $\ln x$ and its derivative using a TI-83/84 Plus calculator, first ensure your calculator mode is set to CLASSIC and then perform the following steps:

1. Enter the function $\ln x$ into Y1.

2. Enter the derivative of $\ln x$ into Y2 by first pressing MATH and selecting item 8:nDeriv(. At the blinking cursor, type in Y1,X,X).

3. Press 2nd window to access TABLE SETUP.

4. For this table we will input specific values of x, so TblStart, ΔTbl, and Depend can be ignored.

5. For Indpnt, move the blinking cursor over Ask. Press enter.

6. Press 2nd graph to access the table.

7. With the cursor in the **X** column, enter the first x-value you would like to find y for (use 0.1). Press `enter`. Then -2.303 will appear in the **Y1** column and 10 will appear in the **Y2** column.

8. Continue entering the remaining x-values.

X	Y1	Y2
.1	-2.303	10
.25	-1.386	4
.5	-.6931	2
1	0	1
2	.69315	.5
3	1.0986	.33333
4	1.3863	.25

X=.1

We are ready to give a simple and clear derivation now for one of the most amazing formulas in calculus. It is not expected that the student would carry out the steps in the proof or would even anticipate any of them, but each is straightforward and all the preliminaries have been given. Before looking at the details, the student might review the definition of the number e, as it will be needed in the following proof.

To find the formula for the derivative of $f(x) = \ln x$, for $x > 0$, we go directly to the definition of the derivative.

$$f'(x) = \lim_{h \to 0} \frac{f(x+h) - f(x)}{h}$$

Step 1: Form the difference quotient.

$$\frac{f(x+h) - f(x)}{h} = \frac{\ln(x+h) - \ln x}{h}$$

Step 2: Simplify the difference quotient. Here we use properties of logarithms.

$$\frac{\ln(x+h) - \ln x}{h} = \frac{\ln\left(\dfrac{x+h}{x}\right)}{h} \qquad \text{By Property 2}$$

$$= \frac{1}{h} \cdot \ln\left(1 + \frac{h}{x}\right)$$

$$= \ln\left(1 + \frac{h}{x}\right)^{\frac{1}{h}} \qquad \text{By Property 3}$$

$$= \frac{1}{x} \cdot x \cdot \ln\left(1 + \frac{h}{x}\right)^{\frac{1}{h}} \qquad \frac{1}{x} \cdot x = 1$$

$$= \frac{1}{x} \ln\left(1 + \frac{h}{x}\right)^{\frac{x}{h}} \qquad \text{By Property 3}$$

Step 3: Find the limit. Here we substitute $u = \dfrac{x}{h}$ and $\dfrac{1}{u} = \dfrac{h}{x}$. Then, since x is some fixed positive value, as $h \to 0^+$, we have $u \to +\infty$. (We omit the discussion of what happens as $h \to 0^-$. In effect, the result will be the same as when $h \to 0^+$.)

$$\lim_{h \to 0}\left[\frac{1}{x}\cdot \ln\left(1+\frac{h}{x}\right)^{\frac{x}{h}}\right] = \lim_{u \to \infty}\left[\frac{1}{x}\cdot \ln\left(1+\frac{1}{u}\right)^{u}\right] \qquad \text{Substitute } u = \frac{x}{h}.$$

$$= \frac{1}{x}\cdot \lim_{u \to \infty}\left[\ln\left(1+\frac{1}{u}\right)^{u}\right] \qquad \text{Since } x \text{ is constant}$$

$$= \frac{1}{x}\ln\left(\lim_{u \to \infty}\left[1+\frac{1}{u}\right]^{u}\right) \qquad \text{Since the expression is continuous}$$

$$= \frac{1}{x}\cdot \ln(e) \qquad \text{Using the definition of } e$$

$$= \frac{1}{x}\cdot 1 \qquad \ln e = 1$$

$$= \frac{1}{x}$$

Thus, the derivative of $\ln x$ has been proved.

Derivative of In x

If $f(x) = \ln x$, then

$$f'(x) = \frac{1}{x}.$$

Example 1: Finding the Derivative

Find $\dfrac{dy}{dx}$ for $y = x^3 + \ln x$.

Solution

$$\frac{dy}{dx} = 3x^2 + \frac{1}{x} \qquad \text{By the Sum and Difference Rule}$$

Example 2: Finding the Derivative

Find $\dfrac{dy}{dx}$ for $y = x^2 \ln x$.

Solution

$$\frac{dy}{dx} = x^2 \cdot \frac{d}{dx}\left[\ln x\right] + \ln x \cdot \frac{d}{dx}\left[x^2\right] \qquad \text{By the Product Rule}$$

$$= x^2 \cdot \frac{1}{x} + \ln x \cdot (2x)$$

$$= x + 2x \ln x$$

Example 3: Equation of a Tangent Line

Find the equation of the tangent line to $y = \ln x$ at the point $(5, 1.61)$. (Note that $\ln 5$ is actually equal to $1.609437912\ldots$, but we round to two decimal places here.)

Solution

For $f(x) = \ln x$, the slope we need is $f'(5) = \dfrac{1}{5}$. In point-slope form, we write the equation of the tangent line as follows.

$$y - 1.61 = \frac{1}{5}(x - 5)$$

$$y - 1.61 = 0.2x - 1$$

$$y = 0.2x + 0.61$$

Use the given (x, y)-coordinate $(5, 1.61)$ in the point-slope formula as well as

$$f'(5) = \frac{1}{5} = m.$$

The Derivative of $y = \ln(g(x))$

Suppose we want to find $\dfrac{dy}{dx}$ given that $y = \ln(x^2 + 1)$.

We can let

$$y = \ln u \quad \text{and} \quad u = x^2 + 1.$$

Then

$$\frac{dy}{du} = \frac{1}{u} \quad \text{and} \quad \frac{du}{dx} = 2x.$$

So, by the Chain Rule, we have

$$\frac{dy}{dx} = \frac{dy}{du} \cdot \frac{du}{dx}$$

$$= \frac{1}{u} \cdot 2x$$

$$= \frac{1}{x^2 + 1} \cdot 2x \qquad \text{Substitute } u = x^2 + 1.$$

Use the same approach for the general case:

$$y = \ln(g(x)),$$

where $g(x)$ is positive and differentiable.

Let

$$y = \ln u \quad \text{and} \quad u = g(x).$$

Then

$$\frac{dy}{du} = \frac{1}{u} \quad \text{and} \quad \frac{du}{dx} = g'(x).$$

Applying the Chain Rule, we get

$$\frac{dy}{dx} = \frac{dy}{du} \cdot \frac{du}{dx}$$

$$= \frac{1}{u} \cdot g'(x)$$

$$= \frac{1}{g(x)} \cdot g'(x) \qquad\qquad \text{Substitute } u = g(x).$$

$$= \frac{g'(x)}{g(x)}.$$

Derivative of ln(g(x))

If $y = \ln(g(x))$, then

$$y' = \frac{1}{g(x)} \cdot g'(x) = \frac{g'(x)}{g(x)}.$$

Example 4: Finding the Derivative

a. Find $f'(x)$ for $f(x) = \ln(x^2 + 2x + 10)$.

Solution

Let $g(x) = x^2 + 2x + 10$. Then $g'(x) = 2x + 2$.

So

$$f'(x) = \frac{g'(x)}{g(x)} = \frac{2x+2}{x^2+2x+10}.$$

b. Find $f'(x)$ for $f(x) = \ln \sqrt{x^2 + 3}$.

Solution

We simplify first to obtain

$$f(x) = \ln \sqrt{x^2+3} = \ln\left(x^2+3\right)^{\frac{1}{2}} = \frac{1}{2}\ln\left(x^2+3\right).$$

Now, differentiating with $g(x) = x^2 + 3$ and $g'(x) = 2x$, we have

$$f'(x) = \frac{1}{2} \cdot \frac{g'(x)}{g(x)} = \frac{1}{2} \cdot \frac{2x}{x^2+3} = \frac{x}{x^2+3}.$$

Logarithmic Differentiation

Functions that involve several products and quotients can be very difficult to differentiate directly. However, we know that logarithms of products and quotients can be written as sums and differences. And sums and differences are generally easier to differentiate than products and quotients. Taking the logarithm of both sides of an equation and then differentiating is called **logarithmic differentiation**. The only condition is that both sides must represent positive values.

Example 5: Using Logarithmic Differentiation

Use logarithmic differentiation to find $f'(x)$, given that $f(x) = \dfrac{\left(x^2+1\right)^3 \sqrt{x^2-2x}}{x-5}$.

Solution

Take the natural logarithm of both sides and simplify.

$$\ln f(x) = \ln\left(\frac{\left(x^2+1\right)^3 \sqrt{x^2-2x}}{x-5}\right)$$

$$= \ln\left(x^2+1\right)^3 + \ln\left(x^2-2x\right)^{\frac{1}{2}} - \ln\left(x-5\right)$$

$$= 3\ln\left(x^2+1\right) + \frac{1}{2}\ln\left(x^2-2x\right) - \ln\left(x-5\right)$$

Now differentiate both sides.

$$\frac{f'(x)}{f(x)} = 3\cdot\frac{1}{x^2+1}\cdot 2x + \frac{1}{2}\cdot\frac{1}{x^2-2x}\cdot(2x-2) - \frac{1}{x-5}$$

$$f'(x) = f(x)\cdot\left(\frac{6x}{x^2+1} + \frac{x-1}{x^2-2x} - \frac{1}{x-5}\right)$$

$$= \frac{\left(x^2+1\right)^3 \sqrt{x^2-2x}}{x-5}\cdot\left(\frac{6x}{x^2+1} + \frac{x-1}{x^2-2x} - \frac{1}{x-5}\right)$$

We complete this section with two examples that apply logarithmic functions and their derivatives. In Example 6 we find the absolute extrema of a function on a closed interval. In Example 7 we use curve sketching techniques for graphing functions.

Example 6: Locating Absolute Extrema

Find the absolute extrema for the function $f(x) = x^2\ln x$ on the interval $[0.1, 2.0]$.

Solution

The function $f(x) = x^2\ln x$ is a continuous function over the interval $(0, +\infty)$, so it has an absolute maximum and an absolute minimum on the closed interval $[0.1, 2.0]$.

We know from Example 2 that

$$f'(x) = x + 2x\ln x$$

$$= x\left(1 + 2\ln x\right).$$

Setting $f'(x) = 0$ gives

$$x(1+2\ln x)=0$$

$$\cancel{x=0} \text{ or } 1+2\ln x=0 \qquad\qquad x \text{ cannot be 0.}$$

$$2\ln x=-1 \qquad\qquad \text{Solve } 1+2\ln x=0 \text{ for } x \text{ to find the}$$

$$\ln x=-\frac{1}{2}=-0.5 \qquad\qquad \text{critical value.}$$

$$x=e^{-0.5}\approx 0.61.$$

Evaluate $f(x)$ at the critical values. (Remember, these include the endpoints of the function's interval.)

x	$f(x)=x^2 \ln x$	
0.1	$f(0.1)=(0.1)^2\ln 0.1 \approx -0.02$	
0.61	$f(0.61)=(0.61)^2\ln 0.61 \approx -0.18$	Absolute min
2.0	$f(2.0)=(2.0)^2\ln 2.0 \approx 2.77$	Absolute max

The absolute minimum is -0.18 and occurs at $x=0.61$. The absolute maximum is 2.77 and occurs at $x=2.0$.

Example 7: Graphing Logarithmic Functions

Using curve sketching techniques, sketch the graph of the function $f(x)=x\ln x$.

Solution

The domain is the interval $(0, +\infty)$ because $\ln x$ is defined only for positive values of x.

$$f'(x)=x\cdot\frac{1}{x}+\ln x\cdot 1 \qquad\qquad \text{Find the critical value(s) by setting } f'(x)$$

$$=1+\ln x \qquad\qquad \text{equal to 0 and solving for } x.$$

$$0=1+\ln x$$

$$\ln x=-1$$

$$x=e^{-1}\approx 0.37$$

Using the critical value $x=e^{-1}$ as well as the endpoint $x=0$, we have two intervals to test.

By testing values, we find that $f(x)$ is decreasing on the interval $(0, e^{-1})$ and increasing on the interval $(e^{-1}, +\infty)$.

Taking the second derivative of $f(x)$, we find that $f''(x)=\dfrac{1}{x}>0$ for all x on the interval $(0, +\infty)$. Therefore, $f(x)$ is concave upward on the interval $(0, +\infty)$.

A local (and absolute) minimum occurs at the critical value $x = e^{-1}$, and

$$f\left(e^{-1}\right) = e^{-1} \ln e^{-1}$$
$$= e^{-1}(-1) = -e^{-1}.$$

Note: One easy point to get without a calculator is $(1, 0)$ (recall that $\ln 1 = 0$). Therefore, it is possible to notice that $1 \cdot \ln 1 = 0$, and $(1, 0)$ is an x-intercept.

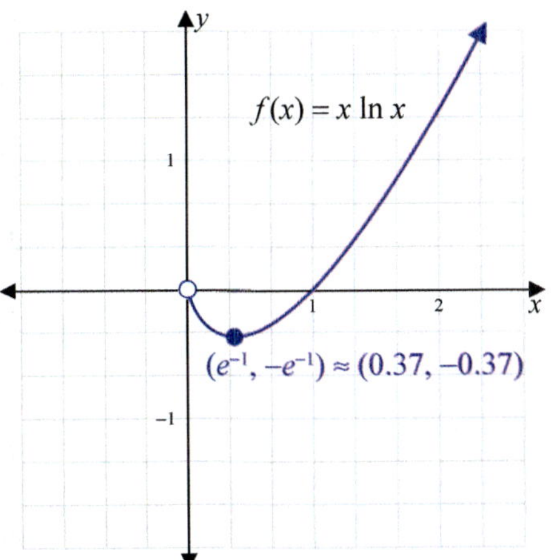

13.1 EXERCISES

PRACTICE

Find the derivative of each of the functions in Exercises 1–28.

1. $f(x) = x^2 - \ln x$

2. $f(x) = 4x^2 + \ln x^2$

3. $f(x) = 25 + x \ln x$

4. $f(x) = (x^2 + 1) \ln x$

5. $y = \dfrac{\ln x}{x}$

6. $y = \dfrac{x^2}{\ln x}$

7. $y = (\ln x)^3$

8. $y = \sqrt{\ln x}$

9. $f(x) = \ln(5x + 3)$

10. $f(x) = \ln(7x - 2)$

11. $f(x) = \ln(x^2 + 2)$

12. $f(x) = \ln(x^2 + 3x)$

13. $f(x) = \ln \sqrt{2x^2 - 1}$

14. $f(x) = \ln \sqrt{x^3 + 4}$

15. $y = \ln(4x + 3)^2$

16. $f(x) = \ln(x^2 - 4)^3$

17. $y = \sqrt{x} \ln(x^2 + 2)$

18. $y = \dfrac{\ln(5x + 2)}{x^3}$

19. $y = \dfrac{\ln(x^2 + 2x - 1)}{x}$

20. $y = x^{-2} \ln(x^2 - 3x + 4)$

21. $f(x) = \ln\big((3x + 1)(x^2 + 3)\big)$

22. $f(x) = \ln\big(x^2(4x - 1)\big)$

23. $f(x) = \ln\big((2x - 1)^2 (x^2 - 2)\big)$

24. $f(x) = \ln\big(\sqrt{4x - 7}(6x + 7)\big)$

25. $y = \ln\left(\dfrac{x + 1}{x - 2}\right)$

26. $y = \ln\left(\dfrac{3x - 1}{x + 5}\right)$

27. $f(x) = \ln\left(\dfrac{x^2 - 5}{2x + 9}\right)$

28. $f(x) = \ln\sqrt{\dfrac{4x + 3}{x^2 - 6}}$

For Exercises 29–34, determine a formula for $f''(x)$. Use your calculator (if necessary) to solve the equation $f''(x) = 0$ and locate any possible inflection points.

29. $f(x) = \dfrac{x}{\ln x}$

30. $f(x) = (2x + x^2)\ln x$

31. $f(x) = 3x^2 \ln x$

32. $f(x) = \ln(x^3)$

33. $f(x) = (\ln x)^3$

34. $f(x) = \ln x + x^2 + 3x + 2$

Use logarithmic differentiation to find the derivatives of the functions in Exercises 35–42.

35. $y = (2x - 5)^3 \sqrt{x^2 - 2x}$

36. $y = (4 - 5x)^4 (7x + 2)^{-\frac{2}{3}}$

37. $y = (x^2 + 2)^4 (x^2 - 1)^{-\frac{1}{3}}$

38. $y = (3x - 2)^5 \sqrt{x^2 + x}$

39. $y = \dfrac{(2x + x^2)^3}{(4x - 9)^2}$

40. $y = \dfrac{\sqrt[3]{x^2 + 4}}{(2 - 5x)^3}$

41. $y = \dfrac{(x + 2)^2 (3x + 4)^2}{\sqrt{x - 6}}$

42. $y = \dfrac{(x^2 - 3)^2 \sqrt{x^2 + 3x}}{x + 7}$

In Exercises 43–48, find the absolute extrema for each of the functions on the indicated interval.

43. $f(x) = x - \ln x;\ [0.5, 2]$

44. $f(x) = x^2 - 8\ln x;\ [0.3, 4]$

45. $f(x) = \dfrac{x}{\ln x};\ [1.2, 3]$

46. $f(x) = \dfrac{\ln x}{x^2};\ [1, 2]$

47. $f(x) = x^2 - \ln x^3;\ [1, e]$

48. $f(x) = \ln(3 + 2x - x^2);\ [0, 2.5]$

Using the graph-sketching techniques discussed in Chapter 12, sketch the graph of each function in Exercises 49–54.

49. $f(x) = 4x \ln x$

50. $f(x) = x^2 \ln x$

51. $f(x) = x - 3\ln x$

52. $f(x) = 4x - \ln x^2$

53. $f(x) = \dfrac{\ln x}{x}$

54. $f(x) = \dfrac{\ln x}{x^2}$

APPLICATIONS

55. **Marginal revenue:** A retailer has determined that the revenue from the sale of x end tables is given by the function $R(x) = 96x + \ln(4x^2 + 15)$ dollars. Find the marginal revenue.

56. **Advertising:** An automobile dealer has estimated that he can sell $N(x) = 420 + 72\ln(1 + 0.5x)$ cars annually, where x (in thousands of dollars) is the amount spent on advertising.
    a. Find the number of cars sold if $6000 is spent on advertising.
    b. Find the rate of change in number of cars sold if $6000 is spent on advertising.

57. **Revenue:** The daily demand for a product is given by $p = -8\ln(0.01x)$, where p is the price in dollars when x units are sold and $0 < x \le 100$. Find the maximum daily revenue.

58. **Revenue:** The demand for a product is given by $p = 14 - 6.5\ln x$, where x is the number of units (in thousands) that can be sold at a price p dollars and $2 \le x \le 8$. Find the maximum revenue.

59. **Insect population:** Mediterranean fruit flies are discovered in a citrus orchard. The Department of Agriculture has determined that the population of flies t hours after the orchard has been sprayed with pesticide is approximated by $N(t) = 25 - 5t\ln(0.04t) - t$, where $0 < t \le 25$.
    a. Find $N(3)$, $N(10)$, $N(25)$.
    b. What will be the maximum number of flies in the orchard?

60. **Air quality:** The Air Quality Management Board estimates that t hours after midnight in a major city the level of ozone in the air is about $N(t) = 0.013 - 0.007t\ln(0.026t)$ parts per million, where $0 < t \le 18$.
    a. Find $N(6)$, $N(10)$, $N(18)$.
    b. Find the maximum level of pollution.

13.2 DERIVATIVES OF EXPONENTIAL FUNCTIONS

■ TOPICS

- ■ Slopes and Exponential Functions
- ■ Chain Rule for Exponential Functions

Slopes and Exponential Functions

For the function $y = 2^x$, let us approximate the slope at $x = 0$ by considering the line through $(0, 2^0)$ and $(0.0001, 2^{0.0001})$. We determine the slope is

$$\frac{2^{0.0001} - 2^0}{0.0001 - 0} = \frac{0.00006931\ldots}{0.0001} = 0.6931\ldots.$$

Here we want to consider Table 1. Using the method above or a graphing calculator, we calculate y-values and slopes for $y = 2^x$.

Because of the exponential character of the function, as x increases by 1, y doubles. This is not surprising, but what is surprising is that y' also doubles! Put another way, $f'(0) = 0.6931$ is a number which specifies the slope not only at $x = 0$ but also for other x-values. This is revealed if we add another column to Table 1 (see Table 2).

$f(x) = 2^x$

x	y	y'
0	1	0.6931
1	2	1.386
2	4	2.773
3	8	5.545

TABLE 1

$f(x) = 2^x$

x	y	y'	y'
0	1	0.6931	1(0.6931)
1	2	1.386	2(0.6931)
2	4	2.773	4(0.6931)
3	8	5.545	8(0.6931)

TABLE 2

Table 2 shows, numerically, that for $y = 2^x$, we have

$$y' = (0.6931)2^x = 0.6931y.$$

We can restate this as:

> The slope of $f(x) = 2^x$ is linearly related (directly proportional) to $f(x)$ itself.

This linear relationship between y-values of exponential functions and the corresponding slopes was understood by the mathematician John Napier (1550–1617) and is exactly what he tried to capture with his logarithm tables.

Suppose we let $f(x) = A^x$ and we look at other values of A besides 2. For example, if $A = 3$, we get the numbers shown in Table 3.

$f(x) = 3^x$

x	$y = 3^x$	$y' = 1.0987y$
0	1	1.0987
1	3	3.296 = 3(1.0987)
2	9	9.888 = 9(1.0987)
3	27	29.665 = 27(1.0987)

TABLE 3

There are two important facts which we want to infer from these tables. First, that for any exponential function $y = A^x$, the slopes are directly proportional to the y-values, that is,

$$\frac{d}{dx} A^x = k \cdot A^x$$

or, $y' = ky$ for some constant k. Secondly, the constant k, called the **constant of proportionality**, is given by

$$k = f'(0).$$

We will eventually prove these facts. Before that, however, we want to consider the set of possible k-values. Notice as the base A increases, for the function $y = A^x$, the value of k increases. Apparently, however, between $A = 2$ and $A = 3$, there is a value of A which makes $k = 1$. This value $k = 1$ is very nice since k is used to multiply in the derivative formula $y' = ky$.

When $k = 1$, the formula for slope is just $y' = y$. That is, the function has the remarkable property that the slope exactly equals the corresponding y-value, regardless of x. In a real sense this base, for which the slope equals y-values, is a "perfect" base for calculus.

We can confirm analytically that e is the perfect base; that is, for $y = e^x$, the slopes are equal to the y-values. The key is to use the results for $\ln x$ and the fact that e^x and $\ln x$ are inverses of each other. The details are simple—use the fact that e^x is always positive as well as your knowledge of logarithmic differentiation.

$$y = e^x$$
$$\ln y = \ln e^x$$
$$\ln y = x \qquad \text{\textcolor{teal}{$\ln e^x = x$}}$$
$$\frac{d}{dx}[\ln y] = \frac{d}{dx} x \qquad \text{\textcolor{teal}{Differentiate both sides and treat y as}}$$
$$\qquad\qquad\qquad \text{\textcolor{teal}{$g(x)$ in the expression $\ln(g(x))$.}}$$
$$\frac{1}{y} \cdot \frac{dy}{dx} = 1$$
$$\frac{dy}{dx} = y = e^x$$

This proves that the derivative of $y = e^x$ is the function itself.

Derivative of the Exponential Function

If $f(x) = e^x$, then

$$f'(x) = e^x.$$

This derivative formula and the corresponding one for $\ln x$ are probably the two most amazing formulas in calculus.

Example 1: Finding the Derivative

a. Find $\dfrac{dy}{dx}$ for $y = x^3 + e^x$.

Solution

$$\frac{dy}{dx} = 3x^2 + e^x \qquad\qquad \text{By the Sum and Difference Rule}$$

b. Find $\dfrac{dy}{dx}$ for $y = x^2 e^x$.

Solution

$$\frac{dy}{dx} = x^2 \cdot \frac{d}{dx}\left[e^x\right] + e^x \cdot \frac{d}{dx}\left[x^2\right] \qquad \text{By the Product Rule}$$

$$= x^2 e^x + e^x \cdot 2x$$

$$= xe^x\left(x+2\right)$$

Chain Rule for Exponential Functions

Now consider the function

$$y = e^{-x^2},$$

where the exponent is a function of x other than x itself. To find $\dfrac{dy}{dx}$, we can proceed as before and take the natural logarithm of both sides.

$$\ln y = \ln e^{-x^2}$$

$$\ln y = -x^2 \ln e$$

$$\ln y = -x^2$$

$$\frac{1}{y} \cdot \frac{dy}{dx} = -2x \qquad\qquad \text{Differentiate both sides.}$$

$$\frac{dy}{dx} = -2x \cdot e^{-x^2} \qquad\qquad \text{Substitute } e^{-x^2} \text{ for } y.$$

While logarithmic differentiation works well here and is a good technique, another approach is to use the Chain Rule with

$$y = e^u \quad \text{and} \quad u = -x^2.$$

Thus

$$\frac{dy}{du} = e^u \quad \text{and} \quad \frac{du}{dx} = -2x.$$

Therefore,

$$\frac{dy}{dx} = \frac{dy}{du} \cdot \frac{du}{dx} = e^u\left(-2x\right) = -2xe^{-x^2}.$$

Applying either technique to the general case $y = e^{g(x)}$, where $g(x)$ is differentiable, we get the following result.

Chain Rule for Exponential Functions

If $f(x) = e^{g(x)}$, then

$$f'(x) = e^{g(x)} \cdot g'(x).$$

Example 2: Using the Chain Rule

Find $\dfrac{dy}{dx}$ if $y = e^{x^2 + 3x}$.

Solution

Let $g(x) = x^2 + 3x$. Then $g'(x) = 2x + 3$.

Thus

$$\frac{dy}{dx} = e^{g(x)} \cdot g'(x) = e^{x^2 + 3x}(2x + 3) = (2x + 3)e^{x^2 + 3x} \, .$$

Example 3: Using the Chain Rule

For $f(x) = xe^{-x}$, find

a. $f'(x)$,

b. $f''(x)$, and

c. the equation of the tangent line to $f(x)$ at the point where $x = -1$.

Solution

a. $\displaystyle f'(x) = x \cdot \frac{d}{dx}\left[e^{-x}\right] + e^{-x} \cdot \frac{d}{dx}[x]$ By the Product Rule

$\displaystyle \quad = x \cdot e^{-x}(-1) + e^{-x}(1)$

$\displaystyle \quad = e^{-x}(-x + 1)$ Factor out e^{-x}.

$\displaystyle \quad = e^{-x}(1 - x)$

b. $\displaystyle f''(x) = e^{-x} \cdot \frac{d}{dx}[1 - x] + (1 - x) \cdot \frac{d}{dx}\left[e^{-x}\right]$ By the Product Rule applied to $f'(x)$

$\displaystyle \quad = e^{-x}(-1) + (1 - x)(-1)e^{-x}$

$\displaystyle \quad = e^{-x}(-1 - 1 + x)$ Factor out e^{-x}.

$\displaystyle \quad = e^{-x}(x - 2)$

c. Since $f(-1) = (-1)e^{-(-1)} = -e$, we need the equation of the tangent line at the point $(-1, -e)$.

To determine the slope of the tangent line at the point $(-1, -e)$, we need to find $f'(-1)$.

$$f'(-1) = e^{-(-1)}(1 - (-1)) = 2e$$

Using the point-slope formula, the equation of the tangent line is

$$y-(-e)=2e\big(x-(-1)\big)$$
$$y+e=2ex+2e$$
$$y=2ex+e$$
$$y\approx 5.437x+2.718.$$

In Example 4 we will use the results found in Example 3 to analyze the function $f(x)=xe^{-x}$ and sketch its graph.

Example 4: Graphing Exponential Functions

Let $f(x)=xe^{-x}$.

a. Find any critical values.

b. Find any hypercritical values.

c. Sketch the graph of the function.

Solution

a. To find critical values, set $f'(x)=0$.

$$f'(x)=e^{-x}(1-x)$$
$$0=e^{-x}(1-x)$$
$$0=1-x$$
$$x=1$$

We know $f'(x)$ from Example 3.

Note that e^{-x} is never equal to 0, so we can divide both sides by e^{-x} (that is, cancel e^{-x}).

The only critical value is $x=1$, since there are no values where $f'(x)$ is undefined.

b. To find hypercritical values, set $f''(x)=0$.

$$f''(x)=e^{-x}(x-2)$$
$$0=e^{-x}(x-2)$$
$$0=x-2$$
$$x=2$$

We know $f''(x)$ from Example 3.

The only hypercritical value is $x=2$.

c. Using the critical value and the hypercritical value found in parts **a.** and **b.**, we can find the local extrema and the points of inflection.

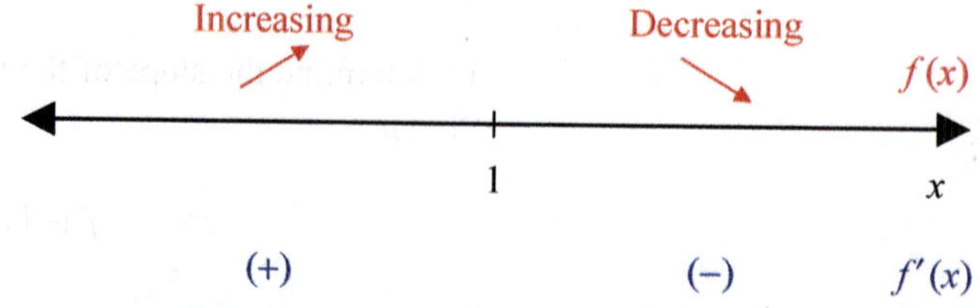

So there is a local max at $(1, e^{-1})$.

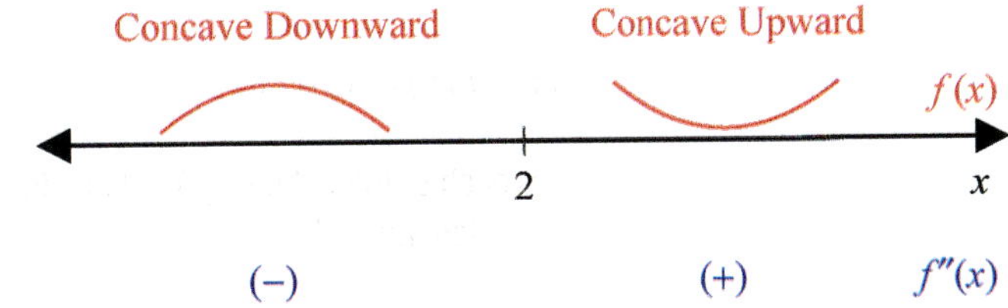

There is a point of inflection at $(2, 2e^{-2})$.

13.2 EXERCISES

💡 PRACTICE

Find the first and second derivative of each of the functions in Exercises 1–6.

1. $f(x) = 3e^x$

2. $f(x) = -6e^x$

3. $f(x) = x^2 + 5e^x$

4. $f(x) = 4x^2 - 2e^x$

5. $f(x) = xe^x$

6. $f(x) = -7x^2e^x$

For Exercises 7–14, find a formula for $f'(x)$ and determine the slope $f'(a)$ at the point where $x = a$ is given.

7. $f(x) = e^x \ln x;\ \ x = 1$

8. $f(x) = e^x \ln(x + 4);\ \ x = 1$

9. $f(x) = \dfrac{e^x}{e^x - 1};\ \ x = 2$

10. $f(x) = \dfrac{e^x}{\ln x};\ \ x = e$

11. $f(x) = 2e^{4x};\ \ x = -1$

12. $f(x) = 8e^{-3x};\ \ x = 4$

13. $f(x) = 3e^{2x+1};\ \ x = 0$

14. $f(x) = 5e^{\frac{x}{2}};\ \ x = 0$

For Exercises 15–18, find a formula for $f'(x)$ and use it to determine the intervals on which $f(x)$ is increasing or decreasing.

15. $f(x) = e^{1-x^2}$

16. $f(x) = e^{-0.04x^2}$

17. $f(x) = \left(e^{2x} - 4\right)^2$

18. $f(x) = \left(e^{4x} + 2\right)^3$

For Exercises 19 and 20, determine $f'(x)$ and use it to determine the intervals on which $f(x)$ is increasing or decreasing. Determine for each function if there is a horizontal asymptote. Confirm your results with a graphing calculator.

19. $f(x) = \sqrt{e^{-0.2x} + 11}$

20. $f(x) = \dfrac{1}{\sqrt{3e^x + 1}}$

Find $f'(x)$ and use it to argue whether or not there is an oblique asymptote for each of the functions in Exercises 21 and 22.

21. $f(x) = \ln\left(e^x + 1\right)$

22. $f(x) = \ln\sqrt{5 + e^{2x}}$

For each of the following functions in Exercises 23–30, find the absolute extrema on the indicated interval.

23. $f(x) = xe^{2x};\ [-2, 1]$

24. $f(x) = xe^{\frac{x}{3}};\ [-4, 0]$

25. $f(x) = x^2 e^{-x};\ [-1, 2]$

26. $f(x) = 2x^2 e^{-x};\ [-2, 3]$

27. $f(x) = 5e^{1-x^2};\ [-2, 1]$

28. $f(x) = 3xe^{-x^2};\ [-2, 2]$

29. $f(x) = (2x + 3)e^{-0.2x};\ [1, 4]$

30. $f(x) = (4x - 1)e^{-0.5x};\ [0, 3]$

For each of the functions in Exercises 31–36, **a.** find any critical values, **b.** find any hypercritical values, **c.** find all intervals of concavity, and **d.** sketch the graph of the function. If available, confirm your results with a graphing utility.

31. $f(x) = xe^{-0.4x}$

32. $f(x) = 2xe^{-0.5x}$

33. $f(x) = 4x^2 e^{-x}$

34. $f(x) = 3x^2 e^{-x}$

35. $f(x) = e^x + e^{-x}$

36. $f(x) = \dfrac{e^x}{x}$

🚀 APPLICATIONS

37. Revenue: The demand for a product is given by $D(x) = 140e^{-0.05x}$, where x is the number of units sold each week and $0 \le x \le 30$.
  a. Find the number of units sold that will yield maximum revenue.
  b. Find the price per unit that will yield maximum revenue.

38. Revenue: The demand equation for a certain product is given by $D(x) = 210e^{-0.025x}$, where x is the number of units sold each week and $0 \le x \le 60$.
  a. Find the number of units sold that will yield the maximum revenue.
  b. Find the price per unit that will yield maximum revenue.

39. Advertising: An automobile manufacturer is planning a television advertisement campaign to introduce a new model for their truck. It is estimated that $N(t) = 600\left(1 - e^{-0.02t}\right)$ people (in thousands) will have seen the advertisement after t days of advertising. How fast is N increasing at the end of 7 days?

40. Insect population: The mosquito population of a pool of water is estimated to be $P(t) = 400 + 1400e^{-0.3t}$, where t is the number of hours after the pool has been treated. Find the rate of change in the population at the end of 5 hours.

41. Bacterial population: The population of bacteria in an experimental culture is estimated by $N(t) = \dfrac{10,000}{1 + 9e^{-0.14t}}$, where t is the number of hours after the experiment begins. How fast is the population changing at the end of 5 hours?

42. **Disease control:** The elk herd in a national park has been infected by a contagious disease. The number of infected animals is estimated by $N(t) = \dfrac{600}{1 + 49e^{-0.36t}}$, where t is the number of days after the disease was discovered. How fast is the disease spreading after 4 days?

43. Suppose the value of the inventory of original Winchester rifles at Bill's Antique Firearms Company has increased according to the formula $r(t) = \dfrac{8500}{1 + 10e^{-0.6t}}$, where r is the average value of one of their rifles and t is the number of years since 2000.
    a. What was the average value of a rifle in 2000? In 2005?
    b. At what rate was r changing in 2005? In 2006?
    c. If there is an inflection point for $r(t)$, locate it and explain its significance in the application.
    d. When is the rate of increase of r at a maximum?

44. Suppose an advertising campaign for the sale of a new magazine, Dungeons and Creeps, causes sales to vary according to the formula $S(t) = 8\left(1 - 0.3e^{-0.2t+1}\right)$, where S is sales in thousands and t is time in months since the ad campaign started.
    a. What were the monthly sales when the ad campaign started?
    b. What was the rate at which sales were changing after 4 months into the campaign?
    c. What are the long-term sales expectations?

45. The growth cycle of a mass of algae in a pond grows according to $A(t) = 1 + 2te^{-0.5t}$, where A represents the mass-density of algae in the pond in suitable units and t is the time in months ($t = 0$ corresponds to April 1st).
    a. What day of the year corresponds to a maximum A value?
    b. When does the rate of decline in algae reach its maximum?

46. A certain calculus student recalls information according to the formula $p = 70e^{-0.6x} + 30$, where p is the percentage of information retained after x weeks.
    a. After 4 weeks, what percentage of a lesson is retained?
    b. After 4 weeks, at what rate is the percentage changing?
    c. What does the model predict a few months after the Calculus course is over?

47. Inexpensive videos detailing the championship basketball season of Castle High School are sold locally by a civic club to raise money for next year's team. The total sales are given by $S = \dfrac{12{,}500}{1 + 15e^{-0.5x}}$, where S is the total sales after x weeks.
    a. What are the total sales after 3 weeks?
    b. What is the rate of change of sales after 3 weeks?
    c. After about how many weeks will the total sales begin to level off?
    d. When is the sales rate increasing fastest? Illustrate this point graphically.

✏ WRITING & THINKING

48. **a.** Find the equation of the tangent line to $f(x) = 2e^{-x^2+1} + 4$ at the point where $x = 2$.

  b. Discuss the advantages and disadvantages of using the tangent line to get values of $f(x)$ for $x \geq 2$ rather than the function itself.

49. Determine k in the equations that follow by finding $f'(0)$. Use logarithmic differentiation.

  a. Let $f(x) = 10^x$. Determine the value of k in the formula $f'(x) = k \cdot 10^x$.

  b. Let $y = f(x) = \pi^x$. Determine k in the formula $f'(x) = ky$.

⬚ TECHNOLOGY

Use a graphing calculator to graph $f(x)$ and $f'(x)$. Then locate all extrema and all inflection points, if any.

50. $f(x) = e^{-x^2} \ln\left(x^2 + 2\right)$

51. $f(x) = \ln \dfrac{2 + 3e^{-x}}{x+2}$

13.3 GROWTH AND DECAY

■ TOPICS

- ■ Exponential Growth
- ■ Exponential Decay
- ■ Limited Growth
- ■ Half-Life
- ■ Logarithmic Differentiation and the General Exponential and Logarithmic Derivatives

Exponential Growth

Exponential and logarithmic functions can be used to describe (or model) the behavior of real-life phenomena. A quantity that increases at a rate of change that is proportional to the amount present is said to **increase exponentially** (or **grow exponentially**). For example, in the study of bacterial cultures in biology and medicine, the number of bacteria increases as time goes on and the rate of growth often increases at any time t in proportion to the number of bacteria present at that time.

Another example of exponential growth is continuously compounded interest. The related function is the exponential function

$$A = Pe^{rt}.$$

Now suppose that $1000 is invested in an account that earns 6 percent interest compounded continuously. Then

$$P = \$1000, \quad r = 0.06, \quad \text{and} \quad A = 1000e^{0.06t}.$$

To find the rate of change of the amount A at time t, we differentiate with respect to t.

$$\frac{dA}{dt} = 1000(0.06)e^{0.06t}$$
$$= 0.06(1000e^{0.06t})$$
$$= 0.06A$$

Thus we see that the rate of change of A, $\dfrac{dA}{dt}$, is directly proportional to A. The constant 0.06 is called the **constant of proportionality**. For continuous compounding, $r = 0.06 = 6\%$ is the **interest rate** or the **growth rate**. The rate of change of A is found by multiplying A by its growth rate.

More generally, suppose that the mathematical model of a situation is an exponential function of the form

$$y = ce^{kt},$$

where c and k are positive constants.

Then

$$\frac{dy}{dt} = c \cdot k e^{kt}$$
$$= k \cdot c e^{kt}$$
$$= ky$$

and the rate of change in y is directly proportional to y. Therefore, the function

$$y = c e^{kt}$$

is a mathematical model for exponential growth.

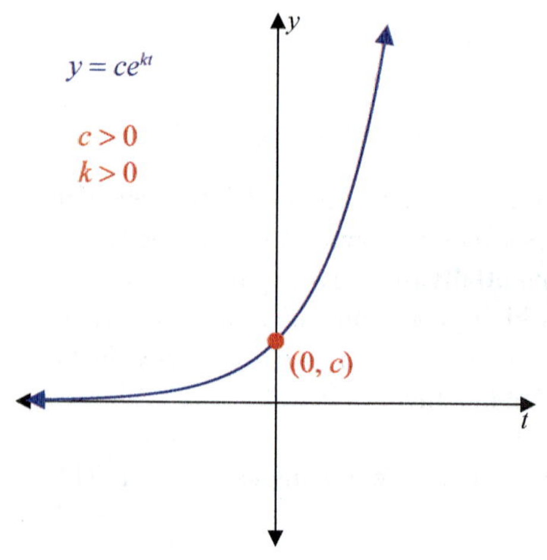

FIGURE 1: Exponential Growth

Exponential Growth

A function $y = f(t)$ satisfies the equation

$$\frac{dy}{dt} = ky,$$

if and only if

$$y = c e^{kt}.$$

If k and c are positive, we say that y **grows** (or **increases**) **exponentially**.

See Figure 1 for a general graph of exponential growth.

Example 1: Population Growth

Suppose that a population of rabbits grows exponentially; that is, the rate of population growth depends on the size of the population. Further, suppose that on Monday at 8:00 a.m. there are 100 rabbits and that at 8:00 a.m. on Wednesday there are 130 rabbits.

a. Find an exponential function that represents the rabbit population at time t, where t is measured in days from Monday at 8:00 a.m.

b. Find the time that the population will be 200 rabbits (double the original 100).

Solution

a. The general form of the function is

$$y = c e^{kt}.$$

When $t = 0$ (Monday at 8:00 a.m.), $y = 100$.

$$100 = c \cdot e^{k \cdot 0} = c \cdot 1$$
$$100 = c$$

Substitute the given values into the general form of the function and solve for c.

This gives

$$y = 100 e^{kt}.$$

Substitute $c = 100$ into the general form of the function.

Now, when $t = 2$ (Wednesday at 8:00 a.m.), $y = 130$.

$$130 = 100e^{k \cdot 2}$$
$$1.30 = e^{2k}$$
$$2k = \ln 1.30$$
$$k = \frac{\ln 1.30}{2} \approx 0.13$$

Substitute the given values into our formula for y.

Use the definition of natural logarithm.

Therefore, the exponential function is $y = 100e^{0.13t}$, where t is measured in days from Monday at 8:00 a.m.

b. Setting $y = 200$ and solving for t, we have

$$200 = 100e^{0.13t}$$
$$2 = e^{0.13t}$$
$$0.13t = \ln 2$$
$$t = \frac{\ln 2}{0.13} \approx 5.3 \, \text{days},$$

which is approximately 3:00 p.m. on Saturday.

Example 2: Continuous Compounding Interest

Suppose that \$2000 is invested in an account that earns **8** percent compounded continuously.

a. What will be the balance in 1 year?

b. When will the original investment be doubled?

Solution

a. We know that the growth is exponential because the interest is compounded continuously. Thus, the account balance after t years is determined by the following formula.

$$A = Pe^{rt} = 2000e^{0.08t}$$

Substitute $P = 2000$ and $r = 0.08$ into the formula for continuously compounded interest.

Now we substitute $t = 1$.

$$A = 2000e^{0.08 \cdot 1}$$
$$\approx \$2166.57$$

\$2166.57 is the balance at the end of 1 year.

b. Setting $A = 4000$ and solving for t, we obtain

$$4000 = 2000e^{0.08t}$$

$$2 = e^{0.08t}$$

$$0.08t = \ln 2$$

$$t = \frac{\ln 2}{0.08} \approx 8.7 \text{ years.}$$

The original investment of \$2000 will double in approximately 8.7 years.

Exponential Decay

If, instead of growing at a rate proportional to the amount present, a quantity declines (or decays) at a rate proportional to the amount present, then the quantity is said to **decline exponentially** (or **decay exponentially**). The function that models this situation is

$$y = ce^{-kt},$$

where c and k are positive constants.

Exponential decline is associated with the study of radioactive substances, food production under adverse conditions, amounts of medicine remaining in the blood stream after an injection, and populations attacked by disease.

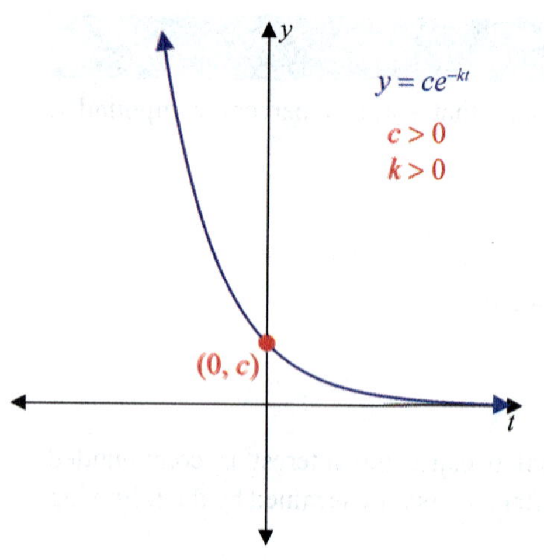

FIGURE 2: Exponential Decline

Exponential Decline

A function $y = f(t)$ satisfies the equation

$$\frac{dy}{dt} = -ky,$$

if and only if

$$y = ce^{-kt}.$$

If k and c are positive, we say that y **declines** (or **decays**) **exponentially**.

See Figure 2 for a general graph of exponential decline.

Example 3: Carbon-14

The radioactive substance carbon-14 is known to maintain a constant level in living plants and animals. When a plant or animal dies, the level of carbon-14 decays at a rate proportional to the amount present. That is, carbon-14 decays exponentially. For carbon-14, the constant of proportionality is approximately $k = 0.000124$. Suppose that an animal bone is found at an archaeological dig and that it contains 40 percent of its original amount of carbon-14. How old is the bone?

Solution

Using the function

$$y = ce^{-0.000124t}$$

to represent the amount of carbon-14 present, we find that if $t = 0$, then

$$y = ce^0 = c.$$

That is, c is the amount of carbon-14 present when the animal died. We know that t years after the animal's death there is 40 percent of the original amount of carbon-14 left. The amount of carbon-14 currently present is thus

$$y = 0.40c.$$

Now, to find the age of the bone t, we substitute $y = 0.40c$ into our original equation and solve for t.

$$y = ce^{-0.000124t}$$
$$0.40c = ce^{-0.000124t}$$
$$0.40 = e^{-0.000124t}$$
$$-0.000124t = \ln 0.40$$
$$t = \frac{\ln 0.40}{-0.000124} \approx 7400 \text{ years}$$

The bone is approximately 7400 years old.

Limited Growth

Even when environmental conditions are good, the growth of populations of wild animals is limited by such factors as a finite food supply and natural enemies or predators. Another example of limited growth occurs in industry. As workers become more efficient, their production level increases, but not more than the capacity of the company as restricted by money, materials, and machinery. This type of limited growth in productivity is called a **learning curve**.

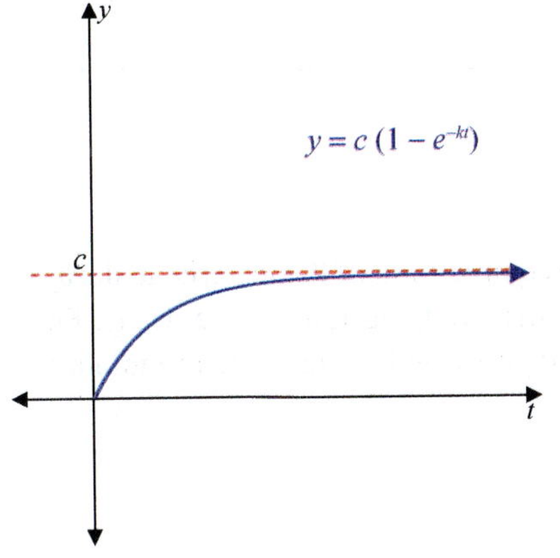

FIGURE 3: Limited Growth

Limited Growth

The mathematical model for limited growth is

$$y = c(1 - e^{-kt}),$$

where c and k are positive constants.

Note the following characteristics of the limited growth model.

1. As t increases, $e^{-kt} = \dfrac{1}{e^{kt}}$ decreases and approaches 0.

2. As t increases, $(1 - e^{-kt})$ approaches 1.

3. As t increases, y approaches c, as shown in Figure 3.

Example 4: Limited Growth

Suppose that a man learning to make widgets can make $N = 100(1 - e^{-0.3t})$ widgets per week after t weeks on the job.

a. About how many widgets per week can this man produce after 4 weeks on the job?

b. At what rate will his widget-making skills be improving at the end of the 4th week? the 10th week?

Solution

a. Substitute $t = 4$ into the formula for N.

$$N = 100\left(1 - e^{-0.3 \cdot 4}\right)$$
$$= 100\left(1 - e^{-1.2}\right)$$
$$\approx 100\left(1 - 0.30\right)$$
$$= 100\left(0.70\right) = 70$$

He can make approximately 70 widgets per week after 4 weeks on the job.

b. To find his rate of learning (or rate of improvement), we need $\dfrac{dN}{dt}$.

$$N = 100\left(1 - e^{-0.3t}\right) = 100 - 100e^{-0.3t}$$
$$\frac{dN}{dt} = -100\left(-0.3\right)e^{-0.3t}$$
$$= 30e^{-0.3t}$$

At $t = 4$,

$$\left.\frac{dN}{dt}\right|_{t=4} = 30e^{-0.3 \cdot 4} = 30e^{-1.2} \approx 9 .$$

At $t = 10$,

$$\left.\frac{dN}{dt}\right|_{t=10} = 30e^{-0.3 \cdot 10} = 30e^{-3} \approx 1.5 .$$

At the end of 4 weeks, his rate of improvement is about 9 widgets per week. After 10 weeks, he is getting closer to the top of his learning curve as his rate of improvement is slowing to 1.5 widgets per week.

Half-Life

The half-life of a material is the amount of time required for half the material to decay. The concept of half-life is particularly useful when studying radioactive decay. For example, the half-life of carbon-14 is about 5700 years, and the half-life of radium is about 1600 years.

Example 5: Half-Life

Suppose that the decay of a particular radioactive isotope is described by the equation

$$y = 200e^{-0.05t},$$

where y is the amount of the isotope in grams and t is time in years. Find the half-life of this isotope.

Solution

When $t = 0$, we have

$$y = 200e^0 = 200.$$

Thus, initially, there are 200 grams of the isotope present. We want to find the time it takes for there to be 100 grams left (half of the initial amount of 200 grams).

Therefore, we want to find t in the equation

$$100 = 200e^{-0.05t}.$$

We divide both sides of the equation by 200 and then use the definition of natural logarithm:

$$\frac{100}{200} = \frac{200e^{-0.05t}}{200}$$

$$\frac{1}{2} = e^{-0.05t}$$

$$-0.05t = \ln\frac{1}{2}.$$

Thus

$$t = \frac{\ln\frac{1}{2}}{-0.05} \approx 13.86.$$

The half-life of the isotope is about 13.86 years.

Logarithmic Differentiation and the General Exponential and Logarithmic Derivatives

In general, when we consider any general exponential function $f(x) = B^x$, for some $B > 0$, we discover that there is a constant, say b, such that $f'(x) = b \cdot B^x$. When $B = e$, then $b = 1$. We can use the technique of logarithmic differentiation to easily discover the nature of the constant b.

Starting with

$$f(x) = B^x$$

we can take the natural logarithm of both sides. This gives

$$\ln(f(x)) = \ln(B^x) = \ln(B) \cdot x.$$

Then we differentiate both sides.

$$\frac{d}{dx}\ln\big(f(x)\big) = \frac{d}{dx}\big(\ln(B)\cdot x\big) = \ln B \cdot \frac{d}{dx}x = \ln B \cdot 1$$

$$\frac{1}{f(x)}\cdot\frac{d}{dx}f(x) = \ln B \cdot 1$$

General Exponential Derivative

1. If $f(x) = B^x$, then $f'(x) = \ln B \cdot f(x) = (\ln B)\cdot B^x$.

2. Chain Rule: If $f(x) = B^{g(x)}$,

   then $f'(x) = (\ln B)\cdot B^{g(x)}\cdot g'(x)$.

The mysterious constant is the natural logarithm of the base B. If $x = 0$, then we see that $f'(0) = \ln B$.

Restated, if $f(x) = B^x$ for some $B > 0$, then $f'(x) = \ln B \cdot B^x$. Moreover, $f'(0) = \ln B$.

We can use the same method and the result just obtained to get a general formula for the derivative of any logarithmic function.

Suppose $y = \log(x) = \log_{10}(x)$. Then $10^y = x$.

Then we use logarithmic differentiation as follows.

$$\ln\big(10^y\big) = \ln x$$

$$y\cdot\ln(10) = \ln x$$

$$y = \frac{1}{\ln 10}\ln x$$

$$y' = \frac{1}{\ln 10}\cdot\frac{1}{x}$$

General Logarithmic Derivative

1. If $y = \log_B(x)$ then $y' = \dfrac{1}{\ln B}\cdot\dfrac{1}{x}$.

2. Chain Rule: If $y = \log_B(g(x))$, then $y' = \dfrac{1}{\ln B}\cdot\dfrac{1}{g(x)}\cdot g'(x)$.

Example 6: Using the Derivatives

Given $f(x) = (x^2 - 3x + 6)\log_2(x)$, determine $f'(x)$.

Solution

Use the Product Rule.

$$f'(x) = (x^2 - 3x + 6)\cdot\frac{d}{dx}\log_2(x) + \log_2(x)\cdot\frac{d}{dx}(x^2 - 3x + 6)$$

$$= (x^2 - 3x + 6)\cdot\frac{1}{\ln 2}\cdot\frac{1}{x} + \log_2(x)\cdot(2x - 3)$$

13.3 EXERCISES

🚀 APPLICATIONS

1. **Population:** The population of a city is growing exponentially at a rate of 3.5 percent per year. The population was 8400 in 2000.
   a. Find an exponential function that represents the population t years after 2000.
   b. What was the population in the year 2010?
   c. When was the population 12,800?

2. **Bee population:** A swarm of bees grows exponentially at a rate of 4 percent hourly. Initially, there were 900 bees in the swarm.
   a. Find an exponential function for the number of bees in the swarm after t hours.
   b. How many bees are in the swarm after 6 hours?
   c. How many hours will it take for the swarm to double in size? Round your answer to the nearest tenth.

3. **Cost:** In 2018, the cost of a medium pizza was about $9.00. In 2021, the cost was $12.00. If the cost is growing exponentially, predict the cost of a medium pizza in 2027?

4. **Ant colony:** A colony of ants is growing exponentially. When first observed, the colony contained about 400 ants. If at the end of 9 days there are about 700 ants, approximately how many ants will be present at the end of 15 days?

5. **Bacterial population:** A bacteria culture grows at a rate proportional to its size. If the population doubles every 6 hours, how long will it take for the population to be three times its initial size?

6. **Demand for oil:** The demand for oil in the United States doubles every 8 years. How long will it take for the demand to triple?

7. **Inflation:** The amount of goods and services that costs $100 on January 1, 2015 costs $139.10 on January 1, 2018. Estimate the cost of the same goods and services on January 1, 2025. Assume the cost is growing exponentially.

8. **Interest compounded continuously:** One thousand dollars is deposited in a savings account where the interest is compounded continuously. After 4 years, the balance will be $1366.15. When will the balance be $1870.00?

9. **Half-life:** The decay rate for a radioactive isotope is 2.6 percent per year. Find its half-life.

10. **Half-life:** The decay rate of a radioactive isotope is 6.5 percent per year. Find its half-life.

11. **Archaeological dating:** A wooden carving found at an archaeological dig contains about 34 percent of its carbon-14. Approximately how old is the carving?

12. **Archaeological dating:** Bones from the skeleton of an animal have lost 62 percent of their carbon-14. Estimate the age of the bones.

13. **Atmospheric pressure:** As the elevation above sea level is increased, the atmospheric pressure declines exponentially. The pressure at sea level is approximately 15 lb/in.² and the pressure at 3000 feet of elevation is about 13 lb/in.² Find the pressure at 5000 ft.

14. **Drug concentration:** The concentration of a drug in the body fluids is known to decline exponentially. If 20 mg of a drug is administered and 8 mg remains after 3 hours, how much will remain after 5 hours?

15. **Depreciation:** It is determined that the value of a piece of machinery declines exponentially. A machine that was purchased 5 years ago for $65,000 is worth $35,000 today. What will be the value of the machine 5 years from now?

16. **Population:** The population of a certain economically depressed union is declining exponentially at a rate of 1.5 percent. If the population in 2010 was 30,000, estimate the population in 2030.

17. **Reliability:** Studies show that the fractional part P of light bulbs that has burned out after t hours of use is given by $P = 1 - e^{-0.03t}$. What fractional part of the bulbs has burned out after 50 hours? How long will it be before half of the bulbs have burned out?

18. **Advertising:** A radio station estimates that during an intense advertising campaign, the number of people N who will hear a commercial is given by $N = A\left(1 - e^{-0.02x}\right)$, where A is the number of people in the broadcasting area and x is the number of times the commercial is run. If there are 60,000 people in the area,
    a. How many people will hear the commercial if it is run 20 times?
    b. How many times should the station plan to run the commercial to be certain that at least 30,000 people hear it?

19. **Ecology:** The Department of Fisheries has begun a reclamation project at a lake where the fish population was nearly destroyed by agricultural chemicals. They estimate that the population of fish in t years will be $P = 6000 - 5200e^{-0.28t}$.
    a. What was the initial population?
    b. What will be the population after 4 years?
    c. How long will it take for the population to be 5000 fish?

20. **Advertising:** The manager of The Sound Lab has determined that after an intense advertising campaign, the monthly sales of a particular wireless speaker can be approximated by $N = 300 + 180e^{-0.04t}$ units, where t is the number of months after the campaign.
    a. Find the monthly sales initially.
    b. Find the monthly sales when $t = 6$.
    c. When will the monthly sales be 400 units?

21. **Skills development:** Beverly is making a small souvenir to give to each person attending her family reunion. The length of time, in minutes, she takes to make the nth one is given by the function $T(n) = 12 + 30e^{-0.1n}$. How long will it take her to make the 30th souvenir?

22. **Dairy farming:** The number of dairy farmers in a particular state who are feeding a new supplement to their milking cows is given by the function $W(t) = 340(1 - e^{-0.09t})$, where t is the number of months the supplement has been available. How long will it be before 200 farmers are feeding the supplement to their cows?

23. **Cost:** The total cost function for a local company is given by $C(t) = 12 - ce^{-kt}$ in thousands of dollars, where t is the time in months. The fixed costs are \$5000 and the total cost after 2 months is \$10,200. Find the total cost at the end of 6 months.

24. **Skills development:** The time that it takes a service attendant to change a tire is given by the function $T(x) = 4.4 + Ce^{-kx}$ minutes, where x is the number of tires the attendant has changed before. It takes Patrick 15 minutes to change the first tire $(x = 0)$ and 9.3 minutes to change the seventh tire. How long will it take him to change the eleventh tire?

13.4 ELASTICITY OF DEMAND

Previously we introduced the demand function $p = D(x)$, where p is the price per item at which consumers are willing to buy x items. Recall that revenue is the product of the number of items sold times the price per item,

$$R(x) = x \cdot p = x \cdot D(x).$$

We know that the demand functions are always decreasing, indicating that an increase in supply (items demanded or sold) will result in a decrease in price. Since revenue depends on both price and supply, a natural concern is whether an increase in supply will result in a percent price change so small that revenue increases or a percent price change so great that revenue decreases. Economists have developed a concept called **elasticity of demand** for determining just what effect can be expected.

Intuitively, if demand is "elastic," then an increase in sales will be accompanied by a relatively small decrease in price and will result in an increase in revenue. If demand is "inelastic," then an increase in sales will be accompanied by a relatively large decrease in price and will result in a decrease in revenue.

For instance, if computers are currently selling about 10,000 units a year, then an increase of sales of 200 units will result in a **relative rate of change** in demand of $\frac{200}{10,000} = 0.02 = 2$ percent per year. The calculation of elasticity of demand involves a comparison of the *relative rate of change of the quantity demanded with the relative rate of change in price.*

Consider the following analysis:

1. If Δx is the change in quantity demanded, then $\dfrac{\Delta x}{x}$ is the percent change in quantity.

2. If Δp is the change in price, then $\dfrac{\Delta p}{p}$ is the percent change in price.

3. The comparison we want is the ratio

$$\frac{\text{Percent change in } x}{\text{Percent change in } p} = \frac{\dfrac{\Delta x}{x}}{\dfrac{\Delta p}{p}} = \frac{p}{x} \cdot \frac{\Delta x}{\Delta p} = \frac{p}{x} \cdot \frac{1}{\dfrac{\Delta p}{\Delta x}}.$$

4. Now, if we let $\Delta x \to 0$ and assume that $p = D(x)$ is continuous, we have

$$\lim_{\Delta x \to 0} \left[\frac{p}{x} \cdot \frac{1}{\dfrac{\Delta p}{\Delta x}} \right] = \frac{p}{x} \cdot \frac{1}{\dfrac{dp}{dx}}.$$

For simplicity, we make the following adjustments in the notation:

$$\frac{p}{x} \cdot \frac{1}{\dfrac{dp}{dx}} = \frac{1}{x} \cdot \frac{p}{\dfrac{dp}{dx}} = \frac{1}{x} \cdot \frac{D(x)}{D'(x)}.$$

This last expression will be negative because $D'(x)$ is negative and x and $D(x)$ are positive. So that the measure of elasticity of demand will be a positive number, this expression is multiplied by -1.

Elasticity of Demand

If $p = D(x)$ is the demand function for a product, then the **elasticity of demand** for that product is

$$E = -\frac{1}{x} \cdot \frac{D(x)}{D'(x)}.$$

Since total revenue, $R(x)$, equals quantity times unit price, we have $R(x) = x \cdot D(x)$.

Differentiating both sides gives (with the product rule) $R'(x) = D(x) + xD'(x)$. Using the definition of elasticity of demand, $xD'(x) = \dfrac{-D(x)}{E}$. Now, substituting in the expression for $R'(x)$,

$$R'(x) = D(x) - \frac{D(x)}{E}$$

$$R'(x) = D(x)\left(1 - \frac{1}{E}\right).$$

This relates the rate of change of revenue to elasticity of demand and helps one understand the analysis used in economics.

Properties of Elasticity of Demand

1. If $E > 1$, then $R'(x) = D(x)\left(1 - \dfrac{1}{E}\right) > 0$, the revenue is increasing, and we say that the demand is **elastic**.

2. If $E < 1$, then $R'(x) = D(x)\left(1 - \dfrac{1}{E}\right) < 0$, the revenue is decreasing, and we say that the demand is **inelastic**.

3. If $E = 1$, then $R'(x) = 0$, the revenue is at its maximum, and we say that the demand has **unit elasticity**.

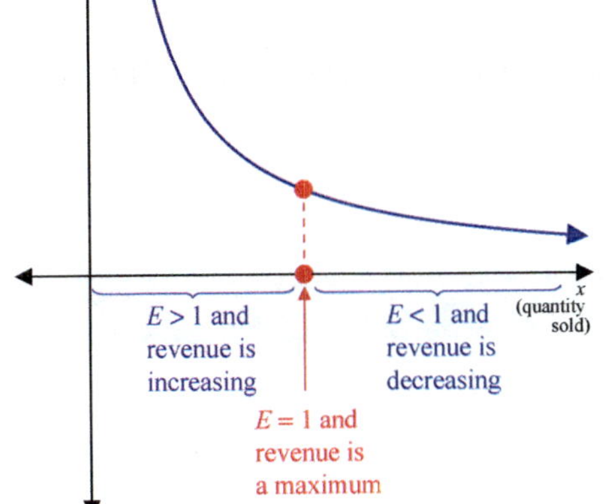

FIGURE 1: Elasticity of Demand

Figure 1 illustrates the various possibilities for E.

Example 1: Elasticity of Demand

Suppose that the demand function for a product is $p = D(x) = 300 - 2x$.

a. What is the price per unit if 50 units are sold?

b. Find the function describing the elasticity of demand E.

c. Find $E(50)$ and $E(110)$.

d. What quantity x maximizes revenue, R?

Solution

a. For $x = 50$,

$$p = D(50) = 300 - 2(50) = 200.$$

If the demand is 50 units, the price per unit is \$200.

b. For $D(x) = 300 - 2x$,

$$D'(x) = -2. \qquad \text{Find } D'(x).$$

Substituting $D(x) = 300 - 2x$ and $D'(x) = -2$ in the formula for $E(x)$ gives

$$E(x) = -\frac{1}{x} \cdot \frac{300 - 2x}{-2} = \frac{150 - x}{x}.$$

c. $E(50) = \dfrac{150 - 50}{50} = \dfrac{100}{50} = 2 > 1$ \qquad Demand is elastic.

$E(110) = \dfrac{150 - 110}{110} = \dfrac{40}{110} \approx 0.36 < 1$ \quad Demand is inelastic.

When the demand level is at 50 units, a change in the quantity sold will result in a relatively small change in price, and revenue will be increasing. When the demand level reaches 110, the revenue will be decreasing.

d. $R = xD(x) = x(300 - 2x) = 300x - 2x^2$
$R' = 300 - 4x$

If $R' = 0, x = 75$.

Since $R'' = -4$, R is concave down so $x = 75$ gives a maximum for R. This is consistent with part **c.**

Example 2: Elasticity of Demand

A product is known to have a demand function $p = D(x) = 100e^{-0.5x}$.

a. Find the value of x for which $E = 1$.

b. Find the value for x that maximizes the revenue.

The answers for parts a. and b. should be the same x-value.

Solution

a. $D(x) = 100e^{-0.5x}$

$D'(x) = 100(-0.5)e^{-0.5x} = -50e^{-0.5x}$ \qquad Find $D'(x)$.

Thus

$$E(x) = -\frac{1}{x} \cdot \frac{100e^{-0.5x}}{-50e^{-0.5x}} = \frac{2}{x}. \qquad \text{Substitute } D(x) \text{ and } D'(x) \text{ into } E(x).$$

Setting $E(x) = 1$ gives

$$\frac{2}{x} = 1$$

$$x = 2.$$

To find the value of x that maximizes the revenue in Example 2 using a TI-83/84 Plus calculator, perform the following steps:

1. Enter the revenue function into Y1.

2. Graph Y1 in an appropriate window (0 by 20 for x and 0 by 100 for y).

3. With the graph displayed, press **2nd** **trace** and select item 4:maximum.

4. When asked LeftBound? on the graphing screen, use the arrows to move the blinking tracer to the left side of the maximum point. Press **enter**.

5. For RightBound?, move the tracer to the right of the maximum and press **enter**.

6. When asked Guess?, move the tracer in the middle of the left and right bound points and again press **enter**.

7. The coordinates of the maximum point will appear in decimal form at the bottom of the screen.

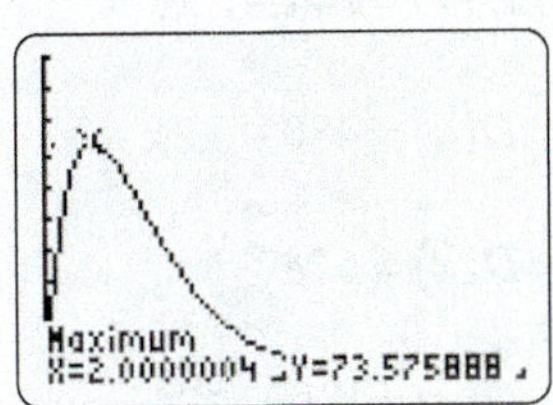

b. The revenue function is $R(x) = x \cdot p = x \cdot 100e^{-0.5x} = 100xe^{-0.5x}$.

Find $R'(x)$ and set $R'(x) = 0$ to find the value of x that gives the maximum revenue.

$$R'(x) = 100x \cdot (-0.5)e^{-0.5x} + e^{-0.5x}(100)$$
$$= e^{-0.5x}(-50x + 100)$$
$$0 = e^{-0.5x}(-50x + 100)$$
$$0 = -50x + 100 \text{ or } \cancel{0 = e^{-0.5x}}$$
$$50x = 100$$
$$x = 2$$

Thus $x = 2$ gives the maximum revenue.

There is also an alternative approach to elasticity of demand, which we briefly outline here. The slope of the demand curve is negative, and as a consequence, the demand function is one-to-one and thus is invertible. Let $F = D^{-1}$ and write $F(p) = x$ whenever $p = D(x)$. If p is the unit price, then x is the quantity demanded at that price. We can differentiate $F(p)$ with respect to x as follows.

$$\frac{d}{dx}F(p) = \frac{d}{dx}(x)$$

Note: $F = F(p)$ is the quantity demanded when the unit price is p.

$$F'(p) \cdot \frac{dp}{dx} = 1$$

Now we can use substitution with $\dfrac{dp}{dx} = D'(x)$, $x = F(p)$, and $p = D(x)$ in order to rewrite the equation for elasticity of demand.

$$E = -\frac{1}{x} \cdot \frac{D(x)}{D'(x)} = -\frac{1}{F(p)} \cdot \frac{p}{\dfrac{dp}{dx}}$$

$$E = -\frac{pF'(p)}{F(p)} = \frac{-p}{F(p) \cdot \dfrac{1}{F'(p)}}$$

This gives a useful equivalent formulation of elasticity of demand.

A grocery store determines that the demand function for its bakery's bread is $x = 180 - 30p$, where x is the number of loaves of bread it sells daily and p is the unit price.

a. Find the quantity demanded when the price is \$1.70.

b. Find the function describing the elasticity as a function of p.

c. Find the elasticity at $p = \$1.70$.

d. Interpret the resulting elasticity.

e. Determine the revenue function R and find p so that R is a maximum.

Solution

a. The quantity demanded is x and $x = F(p)$.

$$x = F(1.70)$$
$$= 180 - 30(1.70)$$
$$= 129$$

b. $E = -\dfrac{pF'(p)}{F(p)} = -\dfrac{p(-30)}{180 - 30p} = \dfrac{p}{6 - p}$

c. $E(1.70) = \dfrac{30(1.70)}{180 - 30(1.70)} = \dfrac{51}{129} = \dfrac{17}{43}$

d. Since $E < 1$, the demand is inelastic and so an increase in price will bring a decrease in demand and an increase in revenue.

e. $R = xp = (180 - 30p)p$

$$= 180p - 30p^2$$

Thus $R' = 180 - 60p$. Setting $R' = 0$, we obtain $60p = 180$, so $p = \$3.00$ per loaf.

Since $R'' = -60$, the function R is concave down at $p = 3$ and this price gives a maximum for the revenue.

13.4 EXERCISES

💡 PRACTICE

For each of the demand functions in Exercises 1–16, find **a.** the function describing the elasticity of demand and **b.** the value of x that maximizes the revenue.

1. $p = D(x) = 84 - 3x$

2. $p = D(x) = 144 - 1.5x$

3. $p = D(x) = 520 - 2.6x$

4. $p = D(x) = 480 - 3.2x$

5. $p = D(x) = 200e^{-0.2x}$

6. $p = D(x) = 67e^{-0.1x}$

7. $p = D(x) = 88e^{-0.025x}$

8. $p = D(x) = 130e^{-0.04x}$

9. $p = D(x) = \sqrt{150 - x}$

10. $p = D(x) = \sqrt{180 - 2x}$

11. $p = D(x) = \sqrt{162 - 3x}$

12. $p = D(x) = \sqrt{255 - 2.5x}$

13. $p = D(x) = 18 - \sqrt{x}$

14. $p = D(x) = 21 - 2\sqrt{x}$

15. $p = D(x) = 363 - x^2, \ x \le 18$

16. $p = D(x) = 600 - 0.5x^2, \ x \le 34$

APPLICATIONS

17. Maximum revenue: The demand function for an electric pencil sharpener is given by $p = D(x) = 19.2 - 0.4x$ dollars. Find the level of production for which the revenue is maximized.

18. Maximum revenue: The demand function for a popular stereo receiver is given by $p = D(x) = 540 - 0.05x^2$ dollars. Find the level of production for which the revenue is maximized.

19. Elastic demand: The demand function for an exclusive wool blanket is given by $p = D(x) = 33 - 2\sqrt{x}$ dollars, where x is in thousands of blankets. Find the level of production for which the demand is elastic.

20. Elastic demand: Find the levels of production for which the demand is elastic if the demand is given by $p = D(x) = \sqrt{207 - 3x}$ dollars.

21. Elastic demand: An arcade sells video games and determines that $x = 30\left(1 - e^{-\frac{p}{10}}\right)$, where x is the number of video games demanded for a unit price p.
   a. Determine the quantity demanded when $p = \$10$ per game.
   b. Determine E and interpret the result at $p = \$10$.
   c. What revenue is generated at $p = \$10$?

22. Elastic demand: Lucky Blooms sells a new rose variety which has established a demand of $x = f(p) = \dfrac{e^{\frac{p}{3}} + 350}{e^{\frac{p}{2}}}$.
   a. Determine the quantity demanded when $p = \$3$.
   b. Determine E and interpret the result at $p = \$3$.

23. Elastic demand: The demand for a product is given by $x = F(p) = \dfrac{1800}{10 + \ln(1 + p)}$.
   a. If $p = 20$, determine E and interpret the results.
   b. What is the revenue function R?
   c. Use R' to determine if R is increasing at $p = 20$. Is your answer consistent with part **a.**?

24. Elastic demand: Suppose a product has a demand function $x = F(p) = 300e^{-\frac{p}{10}}$.
   a. Find the elasticity function.
   b. Is the demand elastic or inelastic at $p = \$20$?
   c. Determine the unit price which maximizes revenue.
   d. Discuss whether or not your answers to **b.** and **c.** are consistent.

25. Elastic demand: Suppose the demand for a product is $p = D(x) = 300e^{-\frac{x^2}{200}}$.
   a. Determine the unit price if the quantity $x = 5$.
   b. What is formula for $D'(x)$?
   c. What is the elasticity at $x = 5$?
   d. Determine the value of x which maximizes revenue.

13.5 L'HÔPITAL'S RULE

■ TOPICS

- Limits of Indeterminate Forms 0/0 and ∞/∞
- Limits of Other Indeterminate Forms

In this section, we develop a calculus-based tool that often allows us to evaluate, fairly easily, limits that would otherwise pose a considerable challenge. The tool is called *l'Hôpital's Rule*, and the limits it applies to are called *limits of indeterminate form*.

Limits of Indeterminate Forms 0/0 and ∞/∞

As motivation, consider a limit of the form

$$\lim_{x \to c} \frac{f(x)}{g(x)},$$

where $f(c) = g(c) = 0$, $f'(c)$ and $g'(c)$ both exist, and $g'(c) \neq 0$. Such a limit is said to be of **indeterminate form** $\dfrac{0}{0}$, and the limit cannot be determined by simply evaluating $\dfrac{f(c)}{g(c)}$. However, we can rewrite the limit in a form that can be evaluated, as follows.

$$\lim_{x \to c} \frac{f(x)}{g(x)} = \lim_{x \to c} \frac{f(x) - 0}{g(x) - 0}$$

$$= \lim_{x \to c} \frac{f(x) - f(c)}{g(x) - g(c)} \qquad \color{blue}{f(c) = g(c) = 0}$$

$$= \lim_{x \to c} \frac{\dfrac{f(x) - f(c)}{x - c}}{\dfrac{g(x) - g(c)}{x - c}} \qquad \color{blue}{\text{Divide top and bottom by } x - c.}$$

$$= \frac{\displaystyle\lim_{x \to c} \frac{f(x) - f(c)}{x - c}}{\displaystyle\lim_{x \to c} \frac{g(x) - g(c)}{x - c}}$$

$$= \frac{f'(c)}{g'(c)}$$

So if the values of these derivatives are known (and $g'(c) \neq 0$) the evaluation of the limit *can* be accomplished with a different substitution. This observation is a simple form of l'Hôpital's Rule, named after the French nobleman Guillaume François Antoine de l'Hôpital (1661–1704) in whose introductory calculus textbook it first appeared—the result is actually due to the Swiss mathematician Johann Bernoulli (1667–1748).

Determine $\displaystyle\lim_{x \to 0} \frac{5x + 1 - e^{2x}}{x}$.

Solution

The limit is of indeterminate form $\dfrac{0}{0}$, so we apply l'Hôpital's Rule.

$$\lim_{x \to 0} \frac{5x + 1 - e^{2x}}{x} = \lim_{x \to 0} \frac{5 - 2e^{2x}}{1} \qquad \text{Differentiate top and bottom.}$$
$$= 5 - 2(1) \qquad \text{Substitute } x = 0.$$
$$= 3$$

Before presenting the full form of l'Hôpital's Rule, we will state one other useful theorem—it is a formulation of the Mean Value Theorem due to Augustin-Louis Cauchy (1789–1857). Cauchy's MVT is the basis for one of the more elegant proofs of l'Hôpital's Rule.

Suppose that f and g are continuous on $[a, b]$ and differentiable on (a, b), $g'(x) \neq 0$ on (a, b), and $g(a) \neq g(b)$. Then there is a point $c \in (a, b)$ such that

$$\frac{f'(c)}{g'(c)} = \frac{f(b) - f(a)}{g(b) - g(a)}.$$

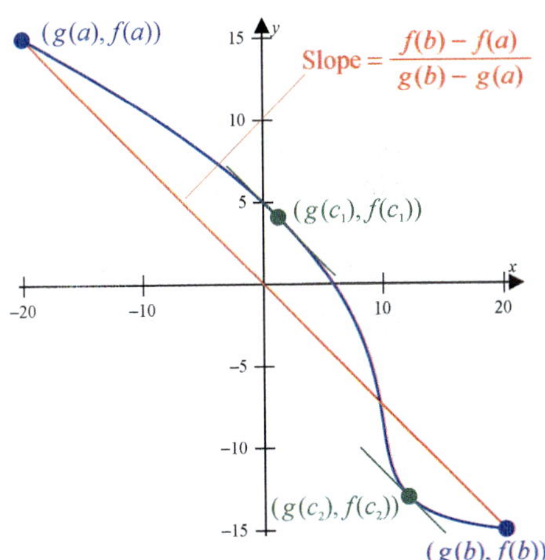

FIGURE 1: Cauchy's Mean Value Theorem

Note that Cauchy's Mean Value Theorem reduces to the simpler Mean Value Theorem if $g(x) = x$.

The simpler version of the MVT guarantees the existence of a point where the tangent to a function is parallel to a secant line, and Cauchy's MVT does something similar. Given two functions f and g with the properties above, the collection of ordered pairs $\{(g(x), f(x)) \,|\, x \in [a, b]\}$ defines a curve in $\mathbb{R}^2$ such as the one depicted in Figure 1, and the red line segment connecting $(g(a), f(a))$ and $(g(b), f(b))$ is also called a secant line. Specifically, the blue curve in Figure 1 is defined by the two functions $f(x) = x^2 - 5x - 9$ and $g(x) = x^3 + x + 10$, and the interval $[a, b]$ is $[-3, 2]$; the slope of the secant line is $-\dfrac{3}{4}$. Cauchy's MVT tells us that for at least one number $c \in (a, b)$, the line tangent to the curve at $(g(c), f(c))$ is parallel to the secant line. (Curves defined in this manner are called *parametric curves*.)

We are now ready for a stronger form of l'Hôpital's Rule.

L'Hôpital's Rule

Suppose f and g are differentiable at all points of an open interval I containing c, and that $g'(x) \neq 0$ for all $x \in I$ except possibly at $x = c$. Suppose further that either

$$\lim_{x \to c} f(x) = 0 \quad \text{and} \quad \lim_{x \to c} g(x) = 0$$

or

$$\lim_{x \to c} f(x) = \pm\infty \quad \text{and} \quad \lim_{x \to c} g(x) = \pm\infty.$$

Then

$$\lim_{x \to c} \frac{f(x)}{g(x)} = \lim_{x \to c} \frac{f'(x)}{g'(x)}$$

assuming the limit on the right is a real number or ∞ or $-\infty$.

Further, the rule is true for one-sided limits at c and for limits at infinity; that is, $x \to c$ can be replaced with $x \to c^+$, $x \to c^-$, $x \to -\infty$, or $x \to \infty$, assuming always that the limit on the right is a real number or ∞ or $-\infty$.

We have already mentioned limits of indeterminate form $\dfrac{0}{0}$ in passing; the other type of limit described in l'Hôpital's Rule is of **indeterminate form** $\dfrac{\infty}{\infty}$ and we will discuss other variations soon.

Proof

We will prove only the case in which

$$\lim_{x \to c} f(x) = 0 \quad \text{and} \quad \lim_{x \to c} g(x) = 0,$$

and we will prove that the claim is true as $x \to c^-$; the corresponding result for $x \to c^+$ is nearly identical, and the two one-sided limits together prove the theorem.

Suppose that $x \in I$ is a number lying to the left of c. Then $g'(x) \neq 0$ and we can apply Cauchy's MVT to the interval $[x, c]$. Thus, there is a point $\tilde{c} \in (x, c)$ such that

$$\frac{f'(\tilde{c})}{g'(\tilde{c})} = \frac{f(c) - f(x)}{g(c) - g(x)}$$

$$= \frac{f(x)}{g(x)} \qquad \qquad f(c) = g(c) = 0$$

As we let $x \to c^-$, $\tilde{c} \to c^-$ as well since $\tilde{c}$ always lies between x and c. Hence,

$$\lim_{x \to c^-} \frac{f(x)}{g(x)} = \lim_{\tilde{c} \to c^-} \frac{f'(\tilde{c})}{g'(\tilde{c})} = \lim_{x \to c^-} \frac{f'(x)}{g'(x)}.$$

The value of l'Hôpital's Rule comes from the fact that differentiating the numerator and denominator of a fraction of indeterminate form often results in a fraction that is not, making the limit easier to determine.

Example 2: L'Hôpital's Rule and the Indeterminate Form ∞/∞

Determine $\lim\limits_{x \to \infty} \dfrac{\ln x}{2\sqrt{x}}$.

Solution

First, note that l'Hôpital's Rule is indeed applicable: this limit at infinity is of indeterminate form $\dfrac{\infty}{\infty}$. So we evaluate the limit as follows.

$$\lim_{x \to \infty} \frac{\ln x}{2\sqrt{x}} = \lim_{x \to \infty} \frac{\dfrac{d}{dx}(\ln x)}{\dfrac{d}{dx}(2\sqrt{x})} = \lim_{x \to \infty} \frac{\dfrac{1}{x}}{\dfrac{1}{\sqrt{x}}} = \lim_{x \to \infty} \frac{1}{\sqrt{x}} = 0$$

Example 3: L'Hôpital's Rule and the Indeterminate Form 0/0

Determine $\lim\limits_{x \to 0} \dfrac{a^x - 1}{x}$, where $a > 0$ is a constant.

Solution

The limit is of indeterminate form $\dfrac{0}{0}$, so we proceed using l'Hôpital's Rule.

$$\lim_{x \to 0} \frac{a^x - 1}{x} = \lim_{x \to 0} \frac{(\ln a)a^x}{1} = \ln a$$

> **⚠ CAUTION**
>
> L'Hôpital's Rule says the limit of a quotient of two functions is equal to the limit of the quotient of their derivatives—don't mistakenly apply the Quotient Rule of differentiation! After verifying that the conditions of l'Hôpital's Rule are satisfied, proceed by differentiating the numerator and denominator individually.

L'Hôpital's Rule is obviously useful, but there is even more power in it than might appear at first glance. Suppose the rule is applied to a limit of indeterminate form, and the resulting limit is again of indeterminate form. Does that mean the rule fails to tell us anything? Often, no—if the new limit is also indeterminate, then we can again apply l'Hôpital's Rule. This process can be repeated as often as necessary, *as long as we stop as soon as we reach a limit that is not indeterminate*. The next example illustrates how repeated application of the rule works, and the sort of mistake that can arise if we misuse it.

Example 4: Repeated Applications of L'Hôpital's Rule

Evaluate the following limits.

a. $\lim\limits_{x \to \infty} \dfrac{e^x}{5x^2 - 3x + 2}$

b. $\lim\limits_{x \to 0} \dfrac{x - \ln(x + 1)}{3x^2 + 7x}$

Solution

a. The limit is of indeterminate form $\dfrac{\infty}{\infty}$ so we can apply l'Hôpital's Rule.

$$\lim_{x \to \infty} \frac{e^x}{5x^2 - 3x + 2} = \lim_{x \to \infty} \frac{e^x}{10x - 3}$$

Limit is still of indeterminate form $\dfrac{\infty}{\infty}$. Apply l'Hôpital's Rule again.

$$= \lim_{x \to \infty} \frac{e^x}{10}$$

Limit can now be determined; $e^x \to \infty$ as $x \to \infty$.

$$= \infty$$

b. The limit is of indeterminate form $\dfrac{0}{0}$ and we apply l'Hôpital's Rule.

$$\lim_{x\to 0}\frac{x-\ln(x+1)}{3x^2+7x}=\lim_{x\to 0}\frac{1-\dfrac{1}{x+1}}{6x+7}=\frac{0}{7}=0$$

If we mistakenly continued to apply the rule, we would have obtained

$$\lim_{x\to 0}\frac{1-\dfrac{1}{x+1}}{6x+7}=\lim_{x\to 0}\frac{\dfrac{1}{(x+1)^2}}{6}=\frac{1}{6},$$

which is incorrect.

Limits of Other Indeterminate Forms

In addition to the two indeterminate forms we have already seen, l'Hôpital's Rule can be used to evaluate other potentially challenging limits. The remaining examples illustrate how indeterminate products, differences, and powers can be rewritten in such a way that l'Hôpital's Rule applies.

Example 5: L'Hôpital's Rule and the Indeterminate Form $0 \cdot \infty$

Determine $\lim\limits_{x\to 0^+} \sqrt{x}\ln x$.

Solution

We say that a limit of a product fg is of **indeterminate form** $0 \cdot \infty$ if one of the functions approaches 0 and the other approaches ∞ or $-\infty$. The limit, if it exists, depends on which function dominates—the product could tend toward 0, could grow unbounded, or could approach some nonzero real number if the two functions balance one another just right.

We can apply l'Hôpital's Rule if we can rewrite the product as a quotient in either the indeterminate form $\dfrac{0}{0}$ or $\dfrac{\infty}{\infty}$. That is, we rewrite fg as either

$$\frac{f}{\dfrac{1}{g}} \quad \text{or} \quad \frac{g}{\dfrac{1}{f}},$$

whichever is easier to work with. In this example, we rewrite the limit and then apply l'Hôpital's Rule as follows.

$$\lim_{x \to 0^+} \sqrt{x}\,\ln x = \lim_{x \to 0^+} \frac{\ln x}{\dfrac{1}{\sqrt{x}}}$$

Rewrite to obtain the indeterminate form $\dfrac{\infty}{\infty}$.

$$= \lim_{x \to 0^+} \frac{\dfrac{1}{x}}{-\dfrac{1}{2}x^{-\frac{3}{2}}} = \lim_{x \to 0^+}\left(-2\sqrt{x}\right) = 0$$

Example 6: L'Hôpital's Rule and the Indeterminate Form ∞−∞

Determine $\displaystyle\lim_{x \to 1}\left(\frac{x}{x-1} - \frac{1}{\ln x}\right)$.

Solution

A limit of a difference $f - g$ is of **indeterminate form** $\infty - \infty$ if $f \to \infty$ and $g \to \infty$. Again, such limits are usually not trivial: if f dominates, the difference will tend to ∞; and if g dominates, the difference will tend to $-\infty$; but it is also possible for the two to balance out and result in a finite limit.

To use l'Hôpital's Rule, we need to rewrite the difference as a quotient. In this case, we can do so by combining the two fractions using a common denominator; in other such problems, rationalization or factoring out a common factor may be helpful.

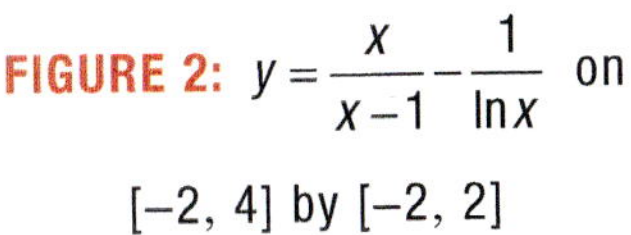

FIGURE 2: $y = \dfrac{x}{x-1} - \dfrac{1}{\ln x}$ on

$[-2, 4]$ by $[-2, 2]$

$$\lim_{x \to 1}\left(\frac{x}{x-1} - \frac{1}{\ln x}\right) = \lim_{x \to 1}\frac{x\ln x - x + 1}{(x-1)\ln x}$$

$$= \lim_{x \to 1}\frac{\ln x + 1 - 1}{\ln x + \dfrac{x-1}{x}} \qquad \text{Apply l'Hôpital's Rule once.}$$

$$= \lim_{x \to 1}\frac{x\ln x}{x\ln x + x - 1} \qquad \text{Simplify the fraction.}$$

$$= \lim_{x \to 1}\frac{\ln x + 1}{\ln x + 1 + 1} \qquad \text{Apply l'Hôpital's Rule again.}$$

$$= \frac{1}{2}$$

Figure 2 visually confirms the limit we found.

The next three limits are all of the form f^g, and constitute limits of **indeterminate forms** 1^∞, 0^0, **and** ∞^0. As with the other indeterminate forms, there is competition between the effects of the two functions in the limit. In order to use l'Hôpital's Rule, we set $y = f^g$ and take the natural logarithm of both sides to obtain $\ln y = g\ln f$. If we can determine the limit of $g\ln f$, we can determine the limit of $y = e^{\ln y} = e^{g\ln f}$.

Example 7: L'Hôpital's Rule and the Indeterminate Form 1^∞

Determine $\lim\limits_{x \to 0^+} (1+x)^{\frac{1}{x}}$.

Solution

Note that the base of the expression $(1+x)^{\frac{1}{x}}$ goes to 1 and the exponent goes to ∞ as $x \to 0^+$. We let $y = (1+x)^{\frac{1}{x}}$.

$$\ln y = \left(\frac{1}{x}\right) \ln(1+x) = \frac{\ln(1+x)}{x}$$

The limit of this last expression is of the form $\dfrac{0}{0}$, and we can apply l'Hôpital's Rule.

$$\lim_{x \to 0^+} \frac{\ln(1+x)}{x} = \lim_{x \to 0^+} \frac{\frac{1}{1+x}}{1} = 1$$

Since $\lim\limits_{x \to 0^+} \ln y = 1$, $\lim\limits_{x \to 0^+} y = \lim\limits_{x \to 0^+} e^{\ln y} = e^1 = e$. Hence, $\lim\limits_{x \to 0^+} (1+x)^{\frac{1}{x}} = e$.

Example 8: L'Hôpital's Rule and the Indeterminate Form 0^0

Determine $\lim\limits_{x \to 0^+} x^x$.

Solution

Both the base and the exponent approach 0. Letting $y = x^x$, we arrive at

$$\ln y = x \ln x = \frac{\ln x}{\frac{1}{x}},$$

a limit of indeterminate form $\dfrac{\infty}{\infty}$. So,

$$\lim_{x \to 0^+} \frac{\ln x}{\frac{1}{x}} = \lim_{x \to 0^+} \frac{\frac{1}{x}}{-\frac{1}{x^2}} = \lim_{x \to 0^+} (-x) = 0$$

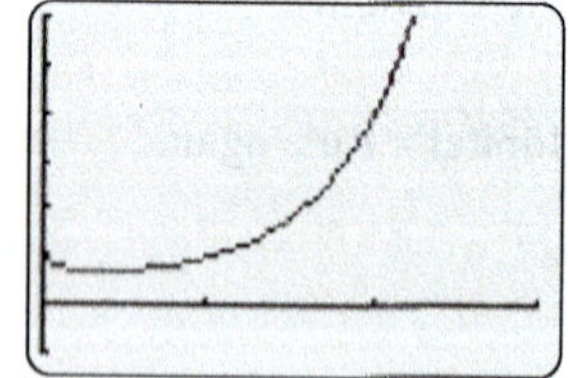

FIGURE 3: $y = x^x$ on $[0, 3]$ by $[-1, 6]$

and hence $x^x \to e^0 = 1$ as $x \to 0^+$ (don't forget this last step!). Figure 3 visually confirms our result.

Example 9: L'Hôpital's Rule and the Indeterminte Form ∞^0

Determine $\lim\limits_{x\to\infty} x^{\frac{1}{x}}$.

Solution

The base has a limit of ∞ and the exponent has a limit of 0. We proceed as in the last two examples.

$$y = x^{\frac{1}{x}}$$

$$\ln y = \frac{1}{x}\ln x = \frac{\ln x}{x} \qquad \text{Indeterminate form } \frac{\infty}{\infty}$$

Applying l'Hôpital's Rule,

$$\lim_{x\to\infty}\ln y = \lim_{x\to\infty}\frac{\ln x}{x} = \lim_{x\to\infty}\frac{\dfrac{1}{x}}{1} = 0$$

and therefore $\lim\limits_{x\to\infty} x^{\frac{1}{x}} = \lim\limits_{x\to\infty} y = e^0 = 1$.

13.5 EXERCISES

💡 PRACTICE

Evaluate the limit using previous techniques. Then decide whether l'Hôpital's Rule is applicable and, if so, use it to check your answer.

1. $\lim\limits_{x\to 3}\dfrac{2x^2 - 18}{x - 3}$

2. $\lim\limits_{x\to -2}\dfrac{x^3 + 8}{x + 2}$

3. $\lim\limits_{x\to\infty}\dfrac{6x^2 - x + 7}{x - 3x^2}$

4. $\lim\limits_{x\to -\infty}\dfrac{5x^2 - 2x + 1}{2.5x^3 - 3x^2 + 6}$

5. $\lim\limits_{x\to 0}\dfrac{2x}{\sqrt{x+3} - \sqrt{3}}$

6. $\lim\limits_{x\to 0^+}\left(\sqrt{x}\right)^{1/x}$

7. $\lim\limits_{x\to 0}\left(\dfrac{1}{x} - \dfrac{1}{x\sqrt{x+1}}\right)$

Two functions are in competition to determine the indicated limit. Identify the type of the indeterminate form, and fill out the table to decide which function dominates.

8. $\lim\limits_{x\to\infty} f(x)$, where $f(x) = \dfrac{\sqrt{5x^3 + 7}}{0.2x^2 + 1}$

x	1	10	100	1000	10,000	100,000
$f(x)$						

9. $\lim\limits_{x \to \infty} g(x)$, where $g(x) = \dfrac{0.5\sqrt{x}}{\ln(x+1)}$

x	1	10	100	1000	10,000	100,000
$g(x)$						

10. $\lim\limits_{x \to \infty} h(x)$, where $h(x) = x^{100} e^{-x}$

x	1	10	100	1000	10,000	100,000
$h(x)$						

Check whether l'Hôpital's Rule applies to the given limit. If it does, use it to determine the value of the limit. If it does not, find the limit some other way. (When necessary, apply l'Hôpital's Rule several times.)

11. $\lim\limits_{x \to \infty} \dfrac{2x+5}{x^2-7}$

12. $\lim\limits_{x \to \infty} \dfrac{4-2.5x}{x+3}$

13. $\lim\limits_{x \to -\infty} \dfrac{1.5x^3 - 2x^2 + x + 9}{x^2 + 2.1x - 4}$

14. $\lim\limits_{x \to -\infty} \dfrac{4.5x^4 + x^3 - 2}{3 - 1.5x^4}$

15. $\lim\limits_{x \to 0^+} \dfrac{\sqrt{x}}{\ln x}$

16. $\lim\limits_{x \to \infty} \dfrac{\dfrac{1}{x}+2}{2x+1}$

17. $\lim\limits_{t \to 0} \dfrac{t}{\sqrt{2t+9}-3}$

18. $\lim\limits_{x \to \infty} \dfrac{\ln x}{\ln\left(x^2+3x\right)}$

19. $\lim\limits_{x \to 0} \dfrac{\log_{10}\left(x^2+2x+1\right)}{\log_{10}(x+1)}$

20. $\lim\limits_{x \to 0} \dfrac{x}{3^{x/2}-1}$

21. $\lim\limits_{x \to \infty} \dfrac{2^x}{x^2-3x+4}$

22. $\lim\limits_{x \to \infty} \dfrac{x+2^x}{5^x-x}$

23. $\lim\limits_{x \to \infty} \dfrac{4^x+x^2}{3^x-x}$

24. $\lim\limits_{x \to \infty} \dfrac{\ln(\ln x)}{x \ln x}$

25. $\lim\limits_{x \to \infty} \dfrac{\log_4(2x+1)}{\log_5(x-4)}$

26. $\lim\limits_{x \to 0^+} \dfrac{\log_4(x+1)}{\log_3 x}$

27. $\lim\limits_{x \to 0} \dfrac{3^x-1}{x 3^x}$

Identify the indeterminate product, quotient, difference, or power, and use l'Hôpital's Rule to find the limit. If the limit is not of indeterminate form, say so and find it by other means.

28. $\lim\limits_{x \to 0^+} x \ln x$

29. $\lim\limits_{x \to \infty} \dfrac{\sqrt{2x^2+1}}{x+3}$

30. $\lim\limits_{x \to \infty} (\ln x)^{-1/x}$

31. $\lim\limits_{x \to 0^+} \left(\dfrac{1}{x}\right)^x$

32. $\lim\limits_{x \to 0^+} (-\ln x)^x$

33. $\lim\limits_{x \to 1^+} \left(\dfrac{1}{\ln x} - \dfrac{2}{x-1}\right)$

34. $\lim\limits_{x \to 0^+} \left(\dfrac{1}{x} + \ln x\right)$

35. $\lim\limits_{x \to 4^+} \left(\dfrac{32}{x^2-16} - \dfrac{x}{x-4}\right)$

36. $\lim\limits_{x \to 0^+} x^{\left(x^2\right)}$

37. $\lim\limits_{x \to 0^+} \left(2^x - x\right)^{1/x}$

38. $\lim\limits_{x \to 0^+} (1-x)^{1/x}$

39. $\lim\limits_{x \to \infty} \left(\sqrt{x^2-3x} - \dfrac{3}{x^2+1}\right)$

40. $\lim\limits_{x\to\infty}(x-1)^{1/x}$

41. $\lim\limits_{x\to\infty}\dfrac{\ln x}{x^{7/5}}$

42. $\lim\limits_{x\to\infty}\dfrac{x^{100}}{3^x}$

43. $\lim\limits_{x\to\infty}\dfrac{\ln\left(100x^2+e^x\right)}{100x}$

44. $\lim\limits_{x\to0}(1+2x)^{1/x}$

45. $\lim\limits_{x\to1}x^{1/(1-x)}$

Find the limit. If applicable, use l'Hôpital's Rule (as many times as appropriate).

46. $\lim\limits_{x\to\infty}\dfrac{2x^5+x^3-4}{e^x}$

47. $\lim\limits_{x\to\infty}x^{1/x^3}$

48. $\lim\limits_{x\to0^+}x^{x^x}$

49. $\lim\limits_{x\to0^+}\left(x^x\right)^x$

50. $\lim\limits_{x\to\infty}x^{1/x^n},\quad n\in\mathbb{Z}^+$

51. $\lim\limits_{x\to\infty}\dfrac{(\ln x)^3}{x^2}$

52. $\lim\limits_{x\to\infty}\dfrac{x^2+1}{2^x}$

53. $\lim\limits_{x\to\infty}\left(1+\dfrac{1}{x}\right)^x$

54. $\lim\limits_{x\to\infty}\sqrt[x]{x}$

55. $\lim\limits_{x\to\infty}\dfrac{2^x+5^x}{6^x}$

Convince yourself that the initial use of l'Hôpital's Rule is not helpful in finding the limit. If possible, try to find a way to make use of the theorem, or evaluate the limit in some other way.

56. $\lim\limits_{x\to\infty}\dfrac{\sqrt{x+2}}{\sqrt{x}}$

57. $\lim\limits_{x\to\infty}\dfrac{\sqrt[3]{x+1}-2}{\sqrt{x^2+2}}$

58. $\lim\limits_{x\to\infty}\dfrac{2^x+3^x}{5^x}$

59. $\lim\limits_{x\to\infty}\dfrac{5^x-6^x}{7^x+8^x}$

60. $\lim\limits_{x\to\infty}\dfrac{2^{-x}}{x^{-1}}$

61. $\lim\limits_{x\to\infty}\left(\dfrac{1}{x+1}\right)^{-x^3}$

62. $\lim\limits_{x\to\infty}\left(\dfrac{1}{x^2}\right)^{e^{-x}}$

63. $\lim\limits_{x\to\infty}\dfrac{x}{\sqrt{x^2+1}}$

64. $\lim\limits_{x\to1^+}\left(\dfrac{1}{x-1}-\dfrac{1}{\ln x}\right)$

65. $\lim\limits_{x\to\infty}2^{-x}x\ln x$

✎ WRITING & THINKING

Find the error(s) in the limit calculation.

66. $\lim\limits_{x\to2}\dfrac{x^2-2}{x-2}=\lim\limits_{x\to2}\dfrac{2x}{1}=4$ (Incorrect!)

67. $\lim\limits_{x\to-\infty}\dfrac{5^x+1}{5^x}=\lim\limits_{x\to-\infty}\dfrac{(\ln5)5^x}{(\ln5)5^x}=1$ (Incorrect!)

Use l'Hôpital's Rule to prove the assertion.

68. $\lim\limits_{x\to\infty}\dfrac{p(x)}{e^{kx}}=0$ ($p(x)$ is a polynomial, $k>0$)

69. $\lim\limits_{x\to\infty}\dfrac{(\ln x)^n}{x^k}=0$ ($n\in\mathbb{N},\,k>0$)

70. $\lim\limits_{x\to\infty}\dfrac{a^x}{x^n}=\infty$ ($a>1,\,n\in\mathbb{N}$)

Find the value(s) of c satisfying the conclusion of Cauchy's Mean Value Theorem. If the theorem doesn't apply, explain why.

71. $f(x) = x, \quad g(x) = x^2 + 1; \quad [0,1]$

72. $f(x) = x^3 - 1, \quad g(x) = x^2 + 2x; \quad [-1,1]$

73. $f(x) = x^3 - x, \quad g(x) = -x^2 + 2x + 3; \quad [-1,3]$

74. $f(x) = x^3, \quad g(x) = -x^2; \quad [-2,3]$

75. $f(x) = x^2 + 3x, \quad g(x) = 3x^2 - 5x + 3; \quad [-1,3]$

76. $f(x) = \dfrac{1}{x}, \quad g(x) = \ln x; \quad [1,2]$

77. $f(x) = x^2 - 5x - 9, \quad g(x) = x^3 + x + 10; \quad [-3,2]$

78. Recall the compound interest formula for the value of an investment of P dollars after t years, compounded n times a year at an annual interest rate of r:

$$A = P\left(1 + \frac{r}{n}\right)^{nt}$$

Use l'Hôpital's Rule to prove that if we let $n \to \infty$, we obtain the continuous compounding formula $A = Pe^{rt}$.

79. The strength of an electric field due to a disk charge is obtained from the formula

$$E(x) = \frac{\sigma}{2\varepsilon_0}\left(1 - \frac{x}{\sqrt{x^2 + R^2}}\right)$$

where σ is the electric charge per unit area (in C/m^2), $\varepsilon_0 = 8.85 \cdot 10^{-12} \ C^2/Nm^2$, R is the radius of the ring, and x is the distance to the charge in meters. Use l'Hôpital's Rule to confirm that $E(x) \to 0$ as $x \to \infty$. How is E affected by σ and R at a given distance? What happens to the rate of change of E as x increases? (**Hint:** Apply l'Hôpital's Rule to dE/dx as $x \to \infty$.)

80. Marquis de l'Hôpital first illustrated the rule named after him in his 1696 textbook, *Analyse des Infiniment Petits*. He used an example where the objective was to find

$$\lim_{x \to a} \frac{\sqrt{2a^3 x - x^4} - a\sqrt[3]{a^2 x}}{a - \sqrt[4]{ax^3}}$$

for $a > 0$. Determine the above limit.

Use a graphing utility to graph the function for different values of the parameter c. Examine how the values of the parameter affect the indicated limit.

81. $\lim\limits_{x \to \infty}\left(1 + \dfrac{1}{cx}\right)^x; \quad$ What happens to the limit when $|c| \to \infty$?

82. $\lim\limits_{x \to 0^+} \dfrac{1 - c^x}{cx}; \quad$ What happens to the limit when $c \to \infty$?

13.6 DIFFERENTIALS

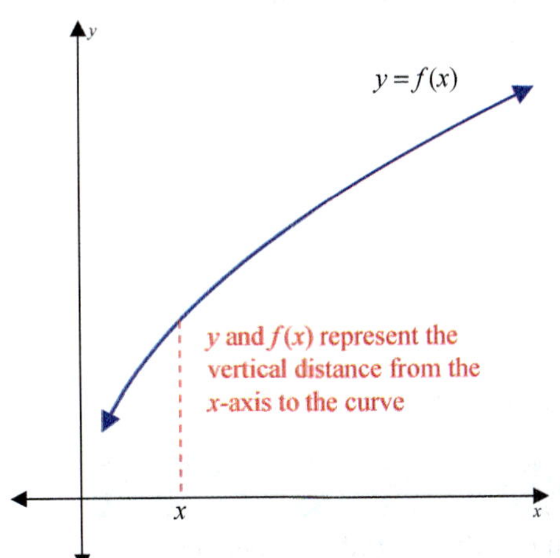

FIGURE 1

For the function $y = f(x)$, the expressions y and $f(x)$ both represent the distance from the x-axis to the point $(x, f(x))$ on the curve. (See Figure 1.)

If x is changed by a small amount h, then the corresponding y is also changed. The value of y becomes $f(x + h)$, and the change in y becomes $f(x + h) - f(x)$. (See Figure 2.)

The change in y can be denoted by Δy (delta y):

$$\Delta y = f(x + h) - f(x).$$

Now $\Delta x = h$ and the **difference quotient** is given by

$$\frac{\Delta y}{\Delta x} = \frac{f(x+h) - f(x)}{h} \quad \text{or} \quad \frac{\Delta y}{\Delta x} = \frac{f(x+\Delta x) - f(x)}{\Delta x}.$$

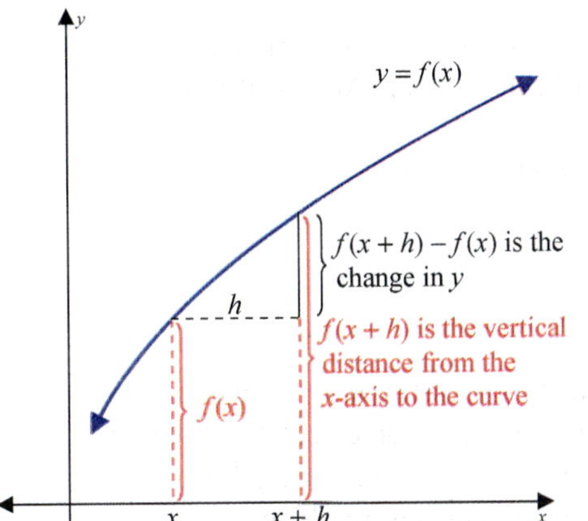

FIGURE 2

Geometrically, the difference quotient is the slope of the secant line, as illustrated in Figure 3.

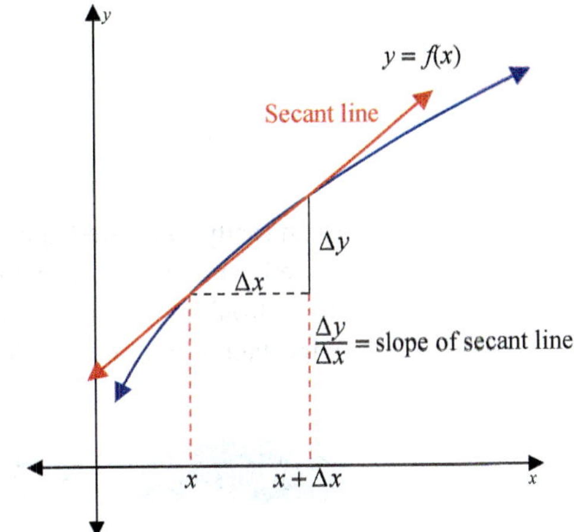

FIGURE 3

The slope of the tangent is the first derivative $\dfrac{dy}{dx}$. Now we take some liberty with the notation $\dfrac{dy}{dx}$ and think of it as being composed of two parts, dy and dx. (See Figure 4.)

The symbols dy and dx are called **differentials**. We let $h = \Delta x = dx$. Then, as illustrated in Figure 5, the difference quotient and the derivative $\dfrac{dy}{dx}$ (quotient of the differentials) are approximately equal for small values of Δx.

$$\frac{dy}{dx} \approx \frac{\Delta y}{\Delta x}$$

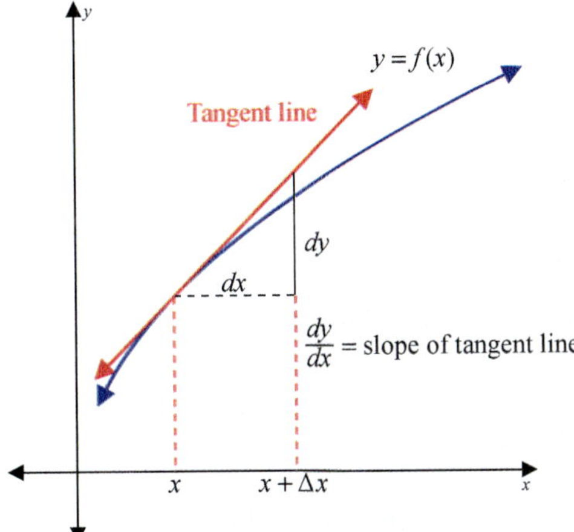

FIGURE 4

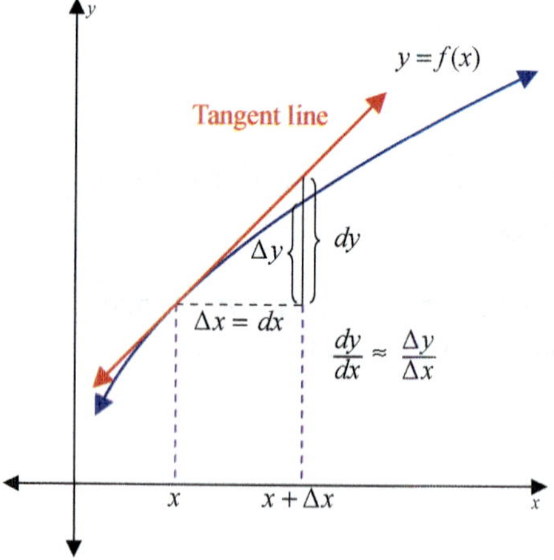

FIGURE 5

From Figure 5, we can see that the differential dy and the change in y, Δy, are almost equal. In fact, the smaller Δx is, the closer dy and Δy will be to each other. We make the following observations.

$$\frac{dy}{dx} \approx \frac{\Delta y}{\Delta x}$$

$$dy \approx \frac{\Delta y}{\Delta x}\,dx$$

$$dy = \lim_{\Delta x \to 0}\left(\frac{\Delta y}{\Delta x}\right) \cdot dx = f'(x)\,dx$$

This last expression becomes the definition for the differential dy.

Differential

If $y = f(x)$ is a function and $y' = f'(x)$ exists, then the **differential** dy is defined as

$$dy = f'(x) \cdot dx.$$

As we have seen in the previous development, the change in y (represented by Δy) can be approximated by the differential dy. That is, we can calculate Δy directly as

$$\Delta y = f(x + \Delta x) - f(x)$$

or use

$$dy = f'(x)\,dx$$

as an easily calculated approximation to Δy. While this use of dy may seem somewhat forced, particularly since most students have calculators and can calculate Δy without much trouble, we will find other uses for differentials later. At this stage, though, it is important to become familiar with both the notation and the terminology.

Example 1: Finding Differentials

a. Find dy for $y = x^2$.   **b.** Find du for $u = (x^2 + 5)^3$.

c. Find dv if $v = \dfrac{x^2}{x^2 + 9}$.

Solution

a. $f(x) = x^2$ and $f'(x) = 2x$ Find the derivative of y.

So

$$dy = f'(x) \cdot dx \quad \text{Substitute } f'(x) = 2x \text{ into the definition of differential.}$$
$$= 2x \cdot dx.$$

b. The differential du is found in the same way as dy.

$$f(x) = (x^2 + 5)^3 \quad \text{and} \quad f'(x) = 3(x^2 + 5)^2 \cdot 2x \quad \text{Find the derivative of } u.$$
$$= 6x(x^2 + 5)^2$$

So

$$du = f'(x)\,dx$$
$$= 6x\left(x^2 + 5\right)^2 dx.$$

Substitute $f'(x) = 6x(x^2 + 5)^2$ into the definition of differential.

c. $f(x) = \dfrac{x^2}{x^2 + 9}$ and $f'(x) = \dfrac{\left(x^2 + 9\right)(2x) - x^2(2x)}{\left(x^2 + 9\right)^2}$ Find the derivative of v.

$$= \dfrac{18x}{\left(x^2 + 9\right)^2}$$

So

$$dv = \dfrac{18x}{\left(x^2 + 9\right)^2}\,dx.$$

Substitute $f'(x) = \dfrac{18x}{\left(x^2 + 9\right)^2}$ into the definition of differential.

Example 2: Using Differentials

For $y = x^2$, $x = 9$, $\Delta x = 0.1$, find

a. Δy,   **b.** dy, and   **c.** $\Delta y - dy$.

Solution

a. $\Delta y = f\left(x + \Delta x\right) - f\left(x\right)$

$\quad = f\left(9 + 0.1\right) - f\left(9\right)$ Substitute $x = 9$ and $\Delta x = 0.1$.

$\quad = \left(9.1\right)^2 - \left(9\right)^2$ Find $f(9.1)$ and $f(9)$.

$\quad = 82.81 - 81$

$\quad = 1.81$ Solve the equation.

b. $f\left(x\right) = x^2$ and $f'\left(x\right) = 2x$ Find the derivative of y.

$\quad dy = f'\left(x\right)dx$

$\quad = 2x\,dx$ Substitute $f'(x) = 2x$ into the definition of differential.

$\quad = 2\left(9\right)\left(0.1\right)$ Substitute in $x = 9$ and $dx = 0.1$.

$\quad = 1.8$

c. $\Delta y - dy = 1.81 - 1.8 = 0.01$ Substitute the values found in parts **a.** and **b.** into the equation and solve.

Example 3: Estimating Square Roots

Using differentials, estimate $\sqrt{15}$.

Solution

Here we let $f(x) = \sqrt{x}$, $x = 16$, and $\Delta x = -1$.

We have chosen $x = 16$ because 16 is the closest square number to 15. This gives $15 = 16 + \Delta x$, which leads to $\Delta x = -1$.

$$f(x) = \sqrt{x} = x^{\frac{1}{2}}$$ Rewrite $f(x)$ using exponents.

$$f'(x) = \frac{1}{2}x^{-\frac{1}{2}} = \frac{1}{2\sqrt{x}}$$ Find the derivative of $f(x)$.

$$dy = f'(x)\,dx$$

$$= \frac{1}{2\sqrt{x}}\,dx$$ Substitute $f'(x)$ into the definition of differential.

$$dy\big|_{x=16} = \frac{1}{2\sqrt{16}}(-1)$$ Find dy for $x = 16$: substitute $x = 16$ and $\Delta x = -1$.

$$= -\frac{1}{8}$$

$$= -0.125$$

Now using $dy \approx \Delta y = f(x + \Delta x) - f(x)$, we obtain that $f(x + \Delta x) \approx f(x) + dy$ or $\sqrt{x + \Delta x} \approx \sqrt{x} + dy$.

$$\sqrt{16 + (-1)} \approx \sqrt{16} + dy$$

$$\sqrt{15} \approx \sqrt{16} + dy$$

$$= 4 - 0.125$$

$$= 3.875$$

Your calculator will give $\sqrt{15} \approx 3.872983346$.

Example 4: Change in Volume

The volume of a cube with edge x is given by $V = x^3$. Using differentials, approximate the change in V if x is

a. changed from 5 cm to 5.01 cm; **b.** changed from 5 cm to 4.99 cm.

Solution

a.
$$V = f(x) = x^3$$

$$f'(x) = 3x^2$$ Find the derivative of $f(x)$.

$$dV = f'(x)\,dx = 3x^2\,dx$$ Find the differential.

If $x = 5$ (the original length of an edge) and $dx = 0.01$ (the change in length), then

$$dV = 3(5)^2(0.01) = 0.75 \text{ cm}^3.$$

b. Using $x = 5$ and $dx = -0.01$,

$$dV = 3(5)^2(-0.01) = -0.75 \text{ cm}^3.$$

Note that with a calculator, we can find the volume of each cube.

$$V(5) = 5^3 = 125 \text{ cm}^3$$
$$V(5.01) = (5.01)^3 = 125.751501 \text{ cm}^3$$
$$V(4.99) = (4.99)^3 = 124.251499 \text{ cm}^3$$

So our values of $dV = \pm 0.75$ (calculated in parts **a.** and **b.**) are very close to the actual changes in volume.

Example 5: Change in Revenue

Suppose that a company makes and sells x tennis rackets per week, and the corresponding revenue function (in hundreds of dollars) is $R(x) = 20x - \dfrac{x^2}{30}$, where x is between 0 and 600. Use dR to estimate the approximate change in revenue if production is

a. increased from 150 to 160 rackets;   **b.** increased from 480 to 490 rackets.

Solution

a. $R(x) = 20x - \dfrac{x^2}{30}$

$R'(x) = 20 - \dfrac{x}{15}$ Find the derivative of $R(x)$.

$dR = R'(x)\,dx = \left(20 - \dfrac{x}{15}\right)dx$ Find the differential.

Let $x = 150$ and $\Delta x = dx = 160 - 150 = 10$.

$$dR = \left(20 - \frac{150}{15}\right)10$$
$$= (20 - 10)10$$
$$= 100$$

The change in revenue is approximately 100 hundred dollars or $10,000.

b. Let $x = 480$ and $\Delta x = dx = 490 - 480 = 10$.

$$dR = \left(20 - \frac{480}{15}\right)10$$
$$= (20 - 32)10$$
$$= -120$$

The change in revenue is negative 120 hundred dollars, or a loss of $12,000. In this case, the revenue is actually decreasing at a high level of production and sales because the corresponding price of rackets has been lowered to maintain sales.

13.6 EXERCISES

PRACTICE

Find the differential for each of the functions in Exercises 1–14.

1. $y = x^3 + 5$

2. $y = 4x^3 + x - 7$

3. $u = \left(2t^2 + 1\right)^2$

4. $u = \left(5t + 9\right)^2$

5. $A = \pi r^2$

6. $A = x\left(44 - 2x\right)$

7. $V = \dfrac{4}{3}\pi r^3$

8. $V = x^3$

9. $S = 4x^2 + \dfrac{1350}{x}$

10. $C = 75 + 10x - 0.6\sqrt{2x}$

11. $C = 40 + 3x + 0.4\sqrt{x}$

12. $S = 2\pi r^2 + \dfrac{90\pi}{r}$

13. $P = -0.2x^2 + 75x - 2400$

14. $P = -0.3x^2 + 84x - 870$

For each of the functions given in Exercises 15–22, use the given values for x and Δx to find
a. Δy, **b.** dy, and **c.** $\Delta y - dy$.

15. $y = x^2 - 3x + 4$, $x = 3$, $\Delta x = 0.2$

16. $y = x^2 + 5x - 9$, $x = 2$, $\Delta x = 0.15$

17. $y = \left(2x^2 - 4\right)^3$, $x = -2$, $\Delta x = 0.04$

18. $y = \left(x^2 + x - 1\right)^3$, $x = -3$, $\Delta x = 0.05$

19. $y = 20\left(x - \dfrac{24}{x}\right)$, $x = 2$, $\Delta x = -0.12$

20. $y = 2x^2 + \dfrac{125}{x^2}$, $x = 5$, $\Delta x = -0.15$

21. $y = \sqrt{3x + 4}$, $x = 7$, $\Delta x = 0.5$

22. $y = \sqrt{12 - 5x}$, $x = 2$, $\Delta x = 0.07$

Use differentials to approximate the indicated roots in Exercise 23–30. Express your answer as $a \pm \dfrac{b}{c}$, where a is the nearest integer.

23. $\sqrt{37}$

24. $\sqrt{65}$

25. $\sqrt[3]{26}$

26. $\sqrt[3]{126}$

27. $\sqrt{50.4}$

28. $\sqrt{79.5}$

29. $\sqrt[3]{62.3}$

30. $\sqrt[3]{218.3}$

APPLICATIONS

31. **Cost:** A total cost function (in dollars) is given by $C(x) = 375 + 9x + 0.01x^2$. Use differentials to estimate the change in cost when the level of production is increased from 60 to 62 units.

32. **Cost:** A total cost function (in dollars) is given by $C(x) = 930 + 15x + 0.2x^2$. Using differentials, estimate the change in cost from $x = 100$ to $x = 101$.

33. **Profit:** The weekly revenue (in dollars) from the sale of x coffee makers is given by $R(x) = 40x$. The total cost function is given by $C(x) = 370 + 16x + 0.2x^2$. Use differentials to approximate the change in profit if the weekly sales are increased from 25 to 28 coffee makers.

34. **Profit:** The monthly revenue from the sale of x 50-gallon aquariums is given by $R(x) = 54x - 0.3x^2$ dollars. The total cost function is given by $C(x) = 0.1x^2 + 4x + 200$ dollars. Using differentials, find the approximate change in profit if the monthly sales are increased from 40 to 44 aquariums.

35. **Population:** It is estimated that t years from now the population of a city will be $P(t) = 10(4000 + 2t^2) - 1600t$. Use differentials to estimate the change in population as t changes from 6 to 6.25 years.

36. **Bacterial population:** It is estimated that t hours from now the population of bacteria in a culture will be $P(t) = \dfrac{8000}{\sqrt{8 - 0.5t}}$. Use differentials to estimate the change in population as t changes from 8 to 8.3 hours.

37. **Volume:** The edge of a cube measures 18 in. with a possible error in measurement of 0.02 in. Use differentials to estimate the possible error in computing the volume.

38. **Fiberglass coating:** A cube is 12 in. on a side. It is to be covered with a fiberglass coating 0.25 in. thick. Use differentials to estimate the volume of the fiberglass coating.

39. **Measurement error:** A manufacturer of cargo containers receives an order for a cube-shaped container. The specifications state that the volume should be 125 ft³. with a maximum error of no more than 1 ft³. Using differentials, find the possible error in the length of the edges.

40. **Melting ice:** A block of ice is in the form of a 10 in. cube. If it melts uniformly until the volume changes to 972 in.³, approximate the change in the length of each edge by using differentials.

41. **Volume of a weather balloon:** A spherical weather balloon is being inflated. Use differentials to find the approximate change in the volume if the radius changes from 20 to 21.5 inches. (**Hint:** $V = \dfrac{4}{3}\pi r^3$.)

42. **Volume of a tumor:** A spherical cancer tumor is being treated with an experimental drug. The radius of the tumor has been reduced from 1.6 to 1.4 cm. Use differentials to estimate the change in the volume of the tumor. (**Hint:** $V = \dfrac{4}{3}\pi r^3$.)

Chapter 14

INTEGRATION WITH APPLICATIONS

14.1 THE INDEFINITE INTEGRAL

■ TOPICS

- The Antiderivative
- Formulas of Integration
- Distance, Velocity, and Acceleration

The Antiderivative

We have developed several formulas for differentiation and used those formulas to find rates of change, slopes of tangent lines, marginal cost, maxima and minima, and so on. Now we want to consider reversing the process of differentiation. That is, we want to **antidifferentiate**.

If the management of a company knows the current rate of change of cost (i.e., they know the marginal cost function), they can use this knowledge to find the related cost function and proceed to make a reasonably accurate budget projection. With special instruments, scientists can measure the speed of particles along straight lines. That is, they know the derivative of the position function (or velocity), and with this information, they can find the function that will tell the location of the particle at any particular time. In such situations, a derivative is known and the objective is to find the function (or functions) with this derivative.

Antiderivative

If F and f are two functions and $F'(x) = f(x)$ for all x in the domain of f, then F is an **antiderivative** of f.

Note: We also say that $F(x)$ is an antiderivative of $f(x)$.

For example, suppose that

$$P(x) = 4x^2, \quad Q(x) = 4x^2 + 5, \quad \text{and} \quad R(x) = 4x^2 - 100.$$

In each case, we have

$$P'(x) = Q'(x) = R'(x) = 8x.$$

Thus $P(x)$, $Q(x)$, and $R(x)$ are all antiderivatives of $8x$. Any two of these antiderivatives differ from each other by a constant; in fact, all the antiderivatives of $8x$ belong to the family of functions $4x^2 + C$, where C is an arbitrary constant.

Consider the following general analysis. Suppose that $G(x)$ and $F(x)$ are both antiderivatives of $f(x)$. This means that $G'(x) = f(x)$ and $F'(x) = f(x)$. Now let

$$H(x) = G(x) - F(x).$$

Then

$$H'(x) = G'(x) - F'(x) = f(x) - f(x) = 0.$$

If the derivative of $H(x)$ is zero for every x, then we know that $H(x)$ is some constant C because only constant functions have 0 for a derivative. Therefore,

$$G(x) - F(x) = C$$

or

$$G(x) = F(x) + C.$$

Antiderivatives Differ by a Constant

If G and F are both antiderivatives of the same function f, then

$$G(x) = F(x) + C,$$

where C is a constant.

We emphasize that "any two antiderivatives of the same function differ by a constant." Geometrically, this means that the graphs of $y = G(x)$ and $y = F(x)$ are congruent, and the vertical distance between any two points on their graphs is C.

Example 1: Antiderivatives

Show that both $F(x) = x^3 + x^2 + 50$ and $G(x) = x^3 + x^2 - 80$ are antiderivatives of $f(x) = 3x^2 + 2x$.

Solution

We start with differentiating both F and G.

$$F(x) = x^3 + x^2 + 50 \quad \text{and} \quad G(x) = x^3 + x^2 - 80$$
$$F'(x) = 3x^2 + 2x \qquad\qquad G'(x) = 3x^2 + 2x$$

Thus

$$F'(x) = G'(x) = f(x),$$

and F and G are both antiderivatives of f. Note that $F(x) - G(x)$ is a constant.

Formulas of Integration

Now we want to develop a general approach to find antiderivatives. The **indefinite integral** of a function represents the set of all its antiderivatives. This integral is symbolized by an elongated s in the form $\int$.

NOTE

The differential dx indicates that the integral is taken with respect to the variable x. At this time it plays a minor role, but it is a necessary part of the notation.

The Indefinite Integral

If $F(x)$ is any antiderivative of $f(x)$, then the **indefinite integral** of $f(x)$, symbolized by $\int f(x)\,dx$, is defined as

$$\int f(x)\,dx = F(x) + C,$$

where C is an arbitrary constant. Note that $f(x)$ is called the **integrand**, dx is called the **differential**, and C is called the **constant of integration**.

To **integrate** (or **to find the indefinite integral of**) a function, we develop formulas based on reversing the differentiation formulas that we already know. Now we are interested in only four basic integrals and combinations of these integrals. We develop these integrals on an informal basis with the knowledge that their proofs lie in the fact that their derivatives do give the integrands.

Formulas of Integration

I. $\int k\,dx = kx + C$

II. $\int x^r\,dx = \dfrac{1}{r+1} \cdot x^{r+1} + C$, where $r \neq -1$

III. $\int \dfrac{1}{x}\,dx = \int x^{-1}\,dx = \ln|x| + C$ **Note:** The power rule, Formula II, does not apply.

IV. $\int e^x\,dx = e^x + C$

I. $\int k\,dx = kx + C$

Consider the following pattern:

a. If $F(x) = 8x$, then $F'(x) = 8$. Therefore, $\int 8\,dx = 8x + C$.

b. If $F(x) = 3x + 5$, then $F'(x) = 3$. Therefore, $\int 3\,dx = 3x + C$.

c. If $F(x) = 2x - 7$, then $F'(x) = 2$. Therefore, $\int 2\,dx = 2x + C$.

These results imply the following: for any constant k,

$$\int k\,dx = kx + C.$$

II. $\int x^r\,dx = \dfrac{1}{r+1} \cdot x^{r+1} + C$, where $r \neq -1$

a. If $F(x) = \dfrac{1}{2}x^2$, then $F'(x) = \dfrac{1}{\cancel{2}} \cdot \cancel{2}x = x$.

Therefore, $\int x\,dx = \dfrac{1}{2}x^2 + C$.

b. If $F(x) = \dfrac{1}{3}x^3 + 14$, then $F'(x) = \dfrac{1}{\cancel{3}} \cdot \cancel{3}x^2 = x^2$.

Therefore, $\int x^2\,dx = \dfrac{1}{3}x^3 + C$.

c. If $F(x) = -\dfrac{1}{3}x^{-3}$, then $F'(x) = -\dfrac{1}{3}(-3)x^{-4} = x^{-4}$.

Therefore, $\displaystyle\int x^{-4}\,dx = -\dfrac{1}{3}x^{-3} + C$.

Thus, for $r \neq -1$,

$$\int x^r\,dx = \frac{1}{r+1}\cdot x^{r+1} + C.$$

III. $\displaystyle\int \frac{1}{x}\,dx = \int x^{-1}\,dx = \ln|x| + C$

We know that if $x > 0$ and $F(x) = \ln x$, then

$$F'(x) = \frac{1}{x}.$$

However, suppose that $x < 0$ and

$$G(x) = \ln(-x).$$

Then, by the Chain Rule,

$$G'(x) = \frac{1}{-x}\cdot\frac{d}{dx}[-x] = -\frac{1}{x}(-1) = \frac{1}{x}.$$

This means that both

$$F(x) = \ln x \ (\text{for } x > 0)$$

and

$$G(x) = \ln(-x) \ (\text{for } x < 0)$$

have the same derivative, namely, $f(x) = \dfrac{1}{x} = x^{-1}$. Since logarithms are defined only for positive numbers, we use $|x|$ in the following integral. This means that the integral $\displaystyle\int \frac{1}{x}\,dx$ is defined when x is positive or when x is negative.

$$\int \frac{1}{x}\,dx = \int x^{-1}\,dx = \ln|x| + C$$

IV. $\displaystyle\int e^x\,dx = e^x + C$

Since the function $f(x) = e^x$ is its own derivative, a reasonable conclusion is that it is also its own integral. For example, if

$$F(x) = e^x + 7, \text{ then } F'(x) = e^x.$$

Thus we see that

$$\int e^x\,dx = e^x + C.$$

Individually, the four integral formulas we have just discussed are somewhat limiting in their applications. However, the following two rules, essentially reversing two familiar rules of differentiation, allow us to integrate a wide variety of functions.

> ### Constant Multiple Rule
>
> $$\int kf(x)\,dx = k\int f(x)\,dx$$
>
> In words, the integral of a constant times a function is equal to that constant times the integral of the function.

> ### Sum and Difference Rule
>
> $$\int \left[f(x) \pm g(x) \right] dx = \int f(x)\,dx \pm \int g(x)\,dx$$
>
> In words, the integral of a sum (or difference) of two functions is the sum (or difference) of their individual integrals.

The following examples illustrate a variety of applications of the formulas and rules that we have discussed for finding indefinite integrals.

Example 2: Finding the Indefinite Integral

Find $\int 3x^5\,dx.$

Solution

> ### NOTE
>
> Just as the derivative rules apply regardless of the variables used, it is also true that the variable used in integration is a matter of choice.
>
> For example, $\int 3w^5\,dw = \dfrac{1}{2}w^6 + C.$

$$\int 3x^5\,dx = 3\int x^5\,dx \qquad \text{By the Constant Multiple Rule}$$

$$= 3 \cdot \frac{1}{5+1} \cdot x^{5+1} + C \qquad \text{By Formula II}$$

$$= \frac{\cancel{3}}{\cancel{6}\,2}x^6 + C$$

$$= \frac{1}{2}x^6 + C$$

Check

Each integral can be checked by differentiation. Its derivative must be the integrand.

$$\frac{d}{dx}\left[\frac{1}{2}x^6 + C\right] = \frac{1}{2}\cdot 6x^5 + 0 = 3x^5$$

Example 3: Finding the Indefinite Integral of Polynomial Expressions

Find $\int \left(6t^5 - 4t^3 + 10t^2 + 1\right) dt.$

Solution

$$\int \left(6t^5 - 4t^3 + 10t^2 + 1\right) dt = 6\int t^5\, dt - 4\int t^3\, dt + 10\int t^2\, dt + 1\int dt$$

$$= 6\left(\frac{1}{6}t^6\right) - 4\left(\frac{1}{4}t^4\right) + 10\left(\frac{1}{3}t^3\right) + 1(t) + C$$

$$= t^6 - t^4 + \frac{10}{3}t^3 + t + C$$

Check

$$\frac{d}{dt}\left(t^6 - t^4 + \frac{10}{3}t^3 + t + C\right) = 6t^5 - 4t^3 + \frac{10}{3}\left(3t^2\right) + 1 + 0$$

$$= 6t^5 - 4t^3 + 10t^2 + 1 \qquad \text{This is the original integrand.}$$

Example 4: Finding the Indefinite Integral

Find $\displaystyle\int \frac{5}{x}\, dx$.

Solution

$$\int \frac{5}{x}\, dx = 5\int \frac{1}{x}\, dx \qquad \text{By the Constant Multiple Rule}$$

$$= 5\ln|x| + C \qquad \text{By Formula III}$$

Example 5: Finding the Indefinite Integral

Find $\displaystyle\int \frac{1}{\sqrt{y}}\, dy$.

Solution

$$\int \frac{1}{\sqrt{y}}\, dy = \int \frac{1}{y^{\frac{1}{2}}}\, dy = \int y^{-\frac{1}{2}}\, dy \qquad \text{Rewrite using exponents.}$$

$$= \frac{1}{-\frac{1}{2}+1} \cdot y^{-\frac{1}{2}+1} + C \qquad \text{By Formula II}$$

$$= \frac{1}{\frac{1}{2}} y^{\frac{1}{2}} + C$$

$$= 2y^{\frac{1}{2}} + C$$

$$= 2\sqrt{y} + C$$

Example 6: Using Several Formulas for Integration

Find $\int \left(e^x - x^{\frac{2}{5}} + 3\right) dx$.

Solution

$$\int \left(e^x - x^{\frac{2}{5}} + 3\right) dx = \int e^x dx - \int x^{\frac{2}{5}} dx + \int 3 dx$$

By the Sum and Difference Rule

$$= \left(e^x + C_1\right) - \left(\frac{1}{\frac{2}{5}+1} \cdot x^{\frac{2}{5}+1} + C_2\right) + \left(3x + C_3\right)$$

By Formulas IV, II, and I

$$= e^x - \frac{5}{7} x^{\frac{7}{5}} + 3x + C$$

One use of the constant representative C is sufficient since the sum of three constants is simply another constant.

Example 7: Finding the Indefinite Integral of a Rational Function

Find $\int \dfrac{t^3 + t + 1}{t^2} dt$.

Solution

The general approach is to replace a quotient with a polynomial-like expression whenever the denominator is a single term.

$$\int \frac{t^3 + t + 1}{t^2} dt = \int \left(t + \frac{1}{t} + \frac{1}{t^2}\right) dt$$

$$= \int t\, dt + \int \frac{1}{t} dt + \int \frac{1}{t^2} dt$$

$$= \int t\, dt + \int t^{-1} dt + \int t^{-2} dt$$

$$= \frac{1}{2} t^2 + \ln|t| + \frac{1}{-1} t^{-1} + C \quad \text{By Formulas II and III}$$

$$= \frac{t^2}{2} + \ln|t| - \frac{1}{t} + C$$

Example 8: Finding the Constant of Integration

If $F'(x) = x - \dfrac{1}{x}$ and $F(1) = \dfrac{3}{2}$, find $F(x)$.

Solution

Since $F'(x) = x - \dfrac{1}{x}$, we know that

$$F(x) = \int \left(x - \frac{1}{x}\right) dx.$$

Thus

$$F(x) = \int\left(x - \frac{1}{x}\right)dx$$

$$= \int x\,dx - \int \frac{1}{x}\,dx \qquad \text{By the Sum and Difference Rule}$$

$$= \frac{x^2}{2} - \ln|x| + C \qquad \text{By Formulas II and III}$$

Now we use the fact that $F(1) = \frac{3}{2}$ to find the value of C.

$$F(1) = \frac{(1)^2}{2} - \ln|1| + C = \frac{3}{2} \qquad \text{Substitute the given values into } F(x).$$

$$\frac{1}{2} - 0 + C = \frac{3}{2}$$

$$C = 1 \qquad \text{Solve for } C.$$

Therefore,

$$F(x) = \frac{x^2}{2} - \ln|x| + 1.$$

Example 9: Finding the Indefinite Integral

Find $\int 25x^4(6x - 1)\,dx$.

Solution

Multiply the integrand first; then integrate.

$$\int 25x^4(6x - 1)\,dx = \int\left(150x^5 - 25x^4\right)dx$$

$$= \frac{150x^6}{6} - \frac{25x^5}{5} + C = 25x^6 - 5x^5 + C$$

Example 10: Cost Function

A company's marginal cost function is given as $f(x) = 3x + 4$, and fixed costs are known to be $5000. What is the cost function?

Solution

We find the cost function by integrating the marginal cost function, since marginal cost is the derivative of cost. That is, in this case, $f(x) = C'(x)$.

$$C(x) = \int(3x + 4)\,dx = \frac{3}{2}x^2 + 4x + K \qquad \text{In this example, we are using } K \text{ for the constant of integration.}$$

Since fixed costs are $5000, we know that

$$C(0) = 5000,$$

and this allows us to determine a specific value for K, the constant of integration.

$$C(0) = \frac{3}{2}(0)^2 + 4 \cdot 0 + K = 5000$$

Therefore,

$$K = 5000$$

and

$$C(x) = \frac{3}{2}x^2 + 4x + 5000.$$

Distance, Velocity, and Acceleration

Suppose the distance s of a particle from an origin is given by a function of time t. We indicate this by $s = s(t)$.

In this case the velocity v of the particle is given by $v = \dfrac{ds}{dt}$, and the acceleration a is $a = s''(t) = \dfrac{d^2 s}{dt^2}$. Suppose only the acceleration function is known. How would we obtain the distance function $s(t)$?

For example, a meteor 50,000 kilometers away is traveling toward Earth with an acceleration of $-60t + 64$ meters per second per second. Since $a = \dfrac{dv}{dt}$,

$$dv = a \cdot dt.$$

The term dv is the differential of v and we refer to this step as "multiplying" by dt on both sides.

Now, integrating on both sides,

$$\int dv = \int a \cdot dt.$$

Since $v = \int dv$ and $a = -60t + 64$, by substitution, we have

$$v = \int (-60t + 64)\, dt$$

$$= -\overset{30}{\cancel{60}}\left(\frac{1}{\underset{2}{\cancel{2}}}t^2\right) + 64t + C_1$$

$$= -30t^2 + 64t + C_1.$$

Since v is also a function of time t, we see $v(t) = -30t^2 + 64t + C_1$. The constant C_1 can be evaluated only if there is more information. Suppose, with careful observation we measure the speed of the meteor and determine that it is 200 m/s. It is customary and convenient to refer to the time of the initial observation as $t = 0$. Substituting 0 for t, we have $v(0) = -30(0)^2 + 64(0) + C_1$. So $v(0) = C_1$. The constant of integration C_1 is the speed at $t = 0$. Thus $C_1 = -200$ m/s. We use -200 rather than $+200$ since the direction of the meteor is toward the origin (so s is decreasing due to the speed). Therefore,

$$v(t) = -30t^2 + 64t - 200.$$

Since $v = \dfrac{ds}{dt}$, we can say $ds = v \cdot dt$. Integrating again, we have

$$\int ds = \int v \cdot dt = \int \left(-30t^2 + 64t - 200\right) \cdot dt.$$

Integrating a second time gives

$$s = -30\left(\frac{1}{3}t^3\right) + 64\left(\frac{1}{2}t^2\right) - 200t + C_2$$

$$s = -10t^3 + 32t^2 - 200t + C_2.$$

Once again there is a constant of integration to evaluate. If at the time of initial measurement the distance of the meteor from Earth is 50,000 kilometers, then $s(0) = 50,000,000$ meters.

$$50,000,000 = s(0) = -10(0)^3 + 32(0)^2 - 200(0) + C_2 = C_2$$

Here we see $C_2 = 50,000,000$ meters, the initial distance of the meteor from Earth.

It is typical of "acceleration" problems that there are two integrations, two constants of integration to evaluate, and two additional pieces of data necessary for this evaluation. It is common to use the notation v_0 and s_0 to denote initial ($t = 0$) values of velocity and distance.

One case of special interest is a body falling due to Earth's gravity. In this case the acceleration a is a constant. As before, $a = \dfrac{dv}{dt}$ so $dv = a \cdot dt$. Integrating both sides gives

$$v = at + C.$$

The constant C is v_0, the initial velocity. That is, $v = at + v_0$. Since $\dfrac{ds}{dt} = v$, $ds = v \cdot dt$. One more integration gives

$$s = \int ds = \int \left(at + v_0\right) dt = a\left(\frac{1}{2}t^2\right) + v_0 t + s_0$$

$$s = \frac{1}{2}at^2 + v_0 t + s_0.$$

14.1 EXERCISES

💡 PRACTICE

In Exercises 1–12, show that the function $F(x)$ is an antiderivative of the function $f(x)$ by differentiating F.

1. $F(x) = 4x - 1,\ \ f(x) = 4$

2. $F(x) = 6x,\ \ f(x) = 6$

3. $F(x) = 3x^2 + 5x + 2,\ \ f(x) = 6x + 5$

4. $F(x) = \dfrac{1}{2}x^2 - 4x + e^{2x} - 1, \quad f(x) = x - 4 + 2e^{2x}$

5. $F(x) = \ln x - \dfrac{1}{x} - 4e^{x^2}, \quad f(x) = \dfrac{1}{x} + \dfrac{1}{x^2} - 8xe^{x^2}$

6. $F(x) = \ln x^3 + \dfrac{1}{x^2} + 6, \quad f(x) = \dfrac{3}{x} - \dfrac{2}{x^3}$

7. $F(x) = \left(x^2 + 3\right)^4 - 1, \quad f(x) = 8x\left(x^2 + 3\right)^3$

8. $F(x) = 3\left(5x - 1\right)^{\frac{2}{3}} + 8, \quad f(x) = \dfrac{10}{\sqrt[3]{5x - 1}}$

9. $F(x) = \dfrac{5}{3}\left(e^x - 4\right)^3 + e, \quad f(x) = 5e^x\left(e^x - 4\right)^2$

10. $F(x) = 3e^{x^2 - 1} - 7, \quad f(x) = 6xe^{x^2 - 1}$

11. $F(x) = \ln\left(x^2 + 5x - 3\right) - \sqrt{5}, \quad f(x) = \dfrac{2x + 5}{x^2 + 5x - 3}$

12. $F(x) = \ln\left(e^{3x} - x\right) + \sqrt{11}, \quad f(x) = \dfrac{3e^{3x} - 1}{e^{3x} - x}$

Find the indefinite integrals in Exercises 13–32.

13. $\displaystyle\int 7\,dx$

14. $\displaystyle\int \dfrac{2}{3}\,dx$

15. $\displaystyle\int 5x^4\,dx$

16. $\displaystyle\int -2x^{-3}\,dx$

17. $\displaystyle\int \left(x^2 - 3\right)dx$

18. $\displaystyle\int \left(x^4 + 5\right)dx$

19. $\displaystyle\int \left(\dfrac{1}{3} - e^t\right)dt$

20. $\displaystyle\int \left(e^t + t\right)dt$

21. $\displaystyle\int \left(\dfrac{1}{y} + y^3\right)dy$

22. $\displaystyle\int \left(\dfrac{1}{\sqrt{y}} - \dfrac{1}{y}\right)dy$

23. $\displaystyle\int \left(4x^2 + \dfrac{2}{x} + \dfrac{1}{x^2}\right)dx$

24. $\displaystyle\int \left(9x - \dfrac{3}{x} - \dfrac{1}{\sqrt{x}}\right)dx$

25. $\displaystyle\int \left(2\sqrt[3]{x} + 5\sqrt{x}\right)dx$

26. $\displaystyle\int \left(\sqrt{x} + 6e^x - \dfrac{5}{x}\right)dx$

27. $\displaystyle\int \left(4e^y - 2y^5 - \dfrac{1}{5}\right)dy$

28. $\displaystyle\int \left(y^{\frac{3}{2}} + 5y^{-\frac{2}{3}} - y^{-1}\right)dy$

29. $\displaystyle\int \left(\dfrac{2}{\sqrt[3]{t}} + \dfrac{7}{t^3}\right)dt$

30. $\displaystyle\int \left(2t^{\frac{5}{2}} + \dfrac{4}{t} - \sqrt{3}\right)dt$

31. $\displaystyle\int \left(\dfrac{2}{3}e^x + x^{-\frac{3}{2}} - 7x^{-1}\right)dx$

32. $\displaystyle\int \left(0.25e^x + 4x^{-\frac{1}{4}}\right)dx$

In Exercises 33–38, perform the indicated multiplication and then integrate.

33. $\int x^3 (2x-1)\, dx$

34. $\int x^2 (3x-5)\, dx$

35. $\int (3t+2)^2\, dt$

36. $\int (5x+6)^2\, dx$

37. $\int \sqrt{y}\,(y^2 + 2y - 1)\, dy$

38. $\int \sqrt{y}\,(4 - 3y - 2y^2)\, dy$

In Exercises 39–44, simplify the indicated quotient and then integrate.

39. $\int \dfrac{3x^2 + 5x - 4}{x^2}\, dx$

40. $\int \dfrac{x^3 - 6x^2 + x}{x^2}\, dx$

41. $\int \dfrac{4 + \sqrt{x} - 3x}{x}\, dx$

42. $\int \dfrac{5x^2 - 2x + 3}{\sqrt{x}}\, dx$

43. $\int \dfrac{x^{\frac{3}{2}} + 6 - 2xe^x}{x}\, dx$

44. $\int \dfrac{4x^2 + 4\sqrt{x} - 7x}{x^2}\, dx$

45. Find the antiderivative $F(x)$ that satisfies the given condition.

    a. $F'(x) = x^2 - e^x,\ \ F(0) = 1$

    b. $F'(x) = 6x^2 + x - 10,\ \ F(0) = 0$

    c. $\dfrac{dF}{dx} = \dfrac{10}{\sqrt{x}},\ \ F(1) = 20$

    d. $\dfrac{dF}{dx} = 6e^x - 2,\ \ F(0) = -10$

46. Given that $f'(x) = x^2 - 2$, determine the function $f(x)$ with the given constant of integration C. Draw all three functions on the same coordinate system.
    a. $C = -1$
    b. $C = 1$
    c. $C = 3$

47. Given $f'(x) = 6x^2 - 24x$.
    a. Determine the x-values at which $f(x)$ has a local maximum or minimum.
    b. Determine whether there is an inflection point.
    c. Given $f(1) = -9$, sketch f and determine if the answers to part a. are correct.

🚀 APPLICATIONS

48. Cost: The weekly marginal cost of producing x ice cream makers is $28 + 0.05x$ dollars per ice cream maker. Find the cost function if the fixed costs are \$2400.

49. Cost: The marginal cost of producing x clock radios is $0.3x^2 - 0.8x + 24$ dollars per clock radio. The fixed costs are \$1500. Find the cost function.

50. Revenue: The marginal revenue from selling x irons is $94 - 0.06x$ dollars per iron. Find the revenue function. (**Hint:** $R(0) = 0$.)

51. Revenue: The marginal revenue from selling x floor lamps is $100 - 0.2x$ dollars per lamp. Find the revenue function. (**Hint:** $R(0) = 0$.)

52. **Profit:** The marginal profit from the production and sale of x cameras is estimated to be $24 - 0.4x$ dollars per unit.
    a. Find the firm's profit function if the profit from the production and sale of 80 units is $240.
    b. What is the profit from the sale of 90 units?

53. **Profit:** A manufacturer has determined that the marginal profit from the production and sale of x wireless speakers is approximately $120 - 3x$ dollars per speaker.
    a. Find the profit function if the profit from the production and sale of 30 speakers is $1200.
    b. What is the profit from the sale of 40 speakers?

54. **Population:** The population of a community is growing at a rate given by $\dfrac{dP}{dt} = 120 - 15t^{\frac{1}{2}}$ people per year. Find a function to describe the population t years from now if the present population is 8600 people.

55. **Rodent control:** Animal control officers have implemented a program to eliminate rats in a community. They estimate that the population of rats is changing at a rate of $\dfrac{dP}{dt} = 24t^{\frac{1}{2}} - 40t$ rats per month. Find a function for the rat population t months from now if the current population is estimated to be 6300 rats.

56. **Air quality:** The air quality control office estimates that for a population of x thousand people, the level of pollution in the air is increasing at a rate of $\dfrac{dL}{dx} = 0.2 + 0.002x$ parts per million per thousand people. Find a function to estimate the level of the pollutants if the level is 5.4 parts per million when the population is 20,000 people.

57. **Ecology:** Biologists are treating a stream contaminated with bacteria. The level of contamination is changing at a rate of $\dfrac{dN}{dt} = -\dfrac{960}{t^2} - 240$ bacteria per cubic centimeter per day, where t is the number of days since the treatment began. Find a function $N(t)$ to estimate the level of contamination if the level after 1 day was about 5720 bacteria per cubic centimeter.

58. **Height:** An object is projected vertically so that the velocity after t seconds is given by $v(t) = 96 - 32t$ feet per second.
    a. Find the height function $s(t)$ if $s(0) = 18$ feet.
    b. What will be the height after 3 seconds?

59. **Distance:** A vehicle travels in a straight line for t minutes with a velocity $v(t) = 72t - 6t^2$ feet per minute, for $0 \le t \le 10$.
    a. Find the distance function $s(t)$ if $s(0) = 0$.
    b. How far will the vehicle travel in 5 minutes?
    c. How far will the vehicle travel in 10 minutes?

60. Distance: A particle moves along an axis with velocity given by $v(t) = 3t - 1$, where t is in seconds and v is in ft/s.

   a. Determine the acceleration, $a(t)$.

   b. What is the distance function, $s(t)$, if $s(0) = 5$?

61. Distance: A meteor falls partly under the influence of Earth's gravity at a velocity given by $v(t) = 200 + 30t + 24t^{\frac{1}{2}}$ for $0 \le t \le 24$, where t is in hours and v is in miles per hour.

   a. Determine the acceleration.

   b. Determine the distance function if $s(0) = 5000$ miles.

✎ WRITING & THINKING

62. a. Compute the derivative of $y = e^{mx+b}$ where m and b are constants. Use your answer to determine an integration formula for $\int e^{mx+b}\, dx$.

   b. Compute the derivative of $y = \dfrac{1}{e^{mx+b}}$ where m and b are constants. Use your answer to determine an integration formula for $\int \dfrac{1}{e^{mx+b}}\, dx$.

14.2 INTEGRATION BY SUBSTITUTION

In this section we will develop a method of integration for finding integrals for which integration formulas or rules can not be used directly, such as the following:

$$\int (x^2 + 4)^6 \, 2x \, dx, \quad \int e^{x^2} \, 2x \, dx, \quad \text{or} \quad \int \frac{2t}{t^2 + 1} \, dt.$$

This method is called **integration by substitution**.

The concept of substitution can be used in formulas for differentiation. For example, if

$$f(x) = e^{g(x)}, \text{ then } f'(x) = e^{g(x)} g'(x).$$

Therefore, for

$$f(x) = e^{x^2 + 3x}, \text{ we have } f'(x) = e^{x^2 + 3x}(2x + 3).$$

In this case we have "substituted" $x^2 + 3x$ for $g(x)$ and $(2x + 3)$ for $g'(x)$.

Similarly, the General Power Rule states that if

$$f(x) = [g(x)]^n, \text{ then } f'(x) = n[g(x)]^{n-1} g'(x).$$

Thus, for

$$f(x) = (x^2 + 4)^7, \text{ we have } f'(x) = 7(x^2 + 4)^6 2x.$$

In this case we have substituted $x^2 + 4$ for $g(x)$ and $2x$ for $g'(x)$.

The four formulas for integration with function notation can be restated with the variable x replaced by $g(x)$ and the differential dx replaced by $g'(x)dx$. These formulas can be verified by using the Chain Rule to show that the derivative of each function on the right is equal to the integrand.

Formulas of Integration with Function Notation

1. $\int kg'(x)\,dx = kg(x) + C$

2. $\int [g(x)]^r g'(x)\,dx = \dfrac{1}{r+1}[g(x)]^{r+1} + C, \text{ where } r \neq -1$

3. $\int \dfrac{g'(x)\,dx}{g(x)} = \int \dfrac{1}{g(x)} g'(x)\,dx = \int [g(x)]^{-1} g'(x)\,dx = \ln|g(x)| + C$

4. $\int e^{g(x)} g'(x)\,dx = e^{g(x)} + C$

These same formulas for integration can be restated again where substitutions (called **u-substitutions**) are made as follows:

$$u = g(x) \quad \text{and} \quad du = g'(x)dx.$$

These forms indicate three particularly informative ideas about integration formulas:

1. The letter x that we used previously in integration formulas is a placeholder. Any other letter, such as u, v, y, or t, can be used.

2. Any substitution made in an integrand requires a related substitution for the differential such as du, dv, dy, or dt.

3. The u-substitution method will be appropriate to use when, in order to check the correctness of the antiderivative found, one must use the Chain Rule in some form.

There is no product rule for integration.

Formulas of Integration with u-Substitution

1. $\int k\,du = ku + C$

2. $\int u^r\,du = \dfrac{1}{r+1}u^{r+1} + C$, where $r \neq -1$

3. $\int \dfrac{du}{u} = \int \dfrac{1}{u}\,du = \int u^{-1}\,du = \ln|u| + C$

4. $\int e^u\,du = e^u + C$

Now consider the integral listed first in the introduction to this section:

$$\int \left(x^2 + 4\right)^6 2x\,dx.$$

Our goal is to make substitutions in the integrand and differential so that the resulting integral will be one that we recognize in the list of integration formulas. Our choice (for it is a choice) is based on this goal. If this substitution does not lead to a recognizable integral form, then we make another choice.

If we let $u = x^2 + 4$, then $du = 2x\,dx$.   We know that if $u = g(x)$, then $du = g'(x)dx$.

If we make these substitutions, the integral takes the form of one of the formulas we know.

$$\int \left(x^2 + 4\right)^6 2x\,dx = \int u^6\,du \qquad \text{Substituting } u \text{ and } du$$

$$= \frac{1}{7}u^7 + C \qquad \text{Using Formula II}$$

$$= \frac{1}{7}\left(x^2 + 4\right)^7 + C \qquad \text{Substitute } x^2 + 4 \text{ for } u \text{ so that the answer is in the terms of the original variable } x.$$

We have just performed **integration by substitution**, which involves the following two substitutions:

1. If $u = g(x)$,   In the discussion above, we let $u = x^2 + 4$ and $du = 2x\,dx$.

2. then $du = g'(x)dx$.

The following examples illustrate integration by substitution in a variety of situations. Study these examples carefully.

Example 1: Integration by Substitution with Rational Expressions

Find $\int \dfrac{2t}{t^2 + 1}\,dt$.

Solution

Let $u = t^2 + 1$. Then $du = 2t\,dt$.

In this case $2t\,dt$ is in the integral expression just as we need it. So no adjustments are necessary.

$$\int \frac{2t}{t^2+1}\,dt = \int \frac{du}{u}$$ Make the substitutions.

$$= \int u^{-1}\,du$$ Rewriting the integrand emphasizes that the exponent is -1.

$$= \ln|u| + C$$ By Formula III

$$= \ln|t^2+1| + C$$ Substitute in $u = t^2 + 1$.

$$= \ln(t^2+1) + C$$ In this problem, the absolute value symbols are not necessary because $t^2 + 1 > 0$ for all t.

Example 2: Integration by Substitution with Radicals

Find $\int \sqrt{5x+1}\,dx$.

Solution

Choose $u = 5x + 1$. Then $\dfrac{du}{dx} = 5$ and, multiplying by dx on both sides, we get $du = 5\,dx$.

Since $du = 5\,dx$ and the constant factor 5 is not part of the original integrand, multiply the integrand by 1 in the form $\dfrac{1}{5}\cdot 5$ as follows.

$$\int \sqrt{5x+1}\,dx = \int (5x+1)^{\frac{1}{2}}\cdot \frac{1}{5}\cdot 5\,dx$$ Use fractional exponents.

$$= \int u^{\frac{1}{2}}\cdot \frac{1}{5}\cdot du$$ Substitute.

$$= \frac{1}{5}\int u^{\frac{1}{2}}\,du$$ By the Constant Multiple Rule

$$= \frac{1}{5}\cdot \frac{1}{\frac{1}{2}+1}\cdot u^{\frac{1}{2}+1} + C$$ By Formula II

$$= \frac{1}{5}\cdot \frac{2}{3}\cdot u^{\frac{3}{2}} + C$$ Simplify.

$$= \frac{2}{15}(5x+1)^{\frac{3}{2}} + C$$ Substitute $5x + 1$ for u.

A slightly different approach: "Solve" $du = 5\,dx$ for dx; we get $dx = \dfrac{1}{5}\,du$. Then substitute this with $u = (5x + 1)$, resulting directly in the integral $\int\left(u^{\frac{1}{2}}\cdot \dfrac{1}{5}\,du\right)$.

⚠ **CAUTION**

The technique used in Example 2, multiplying by 1 in the form $\dfrac{1}{5}\cdot 5$ and then factoring out $\dfrac{1}{5}$ from the integral, is valid because $\dfrac{1}{5}$ is a constant. The same technique is not valid with variables. For example, inserting $\dfrac{1}{x}\cdot x$ is valid (if $x \neq 0$), but then factoring out $\dfrac{1}{x}$ from the integral is not valid.

Example 3: Integration by Substitution with e

Find $\int 7xe^{x^2+3}\,dx$.

Solution

Let $u = x^2 + 3$. Then $du = 2x\,dx$.

We adjust for the 2 by introducing $\frac{1}{2} \cdot 2$ into the integrand.

$$\int 7xe^{x^2+3}\,dx = 7\int e^{x^2+3}\cdot\frac{1}{2}\cdot 2x\,dx$$

Use the Constant Multiple Rule with 7, and multiply by $\frac{1}{2}\cdot 2 = 1$ since we need $2x\,dx$ for du.

$$= \frac{7}{2}\int e^u\,du$$

Substitute and use the Constant Multiple Rule with $\frac{7}{2}$.

$$= \frac{7}{2}e^u + C$$

By Formula IV

$$= \frac{7}{2}e^{x^2+3} + C$$

Example 4: Integration by Substitution

Find $\displaystyle\int \frac{x^2}{\left(x^3-1\right)^5}\,dx$.

Solution

Choose $u = x^3 - 1$. Then $du = 3x^2\,dx$.

$$\int \frac{x^2}{\left(x^3-1\right)^5}\,dx = \int \left(\frac{\frac{1}{3}\cdot 3x^2\,dx}{\left(x^3-1\right)^5}\right)$$

Multiply the integrand by $\frac{1}{3}\cdot 3 = 1$ to get $3x^2\,dx$.

$$= \frac{1}{3}\int \frac{du}{u^5}$$

By the Constant Multiple Rule

$$= \frac{1}{3}\int u^{-5}\,du$$

$$= \frac{1}{3}\cdot\frac{1}{(-5+1)}\cdot u^{-5+1} + C$$

By Formula II

$$= -\frac{1}{12}u^{-4} + C$$

$$= -\frac{1}{12}\left(x^3-1\right)^{-4} + C$$

$$= -\frac{1}{12\left(x^3-1\right)^4} + C$$

> **✎ NOTE**
>
> We have used the letter u throughout this section. This particular letter is commonly used in most calculus courses, and the substitution technique is sometimes known as the **u-substitution** technique.
>
> A rule of thumb that you may have observed while using this technique is to substitute u for
>
> 1. the exponent in an exponential function,
>
> 2. an expression in parentheses,
>
> 3. a denominator, or
>
> 4. an expression under a radical sign.

The following example illustrates *an incorrect choice* for substitution. Remember, if a substitution does not result in an integral that matches one of the basic formulas we know, then try another substitution.

Example 5: Incorrect Integration by Substitution

Find $\displaystyle\int 3x^2\sqrt{x^3+1}\,dx$.

Solution

Let $u = 3x^2$. Then $du = 6xdx$. This substitution will not work because the expression $\sqrt{x^3 + 1}\,dx$ is not du.

A better substitution is to let u represent the expression under the radical sign. Let $u = x^3 + 1$. Then $du = 3x^2dx$. Now substitution gives a familiar form.

$$\int 3x^2\sqrt{x^3+1}\,dx = \int \left(x^3+1\right)^{\frac{1}{2}} \cdot 3x^2\,dx \quad \text{Rewrite using exponents.}$$

$$= \int u^{\frac{1}{2}}\,du \qquad\qquad \text{Substitute.}$$

$$= \frac{1}{\frac{1}{2}+1}u^{\frac{1}{2}+1} + C \qquad \text{By Formula II}$$

$$= \frac{2}{3}u^{\frac{3}{2}} + C \qquad\qquad \text{Simplify.}$$

$$= \frac{2}{3}\left(x^3+1\right)^{\frac{3}{2}} + C \qquad \text{Substitute } x^3 + 1 \text{ for } u.$$

Example 6: Incorrect Integration by Substitution

Find $\int \left(x^2+5\right)^{10} dx$.

Solution

Let $u = x^2 + 5$. Then $du = 2xdx$ and $u^{10} = (x^2 + 5)^{10}$.

This substitution will not work. The differential is $dx = \dfrac{du}{2x}$. The factor $2x$ is not present in the original problem.

Actually, in this example, the method cannot be made to work by any u-substitution.

Integrals of the particular form in Example 6 can be worked by a method called trigonometric substitution, which is not discussed in this course. We do not suggest expanding the polynomial and integrating term by term!

There are many problems that require other methods, and there are some functions that cannot be integrated in "closed form." This means there is no simple formula for the antiderivative.

14.2 EXERCISES

PRACTICE

In Exercises 1–36, use the technique of substitution to perform each integration.

1. $\displaystyle\int (x+4)^7\, dx$

2. $\displaystyle\int (y-6)^{-3}\, dy$

3. $\displaystyle\int 2x\left(x^2-1\right)^{\frac{1}{2}}\, dx$

4. $\displaystyle\int 3x^2\left(x^3-5\right)^{\frac{1}{3}}\, dx$

5. $\displaystyle\int \frac{1}{t+2}\, dt$

6. $\displaystyle\int \frac{1}{x-11}\, dx$

7. $\displaystyle\int \frac{3t^2}{t^3+4}\, dt$

8. $\displaystyle\int \frac{1}{y^2-8}\cdot 2y\, dy$

9. $\displaystyle\int e^{y+5}\, dy$

10. $\displaystyle\int e^{x-9}\, dx$

11. $\displaystyle\int e^{-0.2x}\left(-0.2\right)\, dx$

12. $\displaystyle\int e^{0.5t}\left(0.5\right)\, dt$

13. $\displaystyle\int \frac{1}{5x+3}\, dx$

14. $\displaystyle\int (3y-2)^{-2}\, dy$

15. $\displaystyle\int \frac{1}{\sqrt{4x-1}}\, dx$

16. $\displaystyle\int e^{-4x}\, dx$

17. $\displaystyle\int xe^{2x^2}\, dx$

18. $\displaystyle\int \frac{x}{2x^2+5}\, dx$

19. $\displaystyle\int \frac{x}{\left(3x^2-1\right)^2}\, dx$

20. $\displaystyle\int y^2 e^{-2y^3}\, dy$

21. $\displaystyle\int \frac{2t+1}{t^2+t-4}\, dt$

22. $\displaystyle\int \left(x^2+3x-1\right)^4 (2x+3)\, dx$

23. $\displaystyle\int 5ye^{-y^2}\, dy$

24. $\displaystyle\int \frac{e^{\sqrt{t}}}{\sqrt{t}}\, dt$

25. $\displaystyle\int 6x\sqrt{5+2x^2}\, dx$

26. $\displaystyle\int \frac{e^t}{e^t-1}\, dt$

27. $\displaystyle\int \frac{4}{x^2}e^{\frac{1}{x}}\, dx$

28. $\displaystyle\int 4y\sqrt[3]{3y^2+7}\, dy$

29. $\displaystyle\int e^{2x}\left(1-3e^{2x}\right)^2\, dx$

30. $\displaystyle\int \frac{4x+10}{x^2+5x+2}\, dx$

31. $\displaystyle\int \frac{\ln x}{x}\, dx$

32. $\displaystyle\int \frac{\ln 4x}{x}\, dx$

33. $\displaystyle\int \frac{1}{x \ln x}\,dx$

34. $\displaystyle\int \frac{(\ln x)^2}{x}\,dx$

35. $\displaystyle\int \frac{e^x - e^{-x}}{e^x + e^{-x}}\,dx$

36. $\displaystyle\int \frac{7x}{e^{x^2}}\,dx$

In Exercises 37–40, divide first and then integrate.

37. a. $\displaystyle\int \frac{x-1}{x-2}\,dx$ $\qquad \left(\textbf{Hint: } \dfrac{x-1}{x-2} = 1 + \dfrac{1}{x-2}\right)$

b. $\displaystyle\int \frac{x+3}{x+1}\,dx$ $\qquad \left(\textbf{Hint: } \dfrac{x+3}{x+1} = 1 + \dfrac{2}{x+1}\right)$

38. a. Rework Exercise 37a and substitute $u = x - 2$.
b. Rework Exercise 37b and substitute $u = x + 1$.

39. $\displaystyle\int \frac{x+2}{x+4}\,dx$

40. $\displaystyle\int \frac{x+5}{x-3}\,dx$

In Exercises 41–44, find a function $f(x)$ given $f'(x)$ and one (x, y)-value.

41. $f'(x) = 10xe^{5x^2}$; $(0, 11)$

42. $f'(x) = \dfrac{3}{x-5}$; $(6, 5)$

43. $f'(x) = (2x+2)^5$; $(-1, 10)$

44. $f'(x) = \dfrac{12x^2}{\left(x^3 + 6\right)^2}$; $\left(0, -\dfrac{2}{3}\right)$

🚀 APPLICATIONS

45. Appreciation: The value V of a painting is increasing at a rate of $4500(25 - 1.8t)^{-\frac{3}{2}}$ dollars per year. The painting originally sold for \$1000.
a. Write a function for its value t years after the original sale.
b. What will the painting be worth 5 years after the original sale?

46. Skills development: It is estimated that after t weeks of practice, students in a typing class can increase their speed $24e^{-0.2t}$ words per minute for each additional week of practice.
a. If their initial speed was 0 words per minute ($S(0) = 0$), write a function to represent their speed after t weeks of practice.
b. How fast can they type after 10 weeks (to the nearest word)?

47. Price: After the NBA championship basketball game, the price of a souvenir cap changes at the rate of $-\dfrac{1}{2x}$ dollars per cap, where x is the number of caps sold (in thousands). Write a function for the price p if $p(10) = 12.85$.

48. Profit: The marginal profit from the production and sale of x units of a product is estimated to be $\dfrac{100}{1+0.5x} - 4.5$ dollars per unit. If $P(0) = -60$, find the profit function $P(x)$.

49. Daily production: Records show that t hours after starting work on a typical day, an employee can assemble bikes at a rate of $\dfrac{6t+15}{\sqrt{t^2+5t}}$ per hour. Find the daily production function $N(t)$ if $N(0)=0$.

50. Position of a particle: A particle is moving in a straight line with the velocity $v(t)=\sqrt{2t+7}$ feet per second, where t represents time in seconds. Find the position function $s(t)$ if $s(0)=0$.

51. Position of a projectile: The velocity of a projectile moving in a straight line t seconds after it is fired is given by $v(t)=36+\dfrac{60}{(t+1)^2}$ feet per second. Find the position function $s(t)$ if $s(1)=10$.

52. The marginal value for a tract of land is $V'=\dfrac{20}{2t+1}$, where t is time in years since 2000 and V is in thousands of dollars.
  a. Determine the function V in terms of t if $V(0)=20$.
  b. What was the value of the land in 2015?

53. A point mass moving on a horizontal axis has a deceleration given by $a(t)=-48(2t+1)^2$, where t is time in seconds and a is in feet per second per second.
  a. If $v(0)=6$ ft/s, determine a velocity function $v(t)$.
  b. If $s(0)=-\dfrac{3}{4}$ feet, determine a distance function $s(t)$.

54. The number of viewers of a new TV show grew at a rate $V'(t)=900e^{0.3t-4}$ all summer long, where V is the total number of viewers in thousands and t is the number of weeks since June 1st.
  a. Determine $V(t)$ if $V(0)=100$.
  b. At what rate is V changing when $t=5$?
  c. When will the show hit 1,000,000 viewers ($V=1000$)?

55. Suppose in a medieval country, from 1200 to 1300, the life expectancy of a female serf changed at the rate $f'(t)=\dfrac{0.3}{1+0.01t}$, where t is time in years after 1200 and $f(t)$ is the average age of death.
  a. What are the units of $f'(t)$?
  b. Determine $f(t)$ if $f(0)=30$.
  c. What was the life expectancy of a female serf in 1300?

56. A new evening school is growing at the rate of $p'(t)=\dfrac{150}{\sqrt{1+0.2t}}$, where $p(t)$ is the total evening school student population and t is the time in years after 2000. The initial enrollment was 1500 students.
  a. How fast was enrollment changing in 2002?
  b. What was the expected enrollment in 2005?

14.3 AREA AND RIEMANN SUMS

■ TOPICS

- ■ Area under a Curve
- ■ Riemann Sums

Area under a Curve

We remind the reader that the area of a planar region is defined simply as the number of unit squares (1-by-1 squares) contained inside the region. It follows that the area of a rectangle is the product of its base and height. In this section we will try to calculate, using calculus, areas bounded by continuous curves. Our general method is to consider the problem of "filling" a region in a coordinate plane with rectangles. The idea of using polygons to calculate areas bounded by curves is very old. Archimedes made very accurate estimates of π by filling circles with regular polygons with ever increasing numbers of sides, and he calculated the exact area of a section of a parabola by filling the area with infinitely many triangles (of uniformly decreasing size). Kepler used triangles to determine his famous Second Law of Planetary Motion; the ray from the sun to a planet sweeps out equal areas in equal times. Amazingly, he did this 50 years before the invention of calculus.

There is another very direct way to approach areas. We now consider the problem from another point of view. Let us consider what the area might represent. Suppose $f'(t)$ is the velocity of an object at time t and the graph of $f'(t)$ is that in Figure 1. We are interested in the area in quadrant I under the curve from $t = a$ to $t = b$. Imagine a rectangle with base Δt and with height $f'(\hat{t})$ where $\hat{t}$ is some t-value in the interval (a, b). The area of this rectangle is $f'(\hat{t})\Delta t$. Suppose $f'(\hat{t}) = 30$ meters/second and that $\Delta t = 2$ seconds. The area is then $\left(30 \ \dfrac{\text{meters}}{\text{second}}\right)(2 \text{ seconds}) = 60$ meters.

That is, the numerical value of the rectangular area represents a close approximation to the distance traveled during the 2 seconds.

The exact area represents the exact distance traveled. The distance function $f(t)$ gives the distance from the origin at time t, and the number $f(b) - f(a)$ is the net distance traveled from time $t = a$ to $t = b$. This shows that, from the numerical point of view, this net distance is exactly the numerical value of the area under the graph of $f'(t)$.

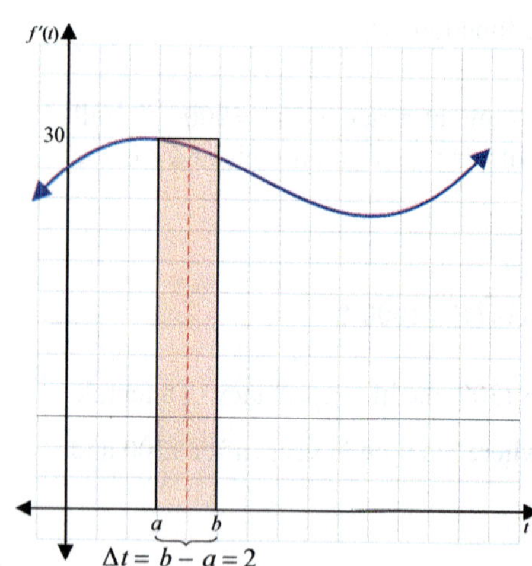

FIGURE 1

Having discovered what the area represents, we have also discovered a way to calculate it! Just use $f(t)$. This is part of what is called the **Fundamental Theorem of Calculus**, namely that the area under any "nice" curve, say $g(x)$, from $x = a$ to $x = b$, can be calculated using any antiderivative $G(x)$ of $g(x)$. The area will be $G(b) - G(a)$.

Determining which curves this applies to and proving the details in a formal way was not worked out satisfactorily until in the early 1800s—after 250 years of using calculus! Example 1 demonstrates how the theorem is used to find the area under a curve.

Use the Fundamental Theorem of Calculus to determine the area under the graph of $f(x) = 30 - 20x + x^5$ and over the x-axis interval $[1, 2]$.

Solution

First, we must find the antiderivative, $F(x)$.

$$F(x) = 30x - 10x^2 + \frac{1}{6}x^6 \quad \text{Note: } \frac{d}{dx}F(x) = \frac{d}{dx}\left(30x - 10x^2 + \frac{1}{6}x^6\right)$$

$$= 30 - 20x + x^5$$

$$= f(x)$$

Using the antiderivative, solve for $F(b) - F(a)$. Use $a = 1$ and $b = 2$.

$$F(2) - F(1) = \left[30(2) - 10(2)^2 + \frac{(2)^6}{6}\right] - \left[30(1) - 10(1)^2 + \frac{(1)^6}{6}\right]$$

$$= [60 - 40 + 10.667] - [20.167]$$

$$= 10.5$$

The area is 10.5 square units.

Calculations of other areas, the use of graphing utilities, and different applications of the Fundamental Theorem are deferred to the next section. In the next subsection we digress to present an introduction to the modern approach to integral calculus using the Fundamental Theorem of Calculus. An overall view of this approach is shown in Figure 2.

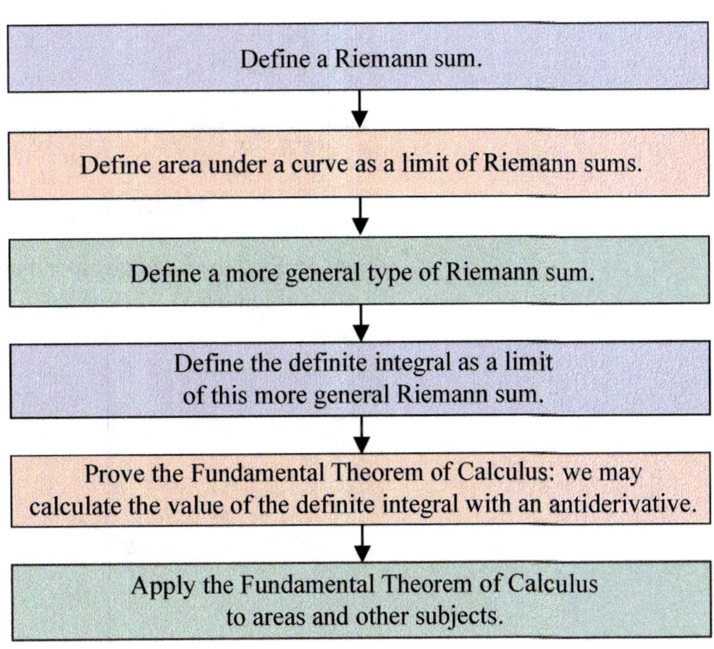

FIGURE 2

We return to the problem of using rectangles to estimate the area between a curve and the x-axis when the curve is above the x-axis and x is restricted to some closed interval $[a, b]$. This approach to the theoretical development of definite integrals is credited to Bernhard Riemann (1826–1866).

Riemann Sums

Suppose that $y = f(x)$ is a nonnegative continuous function on the interval $[a, b]$ and our objective is to estimate the area under the curve (i.e., the area between the curve and the x-axis). (See Figure 3.)

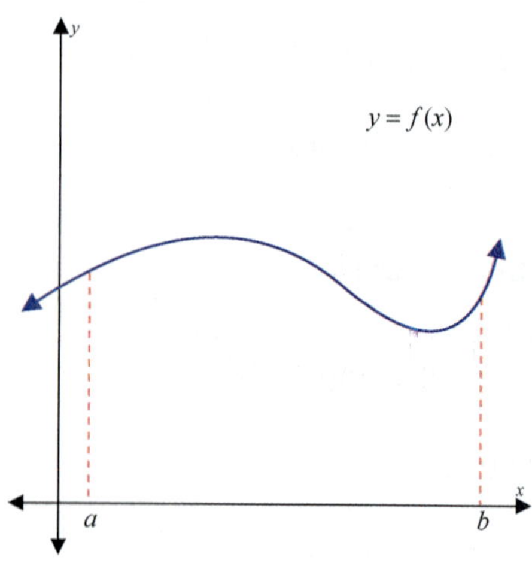

FIGURE 3

We estimate the area by summing the areas of adjacent rectangles as illustrated in the following discussion. In this step-by-step development, we use only five rectangles for simplicity.

Step 1: Partition the interval $[a, b]$ into $n = 5$ subintervals, each of width $\Delta x = \dfrac{b - a}{n} = \dfrac{b - a}{5}$ with endpoints $a = x_0,\ x_1,\ x_2,\ x_3,\ x_4,$ and $x_5 = b$. (See Figure 4.)

Step 2: Choose one value for x in each subinterval. In Figure 5, the chosen values are labeled $c_1, c_2, c_3, c_4,$ and c_5.

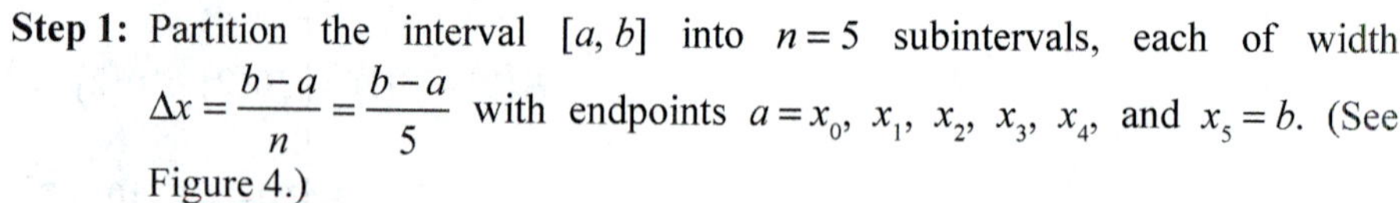
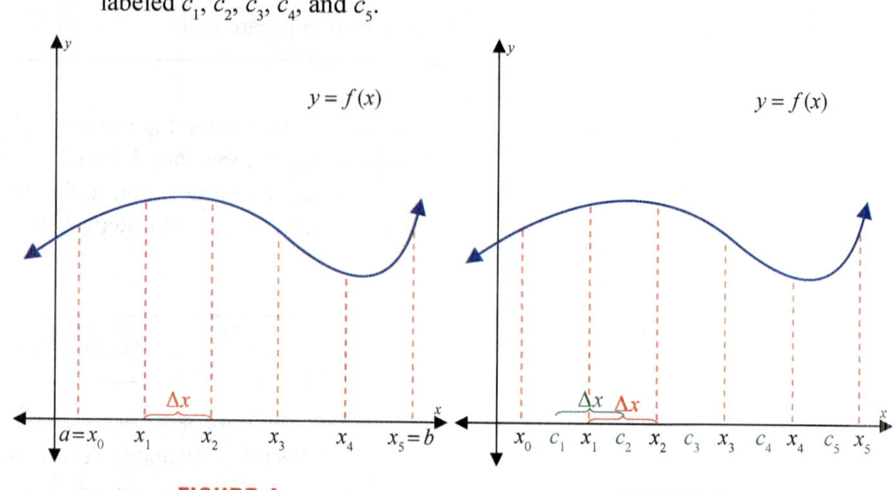

FIGURE 4   **FIGURE 5**

Step 3: Find the five corresponding y-values: $f(c_1), f(c_2), f(c_3), f(c_4),$ and $f(c_5)$. (See Figure 6.)

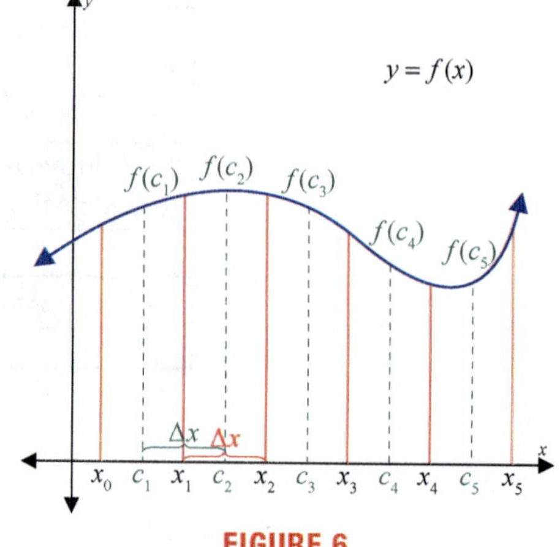

FIGURE 6

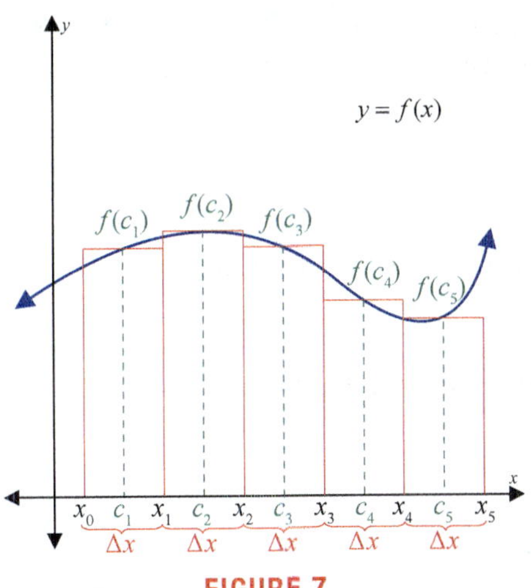

FIGURE 7

Step 4: Now, as illustrated in Figure 7, we treat each of these y-values as the height of a rectangle of width Δx. We multiply each y-value by Δx to find the areas of the rectangles and then form the sum of the products. This sum is called a **Riemann sum**, and it is an approximation of the area between the curve and the x-axis.

Here area $= f(c_1)\Delta x + f(c_2)\Delta x + \cdots + f(c_5)\Delta x$. The use of $n = 5$ subintervals in the previous discussion was purely arbitrary. Any finite number of subintervals could have been used, as we indicate in the following statement of the general form of a Riemann sum.

General Form of a Riemann Sum

The **general form of a Riemann sum** for a function $y = f(x)$, continuous on $[a, b]$, is

$$S_n = [f(c_1) + f(c_2) + \cdots + f(c_n)]\Delta x,$$

where $\Delta x = \dfrac{b-a}{n}$, n is the number of subintervals of $[a, b]$, and each of the numbers $c_1, c_2, \ldots, c_n$ represents one x-value from each subinterval.

Example 2: Riemann Sum

Use a Riemann sum to estimate the area between $f(x) = \dfrac{1}{x}$ and the x-axis on the interval $[a, b] = [1, 5]$ with $n = 4$.

Solution

$$\Delta x = \frac{b-a}{n} = \frac{5-1}{4} = \frac{4}{4} = 1 \qquad \text{First, we find } \Delta x \text{ by substituting the given values } a = 1, b = 5, \text{ and } n = 4.$$

In this example we use the midpoint of each subinterval for the values of c_1, c_2, c_3, and c_4.

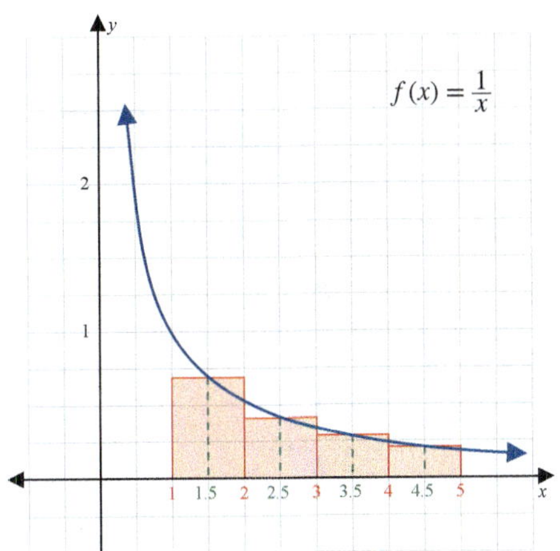

$$S_4 = \left[f(c_1) + f(c_2) + f(c_3) + f(c_4)\right]\Delta x$$

$$= \left(\frac{1}{1.5} + \frac{1}{2.5} + \frac{1}{3.5} + \frac{1}{4.5}\right) \cdot 1 \qquad \text{Substitute the appropriate values into the general form of the Riemann sum.}$$

$$\approx 0.666667 + 0.4 + 0.285714 + 0.222222$$

$$\approx 1.57$$

Use a Riemann sum to estimate the area between $y = x^2 + 1$ and the x-axis on the interval $[a, b] = [-1, 2]$ with $n = 6$.

Solution

$$\Delta x = \frac{2 - (-1)}{6} = \frac{3}{6} = 0.5$$

First, we find Δx by substituting the given values $a = -1$, $b = 2$, and $n = 6$.

In this example we use the left endpoint of each subinterval for the values of $c_1, c_2, c_3, c_4, c_5,$ and c_6.

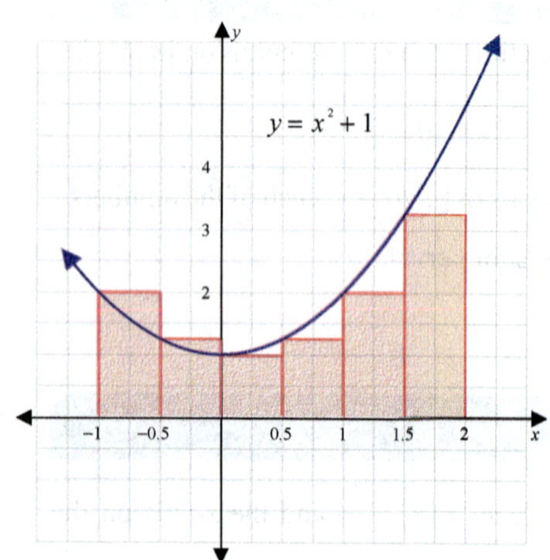

$$S_6 = \left[\begin{array}{l} \left((-1)^2 + 1\right) + \left((-0.5)^2 + 1\right) + \left((0)^2 + 1\right) \\ + \left((0.5)^2 + 1\right) + \left((1)^2 + 1\right) + \left((1.5)^2 + 1\right) \end{array} \right] \cdot (0.5)$$

Substitute the appropriate values into the general form of the Riemann sum.

$$= (2 + 1.25 + 1 + 1.25 + 2 + 3.25)(0.5)$$

$$= (10.75)(0.5)$$

$$= 5.375$$

> **📝 NOTE**
>
> Programmable graphing calculators (like the TI-84 Plus) will calculate various types of Riemann sums directly using a menu. We do not include such problems in this course.

In a Riemann sum, the more rectangles involved (and the smaller the value for Δx), the closer the sum is to the area under the curve. (See Figure 8.)

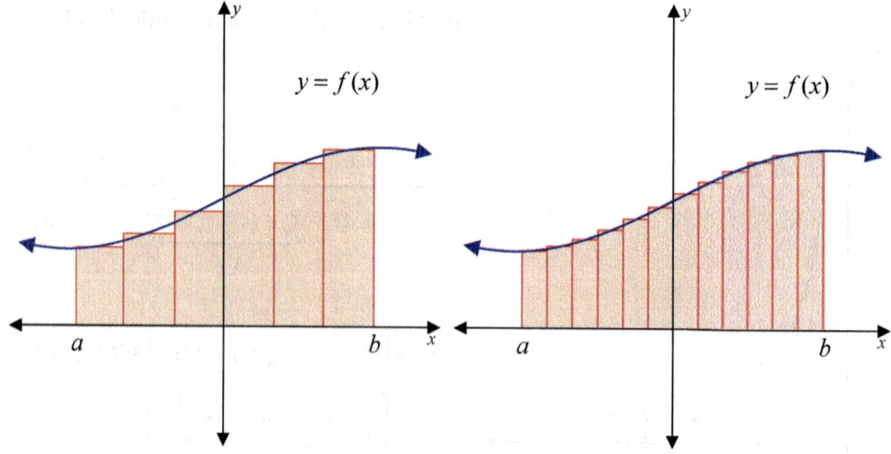

FIGURE 8

14.3 EXERCISES

PRACTICE

In Exercises 1–2, find the area of the shaded region.

1. 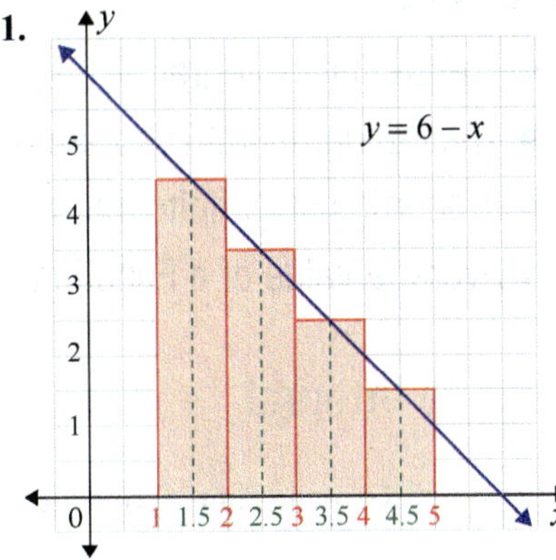

2. 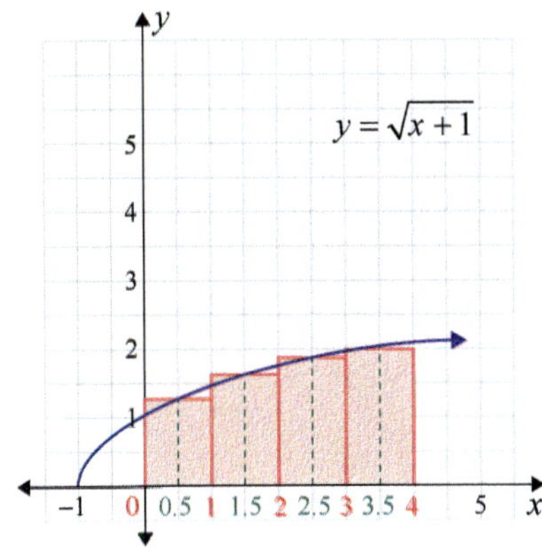

In Exercises 3–6, find the Riemann sum S_n for the given function, interval, and value of n.

3. $f(x) = x^2$; $[a, b] = [0, 4]$; $n = 4$; $c_1 = 0.5$, $c_2 = 1.5$, $c_3 = 2.5$, $c_4 = 3.5$

4. $f(x) = 9 - x^2$; $[a, b] = [-3, 2]$; $n = 5$; $c_1 = -2.5$, $c_2 = -1.5$, $c_3 = -0.5$, $c_4 = 0.5$, $c_5 = 1.5$

5. $f(x) = \sqrt{2 - x}$; $[a, b] = [-2, 2]$; $n = 4$; Use the midpoint of each subinterval for the value of each c_k.

6. $f(x) = \dfrac{1}{x + 2}$; $[a, b] = [-1, 3]$; $n = 4$; Use the midpoint of each subinterval for the value of each c_k.

WRITING & THINKING

Use four rectangles to estimate the area between the graph of the given function and the x-axis on the given interval. Construct three estimates for the function: the first using the left endpoints of the subintervals as the sample points, the second using the right endpoints of the subintervals, and the third using the midpoints of the subintervals. Can you tell which are guaranteed to be underestimates or overestimates? (**Hint:** Consider the increasing/decreasing and concavity features of the graph. It is helpful to make a sketch.)

7. $f(x) = \sqrt{x}$ on $[0, 4]$

8. $f(x) = \dfrac{x^3}{16}$ on $[0, 4]$

9. $f(x) = \dfrac{1}{x}$ on $[1, 5]$

10. $f(x) = \sqrt{4 - x^2}$ on $[-2, 2]$

11. $f(x) = e^{2-x}$ on $[0, 2]$

14.4 THE DEFINITE INTEGRAL AND THE FUNDAMENTAL THEOREM OF CALCULUS

▇ TOPICS

- The Definite Integral
- The Fundamental Theorem of Calculus
- Properties of Definite Integrals
- Average Value of a Function

The Definite Integral

Our discussion of Riemann sums forms the basis for the following definition.

Area under a Curve

If a function $y = f(x)$ is nonnegative and continuous on the interval $[a, b]$, then the **area under the curve** is defined to be $A = \lim_{n \to +\infty} S_n$, where S_n is the general form of a Riemann sum for the function f.

We have made two implicit restrictions in our discussions of Riemann sums that should be explained.

1. Since our initial objective was to find the area under a curve, the discussion was restricted to nonnegative functions (i.e., functions whose graphs lie on or above the x-axis.) In general, Riemann sums can be defined for any continuous function. For those functions that are negative on some intervals, some of the corresponding rectangles may be below the x-axis.

2. Also, for convenience, we required each subinterval to be the same width. Again, Riemann sums may be defined when the subintervals are of varying widths.

Now we define the **definite integral** of any continuous function in terms of Riemann sums.

The Definite Integral

If a function $y = f(x)$ is continuous on the interval $[a, b]$, then the **definite integral** of f from a to b, symbolized as $\int_a^b f(x)\,dx$, is defined to be

$$\int_a^b f(x)\,dx = \lim_{n \to +\infty} S_n,$$

where S_n is the general form of a Riemann sum for the function f.

The number a is called the **lower limit of integration** and the number b is called the **upper limit of integration**.

The two previous definitions lead to the following statement connecting area and the definite integral of a nonnegative function.

Formula for Area under a Curve

For a function $y = f(x)$, nonnegative and continuous on $[a, b]$, the area under the curve is given by

$$A = \int_a^b f(x)\,dx.$$

The Fundamental Theorem of Calculus

We will use the handy notational convenience (the use of the labeled bracket notation) to indicate $F(b) - F(a)$ as illustrated below:

$$F(b) - F(a) = F(x)\Big]_a^b$$

or even

$$\int_a^b f(x)\,dx = \left[\int f(x)\,dx\right]_a^b.$$

In some cases, other similar notations may be used, such as the following:

$$F(b) - F(a) = \left[F(x)\right]_a^b$$
$$= F(x)\Big]_a^b$$
$$= F(x)\Big|_a^b$$
$$= \left[F(x)\right]_{x=a}^{x=b}.$$

One way we have of finding the value indicated by a definite integral of a function is to find a general Riemann sum S_n for the function and then take the limit of S_n as $n \to +\infty$. This approach takes time and can involve techniques and formulas not covered in this course. However, it is essential in many applied areas of physics and engineering to construct a Riemann sum expression from theoretical considerations and only then consider the best way to calculate its value. We do exactly this later in this section when calculating the average value of a function. But now, we turn directly to the Fundamental Theorem of Calculus to provide the necessary tools for evaluating definite integrals. The full proof of the theorem, which involves Riemann sums, is beyond the scope of this course and therefore is not included.

The Fundamental Theorem of Calculus

If a function $y = f(x)$ is continuous on the interval $[a, b]$ and $F(x)$ is any antiderivative of $f(x)$, then

$$\int_a^b f(x)\,dx = \left[F(x)\right]_a^b = F(b) - F(a).$$

As the Fundamental Theorem implies, the constant of integration, C, can be omitted when definite integrals are evaluated. Consider the following analysis.

Suppose that $F(x)$ is any antiderivative of $f(x)$. Then

$$\int_a^b f(x)\,dx = F(x) + C\Big]_a^b$$
$$= \left[F(b) + C\right] - \left[F(a) + C\right]$$
$$= F(b) - F(a) + C - C$$
$$= F(b) - F(a).$$

Thus the constant of integration is subtracted out when a definite integral is evaluated.

Example 1: Definite Integrals

a. Evaluate $\int_1^2 2x\,dx.$ **b.** Find the value of $\int_1^4 \dfrac{1}{\sqrt{x}}\,dx.$ **c.** Evaluate $\int_0^1 e^x\,dx.$

Solution

a. $\displaystyle\int_1^2 2x\,dx = x^2\Big]_1^2$ Determine the antiderivative $F(x)$.

$$= (2)^2 - (1)^2$$ Use the Fundamental Theorem of
$$= 4 - 1$$ Calculus with $a = 1$ and $b = 2$.
$$= 3$$

b. $\displaystyle\int_1^4 \frac{1}{\sqrt{x}}\,dx = \int_1^4 x^{-\frac{1}{2}}\,dx = \frac{2}{1}x^{\frac{1}{2}}\Big]_1^4 = 2\sqrt{x}\Big]_1^4$ Determine the antiderivative $F(x)$.

$$= 2\sqrt{4} - 2\sqrt{1}$$ Use the Fundamental Theorem of
$$= 4 - 2$$ Calculus with $a = 1$ and $b = 4$.
$$= 2$$

c. $\displaystyle\int_0^1 e^x\,dx = e^x\Big]_0^1$ Determine the antiderivative $F(x)$.

$$= e^1 - e^0$$ Use the Fundamental Theorem of
$$= e - 1$$ Calculus with $a = 0$ and $b = 1$.

Properties of Definite Integrals

The properties of definite integrals listed here follow directly from the Fundamental
Theorem of Calculus and provide the basis for expansion of the techniques used to
evaluate definite integrals.

Properties of Definite Integrals

1. $\displaystyle\int_a^a f(x)\,dx = 0$

2. $\displaystyle\int_a^b kf(x)\,dx = k\int_a^b f(x)\,dx$

3. $\displaystyle\int_a^b \big[f(x) \pm g(x)\big]\,dx = \int_a^b f(x)\,dx \pm \int_a^b g(x)\,dx$

Example 2: Using Properties of Definite Integrals

a. Evaluate $\displaystyle\int_1^2 \left(x^2 + \frac{1}{x} - 3e^x\right)dx$.

b. Find the value of $\displaystyle\int_0^1 x\sqrt{x^2 + 1}\,dx$.

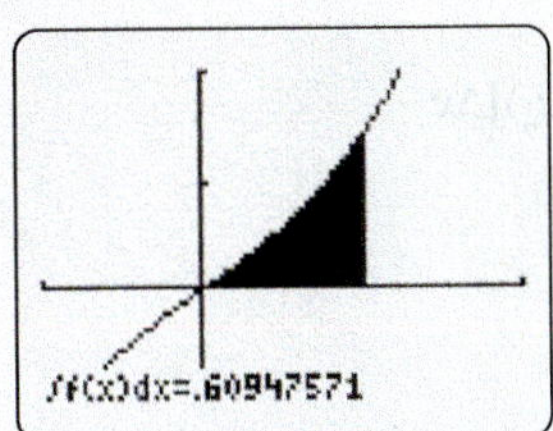

Solution

a. $\displaystyle\int_1^2\left(x^2+\frac{1}{x}-3e^x\right)dx = \frac{x^3}{3}+\ln|x|-3e^x\,\Bigg]_1^2$ Determine the antiderivative $F(x)$.

$$= \left(\frac{(2)^3}{3}+\ln|2|-3e^2\right)-\left(\frac{(1)^3}{3}+\ln|1|-3e^1\right)$$

Use the Fundamental Theorem of Calculus with $a = 1$ and $b = 2$.

$$= \frac{8}{3}+\ln 2-3e^2-\frac{1}{3}-0+3e$$

$$= \frac{7}{3}+\ln 2-3e^2+3e$$

b. Use the u-substitution technique with $u = x^2+1$ and $du = 2x\,dx$ to find the antiderivative. Then replace u with x^2+1 before evaluating the definite integral.

$$\int x\sqrt{x^2+1}\,dx = \frac{1}{2}\int(x^2+1)^{\frac{1}{2}}\,2x\,dx$$ Use u-substitution to find the antiderivative.

$$= \frac{1}{2}\int u^{\frac{1}{2}}\,du$$

$$= \frac{1}{2}\cdot\frac{2}{3}u^{\frac{3}{2}}+C$$

$$= \frac{1}{3}(x^2+1)^{\frac{3}{2}}+C$$

$$\int_0^1 x\sqrt{x^2+1}\,dx = \frac{1}{3}(x^2+1)^{\frac{3}{2}}\,\Bigg]_0^1$$ Evaluate the definite integral.

$$= \left[\frac{1}{3}\big((1)^2+1\big)^{\frac{3}{2}}\right]-\left[\frac{1}{3}\big((0)^2+1\big)^{\frac{3}{2}}\right]$$

$$= \left(\frac{1}{3}\cdot 2^{\frac{3}{2}}\right)-\left(\frac{1}{3}\right)$$

$$= \frac{1}{3}\cdot 2\sqrt{2}-\frac{1}{3}$$

$$= \frac{1}{3}\left(2\sqrt{2}-1\right)$$

Average Value of a Function

Suppose that the last three cars you bought cost $7500, $8200, and $10,400. To find the average cost, you would add these numbers and divide by 3:

$$\frac{7500+8200+10,400}{3} = \frac{26,100}{3} = \$8700.$$

To find the average value of a continuous function $y = f(x)$ on an interval $[a, b]$ is somewhat more difficult because $f(x)$ has an infinite number of values on $[a, b]$.

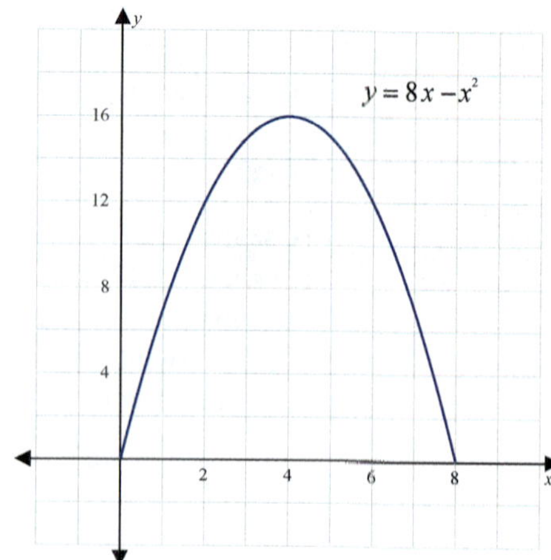

$$y = 8x - x^2$$

FIGURE 1

For example, let's numerically estimate an average y-value for $y = 8x - x^2$ considered over the x-axis interval $[0, 8]$. (See Figure 1.)

We could average the highest y-value, 16, with the lowest y-value, 0, and get 8. If we average $f(0)$ and $f(8)$, we get 0. Perhaps, it is better if we pick a few evenly spaced x-values and average the corresponding y-values. Choosing four values, we get

$$\frac{f(0) + f(2) + f(4) + f(6)}{4} = \frac{0 + 12 + 16 + 12}{4} = \frac{40}{4} = 10$$

or we might get

$$\frac{f(1) + f(3) + f(5) + f(7)}{4} = \frac{7 + 15 + 15 + 7}{4} = \frac{44}{4} = 11.$$

Evidently, the average y-value is somewhere near 10 or 11.

In general, one may estimate an average value of $f(x)$ on the interval $[a, b]$ by taking n evenly spaced x-values and computing the following expression.

$$\frac{f(x_1) + f(x_2) + \cdots + f(x_n)}{n}$$

Presumably, we get a better estimate as n gets bigger (i.e., if we use more points). If we could get a convenient limit expression, which could be seen to converge, and is easy to calculate, then we could always estimate an average y-value. Actually, this expression is very suggestive of a Riemann sum since a Riemann sum involves the sum of y-values. We proceed by analyzing a general Riemann sum; and, not surprisingly, we find a relationship between the concept of average value and the definite integral.

Step 1: Form the general Riemann sum S_n for the function f.

$$S_n = [f(c_1) + f(c_2) + \cdots + f(c_n)]\Delta x$$

Step 2: Substitute $\Delta x = \dfrac{b - a}{n}$.

$$S_n = \left[f(c_1) + f(c_2) + \cdots + f(c_n) \right]\frac{b - a}{n}$$

$$= (b - a)\left[\frac{f(c_1) + f(c_2) + \cdots + f(c_n)}{n} \right] \qquad \text{Observe that the factor in brackets is exactly what we need.}$$

Step 3: We know that $\displaystyle\lim_{n \to +\infty} S_n = \int_a^b f(x)\,dx$.

So

$$\int_a^b f(x)\,dx = (b - a) \cdot \lim_{n \to +\infty} \left[\frac{f(c_1) + f(c_2) + \cdots + f(c_n)}{n} \right],$$

or

$$\frac{1}{b - a}\int_a^b f(x)\,dx = \lim_{n \to +\infty} \left[\frac{f(c_1) + f(c_2) + \cdots + f(c_n)}{n} \right].$$

Thus the very limit we wanted exists! We define the limit on the right to be the **average value (AV)** of f on the interval $[a, b]$.

Average Value

For a function $y = f(x)$, continuous on the interval $[a, b]$, the **average value** is

$$AV = \frac{1}{b-a} \int_a^b f(x)\, dx.$$

Geometrically, the average value for a nonnegative continuous function is the height of a rectangle with base $(b - a)$ that has area exactly equal to the area under the curve. (See Figure 2.)

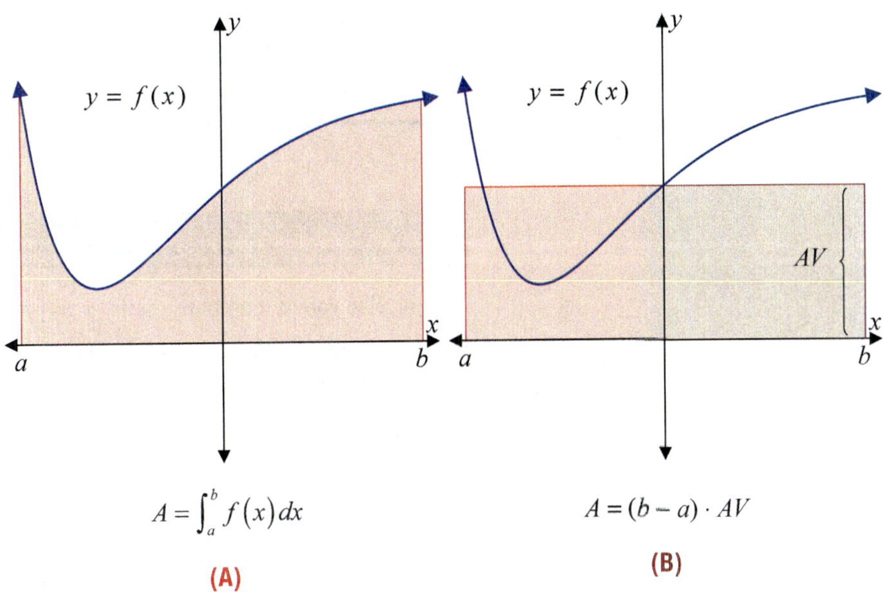

The shaded area in (A) exactly equals the rectangular area in (B).

FIGURE 2

Example 3: Average Value

a. Find the average value of $f(x) = 8x - x^2$ on the interval $[0, 8]$.

Solution

$$AV = \frac{1}{8-0} \cdot \int_0^8 \left(8x - x^2\right) dx$$

Substitute $a = 0$, $b = 8$, and the given $f(x)$ into the formula for average value.

$$= \frac{1}{8} \cdot \left[\frac{8x^2}{2} - \frac{x^3}{3} \right]_0^8$$

Solve.

$$= \frac{1}{8} \cdot \left[\frac{8(8)^2}{2} - \frac{(8)^3}{3} \right] - \frac{1}{8} \cdot \left[\frac{8(0)^2}{2} - \frac{(0)^3}{3} \right]$$

$$= 32 - \frac{64}{3} - 0$$

$$= 10.\overline{6}$$

Note that this is between 10 and 11 as estimated earlier.

b. Find the average value of $f(x) = x^2 + 2$ on the interval $[0, 3]$.

Solution

$$AV = \frac{1}{3-0}\int_0^3 \left(x^2 + 2\right)dx$$

Substitute $a = 0$, $b = 3$, and the given $f(x)$ into the formula for average value.

$$= \frac{1}{3}\left(\frac{x^3}{3} + 2x\right)\Bigg]_0^3$$

Solve.

$$= \frac{1}{3}\left[\left(\frac{(3)^3}{3} + 2\cdot 3\right) - \left(\frac{(0)^3}{3} + 2\cdot 0\right)\right]$$

$$= \frac{1}{3}(9 + 6)$$

$$= 5$$

Example 4: Average Sales

A new electronics company sells $y = \dfrac{1}{2}t^2 + t$ wireless speakers (in hundreds) per month, where t is the number of months the company has been in business. Find the average sales per month during the company's first 6 months in business.

Solution

$$AV = \frac{1}{6-0}\int_0^6 \left(\frac{1}{2}t^2 + t\right)dt$$

Substitute $a = 0$, $b = 6$, and the given $f(x)$ into the formula for average value.

$$= \frac{1}{6}\left(\frac{t^3}{6} + \frac{t^2}{2}\right)\Bigg]_0^6$$

Solve.

$$= \frac{1}{6}\left[\left(\frac{(6)^3}{6} + \frac{(6)^2}{2}\right) - \left(\frac{(0)^3}{6} + \frac{(0)^2}{2}\right)\right]$$

$$= \frac{1}{6}(36 + 18 - 0)$$

$$= \frac{1}{\cancel{6}}\cancel{(54)}^{9}$$

$$= 9$$

The company averaged sales of 900 wireless speakers per month during its first 6 months in business.

14.4 EXERCISES

🔆 PRACTICE

In Exercises 1–32, evaluate each definite integral.

1. $\displaystyle\int_{-1}^{0} 5x^2\, dx$

2. $\displaystyle\int_{1}^{4} \frac{1}{x}\, dx$

3. $\displaystyle\int_{0}^{3} \left(1+e^x\right) dx$

4. $\displaystyle\int_{0}^{2} 6e^x\, dx$

5. $\displaystyle\int_{3}^{5} \frac{1}{x-2}\, dx$

6. $\displaystyle\int_{1}^{9} \sqrt{x}\, dx$

7. $\displaystyle\int_{-4}^{-2} \frac{1}{x^2}\, dx$

8. $\displaystyle\int_{2}^{4} \left(7x+2\right) dx$

9. $\displaystyle\int_{-1}^{3} \left(4x+1\right) dx$

10. $\displaystyle\int_{-2}^{1} \frac{4}{x+3}\, dx$

11. $\displaystyle\int_{-1}^{3} e^{x+1}\, dx$

12. $\displaystyle\int_{2}^{3} \left(x^2+2x-4\right) dx$

13. $\displaystyle\int_{2}^{4} x\left(x^2-3\right) dx$

14. $\displaystyle\int_{1}^{8} \left(1+\frac{1}{\sqrt[3]{x}}\right) dx$

15. $\displaystyle\int_{0}^{3} \frac{1}{3x+1}\, dx$

16. $\displaystyle\int_{1}^{3} 2e^{-1.5x}\, dx$

17. $\displaystyle\int_{3}^{5} \frac{1}{(3x+1)^2}\, dx$

18. $\displaystyle\int_{-1}^{0} \sqrt{3x+4}\, dx$

19. $\displaystyle\int_{2}^{3} (3-2x)^4\, dx$

20. $\displaystyle\int_{0}^{2} \frac{x}{\sqrt[3]{x^2+4}}\, dx$

21. $\displaystyle\int_{2}^{6} \frac{3x}{x^2-3}\, dx$

22. $\displaystyle\int_{3}^{5} \frac{x+2}{x^2+4x+3}\, dx$

23. $\displaystyle\int_{0}^{1} e^x\left(e^x+1\right) dx$

24. $\displaystyle\int_{0}^{5} xe^{-0.24x^2}\, dx$

25. $\displaystyle\int_{1}^{3} xe^{x^2-1}\, dx$

26. $\displaystyle\int_{1}^{2} (x-1)\left(2x^2-4x+1\right)^2 dx$

27. $\displaystyle\int_{6}^{7} \frac{x-3}{\sqrt{x^2-6x+4}}\, dx$

28. $\displaystyle\int_{1}^{8} \left(2x^{\frac{2}{3}}-x^{-2}\right) dx$

29. $\displaystyle\int_{0}^{1} \frac{e^x}{e^x+1}\, dx$

30. $\displaystyle\int_{1}^{4} \frac{\ln x}{x}\, dx$

31. $\displaystyle\int_{1}^{3} \frac{1+\ln x}{x}\, dx$

32. $\displaystyle\int_{2}^{4} \frac{1}{x^2} e^{\frac{1}{x}}\, dx$

For Exercises 33–38, find the average value of the function on the given interval.

33. $f(x)=x^2+6;\ [1,\,4]$

34. $f(x)=4x^2-3x+1;\ [-1,\,3]$

35. $f(x)=\sqrt{x+1};\ [3,\,8]$

36. $f(x)=\sqrt[3]{2x+1};\ [0,\,13]$

37. $f(x)=2e^{-0.25x};\ [0,\,4]$

38. $f(x)=1+e^{-0.4x};\ [0,\,5]$

APPLICATIONS

39. Pollution: The level of pollution in San Felipe Bay, due to an oil spill, is estimated to be $f(t) = \dfrac{1800t}{\sqrt{t^2 + 11}}$ parts per million, where t is the time in days since the spill occurred. Find the average level of pollution during the first 5 days after the spill occurred.

40. Bacterial population: It is estimated that the number of bacteria present in a culture t hours after bacteria are introduced to the culture is given by $N(t) = \dfrac{8000}{\sqrt{8 - 0.5t}}$. Find the average number of bacteria present during the first 8 hours.

41. Average production: The daily production level for a product is given by $N(t) = 240 - 240e^{-0.2t}$ units, where t is the time in hours after production begins. Find the average production during the first 4 hours.

42. Average marginal profit: The marginal profit from the production and the sale of x barbecue grills is given by $P'(x) = 52 - 0.8x$ dollars per grill. Find the average marginal profit for the first 40 grills produced and sold.

14.5 AREA UNDER A CURVE (WITH APPLICATIONS)

■ TOPICS

- ■ Area Bounded by a Curve and the *x*-Axis
- ■ Marginal Analysis

Previously, we discussed the fact that when a function is nonnegative, the definite integral of the function represents the area between the graph of the function and the *x*-axis on the interval $[a, b]$. In this section, we will continue that discussion and relate definite integrals to area when a function has negative values.

Area Bounded by a Curve and the *x*-Axis

A Riemann sum can involve rectangles both above and below the *x*-axis, as illustrated in Figure 1.

In Figure 1

$$S_4 = f\left(c_1\right)\Delta x + f\left(c_2\right)\Delta x + f\left(c_3\right)\Delta x + f\left(c_4\right)\Delta x$$
$$= A_1 + A_2 - A_3 - A_4.$$

Since $f(c_3)$ is negative, $f(c_3)\Delta x$ is **not** an area. It is instead the negative of an area, $f(c_3)\Delta x = -A_3$ (sometimes called a "signed area"). Similarly, $f(c_4)\Delta x = -A_4$. Thus, in this figure, the Riemann sum is not a total area; instead, it is the difference between areas. This concept leads to the following related statement about computing areas.

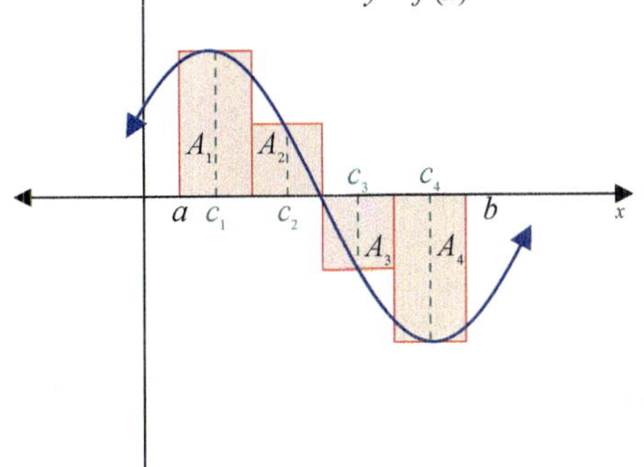

FIGURE 1

The Integral as Area

For $y = f(x)$, a continuous function on the interval $[a, b]$, the integral $\displaystyle\int_a^b f(x)\,dx$ represents

1. the total area bounded by the curve and the *x*-axis from $x = a$ to $x = b$ if $f(x)$ is nonnegative for all x in $[a, b]$;

2. the difference between the areas above the *x*-axis and below the *x*-axis that are bounded by the curve and the *x*-axis from $x = a$ to $x = b$ if $f(x)$ is negative for any x in $[a, b]$.

Example 1: Interpreting the Integral

a. Evaluate $\displaystyle\int_0^2 (x+1)\,dx$ and interpret the integral geometrically.

Solution

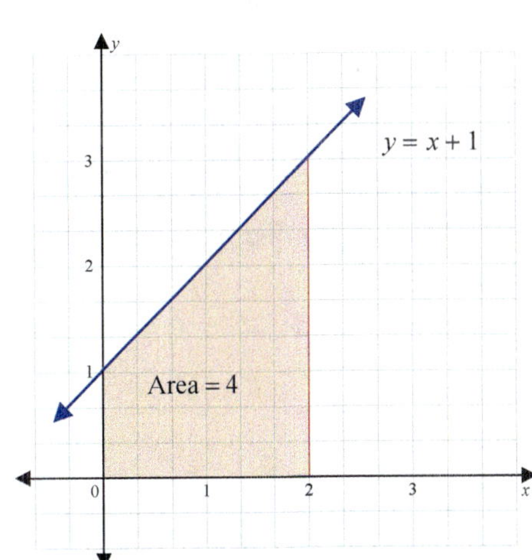

$$\int_0^2 (x+1)\,dx = \frac{x^2}{2} + x \Bigg]_0^2 = \left(\frac{(2)^2}{2} + 2\right) - \left(\frac{(0)^2}{2} + 0\right) = 4$$

Geometrically, 4 is the area between the curve $y = x + 1$ and the *x*-axis on the interval $[0, 2]$.

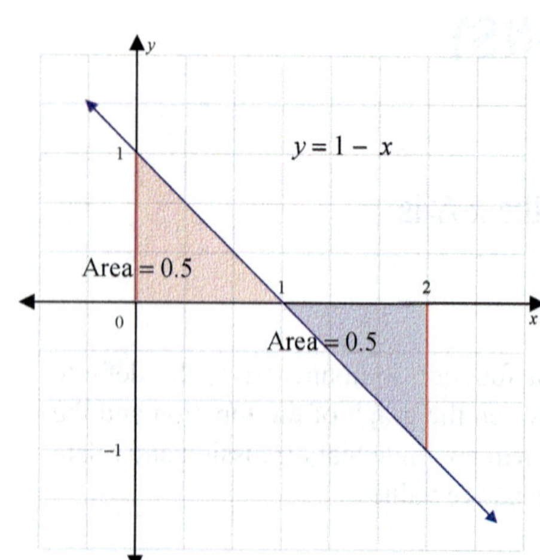

b. Evaluate $\int_0^2 (1-x)\,dx$ and interpret the integral geometrically.

Solution

$$\int_0^2 (1-x)\,dx = x - \frac{x^2}{2}\bigg]_0^2$$

$$= \left(2 - \frac{(2)^2}{2}\right) - \left(0 - \frac{(0)^2}{2}\right)$$

$$= 0$$

Geometrically, the 0 means that the area above the x-axis is equal to the area below the x-axis on the interval $[0, 2]$.

In Example 1b, the triangular region above the x-axis has the same area (0.5) as the triangular region below the x-axis. Thus, if we wanted to find the total of the areas, we could, because of the symmetry in the function, integrate the function from $x = 0$ to $x = 1$ and double the result.

$$\text{Total area} = 2\int_0^1 (1-x)\,dx$$

$$= 2\left(x - \frac{x^2}{2}\right)\bigg]_0^1$$

$$= 2\left(1 - \frac{1}{2}\right)$$

$$= 2\left(\frac{1}{2}\right) = 1$$

Example 2 shows a case where the regions above and below the x-axis do not have the same area. In this case we must separate the integral into two parts to find the total area.

Example 2: Total Area

Find the total area bounded by the x-axis and the curve $y = 4 - x^2$ on the interval $[-2, 3]$.

Solution

The curve $y = 4 - x^2$ is a parabola that crosses the x-axis at $x = -2$ and $x = 2$. We integrate the function from $x = -2$ to $x = 2$ to find A_1. Then, to find A_2, we take the additive inverse (negative) of the integral from $x = 2$ to $x = 3$.

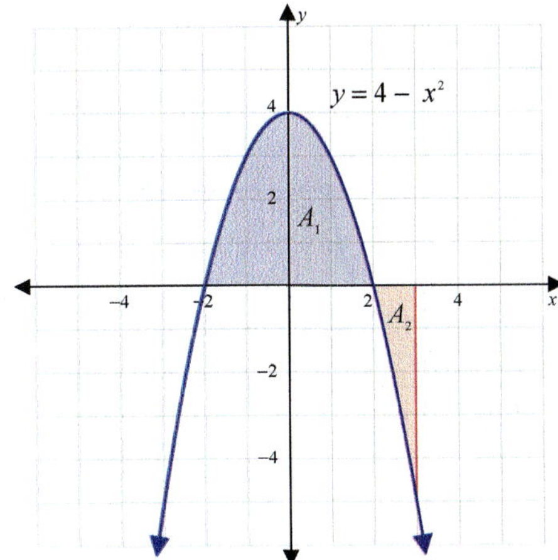

$$A_1 = \int_{-2}^{2}\left(4-x^2\right)dx$$

$$= 4x - \frac{x^3}{3}\Bigg]_{-2}^{2}$$

$$= \left[4(2) - \frac{(2)^3}{3}\right] - \left[4(-2) - \frac{(-2)^3}{3}\right]$$

$$= \left(8 - \frac{8}{3}\right) - \left(-8 + \frac{8}{3}\right)$$

$$= 8 - \frac{8}{3} + 8 - \frac{8}{3}$$

$$= 16 - \frac{16}{3} = \frac{32}{3}$$

Note: In general, we don't add a minus sign in front of an integral unless we are finding the area of a region below the x-axis.

$$A_2 = -\int_{2}^{3}\left(4-x^2\right)dx$$

$$= -\left(4x - \frac{x^3}{3}\right)\Bigg]_{2}^{3}$$

$$= -\left[\left(4(3) - \frac{(3)^3}{3}\right) - \left(4(2) - \frac{(2)^3}{3}\right)\right]$$

$$= -\left[(12-9) - \left(8 - \frac{8}{3}\right)\right]$$

$$= -\left(12 - 9 - 8 + \frac{8}{3}\right)$$

$$= -\left(-5 + \frac{8}{3}\right) = -\left(-\frac{15}{3} + \frac{8}{3}\right)$$

$$= -\left(-\frac{7}{3}\right) = \frac{7}{3}$$

Graphs, as shown in Examples 1 and 2, are valuable visual aids in understanding definite integrals. A sketch of the graph of the function being integrated should be made whenever possible.

$$\text{Total Area} = A_1 + A_2 = \frac{32}{3} + \frac{7}{3} = \frac{39}{3} = 13$$

Note that $\int_{-2}^{3}\left(4-x^2\right)dx = A_1 - A_2 = \frac{32}{3} - \frac{7}{3} = \frac{25}{3}$, which is not the total area.

The following two properties of the definite integral can be added to the list of three properties stated in the previous section.

Additional Properties of the Definite Integral

$$\int_{a}^{b} f(x)\,dx = -\int_{b}^{a} f(x)\,dx$$

$$\int_{a}^{b} f(x)\,dx = \int_{a}^{c} f(x)\,dx + \int_{c}^{b} f(x)\,dx \quad \text{where } c \text{ is any point with } a \le c \le b.$$

The latter of the previous properties is particularly useful if a function is defined in pieces and we want to find the area bounded by the curve and the x-axis. (See Figure 2.)

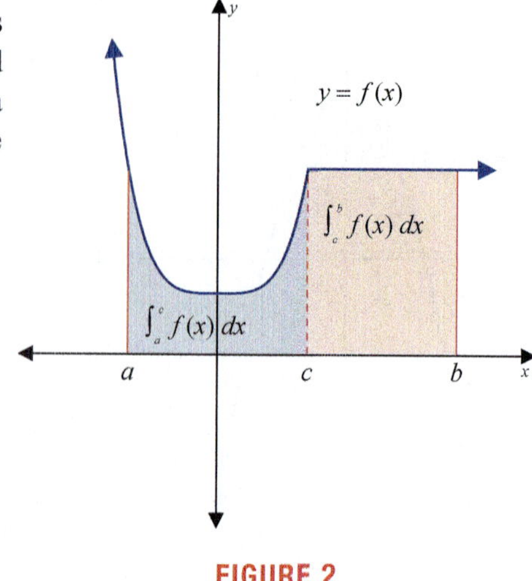

FIGURE 2

Example 3: Bounded Area

Find the area of the region bounded by the function

$$f(x) = \begin{cases} x+2 & \text{if } x \le 1 \\ 5-2x & \text{if } x > 1 \end{cases}$$

and the x-axis from $x = -1$ to $x = 2$.

Solution

The graph of the function consists of two rays.

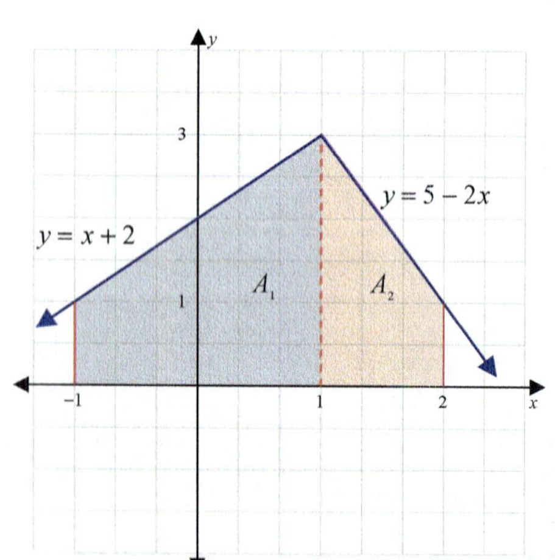

The area of the bounded region from $x = -1$ to $x = 2$ is found as follows.

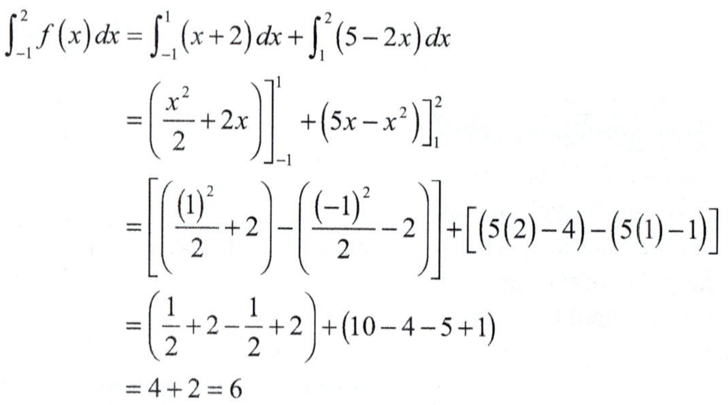

$$\int_{-1}^{2} f(x)\,dx = \int_{-1}^{1}(x+2)\,dx + \int_{1}^{2}(5-2x)\,dx$$

$$= \left(\frac{x^2}{2}+2x\right)\Bigg]_{-1}^{1} + \left(5x - x^2\right)\Bigg]_{1}^{2}$$

$$= \left[\left(\frac{(1)^2}{2}+2\right)-\left(\frac{(-1)^2}{2}-2\right)\right]+\left[\left(5(2)-4\right)-\left(5(1)-1\right)\right]$$

$$= \left(\frac{1}{2}+2-\frac{1}{2}+2\right)+(10-4-5+1)$$

$$= 4+2 = 6$$

Marginal Analysis

We know that marginal cost is the derivative $C'(x)$, marginal revenue is the derivative $R'(x)$, and marginal profit is the derivative $P'(x)$. Therefore, the integrals of each of these functions will yield the antiderivatives, which are, respectively, the cost function $C(x)$, the revenue function $R(x)$, and the profit function $P(x)$. We want to show how the definite integral can be used to measure changes in cost, revenue, and profit. The integrals do not take into account fixed costs such as rent. Such costs must be treated separately and are ignored here.

Suppose that a marginal profit function $P'(x) = 20 - 0.05x$ is known, where x is the number of items sold and the marginal profit is in dollars per item. Since P' represents the change in profit per item, we note that the profit is decreasing by 5 cents per item. The area under the curve corresponding to P' on an interval $[0, a]$ represents the profit when a items are produced and sold. (See Figure 3.) These ideas are illustrated in Example 4.

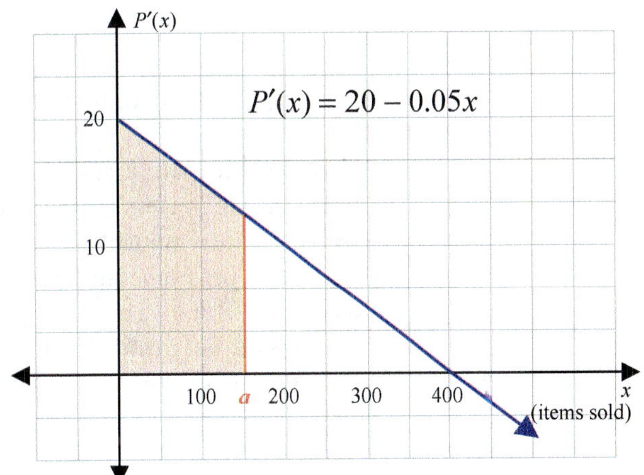

Figure 3 shows that if more than 400 items are sold, the company's profits will decline (the area will be below the x-axis). Note that at $x = 100$ (say), marginal profit is decreasing, but profit (measured by the area) is increasing at $x = 100$. If marginal profit is decreasing, it might still be worthwhile to continue producing and selling the product.

FIGURE 3

Example 4: Marginal Analysis

A picture frame maker knows that his profit (in dollars) is changing at a rate given by the function $P'(x) = 20 - 0.05x$, where x is the number of frames he makes and sells.

a. Find the profit from selling the first 200 frames.

b. Find the change in his profit when sales increase from 200 to 400 frames.

Solution

a. Integrate the marginal profit function from $x = 0$ to $x = 200$.

$$\int_0^{200} (20 - 0.05x)\, dx = 20x - 0.025x^2 \Big]_0^{200}$$

$$= \left[20(200) - 0.025(200)^2 \right] - \left[20(0) - 0.025(0)^2 \right]$$

$$= (4000 - 1000) - 0$$

$$= 3000$$

His profit is \$3000 on the first 200 frames if we ignore any fixed costs.

b. Integrate from $x = 200$ to $x = 400$.

$$\int_{200}^{400} (20 - 0.05x)\, dx = 20x - 0.025x^2 \Big]_{200}^{400}$$

$$= \left[20(400) - 0.025(400)^2 \right] - \left[20(200) - 0.025(200)^2 \right]$$

$$= (8000 - 4000) - (4000 - 1000)$$

$$= 1000$$

His profit changes by $1000 when sales increase from 200 to 400 frames.

Note that from parts a. and b. we see that the profit on sales for the first 200 frames ($x = 0$ to $x = 200$) is greater than the profit for the second 200 frames ($x = 200$ to $x = 400$). This result is quite reasonable because the marginal profit, $P'(x) = 20 - 0.05x$, is decreasing by 5 cents per frame. In this problem, the fixed costs are relevant. Suppose the fixed costs are $1000. Then $P(x) = 20x - 0.025x^2 - 1000$ dollars, and $P(200) = \$2000$, which is the **net profit**. The integral from $x = 0$ to $x = 200$ gives the increase in profits, $P(200) - P(0) = 2000 - (-1000) = 3000$.

14.5 EXERCISES

💡 PRACTICE

For Exercises 1–18, find the total area bounded by the x-axis and the curve $y = f(x)$ on the indicated interval.

1. $f(x) = 3x + 1,\ \ [0, 5]$

2. $f(x) = 7 - 2x,\ \ [-1, 3]$

3. $f(x) = x^2 + 1,\ \ [-2, 2]$

4. $f(x) = 0.5x^2 + 2,\ \ [1, 4]$

5. $f(x) = x^3 + 2,\ \ [-1, 1]$

6. $f(x) = 2x^3 - 1,\ \ [1, 2]$

7. $f(x) = x^2 + x + 1,\ \ [-1, 3]$

8. $f(x) = x^2 + 2x - 3,\ \ [1, 3]$

9. $f(x) = \dfrac{4}{x+1},\ \ [0, 3]$

10. $f(x) = \dfrac{3}{2x+1},\ \ [0, 2]$

11. $f(x) = 3e^{0.6x},\ \ [0, 5]$

12. $f(x) = 1 + e^{-0.3x},\ \ [0, 4]$

13. $f(x) = x^2 - 2x - 8,\ \ [2, 5]$

14. $f(x) = x^2 + 3x - 4,\ \ [0, 4]$

15. $f(x) = \begin{cases} 2 - x & \text{if } -1 \le x \le 2 \\ x^2 - 4 & \text{if } 2 \le x \le 3 \end{cases}$

16. $f(x) = \begin{cases} x^2 & \text{if } -2 \le x \le 1 \\ 2x - 1 & \text{if } 1 \le x \le 2 \end{cases}$

17. $f(x) = \begin{cases} x + 2 & \text{if } -2 \le x \le 0 \\ \sqrt{x+4} & \text{if } 0 \le x \le 5 \end{cases}$

18. $f(x) = \begin{cases} 1 - 2x & \text{if } -2 \le x \le 0 \\ e^{2x} & \text{if } 0 \le x \le 1.5 \end{cases}$

🚀 APPLICATIONS

19. **Profit:** The marginal profit for a certain style of sports jacket is given by $P'(x) = 56 - 0.8x$ dollars per jacket, where x is the number of jackets produced and sold weekly. Find the profit for the first 50 jackets that are produced and sold. (Ignore any fixed costs.)

20. **Profit:** The marginal profit of an important product is given by $P'(x) = 10 - 0.015e^{0.6x}$ dollars per item, where x is the number of items produced and sold. Find the profit for the first 8 items. (Ignore any fixed costs.)

21. **Cost:** The marginal cost of a product is given by $15 + \dfrac{4}{\sqrt{x}}$ dollars per unit, where x is the number of units produced. The current level of production is 100 units weekly. If the level of production is increased to 169 units weekly, find the increase in the total costs.

22. **Revenue:** The marginal revenue from the sale of x bottles of a wine is given by $8.4 - 0.3\sqrt{x}$ dollars per bottle. Find the increase in total revenue if the number of bottles sold is increased from 225 to 350.

23. **Wildlife management:** The manager of a wildlife preserve has started a management program to control the population of the preserve's bison herd. It is estimated that the population will continue to grow according to the function $N'(t) = 15 - 6t^{\frac{1}{2}}$ bison per year, where t is the number of years after implementation of the plan and $0 \le t \le 5$. Find the increase in the population during the first 4 years of the program.

24. **Bacterial population:** It is estimated that t hours after some particular bacteria are introduced into a culture, the population will be increasing at a rate of $P'(t) = \dfrac{1200}{(12 - 0.5t)^{\frac{1}{2}}}$ bacteria per hour. Find the increase in the population during the first 6 hours.

✏️ WRITING & THINKING

In Exercises 25 and 26, explain the meaning of the shaded region in each graph.

25.   26.

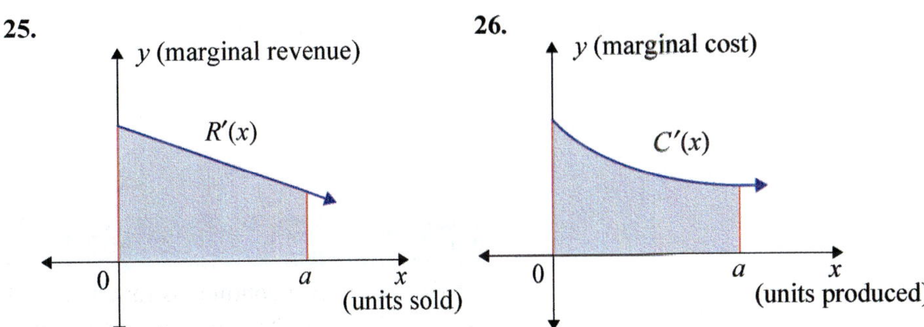

14.6 AREA BETWEEN TWO CURVES (WITH APPLICATIONS)

■ TOPICS

- Area between Two Curves
- Consumers' Surplus and Producers' Surplus
- Lorenz Curves

Previously we studied the area between a curve and the x-axis. The x-axis itself is the line (or curve) corresponding to the equation $y = 0$. Thus the area between a curve and the x-axis can be thought of as the area between two curves. In this section we will expand on this idea by developing general techniques for finding the area bounded by two curves and by discussing some interesting applications.

Area between Two Curves

Suppose that $y = f(x)$ and $y = g(x)$ are two continuous functions and $f(x) \geq g(x)$ on some interval $[a, b]$. Then the area of the region between the two curves can be found by the following definite integral. (See Figure 1.)

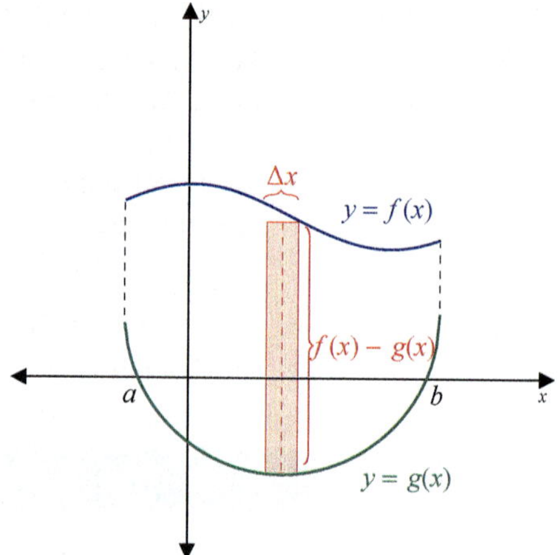

The area of the rectangle is $[f(x) - g(x)]\Delta x$ where x is a point in the interval Δx.

The integral $A = \int_a^b \left[f(x) - g(x) \right] dx$ represents the limit of a Riemann sum of such rectangles.

FIGURE 1

Area between Two Curves

If f and g are two continuous functions and $f(x) \geq g(x)$ on the interval $[a, b]$, then the area between the two curves on this interval is

$$A = \int_a^b \left[f(x) - g(x) \right] dx$$

Since $f(x) \geq g(x)$, we are assured that $[f(x) - g(x)]$ is nonnegative. Thus whether or not $f(x)$ and $g(x)$ are themselves positive or negative is not of concern when we are finding the area between the curves. The graph, though, is still a valuable tool, as we shall see in Examples 1 and 2.

Example 1: Finding the Area between Two Curves

Find the area enclosed by the curves $y = x^2 - 1$ and $y = 1 - x$.

Solution

Sketch the graphs of both functions to help determine which function is larger. Now set the two y-values equal to each other to find the points of intersection. These x-values will be the limits of integration.

$$x^2 - 1 = 1 - x$$
$$x^2 + x - 2 = 0$$
$$(x + 2)(x - 1) = 0$$
$$x = -2, 1$$

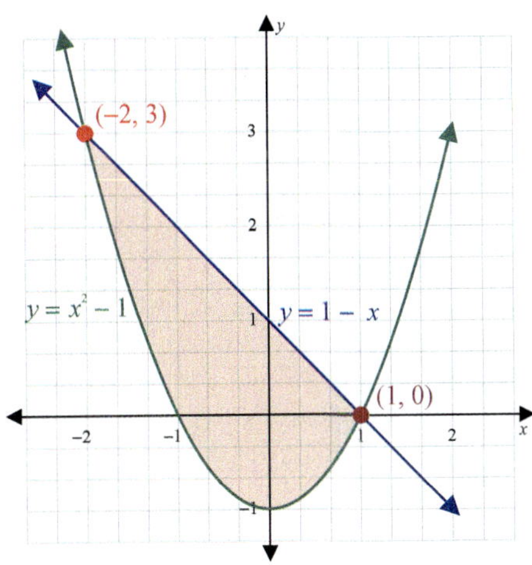

$$A = \int_{-2}^{1} \left[(1 - x) - (x^2 - 1) \right] dx$$
$$= \int_{-2}^{1} \left(2 - x - x^2 \right) dx$$
$$= 2x - \frac{x^2}{2} - \frac{x^3}{3} \Bigg]_{-2}^{1}$$
$$= \left(2(1) - \frac{(1)^2}{2} - \frac{(1)^3}{3} \right) - \left(2(-2) - \frac{(-2)^2}{2} - \frac{(-2)^3}{3} \right)$$
$$= \left(2 - \frac{1}{2} - \frac{1}{3} \right) - \left(-4 - \frac{4}{2} - \frac{-8}{3} \right)$$
$$= 2 - \frac{1}{2} - \frac{1}{3} + 4 + 2 - \frac{8}{3}$$
$$= \frac{9}{2}$$
$$= 4.5$$

Find the area between the curves $y = x^3$ and $y = x^2 + 1$ on the interval $[-1, 1]$.

Solution

Sketch the graphs of both functions to help determine whether or not the curves intersect on the interval $[-1, 1]$.

These curves do not intersect on the interval $[-1, 1]$ and $y = x^2 + 1$ is larger on the entire interval.

$$A = \int_{-1}^{1} \left[\left(x^2 + 1 \right) - \left(x^3 \right) \right] dx$$

$$= \int_{-1}^{1} \left(x^2 + 1 - x^3 \right) dx$$

$$= \frac{x^3}{3} + x - \frac{x^4}{4} \Bigg]_{-1}^{1}$$

$$= \left(\frac{(1)^3}{3} + 1 - \frac{(1)^4}{4} \right) - \left(\frac{(-1)^3}{3} + (-1) - \frac{(-1)^4}{4} \right)$$

$$= \left(\frac{1}{3} + 1 - \frac{1}{4} \right) - \left(-\frac{1}{3} - 1 - \frac{1}{4} \right)$$

$$= \frac{1}{3} + 1 - \frac{1}{4} + \frac{1}{3} + 1 + \frac{1}{4}$$

$$= \frac{8}{3}$$

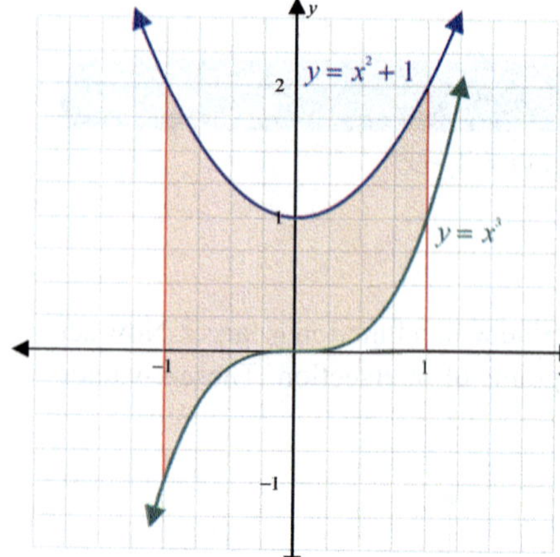

Now we will show how the definite integral, along with the concept of area between two curves, can be applied to two topics from the realm of business and economics: consumers' and producers' surplus and Lorenz curves.

Consumers' Surplus and Producers' Surplus

For consumers, the demand curve $p = D(x)$ represents the price they are willing to pay per item when x items are available in the marketplace. For producers, the supply curve $p = S(x)$ represents the price per item at which they are willing to produce and sell x items. The equilibrium point (x_E, p_E) is the point where the two curves intersect (see Figure 2).

Some consumers are happy to pay the equilibrium price p_E because, as Figure 2 shows, they would have been willing to pay a higher price if there had been fewer than x_E items on the market. Thus these people have saved money by buying at a lower price. This savings, called the **consumers' surplus (CS)**, is the difference between the total value to the consumers and the actual total cost to the consumers, as illustrated in Figure 3.

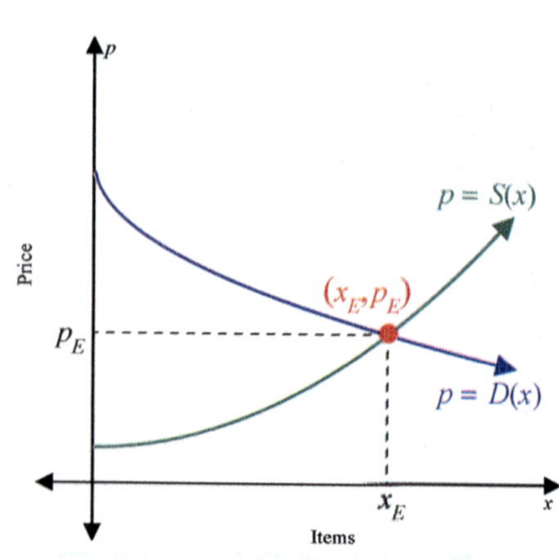

FIGURE 2

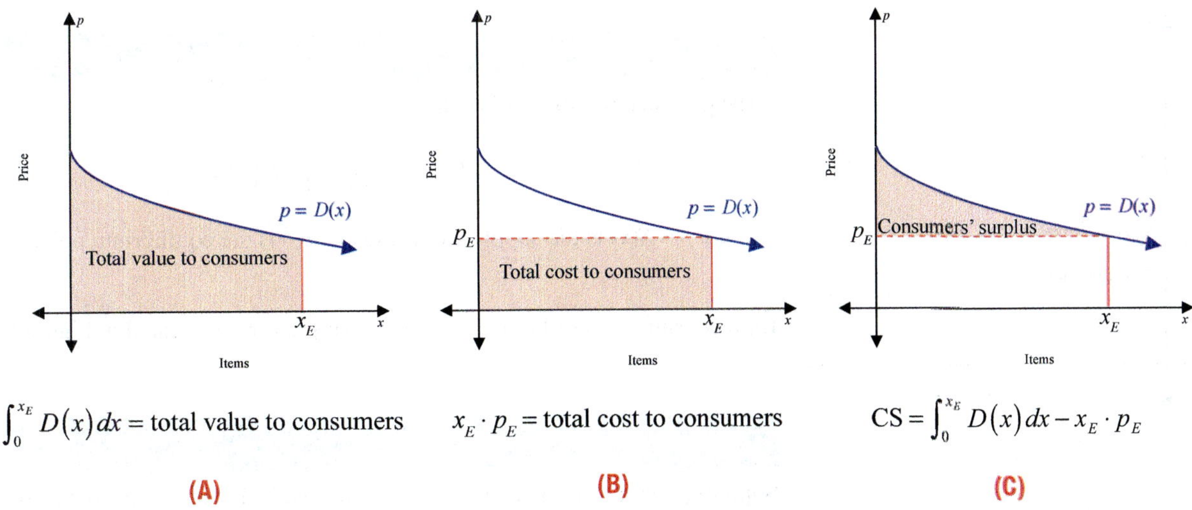

FIGURE 3

Consumers' Surplus

The **consumers' surplus** is defined to be

$$CS = \int_0^{x_E} D(x)\,dx - x_E \cdot p_E,$$

where $p = D(x)$ is the demand curve and (x_E, p_E) is the equilibrium point.

Some producers are pleased to sell at price p_E. They would have been willing to sell fewer items at a lower price. Thus the consumers have spent more than these producers anticipated. This extra income, called the **producers' surplus (PS)**, is the difference between the total cost to the consumers and the total value to the producers, as illustrated in Figure 4.

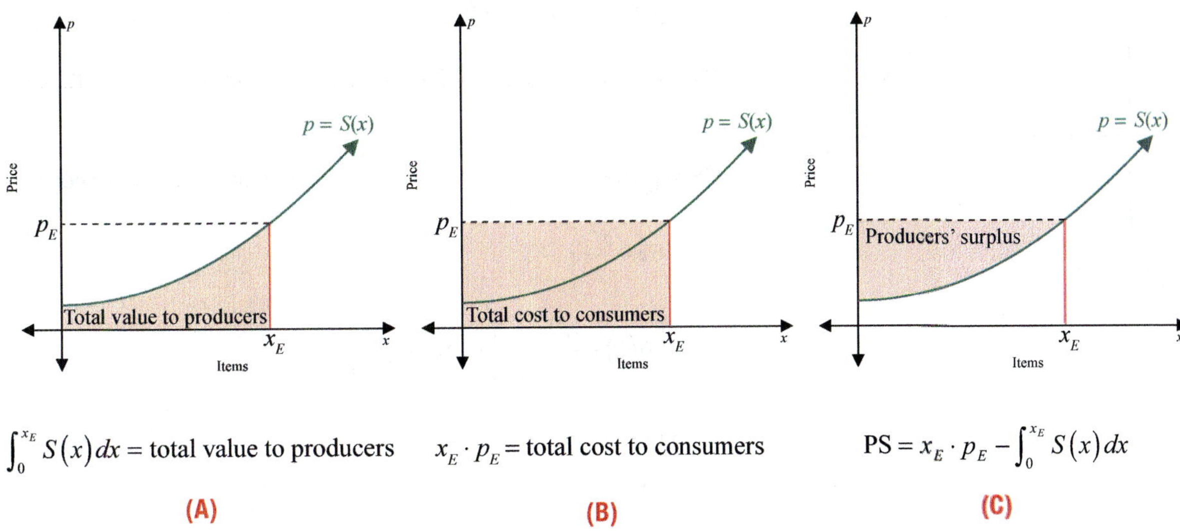

FIGURE 4

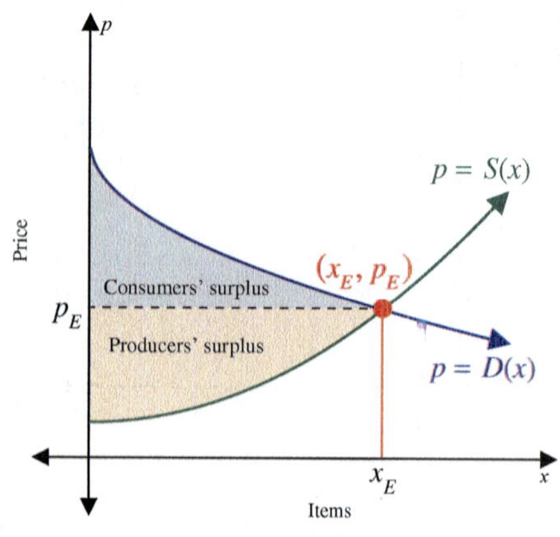

FIGURE 5

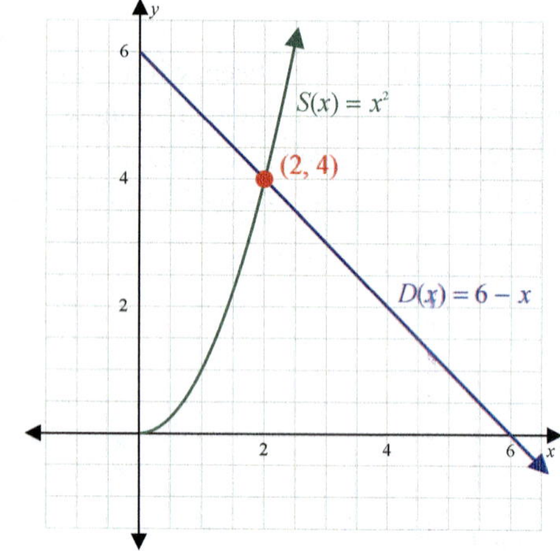

Producers' Surplus

The **producers' surplus** is defined to be

$$PS = x_E \cdot p_E - \int_0^{x_E} S(x)\,dx,$$

where $p = S(x)$ is the supply curve and (x_E, p_E) is the equilibrium point.

Both consumers' surplus and producers' surplus are illustrated in Figure 5.

Example 3: Determining Surpluses

Suppose, for a certain new brand of mechanical pencil, the demand function is $D(x) = 6 - x$ and the supply function is $S(x) = x^2$, where x is in thousands of pencils.

a. Find the equilibrium point.

b. Find the consumers' surplus.

c. Find the producers' surplus.

Solution

a. Set $S(x) = D(x)$ and solve for x.

$$x^2 = 6 - x$$
$$x^2 + x - 6 = 0$$
$$(x+3)(x-2) = 0$$
$$x = -3 \quad \text{or} \quad x = 2 \qquad \text{Note that } x \text{ cannot be negative.}$$

Therefore, $x_E = 2$ and $p_E = 6 - 2 = 4$. The equilibrium point is (2, 4).

b. $\text{CS} = \int_0^2 (6 - x)\,dx - (2 \cdot 4)$ Use the formula for consumers' surplus.

$$= 6x - \frac{x^2}{2}\Bigg]_0^2 - 8$$

$$= \left(6(2) - \frac{(2)^2}{2}\right) - \left(6(0) - \frac{(0)^2}{2}\right) - 8$$

$$= 12 - \frac{4}{2} - 8$$

$$= \$2$$

c. $\text{PS} = (2 \cdot 4) - \int_0^2 x^2 \, dx$ Use the formula for producers' surplus.

$$= 8 - \left(\frac{x^3}{3} \bigg]_0^2 \right)$$

$$= 8 - \left[\left(\frac{(2)^3}{3} \right) - \left(\frac{(0)^3}{3} \right) \right]$$

$$= 8 - \frac{8}{3}$$

$$= \frac{16}{3} \approx \$5.33$$

Lorenz Curves

Economists and sociologists use curves known as **Lorenz curves** to study the distribution of income within a society or country. In the graph of a Lorenz curve (see Figure 6), the x-axis shows the cumulative percentage of a country's families, starting from the lowest income families. The y-axis shows the cumulative percentage of the country's total income. Thus, for every number x in the interval $[0, 1]$, $f(x)$ is defined to be the percentage of the country's total income earned by the lower $100x$ percent of the families.

For example, in Figure 6, the point $(0.3, 0.1)$ indicates that the bottom 30 percent of the families earn 10 percent of the country's income. Similarly, the point $(0.7, 0.5)$ indicates that the bottom 70 percent earn 50 percent of the income.

If every family had the same income, then the distribution of income would be represented by $y = x$, the **line of complete equality**. The area between the line $y = x$ and the Lorenz curve $y = f(x)$ is used to measure how much the distribution of income differs from complete equality. The ratio of the area between $y = x$ and $y = f(x)$ to the area under the line $y = x$ is called the **coefficient of inequality**. Since the square with upper right corner at $(1, 1)$ in Figure 6 has area 1, we see that the area under the line $y = x$ is $\frac{1}{2}$. Thus

$$\text{Coefficient of inequality} = \frac{\text{area between curves}}{\frac{1}{2}} = 2 \cdot (\text{area between curves}).$$

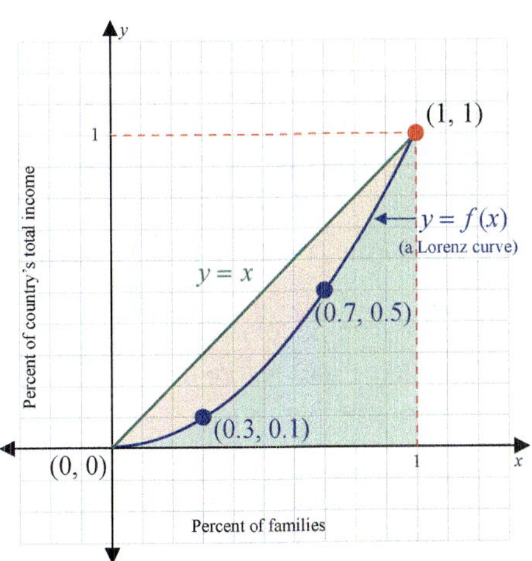

In this graph, percent is represented in decimal form.

FIGURE 6

Coefficient of Inequality for a Lorenz Curve

If $y = f(x)$ represents a Lorenz curve, then

$$\text{Coefficient of inequality} = 2 \int_0^1 \left[x - f(x) \right] dx.$$

A coefficient of 0 corresponds to a distribution along the line of complete equality. A coefficient of 1 corresponds to a maximum of "inequality" in which $f(x) = 0$. Any function that describes a Lorenz curve must have the following properties:

1. The domain is $[0, 1]$.

2. The range is $[0, 1]$.

3. $f(0) = 0$ and $f(1) = 1$.

4. $f(x) \le x$ for all x in the interval $[0, 1]$.

Example 4: Using a Lorenz Curve

Suppose that $f(x) = 0.6x^2 + 0.4x$ represents a Lorenz curve for some country.

a. What percent of the country's total income is earned by the lower 50 percent of the families in this country?

b. Find the coefficient of inequality.

Solution

a. $f(0.5) = 0.6(0.5)^2 + 0.4(0.5) = 0.35$ Since we want the lower 50% of families, set $x = 0.5$.

The lower 50 percent of the families earn 35 percent of the country's total income.

b. $2\int_0^1 \left[x - \left(0.6x^2 + 0.4x \right) \right] dx$ Combine terms in the integrand before finding the antiderivative.

$$= 2\int_0^1 \left[\left(0.6x - 0.6x^2 \right) \right] dx$$

$$= 2 \left(\frac{0.6x^2}{2} - \frac{0.6x^3}{3} \right) \Bigg]_0^1$$

$$= 2 \left[\left(\frac{0.6(1)^2}{2} - \frac{0.6(1)^3}{3} \right) - \left(\frac{0.6(0)^2}{2} - \frac{0.6(0)^3}{3} \right) \right]$$

$$= 2 \left[(0.3 - 0.2) - 0 \right]$$

$$= 0.2$$

The coefficient of inequality is 0.2.

14.6 EXERCISES

💡 PRACTICE

In Exercises 1–16, find the area of the region bounded by the graphs of the given equations.

1. $y = x^2$, $y = x - 1$, $x = -1$, $x = 4$ 2. $y = x^2 + 2$, $y = x$, $x = 2$, $x = 5$

3. $y = x^3$, $y = x^2$, $x = 0$, $x = 1$ 4. $y = x^2 + 1$, $y = 1 - 2x$, $x = 0$, $x = 3$

5. $y = \sqrt{2x+1}$, $y = 3x+2$, $x = 0$, $x = 2$ **6.** $y = e^{x-1}$, $y = x$, $x = 1$, $x = 4$

7. $y = e^{-x}$, $y = x+1$, $x = 0$, $x = 3$ **8.** $y = x^2 +1$, $y = e^{-0.2x}$, $x = 0$, $x = 4$

9. $y = \dfrac{1}{x}$, $y = \dfrac{5}{2} - x$, $x = \dfrac{1}{2}$, $x = 2$ **10.** $y = \dfrac{1}{x+1}$, $y = e^{0.7x}$, $x = 0$, $x = 2$

11. $y = x+1$, $y = x^2 + x$ **12.** $y = x^2 +1$, $y = 6-x$

13. $y = \sqrt{x}$, $y = x^2$ **14.** $y = x^2 - 6x$, $y = -x^2$

15. $y = x^2 - 2x - 3$, $y = 2x+2$ **16.** $y = x^2 + 5x - 1$, $y = 2 - x^2$

For Exercises 17–21, determine the area pictured (check each answer using a graphing utility if possibe). In each case you must determine the limits of integration if necessary.

17.

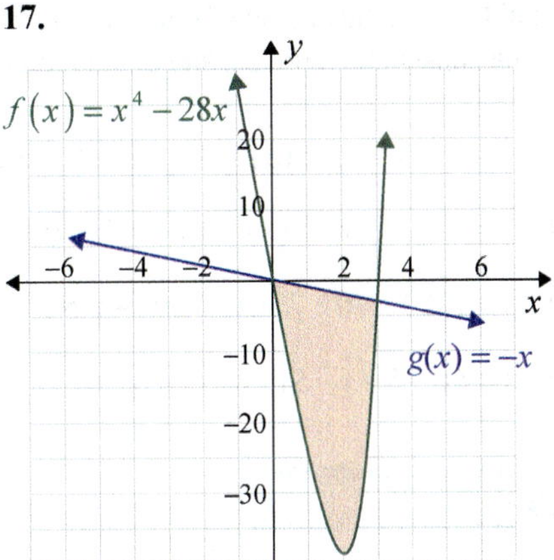

18.

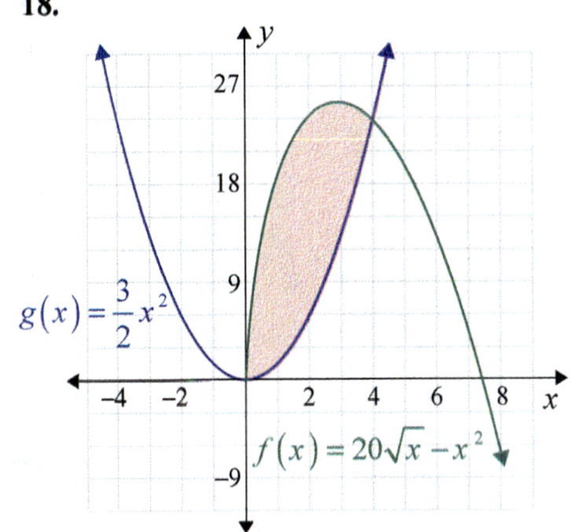

19.

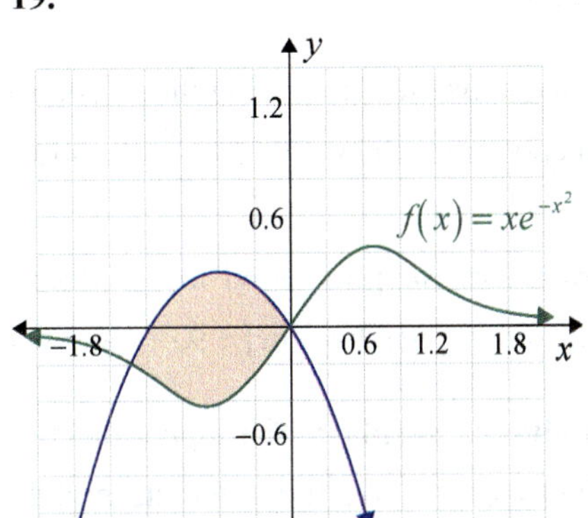

20.

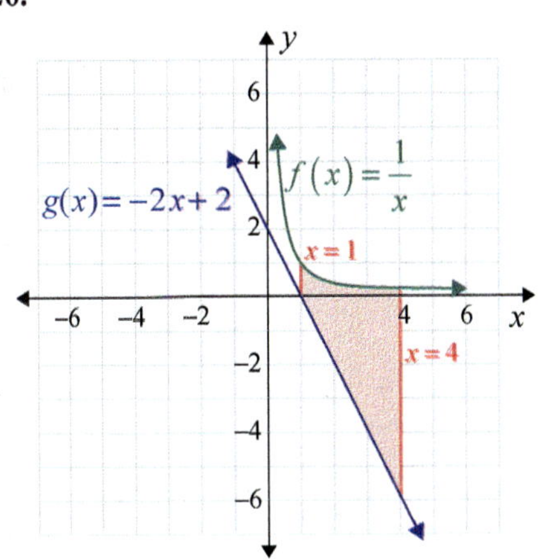

21.

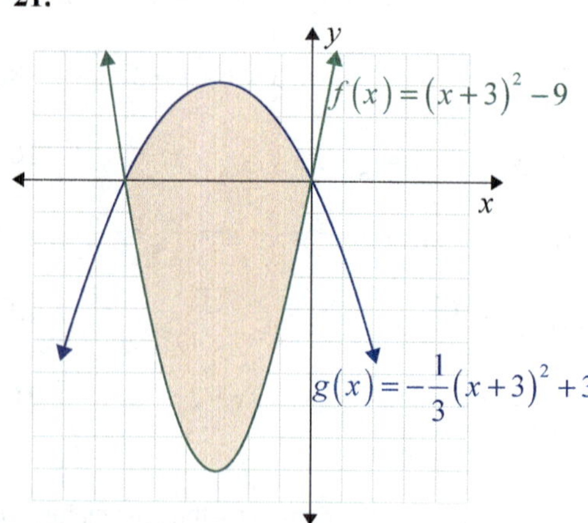

For each of the demand and supply functions in Exercises 22–25, find **a.** the equilibrium point, **b.** the consumers' surplus, and **c.** the producers' surplus.

22. $D(x) = 18 - 0.4x, \ S(x) = 3 + 0.1x, \ 0 \le x \le 40$

23. $D(x) = 24 - 0.2x, \ S(x) = 10 + 0.5x, \ 0 \le x \le 100$

24. $D(x) = 1000 - 30x, \ S(x) = 200 + 0.5x^2, \ 0 \le x \le 30$

25. $D(x) = 66 - 5\sqrt{x}, \ S(x) = 16 + x, \ 0 \le x \le 120$

🚀 **APPLICATIONS**

26. Consumers' surplus: The demand function for a particular product is given by the function $D(x) = 24 - 0.6x - 0.03x^2$. If $x_E = 10$ units, find the consumers' surplus.

27. Consumers' surplus: Find the consumers' surplus for a product if the demand function is given by $D(x) = \dfrac{800}{x+4}$ and $x_E = 4$ units.

28. Producers' surplus: Find the producers' surplus for a product if the supply function is given by $S(x) = 9e^{0.4x}$ and $x_E = 5$ units.

29. Producers' surplus: The supply function for a product is given by the function $S(x) = \sqrt{16 + 1.5x}$. If $x_E = 6$ units, find the producers' surplus.

30. Consumers' and producers' surplus: The demand curve for a product is given by $D(x) = 18 - 3x$ and the corresponding supply curve is $S(x) = 3x + 6$. Find the consumers' surplus and the producers' surplus.

31. Consumers' and producers' surplus: The demand curve for a product is given by $D(x) = 125 - 15x$ and the corresponding supply curve is $S(x) = 50 + 10x$. Find the equilibrium point, the CS, and the PS.

32. Lorenz curve: The income distribution of a small country is estimated by the Lorenz curve $f(x) = \dfrac{13}{18}x^2 + \dfrac{5}{18}x$.

  a. What percentage of the country's total income is earned by the lower 80 percent of its families? Round to the nearest percentage point.

  b. Find the coefficient of inequality.

33. Lorenz curve: The Lorenz curve for estimating the income distribution of a country is given by $f(x) = \dfrac{7}{16}x^2 + \dfrac{9}{16}x$.

  a. What percentage of the country's total income is earned by the lower 70 percent of its families? Round to the nearest percentage point.

  b. Find the coefficient of inequality.

34. Lorenz curve: A study shows that the income distribution of farmers in a certain state is estimated by $f(x) = 0.47x^3 + 0.24x^2 + 0.29x$.

  a. What percentage of the state's farming income is earned by the lower 60 percent of the state's farmers? Round to the nearest percentage point.

  b. Find the coefficient of inequality.

35. Lorenz curve: In a certain state the income distribution for the lumber and the logging industry is estimated by $f(x) = 1.16x^3 - 0.82x^2 + 0.66x$.

  a. What percentage of the state's lumber and logging income is earned by the lower 50 percent of the companies? Round to the nearest percentage point.

  b. Find the coefficient of inequality.

36. Lorenz curve: The income distribution for a certain country in 1996 was estimated by the function $f(x) = 0.34x + 0.66x^2$. In 2000 the income distribution was estimated by the function $f(x) = 0.3x + 0.72x^2 - 0.02x^3$.

  a. Find the coefficient of inequality for each of the years.

  b. Which year had a more equitable income distribution?

37. Lorenz curve: The income distribution for country A is estimated by the function $f(x) = 0.24x + 0.72x^2 + 0.04x^3$. The income distribution for country B is estimated by the function $f(x) = 0.28x + 0.69x^2 + 0.03x^3$.

  a. Find the coefficient of inequality for each of the two countries.

  b. Which country has a more equitable income distribution?

For Exercises 38–41, use a graphing utility to determine the area.

38.

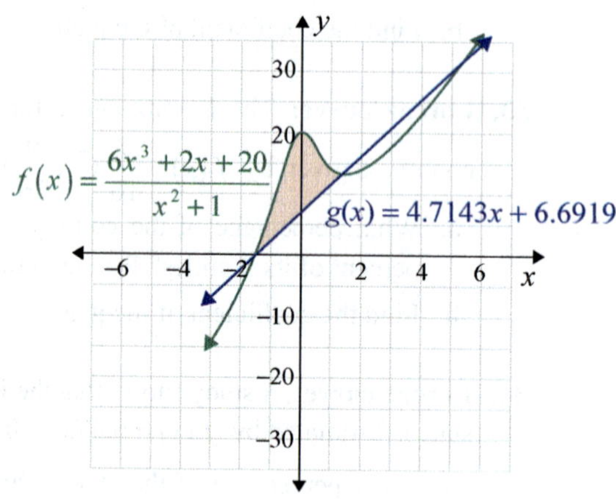

39.

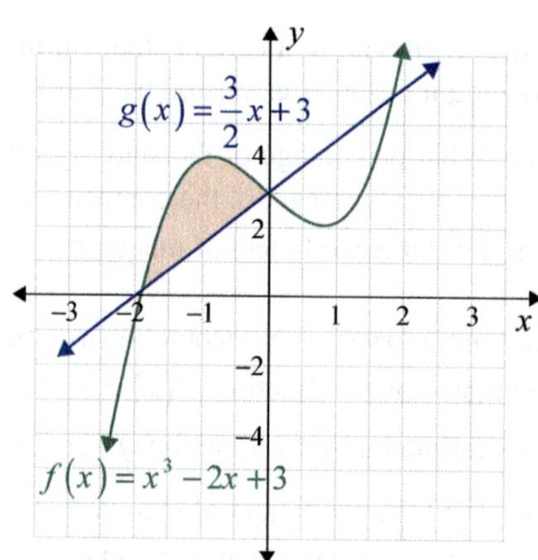

40.

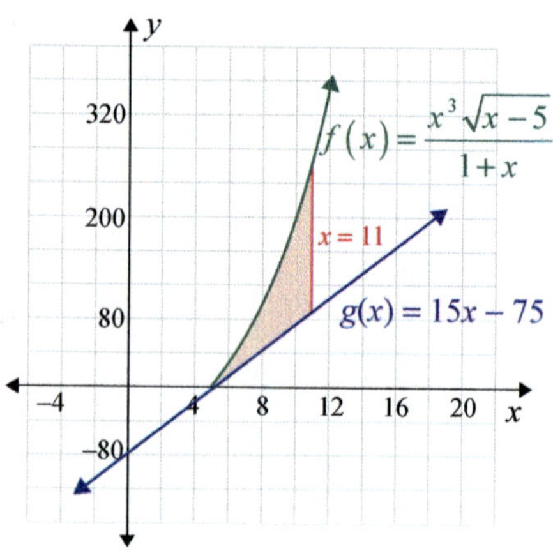

41.

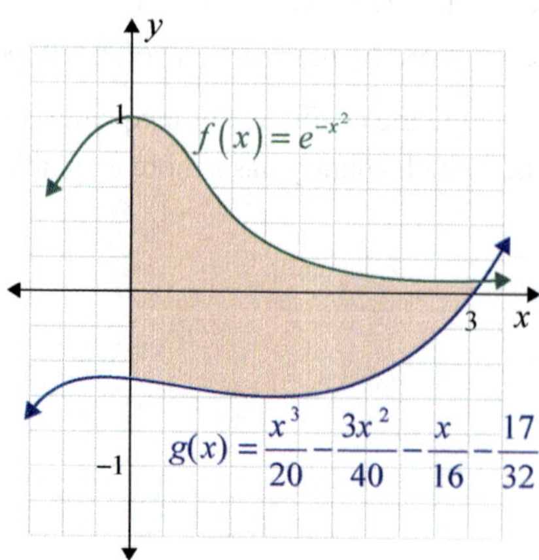

14.7 DIFFERENTIAL EQUATIONS

■ TOPICS

- ■ Verification of Solutions
- ■ Separation of Variables
- ■ Applications

A **differential equation** is an equation that involves differentials or derivatives. Previously, our discussion of exponential growth and decline and continuously compounded interest involved one type of differential equation,

$$\frac{dA}{dt} = kA,$$

and the general solution,

$$A = ce^{kt}.$$

Other examples of differential equations are

1. $\dfrac{dy}{dx} = -\dfrac{x}{y}$,

2. $\dfrac{d^2 y}{dt^2} + 2\dfrac{dy}{dt} = 0$,

3. $y' - 2y = 4x$, and

4. $y'' - y' - 6y = 0$.

Applications of differential equations appear in such fields as engineering, physics, business, economics, biology, and psychology in problems involving motion or measurement of rates of change. In general, the techniques and integrals used in solving differential equations are not elementary in nature. In fact, entire courses are devoted to the study of differential equations. Our presentation is limited to a few important types of problems.

Verification of Solutions

The order of a differential equation is the order of the highest derivative that appears in the equation. In the examples shown previously, equations 1 and 3 are *first-order* differential equations and equations 2 and 4 are *second-order* differential equations.

> ### Solution of a Differential Equation
>
> A function $y = f(x)$ is a **solution of a differential equation** if the function and its appropriate derivatives satisfy the equation.

Solutions of differential equations are classified as

1. **general solutions** if arbitrary constants are involved,

2. **particular solutions** if no arbitrary constants are involved, or

3. the **trivial solution** $y = 0$.

For example, $y = ce^{3t}$ is a general solution of the equation $y' - 3y = 0$, and $y = 4e^{3t}$ is a particular solution of the same equation.

The following examples illustrate how to verify that a given function is a solution of a differential equation.

Example 1: Solutions of Differential Equations

Verify that the function $y = cx^3 - 4$ is a general solution of the differential equation $xy' - 3y = 12$.

Solution

Step 1: Find y'.

$$y = cx^3 - 4$$
$$y' = 3cx^2$$

Step 2: Substitute for y and y'.

$$xy' - 3y = x\left(3cx^2\right) - 3\left(cx^3 - 4\right)$$
$$= 3cx^3 - 3cx^3 + 12$$
$$= 12$$

Therefore, since the function contains an arbitrary constant, c, and is a solution, it is a general solution.

Example 2: Solutions of Differential Equations

Verify that the function $y = e^{3x} + e^{-2x}$ satisfies the second-order differential equation $y'' - y' - 6y = 0$ and is, therefore, a particular solution.

Solution

Step 1: Find y' and y''.

$$y = e^{3x} + e^{-2x}$$
$$y' = 3e^{3x} - 2e^{-2x}$$
$$y'' = 9e^{3x} + 4e^{-2x}$$

Step 2: Substitute for y, y', and y''.

$$y'' - y' - 6y = \left(9e^{3x} + 4e^{-2x}\right) - \left(3e^{3x} - 2e^{-2x}\right) - 6\left(e^{3x} + e^{-2x}\right)$$
$$= 9e^{3x} + 4e^{-2x} - 3e^{3x} + 2e^{-2x} - 6e^{3x} - 6e^{-2x}$$
$$= (9 - 3 - 6)e^{3x} + (4 + 2 - 6)e^{-2x}$$
$$= 0$$

Therefore, the given function is a solution, and since it contains no arbitrary constants, it is a particular solution.

Separation of Variables

Sometimes a first-order differential equation can be written in the form

$$g(y)y' = f(x).$$

In such cases we can treat $y' = \dfrac{dy}{dx}$ as a ratio of two differentials dy and dx and rewrite the equation in the form $g(y)dy = f(x)dx$.

This equation is called a **separable equation** because the variables x and y and their corresponding differentials are separated and appear on opposite sides of the equation. If $g(y)$ and $f(x)$ are continuous functions, the general solution can be found by integrating both sides:

$$\int g(y)\,dy = \int f(x)\,dx.$$

Only one constant of integration is necessary and is added after this step. This technique of solving first-order differential equations is called the method of **separation of variables** and is essentially a three-step process.

To Solve a First-Order Separable Differential Equation

Step 1: Write y' as $\dfrac{dy}{dx}$.

Step 2: Separate variables by writing dx with all terms involving x on one side of the equation and dy with all terms involving y on the other side.

Step 3: Integrate both sides of the new equation.

The method of solving differential equations by separating variables is illustrated in Examples 3 and 4.

Example 3: Using the Method of Separating Variables

Solve the differential equation $y' = \dfrac{y}{x-1}$ where $x > 1$ and $y > 0$.

Solution

Step 1: $\quad \dfrac{dy}{dx} = \dfrac{y}{x-1}$ $\qquad\qquad$ Write y' as $\dfrac{dy}{dx}$.

Step 2: $\quad \dfrac{dy}{y} = \dfrac{dx}{x-1}$ $\qquad\qquad$ Separate the variables.

Step 3: $\quad \displaystyle\int \dfrac{1}{y}\,dy = \int \dfrac{1}{x-1}\,dx$ $\qquad$ Integrate both sides of the equation.

$$\ln y = \ln(x-1) + C$$

$$\ln y = \ln(x-1) + \ln k$$

$$\ln y = \ln k(x-1)$$

$$y = k(x-1)$$

We rewrite the constant C as the constant $\ln k$ to help simplify the expression.

Example 4: Using the Method of Separating Variables

Solve the differential equation $y' = 2xy$, where $y > 0$.

Solution

Step 1: $\dfrac{dy}{dx} = 2xy$ Write y' as $\dfrac{dy}{dx}$.

Step 2: $\dfrac{dy}{y} = 2xdx$ Separate the variables.

Step 3: $\displaystyle\int \dfrac{1}{y}dy = \int 2xdx$ Integrate both sides of the equation.

$$\ln y = x^2 + C$$

$$y = e^{x^2 + C}$$

$$y = e^C e^{x^2}$$

$$y = ke^{x^2}$$

Solve for y by rewriting the equation in exponential form. e^C is a positive constant and is represented by the single letter k.

The following example is called an **initial-value problem** because an initial condition is given on x and y that allows the evaluation of the arbitrary constant C in the solution of the differential equation. Thus an initial-value problem results in a particular solution.

Example 5: Initial-Value Problem

Solve the initial-value problem $y' = x$ with initial condition $f(0) = 1$. (That is, $y = 1$ when $x = 0$.)

Solution

$$\dfrac{dy}{dx} = x \qquad \text{Step 1}$$

$$dy = xdx \qquad \text{Step 2}$$

$$\int dy = \int xdx \qquad \text{Step 3}$$

$$y = \dfrac{1}{2}x^2 + C \qquad \text{This is the general solution.}$$

Since $f(0) = 1$, we have

$$1 = \dfrac{1}{2}\cdot 0^2 + C$$

$$1 = C$$

Therefore, the particular solution is $y = \dfrac{1}{2}x^2 + 1$.

Applications

We have discussed elasticity of demand as represented by the formula,

$$E = -\frac{1}{x} \cdot \frac{D(x)}{D'(x)},$$

where $p = D(x)$ is a demand function and $D'(x) = \dfrac{dp}{dx}$. If E is a known constant value, then the formula can be treated as a differential equation that has the demand function as its solution.

Example 6: Elasticity of Demand

Find the demand function $p = D(x)$ given that the elasticity of demand $E = 3$ for all positive x.

Solution

$$3 = -\frac{1}{x} \cdot \frac{p}{\frac{dp}{dx}}$$
Use the formula for E.

$$\frac{1}{p}\,dp = -\frac{1}{3x}\,dx$$
Separate the variables.

$$\int \frac{1}{p}\,dp = -\frac{1}{3}\int \frac{1}{x}\,dx$$
Integrate both sides of the equation.

$$\ln p = -\frac{1}{3}\ln x + C$$
Since $x > 0$, no absolute value signs are needed for $\ln x$.

$$\ln p = -\frac{1}{3}\ln x + \ln k$$
We rewrite the constant C as the constant $\ln k$ to help simplify the expression.

$$\ln p = \ln\left(k \cdot x^{\frac{-1}{3}}\right)$$

$$p = k \cdot x^{\frac{-1}{3}}$$

$$p = \frac{k}{\sqrt[3]{x}}$$

Another interesting application involves an equation known as the **logistic equation**. This equation (and its corresponding curve, known as a **logistic curve** or **S curve**) represents a type of modified exponential growth. The rate of growth is slow at first, increases to a maximum rate, and then tapers off as the population approaches some upper limit (or upper bound). (See Figure 1.)

We know that the exponential growth of a population P can be described by the differential equation

$$\frac{dP}{dt} = kP.$$

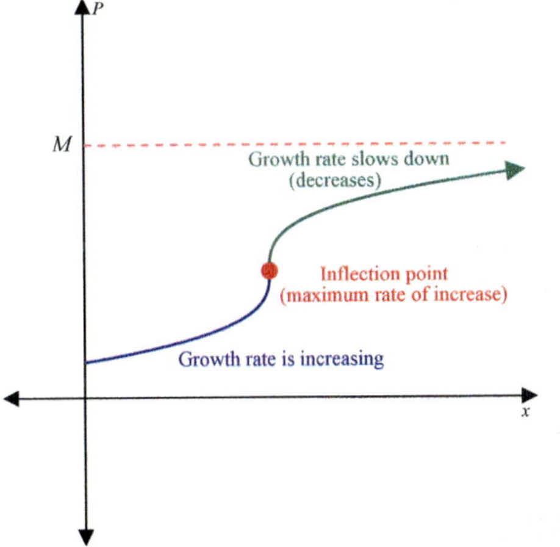

FIGURE 1: The Graph of a Logistic Curve

Prolonged exponential growth is unrealistic in many applications. In 1844, the Belgian mathematician Pierre François Verhulst introduced an inhibiting factor proportional to $-P^2$ resulting in the **logistic equation**

$$\frac{dP}{dt} = k\left(MP - P^2\right).$$

We can solve this equation by separating variables in the form

$$\frac{dP}{P(M-P)} = kdt$$

and using the table of solutions (Table 1) given at the end of this section. The solution (where $P \ne M$ and $P \ne 0$) is

$$P(t) = \frac{M}{1 + Ce^{-Mkt}}.$$

The constant M is the limiting value (or upper bound) of P as $t \to +\infty$.

Example 7: Population of Bacteria

At the beginning of an experiment, a culture of 100 bacteria is growing in a medium that will allow a maximum of 10,000 bacteria to survive. If the population is 200 after 5 hours, use the logistic equation to represent the number of bacteria present t hours after the experiment has begun.

Solution

We know that the solution function has the form

$$P(t) = \frac{10,000}{1 + Ce^{-10,000kt}}.$$

We also know that $P(0) = 100$ and $P(5) = 200$. Substituting $t = 0$ and $P = 100$ gives

$$100 = \frac{10,000}{1 + Ce^0}$$

$$100 = \frac{10,000}{1 + C}$$

$$C = 99.$$

Substituting $t = 5$ and $P = 200$ gives

$$200 = \frac{10,000}{1 + 99e^{-10,000k(5)}}$$

$$200 = \frac{10,000}{1 + 99e^{-50,000k}}$$

$$1 + 99e^{-50,000k} = 50$$

$$e^{-50,000k} = \frac{49}{99}$$

$$-50,000k = \ln\left(\frac{49}{99}\right)$$

$$-50,000k \approx -0.7033$$

$$k \approx 0.000014.$$

Thus the final form for $P(t)$ is $P(t) = \dfrac{10,000}{1+99e^{-0.14t}}$.

Table 1 contains the general solutions of differential equations that arise in three basic growth applications.

Application	Differential Equation	General Solution
Unbounded growth	$\dfrac{dy}{dt} = ky$	$y = Ce^{kt}$
Bounded growth	$\dfrac{dy}{dt} = k(M-y)$	$y = M + Ce^{-kt}$
Logistic curve	$\dfrac{dy}{dt} = ky(M-y)$	$y = \dfrac{M}{1+Ce^{-Mkt}}$

TABLE 1: General Solutions for Growth Applications

14.7 EXERCISES

💡 PRACTICE

In Exercises 1–12, verify that the differential equation has the given function as a particular solution.

1. $\dfrac{dy}{dx} = 6, \ y = 6x - 1$

2. $\dfrac{dy}{dx} = 3x - 2, \ y = \dfrac{3}{2}x^2 - 2x - 4$

3. $\dfrac{dy}{dx} = 3 + y, \ y = e^x - 3$

4. $x\dfrac{dy}{dx} - x + 2 = 0, \ y = x - 2\ln x + 7$

5. $\dfrac{dy}{dx} = y^{\frac{3}{2}}, \ y = \dfrac{4}{(5-x)^2}$

6. $\dfrac{dy}{dx} = -0.5y, \ y = 3e^{-0.5x}$

7. $2\dfrac{dy}{dx} + 3y = 1, \ y = \dfrac{1}{3} - 2e^{-1.5x}$

8. $x\dfrac{dy}{dx} = xy + y, \ y = 4xe^x$

9. $2x^2 y'' - xy' - 2y = 4 - 15x, \ y = 3x^2 + 5x - 2$

10. $x^2 y'' + xy' - y + \ln x = 0, \ y = x + \ln x$

11. $x^2 y'' - xy' + y = 0, \ y = x\ln x$

12. $y'' - 2y' + y = 0, \ y = e^x(x+2)$

In Exercises 13–20, find the solution of each separable differential equation.

13. $\dfrac{dy}{dx} = 3x + \dfrac{1}{x}$

14. $\dfrac{dy}{dx} = -3xy$

15. $\dfrac{dy}{dx} = \dfrac{2y-1}{x+1}$

16. $x\dfrac{dy}{dx} = \dfrac{x^2 + 1}{y^2}$

17. $\dfrac{dy}{dx} = -0.4y$

18. $\dfrac{dy}{dx} = -2(26 - y)$

19. $\dfrac{dy}{dx} = (1 - 5x)y^2$

20. $\dfrac{dy}{dx} = (x^2 + 1)e^{-y}$

In Exercises 21–32, solve each initial-value problem or obtain a general solution as indicated. (Refer to Table 1 in the text if necessary.)

21. $\dfrac{dy}{dx} = 0.6y(20 - y)$

22. $\dfrac{dy}{dx} = 0.3y(50 - y)$

23. $\dfrac{dy}{dx} = 0.25y, \ y = 5$ when $x = 0$

24. $\dfrac{dy}{dx} = -3x, \ y = 10$ when $x = 2$

25. $x\dfrac{dy}{dx} = y + 1, \ y = 14$ when $x = 3$

26. $\dfrac{dy}{dx} = -2xy, \ y = 18$ when $x = 0$

27. $\dfrac{dy}{dx} = -6x^2 y^2, \ y = 2$ when $x = 1$

28. $\dfrac{dy}{dx} = \dfrac{xy}{x^2 + 1}, \ y = 7$ when $x = 0$

29. $\dfrac{dy}{dx} = 8x + 2xy, \ y = 10$ when $x = 0$

30. $\dfrac{dy}{dx} = 0.3(80 - y), \ y = 60$ when $x = 0$

31. $\dfrac{dy}{dx} = 0.8y(40 - y), \ y = 30$ when $x = 0$

32. $\dfrac{dy}{dx} = 0.04y(60 - y), \ y = 10$ when $x = 0$

🚀 APPLICATIONS

33. Elasticity of demand: The elasticity of demand for a product is given by $E = 1.5$. Find the demand function $p = D(x)$ if $D(8) = 24$.

34. Elasticity of demand: The elasticity of demand for a product is given by $E = 2$. Find the demand function $p = D(x)$ if $D(25) = 30$.

35. Elasticity of demand: The elasticity of demand for a product is given by $E = \dfrac{2(120 - x)}{x}$. Find the demand function $p = D(x)$ if $D(20) = 180$.

36. Elasticity of demand: The elasticity of demand for a product is given by $E = \dfrac{60 - 0.4x}{0.2x}$. Find the demand function $p = D(x)$ if $D(70) = 28$.

37. **Resale value:** The resale or salvage value V of a machine decreases at a rate proportional to its value. Thus $\dfrac{dV}{dt} = -kV$, where t is the machine's age in years and k is its rate of decrease in value.

    a. Find the expression for the value when the machine is t years old if the original value was \$24,000 and the rate of decrease is 6 percent.
    b. Find the value of the machine when it is 7 years old.

38. **Drug concentration:** The amount A of a drug remaining in a body t hours after an injection decreases at a rate proportional to the amount present. This suggests the differential equation $\dfrac{dA}{dt} = -kA$. The amount of a certain drug decreases at a rate of 3 percent per hour. Find the amount of the drug remaining in the body 4 hours after an injection of 20 cc of the drug.

39. **Newton's Law of Cooling:** Newton's Law of Cooling states that the rate at which the temperature T of an object changes is proportional to the difference between the temperature of the object and the temperature of the surrounding medium. That is $\dfrac{dT}{dt} = -k(T - M)$, where k is the constant of proportionality, t is time, and M is the constant temperature of the medium.

    a. Solve the differential equation for $T(t)$.
    b. Find $T(5)$, if $T(0) = 78°$, $M = 26°$, and $k = 0.3$.

40. **Newton's Law of Cooling:** The temperature of a roast was $160°$ when it was removed from an oven and placed in a room with constant temperature of $76°$. After 10 minutes, the temperature of the roast was $152°$. Find the temperature 20 minutes after the roast was removed from the oven. (See Exercise 39.)

41. **Spread of a rumor:** In a small community with a population of 2800, a rumor about the mayor was started. The rate at which the rumor spread was approximated by $\dfrac{dN}{dt} = 0.0003N(2800 - N)$ people per day, where N is the number of people who have heard the rumor t days after the rumor was started.

    a. Write an equation for $N(t)$, assuming that 20 people have heard the rumor at $t = 0$.
    b. How many days will it take for 1500 people to hear the rumor? Round to the nearest day.

42. **Spread of a disease:** The population of seals on an island is about 600. Biologists estimate that there are 12 seals with a very infectious disease. The disease will spread at a rate $\dfrac{dN}{dt} = 0.006N(600 - N)$ seals per day, where N is the number of infected seals t days after the discovery of the disease.

    a. Write a function $N(t)$ for the number of seals infected t days after the discovery of the disease.
    b. At what time t will 300 seals will be infected? Round to the nearest day.

15

Chapter 15

ADDITIONAL INTEGRATION TOPICS

15.1 INTEGRATION BY PARTS

We previously introduced four basic formulas and two rules for integration. We then expanded the applications of these same four formulas by using the technique of substitution. In this section we will develop another substitution technique that will allow us to evaluate an even wider variety of integrals. This new technique, called **integration by parts**, is based on the Product Rule for Differentiation:

$$\frac{d}{dx}\left[f(x)\cdot g(x)\right]=f(x)\cdot g'(x)+g(x)\cdot f'(x).$$

If we substitute

$$u=f(x),\quad v=g(x),\quad \frac{du}{dx}=f'(x),\quad \text{and}\quad \frac{dv}{dx}=g'(x),$$

the formula can be written as

$$\frac{d}{dx}[u\cdot v]=u\cdot\frac{dv}{dx}+v\cdot\frac{du}{dx}.$$

Now, thinking in terms of differentials, we multiply both sides of the equation by dx and get

$$d[u\cdot v]=u\cdot dv+v\cdot du.$$

Integrating both sides of the equation gives

$$\int d[u\cdot v]=\int u\cdot dv+\int v\cdot du$$

and

$$u\cdot v=\int u\cdot dv+\int v\cdot du.$$

Solving for $\int u\cdot dv$ gives the desired form,

$$\int u\cdot dv=u\cdot v-\int v\cdot du.$$

This equation is known as the formula for **integration by parts** or just as the **parts formula**. The given integral is $\int u\cdot dv$, and the two parts are u and dv. The objective is to choose these two parts in such a way that the resulting integral on the right, $\int v\cdot du$, is easier to evaluate than the original integral.

> **Formula for Integration by Parts**
>
> $$\int u\cdot dv=u\cdot v-\int v\cdot du$$

> **✓ NOTE**
>
> The constant of integration C is not an explicit part of the formula. It will appear later when the last integration is performed.

To see how this formula works, consider the integral

$$\int xe^x dx$$

and let

$$u=x\quad \text{and}\quad dv=e^x dx.$$

Then

$$du=dx\quad \text{and}\quad v=\int dv=\int e^x dx=e^x.$$

Now, by using the formula for integration by parts, we obtain

$$\int u \cdot dv = u \cdot v - \int v \cdot du$$

$$\int x \cdot e^x dx = x \cdot e^x - \int e^x dx \qquad \text{This last integral, } \int xe^x dx, \text{ is one we are familiar with.}$$

$$= xe^x - e^x + C.$$

Integration by parts can involve somewhat of a trial-and-error approach in the choice of u and dv. Suppose, for example, that we had chosen the parts as follows:

$$u = e^x \quad \text{and} \quad dv = x dx.$$

Then

$$du = e^x dx \quad \text{and} \quad v = \int dv = \int x dx = \frac{1}{2} x^2.$$

Integrating by parts gives

$$\int xe^x dx = \frac{1}{2} x^2 e^x - \int \frac{1}{2} x^2 e^x dx$$

and the last integral on the right is more difficult to evaluate than the original integral. In such a case, we simply start over and make new choices for u and dv.

As an organizational aid, we will use the following box-type format when we integrate by parts.

$u = $ (first part)	$dv = $ (second part)
$du = $ (differential of first part)	$v = $ (integral of second part)

The outline emphasizes that we have separated the original problem into two parts, one of which (the "u") needs to be differentiated (usually very easily) and the other of which (the "dv") needs to be integrated (also without difficulty, if we have chosen well). The following integrals are evaluated with the technique of integration by parts.

Example 1: Integration by Parts

Find $\int \ln x dx$.

Solution

The differential dx will always be part of the differential dv. In this case, $dv = \ln x dx$ is not appropriate since the integration of $\ln x dx$ is the original problem. Thus, we choose $dv = dx$ (a very easy integration) and $u = \ln x$.

$u = \ln x$	$dv = dx$
$du = \frac{1}{x} dx$	$v = \int dv = \int dx = x$

Now we substitute in the parts for our formula.

$$\int u dv = uv - \int v du$$

$$\int \ln x dx = (\ln x) x - \int x \frac{1}{x} dx$$

$$= x \ln x - x + C$$

As a check, we differentiate.

$$\frac{d}{dx}\left(x\ln x - x + C\right) = x \cdot \frac{d}{dx}\ln x + \ln x \cdot \frac{d}{dx}x - 1 + 0$$

$$= x \cdot \frac{1}{x} + \ln x(1) - 1 + 0$$

$$= 1 + \ln x - 1$$

$$= \ln x$$

This shows the solution is correct.

Example 2: Integration by Parts

Find $\int x\ln x\, dx$.

Solution

Since we differentiate the choice for u, we could let $u = x$ (with $dv = \ln x\, dx$) or $u = \ln x$ (with $dv = x\, dx$). However, we integrate the choice for dv, so we do **not** choose $dv = \ln x\, dx$ because we do not have a standard way to integrate this choice of dv directly (in theory we could use Example 1, but this is not as easy as the choice we will make).

$u = \ln x$	$dv = x\, dx$
$du = \dfrac{1}{x}\, dx$	$v = \int dv = \int x\, dx = \dfrac{1}{2}x^2$

Now we apply the formula for integration by parts.

$$\int u\, dv = uv - \int v\, du$$

$$\int \ln x \cdot x\, dx = \left(\ln x\right)\frac{1}{2}x^2 - \int \frac{1}{2}x^2 \cdot \frac{1}{x}\, dx \qquad \text{Rewrite } \int x\ln x\, dx \text{ as } \int \ln x \cdot x\, dx.$$

$$= \ln x \cdot \frac{1}{2}x^2 - \int \frac{1}{2}x\, dx$$

$$= \frac{1}{2}x^2 \ln x - \frac{1}{2}\int x\, dx$$

$$= \frac{1}{2}x^2 \ln x - \frac{1}{2}\cdot\frac{1}{2}x^2 + C$$

$$= \frac{1}{2}x^2 \ln x - \frac{1}{4}x^2 + C$$

The cancellation which occurs when we multiply v times du tells us that the choice for u and dv was excellent!

Example 3: Integration by Parts

Find $\int x^2 e^{-x} dx$.

Solution

In this problem we apply integration by parts **twice**.

$u = x^2$	$dv = e^{-x} dx$
$du = 2x\,dx$	$v = \int e^{-x} dx = -e^{-x}$

Therefore,

$$\int u\,dv = uv - \int v\,du$$

$$\int x^2 e^{-x} dx = x^2 \cdot \left(-e^{-x}\right) - \int -e^{-x} \cdot 2x\,dx$$

$$= -x^2 e^{-x} + \int 2x e^{-x} dx.$$

Now integrate $\int 2x e^{-x} dx$ by parts.

$u = 2x$	$dv = e^{-x} dx$
$du = 2\,dx$	$v = \int e^{-x} dx = -e^{-x}$

Therefore,

$$\int u\,dv = uv - \int v\,du$$

$$\int 2x e^{-x} dx = 2x \cdot \left(-e^{-x}\right) - \int -e^{-x} \cdot 2\,dx$$

$$= -2x e^{-x} + 2\int e^{-x} dx$$

$$= -2x e^{-x} - 2e^{-x} + C.$$

Putting all the results together, we have

$$\int x^2 e^{-x} dx = -x^2 e^{-x} - 2x e^{-x} - 2e^{-x} + C.$$

Example 4: Integration by Parts

Evaluate the definite integral $\int_1^5 x\sqrt{x-1}\,dx$.

Solution

$u = x$	$dv = \sqrt{x-1}\,dx$
$du = dx$	$v = \int (x-1)^{\frac{1}{2}} dx = \dfrac{2}{3}(x-1)^{\frac{3}{2}}$

Now we evaluate the definite integral using the formula for integration by parts.

$$\int u\,dv = uv - \int v\,du$$

$$\int_1^5 x\sqrt{x-1}\,dx = x\cdot\frac{2}{3}(x-1)^{\frac{3}{2}}\Big]_1^5 - \int_1^5 \frac{2}{3}(x-1)^{\frac{3}{2}}\,dx$$

$$= \frac{2}{3}x(x-1)^{\frac{3}{2}}\Big]_1^5 - \frac{2}{3}\cdot\frac{2}{5}(x-1)^{\frac{5}{2}}\Big]_1^5$$

$$= \frac{2}{3}x(x-1)^{\frac{3}{2}} - \frac{4}{15}(x-1)^{\frac{5}{2}}\Big]_1^5$$

$$= \left[\frac{10}{3}(4)^{\frac{3}{2}} - \frac{4}{15}(4)^{\frac{5}{2}}\right] - (0)$$

$$= \frac{80}{3} - \frac{128}{15}$$

$$= \frac{400}{15} - \frac{128}{15}$$

$$= \frac{272}{15}$$

15.1 EXERCISES

💡 PRACTICE

In Exercises 1–16, use the technique of integration by parts to evaluate the integrals.

1. $\displaystyle\int xe^{2x}\,dx$

2. $\displaystyle\int 3xe^{-x}\,dx$

3. $\displaystyle\int 2ye^{0.5y}\,dy$

4. $\displaystyle\int 5te^{0.4t}\,dt$

5. $\displaystyle\int \ln t\,dt$

6. $\displaystyle\int y^2 \ln y\,dy$

7. $\displaystyle\int x^3 \ln 5x\,dx$

8. $\displaystyle\int 8x\ln 3x\,dx$

9. $\displaystyle\int x\sqrt{x+2}\,dx$

10. $\displaystyle\int x\sqrt{x-3}\,dx$

11. $\displaystyle\int x(x+4)^{-2}\,dx$

12. $\displaystyle\int x(x-1)^{-3}\,dx$

13. $\displaystyle\int \frac{t}{2e^{0.6t}}\,dt$

14. $\displaystyle\int y^2 e^{3y}\,dy$

15. $\displaystyle\int \sqrt{x}\,\ln 7x\,dx$

16. $\displaystyle\int 3x(x-6)^{-\frac{2}{3}}\,dx$

In Exercises 17–22, use the technique of integration by parts to evaluate each definite integral. Round your answer to the nearest hundredth.

17. $\displaystyle\int_0^2 xe^{-2x}\,dx$

18. $\displaystyle\int_0^3 (x+1)e^{-0.5x}\,dx$

19. $\displaystyle\int_0^1 (x+2)e^{-4x}\,dx$

20. $\displaystyle\int_0^4 (1-2x)e^{1.2x}\,dx$

21. $\displaystyle\int_{-2}^3 \frac{x}{\sqrt{6+x}}\,dx$

22. $\displaystyle\int_0^4 x\sqrt{1+2x}\,dx$

In each of Exercises 23–30, identify the *u* and *dv* which would solve the integral using integration by parts. Then evaluate the integral and round your answer to the nearest hundredth.

23. $\displaystyle\int_0^1 4x(3x+1)^5\,dx$

24. $\displaystyle\int_1^2 \frac{x}{\sqrt{2x+5}}\,dx$

25. $\displaystyle\int_{-1}^2 (x+1)(x+2)^{\frac{3}{2}}\,dx$

26. $\displaystyle\int_1^4 \sqrt{x}\,\ln x\,dx$

27. $\displaystyle\int_1^5 x^2 \ln x\,dx$

28. $\displaystyle\int_1^3 \frac{\ln t}{t^2}\,dt$

29. $\displaystyle\int_0^6 \ln(x+1)\,dx$

30. $\displaystyle\int_1^2 (2x+1)\ln x\,dx$

In Exercises 31–40, use the technique of substitution or integration by parts to evaluate the integrals.

31. $\displaystyle\int 5te^{-2t}\,dt$

32. $\displaystyle\int 5te^{-2t^2}\,dt$

33. $\displaystyle\int \sqrt{3x}\,\ln x\,dx$

34. $\displaystyle\int \frac{\ln x}{x}\,dx$

35. $\displaystyle\int 3x(2x^2-1)^{\frac{3}{2}}\,dx$

36. $\displaystyle\int 3x(2x-1)^{\frac{3}{2}}\,dx$

37. $\displaystyle\int \frac{(\ln x)^2}{x}\,dx$

38. $\displaystyle\int x\ln x^2\,dx$

39. $\displaystyle\int \frac{e^x}{1-e^x}\,dx$

40. $\displaystyle\int \frac{x}{\sqrt{5x^2-3}}\,dx$

🚀 APPLICATIONS

41. Demand for a natural resource: The demand for a natural resource t years from now will be increasing at a rate of $te^{0.01t}$ million units per year. If the current demand is 80 million units, write a function for the demand t years from now.

42. Revenue: The marginal revenue for x units of a product is given by $R'(x)=(200-30x)e^{-0.15x}$ dollars per unit. Find the revenue function $R(x)$ if $R(0)=0$.

43. Revenue: The marginal revenue for x units of a product is given by $R'(x)=18-0.4\ln x$ dollars per unit, where $x \ge 1$. Find the revenue function if $R(1)=\$18.40$.

44. Resale value: The value of a machine depreciates at a rate of $-200t(t+1)^{-2}$ dollars per year, where t is the age (in years) of the machine. If the original cost of the machine is $\$540$, find a function for the value of the machine when it is t years old.

15.2 ANNUITIES AND INCOME STREAMS

■ TOPICS

- ■ Annuities
- ■ Income Streams

We are, in general, familiar with periodic payments and receipts of money such as rent payments, loan payments, and interest on savings accounts. In some large businesses, such as hotel and restaurant chains, banks, and department stores, money appears to flow continuously. We will find that even though this flow is not exactly continuous, we can represent the flow of money with a continuous function and obtain practical results. In this section we will discuss two topics related to the flow of money: annuities and income streams.

Annuities

An **annuity** consists of a series of equal payments made at equal time intervals. The **amount** (or **future value**) of an annuity is its total value, that is, the sum of the payments and interest earned on each payment. In the case of an **ordinary annuity**, the payments are made at the end of each time period. For an **annuity due**, the payments are made at the beginning of each time period. Mortgage payments and rent payments are examples of annuities due.

Suppose that you deposit $100 into a savings account at the end of each month for 6 months. Each $100 earns interest over a different period of time. (In fact, the last $100 does not earn any interest during the 6 months.) If interest on each deposit is compounded continuously at 12 percent, then the amount of the annuity at the end of the sixth month is the following sum. Note that since 12 percent $= 0.12$ is an annual rate of interest, $\dfrac{0.12}{12} = 0.01$ is the monthly rate of interest.

$$S_6 = 100e^{0.01(5)} + 100e^{0.01(4)} + 100e^{0.01(3)} + 100e^{0.01(2)} + 100e^{0.01(1)} + 100e^{0.01(0)}$$

$$= 100\left(e^{0.01(5)} + e^{0.01(4)} + e^{0.01(3)} + e^{0.01(2)} + e^{0.01(1)} + e^{0.01(0)}\right)$$

$$\approx 100(1.0513 + 1.0408 + 1.0305 + 1.0202 + 1.0101 + 1.00)$$

$$= 100(6.1529)$$

$$= \$615.29$$

The process is illustrated in Figure 1.

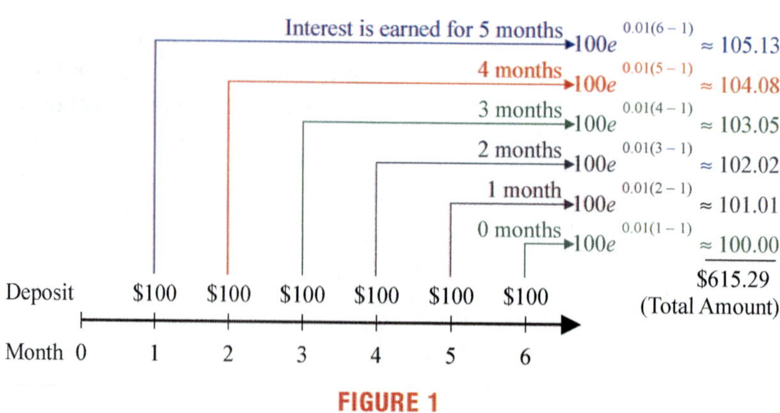

FIGURE 1

The sum S_6 is a Riemann sum with $\Delta t = 1$ month. Thus the integral

$$\int_0^6 100e^{0.01(6-t)}\,dt$$

gives a good approximation of the amount of the annuity at the end of the 6 months.

$$\int_0^6 100e^{0.01(6-t)}\,dt = -\frac{100}{0.01}e^{0.01(6-t)}\Big]_0^6$$

$$= -10,000\left(e^{0.01(6-6)} - e^{0.01(6-0)}\right)$$

$$\approx -10,000\left(1 - 1.061836547\right)$$

$$\approx \$618.37$$

Future Value of an Annuity

The **amount** (or **future value**) **of an annuity** at the end of N time periods is approximated by the integral

$$\int_0^N Pe^{r(N-t)}\,dt = \frac{P}{r}\left(e^{rN} - 1\right),$$

where P is the number of dollars invested each time period, and r is the interest rate (as a decimal) per time period.

For an **annual annuity**, r is the annual rate of interest and N is the number of years that payments are made. The preceding formula gives a good approximation whether the annuity is an ordinary annuity or an annuity due.

Example 1: Future Value of an Annuity

Suppose that the parents of a child set up an annuity account paying 10 percent compounded continuously for the child's college education. They deposit $500 each year for 20 years. What will be the approximate amount of the annuity in 20 years?

Solution

Since the deposits are made yearly, $r = 0.10$, $N = 20$, and $P = 500$.

$$\int_0^{20} 500e^{0.10(20-t)}\,dt = -\frac{500}{0.10}\left(e^{0.10(20-t)}\right)\Big]_0^{20}$$

$$= -\frac{500}{0.10}\left(1 - e^{0.10(20)}\right)$$

$$= -5000\left(1 - e^2\right)$$

$$\approx -5000\left(1 - 7.389056\right)$$

$$= 31,945.28$$

The value of the annuity will be approximately $31,945.28 in 20 years.

Income Streams

Suppose that money is not paid in equal amounts at regular time intervals, but instead, income flows almost continuously, as does the **income stream** from a video arcade game or from airline ticket sales. If $R(t)$ represents the rate of flow of revenue, we can use a definite integral to find the amount of income over a period of time.

Future Value of an Income Stream

The **amount** (or **future value**) **of an income stream** at the end of T years is given by the integral

$$A = \int_0^T R(t)e^{r(T-t)}dt,$$

where $R(t)$ is the rate of flow of revenue at time t, r is the annual interest rate (as a decimal), and interest is compounded continuously.

Example 2: Future Value of an Income Stream

A travel agency expects income from airline ticket sales to increase at a continuous rate represented by $R(t) = 210 + 0.1t$ in hundreds of dollars per year over the next 3 years. With interest at 5 percent compounded continuously, what income does the agency expect from airline ticket sales over the next 3 years?

Solution

Using the formula $\int_0^T R(t)e^{r(T-t)}dt$ with $R(t) = 210 + 0.1t$, $T = 3$, and $r = 0.05$, we have

$$A = \int_0^3 (210 + 0.1t)e^{0.05(3-t)}dt.$$

Integrating by parts, we obtain the following results.

$u = 210 + 0.1t$	$dv = e^{0.05(3-t)}dt$
$du = 0.1\,dt$	$v = \int e^{0.05(3-t)}dt = -\dfrac{1}{0.05}e^{0.05(3-t)} = -20e^{0.05(3-t)}$

$$A = (210 + 0.1t)\left(-20e^{0.05(3-t)}\right)\Big]_0^3 - \int_0^3 \left(-20e^{0.05(3-t)}\right)(0.1\,dt)$$

Using the parts formula, $uv - \int v\,du$

$$= (210 + 0.1t)\left(-20e^{0.05(3-t)}\right)\Big]_0^3 + 2\int_0^3 e^{0.05(3-t)}dt$$

$$= (210 + 0.1t)\left(-20e^{0.05(3-t)}\right)\Big]_0^3 - 2\left(\frac{1}{0.05}e^{0.05(3-t)}\right)\Big]_0^3$$

$$= 210.3(-20)\left(e^0\right) - 210(-20)\left(e^{0.15}\right) - 2(20)\left(e^0\right) + 2(20)\left(e^{0.15}\right)$$

$$\approx -4206 - 210(-23.23668485) - 40 + 2(23.23668485)$$

$$= 210(23.23668485) + 2(23.23668485) - 4246$$

$$= 212(23.23668485) - 4246$$

$$\approx 4926 - 4246 = 680$$

The units are hundreds of dollars.

The agency expects approximately \$68,000 from airline ticket sales over the next 3 years.

We know that if P dollars are invested with interest compounded continuously at a rate r for t years, then the future value A is given by the formula

$$A = Pe^{rt}.$$

If this formula is solved for P, we have

$$P = Ae^{-rt}.$$

We call P the **present value of A**. That is, P is the amount to be invested now so that the amount A will be accumulated if interest is compounded continuously at a rate r for t years. Reasoning in a similar manner, we can relate the concept of the future value of an income stream to the **present value of an income stream**.

Present Value of an Income Stream

The **present value of an income stream** over T years is given by the integral

$$PV = \int_0^T R(t)e^{-rt}\,dt,$$

where $R(t)$ is the rate of flow of revenue at time t, r is the annual interest rate (as a decimal), and interest is compounded continuously.

Example 3: Present Value of an Income Stream

A new department store is expected to generate income at a continuous rate described by the function $R(t) = 50{,}000 + 2000t$ dollars per year over the next 5 years. Find the present value of the store's income stream if the current interest rate is 10 percent compounded continuously.

Solution

Using the formula $\int_0^T R(t)e^{-rt}\,dt$ with $R(t) = 50{,}000 + 2000t$, $T = 5$, and $r = 0.10$, we have

$$PV = \int_0^5 (50{,}000 + 2000t)e^{-0.10t}\,dt.$$

Integrating by parts gives the following results.

$u = 50{,}000 + 2000t$	$dv = e^{-0.10t}\,dt$
$du = 2000\,dt$	$v = \int e^{-0.10t}\,dt = -\dfrac{1}{0.10}e^{-0.10t} = -10e^{-0.10t}$

$$PV = \left(50,000 + 2000t\right)\left(-10e^{-0.10t}\right)\Big]_0^5 - \int_0^5 \left(-10e^{-0.10t}\right)\left(2000dt\right) \quad \text{Using the parts}$$

$$\text{formula, } uv - \int v\,du$$

$$= \left(50,000 + 2000t\right)\left(-10e^{-0.10t}\right)\Big]_0^5 + 20,000\int_0^5 e^{-0.10t}\,dt$$

$$= \left(50,000 + 2000t\right)\left(-10e^{-0.10t}\right)\Big]_0^5 + \left(-200,000e^{-0.10t}\right)\Big]_0^5$$

$$= 60,000\left(-10e^{-0.5}\right) + 500,000 - 200,000\left(e^{-0.5}\right) + 200,000$$

$$= 700,000 - 800,000e^{-0.5}$$

$$\approx 700,000 - 800,000\left(0.6065307\right)$$

$$\approx 700,000 - 485,225 = 214,775$$

The present value of the new store's income stream is approximately \$214,775.

15.2 EXERCISES

🚀 APPLICATIONS

1. **Annuity:** Estimate the amount of an annuity if \$1000 is deposited annually for 10 years at a rate of 8 percent compounded continuously.

2. **Annuity:** An amount of \$6000 is invested in an account each year for 8 years. Find the approximate balance at the end of the 8 years if the account pays interest at a rate of 7 percent compounded continuously.

3. **Annuity:** Christine has decided to invest \$2000 each year into an IRA account that pays interest at the rate of 9 percent compounded continuously. Find the amount in the account at the end of 15 years.

4. **Annuity:** Bob and Ann plan to deposit \$4000 per year into their retirement account. If the account pays interest at a rate of 8.4 percent compounded continuously, approximately how much will be in their account after 12 years?

5. **Annuity:** Bryan plans to deposit \$1200 each year into an annuity. If the account pays interest at a rate of 7.5 percent compounded continuously, find the approximate balance of his account after 10 years.

6. **Income stream:** Find the value of an income stream after 7 years if the rate of flow is estimated to be \$200,000 annually and the income is invested at a rate of 8 percent compounded continuously.

7. **Income stream:** The owner of a local convenience store estimates that the store will generate an annual income of \$340,000 for the next 4 years. If the rate of interest is 9 percent compounded continuously, find the value of the income stream.

8. **Income stream:** A real estate investment is expected to generate an income flow of \$12,000 annually for the next 6 years. Find the amount of the income stream if the interest rate is 7.8 percent compounded continuously.

9. **Income stream:** Find the value of an income stream after 5 years if $R(t) = 3600e^{0.02t}$ is the rate of flow of revenue and the income is deposited at a rate of 7 percent compounded continuously.

10. **Income stream:** A certain investment has a continuous flow of money at a rate of $R(t) = 7200e^{0.01t}$. Find the value of this flow after 4 years if the interest rate is 8.2 percent compounded continuously.

11. **Income stream:** Find the value of an income stream if $R(t) = 50 + 0.2t$ is the rate of flow of revenue reinvested at 6 percent compounded continuously for 8 years.

12. **Income stream:** Find the value of an income stream if $R(t) = 80 + 1.2t$ is the rate of flow of revenue reinvested at 6.4 percent compounded continuously over the next 6 years.

13. **Income stream:** The profit from a number of soft drink machines is estimated to be at the rate of $R(t) = 15 + 0.8t$ thousand dollars per year. If the profits are deposited into an account paying 6.5 percent compounded continuously, find the amount of the income stream after 7 years.

14. **Income stream:** It is estimated that a computer will save accounting fees at a small company at a rate of $R(t) = 4 + 0.6t$ thousand dollars per year. If the savings are reinvested at 5 percent compounded continuously, find the amount of the income stream after 4 years.

15. **Income stream:** Find the present value of an income stream with $R(t) = 60 - 0.4t$, $r = 8$ percent, and $T = 20$.

16. **Income stream:** Find the present value of an income stream with $R(t) = 150 - t$, $r = 12$ percent, and $T = 10$.

17. **Income stream:** The rate of flow of an income stream is estimated by $R(t) = 6000e^{0.015t}$ for the next 4 years. Find the present value of this flow if the interest rate is 6 percent compounded continuously.

18. **Income stream:** The rate of flow of an income stream for the next 6 years is estimated by $R(t) = 10{,}000e^{-0.01t}$. Find the present value of this flow if the interest rate is 8.5 percent compounded continuously.

19. **Income stream:** Sandy estimates that the profits from his ice cream store will be $R(t) = 24 + 3.6t$ thousand dollars per year for the next 5 years. Find the present value of the store if the current interest rate is 10 percent compounded continuously.

20. **Income stream:** Elco Grain Company expects their profits to be $R(t) = 30 + 12e^{0.02t}$ thousand dollars per year for the next 4 years. If the current interest rate is 8 percent compounded continuously, find the present value of the company.

✐ WRITING & THINKING

21. In Figure 1, replace the column information with $100e^{0.01(5)}$, $100e^{0.01(4)}$, $100e^{0.01(3)}$, $100e^{0.01(2)}$, $100e^{0.01(1)}$, $100e^{0.01(0)}$. This suggests that the future value of an annuity could be given by $\int_0^N Pe^{rt}\,dt$. Does this give the same result as the formula for the future value of an annuity shown in the lesson? Explain why or why not.

15.3 TABLES OF INTEGRALS

Extensive tables of integrals contain hundreds of integral formulas listed according to the form that fits the type of function in the integrand. To use such a table to evaluate an integral, we simply look through the table until we find a formula in which the integrand matches the form of the expression we are to integrate. The answer is given by the formula. Table 1 contains a selection of formulas that will be used in this course. In general, the formulas in these tables are categorized according to the types of expressions contained in the integrands.

In Table 1 we have listed elementary forms of integrals (all of which we have seen previously), forms involving the expressions $ax+b$, $\sqrt{ax+b}$, $(ax+b)(cx+d)$, and $x^2 - a^2$, as well as forms involving exponential and logarithmic expressions. Thus, to evaluate an integral such as $\int \dfrac{1}{x(3x+4)}\,dx$, we first note that the integrand contains an expression of the form $ax + b$ with $a = 3$ and $b = 4$. Then we find this category in Table 1 and determine that Formula 10 is the correct form (see Example 1).

If an integrand contains an expression of more than one type, the appropriate integration formula may be difficult to find. In fact, some tables may not contain the correct formula for a specific problem. Before using a formula, we must be sure that the integrand in the problem precisely matches the integrand in the formula.

	Elementary Forms
1.	$\int k\,dx = kx + C$
2.	$\int x^r\,dx = \dfrac{1}{r+1}x^{r+1} + C\ (r \neq -1)$
3.	$\int e^x\,dx = e^x + C$
4.	$\int \dfrac{1}{x}\,dx = \ln|x| + C$
5.	$\int kf(x)\,dx = k\int f(x)\,dx$
6.	$\int \left[f(x) \pm g(x)\right]dx = \int f(x)\,dx \pm \int g(x)\,dx$
7.	$\int u\,dv = uv - \int v\,du$
	Forms Involving $(ax + b)$
8.	$\int \dfrac{1}{ax+b}\,dx = \dfrac{1}{a}\ln|ax+b| + C$
9.	$\int \dfrac{1}{(ax+b)^2}\,dx = -\dfrac{1}{a(ax+b)} + C$
10.	$\int \dfrac{1}{x(ax+b)}\,dx = \dfrac{1}{b}\ln\left|\dfrac{x}{ax+b}\right| + C$

11.	$\displaystyle\int \frac{x}{ax+b}\,dx = \frac{1}{a^2}\left(ax - b\ln\left|ax+b\right|\right) + C$
Forms Involving $(ax+b)(cx+d)$	
12.	$\displaystyle\int \frac{1}{(ax+b)(cx+d)}\,dx = \frac{1}{ad-bc}\ln\left|\frac{ax+b}{cx+d}\right| + C$
13.	$\displaystyle\int \frac{x}{(ax+b)(cx+d)}\,dx = \frac{1}{ad-bc}\left(\frac{d}{c}\ln\left|cx+d\right| - \frac{b}{a}\ln\left|ax+b\right|\right) + C$
Forms Involving $\sqrt{ax+b}$	
14.	$\displaystyle\int \sqrt{ax+b}\,dx = \frac{2}{3a}(ax+b)^{\frac{3}{2}} + C$
15.	$\displaystyle\int x\sqrt{ax+b}\,dx = \frac{2(3ax-2b)}{15a^2}(ax+b)^{\frac{3}{2}} + C$
Forms Involving $(x^2 - a^2)$	
16.	$\displaystyle\int \frac{1}{x^2-a^2}\,dx = \frac{1}{2a}\ln\left|\frac{x-a}{x+a}\right| + C \;\left(x^2 > a^2\right)$
Forms Involving $\sqrt{x^2 \pm a^2}$	
17.	$\displaystyle\int \sqrt{x^2 \pm a^2}\,dx = \frac{x}{2}\sqrt{x^2 \pm a^2} \pm \frac{a^2}{2}\ln\left|x + \sqrt{x^2 \pm a^2}\right| + C$
18.	$\displaystyle\int \frac{1}{\sqrt{x^2 \pm a^2}}\,dx = \ln\left|x + \sqrt{x^2 \pm a^2}\right| + C$
Forms Involving e^{kx}	
19.	$\displaystyle\int e^{kx}\,dx = \frac{1}{k}e^{kx} + C$
20.	$\displaystyle\int x^n e^{kx}\,dx = \frac{x^n e^{kx}}{k} - \frac{n}{k}\int x^{n-1}e^{kx}\,dx$
21.	$\displaystyle\int \frac{1}{a+be^{kx}}\,dx = \frac{x}{a} - \frac{1}{ak}\ln\left|a+be^{kx}\right| + C$
Forms Involving $\ln x$	
22.	$\displaystyle\int \ln x\,dx = x\ln x - x + C$
23.	$\displaystyle\int x^n \ln x\,dx = \frac{x^{n+1}\ln x}{n+1} - \frac{x^{n+1}}{(n+1)^2} + C$

TABLE 1: Integral Formulas

Comments about Integral Tables

1. In this course, as well as in general, the letters in the beginning of the alphabet, such as a, b, c, and d, represent constants. The letter d is used to emphasize certain patterns or symmetry even though it also appears as part of the differential dx. The two appearances of d should not cause confusion. The letter k is often used to represent a constant in an exponential function.

2. The differential might be in the numerator, so the integral $\int \dfrac{1}{ax+b}\,dx$ might look like $\int \dfrac{dx}{ax+b}$.

3. The natural logarithm might be written as $\log x$ instead of $\ln x$.

4. The constant C might be omitted.

The user of the table should be aware of the notation used and its implications.

The following examples illustrate the use of the formulas in Table 1.

Example 1: Using the Integral Table

Find $\displaystyle\int \dfrac{1}{x(3x+4)}\,dx$.

Solution

Using Formula 10 with $a = 3$ and $b = 4$, we have

$$\int \frac{1}{x(3x+4)}\,dx = \frac{1}{4}\ln\left|\frac{x}{3x+4}\right| + C.$$

Example 2: Using the Integral Table

Evaluate $\displaystyle\int \dfrac{9}{x^2-25}\,dx$.

Solution

$$\int \frac{9}{x^2-25}\,dx = 9\int \frac{1}{x^2-25}\,dx \qquad \text{Factor out 9 by Formula 5 with } k=9.$$

$$= 9\int \frac{1}{x^2-5^2}\,dx \qquad \text{Rewrite } x^2-25 \text{ as } x^2-5^2 \text{ and find an integral}$$
$$\text{formula that contains the form } x^2-a^2.$$

$$= 9\cdot\frac{1}{2(5)}\ln\left|\frac{x-5}{x+5}\right| + C \qquad \text{Apply Formula 16 with } a=5.$$

$$= \frac{9}{10}\ln\left|\frac{x-5}{x+5}\right| + C \qquad \text{Simplify.}$$

In Example 3 we use a **reduction formula**, a formula that yields another simpler integral. In some cases, this simpler integral will result in yet another integral to be evaluated. This process continues until no integrals remain. Reduction formulas are said to be **iterative** in nature because of the repeated pattern of the steps involved.

Example 3: Using a Reduction Formula

Find $\int x^2 e^{-3x}\, dx$.

Solution

We will use Formula 20 twice and then use Formula 19.

$$\int x^2 e^{-3x}\, dx = \frac{x^2 e^{-3x}}{-3} - \frac{2}{-3}\int x^{2-1} e^{-3x}\, dx \qquad \text{Use Formula 20 with } n = 2 \text{ and } k = -3.$$

$$= -\frac{1}{3} x^2 e^{-3x} + \frac{2}{3}\left(\frac{x^1 e^{-3x}}{-3} - \frac{1}{-3}\int x^{1-1} e^{-3x}\, dx \right) \qquad \text{Use Formula 20 again on the rightmost part of the expression with } n = 1 \text{ and } k = -3.$$

$$= -\frac{1}{3} x^2 e^{-3x} - \frac{2}{9} x e^{-3x} + \frac{2}{9}\int e^{-3x}\, dx \qquad \text{Simplify.}$$

$$= -\frac{1}{3} x^2 e^{-3x} - \frac{2}{9} x e^{-3x} + \frac{2}{9}\cdot \frac{e^{-3x}}{-3} + C \qquad \text{Use Formula 19 with } k = -3.$$

$$= -\frac{1}{3} x^2 e^{-3x} - \frac{2}{9} x e^{-3x} - \frac{2}{27} e^{-3x} + C \qquad \text{Simplify.}$$

In Example 4 we show how an integrand can be rewritten algebraically as a sum so that each part of the sum fits a formula in the table.

Example 4: Rewriting the Integrand

Find $\int \dfrac{x+1}{x^2 + 5x + 6}\, dx$.

Solution

$$\int \frac{x+1}{x^2 + 5x + 6}\, dx = \int \frac{x+1}{(x+2)(x+3)}\, dx \qquad \text{Factor the denominator.}$$

$$= \int \frac{x}{(x+2)(x+3)}\, dx + \int \frac{1}{(x+2)(x+3)}\, dx \qquad \text{By Formula 6}$$

Now we apply Formula 13 to the first integral and Formula 12 to the second integral.

$$\int \frac{x}{(x+2)(x+3)}\, dx = \frac{1}{3-2}\left(\frac{3}{1}\ln|x+3| - \frac{2}{1}\ln|x+2| \right) + C_1$$

$$= 3\ln|x+3| - 2\ln|x+2| + C_1$$

$$\int \frac{1}{(x+2)(x+3)}\,dx = \frac{1}{3-2}\ln\left|\frac{x+2}{x+3}\right| + C_2$$

$$= \ln\left|\frac{x+2}{x+3}\right| + C_2$$

$$= \ln|x+2| - \ln|x+3| + C_2$$

Now combining the parts gives the result.

$$\int \frac{x+1}{x^2+5x+6}\,dx = 3\ln|x+3| - 2\ln|x+2| + C_1 + \ln|x+2| - \ln|x+3| + C_2$$

$$= 2\ln|x+3| - \ln|x+2| + C \qquad\qquad \text{Where } C = C_1 + C_2$$

15.3 EXERCISES

💡 PRACTICE

Use Table 1 to find the following integrals.

1. $\displaystyle\int \frac{1}{4x+3}\,dx$

2. $\displaystyle\int \sqrt{9x+2}\,dx$

3. $\displaystyle\int e^{-0.15x}\,dx$

4. $\displaystyle\int \ln x\,dx$

5. $\displaystyle\int \frac{1}{(2x-5)^2}\,dx$

6. $\displaystyle\int \frac{x}{x+6}\,dx$

7. $\displaystyle\int x\sqrt{3x-4}\,dx$

8. $\displaystyle\int \frac{x}{(2x+1)(x-2)}\,dx$

9. $\displaystyle\int \sqrt{x^2+36}\,dx$

10. $\displaystyle\int x^4 \ln x\,dx$

11. $\displaystyle\int \frac{1}{x^2-16}\,dx$

12. $\displaystyle\int \frac{1}{x(4x-3)}\,dx$

13. $\displaystyle\int \frac{1}{(x+8)(5x-1)}\,dx$

14. $\displaystyle\int \frac{1}{2+e^{3x}}\,dx$

15. $\displaystyle\int 7x^5 \ln x\,dx$

16. $\displaystyle\int x^4 e^{-2x}\,dx$

17. $\displaystyle\int \frac{1}{8-5e^{-0.7x}}\,dx$

18. $\displaystyle\int \frac{1}{(0.3x+2)^2}\,dx$

19. $\displaystyle\int \frac{2}{x(3x-1)}\,dx$

20. $\displaystyle\int \frac{4}{x^2-8}\,dx$

21. $\displaystyle\int 14\sqrt{6x-5}\,dx$

22. $\displaystyle\int \frac{13}{(4x-1)(2x+3)}\,dx$

23. $\displaystyle\int \frac{8}{x(0.4x+1)}\,dx$

24. $\displaystyle\int 2x\sqrt{3x-4}\,dx$

25. $\displaystyle\int x^3 e^{1.5x}\,dx$

26. $\displaystyle\int \sqrt{x^2+9}\,dx$

27. $\displaystyle\int \frac{x}{(x-7)(5x+2)}\,dx$

28. $\displaystyle\int \sqrt{x^2-15}\,dx$

29. $\displaystyle\int \frac{1}{\sqrt{x^2-12}}\,dx$

30. $\displaystyle\int \frac{6}{24-9e^{3.1x}}\,dx$

15.4 NUMERICAL INTEGRATION

■ TOPICS

- The Trapezoidal Rule
- Simpson's Rule

If we can find an antiderivative of a given function f, evaluating $\int_a^b f(x)\,dx$ is straightforward. But there are two fairly common circumstances under which this is difficult or impossible. The first is when the integrand f has no antiderivative expressible in terms of *elementary functions*, which are those we work with most frequently: polynomial functions, power functions, exponential and logarithmic functions, and all other functions that can be expressed as a combination of these through the operations of addition, subtraction, multiplication, division, and function composition. Such integrals are termed **nonelementary**, and include the following examples: $\int e^{x^2}\,dx, \int \sqrt{1+x^4}\,dx, \int \dfrac{e^x}{x}\,dx, \int \dfrac{1}{\ln x}\,dx, \int \ln(\ln x)\,dx, \int e^{e^x}\,dx.$ The other circumstance is when the integral in question is defined by a function for which we know only a few values, such as a collection of data points from an experiment or a set of measurements of some physical object. *Numerical integration* is the name given to a set of techniques that can be used to provide approximations in these cases, along with estimates of the errors in the approximations. We study two such techniques in this section.

The Trapezoidal Rule

We have actually already gained considerable experience with the most basic numerical integration technique. When we compute a Riemann sum with either left endpoints or right endpoints of subintervals as the sample points, the result is, of course, an approximation to the integral. The Trapezoidal Rule is a slight modification of such an approach that, in general, improves on the accuracy of the approximation.

Recall that a Riemann sum approximation of the integral $\int_a^b f(x)\,dx$, assuming equal-width subintervals, has the form

$$\int_a^b f(x)\,dx \approx \left[\, f(c_1) + f(c_2) + \cdots + f(c_n)\,\right]\Delta x,$$

where $\Delta x = \dfrac{b-a}{n}$ and $[a, b]$ is partitioned into n subintervals by the points $x_0 = a, x_1, \ldots, x_{n-1}, x_n = b$. The x-value selected from the ith subinterval, c_i, is equal to x_{i-1} if we choose to let each one be the left subinterval endpoint and is equal to x_i if we choose to use right endpoints. Since each Riemann sum is an approximation of the integral, we might hope that the average of the two approximations would be a better approximation. That average is calculated as follows.

$$\int_a^b f(x)\,dx \approx \frac{1}{2}\left(\left[\, f(x_0) + f(x_1) + \cdots + f(x_{n-1})\,\right]\Delta x + \left[\, f(x_1) + f(x_2) + \cdots + f(x_n)\,\right]\Delta x \right)$$

$$= \frac{\Delta x}{2}\left[\left(f(x_0) + f(x_1) \right) + \left(f(x_1) + f(x_2) \right) + \cdots + \left(f(x_{n-1}) + f(x_n) \right) \right]$$

$$= \frac{\Delta x}{2}\left[f(x_0) + 2f(x_1) + 2f(x_2) + \cdots + 2f(x_{n-1}) + f(x_n) \right]$$

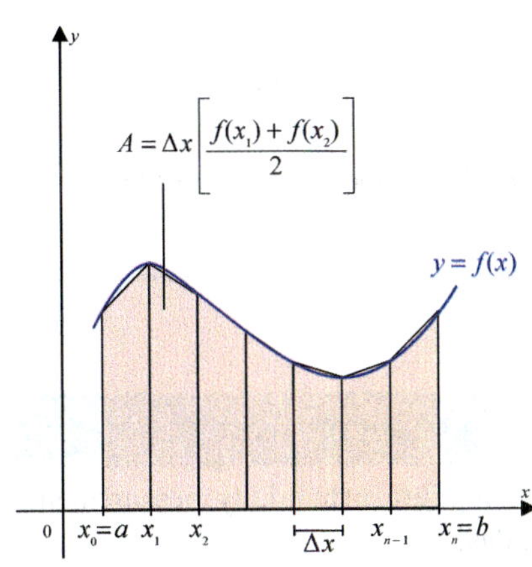

FIGURE 1: Areas of Trapezoids

The sum above is referred to as the trapezoidal approximation to the integral, because each expression

$$\frac{\Delta x}{2}\left[f\left(x_{i-1}\right)+f\left(x_i\right)\right]=\Delta x\left[\frac{f\left(x_{i-1}\right)+f\left(x_i\right)}{2}\right]$$

corresponds to the area of the ith trapezoid shown in Figure 1 (recall that the area of a trapezoid is the width Δx times the mean of the lengths of the bases, which are $f(x_{i-1})$ and $f(x_i)$ in this context).

Trapezoidal Rule

An approximation of $\int_a^b f(x)\,dx$ using the Trapezoidal Rule is the sum

$$T_n = \frac{\Delta x}{2}\left[f\left(x_0\right)+2f\left(x_1\right)+2f\left(x_2\right)+\cdots+2f\left(x_{n-1}\right)+f\left(x_n\right)\right],$$

where $\Delta x = \dfrac{b-a}{n}$ and $x_i = a+i\Delta x$ for $i = 0,\ 1,\ \ldots,\ n.$

Example 1: The Trapezoidal Rule

Use the Trapezoidal Rule to approximate $\int_1^2 \ln x\,dx$ with $n = 10$, and compare the result to the exact value of this integral.

Solution

Given that $a = 1$, $b = 2$, and $n = 10$, we note that $\Delta x = \dfrac{1}{10}$ and $x_i = 1+\dfrac{i}{10}$ for i from 0 to 10.

$$T_{10} = \frac{1}{20}\left[\ln 1 + 2\ln\frac{11}{10} + 2\ln\frac{12}{10} + \cdots + 2\ln\frac{19}{10} + \ln 2\right] \approx 0.385878$$

Using integration by parts, we know that $\int \ln x\,dx = x\ln x - x$, so

$$\int_1^2 \ln x\,dx = \left[x\ln x - x\right]_1^2 = (2\ln 2 - 2)-(-1) = 2\ln 2 - 1 \approx 0.386294.$$

So $\int_1^2 \ln x\,dx - T_{10} \approx |0.386294 - 0.385878| = 0.000416,$ approximately 0.1% of the exact value.

Example 2: The Trapezoidal Rule

A hygrometer at a weather station records the following percent humidity measurements at the top of the hour from midnight to noon one day.

Time	Humidity (%)
12 a.m.	55
1 a.m.	57
2 a.m.	60
3 a.m.	65
4 a.m.	72
5 a.m.	75
6 a.m.	74
7 a.m.	70
8 a.m.	62
9 a.m.	50
10 a.m.	48
11 a.m.	48
12 p.m.	47

Use the Trapezoidal Rule to find the approximate average percent humidity over this 12-hour period.

Solution

Taking this approach to finding the average humidity, we are implicitly making the assumption that humidity h is a continuous function of time—under this assumption, we know that $\int_0^{12} h(t)\,dt$ exists, even if we don't actually know a formula for h. Assumptions like this are reasonable, because physical attributes like humidity don't typically exhibit abrupt discontinuous changes. If we simply plot the 13 data points we have and connect them with straight lines, the result is the graph in Figure 2. Note that the data indicate sharper changes in humidity over some one-hour periods than others.

Using our formula for the average value of a function over an interval we have

$$\frac{1}{b-a}\int_a^b h(t)\,dt = \frac{1}{12}\int_0^{12} h(t)\,dt.$$

Since we don't have a formula for h, we use the Trapezoidal Rule with $\Delta t = 1$ hour.

$$\frac{1}{12}\int_0^{12} h(t)\,dt \approx \left(\frac{1}{12}\right)\left(\frac{1}{2}\right)\left(55 + 2\cdot 57 + 2\cdot 60 + \cdots + 2\cdot 48 + 47\right) = 61\%$$

Compare this to the less sophisticated approach of simply taking the average of the 13 data points, which gives us an answer of approximately 60.2% humidity. The difference comes from the fact that a simple average does not take into account differing rates of change of the humidity over the interval.

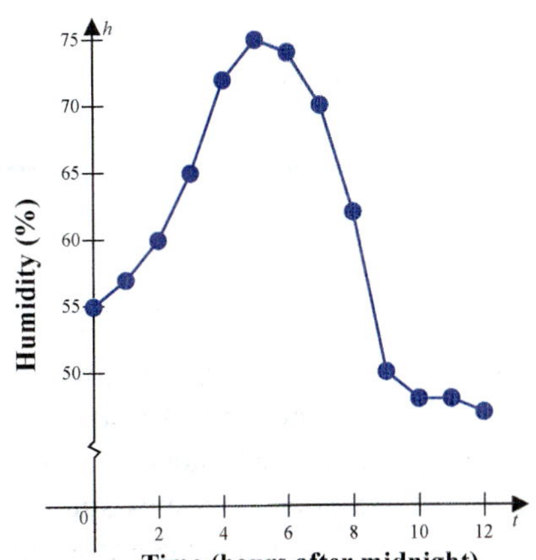

FIGURE 2: Humidity Measurements

There are situations where it can be helpful to estimate the error between the true value of an integral and the Trapezoidal Rule approximation of that integral. Such discussions are left for a different course, but we provide the following theorem for curiosity's sake.

Error Estimate for the Trapezoidal Rule

If f'' is continuous on $[a, b]$ and M is an upper bound for $|f''(x)|$, then the error E_T between the exact value of $\int_a^b f(x)\,dx$ and the Trapezoidal Rule approximation T_n satisfies

$$|E_T| \le \frac{M(b-a)^3}{12n^2}$$

where n is the number of subintervals in the partition.

Example 3: The Trapezoidal Rule

Use the Trapezoidal Rule with $n = 4$ to approximate the nonelementary integral $\int_0^1 e^{x^2}\,dx$.

Solution

With $a = 0$, $b = 1$, and $n = 4$, we have $\Delta x = \dfrac{1}{4}$ and $x_i = \dfrac{i}{4}$ for $i = 0, \ldots, 4$.

$$T_4 = \frac{1}{8}\left[e^0 + 2e^{\frac{1}{16}} + 2e^{\frac{1}{4}} + 2e^{\frac{9}{16}} + e^1 \right] \approx 1.4907$$

Simpson's Rule

Although we derived the Trapezoidal Rule by averaging two Riemann sums, we can think of the resulting formula as the integral of a function composed of a sequence of straight lines, with each straight line approximating the original integrand over a subinterval. Simpson's Rule, named for the English mathematician Thomas Simpson (1710–1761), takes a similar approach, but approximates the integrand with a sequence of parabolas.

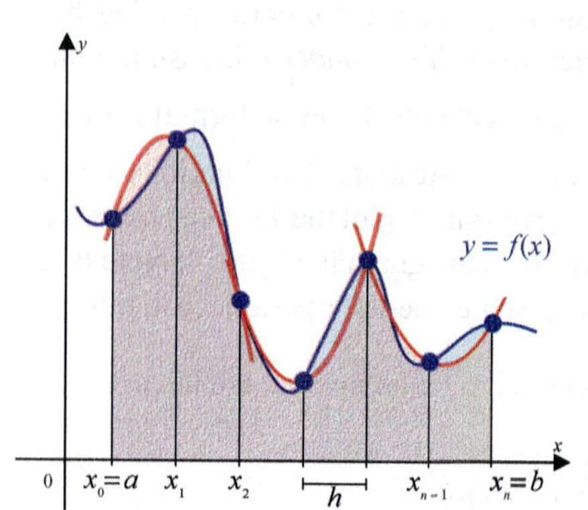

FIGURE 3: Simpson's Rule

There are several variations of Simpson's Rule, but the simplest assumes that the number of subintervals n is even; the resulting formula is then based on parabolas fitted through three points on the graph of the integrand f for each successive pair of subintervals. A representative illustration is shown in Figure 3.

To construct the formula, let $y_i = f(x_i)$ for each $i = 0, \ldots, n$ and let $h = \Delta x = \dfrac{b-a}{n}$. For ease of calculation, we shift the graph of f left or right so that the first two subintervals $[x_0, x_1]$ and $[x_1, x_2]$ are located on either side of 0 as shown in Figure 4; doing so doesn't affect the area between the graph of f and the x-axis over the interval $[x_0, x_2]$. After shifting, note that $x_0 = -h$, $x_1 = 0$, and $x_2 = h$. So the parabola we are looking for must pass through the three points $(-h, y_0)$, $(0, y_1)$, and (h, y_2). If we assume the parabola is the graph of the quadratic $y = Ax^2 + Bx + C$, then we know the following:

$$y_0 = Ah^2 - Bh + C,$$
$$y_1 = C,$$
$$\text{and } y_2 = Ah^2 + Bh + C.$$

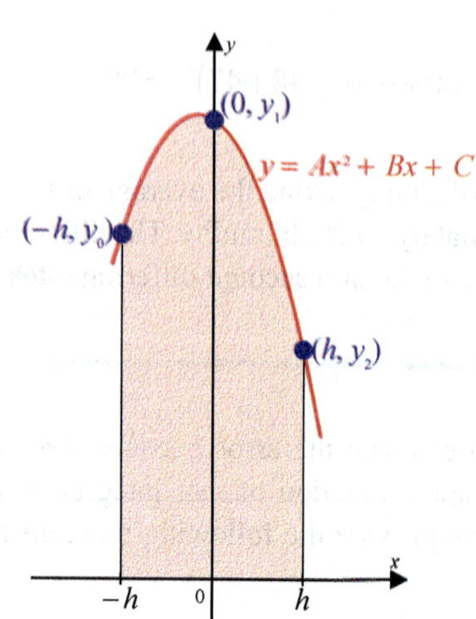

FIGURE 4: Fitting a Parabola

The integral of the quadratic over the interval $[-h, h]$ is

$$I = \int_{-h}^{h} \left(Ax^2 + Bx + C \right) dx = \left[\frac{Ax^3}{3} + \frac{Bx^2}{2} + Cx \right]_{-h}^{h} = \frac{h}{3}\left(2Ah^2 + 6C \right),$$

which can be expressed in terms of y_0, y_1, and y_2 by noting that $y_0 + y_2 = 2Ah^2 + 2C$ and $4y_1 = 4C$. So $I = \left(\frac{h}{3} \right)\left(y_0 + 4y_1 + y_2 \right).$

Similarly, shifting the graph of f so that the subintervals $[x_2, x_3]$ and $[x_3, x_4]$ are located on either side of 0 (so that $x_3 = 0$), the integral of the parabola fitted through $(-h, y_2)$, $(0, y_3)$, and (h, y_4) is equal to $\left(\frac{h}{3} \right)\left(y_2 + 4y_3 + y_4 \right).$ Our approximation to $\int_a^b f(x)\,dx$ is the sum of the integrals of the parabolas.

$$\int_a^b f(x)\,dx \approx \frac{h}{3}\left(y_0 + 4y_1 + y_2 \right) + \frac{h}{3}\left(y_2 + 4y_3 + y_4 \right) + \cdots + \frac{h}{3}\left(y_{n-2} + 4y_{n-1} + y_n \right)$$

$$= \frac{h}{3}\left(y_0 + 4y_1 + 2y_2 + 4y_3 + 2y_4 + \cdots + 2y_{n-2} + 4y_{n-1} + y_n \right)$$

Rephrasing the formula back in terms of the integrand function f gives us Simpson's Rule.

Simpson's Rule

An approximation of $\int_a^b f(x)\,dx$ using Simpson's Rule is the sum

$$S_n = \frac{\Delta x}{3}\left[\begin{array}{l} f(x_0) + 4f(x_1) + 2f(x_2) + 4f(x_3) + 2f(x_4) + \cdots \\ + 2f(x_{n-2}) + 4f(x_{n-1}) + f(x_n) \end{array} \right]$$

where n is even, $\Delta x = \dfrac{b-a}{n}$, and $x_i = a + i\Delta x$ for $i = 0, 1, \ldots, n$. (Note the pattern of the coefficients: 1, 4, 2, 4, 2, $\ldots$, 2, 4, 1.)

Example 4: Simpson's Rule

Use Simpson's Rule to approximate $\int_1^2 \ln x\,dx$ with $n = 10$.

Solution

As in Example 1, $a = 1$, $b = 2$, $\Delta x = \dfrac{1}{10}$, and $x_i = 1 + \dfrac{i}{10}$ for i from 0 to 10.

$$S_{10} = \frac{1}{30}\left[\ln 1 + 4\ln\frac{11}{10} + 2\ln\frac{12}{10} + \cdots + 2\ln\frac{18}{10} + 4\ln\frac{19}{10} + \ln 2 \right] \approx 0.386293$$

This is a considerable improvement over the Trapezoidal Rule approximation, and only differs in the 6th decimal place from the exact answer.

As with the Trapezoidal Rule, there are situations where it can be helpful to estimate the error between the true value of an integral and the Simpson's Rule approximation of that integral. Such discussions are left for a different course, but we provide the following theorem for curiosity's sake.

Error Estimate for Simpson's Rule

If $f^{(4)}$ is continuous on $[a, b]$ and M is an upper bound for $|f^{(4)}(x)|$, then the error E_S between the exact value of $\int_a^b f(x)\,dx$ and the Simpson's Rule approximation S_n satisfies

$$|E_S| \le \frac{M(b-a)^5}{180n^4}$$

where n is the number of subintervals in the partition.

Example 5: Simpson's Rule

Use Simpson's Rule with $n = 4$ to approximate the nonelementary integral $\int_0^1 e^{x^2}\,dx$.

Solution

With $a = 0$, $b = 1$, and $n = 4$, we have $\Delta x = \dfrac{1}{4}$ and $x_i = \dfrac{i}{4}$ for $i = 0, \ldots, 4$.

$$S_4 = \frac{1}{12}\left[e^0 + 4e^{\frac{1}{16}} + 2e^{\frac{1}{4}} + 4e^{\frac{9}{16}} + e^1 \right] \approx 1.4637$$

Example 6: Simpson's Rule

A landscape designer has planned a free-form garden pond with the shape shown in Figure 5 and needs to estimate its volume in cubic feet. The pond will have a uniform depth of 2 feet. At 1-foot intervals, the distances across the pond are to be as indicated in the diagram. Use Simpson's Rule to estimate the volume of the pond.

Solution

To use Simpson's Rule, we let $\Delta x = 1$ and take the distances as shown to be the values of a function. Note that $n = 6$.

$$S_6 = \frac{1}{3}\left[1 + 4\cdot 2 + 2\cdot 3 + 4\cdot 4 + 2\cdot 6 + 4\cdot 7 + 5\right] = 25\frac{1}{3}$$

So the estimated volume of the pond is this surface area times the depth, or $50\frac{2}{3}\,\text{ft}^3$.

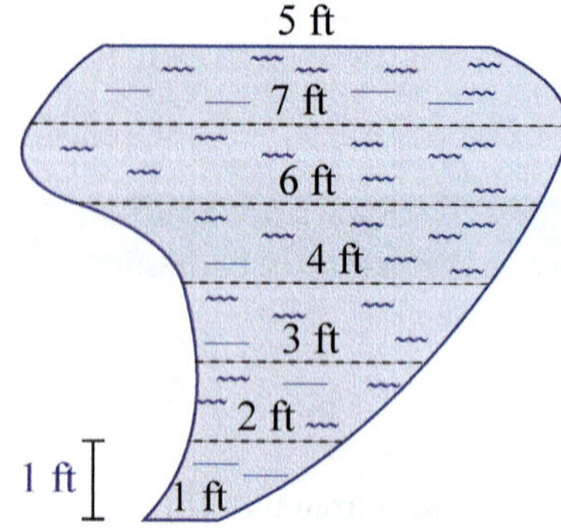

FIGURE 5: Garden Pond

15.4 EXERCISES

PRACTICE

The function $f(x)$ is given by its graph. Use the Trapezoidal Rule and Simpson's Rule, respectively, to approximate the shaded area $\int_a^b f(x)\,dx$ by **a.** T_6 and **b.** S_6.

1.

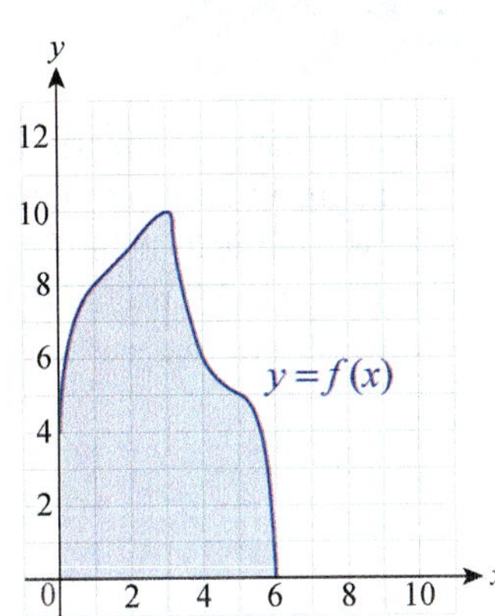

2.

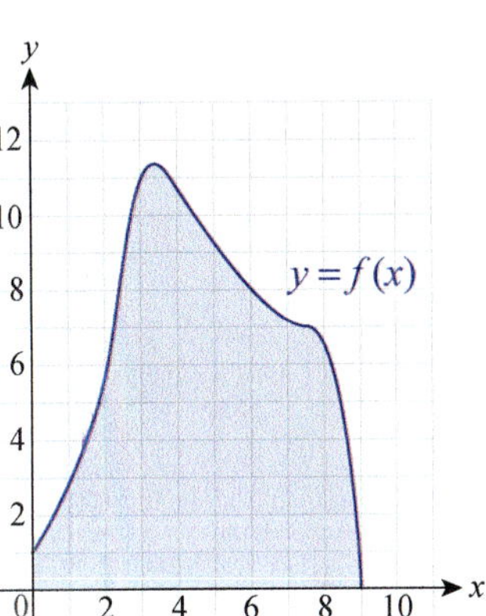

Use the Trapezoidal Rule and Simpson's Rule with $n = 8$ to approximate the integral. Then find the exact value and compare your answers.

3. $\displaystyle\int_0^8 x^4\,dx$

4. $\displaystyle\int_1^5 \frac{1}{x}\,dx$

5. $\displaystyle\int_1^5 \frac{1}{x^2}\,dx$

6. $\displaystyle\int_0^4 \sqrt{x}\,dx$

7. $\displaystyle\int_0^4 x^3\,dx$

8. $\displaystyle\int_{-2}^6 \sqrt[3]{x+2}\,dx$

9. $\displaystyle\int_0^2 e^x\,dx$

10. $\displaystyle\int_1^5 \ln x\,dx$

11. $\displaystyle\int_{-2}^6 \left(4 - \frac{1}{2}x\right)dx$

12. $\displaystyle\int_{-4}^4 \left(16 - x^2\right)dx$

13. $\displaystyle\int_0^4 x\sqrt{x^2+2}\,dx$

14. $\displaystyle\int_0^{16} \frac{1}{\sqrt{x+1}}\,dx$

15. $\displaystyle\int_0^8 \frac{x}{\sqrt{x^2+1}}\,dx$

Use **a.** the Trapezoidal Rule and **b.** Simpson's Rule to approximate the definite integral for the indicated value of n.

16. $\displaystyle\int_0^4 \sqrt[4]{x}\,dx; \quad n = 4$

17. $\displaystyle\int_0^5 \sqrt{x^4+4}\,dx; \quad n = 10$

Some texts discuss the "Midpoint Rule" as a numerical integration method. The idea is simply forming a Riemann sum by choosing the midpoint of each subinterval as the sample point.

Use the "Midpoint Rule" with $n = 8$ to approximate the integral and compare your answers to those in Exercises 3–5.

18. $\displaystyle\int_0^8 x^4\,dx$

19. $\displaystyle\int_1^5 \frac{1}{x}\,dx$

20. $\displaystyle\int_1^5 \frac{1}{x^2}\,dx$

21. Use Simpson's Rule with $n = 6$ to approximate $\ln 2 = \int_1^2 \frac{1}{x}\, dx.$

22. Use Simpson's Rule with $n = 6$ to approximate $\pi = \int_0^1 \frac{4}{x^2 + 1}\, dx.$

23. Use Simpson's Rule with $n = 24$ to approximate the area of the region bounded by the graphs of $y = \sqrt{1 + x^4}$, $x = -6$, $x = 6$, and the x-axis. (Notice that this problem leads to a nonelementary integral.)

🚀 APPLICATIONS

24. The following table summarizes acceleration data for the 2012 Ford Mustang Boss 302 Laguna Seca. Use Simpson's Rule to estimate the total distance traveled by "the Boss" during its timed 0–120 mph run. (**Hint:** Sketching a graph similar to the one in Example 2 is useful. Be sure to identify which area you can approximate and how it yields the answer to the problem.)

Time to Speed	
Miles per Hour	Seconds
0–120	13.0
0–110	10.9
0–100	9.1
0–90	7.6
0–80	6.3
0–70	5.2
0–60	4.1
0–50	3.3
0–40	2.4
0–30	1.7
0–20	1.1
0–10	0.4

Source: *Road & Track*

25. Use the Trapezoidal Rule to estimate the amount of water needed to raise the water level by two inches in a pool with the shape shown in the figure. At 2-foot intervals, the distances across the pool (in feet) are as indicated in the diagram.

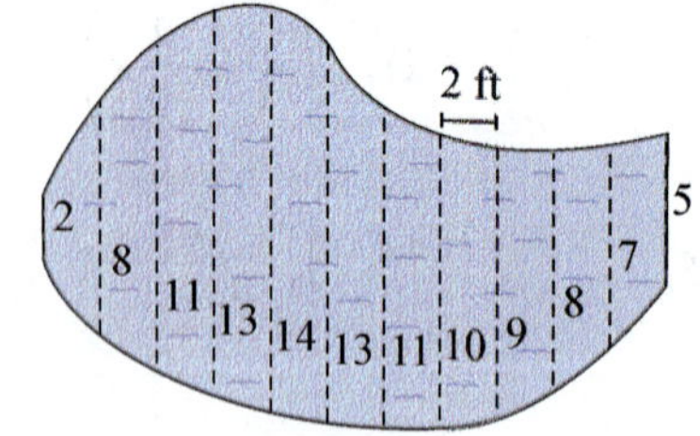

26. The figure shows the vertical cross-section of the Lazee river where the Dinkatown ferry docks. The depth of the river is indicated at 5-foot intervals in the diagram. If the river flows at $5\,\text{ft/s}$, use Simpson's Rule to estimate the amount of water passing by the dock every second.

✏️ WRITING & THINKING

27. Prove that Simpson's Rule actually gives the exact answer for definite integrals of all polynomials of degree 3 or less.

15.5 IMPROPER INTEGRALS

■ TOPICS

■ Improper Integrals Using Technology

There are many practical situations where an integral on an unbounded interval has meaning. An integral on an unbounded interval is called an **improper integral**.

Improper Integral

The integral $\int_a^{+\infty} f(x)\,dx$ is called an **improper integral**. This integral is defined as the following limit:

$$\int_a^{+\infty} f(x)\,dx = \lim_{b \to +\infty} \int_a^b f(x)\,dx.$$

If $\lim\limits_{b \to +\infty} \int_a^b f(x)\,dx$ exists, then the improper integral is said to be **convergent**.

If $\lim\limits_{b \to +\infty} \int_a^b f(x)\,dx$ does not exist, then the improper integral is said to be **divergent**.

It is a curiosity that an area can be finite yet have an infinite boundary. It is like having a backyard fence infinitely long but only a finite yard.

Example 1: Evaluating an Improper Integral

Evaluate $\int_1^{+\infty} \dfrac{1}{x^2}\,dx.$

Solution

First, evaluate the integral from $x = 1$ to $x = b$.

$$\int_1^b \frac{1}{x^2}\,dx = \int_1^b x^{-2}\,dx$$

$$= \frac{x^{-1}}{-1}\Bigg]_1^b = -\frac{1}{x}\Bigg]_1^b = -\frac{1}{b} + 1$$

Next, find the limit as $b \to +\infty$.

$$\lim_{b \to +\infty}\left(-\frac{1}{b} + 1\right) = 0 + 1 = 1 \qquad \text{Note that } \lim_{b \to +\infty} \frac{1}{b} = 0.$$

Therefore,

$$\int_1^{+\infty} \frac{1}{x^2}\,dx = 1$$

We say that the integral **converges** to 1.

The result from Example 1 can be interpreted as the area under the curve $f(x) = \dfrac{1}{x^2}$ on the interval $[1, +\infty]$. As illustrated in Figure 1, the larger b gets, the closer the value of the integral $\displaystyle\int_1^b \dfrac{1}{x^2}\,dx$ gets to 1.

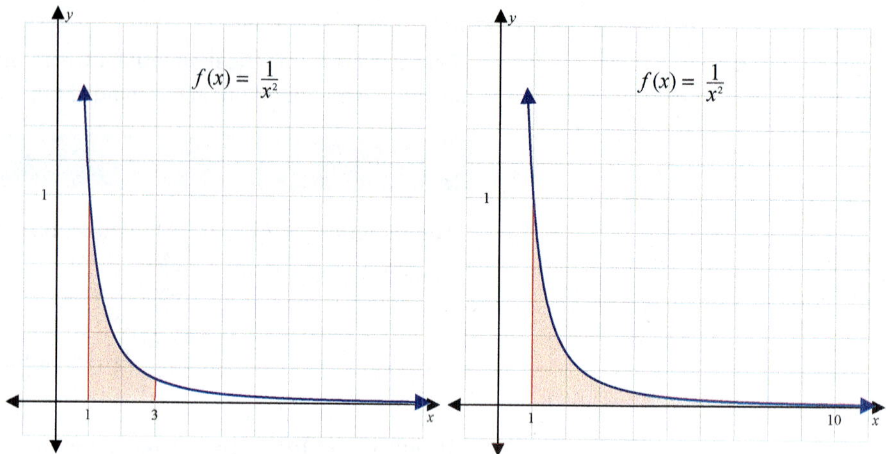

$$\int_1^3 \frac{1}{x^2}\,dx = \frac{2}{3} \qquad\qquad \int_1^{10} \frac{1}{x^2}\,dx = \frac{9}{10}$$

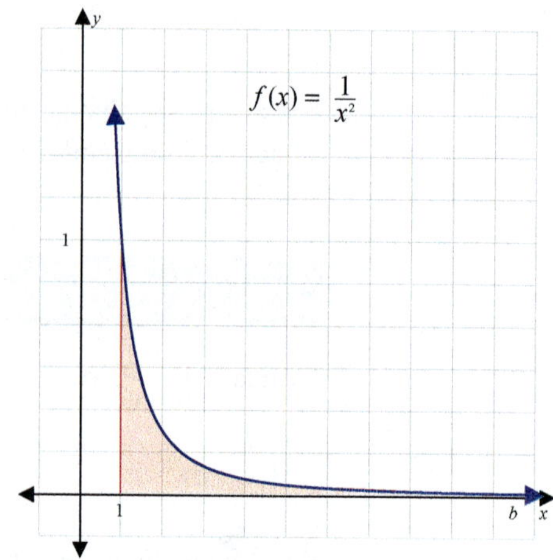

$$\lim_{b \to +\infty} \int_1^b \frac{1}{x^2}\,dx = 1$$

FIGURE 1

In evaluating improper integrals, we must be able to find limits as $b \to +\infty$. These ideas were discussed previously, and we list the following results for use in the examples and exercises that follow.

$$\lim_{b \to +\infty} \frac{1}{b} = 0, \qquad \lim_{b \to +\infty} \frac{1}{e^b} = 0, \qquad \lim_{b \to +\infty} \frac{b}{e^b} = 0,$$

$$\lim_{b \to +\infty} \left(\ln b\right) = +\infty, \qquad \text{and} \qquad \lim_{b \to +\infty} b^{\frac{1}{n}} = +\infty \ \text{(for } n > 0)$$

Example 2: Determining Integral Convergence

Determine whether the improper integral $\int_0^{+\infty} e^{-3x}\, dx$ is convergent or divergent. Evaluate it if it is convergent.

Solution

First, evaluate the integral from 0 to b.

$$\int_0^b e^{-3x}\, dx = -\frac{1}{3}e^{-3x}\bigg]_0^b = -\frac{1}{3e^{3b}} + \frac{1}{3}$$

Next, find the limit as $b \to +\infty$.

$$\lim_{b \to +\infty}\left(-\frac{1}{3e^{3b}} + \frac{1}{3}\right) = 0 + \frac{1}{3} = \frac{1}{3}$$

This integral converges to $\dfrac{1}{3}$.

Example 3: Determining Integral Convergence

Determine whether the improper integral $\int_1^{+\infty} \frac{1}{x}\, dx$ is convergent or divergent. Evaluate it if it is convergent.

Solution

First, evaluate the integral from 1 to b.

$$\int_1^b \frac{1}{x}\, dx = \ln|x|\bigg]_1^b = \ln b - \ln 1 = \ln b$$

Next, find the limit as $b \to +\infty$.

$$\lim_{b \to +\infty} \ln b = +\infty$$

The integral is divergent.

Example 3 illustrates the importance of the limit as well as the unusual and sometimes unexpected nature of improper integrals. In contrast to Example 1, where the area under the curve $f(x) = \dfrac{1}{x^2}$ equals 1, the area under the curve $f(x) = \dfrac{1}{x}$ is infinite. The curves seem similar on an intuitive basis but are quite different in terms of improper integrals and limits. (See Figure 2.)

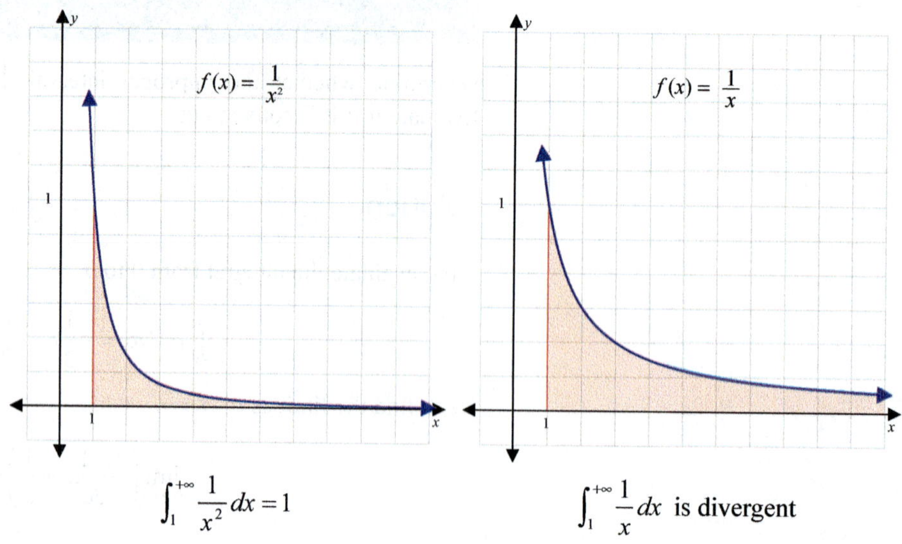

$$\int_1^{+\infty} \frac{1}{x^2}\,dx = 1$$

$$\int_1^{+\infty} \frac{1}{x}\,dx \text{ is divergent}$$

FIGURE 2

Example 4: Determining Integral Convergence

Find the value of the integral $\int_4^{+\infty} \dfrac{1}{\sqrt{x}}\,dx$ if it is convergent.

Solution

First, evaluate the integral from 4 to b.

$$\int_4^b \frac{1}{\sqrt{x}}\,dx = \int_4^b x^{-\frac{1}{2}}\,dx = 2x^{\frac{1}{2}}\Big]_4^b$$

$$= 2b^{\frac{1}{2}} - 4$$

Next, find the limit as $b \to +\infty$.

$$\lim_{b \to +\infty}\left(2b^{\frac{1}{2}} - 4\right) = +\infty$$

The integral is divergent.

Example 5: Determining Integral Convergence

Find the value of the integral $\int_1^{+\infty} \dfrac{1}{x^{\frac{3}{2}}}\,dx$ if it is convergent.

Solution

First, evaluate the integral from 1 to b.

$$\int_1^b \frac{1}{x^{\frac{3}{2}}}\,dx = \int_1^b x^{-\frac{3}{2}}\,dx = -2x^{-\frac{1}{2}}\Big]_1^b$$

$$= -2b^{-\frac{1}{2}} + 2 = -\frac{2}{b^{\frac{1}{2}}} + 2$$

Next, find the limit as $b \to +\infty$.

$$\lim_{b \to +\infty}\left(-\frac{2}{b^{\frac{1}{2}}} + 2\right) = 0 + 2 = 2$$

The integral converges to 2.

There are other types of improper integrals. We consider one more in this course. We consider the case of an area bounded by an asymptote.

Example 6: Area Bounded by an Asymptote

Find the value of $\int_0^1 \frac{1}{x^2}\,dx$.

Solution

Since the integrand is not defined at the lower limit, we must use a limit to evaluate the integral.

$$\int_0^1 \frac{1}{x^2}\,dx = \lim_{a \to 0^+}\int_a^1 \frac{1}{x^2}\,dx$$

$$= \lim_{a \to 0^+}\left[-x^{-1}\right]_a^1$$

$$= \lim_{a \to 0^+}\left[-1 - \left(-a^{-1}\right)\right]$$

$$= \lim_{a \to 0^+}\left[-1 + \frac{1}{a}\right] = -1 + \lim_{a \to 0^+}\frac{1}{a}$$

Since this last limit does not exist, the integral diverges.

Improper Integrals Using Technology

Once a convergent improper integral has been set up, it is possible to use the TI-84 Plus or other graphing utility to do the calculating. For example, let us integrate $\int_1^\infty 2xe^{-3x}\,dx$. The following methods are recommended.

Method 1

Step 1: Graph the function in a suitable window, say $[-1, 3]$ by $[-1, 1]$.

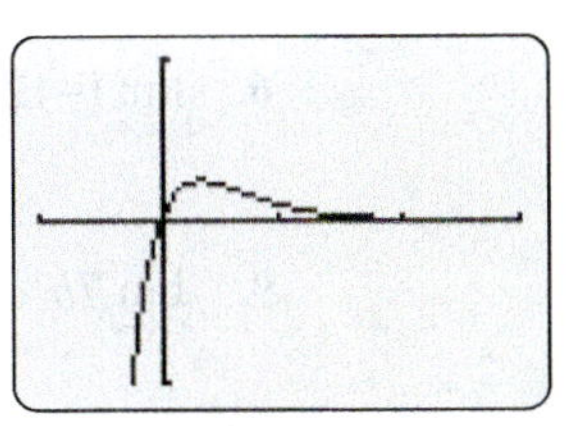

FIGURE 3

The displayed portion of the graph is large enough to easily see that there is very little area beyond $x = 3$. (See Figure 3.)

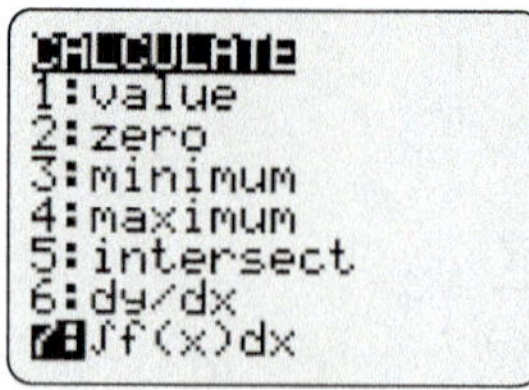

FIGURE 4

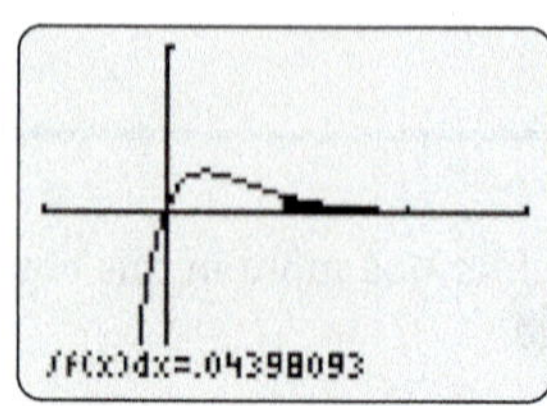

FIGURE 5

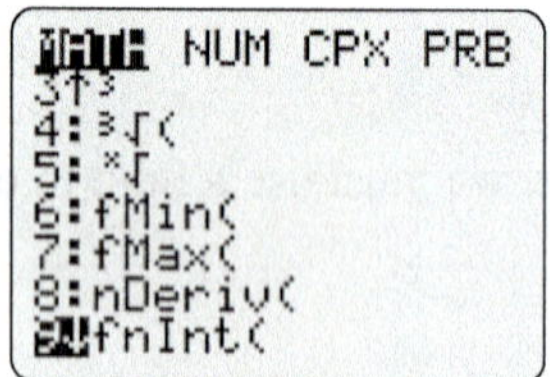

FIGURE 6

Step 2: Press ⟨ 2nd ⟩ ⟨ trace ⟩ and select item $7:\int f(x)dx$ (this is the integration symbol). (See Figure 4.)

At the prompt, type the lower limit 1 and enter. At the next prompt, type the upper limit 3. The area is shaded and the decimal answer $.04398093$ appears at the bottom of the screen (see Figure 5). This is an approximation to the actual value, but when using this method, the upper and lower limits must be in the range of x-values plotted on the screen.

Method 2

Step 1: Press ⟨ mode ⟩ and select CLASSIC.

Step 2: From the home screen, press ⟨ math ⟩ and select item $9:fnInt($ (function integral). (See Figure 6.)

Type the function, the variable of integration x, the lower limit, the upper limit, a right parenthesis, and ⟨ enter ⟩. The four items within the parentheses must be separated by commas. (**Note:** You may use any number for the upper limit; for this function 100 works well.)

The calculator will return $.0442551719$, a more accurate answer than the result from Method 1 (see Figure 7).

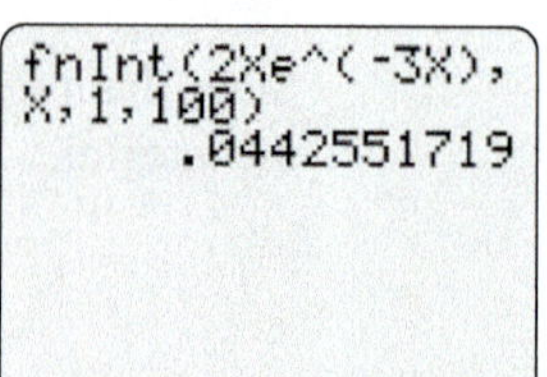

FIGURE 7

15.5 EXERCISES

💡 PRACTICE

In Exercises 1–10, find the limit if it exists.

1. $\displaystyle\lim_{b\to+\infty}\frac{1}{b}$

2. $\displaystyle\lim_{b\to+\infty}\frac{1}{\sqrt[3]{b}}$

3. $\displaystyle\lim_{b\to+\infty}\frac{\sqrt{b}}{20}$

4. $\displaystyle\lim_{b\to+\infty} e^{0.1b}$

5. $\displaystyle\lim_{b\to+\infty} e^{-4b}$

6. $\displaystyle\lim_{b\to+\infty}\left(-12\ln b\right)$

7. $\displaystyle\lim_{b\to+\infty}\left(2+\frac{9}{\sqrt{3b+1}}\right)$

8. $\displaystyle\lim_{b\to+\infty}\left(5+e^{-2b}\right)$

9. $\displaystyle\lim_{b\to+\infty} 7b^4 e^{-b}$

10. $\displaystyle\lim_{b\to+\infty}\left(7b+2\right)^{-\frac{2}{3}}$

In Exercises 11–34, determine whether the improper integrals are convergent or divergent, and evaluate those which are convergent.

11. $\displaystyle\int_{2}^{+\infty} \frac{4}{x^3}\,dx$

12. $\displaystyle\int_{1}^{+\infty} \frac{1}{\sqrt[3]{x}}\,dx$

13. $\displaystyle\int_{8}^{+\infty} x^{-\frac{2}{3}}\,dx$

14. $\displaystyle\int_{4}^{+\infty} 5x^{-\frac{3}{2}}\,dx$

15. $\displaystyle\int_{20}^{+\infty} 3e^{-x}\,dx$

16. $\displaystyle\int_{4}^{+\infty} e^{-2x}\,dx$

17. $\displaystyle\int_{2}^{+\infty} e^{-\frac{x}{3}}\,dx$

18. $\displaystyle\int_{2}^{+\infty} 4e^{-0.5x}\,dx$

19. $\displaystyle\int_{2}^{+\infty} e^{1.5x}\,dx$

20. $\displaystyle\int_{-1}^{+\infty} \frac{1}{80}e^{0.16x}\,dx$

21. $\displaystyle\int_{0}^{+\infty} \frac{1}{(x+3)^2}\,dx$

22. $\displaystyle\int_{0}^{+\infty} \frac{4}{\sqrt{3x+1}}\,dx$

23. $\displaystyle\int_{-1}^{+\infty} \frac{2}{\sqrt[3]{2x+3}}\,dx$

24. $\displaystyle\int_{2}^{+\infty} (3x+2)^{-\frac{4}{3}}\,dx$

25. $\displaystyle\int_{0}^{+\infty} \frac{5}{x+1}\,dx$

26. $\displaystyle\int_{0}^{+\infty} (5x+4)^{-\frac{3}{2}}\,dx$

27. $\displaystyle\int_{0}^{+\infty} x^2 e^{-x^3}\,dx$

28. $\displaystyle\int_{0}^{+\infty} -4xe^{x^2}\,dx$

29. $\displaystyle\int_{1}^{+\infty} xe^{1-x^2}\,dx$

30. $\displaystyle\int_{0}^{+\infty} 7xe^{-x^2}\,dx$

31. $\displaystyle\int_{2}^{+\infty} \frac{1}{x(\ln x)^3}\,dx$

32. $\displaystyle\int_{e}^{+\infty} \frac{1}{x\ln x}\,dx$

33. $\displaystyle\int_{0}^{+\infty} xe^{-x}\,dx$

34. $\displaystyle\int_{0}^{+\infty} xe^{-0.2x}\,dx$

In Exercises 35–38, find the area, if it exists, of the region under the curve $y = f(x)$ on the given interval of the x-axis.

35. $f(x) = \dfrac{4}{x^2},\ \ x \geq 2$

36. $f(x) = 3e^{-x},\ \ x \geq 0$

37. $f(x) = \dfrac{3}{x},\ \ x \geq 6$

38. $f(x) = 2e^{0.8x},\ \ x \geq 0$

✏ WRITING & THINKING

39. The integral $\displaystyle\int_{1}^{\infty} \frac{1}{x^p}\,dx$ converges if and only if (choose all that apply):

    a. $0 < p < 1$
    b. $p \neq 1$
    c. p is an integer greater than or equal to 2
    d. $p > 1$
    e. p is positive
    f. none of the above

〽 TECHNOLOGY

40. Integrate $\displaystyle\int_{1}^{\infty} 2xe^{-3x}\,dx$ by evaluating the limit and compare your answer to the calculator values obtained at the end of the section.

15.6 VOLUME

■ TOPICS

- ■ Solid of Revolution

Solid of Revolution

We use definite integration to find the area of the region between the graph of a continuous function and the x-axis. If such a region is revolved about the x-axis, it sweeps through a **solid of revolution**. The objective of this section is to develop a technique for finding the volume of a solid of revolution. (See Figure 1.)

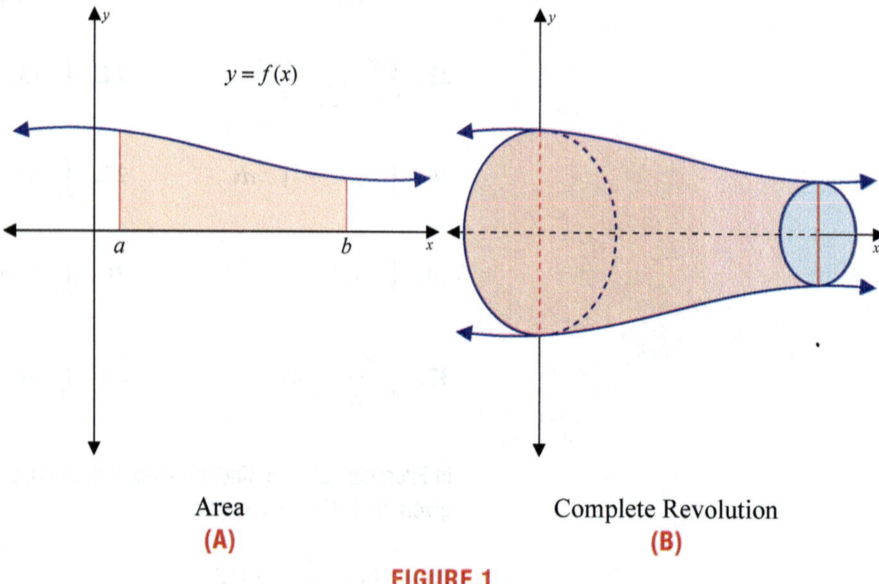

Area

(A)

Complete Revolution

(B)

FIGURE 1

From geometry, we know that the area of a circle with radius r is given by $A = \pi r^2$ and that the volume of a cylinder with a radius r and height h is given by $V = \pi r^2 h$. (See Figure 2.)

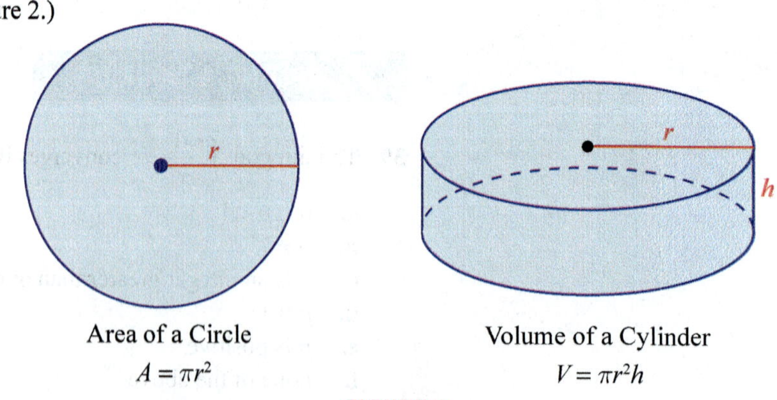

Area of a Circle

$A = \pi r^2$

Volume of a Cylinder

$V = \pi r^2 h$

FIGURE 2

For a solid of revolution, each cross section is a circle and each circle has a radius $f(x)$ that depends on x. The area of such a circular cross section is $A = \pi[f(x)]^2$. (See Figure 3(A).)

In Figure 3(B) the cylindrical cross section with radius $f(x)$ and height Δx is shown to have volume $V = \pi[f(x)]^2 \Delta x$.

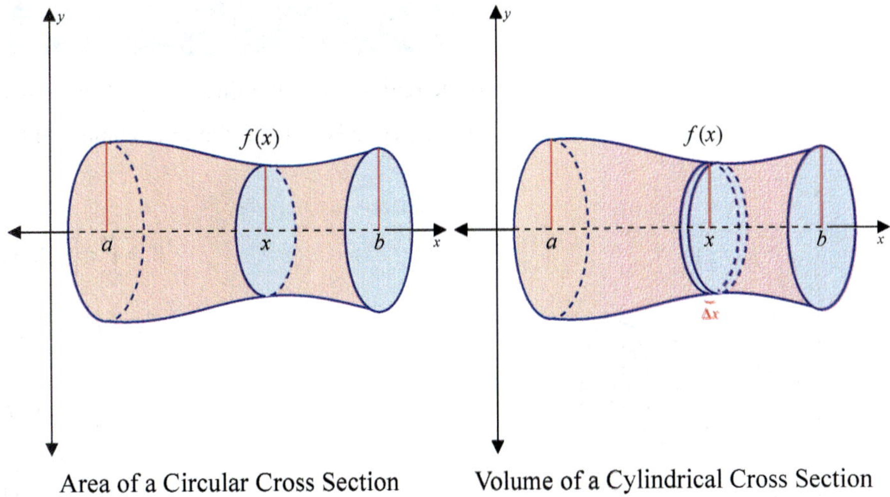

Area of a Circular Cross Section
$$A = \pi[f(x)]^2$$
(A)

Volume of a Cylindrical Cross Section
$$V = \pi[f(x)]^2 \Delta x$$
(B)

FIGURE 3

Now we partition the interval $[a, b]$ into n subintervals, each of width Δx, and choose a value in each subinterval: $c_1, c_2, \ldots, c_n$. The Riemann sum of the volumes of the cylinders with radii $f(c_1), f(c_2), \ldots, f(c_n)$ and height Δx is given by

$$S_n = \pi[f(c_1)]^2\Delta x + \pi[f(c_2)]^2\Delta x + \cdots + \pi[f(c_n)]^2\Delta x.$$

The limit of this sum as $n \to +\infty$ is defined to be the definite integral that is the **volume of the solid of revolution**.

Volume of a Solid of Revolution

If $y = f(x)$ is a nonnegative continuous function on the interval $[a, b]$, then the volume of the solid formed by revolving the region bounded by the graph of the function and the x-axis ($a \leq x \leq b$) about the x-axis is given by

$$V = \int_a^b \pi\left[f(x)\right]^2 dx.$$

Example 1: Finding the Volume of a Solid of Revolution

Find the volume of the solid of revolution generated by revolving the region under the curve $y = \sqrt{x}$ from $x = 0$ to $x = 4$ about the x-axis.

Solution

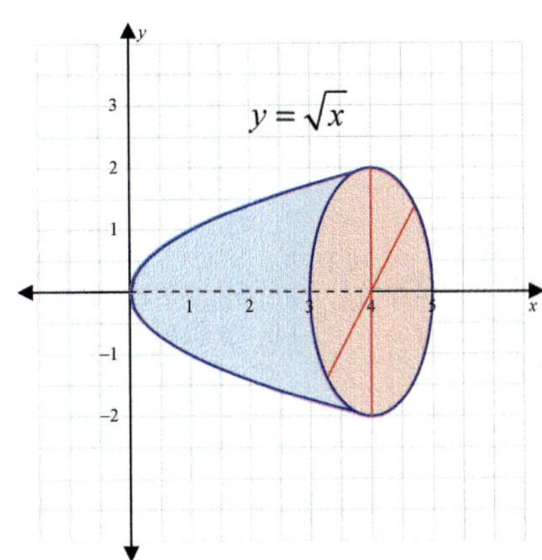

$$V = \int_0^4 \pi\left[f(x)\right]^2 dx$$

$$= \int_0^4 \pi\left(\sqrt{x}\right)^2 dx$$

$$= \pi\int_0^4 x\,dx = \pi\left(\frac{x^2}{2}\right)\Bigg]_0^4 = 8\pi$$

The volume is 8π cubic units.

If the region under the line $y = \dfrac{r}{h}x$ from $x = 0$ to $x = h$ is revolved about the x-axis, a circular cone is formed. Find the volume of this cone.

Solution

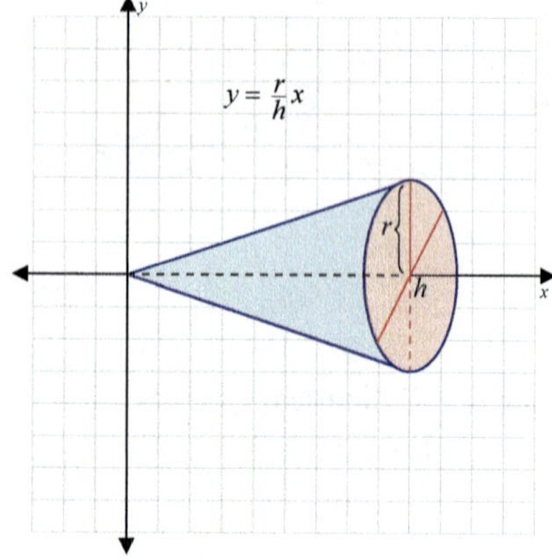

$$V = \int_0^h \pi \left[f(x) \right]^2 dx$$

$$= \pi \int_0^h \left(\frac{r}{h}x \right)^2 dx$$

$$= \pi \frac{r^2}{h^2} \int_0^h x^2 dx$$

$$= \pi \frac{r^2}{h^2} \left(\frac{x^3}{3} \right) \Bigg]_0^h$$

$$= \pi \frac{r^2}{h^2} \left(\frac{h^3}{3} \right) = \frac{1}{3} \pi r^2 h$$

Thus the volume is given by the standard formula $\dfrac{1}{3} \pi r^2 h$.

The region bounded by the parabola $y = 9 - x^2$ and the x-axis is revolved about the x-axis. Find the volume of the solid of revolution that is generated.

Solution

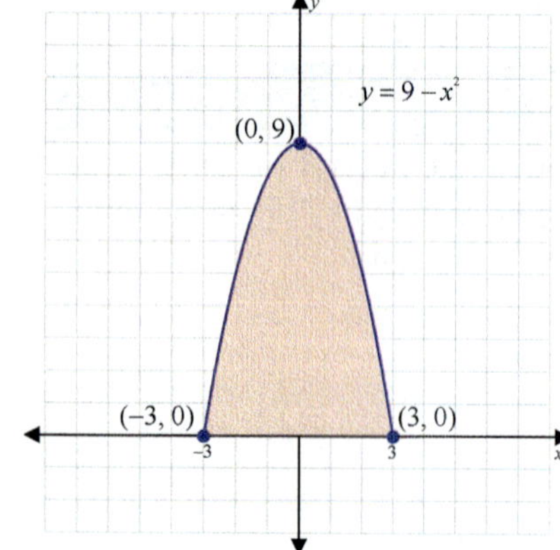

Set $y = 0$ to find the points where the parabola intersects the x-axis.

$$9 - x^2 = 0$$

$$x^2 = 9$$

$$x = \pm 3$$

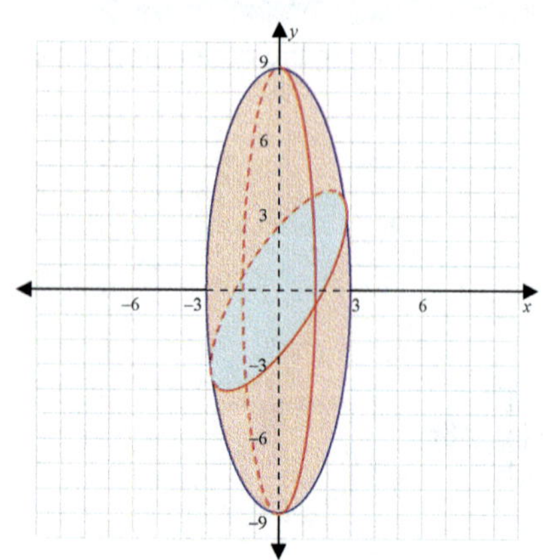

$$V = \int_{-3}^{3} \pi \left[f(x) \right]^2 dx$$

$$= \pi \int_{-3}^{3} \left(9 - x^2 \right)^2 dx$$

$$= \pi \int_{-3}^{3} \left(81 - 18x^2 + x^4 \right) dx$$

$$= \pi \left(81x - \frac{18x^3}{3} + \frac{x^5}{5} \right) \Bigg]_{-3}^{3}$$

$$= \pi \left[\left(243 - 162 + \frac{243}{5} \right) - \left(-243 + 162 - \frac{243}{5} \right) \right]$$

$$= \pi \left(486 - 324 + \frac{486}{5} \right) = \frac{1296\pi}{5}$$

The volume is $\dfrac{1296\pi}{5}$ cubic units.

We have considered only solids of revolution in which a region is rotated about the x-axis. The formula we used would need some adjustment if the region were rotated about some other horizontal line, since the radius of revolution would be represented by some expression other than $f(x)$. Such solids of revolution will not be considered in this course.

15.6 EXERCISES

💡 PRACTICE

Find the volume of the solid generated when the regions bounded by the graphs of the given equations and the x-axis are rotated about the x-axis.

1. $y = x$, $x = 0$, $x = 2$

2. $y = 3x$, $x = 1$, $x = 3$

3. $y = 2\sqrt{x}$, $x = 0$, $x = 4$

4. $y = \sqrt[3]{x}$, $x = 0$, $x = 8$

5. $y = e^x$, $x = -1$, $x = 2$

6. $y = e^{-x}$, $x = -1$, $x = 2$

7. $y = 1 - x^2$, $x = -1$, $x = 1$

8. $y = 4 - x^2$, $x = -2$, $x = 2$

9. $y = \left(16 - x^2\right)^{\frac{1}{2}}$, $x = 0$, $x = 4$

10. $y = \sqrt{3 - x^2}$, $x = 0$, $x = 1$

11. $y = \dfrac{4}{x}$, $x = 1$, $x = 3$

12. $y = \dfrac{2}{x}$, $x = 1$, $x = 2$

13. $y = \dfrac{1}{\sqrt{x}}$, $x = 1$, $x = 6$

14. $y = \dfrac{2}{\sqrt{x}}$, $x = 1$, $x = 5$

15. $y = x + \sqrt{x}$, $x = 1$, $x = 4$

16. $y = \sqrt{x} - x$, $x = 0$, $x = 1$

16

# Chapter 16

MULTIVARIABLE CALCULUS

16.1 FUNCTIONS OF SEVERAL VARIABLES

◼ TOPICS

- ◼ Cobb-Douglas Production Formula
- ◼ Graphs in Three Dimensions

Up to this point we have studied functions of one variable. For example, the cost function

$$C(x) = x^2 + 3x + 100$$

depends only on the single variable x, where x represents the number of items produced. Similarly, the distance function

$$s(t) = 180t - 16t^2$$

depends only on the single variable t, where t represents time. With functions of one variable such as these, we have discussed concepts related to differentiation such as maxima and minima, developed integration techniques, worked with a wide variety of applications, and learned detailed graphing techniques.

Now we want to consider several of these same ideas as they relate to functions that have more than one variable. For example, suppose that a company produces two products, x units of one and y units of the other. Then the cost function depends on both x and y, and we might write

$$C(x, y) = 20,000 + 10x + 30y,$$

where \$20,000 represent the fixed costs.

If another company produces three products, the cost function might be in the form

$$C(x, y, z) = 500 + x + e^{0.2}y + 10.6z^2,$$

where x, y, and z represent the numbers produced of each of the three types of items. Both cost functions are **functions of several variables**.

In the current discussion we will deal primarily with functions of two variables, although the basic ideas can easily be expanded to include functions of more than two variables.

Domain and Range of $f(x, y)$

If D is a set of ordered pairs of real numbers and for each ordered pair (x, y) in D there corresponds a unique real number $f(x, y)$, then f is called a **function of x and y**. The set D is the **domain** of the function. The set of all the values of $f(x, y)$ is called the **range** of the function.

We sometimes write

$$z = f(x, y)$$

and call z the **dependent variable** and x and y the **independent variables**.

Example 1: Evaluating $f(x, y)$

For $f(x, y) = 3x + y^2$ find the following.

a. $f(2, 3)$ **b.** $f\left(2, \sqrt{2}\right)$ **c.** $f(0, 0)$

Solution

a. $f(2, 3) = 3 \cdot 2 + 3^2 = 15$ Substitute $x = 2$ and $y = 3$.

b. $f\left(2, \sqrt{2}\right) = 3 \cdot 2 + \left(\sqrt{2}\right)^2 = 8$ Substitute $x = 2$ and $y = \sqrt{2}$.

c. $f(0, 0) = 3 \cdot 0 + 0^2 = 0$ Substitute $x = 0$ and $y = 0$.

Example 2: Evaluating $f(x, y)$

For $f(x, y) = e^{\sqrt{x}} + \ln y$,

a. find the domain and   **b.** find $f(0, 1)$.

Solution

a. For $\sqrt{x}$ to be defined, we must have $x \geq 0$.
For $\ln y$ to be defined, we must have $y > 0$.
So, the domain is $\{(x, y) \mid x \geq 0 \text{ and } y > 0\}$.

b. $f(0, 1) = e^{\sqrt{0}} + \ln 1 = e^0 + 0 = 1$ Substitute $x = 0$ and $y = 1$.

Example 3: Evaluating Revenue Influenced by Two Variables

A store sells two brands of soda. Brand A sells for \$1.25 per bottle and Brand B sells for \$1.50 per bottle.

a. What is the revenue function for soda?

b. What is the revenue for soda if 100 bottles of Brand A and 150 bottles of Brand B are sold?

Solution

a. Let $x =$ the number of bottles of Brand A sold, and
$y =$ the number of bottles of Brand B sold.

Then the revenue function is

$$R(x, y) = 1.25x + 1.50y.$$

b. $R(100, 150) = 1.25(100) + 1.50(150)$
$$= 125 + 225$$
$$= \$350$$

Example 4: Characteristics of a Box

Suppose that a box has a square base and an open top.

a. Write a function of two variables that represents the volume of the box.

b. Write a function of two variables that represents the surface area of the box.

c. Determine the surface area if the dimensions of the box are 4 by 4 by 5.

Solution

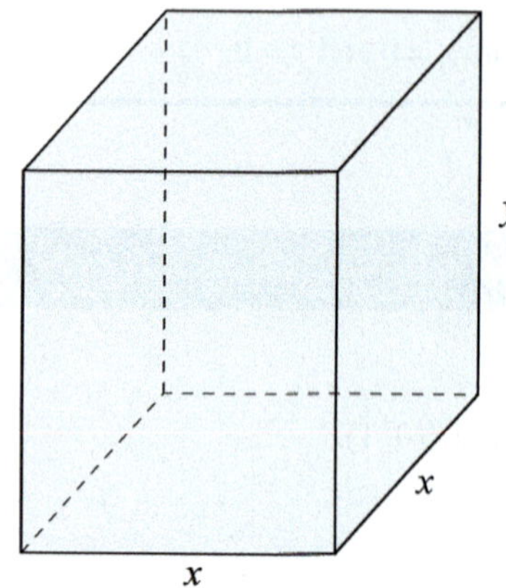

Since the base is square, the length and width of the box are equal.

Let x = length of the box,

$\quad x$ = width of the box, and

$\quad y$ = height of the box.

a. The formula for the volume of a rectangular solid is $V = l \cdot w \cdot h$ where l = length, w = width, and h = height. Thus, the function that represents the volume of this box is as follows.

$$V(x, y) = x \cdot x \cdot y = x^2 y$$

b. The surface area consists of the sum of the areas of the bottom surface, the front and back faces, and the left and the right faces. For this box, we have

$$x^2 = \text{area of bottom surface and}$$

$$xy = \text{area of each of the other four faces.}$$

Thus the total surface area is represented by the function

$$S(x, y) = x^2 + 4xy.$$

c. Evaluate the surface area function at $x = 4$ and $y = 5$.

$$S(4, 5) = 4^2 + 4(4)(5) = 16 + 80 = 96$$

Cobb-Douglas Production Formula

Economists use a formula called the **Cobb-Douglas Production Formula** to model the production levels of a company (or a country). The formula is

$$P(x, y) = kx^a y^{1-a},$$

where P is the total units produced, x is a measure of labor units, y is a measure of capital invested, and k is a constant that varies from product to product. Notice that the sum of the exponents on x and y is 1. That is, $a + (1 - a) = 1$.

Example 5: Using the Cobb-Douglas Production Formula

Suppose that the function $P(x, y) = 500x^{0.3}y^{0.7}$ represents the number of units produced by a company with x units of labor and y units of capital.

a. How many units of a product will be manufactured if 300 units of labor and 50 units of capital are used?

b. How many units will be produced if twice the numbers of units of labor and capital are used?

Solution

a. $P(300, 50) = 500(300)^{0.3}(50)^{0.7}$

$\approx 42,794$ units produced   Fractional units are not counted.

b. If the numbers of units of labor and capital are both doubled, then

$$x = 2 \cdot 300 = 600 \text{ and } y = 2 \cdot 50 = 100.$$

So

$$P(600, 100) = 500(600)^{0.3}(100)^{0.7} \qquad \text{Substitute } x = 600 \text{ and } y = 100.$$

$\approx 85,588$ units produced.

Thus we see that production is doubled if both labor and capital are doubled.

Graphs in Three Dimensions

Graphs of functions of two variables require a *three-dimensional coordinate system*. We will use the Cartesian system with three axes, x-axis, y-axis, and z-axis, each perpendicular to the other two. Points are represented by **ordered triples** in the form (x, y, z). Figure 1 illustrates the location of the point $(1, 2, 3)$.

Intuitively, picture a corner of your classroom where two walls and the floor come together. The x-axis is where the floor meets one wall, the y-axis is where the floor meets the other wall, and the z-axis is where the two walls meet. If several students stand, their feet are in the xy-plane, and the tops of their heads can be considered to be points in the space of the room represented by ordered triples depending on how far they are standing from each wall and their height. (See Figure 2.)

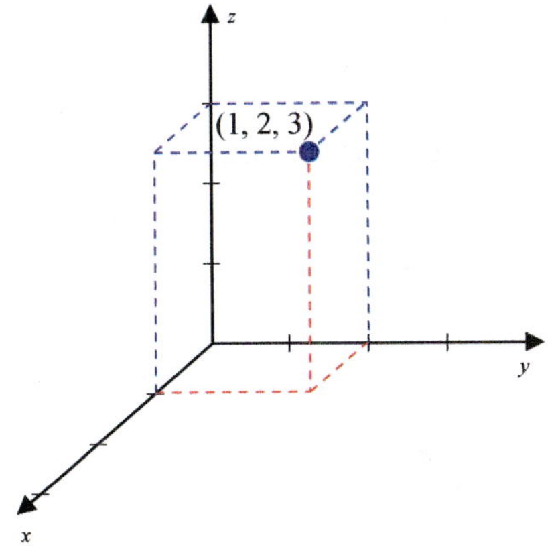

FIGURE 1

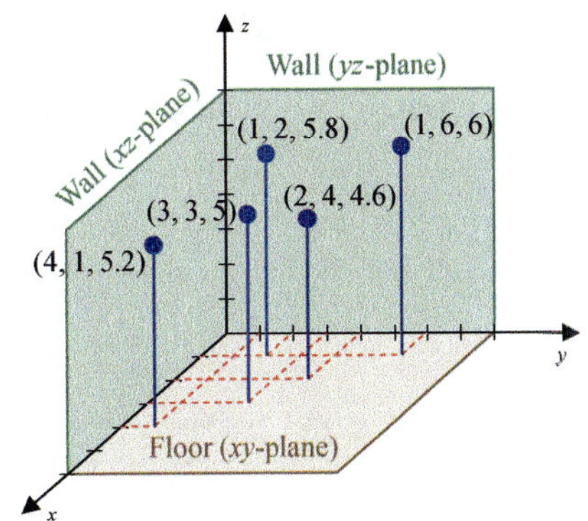

FIGURE 2

Draw a three-dimensional coordinate system; then graph and label the following points.

$$A(1, -2, 3), \quad B(0, 0, 4), \quad C(0, 3, 2), \quad \text{and} \quad D(2, 2, -1)$$

Solution

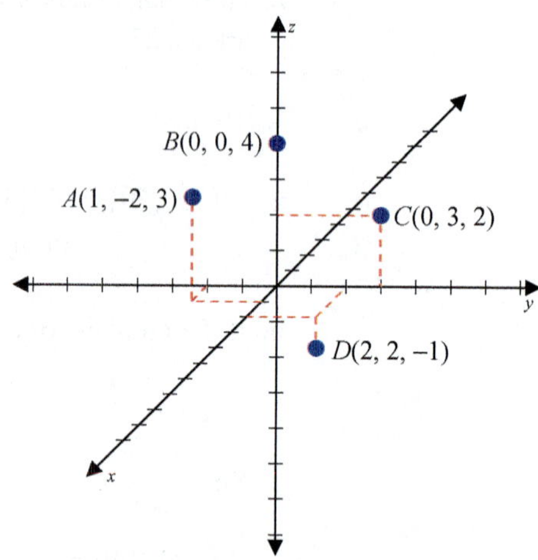

A point in space with coordinates (a, b, c) can be thought of as the intersection of three planes, each plane perpendicular to the other two. The equations of these planes are $x = a$ (where y and z can have all real values), $y = b$ (where x and z can have all real values), and $z = c$ (where x and y can have all real values). Portions of the graphs of these planes are illustrated in Figure 3.

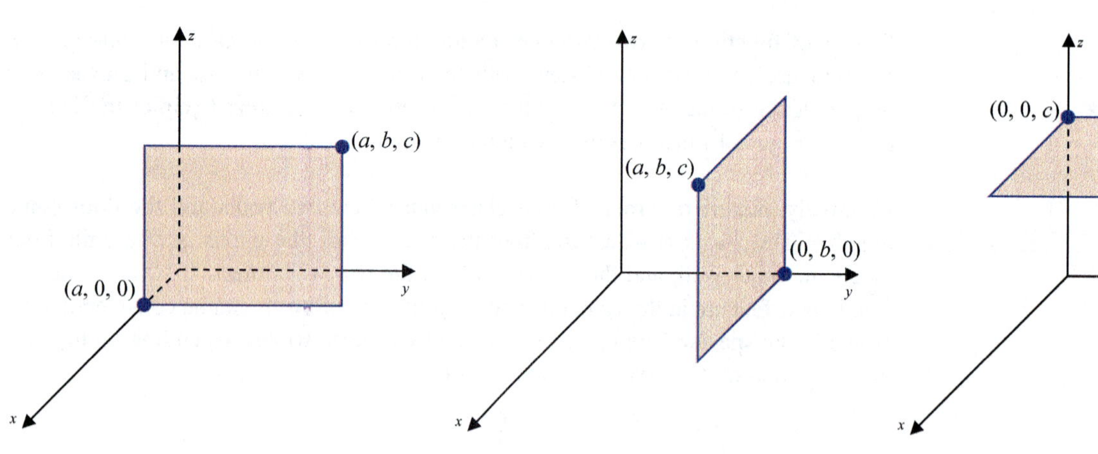

Plane $x = a$ (parallel to the yz-plane)
(A)

Plane $y = b$ (parallel to the xz-plane)
(B)

Plane $z = c$ (parallel to the xy-plane)
(C)

FIGURE 3

Set up a three-dimensional coordinate system to graph each of the following planes.

a. $x = 5$ **b.** $y = -2$

Solution

a.

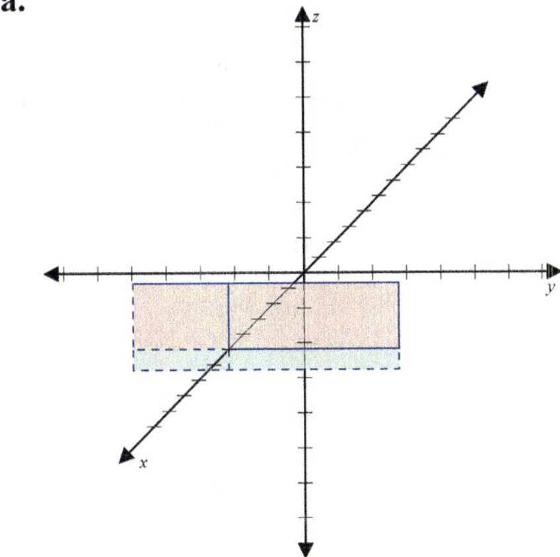

Note that $x = 5$ is a plane parallel to the yz-plane.

b.

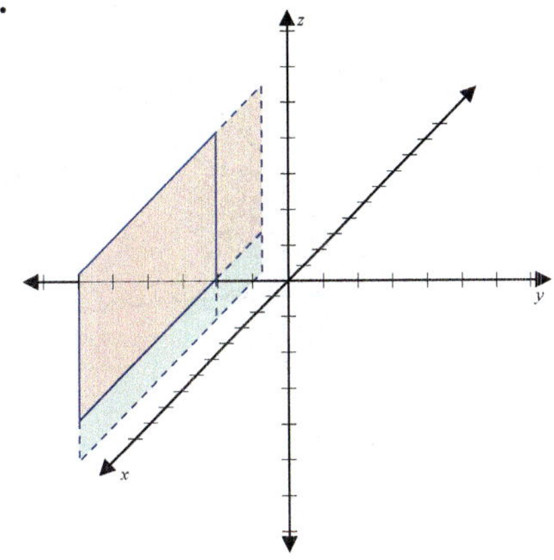

Note that $y = -2$ is a plane parallel to the xz-plane.

In each graph, the contrasting color corresponds to negative z-values and only a portion of the plane is graphed. You might have chosen to sketch a different portion. For your sketch to have a proper perspective, the lines that are shown should be drawn parallel to the coordinate axes.

In general, the graph of a function of two variables is a **surface** and can be difficult to draw. In this course, the student is not expected to draw such surfaces; however, the graphs of several surfaces and their corresponding equations will be presented as aids in understanding certain concepts such as local maxima, local minima, and partial derivatives. Portions of four such surfaces are illustrated in Figure 4.

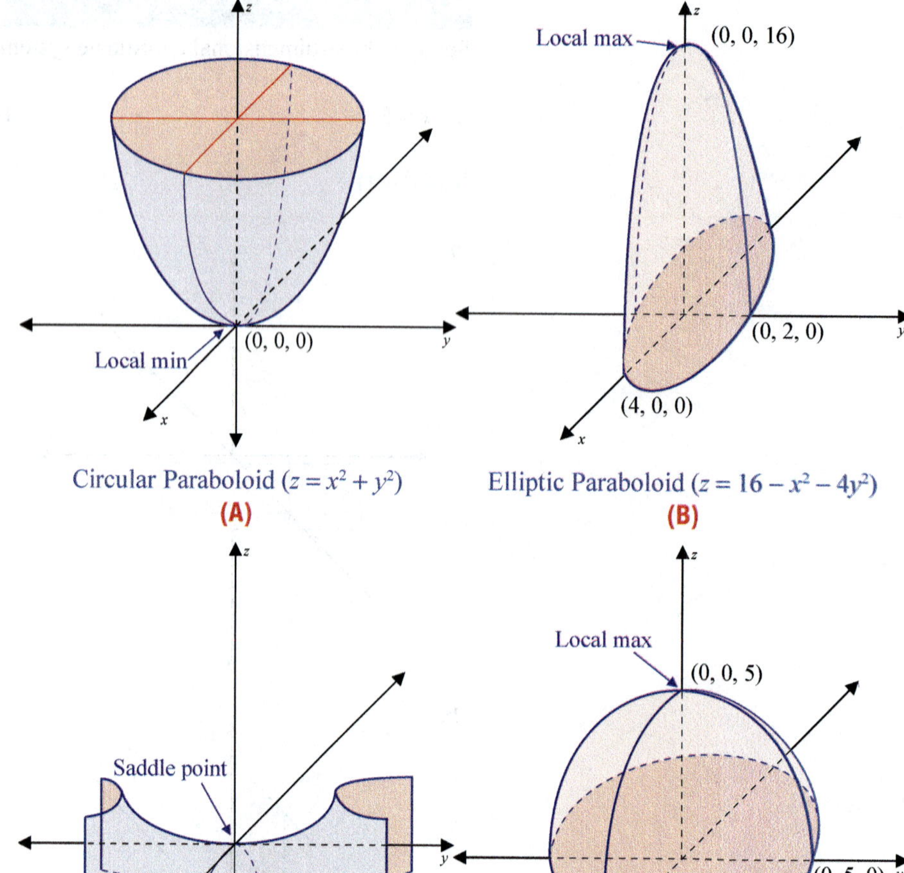

Circular Paraboloid ($z = x^2 + y^2$)

(A)

Elliptic Paraboloid ($z = 16 - x^2 - 4y^2$)

(B)

Hyperbolic Paraboloid ($z = y^2 - x^2$)

(C)

Hemisphere ($z = \sqrt{25 - x^2 - y^2}$)

(D)

FIGURE 4

16.1 EXERCISES

💡 PRACTICE

In Exercises 1–15, find the indicated function values, if possible.

1. $f(x, y) = 12x - 3y + xy$
  a. $f(1, 3)$
  b. $f(0, 4)$

2. $f(x, y) = 7xy - 11x + 9y$
  a. $f(-2, 1)$
  b. $f(3, -2)$

3. $f(x, y) = 4x^2 - 3xy + y^2$
  a. $f(2, 5)$
  b. $f(0, 3)$

4. $g(x, y) = 2xy + 5x^2y + y^3$
  a. $g(-2, 2)$
  b. $g(-1, -1)$

5. $g(x, y) = \dfrac{4x + y}{x - y}$

   a. $g(3, -1)$
   b. $g(2, 2)$

6. $f(x, y) = \dfrac{2x - y}{x^2 + y}$

   a. $f(6, 2)$
   b. $f(3, -9)$

7. $g(x, y) = \dfrac{8x^2 y}{\sqrt{2x + y}}$

   a. $g(1, -4)$
   b. $g(-2, 6)$

8. $f(x, y) = \dfrac{5x^2 + y^3}{\sqrt{4x + y^2}}$

   a. $f(-1, 3)$
   b. $f(3, 2)$

9. $g(x, y) = 3xe^{x+y}$
   a. $g(2, 1)$
   b. $g(-1, 0)$

10. $g(x, y) = ye^{4x} + 2xy$
   a. $g(1, 5)$
   b. $g(-1, 3)$

11. $f(x, y) = x \ln xy + y \ln x$
   a. $f(1, e)$
   b. $f(e^2, 1)$

12. $f(x, y) = 2x^3 y + \ln y$
   a. $f(-2, e)$
   b. $f(3, 1)$

13. $A(P, r, t) = Pe^{rt}$

   a. $A(1000, 0.08, 4)$
   b. $A(500, 0.06, 5)$

14. $A(P, r, t) = P\left(1 + \dfrac{r}{4}\right)^{4t}$

   a. $A(1500, 0.08, 6)$
   b. $A(800, 0.06, 10)$

15. $S(l, w, h) = 2lw + 2lh + 2wh$

   a. $S(18, 15, 9)$
   b. $S(14, 9, 11)$

In Exercises 16–19, draw a three-dimensional coordinate system, and graph and label the given points.

16. $A(0, 0, 3)$; $B(2, -1, 0)$; $C(4, 2, 1)$; $D(3, -2, -2)$

17. $A(0, 1, 0)$; $B(3, 0, -4)$; $C(1, 1, -3)$; $D(-2, 3, 0)$

18. $A(-3, 0, 2)$; $B(0, 0, 2)$; $C(2, -2, 2)$; $D(0, -2, 2)$

19. $A(-3, -2, 1)$; $B(4, 0, -1)$; $C(0, 0, 4)$; $D(-1, 4, -1)$

In Exercises 20–25, draw a three-dimensional coordinate system, and graph the given planes.

20. $x = 0$ **21.** $y = 0$ **22.** $z = 0$

23. $x = 3$ **24.** $y = 2$ **25.** $z = 4$

✦ APPLICATIONS

26. Stock yield: The yield of a stock is given by the function $Y(d, p) = \dfrac{d}{p}$, where d is the dividend per share of stock and p is the price per share. Find the yield of a stock that sells for \$5.88 if the dividend is \$1.00.

27. Intelligence quotient: The intelligence quotient (IQ) of a person is determined by $f(M, C) = 100 \cdot \dfrac{M}{C}$, where M is the mental age (determined by tests) and C is the actual or chronological age. Find the IQ of a child who is 13 years old and has a mental age of 15.4 years. (Round to the nearest integer.)

28. Cobb-Douglas production: The number of units of a product that are manufactured by a company is given by $f(L, K) = 300L^{0.4}K^{0.6}$, where L is the units of labor and K is the units of capital.

    a. How many units of a product will be manufactured by utilizing 30 units of labor and 24 units of capital? (Round to the nearest unit.)

    b. How many units will be produced if the number of units of labor and capital are doubled? (Round to the nearest unit.)

29. Cost: A company manufactures two lawn mower models, standard and self-propelled. The cost of producing each standard mower is $80, and the cost of producing each self-propelled mower is $140. If the fixed costs are $5200, the total cost function is given by $C(x, y) = 5200 + 80x + 140y$, where x is the number of standard and y is the number of self-propelled mowers.

    a. Find $C(30, 20)$.   **b.** Find $C(36, 25)$.

30. Cost: The cost function for producing two models of a product is found to be $C(x, y) = 850 + 32x + 20y$, where x is the number of model A and y is the number of model B. The cost for model A is $32, the cost for model B is $20, and the fixed costs are $850 per week.

    a. Find $C(40, 24)$.   **b.** Find $C(60, 38)$.

31. Cost: The cost of producing the standard model of a video camera is $160. The cost of producing the deluxe model is $220.

    a. If a company has weekly fixed costs of $1360, find the cost function $C(x, y)$, where x is the number of standard models and y is the number of deluxe models.

    b. Find $C(15, 12)$.

32. Cost: A company makes two grades of paint, grade I, guaranteed for 5 years, and grade II, guaranteed for 10 years. A gallon of grade I costs $3.20 to make, while a gallon of grade II costs $3.90 to make. The weekly fixed costs are $4500.

    a. Find the cost function $C(x, y)$ for making x gallons of grade I and y gallons of grade II.

    b. What is the cost of making 200 gallons of grade I and 140 gallons of grade II?

33. Revenue: A grocery store sells two brands of a product, the store brand and a name brand. The manager estimates that if she sells the store brand for x dollars and the name brand for y dollars, she will be able to sell $64 - 20x + 18y$ units of the store brand and $52 + 16x - 22y$ units of the name brand.

    a. Find the revenue function $R(x, y)$.

    b. What is the revenue if she sells the store brand for $4.00 and the name brand for $4.50?

34. Revenue: A pharmacy sells two cold remedies, one a generic remedy and the other a name brand. The store manager has determined that he can sell $26 - 6x + 8y$ bottles of the generic remedy and $22 + 5x - 9y$ bottles of the name brand if the prices are x dollars per bottle and y dollars per bottle, respectively.

    a. Find the revenue function $R(x, y)$.

    b. What is the revenue if the generic remedy is priced at $6.20 per bottle and the name brand is priced at $7.00 per bottle?

35. Volume and surface area: A rectangular box has no top and one partition (see diagram).

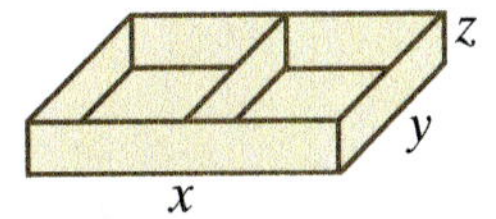

    a. Write a function of three variables for the number of cubic units in the volume of the box.

    b. Write a function of three variables for the number of square units of material needed to construct the box.

36. Volume and surface area: A rectangular box has no top and two intersecting partitions (see diagram).

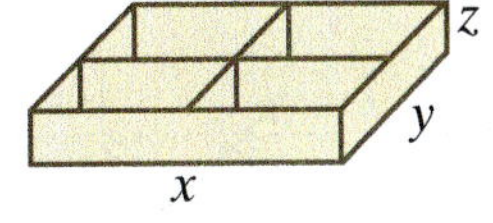

    a. Write a function of three variables for the number of cubic units in the volume of the box.

    b. Write a function of three variables for the number of square units of material needed to construct the box.

37. Compound interest: A deposit of $1000 is made into a savings account earning interest compounded quarterly. The amount $A(r, t)$ after t years is given by

$$A(r,t) = 1000\left(1 + \frac{r}{4}\right)^{4t}, \text{ where } r \text{ is the interest rate in decimal form. Use this}$$

function of two variables to complete the following table.

		Number of Years (t)		
		3	5	10
Rate (r)	0.06			
	0.08			
	0.10			

38. Interest compounded continuously: A deposit of $1000 is made into a savings account earning interest compounded continuously. The amount $A(r, t)$ after t years is given by $A(r, t) = 1000e^{rt}$, where r is the interest rate in decimal form. Use this function of two variables to complete the following table.

		Number of Years (t)		
		5	8	12
Rate (r)	0.080			
	0.085			
	0.100			

16.2 PARTIAL DERIVATIVES

■ TOPICS

- First-Order Partial Derivatives
- Marginal Productivity of Labor and Capital
- Second-Order Partial Derivatives
- Partial Derivatives of Functions of Three Variables

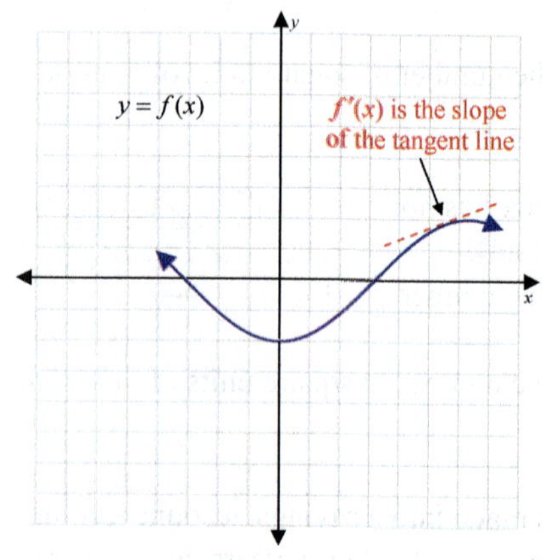

FIGURE 1

For a function of one variable, $y = f(x)$, the first derivative $\dfrac{dy}{dx} = f'(x)$, if it exists, can be interpreted as the instantaneous rate of change of y with respect to x. Geometrically $f'(x)$ is the slope of a line tangent to the graph of the function, as shown in Figure 1.

First-Order Partial Derivatives

Now we will consider the same ideas for a function of two variables, $z = f(x, y)$. We want to find the instantaneous rate of change of z with respect to each variable, one at a time. That is, we find a derivative with respect to x by temporarily treating y as a constant and a derivative with respect to y by temporarily treating x as a constant. Derivatives of this type (if they exist) are called **partial derivatives**.

First-Order Partial Derivatives

For a function of two variables, $z = f(x, y)$, the two **first-order partial derivatives** (or simply **first partial derivatives**) are denoted and defined as follows.

a. The **first partial derivative of f with respect to x** (if it exists) is

$$\frac{\partial f}{\partial x} = \lim_{h \to 0} \frac{f(x + h, y) - f(x, y)}{h}.$$

b. The **first partial derivative of f with respect to y** (if it exists) is

$$\frac{\partial f}{\partial y} = \lim_{k \to 0} \frac{f(x, y + k) - f(x, y)}{k}.$$

c. In our notation, $\dfrac{\partial f}{\partial x} = \dfrac{\partial z}{\partial x}$ and $\dfrac{\partial f}{\partial y} = \dfrac{\partial z}{\partial y}.$

Notice that in the definition of $\dfrac{\partial f}{\partial x}$, x is changed by h and y is unchanged (treated as a constant).

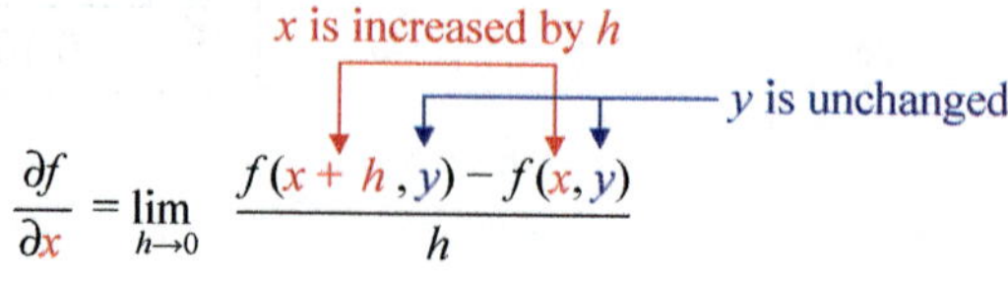

$$\frac{\partial f}{\partial x} = \lim_{h \to 0} \frac{f(x + h, y) - f(x, y)}{h}$$

Similarly in the definition of $\dfrac{\partial f}{\partial y}$, y is changed by k and x is unchanged (treated as a constant).

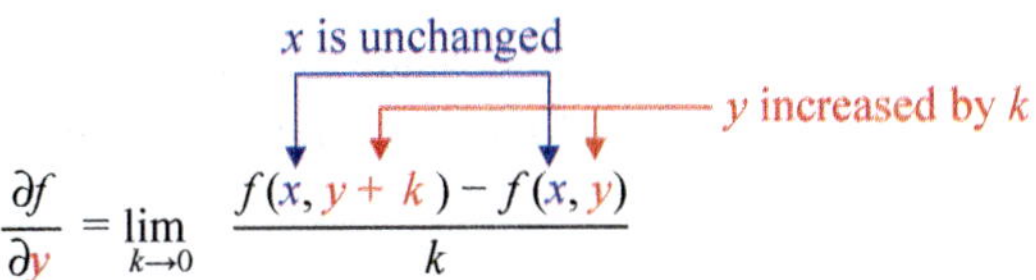

$$\frac{\partial f}{\partial y} = \lim_{k \to 0} \frac{f(x, y + k) - f(x, y)}{k}$$

The two symbols $\dfrac{\partial}{\partial x}$ and $\dfrac{\partial}{\partial y}$ are called **operators**. These operators are similar to $\dfrac{d}{dx}$ but are used to indicate that the operation of partial differentiation is to be performed relative to the variable in the denominator with other variables held constant.

Geometrically, if $\dfrac{\partial f}{\partial x}$ or $\dfrac{\partial z}{\partial x}$ exists, it can be interpreted as the slope of a line tangent to the surface represented by the function $z = f(x, y)$. This line is in a plane where y is constant (a plane parallel to the xz-plane) and is also tangent to the curve formed by the intersection of the surface and the plane. Similarly, $\dfrac{\partial f}{\partial y}$ can be interpreted as the slope of a line tangent to the surface in a plane where x is constant (a plane parallel to the yz-plane). Both situations are illustrated in Figure 2.

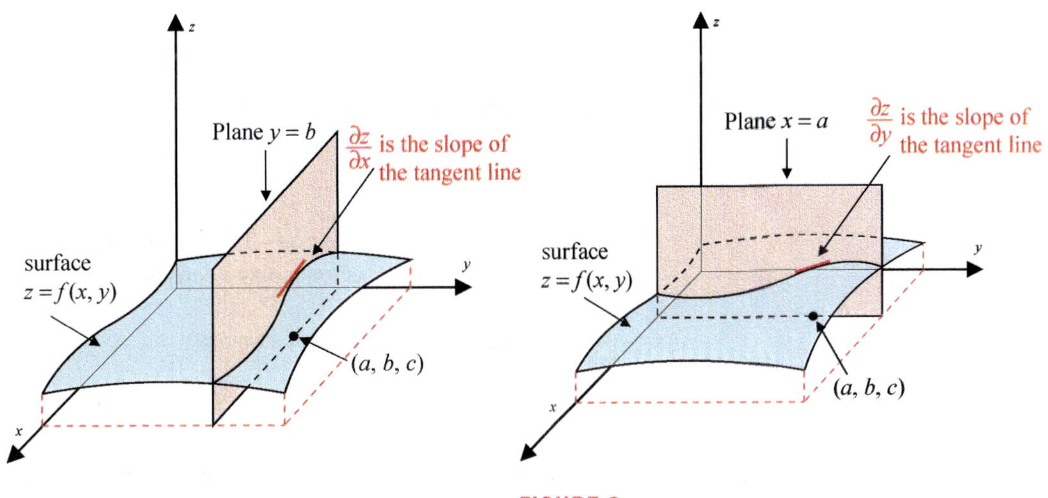

FIGURE 2

In effect, to find $\dfrac{\partial f}{\partial x}$, we simply treat y and any function of y as any other constant and follow the rules of differentiation with one variable. For example, the expressions

$$\ln y, \quad -5y, \quad y^4, \quad \text{and} \quad e^y$$

are all treated as constants when finding $\dfrac{\partial f}{\partial x}$.

Similarly, x and any function of x are treated as constants when finding $\dfrac{\partial f}{\partial y}$.

The following notations are also used to indicate partial derivatives:

$$\frac{\partial f}{\partial x} = \frac{\partial z}{\partial x} = f_x(x, y) = f_x$$

and

$$\frac{\partial f}{\partial y} = \frac{\partial z}{\partial y} = f_y(x, y) = f_y.$$

The value of the partial derivative of $z = f(x, y)$ with respect to x at the point $(a, b, f(a, b))$ can be denoted by any of the following forms:

$$\frac{\partial f}{\partial x}\bigg|_{(a,b)}, \qquad \frac{\partial z}{\partial x}\bigg|_{(a,b)}, \qquad \text{and} \quad f_x(a, b).$$

Similar notations denote the value of the partial derivative with respect to y at that point.

Study the following examples carefully so that you become accustomed to treating y (or x, as the case may be) as a constant.

Example 1: Finding a Partial Derivative

For the function $f(x, y) = 4x^2 - 3xy + 5y^2$, find

a. $\dfrac{\partial f}{\partial x}$ and

b. $\dfrac{\partial f}{\partial y}$.

Solution

a. Treating y as a constant, we obtain

$$\frac{\partial f}{\partial x} = 8x - 3y.$$

Note that in the expression $-3xy$, we treat $-3y$ as a constant coefficient of x. Also, the expression $5y^2$ is treated as a constant.

b. Treating x as a constant, we obtain

$$\frac{\partial f}{\partial y} = -3x + 10y.$$

In this case $4x^2$ is treated as a constant and $-3x$ is treated as a constant coefficient of y.

Example 2: Finding a Partial Derivative

For the function $f(x, y) = e^{xy} - \ln x + y^3$, find the following.

a. f_x

b. f_y

Solution

a. To differentiate e^{xy} with respect to x, we use the Chain Rule and differentiate the exponent xy by treating y as a constant. Thus we have

$$f_x = e^{xy} \cdot y - \frac{1}{x} = ye^{xy} - \frac{1}{x}.$$

b. To find f_y, we treat x as a constant. (This means that $\ln x$ is treated as a constant, too.)

$$f_y = x \cdot e^{xy} + 3y^2$$

Example 3: Finding a Partial Derivative

For the function $f(x, y) = xe^{xy} + y^2$, find

a. $f_x(1, 2)$ and

b. $f_y(1, 2)$

Solution

a. To find $f_x(x, y)$, we treat y^2 as a constant. However, we treat xe^{xy} as a product of two functions of x. The two functions are x and e^{xy}.

$$f_x(x, y) = x \cdot \frac{\partial}{\partial x}\left(e^{xy}\right) + e^{xy} \cdot \frac{\partial}{\partial x}(x) + \frac{\partial}{\partial x}\left(y^2\right)$$
$$= x \cdot e^{xy} \cdot y + e^{xy} \cdot 1 + 0$$
$$= xye^{xy} + e^{xy}$$

Now we evaluate $f_x(1, 2)$.

$$f_x(1, 2) = 1 \cdot 2 \cdot e^{1 \cdot 2} + e^{1 \cdot 2}$$
$$= 2e^2 + e^2$$
$$= 3e^2$$

b. To find $f_y(x, y)$, we do **not** use the Product Rule to differentiate xe^{xy} with respect to y, since x is treated as a constant.

$$f_y(x, y) = x \cdot e^{xy} \cdot x + 2y$$
$$= x^2 e^{xy} + 2y$$

Now we evaluate $f_y(1, 2)$.

$$f_y(1, 2) = 1^2 \cdot e^{1 \cdot 2} + 2 \cdot 2$$
$$= e^2 + 4$$

Example 4: Using the Partial Derivatives

Given the function $z = x^2 + xy + y^2 - 7x - 8y + 1$, find all the ordered pairs (x, y) where both $\dfrac{\partial z}{\partial x} = 0$ and $\dfrac{\partial z}{\partial y} = 0$.

Solution

First, we find the partial derivatives.

$$\frac{\partial z}{\partial x} = 2x + y - 7 \quad \text{and} \quad \frac{\partial z}{\partial y} = x + 2y - 8$$

We want to solve the following system of equations.

$$\begin{cases} 2x + y - 7 = 0 \\ x + 2y - 8 = 0 \end{cases}$$

Multiply each term in the first equation by −2, add to eliminate y, and then solve the resulting equation for x.

$$-4x - 2y = -14 \quad \text{Multiply each term by } -2.$$
$$\underline{\hphantom{-4}x + 2y = 8\hphantom{-14}}$$
$$-3x \hphantom{+2y} = -6 \quad \text{Add the equations to eliminate } y.$$
$$x = 2 \quad \text{Solve for } x.$$
$$2 + 2y - 8 = 0 \quad \text{Substitute } x = 2 \text{ in one of the original equations and solve for } y.$$
$$2y = 6$$
$$y = 3$$

Thus both partial derivatives are 0 at (2, 3).

Marginal Productivity of Labor and Capital

We previously introduced the Cobb-Douglas Production Formula $P(x, y) = kx^a y^{1-a}$ where x is a measure of units of labor and y is a measure of units of capital invested.

The partial derivative $\dfrac{\partial P}{\partial x}$ is called the **marginal productivity of labor**, and the partial derivative $\dfrac{\partial P}{\partial y}$ is called the **marginal productivity of capital**. Thus, when $x = x_1$ and $y = y_1$,

$$\left.\frac{\partial P}{\partial x}\right|_{(x_1, y_1)}$$

is the approximate increase in production for one unit of increase in labor,

and

$$\left.\frac{\partial P}{\partial y}\right|_{(x_1, y_1)}$$

is the approximate increase in production for one unit of increase in capital.

Example 5: Using the Cobb-Douglas Production Formula

Suppose that the production function $P(x, y) = 2000x^{0.5}y^{0.5}$ is known. Determine the marginal productivity of labor and the marginal productivity of capital when 16 units of labor and 144 units of capital are used.

Solution

$$\frac{\partial P}{\partial x} = 2000(0.5)x^{-0.5}y^{0.5} = \frac{1000y^{0.5}}{x^{0.5}}$$

$$\frac{\partial P}{\partial y} = 2000(0.5)x^{0.5}y^{-0.5} = \frac{1000x^{0.5}}{y^{0.5}}$$

Substituting $x = 16$ and $y = 144$, we have

$$\left.\frac{\partial P}{\partial x}\right|_{(16,144)} = \frac{1000(144)^{0.5}}{(16)^{0.5}} = \frac{1000(12)}{4} = 3000 \text{ units}$$

and

$$\left.\frac{\partial P}{\partial y}\right|_{(16,144)} = \frac{1000(16)^{0.5}}{(144)^{0.5}} = \frac{1000(4)}{12} = 333 \text{ units.} \quad \text{Only whole units are counted.}$$

Thus we see that adding one unit of labor will increase production by about 3000 units and adding one unit of capital will increase production by about 333 units.

Second-Order Partial Derivatives

For a function of two variables, $z = f(x, y)$, each of the first partial derivatives $f_x(x, y)$ and $f_y(x, y)$ is also a function of two variables. This means that each of the functions f_x and f_y has two partial derivatives, provided that they exist. These four partial derivatives of partial derivatives are called **second-order partial derivatives** (or simply **second partial derivatives**) and can be denoted by the following forms:

$$\frac{\partial}{\partial x}\left(\frac{\partial z}{\partial x}\right) = \frac{\partial^2 z}{\partial x^2} = \frac{\partial^2 f}{\partial x^2} = f_{xx},$$

$$\frac{\partial}{\partial y}\left(\frac{\partial z}{\partial x}\right) = \frac{\partial^2 z}{\partial y \partial x} = \frac{\partial^2 f}{\partial y \partial x} = f_{xy},$$

$$\frac{\partial}{\partial y}\left(\frac{\partial z}{\partial y}\right) = \frac{\partial^2 z}{\partial y^2} = \frac{\partial^2 f}{\partial y^2} = f_{yy}, \quad \text{and}$$

$$\frac{\partial}{\partial x}\left(\frac{\partial z}{\partial y}\right) = \frac{\partial^2 z}{\partial x \partial y} = \frac{\partial^2 f}{\partial x \partial y} = f_{yx}.$$

The notation indicates the specific order in which the partial differentiation is to take place. For example, f_{xy} indicates that the partial differentiation is to be first with respect to x and then with respect to y. When the notation ∂ is used, this same order of partial differentiation is shown with the positions of x and y reversed.

$$f_{xy} = \frac{\partial^2 f}{\partial y \, \partial x}$$

Left-to-right order   Right-to-left order

Later we will show how second-order partial derivatives for a function of two variables can be used to determine whether a local extremum is a local maximum or a local minimum.

Example 6: Finding Second Partial Derivatives

a. Find all four second partial derivatives of $f(x, y) = \ln(x^2 + 4y)$.

b. Find $f_{xx}\left(2, \frac{1}{2}\right)$.

Solution

a. We must find the first partial derivatives f_x and f_y before we can find the second partial derivatives.

$$f_x = \frac{1}{x^2 + 4y}(2x) = \frac{2x}{x^2 + 4y}$$

$$f_y = \frac{1}{x^2 + 4y}(4) = \frac{4}{x^2 + 4y}$$

Now we can find the second partial derivatives.

$$f_{xx} = \frac{(x^2 + 4y) \cdot 2 - 2x(2x)}{(x^2 + 4y)^2} = \frac{-2x^2 + 8y}{(x^2 + 4y)^2}$$

$$f_{xy} = \frac{(x^2 + 4y) \cdot 0 - 2x \cdot 4}{(x^2 + 4y)^2} = \frac{-8x}{(x^2 + 4y)^2}$$

$$f_{yx} = \frac{(x^2 + 4y) \cdot 0 - 4 \cdot 2x}{(x^2 + 4y)^2} = \frac{-8x}{(x^2 + 4y)^2}$$

$$f_{yy} = \frac{(x^2 + 4y) \cdot 0 - 4 \cdot 4}{(x^2 + 4y)^2} = \frac{-16}{(x^2 + 4y)^2}$$

b. $f_{xx}\left(2, \frac{1}{2}\right) = \dfrac{-2 \cdot 2^2 + 8\left(\frac{1}{2}\right)}{\left[2^2 + 4\left(\frac{1}{2}\right)\right]^2} = \dfrac{-8 + 4}{(4 + 2)^2} = \dfrac{-4}{36} = -\dfrac{1}{9}$

Partial Derivatives of Functions of Three Variables

The concept of partial derivatives is easily extended to include functions of three or more variables. If, for example, $w = f(x, y, z)$, then there are three first partial derivatives:

$$\frac{\partial w}{\partial x} = f_x(x, y, z), \text{ found by treating } y \text{ and } z \text{ as constants,}$$

$$\frac{\partial w}{\partial y} = f_y(x, y, z), \text{ found by treating } x \text{ and } z \text{ as constants, and}$$

$$\frac{\partial w}{\partial z} = f_z(x, y, z), \text{ found by treating } x \text{ and } y \text{ as constants.}$$

Later we use partial derivatives of functions of three variables to determine the maximum or minimum value of a function of two variables with restrictions on the variables.

Example 7: Finding Partial Derivatives with Three Variables

For $w = 2ye^{xy} + z^2$, find

a. $\dfrac{\partial w}{\partial x}$,   **b.** $\dfrac{\partial w}{\partial y}$, and   **c.** $\dfrac{\partial w}{\partial z}$.

Solution

a. Treat y and z as constants.

$$\frac{\partial w}{\partial x} = 2y \cdot e^{xy} \cdot y + 0 = 2y^2 e^{xy}$$

b. Treat x and z as constants. Here $2ye^{xy}$ is a product of two functions of y, and we use the Product Rule to differentiate.

$$\frac{\partial w}{\partial y} = 2y \cdot e^{xy} \cdot x + e^{xy} \cdot 2 + 0 = 2e^{xy}(xy + 1)$$

c. Treat x and y as constants. In this case, the entire expression $2ye^{xy}$ is treated as a constant.

$$\frac{\partial w}{\partial z} = 0 + 2z = 2z$$

16.2 EXERCISES

💡 PRACTICE

Find $\dfrac{\partial f}{\partial x}$ and $\dfrac{\partial f}{\partial y}$ for each of the functions in Exercises 1–28.

1. $f(x,y) = 4x + 7y - 10$

2. $f(x,y) = 11x - 19y + 2$

3. $f(x,y) = 2x^2 + 5y^2$

4. $f(x,y) = 5x^3 - 6y^4$

5. $f(x,y) = x^2 y + 4xy^3 + 6$

6. $f(x,y) = x^3 - 90x^2 y^3 - 9$

7. $f(x,y) = y\sqrt{25 + x^2}$

8. $f(x,y) = x^3 \sqrt{y^2 + 5}$

9. $f(x,y) = \sqrt{49 - x^2 - y^2}$

10. $f(x,y) = \sqrt{16 + 2x^2 + y^2}$

11. $f(x,y) = 4e^{x-y}$

12. $f(x,y) = 7e^{xy}$

13. $f(x,y) = x \ln y$

14. $f(x,y) = \ln(x^2 + y^2)$

15. $f(x,y) = \ln(x^2 + 3xy)$

16. $f(x,y) = \dfrac{y}{\ln x}$

17. $f(x,y) = \dfrac{2x^2}{y^2 + 1}$

18. $f(x,y) = \dfrac{x^2 + 3}{5y - 9}$

19. $f(x,y) = \dfrac{x^2}{xy + 3}$

20. $f(x,y) = \dfrac{x + y}{4x - y}$

21. $f(x,y) = x^3 e^y + ye^{x^2}$

22. $f(x,y) = xe^{y^3} - y^2 e^x$

23. $f(x,y) = x\sqrt{xy + 2}$

24. $f(x,y) = y\sqrt{x^2 + y^3}$

25. $f(x,y) = x^4 e^{xy}$

26. $f(x,y) = y^3 e^{-xy}$

27. $f(x,y) = y^5 \ln(x^2 - 5y^2)$

28. $f(x,y) = 3x \ln xy^4$

In Exercises 29–32, find $\dfrac{\partial S}{\partial m}$ and $\dfrac{\partial S}{\partial b}$.

29. $S(m,b) = (2m + b - 9)^2 + (4m + b - 13)^2 + (5m + b - 18)^2$

30. $S(m,b) = (8m + b - 17)^2 + (9m + b - 23)^2 + (10m + b - 28)^2$

31. $S(m,b) = (6m + b - 40)^2 + (8m + b - 49)^2 + (9m + b - 55)^2 + (10m + b - 62)^2$

32. $S(m,b) = (12m + b - 81)^2 + (13m + b - 88)^2 + (14m + b - 96)^2 + (15m + b - 101)^2$

In Exercises 33–44, find all second-order partial derivatives f_{xx}, f_{xy}, f_{yx}, and f_{yy}.

33. $f(x,y) = 3xy + x^2 y^3 - 19$

34. $f(x,y) = 5x^3 y - 3x^3 y^2 - 13$

35. $f(x,y) = x^4 y^{\frac{2}{3}}$

36. $f(x,y) = (xy)^{\frac{3}{4}}$

37. $f(x,y) = xe^{2y}$

38. $f(x,y) = \dfrac{e^{x^3}}{y^4}$

39. $f(x,y) = (4x - 3y)^{\frac{5}{3}}$

40. $f(x,y) = \sqrt{7x^3 + y^2}$

41. $f(x,y) = \dfrac{3x+1}{5y+3}$

42. $f(x,y) = \dfrac{6y^2 - 5}{2x + 7}$

43. $f(x,y) = \dfrac{2xy}{x-y}$

44. $f(x,y) = \dfrac{x-y}{xy}$

In Exercises 45–48, find f_x, f_y, and f_z.

45. $f(x,y,z) = xy + 2xz + 9yz$

46. $f(x,y,z) = 3x^2 y + 2xyz + 7xz^2$

47. $f(x,y,z) = (8x^2 + 5y^2 - 2z^2)^2$

48. $f(x,y,z) = \sqrt{x^2 + 2y^2 + 4z^2}$

In Exercises 49–52, find $\dfrac{\partial F}{\partial x}, \dfrac{\partial F}{\partial y}$, and $\dfrac{\partial F}{\partial \lambda}$. (Note that λ is the Greek letter lambda.)

49. $F(x,y,\lambda) = 8x + 15xy - 2y^2 + \lambda(x + y - 60)$

50. $F(x,y,\lambda) = 3x^2 + 12y^2 + \lambda(x + 2y - 84)$

51. $F(x,y,\lambda) = 5x^2 + 3xy - 10y^2 + \lambda(14x + 17y - 49)$

52. $F(x,y,\lambda) = 7x^2 - 2xy + 9y^2 + \lambda(8x + 15y - 120)$

🚀 APPLICATIONS

53. Marginal productivity: The number of units of a product that are manufactured by a company is given by $f(L,K) = 80L^{\frac{2}{3}} K^{\frac{1}{3}}$, where L is the units of labor and K is the units of capital. Find the marginal productivity of labor and the marginal productivity of capital if the company is currently utilizing 27 units of labor and 64 units of capital.

54. Marginal productivity: The productivity of a company is approximated by $f(L,K) = 20L^{\frac{2}{5}} K^{\frac{3}{5}}$, where L is the units of labor and K is the units of capital. Find the marginal productivity of labor and the marginal productivity of capital if the company is currently utilizing 32 units of labor and 32 units of capital.

55. Marginal cost: A company manufactures two products, product A and product B. The cost of producing x units of A and y units of B is $C(x,y) = 3000 + 7x + 5.8y + 0.03x^2 - xy + 0.02y^2$.
   a. Find the marginal cost with respect to x.
   b. Find the marginal cost with respect to y.

56. Marginal profit: The profit from the sale of two products is given by the function $P(x, y) = 88x + 54y - 0.02x^2 - 0.015y^2 - 68$, where x is the number of units of product A sold, and y is the number of units of product B sold.
   a. Find the marginal profit with respect to x.
   b. Find the marginal profit with respect to y.

57. Marginal profit: A company produces two models of a product. The cost function is given by $C(x, y) = x^2 - 2xy + 2y^2 + 4x + 3y + 8$ and the revenue function is given by $R(x, y) = 20x + 15y$, where x is the number of units of model A and y is the number of units of model B produced and sold.
   a. Find the profit function.
   b. Find $P_x(20,14)$ and $P_y(20,14)$ and interpret the results.

58. Marginal profit: A firm produces and sells x units of product A and y units of product B. Its revenue function is given by $R(x, y) = 80x + 100y$ and its cost function is given by $C(x, y) = x^2 + 1.5y^2 - xy + 1500$. Find $P_x(50, 25)$ and $P_y(50, 25)$ and interpret the results.

59. Marginal profit: A marketing manager of a department store has determined that revenue is related to the number of units of television advertising x and the number of units of newspaper advertising y by the function $R(x, y) = 500(20x + 5y + 20xy - x^2)$. Each unit of television advertising costs $5000 and each unit of newspaper advertising costs $2500.
   a. Find the marginal profit with respect to x.
   b. Find the marginal profit with respect to y.

60. Marginal profit: A firm manufactures and sells two models of electric mowers. The standard model of the mower sells for $300, and the self-propelled model of the mower sells for $400. The total cost function is $C(x, y) = 90,000 + 0.05x^2 + 0.1y^2 + 0.125xy$, where x is the number of standard models and y is the number of self-propelled models.
   a. Find the marginal profit with respect to x.
   b. Find the marginal profit with respect to y.

16.3 LOCAL EXTREMA FOR FUNCTIONS OF TWO VARIABLES

Previously, we discussed differentiation of functions of one variable and how the first derivative can be used to help locate and test local extrema. We also tested local extrema with the Second Derivative Test. In this section, we will define local maxima and local minima for functions of two variables and discuss a test involving second-order partial derivatives for determining the nature of these local extrema.

We will find that the conditions and concepts related to local maxima and local minima are quite similar for functions of one variable and functions of two variables. For functions of one variable, we dealt with intervals on real number lines: open intervals (endpoints not included) and closed intervals (endpoints included). For functions of two variables, we deal with regions in planes: **open regions** (boundary points not included) and **closed regions** (boundary points included).

> ### Local Extrema for a Function $z = f(x, y)$
>
> Suppose that $z = f(x, y)$ is a function defined on a region in the xy-plane and (a, b) is a point in that region.
>
> **1.** If there is an open region R containing (a, b) such that
>
> $$f(a, b) \geq f(x, y)$$
>
> for all (x, y) in R, then $f(a, b)$ is called a **local maximum** of f.
>
> **2.** If there is an open region R containing (a, b) such that
>
> $$f(a, b) \leq f(x, y)$$
>
> for all (x, y) in R, then $f(a, b)$ is called a **local minimum** of f.

As with functions of one variable, local extrema for functions of two variables may or may not be absolute extrema. In this course we will not be concerned with determining absolute extrema for functions of two variables. Several local extrema are illustrated in Figure 1.

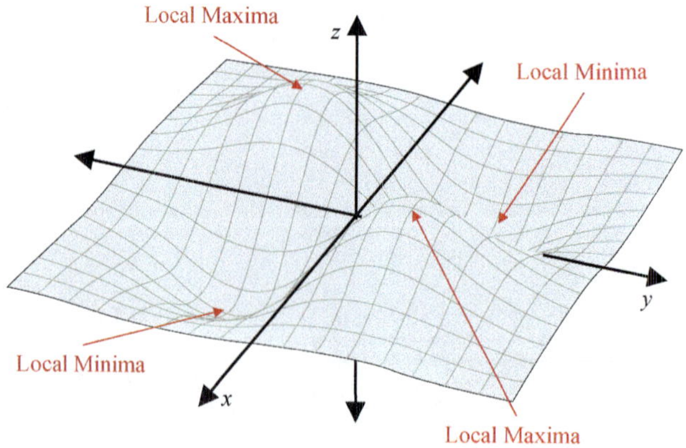

FIGURE 1

Recall that, for $y = f(x)$, a function of one variable, $x = c$ is a critical value if $f'(c) = 0$ or $f'(c)$ does not exist. Similarly, for a function $z = f(x, y)$, a point (a, b) in the domain of f is a **critical point** if $f_x(a, b) = 0$ and $f_y(a, b) = 0$.

If $f(a, b)$ is a local minimum (or a local maximum), then (a, b) is a critical point. However, a critical point does not guarantee a local extremum. In some cases, a critical point gives a **saddle point** where there is neither a local minimum nor a local maximum.

Example 1: Locating Critical Points

Locate the critical points for the function $f(x, y) = 4 + (x - 3)^2 + (y + 5)^2$.

Solution

We find the first partial derivatives, set them equal to zero, and solve the resulting system of equations.

$$f_x(x, y) = 2(x - 3) \quad \text{and} \quad f_y(x, y) = 2(y + 5)$$

The solution of the system

$$\begin{cases} 2(x - 3) = 0 \\ 2(y + 5) = 0 \end{cases}$$

is the point $(3, -5)$. Thus the only critical point is $(3, -5)$. Therefore, if f has a local minimum or a local maximum, then it must occur at $(3, -5)$.

To determine whether or not a critical point is a local minimum, a local maximum, or a saddle point, we can use the following test, known as the **Second Partials Test** (or the **D-Test**).

Second Partials Test (or D-Test)

Suppose that a function $z = f(x, y)$ and the first partial derivatives and the second partial derivatives are all defined in an open region R and that (a, b) is a critical point in R such that

$$f_x(a, b) = 0 \quad \text{and} \quad f_y(a, b) = 0.$$

Define the quantity D as follows.

$$D = f_{xx}(a, b) \cdot f_{yy}(a, b) - [f_{xy}(a, b)]^2$$

Case 1: If $D > 0$ and $f_{xx}(a, b) > 0$, then $f(a, b)$ is a **local minimum**.

Case 2: If $D > 0$ and $f_{xx}(a, b) < 0$, then $f(a, b)$ is a **local maximum**.

Case 3: If $D < 0$, then $(a, b, f(a, b))$ is a **saddle point**.

Case 4: If $D = 0$, then this test gives no information.

Example 2: Applying the D-Test

Find all the local minima, local maxima, and saddle points for the function

$$f(x, y) = x^2 - xy + y^2 - 9x + 5.$$

Solution

Find the first partial derivatives f_x and f_y.

$$f_x(x, y) = 2x - y - 9 \quad \text{and} \quad f_y(x, y) = -x + 2y$$

Now, to find any critical points, solve the following system.

$$\begin{cases} 2x - y - 9 = 0 \\ -x + 2y = 0 \end{cases}$$

$$\begin{aligned} 2x - y &= 9 \\ \underline{-2x + 4y} &= 0 \qquad \text{Multiply the second equation by 2 and add the two equations.} \\ 3y &= 9 \\ y &= 3 \\ -x + 2 \cdot 3 &= 0 \qquad \text{Substitute } y = 3 \text{ into the second equation and solve for } x. \\ x &= 6 \end{aligned}$$

The only critical point is (6, 3).

The second partials are

$$f_{xx}(x, y) = 2, \quad f_{yy}(x, y) = 2, \quad \text{and} \quad f_{xy}(x, y) = -1.$$

In this case each of the second partials is a constant and will have that constant value at (6, 3).

$$\begin{aligned} D &= f_{xx}(6,3) \cdot f_{yy}(6,3) - \left[f_{xy}(6,3) \right]^2 \\ &= 2 \cdot 2 - (-1)^2 \\ &= 4 - 1 = 3 > 0 \end{aligned}$$

Since $D > 0$, we check the sign of $f_{xx}(6, 3)$ to determine whether (6, 3) yields a local minimum or a local maximum.

$$f_{xx}(6, 3) = 2 > 0$$

Therefore, by Case 1 in the D-Test, the critical point (6, 3) yields a local minimum value. This value is

$$f(6, 3) = 6^2 - 6 \cdot 3 + 3^2 - 9 \cdot 6 + 5 = -22.$$

Example 3: Locating and Classifying Critical Points

Locate and classify all critical points of the function

$$g(x, y) = \frac{1}{3}x^3 - \frac{1}{3}y^3 + xy.$$

Solution

The first partial derivatives are as follows.

$$g_x = x^2 + y \quad \text{and} \quad g_y = -y^2 + x$$

Now solve the following system where $g_x = 0$ and $g_y = 0$.

$$\begin{cases} x^2 + y = 0 \\ -y^2 + x = 0 \end{cases}$$

Solving the second equation for x gives $x = y^2$, and substituting y^2 for x in the first equation gives

$$\left(y^2\right)^2 + y = 0$$
$$y^4 + y = 0$$
$$y\left(y^3 + 1\right) = 0$$
$$y = 0 \quad \text{or} \quad y^3 = -1$$
$$y = -1$$
$$x = 0^2 = 0 \quad x = \left(-1\right)^2 = 1$$

Thus there are two critical points: $(0, 0)$ and $(1, -1)$.

Now find the second partials and calculate D for each critical point.

$$g_{xx} = 2x, \quad g_{yy} = -2y, \quad \text{and} \quad g_{xy} = 1$$

For $(0, 0)$:

$$g_{xx}(0, 0) = 0, \quad g_{yy}(0, 0) = 0, \quad \text{and} \quad g_{xy}(0, 0) = 1$$

$$D = 0 \cdot 0 - (1)^2 = -1 < 0$$

Therefore, by Case 3 of the D-Test, the point $(0, 0, g(0, 0)) = (0, 0, 0)$ is a saddle point.

For $(1, -1)$:

$$g_{xx}(1, -1) = 2, \quad g_{yy}(1, -1) = 2, \quad \text{and} \quad g_{xy}(1, -1) = 1$$

$$D = 2 \cdot 2 - (1)^2 = 4 - 1 = 3 > 0$$

Since $D > 0$ and $g_{xx}(1, -1) = 2 > 0$, by Case 1 of the D-Test, the critical point $(1, -1)$ yields a local minimum. The minimum value is

$$g(1, -1) = \frac{1}{3}(1)^3 - \frac{1}{3}(-1)^3 + 1(-1)$$
$$= \frac{1}{3} + \frac{1}{3} - 1 = -\frac{1}{3}.$$

> ### Example 4: Optimizing an Equation in Two Variables

A company produces and sells two styles of umbrellas. One style sells for $20 each and the other sells for $25 each. The company has determined that if x thousand of the first style and y thousand of the second style are produced, then the total cost in thousands of dollars is given by the function

$$C(x, y) = 3x^2 - 3xy + \frac{3}{2}y^2 + 32x - 29y + 70.$$

How many of each style of umbrella should the company produce and sell in order to maximize profit?

Solution

Since x thousand umbrellas sell for $20 each and y thousand umbrellas sell for $25 each, the revenue function (in thousands of dollars) is given by

$$R(x, y) = 20x + 25y.$$

Next, the profit function can be calculated.

$$P(x, y) = R(x, y) - C(x, y)$$

$$= 20x + 25y - \left(3x^2 - 3xy + \frac{3}{2}y^2 + 32x - 29y + 70 \right)$$

$$= -3x^2 + 3xy - \frac{3}{2}y^2 - 12x + 54y - 70$$

The first partial derivatives are as follows.

$$P_x = -6x + 3y - 12 \quad \text{and} \quad P_y = 3x - 3y + 54$$

Now solve the following system of equations.

$$\begin{cases} -6x + 3y - 12 = 0 \\ 3x - 3y + 54 = 0 \end{cases}$$

$$\begin{aligned} -6x + 3y &= 12 \\ \underline{3x - 3y} &= \underline{-54} \\ -3x &= -42 \\ x &= 14 \end{aligned}$$

$$3(14) - 3y = -54 \qquad \text{Substitute } x = 14 \text{ into the second equation and solve for } y.$$

$$-3y = -96$$

$$y = 32$$

The company will make the maximum profit if it produces and sells 14,000 of the first style of umbrella and 32,000 of the second style. The student should verify (with the D-Test) that the profit is indeed a maximum at $(14, 32)$.

16.3 EXERCISES

💡 PRACTICE

For each of the Exercises 1–24, find all local maxima, local minima, and saddle points.

1. $f(x,y) = x^2 + y^2 - 6x + 2y - 4$

2. $f(x,y) = x^2 + 2y^2 + 8x - 4y + 2$

3. $f(x,y) = x^2 - y^2 - 10x + 2y + 9$

4. $f(x,y) = 12x + 8y - x^2 - y^2 - 7$

5. $f(x,y) = 5x + 8y - x^2 - y^2 + 11$

6. $f(x,y) = y^2 - x^2 + 6x - y - 10$

7. $f(x,y) = x^2 + xy - y^2 - 5y - 8$

8. $f(x,y) = x^2 - 2xy + 4y^2 - 6y + 3$

9. $f(x,y) = 2x^2 - 3xy + 3y^2 + 5x - 13$

10. $f(x,y) = 10x - 2x^2 + 2xy - 3y^2 + 5$

11. $f(x,y) = x^2 - xy + y^2 - 2x - 2y + 1$

12. $f(x,y) = 3x^2 - 2xy + y^2 - 16x + 4y + 14$

13. $f(x,y) = -x^2 + xy - y^2 + 4x - 5y + 6$

14. $f(x,y) = x^2 + 3y^2 + 5xy + 4x - 3y + 15$

15. $f(x,y) = x^3 + y^2 - 3x - 6y + 11$

16. $f(x,y) = x^3 - 3x^2 - 2y^2 - 9x + 8y + 7$

17. $f(x,y) = x^2 - y^3 + 9y^2 - 4x - 15y - 14$

18. $f(x,y) = 2x^2 + y^3 - 3y^2 - 12x - 24y + 21$

19. $f(x,y) = x^3 - 3x^2y + 12y$

20. $f(x,y) = 9x - xy^2 + 2y^3$

21. $f(x,y) = 2x^2 - x^2y + y^2$

22. $f(x,y) = x^2 - 2xy^2 + 4y^2$

23. $f(x,y) = xy + \dfrac{2}{x} + \dfrac{4}{y}$

24. $f(x,y) = xy + \dfrac{9}{x} + \dfrac{3}{y}$

🚀 APPLICATIONS

25. Profit: An automobile agency sells two models of a car. The annual profit is estimated by $P(x,y) = -0.1x^2 - 0.2y^2 + 6x + 10y - 160$ in thousands of dollars. Find the number of each model that should be sold to maximize profit.

26. **Sales:** The owner of a small business advertises in the newspaper and on radio. He has found that the number of units that he sells is approximated by $N(x,y) = -0.5x^2 - y^2 + 8x + 12y + 240$, where x (in thousands of dollars) is the amount spent on newspaper advertising and y (in thousands of dollars) is the amount spent on radio advertising. How much should he spend on each to maximize the number of units sold?

27. **Profit:** A firm produces two kinds of magazine racks, one selling for \$50 and the other for \$45. The total cost of producing x of the \$50 racks and y of the \$45 racks is given by $C(x,y) = 0.15x^2 + 0.1y^2 - 10x - 3y + 4760$ dollars. Find the number of each kind that should be produced and sold to maximize the profit.

28. **Profit:** A company makes two types of work gloves, leather and cloth. The leather gloves sell for \$5.80 and the cloth gloves for \$1.60. The total cost function is $C(x,y) = 0.25x^2 + 0.03y^2 + 1.3x + 0.4y + 14$ in thousands of dollars, where x (in thousand pairs) is the number of leather gloves and y (in thousand pairs) is the number of cloth gloves. How many gloves of each type should be produced and sold to maximize profits?

29. **Revenue:** A department store sells two types of T-shirts, adult and youth. The store manager has determined that he can sell $23 - 6x + 8y$ adult T-shirts and $26 + 5x - 9y$ youth T-shirts if the price is x dollars for the adult and y dollars for the youth. Find the price of each that will yield maximum revenue.

30. **Revenue:** A grocery store sells two brands of a product, a name brand and a store brand. The manager estimates that if she sells the name brand for x dollars per unit and the store brand for y dollars per unit, she will be able to sell $62 - 20x + 18y$ units of the name brand and $53 + 16x - 22y$ units of the store brand. Find the price of each that will yield maximum revenue.

31. **Construction:** A rectangular box is to be constructed without a top and with one partition. The volume of the box must be 162 in.^{3} Find the dimensions that will minimize the material required to construct the box.

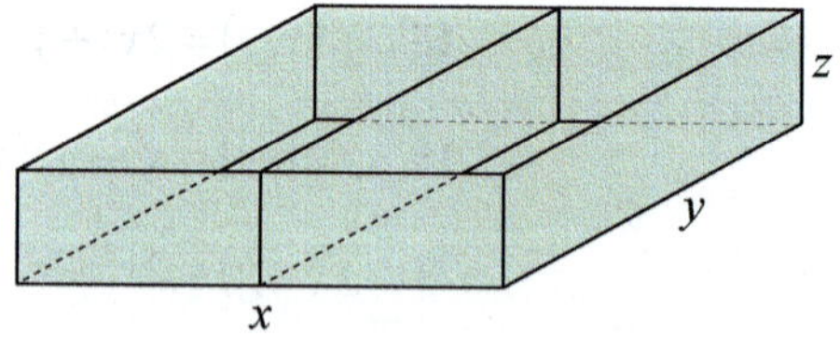

32. **Packaging:** The Postal Service has a limit of 108 inches on the combined length and girth of a rectangular package to be sent by parcel post. Length is the measurement for the longest side and girth is the distance around the package perpendicular to the length. Find the dimensions of the package of maximum volume that can be sent.

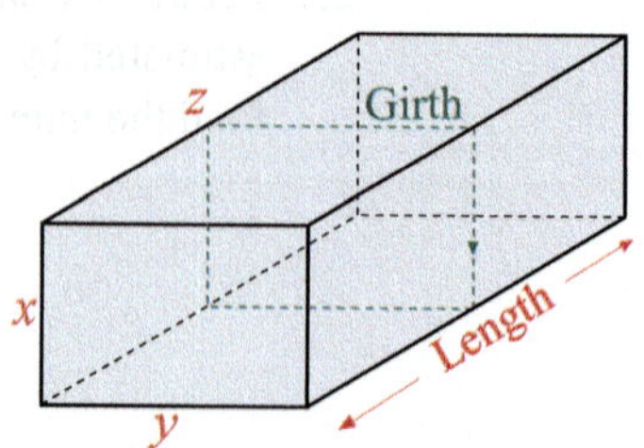

16.4 LAGRANGE MULTIPLIERS

In this section we will consider problems of locating the maxima and minima of functions of two variables in which there are certain restrictions on the variables. The restrictions are also called **side conditions** or **constraints**. The problems themselves are known as **constrained optimization problems**. For example, suppose that we want to find the minimum value of the function

$$f(x, y) = x^2 + y^2 + 2$$

with the constraint that

$$x + y - 3 = 0.$$

We want to find the point (a, b) where $f(a, b)$ is a minimum under the constraint that $a + b - 3 = 0$. One general method for finding maxima or minima of functions when there is a constraint on the variables is the method of **Lagrange multipliers**, named after the French mathematician Joseph-Louis Lagrange (1736–1813). In this method the Greek letter lambda (λ) is used to represent a variable called the **Lagrange multiplier**.

The technique is based on treating λ as an independent variable and forming the **Lagrange function**

$$F(x, y, \lambda) = f(x, y) + \lambda \cdot g(x, y),$$

where $z = f(x, y)$ is the object function and $g(x, y) = 0$ is the constraint.

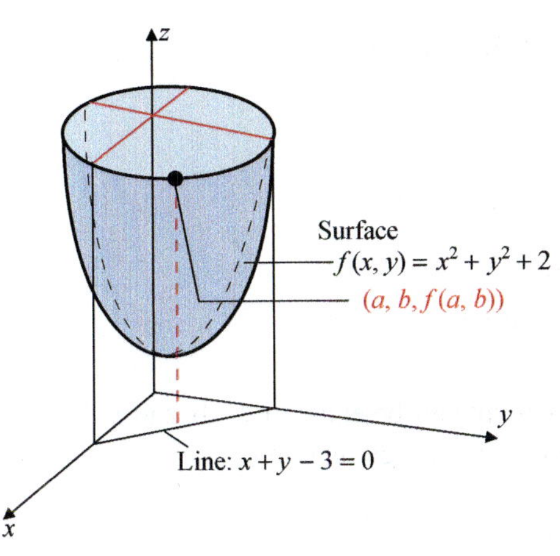

FIGURE 1

The Method of Lagrange Multipliers

To maximize or minimize a function $z = f(x, y)$ subject to the constraint $g(x, y) = 0$:

Step 1: Form the Lagrange function.

$$F(x, y, \lambda) = f(x, y) + \lambda \cdot g(x, y)$$

Step 2: Find each of the partial derivatives F_x, F_y, and F_λ, provided all partials exist.

Step 3: Solve the following system of three equations.

$$\begin{cases} F_x(x, y, \lambda) = 0 \\ F_y(x, y, \lambda) = 0 \\ F_\lambda(x, y, \lambda) = 0 \end{cases}$$

Step 4: If f has a local maximum or local minimum subject to the constraint $g(x, y) = 0$, then the corresponding x- and y-values will be found among the solutions to the system in Step 3.

To determine whether the solution (a, b) found by using the method of Lagrange multipliers gives a maximum or a minimum, we can compare the value $f(a, b)$ with the values $f(x, y)$ where the points (x, y) are near (a, b) and satisfy the constraint $g(x, y) = 0$. For the problems in this course, the values found in the solutions to the system in Step 3 do give the maximum value (or minimum value) as requested in the problems.

Use the method of Lagrange multipliers to find the minimum value of the function $f(x, y) = x^2 + y^2 + 2$ with the constraint $x + y - 3 = 0$.

Solution

Step 1: Form the Lagrange function.

$$F(x, y, \lambda) = x^2 + y^2 + 2 + \lambda(x + y - 3)$$

Step 2: Find each of the partial derivatives.

$$F_x(x, y, \lambda) = 2x + \lambda$$
$$F_y(x, y, \lambda) = 2y + \lambda$$
$$F_\lambda(x, y, \lambda) = x + y - 3$$

Note: The constraint function $g(x, y)$ will always be equal to F_λ because it is treated as a constant coefficient of λ in the Lagrange function.

Step 3: Solve the following system of equations.

$$\begin{cases} 2x + \lambda = 0 & (1) \\ 2y + \lambda = 0 & (2) \\ x + y - 3 = 0 & (3) \end{cases}$$

Isolate λ in equations (1) and (2) and solve the resulting equations.

$$2x + \lambda = 0 \qquad (1)$$
$$\lambda = -2x$$
$$2y + \lambda = 0 \qquad (2)$$
$$\lambda = -2y$$

Thus $-2x = -2y$ and $x = y$. Now substitute x for y in equation (3) and solve for x.

$$x + x - 3 = 0$$
$$2x = 3$$
$$x = 1.5$$

Since $x = y$, we have $y = 1.5$.

Step 4: From the graph in Figure 1, we can see that

$$f(1.5, 1.5) = (1.5)^2 + (1.5)^2 + 2 = 6.50$$

is a minimum. In order to verify this result algebraically, we choose the point $(1.4, 1.6)$, which is close to $(1.5, 1.5)$ and satisfies the constraint $x + y - 3 = 0$, and find that $f(1.4, 1.6) = 6.52 > 6.50$. The D-Test can also be used to verify this.

Example 2: Applying the Method of Lagrange Multipliers

Use the method of Lagrange multipliers to find the maximum value of the function $f(x, y) = xy$ under the restriction $x^2 + y^2 - 200 = 0$, where $x > 0$ and $y > 0$.

Solution

Step 1: Form the Lagrange function.

$$F(x, y, \lambda) = xy + \lambda(x^2 + y^2 - 200)$$

Step 2: Find each of the partial derivatives.

$$F_x = y + \lambda \cdot 2x$$
$$F_y = x + \lambda \cdot 2y$$
$$F_\lambda = x^2 + y^2 - 200$$

Step 3: Solve the following system of equations.

$$\begin{cases} y + \lambda \cdot 2x = 0 & (1) \\ x + \lambda \cdot 2y = 0 & (2) \\ x^2 + y^2 - 200 = 0 & (3) \end{cases}$$

Isolate λ in equations (1) and (2) and solve the resulting equations.

$$y + \lambda \cdot 2x = 0 \quad (1)$$
$$\lambda = \frac{-y}{2x}$$
$$x + \lambda \cdot 2y = 0 \quad (2)$$
$$\lambda = \frac{-x}{2y}$$

We have $\dfrac{-y}{2x} = \dfrac{-x}{2y}$ which gives $-2x^2 = -2y^2$ and $x^2 = y^2$. Substituting x^2 for y^2 in equation (3), we have

$$x^2 + x^2 - 200 = 0$$
$$2x^2 = 200$$
$$x^2 = 100$$
$$x = \pm 10.$$

Since $y^2 = x^2$, $y^2 = 100$, and $y = \pm 10$. Thus there are four points that satisfy the system: $(10, 10)$, $(10, -10)$, $(-10, 10)$, and $(-10, -10)$. However, among these points, only the point $(10, 10)$ satisfies the additional conditions $x > 0$ and $y > 0$.

Step 4: The value $f(10, 10) = (10)(10) = 100$ is a maximum for $f(x, y) = xy$ under the given constraints.

As a check that 100 is indeed a maximum, we find that

$$f(9.88, 10.12) = (9.88)(10.12) = 99.9856 < 100.$$

Note: Under the given constraints, points close to $(10, 10)$ have irrational coordinates; we have chosen 9.88 and 10.12 as rounded values that nearly satisfy the constraints.

Example 3: Optimization Using the Cobb-Douglas Production Formula

Suppose that the Cobb-Douglas Production Formula for a particular product is $f(x, y) = 100x^{0.6}y^{0.4}$, where x represents units of labor at a cost of \$200 per unit and y represents units of capital at a cost of \$300 per unit. If the company's budget allows \$120,000 for labor and capital, then the constraint is represented by the equation $200x + 300y = 120{,}000$. Find the maximum level of production for this product.

Solution

Step 1: Form the Lagrange function.

$$F(x, y, \lambda) = 100x^{0.6}y^{0.4} + \lambda(200x + 300y - 120{,}000)$$

Step 2: Find each of the partial derivatives.

$$F_x = 60x^{-0.4}y^{0.4} + 200\lambda$$
$$F_y = 40x^{0.6}y^{-0.6} + 300\lambda$$
$$F_\lambda = 200x + 300y - 120{,}000$$

Step 3: Solve the following system of equations.

$$\begin{cases} 60x^{-0.4}y^{0.4} + 200\lambda = 0 & (1) \\ 40x^{0.6}y^{-0.6} + 300\lambda = 0 & (2) \\ 200x + 300y - 120{,}000 = 0 & (3) \end{cases}$$

From isolating λ in equations (1) and (2) we have

$$\frac{-60x^{-0.4}y^{0.4}}{200} = \lambda = \frac{-40x^{0.6}y^{-0.6}}{300}.$$

Multiply both sides of the equation by $-600x^{0.4}y^{0.6}$.

$$180x^0y^1 = 80x^1y^0$$
$$180y = 80x$$
$$y = \frac{4}{9}x$$

Substitute $\dfrac{4}{9}x$ for y in equation (3).

$$200x + 300\left(\frac{4}{9}x\right) - 120{,}000 = 0$$
$$3000x = 1{,}080{,}000$$
$$x = 360$$
$$y = \frac{4}{9}(360) = 160$$

Step 4: To maximize production, the company should invest 360 units of labor and 160 units of capital. This investment will result in a production level of approximately

$$f(360,160) = 100(360)^{0.6}(160)^{0.4}$$
$$\approx 100(34.181)(7.615) \approx 26{,}029 \text{ units.}$$

(The student should verify that 26,029 is indeed a maximum.)

NOTE

In economics, the absolute value of the Lagrange multiplier λ used in maximizing a production function is called the **marginal productivity of money**. That is, $|\lambda|$ indicates the increase in production that could be expected for an increased investment of one dollar. In this course we will not be concerned with the marginal productivity of money or any other interpretations of λ. This is why we have not found the values of λ in the examples.

16.4 EXERCISES

◉ PRACTICE

In Exercises 1–8, use the method of Lagrange multipliers to find the minimum value of f subject to the given constraint.

1. $f(x,y) = x^2 + y^2$, subject to $x + y - 4 = 0$

2. $f(x,y) = 4x^2 + 3y^2$, subject to $x + y - 7 = 0$

3. $f(x,y) = 5x^2 + 4y^2 - 2x$, subject to $x - y - 2 = 0$

4. $f(x,y) = 2x^2 + y^2 - 18x$, subject to $3x - y - 8 = 0$

5. $f(x,y) = 6x^2 + 5y^2 - xy$, subject to $2x + y = 24$

6. $f(x,y) = 2x^2 + 3y^2 - 3xy$, subject to $x + y = 16$

7. $f(x,y) = x^3 + y^3$, subject to $x + y = 8$

8. $f(x,y) = x^3 - y^3$, subject to $x - y = 10$

In Exercises 9–16, use the method of Lagrange multipliers to find the maximum value of f subject to the given constraint.

9. $f(x,y) = 2x^2 - 5y^2$, subject to $x - y = 3$

10. $f(x,y) = 5y^2 - 8x^2$, subject to $x + y = 6$

11. $f(x,y) = 8xy - 3x^2$, subject to $x + 2y = 14$

12. $f(x,y) = 6x^2 - 5xy$, subject to $2x - y = 8$

13. $f(x,y) = x^2 + y^2 + 4xy$, subject to $3x + 4y = 23$

14. $f(x,y) = x^2 - 4y^2 + 84xy$, subject to $5x + 2y = 18$

15. $f(x,y) = 15x^{0.4}y^{0.6}$, subject to $10x + 8y = 200$

16. $f(x,y) = 8x^{\frac{1}{2}}y^{\frac{1}{2}}$, subject to $6x + 15y = 450$

In Exercises 17–20, use the method of Lagrange multipliers to find the maximum and minimum values of f subject to the given constraint.

17. $f(x,y) = 4xy$, subject to $x^2 + 4y^2 = 72$

18. $f(x,y) = 5xy$, subject to $9x^2 + y^2 = 162$

19. $f(x,y) = x^3 + 4y^3$, subject to $x + y = 6$

20. $f(x,y) = 3x^3 + y^3$, subject to $3x + y = 8$

APPLICATIONS

21. Cost: A company has a plant in Los Angeles and a plant in Oklahoma City. The firm is committed to produce a total of 40 units of a product each week. The cost function is given by $C(x, y) = 0.3x^2 + 0.2y^2 + 20x + 7y + 200$, where x is the number of units produced in Los Angeles and y is the number of units produced in Oklahoma City. How many units should be produced in each plant to minimize the total weekly costs?

22. Profit: A department store sells two styles of a jacket, lined and unlined. During the month of January, the management expects to sell exactly 250 jackets. The profit function is given by $P(x, y) = -0.3x^2 - 0.4y^2 - 0.3xy + 80x + 65y - 1000$, where x is the number of lined jackets sold and y is the number of unlined jackets sold. How many of each type should be sold to maximize the profit?

23. Production: The production function for a certain product is given by $f(x, y) = 75x^{0.3}y^{0.7}$, where x is the number of units of labor and y is the number of units of capital. Each unit of labor costs $300, and each unit of capital costs $200. If the company's budget allows a total of $20,000 for labor and capital, find the maximum level of production.

24. Production: The management of a company has determined that x units of labor and y units of capital are required to produce $f(x, y) = 130x^{0.4}y^{0.6}$ units of a product. Each unit of labor costs $450, and each unit of capital costs $360. Find the maximum number of units that can be produced if a total of $90,000 is available for labor and capital.

25. Sales: A sales representative for a textbook publishing company estimates her monthly sales for March to be $S(x, y) = 30x + 18y - 1.2x^2 - 0.6y^2$ in thousands of dollars, where x and y represent the number of days spent in each of the two metropolitan areas that comprise her sales territory. If she plans to work 20 days during the month, how many days should she spend in each area to maximize her sales?

26. Revenue: The marketing manager of a department store has determined that revenue, in dollars, is related to the number of units of television advertising x and the number of units of newspaper advertising y by the function $R(x, y) = 500(20x + 5y + 6xy - x^2)$. Each unit of television advertising costs $3000, and each unit of newspaper advertising costs $1500. If the advertising budget is $30,000, find the maximum revenue.

27. Construction: A farmer wants to build a rectangular pen and then divide it with two interior fences. The total area enclosed is to be 2484 ft². The exterior fence costs $18 per foot, and the interior fence costs $16.50 per foot. Find the dimensions of the pen that will minimize the cost of fencing.

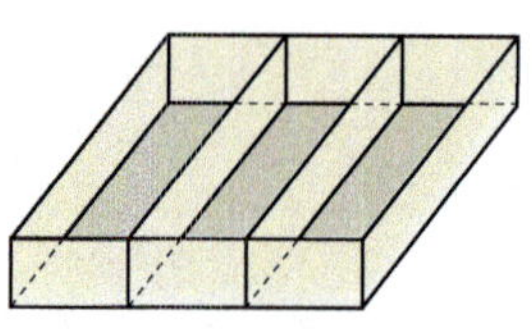

28. Shipping: A container manufacturer is asked to design a closed rectangular shipping crate with a square base. The volume of the crate is 36 ft³. The material for the top costs $1 per square foot, the material for the sides costs $0.90 per square foot, and the material for the bottom costs $1.40 per square foot. Find the dimensions that will minimize the total cost of building the crate.

16.5 THE METHOD OF LEAST SQUARES

■ TOPICS

- ■ Regression Lines
- ■ The Least-Squares Regression Line

Regression Lines

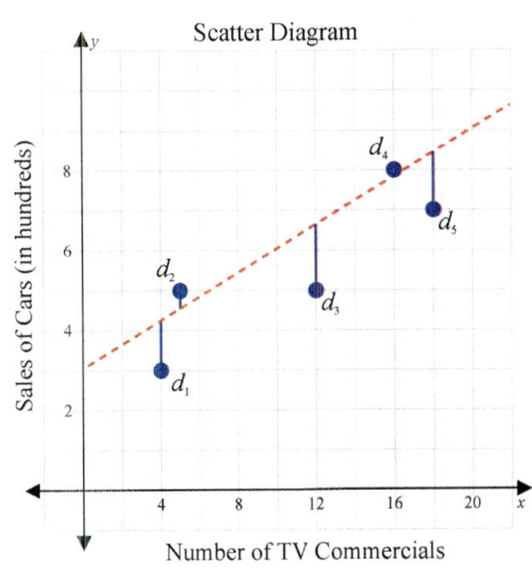

Scatter Diagram

Number of TV Commercials

FIGURE 1

Suppose that an automobile company is running a new television commercial in five cities with approximately the same population. The company tests the success of the commercial by looking at two variables: x (the number of times the commercial is run on TV in each city) and y (the sales in each city). Sales are in hundreds of cars. The results are shown in the table and scatter diagram in Figure 1.

Number of TV Commercials x	Sales of Cars (in hundreds) y
4	3
5	5
12	5
16	8
18	7

The scatter diagram indicates that there is a positive correlation between the two variables x and y. That is, there seems to be a tendency for an increase in sales whenever the number of commercials shown is increased.

With this data, the company would like to know what sales to expect in one of these cities if it uses this commercial a particular number of times. For example, if $x = 10$, what would be the expected value of y (the sales in that city)?

One solution to this problem is to find a **curve of best fit**. In this section, we will be concerned only with **regression lines** (or **lines of best fit**). An approximate position for the regression line for the data given in Figure 1 is shown in Figure 2. Also indicated in Figure 2 are the vertical deviations of the data points from the regression line: d_1, d_2, d_3, d_4, and d_5.

The dashed line in Figure 2 indicates the approximate position of the **regression line** (or **line of best fit**) for the given data.

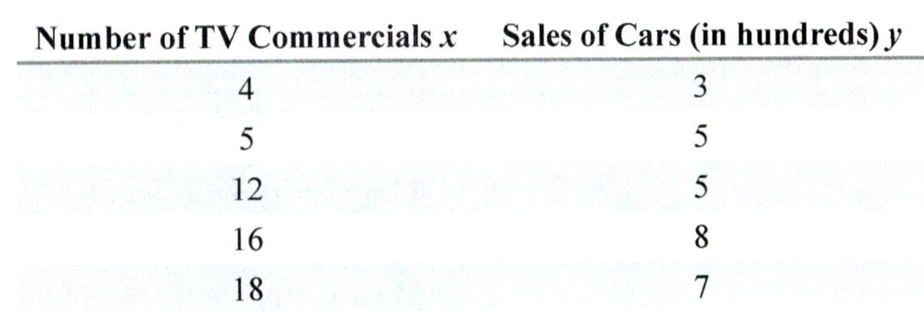

Scatter Diagram

Number of TV Commercials

FIGURE 2

The Least-Squares Regression Line

One algebraic method for finding a regression line is called the **Method of Least Squares**. In this method, the regression line is represented by an equation in the form $y = mx + b$, and the values of the vertical deviations of the given points from this line are squared. The values of m and b in the linear equation are determined by minimizing the sum of these squares. With reference to Figure 2, we want to find the minimum value of the sum

$$\sum_{k=1}^{5} d_k^2 = d_1^2 + d_2^2 + d_3^2 + d_4^2 + d_5^2.$$

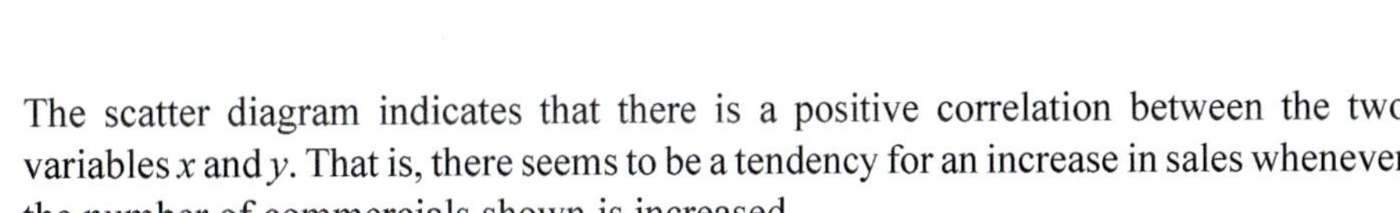

☑ NOTE

The use of the Greek capital letter sigma (Σ) to indicate summations is a standard notation in mathematics.

The vertical deviations of the data points from the dashed line in Figure 2 are calculated as follows:

$$d_1 = (4m + b) - 3 \qquad \text{since } x_1 = 4 \text{ and } y_1 = 3,$$
$$d_2 = (5m + b) - 5 \qquad \text{since } x_2 = 5 \text{ and } y_2 = 5,$$
$$d_3 = (12m + b) - 5 \qquad \text{since } x_3 = 12 \text{ and } y_3 = 5,$$
$$d_4 = (16m + b) - 8 \qquad \text{since } x_4 = 16 \text{ and } y_4 = 8,$$
$$d_5 = (18m + b) - 7 \qquad \text{since } x_5 = 18 \text{ and } y_5 = 7.$$

We see that the value for each d_k depends on the two variables m and b. Thus the sum of the squares is also a function of m and b.

$$f(m,b) = \sum_{k=1}^{5} d_k^2 = (4m + b - 3)^2 + (5m + b - 5)^2 + (12m + b - 5)^2$$
$$+ (16m + b - 8)^2 + (18m + b - 7)^2$$

Our objective is to find the values of m and b that minimize $f(m, b)$ using the D-Test.

$$f_m = 2(4m + b - 3) \cdot 4 + 2(5m + b - 5) \cdot 5 + 2(12m + b - 5) \cdot 12$$
$$+ 2(16m + b - 8) \cdot 16 + 2(18m + b - 7) \cdot 18$$
$$= 1530m + 110b - 702$$
$$f_b = 2(4m + b - 3) + 2(5m + b - 5) + 2(12m + b - 5) + 2(16m + b - 8) + 2(18m + b - 7)$$
$$= 110m + 10b - 56$$

The solutions to the system $f_m = 0$ and $f_b = 0$ are $m = 0.26875$ and $b = 2.64375$.

For our purposes, the rounded values $m \approx 0.27$ and $b \approx 2.64$ are sufficiently accurate. For the D-Test, we have

$$f_{mm} = 1530, \quad f_{bb} = 10, \quad \text{and} \quad f_{mb} = 110.$$

Thus $D = (1530)(10) - (110)^2 = 3200 > 0$ with $f_{mm} = 1530 > 0$ and the values $m \approx 0.27$ and $b \approx 2.64$ do give a minimum value for the sum of the squares represented by the function $f(m, b)$.

Therefore, the regression line is

$$y = 0.27x + 2.64;$$

and for $x = 10$, the estimated value of y given by the regression line is

$$y = 0.27(10) + 2.64 = 5.34.$$

So the company can expect to sell about 534 cars if the commercial is shown 10 times.

For any significant amount of data, finding the representation of the function $f(m, b)$, calculating the partials, and solving the simultaneous equations can be a prohibitive exercise. General formulas for finding the values of m and b for the regression line have been determined. These formulas make use of the following notations, where n is the number of ordered pairs in the data set.

$$\overline{x} = \frac{\sum_{k=1}^{n} x_k}{n} \qquad \text{The average (or mean) of the } x\text{-values}$$

$$\overline{y} = \frac{\sum_{k=1}^{n} y_k}{n} \qquad \text{The average (or mean) of the } y\text{-values}$$

For simplicity, the subscript notation in the summation expressions has been omitted in the following formulas.

The Least-Squares Regression Line

For a set of data points $(x_1, y_1), (x_2, y_2), \ldots, (x_n, y_n)$, the regression line is $y = mx + b$, where (omitting indices in the summations)

1. $m = \dfrac{n \cdot \sum(xy) - \sum x \sum y}{n \cdot \sum x^2 - (\sum x)^2}$,

2. $b = \overline{y} - m\overline{x}$,

and

3. $\overline{y} = \dfrac{1}{n}\sum y$ and $\overline{x} = \dfrac{1}{n}\sum x$.

Example 1: Using Linear Regression on Tabular Data

Use the formulas to find the regression line for the data given in the table in Figure 1.

Solution

The data are shown and calculations are made in Table 1.

	x	y	xy	x^2
	4	3	12	16
	5	5	25	25
	12	5	60	144
	16	8	128	256
	18	7	126	324
Sums	**55**	**28**	**351**	**765**

TABLE 1

We find that

$$\overline{x} = \frac{55}{5} = 11 \quad \text{and} \quad \overline{y} = \frac{28}{5} = 5.6.$$

Thus, from Formula 1,

$$m = \frac{5(351) - (55)(28)}{5(765) - (55)^2} = \frac{1755 - 1540}{3825 - 3025} = \frac{215}{800} \approx 0.27$$

and, from Formula 2,

$$b = \bar{y} - m\bar{x} = 5.6 - (0.27)(11) = 5.6 - 2.97 = 2.63.$$

The line of regression is

$$y = 0.27x + 2.63,$$

which is in agreement (except for slight rounding errors) with the previous calculations.

Example 2: Forecasting Grade Point Averages with Linear Regression

Table 2 shows the high school grade point averages (HS-GPA) and the college grade point averages (C-GPA) after 1 year of college for 10 students.

a. Find the regression line for the data given in Table 2.

b. What would be your best prediction for the C-GPA of a high school student with an HS-GPA of 2.5?

c. Graph the data and the regression line on the same set of axes.

HS-GPA (x)	2.0	2.0	2.2	2.2	2.7	3.2	3.2	3.3	3.5	3.7
C-GPA (y)	1.5	1.8	2.0	1.5	2.0	2.8	3.0	3.5	3.5	3.4

TABLE 2

Solution

a. In Table 3, we calculate the values needed for Formulas 1 and 2.

	x	y	xy	x^2
	2.0	1.5	3.00	4.00
	2.0	1.8	3.60	4.00
	2.2	2.0	4.40	4.84
	2.2	1.5	3.30	4.84
	2.7	2.0	5.40	7.29
	3.2	2.8	8.96	10.24
	3.2	3.0	9.60	10.24
	3.3	3.5	11.55	10.89
	3.5	3.5	12.25	12.25
	3.7	3.4	12.58	13.69
Sums	**28.0**	**25.0**	**74.64**	**82.28**

TABLE 3

Now

$$\bar{x} = \frac{28}{10} = 2.8 \quad \text{and} \quad \bar{y} = \frac{25}{10} = 2.5.$$

From Formula 1, we have

$$m = \frac{10(74.64) - (28.0)(25.0)}{10(82.28) - (28.0)^2} = \frac{46.4}{38.8} \approx 1.20.$$

From Formula 2, we have

$$b = 2.5 - 1.20(2.8) = -0.86.$$

Thus the line of regression is $y = 1.20x - 0.86$.

b. The best prediction for the C-GPA of a student with an HS-GPA of 2.5 is found by substituting $x = 2.5$ in the equation for the line of regression and solving for y:

$$y = 1.20(2.5) - 0.86 = 2.14.$$

c.

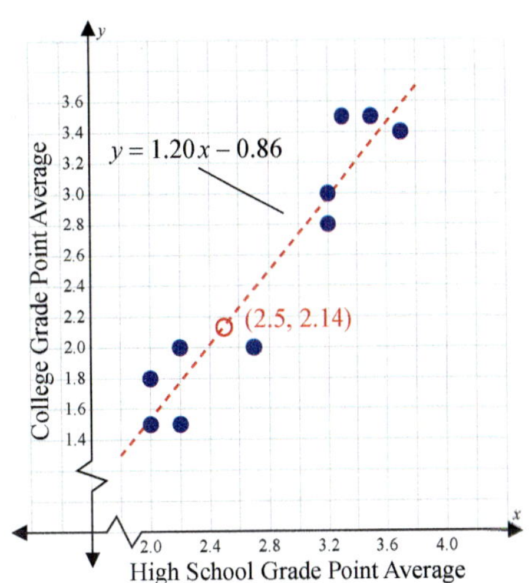

On a subjective basis, a look at a carefully detailed scatter diagram can help determine whether the curve of the best fit is a straight line or some other curve such as a parabola. To deal with a nonlinear relationship between two variables, we would need to modify the model from linear ($y = mx + b$) to quadratic ($y = ax^2 + bx + c$) or some higher-degree function and minimize the sum of the squares by using partial derivatives as before. A nonlinear regression curve is illustrated in Figure 3. We will not analyze nonlinear regression curves in this section.

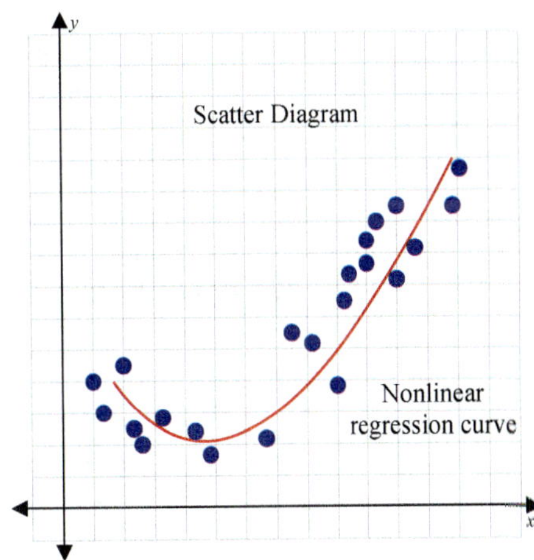

FIGURE 3

16.5 EXERCISES

PRACTICE

In Exercises 1–4, **a.** find the equation of the regression line for the given points, and **b.** draw the scatter diagram and graph the regression line.

1. $(0, 3)$, $(1, 5)$, $(2, 7)$, $(3, 8)$, $(5, 9)$, $(6, 9)$

2. $(1, 10)$, $(2, 8)$, $(3, 7)$, $(4, 6)$, $(5, 5)$, $(6, 5)$, $(7, 4)$

3. $(1, 9.6)$, $(2, 8.7)$, $(3, 7.7)$, $(4, 6.1)$, $(5, 5.0)$

4. $(1, 5.2)$, $(2, 6.4)$, $(3, 8.1)$, $(4, 9.2)$, $(5, 10.6)$

In Exercises 5–10, find the equation of the regression line for the given points.

5. $(10, 6.5)$, $(20, 5.8)$, $(30, 5.6)$, $(40, 3.1)$, $(50, 1.8)$

6. $(1, 0.2)$, $(2, 0.4)$, $(3, 0.3)$, $(4, 0.6)$, $(5, 0.6)$

7. $(1, 236)$, $(2, 248)$, $(3, 270)$, $(4, 285)$, $(5, 291)$

8. $(1, 0.45)$, $(3, 0.71)$, $(4, 0.82)$, $(5, 0.94)$

9. $(0.6, 4.8)$, $(0.8, 5.0)$, $(1.0, 4.8)$, $(1.2, 5.2)$, $(1.4, 5.8)$

10. $(3.2, 0.10)$, $(4.1, 0.15)$, $(4.8, 0.20)$, $(5.1, 0.23)$, $(6.0, 0.29)$

APPLICATIONS

11. **Advertising budget:** During the last 5 years, the advertising manager for a corporation has gathered the following data that shows the relationship between the advertising budget (in millions of dollars) and the total sales (in thousands of units).

Advertising Budget (x) (in millions)	Sales (y) (in thousands)
$4.5	37
$6.5	46
$3.5	42
$3.2	32
$2.6	29

a. Find the regression line for the data.
b. Estimate the sales if $4 million is budgeted for advertising.

12. **Price:** Records at a company for the last 5 years show the following relationship between the units sold (in thousands) and the price of a product.

Price (p)	Quantity Sold (x) (in thousands)
$8.80	3.8
$8.00	5.2
$7.50	7.3
$6.90	8.0
$6.20	9.6

a. Find the regression line for the price in terms of units.
b. Estimate the price that should be charged in order to sell 10,000 units.

13. **Construction:** The following data shows the amount spent on office building construction (in thousands) for a particular county during a six-month period.

Month	Apr	May	June	July	Aug	Sept
Amount (in thousands)	$24	$19	$30	$49	$68	$69

    a. Find the regression line for the data.
    b. Estimate the amount spent in October.

14. **Revenue:** The annual revenue (in millions of dollars) for a corporation is given in the following table.

Year	1999	2000	2001	2002	2003	2004
Revenue (in millions)	$66	$82	$127	$201	$310	$392

    a. Find the line of regression for the data.
    b. Estimate the revenue for 2005.

15. **Livestock futures:** The price of livestock futures is the estimated market price of livestock on the delivery date (end of the indicated month). The cattle futures (in cents per pound) for the months February through July are as follows.

Month	Feb	Mar	Apr	May	June	July
Price (¢ per pound)	79.10	76.02	71.80	71.45	71.45	72.50

    a. Find the line of regression for the data.
    b. Estimate the price for August.

16. **Tourism:** The total number of foreign tourists visiting the United States, as reported by the U.S. Travel and Tourism Administration, is shown in the following table.

Year (x)	2000	2001	2002	2003	2004
Tourists (y) (in millions)	25.7	26.3	29.7	34.2	38.3

    a. Find the regression line for the data.
    b. Estimate the number of foreign tourists that visited the United States in 2005.

16.6 DOUBLE INTEGRALS

Previously, we developed the concept of a definite integral of a continuous function of one variable. A definite integral was defined on a closed interval $[a, b]$ to be the limit of a Riemann sum. The closed interval was partitioned into subintervals of width $\Delta x = \dfrac{b-a}{n}$, where n was the number of subintervals. In an analogous manner, we define the **double integral** of a continuous function of two variables, $z = f(x, y)$, on a closed region R in the xy-plane where R is partitioned into n rectangular subregions each with area $\Delta A = \Delta x \Delta y$. In each of these subregions, a point (x_k, y_k) is chosen and the n products $f(x_k, y_k)\Delta x \Delta y$ are added to form a Riemann sum for the function of two variables. The limit of this sum as $n \to \infty$ is called a **double integral** (or an **iterated integral**) of the function $z = f(x, y)$ on the region R.

Double Integral

For a function $z = f(x, y)$, continuous on a closed region R in the xy-plane, the **double integral** of f on R is denoted and defined as follows.

$$\iint_R f(x,y)\,dA = \iint_R f(x,y)\,dxdy = \lim_{n \to \infty} \sum_{k=1}^{n} f(x_k, y_k)\Delta x \Delta y$$

For a continuous function of one variable, $y = f(x) \geq 0$, the definite integral can be interpreted geometrically as the area between the graph of the function and the x-axis on the closed interval $[a, b]$. Similarly, if a continuous function $z = f(x, y) \geq 0$ for all (x, y) in the region R, then each of the products $f(x_k, y_k)\Delta x \Delta y$ in the Riemann sum can be interpreted as the volume of a rectangular solid. As $n \to \infty$ and the rectangular regions $\Delta x \Delta y$ shrink in size, the sum of the volumes of these rectangular solids approaches the volume of the solid bounded above by f and below by the xy-plane with a vertical boundary passing through the boundary of the region R. (See Figures 1 and 2.)

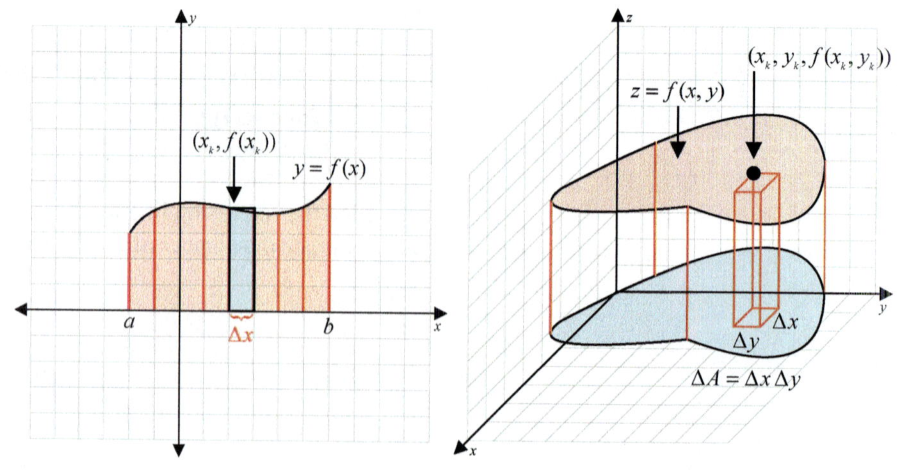

$$\int_a^b f(x)\,dx = \lim_{n \to \infty} S_n$$

$$= \lim_{n \to \infty} \sum_{k=1}^{n} f(x_k)\Delta x$$

$$\iint_R f(x,y)\,dA = \iint_R f(x,y)\,dxdy$$

$$= \lim_{n \to \infty} \sum_{k=1}^{n} f(x_k, y_k)\Delta x \Delta y$$

FIGURE 1   **FIGURE 2**

Thus we define the volume of a solid as depicted in Figure 2 to be the value of the double integral of f on R.

> ### Value of the Double Integral
>
> If $z = f(x, y)$ is a continuous function on a region R in the xy-plane and $f(x, y) \geq 0$ on R, then the volume of the solid bounded above by the graph of f and below by R is the **value of the double integral** of f on R.
>
> $$V = \iint_R f(x, y)\, dx\, dy$$

The actual evaluation of a double integral, whether the interpretation is geometric or not, depends a great deal on the nature of the region R in the xy-plane. We will discuss three types of regions, which we designate **Type I**, **Type II**, and **Type III** for reference purposes only. The general forms of these three types of regions are illustrated in Figures 3, 4, and 5.

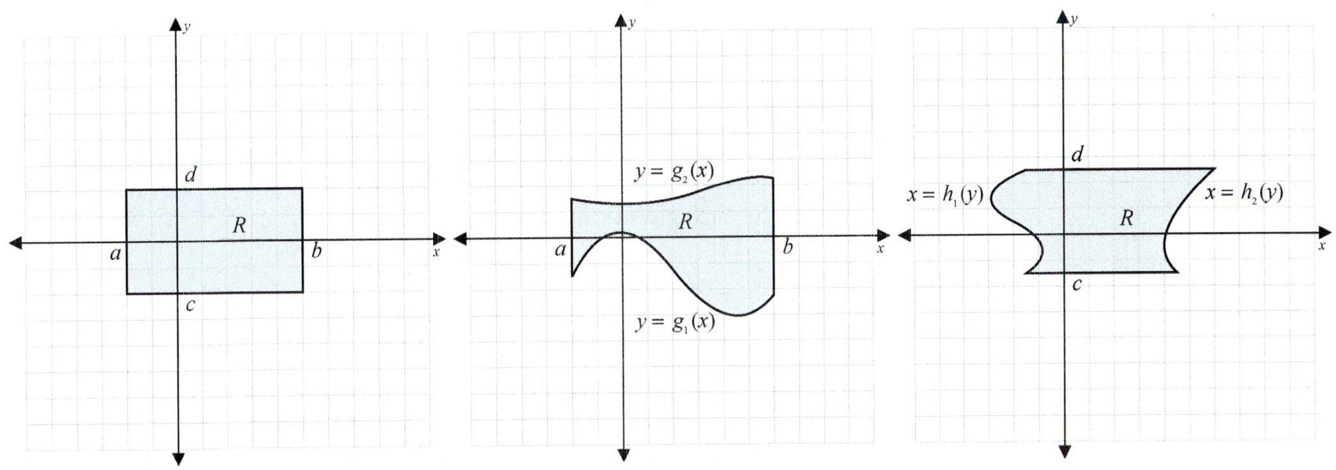

Region R is a rectangle:

$$R = \{(x, y)\mid a \leq x \leq b \text{ and } c \leq y \leq d\}$$

FIGURE 3: Type I Region

Region R is bounded on the left and right by two constant values of x and above and below by two continuous functions of x:

$$R = \{(x, y)\mid a \leq x \leq b \text{ and } g_1(x) \leq y \leq g_2(x)\}$$

FIGURE 4: Type II Region

Region R is bounded on the left and right by two continuous functions of y and above and below by two constant values of y:

$$R = \{(x, y)\mid h_1(y) \leq x \leq h_2(y) \text{ and } c \leq y \leq d\}$$

FIGURE 5: Type III Region

The following theorem, stated here without proof, shows how the evaluation of a double integral depends on the region R and its type.

Evaluating Double Integrals

If $z = f(x, y)$ is a continuous function on a closed region R, then the value of the double integral of f on R is determined as follows, depending on the type of the region R.

Case 1: R is of Type I

$$\iint\limits_R f(x, y)\, dx\, dy = \int_c^d \left(\int_a^b f(x, y)\, dx \right) dy = \int_a^b \left(\int_c^d f(x, y)\, dy \right) dx$$

Case 2: R is of Type II

$$\iint\limits_R f(x, y)\, dx\, dy = \int_a^b \left(\int_{g_1(x)}^{g_2(x)} f(x, y)\, dy \right) dx$$

Case 3: R is of Type III

$$\iint\limits_R f(x, y)\, dx\, dy = \int_c^d \left(\int_{h_1(y)}^{h_2(y)} f(x, y)\, dx \right) dy$$

Note: Only in Case 1 have we shown that the order of integration may be reversed. There are problems of the types in Case 2 and Case 3 in which the order of integration may be reversed; however, we will not discuss reversing the order of integration for such problems.

In each case, the evaluation of a double integral is accomplished in two steps.

Step 1: The "inside" integration is performed with respect to the variable indicated by the inside differential (dy or dx), and the second variable is treated as a constant. Then the integral is evaluated by using the limits on the inside integral sign and substituting for the first variable. The result is a function of the second variable.

Step 2: The resulting function from Step 1 is integrated with respect to the second variable and then evaluated by using the limits on the outside integral sign.

Example 1: Evaluating a Double Integral on a Type I Region

Evaluate the double integral $\displaystyle \int_1^3 \int_1^e \frac{y^2}{x}\, dx\, dy$.

Solution

Since R is of Type I (a rectangular region), we may integrate in either order and obtain the same result. Both integrations are shown here to clarify the idea of reversing the order of integration.

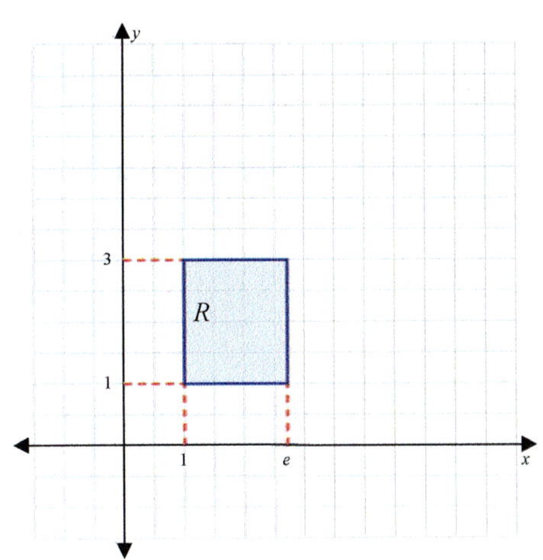

1. $\displaystyle\int_1^3\int_1^e \frac{y^2}{x}\,dx\,dy = \int_1^3\left(\int_1^e \frac{y^2}{x}\,dx\right)dy$

Integrate first with respect to x and evaluate at $x = e$ and $x = 1$. The variable y is treated as constant.

$\displaystyle = \int_1^3\left(y^2\ln|x|\Big]_1^e\right)dy$

$\displaystyle = \int_1^3\left(y^2\ln|e| - y^2\ln|1|\right)dy$

$\displaystyle = \int_1^3\left(y^2 - 0\right)dy$

Now integrate with respect to y and evaluate.

$\displaystyle = \frac{y^3}{3}\Big]_1^3$

$\displaystyle = \frac{3^3}{3} - \frac{1^3}{3}$

$\displaystyle = \frac{27}{3} - \frac{1}{3} = \frac{26}{3}$

2. $\displaystyle\int_1^3\int_1^e \frac{y^2}{x}\,dx\,dy = \int_1^e\left(\int_1^3 \frac{y^2}{x}\,dy\right)dx$

Integrate first with respect to y and evaluate at $y = 3$ and $y = 1$. The variable x is treated as a constant.

$\displaystyle = \int_1^e\left(\frac{1}{x}\cdot\frac{y^3}{3}\Big]_1^3\right)dx$

$\displaystyle = \int_1^e\left(\frac{1}{x}\cdot\frac{3^3}{3} - \frac{1}{x}\cdot\frac{1^3}{3}\right)dx$

$\displaystyle = \int_1^e\left(\frac{27}{3} - \frac{1}{3}\right)\frac{1}{x}\,dx$

$\displaystyle = \frac{26}{3}\int_1^e \frac{1}{x}\,dx$

Now integrate with respect to x and evaluate.

$\displaystyle = \frac{26}{3}\left(\ln|x|\right)\Big]_1^e$

$\displaystyle = \frac{26}{3}\left(\ln|e| - \ln|1|\right)$

$\displaystyle = \frac{26}{3}\left(1 - 0\right) = \frac{26}{3}$

Evaluate the double integral $\displaystyle\int_0^1\int_{x^2}^{x+1} 2x^2 y\,dy\,dx.$

Solution

The region R is of Type II and is illustrated in the graph.

$$\int_0^1 \int_{x^2}^{x+1} 2x^2 y\, dy\, dx = \int_0^1 \left(\int_{x^2}^{x+1} 2x^2 y\, dy \right) dx$$

Integrate first with respect to y and evaluate at $y = x + 1$ and $y = x^2$.

$$= \int_0^1 \left(2x^2 \cdot \frac{y^2}{2} \Big]_{x^2}^{x+1} \right) dx$$

$$= \int_0^1 \left(x^2 y^2 \Big]_{x^2}^{x+1} \right) dx$$

$$= \int_0^1 \left[x^2 (x+1)^2 - x^2 \left(x^2 \right)^2 \right] dx$$

$$= \int_0^1 \left[x^2 \left(x^2 + 2x + 1 \right) - x^6 \right] dx$$

$$= \int_0^1 \left(x^4 + 2x^3 + x^2 - x^6 \right) dx$$

$$= \frac{x^5}{5} + \frac{x^4}{2} + \frac{x^3}{3} - \frac{x^7}{7} \Big]_0^1$$

$$= \left(\frac{1}{5} + \frac{1}{2} + \frac{1}{3} - \frac{1}{7} \right) - (0)$$

$$= \frac{42}{210} + \frac{105}{210} + \frac{70}{210} - \frac{30}{210}$$

$$= \frac{187}{210} \approx 0.89$$

Example 3: Evaluating a Double Integral on a Type III Region

Evaluate the double integral $\int_0^1 \int_0^{\frac{y}{3}} e^{3x+y}\, dx\, dy$.

Solution

$$\int_0^1 \int_0^{\frac{y}{3}} e^{3x+y}\, dx\, dy = \int_0^1 \left(\int_0^{\frac{y}{3}} e^{3x+y}\, dx \right) dy$$

Integrate first with respect to x and evaluate at $x = \frac{y}{3}$ and $x = 0$.

$$= \int_0^1 \left(\frac{1}{3} e^{3x+y} \Big]_0^{\frac{y}{3}} \right) dy$$

$$= \int_0^1 \left(\frac{1}{3} e^{3\left(\frac{y}{3} \right)+y} - \frac{1}{3} e^{3\cdot 0 + y} \right) dy$$

$$= \frac{1}{3} \int_0^1 \left(e^{2y} - e^y \right) dy$$

$$= \frac{1}{3} \left(\frac{1}{2} e^{2y} - e^y \right) \Big]_0^1$$

$$= \frac{1}{3} \left[\left(\frac{1}{2} e^2 - e^1 \right) - \left(\frac{1}{2} e^0 - e^0 \right) \right]$$

$$= \frac{1}{6} e^2 - \frac{1}{3} e + \frac{1}{6}$$

Find the volume of the solid bounded above by the graph of the function $z = 10 - x - y$ and below by the triangular region R in the xy-plane with vertices at $(0, 0, 0)$, $(4, 0, 0)$, and $(0, 8, 0)$.

Solution

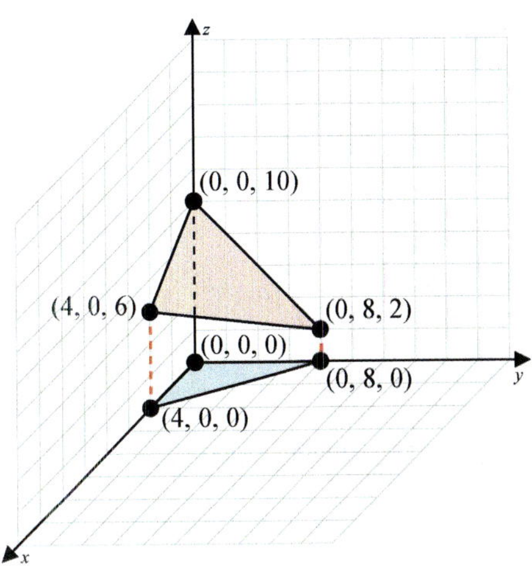

The surface $z = 10 - x - y$ is a plane, and the region R is a triangle in the xy-plane with the given points as vertices. The volume to be found is illustrated in the three-dimensional graph.

To determine the limits of integration, we need to represent the region R in terms of x and y. We can do this by determining the equation of the line through the two points $(4, 0)$ and $(0, 8)$ on an xy-coordinate system and then setting constant restrictions on one of the variables so that the triangular region is described.

The slope of the line is

$$m = \frac{0 - 8}{4 - 0} = -2,$$

and the equation of the line is

$$y - 8 = -2(x - 0)$$
$$y = -2x + 8.$$

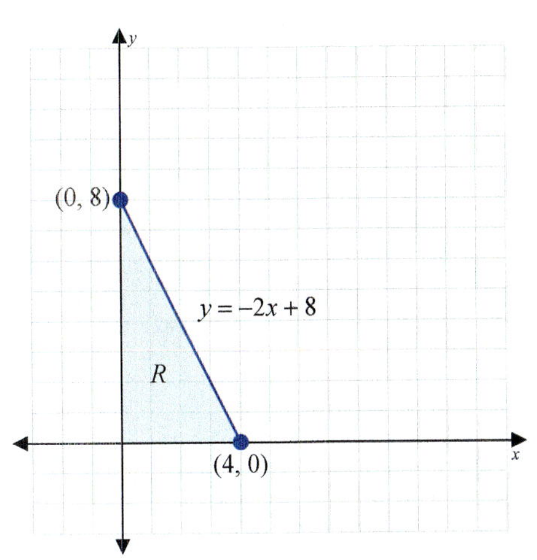

Thus R can be described by the following restrictions on x and y:

$$0 \le x \le 4 \quad \text{and} \quad 0 \le y \le -2x + 8.$$

The described volume is now found by using double integration.

$$\int_0^4 \left(\int_0^{-2x+8} (10 - x - y)\, dy \right) dx = \int_0^4 \left(10y - xy - \frac{y^2}{2} \right]_0^{-2x+8} \right) dx$$

$$= \int_0^4 \left(10(-2x + 8) - x(-2x + 8) - \frac{(-2x + 8)^2}{2} \right) dx$$

$$= \int_0^4 (-12x + 48)\, dx$$

$$= -6x^2 + 48x \Big]_0^4$$

$$= (-96 + 192) - 0 = 96$$

Thus the volume of the solid described is 96 cubic units.

16.6 EXERCISES

PRACTICE

In Exercises 1–10, evaluate the given double integral.

1. $\int_0^1 \int_1^2 (x+1)\,dy\,dx$

2. $\int_0^2 \int_1^3 (4-x)\,dy\,dx$

3. $\int_{-1}^2 \int_1^4 (3x+2y)\,dy\,dx$

4. $\int_{-2}^2 \int_3^4 (2x-y)\,dy\,dx$

5. $\int_1^2 \int_2^3 \frac{2y}{x}\,dy\,dx$

6. $\int_1^3 \int_{-1}^2 \frac{3y}{2x}\,dy\,dx$

7. $\int_1^3 \int_{-2}^2 (x^2+3y^2-1)\,dx\,dy$

8. $\int_2^3 \int_{-1}^1 (2x^2+y^2-x)\,dx\,dy$

9. $\int_0^2 \int_0^1 e^{x+y}\,dx\,dy$

10. $\int_0^1 \int_{-1}^2 ye^{xy}\,dx\,dy$

In Exercises 11–16, evaluate the double integral on the given rectangular region.

11. $\iint\limits_R (x-y^2)\,dA$ $R: 0 \le x \le 2$ and $0 \le y \le 1$

12. $\iint\limits_R (xy+x)\,dA$ $R: 0 \le x \le 3$ and $0 \le y \le 3$

13. $\iint\limits_R y\sqrt{x+1}\,dA$ $R: 0 \le x \le 3$ and $1 \le y \le 5$

14. $\iint\limits_R x^2\sqrt{3+y}\,dA$ $R: -1 \le x \le 4$ and $1 \le y \le 6$

15. $\iint\limits_R e^{x+2y}\,dA$ $R: 0 \le x \le 3$ and $0 \le y \le 4$

16. $\iint\limits_R e^{2x+y}\,dA$ $R: 0 \le x \le 2$ and $0 \le y \le 1$

In Exercises 17–24, evaluate the double integral.

17. $\int_0^2 \int_0^{3x} xy^2\,dy\,dx$

18. $\int_0^1 \int_{2x}^{x^2} x^2 y^2\,dy\,dx$

19. $\int_1^4 \int_0^{x^2} \sqrt{\frac{y}{x}}\,dy\,dx$

20. $\int_1^4 \int_x^{x^2} \sqrt{\frac{x}{y}}\,dy\,dx$

21. $\int_1^3 \int_1^{e^y} \frac{y}{x}\,dx\,dy$

22. $\int_0^2 \int_0^y e^{y^2}\,dx\,dy$

23. $\int_0^4 \int_0^y \sqrt{9+y^2}\,dx\,dy$

24. $\int_0^2 \int_0^{4-y^2} y\sqrt{x}\,dx\,dy$

In Exercises 25–34, evaluate the double integral on the given region.

25. $\iint\limits_R 2xy\,dA$ $R: 0 \le x \le 1$ and $x^2 \le y \le \sqrt{x}$

26. $\iint\limits_R 3xy^2\,dA$ $R: 0 \le x \le 1$ and $x^3 \le y \le \sqrt[3]{x}$

27. $\iint\limits_R (x^2-y)\,dA$ $R: 1 \le x \le 2$ and $x \le y \le x^2$

28. $\displaystyle\iint_R (3-2x-2y)\,dA$ $R:0\le x\le 1$ and $0\le y\le (2-x)$

29. $\displaystyle\iint_R e^y\,dA$ $R:0\le x\le 2$ and $x\le y\le 3x$

30. $\displaystyle\iint_R e^y\,dA$ $R:0\le x\le 1$ and $0\le y\le 2x$

31. $\displaystyle\iint_R (x+y)\,dA$

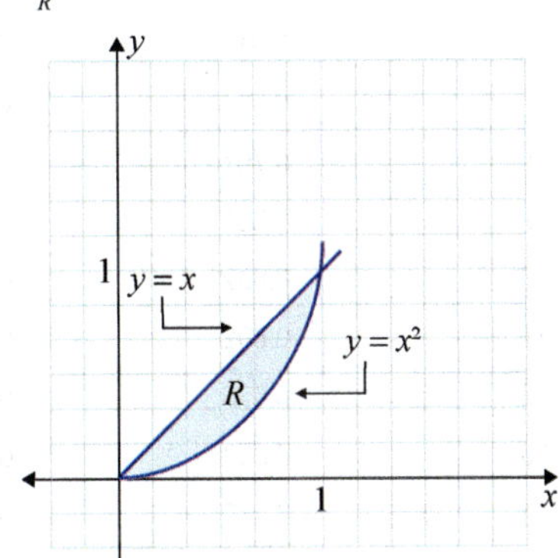

32. $\displaystyle\iint_R (2xy+x)\,dA$

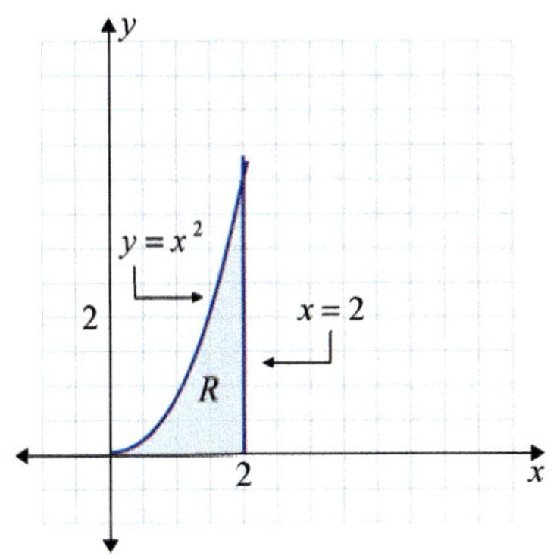

33. $\displaystyle\iint_R (3-2xy)\,dA$

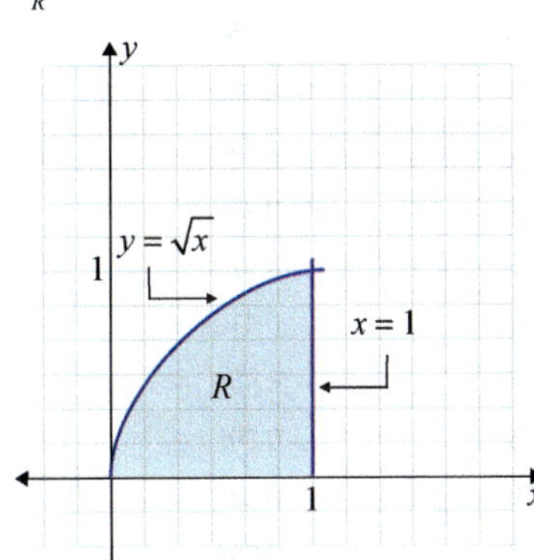

34. $\displaystyle\iint_R \sqrt{4-x^2}\,dA$

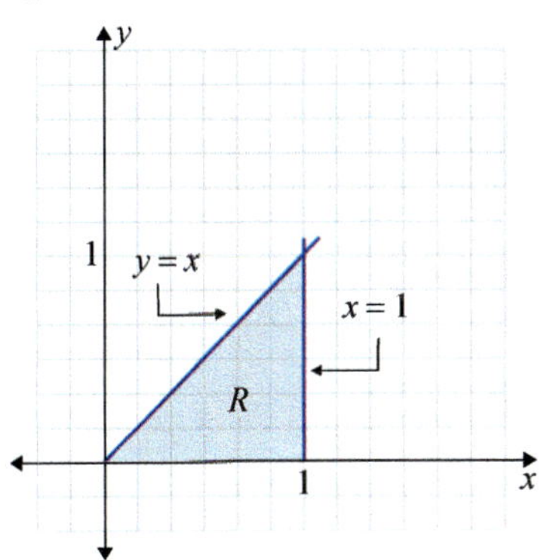

35. Find the volume of the solid bounded above by the graph of $f(x,y)=8-x^2-y^2$ and below by the rectangle $R:-1\le x\le 2$ and $0\le y\le 2$.

36. Find the volume of the solid bounded above by the graph of $f(x,y)=2+x^2+y^2$ and below by the rectangle $R: 0\le x\le 1$ and $0\le y\le 3$.

37. Find the volume of the solid bounded above by the graph of $f(x,y)=8-4x-2y$ and below by the triangle with vertices $(0, 0, 0)$, $(2, 0, 0)$, and $(0, 4, 0)$.

38. Find the volume of the solid bounded above by the graph of $f(x,y)=3+x+2y$ and below by the triangle with vertices $(0, 0, 0)$, $(0, 2, 0)$, and $(2, 0, 0)$.

39. Find the volume of the solid bounded above by the graph of $f(x,y)=2xy$ and below by the region bounded by $y=\sqrt{x}$, $y=0$, and $x=1$.

40. Find the volume of the solid bounded above by the graph of $f(x,y)=4x^2y$ and below by the region bounded by $y=x^2$, $y=0$, and $x=1$.

APPENDIX: CRITICAL VALUES OF THE PEARSON CORRELATION COEFFICIENT

Critical Values of the Pearson Correlation Coefficient

n	$\alpha = 0.05$	$\alpha = 0.01$
4	0.950	0.990
5	0.878	0.959
6	0.811	0.917
7	0.754	0.875
8	0.707	0.834
9	0.666	0.798
10	0.632	0.765
11	0.602	0.735
12	0.576	0.708
13	0.553	0.684
14	0.532	0.661
15	0.514	0.641
16	0.497	0.623
17	0.482	0.606
18	0.468	0.590
19	0.456	0.575
20	0.444	0.561
21	0.433	0.549
22	0.423	0.537
23	0.413	0.526
24	0.404	0.515
25	0.396	0.505
26	0.388	0.496
27	0.381	0.487
28	0.374	0.479
29	0.367	0.471
30	0.361	0.463
35	0.334	0.430
40	0.312	0.403
45	0.294	0.380
50	0.279	0.361
55	0.266	0.345
60	0.254	0.330
65	0.244	0.317
70	0.235	0.306
75	0.227	0.296
80	0.220	0.286
85	0.213	0.278
90	0.207	0.270
95	0.202	0.263
100	0.197	0.256

☑ NOTE

r is statistically significant if $|r|$ is greater than the value given in the table.

Answer Key

Chapter 0: Fundamental Concepts of Algebra

0.1 EXERCISES

1. a. $19,\ 2^5$ **b.** $19,\ \dfrac{0}{15},\ 2^5$

  c. $19,\ \dfrac{0}{15},\ 2^5,\ -33$

  d. $19,\ -4.3,\ \dfrac{0}{15},\ 2^5,\ -33$

  e. $-\sqrt{3}$ **f.** all except $\dfrac{15}{0}$

  g. $\dfrac{15}{0}$

3. a. $|-16|,\ \dfrac{12}{3},\ \sqrt{4}$

  b. $|-16|,\ \dfrac{12}{3},\ 0,\ \sqrt{4}$

  c. $|-16|,\ \dfrac{12}{3},\ 0,\ \sqrt{4}$

  d. all   **e.** none   **f.** all   **g.** none

5.

7.

9. $<,\le$ **11.** $<,\le$

13. $<,\le$ **15.** $>,\ge$

17. $>,\ge$ **19.** $2a+b>c$

21. $9\ge 7$ **23.** $x+5<3$

25. $9\ge 8$

27. $\{3n\,|\,n$ is an integer and $-2\le n\le 3\}$

29. $\{n\,|\,n$ is a prime$\}$

31. $\left\{\dfrac{1}{n}\,\middle|\,n$ is an odd integer$\right\}$

33. $(-\infty,15)$ **35.** $(2.5,3.7]$

37. $(-\infty,4)$ **39.** $\left(-\dfrac{1}{2},\dfrac{2}{5}\right)$

41. $[0,\infty)$

43.

45.

47.

49. 4 **51.** $\sqrt{5}-\sqrt{3}$

53. -15 **55.** 1

57. -1 **59.** -12

61. 8 **63.** 6

65. 11

67. JR > Freddie > Sarah > Aubrey > Elizabeth

69. If age $=x$,
$\{x\,|\,x<2\}=[0,2)\to$ free;
$\{x\,|\,2\le x<12\}=[2,12)\to \$3;$
$\{x\,|\,12\le x<65\}=[12,65)\to \$7;$
$\{x\,|\,x\ge 65\}=[65,\infty)\to \$5$

71. No, because all natural numbers can be expressed as fractions.

0.2 EXERCISES

1. $3x^2y^3,\ -2\sqrt{x+y},\ 7z$

3. $-2,\ \sqrt{x+y}$

5. $1,\ 8.5,\ -14$

7. $\dfrac{-5x}{2yz},\ -8x^5y^3,\ 6.9z$

9. $\dfrac{-5}{2},\dfrac{1}{y},\dfrac{1}{z},x$ **11.** 20

13. 8 **15.** $-\dfrac{\sqrt{2}}{36}+2$

17. 4 **19.** $58+6\pi$

21. $\dfrac{-1}{3}$

23. Commutative   **25.** Associative

27. Associative   **29.** Distributive

31. Commutative

33. Multiplicative cancellation; $\dfrac{1}{5}$

35. Additive cancellation; x

37. Zero-Factor Property

39. Multiplicative cancellation; 6

41. Multiplicative cancellation; $\dfrac{1}{3}$

43. $\dfrac{11}{2}$ **45.** -10

47. 1 **49.** 70

51. $\dfrac{103}{6}$ **53.** $\dfrac{-144}{5}$

55. $\dfrac{37}{2}$ **57.** 23.66

59. 1.64

61. $-\dfrac{1}{5}\left(\sqrt{3(3+7)-5}\right)^3$

63. $\left(\dfrac{\sqrt[3]{x-4}}{2}\right)^2$ **65.** $(-5,4]$

67. $[3,4]$ **69.** $[-\pi,21)$

71. $(3,9]$ **73.** $\mathbb{Z}$

75. $\mathbb{Z}$ **77.** \$66

79. \$102

81. 2.19 square meters

83. It is the same number you began with. Explanations may vary.

85. Answers may vary. (Ex: Please Excuse My Dear Aunt Sally.)

0.3 EXERCISES

1. 16 **3.** -9

5. 81 **7.** 64

9. 1 **11.** $\dfrac{1}{7}$

13. x^3 **15.** $27s^{10}$

17. -2 **19.** x^3

21. $121x^7$ **23.** x

25. $\dfrac{1}{x^2}$ **27.** x^3y^3

29. $\dfrac{16}{s^3}$ **31.** $-\dfrac{y^5}{3x^2}$

33. $\dfrac{1}{3y^2z}$ **35.** $27x^2y^4$

37. 1 **39.** $\dfrac{c^2}{9a^7b^3}$

41. $\dfrac{81y^3z^2}{2x^3}$ **43.** $\dfrac{64a^6}{b^{15}}$

45. $27x^9$ **47.** $\dfrac{1}{5z^6-81x^{12}}$

0.4 EXERCISES

1. -3 **3.** Not real

5. -2 **7.** -5

9. Not real **11.** $-\dfrac{3}{5}$

13. $-\dfrac{1}{2}$ **15.** 2

17. $\dfrac{2}{5}$ **19.** $3|x|$

21. $\dfrac{x^2|z|}{2}$ **23.** $x^2 y^7 z^3$

25. $\dfrac{ab^4}{3c^2}$ **27.** $\dfrac{|x^3|y^2}{2}$

29. $\dfrac{y^6 z^5}{2x^7}$ **31.** $\dfrac{\sqrt[3]{36x^2 y^2}}{3y^2}$

33. $-\sqrt{2}-\sqrt{5}$ **35.** $\sqrt{6}+\sqrt{3}$

37. $\dfrac{x+\sqrt{2x}}{x-2}$

39. $\dfrac{x+2\sqrt{xy}+y}{x-y}$

41. $\dfrac{y-2\sqrt{y}}{y-4}$ **43.** $\dfrac{1}{\sqrt{5}+3}$

45. $\dfrac{9-y}{18-6\sqrt{y}}$ **47.** $\dfrac{1}{\sqrt{13}-\sqrt{t}}$

49. $\dfrac{6-y}{6+y-2\sqrt{6y}}$

51. $bh+bl+hl+l\sqrt{b^2+h^2}$

53. Because a positive number squared is positive and a negative number squared is positive.

0.5 EXERCISES

1. $3x\sqrt[3]{2x}$ **3.** Not possible

5. 0 **7.** $4z\sqrt[3]{2z}$

9. 0 **11.** $\left(3x^2-4\right)^2$

13. 27 **15.** n^2

17. $\dfrac{x^{\frac{4}{5}}}{y^{\frac{5}{3}}}$ **19.** $\dfrac{1}{125}$

21. $y\sqrt[3]{y^2}$ **23.** $\left(ax^2+by\right)^{\frac{1}{12}}$

25. $a^{\frac{15}{4}}$ **27.** $x^{\frac{1}{4}}$

29. $6^{-\frac{1}{3}}$ **31.** $\sqrt[4]{125}$

33. $\sqrt[4]{|y|}$ **35.** x^3

37. $\sqrt[16]{16,807}$

39. $3d^2\sqrt{3};\ 3.326\ \text{cm}^2$

41. Because a root is the same as a fractional exponent.

0.6 EXERCISES

1. $2x^2+x+5$ **3.** $5a^2+14a-2$

5. x^2+8x+2

7. $4x^2+3x+11$

9. $3a^2-8ab$

11. $-7x^2-5xy+5y^2$

13. $6x^4+10x^3-2x^2$

15. $12xy$

17. $x^4,\ x^2y,\ y^2$

19. $14x^2-9x-18$

21. $12x^2-25x+12$

23. $16x^2-24x+9$

25. $9x^2+12xy+4y^2$

27. $36x^2-y^2$

29. $2x^3+6x^2-8x$

31. x^3+27 **33.** x^3+8y^3

35. $-7x+9$ **37.** $7t+7$

39. $4x^2-x$ **41.** $3y,\ 4,\ 6$

43. $(s-7)(s+2)$

45. $(x+13)(x-2)$

47. $4b(b-4)(b+4)$

49. $3(3a-2)(3a+2)$

51. $(x-1)(x+1)\left(x^2+x+1\right)$
$\cdot\left(x^2-x+1\right)$

53. $\left(5x^4-4\right)\left(5x^4+4\right)$

55. $2(t+2y)\left(t^2-2ty+4y^2\right)$

57. $(s+1)(s-1)\left(s^2+1\right)$

59. $100x(y+1)^2$

61. Answers will vary

Chapter 1: Equations and Inequalities in One Variable

1.1 EXERCISES

1. $x=-3$ **3.** $x=2$

5. $x=2$ **7.** $x=2$

9. $y=-1$ **11.** $t=-1$

13. $x=-0.12$ **15.** $x=4$

17. $x=0$ **19.** $y=0$

21. $x=-2$ **23.** $y=-6$

25. $n=6$ **27.** $n=8$

29. $x=0$ **31.** $x=-7$

33. $x=-\dfrac{1}{8}$ **35.** $x=-\dfrac{13}{2}$

37. $x=-\dfrac{21}{5}$ **39.** $x=-\dfrac{8}{15}$

41. $y=\dfrac{28}{5}$ **43.** $x=2$

45. $y=\dfrac{7}{5}$ **47.** $x=-4.5$

49. $x=-44$ **51.** $x=2$

53. $x=-4$ **55.** $y=0.5$

57. $x=1.5$ **59.** $x=0.2$

61. $x=-5$ **63.** $n=3$

65. $y=6$ **67.** $x=3$

69. $n=0$ **71.** $y=0$

73. $z=-1$ **75.** $y=\dfrac{1}{5}$

77. $x=-3$ **79.** $x=-4$

81. $x=-21$ **83.** $y=0$

85. $y=1$ **87.** $x=-\dfrac{3}{2}$

89. $x=\dfrac{1}{4}$ **91.** $x=\dfrac{3}{17}$

93. $x=-\dfrac{1}{4}$ **95.** $x=\dfrac{8}{5}$

97. $x=\dfrac{2}{3}$ **99.** $x=6$

101. $x=-11$ **103.** $x=\dfrac{1}{2}$

105. $x=-5$ **107.** $n=-1.5$

109. $x=0$ **111.** Conditional

113. Conditional **115.** Contradiction

117. Identity   **119.** Conditional

121. a. The 4 should have been multiplied by 3 so that the 3 was distributed over the entire left-hand side of the equation; Correct answer is $x = 15$.

  b. 3 should be subtracted from each side, not from each term, and $5x - 3$ doesn't simplify to $2x$; Correct answer is $x = \dfrac{8}{5}$.

123. a. $x + 6 = 20$; conditional, since there is one age that will make the statement true

  b. $x + 6 = x + 8$; contradiction, since he can't be both 6 years older and 8 years older

  c. $x + 6 = (x + 3) + 3$; identity, since the statement will be true regardless of Ryan's current age

1.2 EXERCISES

1. $r = \dfrac{C}{2\pi}$

3. $a = \dfrac{v^2 - v_0{}^2}{2x}$

5. $F = \dfrac{9}{5}C + 32$

7. $h = \dfrac{A - 2lw}{2w + 2l}$

9. $m = \dfrac{2K}{v^2}$

11. $\dfrac{19}{3}$ hours, or 6 hours and 20 minutes

13. 13.5 miles   **15.** \$390

17. 2 gallons 44%, 1 gallon 50%

19. 24 child tickets, 15 adult tickets

21. 7.5%

23. 26 feet by 26 feet

25. 53, 55, and 57

27. 36.4%

29. $x \approx 0.72$ **31.** $x \approx 13.11$

1.3 EXERCISES

1. $\{-9, 3.14, -2.83, 1, -3, 4\}$

3. $\{-2.83, 1, -3\}$

5. $(-\infty, -3]$

7. $(-\infty, 4.8)$

9. $(-\infty, 2.25)$

11. $\left(-\infty, \dfrac{3}{2}\right)$

13. $\left(-\infty, -\dfrac{3}{11}\right]$

15. $(7, \infty)$

17. $(35, \infty)$

19. $(-3, \infty)$

21. $(-0.11, \infty)$

23. $(1, 5]$

25. $(-10, 6]$

27. $[-8, -2)$

29. $(21, 69]$

31. $\left(\dfrac{23}{7}, \dfrac{25}{7}\right)$

33. $\left[\dfrac{13}{2}, 16\right)$

35. $\left(-\dfrac{5}{3}, 1\right]$

37. $\left(-\infty, -\dfrac{7}{2}\right] \cup \left(\dfrac{15}{2}, \infty\right)$

39. $\left(-\infty, \dfrac{1}{2}\right) \cup \left(\dfrac{5}{2}, \infty\right)$

41. $\varnothing$

43. $(-\infty, 2) \cup (6, \infty)$

45. $\varnothing$ **47.** $\varnothing$

49. $[-4, 0]$

51. $(-\infty, \infty)$

53. $(3, 15)$

55. $(-1, 3]$

57. $(-\infty, \infty)$

59. $[-2, 3)$

61. $[73, 113]$ for an A, $(113, 115)$ for an A+.

63. $(1140, 1600]$

1.4 EXERCISES

1. $\left\{\dfrac{3}{2}, -1\right\}$ **3.** $\{7\}$

5. $\left\{\dfrac{-3}{2}, -3\right\}$ **7.** $\{-3, 1\}$

9. $\{2\}$ **11.** $\{3, 11\}$

13. $\{0, 6\}$ **15.** $\left\{\dfrac{-1 \pm \sqrt{7}}{2}\right\}$

17. $\left\{-\dfrac{5}{3}, \dfrac{1}{3}\right\}$ **19.** $\{-5, 9\}$

21. $\left\{\dfrac{6}{5}, 6\right\}$ **23.** $\{-5, 2\}$

25. $\left\{-\dfrac{7}{2}, \dfrac{9}{2}\right\}$ **27.** $\left\{\dfrac{13}{2}, \dfrac{15}{2}\right\}$

29. $\{-25,-1\}$ **31.** $\left\{-\dfrac{4}{3},1\right\}$

33. $\{-1\}$ **35.** $\left\{\dfrac{-1\pm\sqrt{13}}{6}\right\}$

37. $\left\{-\dfrac{3}{2},5\right\}$ **39.** $\{-14,-6\}$

41. $\{2,14\}$ **43.** $\left\{\dfrac{-1\pm\sqrt{7}}{2}\right\}$

45. $\{-16,12\}$ **47.** $\{-7,-6\}$

49. $\{0,6\}$ **51.** $\{-1,2\}$

53. $\left\{1,2,\dfrac{3\pm\sqrt{17}}{2}\right\}$

55. $\left(3x-1-\sqrt{5}\right)\left(3x-1+\sqrt{5}\right)$

57. $b=-5$ and $c=-24$

1.5 EXERCISES

1. $\{-3,4\}$ **3.** $\{8,13\}$

5. $\left\{\pm\sqrt{2},\,\pm i\sqrt{5}\right\}$

7. $\left\{1\pm 2i,1\pm\sqrt{3}\right\}$

9. $\left\{\dfrac{1}{8},\,27\right\}$ **11.** $\{\pm 2i,\pm 3\}$

13. $\{-1,\pm 2,\,3\}$ **15.** $\left\{1,-\dfrac{8}{27}\right\}$

17. $\{-1,-2,-3\}$ **19.** $\{\pm 1,\,3\}$

21. $\left\{-\dfrac{5}{2},\,0,\,3\right\}$ **23.** $\{\pm 2,\pm 5i\}$

25. $\left\{\pm 2,-\dfrac{6}{5}\right\}$ **27.** $\left\{\pm\dfrac{3}{2},\pm\dfrac{3i}{2}\right\}$

29. $\left\{-\dfrac{5}{2},\,0,\,\dfrac{4}{7}\right\}$

31. $\left\{-\dfrac{4}{3},\dfrac{2\pm 2i\sqrt{3}}{3}\right\}$

33. $\left\{-3,\dfrac{3\pm 3i\sqrt{3}}{2}\right\}$

35. $\left\{\dfrac{5}{2}\right\}$ **37.** $\{1\}$

39. $\{4\}$ **41.** $\{0,2,3\}$

43. $\left\{-\dfrac{1}{5},\dfrac{1}{7}\right\}$ **45.** $\left\{-1,0,\dfrac{2}{5}\right\}$

47. $\left\{\dfrac{8}{3}\right\}$ **49.** $\left\{-3,-\dfrac{13}{4}\right\}$

51. $b=-4$, $c=-12$, and $d=0$

53. $a=1$, $c=-36$, and $d=-144$

55. $a=15$, $b=-16$, and $c=-5$

1.6 EXERCISES

1. $x=7$

3. $x\neq 0,2;\ x=4$

5. $x\neq -3,4;\ x=-10$

7. $x\neq 0;\ x=18$

9. $x\neq 6;\ x=-\dfrac{74}{9}$

11. $x=\dfrac{1}{4}$

13. $x=6$ **15.** $x=4$

17. $x\neq 0;\ x=\dfrac{10}{3}$

19. $x\neq 0;\ x=-\dfrac{3}{4}$

21. $x\neq 0;\ x=-\dfrac{3}{16}$

23. $x\neq -9,-\dfrac{1}{4},0;\ x=-2,1$

25. $x\neq\dfrac{3}{2},0,6;\ x=\dfrac{3}{5},9$

27. $x\neq\dfrac{1}{2},4;\ x=-3$

29. $x\neq -4,-1;\ x=2$

31. $x\neq -1,\dfrac{1}{4};\ x=\dfrac{2}{3}$

33. $x\neq 2,3;\ x=\dfrac{13}{10}$

35. $x\neq -\dfrac{2}{3},2;\ $ no solution

37. $x\neq -1,\dfrac{1}{3},\dfrac{1}{2};\ x=\dfrac{1}{5}$

39. $\{0\}$ **41.** $\{1\}$

43. $\varnothing$ **45.** $\{6\}$

47. $\varnothing$ **49.** $\{10\}$

51. $\{1\}$ **53.** $\{-2,0,2\}$

55. $\varnothing$ **57.** $\{4,44\}$

59. $\{0,5\}$ **61.** $\{4\}$

63. $\{2\}$ **65.** $\{-32\}$

67. $\left\{\pm\dfrac{125}{343}\right\}$ **69.** $\{7,10\}$

71. $\{\pm 3\}$

73. 144 defective bulbs

75. 370 at bats

77. 7.5 inches by 10 inches

79. 8.8 quarts

81. a. 7.5 miles per hour

   b. 55 miles per hour

83. 5 hours

85. a. $x=10$

   b. 8 kilometers per mile

   c. 12 kilometers per mile

   d. 0.5 hour

87. a. $\dfrac{x^2-8x}{4(x-4)(x+4)}$

   b. $x=0,8$

89. a. $\dfrac{14x-16}{5x(x-4)}$ **b.** $x=\dfrac{8}{7}$

Chapter 2: Linear Equations in Two Variables

2.1 EXERCISES

1.

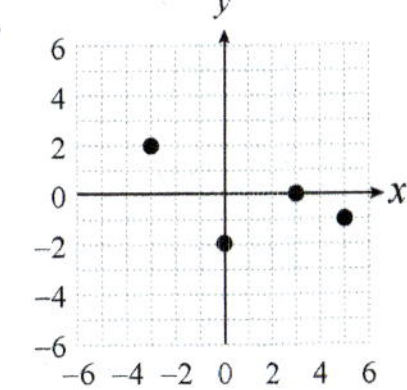

3.

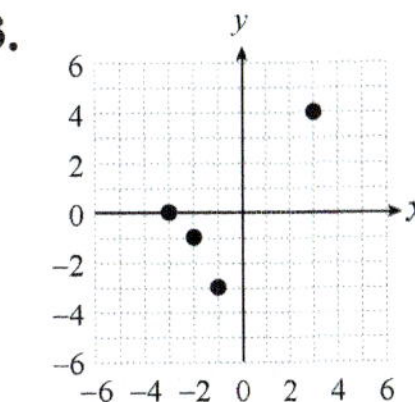

5.

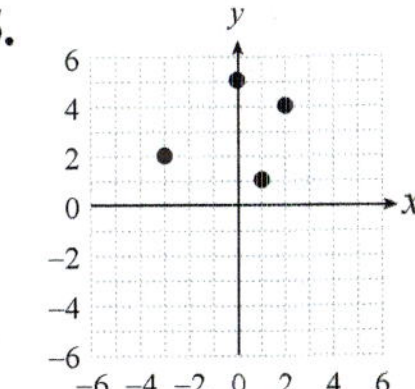

7.

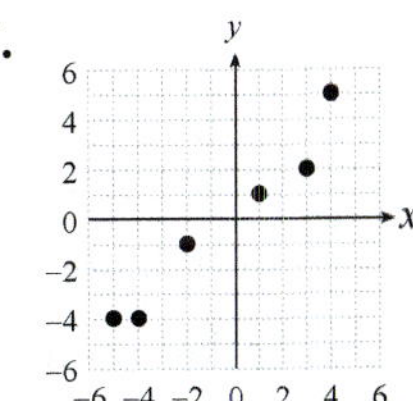

9. III   **11.** IV

13. Positive x-axis

15. III   **17.** IV

19. II   **21.** IV

23. I

25. Negative y-axis

27. $\left\{ (0,-3),(2,0),\left(3,\dfrac{3}{2}\right),(4,3) \right\}$

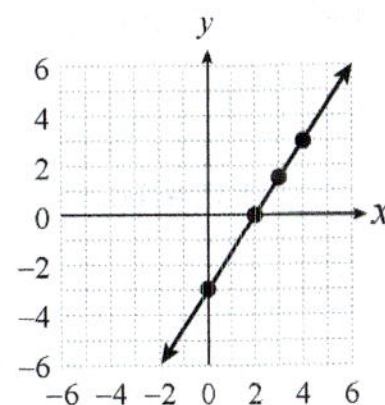

29. $\left\{ (0,0),(1,\pm1),(4,\pm2),(9,\pm3), \left(2,-\sqrt{2}\right) \right\}$

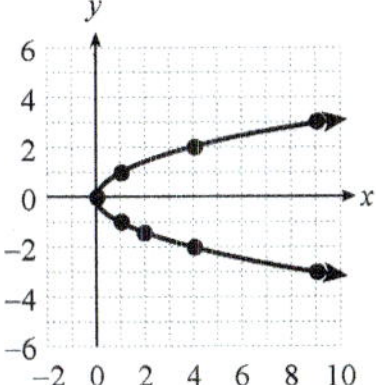

31. $\left\{ (0,\pm3),(\pm3,0),\left(-1,\pm2\sqrt{2}\right), \left(1,\pm2\sqrt{2}\right),\left(\pm\sqrt{5},2\right) \right\}$

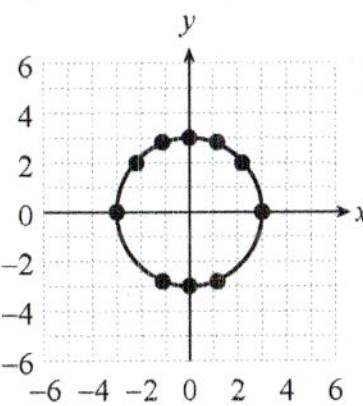

33. $\sqrt{34},\ \left(\dfrac{-7}{2},\dfrac{1}{2}\right)$

35. $\sqrt{58},\ \left(\dfrac{3}{2},\dfrac{7}{2}\right)$

37. $2\sqrt{2},(-1,-1)$

39. $4\sqrt{34},(3,-8)$

41. $10,(1,-6)$

43. $3\sqrt{13},\ \left(2,\dfrac{1}{2}\right)$

45. $10\sqrt{2},\ (3,3)$

47. $x = 2$ or 18

49. $x = 10,\ y = 1$

51. 12

53. $2\sqrt{29} + \sqrt{26} + 5\sqrt{2}$

55. 54

57. 1.25 kilometers

59. a. 249.19 meters

     b. $\left(\dfrac{133}{2},\dfrac{709}{2}\right)$

61. area $= \dfrac{15}{2}$ **63.** area $= 25$

65. area $= 17$ **67.** area $= 48$

69. $x = [-5,6]; y = [-8,9]$

71. $x = [-3,6]; y = [-4,5]$

73. $x = [-6,8]; y = [-9,7]$

2.2 EXERCISES

1. Yes   **3.** No

5. No   **7.** No

9. Yes   **11.** Yes

13. Yes   **15.** No

17. No   **19.** No

21. No   **23.** Yes

25.

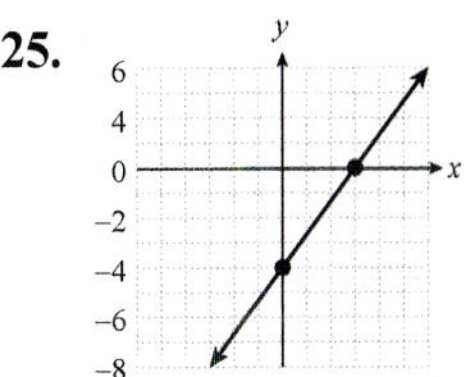

27.

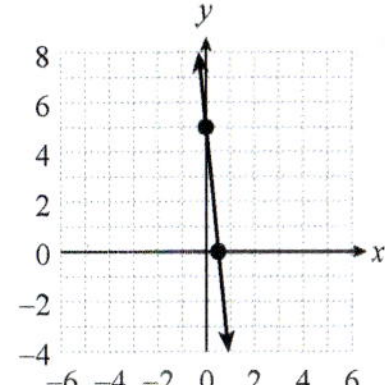

29.

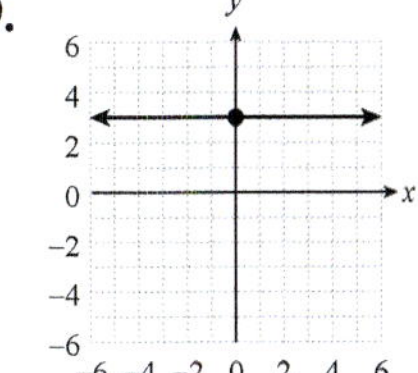

31.

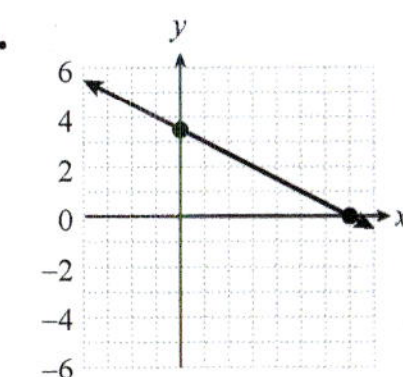

33.

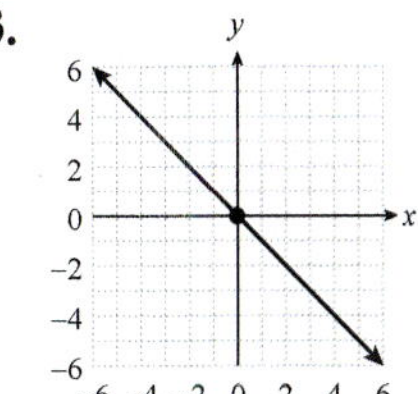

35.

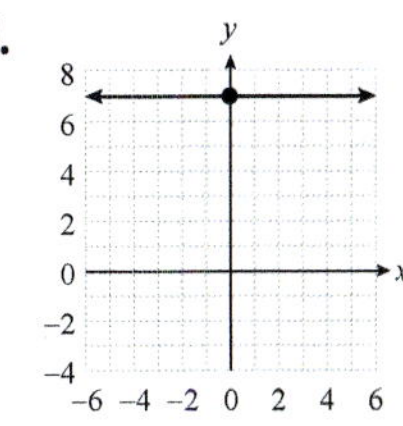

37. 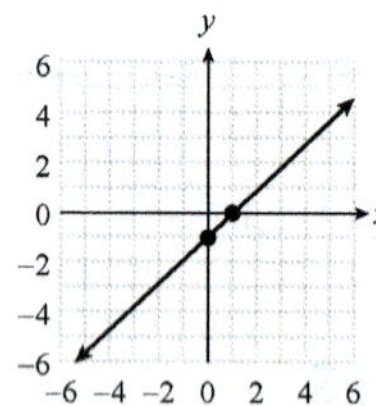

39. 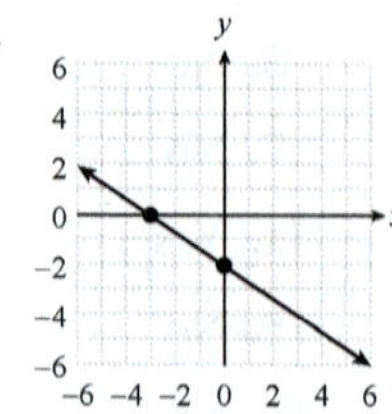

41. e   **43.** c

45. f

47. $a = P - b - c$

49. $j = 24,000 + 9b$;

$b = \dfrac{j - 24,000}{9}$; Yes

2.3 EXERCISES

1. -4 **3.** 0

5. Undefined   **7.** $\dfrac{2}{3}$

9. $\dfrac{1}{6}$ **11.** -7

13. -3 **15.** $-\dfrac{9}{13}$

17. $-\dfrac{1}{4}$ **19.** 0

21. Undefined   **23.** 2

25. $\dfrac{7}{6}$ **27.** $-\dfrac{5}{2}$

29. 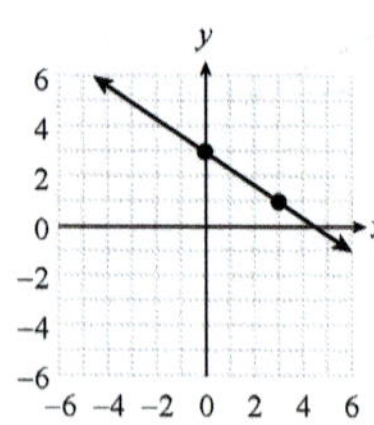

31. 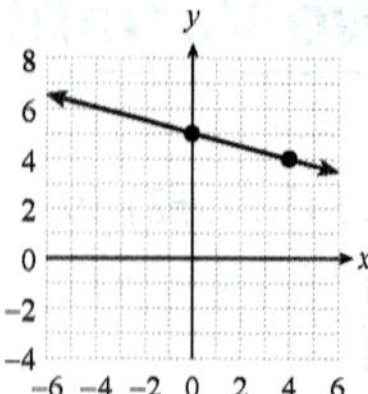

33. 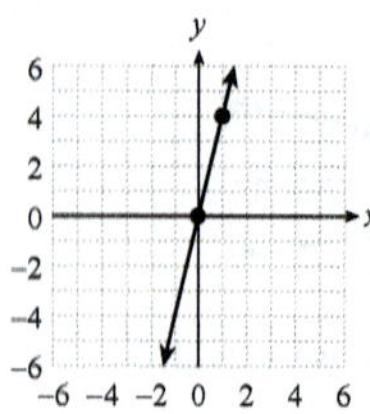

35. 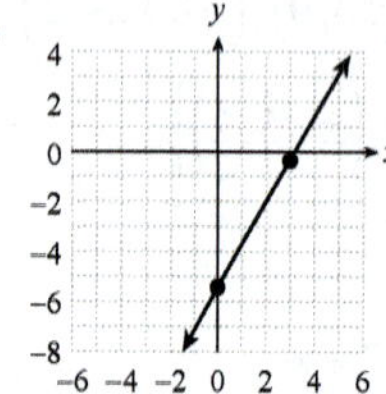

37. $y = \dfrac{3}{4}x - 3$

39. $y = -\dfrac{5}{2}x - 7$

41. $y = -5x - 9$

43. $3x - 2y = 3$

45. $y = 5$

47. $10x - y = 31$

49. $3x + y = 26$

51. $4x + 3y = 5$

53. $x = 2$ **55.** $y = -1$

57. $2x + 7y = 52$

59. $y = 5$

61. $15x - 8y = 0$

63. c   **65.** e

67. d

69. a. $2225   **b.** $2100

c. $0.25

71. $325

2.4 EXERCISES

1. $y = 4x + 9$ **3.** $y = 3x - 11$

5. $y = -9$ **7.** $y = x$

9. $y = \dfrac{7}{6}x + \dfrac{53}{6}$

11. Yes   **13.** Yes

15. Yes   **17.** No

19. No   **21.** No

23. No   **25.** No

27. Yes   **29.** No

31. $y = -\dfrac{1}{3}x - 1$ **33.** $y = 7$

35. $y = -\dfrac{1}{4}x - \dfrac{3}{4}$

37. $y = x + 3$

39. $y = -3x + 28$

41. No   **43.** No

45. No   **47.** No

49. Yes   **51.** No

53. No   **55.** No

57. Yes   **59.** $\dfrac{125}{3}$ ft

2.5 EXERCISES

1. Positive linear correlation

3. Negative linear correlation

5. Strong negative relationship

7. Weak positive relationship

9. No   **11.** Yes

13. a. 606.04   **b.** 886.14

c. 1138.23

15. a. Negative   **b.** $r = -0.538$

c. No

17. a. Positive   **b.** $r = 0.903$

c. Yes

19. a. $\hat{y} = 0.344x + 1.134$

b. Yes, $r = 0.728$

c. 3.886, or about 4 times

21. a. $\hat{y} = 0.429x + 0.4$

b. No, $r = 0.407$

c. N/A

Chapter 3: Functions and Their Graphs

3.1 EXERCISES

1. a. 3   **b.** −11

  c. $2a - 5$ **d.** $2a - 6$

3. a. 9   **b.** 4

  c. a^2 **d.** $a^2 - 2a + 2$

5. a. 4   **b.** −8

  c. $a^3 + 4a^2 + 2a$

  d. $a^3 + a^2 - 3a + 2$

7. a. 35

  b. $4a^2 + 16a + 15$

  c. $4x^2 + 8xh + 4h^2 - 1$

  d. 12

9. a. 2   **b.** $\sqrt{a+7}$

  c. $\sqrt{x+h+5}$ **d.** $3 - \sqrt{6}$

11. a. −5   **b.** −2

  c. 0.25   **d.** 3

13. $3h$ **15.** $h^2 + 2hx$

17. $2h^2 + 4hx$ **19.** $h^2 + 2hx - h$

21. $3h - h^2 - 2hx$

23. $2h^2 + 4xh - 3h$

25. $h^3 + 3h^2x + 3hx^2$

27. $h^3 + 3h^2x + 3hx^2$

29. $x \geq -5$ **31.** $x > -10$

33. $x \neq -1$ **35.** $\mathbb{R}$

37. $x > -\dfrac{5}{2}$ **39.** $\mathbb{R}$

41. Is a function

43. Is a function

45. Is a function

47. Not a function

49. Not a function

51. Is a function

3.2 EXERCISES

1. $x = 3$ **3.** $x = 30, 50$

5. (4, 11)   **7.** (5, 30)

9. a   **11.** c

13. a   **15.** a

17. a

19. a. $C(x) = 135 + 0.5x$

  b. $385

21. a. $P(x) = 4.5x - 135$

  b. $2115

23. $P(x) = 100(4.5x - 135)$

25. 0.75 atm

27. 56.03 atm

29. a. $R(x) = 6.5x$

  b. $C(x) = 1.1x + 378$

  c. $P(x) = 5.4x - 378$

  d. $x = 70$ pies

31. a. $R(x) = 243x$

  b. $C(x) = 73x + 5780$

  c. $P(x) = 170x - 5780$

  d. $x = 34$ sets of clubs

33. a. $0.15/pen

  b. $C(x) = 0.15x + 260$

  c. $260

35. a. $R(x) = 31x - 0.5x^2$

  b. $P(x) = -0.5x^2 + 20x - 500$

37. a. $D(x) = -0.02x + 400$

  b. $R(x) = -0.02x^2 + 400x$

  c. $P(x) = -0.01x^2 + 150x - 3600$

39. a. $I = 0.0625P$

  b. $375

41. a. $p = \dfrac{340}{x}$ **b.** $10

43. The least integer y

  such that $y \geq \dfrac{800}{x}$

45. $C(x) = \begin{cases} 0.65; \text{ for } 0 < x \leq 3 \\ 0.65 + 0.15(x-3); \\ \quad \text{for } x > 3 \end{cases}$

47. $P(x) = 0.2575x$

49. $P(x) = -x^2 + 72x - 720$

51. $A(x) = 138x - x^2$

53. $P(x) = \dfrac{576}{x} + 2x$

55. $A(x) = 360x - \dfrac{3}{2}x^2$

3.3 EXERCISES

1. 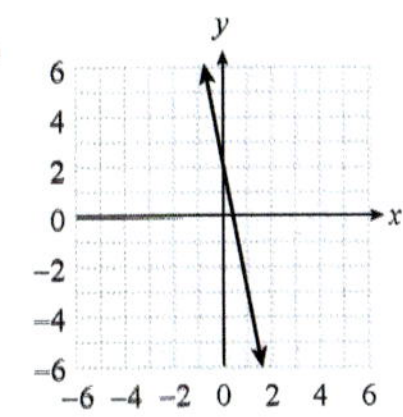

3. 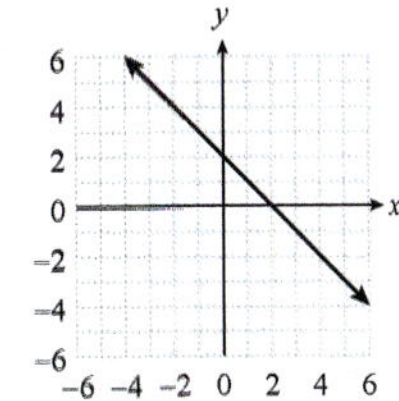

5. 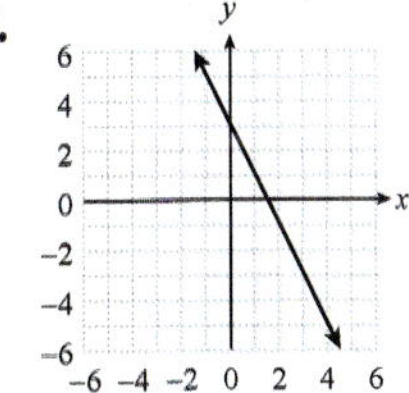

7. 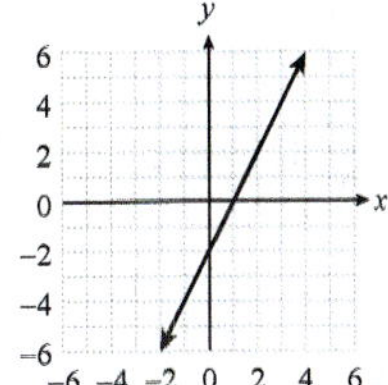

9. 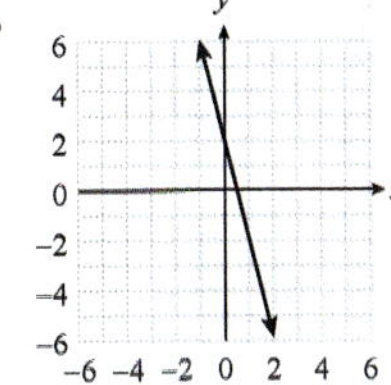

11. 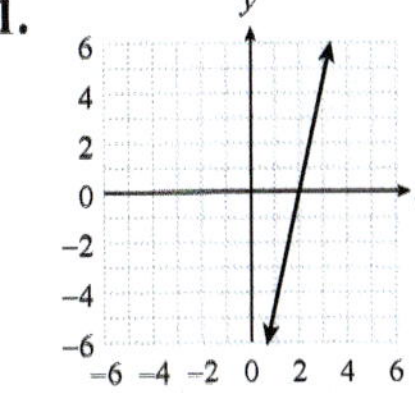

13. 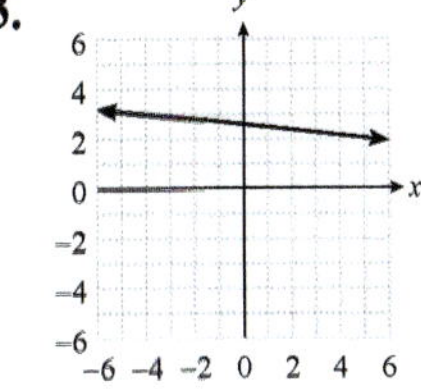

15.
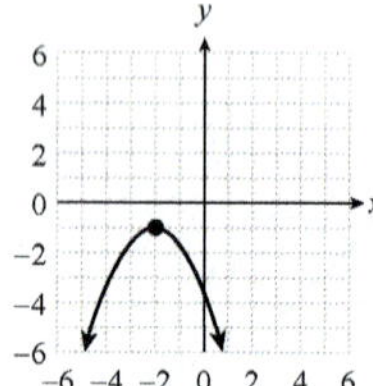

17. b   **19.** a

21. Vertex: $(-2,-1)$; no x-int.
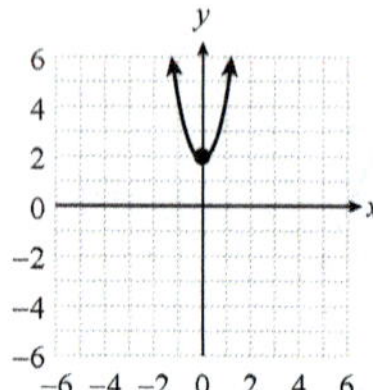

23. Vertex: $(0,2)$; no x-int.
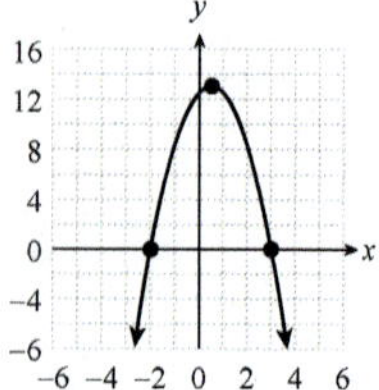

25. Vertex: $\left(\dfrac{1}{2},\dfrac{25}{2}\right)$; x-int.: $x = -2, 3$
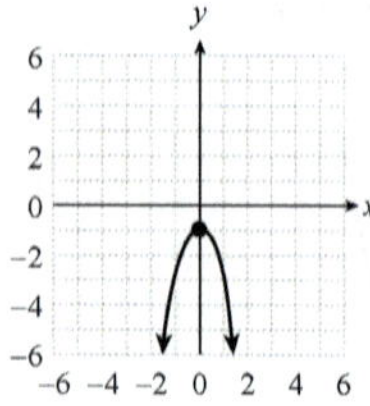

27. Vertex: $(0,-1)$; no x-int.
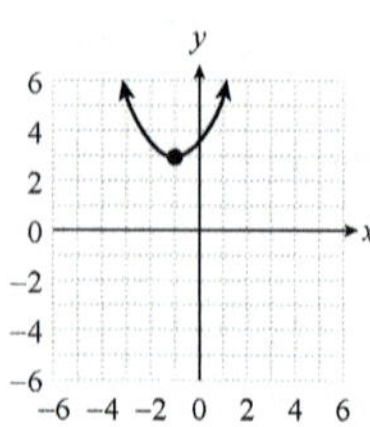

29. Vertex: $(-1,3)$; no x-int.

---

31. Vertex: $(1,-4)$; no x-int.
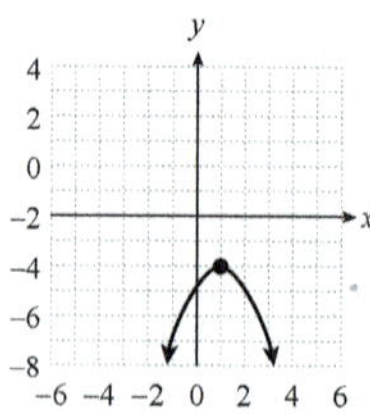

33. Vertex: $(1,-2)$; x-int.: $x = 0, 2$
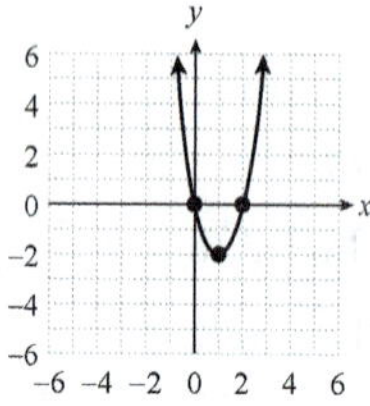

35. c   **37.** a

39. a. 1, on

  b. 2, below

  c. 2, above

  d. 2, above

  e. 2, below

  f. none, below

3.4 EXERCISES

1. Width of 50 feet, length of 100 feet

3. Width and length are 5

5. $(8,4)$

7. 8 and 8

9. 11,250 square feet

11. 500 rooms

13. 1500 cars

15. 6050 square feet

17. 180 feet

19. Vertex: $(4,-1)$;

  x-int.: $x = \dfrac{8 \pm \sqrt{2}}{2}$
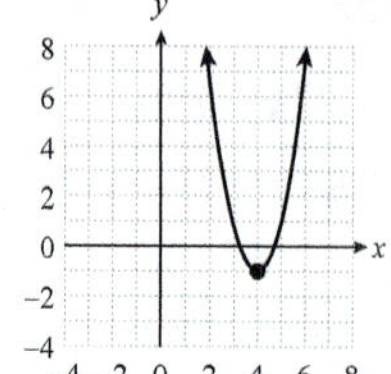

---

21. Vertex: $(4,-36)$;

  x-int.: $x = -2, 10$
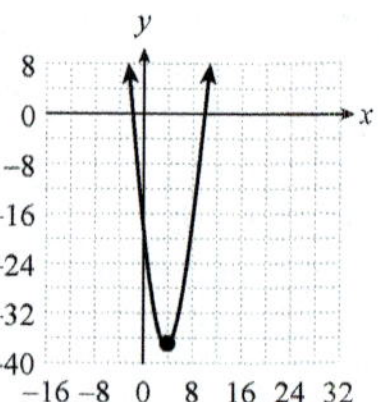

23. Vertex: $(0,25)$; x-int.: $x = -5, 5$
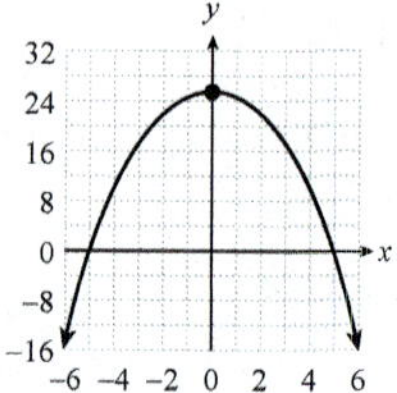

25. Vertex: $(-1,0)$; x-int.: $x = -1$
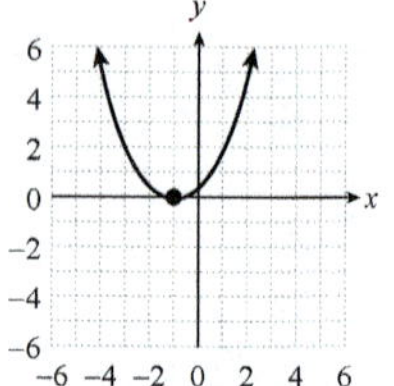

27. Vertex: $(5,21)$;

  x-int.: $x = 5 \pm \sqrt{21}$
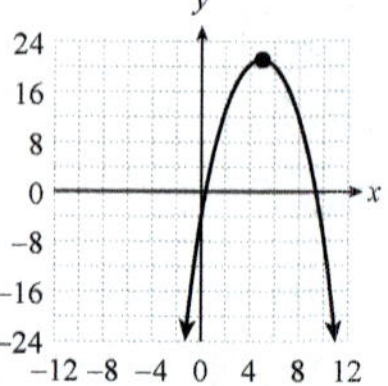

3.5 EXERCISES

1.
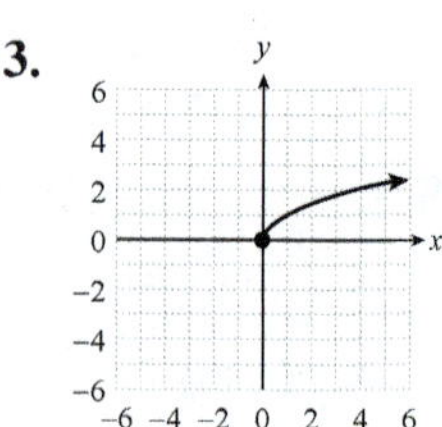

3.

5. **21.**

7. **23.**

9. 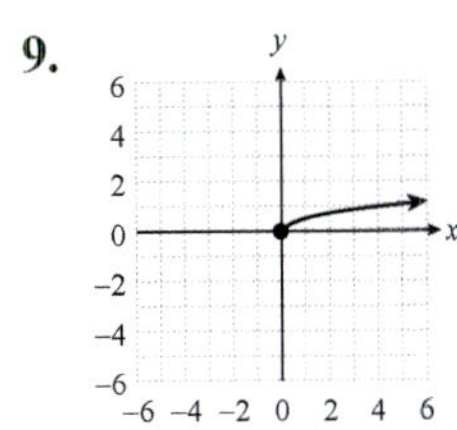 **25.**

11. **27.**

13. 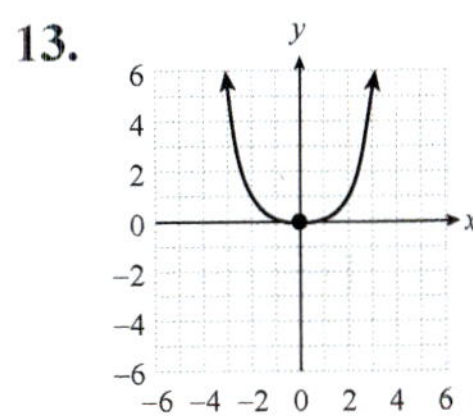 **29.**

15. **31.**

17. 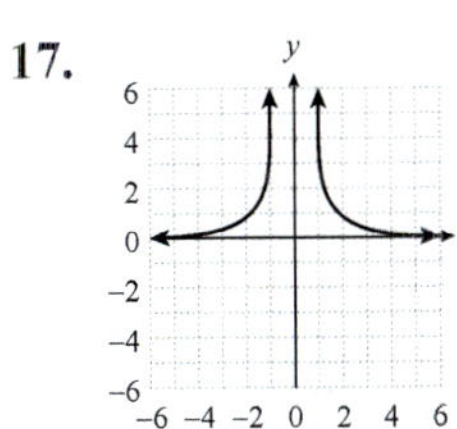 **33.** 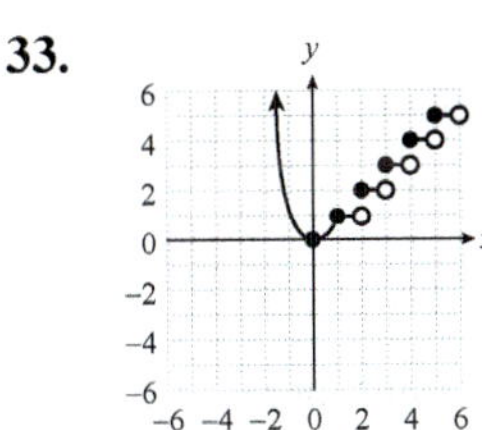

19. **35.**

37. j **39.** a
41. i **43.** e
45. f

3.6 EXERCISES

1. $y = x^2$ **3.** $y = \sqrt[3]{x}$

5. $y = \sqrt{x}$ **7.** $y = \dfrac{1}{x^2}$

9. $y = x^3$ **11.** $y = \sqrt{x}$

13. $y = x^3$

15.

$\text{Dom} = \mathbb{R}, \ \text{Ran} = [0, \infty)$

17.

$\text{Dom} = [-3, \infty), \ \text{Ran} = [-1, \infty)$

19.

$\text{Dom} = \text{Ran} = \mathbb{R}$

21. 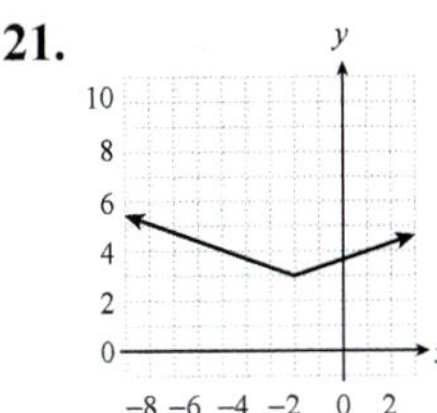

$\text{Dom} = \mathbb{R}, \ \text{Ran} = [3, \infty)$

23. 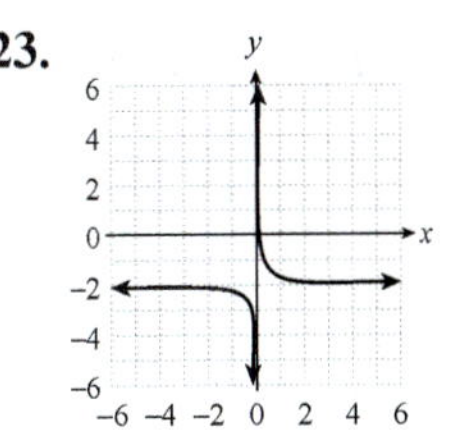

$\text{Dom} = (-\infty, 0) \cup (0, \infty),$
$\text{Ran} = (-\infty, -2) \cup (-2, \infty)$

25. 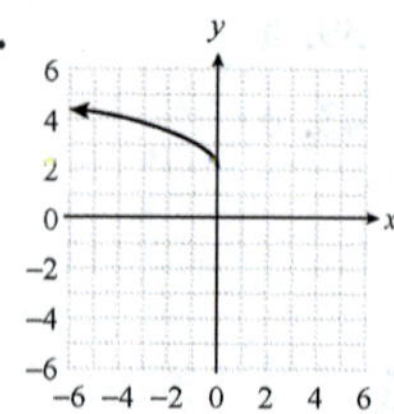

$$\text{Dom} = (-\infty, 0], \ \text{Ran} = [2, \infty)$$

27. 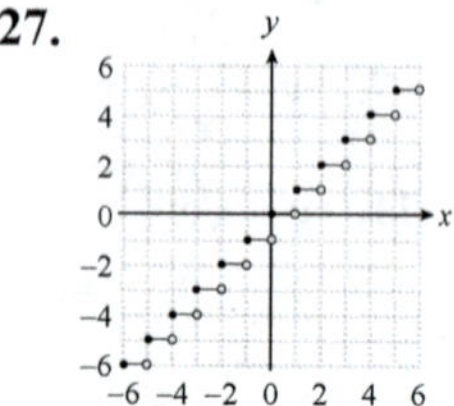

$$\text{Dom} = \mathbb{R}, \ \text{Ran} = \mathbb{Z}$$

29. 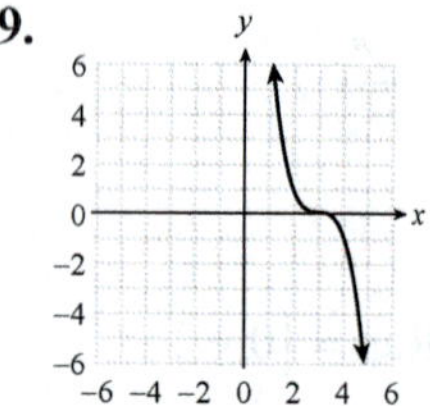

$$\text{Dom} = \text{Ran} = \mathbb{R}$$

31. 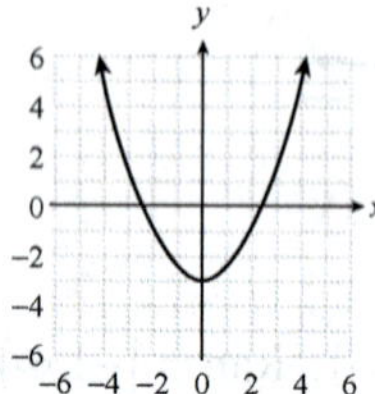

$$\text{Dom} = \mathbb{R}, \ \text{Ran} = [-3, \infty)$$

33. 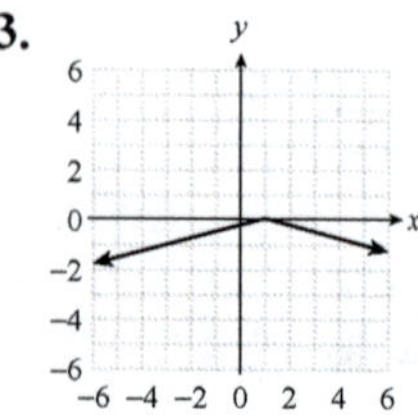

$$\text{Dom} = \mathbb{R}, \ \text{Ran} = (-\infty, 0]$$

35. 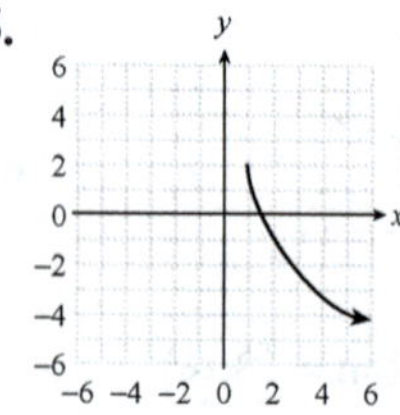

$$\text{Dom} = [1, \infty), \ \text{Ran} = (-\infty, 2]$$

37. 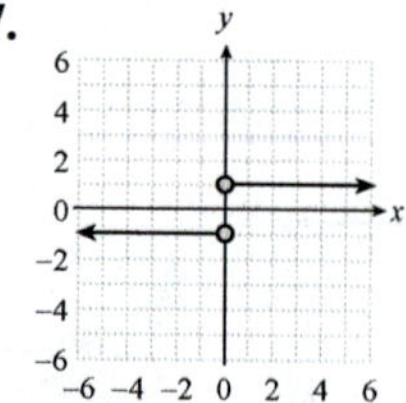

$$\text{Dom} = (-\infty, 0) \cup (0, \infty),$$
$$\text{Ran} = \{-1, 1\}$$

39. 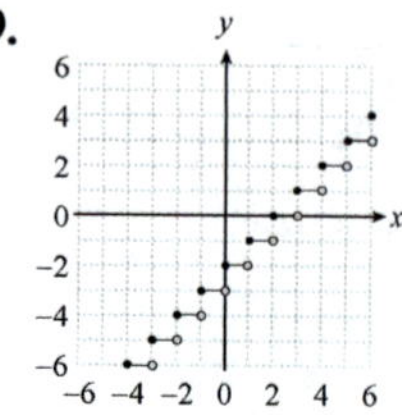

$$\text{Dom} = \mathbb{R}, \ \text{Ran} = \mathbb{Z}$$

41. $f(x) = (x-4)^2 + 2$

43. $f(x) = (-x-2)^2 = (x+2)^2$

45. $f(x) = (x-10)^3 + 4$

47. $f(x) = \sqrt{-x} - 3$

49. $f(x) = -|x-8| - 2$

51. $f(x) = -\sqrt{x+4}$

53. $f(x) = 1 - (x-3)^3$

55. 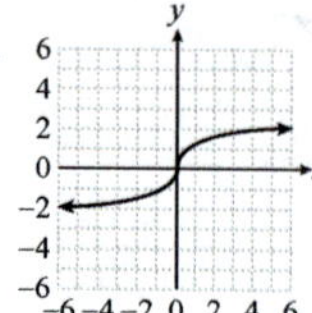

Odd function; origin symmetry

57. 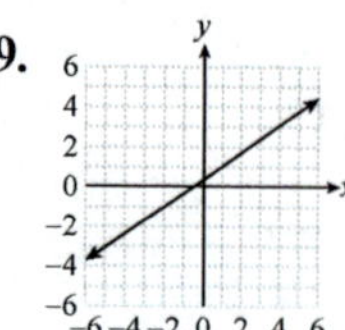

Odd function; origin symmetry

59. 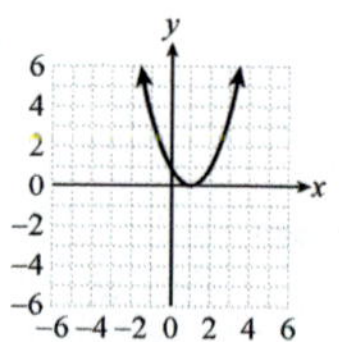

Function; neither; none of the symmetries

61. 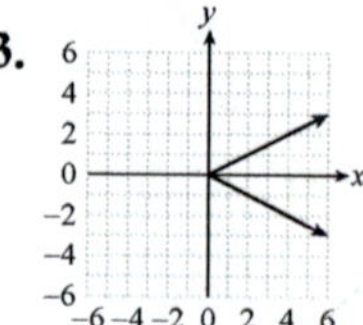

Function; neither; none of the symmetries

63. 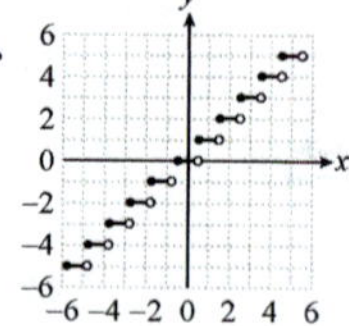

Not a function; x-axis symmetry

65. 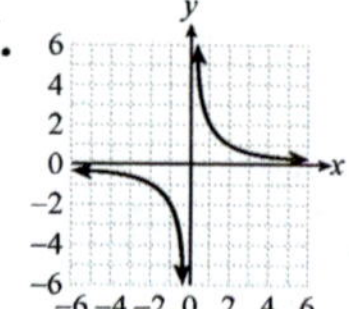

Function; neither; none of the symmetries

67. 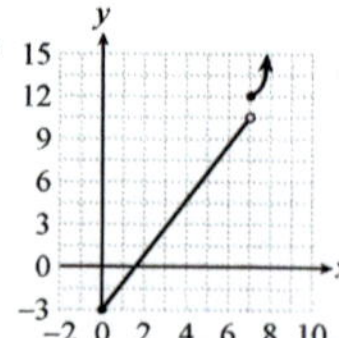

Odd function; Origin symmetry

69. Dec. on $(-\infty, -3)$, Inc. on $(-3, \infty)$

71. Dec. on $(-\infty, 1)$, Dec. on $(1, \infty)$

73. Inc. on $(-1, \infty)$

75. Inc. on $(-\infty, 1)$, Dec. on $(1, \infty)$

77. Inc. on $(-\infty, 7)$, Dec. on $(7, \infty)$

79. Dec. on $(-\infty, 1)$, Inc. on $(1, 3)$, Dec. on $(3, \infty)$

81. Inc. on $(0, \infty)$

89. $f(x) = -(x+3)^2 + 5$

91. $f(x) = -x^3 + 7$

93. $f(x) = (x-2)^3 - 4$

3.7 EXERCISES

19. Yes   **21.** Yes

23. Yes

25. $1 \pm 2i$

27. $-3, \dfrac{1}{2}$

29. $\pm\sqrt{3}, \pm\sqrt{5}$

31. $-\dfrac{5}{2}$

33. $0, 4 \pm 3i$

35. $\pm 1, \pm 2i\sqrt{2}$

37. 7th-degree; lead coef. = 4;
$j(x) \to -\infty$ as $x \to -\infty$
$j(x) \to \infty$ as $x \to \infty$

39. 5th-degree; lead coef. = −6;
$h(x) \to \infty$ as $x \to -\infty$
$h(x) \to -\infty$ as $x \to \infty$

41. 4th-degree; lead coef. = −2;
$f(x) \to -\infty$ as
$x \to -\infty$ and $x \to \infty$

43.

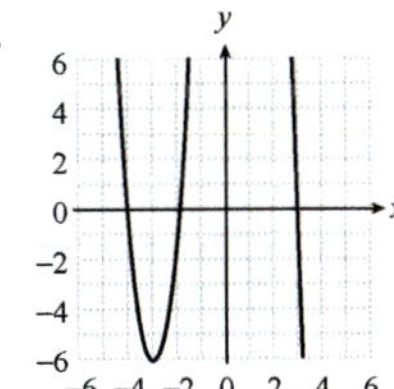

$g(x) \to \infty$ as $x \to -\infty$;
$g(x) \to -\infty$ as $x \to \infty$
x-int : $(-4,0),(-2,0),(3,0)$;
y-int : $(0,24)$

45.

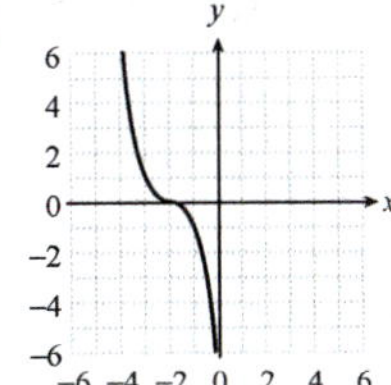

$h(x) \to \infty$ as $x \to -\infty$;
$h(x) \to -\infty$ as $x \to \infty$
x-int : $(-2,0)$;
y-int : $(0,-8)$

47.

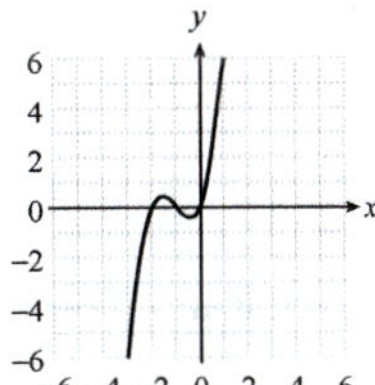

$s(x) \to -\infty$ as $x \to -\infty$;
$s(x) \to \infty$ as $x \to \infty$
x-int : $(-2,0),(-1,0),(0,0)$;
y-int : $(0,0)$

49.

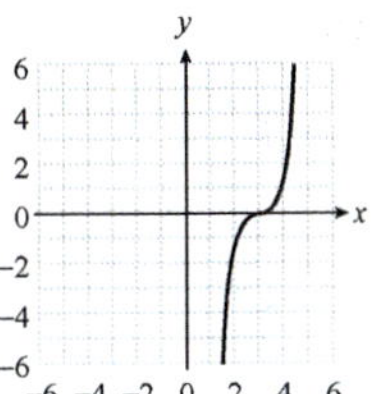

$g(x) \to -\infty$ as $x \to -\infty$;
$g(x) \to \infty$ as $x \to \infty$
x-int : $(3,0)$;
y-int : $(0,-243)$

51. e

53. a

55. f

57. d

59. f

61. b

63. $(-\infty,-2)\cup(3,\infty)$

65. $(-\infty,-2)\cup(-1,0)$

67. $[-2,1]\cup[3,\infty)$

69. $[-5,-1]\cup[1,4]$

71. $\left(-\dfrac{1}{2},2\right)$

73. $(-\infty,-4)\cup(2,3)$

75. All integers between 5 and 27, inclusive

77. All integers between 11 and 23, inclusive

79. Between 3490 and 17,740 phones, inclusive.

81. About 141.4 weeks

3.8 EXERCISES

1. $x = 1$

3. No vertical asymptote

5. $x = 2$

7. $x = 0$

9. $x = -\dfrac{1}{2}$

11. No vertical asymptote

13. $x = 7$

15. $x = -2$

17. $x = -2, x = 2$

19. $y = 0$

21. No horizontal or oblique asymptote

23. $y = 0$

25. $y = 2$

27. $y = 3x + 6$

29. $y = 0$

31. $y = 0$

33. $y = x - 11$

35. $y = 5x + 4$

37.

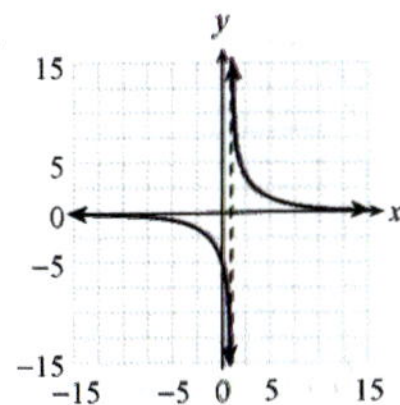

39.

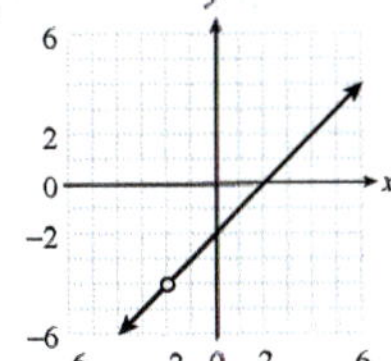

41.

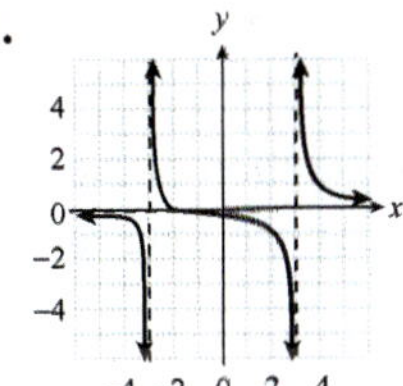

43.

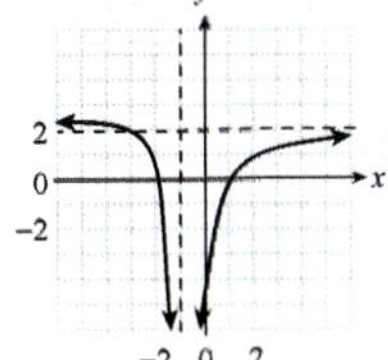

45.

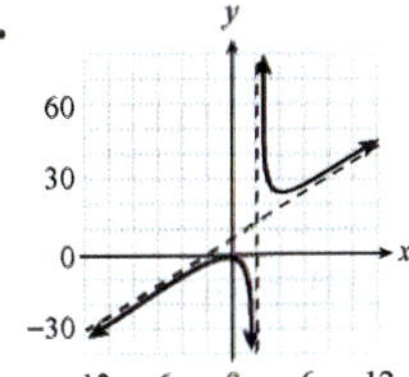

47.

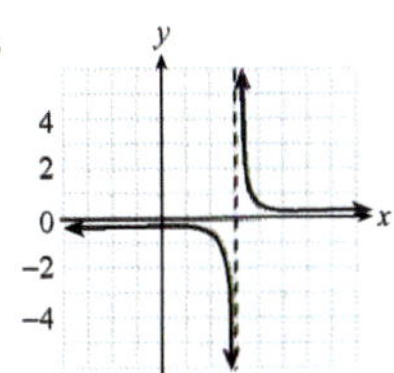

49. a. $x = -2$ **b.** $y = 0$
   c. None   **d.** None
   e. $(0,5)$

51. a. $x = 9$ **b.** $y = 0$
   c. None   **d.** None
   e. $\left(0, -\dfrac{1}{3}\right)$

53. a. $x = -1, x = 1$
   b. None   **c.** $y = x$
   d. $\left(\sqrt[3]{3}, 0\right)$ **e.** $(0,3)$

55. a. $x = 1$ **b.** None
   c. $y = 3x$ **d.** None
   e. $(0,-3)$

57. a.

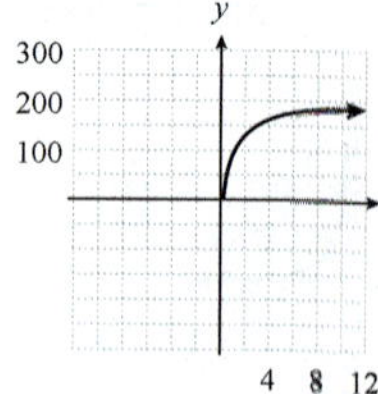

b. April's fish population approaches a maximum of 200 fish.

59. a.

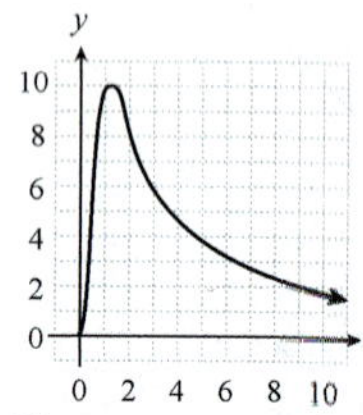

b. The concentration of the drug disappears in the long run.

3.9 EXERCISES

1. $(-\infty, -4] \cup [0, \infty)$

3. $(-\infty, -6)$

5. $(-\infty, -9) \cup (-3, \infty)$

7. $\left(2, \dfrac{5}{2}\right)$ **9.** $(7, \infty)$

11. $\left[\dfrac{7}{5}, 4\right)$ **13.** $\left(-9, -\dfrac{4}{3}\right)$

15. $(-\infty, -4] \cup [0, 3)$

17. $(-2, 0) \cup [5, \infty)$

19. $(-\infty, -2) \cup (-1, 1)$

21. $(-8, -2) \cup (2, \infty)$

23. $(-\infty, -2) \cup (-2, 3)$

25. $(0, 3)$

27. $(-2, -1) \cup (1, \infty)$

29. $(-\infty, -1) \cup \left[-\dfrac{1}{2}, 0\right)$

31. a. $(-4, 0) \cup (1, \infty)$

    b. $(-\infty, -4) \cup (0, 1)$

    c. The function is undefined at $x = 0$.

Chapter 4: Exponential and Logarithmic Functions

4.1 EXERCISES

1.

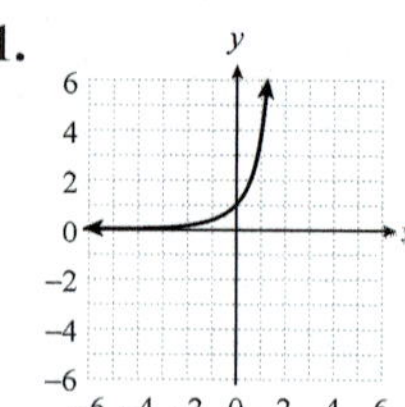

$\text{Dom} = (-\infty, \infty), \text{Ran} = (0, \infty)$

3.

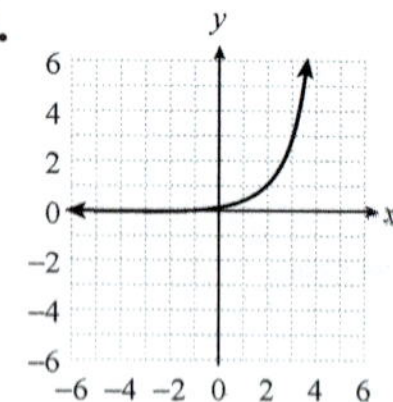

$\text{Dom} = (-\infty, \infty), \text{Ran} = (0, \infty)$

5.

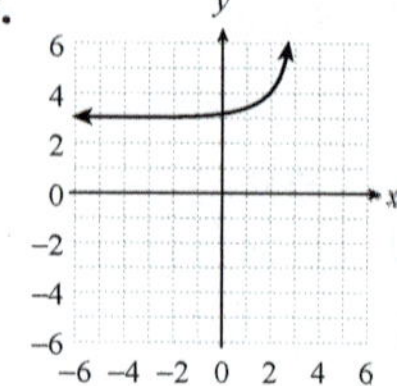

$\text{Dom} = (-\infty, \infty), \text{Ran} = (3, \infty)$

7.

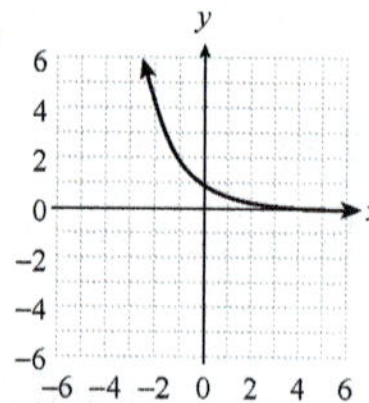

$\text{Dom} = (-\infty, \infty), \text{Ran} = (0, \infty)$

9.

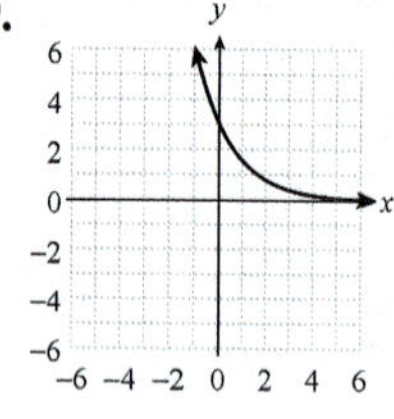

$\text{Dom} = (-\infty, \infty), \text{Ran} = (0, \infty)$

11.

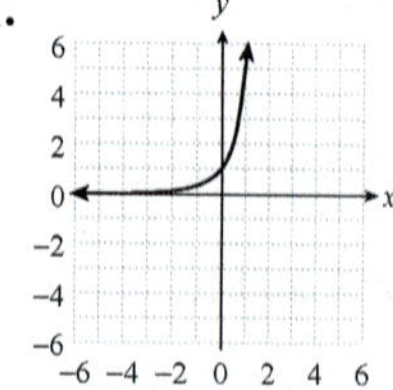

$\text{Dom} = (-\infty, \infty), \text{Ran} = (0, \infty)$

13.

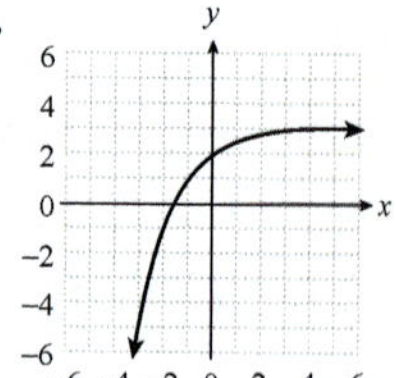

$\text{Dom} = (-\infty, \infty), \text{Ran} = (-\infty, 3)$

15.

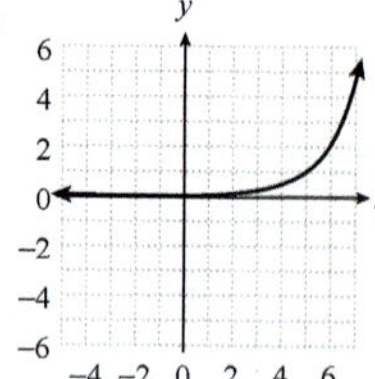

$\text{Dom} = (-\infty, \infty), \text{Ran} = (0, \infty)$

17.

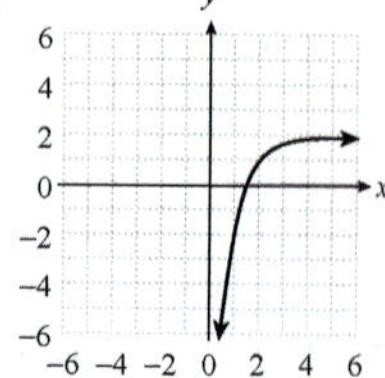

$\text{Dom} = (-\infty, \infty), \text{Ran} = (-\infty, 2)$

19.

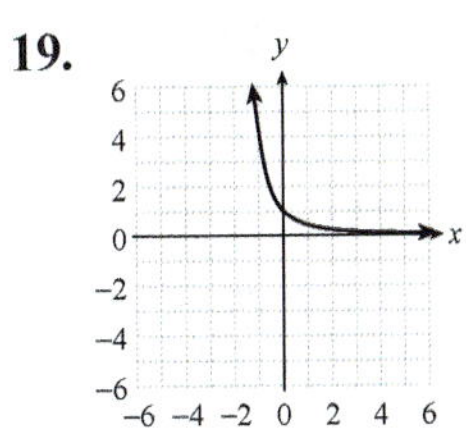

$\text{Dom} = (-\infty, \infty), \text{Ran} = (0, \infty)$

21.

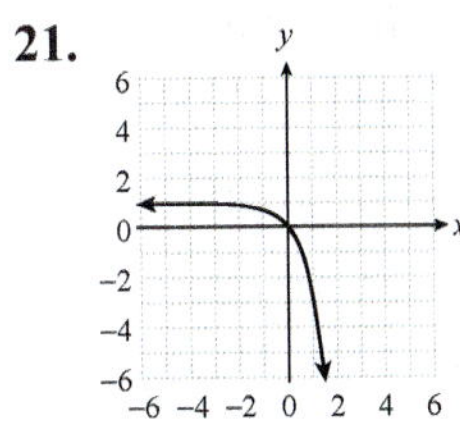

$\text{Dom} = (-\infty, \infty), \text{Ran} = (-\infty, 1)$

23. $\{2\}$

25. $\{-2\}$

27. $\{-13\}$

29. $\{3\}$

31. $\{-2\}$

33. $\{-2, -1\}$

35. $\{7\}$

37. $\{3\}$

39. $\{9\}$

41. $\{-3\}$

43. $\{2\}$

45. $\{-1\}$

47. a

49. i

51. d

53. e

55. h

4.2 EXERCISES

1. $V \approx 178$ people

3. $C \approx \$8526.20$

5. $W \approx 93$ computers

7. a. $a \approx 0.999567$

  b. $A \approx 0.958$ grams

  c. $A \approx 0.648$ grams

9. a. 3 years

  b. 9 years

11. 1118 rabbits

13. The bank offering 2.75% and monthly compounding.

15. Approximately 3.18%

17. \$134,392

19. a. 10

  b. 7490 people

  c. The function approaches 10,000 as time goes on.

21. a. $a \approx 0.965936$

  b. $A \approx 0.707$ kg

  c. $A \approx 7.628$ mg

23. a. \$1521.74 **b.** \$271.74

25. \$9459.48; \$9942.41

27. \$2835.71 **29.** \$20,000

31. \$7318.71

33. a. \$7647.95 **b.** \$7647.57

  c. Yes; daily compounding is a frequency close enough to continuous compounding to make little difference at the hundredths place.

4.3 EXERCISES

1. $4 = \log_5 625$ **3.** $3 = \log_x 27$

5. $3 = \log_{4.2} C$ **7.** $x = \log_4 31$

9. $\sqrt{3} = \log_{4x} 13$ **11.** $e^x = \log_2 11$

13. $81 = 3^4$ **15.** $4 = b^{\frac{1}{2}}$

17. $15 = 2^b$ **19.** $W = 5^{12}$

21. $2x = \pi^4$ **23.** $e^x = 2$

25.

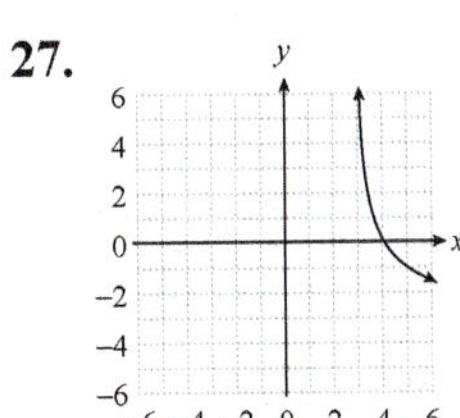

$\text{Dom: } (1, \infty), \text{Ran: } (-\infty, \infty)$

27.

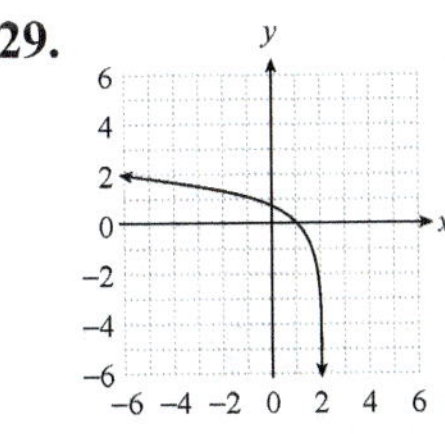

$\text{Dom: } (3, \infty), \text{Ran: } (-\infty, \infty)$

29.

$\text{Dom: } (-\infty, 2), \text{Ran: } (-\infty, \infty)$

31.

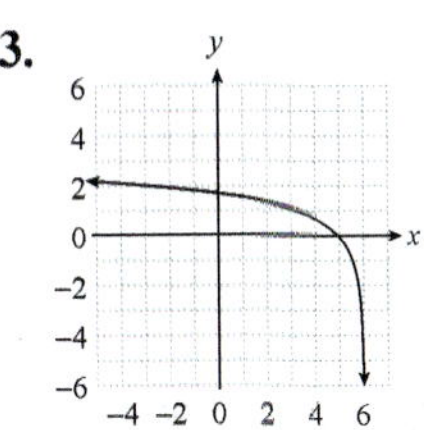

$\text{Dom: } (3, \infty), \text{Ran: } (-\infty, \infty)$

33.

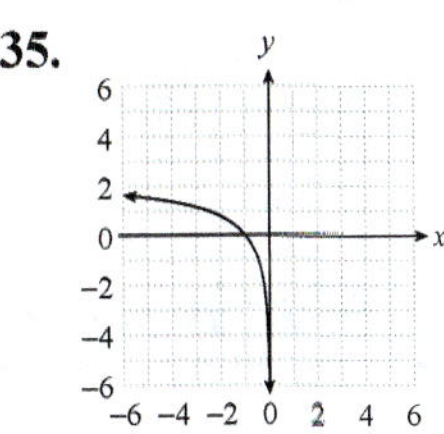

$\text{Dom: } (-\infty, 6), \text{Ran: } (-\infty, \infty)$

35.

$\text{Dom: } (-\infty, 0), \text{Ran: } (-\infty, \infty)$

37. e

39. b

41. h

43. d

45. i

47. -2

49. 3

51. $-\dfrac{1}{2}$

53. $\dfrac{3}{4}$

55. 2.89

57. $\dfrac{5}{3}$

59. 1

61. $\dfrac{1}{5}$

63. 2

65. $\{64\}$

67. $\{9\}$

69. $\left\{-\dfrac{1}{2}\right\}$

71. $\left\{\dfrac{1}{21}\right\}$

73. $\{10\}$

75. $\{36\}$

77. $\left\{\pm\dfrac{1}{10}\right\}$

79. $\{0.18\}$

81. $\{\pm\sqrt{e}\}$ or $\{\pm 1.65\}$

83. $\{12.89\}$

85. $\{10,000,000,002\}$

4.4 EXERCISES

1. $3 + 3\log_5 x$

3. $2 + \ln p - 3\ln q$

5. $1 + \log_9 x - 3\log_9 y$

7. $\dfrac{3}{2}\ln x + \ln p + 5\ln q - 7$

9. $\log(2 + 3\log x)$

11. $1 - \dfrac{1}{2}\log(x + y)$

13. $\log_2(y^2 + z) - 4\log_2 x - 4$

15. $2\log_b x + \dfrac{1}{2}\log_b y - \log_b z$

17. $2 + \log_b a + b\log_b c$

19. $\log\dfrac{x}{y}$ **21.** $\log_5(x + 5)$

23. $\log_2\left(x^{\frac{4}{3}} + 3x^{\frac{1}{3}}\right)$

25. $\ln\left(\dfrac{3p}{q^2}\right)$ **27.** $\log\left(\dfrac{x - 10}{x}\right)$

29. $\ln\left(\dfrac{z^2}{x^3 y^3}\right)$ **31.** $\log_5 4$

33. $\ln 45$ **35.** $\log_3 1 = 0$

37. $\ln 12$ **39.** $\log 11$

41. $\log_8(x^2 - y)$ **43.** x^2

45. $\dfrac{e^2 p}{x}$ **47.** $\dfrac{x^3}{y^4}$

49. x^2 **51.** 4

53. $12x^2$ **55.** 2.04

57. 0.95 **59.** 0.95

61. 2.45 **63.** 3.30

65. 0.74 **67.** 1.20

69. 1.86 **71.** -1

73. 3.85 **75.** 0.77

77. -1.76 **79.** 2

81. 7 **83.** 1

85. $4\sqrt{2} \approx 5.66$ **87.** 9.05

89. 2.08 **91.** 12

93. $1{,}048{,}576$ **95.** 10.25

97. $5{,}011{,}872$ times stronger

99. 133 decibels

101. 7.62; yes

103. a. 15.05 minutes
     b. 7:00 p.m.   **c.** $112\ °F$; no

Chapter 5: Mathematics of Finance

5.1 EXERCISES

1. $7200

3. Taxes are $182.24/week; max. car payment is $292.64/month

5. $3890/month

7. $11,700   **9.** $189.88

11. $117.06   **13.** $164,285.71

15. $312.50   **17.** $6125

5.2 EXERCISES

1. $340   **3.** $585

5. a. $4786.91   **b.** $2286.91

7. a. $4788.57   **b.** $2288.57

9. a. $4692.84   **b.** $2192.84

11. a. $41,425.07   **b.** $28,925.07

13. a. $23,485.22   **b.** $8485.22

15. a. $93,940.86   **b.** $33,940.86

17. $1653.71

19. a. $111,624.25   **b.** $111,945.38

21. a. $6000   **b.** $7373.07
     c. $7358.78

23. a. $1770.36   **b.** $1827.69
     c. $1831.87   **d.** $1832.95
     e. $1833.13

25. a. $3000   **b.** $6000
     c. $12,000   **d.** $24,000

27.

First Bank of Lending Loan APY	
Loan Amount	APY
$< $20{,}000$	11.73%
$20,000–$99,000	9.30%
$\geq $100{,}000$	5.88%

5.3 EXERCISES

1. $11,263.09   **3.** $34,233.00

5. $782.48   **7.** $773.91

9. a. $17.91   **b.** $39.39
     c. $91.19

11. $81,466.12

13. $921.76

15. $978.28

17. a. $257.89
     b. $46,420.20; $I = $28,579.80

19. a. $1,474,258.37
     b. Blake deposited $232,200 and made $1,242,058.37 in interest.

5.4 EXERCISES

1. a. $641.73   **b.** $7700.76

3. a. $274.63   **b.** $9886.68

5. a. $261.70   **b.** $6280.80

7. a. $1240   **b.** $4464

9. 0.9% financing for 48 months will cost $29,492.16 total, and the cash back with 4.75% APR for 48 months will cost $30,752.16 total. The 0.9% financing is the best option.

11. a. $2599.60   **b.** $124,780.80
     c. $10,780.80

13. a. $125.36   **b.** $3008.64
     c. $358.64

15. a. $1266.71   **b.** $206,015.60
     c. $456,015.60   **d.** $1912.48
     e. $94,246.40   **f.** $344,246.40

17. The five-year loan with an APR of 7.5% ($400.76 monthly payment, $4045.60 in interest paid)

19. 20.12 months, so 21 months, or about 1.7 years

21. a. $169,024.96   **b.** $247,502.21

Chapter 6: Systems of Linear Equations; Matrices

6.1 EXERCISES

1. $(-5,2)$ **3.** $(5,3)$

5. $\varnothing$

7. $\left\{\left(\dfrac{y-3}{2},y\right)\middle|y\in\mathbb{R}\right\}$

9. $(-1,7)$ **11.** $(3,11)$

13. $\{(x,4x+1)\,|\,x\in\mathbb{R}\}$

15. $(2,19)$ **17.** $(-5,1)$

19. $(5,6)$

21. $\{(-y-2,y)\,|\,y\in\mathbb{R}\}$

23. $(-5,4)$ **25.** $(-1,1)$

27. $(3,-5)$ **29.** $\varnothing$

31. $(-1,3,0)$ **33.** $(2,2,-1)$

35. $\left\{\left(\dfrac{y-z+2}{3},y,z\right)\middle|y\in\mathbb{R},z\in\mathbb{R}\right\}$

37. $\varnothing$

39. $(1,1,0)$ **41.** $(9,1,1)$

43. $(3,1,-2)$ **45.** $(4,5,5)$

47. $\left(\dfrac{49}{3},\dfrac{-16}{3},\dfrac{5}{4}\right)$

49. $(0,3,2)$

51. 22 pennies, 23 nickels

53. 25 people

55. Eliza is 15 years old.

57. 7 shirts and 4 pairs of shorts

59. 3 quarters, 11 dimes, and 28 pennies

61. Jim is 28 years old.

63. 3 thumb screws

65. Apples: \$0.78, Oranges: \$0.93, Mangos: \$1.05

67. $(0.43,1.28,3.64)$

69. $(-3.42,2.98,2.76)$

71. $(6,8,7)$

6.2 EXERCISES

1. a. 3×2 **b.** -1
  c. None

3. a. 5×2 **b.** None
  c. 10

5. a. 3×4 **b.** None
  c. 286

7. a. 3×2 **b.** 1
  c. None

9. a. 2×5 **b.** 5
  c. 2

11. $\begin{bmatrix} -3 & 1 & -2 & | & -4 \\ \frac{1}{2} & -4 & -1 & | & 1 \\ 0 & -3 & 3 & | & 1 \end{bmatrix}$

13. $\begin{bmatrix} -\frac{3}{2} & -1 & 0 & | & -1 \\ 2 & 2 & 3 & | & 0 \\ 0 & -1 & 6 & | & 0 \end{bmatrix}$

15. $\begin{bmatrix} \frac{12}{5} & \frac{1}{2} & -\frac{3}{2} & | & \frac{1}{5} \\ 1 & 0 & 3 & | & 1 \\ 5 & 2 & 1 & | & -2 \end{bmatrix}$

17. $\begin{bmatrix} \frac{2}{3} & -\frac{4}{3} & -2 & | & 0 \\ 8 & -2 & 6 & | & 7 \\ 3 & -2 & 0 & | & 0 \end{bmatrix}$

19. $\begin{bmatrix} \frac{1}{2} & -14 & -\frac{1}{4} & | & -8 \\ \frac{1}{5} & -\frac{7}{6} & \frac{1}{4} & | & -3 \\ 5 & -5 & \frac{8}{3} & | & -5 \end{bmatrix}$

21. $\begin{cases} x=8 \\ y=3 \end{cases}$

23. $\begin{cases} x+3y+6z=16 \\ \quad\quad y+2z=9 \\ \quad\quad\quad\quad z=4 \end{cases}$

25. $\begin{cases} 9y+13z=27 \\ 2x+21z=19 \\ 7x+18y=32 \end{cases}$

27. $\begin{bmatrix} 2 & -5 & | & 3 \\ 0 & -7 & | & 5 \end{bmatrix}$

29. $\begin{bmatrix} 1 & 3 & | & -2 \\ 9 & -2 & | & 7 \end{bmatrix}$

31. $\begin{bmatrix} 8 & -2 & | & -4 \\ -6 & 2 & | & -14 \end{bmatrix}$

33. $\begin{bmatrix} 4 & 12 & | & -6 \\ 9 & 9 & | & 6 \end{bmatrix}$

35. $\begin{bmatrix} 4 & -1 & | & 5 \\ -6 & 2 & | & 0 \end{bmatrix}$

37. $\begin{bmatrix} 18 & -6 & 15 & | & 42 \\ -7 & 19 & 2 & | & 3 \\ -4.5 & 5.5 & -2 & | & 3.5 \end{bmatrix}$

39. $\begin{bmatrix} 5 & 18 & 22 & | & 5 \\ 32 & -9 & -27 & | & -23 \\ -9 & 21 & 12 & | & 9 \end{bmatrix}$

41. $\begin{bmatrix} 0 & 1 & -9 & | & -3 \\ 1 & 1 & 3 & | & 4 \\ 0 & 0 & 0 & | & 0 \end{bmatrix}$

43. $\begin{bmatrix} -1 & 4 & | & -3 \\ 1 & -6 & | & \frac{5}{2} \end{bmatrix}$

45. $\begin{bmatrix} 1 & 5 & -9 & | & 11 \\ 0 & -1 & 8 & | & -7 \\ 0 & -17 & 41 & | & 1 \end{bmatrix}$

47. Neither   **49.** Neither

51. Neither   **53.** $(3,-1)$

55. $(1,3)$ **57.** $(-7,3)$

59. $\varnothing$ **61.** $(3,2)$

63. $\{(-2y-4,y)\,|\,y\in\mathbb{R}\}$

65. $\varnothing$ **67.** $(4,0,3)$

69. $(15,-21,8)$ **71.** $(3,-5)$

73. $\{(x,-3x-2)\,|\,x\in\mathbb{R}\}$

75. $(-4,1)$ **77.** $(6,4)$

79. $(-11,-5)$ **81.** $(7,3,3)$

83. $(2,2,-1)$

85. $\{(1,y,0)\,|\,y\in\mathbb{R}\}$

87. $(3,-2,3)$ **89.** $(2,3,4)$

91. $(9,-19,7)$ **93.** $(1,-2,-1,3)$

95. 42, 26, 87

97. Small: 10, Medium: 24, Large: 48

6.3 EXERCISES

1. 11 **3.** 15

5. $ab - x^2$ **7.** 8

9. -10 **11.** -39

13. $\{-2,3\}$ **15.** $\{-5,1\}$

17. $\{-5,-4\}$ **19.** $\{-6,4\}$

21. $\{2,5\}$ **23.** 3

25. -9 **27.** 2

29. -2 **31.** 159

33. 78 **35.** -254

37. 404 **39.** 4

41. 120 **43.** 10

45. x^4 **47.** x^8

49. $(76,-53)$

51. $\{(-y-2, y) \mid y \in \mathbb{R}\}$

53. $\varnothing$

55. $\{(-3z-5, -6z-10, z) \mid z \in \mathbb{R}\}$

57. $\left\{\left(\dfrac{-5y-z-5}{2}, \dfrac{-5y+3z-19}{2}, y, z\right) \,\middle|\, y \in \mathbb{R}, z \in \mathbb{R}\right\}$

59. $\left\{(-z+8, -5z+31, -2z+37, z) \mid z \in \mathbb{R}\right\}$

61. $(1647, 2071)$

63. $\varnothing$

65. $(-3, -1, 0, -4)$

67. Candy bars: 5, Ice cream: 6

69. 0.012 **71.** 0.564

73. 1194 **75.** $(1, -1, 2)$

77. $(2, 1, 0, 3)$

6.4 EXERCISES

1. $\begin{bmatrix} 5 & -1 \\ 0 & 0 \\ 2 & 13 \end{bmatrix}$ **3.** $\begin{bmatrix} 6 & -3 \\ 18 & 30 \\ -9 & 21 \end{bmatrix}$

5. Not possible

7. $\begin{bmatrix} 14 & -14 \\ 8 & 0 \\ -4 & 14 \end{bmatrix}$ **9.** $\begin{bmatrix} -7 & 5 \\ 3 & 10 \\ -3 & -8 \end{bmatrix}$

11. Not possible

13. $a = 3$, $b = -1$, $c = 10$

15. $a = 2$, $b = -2$, $c = -1$

17. Not possible

19. $x = 10$, $y = 5$

21. $x = 3$, $y = 1$

23. Not possible

25. $a = 8$, $b = 5$

27. $\begin{bmatrix} 24 & -5 \end{bmatrix}$

29. $\begin{bmatrix} 35 & 18 \end{bmatrix}$

31. Not possible

33. $\begin{bmatrix} -30 & -3 \end{bmatrix}$

35. $\begin{bmatrix} 15 & -3 & -24 \\ 25 & -5 & -40 \\ 30 & -6 & -48 \end{bmatrix}$

37. $\begin{bmatrix} -34 & -7 \end{bmatrix}$

39. $\begin{bmatrix} 11 & 0 \\ 0 & 11 \end{bmatrix}$

41. $\begin{bmatrix} 32 & -20 \\ 56 & -35 \\ -16 & 10 \end{bmatrix}$

43. Not possible

45. $\begin{bmatrix} 14 & -13 \\ -13 & 5 \end{bmatrix}$

47. $\begin{bmatrix} 179 & 76 \end{bmatrix}$

49. $\dfrac{2}{3}$ for store A; $\dfrac{1}{3}$ for store B

51. Solution is incorrect. Explanations may vary.

53. $\begin{bmatrix} 23.94 & -7.56 & 28.98 \\ 21.66 & -6.84 & 26.22 \end{bmatrix}$

55. $\begin{bmatrix} -23.94 & -26.72 \end{bmatrix}$

57. $\begin{bmatrix} -79.59 \\ 39.21 \\ 10.08 \end{bmatrix}$

6.5 EXERCISES

1. $\begin{bmatrix} 14 & -5 \\ 1 & 9 \end{bmatrix}\begin{bmatrix} x \\ y \end{bmatrix} = \begin{bmatrix} 7 \\ 2 \end{bmatrix}$

3. $\begin{bmatrix} 1 & 2 \\ 9 & -3 \end{bmatrix}\begin{bmatrix} x \\ y \end{bmatrix} = \begin{bmatrix} -6 \\ -14 \end{bmatrix}$

5. $\begin{bmatrix} 3 & -7 & 1 \\ 1 & -1 & 0 \\ 0 & 8 & 5 \end{bmatrix}\begin{bmatrix} x_1 \\ x_2 \\ x_3 \end{bmatrix} = \begin{bmatrix} -4 \\ 2 \\ -3 \end{bmatrix}$

7. $\begin{bmatrix} \dfrac{3}{5} & -\dfrac{8}{5} \\ 0 & 1 \end{bmatrix}\begin{bmatrix} x \\ y \end{bmatrix} = \begin{bmatrix} 2 \\ 2 \end{bmatrix}$

9. $\begin{bmatrix} 4 & -3 \\ 2 & -4 \end{bmatrix}\begin{bmatrix} x \\ y \end{bmatrix} = \begin{bmatrix} -9 \\ 13 \end{bmatrix}$

11. $\begin{bmatrix} 2 & -1 & 3 \\ -1 & 1 & 0 \\ 4 & -5 & 1 \end{bmatrix}\begin{bmatrix} x \\ y \\ z \end{bmatrix} = \begin{bmatrix} 0 \\ 17 \\ -2 \end{bmatrix}$

13. $\begin{bmatrix} -\dfrac{1}{20} & -\dfrac{1}{5} \\ \dfrac{1}{4} & 0 \end{bmatrix}$

15. $\begin{bmatrix} -5 & -4 \\ 4 & 3 \end{bmatrix}$

17. $\begin{bmatrix} -5 & 0 \\ 2 & 2 \end{bmatrix}$

19. Not invertible

21. $\begin{bmatrix} 2 & 1 & -4 \\ -4 & -2 & -3 \\ -1 & -1 & -4 \end{bmatrix}$

23. $\begin{bmatrix} -1 & 2 & -1 \\ 0 & -1 & 1 \\ 0 & -4 & 3 \end{bmatrix}$

25. $\begin{bmatrix} -1 & -2 & 1 \\ -2 & 1 & -3 \\ 1 & 2 & 0 \end{bmatrix}$

27. $\begin{bmatrix} -2 & 1 & 1 \\ 2 & 0 & -1 \\ -1 & 0 & 1 \end{bmatrix}$

29. $\begin{bmatrix} 2 & -1 & 2 \\ 0 & 1 & -1 \\ -3 & -2 & -4 \end{bmatrix}$

31. No **33.** Yes

35. No **37.** $\left(-2, -\dfrac{5}{2}\right)$

39. $\left\{\left(\dfrac{3y-1}{2}, y\right) \,\middle|\, y \in \mathbb{R}\right\}$

41. $(-2, 0)$ **43.** $(8, -19)$

45. $(0,5)$ **47.** $(-4,5,-1)$

49. $(-13,19,23);\ (0,0,-1);\ (1,-1,-1)$

51. $(1,-8,7);\ (3,1,1);\ (4,2,0)$

53. $\begin{bmatrix} \dfrac{-1}{4} & \dfrac{1}{10} \\[2mm] \dfrac{-1}{4} & \dfrac{3}{10} \end{bmatrix}$

55. $\begin{bmatrix} 0.053 & -0.258 \\ 0.113 & 0.076 \end{bmatrix}$

57. $\begin{bmatrix} 0.004 & -0.003 & 0.009 \\ 0 & 0.020 & 0.029 \\ 0.012 & 0.014 & 0.013 \end{bmatrix}$

6.6 EXERCISES

1. a. 0.1 units agriculture,
0.6 units manufacturing

b. 2.1 units agriculture,
0.6 units manufacturing

c. $\begin{bmatrix} 0.3 & -0.1 \\ -0.2 & 0.4 \end{bmatrix},\ \begin{bmatrix} 4.0 & 1.0 \\ 2.0 & 3.0 \end{bmatrix}$

d. $\begin{bmatrix} 37,000 \\ 31,000 \end{bmatrix}$

3. a. 0.1 units agriculture,
0.1 units mining,
0.1 units manufacturing

b. 0.4 units agriculture,
0.4 units mining,
0.4 units manufacturing

c. $\begin{bmatrix} 0.9 & -0.2 & -0.3 \\ -0.1 & 0.8 & -0.3 \\ -0.1 & -0.2 & 0.7 \end{bmatrix},$

$\begin{bmatrix} 1.25 & 0.5 & 0.75 \\ 0.25 & 1.5 & 0.75 \\ 0.25 & 0.5 & 1.75 \end{bmatrix}$

d. $\begin{bmatrix} 1700 \\ 2500 \\ 2100 \end{bmatrix}$

5. $\begin{bmatrix} 900 \\ 600 \end{bmatrix}$ **7.** $\begin{bmatrix} 240 \\ 420 \\ 140 \end{bmatrix}$

9. \$800 manufacturing, \$1400 agriculture

11. \$800 computer manufacturing, \$1600 automobile manufacturing

13. \$4750 museums, \$6875 theater, \$4875 sporting events

15. \$2375 curative, \$4125 rehabilitative, \$2875 preventative

17. \$3125 manufacturing, \$2250 agriculture, \$3375 health, \$3750 energy

19. \$180 agriculture, \$80 manufacturing, \$480 energy

21. a. $x_1 = \dfrac{3}{8} x_2$

b. Agriculture needs to produce \$3 for every \$8 of manufacturing produced.

23. a. $x_1 = \dfrac{22}{37} x_3,\ x_2 = \dfrac{43}{37} x_3$

b. For every \$37 of energy produced, agriculture needs to produce \$22 and manufacturing needs to produce \$43.

Chapter 7: Inequalities and Linear Programming

7.1 EXERCISES

1. 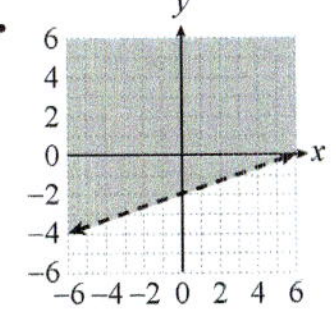

3. 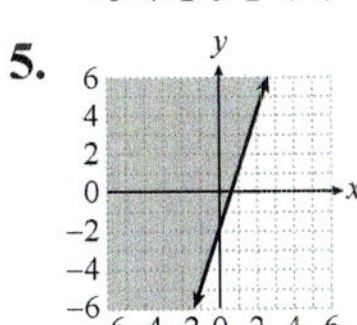

5. 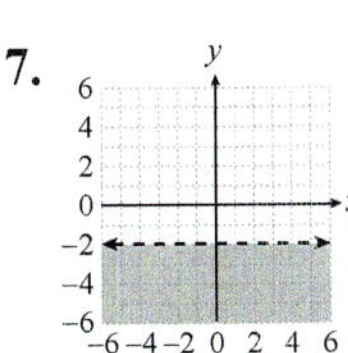

7.

9. 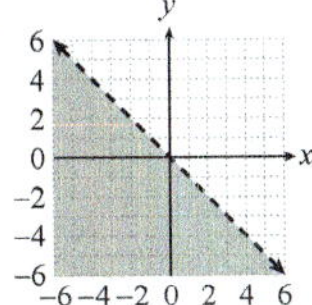

11.

13.

15. 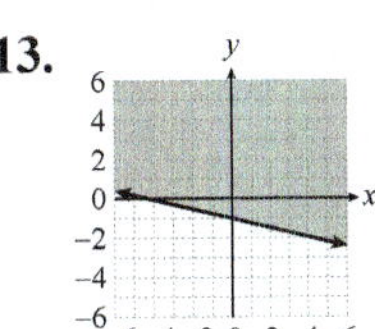

17. 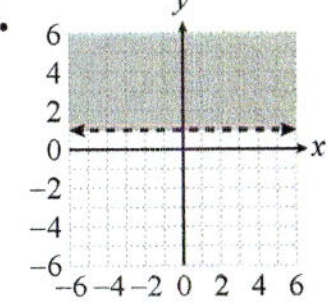

19. 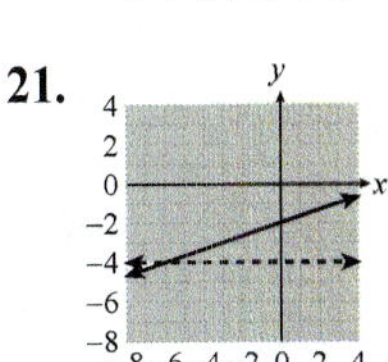

21.

23. 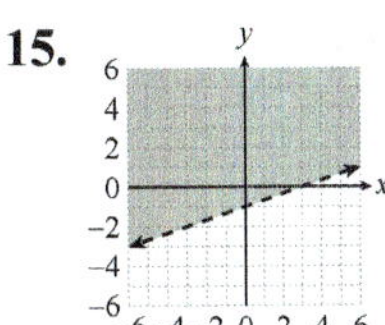

25. 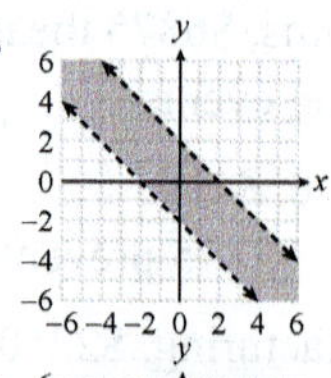

27. 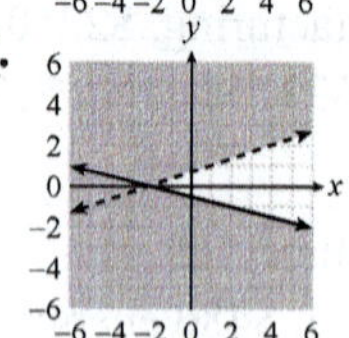

29. 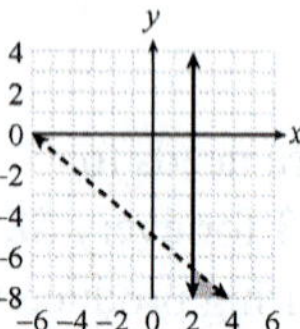

31. 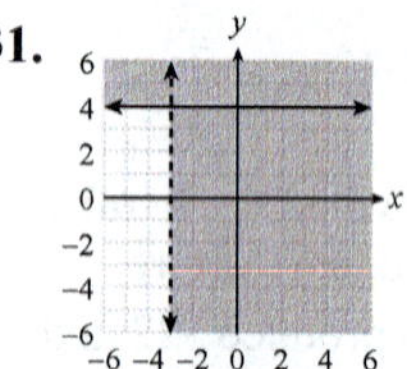

33. 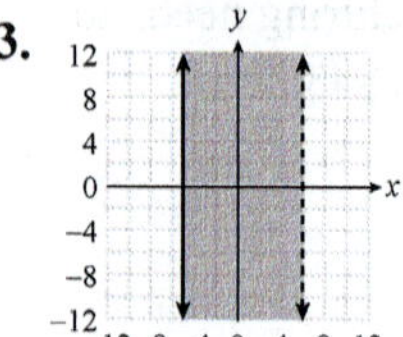

35. 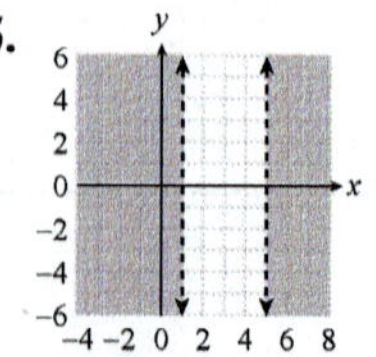

37. 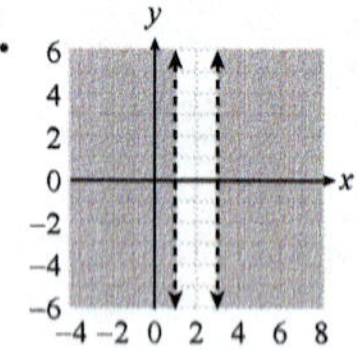

39. 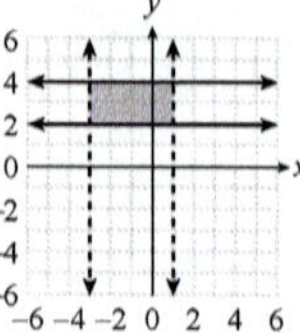

41. 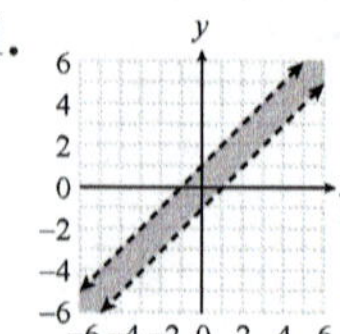

43. 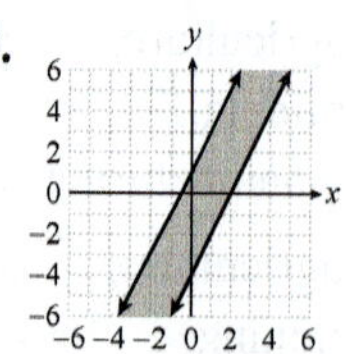

45. 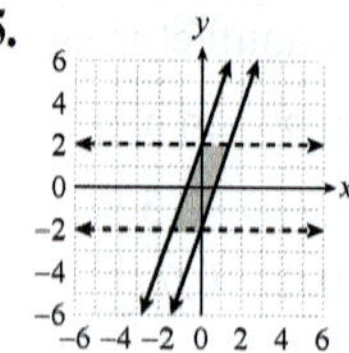

47. h   **49.** b

51. g   **53.** c

55. 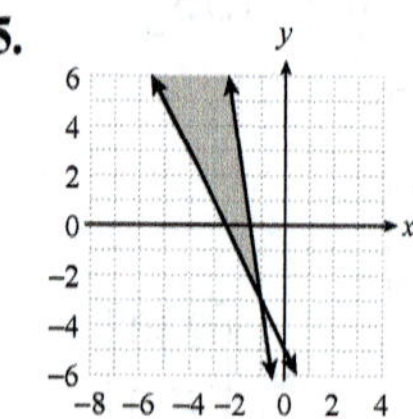

57. 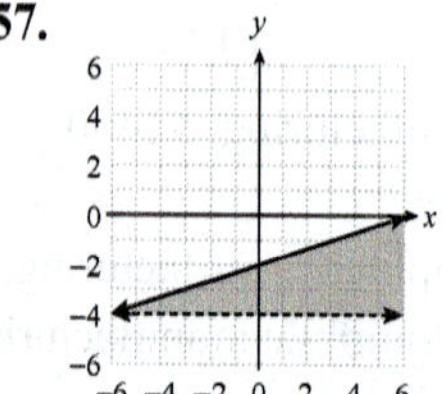

59. 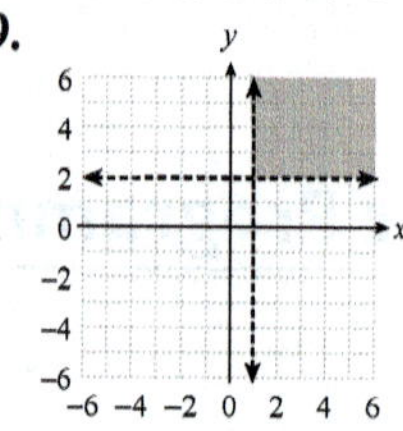

61. 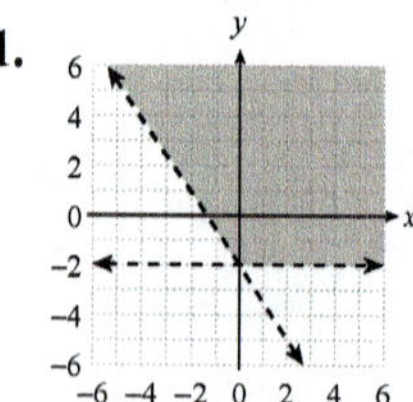

63. 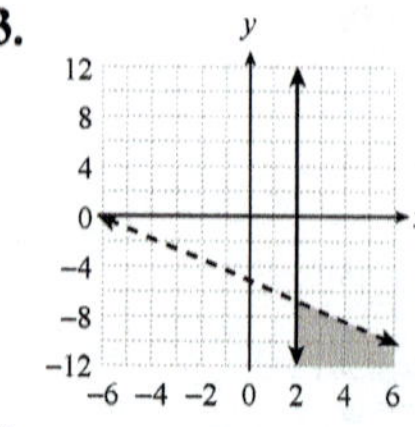

65. 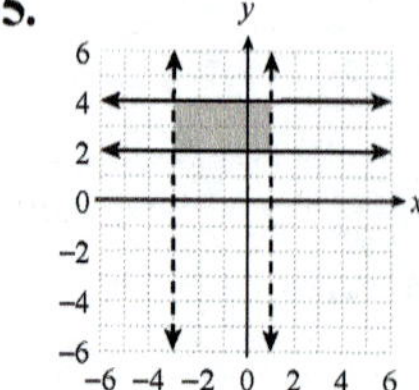

67. $5.5x < 200$

69. $x + y \leq 150$

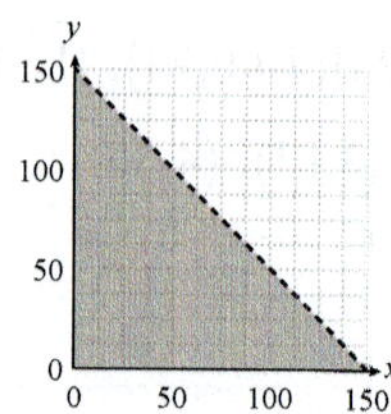

7.2 EXERCISES

1. Min $= 0$ at $(0,0)$;

Max $= 21$ at $(0,7)$

3. Min $= 0$ at $(0,0)$;

Max $= 35$ at $(0,7)$

5. Min $= 25$ at $(5,0)$;

Max $= 95$ at $(7,10)$

7. Min $= \dfrac{40}{3}$ at $\left(0, \dfrac{10}{3}\right)$;

Max $= 44$ at $(4,5)$

9. Min $= 0$ at $(0,0)$;

Max $= 50$ at $(3,4)$

11. Min $= 6$ at $(0,6)$;

Max $= 96$ at $(16,0)$

13. Min $= \dfrac{172}{11}$ at $\left(\dfrac{24}{11}, \dfrac{10}{11}\right)$;

Max $= \dfrac{400}{3}$ at $\left(0, \dfrac{40}{3}\right)$

15. 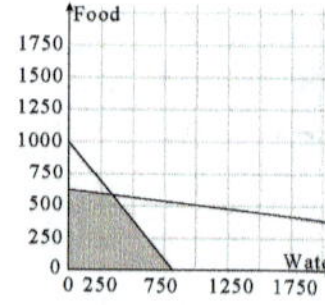

$60x + 50y \leq 50{,}000$;

$x + 10y \leq 6000$;

$x \geq 0;\ y \geq 0$

17. 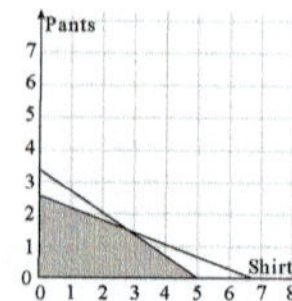

$12x + 32y \leq 80$;

$2x + 3y \leq 10$;

$x \geq 0;\ y \geq 0$

19. Model X: 1000 units;

Model Y: 500 units;

Maximum profit: \$76,000

21. Ashley should buy 3 olive bundles and 2 cranberry bundles. The total cost to make the curtains is \$70.

23. Municipal bonds: \$20,000; Treasury bills: \$5000

7.3 EXERCISES

1. a. s_1, s_2 basic; x_1, x_2 nonbasic

  b. $(x_1, x_2, s_1, s_2, z) = (0, 0, 6, 4, 0)$

  c. 3

  d.
$$\begin{array}{ccccc} x_1 & x_2 & s_1 & s_2 & z \\ \end{array}$$
$$\left[\begin{array}{ccccc|c} \boxed{3} & 1 & 1 & 0 & 0 & 6 \\ 1 & 2 & 0 & 1 & 0 & 4 \\ \hline -7 & -5 & 0 & 0 & 1 & 0 \end{array}\right]$$

$$\xrightarrow{\frac{1}{3}R_1}$$
$$\xrightarrow{-R_1 + R_2}$$
$$\xrightarrow{7R_1 + R_3}$$

$$\begin{array}{ccccc} x_1 & x_2 & s_1 & s_2 & z \\ \end{array}$$
$$\left[\begin{array}{ccccc|c} 1 & \dfrac{1}{3} & \dfrac{1}{3} & 0 & 0 & 2 \\[2mm] 0 & \dfrac{5}{3} & -\dfrac{1}{3} & 1 & 0 & 2 \\[2mm] \hline 0 & -\dfrac{8}{3} & \dfrac{7}{3} & 0 & 1 & 14 \end{array}\right]$$

3. a. x_2, s_3 basic; x_1, s_1, s_2 nonbasic

  b. $(x_1, x_2, s_1, s_2, s_3, z)$
  $= (0, 5, 0, 0, 3, 35)$

  c. 3 (row 2, col 3)

  d.
$$\begin{array}{cccccc} x_1 & x_2 & x_3 & s_1 & s_2 & z \\ \end{array}$$
$$\left[\begin{array}{cccccc|c} 1 & 1 & 1 & 1 & 0 & 0 & 5 \\ 3 & 0 & \boxed{3} & -1 & 1 & 0 & 3 \\ \hline -3 & 0 & -11 & 7 & 0 & 1 & 35 \end{array}\right]$$

$$\xrightarrow{\frac{1}{3}R_2}$$
$$\xrightarrow{-R_2 + R_1}$$
$$\xrightarrow{11R_2 + R_3}$$

$$\begin{array}{cccccc} x_1 & x_2 & x_3 & s_1 & s_2 & z \\ \end{array}$$
$$\left[\begin{array}{cccccc|c} 0 & 1 & 0 & \dfrac{4}{3} & -\dfrac{1}{3} & 0 & 4 \\[2mm] 1 & 0 & 1 & -\dfrac{1}{3} & \dfrac{1}{3} & 0 & 1 \\[2mm] \hline 8 & 0 & 0 & \dfrac{10}{3} & \dfrac{11}{3} & 1 & 46 \end{array}\right]$$

5. max of 38 at $(x_1, x_2) = (4, 3)$

7. max of 16 at multiple points; answers vary: e.g. $(x_1, x_2) = (0, 2)$ and $(x_1, x_2) = \left(\dfrac{8}{5}, \dfrac{6}{5}\right)$

9. max of 48 at $(x_1, x_2, x_3) = (3, 2, 7)$

11. max of 286 at $(x_1, x_2, x_3) = (3, 40, 17)$

13. 70 drills, 30 tables saws; max of \$750

15. 16 pairs A, 8 pairs B; max of \$88

17. 50 Type A, 120 Type B; max of \$8,970,000 (26,250 worker-days of labor)

19. 25 one-bedroom, 50 two-bedroom, 25 three-bedroom; max of \$85,000

21. 20 bookshelves, 15 bed frames, 15 recliners; max of \$21,250

23. 20 newspaper, 14 radio; max of 300,000 women under 24

25. 30 male, 7 female; max of 37 rabbits

27. 3 onions, 3 turnips, 4 radishes; max of 10 plants

29. No; The line $2x_1 + 5x_2 = 80$ doesn't form any boundary of the feasible region, and thus has no candidates for finding a maximum.

31. a. There is a negative entry in the bottom row and $\dfrac{\frac{6840}{7}}{\frac{18}{7}} = 380$ is smaller than $\dfrac{\frac{4800}{7}}{\frac{6}{7}} = 800$ and $\dfrac{\frac{2700}{7}}{\frac{6}{7}} = 450$.

  b. No; It appears the optimal feasible solution is
  $(x_1, x_2, x_3, s_1, s_2, s_3, z)$
  $= (380, 0, 300, 0, 0, 0, 1146),$

  but
  $$f(380, 0, 300)$$
  $$= \frac{3}{2}(380) + \frac{6}{5}(0) + \frac{8}{5}(300)$$
  $$= 1050$$
  is not optimal.

  c. 2; Yes, the number of basic variables doesn't match the number of constraints.

  d. Yes, performing $R_2 + R_1$ will change row 1, column 5 from -1 to 0, and s_2 will become a third basic variable. This yields our desired optimal feasible solution.

$$\begin{array}{ccccccc} x_1 & x_2 & x_3 & s_1 & s_2 & s_3 & z \\ \end{array}$$
$$\left[\begin{array}{ccccccc|c} 0 & 1 & 1 & 2 & -1 & 0 & 0 & 300 \\[2mm] 0 & \dfrac{2}{3} & 0 & -\dfrac{2}{3} & 1 & -\dfrac{1}{3} & 0 & 60 \\[2mm] 1 & -\dfrac{11}{18} & 0 & -\dfrac{8}{9} & 0 & \dfrac{7}{18} & 0 & 380 \\[2mm] \hline 0 & \dfrac{11}{20} & 0 & \dfrac{4}{5} & 0 & \dfrac{1}{20} & 1 & 1146 \end{array}\right]$$

$$\xrightarrow{R_2 + R_1}$$

$$\begin{array}{ccccccc} x_1 & x_2 & x_3 & s_1 & s_2 & s_3 & z \\ \end{array}$$
$$\left[\begin{array}{ccccccc|c} 0 & \dfrac{5}{3} & 1 & \dfrac{4}{3} & 0 & -\dfrac{1}{3} & 0 & 360 \\[2mm] 0 & \dfrac{2}{3} & 0 & -\dfrac{2}{3} & 1 & -\dfrac{1}{3} & 0 & 60 \\[2mm] 0 & -\dfrac{11}{18} & 0 & -\dfrac{8}{9} & 0 & \dfrac{7}{18} & 0 & 380 \\[2mm] \hline 0 & \dfrac{11}{20} & 0 & \dfrac{4}{5} & 0 & \dfrac{1}{20} & 1 & 1146 \end{array}\right]$$

7.4 EXERCISES

1. $A^T = \begin{bmatrix} 2 & 4 \\ 3 & 6 \\ 7 & 2 \end{bmatrix}$

3. $A^T = \begin{bmatrix} 1 & 6 & 7 & -1 \\ 3 & -5 & 2 & 0 \end{bmatrix}$

5. Maximize
$g(y_1, y_2) = 18y_1 + 15y_2$ subject to
constraints $\begin{cases} 3y_1 + 2y_2 \le 7 \\ 5y_1 + 6y_2 \le 2 \end{cases}.$

7. Maximize
$g(y_1, y_2) = 16y_1 + 20y_2$ subject
to constraints $\begin{cases} 2y_1 + 5y_2 \le 3 \\ 3y_1 \le 1 \\ 4y_2 \le 2 \end{cases}.$

9. min of 5 at $(x_1, x_2) = (0, 5)$

11. min of $\dfrac{17}{3}$ at

$$(x_1, x_2, x_3) = \left(5, \frac{1}{3}, 0\right)$$

13. min of 31 at

$$(x_1, x_2, x_3) = \left(\frac{3}{8}, \frac{15}{8}, \frac{11}{8}\right)$$

15. min of 33 at $(x_1, x_2, x_3) = (0, 11, 0)$

17. 25 hours plant A, 10 hours plant B; min of \$45,000

19. 1 bottle Drink I, 2 bottles Drink II; min of \$32

21. 7 days refinery A, 6 days refinery B, 8 days refinery C; min of \$675,000

23. 7 grams food I, 4.5 grams food II; min of 20.5 units of ingredient C

25. 55 minutes radio, 35 minutes television; min of \$19,875

27. 2 servings meal I, 1 serving meal II, 3 servings meal III; min of 1.9 ounces of detrimental substance

29. a. The dual maximization problem has no solution.

b.

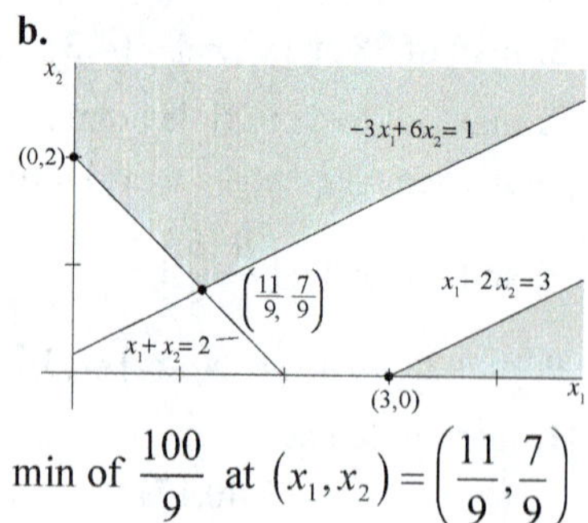

min of $\dfrac{100}{9}$ at $(x_1, x_2) = \left(\dfrac{11}{9}, \dfrac{7}{9}\right)$

7.5 EXERCISES

1. $\begin{cases} 5x_1 + x_2 + s_1 = 15 \\ 3x_1 + 2x_2 - s_2 = 7 \end{cases}$

3. $\begin{cases} 8x_1 + 3x_2 - x_3 - s_2 = 10 \\ 6x_2 - x_3 - s_3 = 7 \\ x_1 + x_2 + x_3 + s_1 = 8 \end{cases}$

5. $\begin{cases} 4x_1 - x_3 - s_3 = 9 \\ x_2 + 2x_3 + s_1 = 13 \\ 5x_1 - x_2 + 4x_3 - s_4 = 14 \\ 3x_1 + x_2 + x_3 + s_2 = 17 \end{cases}$

7. maximize

$$-f(x_1, x_2, x_3) = -4x_1 - 7x_2 - 9x_3;$$

constraints:

$$\begin{cases} x_1 + x_2 + x_3 + s_1 = 9 \\ x_1 - 3x_3 + s_2 = -6 \\ -5x_1 + x_2 + x_3 + s_3 = -7 \end{cases}$$

9. maximize

$$-f(x_1, x_2, x_3) = -4x_1 - 7x_2 - 9x_3;$$

constraints:

$$\begin{cases} x_1 + x_2 + x_3 + s_1 = 9 \\ -x_1 + 3x_3 - s_2 + a_1 = 6 \\ 5x_1 - x_2 - x_3 - s_3 + a_2 = 7 \end{cases};$$

$$z = -4x_1 - 7x_2 - 9x_3 - Ma_1 - Ma_2$$

11. min of 4 at $(x_1, x_2) = (1, 0)$

13. max of 20 at $(x_1, x_2, x_3) = (0, 5, 5)$

15. min of 20 at $(x_1, x_2, x_3) = (1, 2, 3)$

17. max of 56 at $(x_1, x_2, x_3) = (8, 2, 0)$

19. 150 Charlotte to Raleigh, 250 Charlotte to Spartanburg, 200 Charleston to Raleigh, 0 Charleston to Spartanburg; max of \$3150

21. 400 factory A to dealership II, 300 factory B to dealership I, 0 factory A to dealership I, 0 factory B to dealership II; min of \$22,500

23. 2.5 million gallons at plants I and III, none at plant II; min of \$275,000

25. 28 television ads, 5 newspaper ads; max of 57 million voters

27. 30 Tropical Paradise, 45 Arctic Chill, 30 Paradise Blast; max of \$54

29. 52 gallons mix I, 8 gallons mix II, 20 gallons mix III; min of \$4080

31. max of 21 at multiple points, e.g.

$$(x_1, x_2) = \left(0, \frac{7}{3}\right) \text{ and}$$

$$(x_1, x_2) = \left(\frac{8}{5}, \frac{9}{5}\right)$$

33. a. min of 14 at $(x_1, x_2) = (1, 3)$

b. No; the condition $3x_1 + x_2 = 6$ is more restrictive than $3x_1 + x_2 \geq 6$.

Chapter 8: Probability

8.1 EXERCISES

1. False; the set may be an infinite set.

3. True   **5.** True

7. True   **9.** True

11. $B = \{$Nebraska, Nevada, New Hampshire, New Jersey, New Mexico, New York, North Dakota, North Carolina$\}$

13. Answers will vary.

15. $F = \{$Monday, Tuesday, Wednesday, Thursday, Friday$\}$

17. $H = \{x \mid x \in \mathbb{N}, x \leq 50\}$

19. $K = \{x \mid x \in U, x \text{ is an athlete}\}$

21. K is the set of US paper currency that is at most \$100.

23. N is the set of months beginning with the letter J.

25. $B = \{3, 6, 9, 12, 15\}$

27. $D = \{$Wyoming, Nebraska, Kansas, New Mexico, Oklahoma, Utah, Arizona$\}$

29. Yes; they have the same elements, just in a different order.

31. Yes; P, R, and S are all equivalent. They all contain 5 elements. Q is not equivalent to any of them.

33. No; they have different elements.

35. B is equivalent to C since they each have 5 elements. A and D are not equivalent to the other sets.

37. $|X| = 9$

39. $|Y| = 43$, as of printing

41. $A' = \{c, d, f, g, h, i, j, l, m, n, o, p, q, r, u, v, w, x, y, z\}$

43. $A' = \{c, d, f, g, h, i, j, l, m, n, o, p, q, r, u, v, w, x, y, z, A, B, C, D, E, F, G, H, I, J, K, L, M, N, O, P, Q, R, S, T, U, V, W, X, Y, Z\}$

45. 4

47. 6

49. 15

51. $Y = \{$CHN, SWE, ITA, FRA, GER, GBR, RUS, GDR$\}$

53. $G = \{$CHN, ITA, FRA, GER, GBR, URS, USA$\}$

55. No; they have different cardinal numbers, that is, a different number of elements.

57. Answers will vary.

59. Answers will vary. Examples include $U = \{\pi\}$; $U = \mathbb{R}$; $U = \{x \mid x \in \mathbb{R}, x > 0\}$

8.2 EXERCISES

1. $\{1, 2, 3, 4, 5, 6, 7, 8, 10, 12\}$

3. $\varnothing$ or $\{\ \}$

5. $(A \cup B)' = \{9, 11, 13, 14, 15, 16, 17, 18, 19, 20\}$ and $A' = \{9, 10, 11, 12, 13, 14, 15, 16, 17, 18, 19, 20\}$ and $B' = \{1, 3, 5, 7, 9, 11, 13, 14, 15, 16, 17, 18, 19, 20\}$. So $A' \cap B' = \{9, 11, 13, 14, 15, 16, 17, 18, 19, 20\}$.

7. $\{n, u, m, b, e, r, s, l\}$

9. 3

11. $A \cap B = \{r, u, e\}$, $(A \cap B)' = \{a, b, c, d, f, g, h, i, j, k, l, m, n, o, p, q, s, t, v, w, x, y, z\}$, $A' = \{a, c, d, f, g, h, i, j, k, l, o, p, q, t, v, w, x, y, z\}$, $B' = \{a, b, c, d, f, g, h, i, j, k, m, n, o, p, q, s, t, v, w, x, y, z\}$, so $A' \cup B' = \{a, b, c, d, f, g, h, i, j, k, l, m, n, o, p, q, s, t, v, w, x, y, z\}$.

13. $\{C, E\}$

15. $A \cup B = \{I, C, E, U, B\}$, $(A \cup B)' = \{A, D, F, G, H, J, K, L, M, N, O, P, Q, R, S, T, V, W, X, Y, Z\}$, $A' = \{A, B, D, F, G, H, J, K, L, M, N, O, P, Q, R, S, T, U, V, W, X, Y, Z\}$, $B' = \{A, D, F, G, H, I, J, K, L, M, N, O, P, Q, R, S, T, V, W, X, Y, Z\}$, and $A' \cap B' = \{A, D, F, G, H, J, K, L, M, N, O, P, Q, R, S, T, V, W, X, Y, Z\}$

17. $\{A, C, D, F, O, P, R, T, U\}$

19. 4

21. $A \cap B = \{C, O, R, T\}$, $(A \cap B)' = \{A, B, D, E, F, G, H, I, J, K, L, M, N, P, Q, S, U, V, W, X, Y, Z\}$, $A' = \{B, D, E, G, H, I, J, K, L, M, N, P, Q, S, U, V, W, X, Y, Z\}$, $B' = \{A, B, E, F, G, H, I, J, K, L, M, N, Q, S, V, W, X, Y, Z\}$, and $A' \cup B' = \{A, B, D, E, F, G, H, I, J, K, L, M, N, P, Q, S, U, V, W, X, Y, Z\}$

23. $\{g, a, t, o, r, b, i, e\}$

25. $M \cup K = \{b, r, i, d, g, e, a, l, s, t\}$, $(M \cup K)' = \{c, f, h, j, k, m, n, o, p, q, u, v, w, x, y, z\}$, $M' = \{a, c, f, h, j, k, l, m, n, o, p, q, s, t, u, v, w, x, y, z\}$, $K' = \{c, d, f, g, h, j, k, m, n, o, p, q, u, v, w, x, y, z\}$, and $M' \cap K' = \{c, f, h, j, k, m, n, o, p, q, u, v, w, x, y, z\}$.

27. 170 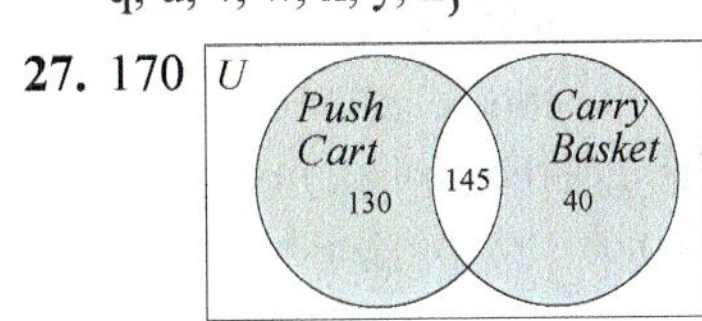

29. Will, David, Kim, Barbara, Alden, Morgan, Ali, Holly, Jessica, Jeff, Kent

31. 11

33. 39

35. 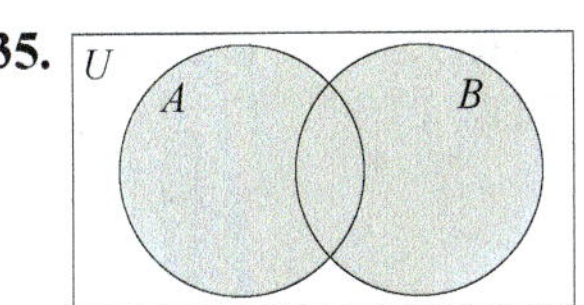

37. 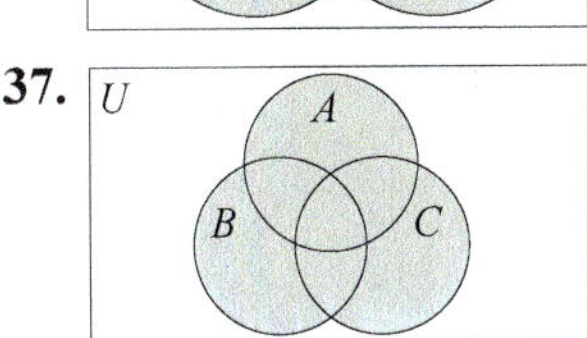

39. $A \cap B'$

8.3 EXERCISES

1. $\{$H1, H2, H3, H4, H5, H6, T1, T2, T3, T4, T5, T6$\}$

3. $\{$BC, BG, BY, BR, CB, CG, CY, CR, GB, GC, GY, GR, YB, YC, YG, YR, RB, RC, RG, RY$\}$

5. Empirical **7.** Classical

9. a. $\dfrac{8}{64} = 0.125$

b. $\dfrac{17}{64} = 0.265625$

c. $\dfrac{49}{64} = 0.765625$

11. $\dfrac{33}{84} \approx 0.392857$

13. $\dfrac{1}{235} \approx 0.004255$

15. $\dfrac{18}{38} \approx 0.473684$

17. $\dfrac{4}{52} \approx 0.076923$

19. $\dfrac{1}{16} = 0.0625$

21. $\dfrac{10}{69} \approx 0.144928$

8.4 EXERCISES

1. $\{$GGG, GGB, GBG, GBB, BGG, BGB, BBG, BBB$\}$

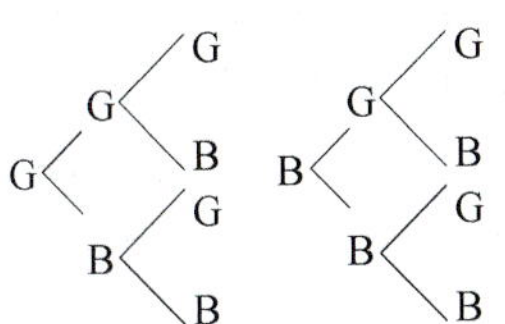

3. $\{$HHH, HHT, HTH, HTT, THH, THT, TTH, TTT$\}$

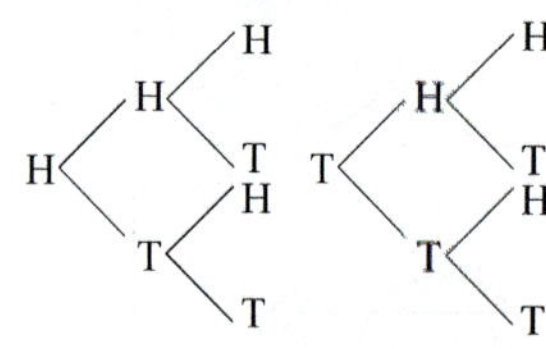

5. 120   **7.** 56

9. 1   **11.** 10

13. 46   **15.** 840

17. 24   **19.** 30

21. $\dfrac{1}{2}$ **23.** 98

25. $10 \cdot 10 \cdot 10 = 1000$

27. $26 + 10 = 36,$
so $36 \cdot 36 \cdot 36 \cdot 36 \cdot 36 \cdot 36$
$= 2{,}176{,}782{,}336$

29. 144 possibilities

31. $3^{11} = 177{,}147$

33. $10^7 = 10{,}000{,}000$

35. $39 \cdot 10 \cdot 8 = 3120$

37. Combination;
$_{120}C_5 = 190{,}578{,}024$ ways

39. Combination; $_{29}C_3 = 3654$ ways

41. Permutation; $_{18}P_3 = 4896$ ways

43. Permutation;

$$\dfrac{10!}{3!3!1!2!1!} = 50{,}400 \text{ ways}$$

45. c.

47. Answers will vary; should be
(number of pants) ·
(number of shirts) ·
(number of pairs of shoes)

8.5 EXERCISES

1. The set of even numbers larger than 0.

3. The complement consists of players aged 8 or 9, so there are nine players in the complement of A.

5. All employees who have worked there for less than or equal to five years.

7. $\dfrac{5}{_{10}C_2} = \dfrac{5}{45} \approx 0.111111$

9. a. $_{16}C_3 = 560$

b. $\dfrac{1}{560} \approx 0.001786$

11. a. $4! = 24$

b. $\dfrac{2 \cdot 3!}{24} = \dfrac{12}{24} = 0.5$

13. a. $_{52}C_5 = 2{,}598{,}960$

b. $\dfrac{4}{2{,}598{,}960} \approx 0.000002$

15. a. $10^4 = 10{,}000$

b. $_{10}P_4 = 10 \cdot 9 \cdot 8 \cdot 7 = 5040$

c. $\dfrac{1}{4!} = \dfrac{1}{24} \approx 0.041667$

d. $\dfrac{1}{3 \cdot \left(\dfrac{4!}{2!1!1!} \right)} = \dfrac{1}{36} \approx 0.027778$

e. Repeat a digit

17. $\dfrac{3}{1000} = 0.003$

19. $\dfrac{1}{26 \cdot 25 \cdot 24} = \dfrac{1}{15{,}600} \approx 0.000064$

21. $\dfrac{_4C_2}{_6C_2} = \dfrac{6}{15} = 0.4$

23. $1 - \dfrac{3}{18} \approx 0.833333$

25. a. $\dfrac{871}{1502} \approx 0.579893$

b. $1 - \dfrac{871}{1502} \approx 0.420107$

27. $1 - \dfrac{1}{2^5} = \dfrac{31}{32} = 0.96875$

8.6 EXERCISES

1. These are independent events because they are choices by three random people.

3. No, the choice of the first person changes the probability of the second person's gender.

5. No, whose card is picked first affects the probabilities of the next pick.

7. $\dfrac{8}{36} = \dfrac{2}{9} \approx 0.222222$

9. $\dfrac{8}{52} = \dfrac{2}{13} \approx 0.153846$

11. a. $\dfrac{102}{137} \approx 0.744526$

b. $\dfrac{76}{137} \approx 0.554745$

13. 0.46

15. $\dfrac{8}{11} \approx 0.727273$

17. a. $\dfrac{21}{113} \approx 0.185841$

b. $\dfrac{76}{113} \approx 0.672566$

c. $\dfrac{40}{113} \approx 0.353982$

19. $\dfrac{6}{11} \approx 0.545455$

21. 0.312

23. $\dfrac{13}{52} \cdot \dfrac{4}{52} = \dfrac{52}{2704} \approx 0.019231$

25. $\dfrac{6}{10} = \dfrac{3}{5} = 0.6$

27. $\dfrac{3}{51} \approx 0.058824$

29. 0.666667

31. $\dfrac{16}{34} = \dfrac{8}{17} \approx 0.470588$

33. $\dfrac{_{18}C_5}{_{20}C_5} = \dfrac{8568}{15{,}504} \approx 0.552632$

8.7 EXERCISES

1. Expected value $= 2.15$

3. Expected value $= 16$

5. 0.65 times per week

7. Expected winnings $= -\$0.03$

9. a. 1 ticket: $-\$8$; 3 tickets: $-\$19$; 5 tickets: $-\$30$

b. 1 ticket

11. 32.25%

13. a. 0   **b.** 0

15. $\dfrac{1}{7}$ **17.** $\dfrac{2}{5}$

19. $\dfrac{5}{7}$

21. a. 0.822   **b.** \$80.24

Chapter 9: Statistics

9.1 EXERCISES

1. census

3. population, population parameters

5. Sample statistic

7. Sample statistic

9. Population: cancer patients; Sample: 675 cancer patients; Sample statistics: 76% and 63%

11. Population: middle-aged women; Sample: 36,000 health records of women; Sample statistics: 18% and 11%; Population parameter: 40%

13. Systematic sampling

15. Cluster sampling

17. Convenience sampling

19. Random sampling

21. Random sample, since there are not clear divisions in the sample group of 1200 patients. Possible biases include age, race, and stage of development of the skin disease in the patients. Answers will vary.

23. Convenience sample, since the population is all children with autism, this is a very large group that may be difficult to study. It may be best to start with a convenience sample to get an idea of where to start. A possible bias to consider is that the effects may be different on children at different levels of functionality, so it may be best to make sure that a range of children are represented. Answers will vary.

25. Answers will vary. No, it is not likely to include all of the American public, since it will not include the section of the public who do not have internet access, and the only people who would respond are people who frequent those news outlet sites. Because some news outlets may have a slight bias one way or the other, each may draw only a certain demographic.

27. Answers will vary. For example, income level, health of residents, happiness of residents, average annual temperature, number of activities in the area.

9.2 EXERCISES

1. histogram   3. line

5. frequency distribution

7. **a.** 6   **b.** 1 kg

   c. 2.00 kg   **d.** 4.99 kg

   e. 89/194 or ≈ 45.9%

9. **a.–b.**

Class	Freq.	Rel. Freq.
0–9	0	0%
10–19	6	17.1%
20–29	8	22.9%
30–39	5	14.3%
40–49	4	11.4%
50–59	3	8.6%
60–69	5	14.3%
70–79	4	11.4%

   c. 11.4%   **d.** 0%

   e. 20–29 calls

11. **a.** 2

   b. The western US

13. **a.** Generally decreasing with brief periods of increase.

   b. June 2011; 9.2%

   c. January 2013; 7.7%

   d. Answers might include making it clearer that these are percentages and that the x-axis is not marked very clearly.

15. **a.** Approximately 6 hours

   b. Approximately 7 hours

   c. There is essentially no difference in the number of hours wives spend sleeping.

   d. A pair of side-by-side bar graphs: one for husbands and one for wives.

9.3 EXERCISES

1. Mean: 24; median: 25; mode: 22, 27; range: 17; standard deviation: 4.8; bimodal

3. Mean: 34.3; median: 35; mode: 26; range: 25; standard deviation: 9.0; unimodal

5. Mean: −1.8; median: −3; mode: −7, −3, 3; range: 18; standard deviation: 5.9; multimodal

7. Mean: 147.3 minutes; median: 149 minutes; mode: no mode; range: 31 minutes; standard deviation: 9.8 minutes

9. Mode   11. Mode

13. 86

15. **a.** 95%   **b.** 16%   **c.** 50%

17. Yes

19. **a.** Min: 35; Q_1: 37; Q_2: 38; Q_3: 40; Max: 42

   b. 38.5

21. 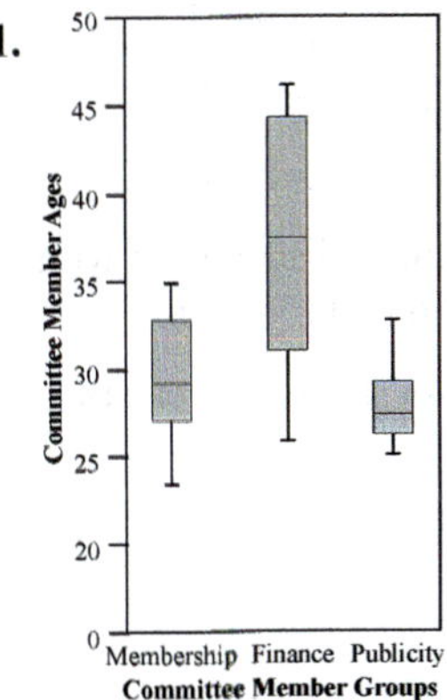

   a. Finance Committee

   b. Publicity Committee

   c. Publicity Committee

23. No, you should not operate with means. We cannot assume that all five classes are the same size. For instance, suppose one section has 100 students with a mean of 72 and another section has only 11 students, but with a mean of 79. We cannot give equal emphasis to the two means. You will not get an accurate answer.

25. No, standard deviations cannot be negative values since they describe a distance.

27. Answers will vary. Possible examples: large variation: salaries of all employees of a Fortune 500 company; small variation: densities of water from different sources.

29. Yes, the five-number summary can be put back together in numerical order: 9, 11, 13.5, 17.5, 19. So, $Q_1 = 11$.

31. No, percentiles give the relative position in terms of percentages. Without knowing more information about the size of the sample, we cannot compute the size of the sample.

9.4 EXERCISES

7. a. 5   **b.** 45   **c.** 15   **d.** 1

9. a. $E(X) = 0.9$ **b.** $\sigma = 0.9$

   c. $P(X = 2) = 0.1722$

   d. $P(X \leq 3) = 0.9917$

   e. $P(X \geq 2) = 0.2252$

   f. $P(X < 5) = 0.9991$

11. a. Binomial distribution with $n = 10$ and $p = 0.10$

   b. $E(X) = 1$ **c.** $\sigma = 0.9487$

   d. $P(X = 1) = 0.3874$

   e. $P(X = 5) = 0.0015$

   f. $P(X \geq 3) = 0.0702$

13. a. Binomial distribution with $n = 7$ and $p = 0.1$

   b. $P(X = 0) = 0.4783$

   There is a 47.83% chance that none of the plants will strike.

   $P(X = 4) = 0.0026$
   There is a 0.26% chance that exactly 4 of the plants will strike.

   $P(X = 7) = 0$

   There is a negligible chance that all 7 plants will strike.

   c. $E(X) = 0.7$

   d. $\sigma = 0.794$. The standard deviation is larger than the expected value. The standard deviation is expressed as the number of plants that strike. Answers may vary.

15. a. $P(X = 2) = 0.375$

   b. $P(X = 4) = 0.0625$

17. a. $P(X \leq 1) = 0.8290$

   b. $E(X) = 0.75$

19. a. $\dfrac{2}{9} = 0.2222$

   b. $P(X = 5) = 0.0389$

   c. $P(X = 0) = 0.0810$

   d. $E(X) = 2.2222$, $\sigma^2 = 1.7284$

9.5 EXERCISES

1. a. $z = 0.55$ **b.** $z = -1.73$

   c. $z = 0.09$

3. a. $z = -1$ **b.** $z = 0.5$

   c. $z = -3.17$

5. -0.45 **7.** 0.41

9. a. $z = -0.87$ **b.** $\mu = 277.12$

   c. $x = 130.30$ **d.** $\sigma = 8.99$

11. 92.79%   **13.** 1.32%

15. 15.15%   **17.** 99.90%

19. 50.00%   **21.** 95%

23. 28.64%   **25.** 27.13%

27. 33.81%   **29.** 98.80%

31. Her second race

33. a. 5.16%   **b.** 95.76%

   c. 99.94%   **d.** 88.88%

35. 1.29

37. $Q_1: -0.67; Q_2: 0; Q_3: 0.67$

9.6 EXERCISES

5. Conditions are not met; $np = 1.5 < 5$.

7. a. 9   **b.** 2.7659

   c. 0.5714   **d.** 0.0000

   e. 0.9390

9. a. 60   **b.** 7.6681

   c. 0.0001   **d.** 0.9924

   e. 0.1075

Chapter 10: Limits and the Derivative

10.1 EXERCISES

1. $\lim = 14$

x	y
6.5	13.5
6.9	13.9
6.99	13.99
6.999	13.999

3. $\lim = 0$

x	y
3.5	0.25
3.1	0.01
3.01	0.0001
3.001	0.000001

5. $\lim = 10$

a	y
1.5	113.33
1.1	15.94
1.01	10.46
1.001	10.045

7. a. $a = 2$ **b.** $\lim = 0.25$

9. a. $a = \pi$ **b.** No limit

11. a. $-2, -1.84, -1.36, -1.19,$
$-1.0199, -1.001999$

  b. -1

13. a. $7, 8.76, 10.84, 11.41, 11.9401,$
11.994001

  b. 12

15. a. $3, 4.\overline{3}, 11, 21, 201, 2001$

  b. $+\infty$

17. a. $0.2, 0.217, 0.238, 0.244,$
$0.249, 0.2499$

  b. 0.25

19. 3 **21.** 0

23. 2 **25.** $-\infty$

27. 2 **29.** 2

31. 0 **33.** $+\infty$

35. 1 **37.** 5

39. -17 **41.** 3

43. 0 **45.** -24

47. $-\infty$ **49.** $+\infty$

51. $\dfrac{1}{4}$ **53.** $-\dfrac{1}{4}$

55. a. 11 **b.** 11

57. a. 8 **b.** 9

59. a. 3 **b.** 3

  c. 3 **d.** $+\infty$

61. yes

10.2 EXERCISES

1. a. 0 **b.** 0 **c.** 0 **d.** 2

3. a. $-\infty$ **b.** 1 **c.** 1 **d.** 1

5. a. 5 **b.** 6
  c. does not exist
  d. 2 **e.** 2 **f.** 2

7. a. $+\infty$ **b.** $-\infty$ **c.** 2 **d.** 0

9. a. $+\infty$ **b.** 3
  c. does not exist **d.** 0

11. 11

13. limit does not exist

15. -3

17. a. 480 **b.** 480 **c.** 480

10.3 EXERCISES

1. 6 **3.** 10

5. $+\infty$ **7.** $\dfrac{6}{7}$

9. $\dfrac{1}{2}$ **11.** 0

13. $-\infty$ **15.** $\dfrac{3}{4}$

17. $-\infty$ **19.** 8

21. 3 **23.** -2

25. a. $\dfrac{1}{4}$ **b.** 6

27. 1

29. a. -3 **b.** -3 **c.** -3 **d.** -2

31. a. 3 **b.** 3 **c.** 3 **d.** 5

33. a. $\$11.88$ **b.** $\$11.88$

  c. $\$11.88$

35. $\$73$

10.4 EXERCISES

1. a. 2 **b.** 2
  c. Does not exist
  d. $f(x)$ is not continuous at $x = 3$ because it has a removable discontinuity ($f(3)$ is undefined).

3. a. -2 **b.** -2 **c.** -2
  d. Yes, because the limit exists, the function exists, and the limit equals the value of the function at $x = 0$, i.e.,
$$\lim_{x \to 0} f(x) = f(0).$$

5. a. -1 **b.** $+\infty$ **c.** -1
  d. $f(x)$ is not continuous at $x = 4$, as the limit doesn't exist at that point.

7. a. No, $f(x)$ is discontinuous at $x = -1$, because
$$\lim_{x \to -1} f(x) \neq f(-1).$$
  b. $f(x)$ is continuous at $x = 3$, as
$$\lim_{x \to 3} f(x) = f(3) = 7.$$

9. At $x = -3$, $f(x)$ is discontinuous; it has a removable discontinuity.

11. At $x = 0$, $f(x)$ is discontinuous; it has a jump discontinuity.

13. $f(x)$ is continuous at $x = 1$.

15. $f(x)$ is discontinuous at $x = 2$.

17. $f(x)$ is continuous at $x = 1$.

19. $f(x)$ is discontinuous at $x = 0$.

21. No points of discontinuity

23. $f(x)$ is discontinuous at $x = -3$; it has an infinite discontinuity.

25. $f(x)$ is discontinuous at $x = 3$; it has an infinite discontinuity.
$f(x)$ is discontinuous at $x = -3$; it has an infinite discontinuity.

27. No points of discontinuity

29. $k = 2$ **31.** $k = 4$

33. a.
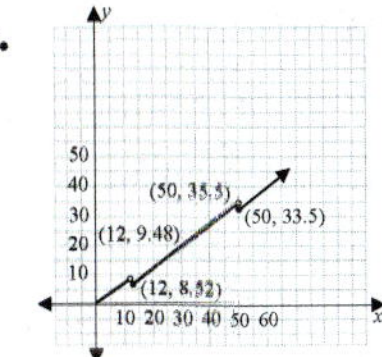

  b. C is discontinuous at $x = 12$ and at $x = 50$. These are jump discontinuities.

35. a.
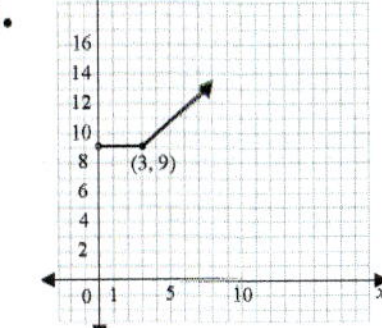

  b. $C(x)$ is continuous for all $x > 0$.

10.5 EXERCISES

1. -1 **3.** 5

5. 8 **7.** 0.1333

9. 0.387

10.6 EXERCISES

1. $f'(15) = 1$ **3.** 12:00 p.m.

5. $f(3) = 14$: 14 is the value of the function at the point $x = 3$.
$f'(3) = 7$: 7 is the slope of the tangent line at $x = 3$.

7. $f'(s)$ is the rate of spending per customer for s customers.

9. $f'(x)$ is the rate of change of cost per toaster.

11. The acceleration (or deceleration) of a race car after t seconds

13. The velocity of the rocket in x seconds

15. $f'(t)$ is the rate of change of cost in dollars per year at t years since 1980.

17. a. $f(500) = 2000$ means 500 man-hours are required to produce 2000 units.

$f'(500) = 3$; 3 more man-hours per unit are required if we increase x from 500 to 501.

b. $f(501) = 2003$

19. a. $f'(x)$ is negative. As x increases, $f(x)$ decreases.

b. $f'(3) = -2500$ means $f(x)$ is changing at the rate $2500/$ year; i.e., the car decreases $2500 in value if x increased from 3 years to 4 years of age.

21. $f(3) = 4$ grams means at the time of 3 weeks after discovery there are 4 grams of cancerous tumor.

$f'(3) = 0.4$ grams/week at 3 weeks, the rate of change in weight of the tumor is 0.4 grams per week.

23. a. $f(20) = 4.25$; it represents the average number of children per woman in 1990.

b. $f'(20) = -0.11$; it represents a decrease in the average number of children per woman in the year 1990.

25. a. $f(30) = 16.57$; it represents the yearly birth rate per 1000 people in 1994.

b. $f'(30) = -0.641$; it represents a decrease in the yearly birth rate per 1000 people in the year 1994.

27. Slope = 27

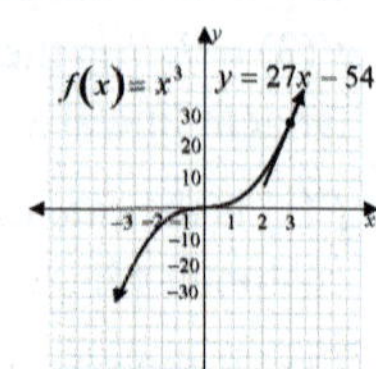

29. Slope = $100 \ln(10) \approx 230.26$

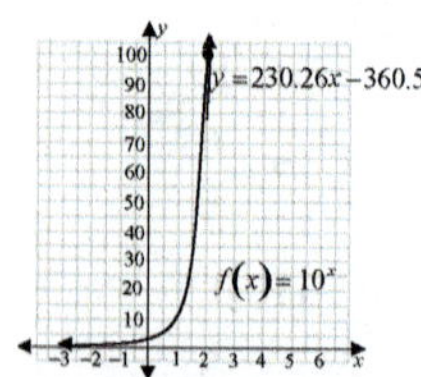

31.

a	$f(a)$	$f'(a)$
-1	-1	3
0	0	0
1	1	3
2	8	12
3	27	27
4	64	48

$f'(x) = 3x^2$

33.

x	y	y-intercept of the tangent line
-1	1	-1
0	0	0
1	1	-1
2	4	-4
3	9	-9
4	16	-16

Tangents at $x = +a$ and $x = -a$ have the same y-intercept.

35. $\left(\dfrac{1}{4}, \dfrac{1}{2}\right)$

37. Window $[-5, 8]$ by $[-110, 800]$

a.

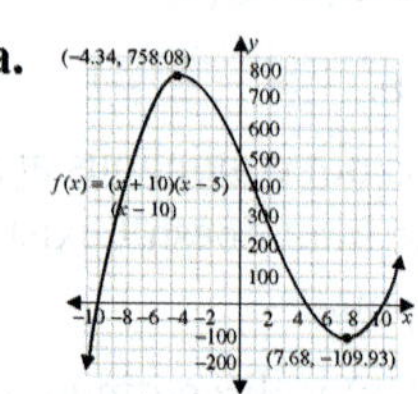

b. $f'(2) = -108$

39. $f(x) = 2x^2$; $f'(x) = 4x$

x	$f(x)$	$f'(x)$
-3	18	-12
-2	8	-8
-1	2	-4
0	0	0
1	2	4
2	8	8
3	18	12

10.7 EXERCISES

1. 0

3. 7

5. $8x$

7. $-\dfrac{7}{x^2}$

9. $-\dfrac{3}{2x^4}$

11. $\dfrac{3}{2\sqrt{x}}$

13. $2.3t^{1.3}$

15. $2.4x^{-0.2}$

17. $-\dfrac{1}{2}u^{-\frac{3}{2}}$

19. $-\dfrac{15}{4}x^{-\frac{1}{4}}$

10.8 EXERCISES

1. $3x^2 - 7$

3. $0.6x - 4$

5. $3x^2 - 12x + 5$

7. $3x^{\frac{1}{2}} + 2x^{-\frac{1}{2}}$

9. $-t^{-\frac{3}{2}} + \dfrac{1}{2}t^{-\frac{1}{2}} + 1$

11. $y = 2x^2 - x - 3$, $y' = 4x - 1$

13. $f(v) = v^{\frac{7}{2}} + 2v^{\frac{5}{2}} - v^{\frac{3}{2}}$, $f'(v) = \dfrac{7}{2}v^{\frac{5}{2}} + 5v^{\frac{3}{2}} - \dfrac{3}{2}v^{\frac{1}{2}}$

15. $f(x) = x^2 + 5x$, $f'(x) = 2x + 5$

17. $g(t) = \sqrt{t} - \dfrac{2}{\sqrt{t}}$, $g'(t) = \dfrac{1}{2}t^{-\frac{1}{2}} + t^{-\frac{3}{2}}$

19. $g(x) = 4\sqrt{x} + 5 - \dfrac{1}{\sqrt{x}}$, $g'(x) = 2x^{-\frac{1}{2}} + \dfrac{1}{2}x^{-\frac{3}{2}}$

21. a. $m = 5$ **b.** $y = 5x - 2$

c.

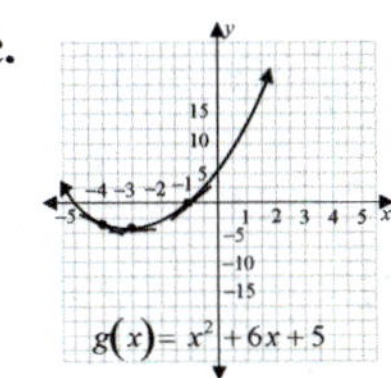

23. a. $m = 4$, $m = 0$, $m = -2$

b. $y = 4x + 4$, $y = -4$,
$y = -2x - 11$

c.

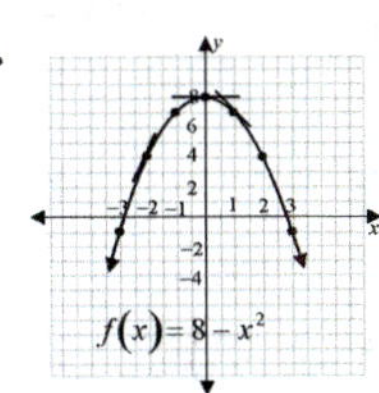

25. a. $m = 4$, $m = 0$, $m = -2$

b. $y = 4x + 12$, $y = 8$,
$y = -2x + 9$

c.

27. a. 192 ft/sec   **b.** 64 ft/sec

c. 6 sec

29. a. 482 people/year

b. 488 people/year

31. a. 51.67 barrels

b. 7.13 barrels/hr

c. 7.09 barrels/hr

33. a. 10 hours

b. -1.3 charge units/hr

c. -0.9 charge units/hr

35. a. 5.585   **b.** 2.015

c. -0.187 births per woman per year

37. a. 7.06%   **b.** 18.82%

c. 0.25% per year

39. a. 13.62 deaths/1000 people

b. -0.37 deaths/1000 people per year

c. 5.18 deaths/1000 people

d. 2008

41. a. $f(30) = 2.66$. This is the number of births per woman in the year 2000.

b. $f'(30) = -0.046$. The number of births per woman was decreasing by 0.046 births per woman in 2000.

10.9 EXERCISES

1. $\overline{C}(x) = \dfrac{220}{x} + 0.4$

3. $D(x) = 3$; $R(x) = 3x$

5. $C(x) = 1.15x + 50$

7. $P(x) = 0.75x + 25$

9. a. $\overline{C}(x) = \dfrac{100}{x} + 150$

b. $D(x) = 450$

c. $R(x) = 450x$

d. $P(x) = 300x - 100$

11. a. $C'(x) = -0.6 - 0.03x^{-0.4}$

b. $S'(x) = 1.6 + 0.03x^{-0.4}$

13. a. $\overline{C}(x) = \dfrac{30}{x} + 2 + 0.003x$

b. $\overline{C}'(x) = -\dfrac{30}{x^2} + 0.003$

c. $x = 100$

15. a. $C(10) = 2705$, $C(20) = 3060$, $C(30) = 3465$

b. $C' = 28 + 0.5x$

c. $C'(10) = 33$, $C'(20) = 38$, $C'(30) = 43$

d. $\overline{C}(x) = \dfrac{2400}{x} + 28 + 0.25x$,
$\overline{C}'(x) = -2400x^{-2} + 0.25$

e. $\overline{C}'(10) = -23.75$,
$\overline{C}'(20) = -5.75$,
$\overline{C}'(30) = -2.42$

17. a. $C(4) = 92$, $C(6) = 102$, $C(9) = 109.5$

b. $C'(x) = 10 - x$

c. $C'(4) = 6$, $C'(6) = 4$, $C'(9) = 1$

d. $\overline{C}(x) = \dfrac{60}{x} + 10 - 0.5x$

e. $\overline{C}'(4) = -4.25$, $\overline{C}'(6) = -2.17$,
$\overline{C}'(9) = -1.24$

19. a. $P(x) = -0.03x^2 + 60x - 10{,}800$

b. $P(200) = \$0$, $P(400) = \$8400$,
$P(600) = \$14{,}400$

c. $P'(x) = -0.06x + 60$

d. $P'(200) = 48$, $P'(400) = 36$,
$P'(600) = 24$

e. $x = 200$, $x = 1800$

21. a. $P(x) = -0.2x^2 + 75x - 2070$

b. $P(60) = \$1710$, $P(80) = \$2650$,
$P(100) = \$3430$

c. $P'(x) = -0.4x + 75$

d. $P'(60) = 51$, $P'(80) = 43$,
$P'(100) = 35$

e. $x = 30$, $x = 345$

23. a. $R(x) = 46x + 0.25x^2$

b. $P(x) = 0.1x^2 + 40x - 190$

c. $P(25) = \$872.50$,
$P(30) = \$1100$, $P(40) = \$1570$

d. $P'(x) = 0.2x + 40$

e. $P'(25) = 45$, $P'(30) = 46$,
$P'(40) = 48$

25. a. $R(x) = 5.5x - 0.0004x^2$

b. $P(x) = -0.0006x^2 + 4.5x - 4650$

c. $P(3000) = 3450$,
$P(3500) = 3750$,
$P(4000) = 3750$

d. $P'(x) = -0.0012x + 4.5$

e. $P'(3000) = 0.9$,
$P'(3500) = 0.3$,
$P'(4000) = -0.3$

27. a. $f(0) = 12{,}519.30$. This year, the value of the car is \$12,519.30.

b. $f'(x) = -1391.10$. As the age of the car increases, the value of the car decreases at the rate of \$1391.10/year.

29. a. $F'(t) = \dfrac{-8.4t^3}{10^6} + \dfrac{4.8t^2}{10^4} - 0.00914t + 0.0994$

b. $F(30) = 7.518$ billion

c. $F'(30) = 0.0304$ billion. In the year 2030, the world population is increasing at the rate of 0.0304 billion/year.

31. a. $H'(t) = \dfrac{-10.4t^3}{10^6} + \dfrac{5.61t^2}{10^4}$
$-0.00792t + 0.116$

b. $H(30) = 8.909$ billion

c. $H(30) = 0.1025$ billion. In the year 2030, the world population is increasing at the rate of 0.1025 billion/year.

33. Argue against; Answers will vary.

Note: $\dfrac{R(x)}{x}$ is also $D(x)$.

Chapter 11: More about the Derivative

11.1 EXERCISES

1. $f'(x) = -8x^3 + 9x^2 + 2x$

3. $f'(x) = \dfrac{15x^2 + 1}{2\sqrt{x}}$

5. $f'(x) = 8x$

7. $g'(x) = -\dfrac{1}{x^{\frac{3}{2}}}$

9. $g'(x) = 6 - 10x^4$

11. $5x^2(x^2 + 3)$ **13.** $\dfrac{3}{2}\left(\dfrac{4t+1}{\sqrt{t}}\right)$

15. $\dfrac{\sqrt{x}(5x+3)}{2}$

17. $-18u^2 + 20u - 9$

19. $\dfrac{1}{5t^2}(50t^3 + 5t^2 - 1)$

21. $18(x+6)^{-2}$

23. $-15(x-7)^{-2}$

25. $\dfrac{x^4 + 3x^2 + 10x}{(x^2+1)^2}$

27. $\dfrac{9-x}{2\sqrt{x}(x+9)^2}$

29. $\dfrac{3u^{\frac{3}{2}} + 4u}{2(\sqrt{u}+1)^2}$

31. $\dfrac{3t^2 - 16t^{\frac{3}{2}} - 3}{2\sqrt{t}(t^2+3)^2}$

33. $\dfrac{6x - 15 + 10x^{\frac{4}{3}} - 20x^{\frac{1}{3}}}{3(1+2\sqrt[3]{x})^2}$

35. $h'(2) = 1$ **37.** $h'(2) = \dfrac{10}{3}$

39. $h'(2) = \dfrac{9}{16}$ **41.** $h'(2) = \dfrac{1}{3}$

43. $h'(2) = -\dfrac{7}{121}$

45. $y = 616.5x - 1766$

47. $y = -\dfrac{19}{25}x + \dfrac{2}{5}$

49. $y = -\dfrac{17}{20}x + \dfrac{37}{20}$

51. $(x, y) = (-2, 36)$, and $\left(\dfrac{8}{3}, -\dfrac{400}{27}\right)$

53. $(x, y) = (10, 100)$

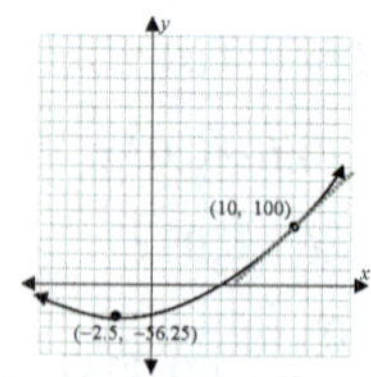

55.

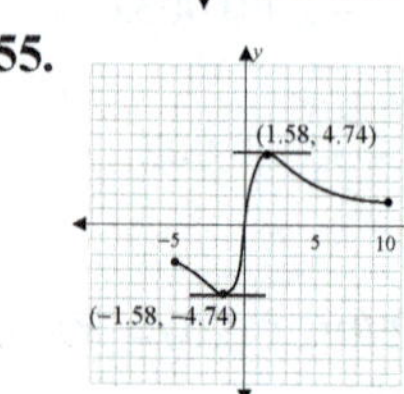

$(-1.58, -4.74), (1.58, 4.74)$

57. \$1.15/item **59.** \$13.20/item

61. 6100 people/year

11.2 EXERCISES

1. $8(2x-5)^3$ **3.** $-12(1-4x)^2$

5. $-4x(x^2+4)^{-3}$

7. $-3(4t+3)(2t^2+3t)^{-4}$

9. $16x^3 + 60x^2 - 6x - 70$

11. $\dfrac{3x^2}{2\sqrt{x^3+1}}$ **13.** $\dfrac{8x}{3(4x^2+1)^{\frac{2}{3}}}$

15. $-\dfrac{3x^2}{2(1-2x^3)^{\frac{3}{4}}}$

17. $5(13t^3 + 3)(t^3 + 3)^3$

19. $6x^2(x^2-8)^2(3x^2-8)$

21. $-\dfrac{x}{(x^2+6)^{\frac{3}{2}}}$ **23.** $\dfrac{8}{(t^2+8)^{\frac{3}{2}}}$

25. $-\dfrac{(10x+18)}{3x^3(2x+3)^{\frac{2}{3}}}$

27. $\dfrac{18x-13}{2(3x-4)^{\frac{1}{2}}}$ **29.** $\dfrac{26(3x-2)}{(5x+1)^3}$

31. $\dfrac{5t-17}{3(t-1)^{\frac{5}{3}}}$

33. $-9(2t+5)^2(t+1)^{-4}$

35. $\dfrac{5}{(4-2x)^{\frac{3}{2}}(x+3)^{\frac{1}{2}}}$

37. $-\dfrac{7}{2(3x-1)^{\frac{3}{2}}(x+2)^{\frac{1}{2}}}$

39. $(7-2x)^{-\frac{3}{2}}(-3x^2+13x+7)$

41. -48 **43.** $-\dfrac{3}{4}$

45. $\dfrac{135}{2}$ **47.** $-\dfrac{7}{8}$

49. 295 **51.** -48

53. -7 **55.** 480

57. $y = 36x - 28$

59. $y = \dfrac{13}{8}x + \dfrac{3}{2}$

61. a. $R(x) = 20x\sqrt{280 - 4x}$

b. \$5880

c. $40\sqrt{70-x} - \dfrac{20x}{\sqrt{70-x}}$

d. \$220 per oven

63. a. 18,560 people

b. 640 people/year

65. 218 units per day

67. a. \$140.83/hour

    b. \$109.48/hour

69. a. 2.5 ppm/year

    b. 30 ppm/year

71. a. 1789 people

    b. $N'(5) = 0.05$

    c. The rate of attendance is increasing at a rate of 50 people per week.

73. a. $P(0) = 2.0$ million students; represents the number of students registered in calculus in 1995.

    b. $P'(t) = -\dfrac{1}{12}\left(1 - \dfrac{t}{30}\right)^{1.5}$

    c. $P'(10) = -0.045$. The rate of registration was decreasing 45,000 students/year in 2005.

11.3 EXERCISES

1. $\dfrac{dy}{dx} = -\dfrac{2x}{y}$ **3.** $\dfrac{dy}{dx} = -\dfrac{2x^2}{y^2}$

5. $\dfrac{dy}{dx} = -\dfrac{\sqrt{y}}{\sqrt{x}}$ **7.** $\dfrac{dy}{dx} = -\dfrac{2y}{x}$

9. $\dfrac{dy}{dx} = -\dfrac{(2x+y)}{(x+2y)}$

11. $\dfrac{dy}{dx} = \dfrac{2-8x-3y}{3x+2y}$

13. $\dfrac{dy}{dx} = \dfrac{x^2 y - 2x^2 y^2 - y^2}{x^3}$

15. $\dfrac{dy}{dx} = \dfrac{4x^{\frac{3}{2}} y^{\frac{1}{2}} + y}{8x^{\frac{1}{2}} y^{\frac{3}{2}} - x}$

17. $\dfrac{dy}{dx} = \dfrac{4x^3 - 2xy^2 - y^3}{2x^2 y + 3xy^2}$

19. $\dfrac{dy}{dx} = \dfrac{x}{2-y}$

21. $\dfrac{dy}{dx} = \dfrac{x}{3y^2}; \; m = \dfrac{2}{3}$

23. $\dfrac{dy}{dx} = \dfrac{4+2xy}{2y-x^2}; \; m = \dfrac{12}{7}$

25. $\dfrac{dy}{dx} = \dfrac{3x^2 + 2y}{2(y-x)}; \; m = -\dfrac{7}{4}$

27. $\dfrac{dy}{dx} = \dfrac{x^2 y - y^3 - 2x^2 y^3}{2x^3}; \; m = 1$

29. $\dfrac{dy}{dx} = \dfrac{20x^{\frac{3}{2}} y^{\frac{1}{2}} - y}{4x^{\frac{1}{2}} y^{\frac{1}{2}} + x}; \; m = \dfrac{76}{9}$

31. $\dfrac{dy}{dt} = \dfrac{x}{4y}\dfrac{dx}{dt}$

33. $\dfrac{dy}{dt} = \dfrac{2-3x^2}{10y}\dfrac{dx}{dt}$

35. $\dfrac{dy}{dt} = -\dfrac{y}{x}\dfrac{dx}{dt}$

37. $\dfrac{dy}{dt} = -\dfrac{(2x+y)}{x+2y}\dfrac{dx}{dt}$

39. $\dfrac{dy}{dt} = \dfrac{(x+y)}{y-x}\dfrac{dx}{dt}$

41. $\dfrac{5}{4}$ speakers per receiver; the number of speakers sold in a week increases by 1.25 for each additional reciever sold

43. $-\dfrac{27}{64}$ units of capital/week

45. \$312/week   **47.** $-\dfrac{4}{8}\text{f} / \sec$

49. 35.4 ft/sec

51. a. 4.5 seconds

    b. $\dfrac{20x}{y}$ **c.** $10\sqrt{2}$ ft/sec

53. a. $p = \$2$ **b.** \$0.38/day

55. a. 800 bears

    b. $\dfrac{dx}{dp} = -\dfrac{x}{p-1}; \; 228.571$

57. a. $(504, 672)$ **b.** $\dfrac{dy}{dx} = -\dfrac{x}{y}$

    c. $\left.\dfrac{dy}{dx}\right|_{t=t_0} = -0.75;$ at this point in time, for every 75 ft the car travels vertically, it also travels 100 ft to the negative horizontal.

    d. $\left.\dfrac{dy}{dt}\right|_{t=t_0} = 37.5$ mph

59. a. $S = 12$

    b. $\dfrac{dp}{dn} = -\dfrac{p-p^2}{n(1-2p)}; \; -0.00049$

11.4 EXERCISES

1. a. $(2, +\infty)$ **b.** $(-\infty, 2)$

3. a. $(-\infty, 2)$ **b.** $(2, +\infty)$

5. a. $(-1, 2)$

    b. $(-\infty, -1), (2, +\infty)$

7. a. $(-2, -1), (2, +\infty)$

    b. $(-\infty, -2), (-1, 2)$

9. a. $(-\infty, -2), (0, 6)$

    b. $(-2, 0), (6, +\infty)$

11. a. Inc.: $(0, +\infty)$; Dec.: $(-\infty, 0)$

    b.

13. a. Inc.: $(-2, 2)$; Dec.: $(-\infty, -2), (2, +\infty)$

    b.

15. a. $x = -3$

    b. Inc.: $(-3, +\infty)$; Dec.: $(-\infty, -3)$

    c.

17. a. $x = \dfrac{7}{4}$

    b. Inc.: $\left(-\infty, \dfrac{7}{4}\right)$; Dec.: $\left(\dfrac{7}{4}, +\infty\right)$

    c.

19. a. $x = \dfrac{2}{3}$

    b. Inc.: $\left(\dfrac{2}{3}, +\infty\right)$;

    Dec.: $\left(-\infty, \dfrac{2}{3}\right)$

    c.

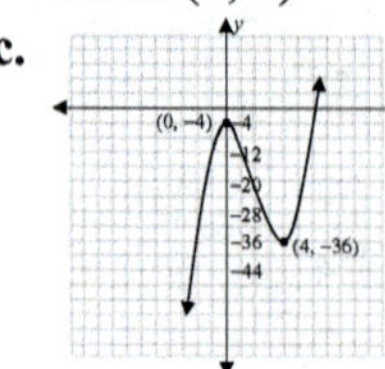

21. a. $x = \dfrac{2}{3}$

    b. Inc.: $\left(\dfrac{2}{3}, +\infty\right)$;

    Dec.: $\left(-\infty, \dfrac{2}{3}\right)$

    c.

23. a. $x = 0$

    b. Inc.: $(-\infty, +\infty)$;
    Dec.: $\varnothing$

    c.

25. a. $x = 0, x = 4$

    b. Inc.: $(-\infty, 0), (4, +\infty)$;
    Dec.: $(0, 4)$

    c.

27. a. $x = -\dfrac{1}{3}, x = 1$

    b. Inc.: $\left(-\infty, -\dfrac{1}{3}\right), (1, +\infty)$;

    Dec.: $\left(-\dfrac{1}{3}, 1\right)$

c.

29. a. $x = -1, x = \dfrac{5}{3}$

    b. Inc.: $(-\infty, -1), \left(\dfrac{5}{3}, +\infty\right)$;

    Dec.: $\left(-1, \dfrac{5}{3}\right)$

    c.

31. a. $x = 1, x = 3$

    b. Inc.: $(-\infty, 1), (3, +\infty)$;
    Dec.: $(1, 3)$

    c.

33. a. $x = -1, x = -4$

    b. Inc.: $(-\infty, -4), (-1, +\infty)$;
    Dec.: $(-4, -1)$

    c.

35. a. None

    b. Inc.: $\varnothing$; Dec.: $(-\infty, 2), (2, +\infty)$

37. a. None

    b. Inc.: $(-\infty, 0), (0, +\infty)$;
    Dec.: $\varnothing$

39. a. $x = -3$

    b. Inc.: $(-\infty, -3), (0, +\infty)$;
    Dec.: $(-3, 0)$

41. a. None

    b. Inc.: $(-\infty, -3), (-3, +\infty)$;
    Dec.: $\varnothing$

43. a. $x = 4$

    b. Inc.: $(4, +\infty)$;
    Dec.: $(-\infty, 0), (0, 4)$

45. Inc.: $(0, 8000)$;
    Dec.: $(8000, 10{,}000)$

47. Inc.: $(0, 62.5)$; Dec.: $(62.5, 100)$

49. 4 years

51. $\overline{C}(x) = \dfrac{250}{x} + 45 - 0.2x$;

$\overline{C}'(x) = -0.2 - \dfrac{250}{x^2} < 0$

11.5 EXERCISES

1. a. $x = 2$

    b. local max: $f(2) = 4$

3. a. $x = -3$

    b. local min: $f(-3) = -11$

5. a. $x = 4$

    b. local min: $f(4) = -5$

7. a. $x = -1, x = \dfrac{1}{3}$

    b. local max: $f(-1) = 4$;

    local min: $f\left(\dfrac{1}{3}\right) = \dfrac{76}{27}$

9. a. $x = -3, x = 2$

    b. local max: $f(2) = 28$;

    local min: $f(-3) = -\dfrac{69}{2}$

11. a. $x = -1, x = 1$

    b. local max: $f(-1) = 8$;
    local min: $f(1) = 4$

13. a. $x = -3, x = 3$

    b. local max: $f(-3) = -6$;
    local min: $f(3) = 6$

15. a. No critical values

    b. No local extrema

17. a. $x = -\dfrac{1}{3}, x = \dfrac{1}{3}$

    b. local max: $f\left(-\dfrac{1}{3}\right) = -6$;

    local min: $f\left(\dfrac{1}{3}\right) = 6$

19. a. $x = \dfrac{1}{2}$

　　b. local min: $f\left(\dfrac{1}{2}\right) = 12$

21. $W'(t) = \dfrac{2100t}{\left(t^2 + 15\right)^2} > 0,$

　　so $W(t)$ is increasing;

　　$\lim\limits_{t \to +\infty} W(t) = 130$

23. $(0, \infty)$ increasing

25. $F'(t) = -\dfrac{8000t}{\left(t^2 + 40\right)^2} < 0,$

　　so $F(t)$ is decreasing;

　　$\lim\limits_{t \to \infty} F(t) = 80$

27. $F'(t) = -\dfrac{3300t}{\left(t^2 + 10\right)^2} < 0,$

　　so $F(t)$ is decreasing;

　　$\lim\limits_{t \to \infty} F(t) = 15$

29. at $t = 50$, -99.5% of light speed

31. $(5.2, 85)$ increasing; maximum
　　efficiency at $v = 5.2$

11.6 EXERCISES

1. Abs. min.: $(5, 3)$;
　　Abs. max.: $(3, 6)$

3. Abs. min.: $(3, 2)$;
　　Abs. max.: $(-5, 8)$

5. Abs. min.: $(4, 4)$;
　　Abs. max.: $(2, 10)$

7. Abs. min.: $(5, -1)$;
　　Abs. max.: $(3, 6)$

9. $f(4) = -16$ min; $f(0) = 0$ max

11. $f(3) = 27$ min; $f(5) = 31$ max

13. $f(4) = 2$ min; $f(0) = 14$ max

15. $f(3) = -19$ min; $f(-1) = 9$ max

17. $f(2) = -16$ min;
　　$f(-2) = f(4) = 16$ max

19. $f\left(\dfrac{1}{3}\right) = -\dfrac{1}{27}$ min;
　　$f(2) = 12$ max

21. $f(-3) = -27$ min; $f(1) = 5$ max

23. $f(2) = -30$ min; $f(4) = 22$ max

25. $f(0) = -4$ min; $f(8) = 0$ max

27. $f(1) = -1$ min; $f\left(\dfrac{1}{8}\right) = 1$ max

29. $f(0) = 2$ min; $f(2) = 2\sqrt{2}$ max

31. $f(0) = -1$ min;
　　$f(-2) = f(2) = \sqrt[3]{3}$ max

33. $f(2) = 4$ min; $f\left(\dfrac{1}{2}\right) = 8.5$ max

35. $f(1.5) = 12$ min; $f(3) = 15$ max

37. $f(2) = 12$ min;
　　$f\left(\dfrac{1}{2}\right) = 32.25$ max

39. 24 units

41. 40 telescopes

43. 120 refrigerators

45. 1:00 p.m.

47. 1600 bacteria at time $t = 6$

49. $F\left(\dfrac{65}{3}\right) = 7041.\overline{6}$ ft

51. $t = 34.32$; 41.77%; yes

53. 500 newtons, $v = 2500$

55. $F(0.8) = 666.\overline{6}$; yes

57. max at $(0, 75)$; min at $(5, 10)$

59. at $t = 25.11$, 0.048 moles per sec.;
　　at $t = 62.01$, 0.025 moles per sec.

61. at $t = 20$ nanosec.; $E(20) = 15$

63. 31st day; 3.298 million

Chapter 12: Applications of the Derivative

12.1 EXERCISES

1. At A: f' is positive;
　　　　f'' is positive
　　At B: f' is positive;
　　　　f'' is zero
　　At C: f' is zero;
　　　　f'' is negative
　　At D: f' is negative;
　　　　f'' is zero
　　At E: f' is negative;
　　　　f'' is positive

3.

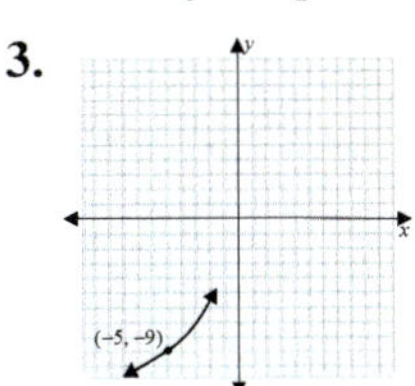

5.

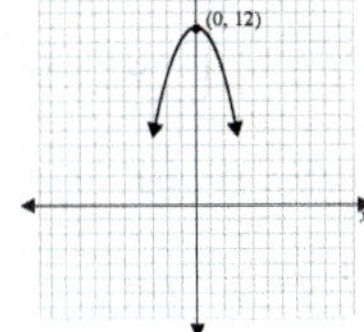

7. $f''(x) = 6x - 2$;
　　$f''(0) = -2$; $f''(1) = 4$;
　　$f''(4) = 22$

9. $f''(x) = 2 - \dfrac{1}{2x^{\frac{3}{2}}}$;

　　$f''(0) =$ does not exist;

　　$f''(1) = \dfrac{3}{2}$; $f''(4) = \dfrac{31}{16}$

11. $f''(x) = -\dfrac{1}{(2x+1)^{\frac{3}{2}}}$;

　　$f''(0) = -1$; $f''(1) = -\dfrac{1}{3\sqrt{3}}$;

　　$f''(4) = -\dfrac{1}{27}$

13. $f''(x) = -\dfrac{12}{(x+4)^3}$;

　　$f''(0) = -\dfrac{3}{16}$; $f''(1) = -\dfrac{12}{125}$;

　　$f''(4) = -\dfrac{3}{128}$

15.

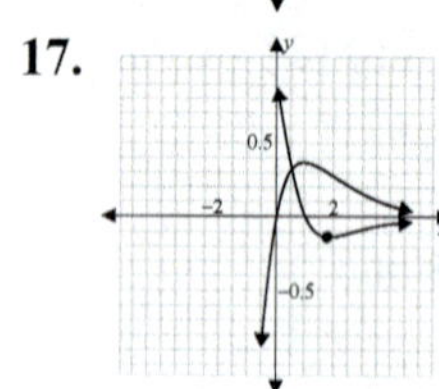

17.

19. a. $(-3, -1), (-1, 1)$
   b. $(-\infty, -3), (1, \infty)$
   c. $(-3, 2), (1, 2)$

21. a. $(-\infty, -3), (1, 5), (5, \infty)$
   b. $(-3, 1)$
   c. $(-3, 3), (1, 4)$

23. a. $(-\infty, +\infty)$ **b.** None
   c. None

25. a. $(-2, +\infty)$ **b.** $(-\infty, -2)$
   c. $(-2, 6)$

27. a. $(0, +\infty)$ **b.** $(-\infty, 0)$
   c. $(0, 2)$

29. a. $(-3, +\infty)$ **b.** $(-\infty, -3)$
   c. $(-3, 7)$

31. a. $\left(-\infty, \dfrac{3}{5}\right)$ **b.** $\left(\dfrac{3}{5}, \infty\right)$
   c. $\left(\dfrac{3}{5}, 0\right)$

33. a. $\left(-\sqrt{5}, 0\right), \left(\sqrt{5}, \infty\right)$
   b. $\left(-\infty, -\sqrt{5}\right), \left(0, \sqrt{5}\right)$
   c. $(0, 0)$

35. $G(x) = x^4$

37. $J(x) = (x - 4)^3$

12.2 EXERCISES

1. $f'(x) = 14x - 28;\ f''(x) = 14;$
   Max: $\varnothing$; Min: $(2, -20)$; I: $\varnothing$;
   Con. up: $\mathbb{R}$; Con. down: $\varnothing$

3. $f'(x) = 6x^2 + 5;\ f''(x) = 12x;$
   Max: $\varnothing$; Min: $\varnothing$; I: $(0, -1)$;
   Con. up: $(0, +\infty)$;
   Con. down: $(-\infty, 0)$

5. $f'(x) = 3x^2 + \dfrac{1}{\sqrt{x}};$
   $f''(x) = 6x - \dfrac{1}{2x^{\frac{3}{2}}};$
   Max: $\varnothing$;   Min: $(0, 5)$;
   I: $(0.3701, 6.2674)$;
   Con. up: $(0.3701, +\infty)$;
   Con. down: $(0, 0.3701)$

7. $f'(x) = 4x\left(x^2 + 7\right);$
   $f''(x) = 12x^2 + 28;$
   Max: $\varnothing$; Min: $(0, 49)$; I: $\varnothing$;
   Con. up: $\mathbb{R}$; Con. down: $\varnothing$

9. $f'(x) = \dfrac{x}{\sqrt{x^2 + 3}};$
   $f''(x) = 3\left(x^2 + 3\right)^{-\frac{3}{2}};$
   Max: $\varnothing$; Min: $\left(0, \sqrt{3}\right)$; I: $\varnothing$;
   Con. up: $\mathbb{R}$; Con. down: $\varnothing$

11. $f'(x) = \dfrac{3\left(1 - x^2\right)}{\left(x^2 + 1\right)^2};$
   $f''(x) = \dfrac{6x\left(x^2 - 3\right)}{\left(x^2 + 1\right)^3};$
   Max: $(1, 1.5)$; Min: $(-1, -1.5)$;
   I: $(0, 0), \left(-\sqrt{3}, \dfrac{-3\sqrt{3}}{4}\right),$
   $\left(\sqrt{3}, \dfrac{3\sqrt{3}}{4}\right);$
   Con. up: $\left(-\sqrt{3}, 0\right), \left(\sqrt{3}, +\infty\right);$
   Con. down: $\left(0, \sqrt{3}\right), \left(-\infty, -\sqrt{3}\right)$

13. POI: $(2.5, 10.179), (0, 0);$
   min: $\left(-\dfrac{5}{4}, -4.04\right)$

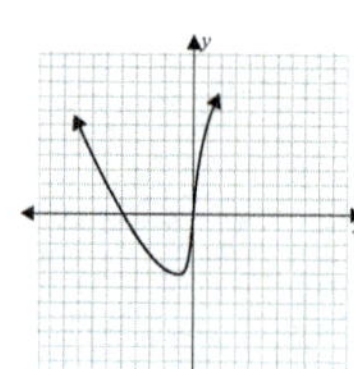

15. POI: $(-1, 0), \left(-1.5, 2.38\right);$
   min: $\left(-\dfrac{3}{4}, -0.94\right)$

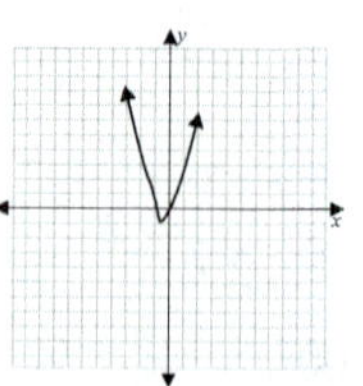

17. Local min: $f\left(\dfrac{3}{2}\right) = \dfrac{11}{4}$

19. Local max: $f(0) = 8;$
   Local min: $f(2) = 4$

21. Local max: $f(-2) = 19;$
   Local min: $f(2) = -13$

23. Local max: $f(-1) = \dfrac{1}{3};$
   Local min: $f(2) = -\dfrac{26}{3}$

25. Local max: $f(0) = 7;$
   Local min: $f(2) = -9, f(-2) = -9$

27. Local min:
   $f\left(-\dfrac{3}{2}\right) = -\dfrac{91}{16}$

29. Local max: $f(-2) = -8;$
   Local min: $f(2) = 8$

31. $x = \$6000$

33. $C'(8) = \$4.80/\text{unit}$

35. a. $N(9) = 104.6$ crimes
   b. $N'(4) = 8$ crimes/month

37. a. Max/min cannot exist;
      POI at $x = 0$
   b. Max/min can exist; POI at
      $x = 0$
   c. No
   d. Answers will vary.

12.3 EXERCISES

1.

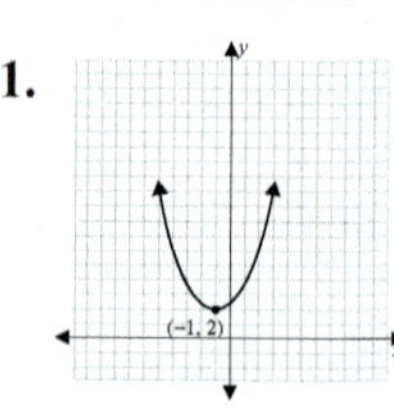

3.

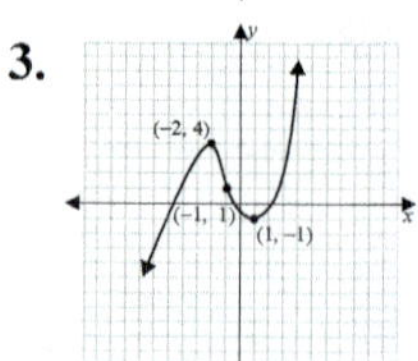

5. 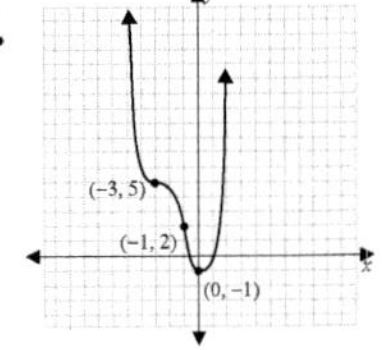

7. 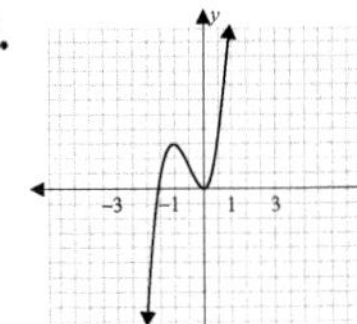

9. 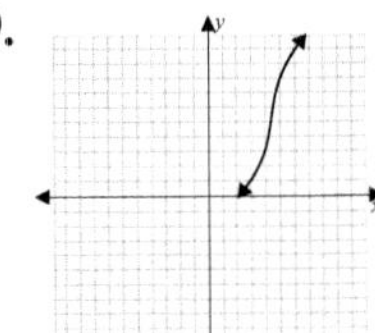

11. 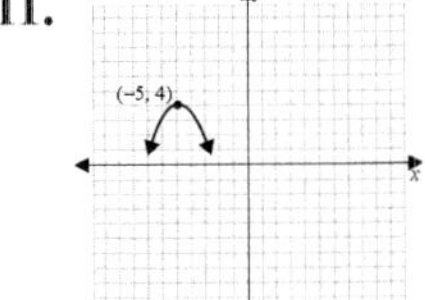

13. 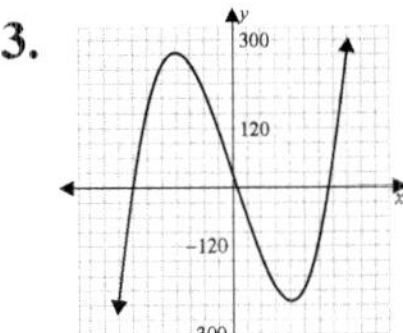

15. 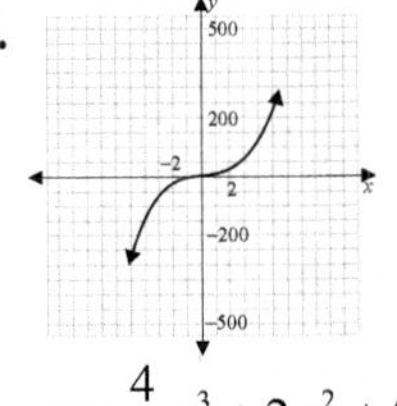

$$y = \frac{4}{3}x^3 + 2x^2 + 5x + 2$$

is implied.

17. $(-\infty, 2)$; $f'(x) < 0$; Decreasing

$(2, +\infty)$; $f'(x) > 0$; Increasing

$x = 2$; $f'(2) = 0$; Local min

$(-\infty, +\infty)$; $f''(x) = 2 > 0$;

Hence $f''(2) = 2 > 0$ and

concave up on $(-\infty, \infty)$.

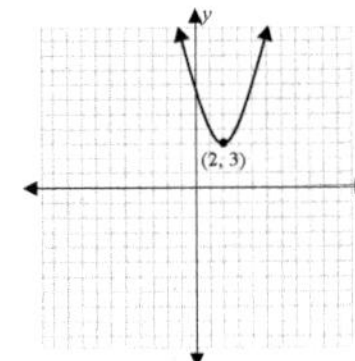

19. $\left(-\infty, \dfrac{5}{2}\right)$; $f'(x) > 0$;

Increasing

$\left(\dfrac{5}{2}, +\infty\right)$; $f'(x) < 0$;

Decreasing

$x = \dfrac{5}{2}$; $f'\left(\dfrac{5}{2}\right) = 0$;

Local max

$(-\infty, +\infty)$; $f''(x) = -2 < 0$;

Concave down on $(-\infty, \infty)$.

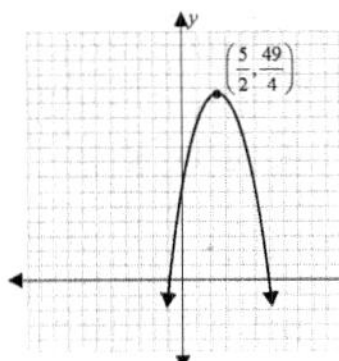

21. $(-\infty, -2), (0, \infty)$;

$f'(x) > 0$; Increasing

$(-2, 0)$; $f'(x) < 0$;

Decreasing

$x = -2$; $f'(-2) = 0$;

Local max

$x = 0$; $f'(0) = 0$;

Local min

$(-\infty, -1)$; $f''(x) < 0$;

Concave down

$(-1, +\infty)$; $f''(x) > 0$;

Concave up

$x = -1$; $f''(-1) = 0$; POI

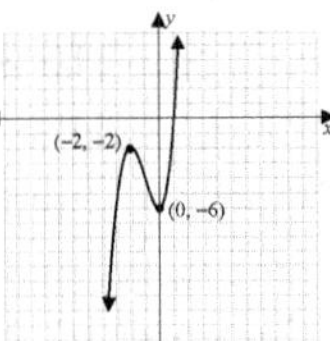

23. $(-\infty, -3), (1, +\infty)$;

$f'(x) > 0$; Increasing

$(-3, 1)$; $f'(x) < 0$;

Decreasing

$x = -3$; $f'(-3) = 0$; Local max

$x = 1$; $f'(1) = 0$; Local min

$(-\infty, -1)$; $f''(x) < 0$;

Concave down

$(-1, +\infty)$; $f''(x) > 0$;

Concave up

$x = -1$; $f''(-1) = 0$; POI

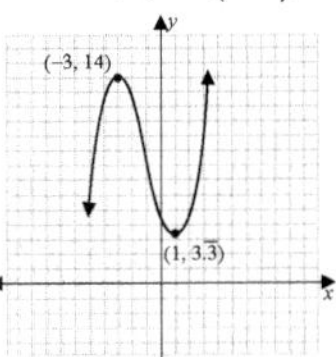

25. $(-\infty, -1), (0, 1)$;

$f'(x) < 0$; Decreasing

$(-1, 0), (1, +\infty)$;

$f'(x) > 0$; Increasing

$x = -1$; $f'(-1) = 0$; Local min

$x = 0$; $f'(0) = 0$; Local max

$x = 1$; $f'(1) = 0$; Local min

$\left(-\infty, -\dfrac{1}{\sqrt{3}}\right), \left(\dfrac{1}{\sqrt{3}}, +\infty\right)$;

$f''(x) > 0$; Concave up

$\left(-\dfrac{1}{\sqrt{3}}, \dfrac{1}{\sqrt{3}}\right)$; $f''(x) < 0$;

Concave down

$x = -\dfrac{1}{\sqrt{3}}$; $f''\left(-\dfrac{1}{\sqrt{3}}\right) = 0$; POI

$x = \dfrac{1}{\sqrt{3}}$; $f''\left(\dfrac{1}{\sqrt{3}}\right) = 0$; POI

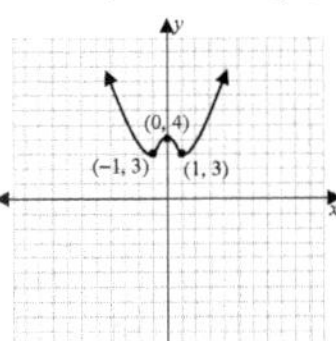

27. $(-\infty, 3)$; $f'(x) < 0$

except $x = 0$; Decreasing

$(3, +\infty)$; $f'(x) > 0$; Increasing

$x = 3$; $f'(3) = 0$; Local min

$(-\infty, 0), (2, +\infty)$; $f''(x) > 0$;

Concave up

$(0, 2)$; $f''(x) < 0$;

Concave down

$x = 0$; $f''(0) = 0$; POI

$x = 2$; $f''(2) = 0$; POI

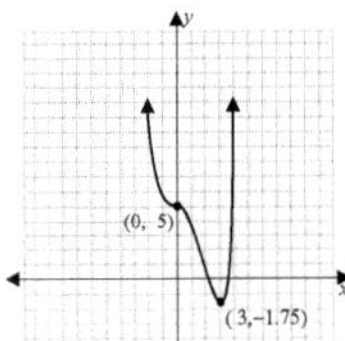

29. $(-\infty, 1); f'(x) < 0;$ Decreasing

$(1, +\infty); f'(x) > 0;$ Increasing

$x = 1; f'(1) = 0;$ Local min

$(-\infty, +\infty); f''(x) > 0;$ Concave up

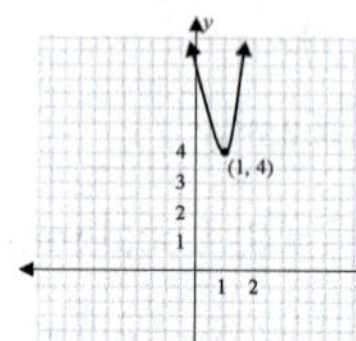

31. $\left(-\infty, -\dfrac{7}{3}\right), (3, +\infty); f'(x) > 0;$ Increasing;

$\left(-\dfrac{7}{3}, 3\right); f'(x) < 0;$ Decreasing;

$x = -\dfrac{7}{3}; f'\left(-\dfrac{7}{3}\right) = 0;$ Local max.;

$x = 3; f'(3) = 0;$ Local min.;

$\left(-\infty, \dfrac{1}{3}\right); f''(x) < 0;$ Concave down;

$\left(\dfrac{1}{3}, +\infty\right); f''(x) > 0;$ Concave up;

$x = \dfrac{1}{3}; f''\left(\dfrac{1}{3}\right) = 0;$ POI

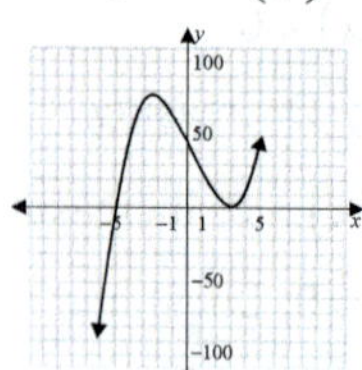

33. $(-\infty, 1.62); f'(x) < 0;$ Decreasing; $(1.62, +\infty); f'(x) > 0$ except $x = 8;$ Increasing; $x = 1.62; f'(1.62) = 0;$ Local min.; $(-\infty, 3.75), (8, +\infty); f''(x) > 0;$ Concave up; $(3.75, 8); f''(x) < 0;$ Concave down; $x = 3.75, f''(3.75) = 0;$ POI;

$x = 8; f''(8) = 0;$ POI

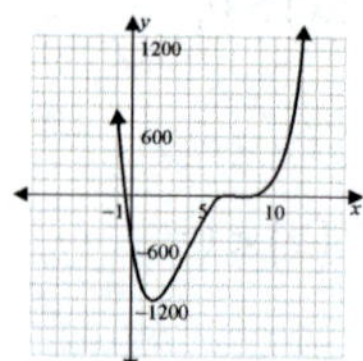

35. $(-\infty, -0.4); f'(x) < 0$ except $x = -1.6;$ Decreasing; $(-0.4, +\infty); f'(x) > 0;$ Increasing; $x = -0.4; f'(-0.4) = 0;$ Local min.; $(-\infty, -1.6), (-0.8, +\infty);$ $f''(x) > 0;$ Concave up; $(-1.6, -0.8); f''(x) < 0;$ Concave down; $x = -1.6; f''(-1.6) = 0;$ POI; $x = -0.8; f''(-0.8) = 0;$ POI

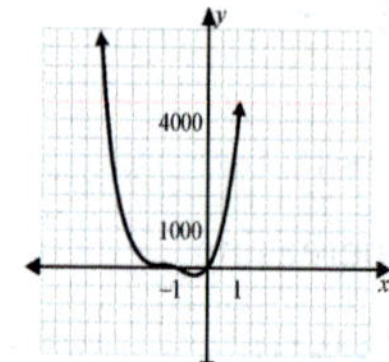

37. 3   **39.** 3

41. a.–b.

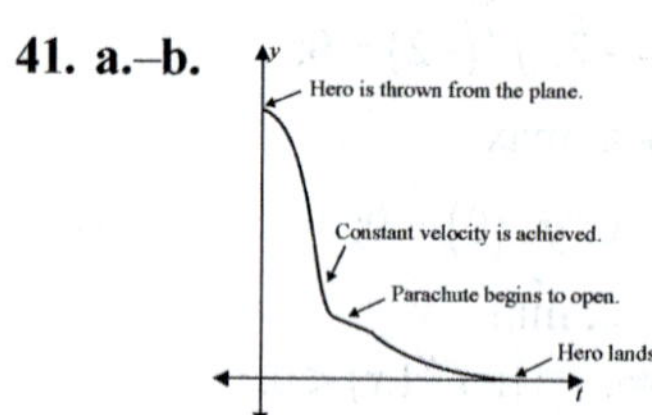

43. a. at $x = 5$

b. The workers should do the most demanding work starting 1 hour and 15 min. after their arrival to 7 hours and 15 min. into their work day. The first 1 hour and 15 min. and the last 45 min. of their day should be spent on the easier tasks. Doing so will achieve a PR of 25.02.

12.4 EXERCISES

1. a. $x = 4$ **b.** $y = 0$ **c.** None

3. a. $x = -8$ **b.** $y = 2$ **c.** None

5. a. None   **b.** $y = 0$ **c.** None

7. a. $x = 1, \ x = -\dfrac{1}{3}$

b. $y = \dfrac{5}{3}$ **c.** None

9. a. $x = 0$ **b.** None   **c.** $y = x$

11. a. $x = -1$ **b.** None

c. $y = x - 1$

13.

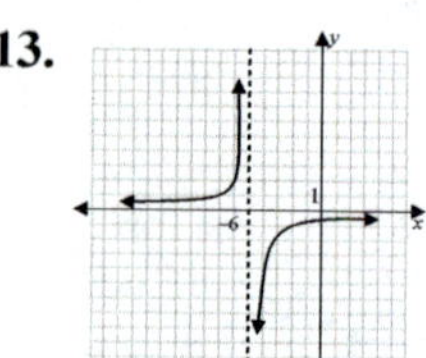

15.

17.

19.

21.

23. a. $F(0) = \$7000$

b. $x = -1$ is a vertical asymptote; $y = 55$ is a horizontal asymptote.

c.

d. \$5500

25. $A(x) = \dfrac{0.03x^2 + 24x + 10}{x}$

a.

b. $x = 0$ is a vertical asymptote; $y = 0.03x + 24$ is an oblique asymptote. Average cost is modeled by the oblique asymptote at large values.

27. a. 32 dinars

b. $x = -3$ is a vertical asymptote (not significant), $y = 20$ is a horizontal asymptote, (Average cost always exceeds 20 dinars.)

c. In the year 2007

29. a. $y = 8$ is a horizontal asymptote.

b. 4 million

12.5 EXERCISES

1. 25 vacuum cleaners per order; 5 orders per year

3. 30 copies per order; 3 orders per year

5. 200 sweatshirts per order; 20 orders per year

7. 120 snowmobiles per order; 8 orders per year

9. $24   **11.** 75 cents

13. 5 salespeople **15.** $6

17. 13 units at $36

19. 1900 microwave ovens at $326

21. a. The company is making $12,000.

b. The profit will rise $2 per cell phone.

c. The rate of profit increase is dropping.

23. a. $R(x) = 12x$

b. $P(x) = -0.3x^2 + 10x - 5$

c. $P(16.67) = 78.33$

25. a. $C(x) = 0.2x^2 + 3x$

b. $A'(x) = 0.2$

c. $C'(x) = 0.4x + 3$

27. a. $26,666.67

b. Value is increasing.

c. Yes. We can use test values with V' to determine if V' is in fact increasing or not (this rate of change is V'').

29.

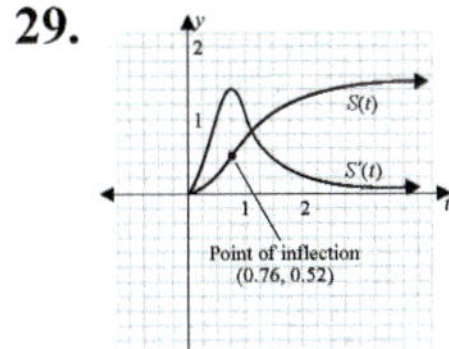

12.6 EXERCISES

1. 4 in. by 4 in.

3. $\dfrac{5}{3}$ in. by $\dfrac{14}{3}$ in. by $\dfrac{35}{3}$ in.

5. a. 12 in. by 12 in. by 6 in.

b. 432 in.^{2}

7. 2 ft by 2 ft by 2.5 ft

9. 1.82 ft

11. 18.5 ft by 37 ft

13. 70 ft by 70 ft; 4900 ft^2

15. 36 ft by 69 ft

17. $\dfrac{65}{3}$ in. by $\dfrac{65}{3}$ in. by $\dfrac{130}{3}$ in.;

$\dfrac{260}{9}$ in. by $\dfrac{260}{9}$ in. by $\dfrac{65}{3}$ in.

yields a package of lesser volume

19. 3 miles from point A

21. 0.8 miles from point A

23. a. 67 ft/sec   **b.** 129 ft

25. a. 324 ft   **b.** 8 sec

c. 144 ft/sec

Chapter 13: Additional Applications of the Derivative

13.1 EXERCISES

1. $f'(x) = 2x - \dfrac{1}{x}$

3. $f'(x) = 1 + \ln x$

5. $\dfrac{dy}{dx} = \dfrac{1 - \ln x}{x^2}$

7. $\dfrac{dy}{dx} = \dfrac{3}{x}(\ln x)^2$

9. $f'(x) = \dfrac{5}{5x + 3}$

11. $f'(x) = \dfrac{2x}{x^2 + 2}$

13. $f'(x) = \dfrac{2x}{2x^2 - 1}$

15. $\dfrac{dy}{dx} = \dfrac{8}{4x + 3}$

17. $\dfrac{dy}{dx} = \dfrac{2x^{\frac{3}{2}}}{x^2 + 2} + \dfrac{\ln(x^2 + 2)}{2\sqrt{x}}$

19. $\dfrac{dy}{dx} = \dfrac{2x + 2}{x(x^2 + 2x - 1)}$

$- \dfrac{\ln(x^2 + 2x - 1)}{x^2}$

21. $f'(x) = \dfrac{9x^2 + 2x + 9}{(3x + 1)(x^2 + 3)}$

23. $f'(x) = \dfrac{8x^2 - 2x - 8}{(2x - 1)(x^2 - 2)}$

25. $\dfrac{dy}{dx} = -\dfrac{3}{(x + 1)(x - 2)}$

27. $f'(x) = \dfrac{2x^2 + 18x + 10}{(x^2 - 5)(2x + 9)}$

29. $f''(x) = \dfrac{2 - \ln x}{x(\ln x)^3}$;

$x = e^2 \approx 7.39$; $(e^2, 3.7)$

31. $f''(x) = 9 + 6\ln x$;

$x = e^{-\frac{3}{2}}$; $\left(e^{-\frac{3}{2}}, -0.224\right)$

33. $f''(x) = \dfrac{3\ln x}{x^2}[2 - \ln x]$;

$x = 1,\ x = e^2 \approx 7.39$;

$(1, 0),\ (e^2, 8)$

35. $\dfrac{dy}{dx} = \dfrac{(2x - 5)^2(8x^2 - 19x + 5)}{\sqrt{x^2 - 2x}}$

37. $\dfrac{dy}{dx} = \dfrac{2x(11x^2 - 14)(x^2 + 2)^3}{3(x^2 - 1)^{\frac{4}{3}}}$

39. $\dfrac{dy}{dx} = \dfrac{\left(16x^2 - 46x - 54\right)\left(2x + x^2\right)^2}{\left(4x - 9\right)^3}$

41. $\dfrac{dy}{dx} = \dfrac{\left[\dfrac{\left(21x^2 - 114x - 248\right)}{(x+2)(3x+4)}\right]}{2(x-6)^{\frac{3}{2}}}$

43. Abs. max.: (2, 1.31);

Abs. min.: (1, 1)

45. Abs. max.: (1.2, 6.58);

Abs. min.: (e, e)

47. Abs. max.: $(e, 4.39)$;

Abs. min.: $\left(\sqrt{1.5}, 0.89\right)$

49.
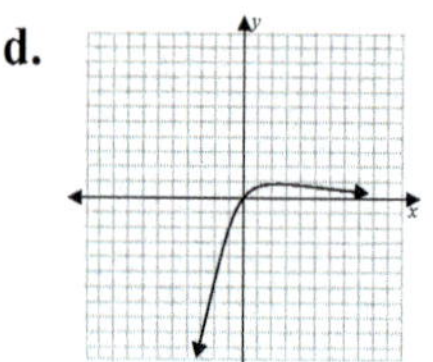

51.

53.

55. $96 + \dfrac{8x}{4x^2 + 15}$ dollars per table

57. \$294.30

59. a. 54 flies, 61 flies, 0 flies

b. 63 flies

13.2 EXERCISES

1. $f'(x) = f''(x) = 3e^x$

3. $f'(x) = 2x + 5e^x$

$f''(x) = 2 + 5e^x$

5. $f'(x) = e^x(x+1)$

$f''(x) = e^x(x+2)$

7. $f'(x) = e^x\left(\dfrac{1}{x} + \ln x\right)$

$f'(1) = e$

9. $f'(x) = -\dfrac{e^x}{\left(e^x - 1\right)^2}$

$f'(2) = -0.181$

11. $f'(x) = 8e^{4x}$

$f'(-1) = 0.147$

13. $f'(x) = 6e^{2x+1}$

$f'(0) = 6e \approx 16.310$

15. $f'(x) = -2xe^{1-x^2}$;

inc. on $(-\infty, 0)$; dec. on $(0, \infty)$

17. $f'(x) = 4e^{2x}\left(e^{2x} - 4\right)$;

dec. on $(-\infty, 0.693)$;

inc. on $(0.693, \infty)$

19. $f'(x) = \dfrac{-0.1e^{-0.2x}}{\left(e^{-0.2x} + 11\right)^{\frac{1}{2}}}$;

always decreasing;

horiz. asym.: $y = \sqrt{11}$

21. $f'(x) = \dfrac{e^x}{e^x + 1}$; There

is an oblique asym. since

$f'(x) \to 1$ as $x \to \infty$.

23. Abs. min.: (−0.5, −0.18);

Abs. max.: (1, 7.39)

25. Abs. min.: (0, 0);

Abs. max.: (−1, e)

27. Abs. min.: (−2, 0.25);

Abs. max.: (0, 13.59)

29. Abs. min.: (1, 4.09);

Abs. max.: (3.5, 4.97)

31. a. $x = 2.5$ **b.** $x = 5$

c. up: $(5, \infty)$; down: $(-\infty, 5)$

d.

33. a. $x = 0, 2$ **b.** $x = 2 \pm \sqrt{2}$

c. up: $(-\infty, 2-\sqrt{2}), (2+\sqrt{2}, \infty)$

down: $(2 - \sqrt{2}, 2 + \sqrt{2})$

d.

35. a. $x = 0$ **b.** None

c. up: $(-\infty, \infty)$

d.

37. a. 20 units   **b.** \$51.50 each

39. 10,432 people per day

41. 209 bacteria per hour

43. a. \$772.73, \$5674.72

b. \$1131.72/yr., \$859.59/yr.

c. (3.8, 4250); The rifle's value begins to increase at a much slower rate and level off.

d. 2004

45. a. June 1st   **b.** Aug 1st

47. a. 2875.58   **b.** 1107.03

c. 39.5 weeks

d. during the 6th week

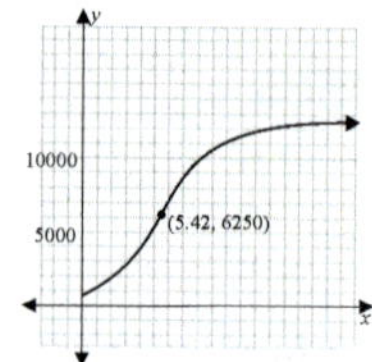

49. a. $k = \ln 10 \approx 2.3$

b. $k = \ln \pi \approx 1.145$

51. $f'(x) = \dfrac{-3e^{-x}(x+3) - 2}{(x+2)\left(2 + 3e^{-x}\right)}$;

No extrema or inflection points

13.3 EXERCISES

1. a. $p(t) = 8400e^{0.035t}$

b. 11,920 people

c. 12 years, or in the year 2012

3. $21.33   **5.** 9.5 hours

7. $300.44   **9.** 26.7 years

11. 8700 years   **13.** 11.8 lb/in.^{2}

15. $18,846

17. 0.777; 23.1 hours

19. a. 800 fish   **b.** 4303 fish

   c. 5.9 years

21. 13.5 minutes **23.** $11,880.94

13.4 EXERCISES

1. a. $E(x) = \dfrac{28 - x}{x}$

   b. $x = 14$

3. a. $E(x) = \dfrac{200 - x}{x}$

   b. $x = 100$

5. a. $E(x) = \dfrac{5}{x}$

   b. $x = 5$

7. a. $E(x) = \dfrac{40}{x}$

   b. $x = 40$

9. a. $E(x) = \dfrac{2(150 - x)}{x}$

   b. $x = 100$

11. a. $E(x) = \dfrac{2(54 - x)}{x}$

   b. $x = 36$

13. a. $E(x) = \dfrac{2(18 - \sqrt{x})}{\sqrt{x}}$

   b. $x = 144$

15. a. $E(x) = \dfrac{363 - x^2}{2x^2}$

   b. $x = 11$

17. 24 sharpeners

19. $x < 121$

21. a. 19 video games

   b. $E = -\dfrac{pe^{-\frac{p}{10}}}{10\left(1 - e^{-\frac{p}{10}}\right)}$

   At $p = 10$, $E = -0.58 < 1$ and revenue is decreasing and demand is inelastic.

   c. $189.64

23. a. $E = 0.073 < 1$, the demand is inelastic.

   b. $R(x) = x\left(e^{\frac{1800}{x} - 10} - 1\right)$

   c. R is increasing as a function of p at $p = 20$ ($x \approx 137.99$); yes.

25. a. $p = 264.75

   b. $D'(x) = -3xe^{-\frac{x^2}{200}}$

   c. $E = 4$ **d.** $x = 10$

13.5 EXERCISES

1. 12   **3.** –2

5. $4\sqrt{3}$ **7.** 1/2

9. ∞/∞; $0.5\sqrt{x}$ dominates

11. 0   **13.** $-\infty$

15. L'Hôpital's Rule does not apply; the limit is 0.

17. 3   **19.** 2

21. ∞ **23.** ∞

25. $\dfrac{\ln 5}{\ln 4}$ **27.** $\ln 3$

29. ∞/∞; the limit is $\sqrt{2}$.

31. ∞^0; the limit is 1.

33. $\infty - \infty$; the limit is $-\infty$.

35. $\infty - \infty$; the limit is $-3/2$.

37. 1^∞; the limit is $2/e$.

39. Not indeterminate form; the limit is ∞.

41. ∞/∞; the limit is 0.

43. ∞/∞; the limit is 1/100.

45. 1^∞; the limit is $1/e$.

47. 1

49. L'Hôpital's Rule does not apply; the limit is 1.

51. 0   **53.** e

55. 0

57. The limit is 0.

59. The limit is 0.

61. The limit is ∞.

63. The limit is 1.

65. The limit is 0.

67. 1/0 is not indeterminate; l'Hôpital's Rule does not apply.

71. $c = 1/2$

73. The theorem does not apply since $g(a) = g(b)$.

75. $c = 1$

77. $c_1 = -17/9$, $c_2 = 1$

79. $\dfrac{dE}{dx} \to 0$

81. The limit of the limits is 1.

13.6 EXERCISES

1. $dy = 3x^2\,dx$

3. $du = 8t(2t^2 + 1)\,dt$

5. $dA = 2\pi r\,dr$

7. $dV = 4\pi r^2\,dr$

9. $dS = \left(8x - \dfrac{1350}{x^2}\right)dx$

11. $dC = \left(3 + \dfrac{0.2}{\sqrt{x}}\right)dx$

13. $dP = (75 - 0.4x)\,dx$

15. a. 0.64   **b.** 0.6

   c. 0.04

17. a. 14.03   **b.** −15.36

   c. 1.33

19. a. −17.7   **b.** −16.8

   c. −0.9

21. a. 0.148   **b.** 0.15

   c. −0.002

23. $6 + \dfrac{1}{12}$ **25.** $3 - \dfrac{1}{27}$

27. $7 + \dfrac{1}{10}$ **29.** $4 - \dfrac{17}{480}$

31. $20.40   **33.** $42

35. −340 people **37.** 19.44 in.^{3}

39. 0.013 ft   **41.** 2400π in.^{3}

Chapter 14: Integration with Applications

14.1 EXERCISES

1. $F'(x) = 4$

3. $F'(x) = 6x + 5$

5. $F'(x) = \dfrac{1}{x} + \dfrac{1}{x^2} - 8xe^{x^2}$

7. $F'(x) = 8x(x^2 + 3)^3$

9. $F'(x) = 5e^x(e^x - 4)^2$

11. $F'(x) = \dfrac{2x + 5}{x^2 + 5x - 3}$

13. $7x + C$ **15.** $x^5 + C$

17. $\dfrac{1}{3}x^3 - 3x + C$

19. $\dfrac{1}{3}t - e^t + C$

21. $\ln|y| + \dfrac{1}{4}y^4 + C$

23. $\dfrac{4}{3}x^3 + 2\ln|x| - \dfrac{1}{x} + C$

25. $\dfrac{3}{2}x^{\frac{4}{3}} + \dfrac{10}{3}x^{\frac{3}{2}} + C$

27. $4e^y - \dfrac{1}{3}y^6 - \dfrac{1}{5}y + C$

29. $3t^{\frac{2}{3}} - \dfrac{7}{2t^2} + C$

31. $\dfrac{2}{3}e^x - 2x^{-\frac{1}{2}} - 7\ln|x| + C$

33. $\dfrac{2}{5}x^5 - \dfrac{1}{4}x^4 + C$

35. $3t^3 + 6t^2 + 4t + C$

37. $\dfrac{2}{7}y^{\frac{7}{2}} + \dfrac{4}{5}y^{\frac{5}{2}} - \dfrac{2}{3}y^{\frac{3}{2}} + C$

39. $3x + 5\ln|x| + \dfrac{4}{x} + C$

41. $4\ln|x| + 2\sqrt{x} - 3x + C$

43. $\dfrac{2}{3}x^{\frac{3}{2}} + 6\ln|x| - 2e^x + C$

45. a. $F(x) = \dfrac{1}{3}x^3 - e^x + 2$

  b. $F(x) = 2x^3 + \dfrac{1}{2}x^2 - 10x$

c. $F(x) = 20\sqrt{x}$

  d. $F(x) = 6e^x - 2x - 16$

47. a. $x = 0$; maximum,

   $x = 4$; minimum

  b. $x = 2$; POI

  c.

100, 70, 40, 10, (4, -63)

49. $C(x) = 0.1x^3 - 0.4x^2 + 24x + 1500$

51. $R(x) = 100x - 0.1x^2$

53. a. $P(x) = 120x - \dfrac{3}{2}x^2 - 1050$

  b. $\$1350$

55. $P(t) = 16t^{\frac{3}{2}} - 20t^2 + 6300$

57. $N(t) = \dfrac{960}{t} - 240t + 5000$

59. a. $s(t) = 36t^2 - 2t^3$

  b. $s(5) = 650$

  c. $s(10) = 1600$

61. a. $a(t) = 30 + \dfrac{12}{\sqrt{t}}$

  b. $s(t) = 200t + 15t^2 + 16t^{\frac{3}{2}}$
     $+ 5000$

14.2 EXERCISES

1. $\dfrac{1}{8}(x + 4)^8 + C$

3. $\dfrac{2}{3}(x^2 - 1)^{\frac{3}{2}} + C$

5. $\ln|t + 2| + C$

7. $\ln|t^3 + 4| + C$

9. $e^{y+5} + C$ **11.** $e^{-0.2x} + C$

13. $\dfrac{1}{5}\ln|5x + 3| + C$

15. $\dfrac{1}{2}\sqrt{4x - 1} + C$

17. $\dfrac{1}{4}e^{2x^2} + C$

19. $-\dfrac{1}{6}(3x^2 - 1)^{-1} + C$

21. $\ln|t^2 + t - 4| + C$

23. $-\dfrac{5}{2}e^{-y^2} + C$

25. $(5 + 2x^2)^{\frac{3}{2}} + C$

27. $-4e^{\frac{1}{x}} + C$

29. $-\dfrac{1}{18}(1 - 3e^{2x})^3 + C$

31. $\dfrac{1}{2}(\ln x)^2 + C$

33. $\ln|\ln x| + C$

35. $\ln(e^x + e^{-x}) + C$

37. a. $x + \ln|x - 2| + C$

  b. $x + 2\ln|x + 1| + C$

39. $x - 2\ln|x + 4| + C$

41. $f(x) = e^{5x^2} + 10$

43. $f(x) = \dfrac{(2x + 2)^6}{12} + 10$

45. a. $V(t) = 5000(25 - 1.8t)^{-\frac{1}{2}}$

  b. $\$1250$

47. $p(x) = 14 - \dfrac{1}{2}\ln x$

49. $N(t) = 6\sqrt{t^2 + 5t}$ bikes

51. $s(t) = 36t - \dfrac{60}{t + 1} + 4$

53. a. $v(t) = -8(2t + 1)^3 + 14$

  b. $s(t) = -(2t + 1)^4 + 14t + \dfrac{1}{4}$

55. a. Average age of death per year

  b. $f(t) = 30\ln|1 + 0.01t| + 30$

  c. 50.8 years

14.3 EXERCISES

1. 12   **3.** 21

5. 5.38

7. Left-endpoint est. ≈ 4.146 (underestimate); right-endpoint est. ≈ 6.146 (overestimate); midpoint est. ≈ 5.384 (overestimate)

9. Left-endpoint est. $= 25/12$ (overestimate); right-endpoint est. $= 77/60$ (underestimate); midpoint est. $= 496/315$ (underestimate)

11. Left-endpoint est. ≈ 8.119 (overestimate); right-endpoint est. ≈ 4.924 (underestimate); midpoint est. ≈ 6.323 (underestimate)

14.4 EXERCISES

1. $\dfrac{5}{3}$ **3.** $2 + e^3 \approx 22.09$

5. $\ln 3 \approx 1.10$ **7.** $\dfrac{1}{4}$

9. 20   **11.** $e^4 - 1 \approx 53.60$

13. 42   **15.** $\dfrac{1}{3}\ln 10 \approx 0.77$

17. $\dfrac{1}{80}$ **19.** $\dfrac{121}{5}$

21. $\dfrac{3}{2}\ln 33 \approx 5.24$

23. $\dfrac{1}{2}\left[(e+1)^2 - 4\right] \approx 4.91$

25. $\dfrac{1}{2}(e^8 - 1) \approx 1489.98$

27. 1.32

29. $\ln(e+1) - \ln 2 \approx 0.62$

31. $\dfrac{1}{2}\left[(1+\ln 3)^2 - 1\right] \approx 1.7$

33. 13   **35.** $\dfrac{38}{15}$

37. $2(1 - e^{-1}) \approx 1.26$

39. $360(6 - \sqrt{11}) \approx 966$ ppm

41. 74.80 units

14.5 EXERCISES

1. $\dfrac{85}{2}$ **3.** $\dfrac{28}{3}$

5. 4   **7.** $\dfrac{52}{3}$

9. $4\ln 4 \approx 5.55$

11. $5(e^3 - 1) \approx 95.43$

13. $\dfrac{38}{3}$ **15.** $\dfrac{41}{6}$

17. $\dfrac{44}{3}$ **19.** \$1800

21. \$1059   **23.** 28 bison

25. Total revenue from the sale of the first a units of a product.

14.6 EXERCISES

1. $\dfrac{115}{6}$ **3.** $\dfrac{1}{12}$

5. $\dfrac{31 - 5\sqrt{5}}{3} \approx 6.61$

7. $6.5 + e^{-3} \approx 6.55$

9. $\dfrac{15}{8} - \ln 2 + \ln\dfrac{1}{2} \approx 0.489$

11. $\dfrac{4}{3}$ **13.** $\dfrac{1}{3}$

15. 36

17. $A = 72.9$ sq. units

19. $A \approx 0.693$ sq. units

21. $A = 48$ sq. units

23. a. $(20, \$20)$

   b. \$40   **c.** \$100

25. a. $(25, \$41)$

   b. \$208.33   **c.** \$312.50

27. \$154.52   **29.** \$2.89

31. $(3, 80)$, CS $= \dfrac{135}{2}$, PS $= 45$

33. a. 61 percent   **b.** $\dfrac{7}{48}$

35. a. 27 percent   **b.** $\dfrac{23}{75}$

37. a. A: 0.26; B: 0.245   **b.** B

39. $A \approx 3.06$ sq. units

41. $A \approx 2.42$ sq. units

14.7 EXERCISES

1. $\dfrac{dy}{dx} = 6$ **3.** $\dfrac{dy}{dx} = e^x = 3 + y$

5. $\dfrac{dy}{dx} = \dfrac{8}{(5-x)^3} = y^{\frac{3}{2}}$

7. $2\dfrac{dy}{dx} + 3y$

$= 2(3e^{-1.5x}) + 3\left(\dfrac{1}{3} - 2e^{-1.5x}\right)$

$= 6e^{-1.5x} + 1 - 6e^{-1.5x} = 1$

9. $2x^2 y'' - xy' - 2y$

$= 2x^2(6) - x(6x + 5)$

$-2(3x^2 + 5x - 2) = 4 - 15x$

11. $x^2 y'' - xy' + y$

$= x^2\left(\dfrac{1}{x}\right) - x(1 + \ln x) + x\ln x = 0$

13. $y = \dfrac{3}{2}x^2 + \ln|x| + C$

15. $y = \dfrac{C(x+1)^2 + 1}{2}$

17. $y = Ce^{-0.4x}$

19. $y = \dfrac{2}{5x^2 - 2x + C}$

21. $y = \dfrac{20}{1 + Ce^{-12x}}$

23. $y = 5e^{0.25x}$ **25.** $y = 5x - 1$

27. $y = \dfrac{2}{4x^3 - 3}$

29. $y = 14e^{x^2} - 4$

31. $y = \dfrac{120}{3 + e^{-32x}}$

33. $D(x) = \dfrac{96}{x^{\frac{2}{3}}}$

35. $D(x) = 18\sqrt{120 - x}$

37. a. $V = 24{,}000e^{-0.06t}$

   b. \$15,769.12

39. a. $T(t) = Ce^{-kt} + M$

   b. 37.6°

41. a. $N(t) = \dfrac{2800}{1 + 139e^{-0.84t}}$

   b. 6 days

Chapter 15: Additional Integration Topics

15.1 EXERCISES

1. $\dfrac{1}{4}e^{2x}(2x-1)+C$

3. $4e^{0.5y}(y-2)+C$

5. $t(\ln t - 1)+C$

7. $\dfrac{1}{16}x^4(4\ln 5x-1)+C$

9. $\dfrac{2}{15}(x+2)^{\frac{3}{2}}(3x-4)+C$

11. $-x(x+4)^{-1}+\ln|x+4|+C$

13. $-\dfrac{5}{18}e^{-0.6t}(3t+5)+C$

15. $\dfrac{2}{9}x^{\frac{3}{2}}(3\ln 7x-2)+C$

17. $\dfrac{1}{4}-\dfrac{5}{4}e^{-4}\approx 0.23$

19. $\dfrac{9}{16}-\dfrac{13}{16}e^{-4}\approx 0.55$

21. $\dfrac{2}{3}\approx 0.67$

23. $u=4x,\ dv=(3x+1)^5\,dx;$

$\dfrac{5158}{7}\approx 736.86$

25. $u=x+1,\ dv=(x+2)^{\frac{3}{2}}\,dx;$

$\dfrac{836}{35}\approx 23.89$

27. $u=\ln x,\ dv=x^2dx;$

$\dfrac{125}{3}\ln 5-\dfrac{124}{9}\approx 53.28$

29. $u=\ln(x+1),\ dv=dx;$

$7\ln 7-6\approx 7.62$

31. $-\dfrac{5}{4}e^{-2t}(2t+1)+C$

33. $\dfrac{2\sqrt{3}}{9}x^{\frac{3}{2}}(3\ln x-2)+C$

35. $\dfrac{3}{10}(2x^2-1)^{\frac{5}{2}}+C$

37. $\dfrac{1}{3}(\ln x)^3+C$

39. $-\ln|1-e^x|+C$

41. $D(t)=100te^{0.01t}-10{,}000e^{0.01t}$
$+10{,}080$ million units

43. $R(x)=18.4x-0.4x\ln x$ dollars

15.2 EXERCISES

1. $15,319.26

3. $63,498.35

5. $17,872.00

7. $1,637,022.23

9. $22,600.56

11. $520.96

13. $155,907.16

15. $568.89

17. $21,963.97

19. $126,906.09

21. Yes, both forms of the integral result in the same formula for the future value of an annuity,

$$\int_0^N Pe^{r(N-t)}dt=\int_0^N Pe^{rt}dt=\dfrac{P}{r}\left(e^{rN}-1\right).$$

15.3 EXERCISES

1. $\dfrac{1}{4}\ln|4x+3|+C$

3. $-\dfrac{20}{3}e^{-0.15x}+C$

5. $-\dfrac{1}{2(2x-5)}+C$

7. $\dfrac{2(9x+8)}{135}\cdot(3x-4)^{\frac{3}{2}}+C$

9. $\dfrac{x}{2}\sqrt{x^2+36}$
$+18\ln\left|x+\sqrt{x^2+36}\right|+C$

11. $\dfrac{1}{8}\ln\left|\dfrac{x-4}{x+4}\right|+C$

13. $-\dfrac{1}{41}\ln\left|\dfrac{x+8}{5x-1}\right|+C$
$=\dfrac{1}{41}\ln\left|\dfrac{5x-1}{x+8}\right|+C$

15. $\dfrac{7x^6}{6}\left(\ln x-\dfrac{1}{6}\right)+C$

17. $\dfrac{x}{8}+\dfrac{1}{5.6}\ln\left|8e^{-0.7x}-5\right|+C$

19. $-2\ln\left|\dfrac{x}{3x-1}\right|+C=2\ln\left|\dfrac{3x-1}{x}\right|+C$

21. $\dfrac{14}{9}(6x-5)^{\frac{3}{2}}+C$

23. $-8\ln\left|\dfrac{2x+5}{x}\right|+C=8\ln\left|\dfrac{x}{2x+5}\right|+C$

25. $\dfrac{x^3e^{1.5x}}{1.5}-\dfrac{3x^2e^{1.5x}}{(1.5)^2}$
$+\dfrac{6xe^{1.5x}}{(1.5)^3}-\dfrac{6e^{1.5x}}{(1.5)^4}+C$

27. $\dfrac{1}{37}\left(\dfrac{2}{5}\ln|5x+2|+7\ln|x-7|\right)+C$

29. $\ln\left|x+\sqrt{x^2-12}\right|+C$

15.4 EXERCISES

1. a. 40   b. 42

3. $T_8=6724;\ S_8\approx 6554.6667;$
exact value: 6553.6

5. $T_8\approx 0.83954;\ S_8\approx 0.8048;$
exact value: 0.8

7. $T_8=65;\ S_8=64;$ exact value: 64

9. $T_8=6.4223;\ S_8=6.3892;$
exact value: $e^2-1=6.389056...$

11. $T_8=24;\ S_8=24;$ exact value: 24

13. $T_8\approx 24.6507;\ S_8=24.5122;$
exact value:
$52\sqrt{2}/3=24.513035...$

15. $T_8\approx 6.9734;\ S_8\approx 7.0665;$
exact value: $\sqrt{65}-1=7.062257...$

17. a. 44.9705   b. 4.7633

19. 1.5998

21. 0.69317

23. 146.30649

25. $35\frac{5}{6}$ cubic feet

15.5 EXERCISES

1. 0

3. $+\infty$

5. 0

7. 2

9. 0

11. $\dfrac{1}{2}$

13. Divergent

15. $3e^{-20}\approx 0.0000000062$

17. $3e^{-\frac{2}{3}}\approx 1.54$

19. Divergent   **21.** $\dfrac{1}{3}$

23. Divergent   **25.** Divergent

27. $\dfrac{1}{3}$ **29.** $\dfrac{1}{2}$

31. $\dfrac{1}{2\left(\ln 2\right)^2} \approx 1.04$

33. 1   **35.** 2

37. Area unbounded, integral divergent

39. d

15.6 EXERCISES

1. $\dfrac{8\pi}{3}$ **3.** 32π

5. $\dfrac{\pi}{2}\left(e^4 - e^{-2}\right) \approx 27.2\pi$

7. $\dfrac{16\pi}{15}$ **9.** $\dfrac{128\pi}{3}$

11. $\dfrac{32\pi}{3}$ **13.** $\pi\ln 6 \approx 1.79\pi$

15. 53.3π

Chapter 16: Multivariable Calculus

16.1 EXERCISES

1. a. 6 **b.** −12

3. a. 11 **b.** 9

5. a. $\dfrac{11}{4}$ **b.** undefined

7. a. undefined on $\mathbb{R}$

    b. $96\sqrt{2}$

9. a. $6e^3 \approx 120.51$

    b. $-3e^{-1} \approx -1.10$

11. a. 1

    b. $2e^2 + 2 \approx 16.78$

13. a. 1377.13 **b.** 674.93

15. a. 1134 **b.** 758

17. 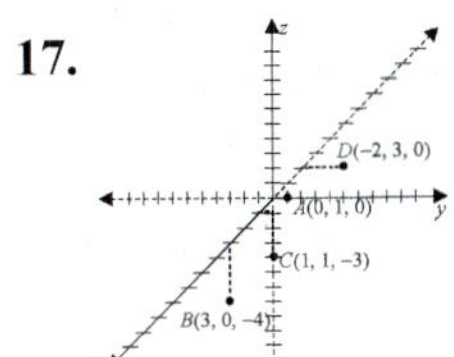

19. 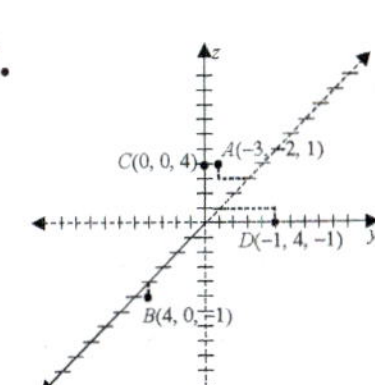

21.

23. 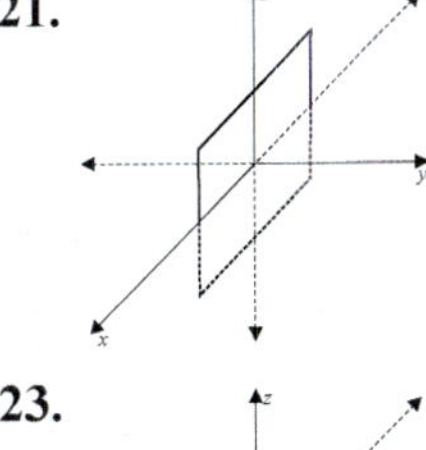

25. 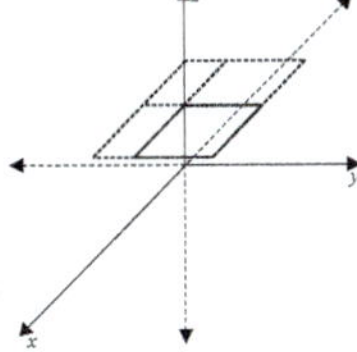

27. 118

29. a. \$10,400 **b.** \$11,580

31. a. $C(x, y) = 1360 + 160x + 220y$

    b. \$6400

33. a. $R(x, y) = -20x^2 - 22y^2 + 64x + 52y + 34xy$

    b. \$336.50

35. a. $V(x, y, z) = xyz$

    b. $S(x, y, z) = xy + 2xz + 3yz$

37.

Number of Years (t)		
3	5	10
1195.62	1346.86	1814.02
1268.24	1485.95	2208.04
1344.89	1638.62	2685.06

16.2 EXERCISES

1. $\dfrac{\partial f}{\partial x} = 4;\ \dfrac{\partial f}{\partial y} = 7$

3. $\dfrac{\partial f}{\partial x} = 4x;\ \dfrac{\partial f}{\partial y} = 10y$

5. $\dfrac{\partial f}{\partial x} = 2xy + 4y^3$

    $\dfrac{\partial f}{\partial y} = x^2 + 12xy^2$

7. $\dfrac{\partial f}{\partial x} = \dfrac{xy}{\sqrt{25 + x^2}}$

    $\dfrac{\partial f}{\partial y} = \sqrt{25 + x^2}$

9. $\dfrac{\partial f}{\partial x} = -\dfrac{x}{\sqrt{49 - x^2 - y^2}}$

    $\dfrac{\partial f}{\partial y} = -\dfrac{y}{\sqrt{49 - x^2 - y^2}}$

11. $\dfrac{\partial f}{\partial x} = 4e^{x-y};\ \dfrac{\partial f}{\partial y} = -4e^{x-y}$

13. $\dfrac{\partial f}{\partial x} = \ln y;\ \dfrac{\partial f}{\partial y} = \dfrac{x}{y}$

15. $\dfrac{\partial f}{\partial x} = \dfrac{2x + 3y}{x^2 + 3xy};\ \dfrac{\partial f}{\partial y} = \dfrac{3x}{x^2 + 3xy}$

17. $\dfrac{\partial f}{\partial x} = \dfrac{4x}{y^2 + 1};\ \dfrac{\partial f}{\partial y} = -\dfrac{4x^2 y}{\left(y^2 + 1\right)^2}$

19. $\dfrac{\partial f}{\partial x} = \dfrac{x^2 y + 6x}{\left(xy + 3\right)^2};\ \dfrac{\partial f}{\partial y} = -\dfrac{x^3}{\left(xy + 3\right)^2}$

21. $\dfrac{\partial f}{\partial x} = 3x^2 e^y + 2xye^{x^2}$

    $\dfrac{\partial f}{\partial y} = x^3 e^y + e^{x^2}$

23. $\dfrac{\partial f}{\partial x} = \dfrac{3xy + 4}{2\sqrt{xy + 2}};\ \dfrac{\partial f}{\partial y} = \dfrac{x^2}{2\sqrt{xy + 2}}$

25. $\dfrac{\partial f}{\partial x} = x^3 e^{xy}\left(xy + 4\right);\ \dfrac{\partial f}{\partial y} = x^5 e^{xy}$

27. $\dfrac{\partial f}{\partial x} = \dfrac{2xy^5}{x^2 - 5y^2}$

    $\dfrac{\partial f}{\partial y} = \dfrac{-10y^6}{x^2 - 5y^2} + 5y^4\ln\left(x^2 - 5y^2\right)$

29. $\dfrac{\partial S}{\partial m} = 90m + 22b - 320$

$\dfrac{\partial S}{\partial b} = 22m + 6b - 80$

31. $\dfrac{\partial S}{\partial m} = 562m + 66b - 3494$

$\dfrac{\partial S}{\partial b} = 66m + 8b - 412$

33. $f_{xx}(x, y) = 2y^3$

$f_{xy}(x, y) = 3 + 6xy^2$

$f_{yx}(x, y) = 3 + 6xy^2$

$f_{yy}(x, y) = 6x^2 y$

35. $f_{xx}(x, y) = 12x^2 y^{\frac{2}{3}}$

$f_{xy}(x, y) = \dfrac{8}{3} x^3 y^{-\frac{1}{3}}$

$f_{yx}(x, y) = \dfrac{8}{3} x^3 y^{-\frac{1}{3}}$

$f_{yy}(x, y) = -\dfrac{2}{9} x^4 y^{-\frac{4}{3}}$

37. $f_{xx}(x, y) = 0$

$f_{xy}(x, y) = 2e^{2y}$

$f_{yx}(x, y) = 2e^{2y}$

$f_{yy}(x, y) = 4xe^{2y}$

39. $f_{xx}(x, y) = \dfrac{160}{9}(4x - 3y)^{-\frac{1}{3}}$

$f_{xy}(x, y) = -\dfrac{40}{3}(4x - 3y)^{-\frac{1}{3}}$

$f_{yx}(x, y) = -\dfrac{40}{3}(4x - 3y)^{-\frac{1}{3}}$

$f_{yy}(x, y) = 10(4x - 3y)^{-\frac{1}{3}}$

41. $f_{xx}(x, y) = 0$

$f_{xy}(x, y) = -\dfrac{15}{(5y + 3)^2}$

$f_{yx}(x, y) = -\dfrac{15}{(5y + 3)^2}$

$f_{yy}(x, y) = \dfrac{50(3x + 1)}{(5y + 3)^3}$

43. $f_{xx}(x, y) = \dfrac{4y^2}{(x - y)^3}$

$f_{xy}(x, y) = -\dfrac{4xy}{(x - y)^3}$

$f_{yx}(x, y) = -\dfrac{4xy}{(x - y)^3}$

$f_{yy}(x, y) = \dfrac{4x^2}{(x - y)^3}$

45. $f_x(x, y, z) = y + 2z$

$f_y(x, y, z) = x + 9z$

$f_z(x, y, z) = 2x + 9y$

47. $f_x(x, y, z)$
$= 32x(8x^2 + 5y^2 - 2z^2)$

$f_y(x, y, z)$
$= 20y(8x^2 + 5y^2 - 2z^2)$

$f_z(x, y, z)$
$= -8z(8x^2 + 5y^2 - 2z^2)$

49. $\dfrac{\partial F}{\partial x} = 8 + 15y + \lambda$

$\dfrac{\partial F}{\partial y} = 15x - 4y + \lambda$

$\dfrac{\partial F}{\partial \lambda} = x + y - 60$

51. $\dfrac{\partial F}{\partial x} = 10x + 3y + 14\lambda$

$\dfrac{\partial F}{\partial y} = 3x - 20y + 17\lambda$

$\dfrac{\partial F}{\partial \lambda} = 14x + 17y - 49$

53. $f_L(27, 64) = \dfrac{640}{9}$;

$f_K(27, 64) = 15$

55. a. $\dfrac{\partial C}{\partial x} = 7 + 0.06x - y$

b. $\dfrac{\partial C}{\partial y} = 5.8 - x + 0.04y$

57. a. $P(x, y) = 16x + 12y - x^2 + 2xy$
$-2y^2 - 8$

b. $P_x(20, 14) = 4$; $P_y(20, 14) = -4$;
The marginal profit with
respect to model A is \$4
and the marginal profit with
respect to model B is −\$4

when 20 units of model A and
14 units of model B are being
produced and sold.

59. a. $P_x(x, y) = 5000 + 10{,}000y$
$- 1000x$

b. $P_y(x, y) = 10{,}000x$

16.3 EXERCISES

1. $f(3, -1) = -14$ is a local
minimum.

3. $(5, 1, -15)$ is a saddle point.

5. $f\left(\dfrac{5}{2}, 4\right) = \dfrac{133}{4}$ is a local
maximum.

7. $(1, -2, -3)$ is a saddle point.

9. $f(-2, -1) = -18$ is a local
minimum.

11. $f(2, 2) = -3$ is a local minimum.

13. $f(1, -2) = 13$ is a local maximum.

15. $f(1, 3) = 0$ is a local minimum;
$(-1, 3, 4)$ is a saddle point.

17. $(2, 5, 7)$ is a saddle point;
$f(2, 1) = -25$ is a local minimum.

19. $(2, 1, 8)$ is a saddle point;
$(-2, -1, -8)$ is a saddle point.

21. $f(0, 0) = 0$ is a local minimum;
$(2, 2, 4)$ is a saddle point;
$(-2, 2, 4)$ is a saddle point.

23. $f(1, 2) = 6$ is a local minimum.

25. 30 of model X, 25 of model Y

27. 200 of the \$50 racks,
240 of the \$45 racks

29. \$16 for the adult T-shirt,
\$13 for the youth T-shirt

31. 9 inches by 6 inches by 3 inches

16.4 EXERCISES

1. $f(2, 2) = 8$ **3.** $f(1, -1) = 7$

5. $f(9, 6) = 612$ **7.** $f(4, 4) = 128$

9. $f(5, 2) = 30$ **11.** $f(4, 5) = 112$

13. $f(5, 2) = 69$ **15.** $f(8, 15) \approx 175$

17. $f(6, 3) = f(-6, -3) = 72$ is a maximum;
$f(6, -3) = f(-6, 3) = -72$ is a minimum

19. $f(12, -6) = 864$ is a maximum; $f(4, 2) = 96$ is a minimum

21. 3 units in LA, 37 units in OKC

23. approximately 3605 units

25. 10 days in each area

27. 69 feet by 36 feet with the interior lanes 36 feet long

16.5 EXERCISES

1. a. $y = 0.96x + 4.1$

b. 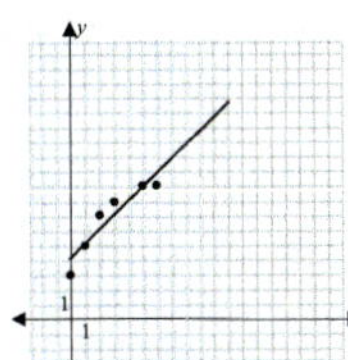

3. a. $y = -1.18x + 10.96$

b. 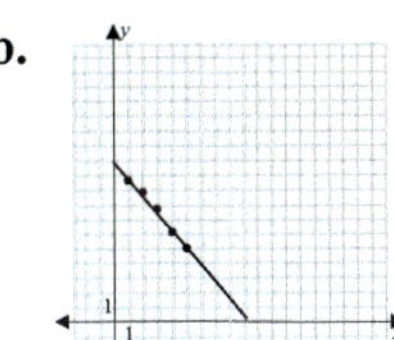

5. $y = -0.12x + 8.19$

7. $y = 14.7x + 221.9$

9. $y = 1.1x + 4.02$

11. a. $y = 3.77x + 21.91$

   b. 37,000 units

13. a. $y = 11.17x + 4.07$

   b. \$82,260

15. a. $y = -0.663x + 76.04$

   b. 71.40 cents per pound

16.6 EXERCISES

1. $\dfrac{3}{2}$ **3.** $\dfrac{117}{2}$

5. $5\ln 2 \approx 3.47$ **7.** $\dfrac{320}{3}$

9. $e^3 - e^2 - e + 1 \approx 10.98$

11. $\dfrac{4}{3}$ **13.** 56

15. $\dfrac{1}{2}\left(e^{11} - e^8 - e^3 + 1\right) \approx 28,437.05$

17. $\dfrac{288}{5}$ **19.** $\dfrac{508}{21}$

21. $\dfrac{26}{3}$ **23.** $\dfrac{98}{3}$

25. $\dfrac{1}{6}$ **27.** $\dfrac{31}{60}$

29. $\dfrac{e^6}{3} - e^2 + \dfrac{2}{3} \approx 127.75$

31. $\dfrac{3}{20}$ **33.** $\dfrac{5}{3}$

35. 34   **37.** $\dfrac{32}{3}$

39. $\dfrac{1}{3}$

Index